Principles of Immunopharmacology

Edited by Frans P. Nijkamp
and Michael J. Parnham

Birkhäuser Verlag
Basel • Boston • Berlin

Frans P. Nijkamp
Faculty of Pharmacy
Department of Pharmacology and Pathophysiology
Sorbonnelaan 16
3584 CA Utrecht
The Netherlands

Michael J. Parnham
PLIVA Research Institute Ltd.
Prilaz baruna Filipovića 29
HR-10 000 Zagreb
Croatia

Library of Congress Cataloging-in-Publication Data
Principles of immunopharmacology / edited by F.P. Nijkamp, M.J. Parnham.-- 2nd ed.
p. ; cm.
Includes bibliographical references and index.
ISBN-13: 978-3-7643-5804-4 (alk. paper)
ISBN-10: 3-7643-5804-1 (alk. paper)
1. Immunopharmacology. I. Nijkamp, Franciscus Petrus, 1947- II. Parnham, Michael J., 1951-
[DNLM: 1. Immunity--drug effects. 2. Adjuvants, Immunologic-pharmacology. 3. Immunologic Tests. 4. Immunosuppressive Agents--pharmacology. 5. Immunotherapy.
QW 540 P957 2005]
RM370.P75 2005
616.07'9--dc22

2005048217

Bibliographic information published by Die Deutsche Bibliothek
Die Deutsche Bibliothek lists this publication in the Deutsche Nationalbibliografie; detailed bibliographic data is available in the internet at http://dnb.ddb.de

ISBN-10: 3-7643-5804-1 Birkhäuser Verlag, Basel – Boston – Berlin
ISBN-13: 978-3-7643-5804-4 Birkhäuser Verlag, Basel – Boston – Berlin

Part of Springer Science+Business Media
Printed on acid-free paper produced from chlorine-free pulp. TCF ∞
Cover design: Micha Lotrovsky, CH-4106 Therwil, Switzerland
Cover illustration: Mechanisms in type I hypersensitivity (allergy)
Printed in Germany
ISBN-10: 3-7643-5804-1
ISBN-13: 978-3-7643-5804-4

9 8 7 6 5 4 3 2 1

www.birkhauser.ch

Contents

Preface to the first edition ... xv

Preface to the second edition ... xvi

Abbreviations ... xvii

A Mechanisms of immunity

A.1 Hematopoiesis, including lymphocyte development and maturation
Introduction: acquisition of specificity ... 3
Hematopoiesis ... 3
T and B lymphocytes: common origin from hematopoietic stem cells ... 6
Specificity: rearrangement of genes encoding antigen receptors ... 7
T lymphocyte maturation in the thymus ... 8
B lymphocyte development in the bone marrow ... 11
Mature T and B lymphocytes: tolerance ... 14
T and B lymphocyte development and maturation: relevance for immunopharmacology ... 14
Summary ... 15
Selected readings ... 16
Recommended website ... 16
References ... 16

A.2 T cell-mediated immunity
Introduction ... 19
Biology of the T lymphocyte immune response ... 19
Composition of the T cell network ... 20
T cell subset markers ... 22
Effectors of T cell-mediated immunity ... 23
Mechanisms of T cell-mediated cytotoxicity ... 24
Mechanisms of T cell activation ... 25
Tolerance ... 26
Summary ... 28
Selected readings ... 28

A.3 Antibody diversity and B cell-mediated immunity
Antibodies and immunoglobulins ... 29
Structure of immunoglobulins ... 29
The generation of antibody diversity ... 30
B cell receptor and signal transduction ... 32
B lymphocyte costimulation ... 34

Cytokine regulation 36
T cell-independent B lymphocyte activation 36
Primary and secondary antibody response 37
Biological functions of antibodies 39
Clinical relevance and future prospects 41
Summary 43
Selected readings 43
Recommended websites 43
References 44

A.4 Cytokines
Introduction 45
Differentiation factors 46
Activation and growth factors of lymphocytes 48
Mediators of inflammation 51
Regulatory factors of immune reactions 54
Chemokines 54
Inhibition of cytokines 58
Summary and outlook 60
Selected readings 60
Recommended website 60
References 60

A.5 Innate immunity – phagocytes, natural killer cells and the complement system
Phagocytes 63
Natural killer cells 72
The complement system 75
Summary 79
Selected readings 79
References 79

A.6 Inflammatory mediators and intracellular signalling
Introduction 81
Eicosanoids 81
Platelet-activating factor (PAF) 87
Toll-like receptors 89
Cytokines 89
Chemokines and their intracellular signalling 92
Neuropeptides 93
Kinins 93
Nitric oxide 95
Reactive oxygen species 98
Amines 99
Summary 101
Selected readings 101
References 101

A.7 Immune response in human pathology: infections caused by bacteria, viruses, fungi, and parasites
Infections 105
Natural resistance 105

Specific resistance ... 107
Defence against bacteria, viruses, fungi, and parasites ... 109
Pathogenesis of shock ... 110
Human immunodeficiency virus infection ... 112
Vaccines and vaccination ... 113
Infections in the new millennium ... 114
Summary ... 115
Acknowledgements ... 115
Selected readings ... 115
Recommended websites ... 115

A.8 Immune response in human pathology. hypersensitivity and autoimmunity
Introduction ... 117
Hypersensitivity ... 117
Autoimmunity ... 123
Conclusion ... 125
Summary ... 126
Selected readings ... 126
Recommended website ... 126
References ... 126

A.9 Cancer immunity
Introduction: cancer immunity from a historical perspective ... 129
Key players in the immune responses against cancer ... 130
Expression of targets for the immune system by cancer cells ... 132
Immunotherapy of cancer ... 136
Conclusions and future developments ... 140
Summary ... 141
Selected readings ... 141
References ... 142

A.10 Neuroimmunoendocrinology
Historical background ... 149
Regulation of the immune system by neuroendocrine hormones ... 150
Neuroendocrine hormone release by immune system cells ... 153
Functions of leukocyte-derived peptide hormones *in vivo* ... 155
Summary and conclusions ... 156
Acknowledgements ... 157
Selected readings ... 157
Recommended websites ... 157
References ... 157

B Immunodiagnosis

B.1 Antibody detection
Introduction ... 163
Basic principle of immunoassays ... 163
Antibody structure ... 164
Clinical relevance of antibody detection ... 164

Antibody detection methods 165
Summary 170
Selected readings 170
References 170

B.2 **Immunoassays**
Introduction 171
Basic principles of assay design 171
Components of immunoassays 173
Data presentation and curve plotting 175
Selected immunoassays 176
Assay performance and validation 178
Summary 180
Selected readings 181
References 181

B.3 **Flow cytometry**
Introduction 183
Mechanistic principles 183
Classification/types of assay 185
Components/construction of assays 187
Examples and their application 192
Summary 195
Selected readings 195
References 195

B.4 **Gene arrays**
Introduction 197
Principle of the technology 197
Production of microarrays 201
Application of microarrays 203
Array data: acquisition, analysis and mining 205
Examples of microarray experiments 207
Summary 210
Selected readings 211
Recommended websites 211
References 211

B.5 **Proteomics and applications within the drug development pipeline**
Introduction 213
Traditional proteomics 213
Array-based proteomics 222
Summary 225
Selected readings 226
Recommended websites 226
References 226

C Immunotherapeutics

C.1 Vaccines

Introduction 231
Historical background 231
Current vaccine categories 232
Pharmacological effects of vaccination 237
New developments 239
Summary 243
Selected readings 244
Recommended websites 244
References 244

C.2 Sera and immunoglobulins

Humoral immunity 247
Antibody structure 247
Effector functions of human Ig 249
The catabolism of Ig 254
Ig preparations for medical use 254
Indications 256
Dosage and administration 258
Side-effects of Ig therapy 258
Monoclonal Ab (mAb) 260
Summary 262
Selected readings 262
Recommended websites 263
References 263

C.3 Anti-allergic drugs

Introduction 265
Disodium cromoglycate and nedocromil sodium (chromones) 265
Histamine receptor antagonists 270
Anti-IgE 275
Summary 277
Selected readings 277
References 278

C.4 Drugs for the treatment of asthma and COPD

Introduction 281
Bronchodilators 281
Controller drugs 300
Mediator antagonists 309
Steroid-sparing therapies 312
New drugs for asthma and COPD 314
Other drugs 328
Summary 329
Selected readings 330
Recommended websites 330
References 330

C.5 **Immunostimulants in cancer therapy**
Introduction . . . 345
Recombinant proteins . . . 345
Natural and synthetic BRMs . . . 357
Combination immunotherapy and cellular therapy . . . 363
Conclusion . . . 365
Summary . . . 366
Selected readings . . . 367
Recommended websites . . . 367
References . . . 368

C.6 **Anti-infective activity of immunomodulators**
Introduction . . . 377
Cytokine immunomodulators . . . 378
Interferons and combinations . . . 379
Colony-stimulating factors . . . 381
Emerging cytokine therapies . . . 381
Synthetic immunomodulators . . . 383
Emerging therapies with synthetic immunomodulators . . . 384
Microbial immunomodulators . . . 386
Summary . . . 388
Recommended websites . . . 388
References . . . 389

C.7 **Mild plant and dietary immunomodulators**
Introduction . . . 391
Plant immunostimulants . . . 391
Zinc . . . 394
Dietary antioxidants . . . 396
Phenolic compounds as immunomodulators . . . 400
Emerging therapies and summary . . . 403
Selected readings . . . 403
Recommended websites . . . 403
References . . . 403

C.8 **Influence of antibacterial drugs on the immune system**
Introduction . . . 407
Antibacterial agents (ABA) and the immune system . . . 407
Class-specific immunomodulatory effects . . . 409
Non-antibacterial effects of antibacterial agents. Therapeutic implications . . . 426
Conclusions . . . 430
Summary . . . 430
Selected readings . . . 432
Recommended website . . . 433
References . . . 433

C.9 **Immunosuppressives in transplant rejection**
Introduction . . . 441
Calcineurin inhibitors . . . 446

Inhibitors of growth factor modulation . . . 450
Cytotoxic drugs . . . 451
Other immunosuppressive drugs . . . 454
Biologicals . . . 455
New approaches and perspectives . . . 457
Summary . . . 460
Selected readings . . . 461
References . . . 461

C.10 Cytotoxic drugs
Background . . . 465
Azathioprine . . . 465
Cyclophosphamide . . . 467
Fludarabine . . . 469
Methotrexate . . . 471
Mycophenolic acid . . . 474
Conclusions . . . 477
Summary . . . 477
Selected readings . . . 479
References . . . 479

C.11 Corticosteroids
Introduction . . . 483
Chemical structures . . . 485
Modes of action . . . 485
Pharmacological effects . . . 487
Pharmacokinetics . . . 490
Clinical indications . . . 490
Side-effects . . . 492
Summary . . . 494
Selected readings . . . 494
Recommended websites . . . 494
References . . . 494

C.12 Non-steroidal anti-inflammatory drugs
Introduction . . . 499
Mode of action of non-steroid anti-inflammatory drugs . . . 499
Cyclo-oxygenase-1 and cyclo-oxygenase-2 . . . 500
Structural basis for COX-2 selectivity . . . 503
Additional uses for selective COX-2 inhibitors . . . 504
COX-2 inhibitors and cancers . . . 504
Premature labour . . . 505
Alzheimer's disease . . . 506
Other cyclo-oxygenases . . . 506
Summary . . . 506
Selected readings . . . 507
References . . . 508

C. 13 Disease-modifying anti-rheumatic drugs
Introduction 511
Pathophysiology of rheumatoid arthritis 511
Corticosteroids 512
Disease-modifying anti-rheumatic drugs (DMARD). 514
Methotrexate 514
Antimalarials (chloroquine and hydroxychloroquine). 520
Azathioprine 521
Cyclosporin 523
Gold complexes. 525
Leflunomide 527
Penicillamine 530
Sulfasalazine 532
Tetracyclines (minocycline) 533
Biological anti-rheumatic therapy 535
Combination DMARD therapies 536
Summary. 537
Selected readings 539
References 540

C.14 Perspectives of immunotherapy in the management of asthma and other allergic conditions
Introduction 545
General aspects of immunotherapy in allergy and asthma. 545
Non-specific immunomodulatory therapies 546
Specific mediator antagonists 546
Anti-cytokine therapy 547
Cytokine therapy. 547
Specific immunotherapy 548
Mechanisms of immunotherapy 548
SIT for asthma. 549
Comparison of SIT with other types of treatment for asthma 549
Effects on natural history of allergic disease 549
Future directions 550
Conclusions. 551
Summary. 552
Selected readings 552
References 552

D Immunotoxicology

D.1 Immunotoxicology
Introduction 559
Mechanisms of immunotoxicity by pharmaceuticals 560
Procedures for preclinical testing of direct immunotoxicity 570
Procedures for preclinical testing of sensitizing capacity 574
Procedures for immunotoxicity testing in humans. 577
Immunotoxicity regulations. 579
Conclusions. 584
Summary. 585
Selected readings 585
References 585

Appendices

Appendix C.1 593
Appendix C.3 598
Appendix C.5 602
Appendix C.6 605
Appendix C.7 606
Appendix C.8 610
Appendix C.12/1 611
Appendix C.12/2 618
Appendix C.13 623
Appendix D.1 625

Glossary 627

List of contributors 645

Index 651

Preface to the first edition

The rapid developments in immunology in recent years have dramatically expanded our knowledge of mammalian host defence mechanisms. The molecular mechanisms of cellular interactions during immune responses have been unravelled, the intracellular responses involved in signal transduction delineated and an ever-increasing number of soluble mediators of immune and inflammatory reponses have been discovered.

The initial result of this explosion of knowledge has been to provide the researcher and the clinician with an arsenal of diagnostic tools with which the immunological bases of disease processes can be investigated. This has made disease diagnosis much more precise, enabling the physician to tailor therapy much more closely to the individual patient's needs. However, better understanding of disease processes only provides a gradual improvement in therapy. This is because the new molecular targets that have been uncovered must first be tested as potential bases for immunomodulatory drug actions and then the new compounds must be subject to extensive development studies. As a result of the molecular unravelling of the immune system, we now understand more precisely the mechanisms of action of some established therapies, such as anti-allergic and anti-asthma agents, including the corticosteroids. New rational treatments based on molecular mechanisms also are now entering clinical practice and are making their mark on cancer, infectious and autoimmune disease therapy.

Concomitantly with these advances in understanding of molecular mechanisms of immunity, immunodiagnosis and immunotherapy, it has become possible to test more accurately the way in which a variety of drug classes interact with the immune system. It is of particular importance to regulatory authorities that the toxic side-effects of immunomodulatory drugs can be distinguished from their beneficial therapeutic effects.

Currently, it is only possible to obtain an overview of these various aspects of immunopharmacology by reading a range of immunological, pharmacological, diagnostic and toxicological literature. Good immunological textbooks are available, while immunopharmacology is covered mainly in terms of the inflammatory response. *Principles of Immunopharmacology* is intended to provide for the first time in a single volume a basic understanding of immunological mechanisms, a review of important immunodiagnostic tools and a description of the main pharmacological agents which modify the immune response, together with an introduction to immunotoxicology. As such we hope that it will be useful as a reference text for physicians, researchers and students with a rudimentary knowledge of immunology.

We, the editors, are grateful to all the authors who have invested their time and effort into this volume. We have received continuous help and encouragement from Petra Gerlach and Katrin Serries of Birkhäuser Verlag and particular thanks are due to Dinij van der Pal for administrative assistance.

Frans P. Nijkamp
Michael J. Parnham
March, 1999

Preface to the second edition

Our knowledge of immunological processes and their modulation has progressed considerably since the first edition of *Principles of Immunopharmacology*. Molecular mechanisms have been elucidated so that we are now in a position to understand many of the complex pathways leading from surface stimulation to cellular responses. We now appreciate much better that the innate immune response is also regulated by far more external and internal stimuli than was previously realised and are starting to understand the role of memory and regulatory cells. Advances in genomics and proteomics have enabled the identification of many genes and new proteins that are intimately involved in the responses of the immune system. We have sought to include the most important of these advances in the first part of this second edition of *Principles of Immunopharmacology*. In addition to including new mechanisms in the section on the immune response, we have also included the new techniques of the genomic and proteomic revolution in the diagnostics section, as methods such as microarrays have now become an essential aspect of cellular analyses.

Inevitably, our increased understanding of immune mechanisms has opened opportunities for the development of novel drugs to treat inflammation and disorders of the immune system. Biologicals are now commanding worldwide interest, both in research and development and in clinical practice. The section on therapy has now been expanded to accommodate these new therapeutic approaches, as well as describing our improved understanding of the mechanisms of action of established agents. The final section on immunotoxicology has also been updated, particularly with regard to new regulatory changes.

While maintaining the unique approach of providing sections on immunology, immunodiagnostics, therapy and immunotoxicology in a single volume, we have also introduced a new double-column format to provide easier access to the text. Important statements are highlighted and instead of giving annotations in the margins, key terms are now indicated in the text and presented in a glossary* at the end of the book. A new appendix summarises important characteristics of commercially available therapeutic agents.

We are very grateful to many of the contributors to the first edition, who have kindly revised and modified their chapters, as well as to the additional authors who have added totally new information. The preparation of this second edition has been the result of close collaboration with Dr. Hans-Detlef Klüber and his colleagues at Birkhäuser Verlag. Thanks for all your help, advice and hard work. We hope you, the reader, will find the new edition useful and informative.

Frans P. Nijkamp
Michael J. Parnham
March, 2005

* Words included in the glossary are highlighted in the text with CAPITAL LETTERS.

Abbreviations

A23187: calcium ionophore
AA: arachidonic acid
AAA: abdominal aortic aneurysm
Ab: antibodies, antibody
ABA: antibacterial agents
ABC transporter: ATP-binding cassette transporter
aca: anticomplementary activity
ACE: angiotensin converting enzyme
ACTH: corticotropin, adrenocorticotrophic hormone
ADCC: antibody-dependent cellular cytotoxicity
ADP: adenosine diphosphate
AFC: antibody-forming cell assay
Ag: antigen
AGP: acid glycoprotein
AgR: antigen receptor
AHA: autoimmune haemolytic anemia
AICD: activation-induced cell death
AICAR: 5-aminoimidazole-4-carboxamide ribonucleotide
AID: activation-induced cytidine deaminase
AIDS: acquired immune deficiency syndrome
ALG: anti-lymphocyte globulin
ALL: acute lymphocytic leukemia
ALS: amyotrophic lateral sclerosis
ALXR: affinity G-protein coupled lipoxin receptors
AML: acute myelogenous leukemia
AML: acute myeloid leukaemia
ANC: absolute neutrophil count
ANLL: acute non-lymphocytic leukemia
ANP: atrial natriuretic peptide
AP-1: activator protein-1
AP50: complement activity that causes 50% haemolysis via the alternative route
APC: allophycocyanin
APC: antigen-presenting cell
βARK: β-adrenergic receptor kinase
ATF-2: activating transcription factor
ATG: anti-thymocyte globulin
ATL: aspirin-triggered lipoxins
ATP: adenosine triphosphate
AVP: arginine vasopressin
AZA: azathioprine

BAGE: B antigen
BAL: bronchoalveolar lavage
BALT: bronchus-associated lymphoid tissue
BCG: bacille Calmette-Guérin
B-CLL: B-chronic lymphocytic leukemia
BCR: B cell receptor
BCR-ABL: breakpoint cluster region-Abelson
BDP: beclomethasone dipropionate
BFU: burst-forming unit
BLNK: B cell linker protein
BM: bone marrow
BMT: bone marrow transplantation
B-NHL: B-non-Hodgkin's lymphoma
BPI: bactericidal permeability increasing protein
BRM: biological response modifier
BT: Buehler test
Btk: Bruton's tyrosine kinase

C/E: cellular/extracellular concentration ratio
C: constant (gene segment)
C3b: large fragment of complement factor 3
C5a: small fragment of complement factor 5
Ca^{2+}: calcium ion
Ca^{2+}_i: intracellular calcium
CAD: coronary artery disease
CAM: cell adhesion molecule
CaM: cytokine-activated monocyte
CCD: charge-coupled device

CCL1: CC chemokine-1
CD: cluster of differentiation
CD40L: CD40 ligand
cDNA: complementary deoxyribonucleic acid
CDR: complementarity determining region
CEA: carcinoembryonic antigen
CF: cystic fibrosis
CFA: complete Freund's adjuvant
CFTR: cystic fibrosis transmembrane conductance regulator
CFU: colony-forming unit
CGD: chronic granulomatous disease
cGMP: cyclic guanosine monophosphate
CGRP: calcitonin gene-related peptide
CH50: complement activity that causes 50% haemolysis via the classical route
CHC: chronic hepatitis C
C_HN: constant domain number N of the heavy chain (N = 1, 2 or 3 for IgA, IgD and IgG and 1, 2, 3 or 4 for IgE and IgM)
Cl^-: chloride ion
CL: class
CML: chronic myelogenous leukemia
CMP: cytidine monophosphate
CNS: central nervous system
COMT: catechol-o-methyl transferase
Con A: concavalin A
COPD: chronic obstructive pulmonary disease
COX: cyclo-oxygenase
CpG DNA: synthetic oligodeoxyribonucleotides containing CpG-dinucleotides
CpG: cytosine-phosphate-guanosine
CPMP: E.U. Committee on Proprietary Medicinal Products
CR: complement receptor
CRAC: Ca^{2+} release-activated Ca^{2+} current
CRH: corticotropin-releasing hormone
CRMO: chronic recurrent multifocal osteomyelitis
CRP: C-reactive protein
CsA: cyclosporin A
CSAID: cytokine suppressive antiinflammatory drugs
CSF: colony-stimulating factor
CT: chemotaxis
CTL: cytotoxic T lymphocyte
CX3CL1: fractalkine
CXC or CXCL: CXC chemokine
CXCR: CXC chemokine receptor
CY: cyclophosphamide
Cy5/Cy3: cyanine dye 5 and 3
CYP isoforms: cytochrome P450 isoforms
CYP: cyclophilin
cysLT: cysteinyl leukotrienes

D: diversity (gene segment)
DAF: decay-accelerating protein
DAG: diacylglycerol
DAO: diamine oxidase
DC: dendritic cell
2DE: two-dimensional gel electrophoresis
DHODH: dihydroorotate dehydrogenase
DISC: death-inducing signaling complexes
DLI: donor leukocyte infusion
DLN: draining lymph nodes
DMARD: disease modifying anti-rheumatic drugs
DN T cells: double negative T cells
DNA: deoxyribonucleic acid
DNAM-1: DNAX accessory molecule 1
dNTP: desoxynucleoside triphosphates
DP: PGD receptor
DPB: diffuse panbronchiolitis
DRESS: Drug Rash with Eosinophilia and Systemic Symptoms
DSF: disease-free survival
DTH: delayed-type hypersensitivity
DTP: diphtheria-tetanus-polio

EAE: encephalomyelitis
EAE: experimental auto-immune encephalomyelitis
EBV: Epstein-Barr virus
EDTA: ethylene diamine tetraacetic acid
EGCG: epigallocatechin-3-gallate
EGFR: epidermal growth factor receptor
eGPx: extracellular glutathione peroxidase
ELISPOT: enzyme-linked immunospot assay
ELK-1: member of ETS oncogene family
EMEA: European Agency for the Evaluation of Medicines
ENA-78: epithelial neutrophil-activating protein-78
ENC: endotoxin neutralizing capacity
eNOS: endothelial nitric oxide synthase
EP: PGE receptor
Ep-CAM: epithelial cell adhesion molecule

EPO: erythropoietin
E-selectin: endothelial selectin
ESI: electrospray ionization
EST: expressed sequence tags

Fab: antigen-binding fragment of immunoglobulins
Fabc: monovalent antigen-binding fragment of immunoglobulins comprising one Fab and Fc part
FasL: Fas ligand
Fc receptor: receptor for the constant binding fragment of immunoglobulins
FcγRI: Fc gamma receptor I
FcεRI: receptor I for IgE
FcγRIIb: receptor IIb for immunoglobulin G
Fc: constant fragment of immunoglobulin
FcRn: neonatal receptor for IgG
FDA: U.S. Food and Drug Administration
FEV_1: forced expiratory volume in 1 second
FGF: fibroblast growth factor
FKBP: FK506-binding proteins
FLAP: five-lipoxygenase-activating-protein
fMLP: formyl-methionyl-leucyl-phenylalanine
FP: PGF receptor
FSH: follicle-stimulating hormone
5-FU: 5-fluorouracil

G-CSF: granulocyte-colony stimulating factor
GAGE: G antigen
GALT: gut-associated lymphoid tissue
G-CSF: granulocyte colony-stimulating factor
GH: growth hormone
GHRH: growth hormone releasing hormone
GI: gastro-intestinal
GlyCAM-1: glycosylation-dependent cell adhesion molecule 1
GM-CSF: granulocyte-monocyte colony-stimulating factor
GMP: Good Manufacturing Practice
GMP: guanosine monophosphate
gp: glycoprotein
GPCR: G protein-coupled receptor
GPI: glycosylphosphatidyl inositol
GPMT: guinea pig maximization test
GR: corticosteroid receptor
GRE: glucocorticoid response element
GSH: glutathione
GSH-Px: glutathione peroxidase
GTP: guanosine triphosphate
GVHD: graft-*versus*-host disease
GVT: graft-*versus*-tumor

H&N: squamous cell carcinoma occurring in the head and neck region
H: histamine
H_2O_2: hydrogen peroxide
HAART: highly active antiretroviral therapy
HAE: hereditary angio-oedema
HAMA: human anti-mouse antibody
HBD: human-β defensin
HBsAg: hepatitis B surface antigen
HBV: hepatitis B virus
HCMV: human cytomegalovirus
HCV: hepatitis C virus
HDAC: histone deacetylase
HDL: high-density lipoprotein
HDT: high-dose chemotherapy
Her-2/neu: human epidermal receptor-2/neurological
HER-2: human epidermal growth factor R 2
HETE: hydroxyeicosatetraenoic acid
HFA: hydrofluoroalkane
HGPRT: hypoxanthine-guanine phosphoribosyltransferase
Hib: *Haemophilus influenzae* type b
HIV: human immunodeficiency virus
HLA: human leukocyte antigen
HMT: histamine *N*-methyltransferase
HMW: high molecular weight
HOCl: hypochlorous acid
HOP: HSP70/HSP90 organizing protein
HPA: hypothalamo-pituitary adrenal axis
HPETE: hydroperoxyeicosatetraenoic acids
HPT: hypothalamic pituitary thyroid
HPV: human papilloma virus
HRF: homologous restriction factor
HSA: human serum albumin
HSC: hematopoietic stem cells
HSP: heat shock protein
HSR: heat shock response
5-HT: serotonin
HTL: helper T lymphocyte
i. d.: intradermal(ly)
i. m.: intramuscular(ly)

i. v.: intravenous(ly)
IAR: immediate asthma response
IC: intracellular
iC3b: inactivated large fragment of complement factor 3
$IC_{50\%}$: inhibitory concentration 50% (concentration which inhibits 50% of activity)
ICAM: intercellular adhesion molecule
ICAM-1: intercellular adhesion molecule-1
ICAT: isotope-coded affinity tagging
ICE: interleukin-1α converting enzyme
ICH: International Conference on Harmonization of Technical Requirements for Registration of Pharmaceuticals for Human Use
IEF: isoelectric focusing
IFN: interferon
IFN-γ: interferon gamma
Ig: immunoglobulin
IgA: immunoglobulin A
IgD: immunoglobulin D
IgE: immunoglobulin E
IGF-1: insulin-like growth factor-1
IgG: immunoglobulin G
IgG Ab: immunoglobulin G antibodies
IgM: immunoglobulin M
IgX: immunoglobulin X (X = A, D, E, G or M)
IL: interleukin
IL-1: interleukin 1
IL-2: interleukin 2
IL-6: interleukin 6
IL-10: interleukin 10
IL-12: interleukin 12
IL-1R: interleukin-1 receptor
IMPDH: inosine monophosphate dehydrogenase
INH: isoniazide
iNOS: inducible nitric oxide synthase
INR: immediate nasal response
IP: inducible protein
IP: PGI_2 receptor
IP_3: inositol 1,4,5-triphosphate
IPC: interferon-producing cell
IPG: immobilized pH gradient
IPV: inactivated polio vaccine
IRAK: interleukin-1 receptor-associated kinase
IRM: immune response modifier
IRS: insulin receptor substrates
IS: immunological synapse
ITAM: immunoreceptor tyrosine-based activation motif
ITIM: immunoreceptor tyrosine-based inhibition motif
ITP: idiopathic thrombocytopenic purpurea
IU: international unit
IVIG: intravenous immunoglobulin
I-κB: inhibitor-κB

J: joining (gene segment)
JAK/STAT: Janus activated kinase/signal transducer activator transcription
JAK: Janus-activated kinase
JC virus: a human polyoma virus
JNK: c-jun N-terminal kinase

K_{Ca}: calcium-activated potassium channel
kDa: kilo Dalton
KIR: killer cell immunoglobulin-like receptor

L: ligand
LABA: long-acting inhaled β_2-agonists
LAD: leukocyte adhesion deficiency
LAF: lymphocyte-activation factor
LAK: lymphokine-activated killer
LAR: late asthma response
LBP: LPS-binding protein
LDL: low-density lipoproteins
LFA-1: leukocyte function-associated antigen-1
LH: luteinizing hormone
LHRH: luteinizing hormone releasing hormone
LLNA: local lymph node assay
LMP: latent membrane protein
LMW: low molecular weight
L-NAME: L-NG-nitro-arginine methyl ester
L-NMMA: L-NG-monomethyl arginine
LNR: late nasal response
5-LO: 5'-lipoxygenase
LOXs: lipoxygenases
LPS: lipopolysaccharide
LRR: leucine-rich repeat
LT: leukotriene
LTB_4: leukotriene B_4

mAb monoclonal antibody(ies)
Mac-1: macrophage adhesion protein-1 (CR3)

MAdCAM-1: mucosal addressin cell adhesion molecule 1
MAGE: melanoma antigen
Mal: MyD88-adapter-like
MALDI: matrix-assisted laser desorption/ionisation
MALT: mucosa-associated lymphoid tissue
MAO: monoamine oxidase
MAPK: mitogen-activated protein kinase
MART: melanoma antigen recognized by T cells
MASP: MBL-associated serine protease
MBL: mannan-binding lectin
MCAT: mass-coded abundance tagging
MCP: membrane cofactor protein
MCP-1: monocyte chemotactic protein 1
M-CSF: macrophage colony-stimulating factor
MDI: metered-dose inhaler
MDP: muramyl dipeptides
MDR: multiple drug resistance
MG: myasthenia gravis
mHAg: minor histocompatibility antigen
MHC: major histocompatibility complex
MHLW: Ministry of Health, Labour and Welfare in Japan
MIC: minimal inhibitory concentration
mIg: membrane-bound immunoglobulin
MIMP: methyl inosine monophosphate
MKK: MAPK kinase kinase
MLV: multilamellar vesicles
MMP: matrix metalloproteinase
MMR: measles-mumps-rubella
MOA: mechanism of action
MODS: multiple organ dysfunction syndrome
MOX: monooxygenase
MPA: mycophenolic acid
MΦ: macrophages
MPL: monophosphoryl lipid A
MPO: myeloperoxidase
MPTP: 1-methyl-4-phenyl 1,2,3,6-tetrahydropyridine
Mr: molecular mass
mRNA: messenger ribonucleic acid
MS/MS: tandem mass spectrometry
MS: mass spectrometry
MS: multiple sclerosis
MSH: melanocyte-stimulating hormone
α-MSH: α-melanocyte stimulating hormone
MTD: maximum tolerated dose
mTOR: mammalian target of rapamycin
MTP-PE: muramyl tripeptide phosphatidylethanolamide
MTX: methotrexate
MudPIT: multidimensional protein identification technology

NADPH: nicotinamide adenine dinucleotide phosphate, reduced form
NDV: Newcastle disease virus
NF-κB: nuclear factor kappa-B
NFAT: nuclear factor of activated T cells
NGF: nerve growth factor
NHL: non-Hodgkin's lymphoma
NIEHS: U.S. National Institute of Environmental Health Sciences
NIOSH: U.S. National Institute for Occupational Safety and Health
NK: natural killer (cells)
NKA: neurokinin A
NKB: neurokinin B
NKR: natural killer receptors
nNOS: neuronal nitric oxide synthase
NO: nitric oxide
NOD: nucleotide-binding oligomerisation domain protein
NOXs: non-phagocytic oxidases
NSAID: non-steroidal anti-inflammatory drugs
(N)SCLC: (non) small-cell lung cancer
NTP: U.S. National Toxicology Program
NY-ESO: New York-esophagus

O_2^-: superoxide
ODN: oligodeoxynucleotide
OECD: Organization for Economic Co-operation and Development
OID: optimal immunomodulatory dose
$ONOO^-$: peroxynitrite
OPV: oral polio vaccine
OS: overall survival

PAF: platelet-activating factor
PALE: post-antibiotic leukocyte enhancement
PALS: periarteriolar lymphoid sheath of spleen
PAMP: pathogen-associated molecular pattern
PBL: peripheral blood leukocyte
PBMCs: peripheral blood mononuclear cells

PBP: penicillin-binding protein
PBS: phosphate buffered saline
PCR: polymerase chain reaction
pDC: plasmacytoid dendritic cell
PDE: phosphodiesterases
PDGF: platelet-derived growth factor
PE: phycoerythrin
PECAM-1: platelet-endothelial-cell adhesion molecule 1
PEG: polyethylene glycol
PerCP: peridinin chlorophyll protein
PFN: perforin
PFS: progression-free survival
PGE_2: prostaglandin E_2
PGI_2: prostacyclin
PGN: peptidoglycan
P-gP: P-glycoprotein
PGs: prostaglandins
PHA: phytohaemagglutinin
pI: isoelectric point
PI3 kinase: phosphatidylinositol-3-kinase
PIP2: phosphatidyl-4,5-inositol bisphosphate
PIP3: phosphatidylinositol (3,4,5) trisphosphate
PKB: protein kinase B
PKC-α: protein kinase C-α
PKC: protein kinase C
PLA_2: phospholipase A_2
PLC: phospholipase C
PLD: phospholipase D
PLNA: popliteal lymph node assay
PMA: phorbol myristate acetate
PMBC: peripheral mononuclear blood cells
PMF: peptide mass fingerprinting
PML-RARα: promyelocytic leukemia-retinoic acid receptor α
PMN: polymorphonuclear leukocyte
PNH: paroxysmal nocturnal haemoglobinurea
POMC: proopiomelanocortin
PPH: phosphatidate phosphohydrolase
PRL: prolactin
PRR: pattern recognition receptor
PSA: prostate-specific antigen
PSCT: peripheral stem cell transplant
PSGL-1: P-selectin glycoprotein ligand-1
PSMA: prostate-specific membrane antigen
PTCA: percutaneous transluminal coronary angioplasty
PTK: protein tyrosine kinase
PTM: post-translational modification
Px: peroxidase

QSAR: quantitative structure-activity relationship
QTOF: quadrupole time-of-flight

RA: reporter antigen
RA: rheumatoid arthritis
RAG: recombinant activation gene protein
RANTES: regulated on activation, normal T cell expressed and secreted
RA-PLNA: reporter antigen popliteal lymph node assay
RBL: rat basophil leukaemia
RDA: recommended daily allowance
r.f.: radio frequency
RFS: recurrence-free survival
RIVM: Rijks Institute voor Volksgezondheit en Milieu (Dutch National Institute for Public Health and the Environment)
RNA: ribonucleic acid
RNCl: secondary N-chloramine
ROS: reactive oxygen species
rRNA: ribosomal RNA
RSV: respiratory syncytial virus
RT: reverse transcription

s.c.: subcutaneous
S: Svedberg coefficient or chemical symbol for sulphur
SAP: serum amyloid protein
SAPHO: synovitis acne pustulosis hyperostosis osteitis
SAR: structure-activity relationship
SRBC: sheep red blood cells
SARS: severe acute respiratory syndrome
SCF: stem cell factor
SCID: severe combined immunodeficient (mouse)
(N)SCLC: (non) small-cell lung cancer
SCT stem cell transplantation
SD: standard deviation or solvent-detergent
SDS-PAGE: sodium dodecyl sulphate polyacryl amid gel electrophoresis
Sel P: selenoprotein P

SEREX: serological identification of antigens by recombinant expression cloning
SG: serglycin
SH2: src-homology 2
SHIP: SH2-containing inositol phosphatase
SHP: SH2-containing protein tyrosine phosphatase
sIgX: surface immunoglobulin of class X (for X see IgX)
SIRP: signal regulatory protein
SIRS: systemic inflammatory response syndrome
SJS: Stevens-Johnson syndrome
SLE: systemic lupus erythematosus
SLP: S-layer protein
SMAC: supramolecular activation complexes
SMX: sulfamethoxazole
SOD: superoxide dismutase
SOM: somatostatin
SP: substance P
SRBC: sheep red blood cells
SRS-A: slow-reacting substance of anaphylaxis
STAT: signal transducers and activators of transcription
STZ: streptozotocin

T-α1: thymosin-α1
TAA: tumor-associated antigen
TACE: TNF-α converting enzyme
TAP: transporter associated with antigen processing
TCC: terminal complement complex
TCD: T cell-depleted
TCGF: T cell growth factor
TCR: T cell receptor (for antigen)
TDI: toluene diisocyanate
TdT: terminal deoxynucleotidyl transferase
TEN: toxic epidermal necrolysis
TGF-β: transforming growth factor-β
TGF: transforming growth factor
Th: T helper cell
TIL: tumor-infiltrating lymphocyte
TIR: Toll/IL-1 receptor
TIRAP: TIR adapter protein
TLR: Toll-like receptor
TLRL: Toll-like receptor ligand
TMP: trimethoprim
TNF: tumor necrosis factor
TNF-α: tumour necrosis factor-α
TNFR: TNF receptor
TNFSFL: TNF-super family ligand
TOF: time-of-flight
Tollip: Toll/IL-1R-interacting protein
TP: TXA_2 receptor
TPEN: (N,N,N,N'-teterakis (2-pyridilmethyl) ethylenediamine
T-PLL: T-prolymphocytic leukemia
TR: thioredoxin reductase
TRAF: TNF receptor associated factor
TRAIL: TNF-related apoptosis-inducing ligand
TREC: T-cell receptor excision circle
Treg: regulatory T cell
TRH: thyrotropin-releasing hormone
TRIF: TIR domain-containing adapter inducing IFN-β
TSH: thyrotropin
TX: thromboxane

UMP: uridine monophosphate

V/Q: ventilation-perfusion
V: variable (gene segment)
VCAM: vascular cell adhesion molecule
VE-cadherin Vascular endothelium cadherin
VEGF: vascular endothelial cell growth factor
VIP: vasoactive intestinal polypeptide
VLA: very late antigen

WBC: white blood cells
WHO: World Health Organization

XLA: X-linked agammaglobulinemia

ZAP70: zeta chain-associated protein 70

Mechanisms of immunity

Hematopoiesis, including lymphocyte development and maturation

Valerie F.J. Quesniaux, Julian D. Down, Henk-Jan Schuurman

Introduction: acquisition of specificity

The immune system has some unique properties which distinguish it from other homeostatic systems in the body. First, it shows an extreme specificity and discrimination in recognition of 'self' and 'nonself' ANTIGENS of potentially pathogenic microorganisms or substances in the external environment. Second, it carries 'memory', resulting in a more rapid and avid response to ANTIGENS upon secondary contact. Third, it responds to antigen in multiple ways, resulting in a variety of potential mechanisms to inactivate or destruct its targets. Related to this, there is a finely-tuned regulatory network of cells and their products maintaining homeostasis in the system or dampening reactions when imbalances arise. LYMPHOCYTES have a central function; for instance, specificity is a characteristic of these cells. This chapter describes their development and maturation, up until the stage at which the cells are able to exert their function in host defence. Development and maturation of lymphoid cells is part of the production of blood cells called HEMATOPOIESIS. HEMATOPOIESIS includes the formation of ERYTHROCYTES (red blood cells), THROMBOCYTES (platelets), and LEUKOCYTES (WBC, white blood cells) in blood; white blood cells can be differentiated into POLYMORPHONUCLEAR neutrophilic GRANULOCYTES, eosinophilic GRANULOCYTES and basophilic GRANULOCYTES, MONOCYTES, and LYMPHOCYTES. In organs, including LYMPHOID ORGANS, the products of hematopoiesis are DENDRITIC CELLS, MACROPHAGES (histiocytes) and mast cells. Although the focus of this chapter is on the lymphoid lineage, some aspects of hematopoiesis will be described as well. Within the lymphoid lineage, the focus is on the acquisition of specificity, because this is the major event during development and maturation. Memory is a typical characteristic of the competent immune system, and is only touched upon in describing affinity maturation of matured B cells. Most of the knowledge described in this chapter is based on studies in mice; although there are many differences between rodents and humans in the complex structure and function of the immune system [1], T and B LYMPHOCYTE maturation in other species relevant for the discipline of immunopharmacology (rat, man) follows similar principles.

Hematopoiesis

A diagram of HEMATOPOIESIS [2] is presented in Figure 1. It starts with the pluripotent hematopoietic STEM CELL. While blood cells in the embryo may first be derived from hematopoietic cells residing in the yolk sac; current evidence strongly suggests that the AORTA-GONAD-MESONEPHROS region is the source of the definitive adult hematopoietic system, which subsequently colonizes the liver and then the BONE MARROW [3]. In adults, hematopoietic STEM CELLS occur almost uniquely in the BONE MARROW; only a very small subset of hematopoietic cells in the peripheral circulation has STEM CELL potential. In rodents, extramedullary hematopoietic activity can also be observed in the red pulp of the spleen. Hematopoietic STEM CELLS are self-renewing and produce a continuous supply of mature cells for the whole life-span. The progeny of the pluripotent hematopoietic STEM CELL comprises cells committed to different lineages in the hematopoietic system; a first differentiation involves the production of myeloid and lymphoid committed progenitor cells. The myeloid progenitor subsequently gives rise to precursors of erythrocytes, thrombocytes, GRANULOCYTES, and MONOCYTES. These precursor cells can be measured *in vitro* in so-called

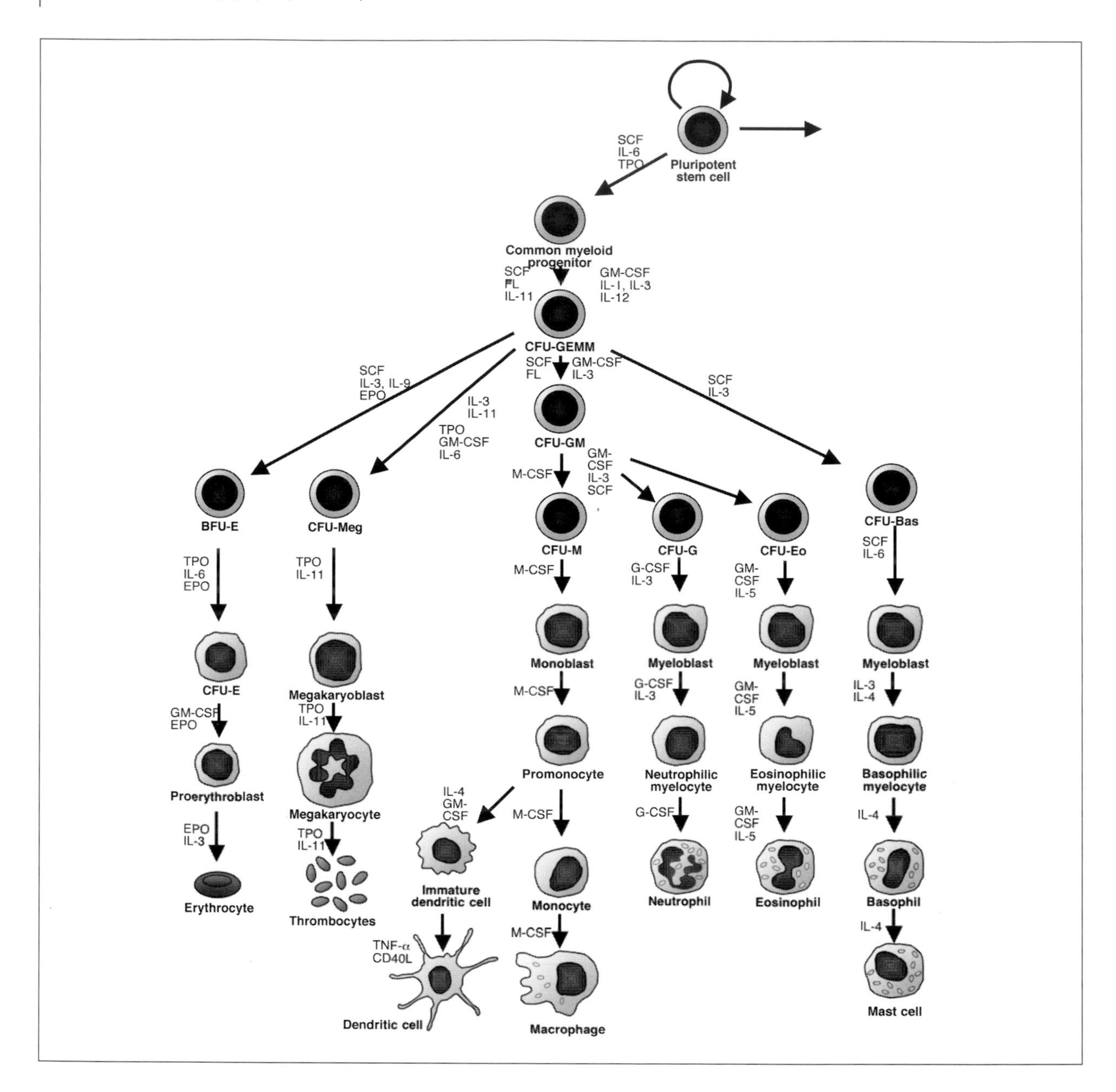

FIGURE 1 **(continued on next page)**

Schematic representation of hematopoiesis, starting with a pluripotent stem cell that develops in a common myeloid progenitor cell or a common lymphoid progenitor cell, and subsequently results in the generation of all blood cell populations: erythrocytes or red blood cells, thrombocytes or platelets, and white blood cells that include macrophages, neutrophilic, eosinophilic and basophilic granulocytes, and mature T or B lymphocytes. The nomenclature of cells in intermittent phases is shown, as well as the cytokines promoting various steps in the development of hematopoietic cells in distinct lineages.

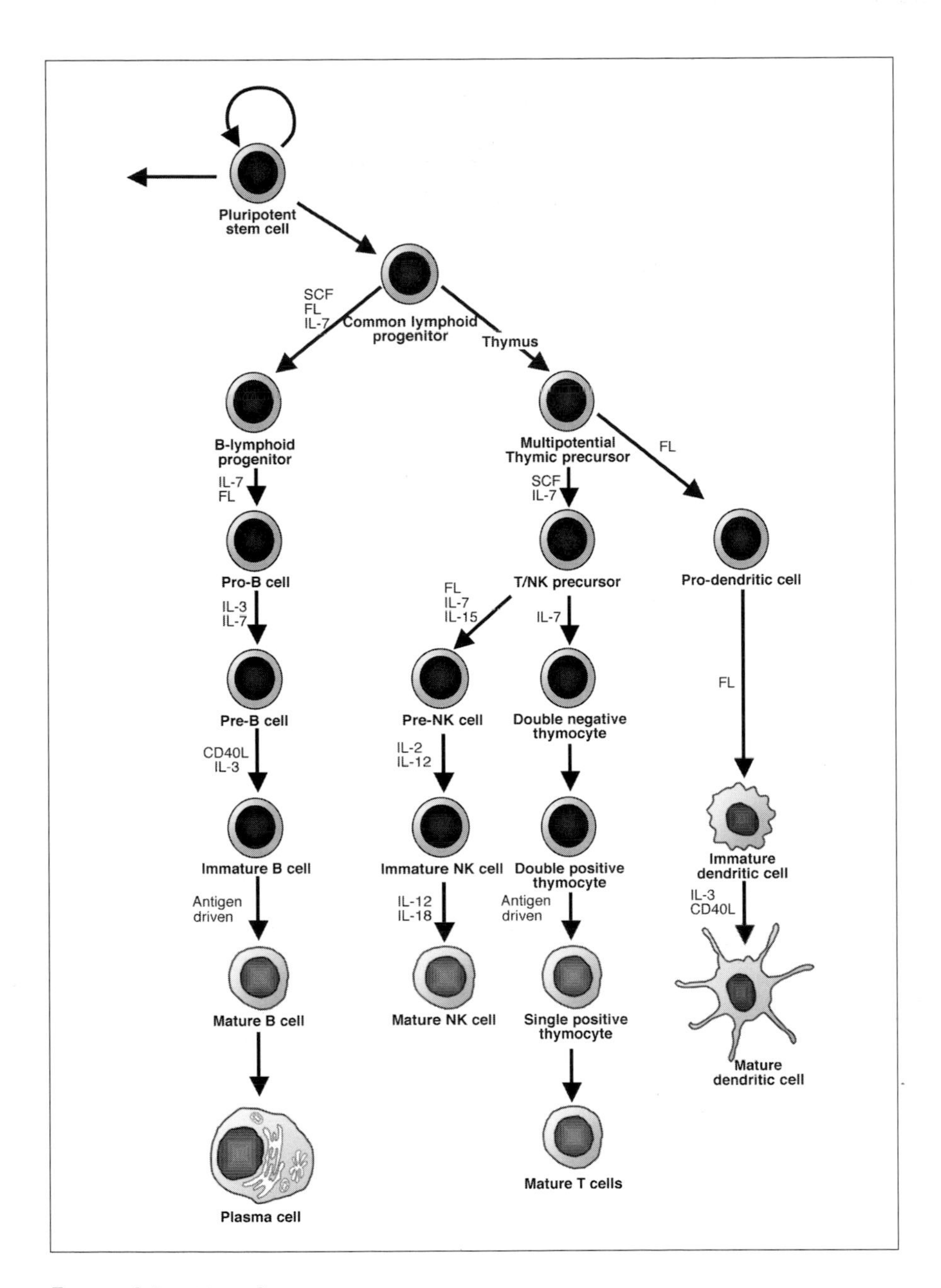

FIGURE 1 (continued)

Abbreviations: BFU-E, burst-forming unit – erythroid; CFU-Bas, colony-stimulating unit – basophilic granulocyte; CFU-E, colony-stimulating unit – erythroid; CFU-Eo, colony-stimulating unit – eosinophilic granulocyte; CFU-G, colony-stimulating unit – granulocyte; CFU-GEMM, colony-stimulating unit – granulocyte-erythroid-macrophage-megakaryocyte; CFU-GM, colony-stimulating unit – granulocyte-macrophage; CFU-M, colony-stimulating unit – monocyte/macrophage; CFU-Meg, colony-stimulating unit – megakaryocyte; EPO, erythropoietin; FL, Flt3 ligand; G-CSF, granulocyte colony-stimulating factor; GM-CSF, granulocyte-macrophage colony-stimulating factor; IL, interleukin; M-CSF, macrophage colony-stimulating factor; NK, natural killer; SCF, stem cell factor; TNF, tumor necrosis factor; TPO, thrombopoietin.

colony-forming assays (colony-forming unit, CFU), in which precursors are cultured in soft agar together with appropriate CYTOKINES (see Fig. 1) and form colonies that can be observed under a microscope; burst-forming units (BFU-E) are precursors for erythrocytes, CFU-Meg are precursors for thrombocytes, CFU-M are precursors of MONOCYTES/MACROPHAGES, CFU-G, CFU-Eo, and CFU-Bas are precursors for neutrophilic, eosinophilic and basophilic GRANULOCYTES, respectively. The existence of a common lymphoid progenitor cell is still controversial, but there is evidence for the presence of progenitor cells that are committed to the T or B cell lineage. A cell surface marker that has been widely used to identify (and enrich) hematopoietic stem and progenitor cells is the CD34 molecule, a molecule with a mucin-like structure whose function is still largely unknown. Other well-recognized antigenic markers of hematopoietic STEM CELLS and progenitors include CD117 (c-kit), CD133 (previously AC133) and, in the case of the mouse, Sca-1.

Besides LYMPHOCYTES, HEMATOPOIESIS results in the production of cells that have relevant functions in the immune system, either in the induction or effector phase of an immune response. Regarding the induction phase, DENDRITIC CELLS deserve attention [4]. Some DENDRITIC CELLS, or precursors thereof, migrate to the thymus where they serve as ANTIGEN-PRESENTING CELLS (APC) and participate in thymic education. Others enter the circulation where they are distributed among tissues and organs to exert their APC function: in skin (epidermis), these cells are the so-called Langerhans cells, and in the draining lymph they are the so-called veiled MACROPHAGES. Subclasses of DENDRITIC CELLS have now been identified and are actively being investigated, but their functions are not yet fully understood. The interstitial tissue in organs harbours so-called 'passenger leukocytes' that, by virtue of their strong expression of MAJOR HISTOCOMPATIBILITY COMPLEX (MHC) class II (see also Chapter A.2), are supposed to have an antigen-presentation function as well. It is noteworthy that in secondary FOLLICLES of lymphoid tissue, a site for B LYMPHOCYTE activation and differentiation, so-called FOLLICULAR DENDRITIC CELLS form part of the stroma with APC function; these cells are not a product of HEMATOPOIESIS but are of mesenchymal origin.

Regarding the effector phase, products of the myeloid lineage like GRANULOCYTES act in concert with ANTIBODIES, for example in PHAGOCYTOSIS of pathogens (see also Chapter A.5). MACROPHAGES in tissue, being the counterpart of MONOCYTES in blood, have a similar function; in the liver, an organ with a huge phagocytic activity, these cells are the so-called Kuppfer cells in sinusoidal spaces. The tissue equivalent of the basophilic granulocyte is the mast cell, that together with IgE-class ANTIBODIES can cause allergic reactions (see also Chapter A.8).

One characteristic typical of both myeloid and lymphoid committed progenitor populations is that many are actively proliferating and are thus susceptible to radiation and many chemotherapeutic agents, including those that are immunosuppressive (e.g., the alkylating drug cyclophosphamide). The loss of myeloid as well as lymphoid progenitors therefore accounts for acute hematological toxicity (leukocytopenia, thrombocytopenia or anemia) as a common adverse side-effect of therapies designed to target the immune system. The rarer and more primitive pluripotent STEM CELLS, however, usually reside in a quiescent non-cycling (G0) state that renders them resistant to most cytotoxic drugs.

T and B lymphocytes: common origin from hematopoietic stem cells

The development of LYMPHOCYTES starts with the common lymphoid progenitor derived from the pluripotent STEM CELL (Fig. 1) [5]. T LYMPHOCYTES gained their name because they develop in the microenvironment of the thymus, an organ located in the mediastinum anterior to the heart; the molecular basis of antigen recognition by these cells is the T CELL RECEPTOR (TCR) complex described in Chapter A.2. B LYMPHOCYTES gained their name because they develop in avian species in the microenvironment of the bursa of Fabricius, an organ located in the intestine near the cloaca, just underneath the epithelium; this organ does not exist in mammalian species, and in these species the BONE MARROW functions as the bursa-equivalent. The molecular basis of antigen recognition of B LYMPHOCYTES is the IMMUNOGLOBULIN

(Ig) molecule described in Chapter A.3. The major B/T types of LYMPHOCYTES are divided into subsets with different functions, e.g., T-helper and T-cytotoxic LYMPHOCYTES, which are further described in Chapters A.2 and A.3.

The differentiation of the common lymphoid progenitor cell into committed progenitors of T or B cell lineage (Fig. 1) is poorly understood. The process seems to involve tight interactions between STEM CELLS and the BONE MARROW microenvironment (stromal cells), and to be under the control of CYTOKINES. VLA-4 is an example of a cell adhesion molecule of the INTEGRIN series for which a pivotal role in differentiation of STEM CELLS into lineage-specific progenitors (lymphoid, myeloid) has been demonstrated. Another example is hyaluronic acid on stromal cells which binds the CD44 molecule on STEM CELL and progenitor populations. Major CYTOKINES promoting the differentiation of pluripotent STEM CELLS into STEM CELLS of the T/B LYMPHOCYTE lineage are STEM CELL factor (SCF, the c-kit ligand) and INTERLEUKIN 7 (IL-7). The adhesion of lymphoid progenitor cells to the stroma probably promotes the binding of SCF from the stromal cells to the c-kit receptor on the progenitor cell. In B-lymphocyte maturation, late pro-B cells (decribed in more detail below, see Figure 3) would then respond to IL-7 for their growth and maturation. Finally, lineage-specific transcription factors promote the development of various cell lineages from hematopoietic STEM CELLS [6, 7]. One of these is the Ikaros gene, which encodes a family of zinc-finger DNA-binding proteins that is crucial for early development into the T or B lymphoid STEM CELL [8].

Specificity: rearrangement of genes encoding antigen receptors

A major event in the differentiation of lineage-specific STEM CELLS into mature T or B LYMPHOCYTES is the synthesis and surface expression of antigen receptors. This involves the generation of all potential specificities required for a functionally active immune system, namely, for T cells, the α and β chain of the TCR, and for B LYMPHOCYTES, the light and heavy chains of the Ig molecule. The germline genome comprises a complex set of gene segments, called V (variable), D (diversity), J (joining) and C (constant) gene segments. A particular combination of one V, D, J and C gene segment has to be generated before transcription and translation into a chain of the antigen-receptor is possible. This is achieved by gene rearrangement: one combination is prepared by excision of intervening genes. As soon as a translocation on one of the two chromosomes encoding a receptor chain is obtained that is in-frame (productive rearrangement), the potential gene rearrangement of the other chromosome is blocked, so that the cell only encodes the product of one single V-D-J-C gene segment leading to a given receptor specificity. Gene rearrangement is an error-prone process, however, and can produce non-productive translocations. Thus, most B LYMPHOCYTES have rearranged the D-J gene segments of the heavy chain of the Ig molecule on both chromosomes, which is the first step in B-lymphopoiesis.

The process of gene rearrangement is unique to T or B LYMPHOCYTES, i.e., the molecular basis of the definition of a B cell is the presence of rearranged genes encoding Ig chains, and that of a T cell is the presence of rearranged genes encoding the TCR α and β chain. Classical T cells use the α and β chain in the TCR complex. A small subset of T cells use two other chains instead, the γ and δ chain. This subset is the so-called γ/δ T cell population. As it expresses a smaller REPERTOIRE of antigen-recognition specificities, it is considered a more 'primitive' (less well-developed) T cell population than α/β T LYMPHOCYTES. The γ/δ T cell population is particularly present during embryonic life. In adults, it apparently has a special function, as a 'sentinel' in initial defence at secretory surfaces [9].

The process of gene rearrangement is essentially random, ocurring under the influence of recombination activation proteins, RAG-1 and RAG-2 [10]. These molecules are upregulated in cells during the rearrangement phase, bind to recombination signal sequences in the DNA and so promote cutting (double-stranded DNA breaks), hairpin formation and splicing out of intervening gene segments. Specific DNA-dependent protein kinases are involved in opening of the coding sequences. In addition to rearrangement of gene segments, diversity is further increased by N-region addition of non-germline encoded nucleotides to V-(D)-J junctions, which is a random process under the influence of the enzyme terminal deoxynucleotidyl transferase (TdT). This nuclear enzyme is present in LYMPHOCYTES during the immature phase of development. It seems plausible that CYTOKINES in the environment, such as IL-7, play a role in initiating gene rearrangement, by upregulation of the DNA-binding proteins and kinases involved. Gene rearrangement occurs in a sequential manner for the different receptor chains, and different stages of development can be distinguished within the rearrangement phase. This is described below separately for T and B LYMPHOCYTES.

Selection of the antigen-recognition repertoire

Essentially, the specificity of the immune system is shaped after receptor gene rearrangement in the contributing T and B cells; the progeny of each distinct cell with a given specificity is considered to be a distinct CLONE, together constituting the whole REPERTOIRE of antigen-recognition specificities. The only way to change specificity within a given CLONE is by SOMATIC MUTATION, a process which occurs later in B cell maturation. However, not all potential rearrangements prove to be useful: the required REPERTOIRE must be selected out of the total potential repertoire. Unwanted (potentially deleterious, e.g., self-reactive) specificities must be deleted and those needed for proper functioning are allowed to expand.

This selection occurs according to various processes that are essentially similar for B and T LYMPHOCYTES, although the requirements for the selection of specificities apparently differ. When the lymphocyte expressing the relevant receptor on its surface is exposed to antigen, and the cell exhibits an unwanted specificity, e.g., an autoreactive lymphocyte with anti-'self' receptor specificity, the cell is activated to initiate cell death by APOPTOSIS (programmed cell death). Should the cell exhibit a desired specificity, it is activated to expand. The first process is called negative selection, the second one positive selection. Apart from environmental factors, such as the nature of the antigen and the APC, intrinsic characteristics of the lymphocyte in a distinct maturation stage determine the outcome of activation, either cell death or expansion. For example, immature T cells undergoing negative selection exhibit an intrinsic purine/pyrimidine metabolism, such that cell activation results in cell death by accumulation of toxic nucleotide triphosphatases. Immature cells undergoing selection also express receptors (Fas/Apo-1) which, upon activation, induce APOPTOSIS in the cell. The hallmark of APOPTOSIS is the endonuclease-induced fragmentation of chromosomes into about 200-basepair fragments.

T lymphocyte maturation in the thymus

The thymus plays a pivotal role in shaping the T cell repertoire, as exemplified by the absence of a functionally active T LYMPHOCYTE system in individuals with a congenital absence of the thymus (humans with the Di-George syndrome, rats and mice with the *rnu* or *nu* mutation). Committed lymphoid progenitors require the thymic microenvironment for differentiation, gene rearrangement and the subsequent selection processes (Fig. 2) [11–14]. The thymus has a unique microenvironment of reticular epithelium which is not observed in other LYMPHOID ORGANS. Other cell populations contributing to T cell development are DENDRITIC CELLS and cells of the monocyte-macrophage cell lineage (Fig. 1) which play a main role in antigen presentation and are mainly found in the medulla of the thymic lobes.

FIGURE 2

T lymphocyte development in the thymus. Various stages in development can be followed by the expression of cell surface markers and T cell receptor α and β chains. During this process recombinant activating gene (RAG) proteins are expressed which mediate gene rearrangement resulting in the capacity to produce the T cell receptor (TCR) complex, first those genes encoding the β chain (Vβ-Dβ-Jβ) and subsequently those encoding the α chain (Vα-Jα). The process

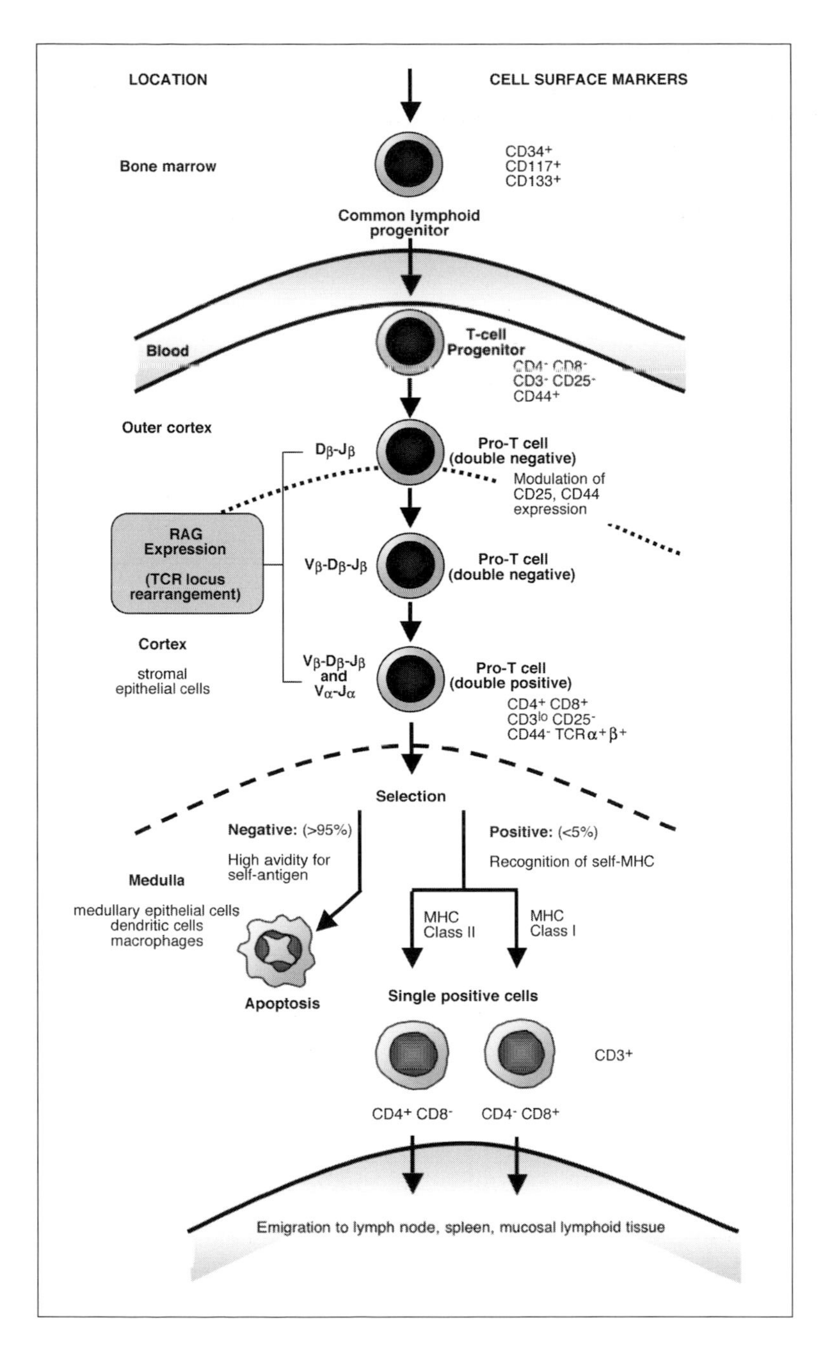

starts after entry of the lymphoid progenitor cell from the bone marrow. 'Double-negative' ($CD4^-CD8^-$) cells in the outer cortex differentiate into 'double-positive' ($CD4^+CD8^+$) cells; during this process there is low expression of the T cell receptor-associated CD3 protein ($CD3^{lo}$). Cells with a high binding (high avidity) for self-antigens including self-MHC undergo negative selection and die by apoptosis. After positive selection (recognition of self-MHC) a small percentage of these cells differentiate into mature 'single-positive' cells expressing either CD4 or CD8. These cells emigrate from the thymus to peripheral lymphoid organs.

Lymphocytes

Four main T LYMPHOCYTE subsets, corresponding to various stages of development, are differentiated, based on their expression of CD3, CD4, CD8 and CD25 antigens. As outlined in other chapters, CD4 and CD8 ANTIGENS are closely linked to the TCR complex and are involved in antigen recognition in association with MHC class I or class II molecules on APCs. CD4 expression in mature T LYMPHOCYTES is linked to helper cell activity and recognition of antigen by MHC class II molecules, and CD8 expression is linked to cytotoxic cell activity and recognition of antigen by MHC class I molecules (see Chapter A.2). The expression of CD25 (IL-2 receptor α chain) on mature T LYMPHOCYTES is associated with a state of activation; together with antigen- or mitogen-induced IL-2 receptor β and γ chains, it confers responsiveness to the T cell growth factor IL-2.

In the thymus, lymphoid progenitors which form a small subset in the outer cortex show a low expression of CD4 and are negative for CD8 and CD25. These cells develop into so-called 'double-negative' cells, being $CD4^-CD8^-$ and $CD3^-CD25^-$. Upon expression of CD25, and presumably under the influence of CYTOKINES (STEM CELL factor, IL-7), cells start to rearrange the genes encoding the TCR complex, first those encoding the β chain (Vβ-Dβ-Jβ) and subsequently the α chain (Vα-Jα). If this rearrangement is successfully completed (e.g., a productive rearrangement from either one of the two chromosomes), the cells develop into the so-called 'double-positive' $CD4^+CD8^+$ stage, which is the major lymphocyte subset in the cortex. At this stage, selection of relevant populations (CLONES) occurs, with first positive and subsequently negative selection. Upon completion of this process, cells develop into so-called 'single-positive' cells, either $CD4^+CD8^-$ or $CD4^-CD8^+$ cells, which form the major populations of LYMPHOCYTES in the medullary areas of thymic lobules. The transmembrane tyrosine phosphatase CD45 isoforms have been implicated in the transition from $CD4^-CD8^-$ to $CD4^+CD8^+$ thymocytes and the further transition into single-positive cells, since these transitions do not occur in mice in which the CD45 gene has been 'knocked-out' [13]. These mice show accumulation of $CD25^+CD44^-$ double-negative cells, which in the absence of CD44 expression are unable to attach to the stroma via the adhesion between CD44 and hyaluronic acid mentioned above.

$CD4^+CD8^-$ or $CD4^-CD8^+$ cells are the main populations emigrating from the thymus into the periphery (e.g., lymph nodes, spleen, mucosal lymphoid tissue). While still considered immature upon migration from the thymus, the cells become fully mature and immunocompetent shortly after arrival in the peripheral immune system. This has been demonstrated in rats using the RT.6 marker, which is a marker for immature cells present on these so-called 'recent thymic emigrants' [15].

Selection

Selection is crucial in intrathymic T cell development [12]. It requires recognition of antigen and MHC molecules by the receptor complex on the developing lymphocyte. Initially, this is an 'immature' type of receptor comprising only the β chain with an α-like chain and the CD3 molecule. After rearrangement of the α chain, the receptor complex is similar to that found on mature LYMPHOCYTES. Positive selection involves the recognition of MHC antigens; in the case of MHC class I linked to CD8, the double-positive cell expands into a $CD4^-CD8^+$ cell population (cytotoxic subset), and in the case of MHC class II linked to CD4, the double-positive cell expands into a $CD4^+CD8^-$ population (helper subset). Epithelial cells represent the major stromal cell population involved in positive selection, which most probably occurs in the thymic cortex. MHC molecules on this cell population are not 'empty' but contain peptides of thymic origin required not only for stabilization of MHC molecules but also for recognition by the TCR. The exact nature of this self-peptide recognition in positive selection remains to be determined.

If not positively selected, the cells die in the thymus cortex, presumably by apoptosis. Negative selection involves the high binding capacity (high avidity) recognition of self antigens, e.g., self-peptides presented by MHC molecules. Should recognition occur with sufficient avidity, the cells are deleted by induction of apoptosis. It has not been completely established which cells in the thymus microenvironment mediate negative selection; originally this property was ascribed to a unique population of DENDRITIC CELLS in the medulla, but a number of studies suggest that epithelial cells themselves are also able to induce negative selection. This means that positive and negative selection do not necessarily have to be performed in strict order; for example, cells can be negatively selected without being first positively selected.

The selection process of immature T LYMPHOCYTES has two relevant consequences for the immunocompetent T cell population. First, it is much reduced in its repertoire. The number of V and J gene segments in the human genome for the β chain is more than 60 and 61, respectively; for the β chain the number of V, D, and J gene segments is more than 106, 2, and 13, respectively. Combined with the potential N-region addition of non-germline encoded nucleotides, the total possible REPERTOIRE of the T cell population is estimated at about 10^{12}. The actual T cell REPERTOIRE is about 10^{6}–10^{7}; this large difference from the total REPERTOIRE is either ascribed to a failure to perform all potential rearrangements, or to the high power of intrathymic selection. Arguments for the latter also come from thymocyte kinetics, i.e., only a very small fraction of cells entering the thymus and generated during development actually emigrate as mature LYMPHOCYTES.

Second, in outbred populations, the REPERTOIRE generated is unique for each individual, as it is biased towards recognition of self-MHC ANTIGENS and various class I and II loci manifest considerable polymorphism.

B lymphocyte development in the bone marrow

The main site of B cell development in mammals is the BONE MARROW (Fig. 3). There are claims that in some species the Peyer's Patches along the intestine are not only a major site for mucosal immune responses, but also contribute to B cell development [16]. Different stages in B cell maturation have been identified [17]. First, hematopoietic STEM CELLS or lymphoid progenitors develop into pro-B cells and rearrange gene segments encoding the V, D and J part of the Ig heavy chain. During this phase the cell expresses RAG-1 and RAG-2 proteins and TdT, which all promote rearrangement processes. In the case of a productive rearrangement, the cell develops into a (large-sized) pre-B cell which is negative for RAG proteins and TdT. The first gene segment adjacent to the VDJ recombinant that encodes the constant part of the heavy chain is the μ-chain segment; in this phase of maturation the cell is able to synthesize heavy chains of the IgM molecule, which are present in the cytoplasm as monomeric proteins.

Upon re-expression of RAG and TdT, the cell starts to rearrange the V and J gene segments encoding the light chain of Igs. The light chain in Ig molecules is exceptional among antigen-receptor chains, because two gene complexes encode this chain, either the κ or the λ chain. However, each individual B cell synthesizes only one light chain ISOTYPE: first the κ gene complex rearranges and if this has resulted in a productive rearrangement, subsequent rearrangement of the λ genome is blocked. After rearrangement of the light chain, the cell ends as a so-called (small-sized) immature B cell with surface expression of IgM.

In a subsequent phase, the mature B cell starts to express IgD on the cell membrane, i.e., the same VDJ transcript is combined with either the Cμ or Cδ gene segment. The simultaneous expression of surface IgM and IgD is possible by alternative RNA processing and termination of transcription. B LYMPHOCYTES showing surface expression of both IgM and IgD are mature B cells capable of responding to antigen stimulation. They form the major resting B cell population found in the blood circulation and peripheral LYMPHOID ORGANS.

During B cell development changes occur in cell surface markers. One of the earliest markers of developing B cells is the tyrosine phosphatase CD45R (B220 in mice), which has an as yet unknown function in B cell development. The expression of CD34 (hematopoietic cell marker) is lost at the (large-sized) pre-B cell stage. Pre-B cells (small-sized) in the BONE MARROW express CD25. The coreceptor CD19, a molecule tightly connected to the B cell antigen

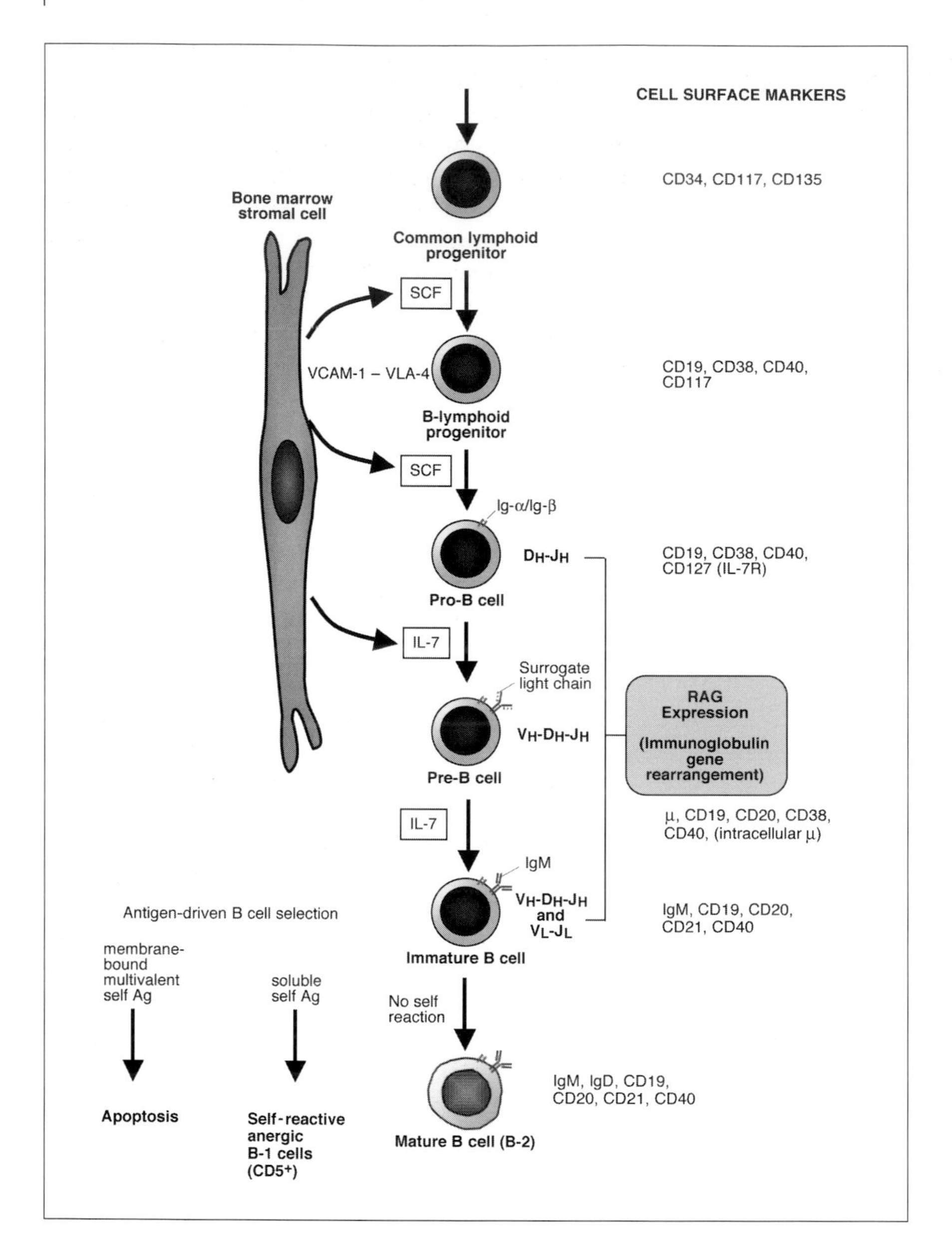

FIGURE 3

Stages of B lymphocyte development. Early B cell development occurs in the bone marrow in close contact with stromal cells through adhesion molecules or interactions between cytokines and their receptors. Prominent in B cell development are stem cell factor (SCF) and IL-7. Various cell surface molecules are expressed during development and can be used as markers. During some stages the cell expresses RAG proteins which mediate gene rearrangement resulting in the capacity to produce the heavy and light chains of the immunoglobulin molecule, first those encoding the heavy chain (V_H-D_H-J_H) and subsequently the light chain (V_L-J_L). MHC class II and CD45R are expressed at all stages. CD45, CD19 and CD21 exhibit signalling functions. CD40 is involved both in B-T cell interaction and signalling; it acts as a receptor for CD40 ligand which is expressed on T cells.

receptor complex with transmembrane signalling function, as well as MHC class II and CD40, are expressed already at the pro-B cell stage. Interactions between CD40 and CD40 ligand are essential for the induction of B LYMPHOCYTE mediated COSTIMULATION in T cell activation.

Selection

Since the generation of the ANTIBODY REPERTOIRE is a random process, it also includes specificities for self components. The restriction and demands for selection between self and non-self are considered less stringent for B LYMPHOCYTES than for the T cell population. A main difference is that B cells and their ANTIBODY products recognize antigen as such (nominal antigen), unlike T cells which recognize antigen after processing to peptides and linking with an MHC molecule. So, there is no apparent need for positive selection.

Self-reactive B cells are allowed to exist in the body, and their ANTIBODY products are assumed to play a regulatory function in immune reactivity. This has been best demonstrated for a particular subset of murine B cells, the so-called B-1 (formerly 'Ly-1' or 'CD5+') B cell population, which differs from conventional (B-2) B LYMPHOCYTES and comprises two populations, the B-1a and B-1b cells [18]. Conventional B (B-2) cells form the bulk of circulating B LYMPHOCYTES and are replenished throughout life from progenitor cells. B-1a cells arise early in embryogenesis, and originate from a fetal liver bipotential precursor cell which has the potential to generate either B cells or MACROPHAGES; these cells constitute a few percent of the total B cells and maintain their numbers by self-replenishment. B-1b cells share many properties with B-1a cells, but in adults can also develop from BONE MARROW progenitors. B-1 cells are most frequent in early embryonic and postnatal life. They preferentially use certain heavy and light chain variable genes, encoding ANTIBODIES reactive with multiple ANTIGENS (polyreactive ANTIBODIES). Their specificities include those with low affinity for self antigens; based on these specificities, a role of these cells in immunoregulation has been proposed. B-1 cells constitutively express the IL-12 receptor β1 subunit and bind IL-12, whereas B-2 cells express IL-12 receptor only after appropriate stimulation.

Selection in the 'classical' B cell population, similar to that of immature T cells, has only been documented in recent years. Immature B cells in the BONE MARROW can be eliminated when they encounter multivalent membrane-bound antigens. This elimination is associated with down-regulation of membrane-bound IgM, an arrest of further maturation and apoptosis. It appears that cells can escape this negative selection by changing their specificity, e.g., by hypermutation or rearrangement of light chain genes [17, 19, 20].

A second process of B LYMPHOCYTE selection occurs at the stage when the cells are mature, in secondary FOLLICLES of lymphoid tissue (so-called GERMINAL CENTRES) [21]. The microenvironment at this location is provided by FOLLICULAR DENDRITIC CELLS, which, unlike the DENDRITIC CELLS mentioned above, do not develop from hematopoietic progenitor cells but originate from mesenchymal fibroblasts. FOLLICULAR DENDRITIC CELLS trap antigen on their surface, presumably in the form of immune complexes, and present it to differentiating mature B cells (so-called centroblasts and centrocytes). At this stage of differentiation, B cells are particularly prone to somatic hypermutation of genes encoding the variable parts of the Ig molecules. Under the influence of T cells and of T cell factors such as IL-2, rearrangements also take place in the gene segments encoding the constant part of the Ig heavy chain, resulting in a loss of IgM or IgD synthesizing capacity and initiation of the synthesis of IgG, IgA or IgE.

These two processes of the T cell-dependent B cell ANTIBODY response, namely Ig class-switch and affinity maturation by SOMATIC MUTATION, are unique to germinal centers and precede the final differentiation of B cells into antibody-producing plasma cells. In binding to the antigen on FOLLICULAR DENDRITIC CELLS, some competition emerges between B cell populations, in that populations with a high affinity for exogenous ANTIGENS survive and those with low affinity (including self-reactive B cells) are deemed to die by apoptosis. In this way, negative selection of self-reactive B cells particularly applies to those B cell populations that are intended to make ANTIBODIES in a T-cell dependent manner. It does not include the B LYMPHOCYTES involved in T cell-independent IgM-class autoANTIBODY synthesis, as these B cells are not subjected to follicular maturation.

Mature T and B lymphocytes: tolerance

The process of development and maturation of T and B cells includes, to some extent, the discrimination between self and non-self. The elimination of self-reactive cells by negative selection soon after cells have expressed their antigen receptors is described as DELETIONAL TOLERANCE and essentially avoids strong reactions to self components [22]. The T cell REPERTOIRE shaped by intrathymic maturation is considered to be rather stable, so that new specificities normally do not emerge in mature populations. SOMATIC MUTATION in the B cell population essentially allows the generation of autoreactive cells. Because the competitive selection in GERMINAL CENTRES mainly involves ANTIGENS from exogenous sources in locally trapped immune complexes, it is unlikely that new autoreactive populations will emerge.

However, to delete for all potential autoANTIGENS in the body would be essentially impossible, first because all autoANTIGENS have to be presented to developing LYMPHOCYTES, and second because this could form exessively large 'holes' in the repertoire. It is, therefore, assumed that the stringent mechanism of DELETIONAL TOLERANCE only holds for those ANTIGENS that are directly exposed to (T) cells of the immune system; otherwise regulatory mechanisms exist in the mature immune system that prevent damage by autoreactive effector cells or their products. This includes the induction of anergy after inappropriate cell stimulation (e.g., for T cells, the absence of CO-STIMULATORY signals in antigenic stimulation), or the presence of immunoregulatory circuits by which potentially autoreactive cells are suppressed. The IDIOTYPE – anti-IDIOTYPE network comprising ANTIBODIES with a specificity for the variable part of other ANTIBODIES is often quoted as an example of such a network; otherwise T cells with suppressor function and/or regulatory T cells with the $CD4^{+}CD25^{+}$ phenotype play a role in this respect. As these mechanisms of tolerance in the strictest sense do not form part of the development and maturation process of T and B LYMPHOCYTES, they are not discussed further in this chapter.

T and B lymphocyte development and maturation: relevance for immunopharmacology

The discipline of immunopharmacology is mainly focussed on mechanisms of and intervention in the process of a competent immune system. The way in which LYMPHOCYTES develop and mature is secondary in this respect. However, cell biological processes during lymphocyte development are in many ways similar to reactions of mature cells upon antigenic stimulation. This not only applies to the basic response of cell proliferation, but also to the involvement of growth factors and intracellular signalling pathways. For instance, besides IL-7, CYTOKINES with a role during intrathymic T cell development include INTERLEUKINS IL-2, IL-4, IL-6 and IL-10, as well as INTERFERON-γ and TUMOR NECROSIS FACTOR (TNF-α). Also with regard to adhesion molecules, interference with the integrin-mediated adhesion of VLA-4 (to VCAM) not only blocks immune responses (e.g., transplant rejection), but also blocks early events in HEMATOPOIESIS in the BONE MARROW.

Some currently used immunosuppressive drugs (see Chapter C.9 for more details) serve to illustrate this principle. Developing progenitor populations may be particularly sensitive to those immunosuppressive agents that damage DNA or interfere with its synthesis (e.g., cyclophosphamide, mycophenolate mofetil and azathioprine). As the myeloid and lymphoid lineages have a common origin, i.e., the pluripotent STEM CELL, and progenitor cells in each lineage are committed to the process of active proliferation, myelosuppression may therefore arise as a significant adverse side-effect. Myelosuppression is deduced from the reduced cellularity in the BONE MARROW, or the selective reduction in one or more leukocyte subsets at this location, resulting in reduced counts of mature cells in the blood circulation; for ERYTHROCYTES this is reflected in anemia or a reduced hematocrit or hemoglobin concentration, for platelets this is called thrombocytopenia, and for WBC leukocytopenia or lymphocytopenia. In this respect, immunosuppressive drugs with an antiproliferative mode of action share the same complications of acute BONE MARROW depression afforded by many 'cytotoxic' drugs used in the treatment of cancer.

The macrolide rapamycin is an immunosuppressive drug that blocks growth factor-induced cell proliferation [23]. Although rodent studies have indicated that its effect on BONE MARROW hematopoietic activity at pharmacological doses is negligible, it induces almost complete atrophy of the thymus at these dosages and hence blocks T cell differentiation. The immunosuppressives cyclosporine A and FK-506 have a more restricted mechanism of action, i.e., inhibition of calcineurin activity with subsequent blockade of intracellular signal transduction leading to synthesis of CYTOKINES, most notably IL-2. These drugs have a peculiar activity on T cell development as they appear to block APOPTOSIS during negative selection of T cells. It is not known whether this bears any relevance for generation of autoreactivity in man. However, under special conditions, namely total BONE MARROW depletion by irradiation and/or antiproliferative drugs followed by autologous BONE MARROW transplantation, which is a treatment procedure in cancer treatment, a short course of cyclosporine treatment results in autoreactive cells [24].

Finally the presumed role of glucocorticosteroids has to be mentioned, in particular its role in T cell development. Stress-induced increases in steroid levels, as well as steroid treatment in experimental animals, result in an immediate atrophy of the thymic cortex which subsequently impairs T cell development. This effect is ascribed to the broad presence of intracellular glucocorticosteroid receptors, although the exact mechanism of action has not been established. Interestingly, the involvement of glucocorticosteroids in T helper cell development has been demonstrated, favoring the generation of T helper type 2 cells [25].

Thus, commonly used immunosuppressive and other drugs aimed to target aberrant immune responses, e.g., autoimmunity and allergy, can affect HEMATOPOIESIS and therefore cause unwanted side-effects. The following example illustrates this phenomenon. Thymic 're-education' is nowadays under consideration as an alternative therapeutic approach to autoimmune diseases, such as severe rheumatoid arthritis, in which the causative agent is not known. Immunoablation would eradicate the pathogenic T LYMPHOCYTES, followed by reintroduction of tolerance (or restoration of T cell control) by transplantation of hematopoietic STEM CELLS which have to undergo re-education leading to an altered REPERTOIRE in view of the ANTIGENS present. Similar to the situation of BONE MARROW transplantation mentioned above, it is evident that such treatment should not be combined with immunosuppressants affecting intrathymic T cell development.

In conclusion, it is of the utmost importance to understand mechanistic similarities between the developing immune system and the responses of the competent immune system when immunopharmacological events require interpretation. In addition, it is likely that immunopharmacology will add to our understanding of lymphocyte development, through the effects of specific pharmacological intervention in the biological pathways for T and B LYMPHOCYTE development and maturation.

Summary

HEMATOPOIESIS is the production of blood cells, that includes the generation of ERYTHROCYTES, THROMBOCYTES and white blood cells (GRANULOCYTES, LYMPHOCYTES and MONOCYTES). The major site of HEMATOPOIESIS is the BONE MARROW, but it can occur at other (extramedullary) locations as well. HEMATOPOIESIS includes the generation of most cells involved in immune reactions, including LYMPHOCYTES carrying the specificity in these reactions, and accessory cells in the induction phase (DENDRITIC CELLS with antigen-presenting function) and effector phase (GRANULOCYTES and MONOCYTES/MACROPHAGES involved in, e.g., PHAGOCYTOSIS). Regarding LYMPHOCYTES, the pluripotent hematopoietic STEM CELL first develops into a common lymphoid progenitor cell, which subsequently develops in either T (thymus-dependent) or B lymphoid progenitors. Subsequently these cells rearrange the genes encoding antigen receptors (TCR, Ig) and express these receptors on the cell surface. In a next phase the cells go through a process of positive and negative selection that has been well established for the T LYMPHOCYTE lineage and occurs within the thymic microenvironment. This selection process is relevant in shaping the antigen-recognition REPERTOIRE and avoids the generation of auto-reactive cells. Since basic cellular processes are similar for HEMATOPOIESIS, including lymphocyte development and reactions of mature LYMPHOCYTES (i.e., cellular proliferation, response to CYTOKINES), pharmacological agents affecting mature LYMPHOCYTES can also affect the development and maturation of these cells. For example, immunosuppressants with an antiproliferative mode of action can show BONE MARROW depression as an adverse side-effect.

Selected readings

Baba Y, Pelayo R, Kincade PW (2004) Relationships between hematopoietic stem cells and lymphocyte progenitors. *Trends Immunol* 25: 645–649

Edry E, Melamed D (2004) Receptor editing in positive and negative selection of B lymphopoiesis. *J Immunol* 173: 4265–4271

Milicevic NM, Milicevic Z (2004) Thymus cell-cell interactions. *Int Rev Cytol* 235: 1–52

Moore MA (2004) The role of cell migration in the ontogeny of the lymphoid system. *Stem Cells Dev* 13: 1–21

Kang J, Der SD (2004) Cytokine functions in the formative stages of a lymphocyte's life. *Curr Opin Immunol* 16: 180–190

Linton PJ, Dorshkind K (2004) Age-related changes in lymphocyte development and function. *Nat Immunol* 5: 133–139

Barthold B, Matthias P (2004) Transcriptional control of B cell development and function. *Gene* 327: 1–23

Paul WE (ed.) (2003) *Fundamental Immunology*, 5th Edition. Philadephia, PA: Lippincot, Williams and Wilkins

Recommended website

Current Opinion of Immunology has a section on lymphocyte development which is reviewed in a number of publications on an annual basis: (*http://www.current-opinion.com/jimm/about.htm?jcode=jimm*) (Accessed March 2005)

References

1 Mestas J, Hughes CCW (2004) Of mice and not men: differences between mouse and human immunology. *J Immunol* 172: 2731–2738

2 Kondo M, Wagers AJ, Manz MG, Prohaska SS, Scherer DC, Beilhack GF, Shizuru JA, Weissman IL (2003) Biology of hematopoietic stem cells and progenitors: implications for clinical application. *Annu Rev Immunol* 21: 759–806

3 Medvinsky A, Dzierzak E (1996) Definitive hematopoiesis is autonomously initiated by the AGM region. *Cell* 86: 897–906

4 Heath WR, Carbone FR (2001) Cross-presentation, dendritic cells, tolerance and immunity. *Annu Rev Immunol* 19: 47–64

5 Storb U, Kruisbeek A (1996) Lymphocyte development (editorial overview). *Curr Opin Immunol* 8: 155–159

6 Clevers HC, Grosschedl R (1996) Transcriptional control of lymphoid development: lessons from gene targeting. *Immunol Today* 17: 336–343

7 Henderson A, Calame K (1998) Transcriptional regulation during B cell development. *Annu Rev Immunol* 16: 163–200

8 Georgopoulos K Winandy S, Avitahl N (1997) The role of the Ikaros gene in lymphocyte development and homeostasis. *Annu Rev Immunol* 15: 155–176

9 Chien YH, Jores R, Crowley MP (1996) Recognition by gamma/delta T cells. *Annu Rev Immunol* 14: 511–532

10 Fugmann SDF, Lee AI, Shockett PE, Villey IJ, Schatz DG (2000) The RAG proteins and V(D)J recombination: complexes, ends, and transposition. *Annu Rev Immunol* 18: 495–527

11 Anderson G, Moore NC, Owen JJT, Jenkinson EJ (1996) Cellular interactions in thymocyte development. *Annu Rev Immunol* 14: 73–99

12 Starr TK, Jameson SC, Hogquist KA (2003) Positive and negative selection of T cells. *Annu Rev Immunol* 21: 139–176

13 Frearson JA, Alexander DR (1996) Protein tyrosine phosphatases in T-cell development, apoptosis and signalling. *Immunol Today* 17: 385–390

14 Zlotnik A, Moore TA (1995) Cytokine production and requirements during T-cell development. *Curr Opin Immunol* 7: 206–213

15 Hosseinzadeh H, Goldschneider I (1993) Recent thymic emigrants in the rat express a unique antigenic

phenotype and undergo post-thymic maturation in peripheral lymphoid tissues. *J Immunol* 150: 1670–1679
16 Griebel PJ, Hein WR (1996) Expanding the role of Peyer's patches in B-cell ontogeny. *Immunol Today* 17: 30–39
17 Melchers F, Rolink A, Grawunder U, Winkler TH, Karasuyama H, Ghia P, Andersson L (1995) Positive and negative selection events during B lymphopoiesis. *Curr Opin Immunol* 7: 214–227
18 Berland R, Wortis HH (2002) Origins and functions of B-1 cells with notes on the role of CD5. *Annu Rev Immunol* 20: 253–300
19 Cornall RJ, Goodnow CC, Cyster JG (1995) The regulation of self-reactive B cells. *Curr Opin Immunol* 7: 804–811
20 Weill J-C, Reynaud CA (1996) Rearrangement/hypermutation/gene conversion: when, where and why? *Immunol Today* 17: 92–97
21 MacLennan ICM (1994) Germinal centres. *Annu Rev Immunol* 12: 117–139
22 Sykes M (1996) Chimerism and central tolerance. *Curr Opin Immunol* 8: 694–703
23 Abraham RT (1998) Mammalian target of rapamycin: immunosuppressive drugs uncover a novel pathway of cytokine receptor signaling. *Curr Opin Immunol* 10: 330–336
24 Hess AD (1993) Autologous graft vs host disease: mechanisms and potential therapeutic effect. Bone Marrow Transpl 12 (Suppl 3): S65–S69
25 Ashwell JD, Lu FW, Vacchio MS (2000) Glucocorticoids in T cell development and function. *Annu Rev Immunol* 18: 309–345

T cell-mediated immunity

Sergey G. Apasov and Michail V. Sitkovsky

Introduction

T cell-mediated immunity is an adaptive process of developing various antigen (Ag)-specific T LYMPHOCYTES to eliminate viral, bacterial, or parasitic infections or malignant cells. Ag specificity of T LYMPHOCYTES is based on high-affinity recognition of unique antigenic peptides derived from a particular pathogen, in contrast to the broad specificity of INNATE IMMUNITY. T cell-mediated immunity includes a primary response by naïve T cells, effector functions by activated T cells, and persistence of Ag-specific memory T cells protecting the organism against a particular pathogen in future encounters (Fig. 1). T cell-mediated immunity is part of a complex and coordinated immune response that includes other effector cells such as MACROPHAGES, NATURAL KILLER CELLS, MAST CELLS, BASOPHILES, EOSINOPHILS, and NEUTROPHILS.

Biology of the T lymphocyte immune response

Each T LYMPHOCYTE expresses a unique T CELL RECEPTOR (TCR) on the surface as the result of developmental selection upon maturation in the thymus (see Chapter A.1 on haematopoiesis). Mature T LYMPHOCYTES, known as naïve T cells, circulate through blood, the lymphatic system, and reside in secondary LYMPHOID ORGANS (Fig. 1). Naïve T cells are those that have not yet encountered foreign Ag and have not yet been activated. Antigenic peptides are presented to the naïve T LYMPHOCYTE in secondary LYMPHOID ORGANS by DENDRITIC CELLS (DC), which are the most efficient "professional" Ag-presenting cells (APC) since they also provide co-stimulatory signals for effective T cell activation. DC acquire Ag in non-lymphoid tissues throughout the body and migrate into secondary LYMPHOID ORGANS guided by inflammatory stimuli and CYTOKINES. APC generate antigenic peptides from a pathogenic agent by ANTIGEN PROCESSING and display them on the cell surface in the context of MHC molecules. The recombinant variability of individual αβTCR, on other hand, ensures that at least a few naïve T cells will have high-affinity binding to an antigenic peptide derived from virtually any pathogen. TCR engagement triggers a cascade of intracellular signaling events resulting in activation of the naïve T cell.

The activated T cells rapidly proliferate (clonal expansion), migrate through the tissues to the sites of Ag presence, and perform effector functions such as cell-mediated cytotoxicity and production of various CYTOKINES (soluble mediators of the immune response). Cytotoxic $CD8^+$ T cells are very effective in direct lysis of infected or malignant cells bearing the Ag, while $CD4^+$ T helper cells produce CYTOKINES that can be directly toxic to the target cells or can stimulate other T cell effector functions, B cell ANTIBODY production, as well as mobilize powerful inflammatory mechanisms (Fig. 1) (see Chapter A.4 for cytokines review).

Most effector T cells will disappear after the antigenic agent is eliminated, although others will remain and form memory T cells. Unlike naïve T cells that live for few months or effector T cells that disappear at the end of the immune response, memory T cells may survive for years in LYMPHOID ORGANS and peripheral tissues. The easily activated memory T cells can perform immediate effector functions in peripheral tissues or undergo activation and clonal expansion in LYMPHOID ORGANS to mount a secondary immune response if the same Ag appears again.

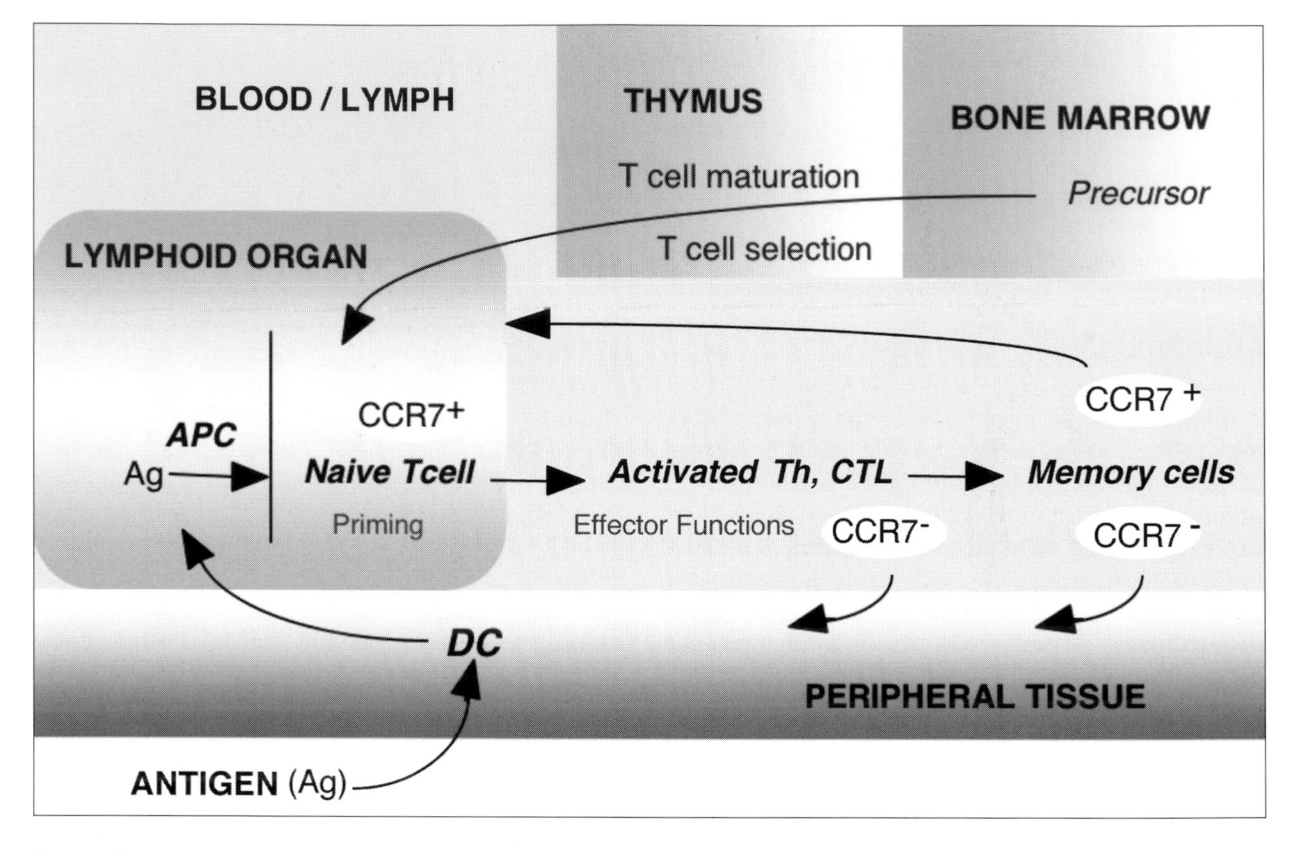

FIGURE 1
Development of T cell-mediated immune response: After thymic selection naïve T lymphocytes become activated by interaction with antigen-presenting cells (APC) in secondary lymphoid organs. Activated T lymphocytes migrate though the body, execute effector functions, and form long-living memory T cells. CCR7 is a chemokine receptor involved in T cell homing into lymphoid tissue.

Memory T cells respond much faster to the Ag than naïve T cells, thus helping to eliminate the source of pathogen infection and effectively preventing the spreading of a disease.

Composition of the T cell network

Lymphoid organs

Primary LYMPHOID ORGANS, – the BONE MARROW, and thymus – are sites of a HEMATOPOIESIS and clonal selection of T cells. The T cell-mediated immune response begins in the secondary LYMPHOID ORGANS: spleen, lymph nodes, and organized lymphoid tissues associated with mucosal surfaces including Peyer's patches, tonsils, bronchial, nasal, and gut–associated lymphoid tissues. The secondary LYMPHOID ORGANS have specialized T cell-rich zones where naïve T LYMPHOCYTES are concentrated; these include the PERIARTERIOLAR LYMPHOID SHEATH of spleen (PALS) and the PARACORTEX of lymph nodes. Naïve T cells reside in the spleen for just a few hours and in the lymph nodes for about one day before they leave via splenic veins or via efferent lymphatic vessels, respectively. Migrating naïve T cells eventually reach the bloodstream and soon after enter new LYMPHOID ORGANS, repeating the cycle until they become activated by antigenic peptides.

T cell subsets

Thymic selection results in the appearance of T cells with two types of TCR. The majority express Ag-binding αβ chains in the TCR, which are disulfide-linked heterodimers of Ig superfamily proteins (Fig. 2) forming unique structures on each T cell. αβTCR T cells have a very diverse REPERTOIRE of Ag recognition and represent mature T cells that circulate through the secondary lymph organs and develop adaptive immune responses. A small fraction of T cells expresses γδ chains in TCR, appears to be much less heterogenic than abTCR T cells, resides in skin and certain mucosal surfaces and may play a role in the initial response to microbial invasion. Although the functions of γδTCR T cells are not fully understood, they are considered as a relatively primitive part of the innate T cell response and will not be reviewed in this chapter.

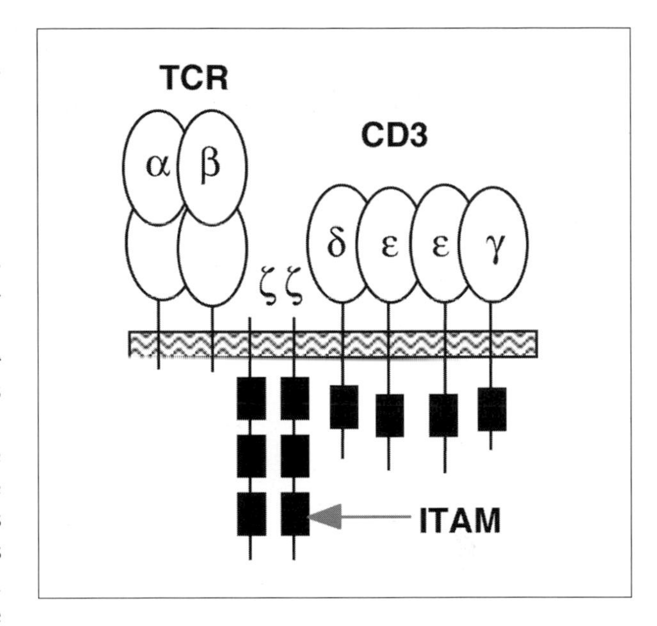

FIGURE 2
T cell receptor complex consists of αβ heterodimers responsible for antigen recognition and CD3 molecules involved in intracellular signaling. Immunoglobulin-like αβ chains are formed upon gene rearrangement and have high variability among individual T cells. Non-polymorphic CD3 chains (ζ, δ, ε, γ) contain intracellular immunoreceptor tyrosine-based activation motifs (ITAMs) initiating cascades of signal transduction.

αβTCR T cells are subdivided into several groups on the basis of lineage markers and functional activities. Two major surface co-receptor molecules, CD4 and CD8, represent two separate T cell lineages with different functions. $CD4^+$ cells recognize Ag in the context of MHC class II molecules and produce CYTOKINES as effector T helper cells. $CD8^+$ T LYMPHOCYTES are activated by Ag-peptide/MHC class I complexes and form effector cytotoxic T LYMPHOCYTES (CTL). On the other hand, the functional status of the T cells allows us to distinguish naïve, effector, and memory cells, as each of these displays extensive diversity in terms of phenotype, function, and anatomic distribution. Naïve T cells are the most homogeneous representatives of $CD4^+$ and $CD8^+$ subsets. Upon activation, however, they can be further distinguished by their cytokine profiles. Thus, activated $CD4^+$ T helper cells can be subdivided into Th1, Th2, and Th0 subsets based on production of signature cytokine IFN-γ (Th1) *versus* IL-4, IL-5 (Th2), or all together (Th0). $CD8^+$ LYMPHOCYTES also could be arranged into Tc1, Tc2, and Tc0 subsets by cytokine profile, although they do not produce as much CYTOKINE as $CD4^+$ T helpers and are not efficient in B cell activation (see Chapter on B cells). Furthermore, these populations can be subdivided into non-lymphoid and lymphoid effectors based on their migration into either secondary lymph organs or peripheral tissues. The patterns of chemokine receptors and adhesion molecules not only determine the migration ability of the effector T cells but also serve as additional surface markers. Memory T cells can be divided on non-lymphoid and lymphoid memory cells by the same principles. Altogether effector and memory LYMPHOCYTES of $CD4^+$ or $CD8^+$ lineage theoretically can be divided into 12 distinct subsets based on the above criteria. In addition, there are subsets of regulatory T cells that make T cell heterogeneity even more complex.

Many cell surface markers can be very useful in the design of the drugs for selective manipulation of the immune response.

T cell subset markers

Naïve T cell markers

Naïve T cells use CD62L ($\alpha 4\beta 7$ integrin), CC chemokine receptor 7 (CCR7), and Leukocyte Function Antigen-1 (LFA-1) for rolling, adhesion, and extravasation through the high endothelial vessels in peripheral lymph nodes and mucosal LYMPHOID ORGANS.

Survival of naïve cells is maintained by low-affinity TCR/self-antigen interaction and signaling as well as by the presence of IL-7. These signals are normally sufficient to maintain homeostasis of naïve T cells for several months.

Effector T cell markers

High-affinity interaction of TCR with foreign Ag-peptide/MHC on mature APC and following activation are reflected in phenotype changes. Activated T cells express CD69 (a very early activation antigen) and CD25 (IL-2Ra). Other important surface receptors of activated T cells are: CD40 ligand, which stimulates APC through CD40, leading to the up-regulation of CD80 (B7-1) and CD86 (B7-2) on APC; CD28, which binds to CD80 and CD86 and propagates co-stimulatory signal enhancing growth factor (IL-2) production and increasing T cell activation. TNF receptor family molecules OX-40, CD27, and 4-1BB, also can be found on primary activated T cells. These receptors were found to sustain T cell proliferation and survival of activated T LYMPHOCYTES upon their binding to the corresponding ligands on the APC. At the peak of proliferation, $CD4^+$ effector cells were also found to change the pattern of adhesion receptors such as CD62L and sPSGL-1 (sialyated form of p-Selectin ligand) and chemokine receptor CXCR5. $CD8^+$ CTL could also be characterized by expression of perforin and granzimes, proteins required for cytolytic functions. The particular set of surface markers may predict homing capacity of the effector T cells; for example, CXCR5 receptor helps $CD4^+$ $CD62L^-$, $sPSGL\text{-}1^-$, $CXCR5^+$ T cells to migrate into B cell-rich FOLLICLES of the lymph nodes and support ANTIBODY production. In contrast, absence of CCR7 and CD62L on CTL allows them to migrate into non-lymphoid tissues such as lung or gut and to clear pathogenic agents in these tissues.

Memory T cell markers

Memory T cells, unlike effector T cells, are not blasts or in the cell cycle; however, they are capable of circulating in lymphoid and non-lymphoid compartments. According to location, memory T cells are divided on central and effector memory cells and express corresponding surface markers. For example, among three phenotypes of $CD8^+$ memory cells that have been identified ($CD45RA^-$, $CCR7^+$; $CD45RA^-$, $CCR7^-$; $CD45RA^+$, $CCR7^-$), the $CCR7^+$ T cells are non-cytotoxic central memory cells, while $CCR7^-$ are effector memory cells. Upon contact with the appropriate Ag, effector memory cells can execute effector functions instantly, whereas central or lymphoid memory cells can rapidly proliferate, expanding and acquiring effector functions. $CD4^+$ memory T cells also appear heterogeneic. At least two subsets of $CD45RA^-$ $CD4^+$ memory cells have been identified in humans. The central memory cells express CCR7 and CD62L and reside in LYMPHOID ORGANS, producing IL-2 upon stimulation. Some of these have been found to migrate into certain inflammation sites depending on the expression of chemokine receptors such as CCR4, CCR6, and CXCR3. The other $CCR7^-$ subset with low CD62L expression produces IFN-γ and IL-4 upon stimulation and apparently represents effector memory cells.

Dendritic cell markers

The marker of DC is CD11c integrin. There are three subsets of DC in spleen and lymph nodes based on the expression of myeloid cell marker CD11b, INTEGRIN CD205, and CD8: myeloid DC ($CD11b^+$, $CD205^-$,

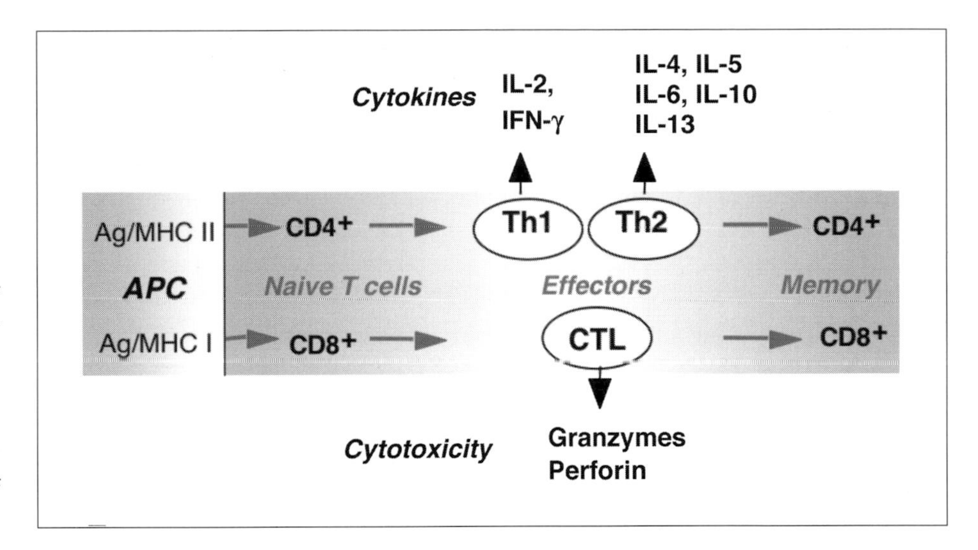

FIGURE 3
The effector phase of immune response includes cytotoxicity by $CD8^+$CTL and cytokine production by $CD4^+$ T helpers subdivided into Th1 and Th2 effectors by the profile of cytokines they produce.

$CD8^-$), lymphoid DC ($CD11b^-$, $CD205^+$, $CD8^+$), and plasmacytoid DC ($CD11b^-$, $CD205^-$, $CD8^-$), the latter of which also express B220 and Gr-1. Two other DC subtypes include "interstitial dendritic cell migrants" ($CD11b^+$, $CD205^+$, $CD8^-$) and Langerhans cells (langerin$^+$) migrating into lymph nodes from interstitial tissue or from skin, respectively.

Effectors of T cell-mediated immunity

$CD4^+$ T helpers

Two major functional T helper subpopulations are distinguished by their cytokine profiles (Fig. 3). Th1 cells produce mainly IFN-γ, but also IL-2, TNF-α, and lymphotoxin. Th1 cells enhance pro-inflammatory cell-mediated immunity and were shown to induce delayed-type (DTH), B cell production of opsonizing ISOTYPES of IgG, and response to some protozoa like *Leishmania* and *Trypanosoma*. Th2 cells secrete IL-4, -5, -6, -10, and -13, promote non-inflammatory immediate immune responses, and were shown to be essential in B cell production of IgG1, IgA, and IgE. Th1 and Th2 development appear to be mutually antagonistic. This gave rise to the model of polarization of immune response in accordance with the nature of the Ag and of surrounding CYTOKINES. For example, IFN-γ and IL-12 are known to support Th1 cells, while IL-4 and IL-10 assist Th2. Although the evidence for the polarized cytokine secretion profiles of Th1 and Th2 is indisputable, several recent studies have shown more complex patterns of CYTOKINE interaction in different models of immune response, including autoimmune models, that are inconsistent with simple dichotomy paradigm.

$CD8^+$ cytotoxic T lymphocytes

The CTL derive from activated naïve $CD8^+$ cells, proliferate in the presence of IL-2, and can expand their number many thousand-fold at the peak of a primary immune response. The dramatic clonal expansion of $CD8^+$ CTL in comparison to $CD4^+$ cells most likely can be attributed to the relatively easy activation by the Ag-MHC class I complex and better survival in circulation. Great expansion and the ability of single $CD8^+$ CTL to destroy more than one target cell while sparing "innocent" bystanders, make CTL very efficient Ag-specific effector cells. Destruction of selected target cells by CTL requires the establishing of cell contact with the target cell and Ag recognition, thus initiating the release of cytolytic granules into the IMMUNOLOGICAL SYNAPSE. CTL, unlike naïve T cells, do not require co-stimulatory signals upon Ag recognition in order to kill; therefore they can destroy a variety of target cells bearing "foreign" Ag.

Mechanisms of T cell-mediated cytotoxicity

Two major pathways of cytotoxicity are described in CTL: Ca^{2+}-dependent perforin/ granzyme-mediated APOPTOSIS, and Ca^{2+}-independent Fas ligand/Fas-mediated APOPTOSIS (Fig. 4). Both pathways are initiated via TCR signaling. Lytic granules (secretory lysosomes containing granzymes, perforin (PFN) and proteoglycan serglycin (SG)) appear to be delivered into target cells as one complex. Granzymes are effector molecules capable of inducing APOPTOSIS in target cells via caspase- dependent and -independent mechanisms. Granzymes enter into the target cell directly via plasma membrane pores formed by PFN or via receptor-mediated endocytosis. In the latter case, PFN provides the translocation of granzymes from endocytic vesicles into the cytosol. Proteoglycan SG presumably serves as a chaperone of PFN until the complex reaches the plasma membrane of the target cells. Lytic granules represent a very efficient natural drug delivery system.

Fas-mediated APOPTOSIS is initiated by binding of Fas molecules on the target cell with Fas ligand on the CTL. The Fas molecule is a member of the TNF receptor superfamily with an intracellular "death" domain initiating caspase-dependent APOPTOSIS upon binding to Fas ligand. TCR cross-linking was shown to induce up-regulation of Fas ligand expression on the cell surface of CTL and in cytolytic granules. Fas-mediated APOPTOSIS appears to be a general phenomenon not restricted to CTL. It was found to be involved in control of cell proliferation and homeostasis among other T cells.

Regulatory T cells

Regulatory T cells (Treg), also referred to as suppressor cells (Ts), include more than one cell type that can fulfill important functions to sustain homeostasis of the immune system, downmodulate the amplitude of an immune response, and prevent autoimmune diseases. There is enough evidence at present to conclude that regulatory T cells participate in all cell-mediated immune responses, directly affecting Th1, Th2, CTL, and B cell reactions against "self" and "foreign" Ag. Although Ts activity was demonstrated decades ago, the major challenge is to find unique T suppressor markers and to reveal the exact mechanisms of suppression.

Although majority of Treg appears within $CD4^+$ T cells, suppressor activity was also reported among $CD8^+$; ($CD8^+$, $CD28^-$); and ($CD4^-$, $CD8^-$, TCR^+) T cells. In the last few years most attention, however, was focused on $CD4^+$ regulatory cells and particularly the $CD4^+CD25^+$ subset. It was found that transfer of T cells depleted of this population into nude mice (athymic animals with no mature T cells of their own) results in various autoimmune diseases, that could be prevented by injection of purified $CD4^+CD25^+$ cells. In addition, $CD4^+CD25^+$ Treg have been implicated in modulation of transplantation systems of long-surviving cardiac and pancreatic islet allografts, inhibition of LPS-induced proliferation of B cells, and down-regulation of T cell-mediated production of self-reactive ANTIBODIES. The inhibitory effect requires stimulation of the TCR on regulatory cells and direct cell contact of the Treg with the populations they inhibit. Further investigations suggested that $CD4^+CD25^+$ Treg do not proliferate upon stimulation but inhibit the proliferation of other $CD4^+$ and $CD8^+$ T cells as well as the B cells response to the Ag. $CD4^+CD25^+$ regulatory cells have been found in mice, rats, and humans. CD25 is not a unique Ts marker; it is the α chain of the IL-2 receptor and is expressed at moderate levels on activated T cells. Several studies in mice and humans suggest that only $CD4^+$ cells with high levels of CD25 behave as Treg. Additional markers found on Treg include CTL associated Ag (CTLA-4) and members of the TNF receptor superfamily: TNFRSF4 (OX40), TNFRSF9 (4-1BB), and TNFRSF18 (GITR). All those molecules, however, are up-regulated on conventional T cells during activation as well, and therefore do not have an advantage over CD25 as specific Treg markers.

Some recent studies in this field indicate that $CD25^-$ cells with high-avidity TCR may play a regulatory role in the absence of $CD25^+$, while others characterized regulatory subsets Tr1 (IL-10-producing $CD4^+$ cells) and Tr3 ($CD4^+$ cells predominantly secreting IL-4, IL-10, and TGF-β), which were capable of amending autoimmune diseases *in vivo*.

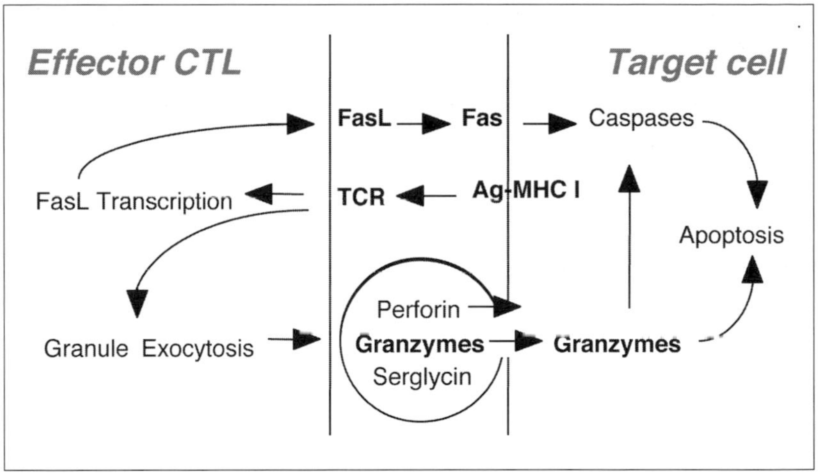

FIGURE 4
Pathways of CTL cytotoxicity involve secretion of cytolytic granules containing perforin and granzymes, and up-regulation of the expression of Fas ligands binding Fas ("death receptor") on the target cells. Both mechanisms are initiated by interaction of CTL with the target cell-bearing antigen, which is recognized by TCR of the CTL.

Mechanisms of T cell activation

Antigen presentation

Antigenic peptides are derived by different molecular mechanisms of Ag processing, from pathogens residing either in the cytosol or in vesicular compartments of the infected cell. MHC class I molecules bind to the antigenic peptides, which are originated in the cytosol of APC by a multimolecular complex of proteases (proteosomes) and transported to the endoplasmic reticulum by TAP-1 and TAP-2 (transporter associated with Ag processing-1 and -2). The newly assembled MHC/ peptide complexes in the endoplasmic reticulum are then translocated through the Golgi to the cell surface. Virtually all cells of the body express MHC class I molecules at different levels and thus could present antigenic peptides to $CD8^+$ CTL and become potential targets of destruction, depending on the Ag.

MHC class II molecules, in contrast, bind peptides deriving from pathogens that appear in intracellular vesicles of the cell or from extracellular proteins internalized by endocytosis. MHC class II molecules are transported from the Golgi to endosomes and lysosomes as a complex bound to the non-polymorphic invariant chain instead of a peptide. Subsequently, the invariant chain is degraded and replaced with peptides generated by vesicular acid proteases at acid pH in the endosomal compartments. MHC II/peptide complexes appear on the surface of only a few types of immune cells, including MACROPHAGES, B cells, and DC.

Another important mechanism is CROSS-PRESENTATION of Ag, a process in which "professional" APC may present an Ag transferred from other cells. Several studies have shown that DC can actually initiate a T cell response against MHC class I-restricted antigenic peptides by a CROSS-PRESENTATION mechanism. CROSS-PRESENTATION may serve as a mechanism for T cell tolerance to self-ANTIGENS in the periphery.

Molecular mechanisms of T lymphocyte activation

Activation of naïve T cells is the most critical step in developing immunity and requires a complex interaction of TCR, co-receptors, and accessory molecules on the surface of the T cell with corresponding ligands on the APC (Fig. 5). TCR-Ag/MHC interaction provides an Ag recognition step and initiates intracellular signaling. Co-receptors such as CD4, CD8, and CTLA-4 modulate the TCR signal, while co-receptors such as CD28 initiate their own intracellular signals that are co-stimulatory to the TCR signal. Accessory molecules such as LFA-1 or CD2 provide adhesion at the cell contact site, strengthening the interaction between the T cell and APC and allowing sustained signal transductions.

The αβ chains of TCR are incovalently associated with invariant chains of the CD3 complex (ζ, δ, ε, and γ) (Fig. 2). Intracellular parts of CD3 chains include one or multiple ITAMs (immunoreceptor tyrosine-based activation motifs). ITAMs provide sites of interaction with protein tyrosine kinases (PTK) that propagate the signaling events.

Src family protein tyrosine kinases *Fyn* and *Lck* phosphorylate ITAMs upon TCR cross-linking by Ag/MHC, and fully phosphorylated ITAMs recruit PTK ZAP-70 to the complex via their SH2 domains. This allows *Lck* to transphosphorylate and to activate ZAP-70. The activated ZAP-70 interacts and phosphorylates Vav, SLP-76, and LAT. Vav/SLP-76 appears to be involved in actine cytoskeleton changes, while LAT is a membrane-associated protein that upon phosphorylation provides binding sites for a number of critical signaling proteins including Grb2, Ras, and PLC-γ. PLC-γ plays a critical role in regulation of Ca^{2+} flux as it cleaves 4,5-bisphosphate (PIP_2) to diacylglycerol (DAG) and inositol 1,4,5-triphosphate (IP_3) upon activation by PI3 kinase. DAG stimulates PKC, while accumulation of IP_3 is the initial trigger for release of intracellular Ca^{2+} that, in turn, triggers the opening of the plasma membrane Ca^{2+} release-activated Ca^{2+} (CRAC) channels. Cascade of the signaling actions eventually results in activation of transcriptional factors including NF-AT, ELK-1, Jun, and ATF-2 and immune gene expression.

Although the first phosphorylation event occurs within a few seconds of TCR cross-linking, the sustained contact and interaction of T cells with APC is required for full T LYMPHOCYTE activation. Recent studies of TCR engagement have focused on IMMUNOLOGICAL SYNAPSE (IS)-dynamic clustering of different surface molecules at the contact point between T cell and APS involving TCR/CD3, co-receptors, and accessory molecules. The latest studies of IS reported a ring-type structure formed by TCR-Ag/MHC complexes around a cluster of LFA-1 and ICAM-1 (INTEGRIN intercellular adhesion molecule-1) followed by inversion of this structure, relocation of TCR/-pMHC in the center, and formation of spatially segregated regions or supramolecular activation complexes (SMAC) (Fig. 6). Mature IS contain central SMAC, a cluster of TCR bound to Ag/MHC, and CD4 or CD8, CD3, CD2, CD2AP, CD28, PKCθ, and PTK *Lck*. sSMAC is surrounded by peripheral SMAC, which contains LFA-1, ICAM-1, and talin. Thus, IS formed on the cell surface may provide prolonged cellular interaction and sustained signaling leading to the Ca^{2+} flux, actin cytoskeleton reorganization, and full-blown T cell activation. It was found that accumulation of cytolytic granules in CTL is directed toward IS and that release of the granules takes place within p-SMAC.

Tolerance

An essential part of T cell-mediated immunity is the development of non-responsiveness toward naturally occurring self-ANTIGENS while mounting effective immune responses against "foreign" Ags. Breakdowns of self-tolerance will result in the development of autoimmune diseases. Self -reactive T cells, both $CD4^+$ and $CD8^+$, were found responsible for initiating and mediating tissue damage in many experimental animal models of organ-specific autoimmunity as well as in human studies.

Immunological tolerance is achieved by different mechanisms on different stages. Initially potential self-reactive T LYMPHOCYTES are deleted during T cell development in the thymus. High-affinity interaction of TCR on immature thymocytes with self-Ag on thymic stroma cells results in apoptosis and elimination of such T cells in the process known as negative selection. T cells with TCR of low to moderate affinity to self-ANTIGENS escape from the thymus and migrate on the periphery. These T cells are normally "ignorant" to self-Ags or develop tolerance after initial activation.

Although the Ag-specific TCR of T cells do not possess an intrinsic mechanism of distinguishing self from non-self peptides, the activation by self-Ag is different than by "foreign" Ag, mainly due to the absence of co-stimulatory signals from non-activated APC. This is in contrast to activated APC that up-regulate co-stimulatory molecules during inflammation, infections, or other pathological conditions.

Partial activation of T cells in the absence of co-stimulatory signals leads, instead of activation, to the

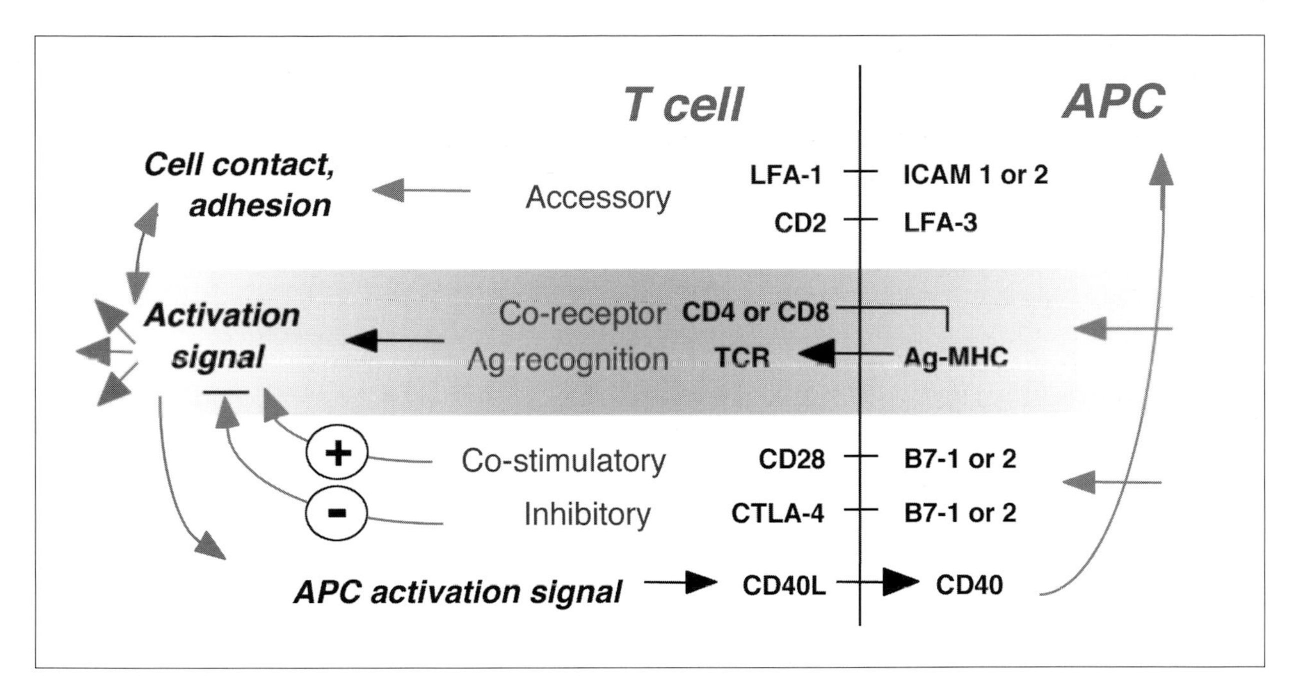

FIGURE 5
Effective T cell activation requires interaction of multiple surface receptors of T lymphocytes with corresponding ligands on the APC. Binding of TCR to the antigen/MHC complex generates an activation signal that is attenuated by co-receptors and adhesion molecules. TCR signals also up-regulate the CD40 ligand which stimulates APC for better expression of antigen-presenting MHC molecules and co-stimulatory receptor ligands improving T cell stimulation and activation.

state of T cell unresponsiveness toward further stimulation, also known as ANERGY.

In most cases co-stimulatory molecules will direct T cell response toward either activation or tolerance. Simple absence of co-stimulatory signals was shown to induce anergy in effector T cells *in vivo* and *in vitro*, while naïve T cells may require a negative signal of CTLA-4 engagement to develop anergy and become tolerant.

Self-reactive cycling T cells may also undergo programmed cell death after re-exposure to the same Ag in the process called activation-induced cell death (AICD). AICD is mediated by death receptors (FAS/FAS-ligands interaction of $CD4^+$ T cells and by TNFRII/TNF interaction of $CD8^+$ T cells) that involve activation of caspase-dependent, death-inducing signaling complexes (DISC).

Peripheral tolerance can be also controlled by immune cytokine divergence and by regulatory T cells. Both natural and adaptive $CD4^+$ regulatory cells have been implicated in the regulation of the autoimmune response. Thymus-derived $CD25^+$ natural regulatory cells suppress other cell activation by largely unknown mechanisms. They require strong co-stimulatory signals for induction and maintenance and express a transcription factor called Foxp3. Adaptive (antigen-induced) regulatory T cells are generated in the periphery by sub-optimal antigenic signals and rely on CYTOKINES such as IL-10 and TGF-β for suppression. These cells of varying phenotype appear often under special conditions such as chronic viral infections. Regulatory T cells present new possibilities for the treatment of autoimmune disorders and for the improvement of transplantations.

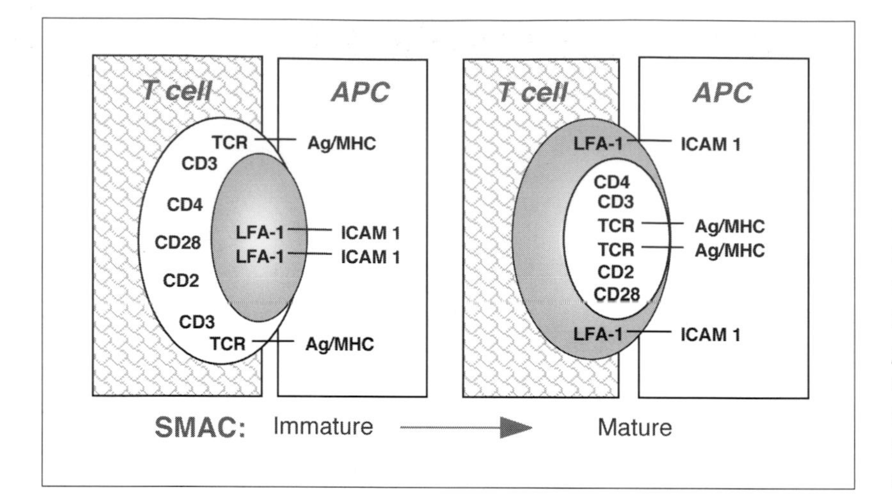

FIGURE 6
Schematic view of receptor clusters of different surface molecules forming supramolecular activation complexes (SMAC) on the membrane of the T cell at the site of interaction with APC. SMAC changes during time of activation inverting central and peripheral compositions.

Summary

T cell-mediated immunity includes priming of naïve T cells, effector functions of activated T helpers and CTL, and long-time persistence of memory T cells. Development of an effective immune response requires proper activation of T LYMPHOCYTES by APC in secondary LYMPHOID ORGANS and migration of the responding T cells to the sites of Ag presence in the body. The efficiency of T cell activation in LYMPHOID ORGANS depends on the concentration of an antigenic peptide and affinity of TCR toward the Ag/MHC complex and is facilitated by inflammatory stimuli, co-stimulatory signals, and CYTOKINES. $CD8^+$ naïve T cells develop into effector CTL after interaction with APC, while $CD4^+$ naïve T cells differentiate into T helper cells of major T helper types: Th1 (producing IL-2, IFN-γ, TNF-α, and LT-α) or Th2 (producing IL-4, IL-5, IL-6, IL-10, and IL-13).

Absence of inflammatory stimuli may induce insufficient activation of DENDRITIC CELLS resulting in induction of anergy and apoptosis among T cells instead of activation and productive response. This may serve as a mechanism of tolerance to self-Ags. Circulation and extravasation of T LYMPHOCYTES are orchestrated by multiple adhesion receptors whose expression and avidity is modulated by CYTOKINES and chemokines. In the process of mediating effector functions, some activated T cells undergo activation-induced cell death (AICD), while other undergo activated T cell autonomous death after the inflammation wanes, thus terminating the immune response. Only a small population of Ag-specific memory cells remains in LYMPHOID ORGANS and through the tissues for a long time after the immune response is over. When exposed to the Ag a second time, memory cells rapidly aquire and mediate effector functions, thereby preventing spread of pathogenic infection.

Selected readings

Paul WE (ed.) (2003) *Fundamental Immunology*, 5th ed., Lippincott Williams and Wilkins, Philadelphia

Janeway CA Jr., Travers P, Walport M, Shlomchik M (2001) *Immunobiology: The Immune System in Health and Disease*, 5th ed, Garland Science, New York

Alt F, Marrack P (eds.) (2003) *Curr Opin Immunol* 15: 82–88, 247–348, 513–565

Paul WE, Fathman CG, Glimcher LH (eds.) (2003) *Ann Rev Immunol* 21: 29–70, 305–334, 659–684, 713–758

Koretzky G, Monroe J (eds.) (2003) *Immunol Rev* 193: 22–38, 58–81; 195: 5–14, 117–135

A3

Antibody diversity and B cell-mediated immunity

Ger T. Rijkers and Lisette van de Corput

Towards the end of the 19th century, Koch and Ehrlich discovered that the serum of immunized animals contained substances (antitoxins) with the ability to neutralize the toxins of diphtheria and tetanus. At Christmas 1891 a group of children received diphtheria antitoxin, which cured them of this otherwise fatal disease. These experiments demonstrated that immunization can induce the formation of humoral substances, which have the ability to protect against infectious diseases. Half a century later, in 1952, Bruton described a patient with (predominantly bacterial) severe and recurrent respiratory tract infections and an agammaglobulinemia. This milestone demonstrated the significant role of immunoglobulins in the defense against infections. Later on, through the pioneering work of Max Cooper and others, it was shown that B LYMPHOCYTES are the cells that produce ANTIBODIES, and that patients such as the one described above (X-linked agammaglobulinemia or XLA) fail to produce ANTIBODIES because they lack B LYMPHOCYTES; B LYMPHOCYTE development in the BONE MARROW stops at the pre-B-cell stage. In 1993 the molecular basis for this disease was found: XLA is caused by structural defects in the gene encoding an enzyme that is now called Bruton's tyrosine kinase (Btk).

Antibodies and immunoglobulins

Host defence against infections with microorganisms depends on the complex interplay between cells and proteins, which together are able to recognize and specifically interact with molecular structures of the microorganism. The proteins involved in this process include TOLL-LIKE RECEPTORS, defensins, collectins such as mannose binding lectin, surfactant proteins, and immunoglobulins. Among these proteins, the immunoglobulins are special because of their extreme diversity in primary structure. It is estimated that up to 10^{12} different IMMUNOGLOBULIN molecules can be formed (see also below). This endows the host with a large spectrum of defence molecules that can bind specifically to virtually any given microorganism. The immunoglobulins in serum comprise approximately 10–20% of total serum proteins; upon electrophoresis they end up in the gamma region, hence the alternative name of gammaglobulins for immunoglobulins.

ANTIBODIES are immunoglobulins and immunoglobulins are antibodies. Still, it can be confusing to mix these terms. The term ANTIBODY should only be used for immunoglobulins with known specificity, such as anti-bloodgroup A antibodies or anti-measles antibodies. Furthermore, the term antibodies is also used when describing the interaction of an IMMUNOGLOBULIN with antigen. The term IMMUNOGLOBULIN is used for the biochemical description of these proteins.

Structure of immunoglobulins

Immunoglobulins are glycoproteins with a basic structure of 4 polypeptide chains, comprising 2 identical heavy chains of 400–500 amino acids and 2 identical light chains of ca. 200 amino acids (Fig. 1). These chains are held together by disulphide bridges and noncovalent protein-protein interactions. On basis of the primary structure of the heavy chains, the immunoglobulins are divided into five classes or ISOTYPES, namely IgG, IgA, IgM, IgD, and IgE. IgG is further subdivided into four subclasses IgG1, IgG2, IgG3, and IgG4, while IgA is subdivided into two subclasses IgA1 and IgA2. There are two types of light chains, kappa (κ) and lambda (λ) chains. Heavy chains and

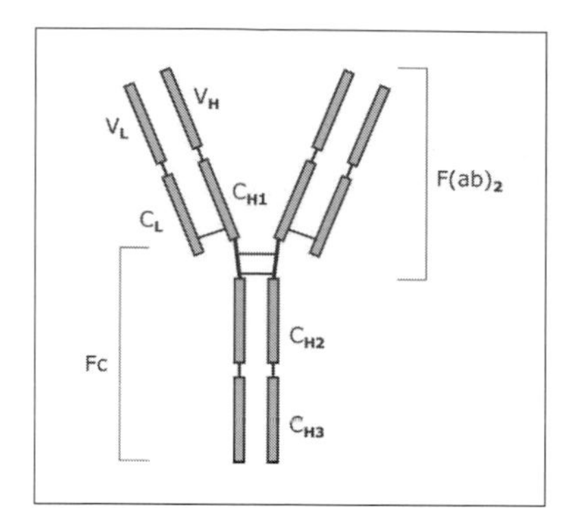

FIGURE 1. SCHEMATIC STRUCTURE OF AN IGG IMMUNOGLOBULIN MOLECULE

The immunoglobulin molecule has 2 identical heavy chains, each composed of 3 constant domains (C_{H1}, C_{H2}, C_{H3}), and a variable domain (V_H). Two identical light chains are connected to the heavy chains by disulphide bonds. Fragments of the molecule that can be obtained after treatment with proteolytic enzymes are indicated with $F(ab)_2$, the antigen-binding fragment, and Fc, the fragment that can be obtained in crystallisable form.

light chains each have a domain structure. Heavy chains consist of four domains, three of them are constant (C_{H1}, C_{H2}, C_{H3}; see Fig. 1) concerning the amino acid structure and one shows sequence variation (V_H). Light chains have a two-domain structure (C_L and V_L). The combination of the variable domains of one heavy and one light chain determines the specific recognition of a microorganism and the binding to it. The constant domains determine the biological activity of the formed antibody-microorganism complex (see below: biological functions of antibodies). In contrast to IgG, the heavy chains of IgM and IgE consist of four constant domains and one variable domain. IgG as shown in Figure 1 is a monomeric structure. Immunoglobulins can also appear as multimeric structures. IgM consists of five monomers (pentamer) and IgA appears in serum predominantly as a monomer but can also appear as a dimer. In secretion fluids and on mucosal surfaces of the respiratory and gastrointestinal tract IgA is present as secretory IgA. Secretory IgA is dimeric IgA coupled to a J chain and a secretory component that is important for the transport of IgA through the epithelial cells to external secretions and the protection of the secreted molecule from proteolytic digestion.

The generation of antibody diversity

The combination of the variable domains of the heavy and the light chain forms the binding site for the antigen [1]. Immunoglobulins are capable of recognizing a wide variety of different ANTIGENS because of their large variation in amino acid sequences within the variable domains. The genes encoding the V_H domain are organized in clusters, each of which encodes parts of the variable domain (gene segments). In precursor B lymphocytes, during the process of gene-rearrangement, three gene segments are joined together, one so-called variable (V) segment, one diversity (D) segment and one joining (J) segment. In humans, there are more than 60 V gene segments, about 30 D and 6 J gene segments. The three segments are joined together at random combination, allowing already more than 10,000 different combinations at this level (Fig. 2). This process requires the activity of two enzymes, the recombinase activating genes RAG-1 and RAG-2. The rearrangements take place in a defined order; a D gene segment joins with a J gene segment and subsequently a V gene segment is joined to the combined DJ sequence. In addition, extra nucleotides can be inserted or deleted by the enzyme terminal-deoxynucleotidyl transferase (TdT) during these gene segment rearrangements, thereby again increasing the variation. Next, the DNA encoding the variable domain and the DNA encoding the constant part are transcribed into mRNA, spliced together and translated into a complete heavy chain. Also during this joining process several nucleotides can be added or deleted. A similar process occurs for the light chain, except that there are only V and J segments for the variable region. The amino acid sequence variability in the V regions is especially pronounced in 3 hypervariable regions. Localized areas of these hypervariable regions of the H and L chains interact to form antigen binding sites (i.e., the COMPLEMENTARITY DETERMINING REGIONS, CDR1, CDR2 and CDR3).

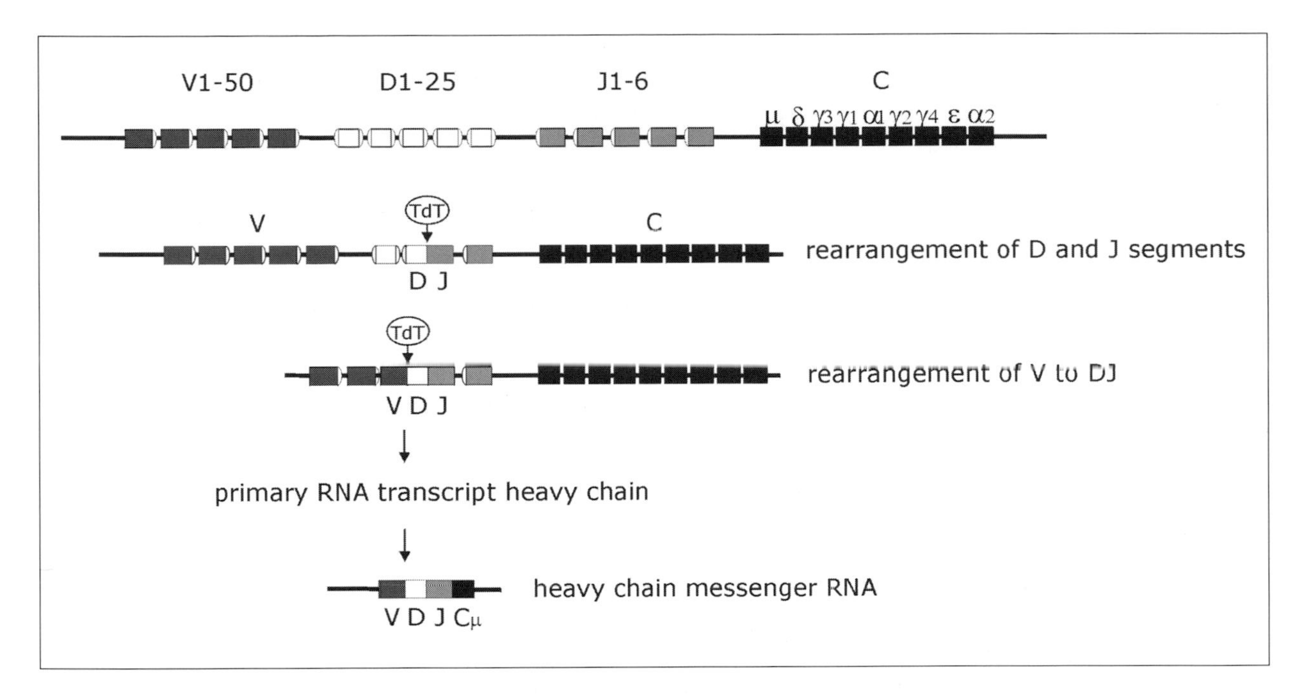

FIGURE 2. REARRANGEMENT OF THE IMMUNOGLOBULIN GENE SEGMENTS
Early during B lymphocyte development, rearrangement of one of the D gene segments to one of the J segments takes place regulated by the recombinase activating genes RAG-1 and RAG-2. Subsequently, rearrangement of one of the V gene segments to the DJ segment occurs. The primary RNA transcript is being processed by splicing the VDJ segment to a constant region gene segment (C).

Early during B LYMPHOCYTE development the rearrangements of the coding segments of the variable regions take place. Daughter cells derived from such a B LYMPHOCYTE form a CLONE of B cells, which all express identical immunoglobulins expressing this particular combination. However, additional diversity can be produced by SOMATIC MUTATIONS and by altering the heavy chain constant region, called class switching. These processes are regulated by the enzyme activation-induced cytidine deaminase (AID). The first transcript to be produced as a B LYMPHOCYTE develops, after VDJ joining, contains the exon for V_H and the exon for C_H of the μ chain, resulting in expression of IgM and production of IgM when stimulated by binding of an antigen. When the B LYMPHOCYTE is further stimulated by an antigen, the class of antibody being produced changes. DNA encoding a different constant domain gene segment is joined to the original V_H exon, while intervening DNA encoding the μ heavy chain gene segment is eliminated. Thus, the B LYMPHOCYTE will produce another IMMUNOGLOBULIN molecule with identical specificity for the antigen. At the same time, but also independent from the class switch, mutations in the variable region may arise, resulting in B lymphocytes with small differences in the affinity of the antibody receptor that is expressed on the membrane. When antigen concentration is low, only the B cells with high affinity for that particular antigen will be activated. In the absence of specific antigen, mature B cells survive in the peripheral circulation for only a few days. Cells which do not encounter antigen within this period of time undergo APOPTOSIS (see Box 1). This is necessary in order to maintain an optimal number of B lymphocytes in the peripheral circulation. However, when immature B cells develop in an

Box 1. Apoptosis

There are two different ways in which cells can die. Cells can die in an uncontrolled manner called necrosis because they are damaged by injurious agents such as toxic agents leading to cell lysis, or they can commit suicide, an active by genes-regulated process. Cells that undergo apoptosis, also called programmed cell death, undergo a characteristic series of changes. The cells shrink, form bubble-like blebs on their surface, and the chromatin (DNA + proteins) in the nucleus are degraded forming a vacuolar nucleus. There are different reasons why cells undergo apoptosis. One is because of proper development, such as formation of fingers and toes of a fetus by apoptosis of the tissue between them. The other reason is for the benefit of the organism. Cells with DNA damage, tumor cells, or cells of the immune system which attack the own body tissues need to be cleared from the organism. Defects in the apoptotic machinery is associated with autoimmune diseases such as rheumatoid arthritis and lupus erythematosus.

environment containing a "self"-antigen, the B cells will also undergo APOPTOSIS and thus be deleted from the repertoire.

B cell receptor and signal transduction

B LYMPHOCYTE activation is initiated by specific recognition of antigen by the antigen receptor, i.e., membrane-bound IMMUNOGLOBULIN (mIg). Resting, primary B lymphocytes express two ISOTYPES of mIg: mIgM and mIgD. Both mIgM and mIgD (as well as other mIg ISOTYPES; see below) are expressed on the cell surface in association with Igα and Igβ molecules; collectively such a complex is called the B-cell receptor (BCR) complex. Igα (CD79a) and Igβ (CD79b) are the protein products of the MB-1 and B29 genes, respectively, and both belong to the Ig superfamily [2]. Igα and Igβ fulfill at least three different functions: they are required for expression of membrane Ig on the B cell surface, they act as transducer elements coupling the antigen receptor to intracellular signalling molecules by virtue of the ITAM motif (see below) and they contain sequences for efficient internalization of antigen.

One of the first signs of cellular activation after antigen-induced ligation of the BCR is the increase in the activity of protein tyrosine kinases (PTKs; see Box 2). Because the cytoplasmic domains of mIgM and mIgD consist of only 3 amino acids, it could be assumed that Igα (cytoplasmic domain of 61 amino acids) and Igβ (48 amino acids) serve a role in signal transduction. Of crucial importance for signal transduction is the ITAM motif (immunoreceptor tyrosine-based activation motif), present in the cytoplasmic domain of Igα and Igβ (Fig. 3). This amino acid motif resides in a 26-amino acid sequence and consists of a tyrosine (Y) followed two residues later by a leucine (L) or isoleucine (I), a sub-motif that is repeated once after six to seven variable residues. The complete ITAM motif also contains two aspartate (D) or glutamate (E) residues at characteristic positions (Fig. 3). The ITAM motif is found in Igα and Igβ, in the CD3γ and CD3δ chains of the T CELL RECEPTOR (TCR) complex, and in the γ chain of the Fcε receptor type I. ζ chains of the TCR contain three copies of the ITAM motif. More or less truncated forms of ITAM are present in CD22 and Fcγ receptor type II. A viral encoded protein, LMP2 from Epstein-Barr virus, also contains an ITAM. The central role of ITAM in cellular signalling through the BCR complex (as well as through the TCR complex) has become apparent from studies in which single amino acid substitution receptor mutants and chimerical receptor molecules have been used.

Upon triggering of mIg, a number of cytoplasmic PTKs become associated with the BCR. These include kinases of the src family, such as lyn, fyn and blk, as well as the syk tyrosine kinase and Bruton's tyrosine kinase (Btk). The binding is mediated by the interaction of the src-homology 2 (SH2) domain within the tyrosine kinase with phosphorylated tyrosine residues within the ITAMs of Igα and Igβ. Note

Box 2. Protein tyrosine kinases (PTK)

Protein tyrosine kinase is an enzyme that catalyzes the phosphorylation of tyrosine residues in proteins with adenosine triphosphate (ATP) or other nucleotides as phosphate donors. The definition of protein tyrosine kinase activity is catalysis of the reaction: ATP + a protein tyrosine = adenosine diphosphate (ADP) + protein tyrosine phosphate. Cell to cell signals concerning growth, differentiation, adhesion, motility, and death, are frequently transmitted through tyrosine kinases. PTKs represent a diverse and rapidly expanding superfamily of proteins, including both transmembrane receptor tyrosine kinases and soluble cytoplasmic enzymes. Activation of the PTK domain of either class of PTK enzymes results in interaction of the protein with other signal transducing molecules and propagation of the signal along a specific signal transduction pathway.

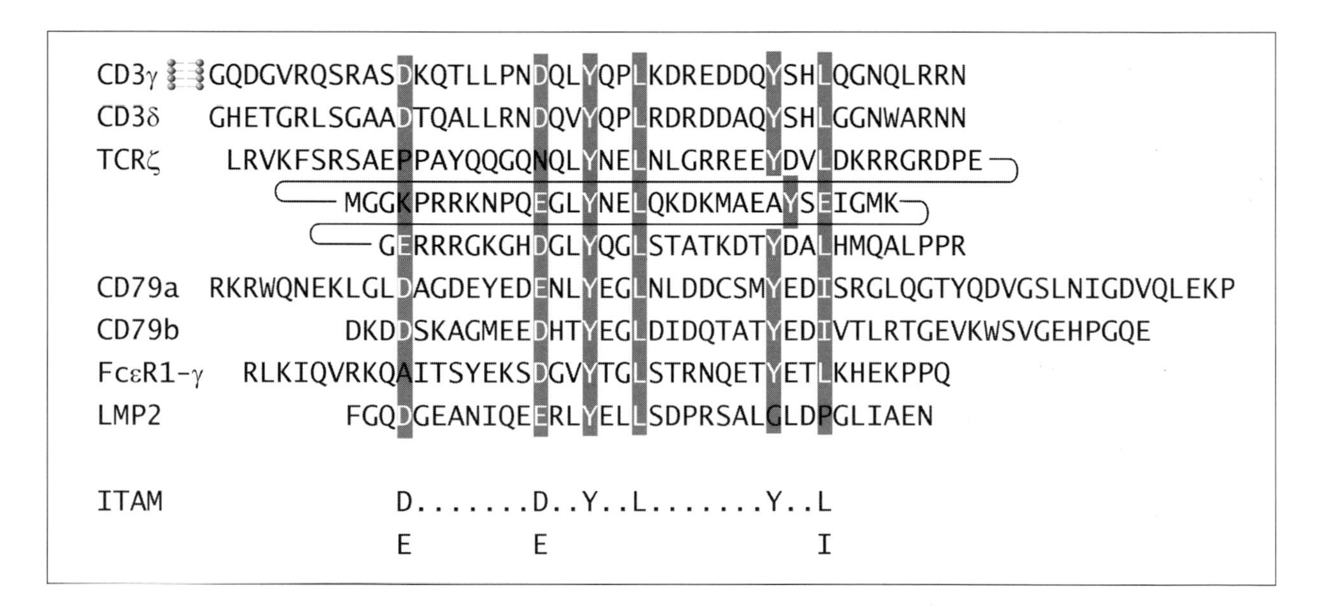

Figure 3. The immunoreceptor tyrosine-based activation motif (ITAM)

The amino acid sequence (given in the single letter code) of the cytoplasmic domains of human CD3γ, CD3δ, the TCRζ chain, Igα (CD79a), Igβ (CD79b), the γ chain of the type I Fcε receptor (FcεRI-γ) and of the EBV encoded LMP2 protein. Note that the cytoplasmic domain of TCRζ is depicted in three interconnected parts in order to allow the alignment of the three copies of the ITAM within the sequence. See text for further explanation.

that this model suffers from a "chicken and egg" problem: if binding of SH2 domains is on phosphorylated ITAM tyrosines, how do ITAM tyrosines become phosphorylated initially? It has been found, however, that an alternative interaction is possible, not depending on phosphotyrosine: the 10 N-terminal residues of src kinases can interact with a specific sequence within the ITAM of Igα (DCSM).

Following phosphorylation of Igα and Igβ by the src family PTKs, syk is recruited and activated [2, 3]. Binding of src and syk kinases to (phosphorylated) ITAMs triggers a series of downstream signalling events, in which adapter proteins are involved. B cell adapter molecules, such as B cell linker (BLNK) and Bam32, function as conduits to effectively channel upstream signals to specific downstream branches. These include activation of phospholipase Cγ2, GTPase-activating protein, MAP kinase (all through the N-terminal regions of lyn, fyn and blk), of phospholipase Cγl (through syk), of the guanine nucleotide-releasing factor Vav, of p85 phosphoinositide-3-kinase (PI-3 kinase, through the SH3 domains of fyn and lyn). Activated PI-3 kinase in turn results in the phosphatidylinositol (3,4,5) trisphosphate (PIP3)-mediated recruitment of the Tec family member Btk to the plasma membrane where it is involved in activation of phospholipase Cγ. Thus, originating from the BCR, several cytosolic PTKs that belong to the src, syk, and Tec families are activated, resulting in the initiation of several distinct cell-signalling pathways (Ras, phospholipase C, PI-3 kinase) via various adapter molecules (e.g., BLNK and Bam32). These signalling pathways result in the activation of a set of protein kinases, which in turn phosphorylate cytoplasmic and nuclear substrates, and ultimately activate transcription (Fig. 4).

B lymphocyte costimulation

Whereas the events described above are causally linked to B LYMPHOCYTE proliferation and differentiation into antibody-secreting plasma cells, in only a few cases is triggering of the BCR by specific antigen sufficient to ensure subsequent B cell activation and differentiation. In all other instances, involving the vast majority of naturally occurring antigens, the process of B cell activation and differentiation depends on activation of additional receptors on the B LYMPHOCYTE. A number of these receptors interact with counter-receptors on T LYMPHOCYTES, thus providing the structural basis for the interaction between these two cell types in the process of antibody formation. Other receptors on B lymphocytes have ligands that are expressed on other cell types (such as MONOCYTES, endothelial cells, etc.) or have soluble ligands.

A major coreceptor on B lymphocytes is CD40 [4]. This 50-kDa glycoprotein is a member of the so-called TNF receptor superfamily and is expressed on all mature B lymphocytes. The counterreceptor for CD40 is the CD40 ligand (CD40L), which is expressed on activated T helper cells. The role of CD40 and CD40L in the process of B LYMPHOCYTE activation is depicted schematically in Figure 5. Antigen that is bound to the BCR is internalized, processed, and peptide fragments derived from the antigen are subsequently presented in MAJOR HISTOCOMPATIBILITY COMPLEX (MHC; see Box 3) class II molecules expressed on the B LYMPHOCYTE surface. Thus presented, peptides can be recognized by specific T cells, leading to T cell activation and thereby expression of CD40L, a 39-kDa cell surface glycoprotein. The interaction of CD40L with CD40 results in progression of the B cell activation process, including acquisition of the capacity to proliferate in response to soluble CYTOKINES produced by the activated T cell. The signal received through CD40 is also important for the process of class switching, the mechanism through which antibodies of IMMUNOGLOBULIN classes other than IgM are produced. The biological significance of the interaction between CD40 and CD40L is illustrated in a human immunodeficiency disease, the so-called X-linked hyper-IgM syndrome. Affected

FIGURE 4. SIGNALLING THROUGH THE BCR COMPLEX

The membrane immunoglobulin is composed of disulphide-linked heavy- and light-chain molecules (only partially shown in the figure) flanked by noncovalently associated dimers of Igα and Igβ. Ovals in membrane immunoglobulin, Igα and Igβ indicate homologous domains of immunoglobulin superfamily members. The cytoplasmic domains of Igα and Igβ contain the ITAM. Phosphorylation (P) of the tyrosine residues (Y) in the ITAMs allows src kinases (lyn, fyn, blk) and syk to associate with the BCR. Activated kinases lead to further phosphorylation of ITAMs, autophosphorylation as well as phosphorylation of a number of cell-signalling molecules. The latter include Shc and Grb2 (initiating the MAP kinase-signalling pathway), Vav (activating Ras and thus also leading to activation of MAP kinase), phospholipase Cγ1 [catalysing the hydrolysis of phosphatidyl inositol bisphosphate (PIP2)], PI-3 kinase [leading to generation

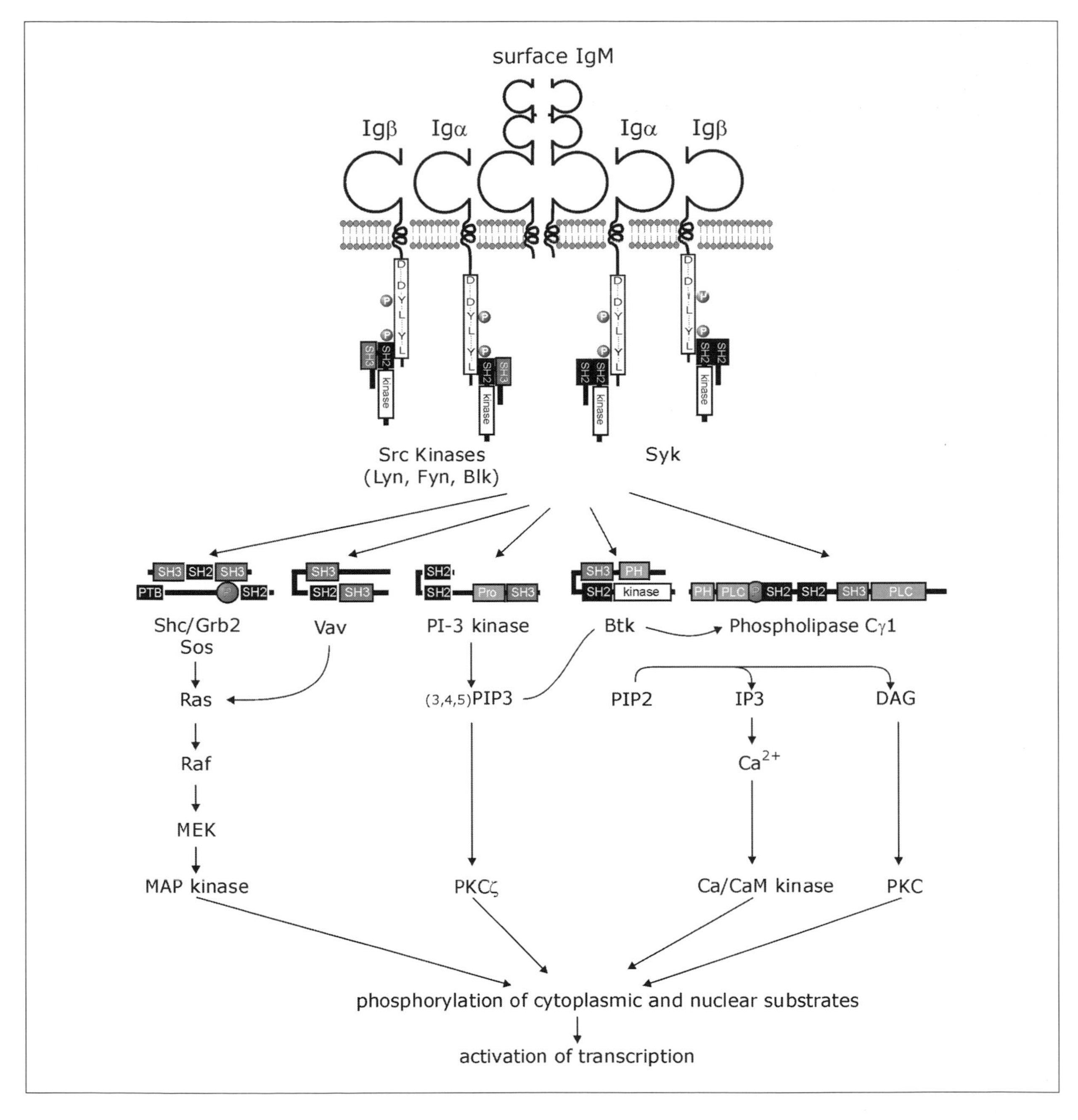

of phosphatidyl inositol trisphosphate (PIP3)] and Btk. PIP2 hydrolysis generates inositol trisphosphate (IP3) and diacylglycerol (DAG). IP3 causes the release of Ca^{2+} from intracellular stores and the subsequent activation of Ca^{2+} /calmodulin-dependent protein kinase (Ca/CaM kinase). DAG, in the presence of phosphatidyl serine and high Ca^{2+}, activates protein kinase C (PKC). PIP3 activates the ζ isoform of PKC. All kinases described above phosphorylate cytoplasmic and nuclear substrates (including transcription factors) leading to activation of transcription and thus B lymphocyte activation.

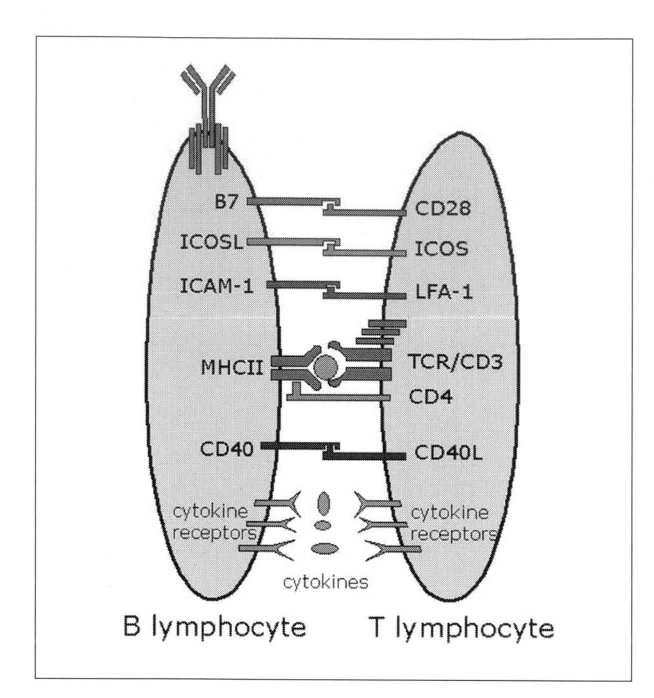

FIGURE 5. INTERACTION BETWEEN B AND T LYMPHOCYTE
A peptide/MHC class II complex expressed on a B lymphocyte can be specifically recognized by a T cell receptor expressed on a CD4⁺ T cell (signal 1). To induce an effective and sustained immune response, an additional signal is needed. CD40L interacts with the CD40 molecule on the B cell, resulting in an increase of expression of B7. Ligation of CD28 by B7 provides a costimulatory signal (signal 2) needed for T cell activation and production of IL-2. Expression of ICOS is induced and the molecule interacts with its ligand. Subsequently, several types of cytokines can be produced and antibody production by B cells is stimulated.

patients carry mutations in the CD40L gene, resulting in the inability to produce antibodies other than IgM.

Additional COSTIMULATORY receptor-counterreceptor pairs contribute to successful interaction between B lymphocytes and T LYMPHOCYTES, such as CD28 and the inducible costimulator (ICOS) on T lymphocytes and B7.1/B7.2 (CD80/CD86) and ICOS ligand (B7h) on B lymphocytes, respectively [5]. As a result of CD40-CD40L interaction, the expression of B7 on the B lymphocyte is upregulated. Ligation of CD28 by B7 provides a COSTIMULATORY signal that is required for T cell activation, proliferation, production of IL-2, and cell survival. Both MHC/peptide complex-T CELL RECEPTOR interaction (signal 1) and CD28-ligand interaction (signal 2) are needed for the induction of a sustained immune response. CD28-deficient mice exhibit defects in ISOTYPE switching and are more susceptible to pathogens that depend on an effective antibody response [6].

Cytokine regulation

The full stimulatory effect of the interaction between T lymphocytes and B lymphocytes depends not only on binding of cell surface receptors and counter-receptors but also on production of T cell CYTOKINES that promote (various stages) of B lymphocyte proliferation and terminal differentiation into plasma cells. The soluble CYTOKINES should not be considered merely as endocrine hormones, because they are secreted at the sites of direct cell-cell contact, and therefore the particular B lymphocyte engaged in cellular interaction with the relevant T lymphocyte benefits most from these growth and differentiation factors.

The CYTOKINES that regulate B lymphocyte growth and differentiation predominantly include INTERLEUKIN (IL)-4, IL-5, IL-6 and IL-10 [7, 8]. IL-4 acts as a costimulator for signals received through the BCR and CD40 in promoting B lymphocyte growth. Both IL-4 and the related cytokine IL-13 can cause switching to IgE and IgG4 production. IL-5 primarily regulates the differentiation of an activated B lymphocyte into an antibody-secreting cell. IL-6 and IL-10 have prominent effects on B lymphocyte differentiation and IMMUNOGLOBULIN secretion.

T cell-independent B lymphocyte activation

In the above section the important role of interaction with T lymphocytes for the process of B lymphocyte activation has been emphasized. There is however a category of ANTIGENS that is unable to activate

Box 3. Major histocompatibilty complex (MHC)

MHC molecules, also known as human leukocyte antigens or HLA, are the products of a cluster of genes in the human DNA known as the major histocompatibility complex (MHC). There are two types of MHC molecules, MHC class I and MHC class II. MHC class I molecules designated as HLA-A, -B, and -C are expressed on all nucleated cells, while MHC class II molecules (HLA-DR, -DQ, and -DP) are primarily expressed on antigen presenting cells (APC) like macrophages, dendritic cells and B cells. MHC molecules bind small protein fragments, called peptides, and form MHC/peptide complexes at the cell surface. Recognition of these MHC/peptide complexes by the T cell receptor are required for T cell activation. B cells, however, are also able to directly recognize antigens via their B cell receptor.

T lymphocytes, whereas B lymphocyte responses and induction of antibodies can be readily demonstrated. These types of ANTIGENS are called T cell-independent antigens, and major representatives are capsular polysaccharides from encapsulated bacteria such as *Streptococcus pneumoniae*, *Haemophilus influenzae* type b and *Neisseria meningitidis*. Until now, neither processing and antigen presentation in the context of MHC class II molecules nor specific T LYMPHOCYTE activation has been demonstrated for polysaccharides. This means that, because CD40/CD40L interaction in the case of T cell-independent ANTIGENS is highly unlikely to take place, polysaccharide-specific B lymphocytes should receive an alternative COSTIMULATORY signal. Indirect evidence points towards a role for the CD19/CD21 receptor complex in this respect [9]. CD19 is a 95-kDa glycoprotein of the IMMUNOGLOBULIN superfamily that is expressed throughout B lymphocyte development. A specific ligand for this molecule has not been identified as yet, although purified CD19 protein does bind to BONE MARROW stromal cells. Activation of CD19 by specific antibodies provides a COSTIMULATORY signal for B lymphocyte activation through the BCR. Indeed, the cytoplasmic domain of the CD19 molecule contains the consensus motifs for binding of src kinases lyn and lck, for binding of PI-3 kinase and for interaction with the protooncogene Vav. On mature B lymphocytes, CD19 is expressed in a molecular complex that includes CD21, the TAPA-1 protein (CD89) and the Leu-13 molecule. The prevailing model is that in this complex CD19 acts as the signal-transducing moiety for CD21. CD21 is a 145-kDa glycoprotein of the complement receptor family, which is expressed on mature B lymphocytes (and also on FOLLICULAR DENDRITIC CELLS and at a low level on a subpopulation of T cells). CD21 is the receptor for the complement component C3 split products iC3b, C3dg and C3d. CD21 also serves as the cellular receptor for the Epstein-Barr virus, and as an INTERFERON α receptor. Furthermore, CD21 can interact with CD23. The (chemical) coupling of C3d to protein ANTIGENS lowers the threshold for antibody induction 100-1000-fold. Bacterial polysaccharides, through the ALTERNATIVE PATHWAY of complement activation, can generate C3 split products, which become deposited on the polysaccharide. Natural complexes of polysaccharide and C3d, thus formed, can cross-link mIg and CD21 on polysaccharide-specific B lymphocytes. This mechanism may bypass the need for engagement of CD40/CD40L in B lymphocyte activation (Fig. 6). Compatible with this mechanism is the finding that children up to the age of 2 years who are unable to respond to polysaccharide ANTIGENS have a reduced expression of CD21 on B lymphocytes [10].

Primary and secondary antibody response

The first contact of the immune system with a given antigen will induce what is called a primary (antibody) response. B lymphocytes become activated and differentiate into plasma cells (along routes as described above). Plasma cells are highly differenti-

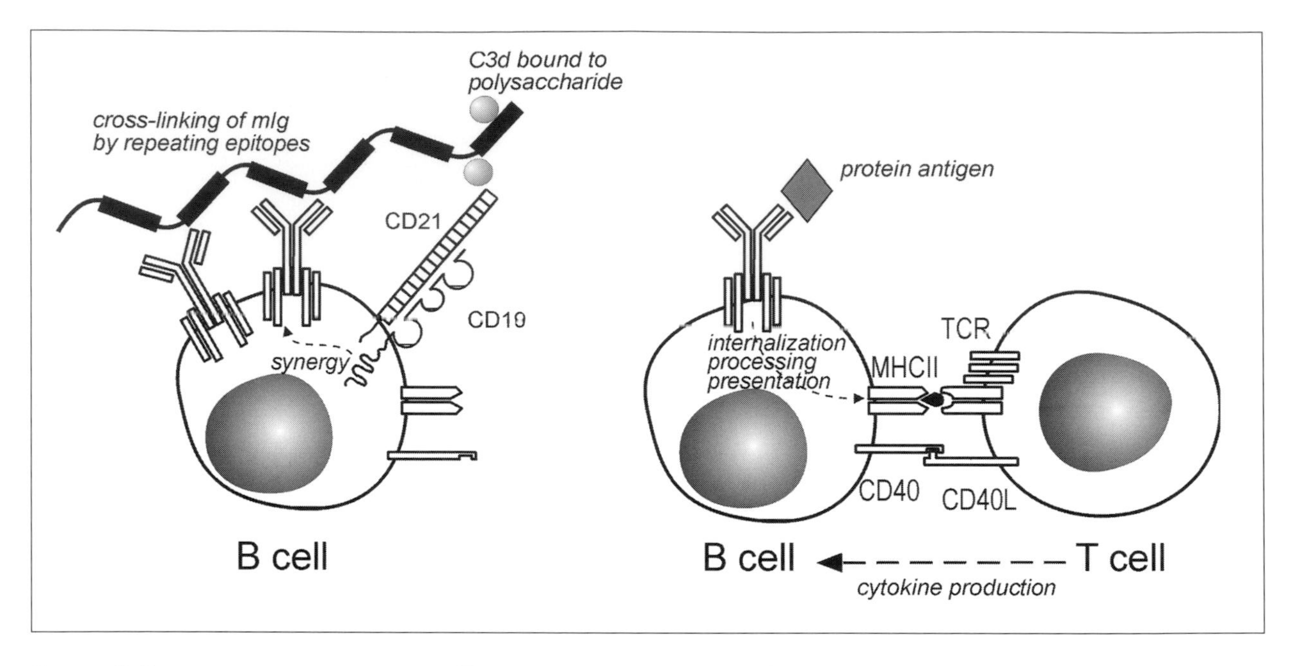

Figure 6. B lymphocyte activation by T cell-independent and by T cell-dependent antigens
Left panel: A polysaccharide antigen with repeating epitopes cross-links mIg on the surface of the B lymphocyte. Deposited C3d is bound by CD21, which provides a synergistic signal for B lymphocyte activation. Right panel: A protein antigen is bound and internalized by mIg; following intracellular processing, peptide fragments are expressed in MHC class II molecules. Specific T lymphocytes recognize these peptides in the context of MHC, become activated and express CD40L. Upon interaction of CD40L with CD40, cytokine production is initiated (see Figure 5 for more details).

ated cells, which maximally produce 10^4 antibody molecules per second, equalling 40% of the total protein-synthesizing capacity of the cell. The lifetime of a plasma cell is 3–4 days in the initial antibody response. These plasma cells are called short-lived plasma cells. A second contact with the same (protein) antigen elicits a secondary antibody response, which is produced by long-lived plasma cells and differs in a number of aspects from a primary response [11]. The latency period (time between contact with antigen and start of antibody production) is shorter in a secondary response (Fig. 7), and antibody levels attained are much higher (1–2 orders of magnitude). While during a primary antibody response predominantly IgM, and to a lesser extent IgG, antibodies are produced, IgG, IgA and IgE antibodies are the major classes produced during a secondary response. The affinity of antibodies produced increases during the response; there may be a 100–1000-fold difference in affinity between antibodies produced by the short-lived plasma cells at the start of a primary response and at the end of a secondary response. This process (affinity maturation) is the combined effect of somatic hyper-mutation of CDR1 and CDR2 regions during B lymphocyte proliferation and the selection of the B lymphocytes with the highest affinity.

A primary antibody response takes place in FOLLICLES and the marginal zone of spleen and lymph nodes. During a secondary response, BONE MARROW is the major site of antibody production. Long-lived plasma cells in the BONE MARROW can survive for at least 90 days in the absence of cell division. Antibodies which are secreted in mucosal tissue of the respiratory and gastrointestinal tract are produced locally by the bronchus-associated lymphoid

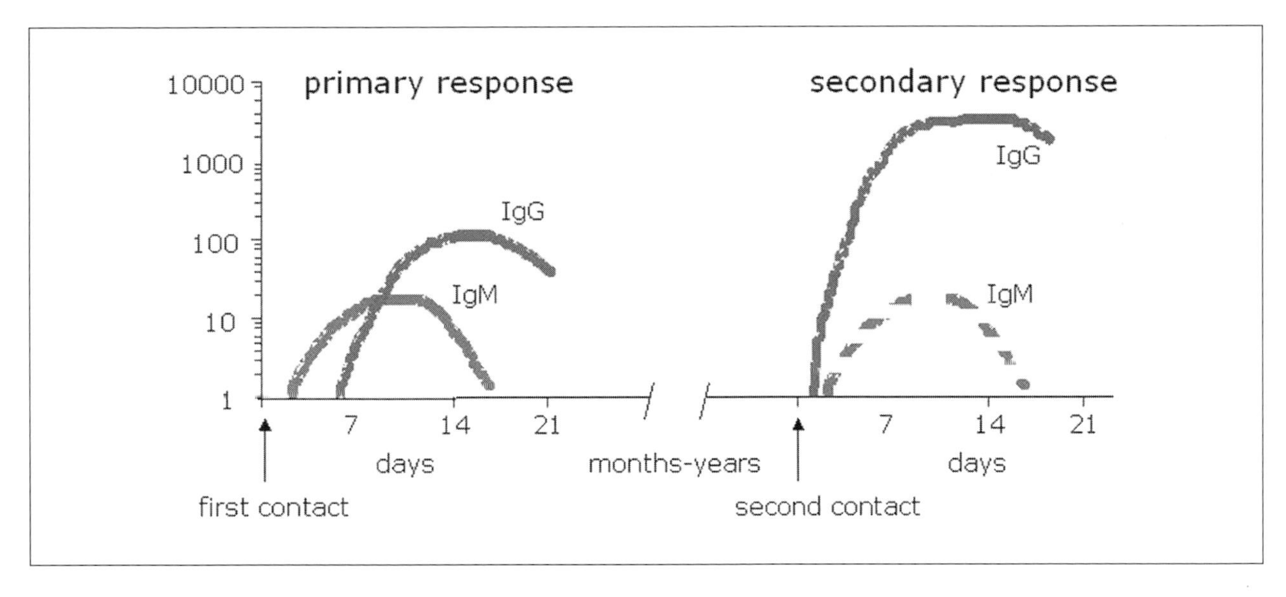

FIGURE 7
The response triggered by the first encounter with a given antigen is called the primary antibody response. During this response IgM appears first, followed by IgG. The most prevalent class of antibody, IgG is produced when a particular antigen is encountered again. This response is called the secondary antibody response. It is faster and results in higher antibody titers than the primary antibody response.

tissue (BALT) and gut-associated lymphoid tissue (GALT), respectively (Fig. 8).

During the primary immune response, a fraction of antigen-specific B lymphocytes do not differentiate into plasma cells but into so-called memory B cells. It should be noted that the term memory cell is largely conceptual; actual differences between naïve and memory B lymphocytes are found in the ISOTYPES of mIg and affinity of mIg. Naïve primary B lymphocytes express mIgM and mIgD; memory B lymphocytes have lost mIgD and express mIgG or mIgA, with or without mIgM. Adult circulating B cells can be separated into three subpopulations on the basis of CD27 and mIgD expression: IgD^+CD27^- naïve B cells, IgD^+CD27^+ non-switched memory B cells, and IgD^-CD27^+ memory B cells (see also Fig. 9). Because of the above-described affinity maturation, the affinity of mIg for antigen on memory B lymphocytes is higher than on primary B lymphocytes.

All characteristics of a primary and secondary antibody response as described above hold true only for protein antigens. For polysaccharide antigens, a second contact with antigen induces an antibody response that is identical in kinetics and magnitude to a primary response. Affinity maturation does not occur, and ISOTYPE distribution of antibodies does not change. Moreover, anti-polysaccharide antibodies use a restricted number of V_H and V_L genes, whereas anti-protein antibodies are more heterogeneous. Finally, IgG anti-protein antibodies to the vast majority of ANTIGENS are of the IgG1 subclass; IgG anti-polysaccharide antibodies in adult individuals are predominantly IgG2.

Biological functions of antibodies

The biological function of antibodies can be discerned in binding of antigen (through variable V_H and V_L domains) and effector functions mediated by the constant domains (in particular C_{H2} and/or C_{H3} domains). The bacterial toxins mentioned at the

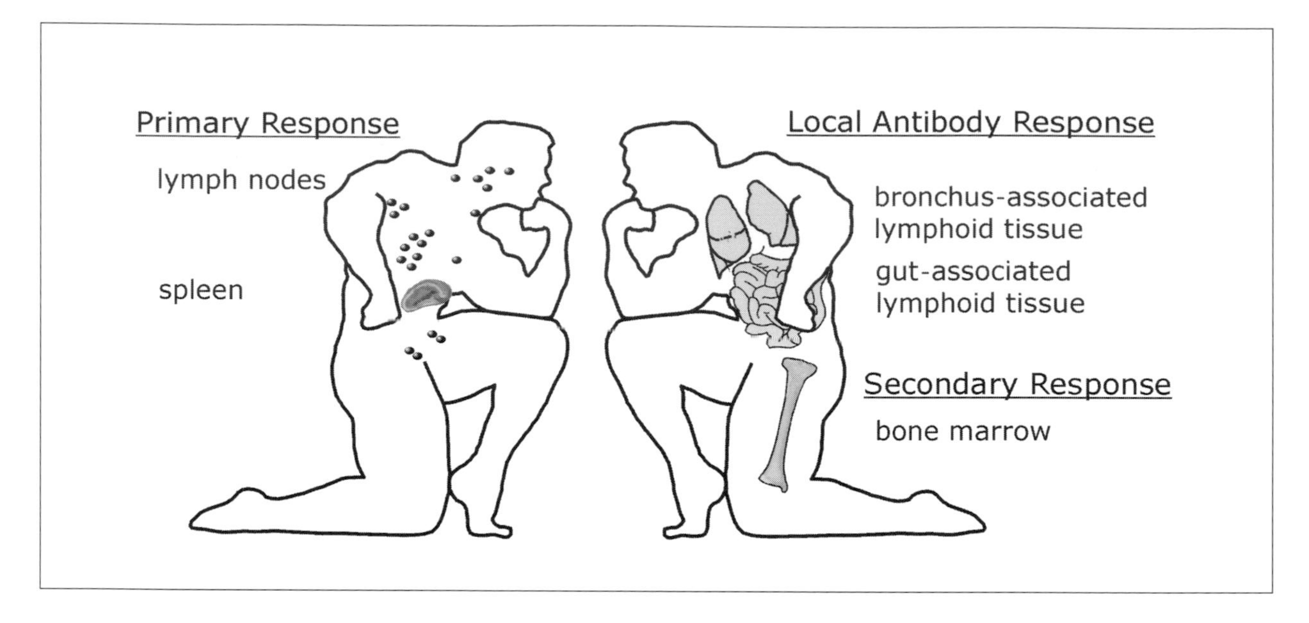

FIGURE 8. SITES OF ANTIBODY PRODUCTION

beginning of this chapter are neutralized when bound by antibodies. Clearance of immune complexes and antibody-opsonized microorganisms depends on the functional integrity of the Fc part of the antibody molecule. The Fc part can either interact with soluble biologically active molecules, such as the COMPLEMENT SYSTEM (see Chapter A.5) or bind to Fc receptors which are expressed on a variety of cells of the immune system [12]. Fc receptors expressed on MONOCYTES, MACROPHAGES and GRANULOCYTES are essential for PHAGOCYTOSIS of immune complexes and opsonized microorganisms. Fc receptors for IgG (Fcγ receptors) are expressed on MONOCYTES and MACROPHAGES; neutrophilic GRANULOCYTES also express receptors for IgA (Fcα receptors). Depending on the class of antibodies in an immune complex, the COMPLEMENT SYSTEM becomes more or less efficiently activated. This will enhance PHAGOCYTOSIS by MONOCYTES, MACROPHAGES and GRANULOCYTES, since these cells also express COMPLEMENT RECEPTORS in addition to Fc receptors.

Fc receptors for IgE (Fcε receptor) are primarily expressed by mast cells. In allergic individuals, Fcε receptors have constitutively bound IgE; exposure to allergens causes cross-linking of Fcε receptors, resulting in mast cell DEGRANULATION and HISTAMINE release. Apart from PHAGOCYTOSIS and DEGRANULATION, Fc receptors also mediate cytotoxicity in a process called ADCC (antibody-dependent cellular cytotoxicity). Target cells (e.g., tumor cells) to which antibodies are bound can be recognized by Fc receptors expressed on cells with cytotoxic potential. The killing process itself is complement-independent. MONOCYTES, neutrophilic and eosinophilic GRANULOCYTES and natural killer (NK) cells display ADCC activity. ADCC can be a mechanism for removal of tumor cells and has been implicated in tissue damage that occurs in autoimmune diseases.

The Fcγ receptor expressed on B lymphocytes (FcγIIb) plays a role in downregulation of B cell activation. When high IgG antibody concentrations are reached during an immune response, antigen-IgG complexes will be formed which can cross-link the BCR and FcγIIb on the surface of the B lymphocyte (see also Fig. 10). The cytoplasmic domain of FcγIIb contains the YSLL motif, which has been termed ITIM for immunoreceptor tyrosine-based inhibitory motif. Tyrosine phosphorylation of this motif causes the association of a protein tyrosine phosphatase. When

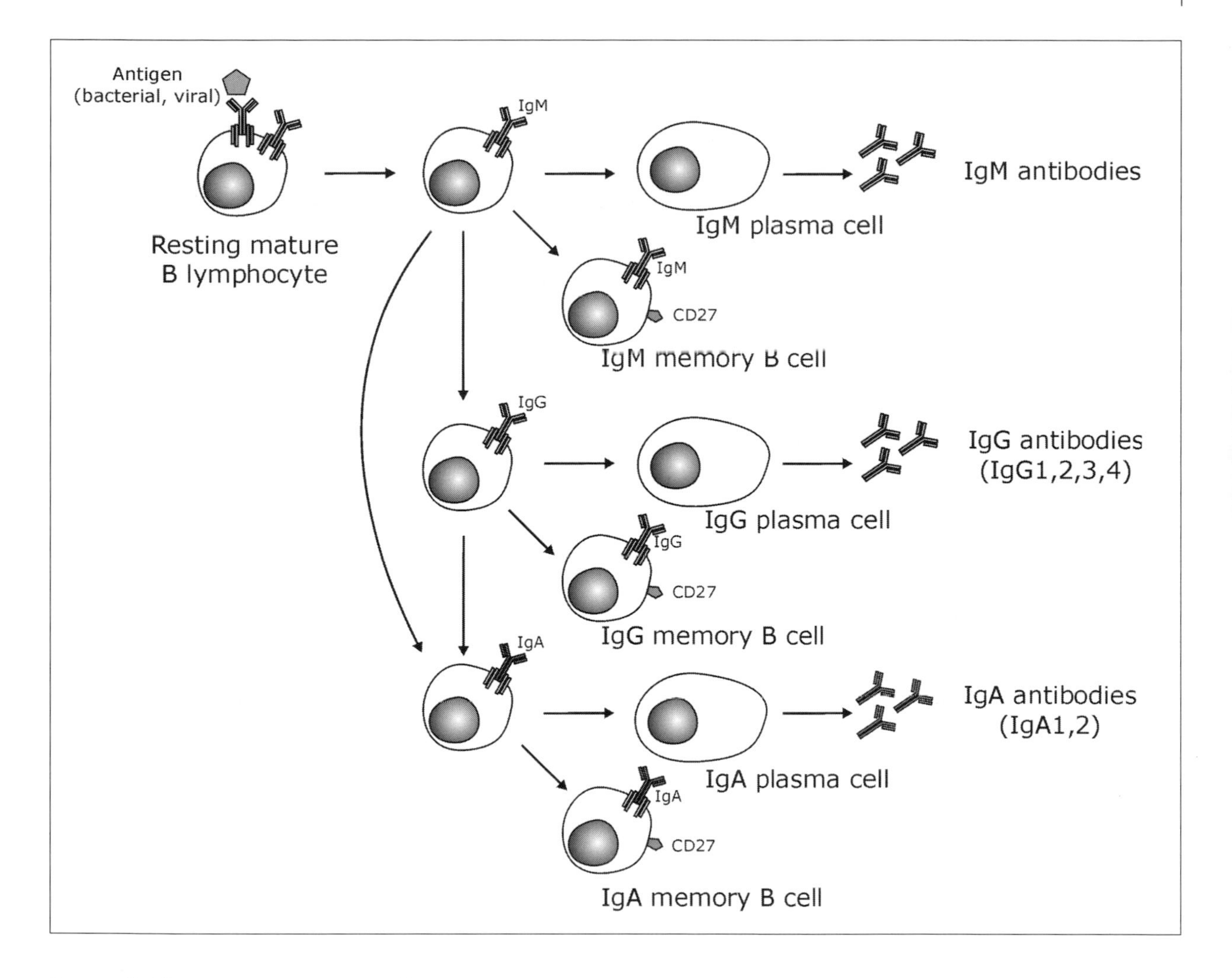

FIGURE 9. GENERATION OF MEMORY B LYMPHOCYTES

brought in close proximity to ITAMs, this enzyme causes tyrosine dephosphorylation and therefore inhibits BCR signaling. This mechanism, by which IgG antibodies interact with FcγIIb and co-crosslink with the BCR, is an example of active downregulation of the antibody response. There are other examples of cell surface receptors with either intrinsic (CD45) or associated protein tyrosine phosphatase activity (CD22), but their cellular ligands, and therefore their role in regulation of B cell activation, are at present unknown.

Clinical relevance and future prospects

The integrity of the humoral immune system is crucial for host defence against bacterial and certain viral infections. Inborn or acquired deficiencies in humoral immunity result in increased susceptibility to potentially life-threatening infections. A dysregulated humoral immune system may result in conditions such as allergy. These and other clinical aspects are discussed elsewhere in this volume (see Chapters A.8, A.9, C.3).

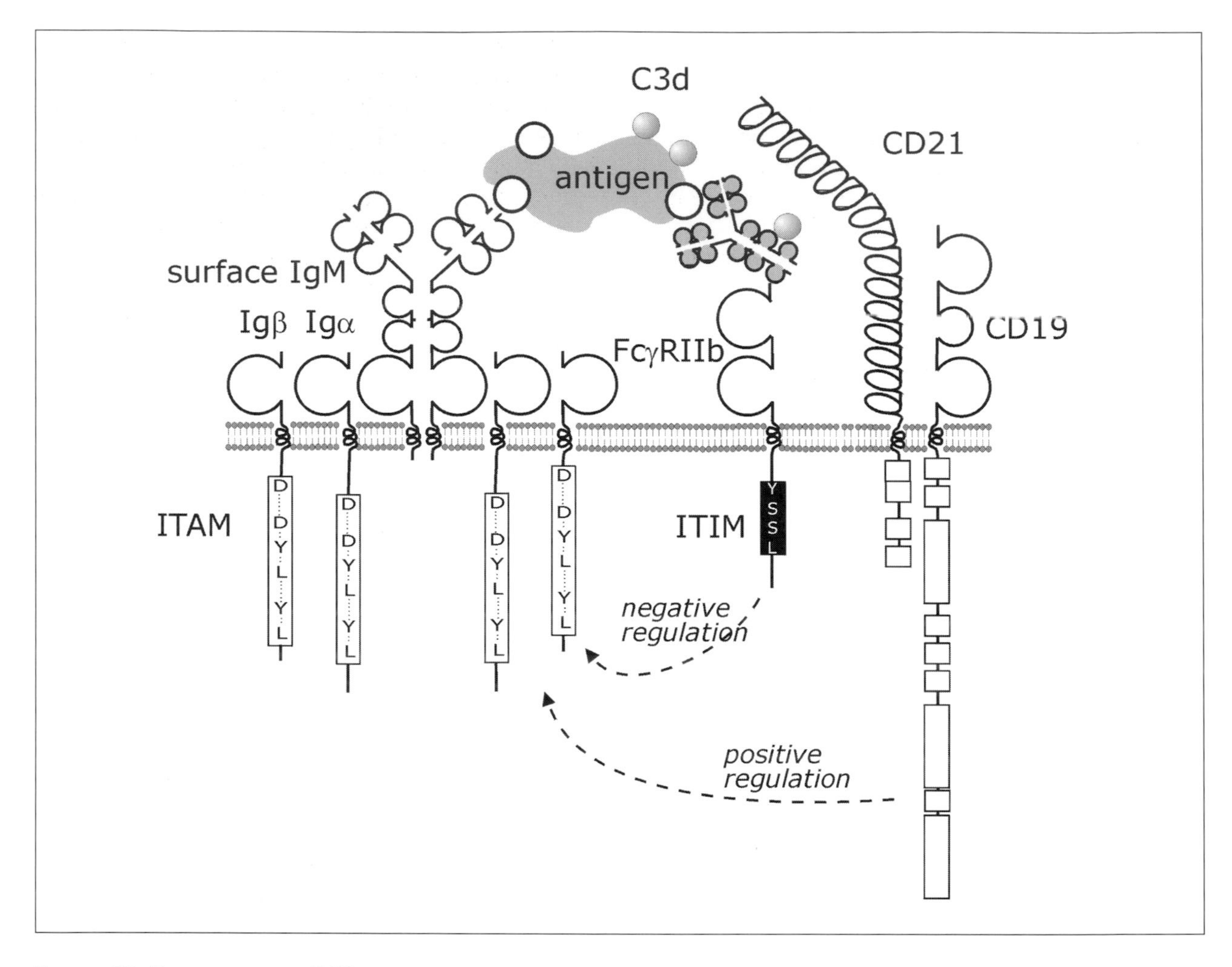

Figure 10. Regulation of BCR activation by coreceptors

Ovals in membrane immunoglobulin (surface IgM), Igα, Igβ, FcγRIIb and CD19 indicate homologous domains of immunoglobulin superfamily members. The cytoplasmic domains of Igα and Igβ contain the ITAM. The Fcγ receptor IIb binds the Fc part of IgG antibodies in immune complexes. The cytoplasmic domain of FcγRIIb contains the immunoreceptor tyrosinase-based inhibitory motif (ITIM) YSLL. CD21 (type 2 complement receptor) binds C3d deposited on the antigen or on IgG antibodies. The cytoplasmic domain of CD19 has a positive regulatory effect on signalling through mIg (see text for further explanation).

Most of what is known about the cellular and molecular aspects of B lymphocyte activation and regulation has been gathered from experiments performed during the last decade. Detailed knowledge of the mechanisms that govern the regulation of expression of cell surface receptors and signalling mechanisms allows pharmacological intervention in antibody-mediated immunity. Intervention is possible at three levels: outside the B lymphocyte, at the cell surface and intracellularly. The coupling of polysaccharides to protein carriers changes the nature of the anti-polysaccharide antibody response from T cell-independent into T cell-dependent. These polysaccharide-protein conjugate vaccines bypass the selective unresponsiveness to T cell-independent ANTIGENS early in life and thus constitute novel and effective tools in prevention of infectious diseases. Conjugate vaccines for *Haemophilus influenzae* type b and *Streptococcus pneumoniae* are able to prevent invasive diseases in otherwise susceptible populations. Specific targeting of ANTIGENS to FcγIIb can be a very efficient way to induce B lymphocyte unresponsiveness. A second level of intervention is regulation of cell surface receptor expression. Substituted guanosines like 8-mercaptoguanosine cause increased expression of CD21 and thus could potentiate B lymphocyte activation. The third level is intracellular intervention. Drugs with the ability to redirect intracellular signalling pathways are potentially powerful tools for enhancement or inhibition of B lymphocyte activation.

A primary antibody response starts with production of IgM antibodies. During a primary response, class switching to IgG and IgA antibodies takes place. Proliferating B lymphocytes undergo somatic hypermutation which can result in antibodies with a higher affinity. During a primary response, part of the B lymphocytes differentiate into long-lived memory B lymphocytes (and express CD27). In a secondary response the expanded CLONE of memory B lymphocytes reacts with a short latency period and high ANTIBODY production.

Upon interaction with antigen, antibodies exert a variety of biological functions: 1) direct neutralization of bacterial toxins; 2) initiate complement activation which, in the case of a cellular antigen, results in cell lysis; 3) augment PHAGOCYTOSIS after interaction with Fc receptors; 4) initiate antibody-dependent cellular cytotoxicity.

Summary

Antibodies are produced by B lymphocytes. Upon binding of antigen to membrane IMMUNOGLOBULIN (mIg), the B LYMPHOCYTE becomes activated and differentiates into antibody -producing plasma cell. mIg is part of the B cell receptor complex, in which Igα and Igβ proteins are signal transduction molecules. Key elements in B lymphocyte signalling are phosphorylation of tyrosine residues within the immunoreceptor tyrosine-based activation motifs (ITAM) of Igα and Igβ and subsequent activation of a series of cytoplasmic tyrosine kinases of the src, syk and tec family. Apart from triggering of mIg, full B LYMPHOCYTE activation requires a number of additional molecular interactions which require cognate cellular interaction with T lymphocytes and/or MONOCYTES (T cell-dependent B cell activation). The binding of CD40 on the B lymphocyte and CD40L on activated T lymphocytes is an example of these co-stimulatory interactions. Furthermore, CYTOKINES, such as IL-4 and IL-5, promote B lymphocyte proliferation and differentiation. For B lymphocyte activation by polysaccharide antigens, T lymphocytes are not required. COSTIMULATION in this case is provided by CD21, the complement receptor on B lymphocytes which is activated by C3d, bound to the polysaccharide.

Selected readings

Bengten E, Wilson M, Miller N, Clem LW, Pilstrom L, Warr GW (2000) immunoglobulin isotypes: structure, function, and genetics. *Curr Top Microbiol Immunol* 248: 189–219

Kurosaki T (2002) Regulation of B-cell signal transduction by adaptor proteins. *Nat Rev Immunol* 2: 354–363

Clark EA, Ledbetter JA (1994) How B and T cells talk to each other. *Nature* 367: 425–428

Zubler RH (2001) Naïve and memory B cells in T-cell-dependent and T-independent responses. *Springer Semin Immunopathol* 23: 405–419

Hardy RR, Hayakawa K (2001) B cell development pathways. *Ann Rev Immunol* 19: 595–562

Recommended websites

http://www.antibodyresource.com/educational.html (portal with many links to relevant websites)

http://www.molbiol.ox.ac.uk/www/pathology/tig/new1/mabth.html (on therapeutic use of antibodies)

http://people.ku.edu/%7Ejbrown/antibody.html (what the heck is an antibody)

(All accessed March 2005)

References

1 Li Z, Woo CJ, Iglesias-Ussel MD, Ronai D, Scharff MD (2004) The generation of antibody diversity through somatic hypermutation and class switch recombination. *Genes Dev* 18: 1–11

2 Campbell KS (1999) Signal transduction from the B cell antigen-receptor. *Curr Opin Immunol* 11: 256–264

3 Fuentes-Pananá EM, Bannish G, Monroe JG (2004) Basal B-cell receptor signaling in B lymphocytes: mechanisms of regulation and role in positive selection, differentiation, and peripheral survival. *Immunol Rev* 197: 26–40

4 Gruss HJ, Herrmann F, Gattei V, Gloghini A, Pinto A, Carbone A (1997) CD40/CD40 ligand interactions in normal, reactive and malignant lympho-hematopoietic tissues. *Leuk Lymphoma* 24: 393–422

5 Sharpe AH, Freeman GJ (2002) The B7-CD28 superfamily. *Nat Rev Immunol* 2: 116–126

6 Ferguson SE, Han S, Kelsoe G, Thompson CB (1996) CD28 is required for germinal center formation. *J Immunol* 156: 4576–4581

7 Cerutti A, Zan H, Schaffer A, Bergsagel L, Harindranath N, Max EE, Casali P (1998) CD40 ligand and appropriate cytokines induce switching to IgG, IgA, and IgE and coordinated germinal center and plasmacytoid phenotypic differentiation in a human monoclonal $IgM^{+}IgD^{+}$ B cell line. *J Immunol* 160: 2145–2157

8 Takatsu K (1998) Interleukin 5 and B cell differentiation. *Cytokine Growth Factor Rev* 9: 25–35

9 Fearon DT, Carroll MC (2000) Regulation of B lymphocyte responses to foreign and self-antigens by the CD19/CD21 complex. *Annu Rev Immunol* 18: 393–422

10 Rijkers GT, Sanders EA, Breukels MA, Zegers BJM (1998) Infant B cell responses to polysaccharide determinants. *Vaccine* 16: 1396–1400

11 Slifka MK, Ahmed R (1998) Long-lived plasma cells: a mechanism for maintaining persistent antibody production. *Curr Opin Immunol* 10: 252–258

12 Radaev S, Sun P (2002) Recognition of immunoglobulins by Fcγ receptors. *Mol Immunol* 38: 1073–1083

Cytokines

Klaus Resch

Introduction

For each immune response several cells have to cooperate. Within the LYMPHOID ORGANS, such as lymph nodes, spleen or Peyer's plaques, the cells of the immune system do not form fixed anatomical structures, but interact in a dynamic fashion. Some cells, especially T LYMPHOCYTES, always leave the LYMPHOID ORGAN to circulate through the body and eventually return. As a consequence, the communication between the cells – LYMPHOCYTES, monocytic cells and GRANULOCYTES – depends on secreted diffusible mediators, the most important of which are the CYTOKINES.

An unambiguous definition for these molecules does not exist, with the exception that they are all proteins. The majority of CYTOKINES is not synthesized exclusively by the cells of the immune system, but by many other cells, and CYTOKINES too, may have effects on many different cells. Therefore they cannot be clearly distinguished from the general growth factors (or hormones), most of which also affect cells of the immune system, although their primary function – e.g., that of the nerve growth factor (NGF) – involves other organs. For practical purposes CYTOKINES can be defined as protein mediators which are (1) primarily synthesized by cells of the immune system, (2) predominantly regulate differentiation and activation of these cells, and (3) are responsible in part for the effector functions of these cells, including inflammation.

The field of cytokine biology as it stands today has three major roots. The first originates in immunology in the 1960s, when it was demonstrated that supernates of stimulated immune cells could regulate function and growth of leukocytes. Partially purified factors were published under function-related names such as lymphocyte activation factor (LAF) or T cell growth factor (TCGF), which, according to their major producer cells, were grouped as monokines or lymphokines. In the seventies the term cytokines was then gradually adapted (upon a suggestion by S. Cohen) as a more general denomination. At the 2nd International Lymphokine Workshop held in 1979, as it became apparent that identical molecules had been described under different names, the term INTERLEUKINS was proposed "as a system of nomenclature... based on the ability to act as communication signals between different populations of leukocytes", and concomitantly the names INTERLEUKIN-1 (IL-1) for LAF and INTERLEUKIN-2 (IL-2) for TCGF were introduced.

The second source of cytokine research stems from virology and classifies the INTERFERONS, originally described in 1957 by A. Isaacs and J. Lindemann, as factors produced by virus-infected cells capable of conferring resistance to infection with homologous or heterologous viruses. Although this activity still defines the groups of INTERFERONS, it soon became apparent that they also regulate immune reactions, which make them a subgroup of cytokines. Regulation of immune and inflammatory reactions is the predominant function of one member, IFN-γ, which is structurally different from the large family of IFN-α/β INTERFERONS.

COLONY-STIMULATING FACTORS (CSFs), which form the third root of cytokines, were first described in 1966 by D. Metcalf, reflecting the observation that they promote the formation of granulocyte or monocyte colonies in a semisolid medium. CSFs predominantly function as proteins which induce growth and differentiation of hematopoietic cells, but can also – as was recognized later – activate fully differentiated cells. A more detailed description of the history of cytokines can be found in the introductory chapter by J. Vilcek in *The Cytokine Handbook* (see "Selected reading").

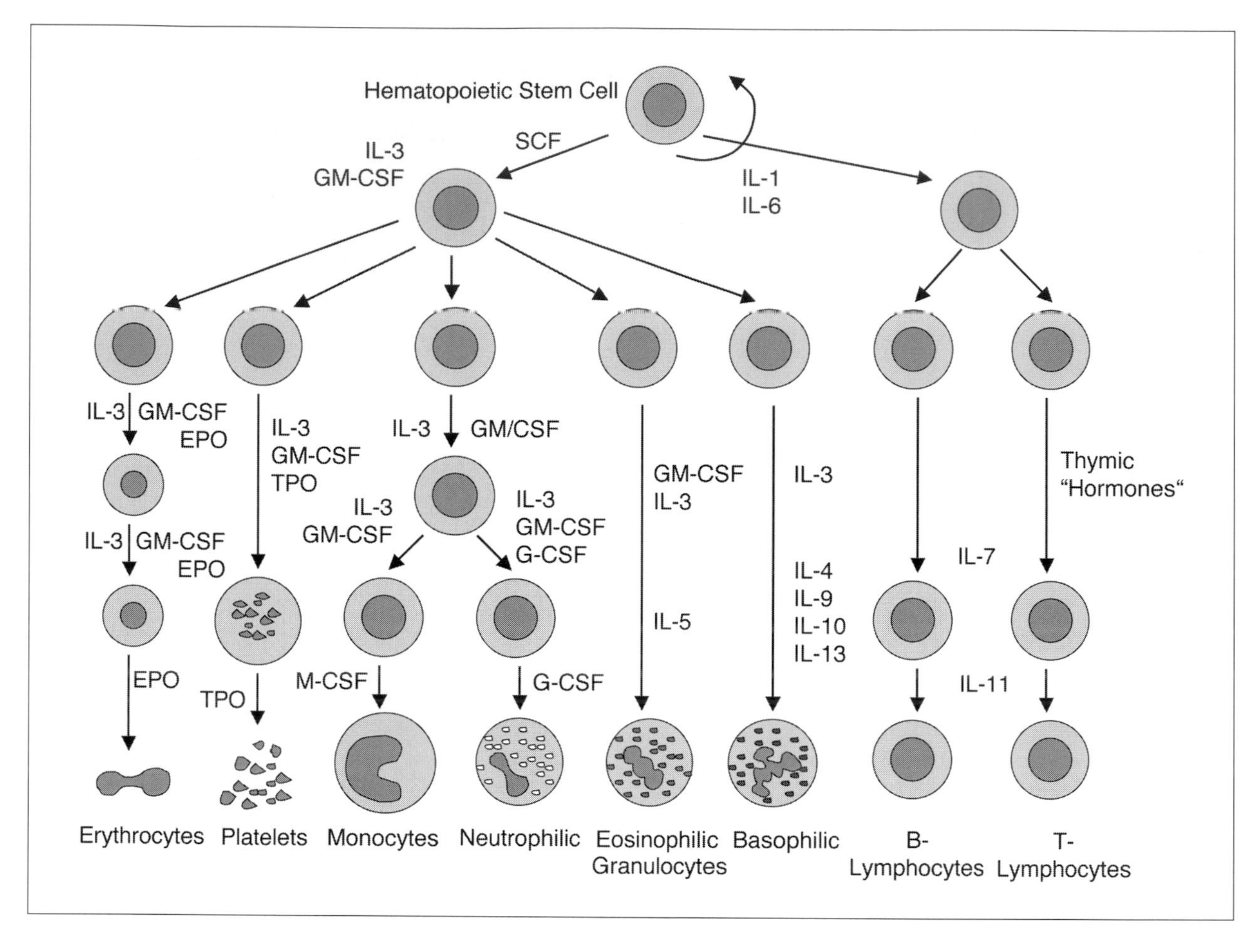

FIGURE 1. CYTOKINES INVOLVED IN THE DIFFERENTIATION OF CELLS OF THE IMMUNE SYSTEM
CSF, colony-stimulating factor; EPO, erythropoietin; G, granulocyte; IL, interleukin; M, monocyte; SCF, stem cell factor; TPO, thrombopoietin

All cytokines exert their biological functions by binding to high-affinity plasma membrane receptors. The structures of the known cytokines, as well as those of most of their respective receptors, have been elucidated by molecular cloning [1–4].

Differentiation factors

The mature cells of the immune system have a finite life span. Their physiological half-life ranges from several hours (neutrophilic GRANULOCYTES) to months (MONOCYTES and most LYMPHOCYTES); some LYMPHOCYTES survive longer, e.g., memory T-helper (Th-) LYMPHOCYTES survive for many years. Therefore the cells must be continuously renewed from a STEM CELL pool, which in adult human beings is located in the BONE MARROW. How the reservoir of STEM CELLS is kept constant by self-renewal is not entirely clear. Besides a direct influence of the stromal microenvironment, the cytokines INTERLEUKIN-1 and INTERLEUKIN-6 (see below) appear to be involved. The hematopoietic STEM CELLS, of which low numbers are also present in blood – carrying the surface marker

Table 1. Myeloid differentiation factors, erythropoietin and thrombopoietin

Cytokine	Molecular mass	Predominant producer cells	Major functions
Stem cell factor	36	stromal cells	differentiation of stem cells
IL-3	14–28	T lymphocytes	differentiation and propagation of myeloid progenitor cells
Granulocyte-macrophage CSF	14–35	T lymphocytes,monocytes, endothelial cells, fibroblasts	differentiation and propagation of myeloid progenitor cells
Granulocyte CSF	18–22	monocytes	propagation and maturation of granulocytes
Macrophage CSF	35–45	endothelial cells	propagation and maturation of monocytes
	18–26	fibroblasts, monocytes	
Erythropoietin	30–32	peritubular renal capillary cells	maturation of erythrocytes
Thrombopoietin	31	liver, kidney	maturation of platelets

CSF, colony-stimulating factor

CD34 – give rise to lymphocytic or myelomonocytic cells, as well as platelets and ERYTHROCYTES (Fig. 1) (see also chapter A.1).

Differentiation of cells of the myelomonocytic cell lineage

The major differentiation factors of the myelomonocytic cell lineage are known best (Tab. 1).

Several of them are named according to the observation which led to their discovery, e.g., the different ways to stimulate outgrowth of colonies from BONE MARROW cell cultures, i.e., CSF. Some of these factors – e.g., STEM CELL factor (SCF), multi-CSF (synonymous with IL-3), regulate early differentiation steps. Others, such as granulocyte/monocyte (GM) CSF, control intermediate steps or selectively induce end differentiation into mature (neutrophilic) GRANULOCYTES (G-CSF) or MONOCYTES (M-CSF). Similarly erythropoietin, synthesized in the kidney, promotes generation of ERYTHROCYTES, and thrombopoietin, which is synthesized in the liver and spleen, promotes formation of platelets. Table 1 summarizes their predominant physiological properties.

Pharmacological implications

As gene technology has facilitated the production of sufficient amounts, CSFs are now exploited as drugs (Tab. 2) [5, 6]. Thus erythropoietin has become established as the drug of choice for the treatment of severe anemias during terminal renal diseases or due to cytostatic therapy [7]. Thrombopoietin has been applied successfully in clinical trials for the treatment of thrombocytopenias [8]. FILGRASTIM (human recombinant G-CSF with an additional methionine, generated from bacteria) was the first CSF approved for the treatment of granulocytopenias. Similarly to LENOGRASTIM (human recombinant G-CSF from eukaryotic cells), it promptly and selectively increases up to 100-fold the number of functionally active neutrophils, for instance in patients with cytotoxic drug-induced granulocytopenias. By treatment with G-CSF the incidence and severity of infections leading to hospital admissions in patients receiving chemotherapy because of malignant tumors could be markedly reduced. So far, however, the therapy has not led to an increase in life expectancy. GM-CSF (MOLGRAMOSTIM) has been approved for similar indications. All other CSFs have been evaluated in clinical trials.

TABLE 2. INDICATION FOR CYTOKINES
I. Reconstitution of a compromised immune system (physiological effects)

Cytokine	Target cell	Indication
Epo	erythroid progenitor cells	anemia
Tpo	megacaryocytic progenitor cells	thrombocytopenia
G-CSF	myeloid progenitor cells	granulocytopenia
M-CSF	monocytic progenitor cells	monocytopenia
GM-CSF	myelomonocytic progenitor cells	leukopenia
IL-3	myelomonocytic progenitor cells	leukopenia
IL-2	lymphocytic progenitor cells	lymphopenia
IL-7	lymphocytic progenitor cells	lymphopenia
IL-11	lymphocytic progenitor cells	lymphopenia
	megacaryocytic progenitor cells	thrombocytopenia
IL-6	hematopoietic stem cells	stem cell deficiency
		expansion of stem cells *in vitro*

Therapeutically administered CSFs are intended to substitute for the loss of a patient's own differentiation factors. Despite this, they – like all drugs – can cause side-effects. For the CSFs these include bone and muscle pain, dysuria, sometimes elevation of liver enzymes and uric acid, and, rarely, a drop in blood pressure, eosinophilia, or allergic reactions [5].

Lymphocyte differentiation

Differentiation of T and B LYMPHOCYTES from STEM CELLS proceeds in a much more complex way. The central process consists of the generation of the huge (>10^8) diversity of antigen-specific receptors. The most important element involved is the free combination of a finite number of gene elements at the level of DNA during the differentiation of the cell lineage. For this gene rearrangement IL-7 appears to be indispensable, in B as well as in T LYMPHOCYTE development. Additionally, less well-characterized cytokines are involved in the maturation of T LYMPHOCYTES in the thymus or of B LYMPHOCYTES in the BONE MARROW, including for instance IL-11 [9]. Both have been tested clinically in patients with lymphopenias.

Activation and growth factors of lymphocytes

Each immune and subsequent inflammatory response requires the interplay of several cell types. The activation of virgin B lymphocytes is initiated by binding of the specific antigen to their surface IgM and IgD receptors. This can lead to secretion of small amounts of IgM. The expansion of the antigen-reactive B cells and, concomitantly, effective IMMUNOGLOBULIN synthesis requires additional stimuli provided by IL-4, IL-5, IL-6, IL-10 and IL-13 [10]. These cytokines are synthesized and secreted by a subpopulation of $CD4^+$ T-helper lymphocytes, the Th 2 cells. These INTERLEUKINS not only promote the maturation of Ig-producing B lymphocytes to the fully secretory plasma cell, but also control switching to synthesis of other Ig-ISOTYPES. This includes IgG, which is impor-

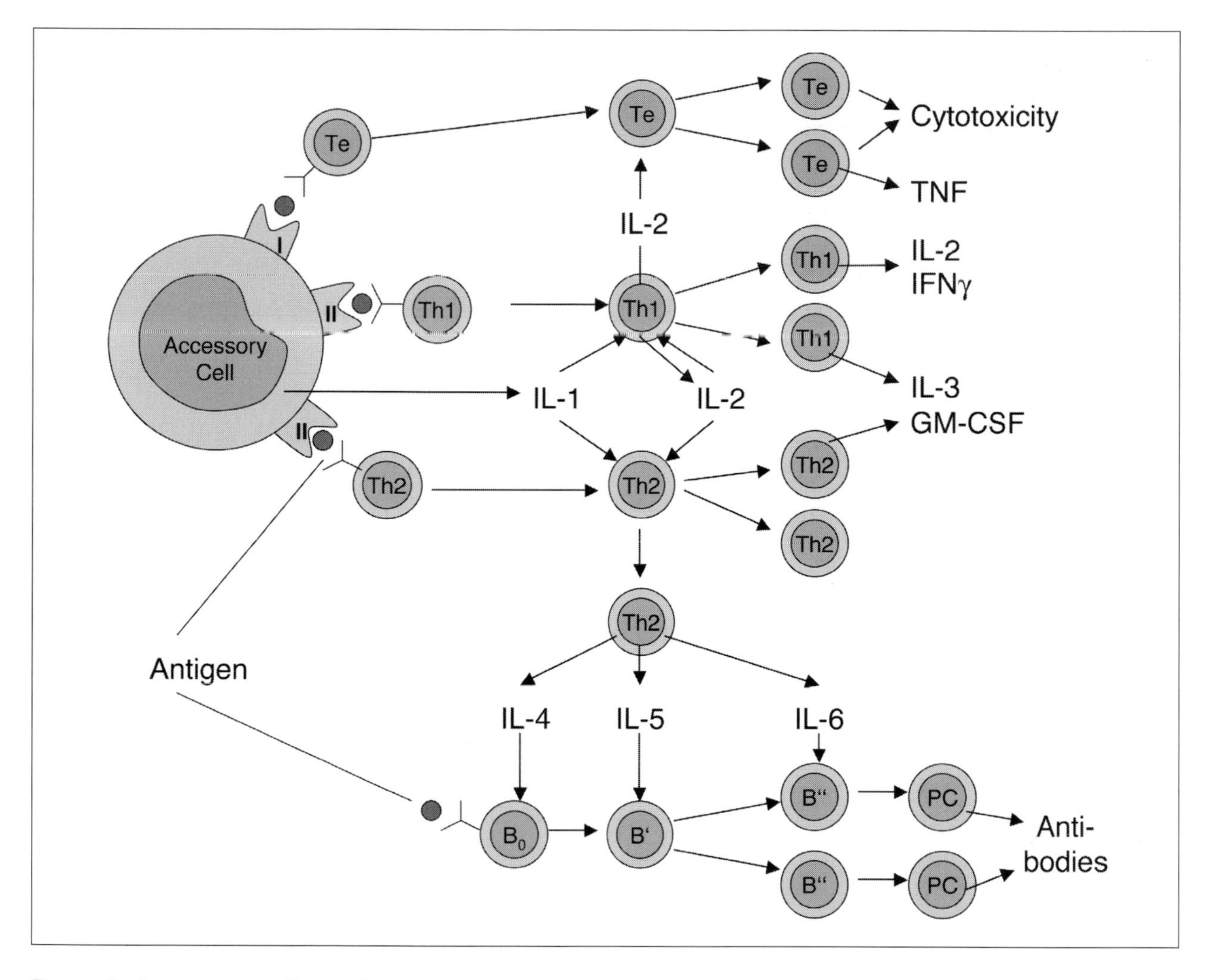

Figure 2. Activation of T and B lymphocytes

B, B lymphocyte; B_0, virgin B lymphocyte; B', B", activated B lymphocytes; GM CSF, granulocyte/monocyte colony-stimulating factor; IFN, interferon; IL, interleukin; PC, plasma cell; Th, T helper lymphocyte; Te, T effector lymphocyte; TNF, tumor necrosis factor; I, MHC class I molecule; II, MHC class II molecule

tant for an effective antibacterial defense, or IgA, which is secreted through mucosal linings and thus exerts an early line of defense before entry of bacteria. For the synthesis of IgE, which is responsible for Type I allergic reactions but also for defense against parasites, IL-4 or IL-13 are of pivotal importance (Fig. 2) (see chapter A.3).

Activation of T lymphocytes, too, requires participation of cytokines [10]. To ensure an effective cellular immune defense, antigen-reactive T lymphocytes proliferate and thereby expand clonally: from a single lymphocyte up to 10^7 descendants may originate in this way. The predominant T lymphocyte growth factor is IL-2, which, if absent, may be substituted by IL-15. IL-2 is formed by the second CD4 positive T-helper subpopulation, the Th-1 cells, and acts on cytotoxic T lymphocytes as well as on all T-helper lymphocytes themselves. Cytotoxic T lymphocytes

Table 3. Interleukins

Cytokine	Molecular mass	Predominant producer cells	Major functions
IL-1α IL-1β	17	monocytes	activation of T lymphocytes and inflammatory cells
IL-2	15	T lymphocytes	activation and proliferation of T lymphocytes, promonocytes, NK cells
IL-3	14–28	T lymphocytes	differentiation and propagation of early myeloid progenitor cells
IL-4	15-20	T lymphocytes	activation and proliferation of Th2 cells, B lymphocytes, inhibition of macrophage activation
IL-5	45-60	T lymphocytes	proliferation of B lymphocytes, maturation of eosinophils
IL-6	26	T lymphocytes, many other cells	activation of B, T lymphocytes and other cells, stem cell expansion
IL-7	25	stromal cells	maturation of T and B lymphocytes
IL-8	10	monocytes	chemotaxis and activation of granulocytes, chemotaxis of T lymphocytes
IL-9	37–40	T lymphocytes	propagation of mast cells, megacaryocytes
IL-10	17–21	T lymphocytes	inhibition of cellular immune and inflammatory reactions, propagation of mast cells
IL-11	23	stromal cells	maturation of lymphocytes, proliferation of myeloid and megacaryocytic progenitor cells
IL-12	p35/p40 dimer	monocytes, dendritic cells	activation of Th1 lymphocytes, NK cells
IL-13	17	T lymphocytes	activation and proliferation of B lymphocytes, inhibition of cellular immune reactions
IL-14	60	T lymphocytes	activation of B lymphocytes,
IL-15	14–15	epithelial cells	activation and proliferation of T lymphocytes, NK cells
IL-16	14	T lymphocytes	chemotaxis of T lymphocytes, macrophages
IL-17	35 homodimer	T lymphocytes	induction of proinflammatory cytokines
IL-18	24	monocytes/dendritic cells	activation and differentiation of Th1 lymphocytes

IL-19, 20, 22, 24 together with IL-10 constitute the IL-10 family
IL-21 related to IL-2, IL-15
IL-23 related to IL-12 (dimer of IL-12, p 40 plus separate p19)

(carrying the surface marker CD8) can kill antigen bearing target cells, most importantly virus-infected cells. Activated Th1 cells, by secreting INTERFERON γ (and other mediators such as TUMOR NECROSIS FACTOR β), initiate inflammatory reactions (see below). It should be noted that by secreting IL-3 and GM-CSF, these T cells also stimulate the formation of monocytic and myeloid cells in the BONE MARROW, which are required for inflammatory reactions. INTERLEUKINS which have been defined so far at a molecular level are summarized in Table 3.

INTERLEUKINS by definition mediate interactions between the different leukocytes. About 30 members have been ascribed so far to this rather heteroge-

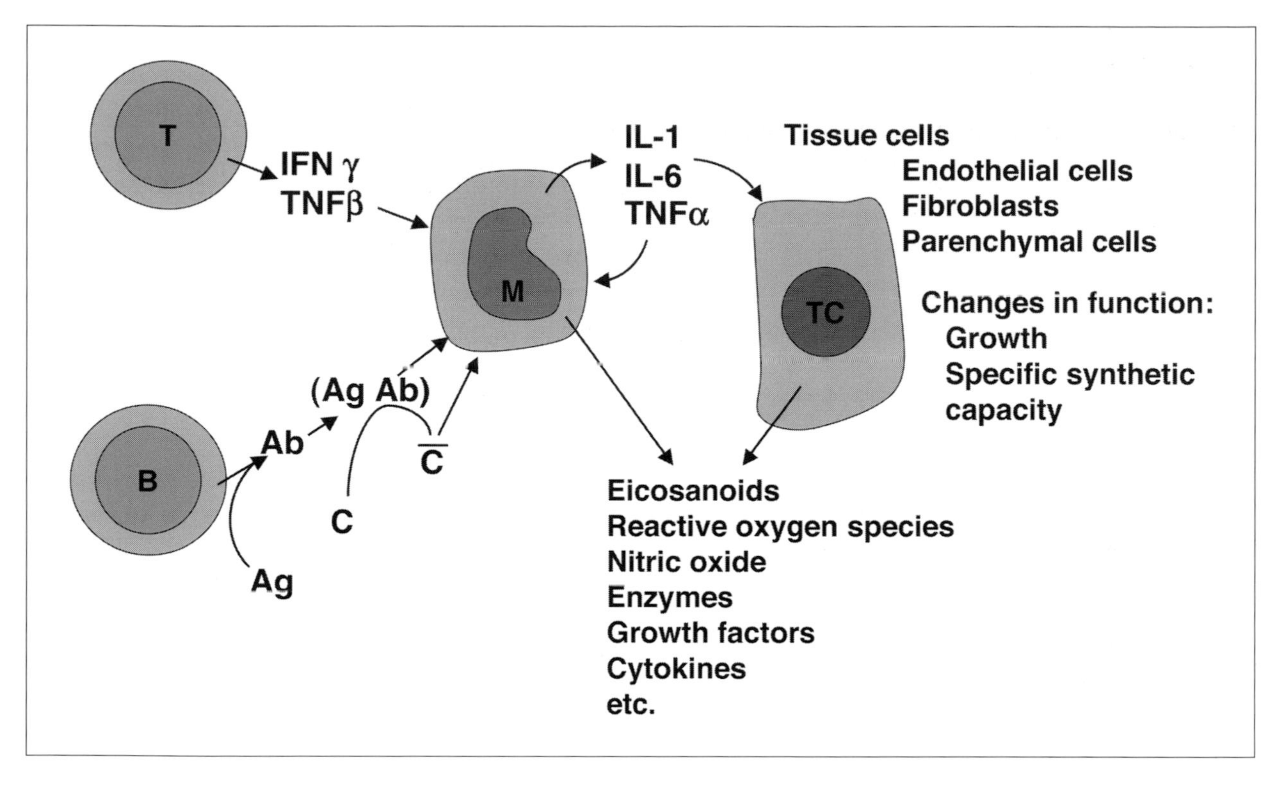

FIGURE 3. IMMUNE REACTIONS AND INFLAMMATION
Ag, antigen; Ab, antibody; B, B lymphocyte; C, complement; C̄, activated complement components; IFN, interferon; IL, interleukin; M, macrophage; TC, tissue cells; TNF, tumor necrosis factor

neous group of proteins and it is likely that their number will further increase in the future. Interestingly, some more "novel" INTERLEUKINS exhibit properties similar to those which had been detected earlier. For example IL-13 exhibits properties similar to IL-4; it even binds to the IL-4 receptor as well as to its own receptor [11]. IL-21 is related to IL-2, and to IL-15. IL-19, 20, 22, 24 together with IL-10 constitute the IL-10 family [12]. This shows that for functionally important INTERLEUKINS, surrogate molecules exist which at least partially can take over the function of the former when it is synthesized in insufficient amounts. The redundancy indicates the importance of a functioning immune system for the survival of a species. Other factors with overlapping functional properties are likely to be responsible for the fine tuning of immune responses.

Mediators of inflammation

Mechanisms and pathophysiology

In response to their specific antigen, B lymphocytes secrete ANTIBODIES and antigen-reactive T lymphocytes expand clonally. ANTIBODIES and cytotoxic T lymphocytes may immediately deal with antigens, or antigen-bearing cells, for example by neutralizing poisons or killing invaded cells. The major proportion of the defense, however, is generally exerted by secondary inflammatory mechanisms. To a large extent, these consist of the infiltration and activation of non-specific leukocytes, mononuclear phagocytes and various GRANULOCYTES [13, 14]. The inflammatory cells are recruited to participate in inflammation by ANTIBODIES as well as by mediators formed by Th1 cells.

Table 4. Interferons

Cytokine	Molecular mass	Predominant producer cells	Major functions
IFN's-α 15 proteins	19–26	monocytes	induction of antiviral activity
IFN-α1			inhibition of tumor cell growth
IFN-α2			activation of cells
... etc.			
IFN-β	23	fibroblasts	induction of antiviral activity
IFN-γ	17–25	T lymphocytes	induction of antiviral activity, activation of macrophages, immunoregulation

As a typical example, the activation of MONOCYTES/MACROPHAGES is depicted in Figure 3. Besides GM-CSF, which not only induces the differentiation of MONOCYTES and GRANULOCYTES, but also activates mature cells, IFN-γ and TNF-β constitute the most important macrophage-activating factors.

IFN-γ, generated by Th1 lymphocytes, is a member of the protein family of INTERFERONS, IFN-α being formed mainly by MONOCYTES and IFN-β by fibroblasts (Tab. 4) [15].

In addition to their antiviral activity, all INTERFERONS can activate cells, most notably the non-specific cells of the immune system, i.e., MACROPHAGES and GRANULOCYTES, to a varying extent.

TNF-β is related to TNF-α, which is synthesized mainly by MACROPHAGES. As suggested by their name, both possess antitumoral activities [16]. TNF-α and β recently were assigned to a large super-family of distantly related proteins, the TNF-super-family ligands (TNFSFL) with more than 20 members. Most of them are type 2 transmembrane proteins and only TNFSFL1 (TNF-β) and TNFSFL2 (TNF-α) usually are referred to as cytokines and are listed in Table 5.

Importantly, IL-1, TNF and to a lesser extent IL-6, effectively activate many tissue cells, including endothelial cells and parenchymal cells [16, 17]. In this way, these tissue cells are recruited to contribute to an inflammatory reaction. In addition, these cytokines also enhance the activities of the mononuclear phagocytes and other leukocytes, in an autocrine and paracrine way, thereby amplifying the inflammatory reaction. These dual properties give the "INFLAMMATORY CYTOKINES" a central role as mediators of inflammation. In addition to their direct effects, the boosting of inflammatory defense mechanisms against infectious agents and tumors constitutes the basis for attempts to positively modulate infectious diseases or malignant tumors by the administration of cytokines (Tab. 6) (see chapters A.9, C.5 and C.6).

Pharmacological relevance

Approved indications for natural INTERFERON-β include severe virus infections, such as recurring *Varicella zoster* infections or *Herpes simplex* infections of the eye. Recombinant INTERFERON-β (INTERFERON-β 1a or 1b) is indicated in some forms of multiple sclerosis.

IFN-α (INTERFERON-α 1a, and INTERFERON-α 2b) is indicated in chronic hepatitis B and hepatitis C (see chapter C.6), as well as in the treatment of some malignant tumors, including hairy cell leukemia or chronic myeloid leukemia. It is effective against some other tumors, such as non-Hodgkin's lymphoma, cutaneous T cell leukemia, malignant melanoma, hypernephroma or bladder carcinoma, but it is ineffective against the majority of carcinomas. Antitumor effects of IFN-β and IFN-γ in humans are uncertain. Recently conjugates of INTERFERON-α 2a or 2b with polyethylene glycol (INTERFERON-α 2a or 2b, PEGYLATED) have been introduced for the treatment of hepatitis C. These conjugates are slowly taken up

TABLE 5. TUMOR NECROSIS FACTORS

Cytokine	Molecular mass	Predominant producer cells	Major functions
TNF-α (TNFSFL2)	17	monocytes	activation of many cells
		many other cells	induction of apoptosis
TNF-β (TNFSFL1)	17	T lymphocytes	cachexia, shock

TABLE 6. INDICATIONS FOR CYTOKINES
II. Activation of normal cells of the immune systems (pharmacodynamic actions)

Cytokine	Target cells	Indication
IFN-α	virus-infected cells, tumor cells	viral infections, malignant tumors
IFN-β	virus-infected cells	viral infections
TNF-α, β	monocytes/macrophages, tumor cells	malignant tumors
IL-1	monocytes/macrophages, tumor cells	malignant tumors (?)
IL-2	T lymphocytes, monocytes, NK-cells	malignant tumors, bacterial, viral infections (AIDS)
IL-6	monocytes/macrophages, tumor cells	some malignant tumors, expansion of stem cells

from a subcutaneous depot and also excreted slowly, thus providing sustained blood levels.

Clinical side-effects of INTERFERONS include the common "flu-like" syndrome (fever, fatigue, shivering, muscle and joint pain), paresthesias, disturbances of the central nervous system, gastrointestinal disturbances, cardiac symptoms, granulocytopenia, thrombopenia or anemia [15].

The cytokines TNF-α, IL-1, IL-2, IL12 and IL18 have been the subjects of clinical trials for treating malignant tumors (see chapter A.9 and C.5). Despite some positive results – such as with IL-2 in metastasizing renal cell carcinoma or melanoma [18] – the overall outcome so far has been rather disappointing. The major reason for this appears to be the limitation by severe toxicity, which arises when these cytokines are administered systemically. This has prompted attempts to increase cytokine concentration locally in the tumor by perfusing it, for instance, with TNF. As experiments in animal models have been encouraging, this strategy is being followed in clinical trials with limb tumors or hepatocarcinomas [19]. Another strategy which is being followed is to transfect either tumor cells or tumor-infiltrating cells in a way that they constitutively secrete high amounts of a cytokine such as IL-2 or TNF. After reinfusion and subsequent redistribution to sites of the tumor, a high local concentration of the cytokines in the tumor may be achieved, which supports sufficient antitumoral activity but is accompanied by tolerable systemic effects [20]. A number of clinical trials of this gene therapy has been initiated. It should be added that the greatest utility so far of cytokines in the treatment of cancer has been for the management of cancer - or cancer therapy – associated side-effects such as decrease in white blood cells, anemia or thrombocytopenia. This includes G-CSF, Epo, thrombopoietin IL-3, IL-7 or IL-11 (see above) [21].

Modern molecular biological methods allow changes in the composition of proteins to yield new or altered functional properties. Thus a desired effect could be enforced, and simultaneously an unwanted side-effect diminished. This concept has been applied to cytokines, and many new functional proteins were created ("designer cytokines", "supercytokines", "MUTEINS"). One example is a fusion protein of IL-6 and the extracellular domain of its receptor ("super IL-6"), which (due to the property that this complex can also activate cells which contain only the second, non-ligand-binding chain of the IL-6 receptor) is more than 100-fold as potent as IL-6. This construct is now used to expand hematopoietic STEM CELLS *in vitro* [22].

Infection with HIV leads to progressive and preferential loss of CD4$^+$ T helper cells. Death results from opportunistic infections or, less often, malignant tumors. In combination with effective antiviral treatment strategies (i.e., at least triple combinations of drugs such as zidovudin, didanosin and sequinavir), IL-2 can dramatically ameliorate immunodeficiency (see chapter C.6).

Regulatory factors of immune reactions

The Th1/Th2 paradigm

While it is desirable during infectious diseases or malignant tumors to augment immune and subsequent inflammatory reactions, in other situations, such as autoimmune, chronic inflammatory or allergic diseases, they are pathogenic. Under physiological conditions, immune reactions are tightly controlled, and immunological diseases may therefore be regarded as failures of immunoregulation. In recent years, the balance between the activity of Th1 and Th2 cells has emerged as the central control mechanism (Fig. 4, see also Fig. 2) [23, 24].

Th1 cells secrete the cytokines IL-2, IL-18 and IFN-γ, which – as described above – promote cellular defence and inflammatory responses. Th2 cells, on the other hand, synthesize those cytokines which predominantly regulate the activation of B lymphocytes, IL-4, IL-5, IL-6 and IL-13. Both Th-subpopulations develop from common CD4$^+$ precursors, Thp. IFN-γ and IL-12 together with IL-18 promote differentiation into Th1 cells, whereas IL-4 and IL-13 are responsible for Th2 differentiation. Simultaneously, the subpopulation-specific cytokines block development of the opposite Th-subpopulation. For IL-10 the suppression of cellular immunoreactions and inflammation is the predominant function. This implies that if one subpopulation gains a developmental advantage, this is reinforced while the corresponding subpopulation is suppressed.

Since no Th1 or Th2 specific ANTIGENS or epitopes have been detected, the initial channeling of preferential differentiation into a specific Th-subpopulation – and thereby the type of immune reaction – must be directed by other cells. For Th1 cells, this is a function of MONOCYTES/ MACROPHAGES and DENDRITIC CELLS which, following interaction with bacteria or their components (e.g., lipopolysaccharide), release IL-12 and IL-18, which in turn initiate differentiation to this subpopulation. For Th2 cells, basophilic GRANULOCYTES or mast cells play a similar role, since they contain vast amounts of IL-4, which can be released in direct contact with a number of known allergens or parasitic antigens.

Pharmacological implications

It is obvious from these data that IL-4, IL-10 or IL-13 should exert antiinflammatory effects in non-allergic situations. Indeed, this is supported by many *in vitro* experiments and more relevant, experimental animal models. In these studies it was found that not only development of Th1 cells was inhibited, but also the activation of MACROPHAGES, the predominant "chronic" inflammatory cells. All these cytokines, therefore, are presently undergoing clinical evaluation in several chronic inflammatory diseases, including rheumatoid arthritis.

Chemokines

Pathophysiology

Nearly all inflammatory – as well as allergic – diseases are confined to certain organs. This implies that all cells of the immune system which participate in the underlying pathomechanisms must escape from the blood stream and invade the pertinent tissue. On the other hand, antigenic material of infec-

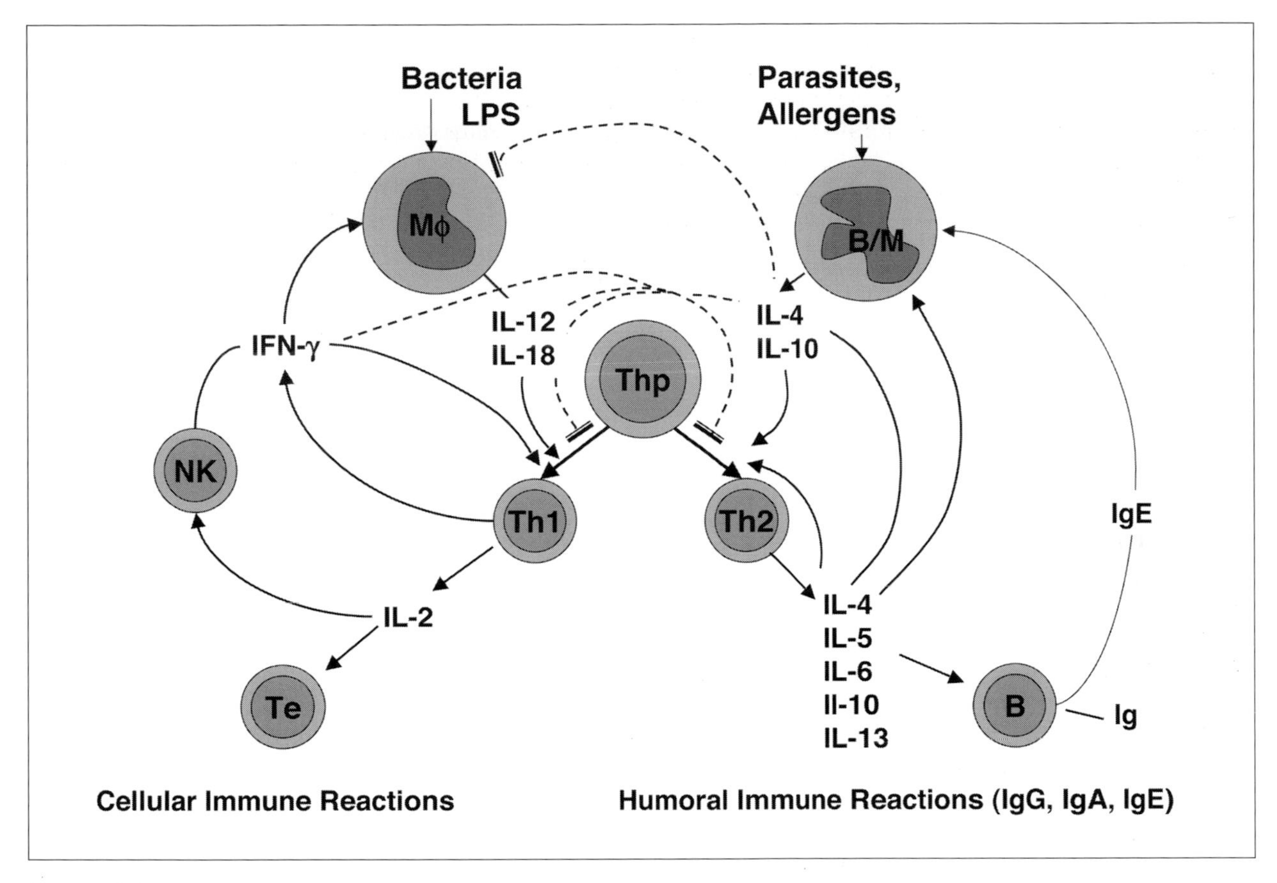

Figure 4. Regulation of immune responses
B/M, basophilic granulocytes/mast cells; B, B lymphocytes; IFN, interferon; Ig, immunoglobulin; IL, interleukin; M, macrophage; NK, natural killer cell; Th, T helper cell; Thp, T helper cell precursor; Te, T effector cell

tive agents penetrating into the body must be taken up by antigen-presenting DENDRITIC CELLS and carried in this form to the adjacent LYMPHOID ORGANS, such as the regional lymph nodes, to initiate an effective immune response. The very complex migration of LEUKOCYTES which proceeds in several subsequent defined steps and, similarly, the migration of DENDRITIC CELLS, are also controlled by various cytokines (Tab. 7).

Among these the protein family of CHEMOKINES plays a pivotal role [25] (Tab. 8).

CHEMOKINES are small proteins with molecular masses between 8 and 10 kDa with certain four characteristic cysteine residues forming intracellular bonds. More than 40 CHEMOKINES have been CLONEd in human beings. According to their structure – which is also represented at the genomic level by gene clusters – three groups can be distinguished. CXC CHEMOKINES, now denominated as CXCL 1 to 15 (L stands for ligand), where the cysteines are separated by an arbitrary amino acid) are predominantly chemotactic for neutrophils and some – interestingly forming a small separate subcluster – for T lymphocyte subsets. CC-CHEMOKINES, CCL1 to 27, mostly attract monocyte MACROPHAGES (again a subcluster of lymphocytes) and XC CHEMOKINES, XCL1 to 2 (or CL1 to 2) lymphocytes. There exists also one member of a CX3C family. The CHEMOKINE families bind with over-

Table 7. Participation of circulating leukocytes in local inflammation

Reaction step	Cytokines	Main target cells
Chemotaxis	IL-8 (CXCL8)	Neutrophils
	Eotaxin (CCL11)	Eosinophils
	MCP-1 (CCL2)	Monocytes
	RANTES (CCL5)	Monocytes
	Lymphotactin (CL1)	Lymphocytes
Emigration from the blood	IFN-γ	Endothelial cells (expression of
	IL-1	cell adhesion molecules such as ELAM-1,
	TNF	ICAM-1, VCAM-1)
Activation	IFN-γ	Monocytes
	IL-1	Monocytes
	TNF	Monocytes
	CSFs	Monocytes
	IL-8 (CXCL8)	Granulocytes
Expansion	CSFs	Myeloid precursor cells

Table 8. Selected chemokines

CXC chemokines	CXCL1	(Gro-α)
	CXCL4	(PF4)
	CXCL8	(IL-8)
	CXCL10	(IP-10)
	CXCL12	(SDF-1α/β)
XC chemokines	XCL1	(Lymphotactin)
	(CL1)	
CX3C chemokines:	CX3CL1	(Fractalkine)
C-C chemokines:	CCL2	(MCP-1)
	CCL3	(MIP-1α)
	CCL4	(MIP-1β)
	CCL5	(RANTES)
	CCL11	(Eotaxin)

lapping patterns to selective receptor families, of which there are five known for CXC Ligands, CXCR1 to 5, 10 for CC Ligands, CCR 1-10, and one XCR1 or CX3R1, which all belong to the seven transmembrane-domain G-protein coupled receptors. CHEMOKINES have fundamental roles in the development, homeostasis and function of the immune system. Within the immune system (which is the scope of this chapter) they can be divided into two categories: homeostatic and inflammatory. The former are those which are constitutively expressed in organs or tissues, suggesting a function involving cell migration.

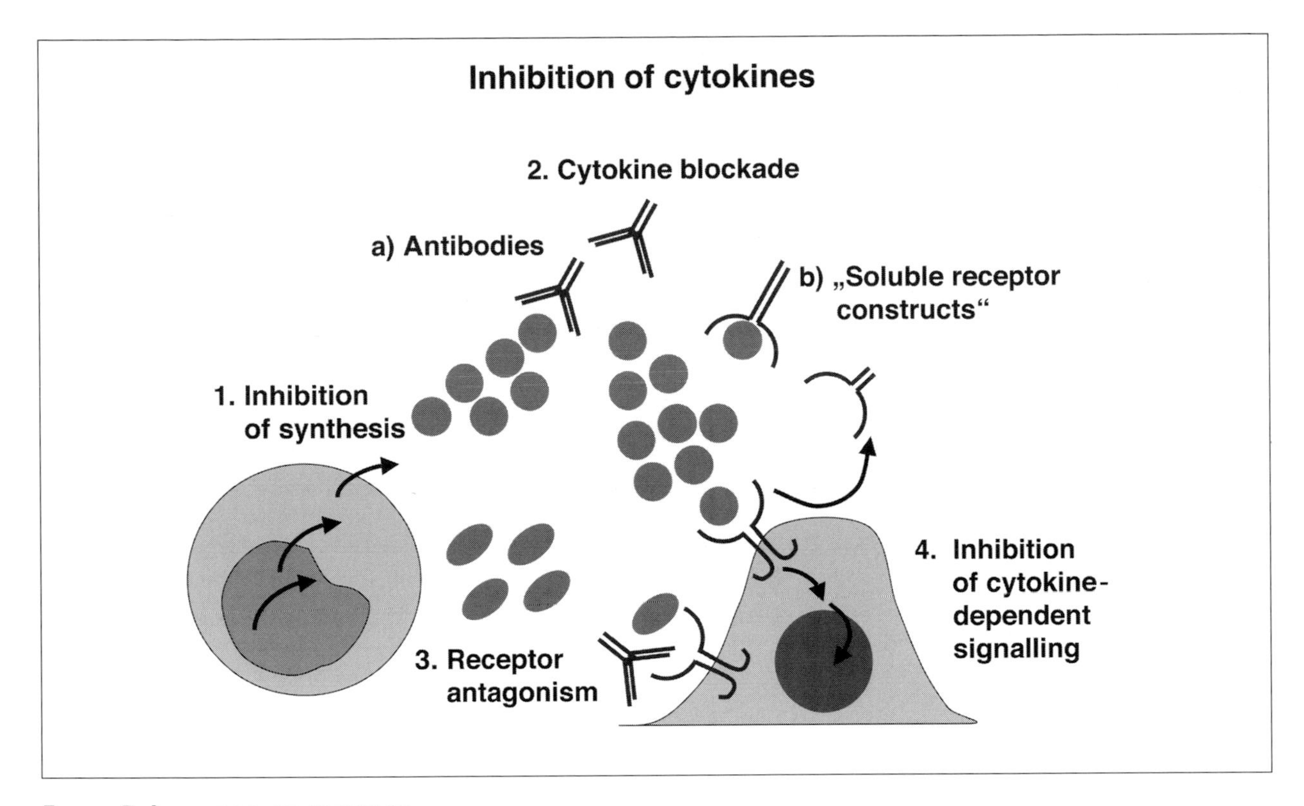

FIGURE 5. INHIBITION OF CYTOKINES

●, cytokine; ⬬, mutated antagonistic cytokine; ⅄, functionally active cytokine receptors; Y, extracellular domains of ("soluble") receptors

The inflammatory chemokines are strongly upregulated by proinflammatory stimuli, predominantly in cells of the immune system (MACROPHAGES, T lymphocytes, but also fibroblasts) and participate in the development of inflammatory and immune reactions. This function is supported by the fact that these CHEMOKINES, in addition to their chemotactic properties, are also potent activators of their target cells.

Pharmacological implications

Their strong proinflammatory properties make CHEMOKINES candidate drugs. Thus quite a number of CHEMOKINES has shown antitumoral effects in various tumor models [26]. Their participation in inflammatory diseases, on the other hand, make them targets for a specific intervention. This is favored by the possibility to generate receptor antagonists of at least some CHEMOKINES by relatively simple procedures (such as deletion of some N-terminal amino acids [27]. Indeed, in some experimental setups encouraging results were found. Progress is hampered, however, by the great redundancy of the CHEMOKINE/CHEMOKINE receptor families.

CCR 3 and CCR 5 have recently raised great interest as coreceptors (in addition to CD4) for the entry of human immunodeficiency virus (HIV) into MACROPHAGES and CXCR4 (also termed fusin) for HIV entry into T lymphocytes. CHEMOKINES and their antagonists effectively blocked *in vitro* entry of macrophage-tropic or lymphotropic HIV strains by binding to their receptor, which raises hope for new therapeutic strategies in AIDS [27] (see chapter C.6).

Table 9. Cytokine inhibitors

Mode of action	Example drugs
1. Inhibition of synthesis	
Reduction of the number of cytokine-producing cells	
cytostatic drugs	azathioprine
monoclonal antibodies to cells	muromonab CD3
regulation of cell activity	
regulatory cytokines	(interleukin-4,-10)*
cyclosporine and related drugs	cyclosporine, tacrolimus
regulation of gene expression	
glucocorticoids	prednisone
2. Decrease of the concentration in active (free) form	
monoclonal antibodies against cytokines	infliximab
soluble cytokine receptors	etanercept
3. Receptor blockade	
monoclonal antibodies against cytokine receptors	basiliximab
cytokine antagonist	anakinra
4. Inhibition of cytokine-dependent signalling	
protein kinase inhibitors	sirolimus

**not yet approved*

Inhibition of cytokines

Pathophysiology

Because of their multiple functions in the inflammatory process, cytokines offer a useful target for therapeutic intervention [27–29]. Inhibition of cytokines may be achieved by several mechanisms[30].

While nearly each feasible possibility is going to be evaluated in clinical trials, quite a number of innovative drugs was approved within the last few years for indications such as rheumatoid arthritis and inflammatory bowel disease, or the prevention of transplant rejection. Suppression of immune and inflammatory reactions can be achieved by inhibition of cytokines in several ways: 1. the inhibition of cytokine synthesis, 2. the decrease of cytokines in free active form, 3. the blocking of the interaction with their receptor, or 4. the inhibition of cytokines-dependent signalling.

Pharmacological implications

For each mechanism, at least one clinically relevant example exists.

Cytostatic drugs such as azathioprine or monoclonal ANTIBODIES against T lymphocyte epitopes (such as CD3 or CD4) decrease the number of (T-) lymphocytes and thereby also cytokine-producing cells. They are effective in preventing transplant rejection and also in the treatment of autoimmune diseases (an example being lupus erythematosus). Without being cytotoxic, by interfering with T CELL RECEPTOR signalling, modern immunosuppressants, including CYCLOSPORIN A or TACROLIMUS, very selectively block the synthesis of T lymphocyte cytokines, predominantly of their growth factor IL-2 (see chapter C.9). In a similar way GLUCOCORTICOIDS (e.g., prednisone) are immunosuppressive by interfering with the gene expression of this cytokine and others synthesized by T lymphocytes. GLUCOCORTICOIDS also represent the most efficient antiinflammatory drugs available. Although they affect multiple proinflammatory mechanisms, their efficiency largely relies on the capacity to block gene expression of most INFLAMMATORY cytokines including IL-1 to IL-8, TNF or IFN-γ. A new group of experimental drugs, termed cytokine suppressive antiinflammatory drugs (CSAID) has been found, which exhibits a striking selectivity for inhibiting the synthesis of IL-1 or TNF by interfering with critical signal transduction steps. None of these has so far been approved for clinical use. Th2 cytokines, such as IL-4 or IL-10, can decrease the synthesis of proinflammatory cytokines by down-regulating the activation of their producer cells, e.g., MACROPHAGES (see Fig. 4). Both – predominantly IL-10 – proved to be effective in clinical studies for the treatment of rheumatoid arthritis and other inflammatory diseases [31].

ANTIBODIES can block the action of secreted cytokines. A humanized version of an ANTIBODY directed to TNF-α (INFLIXIMAB) represented the first example of a specific cytokine blocking antibody, proving efficacious in rheumatoid arthritis or inflammatory bowel disease. INFLIXIMAB resulted – at least in a group of patients – in long-lasting remissions. This ANTIBODY was succeeded by a completely human ANTIBODY (generated by phage display technology) with similar properties (ADALIMUMAB). The major side-effect of the therapy with anti-TNF is an increased risk of infections – as would be expected! – including recurrence of tuberculosis and occasionally septic shock [32]. Several other anticytokine ANTIBODIES, including anti IL-1 and anti IL-6 ANTIBODIES, are being tested in clinical trials [33].

In addition to ANTIBODIES, blocking of free active cytokines can also be achieved in a quite different way. Illustrating the principle of self-limitation, the extracellular domains of many cytokine receptors are released during immune or inflammatory reactions. As these "soluble receptors" contain the full cytokine binding site, they bind the cytokine and thereby dampen its biological effect. A soluble TNF-receptor construct – in which "soluble" TNF-receptors were fused to the constant parts of human IgG to increase affinity and half-life *in vivo* – (ETANERCEPT) has been approved for rheumatoid arthritis and inflammatory bowel disease with similar effectiveness and side-effects as infliximab. As a unique example, a naturally occurring antagonist of IL-1, IL-1ra, has been found and CLONED. For the indication rheumatoid arthritis, this antagonist (ANAKINRA) has been approved recently. Its predominant side-effect, too, is the increased risk of infections.

Several cytokine mutants ("MUTEINS"), which behave as antagonists, have been produced by gene technology. Considerable therapeutic hope has been

TABLE 10. POTENTIAL INDICATIONS FOR INHIBITORS OF CYTOKINES

Cytokine	Indication
IL-1	chronic inflammatory diseases
IL-2	organ transplantation, autoimmune diseases
IL-4, IL-13	allergic (type 1) diseases, esp. allergic asthma
IL-5	allergic asthma (?)
IL-6	chronic inflammatory diseases, some hematological tumors
IL-8	chronic inflammatory diseases, autoimmune diseases and other chemokines
IL-12, IL-18	chronic inflammatory diseases
TNF-α, β	septic shock, chronic inflammatory diseases
IFN-γ	autoimmune diseases

engendered by IL-4 antagonistic MUTEINS for the treatment of allergic asthma which is being tested in several clinical trials. Similarly, several antagonistic CHEMOKINE MUTEINS are being tested clinically in chronic inflammatory diseases, and as described above, in HIV infections [27]. Receptors can also be blocked by specific ANTIBODIES. BASILIXIMAB and DACLIZUMAB are directed against the IL-2 receptor and are effective as immunosuppressants in transplanted organ rejection episodes. Table 10 summarizes indications for inhibitors of cytokines.

Summary and outlook

It was not long after their discovery and subsequent molecular characterization that cytokines were tested for their therapeutic potential. This was only made possible by gene technology, which allowed sufficient amounts to be produced in good quality. Some of them – INTERFERONS or the COLONY-STIMULATING FACTORS – subsequently became established as drugs with great medical and even economic importance. Not all high-flying hopes, however, have been fulfilled, especially with regard to the treatment of malignant tumors. Thus, after a period of set-backs, new strategies have begun to evolve, which allow high local concentrations to be selectively generated, the most sophisticated approach involving the use of genetically altered cells.

On the other hand, cytokines are now known to be crucial participants in the pathogenesis of many diseases. The realization that long-known and valuable drugs, such as the glucocorticosteroids, act predominantly by suppressing the synthesis of certain cytokines, has prompted a search for mechanisms by which the synthesis or function of individual cytokines can be blocked more selectively. Even though cytokines or their inhibitors have developed into indispensable drugs in important indications, it is certain that this is only the beginning. This assumption is based on the growing evidence that these molecules contribute to many more diseases than those anticipated originally; important examples are atherosclerosis, congestive heart failure or neurodegenerative diseases.

Selected readings

Thomson AW, Lotze MT (eds.) (2003) *The Cytokine Handbook*, 4th ed.. Academic Press, London

Oppenheim J, Feldman M (2000) *Cytokine reference, including regularly updated online version*. Academic Press, New York

Recommended website

Horst Ibelgaufts' COPE: Cytokines Online Pathfinder Encyclopaedia: *www.copewithcytokines.de*

References

1 Balkwill FR (ed.) (2000) *Cytokine Molecular Biology. A Practical Approach*. Oxford University Press, Oxford, UK

2 Balkwill FR (2001) *Cytokine Cell Biology. A Practical Approach* 3rd edn. Oxford University Press, Oxford, UK

3 Fitzgerald KA, O'Neill LAJ, Gearing A, Callard RE (2001) *The cytokine facts book*, 2nd edn. Academic Press, San Diego

4 Silvernnoinen O, Ihle JN (1996) *Signalling by the Hematopoietic Cytokine Receptors*. Springer Verlag, Heidelberg

5 Moore MAS (1991) The clinical use of colony stimulating factor. *Annu Rev Immunol* 9: 159–191

6 Metcalf D (1990) The colony stimulating factors. Discovery, development and clinical applications. *Cancer* 10: 2185–2195

7 Jelkmann W (1992) Erythropoietin. Structure, control of production and function. *Physiol Rev* 72: 449–489

8 de Sauvage FJ, Hass PE, Spencer SD, Malloy BE, Gurney AL, Spencer SA, Dabonne WC, Henzel WJ, Wong SC, Kuang WJ et al (1994) Stimulation of megacaryocytopoieses and thrombopoiesis by the c-Mpl ligand. *Nature* 369: 533–538

9 von Boehmer H (1993) The developmental biology of T-lymphocytes. *Annu Rev Immunol* 6: 309–326

10 Paul WE, Seder RA (1994) Lymphocyte responses and cytokines. *Cell* 76: 241–251

11 Zurawski G, de Vries JE (1994) IL-13, an IL-4 like cytokine that acts on monocytes and B-cells, but not on T-cells. *Immunol Today* 15: 19–26

12 Frickenscher H, Hör S, Küpers H, Knappe A, Wittmann S, and Sticht H (2002) The interleukin-10 family of cytokines. *Trends Immunol* 23: 89–96

13 Gallin JJ, Snyderman R (eds.) (1999) *Inflammation. Basic Principles and Clinical Correlates*, 3rd edn.. Lippincott Williams and Wilkins, Philadelphia

14 Kuchroo VK, Sarvetnek N, Hafler DA, Nicholson LB (2001) *Cytokines and autoimmune diseases*. Humana Press, Totowa

15 Stuart-Harris R, Penny R (1996) *Clinical Applications of the Interferons*. Chapman and Hall, London

16 Vasalli P (1992) The pathophysiology of tumor necrosis factor. *Annu Rev Immunol* 10: 411–452

17 Dinarello CA (1996) Biological basis for interleukin-1 in disease. *Blood* 87: 2095–2147

18 Rosenberg SA (2000) Interleukin-2 and the development of immunotherapy for the treatment of patients with cancer. *Cancer J Sci Am* 6: 2–7

19 de Vries MR, ten Hagen TL, Marinelli AW, Eggermont AM (2003) Tumor necrosis factor and isolated hepatic perfusion: from preclinic tumor models to clinical studies. *Anticancer Res* 23: 1811-23

20 Kircheis R, Ostermann E, Wolschek MF, Lichtenberger C, Magin-Lachmann C, Wightman L, Kursa M, Wagner E (2002) Tumor-targeted gene delivery of tumor necrosis factor-alpha induces tumor necrosis and tumor regression without systemic toxicity. *Cancer Gene Ther* 9: 673–680

21 Griffin JD (2001) Hematopoietic growth factors. In: VT DeVita, S Hellmann, SA Rosenberg (eds.): *Cancer Principles and Practice of Oncology*. Lippincott Williams & Wilkins, Philadelphia, 2798–2813

22 Fischer M, Goldschmitt J, Peschel C, Brakenhoff JP, Kallen KJ, Wollmer A, Gortzinger J, Rose-John S (1997) A bioactive designer cytokine for human hematopetic progenitor cell expansion. *Nat Biotech* 15: 142–145

23 Romagnani S (1993) Induction of Th1 and Th2 responses: a key role for the "natural" immune response? *Immunol Today* 13: 379–381

24 Mosmann TR, Sal S (1996) The expanding universe of T-cell subsets: Th1, Th2 and more. *Immunol Today* 17: 138–146

25 Zlotnik A, Yoshie O (2000) Chemokines: A New Classification System and Their Role in Immunity. *Immunity* 12: 121–127

26 Homey B, Müller A, Zlotnik A (2002) Chemokines: Agents for the Immunotherapy of Cancer? *Nat Rev Immunol* 2: 175–184

27 Proudfoot AE (2002) Chemokine Receptors: Multifaceted Therapeutic Targets. *Nat Rev Immunol* 2: 106–107

28 Mantovani A, Dinarello CA, Ghezzi P (2000) Pharmacology of cytokines. Oxford University Press, Oxford

29 Dinarello CA, Moldawer LL (2002) *Proinflammatory and antiinflammatory cytokines in rheumatoid arthritis*. 3rd edn. Amgen Inc. Thousand Oaks, California

30 Ciliberto G, Savino R (2001) Cytokine inhibitors. Marcel Dekker, New York

31 Braat H, Peppelenbosch MP, Hommes DW (2003) Interleukin-10-based therapy for inflammatory bowel disease. Expert Opin Biol Ther 3: 725–731

32 Feldmann M, Ravinder NM (2001) Anti-TNFα Therapy of Rheumatoid Arthritis: What Have We Learned? *Annu Rev Immunol* 19: 163–196

33 Taylor PC (2003) Antibody therapy for rheumatoid arthritis. *Curr Opin Pharmacol* 3: 232–238

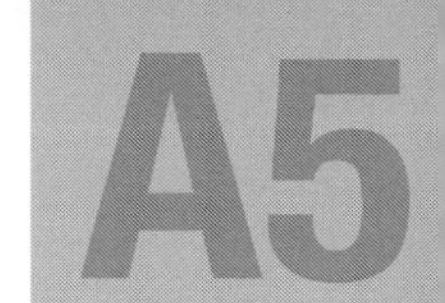

Innate immunity – phagocytes, natural killer cells and the complement system

Dirk Roos, Hergen Spits and C. Erik Hack

In the second half of the 19th century Eli Metchnikoff discovered that bacteria can be ingested (phagocytosed) by leukocytes present in the blood of many different animals, including very primitive ones. At about the same time, Paul Ehrlich found that certain agents dissolved in blood had bactericidal potential. The scientific discussion on the importance of cellular *versus* humoral factors in our defence against bacteria came to an end when it was recognized that both components enforce each other's effect. In 1908, these scientists shared the Nobel Prize for Physiology and Medicine.

Further investigation on the nature and the working mechanism of the cells and the proteins that constitute our immunological defence system showed that each of these components is made up of several different constituents. In its turn, this led to the insight that a functional distinction must be made between the adaptive and the innate branch of the immune system. The adaptive branch is executed by lymphocytes, i.e., white blood cells capable of generating antibodies against structures foreign to the body and killing virus-infected cells. LYMPHOCYTES are able to differentiate between structures that belong to the body and those that are alien. Moreover, these cells display immunological memory: once they have encountered foreign material previously, they will recognize and eliminate this material quicker upon subsequent encounters. The elimination itself is mainly the task of the INNATE IMMUNE SYSTEM. This branch consists of PHAGOCYTES, NATURAL KILLER CELLS and the COMPLEMENT SYSTEM. Phagocytes are white blood cells capable of uptake (phagocytosis) and intracellular killing of microbes, especially after binding of antibodies and complement proteins to the surface of the microbes (Fig. 1). NATURAL KILLER (NK) CELLS are lymphocytes with cytotoxic potential against virus-infected and certain tumour-transformed cells. NK cells differ from cytotoxic T lymphocytes in their HLA-independent manner of target cell recognition. The COMPLEMENT SYSTEM consists of a series of proteolytic enzymes capable of lysing micro-organisms, often in an antibody-accelerated fashion. The activities of these innate systems are tightly regulated, because they are in principle also harmful to the host. This chapter will give a short description of each of the innate systems, their clinical relevance and the potential for therapy in case of failure.

Phagocytes

Many cell types are capable – to some extent – of internalising micro-organisms, which sometimes leads to growth inhibition or even killing of the microbes. However, macrophages and granulocytes are the only 'professional' PHAGOCYTES, because these cells are equipped with a motile apparatus for actively moving to sites of infection (except for organ-localised macrophages, the so-called histiocytes), with surface receptors to bind micro-organisms, with granules filled with cytotoxic proteins and with an enzyme that can generate toxic oxygen radicals. Macrophages and granulocytes are formed in the bone marrow from pluripotent haematopoietic precursor cells, under the influence of growth and differentiation hormones. Macrophages are released into the blood after about ten days of development, as immature monocytes. These cells then move to the various tissues and organs, where they further differentiate into macrophages with site-specific characteristics. Granulocytes take about fourteen days to develop, and are released as mature cells. Most granulocytes differentiate into neutrophilic granulocytes, cells with a high anti-microbial potential. Other granulocyte types are eosinophilic granu-

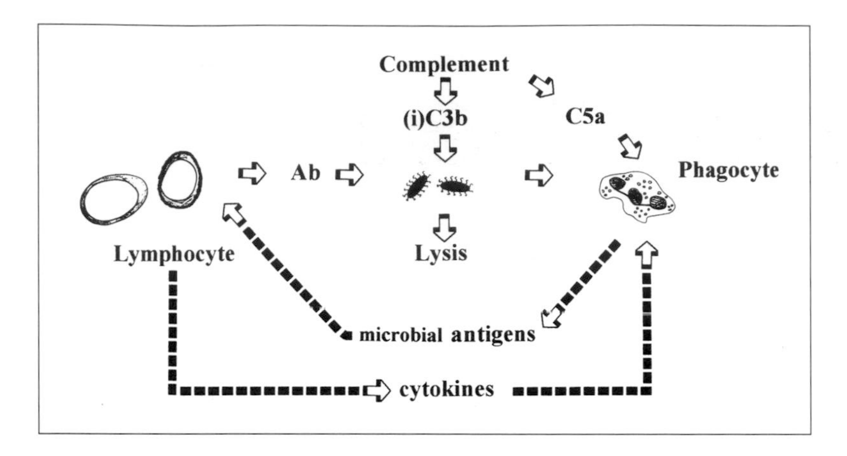

FIGURE 1. INTERACTIONS BETWEEN THE ADAPTIVE AND THE INNATE IMMUNE SYSTEM
Micro-organisms coated with antibodies and/or complement activation products are ingested, killed and degraded by phagocytes. Fragments of microbial proteins are presented to lymphocytes, which may lead to enhanced antibody production and release of cytokines that activate phagocytes. Other complement activation products (see Tab. 4) induce lysis of micro-organisms and attraction of phagocytes to infected areas.

locytes, involved in anti-parasite defence, and basophilic granulocytes, which lack the ability to phagocytose but can release HISTAMINE in inflammatory reactions. These cells also move into the tissues and organs. Macrophages have an estimated life span in the tissues of several months; neutrophilic granulocytes survive only 1–2 days after release from the bone marrow (4–6 days under inflammatory conditions). Thus, neutrophilic granulocytes (neutrophils) need to be formed in much larger numbers than macrophages for efficient surveillance against micro-organisms. Indeed, in healthy adults about 10^{11} neutrophils are released each day from the bone marrow, and this figure can increase ten-fold during infections. Macrophages are formed at not more than 10^9 per day. Phagocytes end their life either through necrosis as a result of phagocytosis and release of toxic mediators (pus formation) or through APOPTOSIS (programmed cell death) and removal by macrophages.

The importance of phagocytes for the host defence against micro-organisms can be concluded from the recurrent, life-threatening infections of patients with a genetic or acquired shortage or deficiency of these cells. Patients with a shortage of phagocytes may be treated with growth factors, such as granulocyte colony-stimulating factor (G-CSF), at least when the receptors for such factors are present on the precursor cells. Complete cure may be obtained by bone-marrow transplantation.

Movement

Neutrophils and monocytes have the ability to actively move to the site of an infection. This is caused by the release in these areas of so-called CHEMOTAXINS, small molecules of bacterial or host origin that diffuse into the surroundings and can bind to specific receptors on the phagocytes. The phagocytes are able to 'sense' the concentration gradient of the CHEMOTAXINS and to move into the direction of the source of these agents until they have reached the site of infection. This process is called CHEMOTAXIS. However, phagocytes in the blood must first pass the blood vessel wall before moving into the tissues [1]. This process of DIAPEDESIS is initiated by reversible interaction of L-SELECTIN, an adhesion protein on the surface of leukocytes, with carbohydrate structures on endothelial blood vessel cells, and by

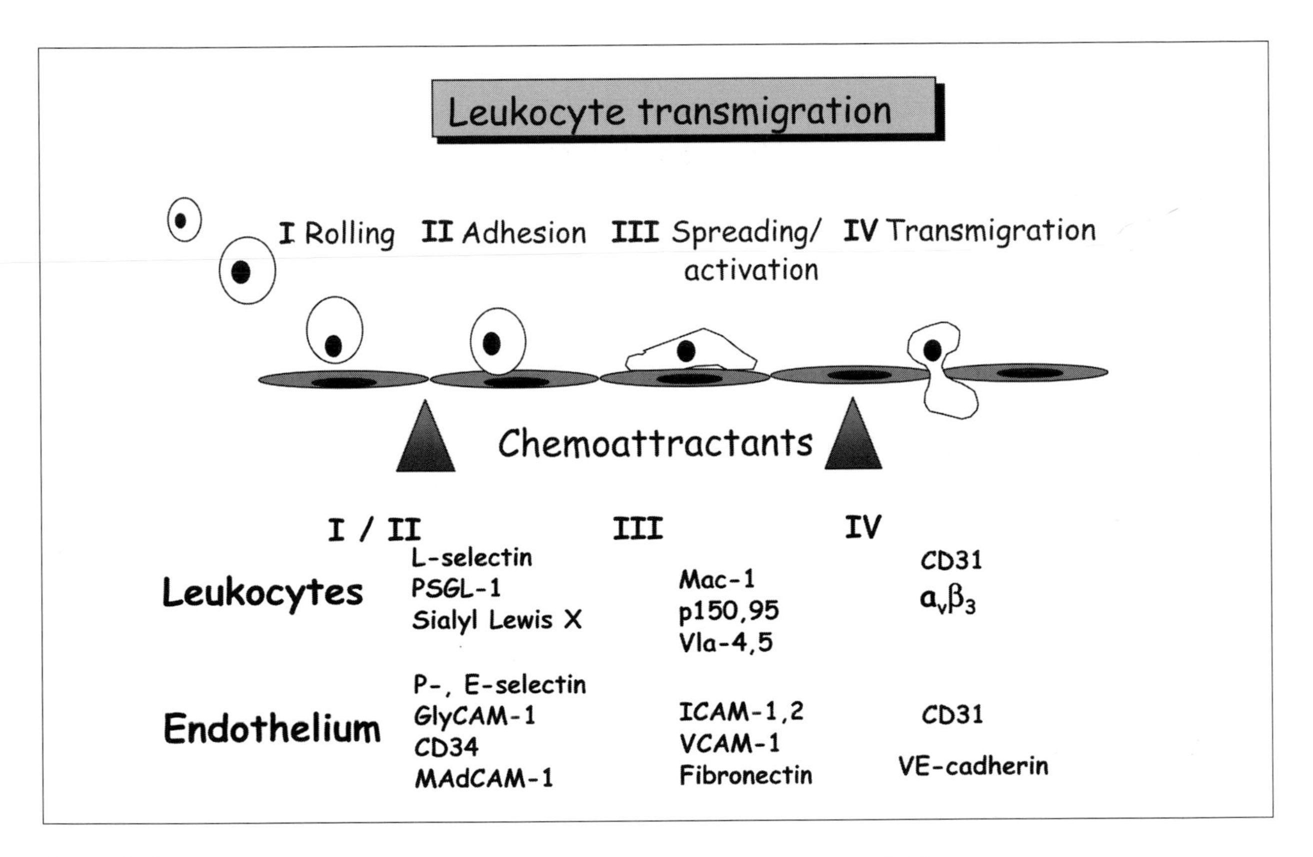

FIGURE 2. SCHEMATIC SURVEY OF PHAGOCYTE INFLUX IN TISSUES DURING INFLAMMATION

Interaction of L-selectin on the phagocytes with carbohydrate structures on the endothelial cells, and of E-selectin on the endothelial cells with carbohydrate structures (sialyl Lewis-X) on the phagocytes, causes "rolling" of the phagocytes over the blood vessel wall (I). Upregulation and activation of integrins (Mac-1, p150,95, Vla-4, Vla-5) on the phagocytes induces stable binding of these molecules to ICAM-1, ICAM-2, VCAM-1 and extracellular matrix proteins (such as fibronectin), which causes stable adhesion (II) and spreading (III) of the phagocytes on the endothelium. PAF and IL-8, produced by the endothelium, then induce diapedesis of the phagocytes between two adjacent endothelial cells into the tissues. Inflammatory mediators and chemokines produced in the infected area cause integrin activation, upregulation of P-selectin, ICAMs and VCAM-1 on the endothelial cells, and increased diapedesis of the phagocytes. The transendothelial migration of the leukoctes (IV) is governed by CD31 (PECAM-1) interactions with CD31 and with the $\alpha_V\beta_3$ integrin, and by VE-cadherin homotypic interactions between adjacent endothelial cells.
PSGL-1, P-selectin glycoprotein ligand 1; GlyCAM-1, glycosylation-dependent cell adhesion molecule 1; MAdCAM-1, mucosal addressin cell adhesion molecule 1

similar interaction between E-selectin on the endothelial cells with carbohydrate structures on the leukocytes. Under normal conditions, this "rolling" of phagocytes over the endothelium leads to stable adhesion and spreading of the phagocytes on the vessel wall, a process in which INTEGRINS play a decisive role (Fig. 2). Finally, DIAPEDESIS (transendothelial migration) and movement of the phagocytes into the tissues takes place. In infected or inflamed areas, these processes are strongly increased by the forma-

TABLE 1. PATTERN RCOGNITION RECEPTORS (PRR) IMPORTANT FOR NEUTROPHIL FUNCTIONS

PRR	Protein family	Ligand	Function
Secreted PRR			
MBL, Ficolin	C-type lectin	Terminal carbohydrate residues	Opsonisation via activation of lectin complement pathway
CRP, SAP	Pentraxins	Phosphorylcholine on microbial membranes	Opsonisation via activation of classical complement pathway
LBP	Lipid-transfer protein	LPS	LPS recognition
Cell-surface PRR			
CD14	Leucine-rich repeats	LPS, peptidoglycan	Co-receptor for TLR
CR3	Integrin	Zymosan	Phagocytosis, activation
Dectin-1	Lectin	β-glucans	Phagocytosis
TLR	Leucine-rich repeats	Various microbial products	Cell activation
Cytosolic PRR			
NOD1	Leucine-rich repeats	Muramyl dipeptide	Cell activation

MBL, mannan-binding lectin; CRP, C-reactive protein; SAP, serum amyloid protein; LBP, LPS-binding protein; LPS, lipopolysaccharide; TLR, Toll-like receptor(s); CR3, complement receptor 3; NOD, Nucleotide-binding oligomerisation domain protein

tion of complement fragment C5a, the bacterial tripeptide formyl-methionyl-leucyl-phenylalanine (fMLP), leukotriene LTB_4 and other chemotaxins. These agents cause an increase in expression of adhesion proteins on the surface of the phagocytes, such as β_1 and β_2 integrins. Moreover, the INTEGRINS are also "activated" by means of a change in their configuration, which causes stronger adhesion to endothelial structures. In addition, endotoxin, also generated in these areas, induces the local macrophages to produce interleukin-1 (IL-1) and TUMOUR NECROSIS FACTOR-α (TNF-α). Endotoxin, IL-1 and TNF-α activate the local endothelial cells to upregulate the expression of intercellular adhesion molecule-1 (ICAM-1), E-selectin and vascular adhesion molecule-1 (VCAM-1), which strongly enhances phagocyte adhesion. Endothelial cells also produce platelet-activating factor (PAF) and IL-8 under these conditions, which remain bound to the endothelial cells and stimulate phagocyte migration.

In the process of diapedesis, the endothelial cells that form the blood vessel wall participate actively. Binding of leukocytes to ICAM-1 or VCAM-1 on the endothelial cells induces signalling in the endothelial cells that leads to a looser cell-cell interaction, especially by disrupting the intercellular VE-cadherin interactions. This enables the leukocytes to squeeze in between two adjacent endothelial cells. The migratory process itself is governed by PECAM-1 (CD31) interactions between the leukocytes and the endothelial cells, in the intercellular cleft between two endothelial cells (Fig. 2).

Changes in the composition of the extracellular matrix, induced by transforming growth factor-β (TGF-β), and generation of additional CHEMOKINES by macrophages, endothelial cells and fibroblasts, add to the influx of phagocytes into inflamed tissues. This influx is phagocyte-specific, because the CHEMOTAXINS of the C-C chemokine family, of which monocyte chemotactic protein-1 (MCP-1) is the prototype, have

TABLE 2. HUMAN TOLL-LIKE RECEPTORS AND THEIR LIGANDS

Toll-like receptor	Ligand
TLR1 (dimer with TLR2)	Bacterial lipopeptides
TLR2	Peptidoglycans, zymosan
TLR3	Double-stranded RNA
TLR4	LPS, heat-shock proteins, taxol
TLR5	Flagellin
TLR6 (dimer with TLR2)	Mycoplasmal lipopeptides
TLR7	Imidazoquinolines (anti-viral compounds)
TLR9	Bacterial DNA (with unmethylated CpG motifs)

specificity for monocytes and macrophages, whereas CHEMOKINES of the C-X-C family, such as IL-8, attract mainly neutrophils and EOSINOPHILS. In the tissues, the phagocytes migrate by local attachment to extracellular matrix proteins, propagation of the cell over this fixed area, attachment at another site and dissociation of the first bonds.

The biological significance of adherence and migration is clearly demonstrated by the clinical symptoms of patients with leukocyte adhesion deficiency (LAD), namely serious, recurrent bacterial infections, retarded wound healing, persistent leukocytosis and a strong deficiency in the generation of inflammatory reactions. The leukocytes from these patients lack β_2-integrins, adhesion proteins involved in spreading of leukocytes on endothelial cells, in DIAPEDESIS and in migration in the tissues. In view of the high incidence of death in LAD patients, aggressive management of infections is indicated. The use of prophylactic treatment with trimethoprim-sulphamethoxazole appears to be beneficial. If a suitable donor is available, bone-marrow transplantation is recommended.

Recognition of pathogens by phagocytes

Phagocytes are specialized in uptake and intracellular killing of a large variety of bacteria, yeasts, fungi and mycoplasmata. Unlike cells of the adaptive immune system, which recognise "microbial non-self" via an immense variety of receptors generated by the variable recombination of a set of germ-line genes, cells of the INNATE IMMUNE SYSTEM depend on the use of products of a limited number of germ-line genes to discriminate "self" from "non-self". The receptors used by cells of the INNATE IMMUNE SYSTEM to recognise "non-self" are collectively termed PATTERN RECOGNITION RECEPTORS (PRRs). These are receptors with the ability to bind patterns specifically expressed by pathogens, so-called PATHOGEN-ASSOCIATED MOLECULAR PATTERNS (PAMPs). Examples of PAMPs are lipopolysaccharide (LPS), β-glucan and peptidoglycan (PGN). PRRs transmit signals that can lead to generation of inflammatory CYTOKINES and CHEMOKINES and to activation of microbicidal systems such as the production of REACTIVE OXYGEN SPECIES (ROS) and the release of antimicrobial peptides. Neutrophils express a number of PRRs (Tab. 1) that can influence the activation status, cytokine secretion and life span of these cells.

A recently identified family of PRRs is that of the TOLL-LIKE RECEPTORS (TLRs). TLRs are transmembrane proteins consisting of an extracellular leucine-rich repeat (LRR) for the binding of PAMPs and a cytoplasmic tail that is responsible for the signal transduction after ligation of these receptors. At least nine different TLRs have been described in the human immune system. For most of these, a ligand in the form of one or more PAMPs has been described (Tab. 2). TLRs are expressed on cells of the INNATE and the ADAPTIVE IMMUNE SYSTEM, but also on cells that do

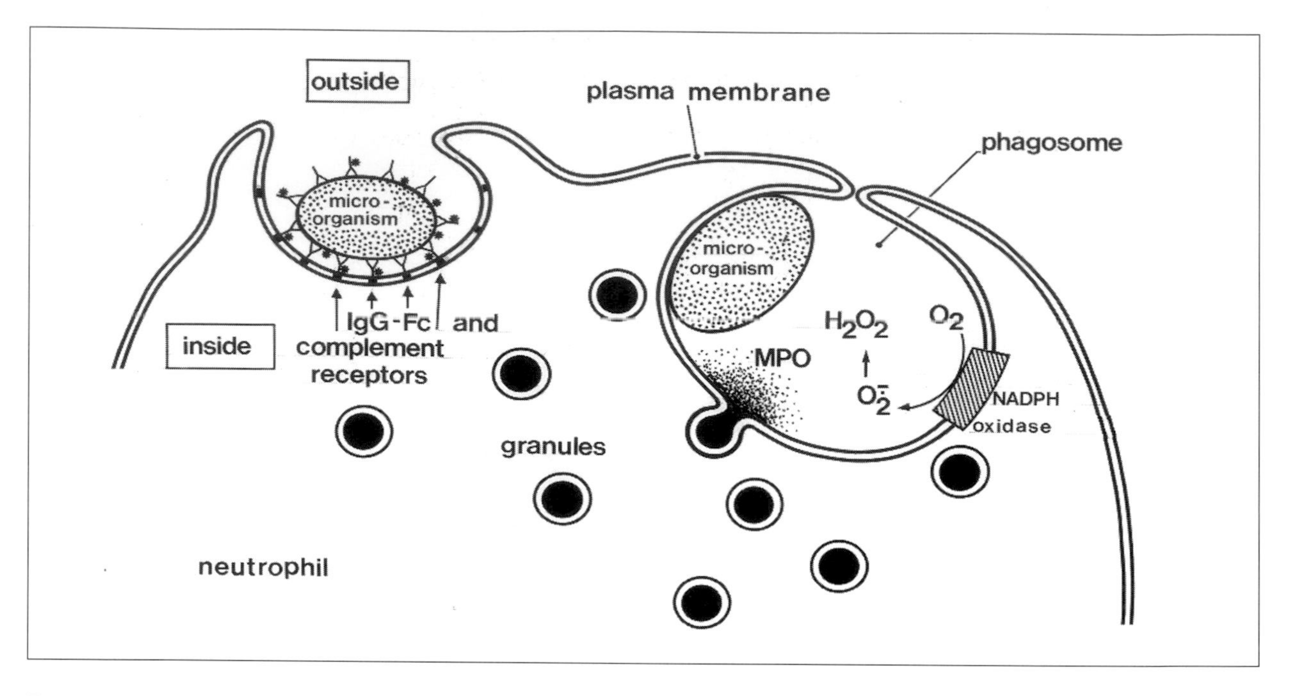

FIGURE 3. SCHEMATIC REPRESENTATION OF PHAGOCYTOSIS, DEGRANULATION AND GENERATION OF REACTIVE OXYGEN PRODUCTS

*MPO, myeloperoxidase; *, complement fragments C3b or iC3b. Figure reproduced with permission from [22]*

not belong to the immune system. TLRs are expressed not only on the surface of phagocytes, enabling these cells to respond to PAMPs present in the extracellular milieu, but also on the phagosomal membrane that surrounds ingested microbes (Fig.3), suggesting that phagocytes can "sense" the contents of the phagosome and react to it.

TLRs share part of their signal transduction pathway with the IL-1 receptor (IL-1R) family. Stimulation of both types of receptor ultimately leads to activation of the transcription factor NF-κB, but also to activation of c-jun N-terminal kinase (JNK) and p38 mitogen-activated protein kinase (MAPK) (Fig. 4). This cascade induces the expression of pro-inflammatory CYTOKINES and the differentiation of various immune cells into effector cells. In addition to this general response to ligation of any TLR, specific responses to each type of TLR also exist, for instance interferon-β (IFN-β) and IFN-γ production after TLR3 and TLR4 ligation. Another example is the activation of the Rac-PI3kinase-PKB pathway as a result of TLR2 signalling. In phagocytes, this pathway is involved in migration, degranulation and NADPH oxidase activation (see next section of this chapter), three functions that are essential for proper finding and killing of microbes. These differential effects of TLR signalling enable the immune system to react differently to various pathogens and thereby orchestrate the immune response specifically for efficient killing of a particular pathogen. TLR are seen as very important links between INNATE and ADAPTIVE IMMUNITY, because the adaptive immune response is highly influenced by the cytokine profile of macrophages and the antigen-presenting capacity of DENDRITIC CELLS, two features that are highly influenced by the outcome of TLR signalling in these cells.

Phagocytosis and killing of micro-organisms

Most micro-organisms can only be efficiently ingested after being covered with specific antibodies

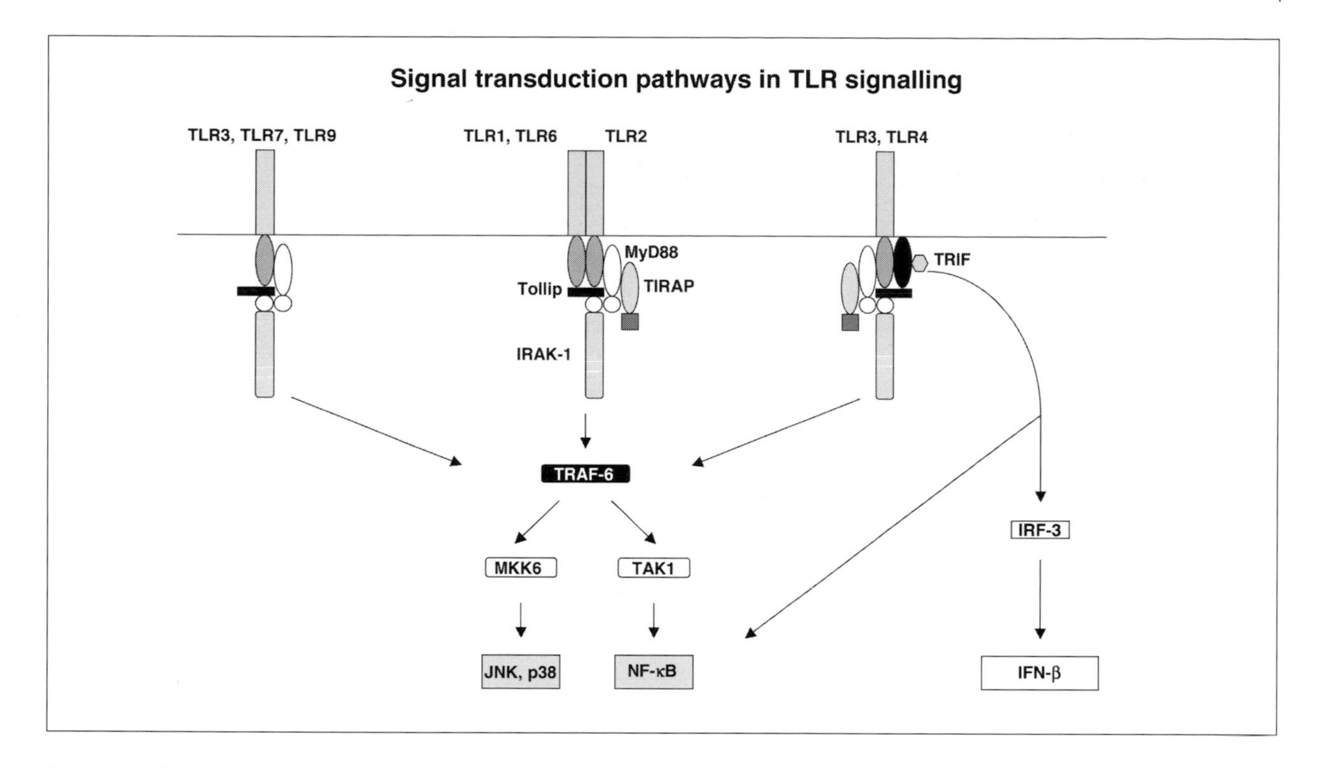

FIGURE 4. SIGNAL TRANSDUCTION PATHWAYS IN TLR SIGNALLING

TLR2 forms heterodimers with either TLR1 or TLR6, the other TLRs act as monomers. The cytoplasmic tails of the TLRs contain a TIR domain also present in the adaptor protein MyD88. Upon ligand association with the TLRs, these TIR domains associate, thus coupling MyD88 to the activated TLRs. This leads to recruitment of the IRAK-1 protein kinase via the death domain in both MyD88 and IRAK-1. This recruitment is mediated by the Tollip protein. IRAK-1 is then phosphorylated, detaches from MyD88 and subsequently binds to and activates a protein called TRAF-6. TRAF-6 triggers the activation of MKK6 and TAK1, which in turn activate JNK, p38 MAPK and NF-κB, respectively. This general signal transduction route induces the expression of a set of genes that cause the induction of inflammatory cytokines and the differentiation of various cell types into effector cells. In addition, the MyD88 adapter protein Mal-TIRAP is involved in signalling via all TLR except TLR3, -7 and -9. In addition, TRIF, another protein with a TIR domain, is involved in IFN production via TLR3 and TLR4. Figure reproduced with permission from [23].

TIR, Toll/IL-1 receptor; IRAK, IL-1 receptor-associated kinase; Tollip, Toll/IL-1R-interacting protein; TRAF, TNF receptor-associated factor; MKK, MAPK kinase kinase; JNK, c-jun N-terminal kinase; Mal, MyD88-adapter-like; TIRAP, TIR adapter protein; TRIF, TIR domain-containing adapter inducing IFN-β.

and/or complement fragments (a process called OPSONISATION). Antibodies bind with their FAB REGIONS to microbial antigens, which results in a spatial arrangement of the FC REGIONS that promotes activation of the classical complement pathway and fixation of fragments such as C3b and iC3b (see below and Fig. 3). The FC REGIONS of the antibodies and the complement fragments can then bind to Fc receptors and COMPLEMENT RECEPTORS, respectively, on the phagocyte surface. This binding of opsonised microorganisms to the phagocytes initiates three reactions in these cells: rearrangement of cytoskeletal elements that result in folding of the plasma membrane around the microbes (the process of PHAGOCYTOSIS), fusion of intracellular granules with this PHAGOSOME (the process of DEGRANULATION) and generation of

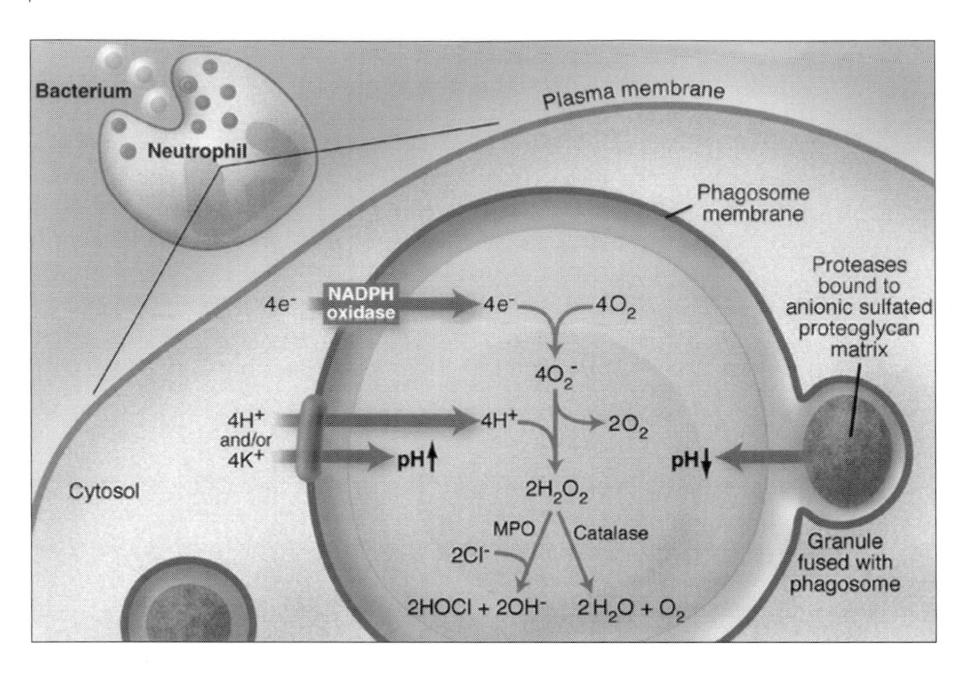

Figure 5. Reactions in the phagosome

The active NADPH oxidase transports electrons over the phagosomal membrane into the phagosome, where the electrons combine with molecular oxygen to generate superoxide (O_2^-). The resulting charge separation is largely compensated by protons (H^+), which are transported by a voltage-gated channel into the phagosome. However, this charge compensation by protons is not total, because the intra-phagosomal pH rises during the first few minutes after phagosome formation. Reeves et al. [3] discovered that potassium ions (K^+) are also entering the phagosome, and that these ions are instrumental in releasing proteases from their proteoglycan matrix in the azurophil granules that have fused with the phagosome. Superoxide combines with protons to form hydrogen peroxide (H_2O_2), which can combine either with chloride ions (Cl^-) in a myeloperoxidase (MPO)-catalysed reaction to hypochlorous acid (HOCl), or with another molecule of H_2O_2 in a catalase-mediated reaction to water and molecular oxygen.
Figure reproduced with permission from [24].

REACTIVE OXYGEN SPECIES within the PHAGOSOME (Fig. 3). The intracellular granules contain an array of microbicidal proteins, such as serine proteases, acid hydrolases, defensins, bactericidal permeability-increasing protein (BPI) and myeloperoxidase, as well as a number of microbistatic proteins, such as metalloproteases, lactoferrin and vitamin B_{12}-binding protein [2]. In neutrophils, these proteins are divided over two distinct types of granule, i.e., azurophil and specific granules, whereas in macrophages only one type of granule seems to be present. Simultaneously with the fusion of the granules with the phagosomal membrane, an NADPH oxidase enzyme in this membrane is activated that pumps electrons donated by NADPH in the cytosol into the phagosome. These electrons combine with molecular oxygen to form superoxide (O_2^-), an anion radical with high reactivity. This sudden increase in oxygen consumption is called the RESPIRATORY BURST. To compensate for the negative charge delivered to the phagosome, protons and other cations are also pumped into the phagosome (Fig. 5). The influx of potassium ions is instrumental in the release of cytotoxic proteins from the proteoglycan matrix of the azurophil granules [3]. The superoxide product of the NADPH oxidase enzyme is spontaneously converted into hydrogen peroxide, which then reacts with chloride anions in a myeloperoxidase-catalyzed reaction to form hypochlorous acid (HOCl). This last product is very toxic to many bacteria, but is rather unstable. However, it can react with primary and secondary amines to form N-chloramines, which are as toxic as HOCl but much more stable. Thus, the NADPH oxidase enzyme is essential in the microbicidal action of phagocytes, both by liberating proteolytic enzymes and by generating reactive oxygen compounds.

The biological significance of the microbicidal apparatus of phagocytes is again illustrated by the consequences of its failure [4]. Patients with chronic granulomatous disease (CGD), whose phagocytes lack an active NADPH oxidase, suffer already at an early age from very serious infections caused by catalase-positive micro-organisms (catalase-negative organisms themselves secrete some hydrogen peroxide, which can be used by CGD phagocytes to kill these organisms). Patients with a deficiency of specific granules (a very rare disorder) suffer from recurrent infections with various microbes. Patients with the syndrome of Chédiak-Higashi are characterized by neutropenia and recurrent infections with purulent micro-organisms. The phagocytes (and many other cell types) of these patients contain aggregated granules, which decrease cell mobility and DEGRANULATION. Infections in CGD and Chédiak-Higashi patients are treated with intravenous antibiotics and surgical drainage or removal of resistant infections. Prophylactic treatment with trimethoprim-sulfamethoxazole is very successful. In addition, prophylaxis with high doses of vitamin C in Chédiak-Higashi patients and with IFN-γ in CGD patients may also be beneficial. Bone-marrow transplantation, although hazardous, is at present the only curative therapy.

Inflammatory reactions

Phagocytes are also involved in many inflammatory reactions, e.g., by presenting microbial antigens to lymphocytes, by releasing inflammatory mediators (chemotactic peptides, LEUKOTRIENES, CYTOKINES), and by removing damaged host cells. Moreover, neutrophils also cause tissue damage; this is usually limited to the infectious period and intended to give the phagocytes access to the infectious agents. However, in chronically inflamed areas, such as those caused by autoimmune reactions, permanent macrophage activation will occur, and neutrophil influx and activation will continue. This will lead to excessive release of proteases from the neutrophils. Under normal conditions, these proteases are quickly inactivated by serine protease inhibitors (SERPINS) and α_2-macroglobulin, which are abundantly present in plasma and tissue fluids. During neutrophil activation, however, ROS and elastase released from these cells will inactivate these protease inhibitors. Moreover, the reactive oxygen compounds will also activate metalloprotease precursors, which will then degrade tissue matrix proteins [5]. Figure 6 shows an overview of these reactions. When this process is not self-limiting, irreversible tissue damage may result. In addition, serpins involved in regulating the complement, the coagulation, the fibrinolytic and the contact system cascades may also be inactivated, which will add to the severity of the clinical symptoms. Well-known clinical conditions in which this may happen are septic shock, gout, rheumatoid arthritis, autoimmune vasculitis, some types of glomerulonephritis, adult respiratory distress syndrome, lung emphysema, acute myocardial infarction, burns, major trauma and pancreatitis.

To limit the extent of these inflammatory reactions, phagocytes, especially neutrophils, have a very short survival time, because they are programmed to die within a few days after leaving the bone marrow by APOPTOSIS. This form of cell death prevents leakage of toxic compounds from the cells into the surroundings but instead leads to surface expression of molecules that induce binding, uptake and degradation of the cells by macrophages. Moreover, to prevent excessive phagocyte activation, nature has equipped these cells with a number of inhibitory mechanisms to dampen the immune response. One of these is a receptor for the glycoprotein CD200. This glycoprotein is found on many tissue cell types, and interaction with the receptor CD200R induces downregulation of the phagocyte reactions by recruiting the inositol phosphatase SHIP. Another is the signal regulatory protein (SIRP)α on macrophages, which upon association with the CD47 protein on haematopoietic cells becomes phosphorylated in its cytoplasmic immunoreceptor tyrosine-based inhibition motifs (ITIM) and then recruits the protein tyrosine phos-

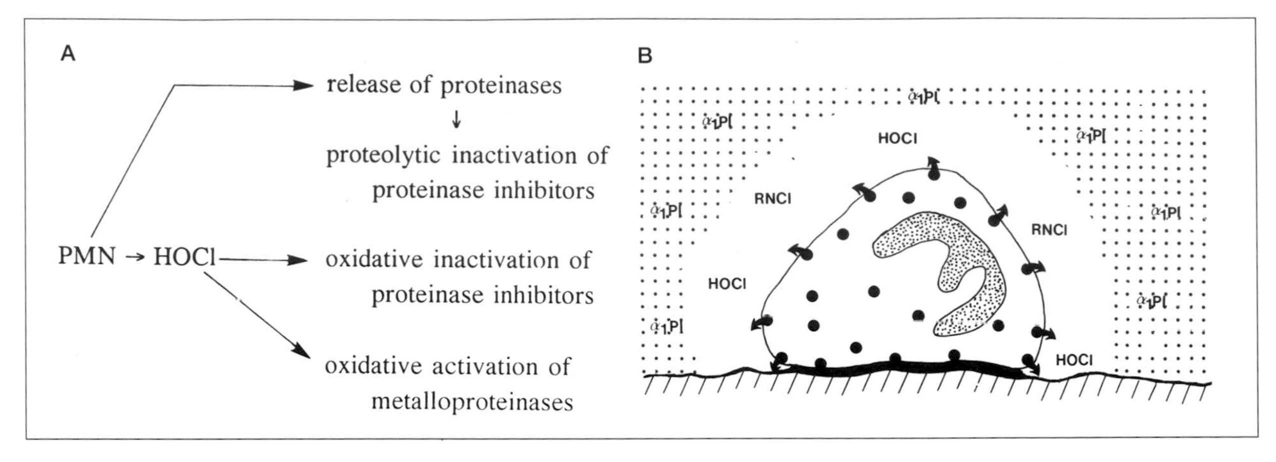

FIGURE 6. OXIDATIVE AND PROTEOLYTIC INFLAMMATORY REACTIONS

A. SCHEMATIC OVERVIEW OF OXIDATIVE AND PROTEOLYTIC INACTIVATION OF PROTEINASE INHIBITORS

During neutrophil activation, for instance by chemokines or phagocytosable material, proteinases are released. Among these are serine proteases from the azurophil granules (elastase, cathepsin G, proteinase 3, urokinase-like plasminogen activator). Normally these proteases are inactivated by serine protease inhibitors (SERPINS) in the extracellular body fluids, such as α_1*-proteinase inhibitor (*α_1*PI),* α_2*-macroglobulin,* α_1*-antichymotrypsin inhibitor and secretory leukoproteinase inhibitor (SLPI). However, neutrophil activation also leads to the generation of reactive oxygen species, which are able to inactivate many of these serpins because they contain the redox-sensitive amino acid methionine in their reactive centre. Other serpins are inactivated by peptide cleavage, especially by elastase. Finally, activated neutrophils also release metalloproteinases, such as collagenase and gelatinase, from their specific granules. These enzymes are secreted in an inactive precursor form, but they can be activated by oxidative processes. Once activated, metalloproteinases can participate in the inactivation of serpins or they can cause tissue degradation by themselves, but they are inhibited by* α_2*-macroglobulin. However, in the presence of oxidants from activated neutrophils, this inhibitor is inactivated and the active metalloproteinases are no longer kept in check.*

B. SCHEMATIC DIAGRAM OF TISSUE INJURY CAUSED BY ACTIVATED NEUTROPHILS

The cells create a zone of inactivated serpins around them (only α_1*PI is depicted) and released proteinases are therefore able to attack the surrounding tissue. Especially when neutrophils are attached to other cells or to basal membranes, tissue destruction may occur. Figure reproduced with permission from [25].*

HOCl, hypochlorous acid; RNCl, secondary chloramine

phatases SHP-1 and SHP-2. Finally, monocytes and macrophages also contain an inhibitory Fcγ receptor, i.e., FcγRIIb, which, in contrast to the activating Fcγ receptors, does not contain or associate with proteins with a cytoplasmic immunoreceptor tyrosine-based activating motif but instead contains again an ITIM. Probably, this FcγRIIb again serves to downmodulate the phagocyte reactions. At present a number of clinical and experimental studies are being conducted to evaluate the benefit of agents that interfere with neutrophil and/or monocyte infiltration, activation or DEGRANULATION. These agents include, amongst others, monoclonal antibodies against adhesion molecules (CD18, CD11b, ICAM-1), oxygen radical scavengers and protease inhibitors.

Natural killer cells

NK cells were described for the first time 28 years ago. Operationally these cells were defined by their ability to kill certain tumour cells *in vitro* without having been in contact with these tumour cells before. Development of NK cells does not require gene rearrangements as is the case for T lymphocytes, but NK cells are nonetheless developmentally closely related to T lymphocytes (reviewed in [6]). The anatomical site of development of NK cells is still unknown. CYTOKINES are critical for development of NK cells both in humans and mice. No NK cells are present in mice with a deficiency of the gamma chain of the IL-2 receptor and in severe combined immunodeficiency patients with mutations in this gamma chain. The IL-2Rγ chain is shared by receptors for several CYTOKINES, including IL-15. This latter cytokine appears to be essential for optimal NK cell development [7]. NK cells are implicated in INNATE IMMUNITY against foreign tissue, tumour cells, and microbes such as parasites, intracellular bacteria and viruses [8]. They appear to be important in early phases of the immune response, in which T lymphocytes do not yet function. There is convincing evidence that, in man, NK cells are involved in the defence against viral infections, in particular against herpes viruses [9]. The mechanisms by which NK cells mediate their effects in infections have not yet been fully elucidated, but it seems likely that CYTOKINES produced by the NK cells are involved. In addition, NK cells can control virus infections by killing virus-infected cells. Excessive activation of NK cells may be deleterious. In animal models for lethal sepsis, it has been shown that elimination of NK cells prevents mortality and improves outcome.

NK cells can mediate acute rejection of bone marrow grafts [10]. This is not the "raison d'être" of NK cells, of course, but this phenomenon has led to the development of the concept that NK cells recognize cells that lack or have modified one or more self-MHC class I antigens, which would explain why normal tissue is protected against NK-cell-mediated lytic activity [10].

Cytokine regulation of NK cells and the role of cytotoxicity mediated by NK cells in immunity against infections

NK cells are intermingled in an intricate cytokine network; they respond to and produce CYTOKINES that play a role in immunity against infections [7, 9, 11]. NK cells respond to IL-15, produced by monocytes, and to IL-12 produced by infected monocytes and dendritic cells. These CYTOKINES induce growth of NK cells, and particularly IL-12 induces NK cells to produce IFN-γ rapidly after infection. IFN-γ not only has anti-viral effects itself but is also a strong inducer of IL-12 production. Moreover, it has been convincingly shown that IFN-γ-activated macrophages are instrumental in the immune response against certain micro-organisms such as *Listeria monocytogenes* [12]. Furthermore, IL-12 plays an essential role in induction of Th1 lymphocytes (producing IFN-γ but not IL-4). Thus, a complex interplay between DENDRITIC CELLS, macrophages, NK cells and T lymphocytes ensures high levels of production of IFN-γ and IL-12, amplified through positive feed-back loops. IL-10, a product of macrophages, lymphocytes and other cell types, is a strong negative regulator of IL-12 production by phagocytes and of IFN-γ production by NK cells. IL-10 may be produced relatively late in an immune response, dampening the strong responses induced by IL-12 and IFN-γ.

Activation of NK cells by viral and microbial infections enhances cytotoxic activity. This is mediated by IFN-α/β, produced by virus- or bacteria-stimulated natural IFN-producing cells (IPC), also called plasmacytoid dendritic cells (pDC) [13]. Bacteria may induce IFN-α/β production through unmethylated CpG motifs, which are prevalent in bacterial but not in vertebrate genomic DNA. Oligodeoxynucleotides (ODN) containing unmethylated CpG motifs activate host defence mechanisms leading to innate and acquired immune responses. The recognition of CpG motifs requires TLR 9 that is expressed on IPC/pDC [13]. Some microbial infections, however, activate NK cytotoxicity without IFN-α/β induction; this appears to be dependent on IL-12 and IFN-γ.

Recognition by NK cells

There are two mechanisms of cell-mediated cytolysis. One is mediated through perforin, a protein secreted by cytotoxic lymphocytes that forms pores in the membranes of target cells. Target cells can also be killed by an interaction of the Fas molecule on the target cell and its ligand on the cytotoxic cell. This interaction activates proteases in the target cell,

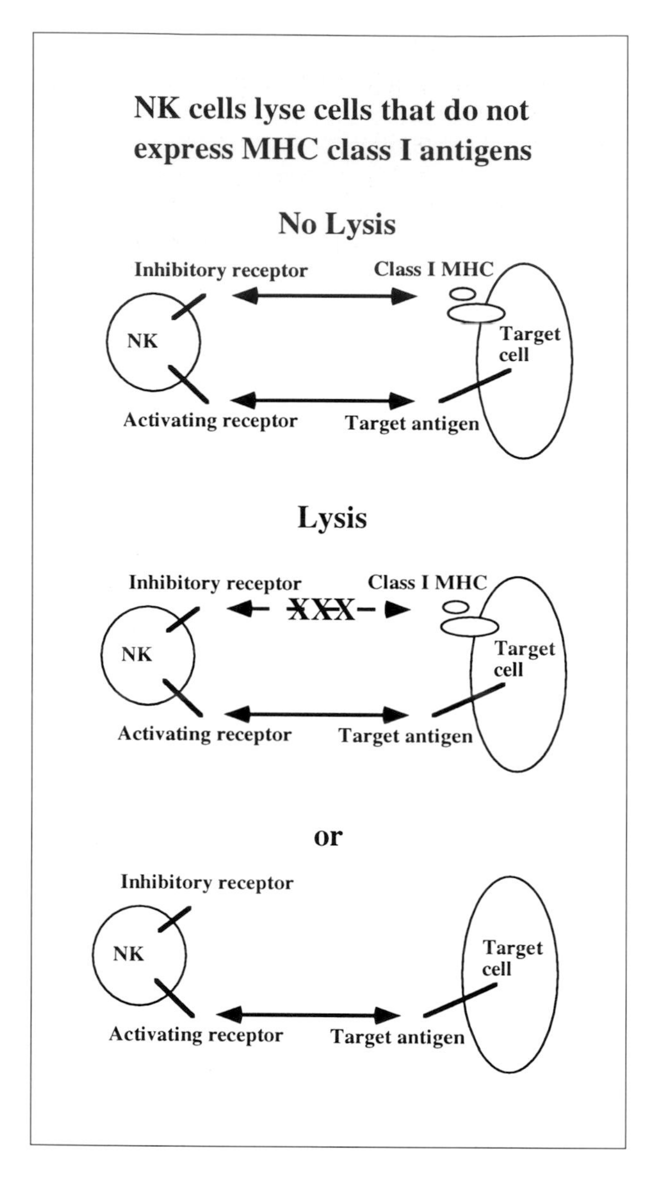

FIGURE 7. MODEL OF REGULATION OF NK ACTIVITY BY CLASS-I MHC-BINDING RECEPTORS

Lysis does not occur when an inhibitory receptor interacts with a class-I MHC antigen on the target cell, despite the fact that a cytolysis-activating receptor also interacts with its ligand. NK cells lyse target cells when the interaction between the inhibitory receptor is not triggered, either because the receptor is not specific for the MHC antigen, when this interaction is blocked by antibodies, or when the target cell does not express class-I MHC antigens at all.

resulting in APOPTOSIS. Clearly NK cells mediate their cytotoxic effects predominantly by a perforin-dependent mechanism, because little NK activity is present in perforin-deficient mice [14]. The remaining NK activity is probably mediated by Fas/FasL interaction [15].

The mechanism of NK cell recognition and the receptors involved have been elusive for a long time. Recent studies, however, have provided some insight into the complex way NK cells recognize their target cells. It is likely that NK cells do not have one single NK receptor that accounts for all biological responses, such as cytokine production and cytotoxic activity. Rather, it appears that NK cells utilize a vast array of receptors that induce their effector functions, which are often counteracted by inhibitory receptors specific for self-MHC class I antigens [8]. The positive and negative signalling pathways used by NK cells share many common features with receptors expressed by T and B lymphocytes. Signals are transmitted by small trans-membrane adaptor proteins that possess the so-called immunoreceptor tyrosine-based activation motifs (ITAM) in their cytoplasmic domains. One of these adaptors, DAP12, associates with numerous NK cell receptors, including Ly49, CD94/NKG2C and CD94/NKG2E in mice, and in humans with several activating killer cell immunoglobulins, with CD94/NKG2C and with NKp44. Other adapters, such as CD3ζ and FcRεRig, associate with the human NK cell receptors NKp30, NKp46 and CD16 [8]. The receptors NKp44 and NKp46 appear to bind the haemagglutinin of influenza virus due to the presence of sialic acid on these receptors [16]. The NKp44 and NKp46 receptors also recognize antigens expressed on tumour cells, but the nature of these ligands is unknown.

Another NK cell receptor, NKG2D, has spurred much interest recently, since it allows NK cells to recognize virus-infected and transformed cells [17]. NKG2D is associated with the adapter DAP10, expressed as a transmembrane-anchored disulfide homodimer. Ligands of NKG2D include cell surface antigens that are upregulated on transformed or virus-infected cells, such as MICA and MICB, two MHC-like stress-dependent cell surface antigens. Monoclonal antibodies against many adhesion, activation or costimulatory molecules on NK cells are able to activate these cells *in vitro*. These antigens include CD2, CD27, CD28, CD44, CD69, LFA-1 and DNAM-1 [16, 18]. However, whether NK cells are activated through one or more of these receptors in the responses against infected cells or in graft rejection in vivo remains to be determined.

The strong cytotoxic activities of activated NK cells raise the question as to how normal tissue is protected from attack by these cells. A solution to this conundrum came from studies on the phenomenon of hybrid resistance [10]. It was recognized in 1979 that NK cells mediate hybrid resistance to bone marrow (BM) or tumour grafts. This is a situation in which BM or tumour grafts of parental origin (either A or B) are rejected by AxB F1 hosts (A and B designate the MHC genotype). This resistance cannot be mediated by T lymphocytes, because these cells are tolerant to the A and B MHC antigens of the parents. It is now clear that NK cells possess a sophisticated system of "inhibiting" receptors that account for their ability to reject BM grafts that lack some MHC antigens present on the NK cells themselves. These inhibiting receptors interact with MHC antigens (Fig. 7). This feature of NK cells allows them to efficiently kill MHC class I-negative tumour cells and to remove infected cells with down-regulated self-MHC. What is more important, this provides for a mechanism by which normal tissue is protected against cytolytic activity by autologous NK cells. One should assume that all NK cells express at least one receptor for self-MHC class I antigens.

Two groups of these inhibitory MHC binding receptors have now been identified (Tab. 3) [8]. One group consists of C-type lectin molecules and is exemplified by the Ly49 gene family in the mouse. Ly49A is the best characterized gene and encodes a disulfide-bonded homodimer that binds to H-2D^k and D^k molecules. As a consequence, target cells that express these MHC antigens are not lysed by Ly49A-positive NK cells. The Ly49 family may comprise around 10 members with different, though overlapping, MHC class I, H-2K and D, specificities (Tab. 3). A second murine inhibitory receptor is CD94/NKG2A, which recognizes the class I MHC-like molecule Qa-1b. Human homologues of Ly49 genes have not yet been identified. However, there is a human homologue of CD94/NKG2A that recognizes HLA-E, an antigen with limited polymorphism.

TABLE 3. INHIBITORY NK CELL RECEPTORS FOR MHC

Species	Receptor	Ligand
Mouse	Ly49	H-2K, H-2D
Mouse	CD94/NK2A	Qa-1b
Human	KIR (KIR2DL, KIR3DL)	HLA-A, HLA-C, HLA-B44
Human	CD94/NKG2A CD159a	HLA-E

In humans, a second group of inhibitory receptors appears to function on NK cells that are designated KILLER CELL IMMUNOGLOBULIN-LIKE RECEPTORS or killer cell inhibitory receptors (KIRs). KIRs are involved in the recognition of HLA-A, HLA-B and HLA-C antigens (Tab. 3). Unlike the Ly49 receptors, KIR are type-I glycoproteins. They are related to the Ig supergene family and are probably encoded by a small number of genes. Three different protein isoforms have been described. The KIR recognizing HLA-Bw4 and HLA-A are proteins with three Ig-like domains (KIR3DL), whereas HLA-C-binding KIRs have two Ig-binding domains (KIR2DL). KIR with Ig gene similarity that bind to MHC class I antigens have not yet been found in mice.

The cytoplasmic domains of all inhibitory NK receptors possess ITIM. These ITIM are phosphorylated upon engagement of the receptors and recruit phosphatases to counteract activating signals that induce phospho-kinase activation. The src homology-containing phosphatases SHP-1 and -2 are the predominant phosphatases, but some Ly49 receptors can recruit the SH2 domain-containing inositol phosphatase SHIP.

The complement system

The COMPLEMENT SYSTEM consists of more than twenty proteins. Most of these are synthesized in the liver and circulate in blood as inactive precursor proteins, also known as complement factors. In addition, some complement proteins are expressed as membrane

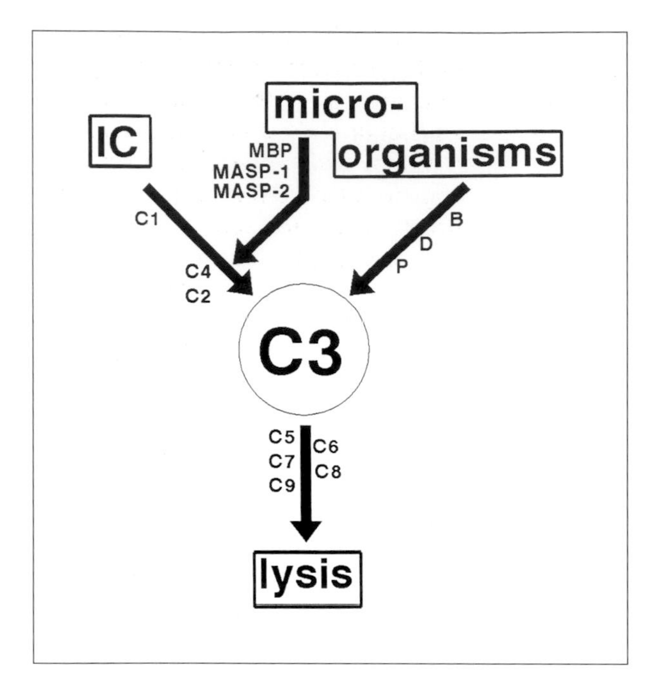

FIGURE 8. COMPLEMENT ACTIVATION PATHWAYS
The pathway starting with C1 is the classical pathway, that with MBP is the mannan-binding lectin pathway, and that with factor B is the alternative pathway. Activation of the classical and mannan-binding lectin pathways is triggered by binding of C1q or mannan-binding protein (MBP), respectively, to an activator (immune complexes [IC] or micro-organisms, respectively). Initiation of alternative pathway activation is more complex and involves interaction of a hydrolyzed form of C3 (not indicated in the figure), factors B and D, factors I and H (not shown in the figure) with the activator. Classical and mannan-binding lectin pathways converge at the level of C4, which in turn converges with the alternative pathway at the level of C3.

proteins, which serve to dampen undesired activation on cell membranes. During activation, one factor activates the subsequent one by limited proteolysis, and so on [19]. Because this activation process resembles a cascade, the complement system is considered as one of the major plasma cascade systems, the others being the coagulation, the fibrinolytic and the contact systems. The physiological role of the complement system is to defend the body against invading micro-organisms and to help in removing immune complexes and tissue debris. Hence, deficiencies of complement may predispose to bacterial infections, immune complex and auto-immune diseases.

Activation

The complement system can be activated via three pathways, namely a CLASSICAL PATHWAY consisting of the proteins C1, C2 and C4, an ALTERNATIVE PATHWAY consisting of factors B, D and P and the inhibitor factors H and I, and a so-called MANNAN-BINDING LECTIN (MBL) pathway, which consists of a mannan-binding protein, the serine proteinases MASP-1 and MASP-2 (MASP stands for mannan-associated serine proteinase), and C4 and C2 [20]. The three pathways converge at the level of the third complement component, C3, to proceed as the common terminal pathway, which consists of C5, C6, C7, C8 and C9 (Fig. 8). Antigen-antibody (immune) complexes are considered to be the main activators of the classical pathway, and microbial surfaces as the main activators of the MBL and of the alternative pathway. In addition, recent evidence suggests that also an acute-phase reactant, C-reactive protein (CRP), may serve as an activator of the classical pathway, in particular when bound to necrotic cells or tissue debris. During activation some complement factors, in particular activated C3, covalently bind to the activator, thereby forming ligands for C3 receptors on phagocytic cells. In the case where activation occurs at a cell membrane, the common pathway proteins (C5 to C9) form macromolecular complexes that insert into the membrane as pores. Under normal conditions, the largest part of these transmembrane pores consists of polymerized C9, in addition to C5b, C6, C7 and C8. Insertion of complement pores into cell membranes will allow the exchange of ions and hence induce lysis of the cell. However, under some conditions complement pores do not lyse cells, but rather lead to signal transduction and hence to an altered activation state of the cell. In addition, complement pores may induce an exchange between phospholipids of outer and inner leaflet of the cell membrane (a so-called flip-flop).

Protection of cells against complement lysis

Membranes of cells near to sites of complement activation may bind activated complement fragments such as C3b or C5b-C9 complexes, thereby becoming targets of complement activation. To prevent lysis of innocent bystander cells, the host cells are equipped with some membrane proteins that inhibit complement activation at various levels. These proteins include the membrane regulatory proteins decay-accelerating factor (DAF; CD55), membrane cofactor protein (MCP; not to be confused with the chemokine MCP) and C3b receptor (CR1), as well as homologous restriction factor (HRF) and membrane inhibitor of reactive lysis (CD59). DAF, MCP and CR1 prevent unwanted activation on membranes by inhibiting the assembly of C3 convertases and accelerating their decay. C3 convertases are macromolecular complexes formed during classical or alternative pathway activation that cleave and activate C3. HRF and CD59 inhibit complement lysis by interfering with the formation of C5-C9 complexes. Several of these proteins, i.e., DAF, CD59 and HRF, are anchored to the cell membrane via a glycan linkage to phosphatidylinositol. This link is defective in the blood cells of patients suffering from paroxysmal nocturnal haemoglobinuria (PNH). Hence, the red cells of these patients have strongly reduced levels of these complement-inhibiting membrane proteins, and are therefore more susceptible to reactive complement lysis, which largely explains the clinical symptoms of PNH. Furthermore, reduction of membrane regulatory proteins has been found locally in the tissues in areas of complement activation, and this is assumed to contribute to complement-mediated tissue damage in inflammation. The mechanism of this reduced expression is not clear.

Biological effects

Activation of complement not only induces fixation of some complement proteins onto the activator, but also results in the generation of biologically active peptides and macromolecular complexes in the fluid phase [19]. Among these are C5a, C3a and C4a, which are released from C5, C3 and C4, respectively, during activation, and – because of their biological effects – are also known as the ANAPHYLATOXINS [21]. For example, C5a, the most potent anaphylatoxin, is chemotactic for neutrophils and able to induce aggregation, activation and DEGRANULATION of these cells. In addition, the ANAPHYLATOXINS may enhance vasopermeability, stimulate adhesion of neutrophils to endothelium, activate platelets and endothelial cells and induce DEGRANULATION of mast cells and the production of the vasoactive eicosanoid thromboxane A_2 and the peptidoleukotrienes LTC_4, LTD_4 and LTE_4 by mononuclear cells. Moreover, they may stimulate or enhance the release of CYTOKINES such as TNF-α and IL-1 and IL-6 by mononuclear cells. The so-called terminal complement complexes (TCC) of C5b, C6, C7, C8 and C9 at sublytic concentrations also induce cells to release mediators, such as CYTOKINES, proteinases and eicosanoids. Finally, complement activation products may induce the expression of tissue factor by cells and thereby initiate and enhance coagulation. Thus, complement activation products have a number of biological effects that may induce and enhance inflammatory reactions (Tab. 4).

Evaluation of the complement system in patients

The functional state of the complement system in patients can be assessed in various ways. The overall activity of the system can be measured by so-called CH50 and AP50 haemolytic assays. In these assays, antibody-sensitized erythrocytes (CH50) or non-sensitized rabbit erythrocytes (AP50) are incubated with dilutions of patient serum. Antibody-sensitized erythrocytes activate the classical pathway, non-sensitized rabbit erythrocytes the alternative pathway. The activity of the serum sample is then expressed in Units, which is the reciprocal of the dilution of serum that lyses 50% of the erythrocytes. The CH50 assay measures the overall activity of classical and common pathways, the AP50 assay that of the alternative and the common pathways. These haemolytic assays were the first to be applied in clinical studies. Decreased haemolytic activity of sera may occur during activation of complement in patients, because activated complement factors are cleared

Table 4. Biological effects of complement activation products

Complement product	Effect
C5a	chemotaxis
C5a; C3a	mast cell degranulation
C5a; C3a	platelet degranulation
C5a	phagocyte degranulation
C5a	stimulation of O_2^- generation by phagocytes
C5a; C5b-9	enhancement of cytokine release
C5b-9; C5a?	expression of tissue factor
C5a; C5b-9	induction of prostaglandin and leukotriene synthesis
C3b; iC3b; C4b	opsonisation of micro-organisms
C5b-9	cell lysis
C3b	enhanced antibody response

from the circulation more rapidly than non-activated (native) complement proteins. However, during an ongoing ACUTE-PHASE REACTION, a decrease in complement protein levels may be masked by increased synthesis. Immunochemical determination of individual complement proteins, for example by nephelometry, has now largely replaced CH50 and AP50 determinations, the more so since the pattern of the relative decreases of complement proteins may provide important diagnostic and prognostic information. Nowadays CH50 and AP50 determinations should only be used to screen for the presence of genetic deficiencies. Deficiencies of the classical pathway will yield decreased activity in the CH50 assay, those of the alternative pathway lead to decreased AP50 activity. Deficiencies of C5 to C9 will yield decreased activity in either assay. Activation of complement in patients can best be assessed by measuring levels of specific complement activation products, such as levels of the anaphylatoxins, in particular C3a, C3b, C4b or circulating C5b-C9 complexes. The availability of monoclonal antibodies specifically reacting with neo-epitopes exposed on activation products and not cross-reacting with the native protein has greatly facilitated the development of specific, sensitive and reproducible immunoassays for these activation products, which are now frequently used in clinical practice.

Novel assays to assess the function and activity of the MBL pathway have also become available. Among these assays is OPSONISATION of mannan, fixed to a solid phase, with complement fragments. These and other assays have revealed that MBL levels may vary widely in the normal population, 15–20% of the people having strongly decreased functional MBL.

Clinical relevance

Genetic deficiencies of various complement proteins have been described. In general, total deficiencies of the classical pathway are associated with an increased risk for systemic lupus erythematosus (SLE). The reason for this association is not clear. It has always been thought that classical pathway deficiencies lead to defective handling of immune complexes, and hence to a greater risk for immune complex diseases such as SLE. Recently, it has become clear that clearance of apoptotic cells is also an important function of the classical pathway of complement. Hence, an alternative explanation for the association of classical pathway deficiencies with SLE is that these deficiencies lead to prolonged exposure of the specific immune system to apoptotic cells and debris, leading to auto-antibody responses. Indeed most, if not all, auto-antibodies in patients with SLE are directed against epitopes exposed by apoptotic cells. Deficiencies of C3 are associated with recurrent infections by pyogenic micro-organisms. Finally, deficiencies of C5 to C8 may lead to an increased risk for *Neisseria* infections. Surprisingly, C9 deficiencies are not associated with an increased risk for infections. Hence, OPSONISATION of micro-organisms by C3 is apparently essential for defence against pyogenic bacteria, whereas the formation of complement pores contributes to defence against *Neisseria*. As stated in the previous paragraph, 15–20% of the people in the Western world have strongly decreased levels of functional MBL, rendering deficiency of this protein the most frequent immune deficiency. A number of studies have shown that individuals with low levels of MBL have an increased risk for infections.

Activation of complement is considered to play an important role in the pathogenesis of a number of inflammatory disorders, including sepsis and septic shock, toxicity induced by the in vivo administration of CYTOKINES or monoclonal antibodies, immune complex diseases such as rheumatoid arthritis, SLE and vasculitis, multiple trauma, ischemia-reperfusion injuries and myocardial infarction. The pathogenetic role of complement activation in these conditions is probably related somehow to the biological effects of its activation products (Tab. 4). Inhibition of complement activation may therefore be beneficial in these conditions, which is substantiated by observations in animal models. The availability of clinically applicable complement inhibitors may help in the treatment of these diseases.

As yet, only C1-esterase inhibitor, a major inhibitor of the classical pathway, is available for clinical use. This is largely due to the fact that a heterozygous deficiency state of this inhibitor is associated with the clinical picture of hereditary angio-oedema (HAE). This disease sometimes leads to the development of life-threatening oedema of the glottis, which must be treated with intubation and the intravenous administration of C1-inhibitor. The pathogenesis of angio-oedema attacks associated with low C1-inhibitor levels is not completely clear but probably involves the generation of C2 peptide and bradykinin. The generation of these peptides results from the unopposed action of activated C1, activated coagulation factor XII and kallikrein of the contact system (C1-inhibitor is the main inhibitor of this system as well). Importantly, low levels of functional C1-inhibitor may be caused by a genetic defect but may also be acquired. Acquired C1-inhibitor deficiency is often associated with the presence of AUTOANTIBODIES against C1-inhibitor, which cause an accelerated consumption of C1-inhibitor. These antibodies are usually produced by a malignant B cell clone. HAE can be treated by attenuated androgens such as Danazol, anti-fibrinolytic agents or intravenously administered C1-inhibitor.

Summary

As indicated in this chapter, the three branches of the INNATE IMMUNE SYSTEM co-operate to protect us against pathogenic micro-organisms and to remove infected, dysregulated, damaged or outdated cells. Phagocytes act by migrating to infected areas and by ingesting and killing micro-organisms. NATURAL KILLER CELLS induce APOPTOSIS in virus-infected or tumour cells. MACROPHAGES remove apoptotic and aged cells. Finally, the COMPLEMENT SYSTEM helps phagocytes to find and ingest micro-organisms but also leads to direct lysis of microbes. Each of these systems is potentially dangerous to the host as well. Therefore, a very tight regulation of their activities exists to protect the host tissue against damage to innocent bystander cells and against excessive and long-lasting activation of these systems. Therapeutic intervention in the case of deficiencies or dysregulation is limited, but the possibilities may increase as we gain more insight into the basic principles.

Selected readings

Janeway CA Jr, Medzhitov R (2002) Innate immune recognition. *Annu Rev Immunol* 20: 197–216

Fearon DT, Locksley RM (1996) The instructive role of innate immunity in the acquired immune response. *Science* 272: 50–53

Worthylake RA, Burridge K (2001) Leukocyte transendothelial migration: orchestrating the underlying molecular machinery. *Curr Opin Cell Biol* 13: 569–577

Cerwenka A, Lanier LL (2003) NKG2D ligands: unconventional MHC class I-like molecules exploited by viruses and cancer. *Tissue Antigens* 61: 335–343

Frank MM (2000) Complement deficiencies. *Pediatr Clin North Am* 47: 1339–1354

References

1 Kuijpers TW, Roos D (1993) Extravasation of leukocytes. *Behring Inst Mitt* 92: 107–137

2 Berton G (1999) Degranulation. In: JI Gallin, R Snyderman (eds): *Inflammation: Basic Principles and Clinical Correlates*. 3rd. Edition. Raven Press, New York, 703–719

3 Reeves EP, Lu H, Jacobs HL, Messina CG, Bolsover S, Gabella G, Potma EO, Warley A, Roes J, Segal AW (2002) Killing activity of neutrophils is mediated through activation of proteases by K^+ flux. *Nature* 416: 275–277
4 Kuijpers TW, Weening RS, Roos D (1999) Clinical and laboratory work-up of patients with neutrophil shortage or dysfunction. *J Immunol Methods* 232: 211–229
5 Weiss SJ (1989) Tissue destruction by neutrophils. *N Engl J Med* 320: 365–376
6 Spits H, Lanier LL, Phillips JH (1995) Development of human T and natural killer cells. *Blood* 85: 2654-2670
7 Liu CC, Perussia B, Young JD (2000) The emerging role of IL-15 in NK-cell development. *Immunol Today* 21: 113–116
8 Cerwenka A, Lanier LL (2001) Natural killer cells, viruses and cancer. *Nat Rev Immunol* 1: 41–49
9 Biron CA, Nguyen KB, Pien GC (2002) Innate immune responses to LCMV infections: natural killer cells and cytokines. *Curr Top Microbiol Immunol* 263: 7–27
10 Kärre K (1997) NK cells, MHC class I antigens and missing self. *Immunol Rev* 155: 5–10
11 Trinchieri G (1997) Cytokines acting on or secreted by macrophages during intracellular infection (IL-10, IL-12, IFN-gamma). *Curr Opin Immunol* 9: 17–23
12 Unanue E (1997) Inter-relationship among macrophages, natural killer cells and neutrophils in early stages of *Listeria* resistance. *Curr Opin Immunol* 9: 35–43
13 Krieg AM (2002) CpG motifs in bacterial DNA and their immune effects. *Annu Rev Immunol* 20: 709–760
14 Kagi D, Ledermann B, Burki K, Seiler P, Odermatt B, Olsen KJ, Podack ER, Zinkernagel RM, Hengartner H (1994) Cytotoxicity mediated by T cells and natural killer cells is greatly impaired in perforin-deficient mice. *Nature* 369: 31–37
15 Oshimi Y, Oda S, Honda Y, Nagata S, Miyazaki S (1996) Involvement of Fas ligand and Fas-mediated pathway in the cytotoxicity of human natural killer cells. *J Immunol* 157: 2909–2915
16 Moretta A, Bottino C, Vitale M, Pende D, Cantoni C, Mingari MC, Biassoni R, Moretta L (2001) Activating receptors and coreceptors involved in human natural killer cell-mediated cytolysis. *Annu Rev Immunol* 19: 197–223
17 Cerwenka A, Lanier LL (2003) NKG2D ligands: unconventional MHC class I-like molecules exploited by viruses and cancer. *Tissue Antigens* 61: 335–343
18 Lanier LL, Corliss B, Phillips JH (1997) Arousal and inhibition of human NK cells. *Immunol Rev* 155: 145–154
19 Cooper NR (1999) Biology of the complement system. In: JI Gallin, R Snyderman (eds): *Inflammation: Basic Principles and Clinical Correlates*. 3rd Ed. Raven Press, New York, 281–315
20 Turner MW (1996) Mannose-binding lectin: the pluripotent molecule of the innate immune system. *Immunol Today* 17: 532–540
21 Vogt W (1986) Anaphylatoxins: possible roles in disease. *Complement* 3: 177–188
22 Roos D (1991) The respiratory burst of phagocytic leukocytes. *Drug Invest* 3 (Suppl 2): 48–53
23 van Bruggen R (2004) *Built for the Kill: Studies on the Neutrophil NADPH Oxidase*. Academic Thesis, University of Amsterdam
24 Roos D, Winterbourn CC (2002) Immunology. Lethal weapons. *Science* 296: 669–671
25 Roos D, Dolman KM (1990) Neutrophil involvement in inflammatory tissue damage. *Neth J Med* 36: 89–94

Inflammatory mediators and intracellular signalling

Richard Korbut and Tomasz J. Guzik

Introduction

Inflammation is a protective response of the macroorganism to injury caused by trauma, noxious chemicals or microbiological toxins. This response is intended to inactivate or destroy invading organisms, remove irritants, and set the stage for tissue repair. The inflammatory response consists of immunological and non-immunological reactions. The latter are triggered by the release from injured tissues and migrating cells of lipid-derived autacoids, such as EICOSANOIDS or "platelet-activating factor" (PAF), large peptides, such as interleukin-1, small peptides, such as bradykinin, and amines, such as HISTAMINE or 5-HYDROXYTRYPTAMINE. These constitute the chemical network of the inflammatory response and result in clinical and pathological manifestations of inflammation (Tab. 1). The concept of the inflammatory response has been introduced over 2000 years ago with the description by Cornelius Celsus as "*rubor et tumor cum calore et dolore*". Centuries later, in 19th century this definition was extended by Rudolph Virchow to incude the loss of function ("*functio lesa*"). It was Virchow and his pupils, like J. Cohnheim, who explained the scientific basis for Celsus' description of inflammation. They found that redness and heat reflected an increased blood flow, swelling is related to the exudation of fluid and accumulation of cells, while pain follows [1]. The first understanding of the mechanism of inflammation was introduced by Elie Metchnikoff, who concluded in his book *Comparative Pathology of Inflammation* published in 1893, that "...inflammation is a local reaction, often beneficial, of living tissue against an irritant substance" [2]. This definition stands until today. For the first time, he observed that this reaction is mainly produced by phagocytic activity of the mesodermic cells, and that it includes "the chemical action of the blood plasma and tissue fluids...", thus introducing the concept of the mediators of inflammation [2]. Further numerous studies since then have identified the roles of individual mediators in inflammation, and we are beginning to understand the genetic molecular aspects of the genesis of inflammatory process.

Eicosanoids

Arachidonic acid (AA) metabolites are formed rapidly from lipids of the cellular membranc, following activation of cells by numerous chemical and physical stimuli (Fig. 1). They exert their effects locally (autacoids), affecting virtually every step of inflammation [3]. EICOSANOIDS encompass cyclic prostanoid structure, i.e., PROSTAGLANDINS (PGs), prostacyclin (PGI2), thromboxane A_2 (TXA_2) and straight chain leukotriene structures (LTs), i.e., chemotactic LTB_4 and pro-inflammatory peptidolipids (LTC_4, LTD_4, LTE_4) (Fig. 2). Recently, a new group of molecules was added to the family of eicosanoids, namely LIPOXINS (LXA_4 and LXB_4), which are products of platelet 12-lipooxygenase metabolism of neutrophil LTA_4 (transcellular biosynthesis). EICOSANOIDS are synthesized by cyclooxygenation (prostanoids) or lipooxygenation (LTs) of a 20-carbon ω-6 polyunsaturated fatty acid (PUFA)- 5,8,11,14-eicosatetraenoic acid (AA) (Fig.1). AA is an important structural constituent of cellular phospholipids, and first must be liberated by acylhydrolases - directly by phospholipase A_2 (PLA_2) or indirectly by PLC before it becomes the substrate for the synthesis of eicosanoids.

TABLE 1. SYMPTOMS OF INFLAMMATION INDUCED BY INFLAMMATORY MEDIATORS

Symptom	Mediators
Vascular permeability	Vasoactive amines Bradykinin Leukotrienes C_4, D_4, E_4 PAF Complement (C3a and C5a) Substance P Nitric oxide
Vasodilation	Nitric oxide PGI_2, PGE_1, PGE_2, PGD_2 Hydrogen peroxide
Vasoconstriction	Thromboxane A_2 Leukotrienes C_4, D_4, E_4 Superoxide
Chemotaxis and leukocyte adhesion	Chemokines LTB_4, HETE, lipoxins Complement (C5a) Bacterial antigens
Pain	Bradykinin Prostaglandins
Fever	IL-1, TNF, IL-6 Prostaglandins
Tissue and endothelial damage	Reactive oxygen species Nitric oxide Lyzosomal enzymes

Prostanoids

Prostanoids are produced by the CYCLOOXYGENASE pathway. Prostaglandin H synthase (PGHS) is a dimeric complex which contains CYCLOOXYGENASE (COX) and peroxidase (Px). COX cyclizes AA to an unstable cyclic 15-hydroperoxy prostaglandin endoperoxide (PGG_2) while Px converts the 15-hydroperoxy to a 15-hydroxy group, in this way yielding PGH_2. Eventually, the end-product of PGHS (the complex which contains either constitutive COX-1, inducible COX-2 or recently discovered COX-3) is an unstable cyclic prostaglandin endoperoxide (PGH_2), which in various types of cells is converted by corresponding isomerases or synthases to stable prostanoids: PGD_2, PGE_2, $PGF_{2\alpha}$, and unstable prostanoids, i.e., PGI_2 or TXA_2. Special biological significance has been ascribed to PGI_2 synthase in vascular endothelial cells and TXA_2 synthase in blood platelets. The transcellular metabolism providing PGH_2 from activated platelets to endothelial cells is the main source of vascular PGI_2 [4]. The biological activity of stable prostanoids is terminated by catabolic enzymes, such as prostaglandin 15-hydroxy dehydrogenase (15-PGDH), D^{13}-reductase or α and ω oxidases which are present in high concentration in the lungs. These enzymes also break down inactive TXB_2 and 6-keto-$PGF_{1\alpha}$.

The role of individual CYCLOOXYGENASE enzymes in the development of inflammation remains unclear. The discovery of the inducible form, COX-2, led to the hypothesis that COX-1 is a constitutive enzyme responsible for physiological activities of PROSTAGLANDINS while COX-2, which is expressed during inflammation, produces "bad" PROSTAGLANDINS that generate pain and fever. This hypothesis quickly turned out to be simplistic and both enzymes show their activities under physiological and pathological conditions [5]. Moreover COX-2 inhibitory drugs possess fewer analgesic properties than non-selective inhibitors. The picture became even more complicated in 2002, with the discovery of COX-3. This isoenzyme is not a separate genetic isoform (like COX-2), but a splice variant of COX-1. In fact, COX-1 mRNA gives rise to four different isoforms including classical COX-1, COX-3 (splice variant including intron 1) and two partially truncated, inactive PCOX-1a and 1b. COX-3, due to the presence of intron 1, which changes its conformational structure, shows significantly diminished activity (25%) [6]. It is expressed mainly in the human brain and the heart. It has been suggested that COX-3 is an isoform particularly involved in the mechanisms of pain and fever during inflammation. Some suggestions exist that this isoform is inhibited by paracetamol, which could explain its analgesic actions.

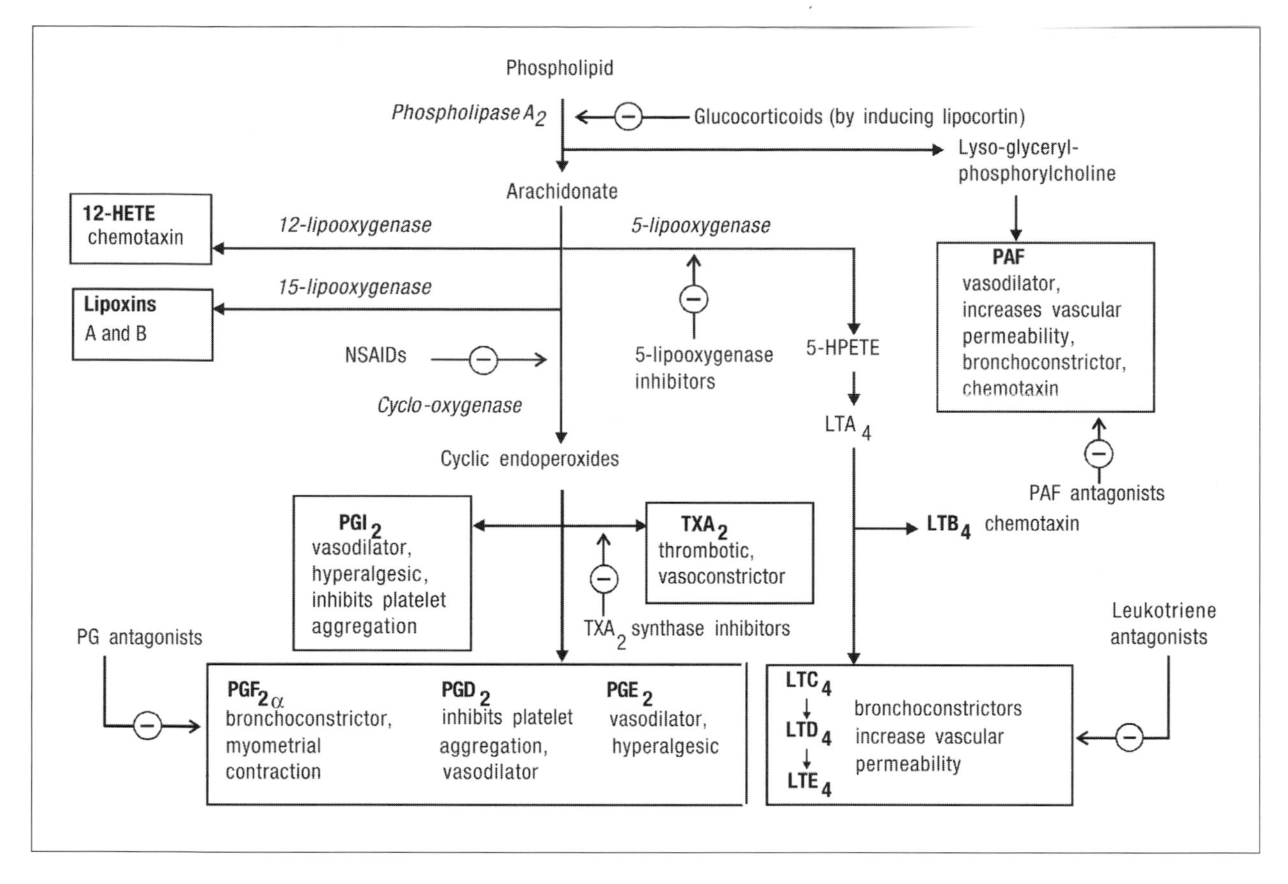

FIGURE 1
Mediators derived from phospholipids and their actions, the sites of action of anti-inflammatory drugs.

Biosynthesis of prostanoids is initiated by transductional mechanisms in an immediate response to activation of various cell membrane receptors or to various physical and chemical stimuli. These lead to an increase in the cytoplasmic levels of calcium ions Ca^{2+}_i and in this way they activate acyl hydrolases, which thereby release free AA for metabolism by PGHS. Alternatively, these enzymes can be induced by delayed transcriptional mechanisms which are usually activated by CYTOKINES or bacterial toxins. The spectrum of prostanoids produced by individual tissues depends on the local expression of individual enzymes. For example, vascular endothelium possesses prostacyclin synthase and COX-2, but lacks TX synthase, present in turn in the platelets. Accordingly, the major prostanoid released by endothelium is PGI_2, while platelets produce TXA_2.

Prostanoids regulate vascular tone and permeability in the development of inflammation. They also (TX) induce platelet aggregation and thrombus formation. Prostaglandins (in particular PGE_2) are also involved in the pathogenesis of pain and fever accompanying inflammation.

Most actions of prostanoids appear to be brought about by activation of the cell surface receptors that are coupled by G proteins to either adenylate cyclase (changes in intracellular c-AMP levels) or PLC (changes in triphosphoinositol – IP_3 and diacylglycerol – DAG levels) [7]. The diversity of the effects of prostanoids is explained by the existence of a

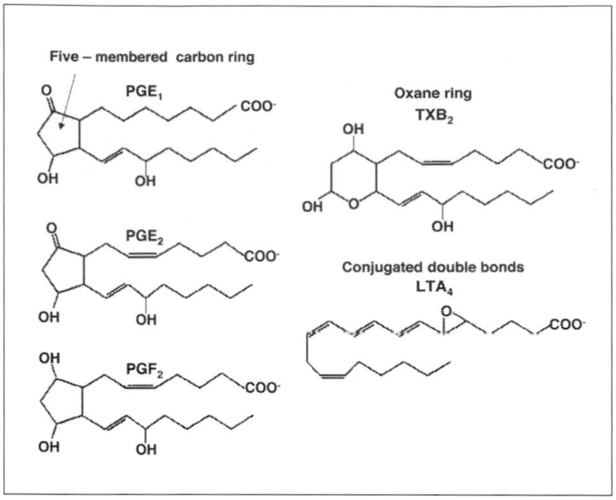

FIGURE 2
Typical eicosanoid structures

number of distinct receptors [8]. The receptors have been divided into five main types, designated DP (PGD), FP (PGF), IP (PGI_2), TP (TXA_2), and EP (PGE). The EP receptors are subdivided further into EP_1 (smooth muscle contraction), EP_2 (smooth muscle relaxation), EP_3 and EP_4, on the basis of physiological and molecular cloning information. Subtype-selective receptor antagonists are under development. Only one gene for TP receptors has been identified, but multiple splice variants exist. PGI_2 binds to IP receptors and activates adenylate cyclase. PGD_2 interacts with a distinct DP receptor that also stimulates adenylate cyclase. PGE_1 acts through IP receptors, PGE_2 activates EP receptors but it may also act on IP and DP receptors.

Products of lipooxygenation of arachidonic acid

AA can be metabolized to straight chain products by lipooxygenases (LOXs) which are a family of cytosolic enzymes that catalyze oxygenation of all polyenic fatty acids with two cis double bonds separated by a methylene group to corresponding lipid hydroperoxides [9] (Fig. 1). As in the case of AA, these hydroperoxides are called hydroperoxyeicosatetraenoic acids (HPETEs). LOXs differ in their specificity for placing the hydroperoxy group, and tissues differ in LOXs that they contain. Platelets have only 12-LOX and synthesize 12-HPETE, whereas LEUKOCYTES contain both 5-LOX and 12-LOX producing both 5-HPETE and 12-HPETE. HPETEs are unstable intermediates, analogous to PGG_2 or PGH_2, and are further transformed by peroxidases or nonenzymatically to their corresponding hydroxy fatty acids (HETEs). 12-HPETE can also undergo catalyzed molecular rearrangement to epoxy-hydroxyeicosatrienoic acids called hepoxillins. 15-HPETE may also be converted by lipooxygenation of LTA_4 to trihydroxylated derivatives called LIPOXINS (Fig. 1).

Leukotrienes

In activated LEUKOCYTES an increase in $Ca^{2+}{}_i$ binds 5-LOX to five-lipoxygenase-activating-protein (FLAP), and this complex converts AA to 5-HPETE, which in turn is the substrate for LTA_4 synthase. In the course of transcellular metabolism between LEUKOCYTES and blood cells or endothelial cells, unstable LTA_4 is converted by corresponding enzymes to stable chemotactic LTB_4 or to cytotoxic cysteinyl-containing LTs – C_4, D_4, E_4 and F_4 (also referred to as sulphidopeptide LTs or peptidolipids) [10] (Fig. 1). Note that the transcellular metabolism of AA can bring about either "protection" as is the case during the platelet/endothelium transfer of PGH_2 to make cytoprotective PGI_2 [1] or "damage" as in the case of the leukocyte/endothelium transfer of LTA_4 to make cytotoxic LTC_4 [6].

Consecutive splicing of amino acids from the glutathione moiety of LTC_4 occurs in the lungs, kidney, and liver. LTE_4 is already substantially deprived of most of the biological activities of LTC_4 and LTD_4. LTC_4 may also be inactivated by oxidation of its cysteinyl sulphur atom to a sulphoxide group. The principal route of inactivation of LTB_4 is by ω-oxidation. LTC_4 and LTD_4 comprise an important endogenous bronchoconstrictor, earlier known as "slow-reacting substance of anaphylaxis" (SRS-A) [11].

Three distinct receptors have been identified for LTs (LTB_4, LTC_4 and LTD_4/LTE_4). Stimulation of all of them appears to activate PLC. LTB_4, acting on specif-

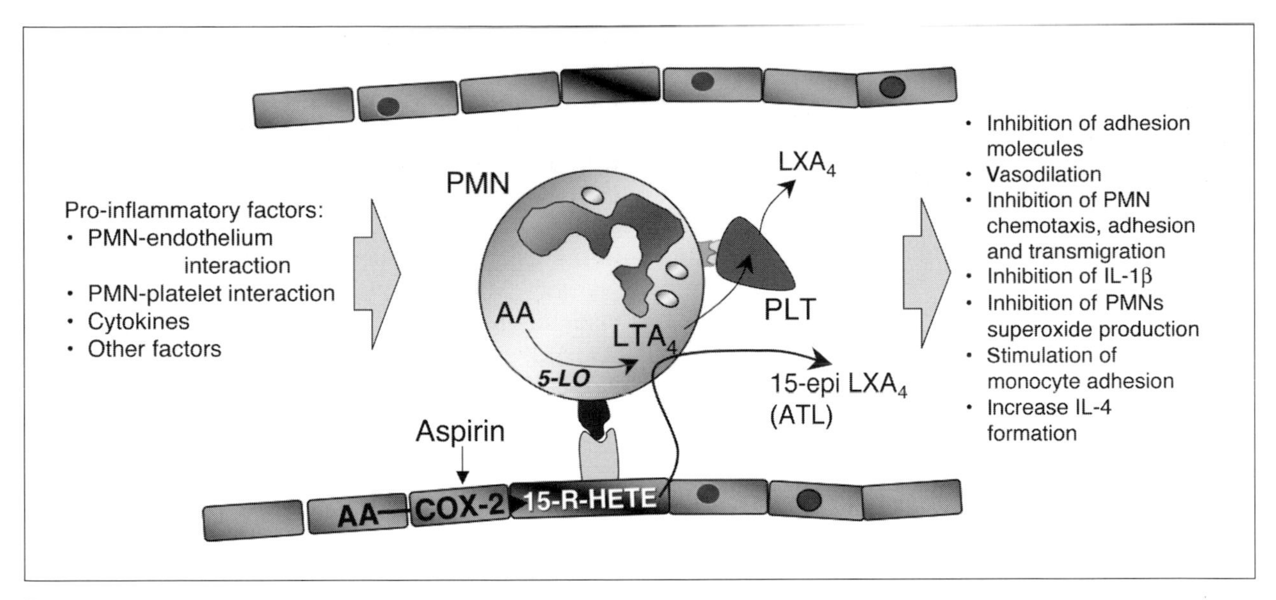

FIGURE 3
Transcellular synthesis of lipoxins and their actions

ic receptors, causes adherence, chemotaxis and activation of POLYMORPHONUCLEAR LEUKOCYTES and MONOCYTES, as well as promoting cytokine production in MACROPHAGES and LYMPHOCYTES. Its potency is comparable with that of various chemotactic peptides and PAF. In higher concentrations, LTB_4 stimulates the aggregation of PMNs and promotes DEGRANULATION and the generation of superoxide. It promotes adhesion of neutrophils to vascular endothelium and their transendothelial migration [12]. The cysteinyl-LTs are strongly cytotoxic, and cause bronchoconstriction and vasodilation in most vessels except the coronary vascular bed.

Lipoxins ("lipooxygenase interaction products")

LIPOXINS are formed by sequential transcellular metabolism of AA by 15- and 5-, or by 5- and 12-LOX [13]. Cellular context is critical for the synthesis of LIPOXINS (Fig. 3). LIPOXINS have several anti-inflammatory properties as well as concomitant proinflammatory actions. LIPOXINS inhibit adhesion molecule expression on endothelium, cause vasodilation and attenuate LTC_4-induced vasoconstriction by antagonism of $cysLT_1$ receptor. They also inhibit chemotaxis, adhesion and transmigration, IL-1β and superoxide production of POLYMORPHONUCLEAR leukocytes. On the other hand, LIPOXINS stimulate monocyte adhesion and increase IL-4 formation [13, 14]. There is an inverse relationship between the amount of LIPOXIN and LT production, which may indicate that LIPOXINS may be "endogenous regulators of leukotriene actions". High-affinity G-protein coupled LIPOXIN receptors (ALXR) have been identified on numerous cells, including MONOCYTES, PMNs, fibroblasts and endothelial and epithelial cells. Its expression may be up-regulated by IFN-γ, IL-13 or even IL-1β. Activation of this receptor modulates phosphatidylinositide 3-kinase (PI3-kinase) activity. LIPOXINS may also competitively bind and block the $cys\text{-}LT_1$ receptor. There are also suggestions that LIPOXINS may also bind within the cell, to ligand-activated transcription factors, therefore regulating gene expression in the nucleus.

A separate group of LIPOXINS was termed aspirin-triggered LIPOXINS (ATLs), as their synthesis is the result of acetylation of CYCLOOXYGENASE-2, which inhibits endothelial cell prostanoid formation and promotes synthesis of 15(R) HETE. These are then converted in PMNs to 15R-enantiomeres: 15-epi LXA_4 or 15-epi-LXB_4. ATLs share many actions of LIPOXINS, albeit with much greater potency [15]. Due to their anti-inflammatory properties, LIPOXIN analogues may find an important place in the treatment of inflammation [13, 14].

Other pathways of arachidonic acid metabolism

AA can be also metabolized by a NADPH-dependent cytochrome P-450-mediated monooxygenase pathway (MOX). The resulting 19-HETE, 20-HETE and a number of epoxyeicosatrienoic and dihydroxyeicosatrienoic acid isomers show vascular, endocrine, renal, and ocular effects, the physiological importance of which remains to be elucidated [16].

Recently, a non-enzymatic, free radical-mediated oxidation of AA, while still embedded in phospholipids, has been discovered. Subsequently, acyl hydrolases gave rise to a novel series of regioisomers of ISOPROSTANES. Formed non-enzymatically, ISOPROSTANES lack the stereospecificity of prostanoids. Highly toxic ISOPROSTANES might contribute to the pathophysiology of inflammatory responses which are insensitive to currently available steroidal and non-steroidal anti-inflammatory drugs. The most thoroughly investigated regioisomer of ISOPROSTANES is 8-epi-$PGF_{2\alpha}$. It has a potent vasoconstrictor action which is mediated by vascular TXA_2/PGH_2 receptors.

Actions and clinical uses of eicosanoids

EICOSANOIDS produce a vast array of biological effects. TXA_2, $PGF_{2\alpha}$ and LTs represent cytotoxic, pro-inflammatory mediators. TXA_2 is strongly thrombogenic through aggregation of blood platelets. LTC_4 injures blood vessels and bronchi subsequent to activation of leukocytes. On a molecular level, their cytotoxicity is frequently mediated by stimulation of PLC or inactivation of adenylate cyclase. Cytoprotective, but not necessarily anti-inflammatory actions are mediated by PGE_2 and PGI_2. They are both naturally occurring vasodilators. PGI_2 is the most comprehensive anti-platelet agent which is responsible for the thromboresistance of the vascular wall. PGE_2, through a similar adenylate cyclase-dependent mechanism, inhibits the activation of leukocytes. PGE_2 is also responsible for protection of the gastric mucosa. PGE_2 and $PGF_{2\alpha}$ may play a physiological role in labor and are sometimes used clinically as abortifacients. Locally generated PGE_2 and PGI_2 modulate vascular tone and the importance of their vascular actions is emphasized by the participation of PGE_2 and PGI_2 in the hypotension associated with septic shock. These PGs also have been implicated in the maintenance of patency of the ductus arteriosus. Various PGs and LTs are prominent components released when sensitized lung tissue is challenged by the appropriate antigen. While both bronchodilator (PGE_2) and bronchoconstrictor ($PGF_{2\alpha}$, TXA_2, LTC_4) substances are released, responses to the peptidoleukotrienes probably dominate during allergic constriction of the airway. The relatively slow metabolism of the LTs in lung tissue contributes to the long-lasting bronchoconstriction that follows challenge with antigen and may be a factor in the high bronchial tone that is observed in asthmatics in periods between attacks. PGs and LTs contribute importantly to the genesis of the signs and symptoms of inflammation. The peptidoleukotrienes have effects on vascular permeability, while LTB_4 is a powerful chemoattractant for POLYMORPHONUCLEAR LEUKOCYTES and can promote exudation of plasma by mobilizing the source of additional inflammatory mediators. PGs do not appear to have direct effect on vascular permeability; however, PGE_2 and PGI_2 markedly enhance oedema formation and leukocyte infiltration by promoting blood flow in the inflamed region. PGEs inhibit the participation of LYMPHOCYTES in delayed reactions. Bradykinin, CYTOKINES (TNF-α, IL-1, IL-8) appear to liberate PGs and probably other mediators that promote hyperalgesia (decreased pain threshold) and the pain of inflammation. Large doses of PGE_2 or $PGF_{2\alpha}$, given to women by intramuscular or subcutaneous injection to induce abortion, cause intense local pain. PGs also can cause headache and vascular pain when infused intravenously. The capacity of PGs to sensitize pain receptors to mechanical and chemical stimulation appears to result from a lowering of the threshold of the polymodal nociceptors of C fibers. Hyperalgesia also is produced by LTB_4. PGE_2, when infused into the cerebral ventricles or when injected into the hypothalamus, produces fever. The mechanism of fever involves the enhanced formation of CYTOKINES that increase the synthesis of PGE_2 in circumventricular organs in and near to the preoptic hypothalamic area, and PGE_2, via increases in c-AMP, triggers the hypothalamus to elevate body temperature by promoting increases in heat generation and decreases in heat loss.

Synthetic PGE_1, acting through IP and EP receptors, is given by infusion to maintain the patency of the ductus arteriosus in infants with transposition of large vessels until surgical correction can be undertaken. PGI_2 (epoprostenol) is occasionally used to prevent platelet aggregation in dialysis machines through inhibition of the thrombocytopenic action of heparin [17]. PGI_2 is also used for the treatment of primary and secondary pulmonary hypertension [18]. Stable analogues of PGI_2 (e.g., iloprost), as well as of PGE_1, are used in selected patients with peripheral vascular disease [17]. The PGE_1 analogue, misoprostol, is approved in the USA for the prevention of peptic ulcers, especially in patients who are required to take high doses of non-steroidal anti-inflammatory drugs (NSAID) for treatment of their arthritis.

Pharmacological interference with eicosanoid synthesis and actions

PLA_2 and COX are inhibited by drugs which are the mainstays in the treatment of inflammation. We discovered that glucocorticosteroids (hydrocortisone, dexamethasone) inhibit the generation of prostanoids *in vivo* through prevention of the release of AA from phospholipids [10]. This effect is mediated by intracellular steroid receptors which, when activated, increase expression of lipocortins which inhibit phospholipases. At present, many other actions of glucocorticosteroids on AA metabolism are known, one of them being inhibition of COX-2 transcription. Aspirin selectively inhibits COX-1, explaining its inhibitory effect on the biosynthesis of TXA_2 in platelets (causing reduced thrombotic tendency), of PGI_2 in endothelial cells and of PGE_2 in gastric mucosa (leading to gastric damage). This action of aspirin is more pronounced than that on the biosynthesis of prostanoids at the site of inflammation, where inducible COX-2 is most active. Consequently, aspirin at low doses seems to be a better anti-thrombotic than anti-inflammatory drug. Aspirin irreversibly acetylates the active centre of COX-1. Unlike endothelial cells, platelets lack the machinery required for *de novo* synthesis of COX-1, and, accordingly, aspirin-induced inhibition of TXA_2 synthesis in platelets is essentially permanent (until new platelets are formed), in contrast to the easily reversible inhibition of PGI_2 synthesis in vascular endothelium. The net effect of aspirin is, therefore, a long-lasting anti-thrombotic action. Unfortunately, most NSAIDs are more effective inhibitors of COX-1 than of COX-2. Meloxicam was the first clinically available drug which is claimed to be a selective COX-2 inhibitor – an anti-inflammatory drug with few side-effects on the gastro-intestinal tract, which causes no bleeding. However, population studies have verified that while protective for gastric mucosa, high doses of COX-2 selective inhibitors may induce cardiovascular (due to inhibition of endothelial COX-2) or renal side-effects [19]. NSAIDs usually are classified as mild analgesics and they are particularly effective in settings in which inflammation has caused sensitization of pain receptors to normally painless mechanical or chemical stimuli. NSAIDs do not inhibit fever caused by direct administration of PGs, but they do inhibit fever caused by agents that enhance the synthesis of IL-1 and other CYTOKINES, which presumably cause fever at least in part by inducing the endogenous synthesis of PGs.

Platelet-activating factor (PAF)

PAF (1-O-alkyl-2-acetyl-*sn*-glycero-3-phosphocholine) is a specialized phospholipid with an alkyl group (12-18C) attached by an ether bond at position 1 of glycerol and acetylated at position 2. PAF is not stored in cells but it is synthesized from 1-O-alkyl-2-

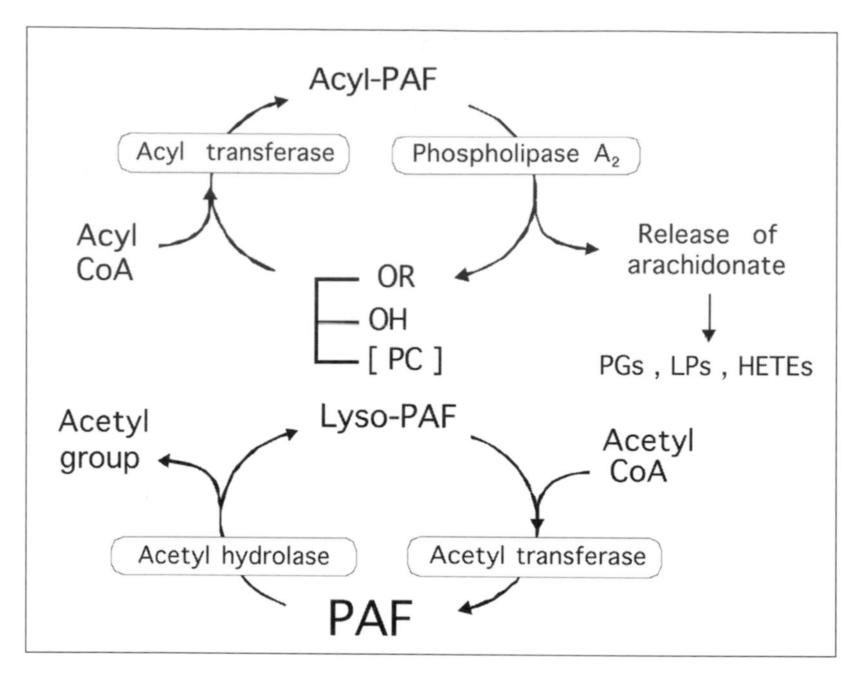

FIGURE 4
The synthesis and metabolism of platelet-activating factor (PAF)

acyl-glycerophosphocholine as required (Fig. 4) [20]. Initially, PLA_2 converts the precursor to the inactive 1-O-alkyl-2-lysoglycerophosphocholine (lyso-PAF) with concomitant release of AA. Incidentally, in GRANULOCYTES, AA produced in this way represents a major source for the synthesis of PGs and LTA_4. In a second step, lyso-PAF is acetylated by acetyl coenzyme A in a reaction catalyzed by lyso-PAF acetyltransferase. This is the rate-limiting step. The synthesis of PAF in different cells is stimulated during antigen-ANTIBODY reactions or by chemotactic peptides (e.g., f-MLP), CYTOKINES, thrombin, collagen, and autacoids. PAF can also stimulate its own formation. Both PLA_2 and lyso-PAF acetyltransferase are calcium-dependent enzymes, and PAF synthesis is regulated by the availability of Ca^{2+}. The anti-inflammatory action of glucocorticosteroids is at least partially dependent on inhibition of the synthesis of PAF by virtue of the inhibitory effect of lipocortin on the activity of PLA_2.

Inactivation of PAF also occurs in two steps (Fig. 4) [20]. Initially, the acetyl group of PAF is removed by PAF acetylhydrolase to form lyso-PAF; this enzyme is present in both cells and plasma. Lyso-PAF is then converted to a 1-O-alkyl-2-acyl-glycerophosphocholine by an acyltransferase. This latter step is inhibited by Ca^{2+}.

PAF is synthesized by platelets, neutrophils, MONOCYTES, basophils and mast cells, EOSINOPHILS, renal mesangial cells, renal medullary cells, and vascular endothelial cells. In most instances, stimulation of the synthesis of PAF results in the release of PAF and lyso-PAF from the cell. However, in some cells (e.g., endothelial cells) PAF is not released and appears to exert its effects intracellularly.

PAF exerts its actions by stimulating a single G protein-coupled, cell-surface receptor [21]. High-affinity binding sites have been detected in the plasma membranes of a number of cell types. Stimulation of these receptors triggers activation of phospholipases C, D, and A_2. and mobilization of $Ca^{2+}{}_i$. Massive direct and indirect release of AA occurs with its subsequent conversion to PGs, TXA_2, or LTs. EICOSANOIDS seem to function as extracellular representatives of the PAF message. As its name suggests, PAF unmasks fibrinogen receptors on platelets, leading directly to platelet aggregation. In endothelial cells the synthesis of PAF may be stimulated by a variety of factors, but PAF is not released extracellularly. Accumulation of

PAF intracellularly is associated with the adhesion of neutrophils to the surface of the endothelial cells and their diapedesis, apparently because it promotes the expression or exposure of surface proteins that recognize and bind neutrophils. Activated endothelial cells play a key role in "targeting" circulating cells to inflammatory sites. Expression of the various adhesion molecules varies among different cell types involved in the inflammatory response. For example, expression of E-selectin is restricted primarily to endothelial cells and is enhanced at sites of inflammation. P-selectin is expressed predominantly on platelets and on endothelial cells. L-SELECTIN is expressed on LEUKOCYTES and is shed when these cells are activated. Cell adhesion appears to occur by recognition of cell surface glycoprotein and carbohydrates on circulating cells by the adhesion molecules whose expression has been enhanced on resident cells. Endothelial activation results in adhesion of LEUKOCYTES by their interaction with newly expressed L-SELECTIN and P-selectin, whereas endothelial-expressed E-selectin interacts with glycoproteins on the leukocyte surface, and endothelial ICAM-1 interacts with leucocyte integrins.

PAF also very strongly increases vascular permeability. As with substances such as HISTAMINE and bradykinin, the increase in permeability is due to contraction of venular endothelial cells, but PAF is 1000–10000-fold more potent than HISTAMINE or bradykinin.

Intradermal injection of PAF duplicates many of the signs and symptoms of inflammation, including vasodilation, increased vascular permeability, hyperalgesia, oedema, and infiltration of neutrophils. Inhaled PAF induces bronchoconstriction, promotes local oedema and accumulation of EOSINOPHILS, and stimulates secretion of mucus. In anaphylactic shock, the plasma concentration of PAF is high and the administration of PAF reproduces many of the signs and symptoms of experimental anaphylactic shock. PAF receptor antagonists prevent the development of pulmonary hypertension in experimental septic shock. Despite the broad implications of these experimental observations, the clinical effects of PAF antagonists in the treatment of bronchial asthma, septic shock and other inflammatory responses have been rather modest.

PAF receptor antagonists include PAF structural analogues, natural products (e.g., ginkgoloids from *Ginkgo biloba*), and interestingly, triazolobenzodiazepines (e.g., triazolam). The development of PAF receptor antagonists is currently at an early stage of clinical development, still leaving the hope that such antagonists may find future therapeutic application in inflammation and sepsis.

Toll-like receptors

The TOLL-LIKE RECEPTORS (TLR1-10) are a part of the innate immune defence, recognizing conserved patterns only on microorganisms, but not mammalian proteins [22]. TLRs and their signalling pathways are present in mammals, fruit flies, and plants. Ten members of the TLR family have been identified in humans, and several of them appear to recognize specific microbial products, including lipopolysaccharide (LPS), bacterial lipoproteins, peptidoglycan, bacterial DNA and viral RNA. Signals initiated by the interaction of TLRs with specific microbial patterns direct the subsequent inflammatory response, including mononuclear phagocytic cell cytokine production. Thus, TLR signalling represents a key component of the innate immune response to microbial infection [22]. These mechanisms are further discussed in Chapter 4.

Cytokines

CYTOKINES are peptides produced by immune cells, which play key roles in regulating virtually all mechanisms of inflammation, including INNATE IMMUNITY, antigen presentation, cellular differentiation, activation and recruitment, as well as repair processes. They are produced primarily by MACROPHAGES and LYMPHOCYTES, but also by other leukocytes, endothelial cells and fibroblasts. Substances considered as CYTOKINES include INTERLEUKINS (IL)-1 to -25, INTERFERONS (IFNs), TUMOUR NECROSIS FACTORS (TNFs), platelet-derived growth factor (PDGF), transforming growth factor- (TGF-)β, CHEMOKINES (which will be discussed separately) and the COLONY-STIMULATING FACTORS. The cytokine production profile in response to immune insult determines the nature of immune response (cell-mediated, humoral, cytotoxic or allergic) [23, 24].

INTERLEUKIN-1 is the term given to a family of four CYTOKINES consisting of two active agonists: IL-1α, IL-1β, an endogenous IL-1-receptor antagonist (IL-1ra) and recently cloned cytokine IL-18, which is structurally related to IL-1. Both IL-1α, IL-1β, as well as a related protein IL-18, are synthesized as a less active precursor. Their secretion in response to various stimuli (antigens, endotoxin, CYTOKINES or microorganisms) depends on the cleavage of the pro-cytokines to their active forms by IL-1 converting enzyme (ICE or caspase 1). IL-1α remains cell-associated and is active mainly during cell-to-cell contact, while the soluble IL-1β is a form predominant in biological fluids. IL-1 is an important inflammatory mediator and it is believed to be implicated in several acute (e.g., systemic inflammatory response syndrome (SIRS) in sepsis) or chronic (e.g., rheumatoid arthritis) inflammatory diseases. IL-1 is also important in immune responses, facilitating interaction of both B or T cells with antigen.

One of the principal actions of IL-1 is activation of T LYMPHOCYTES and B cells by enhancing the production of IL-2 and expression of IL-2 receptors. In IL-1 knockout animals, diminished immune responses or state of tolerance is observed. In vascular endothelial cells, IL-1 increases the synthesis of leukocyte adhesion molecules (VCAM-1, ICAM-1 and E-selectin), stimulates nitric oxide (NO) production, releases PDGF, and activates PLA_2, thus inducing the synthesis of prostanoids and PAF. It stimulates fibroblasts to proliferate, to synthesize collagen and to generate collagenase. It regulates the systemic inflammatory response by stimulating synthesis of acute-phase proteins (C-reactive protein, amyloid and complement), producing neutrophilia, and causing fever by altering a set-point of temperature in the hypothalamus. IL-1 also induces the generation of other CYTOKINES such as the IFNs, IL-3, IL-6, and, in BONE MARROW, the COLONY-STIMULATING FACTORS. It synergises with TNF-α in many of its actions, and its synthesis is stimulated by TNF-α. The therapeutic effects of GLUCOCORTICOIDS in rheumatoid arthritis and other chronic inflammatory and autoimmune diseases may well involve inhibition of both IL-1 production and IL-1 activity. Production of IL-1ra alleviates potentially deleterious effects of IL-1 in the natural course of the disease.

IL-18, although structurally close to the IL-1 family, exerts actions more related to IL-12. It was originally derived from liver, but is produced by numerous cell types (including lung, kidney and smooth muscle cells) apart from LYMPHOCYTES. In contrast to other CYTOKINES, IL-18 procytokine is constitutively expressed and therefore its activity is regulated primarily by caspase-1. It plays a critical role in cellular adhesion, being the final common pathway leading to ICAM-1 expression in response to IL-1, TNF-α and other CYTOKINES. It also synergises with IL-12 in stimulating IFN-γ production. Soluble IL-18 receptor may be particularly interesting from an immunopharmacological point of view as it has lost its signalling domain and may therefore serve as a potent anti-inflammatory molecule.

Tumour necrosis factor-α and -β (TNF-α and -β)

These CYTOKINES are produced primarily in mononuclear phagocytes (TNF-α) and in LYMPHOCYTES (TNF-β), but also by numerous other cells. Activation of TLRs (TLR_2 and TLR_4) by LPS is the most commonly recognised intracellular pathway leading to production of TNF. TNF-α and -β bind with similar affinity to the same cell surface receptors – TNFR 1 (p55) and TNFR 2 (p75), therefore their activities are very close to each other. Their name is based on tumour cytotoxic effects; however, their pharmacological use in the treatment of tumours is limited by severe side-effects. TNF is responsible for severe cachexia during chronic infections and cancer.

In endothelial cells these CYTOKINES induce: expression of adhesion molecules (ICAM-1 and VCAM-1), synthesis of prostacyclin and of CYTOKINES. TNFs act as chemoattractants, as well as potent activators for neutrophils and MACROPHAGES. TNF-α causes fever and releases acute-phase proteins. TNF and IL-1 produce many of the same proinflammatory responses which include induction of CYCLOOXYGENASE and lipooxygenase enzymes as well as the activation of B cells and T cells. It is finally important to point out that TNF is the primary mediator of haemodynamic changes during septic shock through its negative inotropic effects as well increase in vascular permeability.

TGF-α (transforming growth factor-α) is a trophic regulator of cell proliferation and differentiation which is important in repair processes; it is involved in angiogenesis and in the organization of the extracellular matrix, and it is chemotactic for MONOCYTES.

PDGFs cause proliferation of fibroblasts, vascular endothelial cells and smooth muscle; they are implicated in angiogenesis, atherosclerosis and possibly in chronic asthma.

IFNs constitute a group of inducible CYTOKINES which are synthesized in response to viral and other stimuli. There are three classes of IFNs, termed IFN-α, IFN-β and IFN-γ. IFN-α is not a single substance but a family of 15 proteins with similar activities. The three IFNs (α, β and γ) have antiviral activity and IFN-γ has a significant immunoregulatory function and only modest anti-viral activity. Their anti-viral effects are achieved by inhibition of viral replication within infected cells as well as by stimulation of cytotoxic LYMPHOCYTES and NK cells. All IFNs can be induced by other CYTOKINES such as IL-1, IL-2, TNF and COLONY-STIMULATING FACTORS. IFN-α and IFN-β are produced in many cell types – MACROPHAGES, fibroblasts, endothelial cells, osteoblasts, etc.; they can be strongly induced by viruses, and less strongly by other microorganisms and bacterial products. IFNs induce the expression of the major histocompatibility molecules (MHC I and II) that are involved in antigen presentation to T cells. IFNs also stimulate the expression of Fc receptors on GRANULOCYTES, promote the differentiation of myeloid cells and modulate the synthesis of CYTOKINES. IFN-γ is primarily made by T LYMPHOCYTES (T helper type 1), and which may suggest that it is more of an IL than an IFN. Indeed, it functions as an inhibitor of IL-4-dependent expression of low-affinity IgE receptors, therefore inhibiting IgE synthesis.

Colony-stimulating factors. These include IL-3 and GM-CSF (granulocyte macrophage colony-stimulating factor) and several other CYTOKINES. They regulate haematopoiesis and are chemotactic for neutrophils, as well as activating neutrophils and MACROPHAGES.

Anti-inflammatory cytokines. It is important to point out that, apart from pro-inflammatory actions, some CYTOKINES may inhibit inflammatory processes. These include IL-1ra mentioned above, as well as TGF-β or the IL-10 family (includes IL10, 19, 20, 22 and 24).

Intracellular signalling by cytokine receptors [23, 24]

Binding CYTOKINES to their receptors leads to the activation of cytoplasmic tyrosine kinases. Janus kinases (JAKs), a recently described family of four related cytoplasmic protein tyrosine kinases, function in cytokine signalling. There are four JAKs: JAK1, JAK2, JAK3, and TYK2, which transduct signals from cytokine receptors to effector mechanisms. On binding of the cytokine, JAKs bind to the receptor and mediate tyrosine kinase activity. and phosphorylation of the receptor and of receptor-associated JAKs (Fig. 5). The next step in signal transduction involves tyrosine phosphorylation of signal transducers and activators of transcription (STATs) in the cytoplasm. Upon activation, STATs become phosphorylated, form homodimers, and migrate to the nucleus, where they bind to regulatory sequences in the promoters of cytokine-responsive genes, e.g., ICAM-1 or other cytokine genes. Cytokine signalling is based on a relatively small number of redundant tyrosine kinases. For instance, JAK-1 and JAK-3 transduct signals from gc CYTOKINES (i.e., IL-2 or IL-4), while JAK-2 is involved in IL-3, IL-6 and GM-CSF signalling. Similarly, the number of STATs is low when compared to the number of CYTOKINES. Therefore one can conclude that some additional mechanisms will guide different responses to various CYTOKINES. An additional pathway used by many cytokine receptors includes Ras-dependent cascades. In this signal transduction cascade, Ras, Raf-1, Map/Erk kinase kinase (MEKK) and finally mitogen-activated protein kinases (MAPK) are sequentially activated and lead to regulation of cellular proliferation by growth factors and responses to IL-2 or IL-3. The activation of other signalling pathways, like insulin receptor substrates (IRS-1, IRS-2),

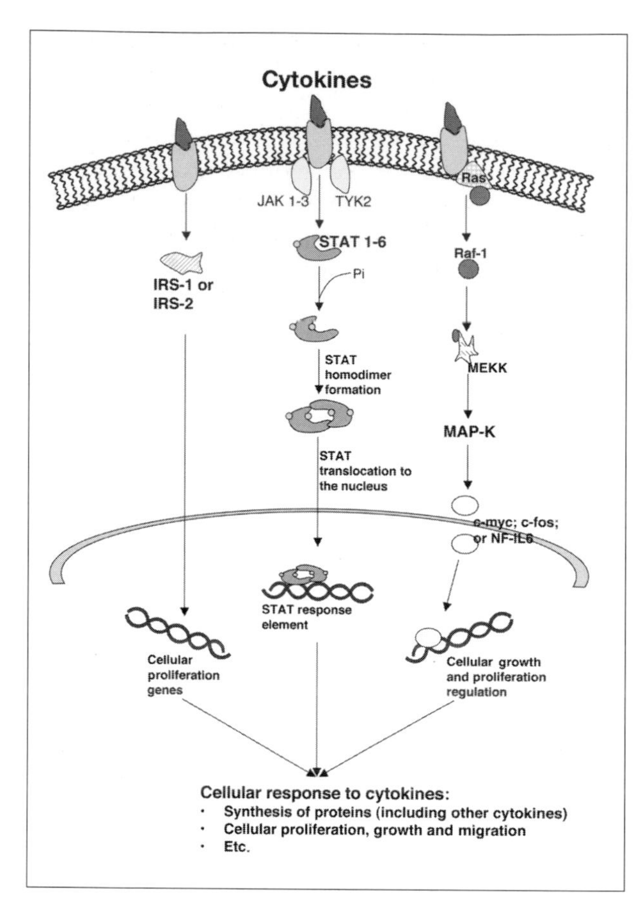

FIGURE 5
Cytokine-induced intracellular signalling

can also mediate some other biological activities of CYTOKINES, including proliferation and regulation of apoptosis. In conclusion, it becomes apparent that combination of the signalling mechanisms described above will lead to many distinct responses to different CYTOKINES.

Chemokines and their intracellular signalling

CHEMOKINES are a family of 8–12 kD molecules, which induce chemotaxis of MONOCYTES, LYMPHOCYTES, neutrophils, and other GRANULOCYTES, as well as vascular smooth muscle cells and a variety of other cells [25]. There are 47 chemokines, sharing 30–60% homology. CHEMOKINES are characterised by the presence of 3-4 conserved cysteine residues. Recently, a new classification of CHEMOKINES has been proposed based on the positioning of the N-terminal cysteine residues (Tab. 2). CHEMOKINES are usually secreted proteins except for fractalkine (CX3CL1), which is the only membrane-bound chemokine. Most CHEMOKINES play roles in recruiting and activating immune cells to and at the site of inflammation while others are important in maintaining homeostasis within the immune system (housekeeping chemokines: CCL5, CCL17-19, 21, 22, 25, 27, 28, CXCL13, CXCL14). Homeostatic CHEMOKINES are expressed in an organ-specific manner while inflammatory CHEMOKINES can be produced by multiple cell types.

Their activities are achieved through interaction with chemokine receptors. There are 18 chemokine receptors currently known; therefore some receptors may bind several ligands, which leads to overlapping functions of known chemokines. Moreover, a single cell may express several chemokine receptors. One of the key features of chemokine receptors, owed to their heptahelical transmembrane structure, is their ability to signal through different intracellular signalling pathways. Binding of the chemokine to the receptor leads to activation of Gα protein and binding of GTP. The Gα subunit activates Src kinases and subsequently MAPKs and protein kinase B (PKB). During activation of Gα protein, a Gβγ complex is liberated and may independently lead to activation of PKB and MAPKs (via PI_3), PKC activation via phospholipase C (PLC) and finally through Pyk-2 [25]. These pathways lead to up-regulation of membrane INTEGRINS and initiate rolling and adhesion of cells as well as their conformational changes. Some of these intracellular pathways (in particular PLC activation) may then lead to an increase in intracellular calcium and its consequences, including DEGRANULATION, NOS activation, etc., within the target cells.

It is difficult to accurately describe the relative importance of individual chemokines. The largest number of studies were conducted looking at actions of IL-8 as the most important chemoattractant for POLYMORPHONUCLEAR leukocytes, although it

Table 2. Classes of chemokines

Subfamily	Chemokines	Characteristics
C-X-C	CXCL 1-16, includes IL-8 (CXCL8)	First two cysteines separated by a variable amino-acid.
C-C	CCL1-28 (include MIP-1 MCP and RANTES)	First two cysteines are adjacent to each other.
C	XCL 1 (lymphotactin)-and XCL 2	Lacks first and third cysteine residue
CX3CL1	CX3CL1 (Fractalkine)	Two N-terminal cysteine residues separated by 3 variable amino acids.

appears late during the inflammatory response. Other well-investigated members of this family include CCL3 (MIP-1a) or RANTES (CCL5).

Apart from effects on chemotaxis, CHEMOKINES have direct and indirect effects on T-cell differentiation into T helper 1 or 2 subclasses, therefore regulating the nature of immune responses [24,25].

Due to the critical role of CHEMOKINES in inflammation, interest has focused on potential therapeutic effects of inhibiting their activity. Both peptide antagonists as well as gene transfer approaches have been successfully used to inhibit inflammation in various animal models (e.g., allergic inflammation models or ApoE-knockout atherosclerosis-prone mice).

Neuropeptides

Neuropeptides are released from sensory neurons and in some tissues they contribute to inflammatory reactions. For example, substance P and other tachykinins produce smooth muscle contraction, mucus secretion, cause vasodilation and increase vascular permeability. "Calcitonin gene-related peptide" (CGRP) is a potent vasodilator, acting on CGRP-receptors leading to activation of adenylate cyclase. The overall pattern of effects of tachykinins is similar, though not identical, to the pattern seen with kinins.

Tachykinins

The mammalian tachykinins comprise three related peptides: substance P (SP), neurokinin A (NKA), also called substance K, and neurokinin B (NKB). They occur mainly in the nervous system, particularly in nociceptive sensory neurons and in enteric neurons. They are released as neurotransmitters, often in combination with other mediators. SP and NKA are encoded by the same gene and they have a similar distribution. Three distinct types of tachykinin receptor are known: NK_1, NK_2, and NK_3. They are selective for three endogenous tachykinins with the following affinity: SP>NKA>NKB for NK_1, NKA>NKB>SP for NK_2 and NKB>NKA>SP for NK_3 receptor. Receptor cloning has shown that tachykinin receptors belong to a family of G-protein-coupled receptors. Several potent antagonists of NK_1 and NK_2 and NK_3-receptors have been discovered [26], and novel therapeutic agents for various disease states (e.g., pain, asthma, arthritis, headache) may be developed.

CGRP differs from other tachykinins. It is coded for by the calcitonin gene which also codes for calcitonin itself. Differential splicing allows cells to produce either procalcitonin (expressed in thyroid cells) or pro-CGRP (expressed in neurons) from the same gene. CGRP is found in non-myelinated sensory neurons and it is a potent inducer of neurogenic inflammation.

Kinins

Kinins are polypeptides with vasodilator/hypotensive, thrombolytic, pro-inflammatory and algesic actions. The two best known kinins are bradykinin and kallidin and they are referred to as plasma

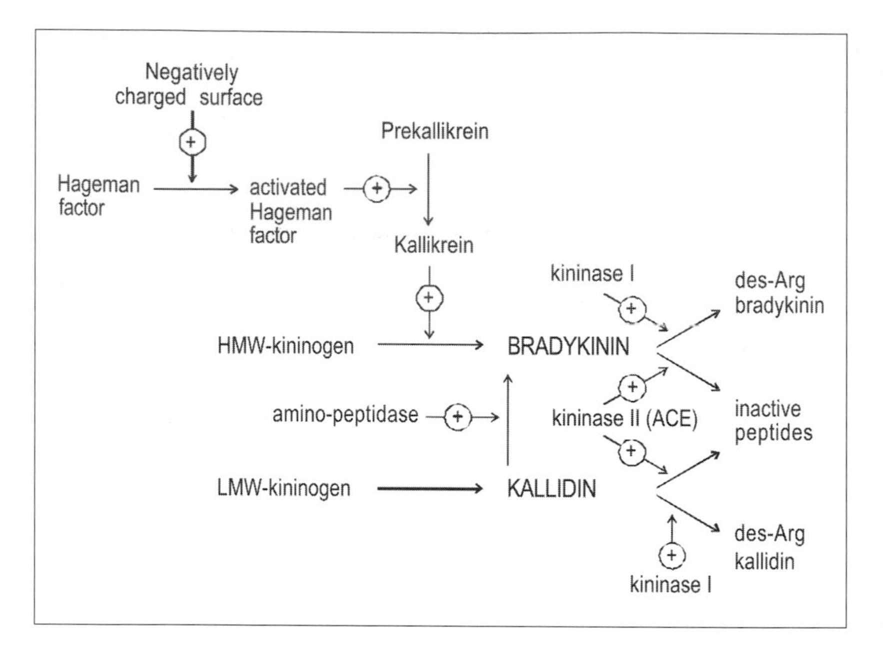

FIGURE 6
The formation and metabolism of kinins

kinins. Since 1980, when Regoli and Barabe divided the kinin receptors into B_1 and B_2 classes, first- and second-generation kinin receptor antagonists have been developed, leading to a much better understanding of the actions of kinins.

Bradykinin is a nonapeptide, kallidin is a decapeptide and has an additional lysine residue at the amino-terminal position. These two peptides are formed from a class of α-2 globulins known as kininogens (Fig. 6). There are two kininogens: high molecular weight (HMW) and low molecular weight (LMW) kininogen, which are products of a single gene that arises by alternative processing of mRNA. The highly specific proteases that release bradykinin and kallidin from the kininogens are termed kallikreins. Two distinct kallikreins, formed by different activation mechanisms from inactive prekallikreins, act on the kininogens. One of these is plasma kallikrein and the other is tissue kallikrein. LMW kininogen is a substrate only for the tissue kallikrein and the product is kallidin, while HMW kininogen is cleaved by plasma and tissue kallikrein to yield bradykinin and kallidin, respectively.

Kallidin is similar in activity to bradykinin and need not be converted to the latter to exert its effects. However, some conversion of kallidin to bradykinin occurs in plasma due to the activity of plasma aminopeptidases.

The half-life of kinins in plasma is about 15 seconds and concentrations of kinins found in the circulation are within the picomolar range. Bradykinin is inactivated by a group of enzymes known as kininases. The major catabolizing enzyme in the lung and in other vascular beds is kininase II, which is identical to peptidyl dipeptidase – known as angiotensin-converting enzyme (ACE). Kininase II is inhibited by captopril, resulting in an increased concentration of circulating bradykinin, which contributes substantially to the antihypertensive effect of captopril. On the other hand, kininase I is arginine carboxypeptidase and it has a slower action than kininase II. It removes the carboxyl-terminal arginine residue producing des-Arg^9-bradykinin or des-Arg^{10}-kallidin, which are themselves potent B_1-kinin receptor agonists.

There are at least two distinct receptors for kinins: B_1 and B_2. The classical, constitutive bradykinin receptor, now designated the B_2 receptor, selectively binds bradykinin and kallidin and mediates a majority of the effects of bradykinin and kallidin in the absence of inflammation, such as the release of PGI_2 and NO from endothelial cells. On the other hand, inducible B_1 receptors are upregulated by inflammation. They bind des-Arg metabolites of bradykinin and kallidin. In contrast to B_1 receptors, the signalling mechanism of B_2 receptors has been well characterized. The B_2 receptor is coupled to G protein and activates both PLA_2 and PLC. While stimulation of the former liberates AA from phospholipids, with its subsequent oxidation to a variety of pro-inflammatory eicosanoids, the activation of PLC through IP_3 and DAG leads directly to pro-inflammatory effects.

During the last decade the existence of other types of kinin receptors (B_3, B_4, B_5) has been suggested. However, recent studies indicate that some of them may actually represent functions of the B_2 receptor [18].

Kinins are among the most potent vasodilators known, acting on arteriolar beds of the heart, liver, skeletal muscle, kidney, intestines, and ovaries. They are claimed to play a minor role in the regulation of blood pressure in healthy individuals, but they play a major vasodepressor regulatory role, most likely mediated by arterial endothelium, in hypertensive patients [27]. Indeed, kinins contract veins and non-vascular smooth muscle, such as gastrointestinal and bronchial muscle. Bradykinin and kallidin have similar contracting properties. At the level of the capillary circulation, kinins increase permeability and produce oedema. Stimulation of B_1 receptors on inflammatory cells such as MACROPHAGES can elicit the production of the inflammatory mediators such as IL-1 and TNF-α [28]. Kinins are also potent pain-inducing agents in both the viscera and skin. In acute pain, B_2 receptors mediate bradykinin algesia. The pain of chronic inflammation appears to involve an increased expression of B_1 receptors.

As in the case of other autacoids, the therapeutic interest in kinins has focused particularly on attempts to modulate their formation or metabolism *in vivo* [29]. Blockade of kinin formation with a kallikrein inhibitor, aprotynin (Trasylol), has been used with some success to treat acute pancreatitis, carcinoid syndrome or Crohn disease. Experimentally, progress has been made in the development of selective antagonists of kinins. Currently, they are not available for clinical use; however, recent studies indicate that kinin receptor antagonists might be useful for the treatment of patients with septic shock, pancreatitis-induced hypotension, bronchial asthma, rhinovirus-induced symptoms and in fighting pain.

Nitric oxide

In animal tissues, nitric oxide (NO) is generated enzymatically by synthases (NOS). The three NOS isoenzymes (neuronal, endothelial and inducible) are flavoproteins which contain tetrahydrobiopterin and haeme and they are homologous with cytochrome p 450 reductase [30]. Isoenzymes of NOS act as dioxygenases using molecular oxygen and NADPH to transform L-arginine to L-citrulline and NO (Fig. 7). NO formed by endothelial constitutive NOS (eNOS) is responsible for maintaining low vascular tone and preventing LEUKOCYTES and platelets from adhering to the vascular wall. eNOS is also found in renal mesangial cells. NO formed by neuronal constitutive NOS (nNOS) acts as a neuromodulator or neuromediator in some central neurons and in peripheral "non-adrenergic non-cholinergic" (NANC) nerve endings. NO formed by inducible NOS (iNOS) in MACROPHAGES and other cells plays a role in the inflammatory response.

NO was discovered by Furchgott and Zawadzki as "endothelium-derived relaxing factor" (EDRF) [31]. It soon became obvious that EDRF, like nitroglycerine, activates soluble guanylate cyclase in vascular smooth muscle by binding to its active haem centre. The rise in cyclic GMP achieved is responsible for vasodilation and for other physiological regulatory functions of NO.

The activities of constitutive nNOS and eNOS are controlled by intracellular calcium/calmodulin levels. For instance, nNOS in central neurons is activated by glutamate binding to NMDA receptors with a subsequent rise in Ca^{2+}_i due to opening of voltage calcium channels, whereas eNOS is activated by blood shear stress or stimulation of endothelial muscarinic, purinergic, kinin, substance P or thrombin receptors. This triggers an increase in Ca^{2+}_i at the expense of the release of Ca^{2+} from endoplasmic reticulum.

Calcium ionophores (e.g., A23187) and polycations (e.g., poly-L-lysine) cause a rise in Ca^{2+}_i and activate eNOS, thereby bypassing the receptor mechanisms.

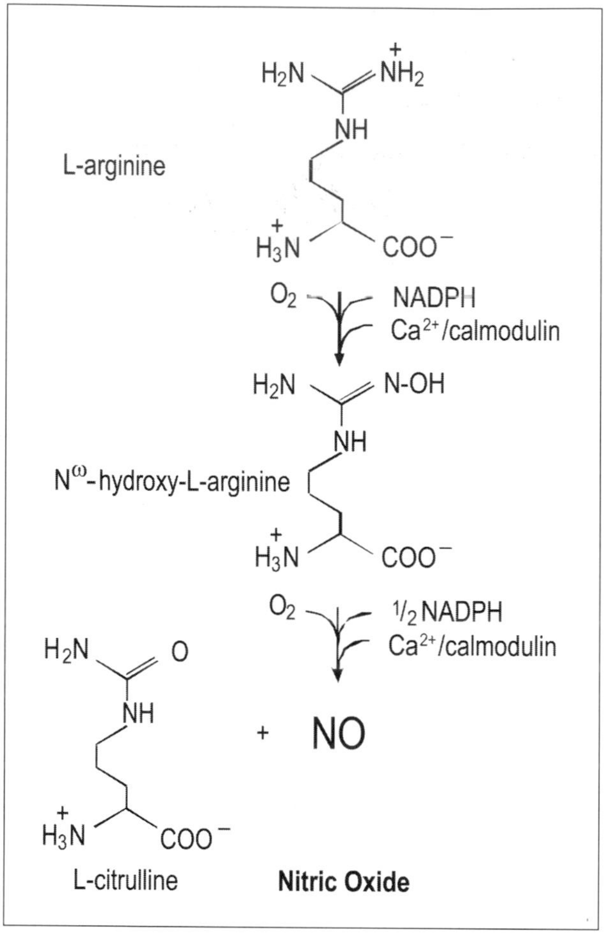

FIGURE 7
The synthesis and metabolism of nitric oxide (NO)

In contrast to the constitutive isoforms of NOS, iNOS does not require a rise in Ca^{2+}_i to initiate its activity. In MACROPHAGES, MONOCYTES and other cells, the induction of iNOS and the presence of L-arginine are sufficient to initiate the generation of NO. Induction of iNOS can be initiated by IFN-γ, TNF–α or IL-1. However, the best recognized inducer is LPS or endotoxin from *Escherichia coli*, which is known to be responsible for the development of SIRS in the course of sepsis due to gram-negative bacteria. Myeloid cells have a receptor for LPS on their cell membrane, m-CD14 protein. LPS, using an "LPS binding protein" (LBP), is anchored to m-CD14 and then triggers a chain of protein phosphorylation which eventually leads to the activation of the major transcription protein NF-k-B. This is responsible for transcription of the message: to make iNOS. In cells which lack m-CD14, the induction of iNOS is completed by a complex of soluble s-CD14 with LBP and LPS itself. In a similar manner, LPS can also induce COX-2. Although NO fulfils more paracrine than autoendocrine functions, yet in the case of iNOS, large amounts of locally formed NO may inhibit iNOS itself as well as COX-2, in a negative feedback reaction. Glucocorticosteroids and some CYTOKINES, such as TGF-β, IL-4 or IL-10, inhibit the induction of iNOS.

Nitric oxide as an effector of inflammation

Kinetics of nitric oxide production by iNOS differ greatly from production by eNOS or nNOS (Fig. 8) [32]. Inducible NOS produces very large, toxic amounts of NO in a sustained manner, whereas consititutive NOS isoforms produce NO within seconds and its activities are direct and short acting. There are multiple intracellular mechanisms through which NO may act as an inflammatory mediator [33]. Low levels of NO produced by constitutive synthases primarily interact directly with positively charged metal ions of guanylate cyclase, cytochrome p450 and NOS itself. Activation of guanylate cyclase leads to an increase in intracellular cyclic guanosine monophosphate (cGMP), which in turn activates cGMP-dependent protein kinases which mediate NO actions including vasorelaxation, increase of vascular permeability, as well as anti-proliferative, antiplatelet and anti-oxidant effects of NO. Recent data have also indicated that NO produced by constitutive NOS enzymes may be involved in immune regulation of T helper cell proliferation and cytokine production. During the course of an inflammatory response, the large amounts of NO formed by iNOS surpass the physiological amounts of NO which are usually made by nNOS or eNOS. The functions of iNOS-derived NO are also different. In immunologically or chemically activated MACROPHAGES, NO kills microorganisms and destroys macromolecules. NO formed by constitutive isoforms of NOS is stored as a

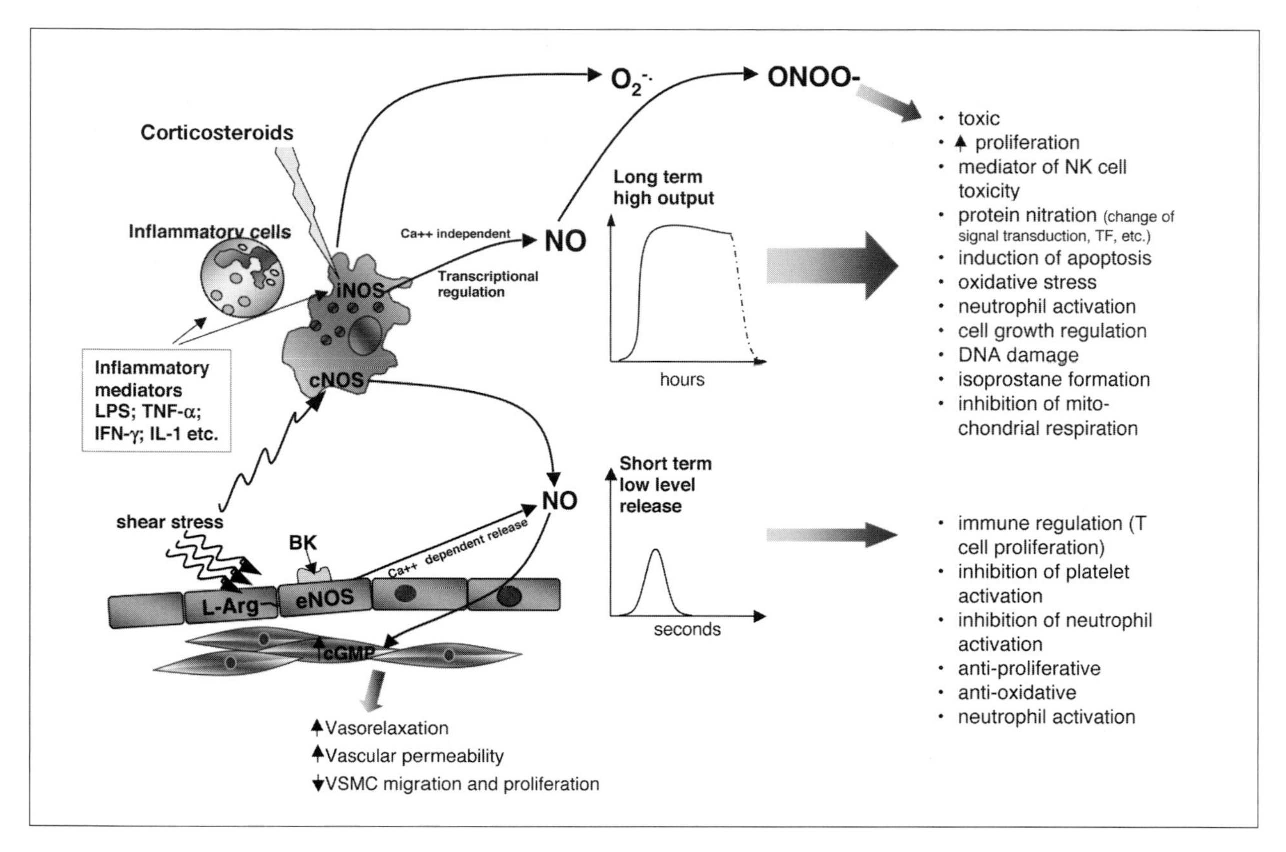

FIGURE 8
Differences between kinetics of nitric oxide generation by eNOS and iNOS

nitrosothiol in albumin and acts physiologically as N-nitrosoglutathione and N-nitrosocysteine. Eventually, within a few seconds, NO is oxidized to nitrites or nitrates. Large amounts of "inflammatory NO" from myeloid cells are usually generated side by side with large amounts of superoxide anion (O_2^-). These two can form peroxynitrite ($ONOO^-$) which mediates the cytotoxic effects of NO, such as DNA damage, LDL oxidation, isoprostane formation, tyrosine nitration, inhibition of aconitase and mitochondrial respiration. The discovery of this reaction opens new possibilities for the therapeutic use of superoxide dismutase (SOD). Indeed superoxide dismutase mimetics have been successfully used to limit the extent of inflammation. Interestingly, over-stimulation of NMDA receptors by glutamate may activate nNOS to such an extent that NO itself exerts neurotoxic properties. NO formed by eNOS seems to be mostly cytoprotective, possibly due to its unusual redox properties.

Large amounts of NO and $ONOO^-$ may target numerous proteins and enzymes critical for cell survival and signalling. These include signalling molecules involved in cytokine signalling like JAK or STAT proteins, NK-κB/IκB pathway as well as MAPK, some G proteins and transcription factors. Nitration of cysteines in these proteins may lead to their activation or inactivation.

NO is scavenged by haemoglobin, methylene blue and pyocyanin from *Pseudomonas coereleus*. These last two are also claimed to be inhibitors of guanylate cyclase. GLUCOCORTICOIDS selectively inhibit the expression of iNOS. Arginine analogues, such as L-N^G-monomethyl arginine (L-NMMA) and L-N^G-nitroarginine methyl ester (L-NAME) inhibit inducible and constitutive NOS isoforms non-selectively. Selective iNOS inhibitors (e.g., alkylisothioureas or aminoguanidines) are being intensively investigated in the hope that selective inhibition of iNOS may prevent development of SIRS or MODS (multiple organ dysfunction syndrome). Indeed, over-production of NO by iNOS during septicaemia is claimed to be responsible for irreversible arterial hypotension, vasoplegia (loss of responses to noradrenaline), lactic acidosis, suffocation of tissues, their necrosis and apoptosis. However, it is important to remember that NO made by iNOS is of benefit to the host defence reaction by contributing to microbial killing.

Moreover, NO generated by eNOS is essential to maintain tissue perfusion with blood, to offer cytoprotection in the pulmonary and coronary circulation against toxic lipids which are released by LPS and to preserve red cell deformability which becomes reduced in septicaemia [34]. Preliminary clinical experience with L-NMMA has been reasonably encouraging, as long as a low dose of the NOS inhibitor is used. In animal models of endotoxic shock, non-selective NOS inhibitors were reported to decrease cardiac output, to increase pulmonary pressure, to decrease nutritional flow to organs, to damage gastric mucosa and to increase mortality rate. On the other hand, inhalation of NO gas (10 ppm) in septic patients has been found to prevent the mismatch of the ventilation/perfusion ratio in their lung. The exact role of NO in various stages of sepsis, SIRS and MODS still awaits further elucidation and evaluation.

Nitric oxide in immune regulation

The exact role of NO in immune regulation is also unclear. Initial mouse studies suggested that antigen-presenting cell-derived NO may inhibit T cell proliferation, particularly of the Th1 subset of T helper cells. Mouse Th1 cells were also shown to produce NO, suggesting that the above mechanism is a part of negative feedback. In this way NO would inhibit Th1 and therefore promote Th2-type cytokine responses leading to humoral and allergic responses. Subsequent studies, however, indicate that both Th1 and Th2 produce similar amounts of NO, and both subsets respond similarly to NO. NO-induced changes in lymphocyte proliferation seem to be dependent more on the effects on the cell cycle proteins than on changes in the cytokine profile [32].

It is also important to discuss mechanisms through which cells that produce NO protect themselves against its toxic actions [35]. Recent studies show that GSH-GSSG anti-oxidative systems protect MACROPHAGES against iNOS-generated large amounts of NO. Similarly, endothelial cells are not the primary responder to eNOS-produced NO. This is possibly due to the fact that increases of intracellular calcium which mediate eNOS activation are also able to inhibit guanylate cyclase activity.

Reactive oxygen species

REACTIVE OXYGEN SPECIES (ROS) production plays an important role in modulation of inflammatory reactions. Major ROS produced within the cell are superoxide anion, hydrogen peroxide and hydroxyl radical [33]. Extracellular release of large amounts of superoxide anion produced by the RESPIRATORY BURST in LEUKOCYTES is an important mechanism of pathogen killing and also leads to endothelial damage, resulting in an increased vascular permeability as well as cellular death. However, vast evidence has recently implicated intracellular ROS production in playing a key role in modulation of release of other mediators of inflammation. This is related mainly to the constitutive expression of NAD(P)H oxidases (termed NOXs- non-phagocytic oxidases) in various tissues [32]. ROS produced by this family of enzymes can regulate adhesion molecule expression on endothelium and inflammatory cells, thus regulating cellular recruitment to the sites of inflammation. They also increase chemokine and cytokine expression. At least part of these effects results from the ability of ROS (in particular H_2O_2) to stimulate MAPK activity which leads to activation of several transcription factors. It is possible that intracellular ROS may act as second messengers in inflammatory signal transduction [32].

Inflammatory CYTOKINES (like TNF-α) may in turn increase NAD(P)H oxidase activity and expression which closes the vicious circle of inflammation.

While loss of NAD(P)H oxidase activity in cells leads to diminished inflammation in the vascular wall, several humoral factors may affect constitutive NAD(P)H oxidase expression in the vascular wall and therefore intracellular ROS production. These include angiotensin II, endothelins, high glucose or high cholesterol levels. Their effects on baseline ROS production may therefore mediate modulatory effects of these factors on inflammation which traditionally were not associated with inflammation.

Accordingly, attempts have been undertaken to inhibit intracellular ROS production in order to limit inflammatory responses. Apocynin, an NAD(P)H oxidase activation inhibitor, has been successfully used in limiting inflammation in animal models of rheumatoid arthritis, while decoy peptides preventing association of NAD(P)H oxidase subunits were shown to be effective in inflammation related to atherosclerosis.

Amines

HISTAMINE, 2-(4-imidazolyl)-ethyl-amine, is an essential biological amine in inflammation and allergy. It is found mostly in the lung, skin and gastro-intestinal tract. It is stored together with macroheparin in granules of mastocytes or basophils (0.01–0.2 pmoles per cell), from which it is released when complement components C3a and C5a interact with specific receptors, or when antigen interacts with cell-fixed IgE. These trigger a secretory process that is initiated by a rise in cytoplasmic Ca^{2+} from intracellular stores. Morphine and tubocurarine release HISTAMINE by a non-receptor action. Agents which increase cAMP formation inhibit HISTAMINE secretion, so it is postulated that, in these cells, c-AMP-dependent protein kinase is an intracellular restraining mechanism. Replenishment of the HISTAMINE content of mast cell or basophil after secretion is a slow process, whereas turnover of HISTAMINE in the gastric histaminocyte is very rapid.

HISTAMINE is synthesized from histidine by a specific decarboxylase and metabolized by histaminases and/or by imidazole N-methyltransferase. HISTAMINE exerts its effects by acting on H_1-, H_2- or H_3-receptors on target cells [36]. It stimulates gastric secretion (H_2), contracts most of the smooth muscle other than that of blood vessels (H_1), causes vasodilation (H_1), and increases vascular permeability by acting on the post-capillary venules [37]. Injected intradermally, HISTAMINE causes the triple response: local vasodilation and wheal by a direct action on blood vessels and the surrounding flare which is due to vasodilation resulting from an axon reflex in sensory nerves, thereby releasing a peptide mediator [37]. Of the many functions of HISTAMINE, stimulation of gastric acid secretion, and mediation of type 1 hypersensitivity, such as urinary and hay fever, are among the most important. The full physiological significance of the H_3-receptor has yet to be established [38]. HISTAMINE may also be involved in T helper cell immune regulation (extensively reviewed in [39]).

5-HYDROXYTRYPTAMINE (5-HT, SEROTONIN) was originally isolated and characterized as a vasoconstrictor released from platelets in clotting blood. 5-HT occurs in chromaffin cells and enteric neurons of the gastrointestinal tract, in platelets and in the central nervous system (CNS). It is often stored together with various peptide hormones, such as somatostatin, substance P or "vasoactive intestinal polypeptide" (VIP). The biosynthesis and metabolism of 5-HT closely parallels that of CATECHOLAMINES, except the precursor for decarboxylase of aromatic amino acids is 5-hydroxytryptophan instead of tyrosine (Fig. 9). 5-HT is inactivated mainly by the monoamine oxidases A or B (MAO A or B) to 5-hydroxyindoleacetic acid (5-HIAA) which is excreted in the urine. Some 5-HT is methylated to 5-methoxytryptamine, which is claimed to be involved in the pathogenesis of affective disorders.

The actions of 5-HT are numerous and complex, showing considerable variation between species [40]. For instance, in the inflammatory response 5-HT seems to be more important in rats than in humans. 5-HT is known to increase gastrointestinal motility, to contract bronchi, uterus and arteries, although 5-HT may also act as a vasodilator through endothelial release of NO. In some species, 5-HT stimulates platelet aggregation, increases microvascular permeability and stimulates peripheral nociceptive nerve endings. A plethora of pathophysiological functions proposed for 5-HT includes control of

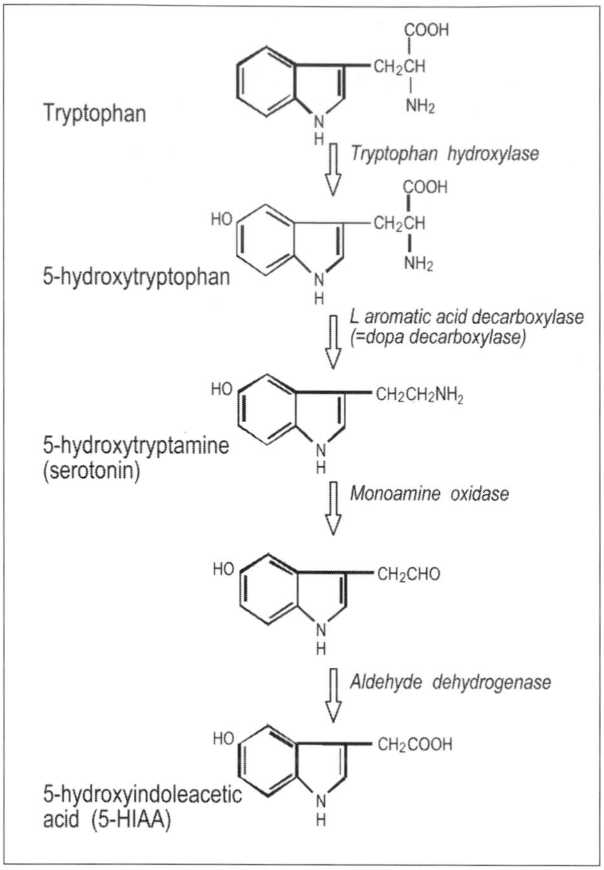

Figure 9
The synthesis and breakdown of 5-HT

peristalsis, vomiting, haemostasis, inflammation and sensitization of nociceptors by peripheral mechanisms or control of appetite, sleep, mood, stereotyped behaviour and pain perception by central mechanisms. Clinically, disturbances in the 5-HT regulation system have been proposed in migraine, carcinoid syndrome, mood disorders and anxiety [40].

These diverse actions of 5-HT are not mediated through one type of receptor. The amino acid sequence for many 5-HT receptor subtypes has been determined by cloning, and the transduction mechanisms to which these receptors are coupled have been explained. The basic four types of receptors are 5-HT_{1-4}. 5-HT_1 and 5-HT_2 receptors are further subdivided into A, B and C subtypes [41]. Types 1, 2 and 4 are G-protein-coupled receptors, type 3 is a ligand-gated cation channel. 5-HT_1 receptors occur mainly in the CNS (all subtypes) and in blood vessels (5-HT_{1D} subtype). 5-HT_{1B} and 5-HT_{1D} receptors appear to be involved, at least in part, in the modulation of neurogenically induced (following electrical, chemical or mechanical depolarization of sensory nerves) vascular inflammation. 5-HT_2 receptors (5-HT_{2A} subtype being functionally the most important) are more distributed in the periphery than in the CNS and they are linked to phospholipase C which catalyses phosphatidylinositol hydrolysis. The role of 5-HT_2 receptors in normal physiological processes is probably a minor one, but it becomes more prominent in pathological conditions, such as asthma, inflammation or vascular thrombosis. 5-HT_3 receptors occur particularly on nociceptive sensory neurons and on autonomic and enteric neurons, on which 5-HT exerts an excitatory effect and evokes pain when injected locally.

Catecholamines. It has become increasingly recognized that the release of catecholamines at autonomic nerve endings and from the adrenal medulla may modulate the function of immunocompetent cells. The major lymphoid organs (spleen, lymph nodes, thymus, and intestinal Peyer's patches) are extensively supplied by noradrenergic sympathetic nerve fibres. Sympathetic nervous system innervation of these lymphoid organs, as well as the presence of adrenergic and dopamine receptors on immune cells, provide the channels for noradrenergic signalling to lymphocytes and macrophages by sympathetic nerves [42]. Catecholamines have a wide range of direct effects on immune cells, particularly on macrophages and lymphocytes. Stimulation of β-adrenergic receptors on LPS-pretreated macrophages prevents the expression and release of pro-inflammatory TNF-α and IL-1, while the release of anti-inflammatory IL-10 is augmented. On the other hand, α-adrenergic stimulation augments phagocytic and tumoricidal activity of macrophages. Catecholamines acting through β-adrenergic and dopaminergic receptors, which are linked to adenylate cyclase through cyclic-AMP, modulate the function of immune cells. An increase in intracellular cyclic-AMP inhibits lymphocyte proliferation and production of proinflammatory CYTOKINES. The demonstration of the presence of α_2-, β-adrenergic, D1 and D2 receptors on various immune cells has recently provided the basis for regulation of cytokine production, specifically ILs and TNF, by these receptors in response to LPS [42]. Vasopressor and inotropic CATECHOLAMINES seem to have potent immunomodulating properties which, as yet, have not been adequately explored and may contribute to the therapeutic effects of dobutamine or dopexamine in the treatment of septic shock and SIRS.

Summary

Inflammation is a protective response of the macroorganism to injury caused by trauma, noxious chemicals or microbiological toxins. This response is intended to inactivate or destroy invading organisms, remove irritants, and set the stage for tissue repair. The inflammatory response consists of immunological and non-immunological reactions. The latter are triggered by the release from injured tissues and migrating cells of lipid-derived autacoids, such as EICOSANOIDS or platelet-activating factor, large peptides, such as INTERLEUKIN-1 and CYTOKINES, small peptides, such as bradykinin, and amines, such as HISTAMINE or 5-HYDROXYTRYPTAMINE. These constitute the chemical network of the inflammatory response and result in clinical and pathological manifestations of inflammation.

Characteristics of each mediator group involved in inflamation are discussed in this chapter.

The role of prostanoids, as autacoids involved in virtually every stage of inflammation, is discussed. They regulate vascular tone and permeability, induce platelet aggregation and thrombus formation and are involved in the pathogenesis of pain and fever accompanying inflammation. Recently discovered LIPOXINS are important regulators of inflammatory reactions. Platelet-activating factor, cytokine and chemokine groups as well as kinins are discussed in the context of their roles in inflammation. Recent studies have brought more light onto our understanding of intracellular signalling mechanisms in response to pro-inflammatory CYTOKINES like IL-1, TUMOUR NECROSIS FACTOR, transforming growth factor and INTERFERONS. TOLL-LIKE RECEPTORS are also briefly discussed.

Apart from the mechanisms mentioned above, NITRIC OXIDE and REACTIVE OXYGEN SPECIES production and interaction appear to play an important role in inflammation. These not only act as important effectors, damaging to invading microorganisms (NO from iNOS or superoxide anion) but may also be very important in immune regulation in part by the redox-sensitive gene regulation. The differences in the biology of these radicals reflected by their contrasting functions are presented. Summarising all of the above mechanisms is very important in the regulation of inflammation. Therefore coordinated pharmacological interventions, which would modify different parallel pathways in the inflammatory cascade, are needed to treat inflammatory diseases.

Selected readings

Cirino G, Fiorucci S, Sessa WC (2003) Endothelial nitric oxide synthase: the Cinderella of inflammation? *Trends Pharmacol Sci* 24: 91–95

Coleman JW (2001) Nitric oxide in immunity and inflammation. *Int Immunopharmacol* 1: 1397–1406

Guzik TJ, Korbut R, Adamek-Guzik T (2003) Nitric oxide and superoxide in inflammation and immune regulation. *J Physiol Pharmacol* 54: 469–487

Salvemini D, Ischiropoulos H, Cuzzocrea S (2003) Roles of nitric oxide and superoxide in inflammation. *Methods Mol Biol* 225: 291–303

References

1 Plytycz B, Seljelid R (2003) From inflammation to sickness: historical perspective. *Arch Immunol Ther Exp* (Warsz) 51: 105–109

2 Metchnikoff E (1893) *Secons sur la pathologie comparee de l'inflammation*. Masson, Paris

3 Harris SG, Padilla J, Koumas L, Ray D, Phipps RP (2002) Prostaglandins as modulators of immunity. *Trends Immunol* 23: 144–150

4 Moncada S, Gryglewski R, Bunting S, Vane JR (1976) An enzyme isolated from arteries transforms prostaglandin endoperoxides to an unstable substance that

inhibits platelet aggregation. *Nature* 263(5579): 663–665

5 Warner TD, Mitchell JA (2002) Cyclooxygenase-3 (COX-3): filling in the gaps toward a COX continuum? *Proc Natl Acad Sci USA* 99: 13371–13373

6 Schwab JM, Schluesener HJ, Laufer S (2003) COX-3: just another COX or the solitary elusive target of paracetamol? *Lancet* 361(9362): 981–982

7 Funk CD (2001) Prostaglandins and LEUKOTRIENES: advances in eicosanoid biology. *Science* 294(5548): 1871–1875

8 Breyer RM, Bagdassarian CK, Myers SA, Breyer MD (2001) Prostanoid receptors: subtypes and signaling. *Annu Rev Pharmacol Toxicol* 41: 661–690

9 Sigal E (1991) The molecular biology of mammalian arachidonic acid metabolism. *Am J Physiol* 260(2 Pt 1): L13–L28

10 Sala A, Aliev GM, Rossoni G, Berti F, Buccellati C, Burnstock G, Folco G, Maclouf J (1996) Morphological and functional changes of coronary vasculature caused by transcellular biosynthesis of sulfidopeptide LEUKOTRIENES in isolated heart of rabbit. *Blood* 87: 1824–1832

11 Feldberg W, Kellaway C (1938) Liberation of HISTAMINE and formation of lysocithin-like substances by cobra venom. *J Physiol* 94: 187–226

12 Yokomizo T, Izumi T, Shimizu T (2001) Leukotriene B4: metabolism and signal transduction. *Arch Biochem Biophys* 385: 231–241

13 Serhan CN (2002) Lipoxins and aspirin-triggered 15-epi-lipoxin biosynthesis: an update and role in anti-inflammation and pro-resolution. *Prostaglandins Other Lipid Mediat* 68–69: 433–455

14 McMahon B, Mitchell S, Brady HR, Godson C (2001) Lipoxins: revelations on resolution. *Trends Pharmacol Sci* 22: 391–395

15 Levy BD, De Sanctis GT, Devchand PR, Kim E, Ackerman K, Schmidt B, Szczeklik W, Drazen JM, Serhan CN (2003) Lipoxins and aspirin-triggered lipoxins in airway responses. *Adv Exp Med Biol* 525: 19–23

16 Carroll MA, McGiff JC, Ferreri NR (2003) Products of arachidonic acid metabolism. *Methods Mol Med* 86: 385–397

17 Kozek-Langenecker SA, Spiss CK, Michalek-Sauberer A, Felfernig M, Zimpfer M (2003) Effect of prostacyclin on platelets, polymorphonuclear cells, and heterotypic cell aggregation during hemofiltration. *Crit Care Med* 31: 864–868

18 Hache M, Denault A, Belisle S, Robitaille D, Couture P, Sheridan P, Pellerin M, Babin D, Noel N, Guertin MC et al (2003) Inhaled epoprostenol (prostacyclin) and pulmonary hypertension before cardiac surgery. *J Thorac Cardiovasc Surg* 125: 642–649

19 Ray WA, Stein CM, Daugherty JR, Hall K, Arbogast PG, Griffin MR (2002) COX-2 selective non-steroidal anti-inflammatory drugs and risk of serious coronary heart disease. *Lancet* 360(9339): 1071–1073

20 Chilton FH, O'Flaherty JT, Ellis JM, Swendsen CL, Wykle RL (1983) Metabolic fate of platelet-activating factor in neutrophils. *J Biol Chem* 258: 6357–6361

21 Chao W, Olson MS (1993) Platelet-activating factor: receptors and signal transduction. *Biochem J* 292 (Pt 3): 617–629

22 Takeda K, Kaisho T, Akira S (2003) Toll-like receptors. *Annu Rev Immunol* 21: 335–376

23 Liew FY (2003) The role of innate cytokines in inflammatory response. *Immunol Lett* 85: 131–134

24 Borish LC, Steinke JW (2003) 2. Cytokines and chemokines. *J Allergy Clin Immunol* 111(2 Suppl): S460–S475

25 Wong MM, Fish EN (2003) Chemokines: attractive mediators of the immune response. *Semin Immunol* 15: 5–14

26 Patacchini R, Maggi CA (2001) Peripheral tachykinin receptors as targets for new drugs. *Eur J Pharmacol* 429: 13–21

27 Pellacani A, Brunner HR, Nussberger J (1992) Antagonizing and measurement: approaches to understanding of hemodynamic effects of kinins. *J Cardiovasc Pharmacol* 20 (Suppl 9): S28–S34

28 Dray A, Perkins M (1993) Bradykinin and inflammatory pain. *Trends Neurosci* 16: 99–104

29 Fein AM, Bernard GR, Criner GJ, Fletcher EC, Good JT Jr, Knaus WA, Levy H, Matuschak GM, Shanies HM, Taylor RW, Rodell TC (1997) Treatment of severe systemic inflammatory response syndrome and sepsis with a novel bradykinin antagonist, deltibant (CP-0127). Results of a randomized, double-blind, placebo-controlled trial. CP-0127 SIRS and Sepsis Study Group. *JAMA* 277: 482–487

30 Cirino G, Fiorucci S, Sessa WC (2003) Endothelial nitric oxide synthase: the Cinderella of inflammation? *Trends Pharmacol Sci* 24: 91–95

31 Furchgott RF, Zawadzki JV (1980) The obligatory role

of endothelial cells in the relaxation of arterial smooth muscle by acetylcholine. *Nature* 288: 373–376

32 Guzik TJ, Korbut R (2003) Nitric oxide and superoxide in inflammation and immune regulation. *J Physiol Pharmacol* 54: 469–487

33 Salvemini D, Ischiropoulos H, Cuzzocrea S (2003) Roles of nitric oxide and superoxide in inflammation. *Methods Mol Biol* 225: 291–303

34 Korbut R, Trabka-Janik E, Gryglewski RJ (1989) Cytoprotection of human polymorphonuclear leukocytes by stimulators of adenylate and guanylate cyclases. *Eur J Pharmacol* 165: 171–172

35 Coleman JW (2001) Nitric oxide in immunity and inflammation. *Int Immunopharmacol* 1: 1397–1406

36 Bakker RA, Timmerman H, Leurs R (2002) Histamine receptors: specific ligands, receptor biochemistry, and signal transduction. *Clin Allergy Immunol* 17: 27–64

37 Repka-Ramirez MS, Baraniuk JN (2002) Histamine in health and disease. *Clin Allergy Immunol* 17: 1–25

38 Dale M, Foreman J, Fan T (1994) *Textbook of immunopharmacology*. Blackwell Scientific Publications, Oxford

39 Akdis CA, Blaser K (2003) Histamine in the immune regulation of allergic inflammation. *J Allergy Clin Immunol* 112: 15–22

40 Mossner R, Lesch KP (1998) Role of serotonin in the immune system and in neuroimmune interactions. *Brain Behav Immun* 12: 249–271

41 Kroeze WK, Kristiansen K, Roth BL (2002) Molecular biology of serotonin receptors structure and function at the molecular level. *Curr Top Med Chem* 2: 507–528

42 Madden KS, Sanders VM, Felten DL (1995) Catecholamine influences and sympathetic neural modulation of immune responsiveness. *Annu Rev Pharmacol Toxicol* 35: 417–448

Immune response in human pathology: infections caused by bacteria, viruses, fungi, and parasites

Jan Verhoef and Harm Snippe

Infections

In the middle of the 19th century, it became clear that micro-organisms could cause disease. Effective treatment, however, was not possible at that time; prevention and spread of infectious diseases depended solely on proper hygienic means. At the beginning of the 20th century, passive and active vaccination procedures were developed against a number of these pathogenic micro-organisms in order to prevent the diseases in question (rabies, diphtheria, tetanus, etc.) and due to the discovery of antimicrobial chemicals (Ehrlich) and antibiotics (Fleming), the threat of infectious diseases seemed to be minimized. Large-scale vaccination programs against childhood diseases (diphtheria, whooping cough and polio) started in the early fifties, giving hope to finally eradicate these diseases from the planet. This approach was successful for smallpox (1980); however, new infectious diseases emerged (e.g., Legionella, HIV, Helicobacter, SARS, etc.). New vaccines and antibiotics are needed. Furthermore, due to the intensive medical treatment with antibiotics and immunosuppressive drugs, hospital infections are a growing problem. Bacteria hitherto deemed harmless are causing opportunistic infections in immunocompromised patients. The pathogens develop multiple resistances to antibiotics and sometimes no effective antibiotics are available to treat those patients.

To make the story evermore serious, man is surrounded and populated by a large number of different non-pathogenic micro-organisms. In the normal – healthy – situation, there is a balance between the offensive capabilities of micro-organisms and the defences of the human body. The body's defences are based on vital non-specific and specific immunological defence mechanisms. An infection means that the microorganism has succeeded in penetrating those lines of defence, signaling a partial or complete breakdown of the body's defence system.

Natural resistance

The body's first line of defence comprises the intact cell layers of skin and mucous membrane, which form a physical barrier. The skin's low pH level and bactericidal fatty acids enhance the protection provided by this physical barrier. The defence in the respiratory tract and the gastrointestinal tract is mucous, the 'ciliary elevator' of the epithelium, and the motility of the small intestine. The presence of normal microbial flora (colonization resistance) in the intestine also plays a role in protection against colonization.

The most important humoral natural resistance factors are complement, lysozyme, INTERFERON, and a number of CYTOKINES. Lysozyme, which is found in almost all body fluids, degrades sections of the cell wall of Gram-positive and – in combination with complement – Gram-negative bacteria. This causes the otherwise sturdy cell wall to leak and the bacterium to burst.

Interferons are glycoproteins which may inhibit the replication of viruses. Within several hours after the onset of a virus infection, INTERFERONS are produced in the infected cell and help protect the neighbouring unaffected cells against infection. This protection is brief, but high concentrations of INTERFERONS are produced at a time when the primary immunological response is relatively ineffective.

CYTOKINES, such as IL-2 (INTERLEUKIN-2), GM-CSF (granulocyte-macrophage colony-stimulating factor), and TNF-α (TUMOR NECROSIS FACTOR α), stimulate non-specifically the proliferation, maturation, and

Box 1. Classification of micro-organisms

Classification of bacteria by

Genotypic characteristics:	chromosomal DNA fragment analysis, nucleic acid sequence analysis, probes
Phenotypic characteristics:	morphology, biotyping, serotyping, antibiotic resistance
Analytic characteristics:	cell-wall analysis, lipid and protein analysis, enzyme typing (catalase)
Gram staining positive or negative	
Aerobic, anaerobic:	fermentation of different sugars

Naming and classification of viruses according to

Structure:	size, morphology (naked, enveloped), nucleic acid (RNA, DNA)
Molecular aspects:	mode of replication, assembly and budding
Disease:	encephalitis, hepatitis
Means of transmission:	droplets, water, blood, insects
Host range:	animal, plant, bacteria

Classification of fungi according to

Structure:	macroscopic morphology of hyphae (mycelium) microscopic morphology of hyphae, conidophores and conidia (spores) shape and size
Cell features:	nucleus, cytosol, plasmalemma (cell membrane which contains cholesterol), physiology, staining properties
Sexual characteristics:	sexual and /or asexual reproduction, extended dikaryotic phase, basidium formation
Genotypic characteristics:	chromosomal DNA fragment analysis, nucleic acid sequence analysis, probes

Diagnosing of parasites by

Macroscopic examination
Concentration of cysts and eggs by microscopic examination
Serological diagnosis: antibody response
Detection of parasite by serology and by nucleic acid hybridization: probes and amplification techniques

function of the cells involved in defence (see Chapter A.4).

Innate immune cells recognize microbes by TOLL-LIKE RECEPTORS (see section Pathogenesis of shock), giving rise to the above production of CYTOKINES in the early phase of the response.

Micro-organisms that succeed in penetrating the first line of defence are ingested, killed, and degraded by phagocytic cells (leukocytes, MONOCYTES, MACROPHAGES), which are attracted to a microbial infection through chemotaxis. The ingestion by phagocytic cells of the micro-organism is enhanced by serum proteins (opsonins), such as ANTIBODIES and the C3b component of complement, for which these phagocytes have a receptor. After ingestion, the particle is surrounded by the membrane of the phagocyte, forming a vacuole known as a PHAGOSOME. The PHAGOSOME then fuses with some of the countless lysosomes in the phagocyte, thus allowing the lysosomal microbicidal agents and enzymes to do their

Box 2. Some examples of important human pathogens

	Species	Disease / location	Treatment / prevention
Bacterium	*Streptococcus pneumoniae*	pneumonia / meningitis	antibiotics / vaccination
	Mycobacterium tuberculosis	lung tuberculosis	antibiotics
	Vibrio cholerae	severe diarrhea	antibiotics / liquid suppletion / sanitation
	Staphylococcus aureus MRSA	wound infection / hospital infection	antibiotics, MRSA not sensitive for standard antibiotics, difficult to treat
	Neisseria meningitidis	meningitis	antibiotics / vaccination
	Bacillus anthracis	systemic infection (sepsis)	antibiotics as early as possible
	Corynebacterium diphteriae	throat / heart	anti-serum, vaccination
	Campylobacter jejuni	intestinal infections	hygiene, especially food (chicken)
	Helicobacter pylori	gastritis, ulcer	antibiotics
DNA Virus	Poxviridae	smallpox	vaccination, eradication
	Herpesviridae	Herpes genitalis	anti-viral agents
	Papavoviridae	warts and cervical carcinoma	surgery
	Hepadnaviridae	hepatitis B	vaccination
RNA Virus	Orthomyxoviridae	influenza	vaccination
	Coronaviridae	SARS	unknown
	Retroviridae	AIDS	anti-viral agents
	Caliciviridae	gastro-intestinal infection	sanitation, hygiene
Parasites	*Plasmodium* species	malaria	prophylactic medication, anti-malarial drugs
	Giardia species	intestinal tract	hygiene
	Trypanosoma cruzi	sleeping sickness	anti-parasitic agents

work. In the case of leukocytes, the formation of toxic oxygen radicals greatly contributes to the killing and elimination of the ingested micro-organism (Fig. 1) (see Chapter A.6).

A special role in cellular natural resistance is reserved for the NK (natural killer) cells, which display considerable cytotoxic activity against virus-infected cells. This NK activity is stimulated by INTERFERONS and, at a very early stage in the infection, serves to reinforce the non-specific defence mechanism.

Specific resistance

In the specific immune response, elements of the natural defence mechanism are directed against a specific enemy. Depending on the micro-organism, either the cellular defence mechanism (tuberculosis) or the humoral antibody-dependent defence mechanism (influenza) is of primary importance. In many cases, a joint cellular and humoral

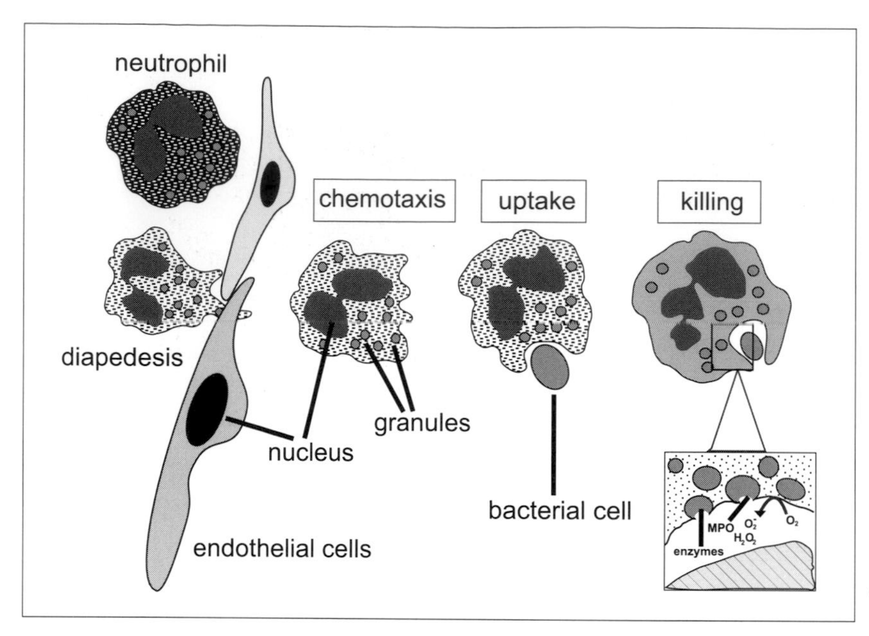

FIGURE 1
Schematic representation of the progressive steps of phagocytic endocytosis.

response is needed to provide an effective defence (typhus).

Both T LYMPHOCYTES and MACROPHAGES play a role in cellular defence. During the first contact with an antigen, MACROPHAGES process the antigen and present its protein fragments (T-cell epitopes) to T cells, which then proliferate and remain present for years in the body as memory cells. When a second encounter occurs, T cells produce lymphokines, which activate the MACROPHAGES. These activated MACROPHAGES grow larger, produce more and better degrading enzymes, and are now able to eliminate micro-organisms which otherwise would have survived intracellularly (tuberculosis, typhoid fever). MACROPHAGES from non-immune animals are not able to eliminate these micro-organisms.

Five different classes of ANTIBODIES can be distinguished in man, namely, IgG, IgA, IgM, IgD, and IgE. They differ from one another in size, charge, amino acid composition, and glycosylation (see Chapter A.3, C.2). In principle, the structure of the ANTIBODIES is the same, i.e., 2 heavy and 2 light chains: it is the variable part of these chains which recognizes the micro-organism. The biological function (see below) is determined by the constant part (Fc) of the heavy chain. With the exception of IgD, all these ANTIBODIES are important in antimicrobial activity.

- IgA, which is found in all external secretions, reacts with the surface of micro-organisms, preventing them from adhering to sensitive cells and mucous membranes.
- IgG neutralizes microbial toxins.
- IgG, IgM, and C3b serve as opsonins, which promote PHAGOCYTOSIS.
- IgG, IgM, and to a lesser extent IgA activate the COMPLEMENT SYSTEM after binding to the micro-organism. Activation products C3a and C5a ensure that the phagocytes are attracted to the inflammatory response.
- IgG and IgM, in combination with complement and lysozyme, have a lytic effect on bacteria and enveloped viruses.
- IgG and IgM inhibit the mobility of micro-organisms by attaching specifically to the flagellum. When this happens the chance of PHAGOCYTOSIS increases and the chance of spreading of disease decreases.
- IgG, together with the killer or K cells, can eliminate infected host cells which carry viral or other foreign ANTIGENS on their surface.

- IgE is of importance in parasite infections. At the site of the infection, mast cells, bearing specific IgE, release large quantities of vasoactive amines, which cause the contraction of smooth muscle tissue and increase the permeability of the blood vessels. In the intestine, this results in worms being detached and eliminated.

Defence against bacteria, viruses, fungi, and parasites

Several non-invasive bacteria, i.e., those that do not invade the body, cause disease through the production of exotoxins (tetanus, diphtheria, cholera). The immune system neutralizes the toxin with the aid of ANTIBODIES (IgG, IgM). If the individual has not been inoculated, the toxin will act on certain cells in the body directly through a receptor. This bond is very strong (i.e., has a high affinity), and is difficult to break by the administration of ANTIBODIES. In practice, if there are clinical symptoms of the disease, then large doses of antitoxins must be administered. If one is trying to prevent the development of the disease, then the presence of small quantities of specific ANTIBODIES (IgG) is sufficient.

The adherence of bacteria to cells is effectively blocked by IgA. Oral vaccination against cholera, for example, is aimed at obtaining sufficient specific IgA in the intestine, so that no colonization of this bacterium can take place, and the cholera toxin can no longer adhere to its receptor.

In general, defence against invasive bacteria is provided by ANTIBODIES (IgG, IgM) that are directed against bacterial surface antigens. In many cases, these bacteria have a capsule which interferes with effective PHAGOCYTOSIS. ANTIBODIES against these capsule ANTIGENS neutralize the interference, with subsequent elimination of the bacteria by phagocytes. ANTIBODIES (IgM, IgG, IgA) in combined action with complement kill bacteria by producing holes in the cell wall of the bacterium.

Although intracellular bacteria (tuberculosis, leprosy, listeriosis, brucellosis, legionellosis, and salmonellosis) are ingested by MACROPHAGES, they are able to survive and multiply. In these cases, cellular immunity alone provides the defence, since ANTIBODIES are not effective. Only activated MACROPHAGES are capable of killing and degrading these bacteria.

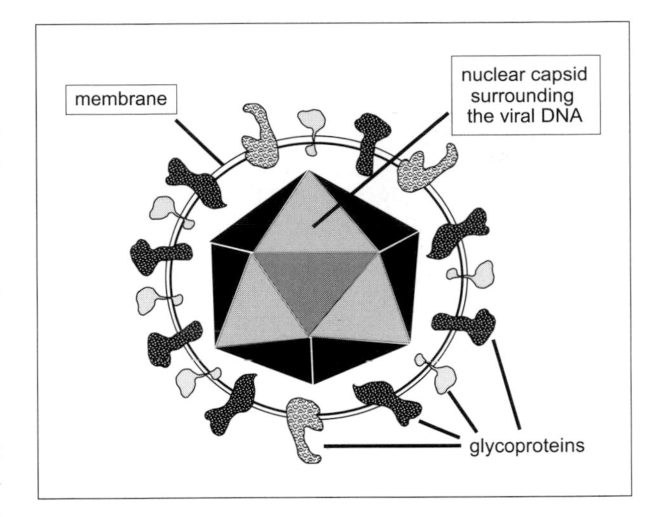

FIGURE 2
*Schematic illustration of an enveloped virus (*Herpes simplex *virus).*

ANTIBODIES neutralize viruses (Fig. 2) directly and/or indirectly by destroying infected cells that carry the virus antigen on their surface. The mechanisms of this defence resemble those of humoral defence against bacterial surfaces. The antibody-dependent cellular cytotoxicity reaction is specific for the defence against viruses. Cells which carry on the surface an antigen encoded by the virus are attacked by cytotoxic K cells, bearing ANTIBODIES which fit the antigen on the target cell (K cells have Fc receptors for IgG).

Not only humoral, but also cellular immunity plays an important role in virus infections. People with a genetic T-cell deficiency are highly susceptible to virus infections. In cellular defence, it is primarily the virus-infected cells which are attacked and eliminated. Cytotoxic T cells recognize MHC-1-presented T-cell epitopes on the surface of virus-infected cells and kill them.

The fungi responsible for human diseases can be divided into two major groups on the basis of their growth forms or on the type of infection they cause. Pathogens exist as branched filamentous forms or as yeasts, although some show both growth forms. The filamentous types (*Trichophyton*) form a 'mycelium'. In asexual reproduction, the fungus is dispersed by means of spores; the spores are a common cause of infection after inhalation. In yeast-like types (*Cryptococcus*), the characteristic form is the single cell, which reproduces by division or budding. Dimorphic types (*Histoplasma*) form a mycelium outside, but occur as yeast cells inside the body. *Candida* shows the reverse condition and forms a mycelium within the body.

In superficial mycoses, the fungus grows on the body surface, for example skin, hair, and nails (*Epidermophyton*, *Trichophyton*), the disease is mild, and the pathogen is spread by direct contact. In deep mycoses (*Aspergillus*, *Candida*, *Cryptococcus*, *Histoplasma*), internal organs are involved and the disease can be life-threatening and is often the result of opportunistic growth in individuals with impaired immunocompetence.

Many of the fungi that cause disease are free-living organisms and are acquired by inhalation or by entry through wounds. Some exist as part of the normal body flora (*Candida*) and are innocuous unless the body's defences are compromised in some way. The filamentous forms grow extracellularly, while yeasts can survive and multiply within phagocytic cells. Neutrophils kill yeasts by means of both intra- and extracellular factors. Some yeasts (*Cryptococcus neoformans*) form a thick polysaccharide capsule in order to prevent phagocytic uptake. In addition, many cell-wall components of yeasts cause suppression of cell-mediated immune responses. The role of humoral and cellular immunity in controlling infections caused by fungi is not yet well defined, but cellular immunity is the cornerstone of host defence against (some) fungal infections. As a consequence, HIV infection, which affects the cellular arm of the immune system, results in previously uncommon infections such as those caused by *C. neoformans*.

The immunological defence systems against *parasites* are considerably more complex than those against bacteria and viruses. This is due to various factors. In the first place, each parasite has its own life cycle, consisting of various stages with specific antigen compositions. Moreover, parasites are able to avoid the host defence system (mimicry), to combat it (immunosuppression), or to mislead it (antigenic variation). Both humoral and cellular immunity are important for the defence against parasites growing intercellularly, as we have seen in the case of bacteria and viruses. ANTIBODY concentrations (IgM, IgG, IgE) are often elevated. IgE also plays a special role in the removal of parasites (especially worm infections) from the intestine (see above).

Pathogenesis of shock

In Gram-negative (Fig. 3) bacterial infections, the interaction between bacterial endotoxin and various host-cell systems has been implicated in the pathogenesis of septic shock. In particular, the release of TNF-α (also called cachectin) and INTERLEUKIN-1 (IL-1), after the activation of host cells by endotoxin, induces hemodynamic shock.

Several lines of evidence support the current hypothesis that the monocyte-macrophage is the principal cellular mediator of endotoxicity. First, C3H/HeJ mice carrying a single gene defect are nonresponsive to lipopolysaccharide (LPS). The transfer of MACROPHAGES of a closely related LPS-sensitive strain makes the mice responsive. Second, when the host is challenged with endotoxin, soluble factors are produced by MACROPHAGES that mediate fever and an acute-phase response. These factors include the proinflammatory CYTOKINES, IL-1, IL-6, IL-8, and TNF-α. Together, TNF-α and IL-1 stimulate endothelial cells to produce and express proteins on their membrane that have adhesive properties for leukocytes, promoting the migration and passage of POLYMORPHONUCLEAR LEUKOCYTES (PMNs) from blood vessels through the endothelial layer, leading to PMN influx into the tissue. Adhesion molecules that mediate the binding of PMNs appear on the endothelium after an inflammatory stimulus, followed by molecules that are specific for adhesion of MONOCYTES or LYMPHOCYTES, which may be why neutrophils enter before mononuclear cells. Molecules that are currently known to be involved in leukocyte-endothelium interactions belong to three structural groups: the IMMUNOGLOBULIN gene superfamily, the INTEGRIN family, and the selectin family.

Concomitant with cytokine release, LPS induces the activation of PMNs, MACROPHAGES, and many other

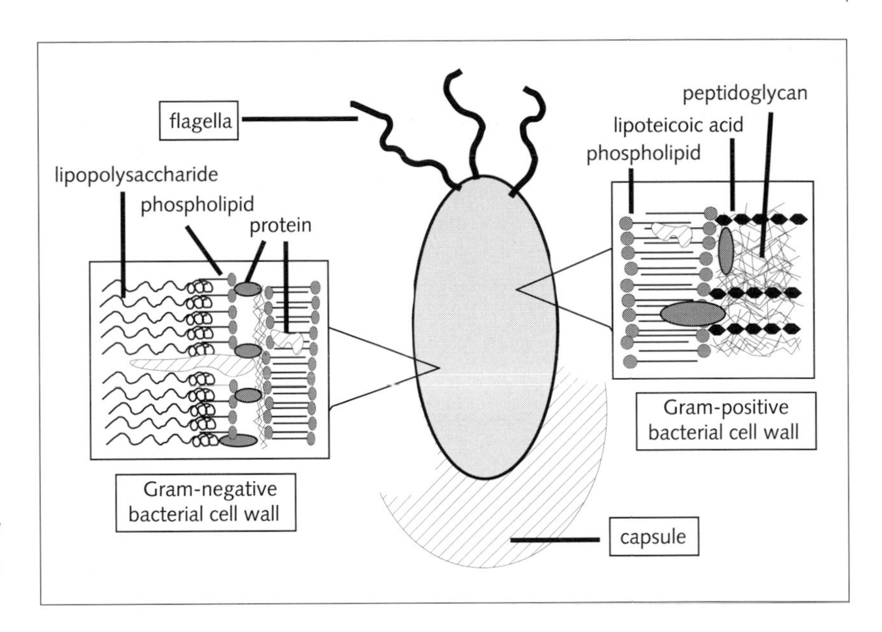

Figure 3
Schematic illustration of the cell envelope of a gram-negative and a gram-positive bacterium.

cells, resulting in the release of toxic oxygen radicals, which lead to tissue damage. At the same time, membrane-associated phospholipases are activated and products of the arachidonic-acid cascade are released through the cyclooxygenase and/or lipooxygenase pathways (see Chapter A.6). Platelet-activating factor (PAF) is also generated, partly in response to the same signals. All these products contribute to a generalized inflammatory state with influx of PMNs, capillary-leak syndrome, disturbances in blood coagulation, and myocardial suppression.

Endotoxin and TNF-α also produce multiple abnormalities in coagulation and fibrinolysis, leading to microvascular clotting and diffuse intravascular coagulation. They also induce endothelial cells to produce plasminogen activator and IL-6, which is an important modulator of the production of acute-phase proteins by the liver. Interestingly, despite having important structural differences, TNF-α and IL-1 have multiple overlapping and few distinct biological activities, act synergistically, and mimic the whole spectrum of toxicity caused by LPS (see Chapter A.4). IL-8 is an important chemoattractant and activator of neutrophils and is crucial in the early stages of inflammation.

Infusion of endotoxin in healthy humans leads to an early and transient increase in plasma levels of TNF-α (detectable after 30 min, peaking after 90–120 min, and undetectable after 4–6 h), which coincides with the development of clinical symptoms and pathophysiological responses encountered in Gram-negative septicemia. TNF-α, IL-1, IL-6, and IL-8 levels are also increased in patients with sepsis syndrome, with high levels of these CYTOKINES being correlated with severity of disease.

All these observations support the concept that endotoxin largely acts by initiating an inflammatory response through the activation of MONOCYTES-MACROPHAGES and the subsequent release of CYTOKINES. It also activates the COMPLEMENT SYSTEM (leading to the generation of C5a, which induces aggregation of PMNs and pulmonary vasoconstriction) and factor XII of the intrinsic coagulation pathway (Hageman factor). Finally, it induces the release of ENDORPHINS, which are also involved in the complex interactions of the inflammatory response in endotoxic septic shock.

Gram-positive bacteria are frequently and increasingly cultured from blood obtained from patients in shock. Unlike the pathophysiology of shock caused by Gram-negative bacteria, not much is

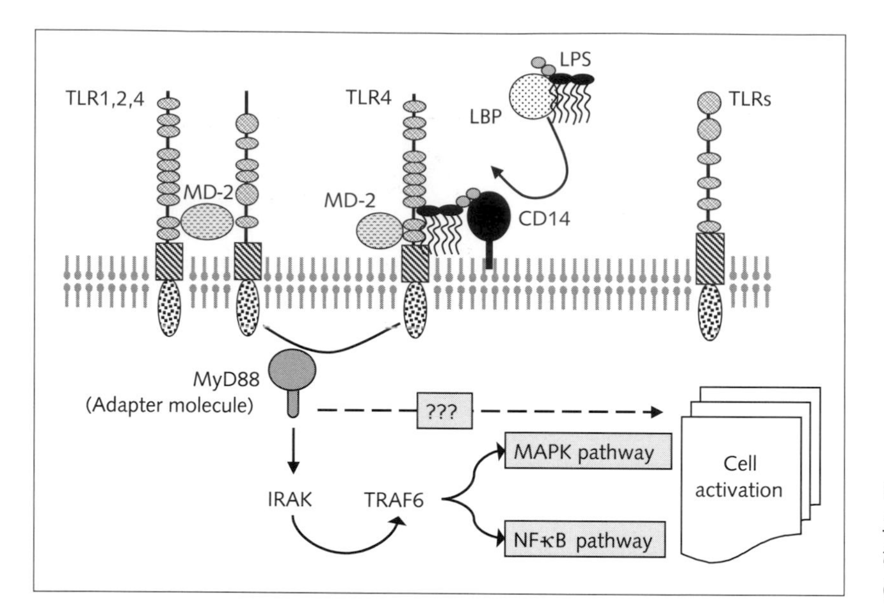

FIGURE 4
Schematic illustration of cell activation through toll-like receptors (TLRs).

known about the sequence of events that controls the signaling of MONOCYTES and MACROPHAGES that leads to the release of CYTOKINES. Cell-wall components, such as peptidoglycan and teichoic acid, are clearly important in the activation of these cells. Exotoxins, however, may also play a role in the pathogenesis of Gram-positive bacterial shock.

CD14 is a cell surface glycoprotein that functions as a binding receptor for LPS. However its membrane anchoring by a glycosylphosphatidyl inositol (GPI) linkage suggests little signaling and suggests the existence of additional coreceptors. Recent studies indicate that innate immune cells recognize conserved pathogen-associated molecular patterns (PAMPs), including LPS, through TOLL-LIKE RECEPTORS (TLRs) (Fig. 4). This family of proteins, that resemble the antimicrobial Toll proteins of *Drosophila*, has been identified in humans and mice. TLR4 was identified as the missing link in LPS-induced cell signal transduction and responsiveness that is associated with MD-2 and CD14. The TLR family members are coupled to a signaling adapter protein (MyD88) and form differential dimers that may explain the discrete responses to TLR ligands such as lipoprotens, heat shock proteins, unmethylated CpG DNA, viral dsRNA and bacterial flagellin. Intracellular signaling involves several kinases depending on the TLRs involved and includes the MAP kinase and NF-κB pathways leading to a cellular response. Recently other protein families have been identified via genetic screening that also participate in direct recognition of pathogens. A new protein was found to be involved in resistance to Gram-positive bacterial infections and recognizes the cell wall component peptidoglycan.

Human immunodeficiency virus infection

The human immunodeficiency virus (HIV) is a retrovirus that infects cells bearing the CD4 antigen, such as T-helper cells (TH), MACROPHAGES, and DENDRITIC CELLS. The CD4 molecule, together with other receptor molecules, like chemokine receptor 5, acts as a binding site for the gp120 envelope glycoprotein of the virus. In an attempt to respond to HIV ANTIGENS and concomitant secondary microbial infections, these cells are activated, thus inducing the replication of HIV in the infected CD4 T cells, which are finally destroyed. In contrast, HIV-1 infection of MACROPHAGES is self-sustained and results in an inexorable growth of chronic active inflammatory processes in

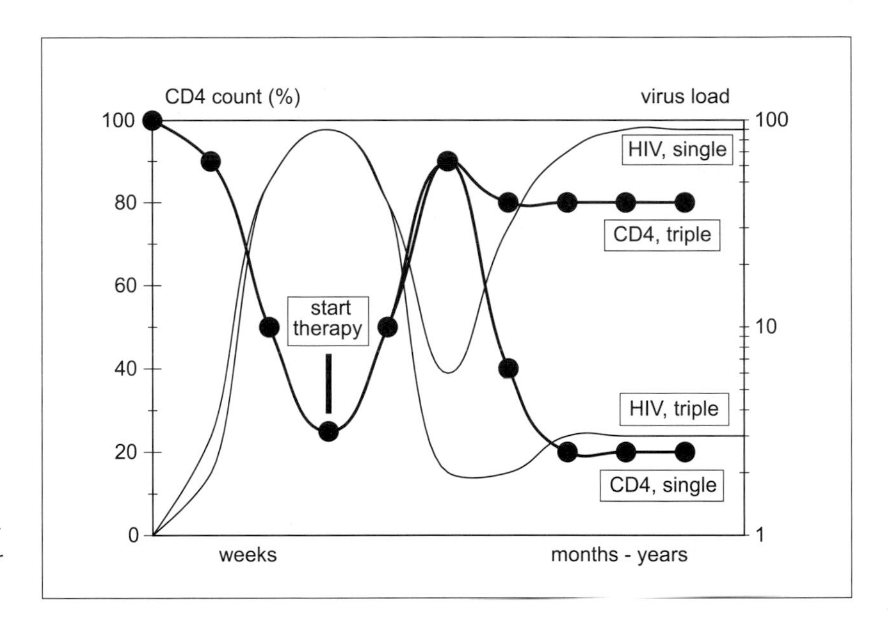

FIGURE 5
The effect of single and triple therapy on viral load and CD4 cells over time in HIV-infected individuals.

many tissue compartments including the central nervous system. Infected cells bear the fusion protein gp41 and may therefore fuse with other infected cells. This helps the virus to spread and accounts for the multinucleated cells seen in lymph nodes and brain. As a result of the decreased numbers of CD4-positive T-helper cells and defects in antigen presentation, depressed immune responses in these patients are observed. During the progression of the disease, opportunistic infections by otherwise harmless micro-organisms can occur. These include *Candida albicans* oesophagitis, mucocutaneous herpes simplex, toxoplasma in the central nervous system, and pneumonia caused by toxoplasma and *Pneumocystis carinii*; Kaposi's sarcoma also occurs frequently in these patients. This has been linked to the presence of a previously unknown type of Herpes virus (HHV-8). This immune deficiency syndrome is called 'acquired immune deficiency syndrome' (AIDS). It has been suggested that infected MONOCYTES/MACROPHAGES carry the HIV virus into the brain where it replicates in microglia and infiltrating MACROPHAGES. As a consequence many AIDS patients develop cognitive and motor brain impairments. However, the picture is complicated by the various persistent infections already present in these patients, which give rise to their own pathology in the brain. These include *Toxoplasma gondii*, *Cryptococcus neoformans* and JC virus.

So far a cure for HIV infection has not been achieved. The main effort in the prevention of HIV infection lies in mass public education programmes. Treatment of infected individuals is possible but expensive. At this moment a triple therapy is being prescribed in the Western countries (two reverse transcriptase inhibitors and one protease inhibitor, Fig. 5), each of which interfere with specific steps in the process of HIV replication. One major problem that has arisen is the increasing resistance to these drugs. Blocking of the chemokine receptor 5, a recently described co-receptor on CD4 cells for HIV, may be an alternative treatment for infected persons. This notion is supported by a recent finding that a homozygous defect in this chemokine receptor accounts for resistance of multiple-exposed individuals to HIV-1 infection.

Vaccines and vaccinations

Pasteur and Koch triggered the stormy development of vaccines (anthrax, rabies, cholera) at the end of the 19th century. While Pasteur remained faithful to the principle of attenuated micro-organisms in preparing his vaccines, Koch employed killed germs (cholera) as a vaccine. Since diphtheria and tetanus cause disease by means of toxins, the next logical step in the development of vaccines was the use of detoxified toxins to induce protection against these diseases (diphtheria, Von Behring and tetanus, Kitasato). Von Behring and Kitasato were the first to demonstrate that the source of the protective activity induced by vaccines was present in blood serum. Von Behring was also the first to prove that protective immunity could be passed on via serum. The development of new vaccines had its ups (yellow fever) and downs (tuberculosis). With the arrival of antibiotics, all work on new bacterial vaccines was suspended or severely curtailed, although some researchers continued to work on viral vaccines, such as rubella, measles, polio, and mumps.

Since it has proved difficult to consistently develop new antibiotics to combat antibiotic-resistant bacteria, interest in vaccines has gradually increased over the last 15 years (see Chapter C.1). Today, thanks to new insights into the immune system and modern molecular biological and chemical techniques of analysis and synthesis, it is possible to produce well-defined vaccines. These contain only those determinants of the pathogenic micro-organism which induce protection (epitopes). These epitopes are usually short peptide or oligosaccharide chains, which can be produced synthetically or by means of recombinant DNA techniques. The immunogenicity of these products can be enhanced by coupling them to a carrier (tetanus toxoid, liposomes) and/or by adding an adjuvant (a substance which strengthens the immune response non-specifically). The recombinant DNA technique can also be used to obtain attenuated strains of micro-organisms, which are fully immunogenic and thus provide protection, but which are no longer virulent. One example of this is the development of a new cholera vaccine based on a bacterium which has all the characteristics of a virulent strain, except the toxin. The bacterium has retained all its adherence factors, which allow it to adhere to the intestinal mucosa; the length of time it spends in the intestine is sufficient to stimulate the local immune system. The newest trend in vaccinology is immunization by introducing plasmid DNA into the host. Success has been attained by this method for hepatitis B vaccination.

Not only are new vaccines being developed, but it is also possible to heighten natural resistance for longer or shorter periods. Various INTERLEUKINS (IL-2, GM-CSF) and INTERFERONS are being studied in order to use them to combat infectious diseases. Monoclonal ANTIBODIES (ANTIBODIES with one specificity) directed against the endotoxin of Gram-negative bacteria are now being administered to patients with severe Gram-negative sepsis (serum therapy). More work is still necessary, however, to refine this technique, as the therapeutic effect is still limited.

Infections in the new millenium

As outlined above for a number of bacteria and viruses, effective vaccines have been developed and applied worldwide. The eradication of smallpox (*Variola major*) virus in the seventies of the last century was a milestone for the World Health Organization. The next goal of the WHO is to eradicate the poliovirus in the coming years. Major problems to be dealt with are the distribution of these vaccines, the costs involved, the registration and the compliance of the vaccinees and molecular techniques to trace the final bug. Meanwhile new unexpected microbiological threats come into focus. Hospital infections caused by multiple resistant micro-organisms due to the abundant use of antibiotics and exchange of genetic material between micro-organisms impose major problems on patients and healthcare workers. New antibiotics and/or vaccines should be developed and new strategies employed to contain these infections. Due to crowding and high mobility of the world population, old and new pathogens, e.g., influenza and SARS, threaten our society. On top of this terrorists might intentionally use micro-organisms (Smallpox, Anthrax, Plague etc.), or bacterial toxins (Botulism) to cause death and disease in humans or animals in a civilian setting. The recognition that an event was caused by a biological weapon presents a severe challenge to be prepared for such an attack, especially for medical care providers and public health officials. Strategies to combat bioterrorism have to be worked out but with the experience of 100 years of combating micro-organisms with hygiene measures, vaccination, antibiotic and anti-viral treatment, there must be a way out.

Summary

Despite the introduction of effective health measurements, vaccination and antimicrobial therapy infectious diseases continue to threaten human life. The reasons are numerous and diverse: antibiotic resistance, hospital-invading pathogens, new emerging infectious diseases, bioterrorism, biological warfare. This chapter is an introduction to several aspects of infectious diseases viewed from the host as well as from the pathogen (bacterium, virus and parasite). Furthermore the basic principles of INNATE and ADAPTIVE IMMUNE RESPONSES, especially in debilitated patients, are described. Detailed information is given on the pathogenesis of septic shock, AIDS and vaccination strategies.

Acknowledgments

We thank Dr. C.P.M. van Kessel of The Eijkman-Winkler Center for the design and layout of the artwork.

Selected readings

1 Jawetz E, Melnick JL, Adelberg EA, Brooks GF, Butel JS, Ornston LN (eds.) (2001) *Medical microbiology*. Prentice-Hall International, London
2 Mims CA, Playfair JHL, Roitt IM, Wakelin D, Williams R, Anderson RM (eds.) (1998) *Medical microbiology*. Mosby, London
3 Roitt I, Brostoff J, Male D (eds.) (2001) *Immunology*. Mosby, London
4 Silverstein AM (ed.) (1988) *A history of immunology*. Academic Press, San Diego
5 Snippe H, Willers JMN (1992) Attack and Defence. In: JJ van Everdingen (ed.): *The beast in man: microbes and macrobes as intimate enemies. Part II. The battle of bugs.* Belvedere, Overveen, The Netherlands
6 Janeway CA, Travers P, Walport M, Shlomchik M (eds.) (2001) *Immunobiology, The immune system in health and disease*. Garland Science Publishing, New York
7 Prescott L, Harley J, Klein D (eds.) (2003) *Microbiology*. Mc Graw Hill, New York

Recommended websites

Scientific Research:
American Society for Microbiologists: *http://www.asm.org*
American Association of Immunology: *http://www.aai.org*
European Federation of Immunological Societies: *http://www.efis.org*
Federation of European Microbiological Societies: *http://www.fems-microbiology.org*
National Library of Medicine: *http://www.ncbi.nlm.nih.gov*

Outbreaks of Infectious Diseases:
Centers for Disease Control and Prevention: *http://www.cdc.gov*
International Society for Infectious Diseases: *http://www.isid.org*
Daily update: *http://www.promedmail.org*
World Health Organization: *http://www.who.int*
(All accessed March 2005)

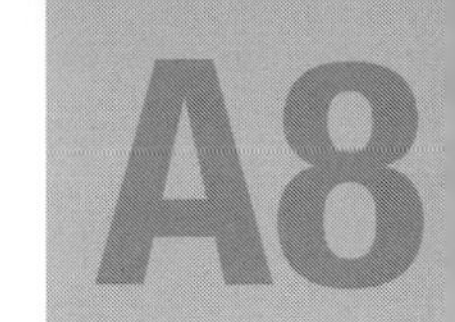

Immune response in human pathology: hypersensitivity and autoimmunity

Jacques Descotes and Thierry Vial

Introduction

The immune system is a complex network of regulatory and effector cells and molecules whose primary function is discrimination of self from non-self to maintain the homeostasis of the body. Closely interacting processes ensure coordinated immune responses. The renewal, activation and differentiation of specialized (immunocompetent) cells are required to achieve a normal level of immunocompetence under the control of many mechanisms with either a redundant or conflicting outcome. It is noteworthy, however, that immune responses are not always beneficial. Thus, inadvertent immunological reactivity against innocent ANTIGENS can lead to HYPERSENSITIVITY reactions while immune responses against self constituents of the host result in autoimmune diseases or reactions. Although the mechanism(s) leading to either type of adverse immune responses can be inhibited, at least to some extent, by pharmacological manipulation, pharmaceutical drugs as well as environmental or industrial chemicals can trigger HYPERSENSITIVITYand autoimmune reactions [1].

Hypersensitivity

Nearly every chemical in our medical, domestic, occupational or natural environment can induce HYPERSENSITIVITY reactions.

Epidemiology of hypersensitivity reactions to drugs

Although HYPERSENSITIVITY is widely held as a major cause of immunotoxic events, there are very few data on the actual incidence of HYPERSENSITIVITY reactions in human beings. Most published data consist of case reports and epidemiological studies are relatively rare. It has been suggested that approximately one-third of all drug-induced adverse effects might have an IMMUNOALLERGIC or PSEUDOALLERGIC origin [2]. More conservative estimates, e.g., 6–10%, have also been proposed and current limitations or uncertainties in the medical diagnosis of drug allergies [3] probably account for these differences. For instance, severe anaphylactic or anaphylactoid reactions may develop in about 1 in 5,000 exposures to antibiotics or radiocontrast media [4], or 1 in 10,000 drug treatments [5]. The same lack of data precludes any reliable estimate of the incidence of HYPERSENSITIVITY reactions in relation to occupational or environmental exposure, as well as to food allergies.

Clinical manifestations of drug-induced hypersensitivity reactions

HYPERSENSITIVITY reactions can affect nearly every organ or tissue of the body, but one organ or tissue is often a predominant target [6].

Anaphylactic shock

Anaphylactic shock is one of the most severe HYPERSENSITIVITY reactions with an estimated death rate of about 1% [7]. ANAPHYLAXIS unexpectedly develops within 1–20 minutes after the last contact with the drug and almost always within the first two hours, depending on the route of administration. Patients usually complain of itching, and then rapidly develop urticaria and angioedema, tachycardia, hypotension progressing to cardiovascular collapse or shock in the most severe cases, and marked respiratory difficulties with cyanosis. Anaphylactic shock is a major

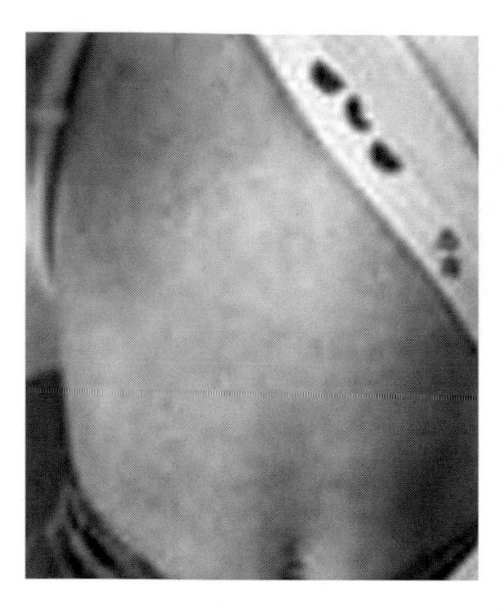

FIGURE 1
Erythematous skin rash due to amoxicillin.

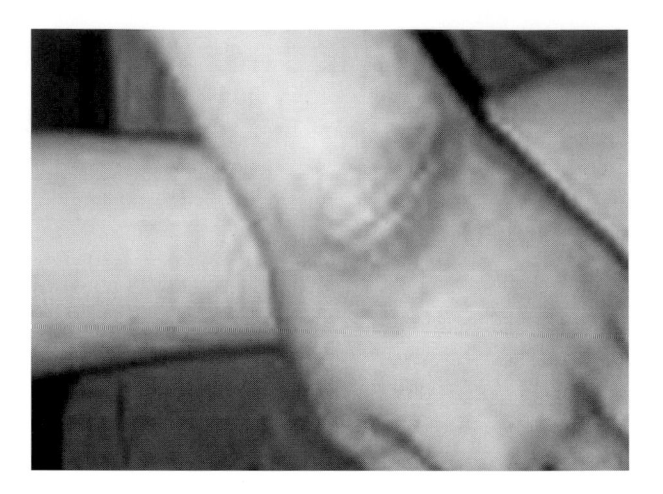

FIGURE 2
Contact dermatitis.

medical emergency. Initial supportive measures tend to maintain or restore normal respiratory and circulatory functions. The key treatment is epinephrine (adrenaline) injected either subcutaneously or intramuscularly.

Skin reactions

Skin reactions are the most frequent immune-mediated adverse effects of drugs [8, 9]. Relatively rare reactions involve specific IgE ANTIBODIES, e.g., urticaria and angioedema, or circulating immune complexes, e.g., vasculitis, but the majority of immune-mediated cutaneous manifestations caused by drugs are presumably due to T cell-mediated mechanisms. The clinical presentation is extremely varied. Morbilliform or exanthematous rash may occur in up to 2% of all treated patients (Fig. 1). The onset varies within 1–2 weeks after the start of treatment. A T cell-mediated cytotoxic mechanism is most likely [10]. Contact dermatitis is a frequent complication of occupational and environmental exposures, but it can also develop following topical drug applications [11]. It is characterized by pruritic vesicles on an erythematous background (Fig. 2). Allergic contact dermatitis is a T LYMPHOCYTE-mediated reaction that should be differentiated from irritant contact dermatitis caused by non antigen-specific mechanism(s). Stevens-Johnson syndrome (SJS) and toxic epidermal necrolysis (TEN) are the most severe and potentially life-threatening cutaneous complications of drug treatments [12]. They affect approximately 0.5-2 in 1 million persons per year [13]. Typically, the first clinical symptoms, i.e., flu-like reaction of variable severity and mucous membrane involvement, appear 7 to 21 days after the start of treatment. Skin lesions extend within 2 to 3 days with purpuric macules and bullae leading to erosion of the epidermis (Fig. 3). Painful erosions of the mucous membranes account for dysphagia, conjunctivitis, keratitis, diarrhea and/or respiratory distress. The prognosis and management are similar to those of severely burnt patients. The mortality rate is about 5% for SJS and 30% for TEN.

The drug HYPERSENSITIVITY syndrome or DRESS – Drug Rash with Eosinophilia and Systemic Symptoms – is characterized by fever and rash, but 50% of patients also present with lymphadenopathy, arthritis or hepatitis, and less frequently kidney, heart, lung, thyroid, or brain involvement (Hyper)eosinophilia is noted in 90% of cases [14].

Immunoallergic cytopenias

Cytopenias generally manifest as antibody-mediated destruction of one or several blood cell lines. Even though drugs are the most likely cause, the incidence is low and probably not more than 1 in 1–300,000 treated patients [15]. Patients with agranulocytosis are either asymptomatic and the diagnosis is then made after a routine blood examination, or they develop clinical symptoms of infection, in particular sore throat. In this latter case, neutropenia is often below 100/mm^3. Hemolytic anemias are due either to direct or antibody-mediated toxicity to the membrane of erythrocytes. Depending whether AUTOANTIBODIES are involved or not, immune-mediated hemolytic anemias are either autoimmune or immunoallergic. However, the distinction may be difficult to make as both AUTOANTIBODIES and drug-dependent ANTIBODIES can be produced by the same drug [16]. Most IMMUNOALLERGIC hemolytic anemias are acute or subacute. Clinical symptoms develop within hours following drug intake and include abdominal and dorsal pain, headache, malaise, fever, nausea and vomiting. Shock and acute renal failure are noted in 30–50% of cases.

Two main mechanisms can be involved. The causative drug, e.g., a third-generation cephalosporin [17], can nonspecifically bind to ERYTHROCYTES and reacts with circulating specific ANTIBODIES. Otherwise, the drug bound to a plasma protein can form an antigenic complex with the resulting production of IgM or IgG against the drug-protein complex that can be passively fixed to erythrocytes. The reintroduction of even a tiny amount of the drug can trigger an antigen-ANTIBODY reaction leading to intravascular hemolysis by activation of the complement cascade. Drug-induced IMMUNOALLERGIC thrombocytopenias are uncommon with the notable exception of those induced by heparin [18]. Most patients with heparin-induced immune thrombocytopenia have detectable ANTIBODIES against the platelet glycoproteins Ib/IX and IIb/IIIa. Clinically, thrombocytopenia leads to bleeding when platelet counts are less than 10–30,000/mm^3.

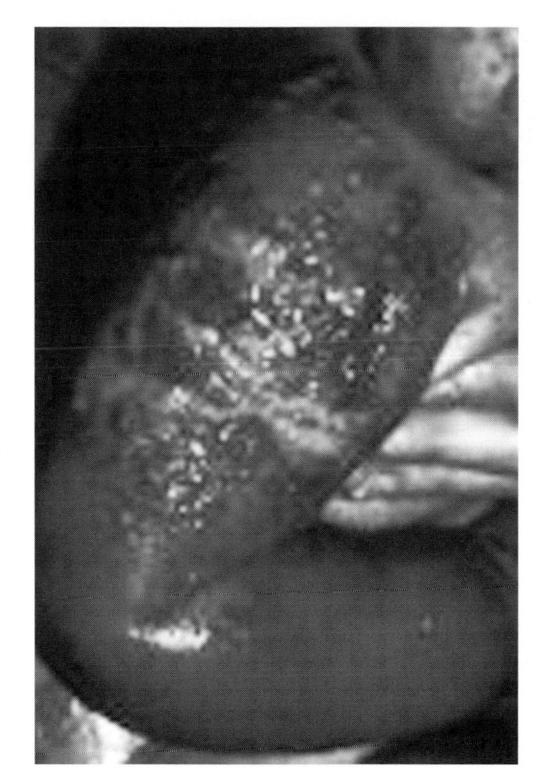

FIGURE 3
Toxic epidermal necrolysis.

Other clinical manifestations

Other clinical manifestations of drug-induced HYPERSENSITIVITY reactions include hepatitis, nephritis and pneumonitis. Severe drug-induced hepatitis is relatively infrequent and considered as an IDIOSYNCRATIC reaction in most instances. An unpredictable adverse immune response against the liver with cytolytic, cholestatic or mixed clinical and biological features, however, can be involved [19]. The causative mechanism is not known in most cases and the association of liver injury with fever, rash and eosinophilia, typically within 1 to 8 weeks after starting drug treatment, is often held as suggestive of an IMMUNOALLERGIC reaction. Drugs are the leading cause of acute interstitial nephritis, which accounts for 1–3% of all cases of acute renal failure [20]. The clinical presentation is nonoliguric renal dysfunction associated with fever, rash and/or eosinophilia in up to 50% of patients. Drugs can cause acute interstitial or eosinophilic pneumonia, and HYPERSENSITIVITY pneumonitis [21]. Clinical manifestations including fever, cough, eosinophilia, and elevated serum IgE levels vary depending on the offending drug.

Mechanisms of drug-induced hypersensitivity reactions

Immune-mediated HYPERSENSITIVITY reactions are the consequence of the exquisite capacity of the immune system to recognize structural elements of non-self molecules or antigens, and to mount specific responses due to the involvement of immunological memory.

Sensitization

An absolute prerequisite for any antigen-specific HYPERSENSITIVITY reaction to develop is that sensitization occurred prior to the eliciting contact. Importantly, it is normally impossible to demonstrate whether a prior contact was actually sensitizing. Identifying a prior and supposedly sensitizing contact is more or less easy. This can be straightforward when a patient who developed an adverse reaction was being treated with the suspected offending drug. At least 5–7 days of treatment are necessary to potentially result in sensitization. The majority of immune-mediated HYPERSENSITIVITY reactions do develop within the first month of treatment, but a longer period of time is nevertheless possible. No difficulty either is expected in patients previously treated with the same drug. An immune-mediated HYPERSENSITIVITY reaction can develop within minutes to several days depending on the route of administration and the underlying mechanism. A prior contact may be far less easy to detect when it is due to exposure via the food chain, or involves a closely related molecule leading to cross-allergenicity. Finally, it is important to bear in mind that even though a prior contact is absolutely required, a clinical reaction often does not develop after a subsequent contact.

Broadly speaking, a molecule can be sensitizing when this is a foreign and large molecule. The majority of drugs are foreign molecules. There is no formally established minimal requirement regarding the size of foreign molecules to be immunogenic (sensitizing). A molecular weight of at least 1,000 or more conservatively 5,000 has been proposed. However, the structural complexity, degree of polymerization and biodegradability also play a major role. The vast majority of drugs are too small to act as direct immunogens. It is widely assumed that low-molecular-weight molecules must therefore play the role of haptens to induce sensitization [22]. Haptens are small molecules that strongly bind to carrier macromolecules so that the formed hapten-carrier complex can mount a specific immune response. Sufficient chemical reactivity is absolutely required for low-molecular-weight molecules to become haptens. As most molecules intended for therapeutic use are devoid of any significant chemical reactivity, metabolites are thought to be involved. However, highly reactive metabolites have very short half-lives so that they often cannot be identified. The consequence is that evidence for the role of metabolites in drug-induced HYPERSENSITIVITY reactions is often indirect and therefore largely assumptive.

Risk factors

Even potent sensitizing drugs, such as penicillin G, induce HYPERSENSITIVITY reactions in only quite a small percentage of treated patients. The involvement of risk factors may account for this finding. Risk factors may be related either to the patient or the drug. Age, gender, atopy and genetic predisposition are the main risk factors related to the patient. Young adults develop more frequent HYPERSENSITIVITY reactions to drugs for unknown reasons. An epidemiological study of severe anaphylactic and anaphylactoid reactions in hospitalized patients found 927 reactions per 1 million in patients less than 20 years of age, 221–276 in patients between 20 and 59 years of age, and only 154 in patients over 60 [23]. Young girls and women seem to develop only slightly more HYPERSENSITIVITY reactions induced by drugs than boys and men. Atopy is characterized by excessive production of IgE associated or not with one or several diseases, such as reaginic asthma, hay fever and constitutional dermatitis. Due to variable definitions over time, conflicting results have been published, but the most recent data support the role of atopy as a risk factor. Multi-generation family and twin studies have demonstrated a genetic component in a number of allergic diseases, in particular IgE-mediated diseases. As drug-induced HYPERSENSITIVITY reactions involving IgE are relatively uncommon, limited evidence supports the role of genetic predisposition. However, as already mentioned, metabolites are widely thought to play a critical role in sensitization to drugs so that the genetic polymorphism of metabolic pathways involved in drug biotransformation is likely to affect the incidence of drug-induced HYPERSENSITIVITY reactions.

Major risk factors related to the drug include the chemical structure, route of administration, and treatment schedule. The role of specific elements of the chemical structure is suspected, but much remains to be done to define reliable structure-immunogenicity relationships [24]. Every route of administration can result in sensitization, but the topical route has a greater potential. The oral route normally leads to tolerance and the mechanism of tolerance breakdown is ill-understood. In sensitized patients, the intravenous route is associated with more rapidly developing and more severe reactions. Finally, intermittent treatment regimens facilitate sensitization.

Pathophysiological mechanisms of drug-induced hypersensitivity reactions

In the early 1960s, Gell and Coombs proposed a classification of IMMUNOALLERGIC reactions, later extended to drugs. This antique classification is still widely used, although it can be misleading as not all mechanisms are covered, several mechanisms can be involved concomitantly, or distinct mechanisms can be involved in different patients treated with the same drug despite clinically similar reactions [25, 26]. Despite major flaws, this classification into 4 types can serve as an introduction to the pathophysiology of drug-induced HYPERSENSITIVITY reactions.

Type I-IV reactions. Type I reactions are immediate HYPERSENSITIVITY (anaphylactic) reactions. Causative ANTIGENS (allergens) induce the production of reaginic ANTIBODIES, namely IgE and to a lesser extent IgG4 in man. IgE bind to high affinity receptors (FcεRI or CD64) on the membrane of mast cells and basophils. After a subsequent contact, the reaction between a divalent drug allergen and specific IgE triggers the DEGRANULATION of mast cells and basophils, which results in the immediate release of preformed mediators including HISTAMINE, neutral proteases (e.g., tryptase), and heparin, that are stored in cytoplasmic granules. Another consequence of DEGRANULATION is de novo synthesis of mediators from membrane phospholipids, including PROSTAGLANDINS and LEUKOTRIENES, and their delayed release (Fig. 4).

Type II reactions are cytotoxicity reactions due to IgM or, less often, IgG. Typically, when the sensitizing drug bound to the surface of blood cells encounters circulating ANTIBODIES, the resulting COMPLEMENT activation provokes the destruction of blood cells as is seen in IMMUNOALLERGIC hemolytic anemias and thrombocytopenias.

Type III reactions are caused by circulating immune complexes that are formed when the antigen is in greater quantity in the serum than IgM or IgG ANTIBODIES. Depending on their size, immune complexes deposit in capillary vessels and activate the COMPLEMENT SYSTEM, platelets, MACROPHAGES and neutrophils. Activated cells release a variety of mediators and free radicals, which damage the endothelial cells. If the antigen is present predominantly at one site, localized damage is seen as in the Arthus reaction. Immune complex deposition is one possible mechanism of cutaneous vasculitis. When the immune complexes are present in the circulation, they may cause serum sickness with fever, arthralgias, cutaneous eruption and proteinuria within 9 to 11 days after the injection of heterologous serum or monoclonal ANTIBODIES. No immune complexes circulating in the blood or deposited in the glomeruli are seen after treatment with low-molecular-weight drugs. The term serum sickness-like disease should be used to avoid confusion.

Type IV or delayed HYPERSENSITIVITY reactions include allergic contact dermatitis and photoallergy. Contact dermatitis can be either non-immune-mediated (i.e., irritant contact dermatitis) or immune-mediated (i.e., allergic contact dermatitis). Allergic contact dermatitis is characterized by the infiltration of T LYMPHOCYTES into the dermis and epidermis. After penetrating into the skin, low-molecular-weight drugs or their metabolites can play the role of haptens that bind to or complex with various cells,

Type I: Immediate hypersensitivity

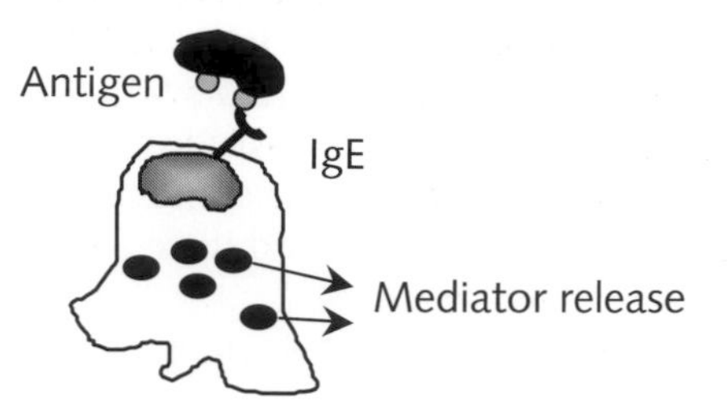

IgE bind FcγRI receptors on mast cells and basophils. Cross-linking of IgE by the antigen results in degranulation with the release of stored (e.g., histamine), then neo-synthesized mediators (e.g., prostaglandins, leukotrienes).

Type II: Cytotoxicity reaction

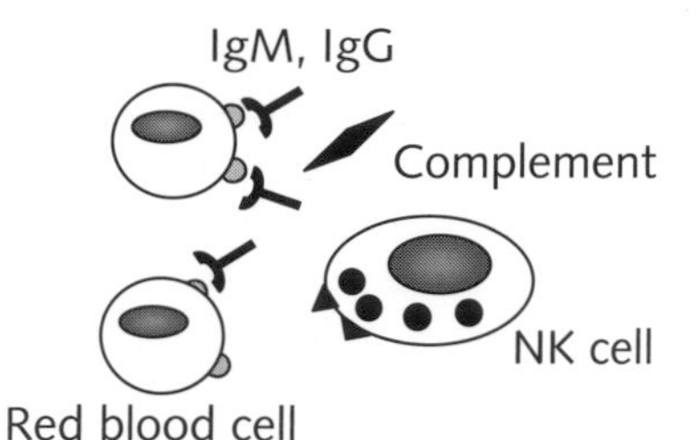

IgM or IgG antibodies directed against antigens on the hosts cells are associated with cytotoxicity by K and NK cells (ADCC) or complement-mediated lysis.

Type III: Immune complex reaction

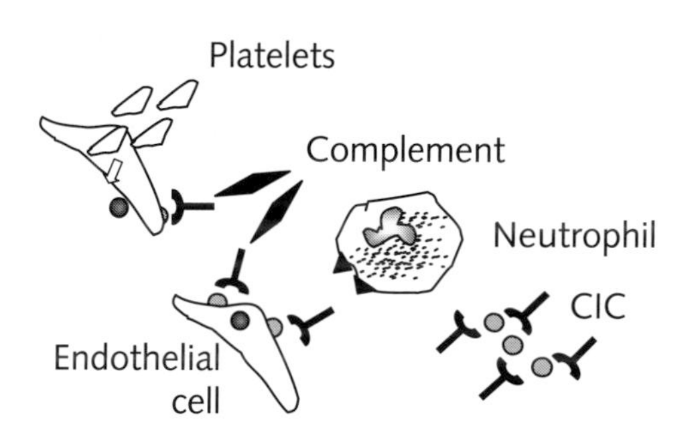

Immune complexes (CIC) are deposited in the walls of small capillaries. Local damage to endothelial cells occurs as a consequence of complement activation and phagocyte attraction to the site of deposition.

Type IV: Delayed-type hypersensitivity

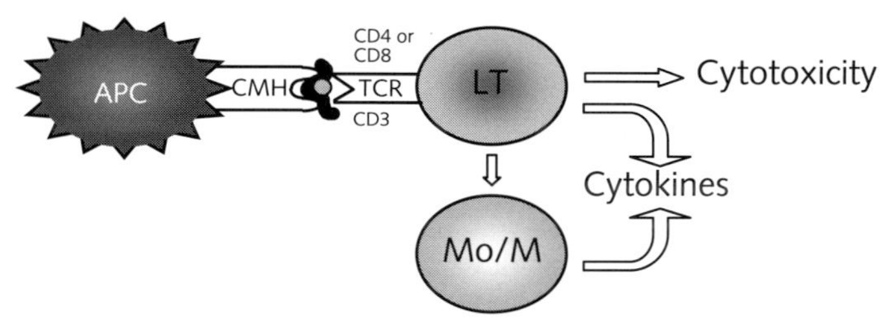

Antigen-sensitized cells release cytokines following a subsequent contact with the same antigen. Cytokines induce an inflammatory response and activate monocytes/macrophages (Mo/M) to release mediators. T lymphocytes can also be directly cytotoxic.

Figure 4
Mechanisms of type I–IV hypersensitivity reactions according to Gell and Coombs classification.

including Langerhans cells and keratinocytes. Langerhans cells process and present the antigen to T LYMPHOCYTES, which leads to the clonal proliferation of sensitized LYMPHOCYTES and to a clinically patent inflammatory reaction.

Delayed hypersensitivity reactions have been proposed to be divided into four distinct sub-categories [27]: type IV-a reactions involve a T_{H1} RESPONSE and closely correspond to the Gell and Coombs type IV reactions. Type IV-b reactions involve a T_{H2} RESPONSE in which the cytokine IL-5 is suspected to play a key role as in the drug HYPERSENSITIVITY syndrome (DRESS) and drug-induced exanthemas. Type IV-c reactions are caused by cytotoxic T LYMPHOCYTES and type IV-d reactions involving IL-8 lead to neutrophilic inflammation. This extended classification offers the advantage to stick more closely to the wide spectrum of drug-induced HYPERSENSITIVITY reactions, but it remains to be established whether it is fully applicable to non-cutaneous drug-induced reactions involving T LYMPHOCYTE-mediated mechanisms.

Pseudo-allergic reactions. An immune-mediated mechanism does not account for all HYPERSENSITIVITY reactions induced by drugs. Clinical manifestations mimicking genuine IgE-mediated reactions have been consistently described in patients exposed for the first time to the same offending drug, hence the proposed term of pseudo-allergy [28]. Several mechanisms have been identified.

HISTAMINE can be released by mast cells and basophils independently of any IgE-mediated mechanism. A cytotoxic or osmotic effect may be involved in direct (non-antigen-specific) HISTAMINE release. Clinical signs and symptoms mimic more or less closely a histaminic reaction with flush, redness of the skin, headache, cough and abdominal pain. The red man syndrome induced by vancomycin is one typical example [29]. Complement activation can be caused by immunological as well as nonimmunological triggers, such as the pharmaceutical solvent Cremophor EL and hydrosoluble radiological contrast media [30]. C3a and C5a are potent byproducts – the ANAPHYLATOXINS – released during complement activation, which induce leukocyte chemotaxis, increased vascular permeability, contraction of bronchial smooth muscle, HISTAMINE release, generation of LEUKOTRIENES, and IL-1 production. Aspirin as well as most NSAIDs can cause acute intolerance reactions [31]. They develop within 1 hour after drug ingestion as an acute asthma attack, often associated with rhinorrhea and conjunctival irritation. Aspirin and the majority of NSAIDs are more potent inhibitors of the COX-1 isoform of the enzyme CYCLOOXYGENASE (COX) than of the COX-2 isoform. Any NSAID with marked COX-1 inhibiting activity can precipitate asthma attacks, presumably by increasing the availability of arachidonic acid and/or the release of LEUKOTRIENES by the enzyme 5-lipooxygenase. That COX-2 inhibitors appear to be safe in patients with a history of aspirin intolerance supports this hypothesis. Finally, angioedema associated with angiotensin-converting enzyme inhibitors probably results from the decreased degradation of bradykinin, which increases vascular permeability, contracts smooth muscles and elicits pain [32].

Autoimmunity

Although autoimmunity is still largely a mystery, autoimmune diseases are relatively common in the general population. Estimates vary widely, but over 1 million new cases may develop every 5 years in the USA [33]. A wide range of xenobiotics have been reported to induce autoimmune diseases [34], but the incidence of drug-induced autoimmunity is seemingly low [35]. It is noteworthy that one given drug can typically induce only one type of autoimmune disease (reaction), e.g., hydralazine and pseudolupus, or α-methyldopa and autoimmune hemolytic anemia. In contrast, treatments with an immunostimulatory drug, such as the therapeutic CYTOKINES rIL-2 and INTERFERONS-α, are associated with a wide range of more frequent autoimmune diseases that cannot be distinguished from spontaneous diseases [36].

Clinical manifestations of autoimmunity

Autoimmune diseases are clinically very diverse and, in many instances, the diagnosis is based on the presence of several clinical signs and symptoms among a predefined set. The clinical presentation of drug-induced autoimmune reactions is more or less variable with respect to the spontaneous disease so that the presence of AUTOANTIBODIES in the sera of patients is a prerequisite. Spontaneous as well as drug-induced autoimmune diseases are divided into systemic and organ-specific.

Systemic autoimmune diseases

Systemic lupus erythematosus (SLE) is estimated to affect 2–10 in 10,000 individuals. The causes of SLE are not known, but endocrine, genetic and environmental factors are likely to be involved. The lupus syndrome or pseudolupus is the most common drug-induced autoimmune reaction [37], and hydralazine and procainamide are by far the most frequent causes. AUTOANTIBODIES have been detected in the sera of up to 25% of patients treated with hydralazine and 50% of those treated with procainamide. However, no clinical signs were associated with AUTOANTIBODIES in the majority of patients. Other drugs seldom reported to induce lupus syndromes include several anti-epileptic drugs, most β-blockers, chlorpromazine and isoniazid. In recent years, minocycline appeared as a leading cause.

Drug-induced lupus syndromes normally bear few clinical and biological similarities to SLE. In contrast to SLE, lupus syndromes are as frequent in men as in women. The most typical clinical signs include arthritis in over 80% of patients, fever, weight loss, and muscular weakness with myalgias. Cutaneous manifestations are often uncharacteristic. Renal involvement is inconsistent and usually mild. No neurological signs are noted. One major distinction is the high incidence of pleural effusion seen in up to 40% of patients and pericardial effusion which can result in cardiac tamponade. No biological abnormalities are typical of a drug-induced lupus syndrome. Antinuclear ANTIBODIES are always present. Anti-ds (double-stranded) or native DNA ANTIBODIES are found in 50–70% of patients with SLE, but in less than 5% of those with the lupus syndrome. In contrast, ANTIBODIES to denatured DNA are relatively common in the lupus syndrome. No AUTOANTIBODIES have so far been identified as markers of drug-induced lupus syndromes. In contrast to SLE, lupus syndromes have a favorable outcome after cessation of the offending drug.

Scleroderma or systemic sclerosis is a relatively rare disease characterized by a more or less diffuse infiltration of the dermis and viscera by collagen with vascular abnormalities including vasospasm and microvascular occlusion. The pathogenesis of scleroderma is not elucidated. There is an overproduction of collagen by fibroblasts. T LYMPHOCYTES are thought to play a pivotal role. Only very few drugs have been reported to induce scleroderma-like diseases. The most severe was the oculo-muco-cutaneous syndrome induced by the beta-blocker practolol, with kerato-conjunctivitis, lesions of the conjunctivae with loss of sight, psoriasis-like eruption, and pleural and/or pericardial effusion [38].

Organ-specific autoimmune diseases

In contrast to systemic autoimmune reactions induced by drugs, organ-specific reactions are characterized by a homogeneous ANTIBODY response against a unique target and clinical symptoms closely mimicking those of the spontaneous autoimmune disease.

Guillain-Barré syndrome usually presents with progressive lower extremity weakness potentially leading to autonomic dysfunction. The mortality rate is 3–5%. A possible link between Guillain-Barré syndrome and vaccination has been suspected [39]. Multiple sclerosis is a multifocal demyelinating disease of the central nervous system that may take a relapsing or progressive course. Early symptoms consist of paresthesias, gait disorders, visual loss and diplopia. Relapses evolve within days and resolve gradually, but incompletely. Multiple sclerosis is characterized by perivascular cuffing of $CD4^+$ T LYMPHOCYTES and myelin destruction. Vaccines, in particular hepatitis B vaccine, have been suspected to induce or facilitate the development of multiple sclerosis, but so far no fully confirmative epidemiological evidence has been published. Myasthenia is characterized by a loss of muscular strength due to impaired neuromuscular transmission. There is a predilection for certain cranial nerves and virtually all patients complain of ocular symptoms. 80–90% of patients with the generalized disease have IgG AUTOANTIBODIES against the nicotinic receptors of acetylcholine in the neuromuscular motor plates. This condition must be differentiated from myasthenia-like syndromes due to a pharmacodynamic interference between the causative drug and acetylcholine. Penicillamine is the most frequent cause of drug-induced myasthenia [40]. Whatever the causative drug, the underlying mechanism is not known.

Thyroiditis is due to a specific autoimmune response involving T cells and AUTOANTIBODIES. Thyroperoxidase is the main autoantigen, but thyroglobulin or the thyrotropin receptor can also be the targets of the autoimmune process [41]. Most often, autoimmune thyroiditis presents as a slowly progressing atrophy of the thyroid gland. Treatments with rIL-2 and the IFNs are associated with a higher incidence of autoimmune thyroiditis [36].

More than 80% of patients with autoimmune hepatitis have hypergammaglobulinemia. The presence of AUTOANTIBODIES, such as antinuclear ANTIBODIES, ANTIBODIES against smooth muscle and liver-kidney microsomal ANTIBODIES, is common, but their diagnostic value is debatable. Several drugs have been reported to cause hepatitis associated with highly specific AUTOANTIBODIES in the sera of affected patients as discussed below [42].

Mechanisms of drug-induced autoimmunity

Overall, our current understanding of the mechanisms involved in drug-induced autoimmunity is very limited.

Cytokine overproduction

Treatments with recombinant CYTOKINES typically induce marked cytokine overproduction [43]. One hypothesis to account for the observed changes in patients treated with recombinant CYTOKINES is an abnormal expression of MHC class II molecules induced by IFN-γ and amplified by IL-1 and TNF-α. Under the influence of IFN-γ, thyroid cells may express MHC class II molecules and act as ANTIGEN-PRESENTING CELLS with the production of antithyroid AUTOANTIBODIES as a consequence.

Formation of neo-antigens

Drugs or their metabolites can bind to cellular constituents with the ensuing formation of neo-autoantigens. This mechanism has been conclusively shown in (autoimmune) hepatitis induced by several drugs [42]. Following biotransformation in the liver, metabolites are formed that bind to CYP isoforms, such as CYP450 1A2 (dihydralazine), CYP450 4E1 (halothane) or CYP450 2C9 (tienilic acid). Structural changes in these hepatocyte constituents trigger a specific immune response that subsequently leads to liver damage.

Molecular mimicry

Basically molecular mimicry means that part of a given protein closely resembles a part of another protein. When a foreign protein penetrates the body, the immune system mounts a specific ANTIBODY response and when the foreign protein closely resembles a self protein of the body, AUTOANTIBODIES are formed that can be pathogenic [43]. Molecular mimicry is the causative mechanism of cardiac changes in rheumatic fever due to cross-reactivity between streptococcal and cardiac myosin. The involvement of molecular mimicry in drug-induced autoimmunity is only assumptive, but a cause for concern during the development of new vaccines or biotechnology-derived therapeutic products.

T cell involvement

Because of their pivotal role in immune responses, T cells are a major focus of current research on autoimmunity [44]. Recognition of closely similar epitopes shared by self and non-self molecules can trigger autoimmune responses due to molecular mimicry or more subtle mechanisms. Activation of T cells could also be due to drugs mimicking co-stimulatory molecules or MHC class II ANTIGENS that are exquisitely involved in the functioning of the IMMUNOLOGICAL SYNAPSE.

Conclusion

HYPERSENSITIVITY and autoimmune reactions are potentially severe immunotoxic consequences of drug treatments. Their clinical features are relatively well known, but much remains to be done to obtain a clear understanding of the underlying mechanisms. This is, however, a crucial step for designing relevant tools that are so much needed [45–47] in order to predict the potential of new drugs for inducing such adverse reactions.

Summary

The immune system is a complex network and inadvertent immunological reactivity can result in HYPERSENSITIVITY or autoimmunity. Nearly every chemical can induce HYPERSENSITIVITY reactions which can affect nearly every organ or tissue of the body. One organ or tissue, however, is often a predominant target. Anaphylactic shock, skin reactions, and cytopenias are the most frequent reactions. HYPERSENSITIVITY reactions can be either immune-mediated or non-immune-mediated (pseudoallergy). A number of different mechanisms are involved and risk factors can serve as triggers. Autoimmunity is still a mystery even though autoimmune diseases are relatively common. Spontaneous autoimmune diseases as well as drug- and chemical-induced autoimmune reactions are divided into systemic and organ-specific. Our current understanding of the mechanisms involved in these adverse reactions is very limited.

Selected readings

Descotes J (ed.) (2004) *Immunotoxicity of Drugs and Chemicals: An Experimental and Clinical Approach. Vol. 1: Principles and Methods of Immunotoxicology*. Elsevier, Amsterdam

Holgate ST, Church MK, Lichtenstein LM (2000) *Allergy* (2nd Edition). Mosby, London

Gorski A, Krotkiewski H, Zimecki M (2001) *Autoimmunity*. Kluwer Academic Publishers, Dordrecht, Boston

Recommended website

National Institutes of Health (NIH) (1998) Understanding Autoimmune Diseases: *http://www.niaid.nih.gov/publications/autoimmune/autoimmune.htm* (Accessed March 2005)

References

1 Descotes J (2004) Health consequences of immunotoxic effects. In: Descotes J (ed): *Immunotoxicity of Drugs and Chemicals*, 3rd edition. Elsevier, Amsterdam, 55–126

2 Demoly P, Bousquet J (2001) Epidemiology of drug allergy. *Curr Opin Allergy Clin Immunol* 1: 305–310

3 Choquet-Kastylevsky G, Vial T, Descotes J (2001) Drug allergy diagnosis in humans: possibilities and pitfalls. *Toxicology* 158: 1–10

4 Bochner BS, Lichtenstein LM (1991) Anaphylaxis. *N Engl J Med* 324: 1785–1790

5 Klein JS, Yocum MN (1995) Underreporting of anaphylaxis in a community emergency room. *J Clin Allergy Immunol* 95: 637–638

6 Gruchalla RS (2003) Drug allergy. *J Allergy Clin Immunol* 111 (Suppl): 548–559

7 Greenberger PA (1999) Allergic emergencies in the physician's office. *Clin Rev Allergy Immunol* 17: 401–412

8 Wolkenstein P, Revuz J (1995) Drug-induced severe skin reactions. *Drug Saf* 13: 56–68

9 Svensson CR, Cowen EW, Gaspari AA (2000) Cutaneous drug reactions. *Pharmacol Rev* 53: 357–379

10 Pichler W, Yawalkar N, Schmid S, Helbling A (2002) Pathogenesis of drug-induced exanthems. *Allergy* 57: 884–893

11 Andersen KE, Maibach HI (1980) Allergic reaction to drugs used topically. *Clin Toxicol* 16: 415–465

12 Fritsch PO, Sidoroff A (2000) Drug-induced Stevens-Johnson syndrome/toxic epidermal necrolysis. *Am J Clin Dermatol* 1: 349–360

13 Roujeau JC, Kelly JP, Naldi L, Rzany B, Stern RS, Anderson T, Auquier A, Bastuji-Garin S, Correia O, Locati F et al (1995) Medication and risk of Stevens-Johnson syndrome or toxic epidermal necrolysis. *N Engl J Med* 333: 1600–1607

14 Bocquet H, Bagot M, Roujeau JC (1996) Drug-induced pseudolymphoma and drug HYPERSENSITIVITY syndrome (Drug Rash with Eosinophilia and Systemic Symptoms: DRESS). *Semin Cutan Med Surg* 15: 250–257

15 Danielson DA, Douglas SW, Herzog J, Jick H, Porter JB (1984) Drug-induced blood disorders. *JAMA* 252: 3257–3260

16 Salama A, Mueller-Eckhardt C (1986) Two types of nomifensine-induced immune haemolytic anaemias: drug-dependent sensitization and/or autoimmunization. *Br J Haematol* 64: 613–620

17 Thompson JW, Jacobs RF (1993) Adverse effects of newer cephalosporins. An update. Drug Saf 9: 132–142

18 Warkentin TE (2002) Heparin-induced thrombocytopenia. *Curr Hematol Rep* 1: 63–72
19 Liu ZX, Kaplowitz N (2002) Immune-mediated drug-induced liver disease. *Clin Liver Dis* 6: 467–486
20 Alexopoulos E (1998) Drug-induced acute interstitial nephritis. *Renal Failure* 20: 809–819
21 Limper AH, Rosenow EC (1996) Drug-induced interstitial lung disease. *Curr Opin Pulm Med* 2: 396–404
22 Park BK, Naisbitt DJ, Gordon SF, Kitteringham NR, Pirmohamed M (2001) Metabolic activation in drug allergies. *Toxicology* 158: 11–23
23 The International Collaborative Study of Severe Anaphylaxis (1998) An epidemiological study of severe anaphylactic and anaphylactoid reactions among hospital patients: methods and overall risks. *Epidemiology* 9: 141–146
24 Baldo BA, Pham NH, Zhao Z (2001) Chemistry of drug allergenicity. *Curr Opin Allergy Clin Immunol* 1: 327–335
25 Descotes J, Choquet-Kastylevsky G (2001) The Gell and Coombs classification: is it still valid? *Toxicology* 158: 43–49
26 Rajan TV (2003) The Gell-Coombs classification of hypersensitivity reactions: a re-interpretation. *Trends Immunol* 24: 376–379
27 Pichler WJ (2003) Delayed drug hypersensitivity reactions. *Ann Intern Med* 139: 683–693
28 Zuberbier T (1999) Pseudoallergy or nonallergic hypersensitivity. *Allergy* 54: 397–398
29 Wallace MR, Mascola JR, Oldfield EC (1991) Red man syndrome: incidence, etiology, and prophylaxis. *J Infect Dis* 164: 1180–1185
30 Szebeni J (2001) Complement activation-related pseudoallergy caused by liposomes, micellar carriers of intravenous drugs, and radiocontrast agents. *Crit Rev Ther Drug Carrier Syst* 18: 567–606
31 Szczeklik A, Stevenson DD (2003) Aspirin-induced asthma: advances in pathogenesis, diagnosis, and management. *J Allergy Clin Immunol* 111: 913–921
32 Agostoni A, Cicardi M, Cugno M, Zingale LC, Gioffre D, Nussberger J (1999) Angioedema due to angiotensin-converting enzyme inhibitors. *Immunopharmacology* 4: 21–25
33 Jacobson DL, Gange SJ, Rose NR, Graham NM (1997) Epidemiology and estimated population burden of selected autoimmune diseases in the United States. *Clin Immunol Immunopathol* 84: 223–243
34 Bigazzi PE (1997) Auto-immunity induced by xenobiotics. *Toxicology* 119: 1–21
35 Vial T, Nicolas B, Descotes J (1997) Drug-induced autoimmunity. Experience of the French Pharmacovigilance system. *Toxicology* 119: 23–27
36 Vial T, Chevrel G, Descotes J (2000) Drugs acting on the immune system. In: MNG Dukes, JK Aronson (eds): Meyler's Side-Effects of Drugs, 14th edition. Elsevier Sciences, Amsterdam, 1246–1337
37 Rubin RL (1999) Etiology and mechanisms of drug-induced lupus. *Curr Opin Rheumatol* 11: 357–363
38 Behan PO, Behan WM, Zacharias FJ, Nicholls JT (1976) Immunological abnormalities in patients who had the oculomucocutaneous syndrome associated with practolol therapy. *Lancet* 2: 984–987
39 Piyasirisilp S, Hemachudha T (2002) Neurological adverse events associated with vaccination. *Curr Opin Neurol* 15: 333–338
40 Penn AS, Low BW, Jaffe IA, Luo L, Jacques JJ (1998) Drug-induced autoimmune myasthenia gravis. *Ann NY Acad Sci* 841: 433–449
41 Pearce EN, Farwell AP, Braverman LE (2003) Thyroiditis. *N Engl J Med* 348: 2646–2655
42 Beaune PH, Lecoeur S (1997) Immunotoxicology of the liver: adverse reactions to drugs. *J Hepatol* 26 (Suppl 2): 37–42
43 Miossec P (1997) Cytokine-induced autoimmune disorders. *Drug Saf* 17: 93–104
44 Fairweather D, Kaya Z, Shellam GR, Lawson CM, Rose NR (2001) From infection to autoimmunity. *J Autoimmun* 16: 175–186
45 Pichler WJ (2003) Drug-induced autoimmunity. *Curr Opin Allergy Clin Immunol* 3: 249–253
46 Choquet-Kastylevsky G, Descotes J (1998) Value of animal models for predicting hypersensitivity reactions to medicinal products. *Toxicology* 129: 27–35
47 Descotes J (2000) Autoimmunity and toxicity testing. *Toxicol Lett* 112–113: 461–465

Cancer immunity

Jan W. Gratama, Reno Debets and Ralph A. Willemsen

Introduction: cancer immunity from a historical perspective

Almost 100 years ago, Ehrlich and coworkers observed the presence of infiltrates of mononuclear cells around or inside tumor lesions [1]. This finding led them to propose that tumors could be recognized and inhibited by the 'magic bullets' of the immune system. At the end of the 19th century, studies were initiated that aimed to actively immunize cancer patients against their own cancerous tissue. During the subsequent decades, cancer patients were nonspecifically immune-stimulated with relatively crude leukocyte extracts such as transfer factor, immune-RNA, bacterial extracts such as bacillus Calmette-Guérain, Coley's toxin or levamisole. These studies were initated in spite of the fact that little was known about the various components of the immune system that could react against cancer and even less was known about the structures on cancer cells that can be recognized by the immune system.

The discrimination between 'self' and 'nonself' by the immune system – the 1960 Nobel Prize-winning concept of Burnet and Medawar – has been pivotal for modern tumor immunology. Subsequently, Thomas and Burnet developed the 'immune surveillance' theory. The core of this theory is that the immune system protects the host from cancer by detecting and destroying newly-formed cancer cells, recognized as nonself [2, 3]. This immune surveillance theory, including its critics and the various experiments supporting and disproving it, has strongly influenced tumor immunological research during the past 40 years.

Ever since, the existence of immunity against cancer has been abundantly demonstrated in animal (mostly murine) models and in man. Interest in the clinical application of immunotherapy to treat cancer has been rekindled in the last two decades by the revival of the immunosurveillance theory, the discovery and structural characterisation of TUMOR-ASSOCIATED ANTIGENS, our progress in understanding the molecular pathways required for the induction and maintenance of immune responses, and methodological advances to generate specific immunological probes in the form of tumor-specific cytotoxic T LYMPHOCYTES (CTL) and monoclonal ANTIBODIES (mAb). Results obtained until now have revealed [4]:

- the increased susceptibility of immunodeficient patients to cancer as compared to immunocompetent individuals, supporting the concept of immunosurveillance against cancer [5–7];
- the molecular characterization of a wide range of various types of TUMOR-ASSOCIATED ANTIGENS on cancer cells, as detailed below [8];
- the feasibility of using TUMOR-ASSOCIATED ANTIGENS as vaccines and DENDRITIC CELLS to induce tumor rejection in tumor-bearing animals and patients [9–14];
- the efficacy of INTERLEUKIN (IL)-2 and adoptively transferred T LYMPHOCYTES, expanded from tumor-infiltrating LYMPHOCYTES (TIL), in some patients with metastatic cancer, in particular melanoma and renal cell carcinoma [15–18];
- the immunosuppressive capacities of tumor cells, i.e., to secrete immunosuppressive CYTOKINES such as transforming growth factor (TGF)-β [19] and IL-10 [20] and to inhibit LEUKOCYTES with antitumor capacities.

Here, we discuss the key elements involved in the generation of antitumor responses: the cellular and humoral components of the immune system, and the target ANTIGENS on the tumor cells. With this basis, we review the various immunotherapeutic approaches to cancer and put the envisaged future developments into the perspective of current experiences.

Key players in the immune responses against cancer

Both innate and adaptive components of the immune system interact to generate antigen-specific immune responses. As detailed in the previous chapters, the INNATE IMMUNE SYSTEM constitutes the body's first line of defense against 'foreign invaders'. INNATE IMMUNITY involves a large number of different cell populations such as epithelial cells, MONOCYTES, MACROPHAGES, DENDRITIC CELLS, POLYMORPHONUCLEAR LEUKOCYTES or GRANULOCYTES, and some lymphocyte subsets that are at the interface between INNATE and ADAPTIVE IMMUNITY (natural killer [NK] LYMPHOCYTES, clusters of differentiation [CD]5$^+$ B LYMPHOCYTES and T cell-receptor [TCR]-$\gamma\delta^+$ T LYMPHOCYTES). The INNATE IMMUNE SYSTEM also comprises a variety of humoral factors such as CYTOKINES, chemokines, enzymes such as lysozyme, metal-binding proteins, integral membrane ion transporters, complex carbohydrates and complement. The phagocytic cells (MACROPHAGES, GRANULOCYTES) and the COMPLEMENT SYSTEM constitute effector mechanisms by which the 'invaders' can be destroyed. The production of CYTOKINES and CHEMOKINES acts in concert with antigen presentation by DENDRITIC CELLS and MONOCYTES to initiate adaptive immune responses.

ADAPTIVE IMMUNITY makes use of a unique mechanism whereby genetic mutations occurring in two specialised cell populations, B and T LYMPHOCYTES, produce numerous molecular 'shapes' that are expressed as ANTIBODIES (Ab) and TCR. Figure 1 provides a simplified overview of how the effector components of the adaptive immune system (T cells and Ab) are regulated and eliminate their targets. Antigen-specific immunity is generated when Ab and TCR are expressed and upregulated through the formation and release of CYTOKINES and chemokines. Thus, ADAPTIVE IMMUNITY involves a wide range of antigen receptors expressed on the surface of T and B LYMPHOCYTES to detect 'foreign' molecules. B LYMPHOCYTES respond to antigen by secreting their own antigen receptors as Ab after having differentiated into plasma cells. Ab interact with effector mechanisms via binding of their constant part (Fc) to complement, Fc-receptor-bearing phagocytes and Fc-receptor-bearing cytotoxic (NK and T) LYMPHOCYTES.

The major T LYMPHOCYTE SUBSETS are characterized by expression of the differentiation markers CD4 or CD8. Whilst Ab mostly react with intact proteins or carbohydrates, T cells do so with peptides expressed on the cellular surface via presentation by scaffolds, i. e., molecules of the MAJOR HISTOCOMPATIBILITY COMPLEX (MHC), on ANTIGEN-PRESENTING CELLS (APC). One of the most potent types of APC is the dendritic cell (DC). DC pick up ANTIGENS in the skin or other peripheral tissues, and migrate to the T-cell zones of LYMPHOID ORGANS where they stimulate naïve CD4$^+$ and CD8$^+$ T cells.

Most CD4$^+$ T cells are 'T-helper (Th) cells' and recognize antigen in the form of 15 to 25-mer peptides presented by class II MHC molecules. These molecules present peptides that are mainly derived from the extracellular compartment, as opposed to peptides presented by class I MHC molecules that capture endogenously processed peptides (see below; [21]). CD4$^+$ T cells are the major regulators of most immune responses. They augment immune responses by secreting CYTOKINES that stimulate cytotoxic CD8$^+$ T cells (so-called Th1-type CYTOKINES such as INTERFERON [IFN]-γ), or B cells to mount Ab responses (so-called Th2-type CYTOKINES such as IL-4 and IL-5).

CD8$^+$ cytotoxic T LYMPHOCYTES (CTL) react with 8 to 10-mer peptides presented on the cellular surface by the scaffolds formed by their class I MHC molecules. These peptides contain 2 to 3 so-called 'anchoring residues' that fit into specific pockets of the class I MHC molecule [22]. CD8$^+$ T cells destroy their target cells, after attachment of (1) their TCR to the appropriate MHC-peptide complex, and (2) their accessory molecules (such as CD8 and CD28) to the corresponding ligands, by perforating their membranes with enzymes (i.e., perforin and granzymes) or by triggering a process of self-destruction (termed apoptosis). In this way, CD8$^+$ T cells can move from one tumor cell to another expressing the same MHC-peptide complexes, and thus can mount a very specific and robust antitumor response.

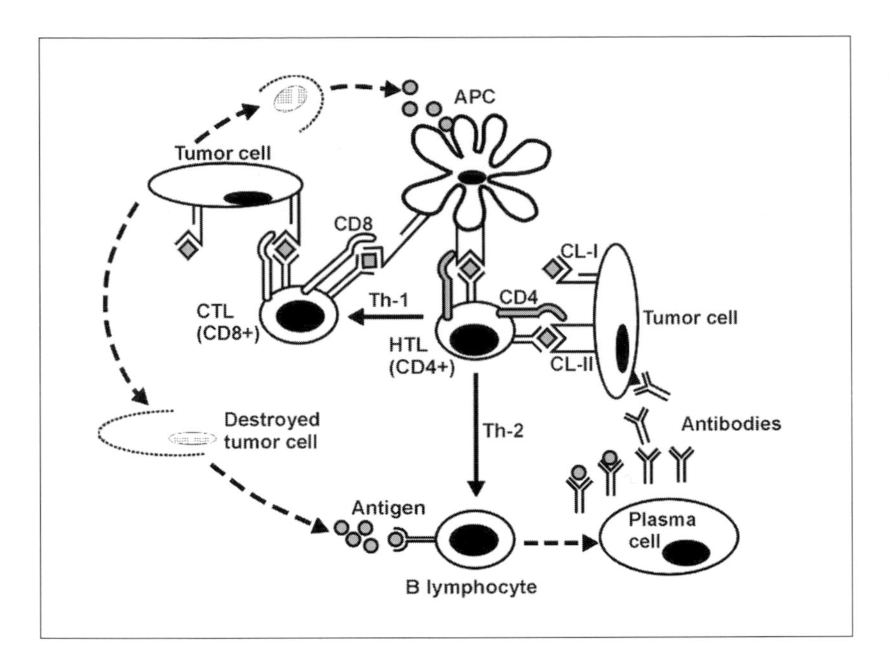

FIGURE 1

Key players of the immune system in mounting an antitumor response. A 'professional' antigen-presenting cell (APC) presents antigenic peptides (depicted as diamonds) to a helper T lymphocyte (HTL) via its Class II MHC molecules (CL-II), and to a cytotoxic T lymphocyte (CTL) via its Class I MHC molecules (CL-I). The T-helper cells may also recognize antigen on a tumor cell itself if the tumor cells express Class II MHC molecules (right side of figure); similarly, a CTL may also recognize antigen on a tumor cell itself if the tumor cells express Class I MHC molecules (upper left). The T-helper cells recognize antigen through their T-cell receptors for antigen (TCR) which are supported in this task by CD4 coreceptor molecules. By the same token, CTL recognize antigen by their TCR supported by CD8 coreceptor molecules. Other important accessory molecules and their receptors, such as CD28 on T cells and CD80/CD86 on APC, as well as downregulatory T cells are not shown here (see text for details). The T-helper cells support CTL activation and proliferation by secreting so-called T-helper 1 cytokines (Th1; e.g., IL-2 and IFN-γ). On the other hand, T-helper cells stimulate B lymphocytes through Th2 cytokines (e.g., IL-4 and IL-5). B cells recognize soluble antigen through their B-cell receptor for antigen; upon activation, B cells differentiate into plasma cells which secrete antibodies specific for that antigen. Upon engagement of their TCR and accessory molecules, CD8 and also CD4$^+$ T cells can destroy tumor cells by secreting granzymes, perforins and cytokines such as TNF-α, or by upregulation of CD95 (Fas ligand) on tumor cells. The remnants of destroyed tumor cells can be taken up by APC, processed and presented to T cells; in addition, they can be specifically recognized by Ag and eliminated via Fc-receptor-expressing phagocytes.

In the setting of ADOPTIVE IMMUNOTHERAPY of cancer (see below), the administration of autologous CD4$^+$ T helper cells concurrently with CD8$^+$ CTL has been shown to prevent exhaustion of the infused CD8$^+$ CTL [23, 24]. CD4$^+$ T cells activate DENDRITIC CELLS through crosslinking of CD40, which thus provide enhanced antigen presentation and co-stimulation, which leads to priming of CD8$^+$ CTL function [25]. Thus, CD4$^+$ T-cell help results in *de novo* generation of tumor-specific CD8$^+$ CTL and concomitant tumor destruction [26]. The exact requirement for CD4$^+$ T-cell help during priming may depend on the nature of the stimulus, but the requirement for CD4$^+$ T cells during memory responses is beyond dispute [27]. We propose that CD4$^+$ T cells have a critical role in anti-tumor immunity that goes beyond the mere induction and maintenance of tumor-specific CD8+ CTL. Specifically, CD4+ T cells activate CD8+ CTL already present at the site of the tumor and enable other CD8+ CTL to migrate into the tumor, induce an anti-tumor Ab response, and otherwise contribute to tumor regression through production of IFN-γ which inhibits tumor-induced angiogenesis [23, 28].

CYTOKINES produced by CD4+ T-helper cells also regulate MACROPHAGES by stopping their migration after their engagement, allowing them to accumulate at a particular site. Thus, more efficient PHAGOCYTOSIS is stimulated, so that increasing numbers of 'invaders' are efficiently destroyed. CD4+ T cells amplify themselves by secreting CYTOKINES and interacting with them; IL-2 is a key example for this 'autocrine loop.' This action enhances the T-helper cell response and thus the entire immune system's response to foreign antigens.

A minor subset of CD4+ T cells, characterized by coexpression of CD25 and in particular the glucocorticoid-induced TUMOR NECROSIS FACTOR receptor, are essential negative regulators of immune function [29]. These T-regulatory cells can suppress both innate and adaptive immune responses. They are capable of robust Ag-driven proliferation as observed for all other adaptive immune responses. Importantly, the presence of tumor-specific CD4+ T-regulatory cells at tumor sites may inhibit T-cell responses against cancer (see below; [30]).

Expression of targets for the immune system by cancer cells

Tumor-associated antigens (TAA) recognized by T lymphocytes

Clinical immunotherapeutic studies of cancer performed using *in vitro* expanded T LYMPHOCYTES isolated from tumors (see below) have demonstrated that CD8+ T cells can specifically lyse, in an MHC-restricted fashion, autologous tumor cells or tumors arising from different tissues, but not normal cells [31, 32]. Class I MHC molecules are expressed on the surface of most nucleated cells. To identify these MHC-restricted tumor antigens, cDNA or DNA cosmid pools derived from tumor cells have been transiently transfected into cell lines expressing the relevant class I MHC molecules. These transfected cell lines were then assayed for their ability to specifically stimulate cytokine production by class I MHC-matched, tumor-reactive T-cell CLONES. The subsequent transfection of individual cDNAs or cosmids from pools that specifically stimulated the T cell CLONES allowed the identification of individual TAA. The first human TAA that was identified in this way by screening a large genomic library from a human leukocyte antigen (HLA)-A*0101+ melanoma patient was termed melanoma antigen (MAGE)-1 [33]. To identify the MAGE-1 peptide presented by HLA-A*0101, selected 9-mer peptides were synthesized and screened for their capacity to specifically stimulate the MAGE-1-specific cytotoxic T-cell CLONE isolated from this patient [34]. Ever since, the use of this 'reverse immunology' approach has allowed the identification of a large number of TAA (see below and Table 1). The definition of class I MHC peptide-binding motifs, i.e., the nature and position of pockets along the peptide-binding groove of the MHC molecule that bind the anchoring residues of the peptides, has allowed the computer-assisted screening of large numbers of potentially immunogenic peptides, including many TAA [22].

The identification of class II MHC-binding peptides has thus far met with more problems than that of class I MHC-binding peptides due to (1) the variable length of the peptides that can be bound by class II molecules, and (2) the availability of less information on peptide-anchoring residues for class II MHC molecules. To date, there is no strategy available for the identification of class II MHC-restricted antigen that matches the successes obtained in the identification of class I MHC-restricted antigens.

Antigens recognized by antibodies (Ab)

In general, ANTIGENS recognized by Ab are mostly intact proteins or carbohydrates, and are not presented by MHC molecules. Therefore, such ANTIGENS cannot be recognized by T LYMPHOCYTES via their TCR which only recognize MHC-presented peptides. Many of these TAA are merely overexpressed by tumor cells in comparison to their normal counterparts. Although these TAA are not 'tumor-specific' in the strict immunological sense, they can effectively be used to selectively target the immune system to tumors resulting in clinical antitumor effects [35]. Using monoclonal Ab, tissue-specific ANTIGENS have been identified that, although not tumor-specific, can serve as target structures for immunotherapeutic interventions (see below).

A highly significant development in the study of the humoral response to cancer is the serological identification of ANTIGENS by recombinant expression cloning (SEREX). This technique uses diluted sera from patients with cancer to screen tumor cDNA libraries expressed in *Escherichia coli* in order to identify ANTIGENS that have elicited high-titer IgG Abs [36]. The first application of SEREX revealed the New York-esophagus 1 (NY-ESO-1) antigen using serum from a patient with squamous cell carcinoma of that organ [37]. Ever since, a multitude of ANTIGENS from tumors of many histologies, including melanoma and colon, lung and renal cancers have been discovered using SEREX. The function of many of the newly identified proteins is still unknown. Importantly, screening for high-titer IgG TAA-specific Ab should identify patients with CD4+ T-cell responses to that TAA as such a response is required for Ig class switching. Indeed, Class II MHC-restricted TAA-specific CD4+ T cells, as well as class I MHC-restricted CD8+ T-cell responses to the same tumor-derived proteins as identified by the high-titer IgG TAA-specific Ab, have been detected in some of these cases [38, 39].

Classification of tumor-associated antigens (TAA)

The identified TAA can be grouped into several general categories [8]. Table 1 provides an overview of these categories, examples of TAA within each category, expression patterns of the listed TAA and information as to whether cellular and humoral immune responses against these TAA have been reported.

The first group is formed by tissue-specific differentiation antigens. These ANTIGENS are expressed by normal and neoplastic cells of the same lineage. Examples are melanoma antigen (Melan)-A recognized by T cells (MART)-1 [40], glycoprotein (gp) 100 [41] and tyrosinase [42] in melanoma, as well as prostate-specific antigen (PSA; [43]) and prostate-specific membrane antigen (PSMA; [44]) in prostate cancer. These ANTIGENS can elicit 'spontaneous' CD8+ CTL responses in patients with melanoma and healthy individuals. In melanoma, increased frequencies of these CTL are seen, but these are functionally inactive (i.e., anergic) [51]. A point of concern with the use of these ANTIGENS as targets for cancer immunotherapy is that normal cells expressing these ANTIGENS also will be destroyed, as illustrated in a clinical trial of melanoma patients treated with *ex vivo* expanded TIL specific for MART-1 and gp100 [15].

The second category comprises the 'cancer-testis antigens,' so-called for their expression by histologically different tumors and, among normal tissues, by spermatogonia and spermatocytes of the testis and occasionally by placental trophoblasts. The expression of cancer/testis ANTIGENS by tumors results from reactivation of genes that are normally silent in adult tissues but are transcriptionally activated in some tumors. Examples of cancer-testis ANTIGENS are the MAGE [48], B antigen (BAGE) [49] and G antigen (GAGE) [50] families of antigens, as well as NY-ESO-1 [37]. These ANTIGENS are less immunogenic than differentiation antigens, and do not elicit 'spontaneous' CTL responses. In this context it is interesting that members of the MAGE-A group are expressed by medullary thymic epithelial cells, where central tolerance for these ANTIGENS may be induced [52].

The third group is constituted by ANTIGENS that are overexpressed by tumors arising from a variety of tissues, and that are also expressed, albeit at much lower levels, by normal tissues. This high level of expression by tumor cells as revealed by microarray or differential display analyses has been instrumental for the identification of many of these ANTIGENS. Examples are carcinoembryonic antigen (CEA) on gastrointestinal, breast and lung cancers [53], human epidermal receptor-2/neurological (Her-2/neu) on melanoma, ovarian and breast cancers [54], and carboxy-anhydrase 9 on renal cell carcinoma [55]. Several mAb have been raised against TAA from this group which are currently used for cancer immunotherapy (see below).

TABLE 1. EXAMPLES OF TUMOR-ASSOCIATED ANTIGENS[1]

Tumor-associated antigen	Observed type of immune response: T-cell-mediated	Antibody-mediated	Tissue distribution by malignant cells and their normal counterparts[2]	Refs.[3]
Tissue-specific differentiation antigens				
Melan-A/MART-1	+	–	*Malignant*: melanoma *Normal*: melanocytes	[40]
gp100	+	–	*Malignant*: melanoma *Normal*: melanocytes	[41]
Tyrosinase	+	+	*Malignant*: melanoma *Normal*: melanocytes	[42, 45]
PSA	+	+	*Malignant*: prostate adenocarcinoma *Normal*: epithelial cells of the prostate gland	[43, 46]
PSMA	+	+	*Malignant*: prostate adenocarcinoma *Normal*: epithelial cells of the prostate gland	[44, 47]
Cancer-testis antigens[4]				
MAGE family	+	+	*Malignant*: melanoma, breast, colon, H&N, (N)SCLC, sarcoma, thyroid[5]	[36, 48]
BAGE	+	+	*Malignant*: melanoma, bladder, breast, H&N, NSCLC, sarcoma	[45, 49]
GAGE family	+	–	*Malignant*: melanoma, bladder, breast, colon, esophagus, leukemias, lymphomas, mesothelioma, (N)SCLC, sarcoma, seminoma	[50]
NY-ESO-1	+	+	*Malignant*: melanoma, bladder, breast, H&N, liver, (N)SCLC, lymphoma, ovary, prostate, sarcoma, thyroid	[37–39]
Overexpressed antigens				
CEA	+	+[6]	*Malignant*: colon, breast, gastric,(N)SCLC, pancreas, rectum *Normal*: embryonic gastro-intestinal tissues	[53, 56]
Her-2/neu	+	+	*Malignant*: breast, gastric, melanoma, ovary, pancreas *Normal*: epithelial cells	[57, 58]
Carboxyanhydrase 9	+	+	*Malignant*: renal cell carcinoma *Normal*: bile duct epithelium, gastric mucosal cells	[55, 59]
Malignancy-associated fusion proteins[4]				
BCR-ABL	+	+	*Malignant*: chronic myelogenous leukemia	[63–65]
PML-RARα	–[7]	–	*Malignant*: acute promyelocytic leukemia	[66]

TABLE 1 (continued)

Tumor-associated antigen	Observed type of immune response T-cell-mediated	 Antibody-mediated	Tissue distribution by malignant cells and their normal counterparts[2]	Refs.[3]
Malignancy-associated point mutations of normal genes[4]				
p53	–	+	*Malignant*: colon, breast, ovary	[58, 67, 68]
β-catenin	+	–	*Malignant*: melanoma	[69]
Viral proteins expressed by malignant cells[4]				
EBV LMP-1	+	+	*Malignant*: EBV$^+$ Hodgkin's lymphoma, nasopharyngeal carcinoma	[73, 74]
HPV E6 & E7	+	+	*Malignant*: HPV (serotypes 16 and 18)$^+$ cervical carcinoma	[75, 76]
Malignancy-associated minor histocompatibility genes				
HA-1	+	–	*Malignant*: hematopoietic malignancies, some solid tumors *Normal*: hematopoietic cells	[80]
BCL2A1	+	–	*Malignant*: hematopoietic malignancies, some solid tumors *Normal*: hematopoietic cells	[81]

[1] *Ref. [8] provides a comprehensive listing of TAA as published until 1 August 2000. Information in this table is an excerpt of this listing, extended with information from some more recent publications.*
[2] *Abbreviations used in this column: (N)SCLC, (non) small-cell lung cancer; GI, gastrointestinal; H&N, squamous cell carcinoma occurring in the head and neck region.*
[3] *References providing evidence for cellular or humoral immune reactivity against the relevant TAA are provided.*
[4] *This group of TAA is not expressed by normal tissues.*
[5] *Expression patterns of MAGE-A1 are shown. The MAGE family includes the MAGE-A, B and C groups; within each group, multiple proteins have been identified [48].*
[6] *Humoral responses against CEA have only been observed by vaccination [56].*
[7] *Cellular immune responses against PML-RARα have only been observed* in vitro *[67].*

A fourth group of truly tumor-specific ANTIGENS arises from the fusion of distant genes resulting from translocation of chromosomes in tumor cells. Therefore, the resulting fusion proteins are highly disease-specific (e.g., breakpoint cluster region-Abelson [BCR-ABL] in chronic myelogenous leukemia [60] and promyelocytic leukemia-retinoic acid receptor α [PML-RARα] in acute promyelocytic leukemia [61]). The candidate ANTIGENS are derived from the region immediately surrounding and containing the fusion site. Although these ANTIGENS may represent the most specific targets for immunotherapy, their use for this purpose is limited, as only a few epitopes have been observed to bind efficiently to a small subset of class I molecules and to be naturally processed from the protein [62].

Another group of unique tumor ANTIGENS arises from point mutations of normal genes whose molecular changes often accompany neoplastic transformation or tumor progression. These mutational antigens are only expressed by tumor cells and not by their normal counterparts. In addition to their limited immunogenicity for T-cell responses (similar to fusion proteins), their therapeutic utility is limited further because induced immune responses are restricted to those individual tumors expressing these antigens. Examples are mutated forms of the p53 tumor suppressor protein in a variety of tumors [58, 67, 68], as well as mutated β-catenin in melanoma [69].

Virus-encoded proteins have been identified as sources of TAA in tumors that emerge from virally transformed cells. These proteins typically contribute to the malignant transformation. Examples of virally encoded TAA are the latent membrane protein (LMP)-1 of Epstein-Barr virus (EBV) that is expressed by nasopharyngeal carcinoma and EBV^+ Hodgkin's lymphoma [70], and the human papilloma virus (HPV)-encoded E6 and E7 proteins expressed by HPV^+ cervical carcinoma [71, 72]. As these ANTIGENS are not expressed by normal cells, their usefulness as targets for immunotherapy of cancer appears to be straightforward.

Last but not least (see below), so-called 'minor histocompatibility antigens' (mHAg) can act as TAA and be instrumental in the rejection of malignant cells after allogeneic hematopoietic STEM CELL transplantation (SCT) [77]. mHAg are Class I MHC-presented peptides from polymorphic, naturally processed intracellular proteins that can be encoded by mitochondrial DNA, the Y chromosome or autosomal chromosomes [78]. mHAg-specific $CD8^+$ T cells have been isolated from SCT recipients, particularly those having developed graft-*versus*-host disease [79]. Twelve mHAg have now been traced back to the proteins from which they originate; interestingly, half of these proteins are expressed by malignant cells and/or appear to have a role in malignant cell transformation [77]. Therefore, these mHAg can be classified as TAA. As mHAg are expressed by normal hematopoietic cells, their use as targets for immunotherapy of cancer is limited to the setting of allogeneic STEM CELL transplantation, in which the patient's hematopoietic system is replaced by that of an HLA-matched, but mHAg-mismatched STEM CELL donor with the capability to recognize the mHAg expressed by the tumor as foreign. In this setting, mHAg, being allo-antigens, are not subject to self-tolerance and likely to be more immunogenic than the above-described MHC-restricted tumor-associated self antigens.

Immunotherapy of cancer

Cytokines

CYTOKINES are endogenous molecules that affect the immune response and include, e.g., IL-1, IFN-α and TUMOR NECROSIS FACTOR (TNF)-α (with proinflammatory properties), IL-2, IL-12 and IFN-γ (with T-cell potentiating properties), IL-8 and macrophage inflammatory protein-1 (with chemotactic properties) and granulocyte-macrophage colony-stimulating factor (GM-CSF, with classical growth-stimulating properties). Recombinant CYTOKINES that have been most widely used against cancer are IFN-α, IL-2 and TNF-α. IFN-α has multiple effects on (tumor) cell proliferation, angiogenesis, immune function and expression of genes coding for Class I MHC molecules, tumor ANTIGENS and adhesion molecules. Its antitumor effects have been well documented in hairy cell leukemia [82], Karposi's sarcoma [83] and metastatic renal cell carcinoma [84,85] whilst its efficacy as adjuvant systemic therapy for stage II-III melanoma is still questionable [86].

IL-2 has never been shown to have direct antitumor activity, but has been used for immuno-stimulation in metastatic renal cell carcinoma and malignant melanoma. Some of these patients obtained longstanding partial (10–15%) or complete (~5%) regression of their tumors; the latter were frequently of long duration (i.e., >7 years), suggesting immunological eradication of cancer [87]. The diverse multisystem toxicity observed with high-dose i.v. IL-2 therapy has prompted major efforts to develop effective regimens using lower doses of IL-2. Whilst low-dose IL-2 regimens in metastatic melanoma do not appear to be effective, response rates similar to those obtained with high-dose IL-2 therapy have been observed in renal cell carcinoma patients [88]. A variety of combination regimens has been tested, especially with low-dose IL-2 and IFN-α for renal cell carcinoma. For example, the combination of IL-2, IFN-α and the cytotoxic drug 5-fluorouracil has yielded response rates of 2%, 16% and 39% in 3 major studies [89–91]. This large variation in response rates indicates that there is no convincing evidence that combination therapy is superior to that with IL-2 alone.

TNF-α is a pleiotropic agent with direct and indirect antitumor effects, and is a mediator of septic shock [92]. Hence, dose-limiting toxicity of systemic TNF-α is already encountered at concentrations 10-50 times lower than needed for antitumor effects in murine models [93]. However, the surgical technique of isolated limb perfusion allows the application of therapeutically effective concentrations of TNF-α; response rates between 64% and 100% have been reported when TNF-α was combined with the cytotoxic drug melphalan for the treatment of in-transit melanoma metastases (reviewed in [94]). The combination of TNF-α and melphalan was similarly effective to treat unresectable soft-tissue sarcoma, where limb amputation could be avoided in 64% to 90% of the patients [94]. Animal studies have revealed that the combined use of TNF-α with melphalan rather than the individual compounds is critical to destroy tumors; TNF-α targets the tumor vasculature, allowing strongly enhanced accumulation of melphalan in the tumor leading to tumor vessel destruction and metabolic shut-down of the tumor [95].

Monoclonal antibodies (mAb)

The early clinical applications of murine mAb against cancer in the 80s were hampered by the development of human antimurine Ab (HAMA) responses that rendered these mAb ineffective by neutralization and shortened their *in vivo* survival. However, the subsequently developed technology to 'humanize' murine mAb in the 90s to avoid the generation of HAMA revived the interest to treat cancer with mAb. The genetic assembly of human IMMUNOGLOBULIN framework regions with the COMPLEMENTARITY DETERMINING REGIONS of murine Ab specific for human cancers allowed the creation of immunoglobulins with the desired specificities, subtypes and affinities for clinical use [96]. Examples of humanized mAb currently in clinical use against cancer are shown in Table 2. The mechanism of action of mAb against tumors is complex and highly dependent on the nature of the target molecule. The effects of Ab can be enhanced by combination with cytoreductive treatment or CYTOKINES. Ab can activate effector functions by their Fc portions: antibody-dependent cellular cytotoxicity (ADCC, in which the mAb interact via their Fc portions with the Fc receptors on phagocytic cells and NK LYMPHOCYTES) and complement-dependent cytotoxicity (in which the Fc receptors bind components of the complement cascade). For example, ADCC appears to be the main effector mechanism for rituximab to eliminate B cells [106]. In addition, Ab may also induce apoptosis, cell cycle arrest, inhibition of cell proliferation as well as angiogenesis and metastatic spread. A case in point is trastuzumab whose mechanism of action includes, besides activation of the immune system, downregulation of the constitutive growth-signalling properties of the Her-2 receptor network on cancer cells [107]. Finally, anticancer mAb may be covalently linked to drugs in order to selectively deliver these to tumors with the aim to improve antitumor efficacy and reduce the systemic toxicity of therapy. An example of such an immunoconjugate is gemtuzumab ozogamicin, a humanized CD33 mAb conjugated with calicheamicin: a highly potent antitumor antibiotic that cleaves double-stranded DNA. This compound has been found effective in the treatment of relapsed acute myelogenous leukemia [102].

Cancer vaccination

The purpose of active antigen-specific immunotherapy, i.e., vaccination, against cancer is to induce polyclonal immune responses against TAA, with the added benefits of (i) achieving more stable levels of these immune responses, and (ii) inducing immunological memory to obtain long-term immunosurveillance against TAA-expressing cancer cells [108]. Early vaccination efforts made use of whole tumor cells or tumor lysates of autologous or allogeneic origin. These efforts were mainly directed at melanoma

TABLE 2. MONOCLONAL ANTITUMOR ANTIBODIES IN CLINICAL USE

Name	Target Ag[1]	Clinical use[2]	Refs.
Alemtuzumab	CD52	(Refractory) B-CLL, T-PLL	[97, 98]
Bevacizumab	VEGF	Colorectal cancer (combined with cytoreductive therapy)	[99]
Cetuximab	EGFR	Head and neck cancer (combined with local radiotherapy)	[100]
Edrecolomab	Ep-CAM	Colorectal cancer	[101]
Gemtuzumabozo-gamicin	CD33	AML in first relapse	[102]
Rituximab	CD20	Various subgroups of B-NHL, B-CLL	[103, 104]
Trastuzumab	Her-2/neu	Breast cancer	[105]

[1] *Abbreviations used in this column: EGFR, epidermal growth factor receptor; Ep-CAM, epithelial cell adhesion molecule; Her-2/neu, human epidermal receptor-2/neurological; VEGF, vascular endothelial growth factor.*
[2] *AML, acute myelogenous leukemia; B-CLL, B-chronic lymphocytic leukemia; B-NHL, B-non-Hodgkin's lymphoma; T-PLL, T-prolymphocytic leukemia.*

and yielded little evidence of consistent efficacy. For example, long-term follow-up of patients with metastatic melanoma, randomized to observation *versus* immunization with an allogeneic melanoma oncolysate, showed no significant differences in overall or disease-free survival between both groups [109]. Meanwhile, the definition of TAA (see above) and the advent of sensitive techniques to detect TAA-specific T-cell responses [110] had greatly augmented the ability to perform immunological assessments in patients, but the few clinical responses to vaccination hindered the ability to interpret these *in vitro* assessments.

Preclinical experiments indicated that a variety of methods could be used to vaccinate against TAA, such as the use of minimal determinant peptides [111], proteins [112], viral oncolysates [113], TAA-encoding DNA [114], heat shock proteins [115], whole tumor cells [116], and DCs [9, 11–15]. The modification and optimization of Class I MHC-binding 'anchor' residues in some minimal determinant peptides was found to improve their immunogenicity *in vitro* and *in vivo*, whilst their ability to generate $CD8^+$ T cells that crossreact with and recognize the native, unmodified peptides remained intact. An example is the melanoma protein gp100-derived peptide ITDQVPFSV, which is presented by HLA-A*0201. $CD8^+$ T cells specific for this peptide have been isolated from tumors, but the MHC binding algorithm [22] predicts that methionine or leucine rather than threonine at position 2 would have been optimal for binding to HLA-A*0201. Indeed, a variant peptide with M substituted for T showed better immunogenicity *in vitro* and *in vivo* [117].

Fully matured DC can be used *in vitro* and *in vivo* to generate T-cell reponses against TAA after they have been loaded with TAA. Myeloid DC, derived from cultures of MONOCYTES or $CD34^+$ hematopoietic precursor cells, have mostly been used in clinical trials. Loading can be done with longer peptides containing both Class I and Class II MHC-presented epitopes, recombinant protein or tumor lysates. In all cases, the MHC molecules will be endogenously loaded with TAA after proteolytic processing. Of interest is that vaccination with DC loaded with a MHC class II-restricted peptide derived from the MAGE-C2 antigen rapidly induced strong and peptide-specific T_{H1} RESPONSES in patients with metastatic melanoma [118]. Also, DNA and RNA encoding TAA can be used to express these ANTIGENS in DC. Early clinical trials have shown the feasibility of DC vaccination, its lack of toxicity and its antitumor potential. Still, many parameters of the preparation of this type of vaccine must be better defined to prove its antitumor efficacy in larger clinical trials [119].

Alternatively, vaccination with cultured tumor cells that have been genetically modified to produce an immunostimulatory cytokine (e.g., IL-2, GM-CSF) or to express an accessory molecule for T cells (e.g., CD80/CD86) on their surface, has been shown to improve antitumor immune reactivity in animal models. The use of GM-CSF yielded the most consistent data in this respect [120]. Clinical application requires that patients must have their tumor excised in order to establish cell lines which then have to be transduced with a highly efficient retroviral vector encoding GM-CSF. Initial clinical trials showed sometimes significant *in vitro* but infrequent clinical antitumor responses [121, 122].

Adoptive autologous cellular immunotherapy

Based on promising animal data, initial approaches to adoptive cellular immunotherapy of cancer have made use of *ex vivo* IL-2-activated peripheral blood mononuclear cells that exhibited so-called lymphokine-activated killer (LAK) activity [123]. LAK activity was defined as the capability to kill fresh and cultured tumor cells, but not normal cells, in an MHC-unrestricted fashion. Patients with advanced melanoma and renal cell cancer were treated with high-dose IL-2 and up to 2×10^{11} autologous LYMPHOCYTES with LAK activity. Although complete and partial tumor regressions were seen in up to one-third of patients [124], subsequent randomized studies attributed these responses to IL-2, whilst no evidence for a benefit from LAK cells was demonstrated [125, 126].

As an alternative to LAK as a source of antitumor activity, TIL were isolated, expanded *ex vivo* using IL-2 to very high numbers (i.e., up to 10^{11}) and reinfused into the patients. Such TIL cultures from melanoma lesions frequently yielded CTL CLONES with HLA-restricted tumor recognition, which proved instrumental for the identification of TAA (see above). Treatment of melanoma patients with polyclonal autologous TIL plus IL-2 yielded complete or partial responses in approximately one-third of the patients [27]. Alternatively, $CD8^+$ CTL CLONES isolated and expanded from cultured peripheral blood mononuclear cells were used and shown to be effective for prevention of herpesvirus reactivations [127] and treatment of melanoma [128]. When this form of treatment was preceded by cytoreductive therapy to induce lymphopenia, an extensive and persistent clonal T-cell repopulation was observed. These T-cells had the original TAA specificity of the TIL, and were able to migrate to the tumor sites and induce partial antitumor responses in almost half of the patients [15]. Such studies clearly indicate the promise of adoptive cellular immunotherapy for cancer, provided that the tumor-specific T cells persist for prolonged periods of time in order to exert their antitumor effects, and do not destroy normal tissues resulting in severe adverse events. For the long-term persistence of adoptively transferred $CD8^+$ T cells, $CD4^+$ T-cell help is needed (see above and [129]), which currently is provided in surrogate form by recombinant CYTOKINES such as IL-2.

However, adoptive transfer of tumor-specific CTL for cancer treatment can be severely hampered by the difficulty of reproducibly isolating and expanding such LYMPHOCYTES. As an alternative, polyclonally activated T cells have been 'retargeted', initially by using bispecific mAb [130] and later by transfer of genes encoding tumor-specific receptors [131]. Bispecific mAb can be directed towards both a TAA and CD3, a part of the TCR complex. By coating polyclonal CTL with such bispecific mAbs, their tumor-cell killing potential can be targeted to tumor cells bearing that TAA. This principle has been tested clinically for the locoregional treatment of advanced ovarian cancer. In spite of objective antitumor responses at the site of treatment (i.e., intraperitoneally) in one-third of patients, this approach proved to be complicated by limited accessibility of solid tumors to the therapy, dissociation of bi-specific mAb from CTL, development of human anti-mouse ANTIBODY responses and the limited capacity of mAb-sensitized T LYMPHOCYTES to recycle their tumoricidal capacity [130]. The transfer of genes encoding TAA-specific receptors into T cells (i.e., genetic retargeting) became feasible as an alternative to bispecific mAb-mediated retargeting after the development of highly efficient protocols for retrovirus-mediated gene transfer into polyclonally activated T cells and for rapid expansion of the transduced cells, all under 'good manufacturing practice' (GMP) conditions [131]. Autologous T LYMPHOCYTES genetically retargeted to carboxyanhydrase 9 by a single-chain Ab-type receptor specific for this TAA [55] are currently being investigated clinically in patients with metastatic renal cell cancer [132].

Allogeneic hematopoietic stem cell transplantation (SCT)

High-dose chemoradiotherapy followed by rescue from the resulting ablation of hematopoietic function with SCT from an HLA-matched donor has become standard therapy for many hematological malignancies. One problem with this treatment is graft-*versus*-host disease (GVHD) due to donor-derived T cells recognizing mismatched mHAg expressed by normal tissues of the host. As the malignant cells that survive chemoradiotherapy are also of host origin, patients who develop GVHD have less frequent recurrence of the original disease due to the associated graft-*versus*-tumor (GVT) effect. T cells mediate this antitumor activity, because (1) infusions of T cells from the SCT donor to treat leukemic relapse after SCT sometimes result in complete remissions, and (2) the complete remissions observed after so-called non-myeloablative SCT must result from GVT effects because the reduced-intensity cytoreductive therapy cannot eliminate all residual disease [133]. This concept has been extended to the treatment of solid tumors. Whilst preliminary data from several groups provide evidence for a GVT effect in renal cell cancer, no such effect has been observed in advanced melanoma and only anecdotal data are available on other solid tumors [134, 135]. Several significant clinical antitumor responses were seen in patients with metastatic renal cell cancer undergoing nonmyeloablative SCT from an HLA-matched sibling after preparation with cyclophosphamide and fludarabine. During the first 100 days post-SCT, patients received additional immunosuppression to allow engraftment and suppress GVHD. During this time most patients showed growth of their tumors. After full engraftment, immunosuppression was tapered (and to some patients CYTOKINES or donor lymphocyte infusions were given) until GVHD was sufficiently suppressed or stabilised at a low grade. Among 33 patients treated in this way, 15 showed evidence of regression of their renal tumors [136]. However, the GVT activity with these regimens is often associated with severe and life-threatening GVHD, which may lead to a treatment-related mortality between 10% and 20%. The characterization of antigenic targets with limited tissue distribution, permitting donor LYMPHOCYTES to preferentially target malignant cells and not critical normal tissues, coupled with methods to select and/or generate T cells with such specificities, should provide a much-needed improvement to this approach.

Conclusions and future developments

It has now been firmly established that the human immune system is capable of recognizing and eliminating spontaneously arising tumors, although cancer cells are generally less immunogenic than microbial pathogens such as bacteria, fungi and viruses. Nevertheless, immunotherapeutic modalities, such as CYTOKINES and mAb, already have become components of several standard treament regimens of human malignancies. Many vaccines have advanced through preliminary testing to efficacy trials, and have shown little toxicity but also limited effects in patients with established tumors. Tumors frequently interfere with the development and function of immune responses. Thus, one of the challenges for cancer immunotherapy is to use advances in cellular and molecular immunology to develop strategies that effectively and safely augment antitumor responses. With our increasing understanding of the requirements for immune cell activation, homing and accumulation at tumor sites, and for the disruption of regulatory mechanisms that inhibit immunological anticancer responses at the sites of the tumors, our abilities to design and engineer immunotherapeutic approaches with antitumor capacities beyond what can be elicited from the normal immune system should advance.

Initial clinical trials of adoptive transfer of large numbers of autologous tumor-reactive T LYMPHOCYTES have shown promising antitumor effects, in particular against melanoma. The ability to transfer genes with high efficiency into polyclonally activated T cells has raised wide interest in the genetic retargeting of T LYMPHOCYTES against tumors with the potential to circumvent the complex and cumbersome procedures to isolate and expand 'spontaneous' tumor-specific T cells. However, these approaches are based on end-stage T cells with probably limited survival times upon administration. The transfer of TAA-specific receptors to hematopoietic STEM CELLS (HSC) in order to retarget their progeny against cancer would appear to be a solution for this problem, but the observation of oncogenic events after transfer of a growth factor receptor-encoding gene into HSC [137] has tempered the enthusiasm for this approach. Meanwhile, transplantation of allogeneic HSC (following pretreatment with a reduced-intensity cytoreductive regimen) has emerged as a promising approach to provide cancer patients with long-term immunosurveillance against their tumors, albeit currently with significant toxicity [136]. The expression of so-called mHAg, constituting TAA, by tumor cells is of key importance for the concept of allo-immunotherapy of cancer. The complexes of MHC and allo-TAA are probably more immunogenic than those of MHC and self-TAA that are the targets of classical, autologous T-cell-based immunotherapy [77]. This characteristic, combined with the fact that a long-term source of antitumor immunosurveillance in the form of HSC is provided, has the potential to make alloimmunotherapy a very powerful form of immunotherapy of cancer.

Summary

The human immune system is capable of recognizing and eliminating spontaneously arising tumors. However, tumors frequently interfere with the development and function of immune responses. With the recent advances in cellular and molecular immunology, strategies are now being developed that effectively and safely augment antitumor responses. These advances have enabled the characterization of so-called TUMOR-ASSOCIATED ANTIGENS (TAA) which can function as targets for the immune system. These TAA include peptides that can be recognized by T LYMPHOCYTES as they are presented by class I or II HLA molecules, or intact proteins or carbohydrates that are not presented by HLA molecules and are recognized by ANTIBODIES. TAA can be classified into (1) tissue-specific differentiation antigens, (2) 'cancer-testis' antigens, (3) normally occurring antigens that are overexpressed, (4) fusion proteins, (5) mutational antigens, (6) virally encoded antigens and (7) minor histocompatibility antigens. Various forms of adoptive tumor immunotherapy have been developed, including (combinations of) CYTOKINES, monoclonal antibodies, and autologous cellular immunotherapy – initially using nonspecifi-cally expanded LYMPHOCYTES, later with expanded TAA-specific T cells derived from tumor infiltrates. The active forms of tumor immunotherapy include vaccination and allogeneic hematopoietic STEM CELL (HSC) transplantation. The latter is currently a standard treatment for hematological malignancies but still experimental for solid tumors. As HSC can be a life-long source of antitumor T cells, we believe that this approach has the potential to become a very powerful immunotherapeutic modality of cancer.

Selected readings

Zinkernagel RM, Doherty PC (1997) The discovery of MHC restriction. *Immunol Today* 18: 14–17

Carter P (2001) Improving the efficacy of antibody-based cancer therapies. *Nat Rev Cancer* 1: 118–129

Eggermont AM, de Wilt JH, ten Hagen TL (2003) Current uses of isolated limb perfusion in the clinic and a model system for new strategies. *Lancet Oncol* 4: 429–437

Berzofsky JA, Terabe M, Oh S, Belyakov IM, Ahlers JD, Janik JE, Morris JC (2004) Progress on new vaccine strategies for the immunotherapy and prevention of cancer. *J Clin Invest* 113: 1515–1525

Childs RW, Barrett J (2004) Nonmyeloablative allogeneic immunotherapy for solid tumors. *Annu Rev Med* 55: 459–475

References

1 Ehrlich P (1909) Über den jetzigen Stand der Karzinomforschung. *Ned Tijdschr Geneesk* 53: 273–290
2 Thomas L (1959) Mechanisms involved in tissue damage by the endotoxins of gram negative bacteria. In: HS Lawrence (ed): *Cellular and Humoral Aspects of the Hypersensitive States*. Hoeber-Harper, New York, 451–468
3 Burnet FM (1970) The concept of immunological surveillance. *Prog Exp Tumor Res* 13: 1–27
4 Vakkila J, Lotze MT (2003) Cellular immunotherapy of cancer. *Blood Ther Med* 3: 84–90
5 Pollock BH, Jenson HB, Leach CT, McClain KL, Hutchison RE, Garzarella L, Joshi VV, Parmley RT, Murphy SB (2003) Risk factors for pediatric human immunodeficiency virus-related malignancy. *JAMA* 289: 2393–2399
6 Yang L, Yamagata N, Yadav R, Brandon S, Courtney RL, Morrow JD, Shyr Y, Boothby M, Joyce S, Carbone DP et al (2003) Cancer-associated immunodeficiency and dendritic cell abnormalities mediated by the prostaglandin EP2 receptor. *J Clin Invest* 111: 727–735
7 Euvrard S, Kanitakis J, Claudy A (2003) Skin cancers after organ transplantation. *N Engl J Med* 348: 1681-1691
8 Renkvist N, Castelli C, Robbins PF, Parmiani G (2001) A listing of human tumor antigens recognized by T cells. *Cancer Immunol Immunother* 50: 3–15
9 Tanaka H, Shimizu K, Hayashi T, Shu S (2002) Therapeutic immune response induced by electrofusion of dendritic and tumor cells. *Cell Immunol* 220: 1-12
10 Muller MR, Grunebach F, Nencioni A, Brossart P (2003) Transfection of dendritic cells with RNA induces CD4- and CD8-mediated T cell immunity against breast carcinomas and reveals the immunodominance of presented T cell epitopes. *J Immunol* 170: 5892–5896
11 Cui Y, Kelleher E, Straley E, Fuchs E, Gorski K, Levitsky H, Borrello I, Civin CI, Schoenberger SP, Cheng L et al (2003) Immunotherapy of established tumors using bone marrow transplantation with antigen gene-modified hematopoietic stem cells. *Nat Med* 9: 952–958
12 Kokhaei P, Rezvany MR, Virving L, Choudhury A, Rabbani H, Osterborg A, Mellstedt H (2003) dendritic cells loaded with apoptotic tumour cells induce a stronger T-cell response than dendritic cell-tumour hybrids in B-CLL. *Leukemia* 17: 894–899
13 Su Z, Dannull J, Heiser A, Yancey D, Pruitt S, Madden J, Coleman D, Niedzwiecki D, Gilboa E, Vieweg J (2003) Immunological and clinical responses in metastatic renal cancer patients vaccinated with tumor RNA-transfected dendritic cells. *Cancer Res* 63: 2127–2133
14 O'Rourke MG, Johnson M, Lanagan C, See J, Yang J, Bell JR, Slater GJ, Kerr BM, Crowe B, Purdie DM et al (2003) Durable complete clinical responses in a phase I/II trial using an autologous melanoma cell/dendritic cell vaccine. *Cancer Immunol Immunother* 52: 387–395
15 Dudley ME, Wunderlich JR, Robbins PF, Yang JC, Hwu P, Schwartzentruber DJ, Topalian SL, Sherry R, Restifo NP, Hubicki AM et al (2002) Cancer regression and autoimmunity in patients after clonal repopulation with antitumor lymphocytes. *Science* 298: 850–854
16 Freedman RS, Kudelka AP, Kavanagh JJ, Verschraegen C, Edwards CL, Nash M, Levy L, Atkinson EN, Zhang HZ, Melichar B et al (2000) Clinical and biological effects of intraperitoneal injections of recombinant interferon-gamma and recombinant interleukin 2 with or without tumor-infiltrating lymphocytes in patients with ovarian or peritoneal carcinoma. *Clin Cancer Res* 6: 2268–2278
17 Figlin RA, Thompson JA, Bukowski RM, Vogelzang NJ, Novick AC, Lange P, Steinberg GD, Belldegrun AS (1999) Multicenter, randomized, phase III trial of CD8(+) tumor-infiltrating lymphocytes in combination with recombinant interleukin-2 in metastatic renal cell carcinoma. *J Clin Oncol* 17: 2521–2529
18 Rosenberg SA, Yannelli JR, Yang JC, Topalian SL, Schwartzentruber DJ, Weber JS, Parkinson DR, Seipp CA, Einhorn JH, White DE (1994) Treatment of patients with metastatic melanoma with autologous tumor-infiltrating lymphocytes and interleukin 2. *J Natl Cancer Inst* 86: 1159–1166
19 Gorelik L, Flavell RA (2001) Immune-mediated eradication of tumors through the blockade of transforming growth factor-beta signaling in T cells. *Nat Med* 7: 1118–1122
20 Chia CS, Ban K, Ithnin H, Singh H, Krishnan R, Mokhtar S, Malihan N, Seow HF (2002) Expression of interleukin-18, INTERFERON-gamma and interleukin-10 in hepatocellular carcinoma. *Immunol Lett* 84: 163–172
21 Pieters J (2000) MHC class II-restricted antigen processing and presentation. *Adv Immunol* 75: 159-208
22 Falk K, Rotzschke O, Stevanovic S, Jung G, Rammensee HG (1991) Allele-specific motifs revealed by sequenc-

ing of self-peptides eluted from MHC molecules. *Nature* 351: 290–296

23 Hunziker L, Klenerman P, Zinkernagel RM, Ehl S (2002) Exhaustion of cytotoxic T cells during adoptive immunotherapy of virus carrier mice can be prevented by B cells or CD4+ T cells. Eur *J Immunol* 2: 374–382

24 Marzo AL, Kinnear BF, Lake RA, Frelinger JJ, Collins EJ, Robinson BW, Scott B (2000) Tumor-specific CD4+ T cells have a major "post-licensing" role in CTL mediated anti-tumor immunity. *J Immunol* 165: 6047–6055

25 Schoenberger SP, Toes RE, van der Voort EI, Offringa R, Melief CJ (1998) T-cell help for cytotoxic T lymphocytes is mediated by CD40-CD40L interactions. *Nature* 393: 480–483

26 Surman DR, Dudley ME, Overwijk WW, Restifo NP (2000) Cutting edge: CD4+ T cell control of CD8+ T cell reactivity to a model tumor antigen. *J Immunol* 164: 562–565

27 Bevan MJ (2004) Helping the CD8(+) T-cell response. *Nat Rev Immunol* 4: 595–602

28 Ibe S, Qin Z, Schuler T, Preiss S, Blankenstein T (2001) Tumor rejection by disturbing tumor stroma cell interactions. *J Exp Med* 194: 1549–1559

29 Wang HY, Lee DA, Peng G, Guo Z, Li Y, Kiniwa Y, Shevach EM, Wang RF (2004) Tumor-specific human CD4+ regulatory T cells and their ligands: implications for immunotherapy. *Immunity* 20: 107–118

30 Gavin M, Rudensky A (2003) Control of immune homeostasis by naturally arising regulatory CD4+ T cells. *Curr Opin Immunol* 15: 690–696

31 Zinkernagel RM, Doherty PC (1997) The discovery of MHC restriction. *Immunol Today* 18: 14–17

32 Carrel S, Johnson JP (1993) Immunologic recognition of malignant melanoma by autologous T lymphocytes. *Curr Opin Oncol* 5: 383–389

33 Van der Bruggen P, Traversari C, Chomez P, Lurquin C, De Plaen E, Van den Eynde B, Knuth A, Boon T (1991) A gene encoding an antigen recognized by cytolytic T lymphocytes on a human melanoma. *Science* 254: 1643–1647

34 Traversari C, van der Bruggen P, Luescher IF, Lurquin C, Chomez P, Van Pel A, De Plaen E, Amar-Costesec A, Boon T (1992) A nonapeptide encoded by human gene MAGE-1 is recognized on HLA-A1 by cytolytic T lymphocytes directed against tumor antigen MZ2-E. *J Exp Med* 176: 1453–1457

35 Carter P (2001) Improving the efficacy of antibody-based cancer therapies. *Nat Rev Cancer* 1: 118–129

36 Sahin U, Tureci O, Schmitt H, Cochlovius B, Johannes T, Schmits R, Stenner F, Luo G, Schobert I, Pfreundschuh M (1995) Human neoplasms elicit multiple specific immune responses in the autologous host. *Proc Natl Acad Sci USA* 92: 11810–11813

37 Chen YT, Scanlan MJ, Sahin U, Tureci O, Gure AO, Tsang S, Williamson B, Stockert E, Pfreundschuh M, Old LJ (1997) A testicular antigen aberrantly expressed in human cancers detected by autologous antibody screening. *Proc Natl Acad Sci USA* 94: 1914–1918

38 Jager E, Jager D, Karbach J, Chen YT, Ritter G, Nagata Y, Gnjatic S, Stockert E, Arand M, Old LJ et al (2000) Identification of NY-ESO-1 epitopes presented by human histocompatibility antigen (HLA)-DRB4*0101-0103 and recognized by CD4(+) T lymphocytes of patients with NY-ESO-1-expressing melanoma. *J Exp Med* 191: 625–630

39 Wang RF, Johnston SL, Zeng G, Topalian SL, Schwartzentruber DJ, Rosenberg SA (1998) A breast and melanoma-shared tumor antigen: T cell responses to antigenic peptides translated from different open reading frames. *J Immunol* 161: 3598–3606

40 Kawakami Y, Eliyahu S, Delgado CH, Robbins PF, Sakaguchi K, Appella E, Yannelli JR, Adema GJ, Miki T, Rosenberg SA (1994) Identification of a human melanoma antigen recognized by tumor-infiltrating lymphocytes associated with *in vivo* tumor rejection. *Proc Natl Acad Sci USA* 91: 6458–6462

41 Bakker AB, Schreurs MW, de Boer AJ, Kawakami Y, Rosenberg SA, Adema GJ, Figdor CG (1994) Melanocyte lineage-specific antigen gp100 is recognized by melanoma-derived tumor-infiltrating lymphocytes. *J Exp Med* 179: 1005–1009

42 Brichard V, Van Pel A, Wolfel T, Wolfel C, De Plaen E, Lethe B, Coulie P, Boon T (1993) The tyrosinase gene codes for an antigen recognized by autologous cytolytic T lymphocytes on HLA-A2 melanomas. *J Exp Med* 178: 489–495

43 Epstein JI (1993) PSA and PAP as immunohistochemical markers in prostate cancer. *Urol Clin North Am* 20: 757–770

44 Israeli RS, Powell CT, Corr JG, Fair WR, Heston WD (1994) Expression of the prostate-specific membrane antigen. *Cancer Res* 54: 1807–1811

45 Cancer Immunome Database. URL: *http://www2.licr.*

org/CancerImmunomeDB/ (Accessed September 2004)

46 Harada M, Kobayashi K, Matsueda S, Nakagawa M, Noguchi M, Itoh K (2003) Prostate-specific antigen-derived epitopes capable of inducing cellular and humoral responses in HLA-A24$^+$ prostate cancer patients. *Prostate* 57: 152–159

47 Kobayashi K, Noguchi M, Itoh K, Harada M (2003) Identification of a prostate-specific membrane antigen-derived peptide capable of eliciting both cellular and humoral immune responses in HLA-A24$^+$ prostate cancer patients. *Cancer Sci* 94: 622–627

48 De Plaen E, Arden K, Traversari C, Gaforio JJ, Szikora JP, De Smet C, Brasseur F, van der Bruggen P, Lethe B, Lurquin C et al (1994) Structure, chromosomal localization, and expression of 12 genes of the MAGE family. *Immunogenetics* 40: 360–369

49 Boël P, Wildmann C, Sensi ML, Brasseur R, Renauld JC, Coulie P, Boon T, van der Bruggen P (1995) BAGE: a new gene encoding an antigen recognized on human melanomas by cytolytic T lymphocytes. *Immunity* 2: 167–175

50 Van den Eynde B, Peeters O, De Backer O, Gaugler B, Lucas S, Boon T (1995) A new family of genes coding for an antigen recognized by autologous cytolytic T lymphocytes on a human melanoma. *J Exp Med* 182: 689–698

51 Zippelius A, Batard P, Rubio-Godoy V, Bioley G, Lienard D, Lejeune F, Rimoldi D, Guillaume P, Meidenbauer N, Mackensen A et al (2004) Effector function of human tumor-specific CD8 T cells in melanoma lesions: a state of local functional tolerance. *Cancer Res* 64: 2865–2873

52 Gotter J, Brors B, Hergenhahn M, Kyewski B (2004) Medullary epithelial cells of the human thymus express a highly diverse selection of tissue-specific genes colocalized in chromosomal clusters. *J Exp Med* 199: 155–166

53 Tsang KY, Zaremba S, Nieroda CA, Zhu MZ, Hamilton JM, Schlom J (1995) Generation of human cytotoxic T cells specific for human carcinoembryonic antigen epitopes from patients immunized with recombinant vaccinia-CEA vaccine. *J Natl Cancer Inst* 87: 982–990

54 Menard S, Pupa SM, Campiglio M, Tagliabue E (2003) Biologic and therapeutic role of HER2 in cancer. *Oncogene* 22: 6570–6578

55 Oosterwijk E, Ruiter DJ, Hoedemaeker PJ, Pauwels EK, Jonas U, Zwartendijk J, Warnaar SO (1986) Monoclonal ANTIBODY G 250 recognizes a determinant present in renal-cell carcinoma and absent from normal kidney. *Int J Cancer* 38: 489–494

56 Berinstein NL (2002) Carcinoembryonic antigen as a target for therapeutic anticancer vaccines: a review. *J Clin Oncol* 20: 2197–2207

57 Fisk B, Blevins TL, Wharton JT, Ioannides CG (1995) Identification of an immunodominant peptide of HER-2/neu protooncogene recognized by ovarian tumor-specific cytotoxic T-lymphocyte lines. *J Exp Med* 181: 2109–2117

58 Scanlan MJ, Gout I, Gordon CM, Williamson B, Stockert E, Gure AO, Jager D, Chen YT, Mackay A, O'Hare MJ et al (2001) Humoral immunity to human breast cancer: antigen definition and quantitative analysis of mRNA expression. *Cancer Immun* 1: 4

59 Vissers JL, De Vries IJ, Engelen LP, Scharenborg NM, Molkenboer J, Figdor CG, Oosterwijk E, Adema GJ (2002) Renal cell carcinoma-associated antigen G250 encodes a naturally processed epitope presented by human leukocyte antigen-DR molecules to CD4(+) T lymphocytes. *Int J Cancer* 100: 441–444

60 Kurzrock R, Gutterman JU, Talpaz M (1988) The molecular genetics of Philadelphia chromosome-positive leukemias. *N Engl J Med* 319: 990–998

61 Ferrucci PF, Grignani F, Pearson M, Fagioli M, Nicoletti I, Pelicci PG (1997) Cell death induction by the acute promyelocytic leukemia-specific PML/RARalpha fusion protein. *Proc Natl Acad Sci USA* 94: 10901–10906

62 Dermime S, Bertazzoli C, Marchesi E, Ravagnani F, Blaser K, Corneo GM, Pogliani E, Parmiani G, Gambacorti-Passerini C (1996) Lack of T-cell-mediated recognition of the fusion region of the pml/RAR-alpha hybrid protein by lymphocytes of acute promyelocytic leukemia patients. *Clin Cancer Res* 2: 593–600

63 Clark RE, Dodi IA, Hill SC, Lill JR, Aubert G, Macintyre AR, Rojas J, Bourdon A, Bonner PL, Wang L et al (2001) Direct evidence that leukemic cells present HLA-associated immunogenic peptides derived from the BCR-ABL b3a2 fusion protein. *Blood* 98: 2887–2893

64 Yasukawa M, Ohminami H, Kojima K, Hato T, Hasegawa A, Takahashi T, Hirai H, Fujita S (2001) HLA class II-restricted antigen presentation of endogenous bcr-abl fusion protein by chronic myelogenous leukemia-derived dendritic cells to CD4(+) T lymphocytes. *Blood* 98: 1498–1505

65 Talpaz M, Qiu X, Cheng K, Cortes JE, Kantarjian H, Kurzrock R (2000) Autoantibodies to Abl and Bcr proteins. *Leukemia* 14: 1661–1666

66 Gambacorti-Passerini C, Grignani F, Arienti F, Pandolfi PP, Pelicci PG, Parmiani G (1993) Human CD4 lymphocytes specifically recognize a peptide representing the fusion region of the hybrid protein pml/RAR alpha present in acute promyelocytic leukemia cells. *Blood* 81: 1369–1375

67 Scanlan MJ, Chen YT, Williamson B, Gure AO, Stockert E, Gordan JD, Tureci O, Sahin U, Pfreundschuh M, Old LJ (1998) Characterization of human colon cancer antigens recognized by autologous antibodies. *Int J Cancer* 76: 652–658

68 Stone B, Schummer M, Paley PJ, Thompson L, Stewart J, Ford M, Crawford M, Urban N, O'Briant K, Nelson BH (2003) Serologic analysis of ovarian tumor antigens reveals a bias toward antigens encoded on 17q. *Int J Cancer* 104: 73–84

69 Robbins PF, El-Gamil M, Li YF, Kawakami Y, Loftus D, Appella E, Rosenberg SA (1996) A mutated beta-catenin gene encodes a melanoma-specific antigen recognized by tumor infiltrating lymphocytes. *J Exp Med* 183: 1185–1192

70 Li HP, Chang YS (2003) Epstein-Barr virus latent membrane protein 1: structure and functions. *J Biomed Sci* 10: 490–504

71 Munger K, Basile JR, Duensing S, Eichten A, Gonzalez SL, Grace M, Zacny VL (2001) Biological activities and molecular targets of the human papillomavirus E7 oncoprotein. *Oncogene* 20: 7888–7898

72 Mantovani F, Banks L (2001) The human papillomavirus E6 protein and its contribution to malignant progression. *Oncogene* 20: 7874–7887

73 Subklewe M, Chahroudi A, Bickham K, Larsson M, Kurilla MG, Bhardwaj N, Steinman RM (1999) Presentation of Epstein-Barr virus latency antigens to CD8(+), interferon-gamma-secreting, T lymphocytes. *Eur J Immunol* 29: 3995– 4001

74 Xu J, Ahmad A, D'Addario M, Knafo L, Jones JF, Prasad U, Dolcetti R, Vaccher E, Menezes J (2000) Analysis and significance of anti-latent membrane protein-1 antibodies in the sera of patients with EBV-associated diseases. *J Immunol* 164: 2815–2822

75 Nakagawa M, Viscidi R, Deshmukh I, Costa MD, Palefsky JM, Farhat S, Moscicki AB (2002) Time course of humoral and cell-mediated immune responses to human papillomavirus type 16 in infected women. *Clin Diagn Lab Immunol* 9: 877–882

76 Baay MF, Duk JM, Burger MP, de Bruijn HW, Stolz E, Herbrink P (1999) Humoral immune response against proteins E6 and E7 in cervical carcinoma patients positive for human papilloma virus type 16 during treatment and follow-up. *Eur J Clin Microbiol Infect Dis* 18: 126–132

77 Spierings E, Wieles B, Goulmy E (2004) Minor histocompatibility antigens big in tumour therapy. *Trends Immunol* 25: 56–60

78 Goulmy E (1997) Minor histocompatibility antigens: from T cell recognition to peptide identification. *Hum Immunol* 54: 8–14

79 Mutis T, Gillespie G, Schrama E, Falkenburg JH, Moss P, Goulmy E (1999) Tetrameric HLA class I minor histocompatibility antigen peptide complexes demonstrate minor histocompatibility antigen-specific cytotoxic T lymphocytes in patients with graft-*versus*-host disease. *Nat Med* 5: 839–842

80 Den Haan JM, Meadows LM, Wang W, Pool J, Blokland E, Bishop TL, Reinhardus C, Shabanowitz J, Offringa R, Hunt DF et al (1998) The minor histocompatibility antigen HA-1: a diallelic gene with a single amino acid polymorphism. *Science* 279: 1054–1057

81 Akatsuka Y, Nishida T, Kondo E, Miyazaki M, Taji H, Iida H, Tsujimura K, Yazaki M, Naoe T, Morishima Y et al (2003) Identification of a polymorphic gene, BCL2A1, encoding two novel hematopoietic lineage-specific minor histocompatibility antigens. *J Exp Med* 197: 1489–1500

82 Grever M, Kopecky K, Foucar MK, Head D, Bennett JM, Hutchison RE, Corbett WE, Cassileth PA, Habermann T, Golomb H et al (1995) Randomized comparison of pentostatin *versus* interferon alfa-2a in previously untreated patients with hairy cell leukemia: an intergroup study. *J Clin Oncol* 13: 974–982

83 Krown SE (2001) Management of Kaposi sarcoma: the role of interferon and thalidomide. *Curr Opin Oncol* 13: 374–381

84 Medical Research Council Renal Cancer Collaborators (1999) interferon alpha and survival in metastatic renal carcinoma: early results of a randomised controlled trial. *Lancet* 353: 14–17

85 Pyrhonen S, Salminen E, Ruutu M, Lehtonen T, Nurmi M, Tammela T, Juusela H, Rintala E, Hietanen P, Kellokumpu-Lehtinen PL (1999) Prospective randomized

trial of interferon alfa-2a plus vinblastine *versus* vinblastine alone in patients with advanced renal cell cancer. *J Clin Oncol* 17: 2859–2867

86 Eggermont AM, Punt CJ (2003) Does adjuvant systemic therapy with interferon-alpha for stage II-III melanoma prolong survival? *Am J Clin Dermatol* 4: 531–536

87 Rosenberg SA, Yang JC, White DE, Steinberg SM (1998) Durability of complete responses in patients with metastatic cancer treated with high-dose interleukin-2: identification of the antigens mediating response. *Ann Surg* 228: 307–319

88 Tourani JM, Lucas V, Mayeur D, Dufour B, DiPalma M, Boaziz C, Grise P, Varette C, Pavlovitch JM, Pujade-Lauraine E et al (1996) Subcutaneous recombinant interleukin-2 (rIL-2) in out-patients with metastatic renal cell carcinoma. Results of a multicenter SCAPP1 trial. *Ann Oncol* 7: 525–528

89 Ravaud A, Audhuy B, Gomez F, Escudier B, Lesimple T, Chevreau C, Douillard JY, Caty A, Geoffrois L, Ferrero JM et al (1998) Subcutaneous interleukin-2, interferon alfa-2a, and continuous infusion of fluorouracil in metastatic renal cell carcinoma: a multicenter phase II trial. Groupe Francais d'Immunotherapie. *J Clin Oncol* 16: 2728–2732

90 Dutcher JP, Atkins M, Fisher R, Weiss G, Margolin K, Aronson F, Sosman J, Lotze M, Gordon M, Logan T et al (1997) Interleukin-2-based therapy for metastatic renal cell cancer: the Cytokine Working Group experience, 1989–1997. *Cancer J Sci Am* 3 (Suppl 1): S73–S78

91 Lopez Hanninen E, Kirchner H, Atzpodien J (1996) Interleukin-2 based home therapy of metastatic renal cell carcinoma: risks and benefits in 215 consecutive single institution patients. *J Urol* 155: 19–25

92 Carswell EA, Old LJ, Kassel RL, Green S, Fiore N, Williamson B (1975) An endotoxin-induced serum factor that causes necrosis of tumors. *Proc Natl Acad Sci USA* 72: 3666–3670

93 Asher A, Mule JJ, Reichert CM, Shiloni E, Rosenberg SA (1987) Studies on the anti-tumor efficacy of systemically administered recombinant tumor necrosis factor against several murine tumors *in vivo*. *J Immunol* 138: 963–974

94 Eggermont AM, de Wilt JH, ten Hagen TL (2003) Current uses of isolated limb perfusion in the clinic and a model system for new strategies. *Lancet Oncol* 4: 429–437

95 De Wilt JH, ten Hagen TL, de Boeck G, van Tiel ST, de Bruijn EA, Eggermont AM (2000) Tumour necrosis factor alpha increases melphalan concentration in tumour tissue after isolated limb perfusion. *Br J Cancer* 82: 1000–1003

96 Presta LG, Chen H, O'Connor SJ, Chisholm V, Meng YG, Krummen L, Winkler M, Ferrara N (1997) Humanization of an anti-vascular endothelial growth factor monoclonal antibody for the therapy of solid tumors and other disorders. *Cancer Res* 57: 4593–4599

97 Keating MJ, Flinn I, Jain V, Binet JL, Hillmen P, Byrd J, Albitar M, Brettman L, Santabarbara P, Wacker B et al (2002) Therapeutic role of alemtuzumab (Campath-1H) in patients who have failed fludarabine: results of a large international study. *Blood* 99: 3554–3561

98 Dearden CE, Matutes E, Catovsky D (2002) Alemtuzumab in T-cell malignancies. *Med Oncol* 19 (Suppl): S27–S32

99 Hurwitz H, Fehrenbacher L, Novotny W, Cartwright T, Hainsworth J, Heim W, Berlin J, Baron A, Griffing S, Holmgren E et al (2004) Bevacizumab plus irinotecan, fluorouracil, and leucovorin for metastatic colorectal cancer. *N Engl J Med* 350: 2335–2342

100 Robert F, Ezekiel MP, Spencer SA, Meredith RF, Bonner JA, Khazaeli MB, Saleh MN, Carey D, LoBuglio AF, Wheeler RH et al (2001) Phase I study of anti-epidermal growth factor receptor antibody cetuximab in combination with radiation therapy in patients with advanced head and neck cancer. *J Clin Oncol* 19: 3234–3243

101 Riethmuller G, Holz E, Schlimok G, Schmiegel W, Raab R, Hoffken K, Gruber R, Funke I, Pichlmaier H, Hirche H et al (1998) Monoclonal ANTIBODY therapy for resected Dukes' C colorectal cancer: seven year outcome of a multicenter randomized trial. *J Clin Oncol* 16: 1788–1794

102 Sievers EL, Larson RA, Stadtmauer EA, Estey E, Lowenberg B, Dombret H, Karanes C, Theobald M, Bennett JM, Sherman ML et al [Mylotarg Study Group] (2001) Efficacy and safety of gemtuzumab ozogamicin in patients with CD33-positive acute myeloid leukemia in first relapse. *J Clin Oncol* 19: 3244–3254

103 McLaughlin P, Grillo-Lopez AJ, Link BK, Levy R, Czuczman MS, Williams ME, Heyman MR, Bence-Bruckler I, White CA, Cabanillas F et al (1998) Rituximab chimeric anti-CD20 monoclonal antibody therapy for relapsed indolent lymphoma: half of patients respond

to a four-dose treatment program. *J Clin Oncol* 16: 2825–2833
104 Lin TS, Lucas MS, Byrd JC (2003) Rituximab in B-cell chronic lymphocytic leukemia. *Semin Oncol* 30: 483–492
105 Cobleigh MA, Vogel CL, Tripathy D, Robert NJ, Scholl S, Fehrenbacher L, Wolter JM, Paton V, Shak S, Lieberman G et al (1999) Multinational study of the efficacy and safety of humanized anti-HER2 monoclonal antibody in women who have HER2-overexpressing metastatic breast cancer that has progressed after chemotherapy for metastatic disease. *J Clin Oncol* 17: 2639–2648
106 Uchida J, Hamaguchi Y, Oliver JA, Ravetch JV, Poe JC, Haas KM, Tedder TF (2004) The innate mononuclear phagocyte network depletes B lymphocytes through Fc receptor-dependent mechanisms during anti-CD20 antibody immunotherapy. *J Exp Med* 199: 1659–1669
107 Baselga J, Albanell J (2001) Mechanism of action of anti-HER2 monoclonal antibodies. *Ann Oncol* 12 (Suppl 1): S35–S41
108 Berzofsky JA, Terabe M, Oh S, Belyakov IM, Ahlers JD, Janik JE, Morris JC (2004) Progress on new vaccine strategies for the immunotherapy and prevention of cancer. *J Clin Invest* 113: 1515–1525
109 Wallack MK, Sivanandham M, Balch CM, Urist MM, Bland KI, Murray D, Robinson WA, Flaherty L, Richards JM, Bartolucci AA et al (1998) Surgical adjuvant active specific immunotherapy for patients with stage III melanoma: the final analysis of data from a phase III, randomized, double-blind, multicenter vaccinia melanoma oncolysate trial. *J Am Coll Surg* 187: 69–77
110 Romero P, Cerottini JC, Speiser DE (2004) Monitoring tumor antigen specific T-cell responses in cancer patients and phase I clinical trials of peptide-based vaccination. *Cancer Immunol Immunother* 53: 249–255
111 Scheibenbogen C, Letsch A, Schmittel A, Asemissen AM, Thiel E, Keilholz U (2003) Rational peptide-based tumour vaccine development and T cell monitoring. *Semin Cancer Biol* 13: 423–429
112 Disis ML, Cheever MA (1996) Oncogenic proteins as tumor antigens. *Curr Opin Immunol* 8: 637–642
113 Kim EM, Sivanandham M, Stavropoulos CI, Bartolucci AA, Wallack MK (2001) Overview analysis of adjuvant therapies for melanoma – a special reference to results from vaccinia melanoma oncolysate adjuvant therapy trials. *Surg Oncol* 10: 53–59
114 Nabel GJ, Nabel EG, Yang ZY, Fox BA, Plautz GE, Gao X, Huang L, Shu S, Gordon D, Chang AE (1993) Direct gene transfer with DNA-liposome complexes in melanoma: expression, biologic activity, and lack of toxicity in humans. *Proc Natl Acad Sci USA* 90: 11307–11311
115 Castelli C, Rivoltini L, Rini F, Belli F, Testori A, Maio M, Mazzaferro V, Coppa J, Srivastava PK, Parmiani G (2004) Heat shock proteins: biological functions and clinical application as personalized vaccines for human cancer. *Cancer Immunol Immunother* 53: 227–233
116 Ward S, Casey D, Labarthe MC, Whelan M, Dalgleish A, Pandha H, Todryk S (2002) Immunotherapeutic potential of whole tumour cells. *Cancer Immunol Immunother* 51: 351–357
117 Parkhurst MR, Salgaller ML, Southwood S, Robbins PF, Sette A, Rosenberg SA, Kawakami Y (1996) Improved induction of melanoma-reactive CTL with peptides from the melanoma antigen gp100 modified at HLA-A*0201-binding residues. *J Immunol* 157: 2539–2548
118 Schuler-Thurner B, Schultz ES, Berger TG, Weinlich G, Ebner S, Woerl P, Bender A, Feuerstein B, Fritsch PO, Romani N et al (2002) Rapid induction of tumor-specific type 1 T helper cells in metastatic melanoma patients by vaccination with mature, cryopreserved, peptide-loaded monocyte-derived dendritic cells. *J Exp Med* 195: 1279–1288
119 Figdor CG, de Vries IJ, Lesterhuis WJ, Melief CJ (2004) Dendritic cell immunotherapy: mapping the way. *Nat Med* 10: 475–480
120 Dranoff G (2003) GM-CSF-secreting melanoma vaccines. *Oncogene* 22: 3188–3192
121 Soiffer R, Lynch T, Mihm M, Jung K, Rhuda C, Schmollinger JC, Hodi FS, Liebster L, Lam P, Mentzer S et al (1998) Vaccination with irradiated autologous melanoma cells engineered to secrete human granulocyte-macrophage colony-stimulating factor generates potent antitumor immunity in patients with metastatic melanoma. *Proc Natl Acad Sci USA* 95: 13141–13146
122 Antonia SJ, Seigne J, Diaz J, Muro-Cacho C, Extermann M, Farmelo MJ, Friberg M, Alsarraj M, Mahany JJ, Pow-Sang J et al (2002) Phase I trial of a B7-1 (CD80) gene modified autologous tumor cell vaccine in combination with systemic interleukin-2 in patients with metastatic renal cell carcinoma. *J Urol* 167: 1995–2000
123 Grimm EA, Mazumder A, Zhang HZ, Rosenberg SA (1982) Lymphokine-activated killer cell phenomenon.

Lysis of natural killer-resistant fresh solid tumor cells by interleukin 2 activated autologous human peripheral blood lymphocytes. *J Exp Med* 155: 1823–1841

124 Kruit WH, Goey SH, Lamers CH, Gratama JW, Visser B, Schmitz PI, Eggermont AM, Bolhuis RL, Stoter G (1997) High-dose regimen of interleukin-2 and interferon-alpha in combination with lymphokine-activated killer cells in patients with metastatic renal cell cancer. *J Immunother* 20: 312–320

125 Rosenberg SA, Lotze MT, Yang JC, Topalian SL, Chang AE, Schwartzentruber DJ, Aebersold P, Leitman S, Linehan WM, Seipp CA et al (1993) Prospective randomized trial of high-dose interleukin-2 alone or in conjunction with lymphokine-activated killer cells for the treatment of patients with advanced cancer. *J Natl Cancer Inst* 85: 622-632

126 Law TM, Motzer RJ, Mazumdar M, Sell KW, Walther PJ, O'Connell M, Khan A, Vlamis V, Vogelzang NJ, Bajorin DF (1995) Phase III randomized trial of interleukin-2 with or without lymphokine-activated killer cells in the treatment of patients with advanced renal cell carcinoma. *Cancer* 76: 824–832

127 Walter EA, Greenberg PD, Gilbert MJ, Finch RJ, Watanabe KS, Thomas ED, Riddell SR (1995) Reconstitution of cellular immunity against cytomegalovirus in recipients of allogeneic bone marrow by transfer of T-cell clones from the donor. *N Engl J Med* 333: 1038–1044

128 Yee C, Thompson JA, Byrd D, Riddell SR, Roche P, Celis E, Greenberg PD (2002) Adoptive T cell therapy using antigen-specific $CD8^+$ T cell clones for the treatment of patients with metastatic melanoma: *in vivo* persistence, migration, and antitumor effect of transferred T cells. *Proc Natl Acad Sci USA* 99: 16168–16173

129 Rooney CM, Smith CA, Ng CY, Loftin SK, Sixbey JW, Gan Y, Srivastava DK, Bowman LC, Krance RA, Brenner MK et al (1998) Infusion of cytotoxic T cells for the prevention and treatment of Epstein-Barr virus-induced lymphoma in allogeneic transplant recipients. *Blood* 92: 1549–1555

130 Canevari S, Stoter G, Arienti F, Bolis G, Colnaghi MI, Di Re EM, Eggermont AM, Goey SH, Gratama JW, Lamers CH et al (1995) Regression of advanced ovarian carcinoma by intraperitoneal treatment with autologous T lymphocytes retargeted by a bispecific monoclonal antibody. *J Natl Cancer Inst* 87: 1463–1469

131 Lamers CH, Willemsen RA, Luider BA, Debets R, Bolhuis RL (2002) Protocol for gene transduction and expansion of human T lymphocytes for clinical immuno-gene therapy of cancer. *Cancer Gene Ther* 9: 613–623

132 Lamers CHJ, Sleijfer S, Willemsen RA, Debets R, Kruit WHJ, Gratama JW, Stoter G (2004) Adoptive immuno-gene therapy of cancer with single chain antibody [scFv(Ig)] gene modified T lymphocytes. *J Biol Reg Homeost Agents* 18: 134–140

133 Slavin S, Morecki S, Weiss L, Shapira MY, Resnick I, Or R (2004) Nonmyeloablative stem cell transplantation: reduced-intensity conditioning for cancer immunotherapy – from bench to patient bedside. *Semin Oncol* 31: 4–21

134 Childs R, Chernoff A, Contentin N, Bahceci E, Schrump D, Leitman S, Read EJ, Tisdale J, Dunbar C, Linehan WM et al (2000) Regression of metastatic renal cell carcinoma after nonmyeloablative allogeneic peripheral-blood stem-cell transplantation. *N Engl J Med* 343: 750–758

135 Hentschke P, Barkholt L, Uzunel M, Mattsson J, Wersall P, Pisa P, Martola J, Albiin N, Wernerson A, Soderberg M et al (2003) Low-intensity conditioning and hematopoietic stem cell transplantation in patients with renal and colon carcinoma. *Bone Marrow Transplant* 31: 253–261

136 Childs RW, Barrett J (2004) Nonmyeloablative allogeneic immunotherapy for solid tumors. *Annu Rev Med* 55: 459–475

137 Hacein-Bey-Abina S, Von Kalle C, Schmidt M, McCormack MP, Wulffraat N, Leboulch P, Lim A, Osborne CS, Pawliuk R, Morillon E et al (2003) LMO2-associated clonal T cell proliferation in two patients after gene therapy for SCID-X1. *Science* 302: 415–419

Neuroimmunoendocrinology

Douglas A. Weigent and J. Edwin Blalock

Historical background

Psychoneuroimmunology refers to the study of the interactions among behavioral, neural, neuroendocrine, and immunological processes of adaptation. Although relationships between the brain and the immune system had been suggested for many years, research during the last 25 years has provided mechanisms for how these systems may interact. The current interest in psychoneuroimmunology is sustained by the now widely held belief that it represents a bidirectional system. The nervous system not only influences immune function but the immune system modifies the nervous system. One of the very early observations related to the effects of stress on immune function, suggesting a psychosomatic-immune axis [1]. Avoidance conditioning was shown to increase virus infection and psychological factors could alter the onset and course of autoimmune disease. Additional results identifying receptors on cells of the immune system, evaluating brain lesions and identifying nerve fibers in compartments of LYMPHOID ORGANS, along with the biological relevance of innervation on immune function, supported the idea of a very dynamic interaction between the immune and nervous system [2]. Over the same time frame, a number of reports appeared, indicating the efficacious effects on the immune system of hormone therapy in hormone-deficient animals [3]. Thus it became apparent that the neuroendocrine system could interact with and modulate the immune system by the release of hormones.

The reciprocal situation in which the stimulation of the immune system altered central nervous system function established the bidirectional nature of the communication between these two systems. It was shown that the immune system as a consequence of antigenic challenge altered the firing rate of hypothalamic neurons [4]. Supernatant fluids from activated LEUKOCYTES could mimic this phenomenon, and it is now clear that a wide range of lymphocyte products influence the synthesis and secretion or release of neuroendocrine hormones and neurotransmitters [5]. Our studies reviewed, in part, below, initially showed that cells of the immune system could be a source of pituitary hormones and that immune-derived CYTOKINES could function as hormones and hypothalamic releasing factors [5]. Our discovery that cells of the immune system produce PROOPIOMELANOCORTIN (POMC) peptides grew out of the original observation that INTERFERON-α (IFN-α) had analgesic effects. We now know that there are distinct domains in the IFN-α molecule that mediate immune and analgesic effects and that the opioid-like analgesic effects of IFN-α are mediated via the μ-opioid receptor [6]. Further studies suggested a mechanism by which the body's two principal recognition organs, the brain and the immune system, may influence each other: that is, they speak the same biochemical language. Collectively, these relationships provide the foundation for behaviorally induced alterations in immune function and for immunologically based changes in behavior. Accumulating *in vitro* and *in vivo* studies indicate that shared ligands and receptors are used as a common chemical language within and between the immune and neuroendocrine systems. Specifically, physical and psychological stimuli stimulate the release of neurotransmitters, hormones and CYTOKINES which bind to receptors on cells of the immune system and alter their function (Fig. 1). In addition, the immune system converts recognition of noncognitive stimuli such as viruses and bacteria into information in the form of CYTOKINES, peptide hormones, and neurotransmitters, which act on

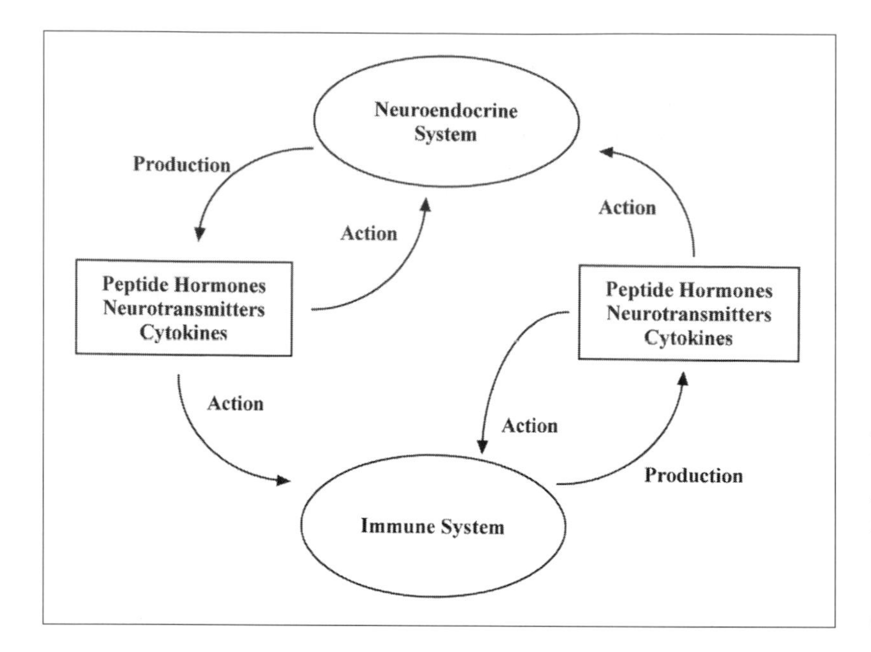

FIGURE 1
A molecular communication circuit within and between the immune and neuroendocrine systems involving shared ligands and their receptors. (Reprinted with permission from Weigent and Blalock [5] by The Society for Leukocyte Biology).

receptors in the neuroendocrine system to alter its function. Soluble products that appear to transmit information from the immune compartment to the central nervous system (CNS) include thymosins, lymphokines (INTERLEUKIN-1 [IL-1], -2, -6, TUMOR NECROSIS FACTOR (TNF-α) and IFN), corticotropin (ACTH), and opioid peptides [7]. The predominant effects of these CYTOKINES is to stimulate the hypothalamic pituitary axis (HPA) and suppress the hypothalamic pituitary thyroid (HPT), hypothalamic pituitary gonadal (HPG) axis, and GROWTH HORMONE (GH) release. The findings suggest an immunoregulatory role for the brain and a sensory function for the immune system [8]. This chapter will briefly describe the basic and clinical evidence for the role neuroendocrine hormones play in communication between the immune and neuroendocrine system.

Regulation of the immune system by neuroendocrine hormones

A large amount of evidence exists to support both the presence of receptors for neuroendocrine hormones on cells of the immune system as well as the ability of these hormones to modulate specific functions of the various immune cell types (Table 1; [5]). Several examples are discussed below and the reader is directed to review articles for a further discussion of the topic.

Actions of corticotropin (ACTH) and endorphins

The receptors for ACTH and ENDORPHINS have been identified on cells of the immune system, as has the ability of these hormones to modulate many aspects of immune reactivity. The findings show that B LYMPHOCYTES contain three times the number of ACTH binding sites compared with T cells and that treatment with a mitogen increases the number of high-affinity sites 2–3-fold on both cell types. Monospecific antiserum to the ACTH receptor on Y-1 adrenal cells recognizes the ACTH receptor on leukocytes. The binding of ACTH initiates a signal transduction pathway that involves both cAMP and mobilization of Ca^{2+}. More recent analysis of the effects of ACTH by patch-clamp methods suggests that this hormone can modulate macrophage functions through the activation of Ca^{2+}-dependent K^+ channels [9]. ACTH

Table 1. Modulation of Immune Responses by Neuroendocrine Hormones

Hormone	Cell type or tissue with hormone receptor	Modulating effects
Corticotropin	Rat spleen T and B cells	Antibody synthesis IFN-γ production B lymphocyte growth
Endorphins	Spleen	Antibody synthesis Mitogenesis Natural killer cell activity
MSH	Lymphocytes, monocytes	Downregulation of proinflammatory cytokines, adhesion molecule expression and NO production
Thyrotropin	Neutrophils, monocytes, B cells	Increased antibody synthesis Comitogenic with ConA
GH	PBL, spleen, thymus	Cytotoxic T cells Mitogenesis
LH and FSH	–	Proliferation Cytokine production
PRL	T and B cells	Comitogenic with ConA Induces IL-2 receptors
CRH	PBL	IL-1 production Enhanced natural killer cell activity Immunosuppressive
TRH	T cell lines	Increased antibody synthesis
GHRH	PBL and spleen	Stimulates proliferation
SOM	PBL	Inhibits natural killer cell activity Inhibits chemotactic response Inhibits proliferation Reduces IFN-γ production

has been shown to suppress MAJOR HISTOCOMPATIBILITY COMPLEX (MHC) class II expression, stimulate natural killer cell activity, suppress INTERFERON-γ (IFN-γ) production, modulate IL-2 production, and function as a late-acting B cell growth factor. The production of opioid peptides in immune cells [10], and lymphocyte receptors for the opioid peptides have also been studied and found to share many of the features, including size, sequence, immunogenicity and intracellular signaling, as those described on neuronal tissue. Many aspects of immunity are modulated by the opiate peptides including: (1) enhancement of the natural cytotoxicity of LYMPHOCYTES and MACROPHAGES toward tumor cells, (2) enhancement or inhibition of T cell mitogenesis, (3) enhancement of T cell rosetting, (4) stimulation of human peripheral blood mononuclear cells, and (5) inhibition of MHC class II antigen expression. The mechanisms by which opiate peptides influence such diverse activities remain unclear. However, it is known that β-ENDORPHIN alters immune cell calcium ion flux while shutting down potassium channel function [11]. Thus, opiates may bind both the classical opioid receptors and K channels to modulate immune cell activity [12].

The anti-inflammatory influences of α-melanocyte stimulating hormone (α-MSH) and other melanocortins are primarily exerted through inhibition of inflammatory mediator production and cell migration [13]. These effects occur through binding of melanocortins to melanocortin receptors on cells of the immune system and via descending anti-inflammatory neural pathways induced by stimulation of α-MSH receptors within the brain [14]. Almost all of the cells responsive to melanocortins, including MACROPHAGES, LYMPHOCYTES, neutrophils, DENDRITIC CELLS, astrocytes and microglia, express the melanocortin type-1 receptors. The *in vitro* and *in vivo* inhibitory effects of α-MSH influence adhesion, production of CYTOKINES and other mediators of inflammation, including IL-1, IL-6, IL-10, TNF-α, chemokines, and nitric oxide [13]. Various skin cells are also the source and target for the anti-inflammatory effects of MSH [15]. The broad effects of α-MSH, on inflammatory mediator production is thought to occur through the participation of G-proteins, the JAK kinase/signal transducer activator transcription (JAK/STAT) pathway, and inhibition of the activation of the nuclear factor NF-κB.

Actions of GH and prolactin (PRL)

It has been clearly shown that cells of the immune system also contain receptors for GH and PRL and that these hormones are potent modulators of the immune response [16]. The PRL and GH receptors have been shown to be members of the superfamily of cytokine receptors involved in the growth and differentiation of hematopoietic cells. A systematic survey of PRL receptor expression by flow cytometry showed that PRL receptors are universally expressed in normal hematopoietic tissues with some differences in density, which could be increased by concanavalin (Con)A treatment *in vitro* and exercise *in vivo*. GH receptors from a number of species have been sequenced and the co-crystallization of human GH with the GH receptor has been achieved. GH binding and cellular processing of the GH receptor have been studied in a cell line of immune origin. In the IM-9 cell line, it has been shown that GH stimulates the proliferation and that the GH receptor can be down-regulated by phorbol esters. Several lines of evidence indicate that activation of the GH receptor increases tyrosine kinase activity and that the GH receptor is associated with a tyrosine kinase in several cell types, including the IM-9 cell. A role for GH in immunoregulation has been demonstrated *in vitro* for numerous immune functions [16], including stimulation of deoxyribonucleic acid (DNA) and ribonucleic acid (RNA) synthesis in the spleen and thymus. GH also affects HEMATOPOIESIS by stimulating neutrophil differentiation, augmenting erythropoiesis, increasing proliferation of BONE MARROW cells, and influences thymic development. GH affects the functional activity of cytolytic cells, including T LYMPHOCYTES and natural killer (NK) cells. GH was necessary for T LYMPHOCYTES to develop cytolytic activity against an allogeneic stimulus in serum-free medium. GH has also been shown to stimulate the production of superoxide anion formation from MACROPHAGES. It is not clear whether GH directly influences intrathymic or extrathymic development or acts indirectly by augmenting the synthesis of thymulin or INSULIN-LIKE GROWTH FACTOR-1 (IGF-1). These observations suggest that GH may stimulate local production of IGF-I, which acts to promote tissue growth and action in a paracrine fashion. The potential effect of GH in tumorigenesis, particularly in acute leukemia, is controversial. An active area of research over the past several years has been the immune-enhancing effects of both synthetic (hexarelin) and natural (GHrelin) GH secretagogues. Their primary effect on cells of the immune system appears to be promoting cell division. A possible role for lymphocyte GH, through an increase in synthesis and secretion, has been suggested in the mechanism of cell proliferation. Likewise, PRL can have modulating effects on the immune system [16]. Data show that suppression of PRL secretion in mice with bromocriptine increases the lethality of a Listeria challenge and abrogates T cell-dependent activation of MACROPHAGES. ANTIBODIES to the PRL receptor have been shown to abolish PRL-induced proliferation of Nb2 cells. More recent studies suggest that PRL may promote survival of certain lymphocyte subsets, modulate the naïve B cell repertoire, and promote antigen-presenting functions [17]. The lymphocyte source of PRL may explain the association of hyperprolactinemia with autoimmune diseases.

Actions of hypothalamic releasing hormones

In addition to pituitary hormones, hypothalamic releasing hormone receptors and their effects have been documented on cells of the immune system. A number of similarities have been identified between the pituitary and spleen binding of CORTICOTROPIN RELEASING HORMONE (CRH) including affinity and apparent subunit molecular weight. The effects of ACTH and ENDORPHINS discussed earlier may be initiated in the immune system via the production of these hormones by cells of the immune system in response to CRH (see below). CRH inhibits lymphocyte proliferation and NK cell activity. The GROWTH HORMONE RELEASING HORMONE (GHRH) receptor has also been identified on cells of the immune system. The GHRH receptor binding sites are saturable and are found on both thymocytes and splenic LYMPHOCYTES. After GHRH binding to its receptor, there is a rapid increase in intracellular Ca^{2+}, which is associated with the stimulation of lymphocyte proliferation. Other *in vitro* findings suggest GHRH may inhibit NK cell activity and chemotaxis and increase IFN-γ secretion. In addition, LEUKOCYTES have been shown to respond to THYROTROPIN RELEASING HORMONE (TRH) treatment by producing THYROTROPIN (TSH) mRNA and protein. Recent work has shown the presence of two receptor types for TRH on T cells. One of these sites satisfies the criteria for a classical TRH receptor and is involved in the release of IFN-γ from T cells. TRH at very low concentrations enhances the *in vitro* plaque-forming cell response via production of TSH. In this instance, T cells were shown to produce TSH while other studies suggest DENDRITIC CELLS and MONOCYTES may also produce biologically active TSH [18]. Interestingly, T LYMPHOCYTES cultured with T3 and T4, but not TSH nor TRH, showed enhanced APOPTOSIS with reduced expression of Bcl-2 protein. The TRH peptide precursor has been reported in the spleen, but the cell type producing this peptide and its mechanism of secretion are yet to be understood. The existence of distinct subsets of SOMATOSTATIN (SOM) receptors on the Jurkat line of human leukemic T cells and U266 IgG-producing human myeloma cells has also been described. The authors speculate that two subsets of receptors may account for the biphasic concentration-dependent nature of the effects of SOM in some systems. Although GH and PRL have immunoenhancing capabilities, SOM has potent inhibitory effects on immune responses. SOM has been shown to significantly inhibit Molt-4 lymphoblast proliferation and phytohemagglutinin (PHA) stimulation of human T LYMPHOCYTES and nanomolar concentrations are able to inhibit the proliferation of both spleen-derived and Peyer's patch-derived LYMPHOCYTES. Other immune responses, such as superantigen-stimulated IFN-γ secretion, endotoxin-induced leukocytosin, and colony-stimulating activity release, are also inhibited by SOM.

Neuroendocrine hormone release by immune system cells

There is now overwhelming evidence that cells of the immune system also produce neuroendocrine hormones. This was first established for ACTH and subsequently for TSH, GH, PRL, luteinizing hormone (LH), follicle-stimulating hormone (FSH) and the hypothalamic hormones SOM, CRH, GHRH, and luteinizing hormone releasing hormone (LHRH) [5]. The evidence strongly supports the notion that neuroendocrine peptides and neurotransmitters, endogenous to the immune system, are used for both intra-immune system regulation, as well as for bidirectional communication between the immune and neuroendocrine systems. The studies show that although the structure of these peptides are identical to those identified in the neuroendocrine system, both similarities and differences exist in the abundance of particular transcripts and in the mechanism of synthesis to the patterns previously described in the neuroendocrine system.

At least two possibilities exist concerning the potential function of these peptide hormones produced by the immune system. First they act on their classic neuroendocrine target tissues. Second, they may serve as endogenous regulators of the immune system. With regard to the latter possibility, it is clear that neuroendocrine peptide hormones can directly modulate immune functions. These studies, however, do not specifically address endogenous as opposed to exogenous regulation by neuroendocrine peptides. A number of investigators have now been able to demonstrate that such regulation is endogenous to the immune system. Specifically, TSH is a pituitary hormone that can be produced by LYMPHOCYTES in response to TRH and, like TSH, TRH enhanced the *in vitro* ANTIBODY response [19]. This enhancement was not observed with GHRH, arginine vasopressin (AVP), or LHRH, and was blocked by ANTIBODIES to the β subunit of TSH. Thus it appears that TRH specifically enhances the *in vitro* ANTIBODY response via production of TSH. This was the first demonstration that a neuroendocrine hormone (TSH) can function as an endogenous regulator within the immune system. A large number of human hematopoietic cell lines and tumors synthesize and release PRL [20]. The evidence suggests a low constitutive level of PRL expression inducible by IL-2 and inhibited by dexamethasone. In T cells, PRL is translocated to the nucleus in an IL-2 dependent P-13 kinase pathway inhibited by rapamycin. Immune cell-derived PRL most likely plays a role in hematopoietic cell differentiation and proliferation. In another study, ANTIBODY to PRL was shown to inhibit mitogenesis through neutralization of the lymphocyte-associated PRL. Furthermore, coordinate gene expression of LHRH and the LHRH-receptor has been shown in the Nb2 T-cell line after PROLACTIN stimulation [21].

Two different approaches have provided convincing evidence that endogenous neuroendocrine peptides have autocrine or paracrine immunoregulatory functions. First, an opiate antagonist was shown to indirectly block CRH enhancement of NK cell activity by inhibiting the action of immunocyte-derived opioid peptides. Second, we have used an antisense oligodeoxynucleotide (ODN) to the translation start site of GH mRNA to specifically inhibit leukocyte production of GH. The ensuing lack of GH resulted in a marked diminution in basal rates of DNA synthesis in such antisense ODN-treated LEUKOCYTES which could be overcome by exogenously added GH [22]. Another group examining SOM found that antisense oligodeoxynucleotides to SOM dramatically increased lymphocyte proliferation in culture [23]. Additionally, LHRH agonists were found to diminish NK cell activity, stimulate T-cell proliferation, and increase IL-2-receptor expression, suggesting an important role for LHRH in the regulation of the immune response [24].

Another major function of GH produced by cells of the immune system is the induction of the synthesis of IGF-1, which, in turn, may inhibit the synthesis of both lymphocyte GH mRNA and protein. Our previous studies also show that both exogenous and endogenous GHRH can stimulate the synthesis of lymphocyte GH. Taken together, these findings support the existence of a complete regulatory loop within cells of the immune system, and they provide a molecular basis whereby GHRH, GH, IGF-1 and their binding proteins may be intimately involved in regulating each other's synthesis. Furthermore, data obtained by immunofluorescence techniques suggest that the cells producing GH also produce IGF-1, which suggests that an intracrine regulatory circuit may be important in the synthesis of these hormones by cells of the immune system [25]. Our most recent findings in a T cell-line show that endogenous GH promotes nitric oxide production, upregulation of IGF-1 receptors and Bcl2 protein along with an inhibition of superoxide formation, clearly establishing a role for lymphocyte GH in APOPTOSIS [26]. A monocyte-derived cytokine, IL-12, has recently been shown to stimulate the synthesis of lymphocyte GH mRNA synthesis and the T-helper cell-1 (Th1) cytokine, IFN-γ [27]. The stress-activated hormones, cortisol and CATECHOLAMINES, decreased lymphocyte GH and the TH1 RESPONSE [27]. Interestingly, LYMPHOCYTES have also been suggested to be important sites of synthesis and action of acetylcholine and CATECHOLAMINES since they contain both the enzymes necessary for biosynthesis of epinephrine and acetylcholine as well as the relevant receptor system [28, 29]. Recent work has identified a neural mechanism involving the vagus nerve and release of acetylcholine that inhibits macrophage activation termed the "cholinergic anti-inflammatory pathway" [30]. The sensory afferent vagus pathway may be activated by low doses of endotoxin, IL1, or products from damaged tissues. The signal is relayed to the brain where activation of the efferent vagus nerve releases acetylcholine. Acetylcholine acts to inhibit macrophage release of the proinflammatory CYTOKINES TNF, IL-1 and IL-18, but not the anti-inflammatory cytokine IL-10. Thus, cholinergic neuron participation in the inhibition of acute inflammation constitutes a "hardwire" neural mechanism of modulation of the immune response. Finally, calcitonin gene-related peptide (CGRP) has also been shown to be produced and secreted by human LYMPHOCYTES and may be involved in inhibition of T-lymphocyte proliferation [31]. In another more recent study, substance P, the potent mediator of neuroimmune regulation, was shown to be upregulated in LYMPHOCYTES by human immunodeficiency virus (HIV) infection, implying it may be involved in immunopathogenesis of HIV infection and acquired immune deficiency syndrome (AIDS) [32]. Neuropeptides, by direct interaction with T cells, induce cytokine secretion and break the commitment to a distinct T helper phenotype [33]. Thus, neurons are not the exclusive source of neurotransmitters and, therefore, provide another instance of shared molecular signals and their receptors between the nervous and immune system.

Functions of leukocyte-derived peptide hormones *in vivo*

Although much work needs to be done to fully establish the clinical relevance of leukocyte-derived neuroendocrine hormones, certain experimental models and clinical observations seem to support the view that leukocyte-derived hormones can also act on their classic neuroendocrine targets.

Actions of ACTH *in vivo*

The finding that cells of the immune system are a source of secreted ACTH suggested that stimuli which elicit the leukocyte-derived hormone should not require a pituitary gland for an ACTH-mediated increase in corticosteroid. This seemed to be the case since Newcastle Disease Virus (NDV) infection of hypophysectomized mice caused a time-dependent increase in corticosterone production which was inhibitable by dexamethasone. Unless such mice were pretreated with dexamethasone, their spleens were positive for ACTH by immunofluorescence [34]. A more recent study has suggested that B LYMPHOCYTES can be responsible for extrapituitary ACTH production. In this report, hypophysectomized chickens were shown to produce ACTH and corticosterone in response to *Brucella abortus*, and the corticosterone response was ablated if B LYMPHOCYTES were deleted by bursectomy prior to hypophysectomy [35]. In children who were pituitary ACTH-deficient and pyrogen tested, an increase in the percentage of ACTH-positive peripheral blood LEUKOCYTES (PBL) was observed. Both the response in hypophysectomized mice and hypopituitary children peaked at approximately 6–8 h after administration of virus and typhoid vaccine, respectively. Such studies have been furthered by a report that CRH administration to pituitary ACTH-deficient individuals results in both an ACTH and cortisol response. In the bovine, both transport stress and pregnancy stimulate ACTH secretion from peripheral LYMPHOCYTES, implying a role for lymphocyte ACTH in stress and probably recognition. Although more work needs to be done, it seems highly likely that locally produced MSH, via cells of the immune system, may affect mast cell function and modulate immediate-type airway and inflammation [13]. The inhibitory effect of MSH on allergic airway inflammation appears to be mediated via enhanced IL-10 production since no effect was observed in IL-10 knockout animals [13]. Finally, LYMPHOCYTES from patients with active systemic lupus erythematosus (SLE) produce increased amounts of PRL [36]. The extrapituitary PRL may initiate or maintain an aberrant immune process in SLE in IL-2 producing LYMPHOCYTES or disturb normal communication between the neuroendocrine system and the immune system in SLE.

Actions of opiate peptides *in vivo*

Gram-negative bacterial infections and endotoxin shock represent another situation in which leukocyte hormones act on the neuroendocrine system. For example, ENDORPHINS have been implicated in the pathophysiology associated with these maladies since the opiate antagonist, naloxone, was shown to improve survival rates and inhibited a number of cardiopulmonary changes associated with these conditions [37]. Furthermore, two separate pools of ENDORPHINS have been observed following bacterial lipopolysaccaride (LPS) administration, and it was suggested that one pool might originate in the immune system. Consistent with this idea is the observation that LYMPHOCYTE depletion, like naloxone treatment, blocked a number of endotoxin-induced cardiopulmonary changes. Our interpretation of these data is that LYMPHOCYTE depletion removed the source of the ENDORPHINS while naloxone blocks their effector function. In a different approach, LPS-resistant inbred mice which have essentially no pathophysiological response to LPS were shown to have a defect in leukocyte processing of POMC to ENDORPHINS. If leukocyte-derived ENDORPHINS were administered to the LPS-resistant mice, they showed much of the pathophysiology associated with LPS administration to sensitive mice [38]. A role for the κ-opioid receptor in immunity has been studied in κ-opioid receptor knockout mice [39]. In these animals, there was an increase in splenocyte number and the humoral response, whereas no changes were observed in μ-opioid receptor and δ-opioid receptor knockout animals. The data suggest that activation of κ-opioid receptors may exert a tonic inhibition of the ANTIBODY response. A similar finding was observed in β-ENDORPHIN-deficient mice [40]. The idea that non-immune tissues may harbor neuroendocrine hormone receptors, produce CYTOKINES, and indirectly influence immunity has also recently been suggested. The tissue-specific expression of INTERLEUKIN-18 (IL-18) in the rat adrenal cortex following ACTH treatment, but not in the spleen, shows that adrenal cells may be the source of IL-18 during stress and not cells of the immune system as expected [41]. Numerous *in vivo* experiments and studies in human disease have shown significant effects and/or changes of α-MSH [14]. Thus, chemokine, TNF-α and adhesion molecule production are reduced in animal models by systemic treatment with α-MSH, while in humans α-MSH plays an important role in fever and the acute-phase response. In the case of lymphocyte GH and IGF, there are now several reports showing the age-related differential expression of GH and IGF molecules and their receptors between LYMPHOCYTES from healthy adults, children, and newborns [42]. In addition, a role for lymphocyte GH *in vivo* has been suggested in HEMATOPOIESIS during fetal development and in the appearance of childhood acute lymphoblastic leukemia.

Antinociception and CRH

Another exciting new development in the opioid field has come with the demonstration that activation of endogenous opioids in rats by a cold water swim results in a local antinociceptive effect in inflamed peripheral tissue. This local antinociception in the inflamed tissue apparently results from production by immune cells of endogenous opioids which interact with opioid receptors on peripheral sensory nerves [43]. Another study strongly suggests that the immune system plays an essential role in pain control [43]. The findings identify locally expressed CRH as the main agent to induce opioid release within inflamed tissue. The opioid receptor-specific anti-nociception in inflamed paws of rats could be blocked by intraplantar α-helical CRH or antiserum to CRH or CRH-antisense oligodeoxynucleotide. This latter treatment reduced the amount of CRH extracted from inflamed paws, as well as the number of CRH-immunostained cells [44]. An upregulation of μ-opioid receptors on sensory nerves and the accumulation of activated/memory T cells containing β-ENDORPHIN in inflamed tissue are consistent with the production of analgesia [45]. A recent work suggests that the molecular pathway mediating the proinflammatory effects of peripheral CRH is through the induction of nuclear factor κB (NF-κB) DNA binding activity [46]. It should also be noted that locally produced neuropeptide Y and its receptor are also involved in the pro-inflammatory responses of paw edema in the rat. In another model, a study employing a GHRH antagonist and GHRH receptor-deficient mice, the data suggest that GHRH plays a crucial role in the development of experimental autoimmune encephalomyelitis (EAE) [47]. Taken together, these observations offer new understanding and suggest new approaches for ameliorating pain and disease in normal, diseased and immunocompromised patients [48].

Summary and conclusions

The activity of cells of the immune system can be strongly influenced by neuroendocrine-derived and leukocyte-derived neuroendocrine peptides. In many respects, the production and regulation of these peptides by LEUKOCYTES is remarkably similar to that observed in neuroendocrine cells. There are, however, a number of noteworthy differences which suggest that rules which apply to pituitary hormone production are not necessarily applicable to the immune system. Once produced, these peptide hormones seem to function in at least two capacities. They are endogenous regulators of the immune system as well as conveyors of information from the immune to the neuroendocrine system. Plasma hormone concentrations do not have to reach the levels required when the pituitary gland is the source because immune cells are not fixed but are mobile and can locally deposit the hormone at the target site and influence the immune response. It is our bias that the transmission of these molecules along with CYTOKINES to the neuroendocrine system represents a sensory function for the immune system wherein LEUKOCYTES recognize stimuli that are not recognizable by the central and peripheral nervous systems [8]. These stimuli have been termed non-cognitive and include bacteria, tumors, viruses, and antigens. The recognition of such non-cognitive stimuli by immunocytes is then converted into information, in the form of peptide hormones and neurotransmitters and CYTOKINES, which is conveyed to the neuroendocrine system and a physiological change occurs.

Acknowledgements

We thank faculty colleagues, students and postdoctoral fellows who were pivotal to the studies reported herein. We are also grateful to Diane Weigent for expert editorial assistance. These studies were supported by NIH grants HL68806 to JEB and AI41651 to DAW.

Selected readings

Weigent DA, Blalock JE (1997) Production of peptide hormones and neurotransmitters by the immune system. *Chem Immunol* 69: 1–30

Torres B, Johnson HM (1997) Neuroendocrine peptide hormone regulation of immunity. *Chem Immunol* 69: 155–184

Recommended websites

Oppenheim JJ, Feldman M (ed.) (2000) Cytokine Reference. Online Database: *http://www.academicpress.com/companions/0122526708* (Accessed November 2004)

Tanase S, Nomiyama H (1995) Cytokine Family Database Home Page: *http://cytokine.medic.kumamoto-u.ac.jp* (Accessed November 2004)

References

1 Ader R (1996) Historical perspectives on psychoneuroimmunology. In: H Friedman, TW Klein, AL Friedman (eds): *Psychoneuroimmunology, stress, and infection*. CRC Press, Boca Raton, 1–24

2 Besedovsky HO, delRey AE, Sorkin E, DaPrada M, Keller HA (1979) Immunoregulation mediated by the sympathetic nervous system. *Cell Immunol* 48: 346

3 Pierpaoli W, Baroni C, Fabris N, Sorkin E (1969) Hormones and immunological capacity. II. Reconstitution of antibody production in hormonally deficient mice by somatotropic hormone, thyrotropic hormone and thyroxin. *J Immunol* 16: 217–230

4 Besedovsky H, Sorkin E, Keller M, Miller J (1977) Hypothalamic changes during the immune response. *Eur J Immunol* 7: 323–325

5 Weigent DA, Blalock JE (1995) Associations between the neuroendocrine and immune systems. *J Leukocyte Biol* 58: 137–150

6 Wang Y, Song L, Chen Y, Jiang C (2002) The analgesic domain of interferon-a2b contains an essential proline residue. Neuroimmunomodulation 10: 5–8

7 Haddad J, Saade N, Safieh-Garabedian B (2002) Cytokines and neuro-immune-endocrine interactions: a role for the hypothalamic-pituitary-adrenal revolving axis. *J Neuroimmunol* 133: 1–19

8 Blalock JE (1984) The immune system as a sensory organ. *J Immunol* 132: 1067–1070

9 Fukushima T, Ichinose M, Shingai R, Sawada M (2001) Adrenocorticotropic hormone activates an outward current in cultured mouse peritoneal macropahges. *J Neuroimmunol* 113: 231–235

10 Smith EM (2003) Opioid peptides in immune cells. In: H Machelska, C Stein (eds): *Immune Mechanisms of*

Pain and Analgesia. Kluwer Academic/Plenum Publishers, New York: 51–68

11 Carr DJ, Bubien JK, Woods WT, Blalock JE (1988) Opioid receptors on murine splenocytes. Possible coupling to K^+ channels. *Ann NY Acad Sci* 540: 694–697

12 Miller D, Mazorow D, Hough C (1997) The K channel blocker, tetraethyl-ammonium, displaces beta-endorphin and naloxone from T-cell binding sites. *J Neuroimmunol* 78: 8–18.

13 Luger TA, Scholzen TE, Brzoska T, Bohm M (2003) New insights into the functions of a-MSH and related peptides in the immune system. *Ann NY Acad Sci* 994: 133–140

14 Lipton J, Catania A (2003) Anti-inflammatory actions of the neuroimmunomodulator a-MSH. *Immunol Today* 18: 140–145

15 Slominiski A, Wortsman J (2000) Neuroendocrinology of the skin. *Endocrine Rev* 21: 457–487

16 Gala RR (1991) Prolactin and growth hormone in the regulation of the immune system. *Proc Soc Exp Biol Med* 198: 513–527

17 Matera L, Mori M, Galetto A (2001) Effect of prolactin on the antigen presenting function of monocyte-derived dendritic cells. *Lupus* 10: 728–734

18 Wang HC, Klein JR (2001) Immune function of thyroid stimulating hormone and receptor. *Crit Rev Immunol* 21: 323–337

19 Kruger TE, Smith LR, Harbour DV, Blalock JE (1989) Thyrotropin: an endogenous regulator of the *in vitro* immune response. *J Immunol* 142: 744–747

20 Montgomery DW (2001) Prolactin production by immune cells. *Lupus* 10: 665–675

21 Wilson T, Yu-Lee L, Kelley M (1995) Coordinate gene expression of luteinizing hormone-releasing hormone (LHRH) and the LHRH receptor after prolactin stimulation in the rat Nb2 T cell line: Implications for a role in immunomodulation and cell cycle gene expression. *Mol Endocrinol* 9: 44–53

22 Weigent DA, Blalock JE, LeBoeuf RD (1991) An antisense oligodeoxynucleotide to growth hormone messenger ribonucleic acid inhibits lymphocyte proliferation. *Endocrinology* 128: 2053–2057

23 Aguila MC, Rodriguez AM, Aguila-Mansila HN, Lee WT (1996) Somatostatin antisense oligodeoxynucleotide-mediated stimulation of lymphocyte proliferation in culture. *Endocrinology* 137: 1585–1590

24 Batticane N, Morale M, Galio F, Farinella Z, Marchetti B (1991) Luteinizing hormone-releasing hormone signalling at the lymphocyte involves stimulation of interleukin-2 receptor expression. *Endocrinol* 129: 277–286

25 Weigent DA, Baxter JB, Blalock JE (1992) The production of growth hormone and insulin-like growth factor-I by the same subpopulation of rat mononuclear leukocytes. *Brain Behav Immun* 6: 365–376

26 Arnold RE, Weigent DA (2003) The production of nitric oxide in EL4 lymphoma cells overexpressing growth hormone. *J Neuroimmunol* 134: 82–94

27 Malarkey WB, Wang J, Cheney C, Glaser R, Nagaraja H (2002) Human lymphocyte growth hormone stimulates interferon gamma production and is inhibited by cortisol and norepinephrine. *J Neuroimmunol* 123: 180–187

28 Tayebati SK, El-Assouad D, Ricci A, Amenta F (2003) Immunochemical and immunocytochemical characterization of cholinergic markers in human peripheral blood lymphocytes. *J Neuroimmunol* 132: 147–155

29 Warthan MD, Freeman J, Loesser K, Lewis C, Hong M, Conway C, Stewart J (2002) Phenylethanolamine N-methyl transferase expression in mouse thymus and spleen. *Brain Behavior Immunity* 16: 493–499

30 Tracey K (2002) The inflammatory reflex. *Nature* 420: 853–859

31 Wang H, Xing L, Li W, Hou L, Guo J, Wang X (2002) Production and secretion of calcitonin gene-related peptide from human lymphocytes. *J Neuroimmunol* 130: 155–162

32 Ho W, Lai J, Li Y, Douglas S (2002) HIV enhances substance P expression in human immune cells. *FASEB J* 16: 616–618

33 Levite M (1998) Neuropeptides, by direct interaction with T cells, induce cytokine secretion and break the commitment to a distinct T helper phenotype. *Proc Natl Acad Sci USA* 95: 12544–12549

34 Smith EM, Meyer WJ, Blalock JE (1982) Virus-induced corticosterone in hypophysectomized mice: a possible lymphoid adrenal axis. *Science* 218: 1311–3112

35 Bayle JE, Guellati M, Ibos F, Roux J (1991) Brucella abortus antigen stimulates the pituitary-adrenal axis through the extrapituitary B lymphoid system. *Prog Neuro Endocrin Immunol.* 4: 99–105

36 Jara LJ, Vera-Lastra O, Miranda JM, Alcala M, Alvarez-Nemegyei J (2001) Prolactin in human systemic lupus erythematosus. *Lupus* 10: 748–756

37 Reynolds DG, Gurll NJ, Vargish T, Lechner RB, Faden AI,

Holaday JW (1980) Blockade of opiate receptors with naloxone improves survival and cardiac performance in canine endotoxic shock. *Circulatory Shock* 7: 39–48

38 Harbour DV, Smith EM, Blalock JE (1987) Splenic lymphocyte production of an endorphin during endotoxic shock. *Brain Behav Immun* 1: 123–133

39 Gaveriaux-Ruff C, Simonin F, Filliol D, Kieffer B (2003) Enhanced humoral response in kappa-opioid receptor knockout mice. *J Neuroimmunol* 134: 72–81

40 Refojo D, Kovalovsky D, Young J, Rubinstein M, Holsboer F, Reul J, Low M, Arzt E (2002) Increased splenocyte proliferative response and cytokine production in b-endorphin deficient mice. *J Neuroimmunol* 131: 126–134

41 Sugama S, Kim Y, Baker H, Tinti C, Kim H, Joh TH, Conti B (2000) Tissue specific expression of rat IL-18 gene and response to ACTH treatment. *J Immunol* 165: 6287–6292

42 Yang Y, Guo L, Liu X (1999) Expression of growth hormone and insulin-like growth factor in the immune system of children. *Horm Metab Res* 31: 380-384

43 Stein C, Hassan AHS, Przewlocki R, Gramsch C, Peter K, Herz A (1990) Opioids from immunocytes interact with receptors on sensory nerves to inhibit nociception in inflammation. *Proc Natl Acad Sci USA* 87: 5935–5939

44 Schafer M, Mousa SA, Zhang Q, Carter L, Stein C (1996) Expression of corticotropin-releasing factor in inflamed tissue is required for intrinsic peripheral opioid anagesia. *Proc Natl Acad Sci USA* 93: 6096–6100

45 Mousa SA, Zhang Q, Ru-Rong J, Stein C (2001) β-endorphin containing memory cells and μ-opioid receptors undergo site-directed transport to peripheral inflamed tissue. J Neuroimmunol 115: 71–78

46 Zhao J, Karalis K (2002) Corticotropin-releasing hormone in mouse thymocytes. *Mol Endocrinol* 16: 2561–2570

47 Ikushima H, Kanaoka M, Kojima S (2003) Requirement for growth hormone-releasing hormone in the development of experimental autoimmune encephalomyelitis. *J Immunol* 171: 2769–2772

48 Stein C, Schafer M, Machelska H (2003) Attacking pain at its source: new perspectives on opioids. *Nature Med* 9: 1003–1008

Immunodiagnosis

B1

Antibody detection

Klaus Hermann, Markus Ollert and Johannes Ring

Introduction

The ACCURATE and reliable determination of biologically active endogenous compounds in plasma, urine and other body fluids for scientific or diagnostic purposes has been a challenge for many decades. In the past, the assessment of such compounds was difficult and tedious because specific and practical analytical tools were not available. In the early days, decisive and convincing conclusions about the significance of a biologically active molecule in disease or health could only be made after purification and isolation of the ANALYTE and identification of its chemical structure. A definite improvement in the analysis of biologically active endogenous compounds was the introduction of bioassays based on an entire animal model or *in vitro* tissue preparations. Although the bioassays possessed sufficient sensitivity, there were problems with their lack of SPECIFICITY. Other analytical procedures such as liquid chromatography, electrophoresis or photometric procedures have also been developed for *in vitro* diagnosis. However, these approaches are either tedious and time-consuming or require expensive equipment and specially trained staff. A landmark in diagnostics was the introduction of immunoassays (IAs) which are inexpensive and easy to perform with high reproducibility, SENSITIVITY and SPECIFICITY.

Basic principle of immunoassays

IAs are based on an antigen-ANTIBODY (Ab) reaction utilizing the exceptional capability of the immune system to produce specific Abs which can recognize and discriminate between a practically infinite number of foreign compounds. The basic requirement for setting up an IA is the production of a specific Ab to a given antigen or hapten by immunizing animals such as mice, rabbits, goats, horses or others. High molecular weight compounds like proteins are usually immunogenic and may serve as an antigen which can be directly injected into the host animal to produce Abs. In contrast, low molecular weight compounds such as drugs, peptides or steroids are haptens which do not induce immune responses. They can be rendered immunogenic after coupling to a carrier. The Abs obtained from the serum after several booster injections with the immunogen are HETEROGENEOUS polyclonal Abs (pAbs). A more sophisticated approach was the generation of monoclonal Abs (mAbs), which was first introduced by Köhler and Milstein in 1975 [1]. This technique allows the production of HOMOGENEOUS Ab species in nearly infinite quantities with high SPECIFICITY for a certain antigen. The method involves the isolation of spleen cells from an immunized animal containing Ab-producing B CELLS. The B CELLS are then fused with myeloma cells. After selection and screening of the desired Ab-secreting cell line (hybridoma), the Ab can be harvested in the supernatant. These hybridomas can be grown in large volumes for the production of huge quantities of the mAb.

Basic essentials for the characterization of the Abs produced are high affinity, SPECIFICITY and sensitivity. The affinity is a measure of the strength of the binding interaction between the antigen and the Ab and can be experimentally determined; according to the law of mass action, it is expressed as equilibrium constant of dissociation, K_d (mol/L), or of association, K_a (L/mol). The lower the K_d the greater the affinity. The SPECIFICITY depends on the recognition of the ANALYTE by the Ab and the extent of crossreaction with closely related or structurally similar ANALYTES. The less the crossreaction, the higher the SPECIFICITY.

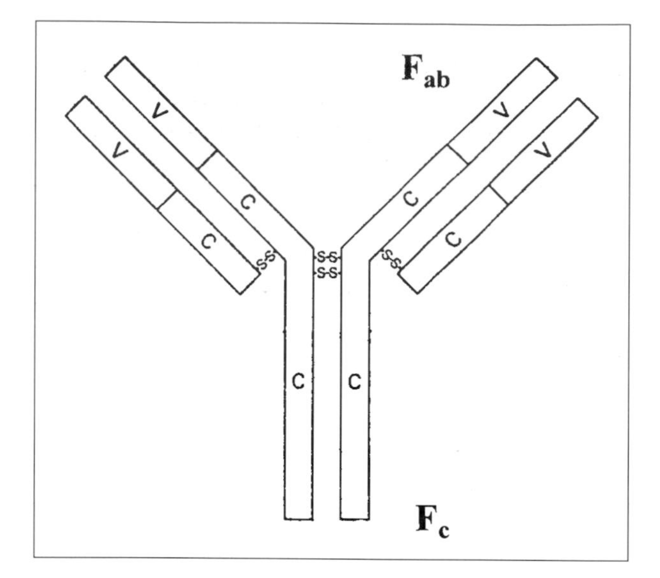

Figure 1. Structure of immunoglobulins
For further details, see text

This in turn is closely related to the ability of the Ab to discriminate between negative and positive samples. The SENSITIVITY describes the detection limit and is defined by the dose-response curve of the antigen to the Ab. The lower the detection limit, the higher the sensitivity.

Antibody structure

Abs, among other serum proteins such as albumins, belong to the gammaglobulin or immunoglobulin (Ig) fraction according to their electrophoretic mobility. They are glycoproteins which are chemically very similar in structure and are constructed of two identical light chains with an approximate molecular weight (MW) of 25 kDa per chain and two identical heavy chains with an approximate MW of 55 kDa per chain. Each of the two light chains is covalently attached to one heavy chain via a single disulfide bond. Likewise, the heavy chains are bound to each other by multiple disulfide bridging. In addition, the three-dimensional Ab structure is stabilized by noncovalent intra- and interchain-interactions. Two locally distinct binding domains are prominent on Ig molecules. One is the antigen binding site (Fab) for the binding and recognition of antigens and the other is the receptor binding site (Fc) for binding to specific receptors on various cells involved in immunological functions such as mononuclear phagocytes, NATURAL KILLER CELLS, mast cells or basophil LEUKOCYTES (Fig. 1). Although Igs share an overall similarity, they can be divided into different classes and subclasses according to their size, charge, solubility and their behavior as antigens. At present, the Ab molecules in humans can be classified into IgA, IgD, IgE, IgG and IgM. IgA and IgG can be further subdivided into their subclasses IgA1 and IgA2 and IgG1, IgG2, IgG3 and IgG4 (see Chapters A.3 and C.2).

Clinical relevance of antibody detection

Initially, the identification and characterization of specific Abs associated with pathophysiological conditions were generally constrained to serve scientific purposes. However, in many instances the measurement of specific Abs evolved into clinically relevant diagnostic markers in health and disease. Medical conditions in which the determination of specific Abs is routinely used are bacterial and viral infections, infestation with parasites, autoimmune diseases and allergies.

Microbial infections

The determination of Abs in infectious diseases has been known and used for a long time. Bacteria-specific Abs of the IgG and IgM class are routinely detected in the serum of patients infected with different bacteria such as *Borrelia burgdorferi* (lyme disease), *Chlamydia trachomatis* (sexually transmitted infection of the urogenital tract), *Legionella pneumophila* (legionnaire disease), *Staphylococcus* and *Streptococcus* and *Treponema pallidum* (lues).

The assessment of IgG and IgM Abs is also a very valuable parameter in the diagnosis of viral infections such as hepatitis, measles, epidemic parotitis and rabies, and infections with the Epstein-Barr virus (mononucleosis), herpesvirus (herpes simplex and herpes zoster) and arborvirus (tick encephalitis). Infestations with parasites leading to diseases such as leishmaniasis, amoebiasis, malaria, toxoplasmosis, schistosomiasis, echinococcosis, tichinellosis, filariasis or others can induce the Ab formation of different Ig classes. In the diagnosis of parasitosis, it is recommended that the identity of the parasite be ascertained first. In addition, serological methods are available to identify circulating antigens, antigen-Ab complexes or circulating Abs.

Autoimmune diseases

Another area of pathophysiological abnormalities in which the measurement of Abs has predictive and diagnostic value is that of the numerous autoimmune diseases (ADs) in which the self-tolerance of the immune system against certain structures of the host is abrogated. Examples of an organ-specific AD are Hashimoto's thyroiditis with circulating Abs against thyroglobulin, and myasthenia gravis with auto-Abs against the acetylcholine receptor. Examples of non-organ-specific ADs are Sjögren's syndrome, rheumatoid arthritis, scleroderma and systemic lupus erythematosus affecting the skin, joints and muscles with Abs against nuclear antigens such as DNA, RNA or histones.

Allergy

The most important immunglobulins for the *in vitro* diagnosis of allergic diseases, either immediate-type HYPERSENSITIVITY reactions such as rhinitis, conjunctivitis, allergic bronchial ASTHMA and ANAPHYLAXIS or LATE PHASE reactions, e.g., allergic contact dermatitis, are Abs of the IgE class [2, 3]. Clinically relevant are the measurement of total IgE or allergen-specific IgE in the patients' serum for the determination of IgE-mediated sensitization. Total serum IgE levels of > 100 (kilo units) KU/l in adults are a good indicator for atopy, a disease characterized by familial HYPERSENSITIVITY to exogenous environmental agents associated with high IgE Ab titers and altered reactivity against various pharmacological stimuli. However, high total IgE Abs can also be induced by parasitic worm infestations. Extremely high values, above 10,000 KU/l, are indicative of IgE-producing myelomas or hyper-IgE syndrome. The measurement of allergen-specific IgG Abs as IgE-blocking Abs for monitoring the success of immune therapy with insect venoms in patients with hymenoptera venom allergy has been used tentatively, but with discrepant results.

Antibody detection methods

The detection of Abs in the circulation or in tissue has become a useful analytical tool for the *in vitro* immunodiagnosis of numerous diseases. Several immunological techniques are available for the routine identification of IgA, IgD, IgE, IgG or IgM Ab classes in the laboratory of clinical chemistry. The most commonly used methods are discussed briefly.

Immunoprecipitation assay

Immunoprecipitation is a very simple and easy to perform *in vitro* assay for the identification and semi-quantification of soluble Abs. The addition of the antigen to the Ab results in the formation of a three-dimensional, insoluble network of precipitating aggregates which can be detected by, e.g., a nephelometer. The assay is very similar to a volumetric acid/base titration. The bulk of precipitate, formed at equivalent concentrations of Ab and antigen, is a measure of the concentration of the Ab. The assay can also be used in reverse to measure the antigen concentration by adding Abs.

A variation of the immunoprecipitation assay is the hemagglutination test and the complement fixation test. The hemagglutination test allows the identification of Abs to red blood cell antigens or the detection of Abs to antigens which are covalently or noncovalently attached to the red cell surface. The complement fixation test is a three-step assay in which the Ab-containing serum is initially incubated with a fixed amount of antigen to form an immune complex. In the second step, COMPLEMENT is added, firmly incorporated by the immune complexes and consumed to some extent. Finally, Ab-coated red blood cells are added as indicator cells for titration of the remaining quantity of COMPLEMENT; the less the red blood cells are lysed the more the immune complexes have been generated before.

Immunocytochemistry

Immunocytochemistry is a technique for the *in situ* detection of an Ab in tissue slices. Frozen tissue or tissue embedded in various embedding media is cut into thin slices and then immobilized on a slide. After fixation of the tissue with formaldehyde, glutaraldehyde, alcohol or acetone, the tissue is incubated with a specific primary Ab directed against the Ab to be detected. In the direct assay, the primary Ab is chemically coupled to a fluorescent dye, which allows the detection of the ANALYTE by fluorescence microscopy. In the indirect assay, excess primary Ab is thoroughly washed off, and the tissue is incubated with a secondary Ab to form a sandwich. The secondary Ab can be fluoresceinated or coupled to an enzyme, e.g., alkaline phosphatase (ALP) or peroxidase. This allows the visualization of the ANALYTE by fluorescence microscopy or by light microscopy after addition of a colorless SUBSTRATE which is enzymatically converted to a colored product. Only cells which contain the ANALYTE will light up under the microscope.

Immunoblotting

The immunoblot or dot blot technique is similar to the immunoprecipitation assay. However, in immunoblotting the antigen-Ab reaction takes place on the solid phase, whereas in the immunoprecipitation assay the Ab reacts with the antigen in solution. The assay utilizes the capability of nitrocellulose or polyvinylidene fluoride (PVDF) membranes to bind antigens. Antigens are applied in small dots, and the membranes are dried. The membranes are treated with blocking buffer (containing e.g. ovalbumin, gelatin or milk proteins) to prevent unspecific adsorption. After blocking, the membranes are incubated with the serum and dilutions of the serum, which contain the Ab. The membranes are washed to remove abundant Ab. Next, the membranes are incubated with a secondary Ab raised against the Ab of interest which is conjugated with an enzyme. The formation of the antigen-Ab-secondary Ab complexes can be visualized by adding a SUBSTRATE which will be converted by the secondary Ab-bound enzyme to yield a colored spot. The use of secondary Ab-bound enzymes represents the basic ELISA (enzyme-linked immunoadsorbent assay) principle. The intensity of the spots is proportional to the amount of Ab present in the serum samples (Fig. 2, left panel). Further characterization of the Ab can be achieved by separating the antigens electrophoretically. The separated components are transferred from the gel to a membrane, a process which is called Western blotting. The membranes are then treated like dot blots as outlined above. This methodology combines the high resolving power of electrophoresis and the discriminating power of an immunological reaction. Components that are recognized by the Ab show up on the Western blot as colored bands (Fig. 2, right panel).

Immunoadsorbent assays

Immunoadsorbent assay (IAA) techniques are widely used for the quantitative measurement of serum IgE and IgG Abs. An antigen, e.g., an allergen extract from chicken meat, grass or tree pollen, or house dust mites, is attached to an inert matrix such as the wall of a reaction vial or microtiter plate wells, or it is chemically coupled to a paper disc. The serum of an allergic patient is incubated in a first-step reaction with the allergen-coated matrix. IgE molecules that recognize the allergen are bound. After removal of excess serum, a secondary Ab, in this case an anti-human IgE Ab raised in mice, rabbits, goats or horses, is added which forms an allergen-IgE-anti-IgE Ab complex (second-step reaction). Excess of the secondary Ab is likewise removed by washing. The formation of the allergen-IgE-anti-IgE-Ab-complex depends on the amount of specific IgE present in the serum sample. Since the secondary Ab, also termed *detecting Ab*, carries a covalently coupled LABEL or tag, the formation of the allergen-IgE-anti-IgE Ab complex, a sandwichlike structure, can be moni-

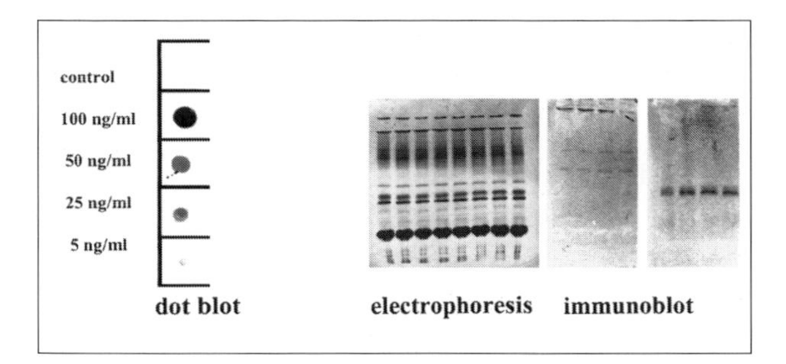

FIGURE 2. DOT BLOTS AND IMMUNOBLOTTING
Specific Abs to an antigen, e.g., an allergen extract, can be detected with the dot blot technique. Further characterization of the Ab or the antigens present in the allergen extract can be achieved by immunoblotting. For further details, see text. The various lanes are different allergen extracts.

tored. Utilizing a standard curve with increasing concentrations of the allergens or, as in most commercial specific IgE assays, with various amounts of the World Health Organization's International Reference Preparation for IgE (2nd IRP75/502), the signal obtained with allergen-IgE-anti-IgE Ab complex in the serum sample can be compared with the signal of the standard curve, which permits the quantification of the IgE Abs. The assay format of an IAA in general is summarized in Figure 3.

The radio-allergo-sorbent test (RAST) is a radioimmunoassay (RIA) version [4] in which the allergens are chemically coupled to a paper disc and the secondary Ab is radioactively labeled with I-125 (Pharmacia Uppsala, Sweden). Similarly, the assay can be performed as an enzyme immunoassay (EIA) in which the secondary Ab is labeled with the enzyme β-galactosidase which can react with a colorless SUBSTRATE to form a colored reaction product (EAST, Sanofi Diagnostics Pasteur, Chaska, MN, USA). In addition to the RAST or EAST, a RAST or EAST inhibition assay can be performed to confirm and validate the results [5]. The serum samples are first incubated *in vitro* with increasing concentrations of the allergens prior to the RAST or EAST. The binding inhibition of the Ab can be illustrated by a dose-response curve which inversely correlates with the concentration of allergens added; low allergen concentrations still give a high signal in the RAST or EAST, whereas the signal vanishes with high concentrations. The concentration of the allergen at 50% inhibition (IC50) can be calculated from the dose-response curve. A low IC_{50} is a good indicator for a high affinity and SPECIFICITY of the Ab to the allergens (Fig. 4).

A further development of the RAST or EAST was the introduction of a three-dimensional cellulose sponge instead of the paper disc. This approach is practiced by the ImmunoCAP system (Pharmacia Uppsala, Sweden). The ImmunoCAP assay is performed in microtiter plates and allows partial to complete automation of the assay. The advantages are higher binding capacity of the cellulose sponges (CAP) and the use of the photometric detection of the allergen-IgE-anti-IgE Ab complex with a micro-

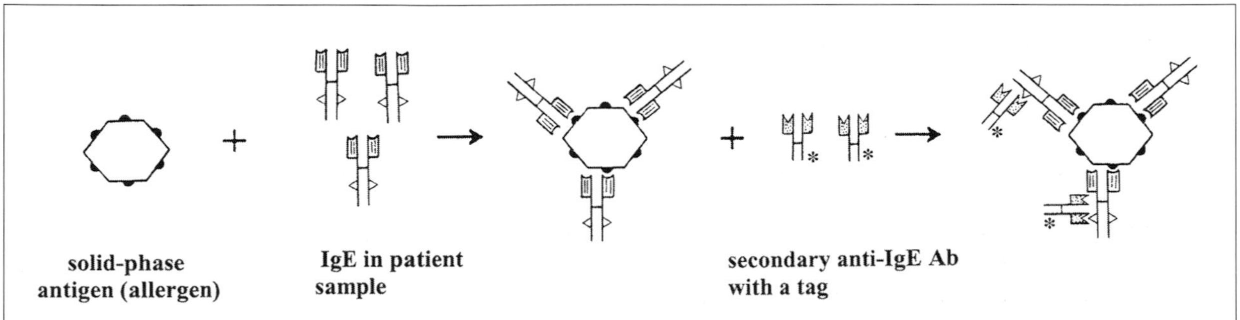

solid-phase antigen (allergen)
IgE in patient sample
secondary anti-IgE Ab with a tag

FIGURE 3. IMMUNO-ABSORBENT ASSAY FORMAT
Reproduced with permission from DPC Biermann, Germany. For further details, see text.

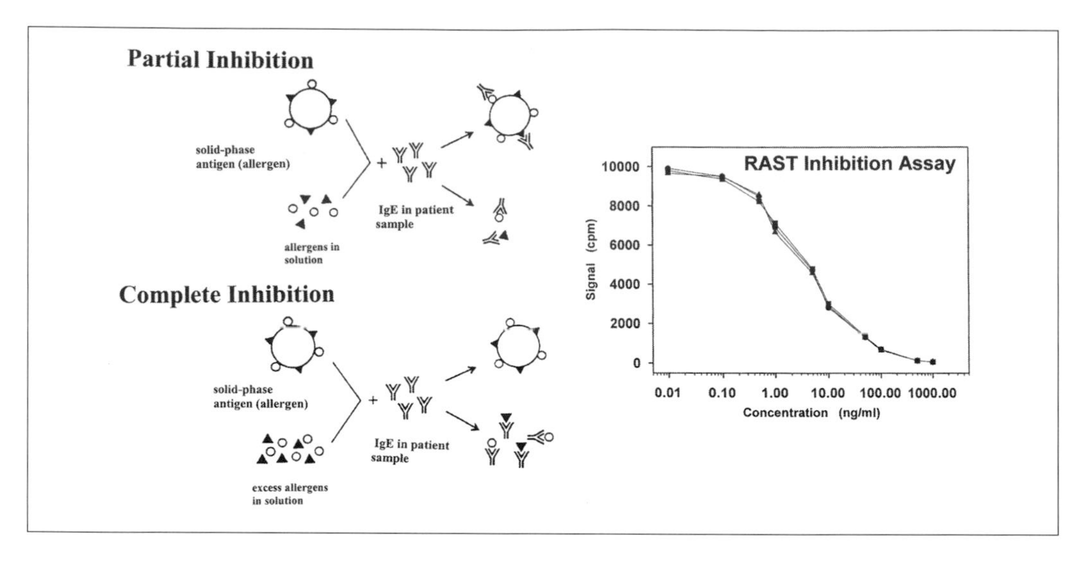

FIGURE 4. RAST INHIBITION ASSAY
Reproduced with permission from DPC Biermann, Germany. For further details, see text.

plate reader. Full automation of the CAP system for IgE detection was achieved with the most recent developments (UniCAP250 and 1000).Variations and modifications of this basic assay format have been made with respect to the characteristics of the allergens, e.g., using allergens in liquid phase and replacing the β-galactosidase-labeled secondary Ab with an Ab that is chemically linked to ALP. This unique concept is realized in several assay formats. The AlaSTAT liquid allergen technology [Diagnostic Product Corporation (DPC), Los Angeles, CA, USA] [6] is based on a four-step reaction. In the first incubation, the biotinylated allergens react with patients' serum IgE. In the second incubation, avidin is added which forms a complex between the biotinylated allergens and IgE. In the third incubation, an enzyme-labeled secondary anti-IgE Ab is added to form a sandwich between the biotin of the reaction vessel, avidin, the biotinylated allergen and the allergen-specific IgE Abs. Finally, the formation of the triple-sandwich complex can be photometrically monitored by the enzymatic conversion of the SUBSTRATE to a colored product (Fig. 5A).

A different assay technology is realized in the Access Allergy Diagnostic System (Sanofi Diagnostics Pasteur). The assay is also a liquid phase assay using paramagnetic polystyrene beads coated with an antibiotin Ab as a capturing Ab. In a first-step reaction, the biotinylated allergens and the allergen-specific IgE form a triple-sandwich complex between the polystyrene beads, the allergen and the IgE. The second reaction is the incubation with an anti-human IgE Ab (secondary Ab) labeled with ALP, which results in the formation of a quadruple sandwich (paramagnetic polystyrene beads, allergens, IgE and anti-IgE). Excess ALP-labeled secondary Ab is also removed in the magnetic field. Finally, dioxetane-phosphate is added as a SUBSTRATE which is converted by ALP to a chemiluminescent product which can be quantified in a luminometer (Fig. 5B).

In the absence of a recognized reference method for *in vitro* sIgE measurement, the Pharmacia's second-generation ImmunoCAP® technology has become a quasi-standard due to its widespread use, analytical reliability and the generally adequate correspondence of its results – on a positive/negative basis – with the results of skin testing [7]. From their inception, however, certain limitations of second-generation test systems for sIgE have been apparent. They are labor-intensive and require specially trained personnel, making them both costly to the laboratory and potentially more prone to human

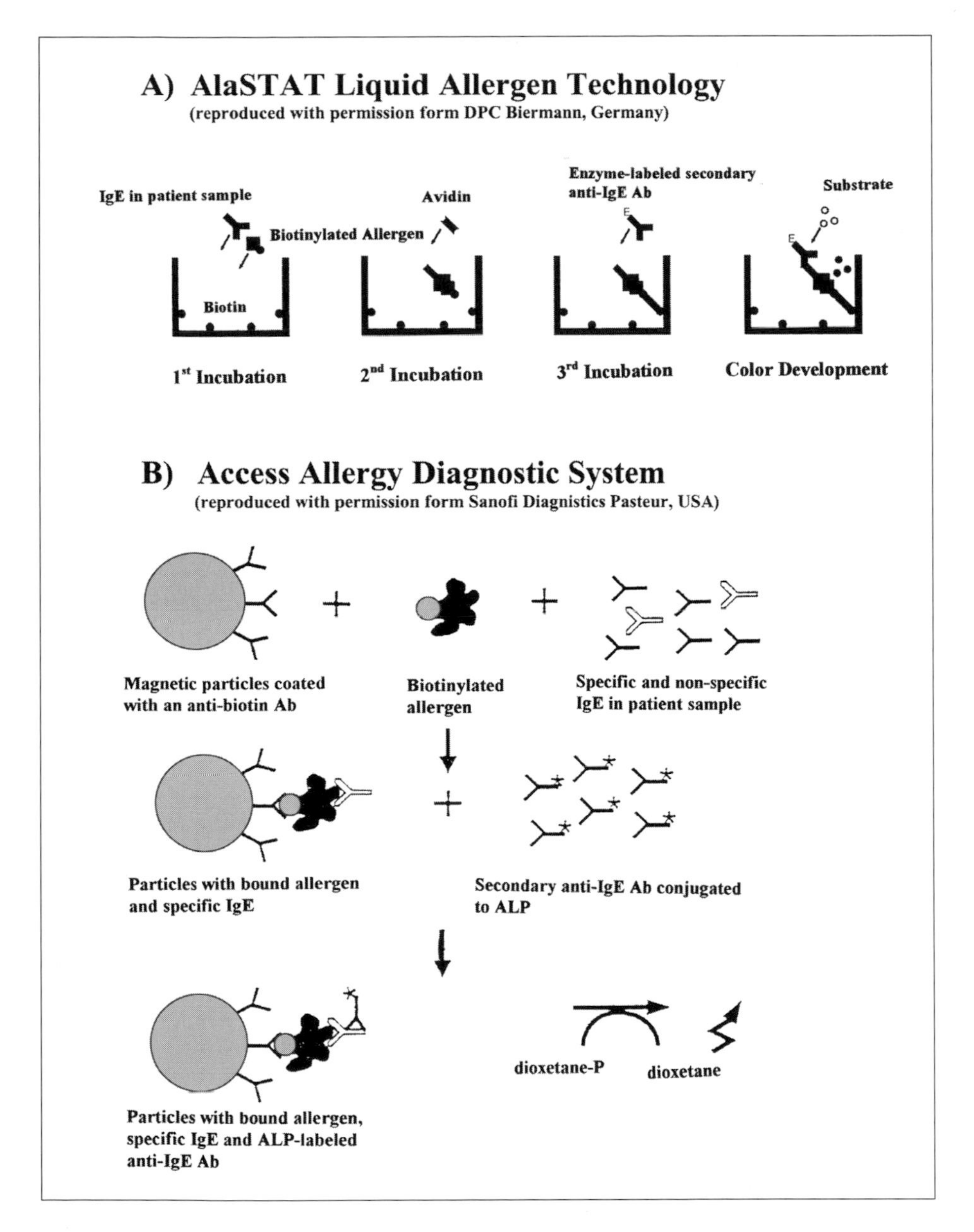

FIGURE 5. ADVANCED IMMUNOASSAY TECHNIQUES

(A) AlaSTAT liquid allergen technology (reproduced with permission from DPC Biermann, Germany). (B) Access Allergy Diagnostics System (reproduced with permission from Sanofi Diagnostics Pasteur, USA). For further details, see text.

error than a fully automated system with barcoding as used in most other fields of clinical chemistry; moreover, turnaround time (≥ 3 hours) is still undesirably slow from the allergy specialist's point of view.

"Third generation" systems for the *in vitro* measurement of sIgE, addressing the major limitations associated with second-generation assays, have finally been developed [8, 9]. These are based on proven technology, have considerably shorter turn-around times, and utilize chemiluminescence. The first fully-automated, random-access assay to become available is DPC's IMMULITE® 2000 Allergy (IML) [9, 10], which represents the implementation of routine sIgE testing on a family of systems already well established in clinical chemistry for performing immunoassays and immunometric assays, in both hospital and reference laboratory settings. IML has a working range of 0.10 to 100 kU/L and uses DPC's liquid allergen chemistry together with the Immulite bead and wash technology and an enzyme-enhanced chemiluminescent detection system [10].

Summary

Antibody detection is crucial for the differential diagnosis of many different pathological conditions. Determination of specific antibodies to bacterial and viral pathogens as well as to parasites enables the correct therapeutic measures to be taken. Antibodies to organ-specific and systemic autoimmune diseases are predictive for prognosis and detection of IgE class antibodies is essential for the diagnosis of immediate HYPERSENSITIVITY reactions.

Methods for detection of antibodies include immunoprecipitation assay, in which Ag-Ab complex aggregates are detected, often by hemagglutination; immunocytochemistry, for *in situ* Ab detection in tissue slices; immunoblotting (dot blot technique) whereby Ag-Ab aggregates are trapped on membranes and then detected with a secondary Ab to yield spots; and immunosorbent assays, which are similar to immunoblotting but, by using a tagged secondary antibody, allow the primary antibody to be quantified. A variety of immunosorbent kits are available which permit rapid, specific, accurate and sensitive detection particularly of IgE antibodies.

Selected readings

Tijssen P (1985) *Practice and theory of enzyme immunoassays*. Elsevier, Amsterdam

Abbas AK, Lichtman AH, Pober JS (eds) (1994) *Cellular and molecular immunology*. W.B. Saunders, Philadelphia

Crowther JR (ed) (1995) *ELISA*. Humana Press, Totowa

Thomas L (ed) (1988) *Labor und Diagnose*. Medizinische Verlagsgesellschaft, Marburg

Roitt IM, Brostoff J, Male DK (1986) *Immunology*. Mosby, St. Louis

References

1 Köhler G, Milstein C (1975) Continous cultures of cells fused secreting antibody of predefined specificity. *Nature* 256: 495–497

2 Ring J (1981) Diagnostic methods in allergy. *Behring Inst Mitt* 68: 141–152

3 Ring J (ed) (1988) *Angewandte Allergologie*. MMV Medizin Verlag, München

4 Wide L, Bennich H, Johansson SGO (1967) Diagnosis of allergy by an *in vitro* test for allergen antibodies. *Lancet* 2: 1105

5 Gleich GJ, Yunginger JW (1981) Variations of the radioallergosorbent test for measurement of IgE antibody levels, allergens and blocking antibody activity. In: J Ring, G Burg (eds): *New trends in allergy. Springer-Verlag*, Heidelberg, 98–107

6 El Shami AS, Alaba O (1988) Liquid-phase *in vitro* allergen-specific IgE assay with *in situ* immobilization. In: *Advances in the Biosciences, Vol. 74: Allergy and Molecular Biology*, Pergamon Press, New York, 191–201

7 Hamilton RG, Adkinson NF Jr (2003) Clinical laboratory assessment of IgE-dependent hypersensitivity. *J Allergy Clin Immunol* 111(Suppl 2): S687–701

8 Petersen AB, Gudmann P, Milvang-Gronager P, Morkeberg R, Bogestrand S, Linneberg A, Johansen N (2004) Performance evaluation of a specific IgE assay developed for the ADVIA centaur® immunoassay system. *Clin Biochem* 37: 882–892

9 Li TM, Chuang T, Tse S, Hovanec-Burns D, El Shami AS (2004) Development and validation of a third generation allergen-specific IgE assay on the continuous random access IMMULITE® 2000 analyzer. *Ann Clin Lab* Sci 34:67–74

10 Ollert MW, Weissenbacher S, Rakoski J, Ring J (2005) Allergen-specific IgE measured by a continuous random-access immunoanalyzer: interassay comparison and agreement with skin testing. *Clin Chem* 51: 1241–1249

Immunoassays

Michael J. O'Sullivan

Introduction

For forty years immunoassays have been the method of choice for quantifying low concentrations of ANALYTES in complex biological fluids. The procedure is equally applicable to the measurement of small molecules such as drugs and large molecules such as proteins. The technique combines SENSITIVITY and SPECIFICITY with ease of use.

Immunoassays are used in basic biological research to investigate the physiological and possible pathological role of a wide range of biologically active substances including cyclic nucleotides, prostaglandins, leukotrienes, growth factors and cytokines [1]. Such research often leads to the identification of new potential targets for therapeutic agents. The assays are also used in the pharmaceutical industry in many aspects of the drug development process. These range from drug screening, toxicological, pharmacological and pharmacokinetic studies through to clinical trials. Immunoassays have perhaps had their greatest impact in the area of clinical diagnostic tests. The technique has been employed for many years in hospital clinical biochemistry laboratories to diagnose disease and metabolic disorders. More recently, applications of this technique have moved out of these core areas into such diverse situations as the biotechnology industry, the food safety industry and even to "over the counter" applications such as home pregnancy testing. In fact it is difficult to think of any area of the biological sciences where immunoassays have not had a significant impact.

The technique was introduced in 1959 by Berson and Yalow [2]. The combination of a signal that can be easily detected and a protein molecule, which binds specifically and with high AFFINITY to the ANALYTE of interest, lies at the heart of all immunoassay procedures. Assay designs have proliferated over the last forty years, as have the different types of signal reagents and detection systems. Sophisticated instruments with associated computer hardware have been developed with the aim of increasing sample throughput. This chapter will discuss and highlight the main elements of the subject but cannot hope to be an in-depth review of the whole field. For the interested reader, *The Immunoassay Handbook, second edition* published in 2001, provides a comprehensive review of the area (see "Selected readings").

Basic principles of assay design

Competitive immunoassays

In the competitive immunoassay (also termed "labelled ANALYTE") approach there is competition between labelled and unlabelled ANALYTE for a limited amount of binding sites on an ANTIBODY. Antibody-bound ANALYTE is separated from unbound ANALYTE and the proportion of LABEL in either fraction is analysed. A curve can then be plotted of the percentage of TRACER bound to the ANTIBODY against a range of known standard concentrations. The concentration of unknown ANALYTE present in the sample can then be determined by interpolation from the standard curve. The principle of the assay is illustrated in Figure 1. In this example there is competition between unlabelled and enzyme-labelled ANALYTE for the capture ANTIBODY on the solid phase. The amount of enzyme-labelled ANALYTE bound to the solid phase is inversely proportional to the concentration of unlabelled ANALYTE. Following a wash step, bound enzyme activity is determined by incubation with the SUBSTRATE. After a timed interval, the reaction is

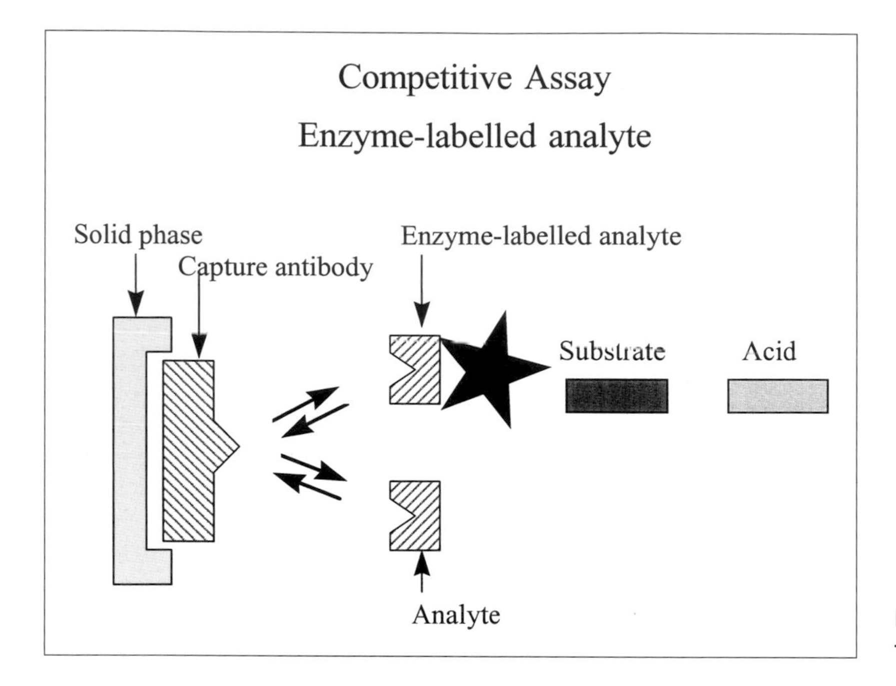

FIGURE 1. PRINCIPLE OF A COMPETITIVE IMMUNOASSAY

terminated with dilute acid and the resultant colour intensity determined in a spectrometer.

Although an ANTIBODY is usually used in these assays, there may be circumstances where it is more appropriate to use a naturally occurring binding protein or receptor. This does not affect the principle of the assay. This assay format has the advantages that only one ANTIBODY is required and it uses relatively small amounts of the sometimes-scarce ANTIBODY reagent. It has the disadvantages that assay SENSITIVITY is limited by ANTIBODY affinity, the assays have a relatively narrow dynamic range and the labelling process may alter the binding characteristics of the labelled ANALYTE. This format tends to be favoured for small ANALYTES.

Immunometric assays

The immunometric approach (also termed "labelled antibody") differs from the competitive approach in a number of ways. In its most common format it involves two antibodies both of which are specific for the ANALYTE. One of these antibodies is labelled; the other is attached to a solid phase. The sample containing the ANALYTE is added and followed by the labelled second antibody. Unbound LABEL is removed by washing. The amount of LABEL bound to the solid phase is related to the amount of ANALYTE in the sample. A standard curve can be constructed using known quantities of ANALYTE and the concentration of ANALYTE in the sample can be determined by interpolation from the curve. This approach is called a two-site immunometric assay. For obvious reasons this format is commonly referred to as a "sandwich assay". The principle of the assay is illustrated in Figure 2. The so-called detection antibody is labelled with an enzyme. Following washing to remove unbound label, the bound enzyme activity is determined by incubation with its SUBSTRATE. The reaction is then terminated with dilute acid and the resultant colour intensity determined in a spectrometer.

This method has the advantages that it tends to be more sensitive and PRECISE than the competitive approach. It also tends to have a wider dynamic range and there is no requirement to LABEL the ANALYTE. The major disadvantages of the technique are the high consumption of antibody and the require-

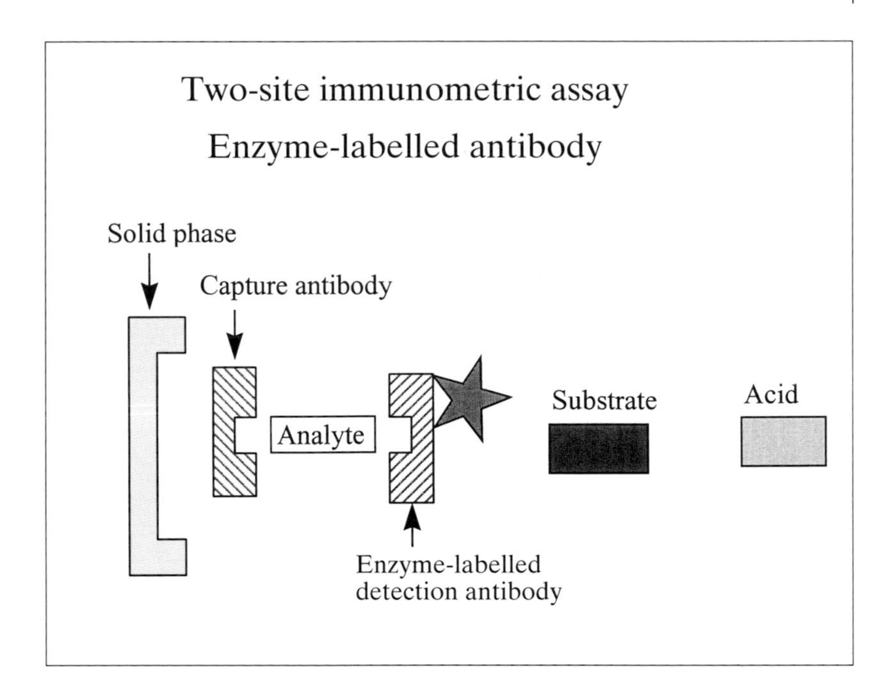

FIGURE 2. PRINCIPLE OF A TWO-SITE IMMUNOMETRIC ASSAY

ment for two antibodies. Immunometric assays are the favoured technique for quantifying large molecules. It cannot be applied to small molecules due to the size restraint on binding two large antibodies to one small molecule at the same time. These two basic approaches have been the subject of endless permutations, some of which will be touched upon in later sections of this chapter.

Homogeneous assays

The assay formats described above suffer from one significant disadvantage in that they are HETEROGENEOUS assays, with a requirement to separate bound from free TRACER. This is a labour-intensive step which is difficult to automate and introduces significant imprecision into the assay. In an attempt to overcome this problem, considerable effort, ingenuity and money have been invested in developing HOMOGENEOUS assays, which do not require a separation step. Several successful methods have been developed for the quantification of small molecular weight ANALYTES, but the methods lack SENSITIVITY and are also not generally applicable to the measurement of large molecules. One exception is the technique termed "scintillation proximity" which will be discussed later in the chapter.

Components of immunoassays

Tracers

Radioisotopes

For many years after the technique was introduced radioisotopes were used, virtually exclusively, as the assay TRACER. Radioactive iodine (iodine-125) was the favoured label: its high specific activity providing good assay SENSITIVITY and a reasonably long half-life giving adequate reagent shelf life. It was also easy in many cases to prepare labelled proteins. The equipment required to measure radioactive decay was also readily available. Finally, the rate and measurement of radioactive decay was not affected by the SAMPLE MATRIX. In most situations iodine-125 remained the LABEL of choice for the next twenty years and is still used even today.

Tritium TRACERS were also widely used for small molecule assays. Such TRACERS are readily commercially available and have a long shelf life. The tritiated molecule is also virtually identical to the non-labelled molecule. Many other labelling techniques change the structure of the labelled molecule, which often results in differences in affinity of the interaction of the antibody with the ANALYTE and the label. This can adversely affect assay performance. For these reasons immunoassays for small molecules can often be set up most quickly using tritium TRACERS. However, tritium TRACERS do have significant disadvantages. In particular, their relatively low specific activity demands long count times and their measurement requires the use of organic scintillant cocktails. For these reasons tritium-based assays have a tendency to be replaced by iodine-125 and non-isotopic TRACERS, if and when labelling problems are overcome.

Non-isotopic tracers

Radioisotopes are perceived as posing a potential health risk and there certainly are regulatory problems associated with their use and disposal. In addition, when non-isotopic assays were first being developed, some researchers believed that there would be advantages associated with the use of non-isotopic TRACERS, although this was hotly disputed at the time by some radioimmunoassay experts. In practice the development of satisfactory labelling techniques and suitable assay designs for non-isotopic TRACERS did prove difficult. A major breakthrough came with the introduction of ninety-six-well microtitre plates with associated washing and measuring equipment [3].

Today enzymes are the most widely used TRACERS [4]. When used in combination with colourimetric end points they provide highly sensitive, robust, PRECISE, ACCURATE and convenient immunoassays [5]. Inexpensive automatic colourimetric multiwell plate readers are readily available. Many commercial kits are on the market, which enable relatively inexperienced workers to measure picogram per ml levels of biologically active compounds in complex biological fluids with inexpensive, readily available laboratory equipment. Horseradish peroxidase in combination with a ready to use formulation of its SUBSTRATE 3,3'5,5'-tetramethly benzidine has proved extremely popular. Many other non-isotopic TRACERS have been used, of which fluorescent and luminescent labels have stood the test of time.

In some assays the detection antibody is labelled with biotin rather than an enzyme. The biotinylated antibody is used in combination with a streptavidin/horseradish peroxidase conjugate. Streptavidin has a very high affinity for, and binds very quickly to biotin, so linking the biotinylated antibody non-covalently to the enzyme. This approach tends to label the antibody more consistently and to give a modest two-to four-fold increase in assay SENSITIVITY. A greater increase in SENSITIVITY can be obtained using macromolecular polymers incorporating many streptavidin and peroxidase molecules.

Binding reagents

Antibodies are used in the vast majority of assays as they can provide the levels of SPECIFICITY and SENSITIVITY required. Binding proteins and receptors are used on occasions when suitable antibodies are not available. Antibodies are either monoclonal or polyclonal. Polyclonal antibodies are produced entirely in animals, particularly rabbits. However an animal's immune system generally produces a rather heterogeneous mixture of antibodies. For this reason and for continuity of supply, monoclonal antibodies are often favoured. Monoclonal production is initiated in mice, but when an antibody response is observed, their spleens are removed and the suspended spleen cells fused with a myeloma cell line. The fused cell hybridomas are grown in culture; if any culture is positive, it is plated out so that each well contains a single cell. This produces cells that are derived from a single progenitor and gives rise to a single species of antibody.

Polyclonal antibodies tend to be of high avidity and can be very specific. However, their exact composition will vary from bleed to bleed, even in the same animal. For this reason it is difficult for commercial kit manufacturers to ensure complete product homogeneity over the lifetime of a commercial immunoassay. Polyclonal antibodies tend to be used in competitive assays, which require high-affinity antibodies and do not consume a lot of antibody. Monoclonal antibodies tend to be of rather lower affinity but provide a more HOMOGENEOUS reagent. They also have a more closely defined SPECIFICITY. These antibodies tend to be used in immunometric assays, often in combination with a polyclonal. Finally it should be admitted that antibody production is more of an art than a science.

Standards

Each time the concentration of an ANALYTE is determined in a sample, it is necessary to prepare a standard curve containing known concentrations of the ANALYTE. The standard is the most important component of an immunoassay. Any error in the standard will produce an error in the estimated ANALYTE concentration. The standard should resemble the ANALYTE as closely as possible. This may seem a rather obvious statement to make, but is often difficult to achieve in practice. For instance, recombinant proteins are often used as standards. Do they have the same conformation and degree of glycosylation as the native molecule? Standards are preferably calibrated against some type of agreed international standard. Commercial companies also have strict internal quality control criteria to ensure that their kit standards do not fall outside tight performance specifications.

Buffers

A multitude of buffers have been employed in immunoassays, although most often phosphate or Tris buffers at near to physiological pH are used. The buffers usually contain a protein additive to reduce non-specific binding to tube or microtitre plate walls. In addition buffers may contain a bacteriostat to prevent bacterial contamination. One difficult problem often encountered in setting up an immunoassay is related to the different composition of the sample and the standard. This can cause problems during assay validation. Buffers often contain additives such as animal proteins in an attempt to minimise such matrix effects.

Separation systems

Many techniques have been employed to separate antibody-bound ANALYTE from free ANALYTE. Activated charcoal is often used with tritium TRACERS. The charcoal selectively adsorbs the free TRACER but is unable to bind the antibody-bound fraction. The charcoal-bound fraction is separated from the free fraction by centrifugation. Precipitation procedures are popular with iodine-125 TRACERS. These methods often employ a second antibody specific for the first to form an immune complex, which can again be separated from the unbound TRACER by centrifugation.

These precipitation techniques have been largely superseded by solid-phase techniques, where either the primary or secondary antibody is bound to a solid phase. Coated particles are widely used with iodine-125 TRACERS. Separation of the bound and free fractions can either be achieved by centrifugation or preferably by magnetisation if a magnetisable component is incorporated into the particle. Coated well techniques have become increasingly popular as the trend away from radioactivity has gained momentum in the immunoassay field. Microtitre plates provide a very convenient format for performing enzyme immunoassays. The antibody is adsorbed to the walls of the plastic wells and separation of the bound from the free fraction is very readily achieved by washing the plates. This is much more convenient than centrifugation methods.

Data presentation and curve plotting

Many approaches have been used for data plotting and standard curve fitting. One approach is to calculate the binding of TRACER in the standard tubes as a percentage of the binding in the absence of standard. This is then plotted against the log of the standard concentration. A log plot spreads out the data points and makes manual calculation of sample concentrations easier (see standard curve one, Figure 3).

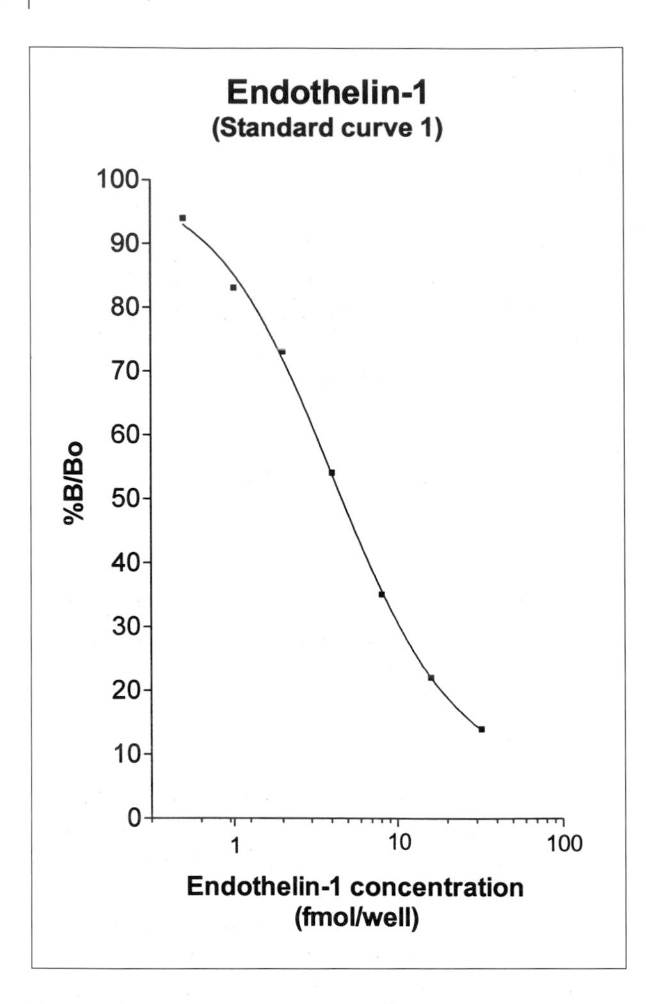

FIGURE 3. A TYPICAL ENDOTHELIN-1 STANDARD CURVE

A number of alternative curve plotting methods are illustrated in standard curves two to four (Figs 5–7). The choice of curve fit software can generate a lot of discussion. Whatever method is chosen, it is important to plot out the data and demonstrate that the curve does actually fit the data points.

Selected immunoassays

Endothelin-1 radioimmunoassay

This assay has been selected as an example of a competitive assay using an iodine-125 TRACER and magnetic separation. Endothelin is a potent vasoconstrictor produced by vascular endothelial cells. It produces a strong and sustained vasoconstriction in most arteries and veins of many mammalian species.

Assay protocol

The assay is performed in polypropylene tubes. Standard (100 µl) or sample (100 µl) and antiserum (100 µl) are added and the tubes incubated at 2–8 °C for 4 hours. TRACER (100 µl of iodine-125 labelled endothelin-1) is then added and the tubes left overnight at 2–8 °C. Amerlex-M (250 µl of magnetizable solid particles) is added and left at room temperature for ten minutes. The antibody-bound fraction is separated by placing the tubes on a magnetic rack for fifteen minutes and then pouring off the supernatant which contains the free phase. The tubes are then counted for one minute in a gamma scintillation counter. In these assays, samples and standards are usually assayed in duplicate.

Cyclic AMP scintillation proximity assay

Cyclic AMP is a member of a biologically important class of molecules termed "second messengers". This is a term for molecules, which are able to transmit intracellularly the biological effects of compounds not able to enter the TARGET cell themselves. The cAMP assay is an example of an HOMOGENEOUS competitive immunoassay, i.e., an assay in which the bound TRACER does not need to be physically separated from the free fraction. This greatly simplifies the assay and makes it more amenable to automation. It is based on the principle that relatively weak beta emitters such as tritium and the iodine-125 Auger electron need to be close to scintillant molecules to produce light, otherwise the energy is dissipated and lost to the solvent. This concept has been used to develop HOMOGENEOUS RIAs by coupling second antibodies onto fluomicrospheres containing scintillant. When a second antibody-coupled fluomicrosphere is added to a RIA tube, any radiolabelled LIGAND that is bound to the primary LIGAND-specific antibody will be immobilized on the fluomicrosphere. This will bring into close proximity the radiolabel

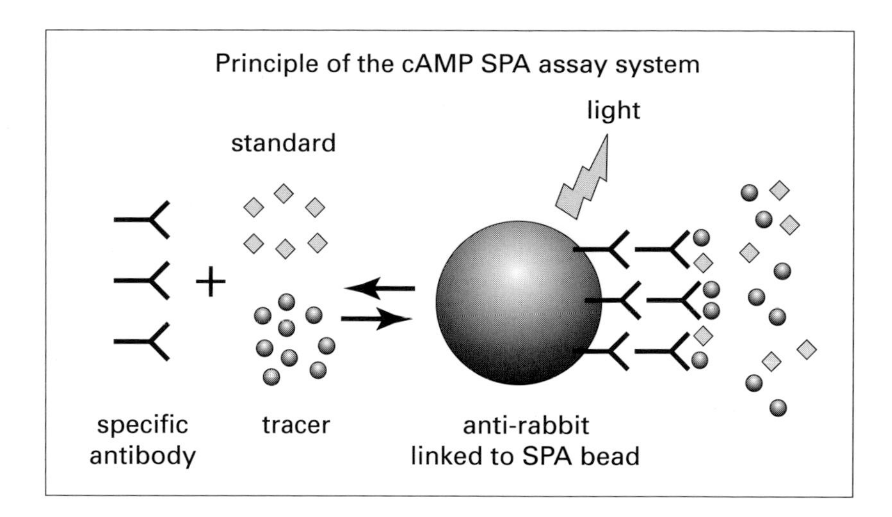

FIGURE 4. THE PRINCIPAL OF A SPA ASSAY

and the scintillant, activating the scintillant to produce light. Any unbound radioligand remains too distant to activate the scintillant (Fig. 4). The signal is measured in a liquid scintillation counter and is inversely proportional to the concentration of LIGAND in the sample or standard [6].

Assay protocol

50 µl of standard or sample followed by 50 µl of iodine-125-labelled cAMP, 50 µl of antiserum and 50 µl of the scintillant beads are pipetted into each assay tube and incubated at room temperature overnight. The amount of TRACER bound to the beads is determined by counting for two minutes in a beta scintillation counter. A typical cAMP SPA plot is shown in standard curve two (Fig. 5). The data are represented as a linear/linear plot.

Leukotriene $C_4/D_4/E_4$ enzyme immunoassay system

The peptido-leukotrienes comprise the slow-reacting substances of anaphylaxis. They are potent mediators of bronchoconstriction, vascular and non-vascular smooth muscle contraction, increase vascular permeability and epithelial mucous secretion. They are widely considered to be important mediators in ASTHMA and antagonists to these compounds are being developed as possible anti-asthma drugs. This assay has been selected as an example of a competitive immunoassay using an enzyme label.

Assay protocol

The assay is performed in a ninety-six-well antibody coated microtitre plate. 50 µl of standard or sample is pipetted into each well and incubated at 4–10 °C for two hours. 50 µl of leukotriene C_4-horseradish peroxidase conjugate is then added and incubated for a further two hours at the same temperature. All wells are washed thoroughly and 150 µl of SUBSTRATE solution added to each. The plate is incubated at room temperature with shaking for thirty minutes and the reaction is then terminated with 100 µl of 1M sulphuric acid. The optical density of each well is determined in an automatic plate reader at 450 nm. A typical leukotriene C_4 plot of optical density against the log concentration is shown in standard curve three (Fig. 6).

Interleukin-10 (mouse) ELISA system

Interleukin-10 is a glycoprotein that inhibits cytokine synthesis by the Th1 sub-population of T cells. The Th1 CYTOKINES are responsible for many aspects of cell mediated immunity, so IL-10 has immunosuppressive activity. There is considerable interest in

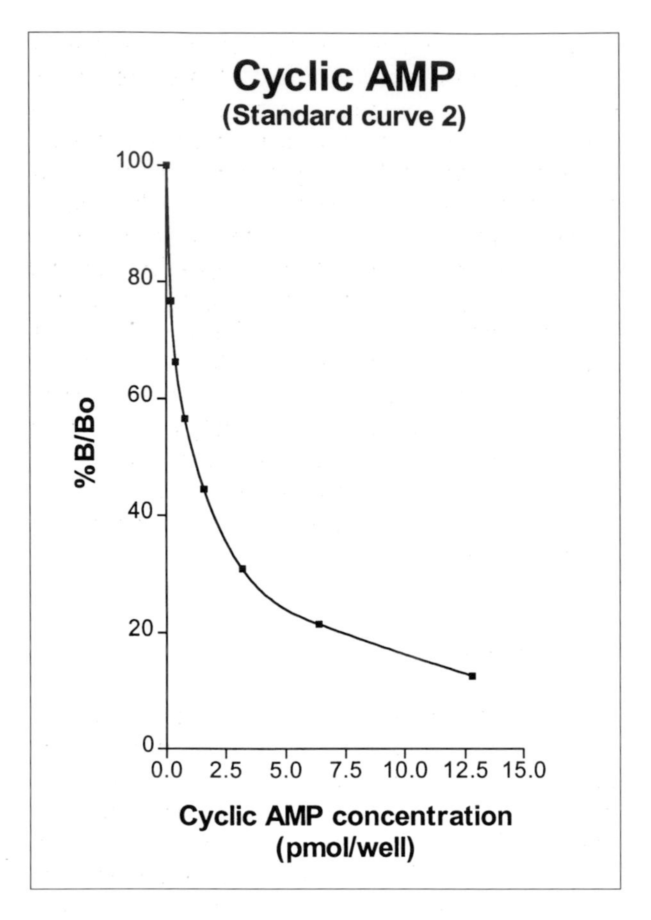

FIGURE 5. A TYPICAL cAMP STANDARD CURVE

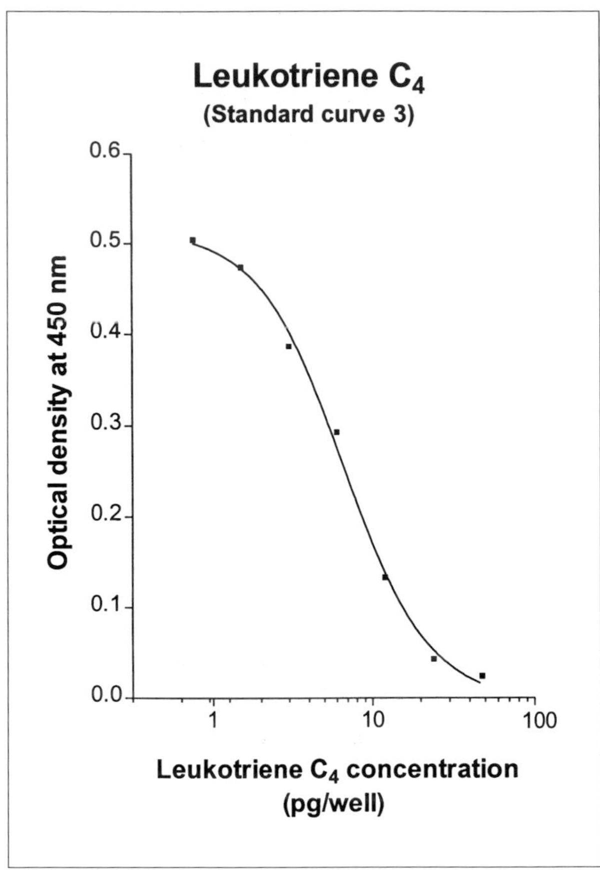

FIGURE 6. A TYPICAL LEUKOTRIENE C_4 STANDARD CURVE

investigating the use of IL-10 in transplantation, rheumatoid arthritis and septic shock. This assay has been chosen as an example of an immunometric assay using a biotin-labelled antibody in combination with a streptavidin/horseradish peroxidase tracer.

Assay protocol

The assay is performed in a ninety-six-well antibody coated microtitre plate. 50 µl of assay buffer and 50 µl of either standard or sample are added to each well. The plate is incubated at room temperature for 3 hours and then washed. 50 µl of biotinylated detection antibody is added to all wells, incubated at room temperature for 1 hour and the plate is washed. 100 µl of streptavidin/horseradish peroxidase is then added, incubated for thirty minutes and the plate is washed again. 100 µl of SUBSTRATE solution is then added and incubated for a further 30 minutes. The reaction is then terminated with 100 µl of dilute sulphuric acid and the optical density measured at 450 nm. A typical IL-10 log/log plot analysed by linear regression is shown in standard curve four (Fig. 7).

Assay performance and validation

When either developing or evaluating an immunoassay, a number of questions relating to the perform-

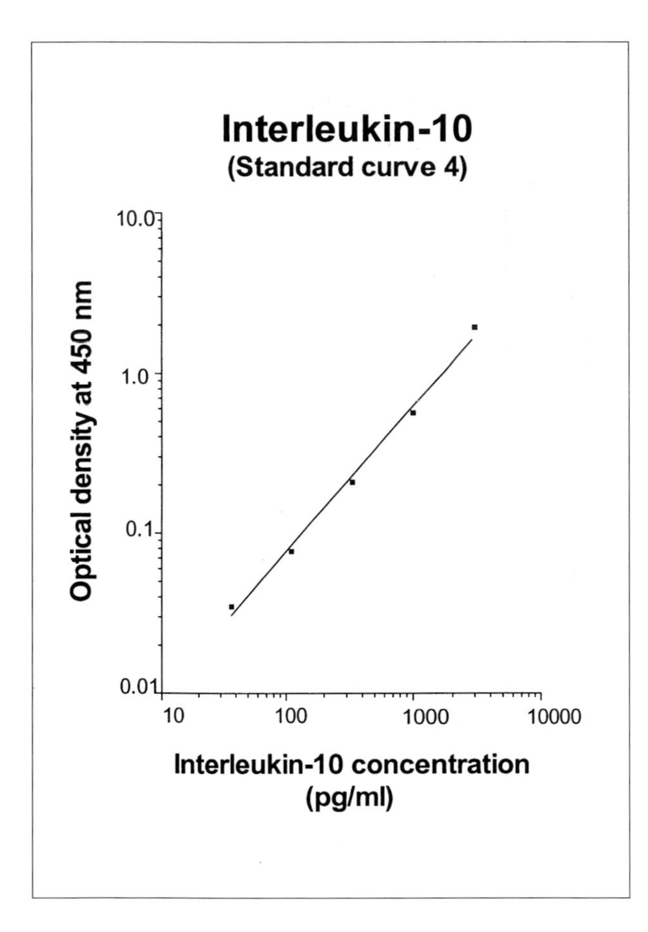

FIGURE 7. A TYPICAL INTERLEUKIN-10 STANDARD CURVE

ance of the assay need to be considered. These include the likely cost of the assay, how easy is it to perform, the equipment required to carry out the assay, what ANALYTE concentration can be measured, how reproducible the assay is and whether or not the assay measures the true ANALYTE concentration. Some of these questions have already been covered. The remainder is discussed in the following sections.

Assay precision

Precision is an index of assay reproducibility and is a guide to how much the determined ANALYTE concentration is likely to vary from measurement to measurement. Within-assay and between-assay precision refers to the reproducibility of measurement in single and multiple assays respectively. Precision is likely to vary throughout the standard curve range. A precision profile of the assay can be constructed by performing multiple measurements at each standard concentration. It is difficult to state what is acceptable with regard to assay precision, as this will vary dependent upon the intended application of the assay. A reasonably well-designed assay will have a within-assay precision of <10% at the extremities of the standard curve and <5% over most of the assay. The between-assay precision is usually a few percent higher than the within-assay precision. In a well-designed assay, the sample concentrations will normally fall within the part of the standard curve having the highest precision.

Assay sensitivity

The SENSITIVITY of an assay is the lowest level of ANALYTE that can be detected. Various ways of calculating assay SENSITIVITY have been used. One common method is to calculate the standard deviation of the zero standard and express the SENSITIVITY as that value corresponding to two standard deviations from the zero standard. However, it is important to be aware that samples cannot be assigned a PRECISE value near to the SENSITIVITY limit of an assay, as the precision at these concentrations will be extremely poor. Samples should only be given values when they fall within the range of the standard curve data points. The range of the data points should be set from the precision profile. Any sample values outside the standard curve range should really be given the value of less than or greater than the lowest or highest standard respectively.

Assay validation

Before using an immunoassay it is important to validate the assay. An assay should be validated for each SAMPLE MATRIX such as plasma, serum or cell culture supernatant that will be used in the assay. The SPECIFICITY of an assay is confirmed by testing against related substances. The analytical recovery is assessed by adding known amounts of the ANALYTE to the SAMPLE MATRIX under evaluation and measuring the percentage recovery of the ANALYTE. Assay linearity is determined by diluting samples and determining whether or not the measured values are in agreement with the non-diluted sample when the dilution factor is taken into account.

In some situations it can prove difficult to develop a valid assay without some kind of sample purification step. Vitamin D metabolite assays, especially the assay for 1-α-25-dihydroxy vitamin D, require such a step. This molecule is present at concentrations ~1000-fold lower than other metabolites such as 25-hydroxy vitamin D. Due to the close similarity of these metabolites, it has proved impossible to produce an antibody with the SPECIFICITY to discriminate sufficiently between the metabolites. In such circumstances it is necessary to remove the 25-hydroxy metabolite, for example on small silica cartridges, prior to the assay. However such procedures are time-consuming and should be avoided if at all possible.

Summary

Immunoassays have remained a core technique in the biological sciences for the last forty years. It is interesting to speculate on future developments in the field. It is easy to predict that the range of ANALYTES measured by immunoassays will continue to grow. For example, immunoassays for CYTOKINES and metalloproteinases will continue to be developed as new molecular species are identified. The way in which the technology will develop is more difficult to foresee. One can predict that there will be two trends, a move towards high-throughput assays and the multiplexing of assays.

Automation will become even more widespread in the clinical field and high throughput pharmacological screening. There is likely to be an increased reliance on fluorescent and luminescent detection systems [7] due to the SENSITIVITY and ease of use of such labels. The format of the assays will move from a 96- to a 384- and possibly 1536-well plate design. These plate designs will increase assay throughput and decrease reagent costs. The drive towards assay miniaturization, high assay throughput and the use of either fluorescent or chemiluminescent detection systems is changing the way researchers are detecting assays. Charged-coupled devices (CCD) such as LEADseeker™ Multimodality Imaging System are capable of imaging any microtitre plate footprint in seconds [8]. The same requirements for high throughput assays are motivating a trend towards the development of HOMOGENEOUS assays. The lack of a separation step means that these systems are much more amenable to automation [9].

A second broad emerging trend is the development of multiplexed assays. In this approach, multiple ANALYTES are measured in one assay [10]. Commercial kits such as the SearchLight™ system are available. This technology resembles a traditional "sandwich ELISA" except that a grid of between nine and twenty-five capture antibodies is arrayed on the base of each well of a 96-well microtitre plate. The assay is performed in a similar fashion to a conventional ELISA except that all the ANALYTES can be simultaneously quantified in each sample. ANALYTE binding is detected using an enzyme LABEL and a chemiluminescent SUBSTRATE. The signal from each spot within the well is quantified using a CCD instrument. The author has used the SearchLight procedure in combination with the LEADseeker imager to quantify nine CYTOKINES simultaneously in each sample. The SENSITIVITY and standard curve range are superior to conventional cytokine ELISAs (unpublished observations).

An alternative approach to multiplexing has been developed by the Luminex Corporation. Luminex internally dye polystyrene microspheres with two spectrally distinct fluorochromes. Using varying ratios of these two fluorochromes, one hundred spectrally distinct microspheres sets have been developed. These microspheres allow the multiplexing of up to one hundred different ANALYTES. A third fluorochrome coupled to a reporter molecule (an adaptation of the assay principle outlined in Figure 2) quantifies the extent of the reactions taking place at the surface of these microspheres. The assay is quantified in a specially designed flow cytometer-like instrument equipped with two lasers. One laser identifies the microsphere (so identifying the ANALYTE); the second quantifies the amount of the fluorochrome bound to each microsphere (so quantifying the ANALYTE). This system has been used to multiplex a variety of ANALYTES [11]. The author has used this system to set up multiplexed assays of four CYTOKINES in both HOMOGENEOUS and HETEROGENEOUS format using a fluorescent phycoerythrin-based detection reagent. The HETEROGENEOUS assays have a similar SENSITIVITY and a wider standard curve range compared to conventional cytokine ELISAs (unpublished data).

Immunoassays have been a core analytical technique in the biological sciences for the last forty years. The continual improvement in the methodology and the development of novel approaches will ensure that the technology will continue to play a centre role for the foreseeable future.

Selected readings

Wild D (ed) (2001) *The Immunoassay Handbook*. Nature Publishing Group, New York

References

1 O'Sullivan MJ, Capper S, Horton JK, Whateley J, Baxendale P (2001) Immunoassay applications in life-science research. In: D Wild (ed): *The Immunoassay Handbook*. Nature Publishing Group, New York, 817–845
2 Yalow RS, Berson SA (1959) Assay of plasma insulin in human subjects by immunologic methods. *Nature* 184: 1684–1689
3 Voller A, Bidwell DE, Huldt G, Engvell E (1978) A microtitre plate method of enzyme linked immunoassay and its application to malaria. *Bull WHO* 51: 209–216
4 O'Sullivan MJ, Marks V (1981) Methods for the preparation of enzyme-antibody-conjugates for use in enzyme immunoassay. In: JJ Langone, HV Vunakis (eds): *Methods in Enzymology, vol. 73, Immunochemical techniques, Part B*. Academic Press, New York, 147–166
5 O'Sullivan MJ, Bridges JW, Marks V (1979) Enzyme immunoassay: A Review. *Annal Clin Biochem* 16: 221–239
6 Baxendale PM, Martin RC, Hughes KT, Lee DY (1990) Development of scintillation proximity assays for prostaglandins and related compounds. *Adv Prost Thromb Leuk Res* 21A: 303–306
7 Kricka LJ, Wild D (2001) Signal generation and detection systems (excluding homogeneous assays). In: D Wild (ed): *The Immunoassay Handbook*. Nature Publishing Group, New York, 159–176
8 Fowler A, Davies I, Norey C (2000) A multi-modality assay platform for ultra-high throughput screening. *Curr Pharma Tech* 1: 265–281
9 Ullman EF (2001) Homogeneous Immunoassay. In: D Wild (ed): *The Immunoassay Handbook*. Nature Publishing Group, New York, 177–210
10 Wiese R, Belosludtsev Y, Powdrill T, Thompson P, Hogan M (2001) Simultaneous multiANALYTE ELISA performed on a microarray platform. *Clin Chem* 47: 1451–1457
11 Carson RT, Vignali DAA (1999) Simultaneous quantification of 15 cytokines using a multiplexed flow cytometric assay. *J Immun Methods* 227: 41–52

B3

Flow cytometry

John F. Dunne and Holden T. Maecker

Introduction

The technologies associated with watching cells flow through an image plane have developed in several important directions over the last 30 years. This chapter will focus on flow cytometers, tools that have optimized fluidics, electronics and optics to generate extraordinary measurement precision and dimensionality on samples of cells in suspension. Outside the scope of this discussion remains the continuum of devices ranging from real optical microscopes that create pictures of cells moving *in situ* like blood cells flowing in a capillary, to simple counting devices that monitor the passage of particles in a sensing region.

While the fundamental fluidics constraints of flow cytometers have not changed, the equipment available to analyze the particles as they pass through a flow cell has been enhanced dramatically with developments in lasers, digital electronics, fluorescence chemistries and computer power. Analysis rates have migrated from hundreds to tens of thousands of cells per second, but more importantly, the number of colours commonly simultaneously measured has gone from one or two in the 1970s, to four becoming common in the 1990s, to the eight to twelve colour experiments being published recently [1]. These polychromatic experiments create a new kind of information about flow samples, as well as a new set of technical challenges. The optimization of instrument and experiment protocols and the efficient analysis of this kind of data requires mind-stretching exercises in multidimensional thinking. Useful highly dimensional commercial tools are beginning to enter research laboratories, and scientists are using them with some satisfaction. As a result, the technology is now bounded more by the complexity of thought than by the availability of sophisticated experimental devices.

Mechanistic principles

Fundamentally, flow cytometers measure fluorescent or scattered light emitted by a cell during its illumination in bright light, typically a highly focused laser beam [2]. In most cases, the cell stream is positioned using hydrodynamic focusing, in which a thin core of cell suspension is limited to the centre of a larger flowing sheath fluid. The cells thus arrive sequentially into the laser beam at thousands of cells per second.

Light is scattered by the cells, and this scattered light is usually measured at narrow angles just above and below the laser beam (commonly called "FORWARD SCATTER"), and by a separate detector at wider angles orthogonal to the beam (commonly called "SIDE SCATTER"). FORWARD SCATTER is descriptive of the size of the cell, while SIDE SCATTER is proportional to size and granularity. Thus in commonly analyzed populations of blood cells, platelets, lymphocytes, monocytes and granulocytes can be distinguished reasonably well simply on the basis of their intrinsic light-scattering properties (Fig. 1).

More valuably, cells can be stained with fluorescent reporter molecules, and the binding of these reporters can describe extremely subtle phenotypes. The most common class of reporters is fluorescently conjugated monoclonal antibodies (mAb). Thousands of these products are commercially available. The antibody protein binds very specifically to particular EPITOPES in proteins present on the cell surface, or if the cell is permeabilized the antibody can enter the cell and bind to EPITOPES within. Since the

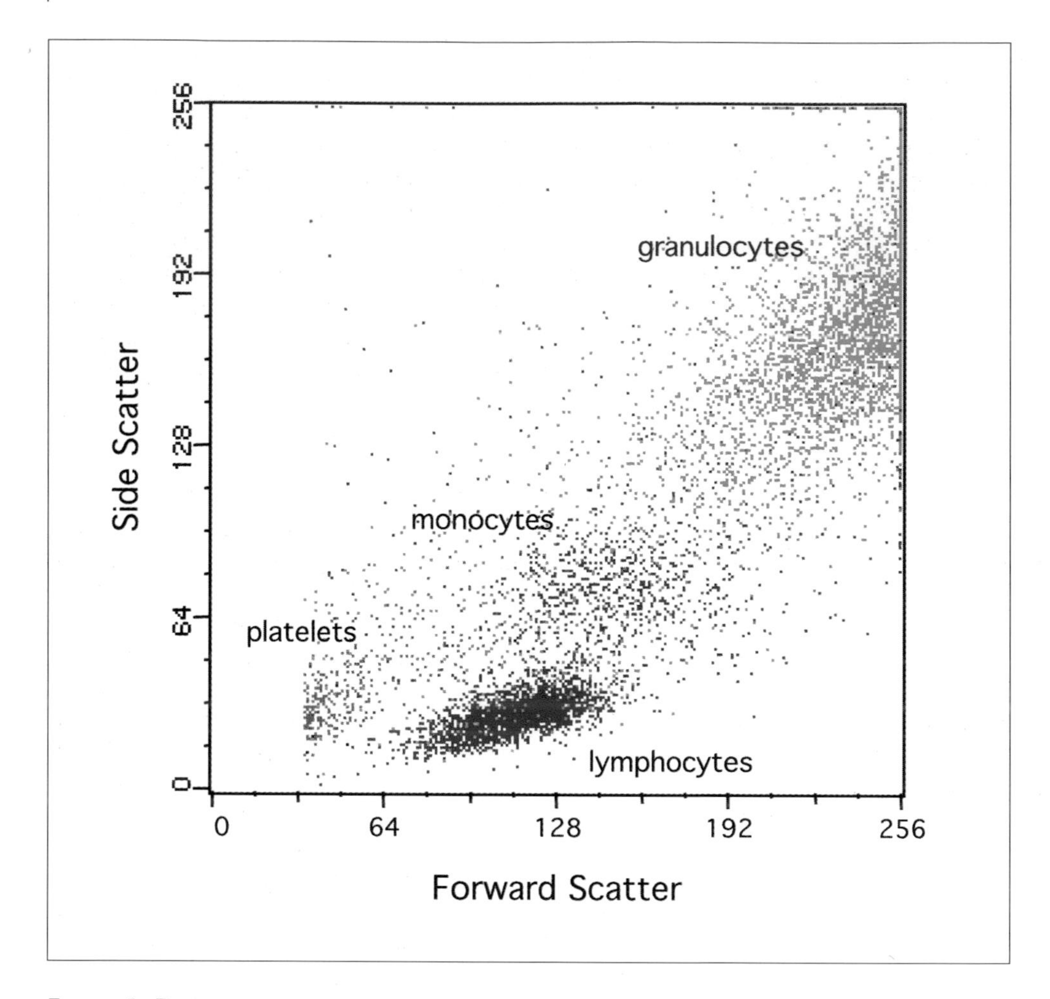

FIGURE 1. DISCRIMINATION OF LEUKOCYTE SUBSETS BY LIGHT SCATTER

The labeled subsets of leukocytes were fluorescently stained with specific mAb (not shown), and the specific mAb-stained populations were coloured as indicated. These populations were then displayed in a dot plot of forward versus side scatter. A threshold was set on forward scatter, excluding much of the platelet population, in order to prevent the collection of small debris that would otherwise interfere with the analysis. Note that the coloured populations are reasonably well resolved from each other. However, the relative positions of these populations will depend upon the treatment of the sample (in this example, formaldehyde-fixed and detergent-permeabilized cells were used).

antibody is covalently linked to a fluorescent molecule, the antibody binding is directly correlated to the fluorescent signal emanating from the cell during laser excitation. In a well-developed assay with binding properties and chemistries well understood, the brightness associated with antibody binding can be a direct measure of the abundance of the relevant antigenic protein (see Box 1). By using several mAb, each conjugated to fluorescent dyes with characteristic colours, combinations of antibody binding can characterize populations of cells that may coexist in suspension. These and other commonly used fluorescence strategies will be discussed in more detail below.

Box 1. Why use flow cytometry as an analytic method?

It is possible to use many different methods for quantifying the proteins expressed by a population of cells. In addition to flow cytometry, Western blotting, immunoprecipitation, ELISA, RIA, enzyme-linked immunospot (ELISPOT), fluorescence microscopy, and immunohistochemical staining are all methods used to quantify cellular proteins. Of these, only flow cytometry, ELISPOT, fluorescence microscopy, and immunohistochemical staining provide information on a single-cell basis. The others are bulk assays that do not quantify the number or phenotype of cells that are expressing the protein of interest (though they may provide other information, like the size of the targeted protein). Of the single-cell assays, flow cytometry is unique in the number of fluorescence parameters that have been combined in a single assay. Thus, it provides the richest information, on a per-cell basis, of the current generation of assays for cellular protein analysis.

Classification/types of assay

Immunophenotype

In the 1980s, with the early propagation of FLOW CYTOMETRY and mAb, there was some expectation that the protein surfaces of hematopoietic cells could be mapped with an ever-increasing complexity to reveal changes in the types of cells present or their relative abundance that might correlate with clinical disease onset or progression. In the intervening 20 years, with thousands of fluorescent antibodies commercially available, the management of only two patient classes is substantially driven by these kinds of flow assays, HIV disease and leukemia/lymphoma.

Flow immunophenotyping in HIV disease

It's now clear that one of the most proximate and dramatic effects of HIV infection is an eventually lethal loss of $CD4^+$ T cells [3]. This was one of the first easily recognized manifestations of infection, indeed the first clinical test that helped identify AIDS patients [4], and has been one of the most useful analytical tools available to characterize mechanisms of HIV pathologies as well as therapeutic benefits of anti-HIV pharmaceuticals and other treatment modalities. More recently augmented by assays that quantify viral load, CD4 counting remains the standard method of monitoring disease progression [4].

The CD4 counting assay is relatively simple among FLOW CYTOMETRY assays. Most commonly, fluorescently conjugated CD4 mAb are added to whole blood, generally as part of a cocktail of 2 to 4 antibodies which together uniquely identify those lymphocytes that bear the CD4 antigen. With specialized software tools and sample preparation procedures that are fully validated and approved as *in vitro* diagnostics, the flow cytometer can report the number of CD4 cells remaining per microlitre of blood. While normal ranges can vary around 1000 cells, AIDS progression commonly results in levels that are half of that, and profound disease is usually correlated with CD4 counts at or below 200 CD4 cells per µl.

While the assay is commonly available to AIDS practitioners in the developed world, technical and financial barriers have thus far prevented the propagation of the current CD4 assays in environments with less well-funded medical infrastructures. Technology providers are now working with governments and other funding organizations to develop viable alternatives and/or equivalents to address this critical world health issue.

Flow immunophenotyping in leukemia/ lymphoma

The general observation that the immunophenotypes commonly found in peripheral blood samples are relatively stable may have been disappointing in some regards, but it was helpful in that it allows the recognition of unusual phenotypes characteristic of transformed cells. Especially in the case of transformed cells of hematopoietic lineages, these unusual phenotypes have now been mapped into more or less broadly recognized categories of malignant disease, and flow cytometric assays are commonly used to help diagnose and monitor leukemias and lymphomas.

Several clinical research consortia have published consensus documents describing the general utility, practical aspects and interpretation of combinations of markers and their distribution on cancer cells [5–7]. Still, no commercially available or otherwise standard kit has been approved for diagnostic use, and common practice is for clinical centres to use their own discretion to implement and validate such assays. Nevertheless, the standard of care of patients suffering from these malignancies has been improved, sometimes dramatically, largely based on practitioners' ability to detect and characterize leukemias and lymphomas by FLOW CYTOMETRY.

Immunotoxicology immunophenotyping

While most common disease states do not change the frequency of normal blood cell components, the toxicity associated with novel pharmacophores or industrial pollutants can have profound effects. Recent work establishing consensus protocols for the evaluation of immunotoxity of such chemical entities documents a growing utility for this class of assays [8]. The foci of these assays are typically rodent, dog or non-human primate models, and mAb reagents are less commonly available or less well characterized. The normal ranges of various lymphoid cell types, other environmental factors that can influence these ranges, and the magnitude of changes in these cell types that should be used as sentinels for concern or to prohibit the use of such chemicals in various exposure scenarios, all represent ongoing work.

Functional assays

An important additional class of flow assays consists of those assays that experimentally induce a functional response in cells of interest, then use FLOW CYTOMETRY to measure the response. One of the most common assays of this type is the measurement of CYTOKINE production in lymphocytes in response to antigenic stimulation. The production of a CYTOKINE can be easily measured using well-established reagents and protocols, and is a hallmark of lymphocyte function. With multicolour FLOW CYTOMETRY, the frequency of responsive cells, the amount and kinetics of CYTOKINE production, and the combination of various CYTOKINES can very richly describe an immunological response qualitatively and quantitatively. These assays will be discussed in detail below, and by way of comparison, other functional assays will be addressed briefly (see Box 2).

Proliferation

Another hallmark of lymphocyte function is cellular proliferation. Several flow assays have been well established in the literature that have been correlated more or less well to traditional ^{3}H-thymidine incorporation associated with DNA synthesis. Perhaps the most directly correlated of these is the use of bromodeoxyuridine (BrdU), an orthologue of thymidine used by DNA synthase and incorporated in the nuclei of proliferating cells, and detected by FLUOROCHROME-conjugated specific mAb [9]. Another proliferation assay utilizes the dilution of a cytosolic dye, commonly carboxyfluorescein diacetate, which results from CYTOKINESIS [10]. One of the earliest flow assays for proliferation, which is still in common use, is the direct measurement of DNA content using a quantitative DNA stain such as propidium iodide or the vital dye Hoescht 33258 [11, 12]. This is especially useful for peripheral blood lymphocytes, which are natively arrested with a 2c DNA content in the G0 phase of the cell cycle, upon stimulation can enter S phase, and progress to the 4c DNA content typical of G2/M. This doubling of DNA is easily recognized flow-cytometrically, and the percentage of cells with more than 2c DNA can be roughly related to traditional metrics like the mitotic index. Another nucleic acid stain, acridine orange, has the unusual property of shifting its emission spectrum depending on whether it is bound to single- or double-stranded nucleic acid (roughly speaking, RNA or

Box 2. Different assays for measuring T cell function

In this chapter, measurement of intracellular cytokines by flow cytometry is described in some detail as a method for determining T cell functional responses after short-term stimulation. However, alternative methods are also in common use. Some are flow cytometry assays, such as MHC-multimer staining [35], or cytokine capture assays [36]; others use different analytical platforms, such as ELISPOT [37]. The main advantage of flow cytometry as an analytical platform is its multiparameter capability (see Box 1). Of the flow cytometry methods available for measuring T cells in a short-term assay, MHC-multimer staining is unique in that it measures T cell specificity rather than function, and thus requires no activation at all. Multimeric forms of MHC molecules with bound antigenic peptides are produced and labeled with a fluorochrome such as PE. These can then be used to stain T cells (via their T cell receptor) in much the same way as a fluorochrome-conjugated antibody. However, their use requires a knowledge of the particular peptide and MHC restriction pattern of a T cell immune response; as such, they are not useful for quantifying the overall T cell response to a pathogen, especially in a MHC-diverse population. Cytokine capture assays, like intracellular cytokine staining, measure responses without regard to MHC restriction. The cytokine capture assay is especially useful if one wishes to maintain the cells in a viable state for sorting and further analysis (since it does not require fixation and permeabilization). Still newer assays, such as measures of granule exocytosis [38], can also be used on viable cells, and add to the armamentarium of tools available for dissecting T cell responses.

DNA) and has been used to simultaneously report DNA doubling and the increase in mRNA characteristic of lymphoid stimulation [13].

Bead matrix immunoassays

Recently bead-based immunoassays have been developed for flow cytometric assessment of various soluble ANALYTES. Each member of a set of beads is commonly identified by specific fluorescence level and/or size, and acts as an ANALYTE capture platform using covalently bound specific mAb. The amount of captured ANALYTE is measured using a second antibody specific for an alternate site of the same ANALYTE, this second antibody being conjugated to a fluorescent reporter. Since the various beads can be recognized independently, and since each captured ANALYTE can be quantified independently, this assay format is well suited to multiplexed analysis. Thus small volumes of biological fluids (as little as 10 µl of tears, for example [14]) can be inspected for quantification of soluble proteins or other LIGANDS for which specific antibody pairs are available. Common implementations of this strategy include the simultaneous quantification of 5–20 different immunomodulatory proteins including CYTOKINES, chemokines, growth factors and hormones. Commercially available kits range from completely integrated systems to basic platforms that can be developed for custom ANALYTE sets, and are used broadly in basic research and drug discovery proteomics projects. No patient management tools in this class have yet been approved for *in vitro* diagnostic use as of this writing.

Components/construction of assays

Traditionally, FLOW CYTOMETRY sample preparation and processing has been constrained by the fact that flow cytometers were designed to accept tubes, and those tubes were loaded onto the instrument manually, one at a time. A second constraint came from instrument set-up and data analysis, which were typically time-consuming and required a certain knowledge base. Both of these constraints are now changing with new instrumentation and software. Current cytometers can often handle racks of tubes that are automatically run in a walk-away mode. Plate loaders are also available that can feed samples from 96-well plates directly to the cytometer, again in a walk-away mode. These developments have been complemented by software that can either: (1) perform data analysis simultaneously with acquisition; or (2) perform a batch analysis routine that analyzes all of the data from an experiment at once. The usefulness of such analysis routines is further augmented by flexible analysis templates that can accommodate changing data without repetitive adjustment of settings by the operator [15].

The impact of these changes in FLOW CYTOMETRY hardware and software have opened up the use of FLOW CYTOMETRY to larger pre-clinical and clinical studies that might involve hundreds of specimens. However, the steps involved in sample preparation can still be complex, as detailed below. Automation of these steps will further allow the use of FLOW CYTOMETRY in high-volume settings where it was previously considered too cumbersome.

Antibodies for cell staining

Since the late 1970s, polyclonal antisera have been increasingly replaced by mAb for most immunological applications, including FLOW CYTOMETRY. MAb can be produced in unlimited quantities, have better lot-to-lot reproducibility, and tend to have lower backgrounds than polyclonal antisera [16].

Another trend in FLOW CYTOMETRY has been the increased use of direct FLUOROCHROME conjugates, rather than indirect fluorescence analysis using second-step antibodies or other reagents that carry the FLUOROCHROME tag. While indirect staining can sometimes amplify the fluorescence signal compared to direct staining, it may be at the expense of increasing background fluorescence [17]. Also, multiparameter FLOW CYTOMETRY is much more difficult using indirect staining methods, due to the potential for interaction between multiple second-step conjugates and the primary antibody TARGETS.

Some of the more common FLUOROCHROMES used in FLOW CYTOMETRY today are listed in Table 1. Not all of these are compatible with all cytometers, or even with all other FLUOROCHROMES, as can be seen by their excitation and emission spectra (see for example www.bdbiosciences.com/spectra). Other considerations of relative brightness and compatibility with certain experimental parameters (intracellular *versus* surface staining, for example) will also steer the choice of certain FLUOROCHROME conjugates over others. For a more complete discussion of the issues of designing multiparameter experiments, see references [18, 19].

Cell types

FLOW CYTOMETRY can be performed on any cell type that can be rendered into a single-cell suspension. Since blood cells already exist in this state, they have been most widely studied by the technique. However, adherent cells or tissues can also be analyzed if they are dissociated from each other and/or their SUBSTRATE using either enzymatic (e.g., trypsin) or non-enzymatic (e.g., EDTA) treatments. Where possible, non-enzymatic dissociation protocols are preferable, because they do not cleave cell-surface proteins that might be TARGETS of the flow cytometric analysis.

Erythrocyte lysis

Whole blood consists of relatively HOMOGENEOUS erythrocytes, and a much more complex collection of leukocytes. The latter can be stained with FLUOROCHROME-conjugated antibodies in the context of whole blood, but analysis of whole blood is difficult because of the light-scattering properties of the erythrocytes, which are so numerous as to obscure the illumination of the leukocytes. Fortunately, erythrocytes are differentially sensitive to hypotonic lysis, and can be removed by incubation of blood with ammonium chloride in water. Leukocytes are more resistant to osmotic damage than erythrocytes, which cannot exclude this salt, and subsequently take up water and are lysed. Typically, whole blood is treated for 10 minutes with a large volume of ammonium chloride solution either before or immediately after staining with FLUOROCHROME-conjugated antibodies. An alternative to ammonium chloride lysis involves hypotonic lysis in the presence of a fixative (e. g., formaldehyde), whereby the erythrocytes are preferentially lysed while LEUKOCYTES are fixed in a single incubation of 10 minutes or so. Solutions for

Table 1. Commonly used fluorochromes for antibody-coupled flow cytometry

Fluorochrome	Type of molecule	Typical excitation laser	Approximate emission peak
Fluorescein isotyocyanate (FITC)	Small organic	488 nm	518 nm
AlexaFluor 488	Small organic	488 nm	518 nm
Phycoerythrin (PE)	Protein	488 or 532 nm	574 nm
PE-Texas Red	Protein tandem	488 or 532 nm	615 nm
PE-Cy5	Protein tandem	488 or 532 nm	665 nm
Peridinin chlorophyll protein (PerCP)	Protein	488 or 532 nm	676 nm
PerCP-Cy5.5	Protein tandem	488 or 532 nm	695 nm
PE-Cy7	Protein tandem	488 or 532 nm	776 nm
Allophycocyanin (APC)	Protein	633 nm	659 nm
AlexaFluor 647	Small organic	633 nm	667 nm
AlexaFluor 700	Small organic	633 nm	718 nm
APC-Cy7	Protein tandem	633 nm	784 nm
Pacific Blue	Small organic	405 nm	454 nm
AmCyan	Protein	405 nm	487 nm

this procedure are commercially available. After lysis/fixation, the cells may be directly analyzed by FLOW CYTOMETRY ("no-wash" assays); or they may be subjected to washing to remove the residual red cell debris and unbound fluorochrome-conjugated antibodies before analysis ("washed" assays). Washed assays typically allow better signal-to-noise discrimination of fluorescently stained populations, and better visualization of lymphocyte scatter properties. However, NO-WASH-ASSAYS have become popular in the clinical marketplace because of their simplicity; and by detecting cells based on threshold staining for a common leukocyte antigen, such as CD45, these assays can adequately resolve subpopulations of lymphocytes.

Density gradient separation

As an alternative to erythrocyte lysis of whole blood, mononuclear LEUKOCYTES (lymphocytes and monocytes) can be directly isolated from whole blood prior to staining. Solutions of high molecular weight carbohydrates, such as FICOLL, are used for this separation, which is accomplished on the basis of density [20]. By underlaying whole blood (usually diluted 1:1 with buffer or tissue culture media) with a similar volume of a FICOLL solution, a density gradient is formed. This biphasic solution is then subjected to centrifugation (typically at about 400 x G for 15-20 minutes). Erythrocytes and granulocytes, which have the greatest buoyant density, are pelleted to the bottom of the FICOLL layer. Lymphocytes and monocytes, which are less dense, collect at the interface of the plasma and FICOLL, where they are collected by pipetting. Platelets, the least dense leukocytes, will remain in the plasma layer. Successive centrifugation of the mononuclear cells collected from the interface is then carried out at 250 × G to remove residual FICOLL and further deplete the mononuclear cells of platelets.

Density gradient separation techniques are a standard, albeit time-consuming method for the isolation of mononuclear cells from small to large volumes of blood. However, simpler alternatives are also available that consist of a gel matrix pre-dispensed into a blood collection tube or centrifuge tube. By adding whole blood and centrifuging at a prescribed speed, the mononuclear cell layer can be collected from the top of the gel interface, while erythrocytes and granulocytes are forced through the gel plug. Such systems are available from commercial vendors, and provide generally equivalent results with greater convenience than the FICOLL separation technique [21].

Activation

As described briefly above, functional assays are those in which the cell types of interest are stimulated *in vitro* and allowed to manifest some kind of response, the quantity and quality of which is measured in the flow cytometer. Lymphocyte activation is a particularly valuable class of functional assays, and typically requires specific antigen or polyclonal MITOGEN as a stimulus. A commonly measured response to this stimulus is CYTOKINE production, which can be detected in a time period of as little as 4 hours (for IFN-γ, IL-2, TNF-α, and IL-4) [22]. Visualization of CYTOKINE production is accomplished with the aid of a secretion inhibitor, such as brefeldin A [23] or monensin [24], which allows for intracellular accumulation of the CYTOKINES where they can be specifically stained after cell permeabilization. Typical antigen-activation protocols involve 6 hours of stimulation in the presence of brefeldin A (for peptides or MITOGENS), or 6 hours of stimulation with protein antigens, the last 4 hours in the presence of brefeldin A [25]. The latter protocol allows for proteins to be processed and presented by MHC molecules prior to inhibition of the secretory pathway by brefeldin A [26]. Activation is carried out at 37 °C, but can be terminated by cooling the cells prior to further processing [22]. Thus, programmeable water baths or similar devices can be used to automatically time the activation process, even if it is initiated at the end of a workday.

Because activation of lymphocytes can induce adhesion molecule expression, a brief treatment with EDTA at the end of the activation period is recommended to dislodge adherent cells from the culture vessel [26].

Fixation/permeabilization

For staining of intracellular EPITOPES such as CYTOKINES, cells are first fixed (usually with a formaldehyde-based buffer) and then permeabilized (usually with a detergent-based buffer). Commercial reagents containing fixatives and detergents are readily available and allow for reproducible fixation and permeabilization protocols.

Cell staining

For activation assays designed to measure CYTOKINE production, intracellular staining is required; but cell-surface staining is usually also performed in the same sample to allow phenotyping of the responding cell population. Depending upon the antibodies and EPITOPES targeted, this surface staining can sometimes be done together with the intracellular CYTOKINE staining, if the targeted EPITOPES are resistant to the conditions of fixation and permeabilization [22, 26]. If they are not, a surface staining step is carried out after activation but before fixation and permeabilization.

For most EPITOPES, surface and/or intracellular staining can be done by incubation with a cocktail of FLUOROCHROME-conjugated antibodies for 30–60 minutes at room temperature. Titration of antibodies is required for optimal staining (although many manufacturers offer pre-titred antibodies and cocktails of antibodies). The optimal titre for surface staining (unfixed cells) is often different from that for intracellular staining (after fixation and permeabilization).

As noted above, simple surface staining of whole blood can be done using a NO-WASH-ASSAY format. But more complex assays, such as intracellular CYTOKINE staining, require washing, both after fixation and permeabilization steps, and after antibody staining.

Increasing automation and throughput

The relative complexity of sample preparation, especially for the more complex functional assays, has inhibited their application to studies of large numbers of animals or large clinical trials. However, robotic sample preparation devices are available and more are in development that would significantly aid in this process. For example, robotic workstations are commercially available that can aliquot whole blood into staining tubes, add antibodies, incubate, and add erythrocyte lysis buffer. There are also workstations that can perform cell washing, and thus can automate most of the steps of a washed assay. Integration of these robotic workstations can not only increase the number of samples handled in these applications, but can also lend considerable standardization to processes that otherwise are highly operator-dependent.

Truly high-throughput sample processing, however, is best accomplished with multiwell plates. Both phenotypic and functional assays can be performed in 96-well plates; in fact, protocols for activation, processing, and analysis of samples for intracellular CYTOKINE staining in a single 96-well plate have been published [15]. With the availability of plate-based loaders for flow cytometers, much larger numbers of samples can be processed in a single run, with minimal incremental technician time. Further automation of plate-based sample processing can also be accomplished using robotic workstations. For example, a robotic workstation has been described that integrates a plate-based liquid handling robot (for dispensing cells and reagents) with an incubation station, plate-based indexing centrifuge, and a flow cytometer equipped with a plate loader [27]. Transfer of plates to and from the various stations is done with a robotic arm, and software controlling the robotic arm also integrates the various components. Further development of such systems will broaden the ability to use FLOW CYTOMETRY for highly parallel studies potentially involving thousands of samples.

Data acquisition and analysis

Although sample preparation can be made parallel using multiwell plates and automation, sample acquisition is still a serial process. Fortunately, tube- and plate-loaders are available that can automate acquisition, and software can be used to annotate data files before they are run.

Data analysis involves the setting of "gates" or regions in one- or two-dimensional data space, followed by further analysis of the fluorescence properties of cells within those gates. Typically, viable cells of interest are first identified by their light scatter properties, using forward and SIDE SCATTER parameters. This may be followed by gating on subsets of lymphocytes, for example, $CD3^+$ (T cells), $CD19^+$ (B CELLS), etc. Subsets of T cells ($CD4^+$, $CD8^+$) may be identified through further gating. In the case of functional assays, a responsive population, e.g., CYTOKINE-positive, is quantified from the subset of interest, e.g., $CD8^+$ T cells.

The PRECISE placement of fluorescence and light scatter gates is historically done by eye, based upon recognition of typical patterns of staining (negative, dim, bright, etc.). However, cluster-finding algorithms are available in some current FLOW CYTOMETRY software programmes that allow these gates to be drawn in a semi-automated way, and to be responsive to changes from one data file to another [15]. Such changes might include slight differences in staining intensities in different donors, between different CYTOKINE antibodies used, etc. By using such cluster-finding algorithms, a gating template can be constructed that can then process all the data of a given experiment without requiring visual checking of each data file. The ungated "raw" data is stored, and the gating template can be applied to the raw data files either contemporaneously with data acquisition, or in a batch analysis mode at the end of data acquisition. If data analysis strategies change as a study unfolds, the raw data can be subsequently reprocessed through revised gating templates. This requires considerably less operator time than traditional methods of data analysis in which the user places gates manually and checks/adjusts them for each data file, then has to re-do all the work manually if strategies change.

A final component of data analysis that is often overlooked is the incorporation of flow cytometric data into a database that may also link it with other types of measurements (patient clinical data, etc.). Fundamentally, this requires the ability to extract measurements of interest from the FLOW CYTOMETRY data files into a spreadsheet along with any annotation associated with the data files. This can be done with modern FLOW CYTOMETRY software packages, such that manual entry of data into a spreadsheet is not necessary. This is a fundamental requirement for

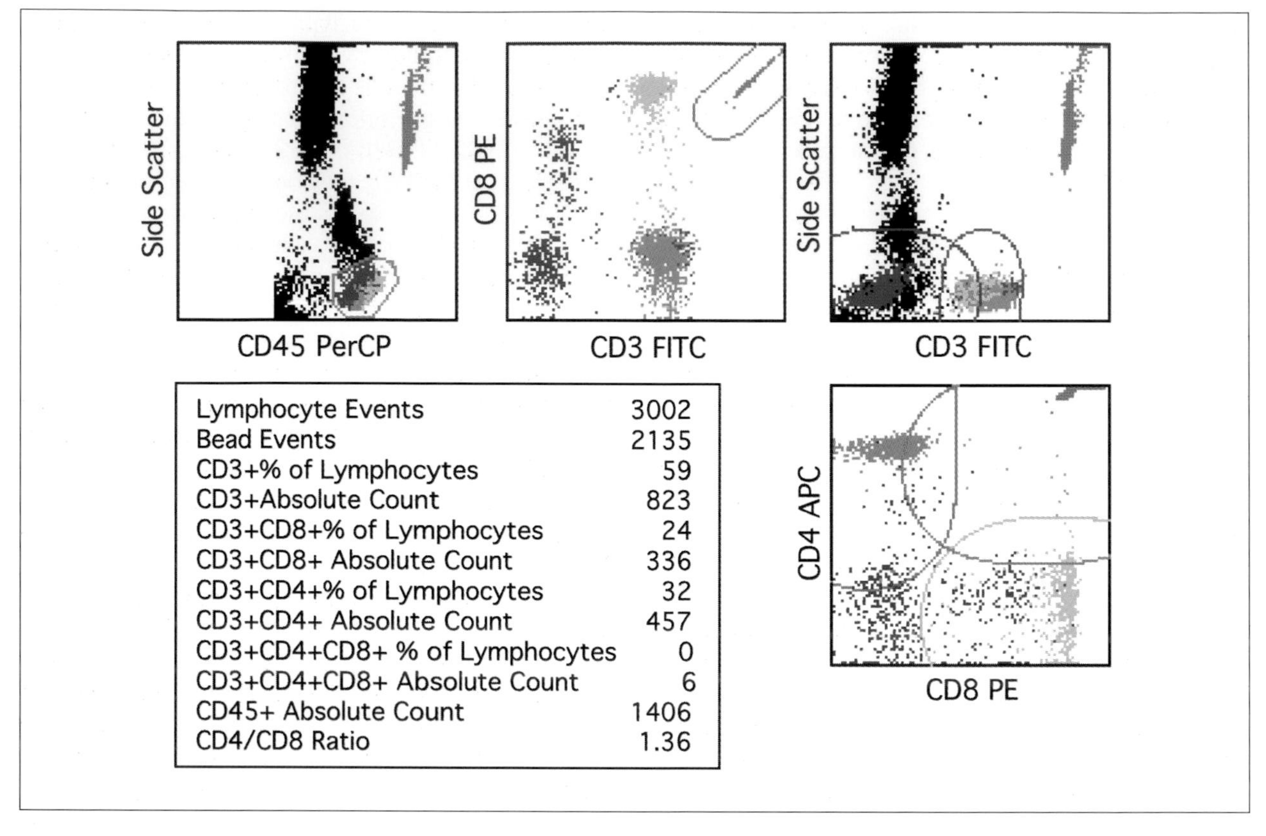

FIGURE 2. CD4 T CELL ENUMERATION ASSAY
Cells and counting beads are successively gated as described in the text, and the percentages and absolute counts of various lymphocyte subsets are automatically reported by the analysis software.

any large studies using FLOW CYTOMETRY as one of their components.

Examples and their application

Immunophenotyping assays

An example of an IMMUNOPHENOTYPING assay to quantify T cell subsets is shown in Figure 2. Note that this was done as a "no-wash" assay in four colours, along with fluorescent beads to allow absolute counting. Fifty microlitres of whole blood were stained with antibodies to CD45, CD3, CD4, and CD8. The blood was subjected to fixation/erythrocyte lysis, then run on a four-colour flow cytometer. An ACQUISITION THRESHOLD was set on CD45 fluorescence, in order to identify leukocytes. All cells above this threshold were collected, with acquisition halted at 20,000 $CD45^+$ events.

Analysis of this assay was done as follows. An initial gate was set on $CD3^+$ cells, in order to identify all T cells. $CD4^+$ and $CD8^+$ T cells were then quantified from a plot gated on $CD3^+$ cells. Note that this sample also contained counting beads which are identified by very high fluorescence in all colours. Because a known number of these beads was added to a known volume of the blood sample, a simple calculation can convert the percentages of each cell subpopulation to an absolute count of cells per microlitre of blood. Since percentages are relative to other populations, absolute counts have become a standard readout for reporting these types of results. Commercial software packages can perform these calculations in an automated fashion for this particular application.

Functional assays

An example of a functional assay (identifying intracellular CYTOKINE production) is shown in Figure 3. This assay was done by stimulation of whole blood with peptides derived from human cytomegalovirus (HCMV), a common herpesvirus that causes chronic, latent infection of various tissues. HCMV causes nonpathological infection in immunocompetent hosts, but is an opportunistic pathogen in immunocompromised hosts [28].

The peptides used for stimulation in this assay were a mixture of 138 15-mers, overlapping by 11 amino acid residues each, and spanning the sequence of the pp65 glycoprotein of HCMV. Pp65 is an immunodominant protein which contains EPITOPES that stimulate CD4 and CD8 T cell responses in a variety of HLA haplotypes [29]. By using such overlapping peptides, both CD4 and CD8 T cell responses can be stimulated with relative efficiency [25]. The blood was stimulated with this peptide mixture for six hours in the presence of brefeldin A, followed by EDTA treatment, fixation/erythrocyte lysis, permeabilization, and surface/intracellular staining. The antibodies used included anti-IFN-γ, CD69, CD4, and CD3. Because this was a washed assay, it was not necessary to use a reagent like CD45 for fluorescence triggering. Instead, a threshold was set on FORWARD SCATTER to eliminate small debris, and an acquisition gate was set around the cluster of small lymphocytes in forward versus SIDE SCATTER. Acquisition was stopped when 40,000 $CD4^+$ T cells were collected.

Analysis of this assay was done as follows. An initial gate was set on $CD3^+$ lymphocytes in a CD3 *versus* SIDE SCATTER plot. Secondary gates were set on $CD3^+CD4^+$ and $CD3^+CD4^-$ cells, respectively. From each of the latter gates, plots of CD69 *versus* IFN-γ expression were displayed, and cells positive for both of these activation markers were quantified. All of these gates were set using a cluster-finding algorithm that allows the gate to automatically shift with the population from sample to sample [15]. An exception is the final region quantifying IFN-γ^+CD69$^+$ cells, which was a user-defined rectangle, but which was "tethered" to the negative population, so that it too would shift in response to changes in the background level of fluorescence from sample to sample. In this way, even very rare populations of IFN-γ^+CD69$^+$ cells could be quantified, whether they formed a recognizable cluster or not.

If desired, one could combine the information from the above two assays, in order to express the HCMV pp65-responsive $CD3^+CD4^+$ or $CD3^+CD4^-$ cells as an absolute number of cells per microlitre. Again, this might provide more standardized results in populations that have varying numbers of CD4 and CD8 T cells (such as HIV patients).

What is the usefulness of identifying and quantifying intracellular CYTOKINE responses? One major application is the development of new vaccines (see Chapter C.1) that are designed to elicit cellular immunity [30, 31]. These include both prophylactic and therapeutic vaccines for HIV, cancer, and certain other viral and intracellular bacterial pathogens. Establishing biomarkers of immunogenicity of vaccines (and eventually surrogate markers of protection) would greatly facilitate the comparison of different vaccine constructs and strategies, and allow more rapid improvement of vaccines for these diseases.

In the area of immunotoxicology (see Chapter D.1), quantification of functional T cell responses, whether to specific antigens or to MITOGENS, could also provide valuable information [30]. While IMMUNOPHENOTYPING can potentially uncover gross changes in cellular subsets in response to a drug, functional assays can uncover much more subtle changes. Suppression (or augmentation) of antigen or MITOGEN responses could give important clues to immunological side-effects of drugs. This could be done as part of the early screening of drug candidates, before expensive animal studies are undertaken. Functional FLOW CYTOMETRY assays could also be used to monitor the dosing of immunomodulatory drugs in settings such as transplantation and autoimmunity [32, 33]. By standardizing, automating, and increasing the ease and throughput of these assays,

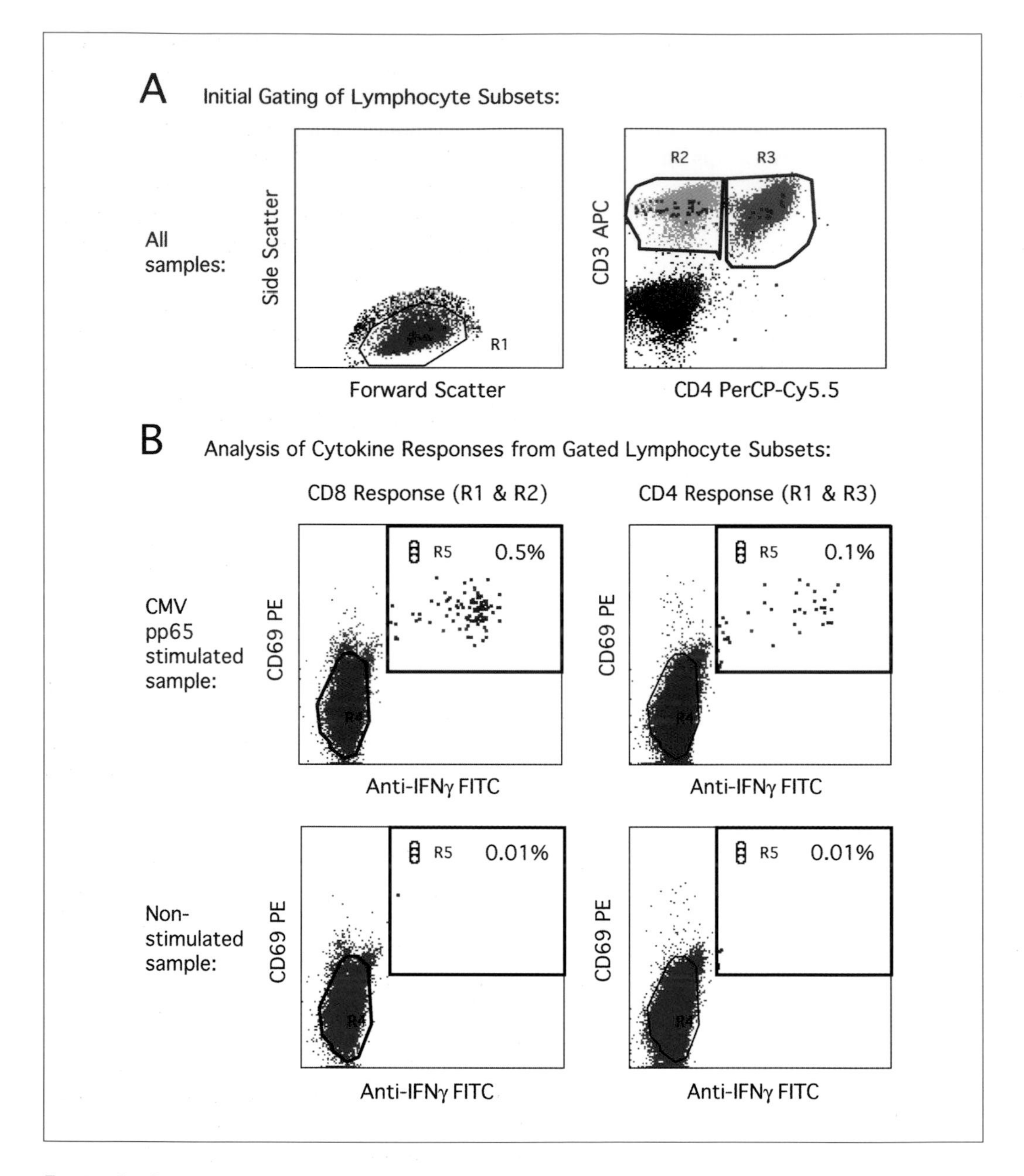

Figure 3. Cytokine flow cytometry assay

(A) Cells are initially gated to identify CD3⁺CD4⁺ and CD3⁺CD4⁻ lymphocytes (the latter population being substantially equivalent to CD8⁺ T cells). (B) Functional responses from each of these populations are then quantified as the percentage of the gated population that is positive for both CD69 (a surface marker of activation) and IFN-γ (a T cell cytokine). The response from an antigen-stimulated sample (top plots) is then compared to that from a control, unstimulated sample (bottom plots). Note that even a very low percentage of positive cells, e.g., 0.1%, can be easily differentiated from the control, when the percentage of positive cells in the control sample is low enough, as in these examples. The statistical significance of the difference between control and test samples can also be calculated [34].

they can become ever more potent tools for immunological research, immunotoxicology, and clinical medicine.

Summary

FLOW CYTOMETRY is a powerful technique for analyzing cells in suspension, and has been extensively applied to the analysis of lymphocytes and their subsets. Flow cytometric assays can be divided into IMMUNOPHENOTYPING assays and functional assays, the latter requiring *in vitro* stimulation of cells in order to read out a response, such as CYTOKINE production. Such assays are being applied to the monitoring of clinical disease states and responses to vaccination or immunomodulation, and they have great potential to be used in measuring immunotoxicology as well.

Selected readings

Givan AL (2001) *Flow cytometry: First Principles*. J. Wiley, New York

Robinson JP (ed) (1999) *Current Protocols in Cytometry* (Current Protocols Series). J. Wiley, New York

Purdue cytometry bulletin board: http://www.cyto.purdue.edu/hmarchiv/cytomail.htm (Accessed November 2004)

References

1 Herzenberg LA, Parks D, Sahaf B, Perez O, Roederer M (2002) The history and future of the fluorescence activated cell sorter and flow cytometry: a view from Stanford. *Clin Chem* 48: 1819–1827

2 Givan AL (2001) *Flow cytometry: First Principles*. J. Wiley, New York

3 Libman H, Makadon HJ (eds) (2003) *HIV* (2nd edn). American College of Physicians, Philadelphia

4 Mandy F, Nicholson J, Autran B, Janossy G (2002) T-cell subset counting and the fight against AIDS: reflections over a 20-year struggle. *Cytometry* 50: 39–45

5 Davis BH, Foucar K, Szczarkowski W, Ball E, Witzig T, Foon KA, Wells D, Kotylo P, Johnson R, Hanson C et al (1997) U.S.-Canadian Consensus recommendations on the immunophenotypic analysis of hematologic neoplasia by flow cytometry: medical indications. *Cytometry* 30: 249–263

6 Stewart CC, Behm FG, Carey JL, Cornbleet J, Duque RE, Hudnall SD, Hurtubise PE, Loken M, Tubbs RR, Wormsley S (1997) U.S.-Canadian Consensus recommendations on the immunophenotypic analysis of hematologic neoplasia by flow cytometry: selection of antibody combinations. *Cytometry* 30: 231–235

7 Rothe G, Schmitz G (1996) Consensus protocol for the flow cytometric immunophenotyping of hematopoietic malignancies. Working Group on Flow Cytometry and Image Analysis. *Leukemia* 10: 877–895

8 Burchiel SW, Kerkvliet NL, Gerberick GF, Lawrence DA, Ladics GS (1997) Assessment of immunotoxicity by multiparameter flow cytometry. *Fundam Appl Toxicol* 38: 38–54

9 Mehta BA, Maino VC (1997) Simultaneous detection of DNA synthesis and cytokine production in staphylococcal enterotoxin B activated CD4+ T lymphocytes by flow cytometry. *J Immunol Methods* 208: 49–59

10 Lyons AB, Parish CR. (1994) Determination of lymphocyte division by flow cytometry. *J Immunol Methods* 171: 131–137

11 Braylan RC, Diamond LW, Powell ML, Harty-Golder B (1980) Percentage of cells in the S phase of the cell cycle in human lymphoma determined by flow cytometry. *Cytometry* 1: 171–174

12 Taylor IW, Milthorpe BK (1980) An evaluation of DNA fluorochromes, staining techniques, and analysis for flow cytometry. I. Unperturbed cell populations. *J Histochem Cytochem* 28: 1224–1232

13 Darzynkiewicz Z (1994) Simultaneous analysis of cellular RNA and DNA content. *Methods Cell Biol* 41: 401–420

14 Cook EB, Stahl JL, Lowe L, Chen R, Morgan E, Wilson J, Varro R, Chan A, Graziano FM, Barney NP (2001) Simultaneous measurement of six cytokines in a single sample of human tears using microparticle-based flow cytometry: allergics vs. non-allergics. *J Immunol Methods* 254: 109–118

15 Suni MA, Dunn HS, Orr PL, deLaat R, Ghanekar SA, Bredt BM, Maino VC, Maecker HT (2003) Performance of plate-based cytokine flow cytometry with automated data analysis. *BMC Immunology* 4: 9

16 Kelley KW, Lewin HA (1986) Monoclonal antibodies:

pragmatic application of immunology and cell biology. *J Anim Sci* 63: 288–309

17 Zolla H (1999) High-sensitivity immunoflourescence/flow cytomtery: Detection of cytokine receptors and other low-abundance membrane molecules. In: JP Robinson (ed): *Current Protocols in Cytometry*. Unit 6.3. J. Wiley, New York

18 Baumgarth N, Roederer M (2000) A practical approach to multicolor flow cytometry for immunophenotyping. J *Immunol Methods* 243: 77–97

19 De Rosa SC, Brenchley JM, Roederer M (2003) Beyond six colors: a new era in flow cytometry. *Nat Med* 9: 112–117

20 Coligan JE, Kruisbeek AM, Margulies DH, Shevach EM, Strober W (eds) (1995) *Current Protocols in Immunology*. Wiley, London

21 Nomura LE, DeHaro ED, Martin LN, Maecker HT (2003) Optimal preparation of rhesus macaque blood for cytokine flow cytometric analysis. *Cytometry* 53A: 28–38

22 Nomura LE, Walker JM, Maecker HT (2000) Optimization of whole blood antigen-specific cytokine assays for CD4(+) T cells. *Cytometry* 40: 60–68

23 Waldrop SL, Pitcher CJ, Peterson DM, Maino VC, Picker LJ (1997) Determination of antigen-specific memory/effector $CD4^+$ T cell frequencies by flow cytometry: evidence for a novel, antigen-specific homeostatic mechanism in HIV-associated immunodeficiency. *J Clin Invest* 99: 1739–1750

24 Prussin C, Metcalfe DD (1995) Detection of intracytoplasmic cytokine using flow cytometry and directly conjugated anti-cytokine antibodies. *J Immunol Methods* 188: 117–128

25 Maecker HT, Dunn HS, Suni MA, Khatamzas E, Pitcher CJ, Bunde T, Persaud N, Trigona W, Fu TM, Sinclair E et al (2001) Use of overlapping peptide mixtures as antigens for cytokine flow cytometry. *J Immunol Methods* 255: 27–40

26 Suni MA, Picker LJ, Maino VC (1998) Detection of antigen-specific T cell cytokine expression in whole blood by flow cytometry. *J Immunol Methods* 212: 89–98

27 Dunne JF, Maecker HT (2004) Automation of cytokine flow cytometry assays. *J Assoc Lab Automation* 9: 5–9

28 Drago F, Aragone MG, Lugani C, Rebora A (2000) Cytomegalovirus infection in normal and immunocompromised humans. A review. *Dermatology* 200: 189–195

29 Kern F, Bunde T, Faulhaber N, Kiecker F, Khatamzas E, Rudawski IM, Pruss A, Gratama JW, Volkmer-Engert R, Ewert R et al (2002) Cytomegalovirus (CMV) phosphoprotein 65 makes a large contribution to shaping the T cell repertoire in CMV-exposed individuals. *J Infect Dis* 185: 1709–1716

30 Maecker HT, Maino VC, Picker LJ (2000) Immunofluorescence analysis of T-cell responses in health and disease. *J Clin Immunol* 20: 391–399

31 Maecker HT, Maino VC (2003) T cell immunity to HIV: defining parameters of protection. *Current HIV Research* 1: 249–259

32 Sindhi R, Allaert J, Gladding D, Koppelman B, Dunne JF (2001) Cytokines and cell surface receptors as target end points of immunosuppression with cyclosporine A. *J Interferon Cytokine Res* 21: 507–514

33 Sindhi R, Allaert J, Gladding D, Haaland P, Koppelman B, Dunne J, Sehgal S (2003) Modeling individual variation in biomarker response to combination immunosuppression with stimulated lymphocyte responses-potential clinical implications. *J Immunol Methods* 272: 257–272

34 Motulsky H. (1995) *Intuitive Biostatistics*. Oxford: Oxford Univ. Press, 197-203.

35 Altman JD, Moss PAH, Goulder PJR, Barouch DH, McHeyzer-Williams MG, Bell JI, McMichael AJ, Davis MM. (1996) Phenotypic analysis of antigen-specific T lymphocytes. *Science* 274: 94–96

36 Brosterhus H, Brings S, Leyendeckers H, Manz RA, Miltenyi S, Radbruch A, Assenmacher M, Schmitz J (1999) Enrichment and detection of live antigen-specific CD4(+) and CD8(+) T cells based on cytokine secretion. *Eur J Immunol* 29: 4053–4059

37 Hutchings PR, Cambridge G, Tite JP, Meager T, Cooke A (1989) The detection and enumeration of cytokine-secreting cells in mice and man and the clinical application of these assays. *J Immunol Methods* 120: 1–8

38 Betts MR, Brenchley JM, Price DA, De Rosa SC, Douek DC, Roederer M, Koup RA (2003) Sensitive and viable identification of antigen-specific $CD8^+$ T cells by a flow cytometric assay for degranulation. *J Immunol Methods* 281: 65–78

Gene arrays

Barbara Schaffrath and Andreas Bosio

Introduction

Function and appearance comprise the phenotype of a cell and this is determined by the amount, the proportion and the condition of proteins which are produced in the cell. Although every cell in an organism possesses the same genetic information, not all genes are active. According to the function and demands of the cell, only certain genes are transcribed into MESSENGER RIBONUCLEIC ACID (mRNA). Based on the messenger RNA, the information is translated into the corresponding protein. After having mapped the whole human genome, questions about function and interaction of the genes and gene products were still open concerning the majority of genes. Until very recently, it was impossible to assess simultaneously the expression of hundreds of genes because of the complex and cumbersome methodologies used in molecular biology. Traditional methods mostly work on the basis of "one gene - one experiment" which makes comprehensive gene expression profiling infeasible. During the 90s a technology was developed which made it possible to study the interaction of thousands of genes simultaneously. The tools of this technology are referred to as DNA MICROARRAYS, DNA CHIPS, or biochips.

The DNA chip technology uses the characteristic of DNA to form helices due to sequence complementarity which is called hybridisation. Since Southern introduced the blotting technique [1], the hybridisation process has been used in a wide range of techniques for the recognition and quantification of DNAs. However, only the reversal of this procedure – the arraying of multiple homogeneous copy DNAs (cDNAs) and testing with a single heterogeneous labeled sample – made it possible to study the GENE EXPRESSION PROFILES of thousands of genes in a given sample.

Principle of the technology

DNA MICROARRAYS are miniaturised devices made up for the analysis of ribonucleic acids by hybridisation. For "classical" hybridisation-based analysis, the extracted genomic DNA (Southern) or RNA (Northern) from the tissue of interest is immobilised on a membrane and a single nucleotide sequence (the probe), which is complementary to a certain sequence, is labeled to detect the corresponding gene or transcript (Fig. 1). For array analysis, several hundreds to thousands of DNA fragments (probes) are immobilised on a SUBSTRATE (a membrane, glass or plastic slide) on defined positions. To answer, e.g. the question of which genes are regulated in certain tumour cells compared to normal cells, RNA from tumour and normal cells is extracted. The extracted RNA is transcribed into its respective copy, the so-called cDNA. cDNAs derived from tumour cells and normal cells are labeled differently, e.g., with nucleotides linked to different cyanine dyes called Cy5 ("red") and Cy3 ("green") (see Fig. 2). Subsequently, the labeled cDNA pools are merged and applied to the DNA array. During this procedure, complementary sequences in the probes and the labeled nucleic acids hybridise. By quantifying the fluorescent intensities (Cy5, Cy3) and downstream image analysis by appropriate software the ratio of bound Cy5- to Cy3-labeled cDNAs for every spot on the array can be determined. This ratio allows conclusions on the activity of single genes in tumour and normal cells to be drawn.

The convenience of DNA MICROARRAYS consists in the opportunity to observe a biological system on the transcriptional level while using minimal amounts of sample material.

DNA microarray formats differ regarding the SUBSTRATES, the probe selection strategies and the way

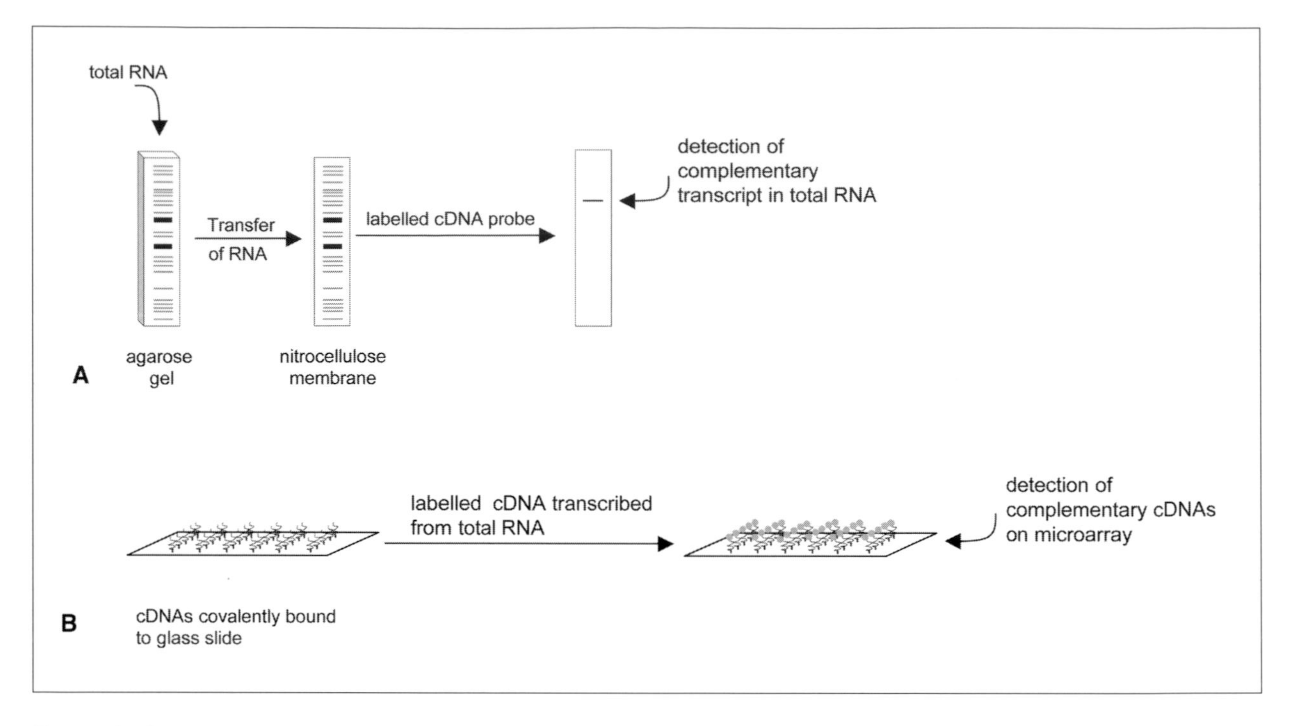

FIGURE 1. COMPARISON OF TRADITIONAL NORTHERN-BLOT AND DNA MICROARRAY
A: Total RNA of the tissue of interest is separated by gel electrophoresis and is blotted to a membrane. A labelled cDNA probe complementary to the transcript of interest is labelled and hybridised to the membrane.
If the transcript is present in the total RNA, a signal can be detected due to hybridisation of probe and transcript.
One experiment – one gene using a single labelled probe.
B: Several cDNAs (hundreds to thousands) complementary to mRNA transcripts of selected genes are covalently bound to a glass slide on a well-known position (spot). Total RNA from the tissue of interest is transcribed into cDNA and labelled by reverse transcription. The labelled cDNA is hybridised to the bound cDNAs. Signals can be detected after hybridisation of two complementary cDNAs.
One experiment – several genes using multiple probes and labelled transcripts.

the probes are immobilised. Basically, the field of DNA MICROARRAYS can be divided into two groups: CDNA ARRAYS and OLIGONUCLEOTIDE ARRAYS. The material of the SUBSTRATE differs independent of the dispensed probes. So does the number of spotted probes, presence of replicates and the presence of controls. Several aspects of the technology will be discussed in detail in the following paragraphs.

There are general criteria, e.g. for probe selection, quality control (of probes, SUBSTRATE and RNA), data acquisition and data analysis, which are relevant for all DNA microarray types. Some criteria have to be considered depending on the microarray types (cDNA, oligonucleotide), manufacturing process (*in situ* hybridisation, non-contact/contact SPOTTING) or labeling procedure (fluorescent dyes, biotin, radioactive labeling, etc.).

The variety of probes

In general, the PROBE SELECTION STRATEGY for MICROARRAYS is subjected to several pre-requisites such as the objective of the experiment, the available sequence information regarding the organism to be investigated and the efforts one is willing to invest. If one is

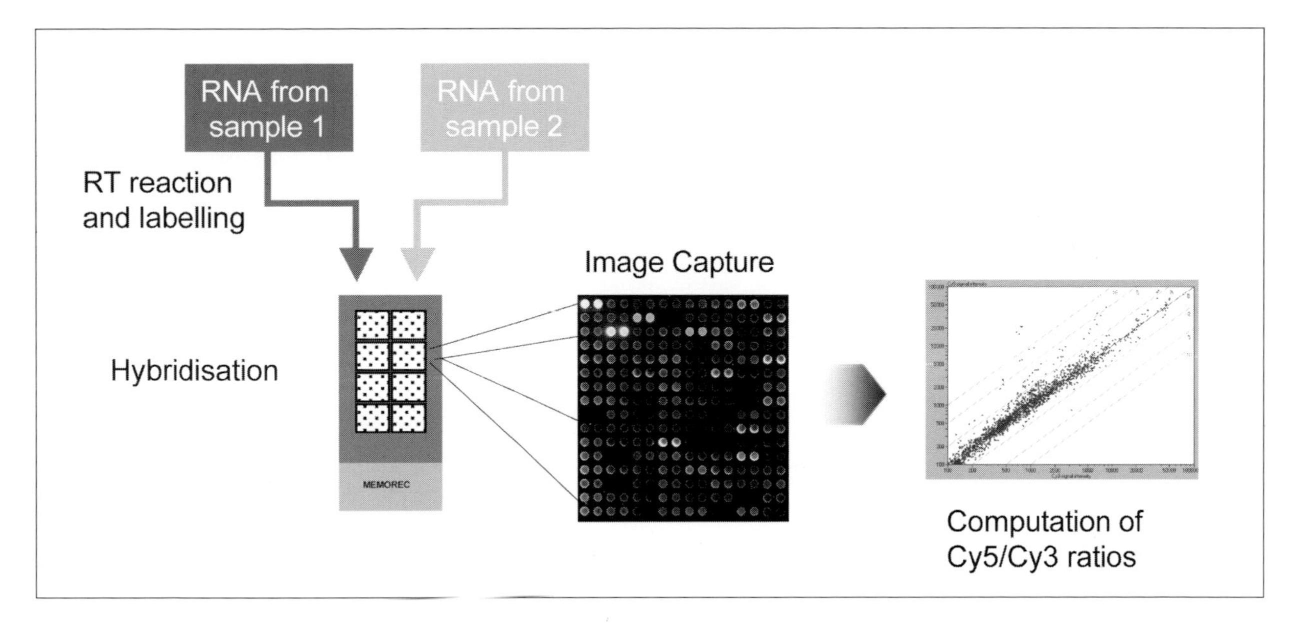

Figure 2: Flow diagram of microarray analysis

interested in a broad overview of the expression levels of as many genes as possible in a small set of experiments with low pre-chip expenses, it would be sufficient to spot cDNA libraries (for def. see Box 1), uncharacterised expressed sequence tags (ESTs) (for def. see Box 2), or oligonucleotides derived therefrom. The in-depth sequence analysis and annotation is thereby postponed to the point when hybridisation has been performed and the resulting 'clones/ genes of interest' have been identified. In contrast, if the experimental setup calls for ACCURATE identification and quantification of particular mRNA species by hybridisation, one can either buy ready-to-spot sets or take the time and effort to clone and sequence-verify the cDNAs (Fig. 3). However, these processes should be accompanied by extensive quality management.

Oligonucleotide probes

Oligonucleotides are generally spotted or synthesised *in situ* directly on the array.

The *in situ* synthesis by PHOTOLITHOGRAPHIC PROCEDURES or inkjet technology allows a parallel production of OLIGONUCLEOTIDE ARRAYS, but is restricted to an oligonucleotide length of 20–60 nucleotides due to a limited coupling efficiency [2, 3]. The direct SPOTTING of pre-synthesised oligonucleotides with a length to up to 70 bp offers the advantage that probes can be quality-controlled before the SPOTTING process. The use of short oligonucleotides (20 to 30 basepairs) is suitable for distinguishing between perfectly matched duplexes and single-base or two-base mismatches [4–6]. When working with short oligonucleotide probes, the use of several different oligonucleotides corresponding to a single gene is typically required to enhance the reliability of the hybridisation signals [7]. Thus, several signals are gained for one gene, which sometimes may be contradictory due to differing hybridisation kinetics or splice variants and therefore hard to interpret.

cDNA probes

As already mentioned, the appropriate PROBE SELECTION STRATEGY depends primarily on the objective of the experiment. If little prior information on relevant genes is available or the prime motivation is an unbiased overview of global changes in gene expression patterns, the SPOTTING of clones from a library with-

Box 1. cDNA libraries

Libraries which store cDNA clones generated from RNA transcripts. Typically these clones represent only the open reading frame (ORF).

Box 2. ESTs

A set of single-pass sequenced cDNAs from an mRNA population derived from a specific cell population (e.g. a specific tissue, organ, developmental state or environmental condition).
- provides a profile of the mRNA population
- quick method for cloning a large number of genes known to be expressed in a cell population

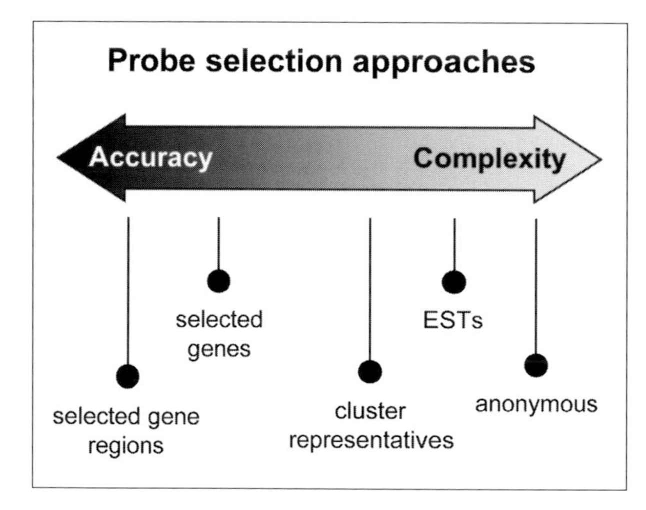

Figure 3. Selection strategies for probes

out prior sequencing is acceptable. Only those clones that show differential expression after hybridisation are submitted to sequencing and further analysis. This approach includes a general lack of reliable sample annotation, shifting some of the necessary work to the post-hybridisation phase.

A more refined strategy relies on available collections of sequenced cDNA clones. The most reliable strategy is the amplification of suitable sequence regions from pre-selected genes by polymerase-chain-reaction (PCR). By doing so, one can adjust important properties of the selected regions like length, orientation or position within the mRNA, which is important to avoid false signals (for more information see Box 3). Beside the careful selection of sequence regions, length and uniformity of the cDNA fragments are of particular interest regarding the robustness of the hybridisation process. Uniformity provides optimal hybridisation conditions (temperature and buffer conditions) for all cDNAs in parallel. With increasing length of the used probes, the hybridisation becomes more stable. A length of 200 bases is adequate to guarantee an efficient hybridisation independent of single nucleotide polymorphisms and varying GC content of the single probes. The sensitivity of cDNA arrays increases the longer the probes are, since more labeled samples may hybridise to the respective immobilised probes.

On the other hand, the length of the probes should be limited to a maximum of about 400 basepairs in order to avoid the probability of cross hybridisation caused by repetitive elements and unspecific interaction. A limitation to ~ 400 basepairs still provides the opportunity to distinguish even genes from highly homologous gene families by choosing fragments from appropriate (e.g. poorly conserved) gene regions.

A determination of single-base or two-base mismatches is not feasible with longer cDNA probes.

Box 3. Selection and annotation of suitable cDNA fragments

To generate suitable fragments for the unambiguous detection of genes, some considerations have to be included:

1. Cross-hybridisation
   Can be reduced by optimizing the experimental conditions if based on unspecific binding. Other reasons for undesirable cross hybridisations are repetitive elements such as Alu repeats, microsatellite repeats, SINEs or LINEs (short or long interspersed elements) within the DNA sequence. Comparing the selected sequence to databases like REPBASE (www.girinst.org) will give some information about this feature.
2. Splice variants/alternative splicing
   Different mRNA species may be transcribed from a single gene. Alternative polyadenylation signals or alternative splicing (the excision of distinct exons) may occur in the sequence. A fragment not detecting all of the possible mRNA variants could lead to contrasting expression profiles when different mRNA species are present.
   On the other hand, the variant gene products of a gene may serve different purposes. It might be interesting to distinguish between the mRNA species.

Depending on the question, fragments are generated either to detect a specific mRNA species, e.g., splice or polyadenylation variants or to detect all possible mRNA species of one gene. Most important with respect to quality management of the array production is a precise documentation of the expected hybridisation properties of the fragment. In addition, each fragment should be completely annotated according to all the names used in public databases like Unigene, Swissprot or trEMBL and other relevant data sources. This alleviates the gene selection for the user by simultaneously avoiding redundancy caused by the probable utilization of different names for an identical gene [8].

Production of microarrays

What are substrates ?

SUBSTRATES represent the solid phase of arrays. The production of suitable SUBSTRATES is as crucial for reliable and reproducible results as the generation of probes. Array SUBSTRATES of various materials (membranes, plastics, glass) with various coatings have flooded the market in the last few years. Generally, SUBSTRATES are composed of a solid phase and one or two layers to hydrophobise the surface by simultaneously providing reactive groups to bind DNA covalently (see Box 4). To guarantee sufficient quality some general features should be taken into account. Surface properties like planarity (which is mainly influenced by the used solid phase), uniformity, mechanical and chemical stability and an optimal DNA binding capacity are crucial. When working with FLUORESCENCE-LABELED SAMPLES, autofluorescence of the SUBSTRATE may strikingly limit the SENSITIVITY of hybridisation experiments and has to be reduced to a minimum. Other points like control of inter- and intra-batch uniformity should be kept in mind while developing a coating protocol with respect to array production.

Production of probes

If oligonucleotides or whole EST clones are used as probes, the SPOTTING process is the next step in the manufacturing of the arrays. This is not the case if selected regions of genes are generated by reverse transcription (RT) PCR (see 1.2.2). Here, the appropriate cDNA fragments are generated with the respective primers, cloned and sequence-verified. When the clones are sequence-verified, the inserts can be amplified by PCR. By using vector-specific, possibly modified primers (see Box 4), the resulting PCR products are prepared for covalent binding to the activated surface of the SUBSTRATE (Fig. 4). This step again should be quality-controlled by checking for the right fragment length and repeating several PCRs using gene-specific primers.

Box 4. Array coatings

In the beginning, most of the coatings which have been developed made use of the reactivity of the nucleobases and lead to an unspecific crosslinking of the DNA to the surface, e.g., silyl (reactive aldehyde), silane (reactive amino-groups), or polylysine [6]. These interactions can reduce the conformational freedom of the cDNAs and hence their affinity for complementary molecules in solution. In contrast, other surfaces, suitable for a covalent binding of the DNA to the surface via a specific linker attached to the 5' or 3' end of the DNA [9, 10], were developed. The end-specific covalent binding offers the opportunity to direct the amount of attached cDNAs to an optimal density at which enough cDNAs are present but charge and steric effects are minimal.

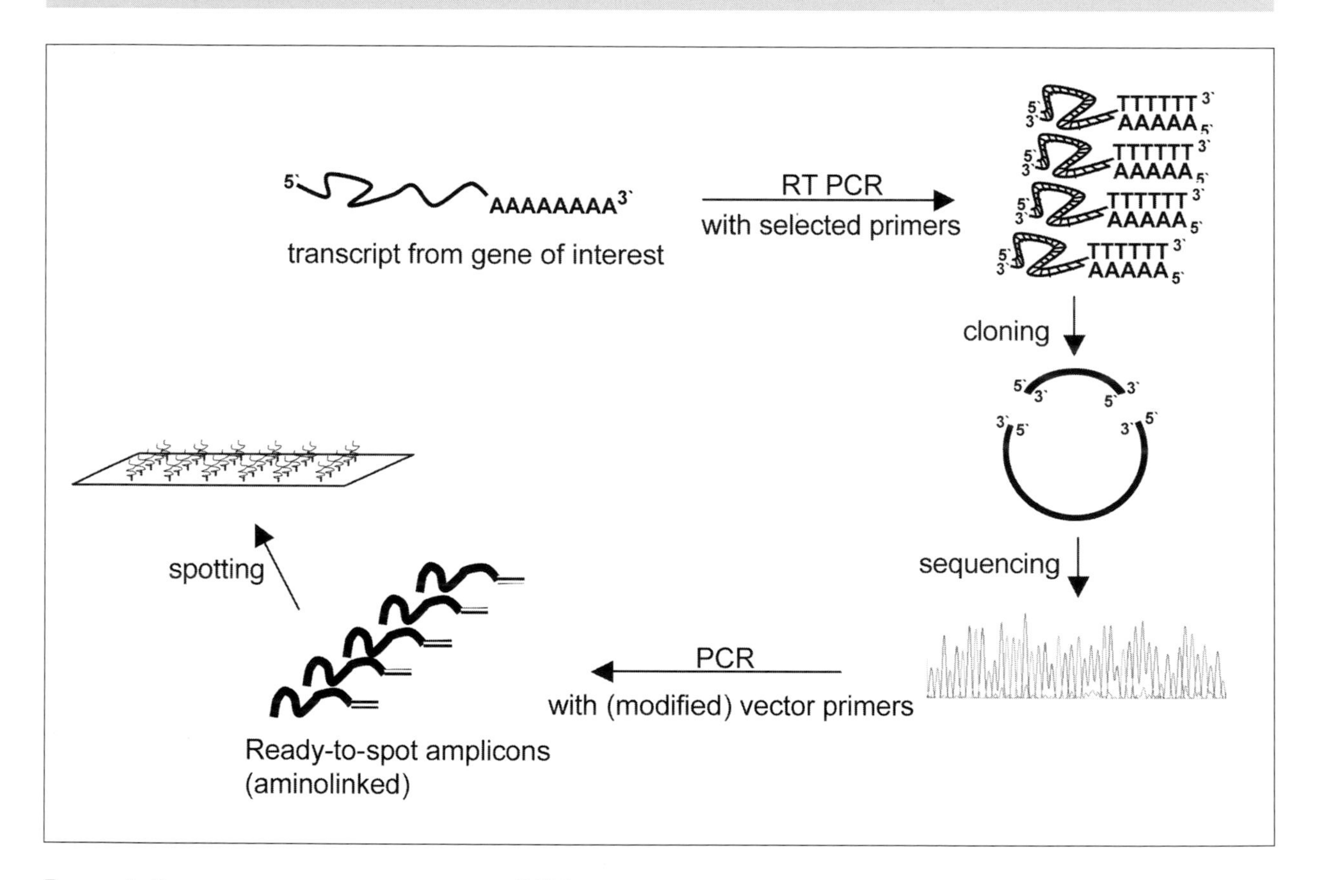

Figure 4. Scheme of the generation of a cDNA microarray starting from a gene of interest

Spotting

Apart from *in situ* synthesis, the SPOTTING of cDNAs or pre-synthesised oligonucleotides is practised. Basically, there are two methods to spot probes: contact printing and non-contact printing. Both technologies are independent of the kind of probes and SUBSTRATES used.

CONTACT PRINTING typically involves rigid pins and is probably the most widely used method for the production of microarrays. There are various types of pins, but independently of the shape, the pins gener-

Box 5

For a highly reliable detection especially of weakly expressed genes, probes should be spotted in replicates. To minimise the impact of probable spatial effects on the measured expression ratio, the replicates should not be arranged in contiguity but be uniformly distributed over the whole spotting area.

Box 6. Quality of total RNA

Integrity and purity are the most critical factors for the quality of RNA.
- Ratio of 28S rRNA and 18S rRNA should be 2.
- Ratio of the extinction 260 nm/280 nm should be between 1.8 and 2.0.
- The sample should preferably be treated with DNAse to avoid contamination of genomic DNA.
- Protocols for RNA extraction has to be adapted according to the analysed tissue (e.g., extreme fatty or fibrous tissue)
- The choice of the preparation protocol may have influence on the range of transcript lengths present in the extracted RNA (e.g., silica filters have a cut-off size of about 50–100 bases and therefore not the whole range of fragment lengths are present in preparations derived therefrom. This will have an impact on the subsequent steps (labelling or amplification).

ally dip into the probes and print them to the surface. NON-CONTACT PRINTING methods are based upon the ink-jet. The sample is dispensed as droplets from the print head.

Independent of the SPOTTING technology the SPOTTING process should be accompanied by an ACCURATE documentation. The positioning of the genes has to be unequivocally traceable (see also Box 5).

Application of microarrays

Preparation and quality of RNA

Even if it sounds trivial, the first crucial step to achieve reliable gene expression results is RNA isolation. If the RNA is slightly degraded or contaminated, the results may be biased and irreproducible (see also Box 6). Cells or tissue from which RNA is extracted are commonly lysed and subsequently extracted using organic solvents or silica filter-based methods. Since RNA extraction protocols may influence the outcome of the expression analysis, it is worthwhile to search in advance for an appropriate protocol in order to use it for all of the samples that should be analysed in one batch of experiments.

Amount of RNA

The necessary amount of total RNA highly depends on the labeling method and the kind of arrays. If any amplification method is used, the required amount of RNA is drastically reduced. Thus, it is hardly possible to draw a general conclusion.

Roughly, amounts varying from 10 to 100 μg total RNA (which resembles 0.2–2 μg mRNA) are required without amplification depending on the labeling method and the kind of array. If the amount of available RNA is limited, e.g. when biopsies or microdissections are analysed, the RNA can be subjected to a linear amplification delivering amplified (a)RNA. Basically, either PCR-based amplification methods or T7-driven *in-vitro* transcriptions from double-stranded cDNA templates are commonly used (Fig. 5) [11].

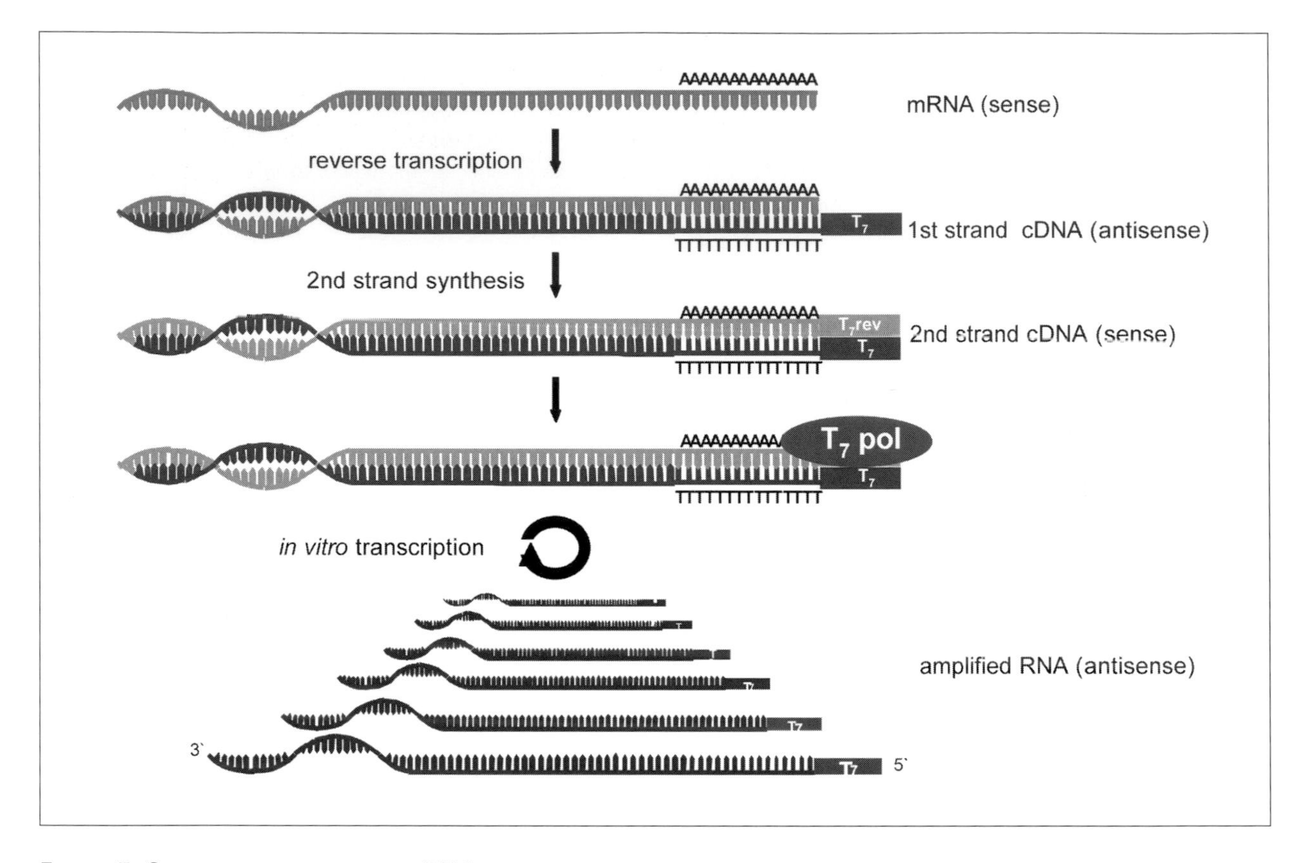

FIGURE 5. SCHEMATIC DIAGRAM OF MRNA AMPLIFICATION

Dyes, labeling and hybridisation methods

Besides radioactive labeling, mainly fluorescent dyes are used (see Box 7). Labeling using silver, gold or platinum particles either as LABEL itself or as enhancer gave good results, but had not been broadly accepted until now. The incorporation of the dyes is either direct or indirect. Performing direct dye incorporation (direct or one-step labeling), the markers are introduced by reverse transcription of mRNA using a dye-labeled desoxyribonucleoside triphosphate (dNTP) (most commonly dCTP) (see Fig. 2).

Since direct integration of the fluorescence dyes is often problematic due to steric effects, two-step labeling protocols (indirect labeling) have been established which first introduce a (smaller) reactive compound into the cDNA, which are then linked in a second step to fluorescent dyes. Some of these protocols increase the SENSITIVITY so that the amount of starting material can be lowered, but most of these indirect protocols are not that highly reproducible, since they require more reactions (not only enzymatic but also chemical ones) and subsequent purification steps.

The hybridisation can be performed as a one-, two- or multi-colour experiment. In fact, mostly one- or two-colour protocols are applied. Using two colours, the sample (e.g. treated cells, pathological tissue) and the respective control (e.g. untreated cells, normal tissue) can be analysed simultaneously. Of course, there is a variety of fluorescent labels on the market, using different dyes with different extinctions. In any case, the use of Cy3 and Cy5 is so far the most common and accepted one.

The most cumbersome and critical step for array applications is the hybridisation step where the

Box 7

The labelling process can be monitored by positive controls, e.g. artificial cDNAs which are present on certain positions on the array. The respective *in vitro* transcripts can be spiked into each RNA sample. Thus, an effective labelling reaction leads to signals on the appropriate positions.

labeled TARGET DNA and the affixed probe DNA are brought together. Therefore, fully or half automated hybridisation machines have been developed during the last years. All systems aim to facilitate the active movement of the labeled TARGET DNA during the hybridisation and therefore be independent from poor diffusion rates.

To ensure that the variance caused by labeling and hybridisation efficiency is minimal, repeated experiments using the same sample should be performed.

Control samples

The measurement of gene expression in a given sample is always referred to the gene expression in other samples, in this text denominated as "control". Alterations in gene expression can only be properly assessed *versus* appropriate control samples. As a consequence, it is very important to choose the right control in order to gain valuable data. The best controls are obviously untreated cells or unaffected tissue of the same origin as treated cells or affected tissue respectively (e.g. peripheral blood mononuclear cells (PBMCs) derived from the same source and treated *in vitro* with phosphate-buffered saline (PBS) (control) or lipopolysaccharide (LPS) (sample)). Before starting a series of experiments, it should be ensured that the control samples are accessible during the whole study. If it is impossible to get samples of the same origin as control samples, which may be the case when dealing with human material, it would be useful to establish an artificial common reference. This can be performed, e.g. by pooling RNA from a set of different cell lines or a (sufficient) number of control samples like normal tissue or untreated cell culture. Which way is the best depends on factors like accessibility, reproducibility and coverage (for example, how many genes are detectable in the control channel). The latter should be determined, of course, before starting the experiments. Accessing several samples from different sources, e.g. from different laboratory animals at one time point for control, the respective RNA should be pooled before starting the labeling, since the biological variation of the control samples could mimic differences between treatments. Pooling of the controls ensures that each individual treatment may be compared to an identical control and that the observed differences can be ascribed to (biological) variation of the treatments.

Array data: acquisition, analysis and mining

Data acquisition

Data acquisition from microarray experiments consists of two parts: the digitalisation of the fluorescence signals and the following image analysis using appropriate software packages.

The scanning process is used to generate digitised images of the array. Two general types of scanners can be used:

Some systems are based on a white light source and a charge-coupled device (CCD) camera detector technology. The array is exposed to the desired excitation wavelengths (filtered from whole UV-VIS spectrum) while the CCD camera is generating one picture at a time from different sections of the array until the whole array is mapped. The other systems are based on different lasers with excitation wave-

Box 8

Among the range of image analysis products, important quality features are:
- The possibilities to discriminate between valid and unwanted signals.
- To flag empty and negative spots as well as spots of irregular shape or with other minor quality features which can be excluded from further analysis.
- Algorithms should be included allowing for automated spot finding which is an invaluable tool to optimise the grid that defines the 'regions of interest' (ROIs = spots) quickly.
- The primary data should include information about the standard deviation of the spots and the background signals.

lengths specific for the used dyes. The images for the separate channels can be generated subsequently or simultaneously.

Additional differences may be due to the way the arrays are scanned. Some devices scan the arrays from the downside which may be a disadvantage because even slight contaminations like dust particles can lead to ghost spots. This problem can be avoided by scanning from the upside of the slides.

For quantification of the signals, appropriate image analysis software should be used to facilitate the production of valid data output ("primary data", see Box 8).

Data analysis and mining

Once having generated the primary data, the ratios of gene expression levels have to be calculated and normalised. In order to include only valid signals in further analyses, a set of rules should be followed before starting the calculation of the expression ratios.

Only 'good quality control (QC) spots' should be processed. Spots of poor quality (empty or negative spots, irregular shape, see Box 8) should be excluded from further analyses. The minimum signal intensity of a spot has to be determined in order to distinguish unreliable spot signals from valid spots. This can be achieved by setting a minimum threshold for signal intensities which is either dependent on the background or on negative controls (e.g. spotting buffer only or fragmented genomic DNA). This way, false positive results can be avoided. The BACKGROUND INTENSITIES should be subtracted from the signal intensities to obtain the net signal. Either the global (the mean of the BACKGROUND INTENSITIES over the whole spotting area) or the local background (defined as an area surrounding the spot) may be used for background correction. But due to local artefacts (contamination or poor hybridisation performance on defined parts of the array) which similarly affects the signal and the background, the local background should be preferred. If a dual labeling is used, ratios of the different dyes are computed for every spot. To a large extent systematic variations like differences in dye integration can be reduced by a process called "NORMALISATION" (see Box 9) and which is performed during data analysis. If replicates are present, the resulting data are averaged.

Because of the multiparametric nature of microarray experiments, bioinformatics and data mining represent essential tools for interpretation of the mass of numerical data produced by (series of) microarray experiments. Starting from relatively simple demands for appropriate visualization of the data, bioinformatics tools are necessary to focus on candidate genes and point out subtle changes in expression over many genes. Such expression patterns have predictive power but are difficult to spot.

Reliable identification of candidate genes by statistical methods often suffers from a limited number of replicate experiments. Since researchers are sometimes overwhelmed by the amount of data produced already by performing one single experiment, they ignore the basic necessity of repeated assays. Biological replicates are essential if dealing with expression profiling, especially if subtle changes of

Box 9. Normalisation

The main idea of normalisation for dual-labelled samples is to adjust differences in the intensity of the two labels. Such differences result from efficiency of dye integration, differences in amount of sample and label used, settings of laser power and photo-multiplier. Normalisation of one-channel arrays mainly corrects spatial heterogeneity. Although normalisation alone cannot control all systematic variations, normalisation plays an important role in the earlier stage of microarray data analysis because expression data can significantly vary under different normalisation procedures. A number of normalisation methods have been proposed, but it is not possible to decide in principle which method performs best. The normalisation method strongly depends on several factors like the number of detectable genes, the number of regulated genes, signal intensities, quality of the hybridisation, etc.

For a rough classification global normalisation can be distinguished from local (signal-intensity-dependent) normalisation.

If global normalisation is used, a single normalisation factor is applied to all detectable genes, leading to a linear shift of all signal intensities. Global normalisation methods are the most widely used methods. The underlying assumption is that constant systematic variations occur, e.g., the lower integration rate of one dye in respect to the second dye. Using the median of the single spot ratios for global normalisation is only advisable if a sufficient number of the detected genes are not regulated. If it is expected that nearly all of the genes will be regulated (which is of special interest regarding small arrays) a set of housekeeping genes should be included in the array configuration. Because housekeeping genes (by definition) are not regulated, the signal intensities of those genes should be the same on dual-labelled arrays. The housekeeping genes have to be picked thoroughly since even so-called housekeeping genes are subject to regulation under particular conditions.

Using local normalisation, a different normalisation factor is calculated for every gene. Local normalisation offers the opportunity of a signal-intensity-dependent normalisation. Some variations (e.g., laser settings) have different impacts on detected genes depending on their signal intensity. Thus, a non-linear shift of the signal intensities can be achieved in reliance on the signal intensity of each single spot.

Several linear and non-linear approaches are applied to normalisation of DNA microarrays. New normalisation methods are still under development. Because there is no general answer to the question of which algorithm best meets requirements, experimenters have to examine their data very carefully to decide which normalisation method to choose.

gene expression shall be used to define e.g. disease states or to distinguish substances by means of their impact on defined cell populations. Based on an appropriate amount of data, classification can be performed by using statistical methods to identify genes characterizing experiment classes.

Additional bioinformatics methods can be used to identify groups of interesting genes. One method commonly used is hierarchical cluster analysis by which genes and arrays can be ordered by similarity in expression behaviour [12] (Fig. 6C). In order to semi-automatically screen these results, bioinformatics infrastructures may be used that integrate the knowledge stored in diverse databases like pathway information, genomic localisation or protein family classification.

Examples of microarray experiments

General considerations

There is no doubt that all changes in cellular processes are reflected by altered gene expression. The measurement of changes in mRNA levels can be paralleled at high standards of reliability and reproducibility, since differences in physicochemical properties of mRNA rising from different genes are negligibly low compared to the respective gene products.

But because alterations in gene expression do not have to meet proportional changes on either the level of protein synthesis or activity, it is necessary to consider whether expression analysis should be accompanied by protein analysis methods after investigation of the expression ratios. Combining such traditional technologies with gene expression analysis will provide a comprehensive view to elucidate molecular mechanisms of actions, e.g. in understanding particular phenotypes, characterising the outcome of treatments or defining disease states.

Some general considerations concerning the experimental design, the configuration of the adequate array and the handling of the data have to be reviewed in advance.

Questions like sample size (to get the appropriate amount of RNA), number of samples and applicable controls (paragraph 3.4) have to be clarified. The preparation and handling of samples and RNA should be accompanied by STANDARD OPERATION PROCEDURES. In this way it is possible to avoid measuring differences in gene expression that are only artefacts caused by comparison of inappropriate samples.

Another often-discussed question is the necessary number of genes per array. In contrast to MEDIUM-TO-LOW-DENSITY ARRAYS, where several hundreds to some thousands of mostly topic-defined probes are spotted, some manufacturers supply 'FULL GENOME ARRAYS' (which have to be updated regularly). Which array format fits best depends on the focus of the experiments, although it has to be taken into account that the higher the number of genes on an array, the more difficult the identification of specific expression patterns characterizing disease states or treatments becomes. The noise caused by genes that can be measured as expressed but are only slightly regulated, hampers, for example, meaningful cluster analysis.

Considering the relatively high numbers of experiments and the number of data, it is necessary to think about how to organise and interpret all the data. The bare expression ratios call for scientific interpretation, which is not possible until additional information at least on the 'history' of each sample is available. In order to provide an optimal tool to assess the meaning of the expression profiling, it is advisable to generate databases holding both the experimental data concerning probe collection and the hybridisation results which are linked to additional data sources (e.g. SwissProt, UniGene, etc.) in a fashion that allows for bi-directional queries.

It is best to consider questions about data storage in advance, since the data that are collected during probe generation (and may be missing if one tries to collect it retrospectively) should match the planned statistical interpretation. General information about the requirements of microarray experiments which assure that microarray data can be easily interpreted and that results derived from its analysis can be independently verified were compiled in the minimum information about a microarray experiment (MIAME) proposal [13].

A practical example for a mode of action study – *in vitro/in vivo* correlation of the causal relationship between cigarette smoking and higher incidence of carcinomas

Little is known in detail about mechanisms leading to a higher incidence of carcinomas as a result of regular inhalation of cigarette smoke. An *in vitro* time course experiment with Swiss3T3 cells exposed to cigarette smoke (CS) was performed. After treatment with an aqueous solution of CS, RNA was isolated at different times spanning 0.5 to 24 hours. RNA from untreated cells was prepared at the same time points and used as control. Expression profiles for every time point was performed using glass arrays representing 513 selected cDNAs. The RNA was labeled either Cy5 (CS exposed cells) or Cy3 (control cells). The labeling proceeds during the RT-reaction of the RNA. Respective samples (Cy5/Cy3 pairs from CS exposed cells and controls) were pooled, cleaned and hybridised on the cDNA array. When hybridisation was finished, the arrays were read out by a laser scanner generating an image for each of the fluorescent dyes. For the following analysis the images were overlayed digitally. Red coloured spots represent spots with a higher intensity in the Cy5 channel ("overexpressed" genes), green spots represent spots with a higher intensity in the Cy3 channel

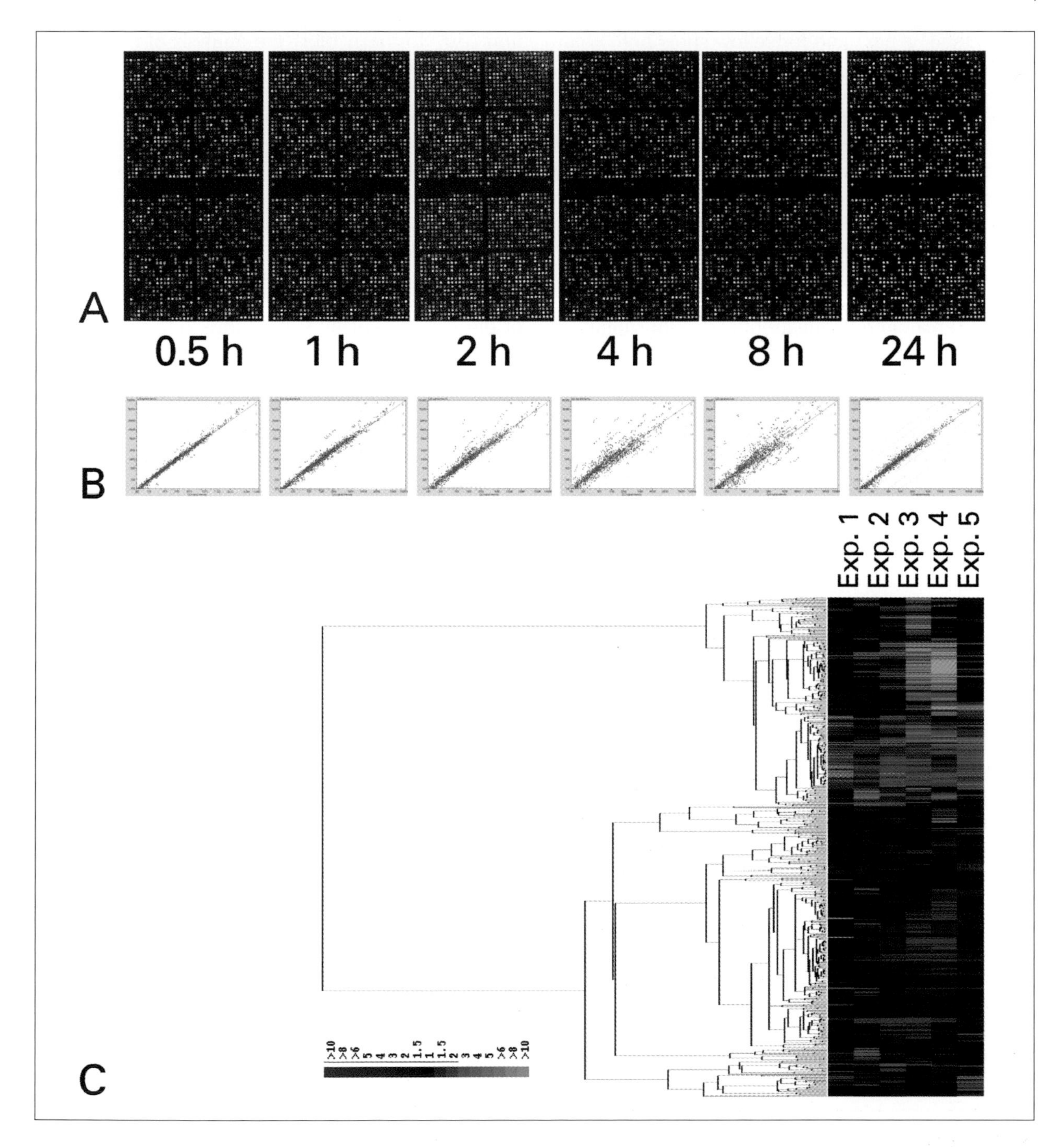

FIGURE 6. GENEXPRESSION ANALYSES OF SWISS 3T3 CELLS EXPOSED TO CIGARETTE SMOKE

A. Digital false colour overlays.

B. Scatterplots of computed signal intensities (Cy5/Cy3).

C. Vertical cluster analysis of similarity of gene expression.

("repressed" genes) and for yellow spots both channels had the same intensity, for example, the gene was expressed similarly in both samples.

Comparing the expression ratios from treated and untreated cells at different time points, there was a pro-inflammatory response that could be tracked for up to eight hours after treatment. After 24 hours, the GENE EXPRESSION PROFILES of the cells had returned to their primary levels. The digital false colour overlay (Fig. 6A) allows for examination of quality in terms of consistency of the hybridisation all over the spotting area or contamination by dust or smear.

Changes in gene expression during the time course can be tracked by scatterplots extracted from the expression ratios of every time point (Fig. 6B). Every detectable gene is represented by a point. Most of the detected genes are not influenced by treatment and therefore are located on the bisector. Genes which are repressed as a result of treatment are located on the right side of the bisector. Induced genes due to treatment can be identified on the left side of the bisector. Replicates of genes are represented in groups of four located nearby, which is also an indication of quality. From a kinetic point of view, the stress response could be characterised by the synchronised upregulation of antioxidant pathways orchestrated by stress-responsive transcription factors. In addition to the regulation of known expected antioxidative and genotoxic/cell-cycle-regulatory genes (e.g., heme-oxygenase-1, methallothionines and a number of heat shock proteins), several genes not yet linked to a CS-evoked stress response were identified [14]. Interestingly, some of the most upregulated genes are known to be responsive to peroxynitrite-induced oxidative stress. Furthermore, genes involved in the inflammatory response and immune modulation were found to be upregulated. These findings – even if derived from *in vitro* studies – contribute to our understanding of the mechanisms leading to CS-related disorders *in vivo*.

Summary

DNA MICROARRAYS are miniaturised devices made for the analysis of ribonucleic acids by hybridisation. In contrast to other technologies based on hybridisation (e.g. Northern blot) the analysis of the expression profiles of several hundred to thousands of genes in parallel is possible. This way, differences in the activity of genes between two different states of cells or tissue, e.g. treated *versus* non-treated, diseased *versus* non-diseased can be described. DNA MICROARRAYS differ regarding the dispensed probes, the SUBSTRATE (solid phase), the labeling procedure and the process of manufacturing. For a rough classification, cDNA ARRAYS can be distinguished from OLIGONUCLEOTIDE ARRAYS. Depending on the objective of the experiment and the efforts one is willing to invest, the sequence information regarding the DNA probes may differ enormously. To get a first overview of the expression levels in a given system it would be sufficient to spot low characterised sequences. The in-depth sequence analysis and annotation is delayed to the point when hybridisation has been performed and the resulting genes of interest have been identified. If ACCURATE identification and quantification of particular mRNA species are desired, time and effort have to be invested to ensure the SPOTTING of sequence-verified cDNAs. Various materials like membranes, plastics or glass may serve as SUBSTRATES for arrays. To guarantee sufficient quality some general features like planarity, uniformity, mechanical and chemical stability and optimal DNA binding capacity are crucial. Apart from *in situ* synthesis of oligonucleotides, the SPOTTING of cDNAs or pre-synthesised oligonucleotides is practised where contact and NON-CONTACT PRINTING are performed. A general declaration about the necessary amount of total RNA for hybridisation is not possible but depends highly on the labeling method and the kind of arrays. Roughly, amounts varying from 10 to 100 µg total RNA are required. If any amplification method is used, the required amount of RNA is drastically reduced.

There is also a broad spectrum of used dyes and labeling methods. Mainly fluorescent dyes are used, mostly Cy5 and Cy3. The incorporation of the dyes is either direct or indirect (two-step labeling). The data acquisition from microarray experiments consists of two parts: the digitalisation of the signals by using scanning devices and the subsequent image analysis using appropriate software packages for quantification of the signals and valid output of primary data.

Having generated the primary data, the ratios of gene expression levels have to be calculated and normalised. Besides the following interpretation of the expression ratios, further bioinformatics methods can be used to identify groups of interesting genes.

Selected readings

Schema M (ed) (1999) *DNA Microarrays: A Practical Approach*. Oxford University Press, Oxford, UK

Berrar DP, Dubitzky W, Granzow M (ed) (2002) *A Practical Approach to Microarray Data Analysis*. Kluwer Academic Publishers, Boston/Dordrecht/London

Nature Genetics (2002) The Chipping Forecast II. *Nat Genet* 32 (issue 4, suppl): 461–552

Recommended websites

Microarray Gene Expression Data Society - MGED Society: Minimum information about a microarray experiment – MIAME: *http://www.mged.org/Workgroups/MIAME/ miame.html* (Accessed April 2005)

National Center for Biotechnology Information: Gene Expression Omnibus: *http://www.ncbi.nlm.nih.gov/geo* (Accessed April 2005)

Unigene: *http://www.ncbi.nlm.ig.gov/UniGene* (Accessed April 2005)

Genetic Information Research Institute: Repbase: *http://www.girinst.org* (Accessed April 2005)

Swiss-Prot: Curated protein sequence database: *http://www.expasy.org/sprot/* (Accessed April 2005)

References

1 Southern EM (1975) Detection of specific sequences among DNA fragments separated by gel electrophoresis. *J Mol Biol* 98: 503–517

2 McGall GH, Fidanza JA (2001) Photolithographic synthesis of high-density oligonucleotide arrays. *Methods Mol Biol* 170: 71–101

3 Pease AC, Solas D, Sullivan EJ, Cronin MT, Holmes CP, Fodor SP (1994) Light-generated oligonucleotide arrays for rapid DNA sequence analysis. *Proc Natl Acad Sci USA* 91: 5022–5026

4 Hacia JG, Collins FS (1999) Mutational analysis using oligonucleotide microarrays. *J Med Genet* 36: 730–736

5 Hacia JG, Fan JB, Ryder O (1999) Determination of ancestral alleles for human single-nucleotide polymorphisms using high-density oligonucleotide arrays. *Nat Genet* 22: 164–167

6 Okamoto T, Suzuki T, Yamamoto N (2000) Microarray fabrication with covalent attachment of DNA using bubble jet technology. *Nat Biotechnol* 18: 438–441

7 Lockhart DJ, Dong H, Byrne MC, Follettie MT, Gallo MV, Chee MS, Mittmann M, Wang C, Kobayashi M, Horton H et al (1996) Expression monitoring by hybridization to high-density oligonucleotide arrays. *Nat Biotechnol* 14: 1675–1680

8 Tomiuk S, Hofmann K (2001) Microarray probe selection strategies. *Brief Bioinform* 2: 329–340

9 Bosio A, Stoffel W, Stoffel M (1999) Device for the parallel identification and quantification of polynucleic acids. In: MEMOREC Stoffel GmbH: EP0965647

10 O'Donnell-Maloney MJ, Smith CL, Cantor CR (1996) The development of microfabricated arrays for DNA sequencing and analysis. *Trends Biotechnol* 14: 401–407

11 Eberwine J (1996) Amplification of mRNA populations using aRNA generated from immobilized oligo(dT)-T7 primed cDNA. *Biotechniques* 20: 584–591

12 Eisen MB, Spellman PT, Brown PO, Botstein D (1998) Cluster analysis and display of genome-wide expression patterns. *Proc Natl Acad Sci USA* 95: 14863–1488

13 Brazma A, Hingamp P, Quackenbush J, Sherlock G, Spellman P, Stoeckert C, Aach J, Ansorge W, Ball CA, Causton HC et al (2001) Minimum information about a microarray experiment (MIAME)-toward standards for microarray data. *Nat Genet* 29: 365–371

14 Bosio A, Knorr C, Janssen U, Gebel S, Haussmann HJ, Muller T (2002) Kinetics of gene expression profiling in Swiss 3T3 cells exposed to aqueous extracts of cigarette smoke. *Carcinogenesis* 23: 741–748

Proteomics and applications within the drug development pipeline

René Houtman and Ian Humphery-Smith

Introduction

Rough estimates at the end of the last century predicted the existence of approximately 100,000 human genes. However, with the completion of the draft of the human genome in early 2001 came what many had already suspected: the actual count reduced this figure by 75%. A shocking conclusion: humans have only twice as many genes as a fruit fly or roundworm, and the gene count is comparable to that of a mouse! The big question that arises is: How in theory do we manage to operate at a higher level of complexity? The answer: It is proteins, and not genes, that are responsible for an organism's overall complexity. The interaction of proteins in a complex network adds up to how an organism functions. This is further complicated by lipids, sugars, the dynamics of molecular shape changes, and the intra-cellular compartmentalization of specific reagents and their respective concentrations.

Indeed, proteins are referred to as the 'molecular workhorses' that determine a phenotype which makes the PROTEOME, the protein complement encoded by a genome, an important object for scientific research, i.e., PROTEOMICS. Post-transcriptional mechanisms that control the rate of synthesis and half-life of proteins [1] are responsible for a poor correlation between mRNA expression levels and protein abundance [2, 3]. Moreover, the importance of CO- and POST-TRANSLATIONAL MODIFICATIONS (PTMs) for the regulation of protein function demands that an improved understanding of molecular systems in health and disease will depend upon access to techniques that provide both quantitative and qualitative information derived from complex protein mixtures.

The complexity and number of samples to be analyzed makes PROTEOMICS a technology-driven science and current approaches can be classified as either traditional or array-based PROTEOMICS. These approaches will be discussed in this chapter.

Traditional proteomics

Traditional PROTEOMICS is based on the separation science and involves sample preparation by extraction of proteins from the tissue or cells of interest followed by quantitative (expression level) and qualitative (identification) analysis. Technologies employed for separation of the proteins include capillary electrophoresis, liquid chromatography, one- and two-dimensional gel electrophoresis (2DE), continuous flow electrophoresis, and variations of mass spectrometry (MS), namely single or tandem MS. The latter can be expanded to MS^n and is based upon repeated ion selection and increasing more refined screening of the breakdown products of a particular polypeptide.

2-Dimensional electrophoresis

Although 2DE was already introduced in the mid 1970s [4], various modifications over the years helped in improving this technique into the powerful tool currently employed in PROTEOMICS research. To date, 2DE has been the mainstay of PROTEOMICS, but is increasingly being complemented by a variety of techniques.

Protein samples are first separated by ISOELECTRIC FOCUSING (IEF, 1st dimension) as illustrated in Figure 1A. This is referred to as charge-driven separation. At a pH above their ISOELECTRIC POINT (pI), amino acids (α) behave as proton donors and hence become negatively charged. When an electric field is applied, these negatively charged residues migrate to the

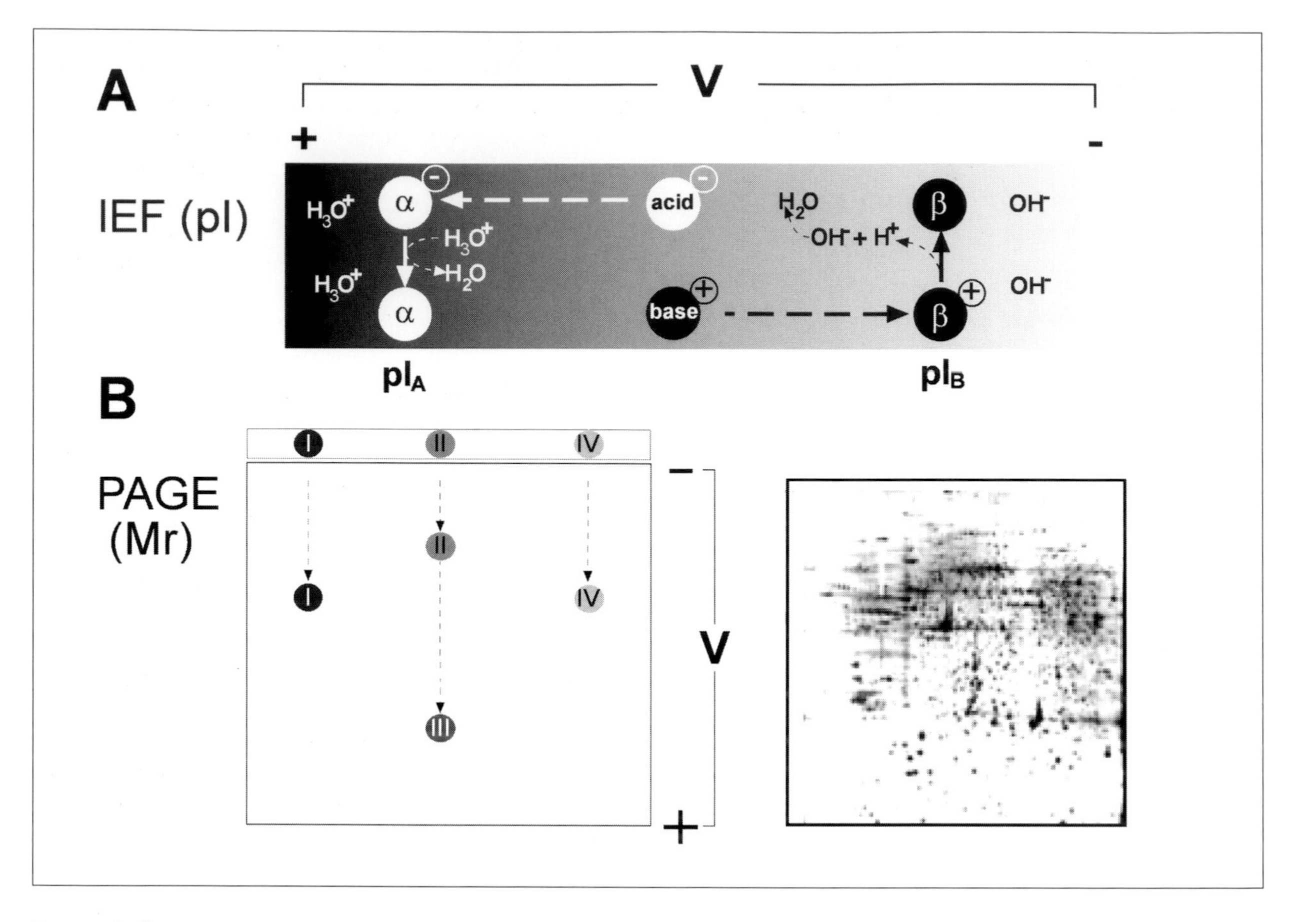

Figure 1. Separation of proteins by two-dimensional electrophoresis
A. Proteins are separated in the 1st dimension based on their isoelectric point (pI), i.e.: isoelectric focusing. Subsequently, B. these proteins are separated in the 2nd dimension based on their molecular mass (Mr) by SDS-PAGE. Separation based on the combination of two protein characteristics enables analysis of complex protein mixtures.

(positive) anode until they reach their pI. At this point the residue's charge is neutralized by proton uptake from its environment resulting in an arrest of migration. Alternatively, at a pH above below their pI, amino acids (β) behave as proton acceptors resulting in a positive charge. Accordingly, in an electric field they move to the cathode until their pI is reached. The pI of a protein is determined by the average of the pIs of the residues of which it consists. Proteins with different amino acid compositions therefore have different pIs. Originally, IEF of proteins was performed in rod-shaped polyacrylamide gels with mixed-in synthetic ampholytes, pH gradient-forming chemicals. However a major improvement of the technique was accomplished with the introduction of immobilized pH gradients (IPG) in the late 1980s. The latter, formed by ampholytes cross-linked to a plastic carrier strip by acryl amide, resulted in enhanced resolution, improved reproducibility, higher loading capacity for preparative gels and possibility of separation of basic proteins under equilibrium conditions [5], but is usually practiced in the presence of ampholytes and thus referred to as 'mixed-bed' electrophoresis.

Following IEF, proteins are separated with respect to mass within a polyacrylamide sieving matrix,

TABLE 1. CLEAVAGE SITE OF SEVERAL PROTEASES COMMONLY USED IN PROTEIN IDENTIFICATION

Enzyme	Cleavage site
Asp-N	X-D[1]
Arg-C	X-R
Glu-C(i)	E-X
Glu-C(ii)	E-X
	D-X
Lys-C	K-X
Trypsin	R-X
	K-X

For amino-acid coding see Table 3; X, any amino acid

namely SDS-PAGE in the 2nd dimension. An anionic detergent sodium dodecyl sulphate (SDS) within the electrophoresis buffer is employed to form charged moieties encompassing the polypeptides needing to be separated. Hence proteins previously separated in an IEF gradient become saturated with a negative charge. Application of an electric field induces migration of these proteins towards the anode from the IPG strip into the PAGE gel where they are separated by means of their molecular mass (Mr), small proteins migrating faster through the maze of acryl polymers then large proteins.

Once separated, proteins can be visualized by various fluorescent and non-fluorescent staining methods.

The power of separation by 2DE, based on a combination of different protein parameters, is illustrated in Figure 1B. Proteins of identical Mr (I and III) can be distinguished with respect to pI due to different amino acid composition, while proteins inseparable by IEF (II and IV) are separated by PAGE due to difference in Mr. In this manner, mixtures containing thousands of proteins can be resolved.

For identification, protein spots are excised from the 2DE gels and subjected to in-gel digestion by an ENDOPEPTIDASE with known specificity as listed in Table 1. PROTEOLYTIC FRAGMENTS diffuse out of the gel and are analyzed by mass spectrometry (MS).

One of the inherent qualities of 2DE is its ability to separate and quantitatively visualize PTMs of proteins, because accompanying changes in mass (see also Table 2) and, most importantly, accompanying charge, result in altered migration and a visible shift on the gel.

Procedures for protein characterization following protein isolation have been reviewed by Humphery-Smith and Ward [6].

Protein identification by mass spectrometry

One of the major breakthroughs for PROTEOMICS research has been the advent of soft-ionization MS-based techniques to facilitate analysis of macromolecules (polypeptides) by mass spectrometry. The instruments used, which enable protein identification from only a few femtomoles of material, can be grouped into two categories: single-stage mass spectrometers producing peptide mass fingerprints (PMF) and tandem MS- (MS/MS) based systems.

Basically, MS is performed by three consecutive events. First, the components of the ANALYTE require ionization/ desorption; second, the ions generated are separated based on their size and charge; and third, the separated ions are detected. Internal resolution of peptides can be obtained by enhancing fragmentation by post-source decay, collision-induced dissociation or electron-coupled dissociation of a parent ion of interest.

Ionization

There are several ways by which the PROTEOLYTIC FRAGMENTS can be ionized, of which matrix-assisted laser desorption/ionization (MALDI) and electrospray ionization (ESI) are most commonly used, illustrated in Figure 2A. In MALDI, the PROTEOLYTIC FRAGMENTS are mixed with a crystalline matrix, spotted onto a solid surface and dried. Next, crystals are targeted by a laser. The energy of the laser is absorbed by the crystal, causing it to vaporize and leading to an 'explosion' and subsequent desorption/ionization of peptides, which then enter the analyzer in gas phase.

Using electrospray ionization (ESI), ionization is achieved by a process referred to as COLOMBATION.

Table 2. Masses of several common post-translational modifications

Modification	Mass
Acetylation	42
Amidation	–1
Biotinylation	226
Carbamylation	41
Carboxylation	44
Deamidation	1
Deoxyhexose	146
Formylation	28
O-GlcNac	203
Glucosylation	162
Hexosamines	161
Hydroxylation	16
Methylation	14
Myristoylation	210
N-acetyl hexosamines	203
Oxidation	16
Palmytoylation	238
Pentose	132
Phosphorylation	80
Sialytation	291
Sodium	22
Sulphation	80

Here, the peptide solution is sprayed under vacuum from a gold-coated glass capillary directly into the analyzer. Before the analyzer is reached, fast evaporation of the solvent induces a rapid increase of charge per droplet size to a critical value, at which time the droplet explodes into naked ions carrying charge.

Separation of ions

Ionized peptides are separated with respect to their mass according to the principle of time-of-flight (TOF). As illustrated in Figure 2B, in an electric field (V) the time (t_D) it takes for ions to travel the distance (d) between the ion source (t_0) and the detector is inversely correlated with their mass-over-charge ratio (m/z). Of two equally charged peptides the one with the highest mass travels more slowly and arrives at the detector later. A typical TOF mass spectrograph is shown in Figure 3 in which each peak represents a parent ion of particular m/z ratio, calculated from t_D.

Peptide mass fingerprinting by MALDI-TOF

Protein identification can be achieved through peptide mass fingerprinting (PMF), usually performed by a single-stage MS instrument using MALDI linked to TOF (MALDI-TOF). The resulting spectrum resolves a peak list of masses corresponding to PROTEOLYTIC FRAGMENTS of the excised protein of interest and autoproteolytic peptides from the protease that was used for the digestion. The characteristic masses of the latter, indicated with * in Figure 2B, are used for internal calibration of the raw mass spectrum, enabling the generation of a high-accuracy list of molecular masses, or peptide mass fingerprint (PMF). The PMF is then compared with those derived from databases of which all proteins have been digested *in silico* with the same protease that was used for digestion of the

Figure 2 (following page)
Protein identification via single-stage mass spectrometry is performed in three subsequent events. A: First, proteolytic fragments are ionized. This is usually achieved by matrix-assisted laser desorption/ionization (MALDI) or electrospray ionization (ESI) B: Second, ions are separated based on their molecular mass by the principle of Time-Of-Flight (TOF) and, finally, detected. This results in a mass-over-charge spectrum (M/Z) spectrum C: A quadrupole can be used as an ion gate so select an ion with a particular M/Z. Tandem-mass spectrometry (MS/MS) is performed for peptide sequencing. The selected ion is fragmented by gas collision. In the resulting M/Z spectrum of all fragments the mass difference between adjacent peaks represents the mass of a terminal amino acid that was lost due to fragmentation. Hence MS/MS is performed for peptide sequencing.

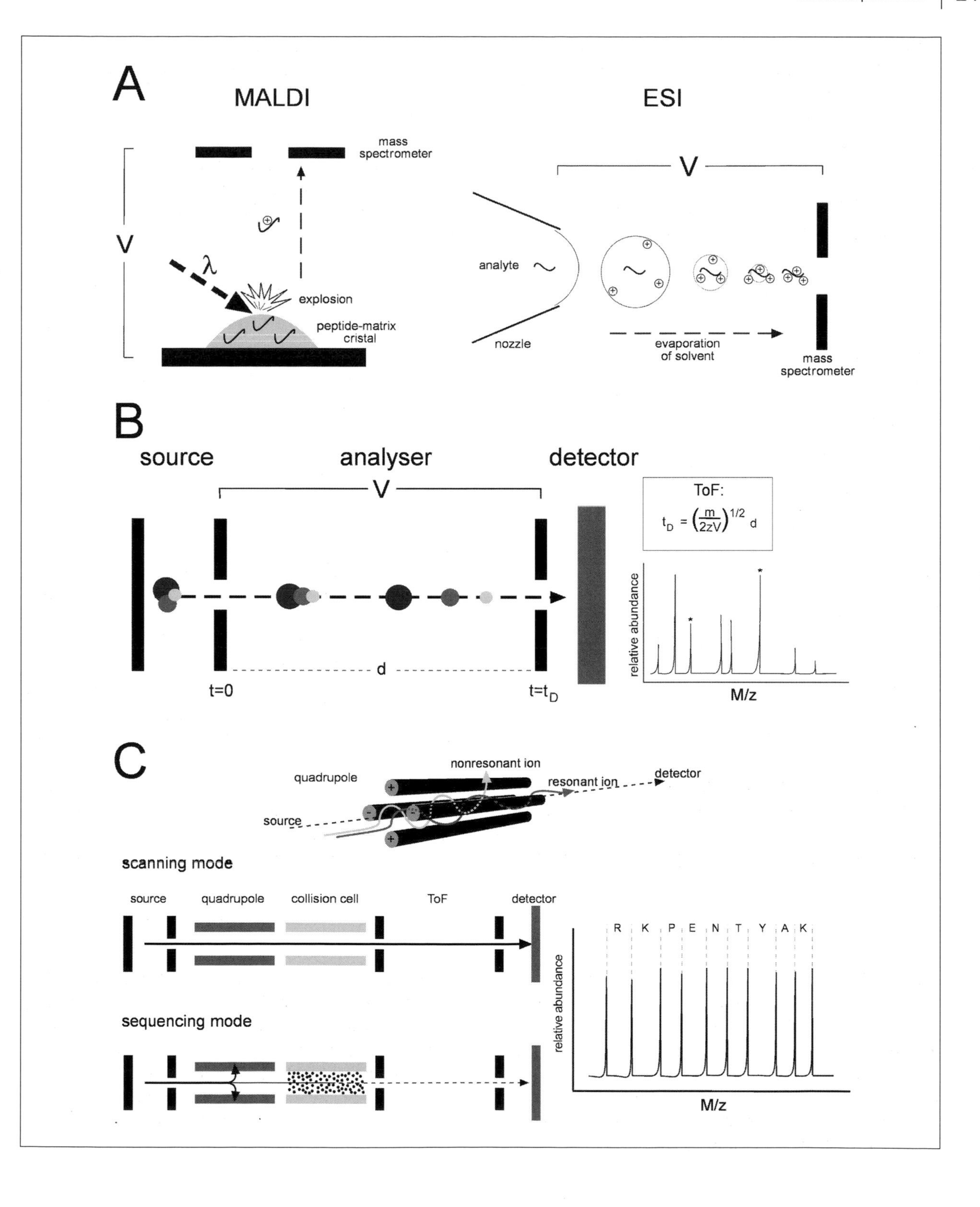
A
MALDI
ESI
mass spectrometer
V
λ
explosion
peptide-matrix cristal
V
analyte
nozzle
evaporation of solvent
mass spectrometer
B
source
analyser
detector
V
ToF:
$t_D = \left(\frac{m}{2zV}\right)^{1/2} d$
relative abundance
M/z
d
t=0
$t=t_D$
C
quadrupole
nonresonant ion
resonant ion
detector
source
scanning mode
source
quadrupole
collision cell
ToF
detector
R K P E N T Y A K
relative abundance
sequencing mode
M/z

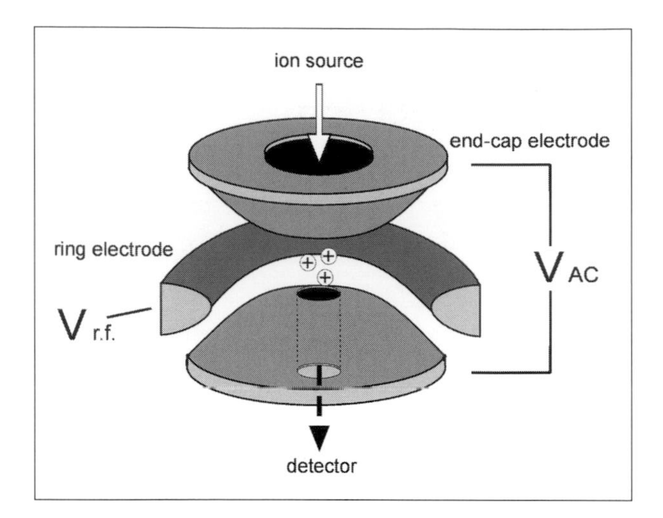

Figure 3

Schematic representation of a quadrupole ion trap, which is used for ion gating. The total mix of ionized peptides is 'trapped' within the space between the ring and end-cap electrodes. A combination of a constant trapping potential (V_{AC}) and alteration of the radio frequency ($V_{r.f.}$) amplitude enables selective elution of ions for analysis by TOF.

excised protein spot. A probability score for protein identity is calculated by an algorithm that takes the percentage of overall peptide sequence coverage taken across the undigested protein and the extent observed: expected mass errors. The success rate of this technique can be improved by the use of more than one protease so that results from more than one PMF can be used to increase the statistical confidence extracted from the search results obtained. [7]. As a general rule in PROTEOMICS, whenever orthogonal techniques and/or orthogonal algorithms are employed on the same sample or experiment, one can enhance the associated statistical confidence in the conclusions through concordance of research findings.

De novo protein sequencing by QTOF

PROTEOLYTIC FRAGMENTS can also be analyzed by tandem MS instruments (MS/MS or MS^n), such as the quadrupole coupled to TOF (QTOF), as illustrated in Figure 2C, by which information pertaining to the amino acid sequence and composition of peptides can be extracted. A quadrupole consists of four parallel rods or poles and serves as a gate for ions, peptides, of interest. The poles have a fixed direct current, DC, while alternating radio frequency voltages are applied to them. Depending on the electric field produced, only ions of a particular m/z will be focused on the detector (resonant ions), while all the others are deflected into the rods (nonresonant ions). Using QTOF, peptides are usually ionized by ESI before entering the instrument, which during the analysis switches back and forth between scanning and MS/MS-(sequencing) mode. During scanning mode, when the quadrupole is switched of, the ANALYTE is separated by TOF. When an ion is detected that meets predetermined criteria, e.g., a proper signal-to-noise ratio and having a mass that does not belong to an autoproteolytic fragment of the applied protease, the instrument switches to scanning mode. When the quadrupole is switched on, it creates an electric field that selects the m/z range for the ion of interest, i.e., parent ion gating. The gated ions are then led through an argon-filled chamber positioned inline with the quadrupole, the collision cell, and are fragmented by collision-induced dissociation prior to entering the TOF. In the resulting spectrograph, as shown in Figure 4, each peak represents a fragment of the parent ion and the mass difference between two adjacent peaks may correspond to that of an amino acid, the masses of which are listed in Table 3. These peaks are derived from the parent ion on the right (Fig. 2C), each having lost varying portions of the intact peptide. The sequence of m/z differences between each consecutive peak in the spectrum can be deconvoluted so as to determine the actual sequence of the selected peptide. After fragmentation of the parent ion and detection of the ion fragments, the instrument switches to scanning mode until a new ion meets the selection criteria. The amino acid sequences obtained by this process can then be employed for use in protein identification by database searching. In addition, MS/MS spectra can be used to study the location and nature of PTMs by focusing on mass differences between adjacent peaks that can only be explained by a net

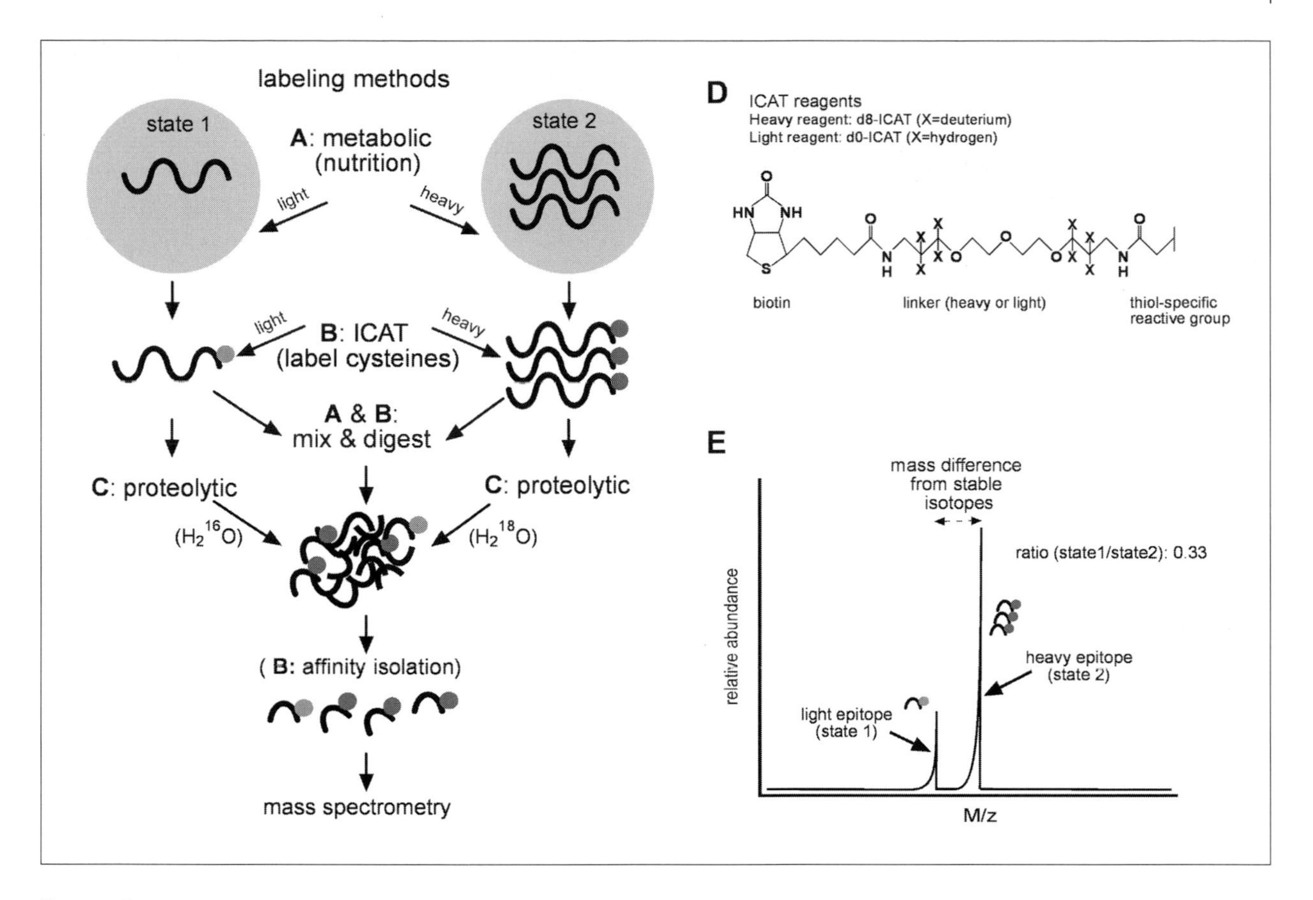

FIGURE 4

Quantitative mass spectrometry-based analysis of protein expression by stable isotope labeling. Comparison of protein abundance between two samples is performed by labeling the proteins of the samples with different mass tags. This labeling is mainly done A. metabolically, B. chemically, e.g., isotope-coded affinity tagging (ICAT, see also D.) or C. proteolytically. Subsequently, labeled samples are mixed, digested and subjected to mass spectrometry. In the resulting spectrum, identical peptides of the different samples are separated by the mass difference of the two isotopes and the ratio of their relative abundance in the spectrum reflects the ratio of protein expression between the two samples.

mass of an amino acid and a PTM as listed in Table 2.

2DE drawbacks

Although 2DE is unsurpassed in its resolving power for separation of complex protein mixtures, it is not without difficulties and shortcomings. Low-abundance proteins are probably the most interesting ones from the point of view of understanding cellular and regulatory proteins; however, only highly expressed proteins are detected and identified with current 2DE protocols [2, 8, 9]. The tendency of 2DE to detect only highly expressed proteins is caused by the intrinsic limitations in dynamic range of various stages of the technique on one hand and the enormous dynamic range of protein expression level across a PROTEOME on the other.

The dynamic range of detection by classical methods such as Coomassie Blue (low sensitivity) or silver staining (non-linear at higher protein concen-

Table 3 Characteristics of amino acids

Residue	1-letter code	Residue mass	Mass of most abundant immonium ion
Alanine	A	71	44
Arginine	R	156	129
Asparagine	N	114	87
Aspartic Acid	D	115	88
Cysteine	C	103	76
Glutamic Acid	E	129	102
Glutamine	G	128	101
Glycine	Q	57	30
Histidine	G	137	110
Isoleucine	H	113	86
Leucine	I	113	86
Lysine	L	128	101
Methionine	M	131	104
Phenylalanine	F	147	120
Proline	P	97	70
Serine	S	87	60
Threonine	T	101	74
Tryptophan	W	186	159
Tyrosine	Y	163	136
Valine	V	99	72

trations) is too narrow and therefore fluorescent staining methods with new probes with enhanced sensitivity and dynamic range have been developed and are becoming increasingly popular. In addition, the number of proteins that can be separated practically by 2DE is limited by the physical dimensions of the gel itself. One solution to enhance resolution is separation of a protein sample on multiple 2DE gels of consecutive partially overlapping narrow-range IPGs. Subsequently, the images of generated gels are combined *in silico* to construct so-called 'zoom gels' [5, 10].

The high variety in protein expression levels is illustrated by the fact that 90% of the proteome of a typical cell is made up of only 10% of the 10,000–20,000 different protein species. An additional factor that adds to this complexity is the large variety of biochemical characteristics (hydrophobicity, charge, pI, etc.) of proteins, which makes solubilization of all proteins during sample preparation virtually impossible. To reduce complexity, samples can be pre-fractionated for specific subsets of proteins by affinity enrichment and application of specific extraction buffers and methods. Still, hydrophobic proteins such as membrane receptors often precipitate during IEF, which prevents transfer into the SDS-PAGE gel and are thus lost during separation.

2DE also suffers from high variation that is introduced during sample preparation, protein loading and the complex nature of the staining procedures employed. In order to obtain statistically significant quantitative data from which valid conclusions regarding biology can be drawn, it is necessary to produce a number of replicate gels from the same sample [11–13]. Furthermore, qualitative and quantitative analysis of 2DE gels performed by dedicated software is still difficult to automate and requires a high degree of human intervention. The latter step is therefore time consuming and currently the rate-limiting factor in 2DE-based PROTEOME analysis. These shortcomings have been ostensibly overcome and fully automated image analysis is becoming a long-awaited reality in the discipline. As with any process, when one bottleneck has been overcome, the goal-posts are transposed elsewhere. Independent of method employed, the major challenge for PROTEOMICS is that of total proteomic coverage, which remains low, except in microbes.

Qualitative MS-based proteomics

Because of the limitations of 2DE, other techniques have been developed for qualitative PROTEOME analysis based on liquid separations coupled to on-line mass spectrometry. In general, complex mixtures of proteins, or peptides, are separated by microcapillary format liquid chromatography, capillary electrophoresis or hybrids thereof prior to ESI and subsequent mass analysis by MS/MS. This approach using one-dimensional separation is sufficient for the analysis of low-complexity samples; however, its resolution is by far inferior to that of 2DE. To date, no studies have been conducted on multiple analyses of the same sample set, but initial indications are that variance due to technology and biology will mean that large numbers of replicates will be necessary to allow valid conclusions due to significant population variance in both test and control groups.

One approach to obtain a higher order of resolution of separation prior to MS is termed multidimensional protein identification technology (MudPIT) [14]. This orthogonal combination of cation exchange and reverse-phase chromatography consists of sequential step-elutions (~15) from global styptic peptides from a strong cation exchange resin with slow ethanol gradients inserted between each salt elution step. After ESI, eluted peptides are analyzed using a quadrupole ion trap (QIT). This instrument consists of two hemispherical end-cap electrodes, spatially separated by ring electrodes, illustrated in Figure 3. The trapping potential, AC voltage, applied to the end-cap electrodes, is used to confine ion species with respect to their respective m/z ratio for subsequent analysis. A mass spectrum is obtained by linearly altering the amplitude of the radio frequency applied to the ring electrode so as to elute a particular ion series from the trap, after which they are analyzed by TOF.

The MudPIT method was shown to be effective in identification of proteins with a wide range of biochemical properties and expression levels, including low-abundance proteins [14, 15]. Although MudPIT is easy to automate and harbors the appropriate resolution and dynamic detection range needed to manage the complexity of global PROTEOME analysis, it will only generate qualitative data. This is a restriction of all MS-based techniques because the relative abundance of a peptide in the spectrum depends on its biochemical make-up and subsequent ability to become ionized and is not necessarily correlated with its concentration in the sample, i.e., relative ion intensity at the detector may be significantly different from that entering the mass spectrometer.

Quantitative MS-based proteomics

Several alternative methods have been proposed for relative quantification of protein expression between paired samples, of which one has been subjected to stable isotope labeling, which can be achieved by several methods (Fig. 4).

Metabolic labeling

In this approach (Fig. 4A), two populations of cells or organisms are grown in parallel, one on a normal source of nutrition and the other on a source depleted or enriched for an isotope of N, C or H [16–18]. This process is referred to as stable isotope labeling with amino acids during culture, or SILAC [19]. Next, protein samples from both populations are mixed, digested and analyzed in parallel by MS. The incorporated isotope induces a mass change reflected by a characteristic shift in the resultant MS spectrum. The mean ratio in relative ion intensity of corresponding peaks is employed as an indication of that of the relative difference in protein expression between the two populations (Fig. 4E). Unfortunately, peptides from the same protein show much variation with respect to the estimates of protein abundance derived from ion intensity measures, due to differences in a peptide's overall charge and associated willingness to enter the gas phase and 'fly' in the MS.

Isotope-coded affinity tagging

Another method is isotope-coded affinity tagging (ICAT) [20] (Fig. 4B) or variants thereof based upon isotope tagging of sulphydryl groups, amino groups, active sites for serine & cysteine hydrolases, phosphate ester groups, and N-linked carbohydrates. The ICAT reagents (Fig. 4D), consist of a protein-(cysteine) reactive group, a linker region and a biotin tag. The linkers are composed of either 8 deuterium (d8, heavy reagent) or 8 hydrogen (d0, light reagent) atoms, resulting in an 8Da mass difference between corresponding peaks in the MS spectrum. A reduced protein sample from one population is derivatized, on cysteine, with the isotopically heavy version of the ICAT reagents, while the other is derivatized with the isotopically light variant. Derivatized samples are mixed and subjected to digestion, after which the mix is enriched for cysteine-containing peptides by affinity purification using the biotin tag of the ICAT linker. This results in an approximate tenfold reduction in sample complexity. One of the weaknesses of this method is the fact that 1 out of 7 proteins does not contain any cysteine and thus does not react with the ICAT reagents. These proteins are lost during affinity purification using biotin and therefore missed in the analysis. Furthermore, the ICAT reagents are relatively large molecules compared with the peptide to which they are attached and thereby influence chromatographic separation prior to ionization and peptide ionization itself, which complicates data interpretation. A detailed comparison of 2DE and ICAT technologies has shown healthy complementarity of these techniques, whereby together enhanced proteomic coverage was achieved with the latter demonstrating a bias for high-Mr proteins and quantification based on the sum of the protein species, while the former showed improved resolution of low-Mr proteins, post-translationally modified polypeptides, processed polypeptides and identified cysteine-free proteins, otherwise invisible to ICAT [21].

Mass coded abundance tagging

To avoid problems associated with poor quantification derived from estimates based on ion intensity measures, mass coded abundance tagging (MCAT) has been introduced [22]. The methodology is based on guanidination of C-terminal lysine residues on tryptic peptides, corresponding elution times, direct peptide identification by MS and a predictable mass change of 21 m/z units. Here, however, ACCURATE protein quantification can be achieved from a reconstructed ion chromatogram for comparison of 'areas under the curve' for differential screening. The latter is somewhat limited by the resolution afforded by the liquid chromatography system. Another variant of MCAT is absolute quantification (AQUA) internal standard peptides [23].

Proteolytic labeling

A third labeling method is performed during protein digestion (Fig. 4C). Two samples are separately digested in either $H_2{}^{16}O$ or $H_2{}^{18}O$. The oxygen atom derived from the aqueous solvent is incorporated into the newly formed C-terminus of each peptide, providing an EPITOPE tag for relative quantification [23]. Differential resolution can be achieved in a similar manner by incorporation of ^{15}N [24].

Array-based proteomics

MICROARRAYS provide the potential for determining thousands of different binding events in a single experiment in a massively parallel fashion. The genomic revolution of the last decade which has delivered information on whole organisms at the level of DNA sequence and transcriptome analysis has been based notably on parallelization, miniaturization and automation – all of which become possible in a protein biochip environment. Usually, a number of different molecules with distinct binding characteristics, CAPTURE MOLECULES or ligands, are immobilized in a grid-like fashion on a solid SUBSTRATE, the array. The latter is incubated with an ANALYTE solution containing the molecules of interest, TARGETS. Detection of binding is usually performed by

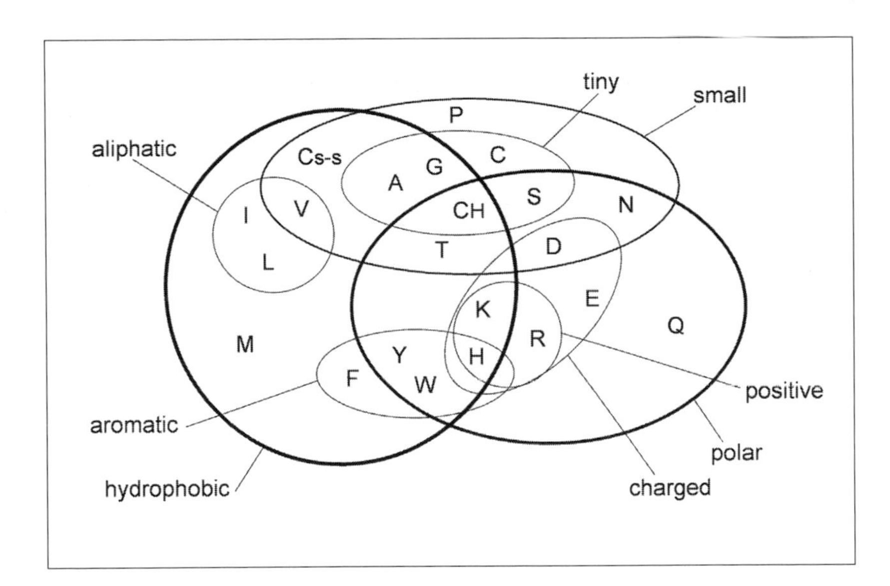

FIGURE 5. BIOCHEMICAL DIVERSITY OF AMINO ACIDS

fluorescent staining methods either before or after TARGET binding.

The microarray format was originally designed for studying DNA dot blots and later mRNA expression levels and is currently being extrapolated for applications to study protein expression and function. The origins and applications of array-based PROTEOMICS have been reviewed by Humphery-Smith [25]. A number of practical problems are associated with the setup of high throughput parallel binding assays involving proteins. Proteins are far more diverse with respect to their biochemical properties than their nucleic acid counterparts, whereby each amino acid combines to rapidly produce an incredibly large potential variety as to properties of the final polypeptide, i.e., the number of amino acid in a polypeptide string to the power of 20 (Fig. 5).

Miniaturization of systems allows for less ANALYTE consumption. But perhaps more importantly, reduced spot size associated with the area devoted to CAPTURE MOLECULES on the array will concomitantly allow for MICROSPOTS with high overall density of binding sites. This results in enhanced SENSITIVITY compared to the use of larger spot areas, i.e., MACROSPOTS. This phenomenon is explained by the theory of 'AMBIENT ANALYTE CONCENTRATION' [26, 27] and is explained below.

Microspot

CAPTURE MOLECULES are immobilized at high density on the solid support and localized within a very small surface area. During the assay, the TARGETS are captured by the MICROSPOT in which the absolute number of capture-TARGET complexes is low owing to its small area. As a result, the capture process does not significantly change the TARGET concentration in the ANALYTE solution, even for TARGETS in low concentration and/or for binding reactions that occur with high affinity. This remains true whenever $<0.1/K$ of the CAPTURE MOLECULES are captured, where K is the initial concentration of potential TARGETS in solution. Under these ambient ANALYTE conditions, the amount of the TARGET captured from solution directly reflects its concentration in the assay system. Such a system is independent of the actual ANALYTE volume and is capable of high-sensitivity concentration measurements with low sample consumption.

The SENSITIVITY that can be obtained is high for two reasons. First, the binding activity occurs at the highest possible TARGET concentration. Second, the CAPTURE MOLECULE-target complex is found only in a small area of the MICROSPOT, resulting in high local signal (Fig. 6). As demonstrated in this diagram, CAPTURE MOLECULES are immobilized in a constant sur-

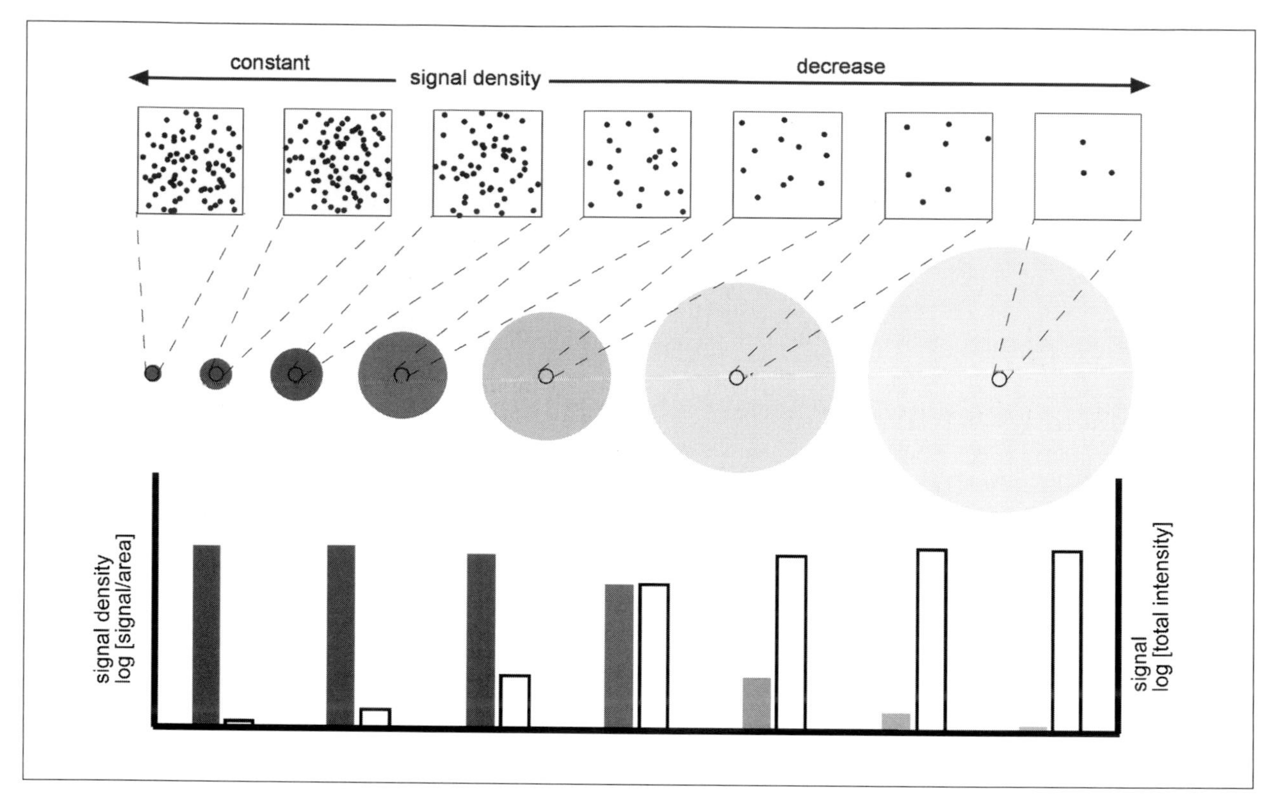

Figure 6. Signal and signal density in microspots (adapted from [28]).
Capture molecules are immobilized in a constant surface density in all spots. The largest spot harbors the highest total amount of capture molecules and overall signal. Here signal density is lowest because of the limiting amount of target, evenly distributed on the area without saturating it. With decreasing spot size the overall signal will decrease but the signal density will increase for smaller spots. Below a certain spot size, the signal density approaches an optimum (ambient analyte conditions, unlimited amount of target) and will stay approximately constant with any further decrease of spot size.

face density onto spots that have increased spot size. With increasing spot size, the total amount of CAPTURE MOLECULES in the assay increases, as does the sum of the obtained signal in the spot. The signal density, however, starts to decrease with increasing spot size because the amount of TARGET starts to become a limiting factor. The capture process leads to a significant reduction of TARGET concentration in solution and at the same time the probe-TARGET complexes are distributed over a larger area. As a result the maximal signal that can be obtained from any spot in the spot is decreased. Decreasing the spot size will decrease the overall signal per MICROSPOT but the signal density will increase for smaller spots. Below a certain spot size, the signal density approaches an optimum (ambient ANALYTE conditions, unlimited amount of TARGET) and will stay approximately constant with any further decrease of spot size. Therefore the highest signal intensities and optimal signal-to-noise ratios can be achieved in small spots. Small spot size also allows for dramatic parallelization on a low-cost biochip format. Furthermore, when conducted in conjunction with grating-coupled surface plasmon resonance, real-time association and dissociation constants can be acquired in parallel so as to provide still greater

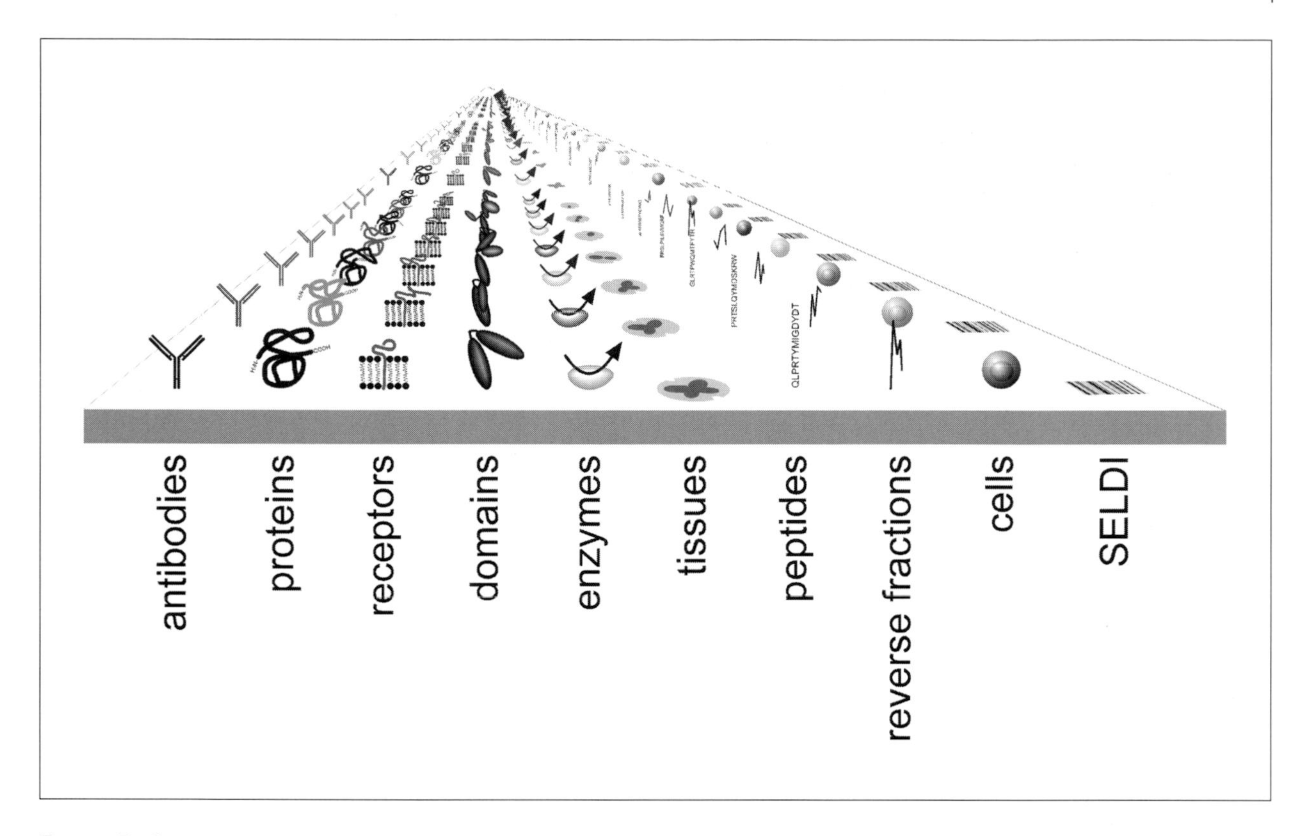

FIGURE 7. A WIDE VARIETY OF ASSAY TYPES IN ARRAY-BASED PROTEOMICS

insight into the dynamics of biomolecular interactions.

To date, protein biochips have been employed to examine a wide variety of assay types, see Figure 7. These include detection of interactions between antibody-antigen; protein-protein; protein-nucleic acid, protein-small molecule, membrane-bound receptors and TARGET, domain screening, and analysis of enzymatic function. In addition, a parallelized format has been adopted for a variety of other applications in PROTEOMICS. These applications include tissue arrays, EPITOPE mapping via peptide arrays, reverse arrays for the examination of naturally occurring protein isoforms found in cellular extracts as eluted from liquid chromatography, cellular arrays for bioassays, and immobilization of various chromatography affinity capture reagents. The latter are now being used extensively to seek out biomarkers associated with the progression, prognosis and early detection of disease and are also being applied to patient cohorting during drug trials, i.e., pharmacoproteomics. Protein arrays hold equally great potential to improved lead development and optimization as a means of verifying TARGET SPECIFICITY in the presence of numerous potential recombinant binders and thereby working towards reducing adverse drug effects as a direct spin-off from increased knowledge of the human genome.

Summary

Here, we have set out to demonstrate that PROTEOMICS in its various forms has demonstrated its utility across the entire drug development pipeline from initial TARGET discovery and validation through to monitoring drug responses during clinical trials. In recent times, proteomic techniques are also increasingly making their impact felt in the very last step in this pipeline, namely process validation during the production phase. This more generalized impact has only become possible as a result of improved reproducibility, sensitivity, user-friendliness and throughput of the techniques employed. The next phase of genomics is now generally hailed as focusing on PROTEOMICS. These techniques are evolving rapidly to meet the demands of the scientific and pharmaceutical communities. Such specialized approaches as chemoproteomics are likely to usher in still further advances in global screening of biomolecule binding to therapeutic agents and high-throughput characterization on line via mass spectrometry of these binders. The results of such analyses can then be fed to *in silico* analysis so as to better understand the shared structural motifs and domains that have given rise to drug binding across a wide spectrum of cellular components. In such a manner, PROTEOMICS data are likely to be seen increasingly to actually drive many aspects of future drug development and improved safety of therapeutic agents, be they traditional small molecules or biomolecules.

Selected Readings

Aebersold R, Mann M (2003) Mass spectrometry-based proteomics. *Nature* 422: 198–207

Phizicky E, Bastiaens PI, Zhu H, Snyder M, Fields S (2003) Protein analysis on a proteomic scale. *Nature* 422: 208–215

Recommended Websites

Web resources for protein scientists: *http://www.proteinsociety.org/docs/WWWResources.html* (Accessed December 2004)

IonSource: *http://ionsource.com/* (Accessed December 2004)

ExPASy Proteomics Server: *http://www.expasy.org/* (Accessed December 2004)

Proteomics Interest Group (ProtIG): *http://proteome.nih.gov/links.html* (Accessed December 2004)

An Introduction to Mass Spectrometry: *http://www.astbury.leeds.ac.uk/Facil/MStut/mstutorial.htm* (Accessed December 2004)

Swiss Proteomics Society (SPS): *http://www.swissproteomicsociety.org/links.html* (Accessed December 2004)

Proteomics – Spectroscopic Applications in Proteomics: *http://www.spectroscopynow.com/Spy/basehtml/SpyH/1,1181,10-4-0-0-0-directories_new-0-0,00.html* (Accessed December 2004)

North Carolina Genomics & Bioinformatics Consortium, LLC: *http://www.ncgbc.org/resources/links/proteomics-links.cfm* (Accessed December 2004)

http://www.hupo.org/ (Accessed December 2004)

References

1 Varshavsky A (1996) The N-end rule: functions, mysteries, uses. *Proc Natl Acad Sci USA* 93: 12142–12149

2 Gygi SP, Rochon Y, Franza BR, Aebersold R (1999) Correlation between protein and mRNA abundance in yeast. *Mol Cell Biol* 19: 1720–1730

3 Futcher B, Latter GI, Monardo P, McLaughlin CS, Garrels JI (1999) A sampling of the yeast proteome. *Mol Cell Biol* 19: 7357–7368

4 O'Farrell PH (1975) High resolution two-dimensional electrophoresis of proteins. *J Biol Chem* 250: 4007–4021

5 Gorg A, Obermaier C, Boguth G, Harder A, Scheibe B, Wildgruber R, Weiss W (2000) The current state of two-dimensional electrophoresis with immobilized pH gradients. *Electrophoresis* 21: 1037–1053

6 Humphery-Smith I, Ward M (2000) Proteome research: methods for protein characterization. In: S Hunt, F Livesey F (eds): *Functional Genomics. Practical Approach*. Oxford University Press, New York, 197–241

7 Cordwell SJ, Wilkins MR, Cerpa-Poljak A, Gooley AA, Duncan M, Williams KL, Humphery-Smith I (1995) Cross-species identification of proteins separated by two-dimensional gel electrophoresis using matrix-assisted laser desorption ionisation/time-of-flight mass spectrometry and amino acid composition. *Electrophoresis* 16: 438–443

8 Gygi SP, Corthals GL, Zhang Y, Rochon Y, Aebersold R (2000) Evaluation of two-dimensional gel elec-

trophoresis-based proteome analysis technology. *Proc Natl Acad Sci USA* 97: 9390–9395

9 Velculescu VE, Zhang L, Zhou W, Vogelstein J, Basrai MA, Bassett DE Jr, Hieter P, Vogelstein B, Kinzler KW (1997) Characterization of the yeast transcriptome. *Cell* 88: 243–251

10 Hoving S, Gerrits B, Voshol H, Muller D, Roberts RC, van Oostrum J (2002) Preparative two-dimensional gel electrophoresis at alkaline pH using narrow range immobilized pH gradients. *Proteomics* 2: 127–134

11 Voss T, Haberl P (2000) Observations on the reproducibility and matching efficiency of two-dimensional electrophoresis gels: consequences for comprehensive data analysis. *Electrophoresis* 21: 3345–3350

12 Quadroni M, James P (1999) Proteomics and automation. Electrophoresis 20: 664–677

13 Houtman R, Krijgsveld J, Kool M, Romijn EP, Redegeld FA, Nijkamp FP, Heck AJ, Humphery-Smith I (2003) Lung proteome alterations in a mouse model for nonallergic asthma. *Proteomics* 3: 2008–2018

14 Washburn MP, Wolters D, Yates JR 3rd (2001) Large-scale analysis of the yeast proteome by multidimensional protein identification technology. *Nat Biotechnol* 19: 242–247

15 Wolters DA, Washburn MP, Yates JR 3rd (2001) An automated multidimensional protein identification technology for shotgun proteomics. *Anal Chem* 73: 5683–5690

16 Weckwerth W, Willmitzer L, Fiehn O (2000) Comparative quantification and identification of phosphoproteins using stable isotope labeling and liquid chromatography/mass spectrometry. *Rapid Commun Mass Spectrom* 14: 1677–1681

17 Berger SJ, Lee SW, Anderson GA, Pasa-Tolic L, Tolic N, Shen Y, Zhao R, Smith RD (2002) High-throughput global peptide proteomic analysis by combining stable isotope amino acid labeling and data-dependent multiplexed-MS/MS. *Anal Chem* 74: 4994–5000

18 Krijgsveld J, Ketting RF, Mahmoudi T, Johansen J, Artal-Sanz M, Verrijzer CP, Plasterk RH, Heck AJ (2003) Metabolic labeling of *C. elegans* and *D. melanogaster* for quantitative proteomics. *Nat Biotechnol* 21: 927–931

19 Ong SE, Blagoev B, Kratchmarova I, Kristensen DB, Steen H, Pandey A, Mann M (2002) Stable isotope labeling by amino acids in cell culture, SILAC, as a simple and accurate approach to expression proteomics. *Mol Cell Proteomics* 1: 376–386

20 Gygi, SP, Rist B, Gerber SA, Turecek F, Gelb MH, Aebersold R (1999) Quantitative analysis of complex protein mixtures using isotope-coded affinity tags. *Nat Biotechnol* 17: 994–999

21 Schmidt F, Donahoe S, Hagens K, Mattow J, Schaible UE, Kaufmann SH, Aebersold R, Jungblut PR (2003) Complementary analysis of the *Mycobacterium tuberculosis* proteome by two-dimensional electrophoresis and isotope coded affinity tag technology. *Mol Cell Proteomics* 3: 24–42

22 Cagney G, Emili A (2002) *De novo* peptide sequencing and quantitative profiling of complex protein mixtures using mass-coded abundance tagging. *Nat Biotechnol* 20: 163–170

23 Gerber SA, Rush J, Stemman O, Kirschner MW, Gygi SP (2003) Absolute quantification of proteins and phosphoproteins from cell lysates by tandem MS. *Proc Natl Acad Sci USA* 100: 6940–6945

24 Munchbach M, Quadroni M, Miotto G, James P (2000) Quantitation and facilitated *de novo* sequencing of proteins by isotopic N-terminal labeling of peptides with a fragmentation-directing moiety. *Anal Chem* 72: 4047–4057

25 Humphery-Smith I (2003) *Protein Biochips and Array-based Proteomics*. Dekker, New York

26 Masseyeff RF (1991) Standardization of immunoassays. *Ann Ist Super Sanita* 27: 427–436

27 Ekins RP (1989) Multi-analyte immunoassay. *J Pharm Biomed Anal* 7: 155–168

28 Templin MF, Stoll D, Schrenk M, Traub PC, Vohringer CF, Joos TO (2002) Protein microarray technology. *Trends Biotechnol* 20: 160–166

Immunotherapeutics

Vaccines

Wim Jiskoot and Gideon F.A. Kersten

Introduction

VACCINES are the most commonly administered immunotherapeutics. Supported by great improvements in sanitation facilities such as safe drinking water, vaccination has been the most effective measure to control a diversity of life-threatening infectious diseases in the 20th century. The most impressive success of vaccination has been the global eradication of smallpox in the 1970s. Moreover, the incidence of many other infectious diseases, such as diphtheria, tetanus, pertussis, poliomyelitis, measles, mumps, and rubella, has been drastically reduced thanks to extensive vaccination programmes.

Upon a natural infection with a pathogen, an unprotected person usually falls ill before the immunological defence system is able to respond adequately. Vaccination aims to stimulate the specific immune response against a pathogen by the administration of attenuated or killed organisms, or fractions thereof. If vaccination is successful and the host comes into contact with the pathogen afterwards, the specific immune response will be immediate and sufficiently strong to kill the invading organism before it has the opportunity to multiply and cause disease. Thus, in a strict sense VACCINES are immunoprophylactics rather than therapeutics.

In most cases, repeated doses of the vaccine are given to boost the immune response. Apart from the number of doses, several other factors determine vaccine EFFICACY, as summarised in Table 1. If a high enough proportion of a population is immunised, vaccination not only protects the immunised individuals, but also may help to protect the community as it decreases the chance that non-immunised persons encounter the pathogen. This is referred to as herd immunity [1].

A brief history of vaccination will be given below. Next, current vaccine categories for human use will be addressed. Finally, new developments in vaccinology will be outlined.

Historical background

Vaccination has a long history [1,2]. The most prominent milestones of vaccinology are listed in Table 2. The first attempts to become immune probably date back to as early as the 7th century, when Indian Buddhists drank snake venom and may thus have become immune against this toxin. Written reports bear witness to the practice of variolation, i.e., the administration of scabs or pustule preparations obtained from patients recovered from smallpox, ever since about 1000 A.D. in various parts of the world, amongst others in China, India, North Africa, and England.

Variolation was widely applied until Edward Jenner introduced cowpox vaccination at the end of the 18th century. His practice was based on the recognition that milkmaids were frequently subjected to mild pox infection acquired from the cows they milked, but were spared from disease during smallpox epidemics. The first demonstration that the principle of immunisation works was Jenner's anecdotal experiment with an 8-year old boy who remained healthy when challenged with smallpox virus after he had been immunised with cowpox virus. It was Jenner who introduced the terms "vaccine" for cowpox preparations (derived from the Latin *vacca* = cow) and "vaccination" for the administration thereof. Later, in honour of Jenner, Louis Pasteur generalised the meaning of vaccination to immunisation with agents other than cowpox. During the 19th century vaccination with live cow-

TABLE 1. FACTORS DETERMINING VACCINE EFFICACY

Pathogen-dependent	Host-dependent	Vaccine-dependent	Vaccination schedule-dependent
Port of entry	Species	Nature of antigenic component(s)	Route of administration
Localisation in host	Age	Antigen content	Number of doses
Antigenic variation	Genetic factors	Delivery systems	Immunisation intervals
Mutation frequency	Physical state	Adjuvants	Simultaneous administration of other vaccines (administered separately)
	Immune status	Combination with other vaccine components (in one vial or syringe)	

pox virus became common practice. In the 20th century, vaccinia virus, which is closely related to cowpox virus [1], became widely used as a live vaccine until smallpox was eradicated (Tab. 2).

Pasteur gave a new impetus to vaccinology in the last quarter of the 19th century. He showed that the virulence (i.e., infectivity) of pathogens could be reduced by successive passage in culture. Vaccination with attenuated strains thus obtained could confer protection without causing disease. The efforts of Louis Pasteur and others led to the development of live ATTENUATED VACCINES against cholera, anthrax, and rabies. Along with the introduction of ATTENUATED VACCINES, it became apparent that infection with live material was not essential to induce immunity. The procedure of killing bacteria by heat and subsequent stabilisation with phenol was developed, resulting in the introduction of heat-inactivated whole-cell VACCINES against cholera, typhoid and plague at the end of the 19th century.

At the beginning of the 20th century the development and introduction of new live (tuberculosis, yellow fever) and killed VACCINES (pertussis, influenza, rickettsia) followed. Moreover, it was being recognised that some components of a micro-organism were more relevant for protection than others, and the concept of SUBUNIT VACCINES was born. This and the discovery of chemical inactivation of bacterial toxins with formaldehyde led to the introduction of SUBUNIT VACCINES against diphtheria (1923) and tetanus (1927).

In the early 1950s tissue-culture techniques for virus propagation were developed. This resulted in the licensing of Salk's inactivated polio vaccine (IPV) in 1955. In the same period Sabin developed an oral polio vaccine (OPV) consisting of live attenuated viruses, which became available in the U.S.A. in 1961. Several other viral VACCINES derived from tissue cultures followed. Furthermore, several bacterial SUBUNIT VACCINES based on purified proteins or polysaccharides were introduced since the 1970s. The first vaccine based on rDNA technology was marketed in 1986.

Current vaccine categories

Classification

The currently available VACCINES for human use are of either bacterial or viral origin and can be divided into several categories (see Table 3). These categories will be discussed below. For a more detailed description of individual VACCINES currently in use, the reader is referred to the textbook of Plotkin & Orenstein [1] (Tab. 3).

TABLE 2. MILESTONES IN VACCINE HISTORY[1]

Year	Event[2]
ca. 1000	Intranasal administration of preparations of scabs from smallpox patients in China
16th–17th century	Parenteral variolation in India by Hindus
17th century	Oral administration of white cow flea pills for smallpox prevention in China
1796	Immunisation of 8-year old boy with cowpox virus and challenging with smallpox virus (Edward Jenner)
1798	Initiation of general cowpox immunisation with Jenner's variola vaccine
1870s	Discovery of attenuation of fowl cholera bacteria (Louis Pasteur)
1884	Attenuated *Vibrio cholerae*: the first bacterial vaccine used in humans (Robert Koch)
1885	First administration to humans of attenuated rabies vaccine (Louis Pasteur)
1896-1897	Introduction of the first heat-inactivated vaccines against typhoid, cholera and plague
1923	Introduction of the first subunit vaccine: formaldehyde-treated diphtheria toxin
1927	Introduction of BCG, attenuated tuberculosis vaccine
1955	Introduction of inactivated poliovirus vaccine (Salk): the first vaccine developed with tissue-culture technique
1961	Attenuated poliovaccine (Sabin) as the first licensed oral vaccine
1980	Declaration of the eradication of smallpox by the WHO
1986	Licensing of the first rDNA vaccine: recombinant HBsAg
1987	Licensing of the first conjugate vaccine against Hib: PRP-T

[1]*Sources: [1, 2]*
[2]*Abbreviations: BCG, bacille Calmette-Guérin; HBsAg, hepatitis B surface antigen; Hib,* Haemophilus influenzae *type b; PRP-T, polyribosylribitol phosphate-tetanus toxoid conjugate vaccine; WHO, World Health Organisation*

Live attenuated vaccines

Attenuation through serial passage and selection of less virulent and less toxic variants has been applied to obtain safe vaccine strains. Once a suitable strain has been obtained, master and working seed lots are prepared. The seed lot system provides the basis for the reproducible production of live (and other) VACCINES. The dose of live VACCINES is determined on the basis of the number of viable organisms.

Live VACCINES have a number of advantages over non-living VACCINES. Although attenuation generally means reduced infectivity, attenuated strains will replicate to some extent in the recipient. This furnishes a sustained Ag dose, inducing strong immune responses even after a single dose. In general, live VACCINES generate higher cell-mediated immune responses than inactivated VACCINES. A single immunisation with a live vaccine can provide lifelong immunity.

The major drawback of live VACCINES is the risk of reversion to pathogenicity. For instance, the occurrence of vaccine-associated paralytic poliomyelitis after the introduction of OPV has been reported [1, 3]. Furthermore, live VACCINES sometimes cause mild symptoms resembling the disease caused by the pathogen. Live VACCINES should never be given to immunosuppressed persons, because they lack the ability to respond even to infections by attenuated organisms.

Attenuated viral vaccines

Examples of live viral VACCINES are polio, measles, mumps and rubella VACCINES. Attenuated polio vaccine is administered orally. Poliovirus is a non-

Table 3. Classification and examples of current vaccines

Category	Example	Vaccine characteristics
Live attenuated organisms		
viral	oral polio vaccine (Sabin)	attenuated viruses, serotypes 1-3; oral vaccine
	measles vaccine	attenuated virus
	mumps vaccine	attenuated virus
	rubella vaccine	attenuated virus
	yellow fever vaccine	attenuated virus
bacterial	tuberculosis vaccine	attenuated *Mycobacterium bovis*
	typhoid fever vaccine	attenuated bacteria, oral vaccine
Killed whole organisms		
viral	Inactivated polio vaccine (Salk)	formaldehyde-inactivated viruses, serotypes 1–3
	rabies vaccine	β-propiolactone-inactivated virus
	hepatitis A vaccine	formaldehyde-inactivated virus
	Japanese encephalitis vaccine	formaldehyde-inactivated virus
bacterial	whooping cough vaccine	heat-inactivated bacteria
	cholera vaccine	phenol-inactivated bacteria
	typhoid fever vaccine	heat-inactivated bacteria
Subunit vaccines		
viral	influenza vaccine	influenza surface antigens
	hepatitis B vaccine	recombinant hepatitis B surface antigen
bacterial	diphtheria vaccine	formaldehyde-treated toxin
	tetanus vaccine	formaldehyde-treated toxin
	whooping cough vaccine	mixture of purified proteins
	meningococcal vaccine	purified capsular polysaccharides
	pneumococcal vaccine	purified capsular polysaccharides
	Hib[1] vaccine	polysaccharide-protein conjugates

[1]Haemophilus influenzae *type b*

enveloped single-strand RNA virus. Its four structural proteins (VP1-4) form a regular three-dimensional structure with a diameter of 28 nm. OPV contains the three existing serotypes (1, 2 and 3), which differ from each other in a number of distinct EPITOPES relevant for protection. The virus is relatively stable and additives such as magnesium chloride or sorbitol further enhance the thermal stability of the vaccine. OPV is included in many childhood immunisation programmes. It is usually administered to infants in a 4-dose scheme and probably provides lifelong protection.

Attenuated mumps, measles and rubella viruses are often combined in a COMBINATION VACCINE (MMR vaccine). These attenuated RNA viruses vary in size and number of structural proteins. Measles, mumps and rubella VACCINES, whether separate or combined, are lyophilised preparations stabilised with sucrose, sorbitol, hydrolysed gelatine and/or amino acids. They contain the antibiotic neomycin and have to be kept refrigerated. The three vaccine components have in common that one single s.c. administration is probably sufficient for lifelong protection. Nevertheless, in some countries the first dose given at 12–15 months of age is followed by a second vaccination at the age of 4–6 or 11–12 years. Both humoral and cell-mediated immunity are important for protection. VACCINE EFFICACY is estimated to be at least 90% and combining the components does not seem to influence their effectiveness. Side-effects are generally mild and usually occur 7–12 days after vaccination. MMR VACCINES are not indicated for infants below the age of one year, because circulating maternal Ab impair vaccine EFFICACY in this age group.

Attenuated bacterial vaccines

The most well-known attenuated bacterial vaccine is tuberculosis vaccine, which has been incorporated in many immunisation programmes as of the 1930s. The vaccine is based on *Mycobacterium bovis* bacteria, which primarily infect cattle but can also infect humans. The vaccine consists of lyophilised attenuated *M. bovis*, known as bacille Calmette-Guérin (BCG), and is administered i.d. to infants and older children. Current vaccine strains vary in the extent of attenuation and the dosage varies among vaccine suppliers. The immunisation schedule varies significantly among nations. The nature of the immune response is not known in detail, but cell-mediated immune mechanisms are probably involved in protection, whereas Ab do not seem to play a substantial role. Besides strain variations, the lack of reliable tests for immunity and the poorly understood MECHANISM OF ACTION contribute to the fact that estimates of vaccine EFFICACY vary from 0 to 80% [1].

Oral attenuated *Salmonella typhi* VACCINES are indicated for high-risk groups, such as children in endemic areas and travellers, to prevent typhoid fever. The only licensed strain is Ty21a, whose attenuation has been stimulated by using nitrosoguanidine, a chemical mutagenic agent. Strain Ty21a lacks the ability to synthesise capsular polysaccharides, which are essential for virulence. In order to protect the bacteria against peptic digestion, the vaccine is formulated as lyophilised bacteria in enteric-coated capsules. Protection is achieved through 3–4 doses administered every other day. The vaccine provides significant protection by inducing relatively strong intestinal IgA and cell-mediated responses, and a weak systemic Ab response. Protective Ab are directed against flagelli and LPS. The duration of the protection is relatively short (3–5 years).

Killed whole organisms

Inactivated bacterial and viral VACCINES are obtained from virulent strains by heat treatment or by chemical inactivation, usually with formaldehyde. Since killed pathogens are not able to propagate after administration, these VACCINES usually are less immunogenic than live VACCINES. An advantage over the latter is the inability to revert to virulence. On the other hand, deficient inactivation has caused vaccine-related accidents. For instance, immunisation with insufficiently inactivated polio vaccine in 1955 resulted in cases of paralytic disease [1,2]. Examples of this category include inactivated polio vaccine (IPV) and whole-cell pertussis vaccine, which are discussed below.

Inactivated polio vaccine

IPV is currently used in several countries. The vaccine consists of formaldehyde-inactivated poliovirus and includes the three serotypes. The dose is determined on the basis of Ag contents. Advantages of IPV over OPV are a better temperature stability, the lower risk of vaccine-related disease and the possibility of combination with diphtheria, tetanus and pertussis components in one formulation (DTP-IPV vaccine). In contrast to OPV, IPV does not elicit substantial amounts of secretory IgA Ab, but its effect relies on the induction of virus-neutralising serum IgG. A vaccination regimen with both IPV and OPV combines the advantages of both VACCINES and is applied in Denmark [1].

Whole-cell pertussis vaccine

Pertussis (or whooping cough) vaccine consists of heat-inactivated *B. pertussis* cells. The dose is determined on the basis of the opacity of the inactivated cell suspension. The vaccine potency is tested by protection assays in mice. The protective EFFICACY of whole-cell pertussis VACCINES is probably based on Ab against several pertussis Ag, such as pertussis toxin, filamentous haemagglutinin and LPS. Whole-cell pertussis VACCINES are notorious for their frequent side-reactions, mostly fever and irritability. Other side reactions include excessive sleeplessness, persistent inconsolable crying and shock-like phenomena. The adverse effects of whole-cell pertussis VACCINES are largely due to the LPS present in *B. pertussis*'s outer membrane. The adverse effects are stronger in older children and adults, so that whole-cell pertussis VACCINES are not indicated for these age groups. Although protection is probably restricted to a period of about 10 years, this is not a significant problem, because whooping cough infections are most dangerous in toddlers and infants. Adolescents and adults in general (but not always) have relatively mild symptoms. The vaccine contains colloidal aluminium salt as ADJUVANT and is usually combined with diphtheria and tetanus vaccine components (DTP vaccine). The vaccine is given in 4-5 i.m. doses.

Subunit vaccines

SUBUNIT VACCINES contain one or more selected Ag (subunits) significant for protection against the pathogen they are derived from. SUBUNIT VACCINES have better defined physicochemical characteristics and show fewer side-effects than VACCINES consisting of attenuated or inactivated organisms. ANTIGENS used for current SUBUNIT VACCINES include viral and bacterial proteins as well as bacterial capsular polysaccharides.

Proteins

Protection against diphtheria or tetanus is mainly based on the presence of Ab directed against the respective toxins. These toxins are water-soluble proteins and form the basis of diphtheria and tetanus VACCINES. In order to eliminate the toxicity of diphtheria and tetanus toxin, they are incubated with formaldehyde. This process is called toxoidation and the resulting products are referred to as toxoids. Formaldehyde forms covalent bonds with the toxin, which is initiated by a reversible reaction of formaldehyde with primary amino groups, followed by an irreversible reaction with other amino acid residues [2]:

$$R\text{-}NH_2 + CH_2O \leftrightarrow R\text{-}NH\text{-}CH_2OH \leftrightarrow R\text{-}N = CH_2 + H_2O \quad (1)$$

$$R\text{-}N = CH_2 + R'\text{-}H \rightarrow R\text{-}NH\text{-}CH_2\text{-}R' \quad (2)$$

where R is the toxin and R' can be a reactive amino acid residue (e.g., lysine, arginine, tryptophan, tyrosine, histidine) in the toxin molecule (or possibly a neighbouring toxin molecule) or a free amino acid present in the medium. Thus, stable crosslinks are formed, yielding a heterogeneous product with respect to number and sites of formaldehyde adducts and molecular weight. The degree of toxoidation is highly dependent on the reaction conditions, including formaldehyde concentration, pH, temperature and the presence of other components. The toxoidation process must be a compromise between sufficient detoxification and preservation of relevant EPITOPES. To enhance the relatively poor immunogenicity of toxoids, they are adsorbed to aluminium salt suspensions.

A solution to the risk of residual toxicity and loss of immunogenic sites has been found by the introduction of genetic toxoidation, which has been applied to pertussis toxin, a crucial component in recently developed SUBUNIT VACCINES against pertussis. Genetic toxoidation of pertussis toxin was achieved by site-directed mutagenesis of the site responsible for toxicity. The resultant toxoid is devoid of toxicity and well defined, does not bear the risk of reversal to toxicity and is more immunogenic than chemically detoxified pertussis toxin [4]. Curiously, genetically obtained pertussis toxoid is stabilised with low concentrations of formaldehyde.

Acellular pertussis VACCINES have been introduced in Japan in the early 1980s as alternatives to the whole-cell VACCINES for the immunisation of older children. Although pertussis toxoid alone may be sufficient for protection, most acellular pertussis VACCINES contain at least two proteins important for the virulence of *B. pertussis*, including (inactivated) pertussis toxin, filamentous haemagglutinin, fimbriae and pertactin. Recent field trials indicate that the EFFICACY of these VACCINES is comparable to that of whole-cell VACCINES, whereas the acellular VACCINES induce virtually no adverse effects [5]. This makes them suitable for immunisation of older children and adults. Acellular pertussis VACCINES have now been introduced in many national immunisation programmes, mostly combined with diphtheria and tetanus components (DTaP vaccine).

Hepatitis B vaccine was the first marketed recombinant vaccine, which has replaced hepatitis B VACCINES obtained from plasma of infected humans. Recombinant hepatitis B VACCINES are composed of hepatitis B surface Ag (HBsAg) derived from yeast or mammalian cells. Introduction of the genes for HBsAg in eukaryotic cells has been accomplished by inserting the protein gene into a plasmid, which was then used to transform the host cells. Purified HBsAg self-assembles to 22-nm particles identical to those excreted by cells infected with the native virus. Advantages of the rDNA vaccine as compared to the plasma-derived product are safety, high yields and consistent quality. The ease of production has made the vaccine available worldwide. Hepatitis B vaccine has been included in several national immunisation programmes [1]. For countries with endemic hepatitis (many developing countries), WHO recommends immunisation of all infants.

Capsular polysaccharides

Many bacteria have a capsule consisting of high-molecular-weight polysaccharides, which act as virulence factors. Capsule-forming species include both Gram-positive (e.g., pneumococci) and Gram-negative bacteria (e.g., meningococci). The polysaccharides of the different species are composed of linear repeat oligosaccharide units that vary in sugar composition and chain length. The host defence against encapsulated bacteria relies on anti-polysaccharide Ab interacting with complement to opsonise the organisms and prepare them for PHAGOCYTOSIS and CLEARANCE. Licensed capsular polysaccharide VACCINES include meningococcal (serogroups A, C, W-135, Y), pneumococcal (up to 23 serotypes) and *Haemophilus influenzae* type b (Hib) VACCINES.

The main disadvantage of capsular polysaccharide VACCINES is their T CELL independency, which implies that they do not elicit immunological memory. Moreover, infants up to 2 years of age show very weak, non-protective immune responses, whereas they belong to the highest risk groups for infections with the encapsulated bacteria mentioned above.

Polysaccharide-protein conjugate vaccines

The poor immunogenicity of plain polysaccharides can be overcome by covalent coupling to carrier proteins containing T CELL EPITOPES. These helper EPITOPES make them T CELL-dependent and enables the induction of strong immune responses and immunological memory in all age groups, including infants. Conjugate VACCINES licensed so far are Hib, meningococcal group C and pneumococcal VACCINES. The Hib polysaccharide consists of repeat units of ribosyl(1-1)ribitol phosphate. An effective *Haemophilus influenzae* vaccine is relatively easy to produce, because – in contrast with the diversity of pathogenic meningococcal and pneumococcal strains – Hib is responsible for about 95% of infections with Haemophilus species, so only one polysaccharide type has to be included in the vaccine. Table 4 shows that the four licensed Hib conjugate VACCINES vary in composition, owing to differences in polysaccharide length, carrier protein, coupling procedure and polysaccharide-to-protein ratio. As a result, these VACCINES differ with respect to immunogenicity and EFFICACY [6]. The VACCINES are incorporated in many childhood immunisation programmes and are normally administered i.m. in a multi-dose schedule with DTP, either separate or as a combined DTP-Hib vaccine. Another example of a licensed conjugate vaccine is a mixture of polysaccharides from seven types of pneumococci conjugated to diphtheria toxin. Finally, several manufacturers produce meningococcus group C conjugate vaccine.

Pharmacological effects of vaccination

The EFFICACY of a vaccine is difficult to estimate, because the relationship between immune response and degree of protection is not straightforward.

TABLE 4. CHARACTERISTICS OF LICENSED *HAEMOPHILUS INFLUENZAE* TYPE B CONJUGATE VACCINES[1]

Property	Vaccine[2] PRP-D	HbOC	PRP-OMP	PRP-T
Polysaccharide size	medium	small	medium	large
Polysaccharide content (μg)	25	10	15	10
Carrier protein	Diphtheria toxoid	Diphtheria toxin mutant	Meningococcal group B outer membrane proteins	Tetanus toxoid
Protein content (μg)	18	20	250	20
Linkage	via spacer	direct	via spacer	via spacer
Formulation	aqueous solution	aqueous solution	lyophilised, reconstituted with alum salt suspension	lyophilised, reconstituted with aqueous buffer

[1]*Source: [1]*
[2]*Abbreviations: PRP, polyribosylribitol phosphate; PRP-D, PRP-diphtheria toxoid conjugate vaccine; HbOC,* Haemophilus *type b oligosaccharide conjugate vaccine; PRP-OMP, PRP-outer membrane protein conjugate vaccine; PRP-T, PRP-tetanus toxoid conjugate vaccine*

Seroconversion, i.e., the increase in the level of specific circulating Ab, is commonly determined as a measure for the immunogenicity. Moreover, the protective quality of these Ab can be measured with assays for bactericidal activity, i.e., their ability to kill bacteria in the presence of complement (e.g., meningococcal VACCINES), virus-neutralising activity (e.g., polio VACCINES) or toxin-neutralising activity (e.g., diphtheria and tetanus VACCINES). However, it is hard to correlate the level and persistence of circulating Ab to protective EFFICACY. Moreover, the extent of cell-mediated immunity may in some instances be a better measure for protection, e.g., against tuberculosis, but is more difficult to measure. The effectiveness of vaccination is most clearly demonstrated by the reduction of disease after introduction of a vaccine in national immunisation programmes. Recent examples are the drastic reduction in incidence of Hib infections observed in those areas where routine vaccination in infants was introduced and a similar effect after the introduction of meningococcus group C vaccination in the UK. There is much indirect evidence of vaccine EFFICACY. For instance, in The Netherlands, where the use of IPV has effectively protected most of the population, two significant outbreaks of poliomyelitis in 1978 and 1992 were restricted to communities which refuse vaccination on religious grounds [3]. Concerns about the safety of whole-cell pertussis VACCINES in the 1970s were the cause of a sharp decline in vaccination levels in Great Britain, which in turn resulted in epidemics in 1978 and 1982; after renewed public acceptance of the vaccine the incidence of disease dropped again [1].

Since the target groups of VACCINES in many cases include healthy infants and young children, vaccine safety is of particular importance. The occurrence of side-effects may be due to the antigenic components (e.g., LPS in whole-cell pertussis vaccine), impurities derived from the production process, (e.g., chick protein from the cell substrate used for measles vaccine production), or to additives used in a vaccine formulation (e.g., neomycin or gelatine in MMR VACCINES, aluminium salts in SUBUNIT VACCINES). Before a new vaccine candidate is licensed, its safety is investigated in phase I, II and III field trials. Phase I trials include a small number of healthy adults and serve to collect preliminary safety data and to assess vaccine dosage. In phase II studies safety and immunogenicity are determined in a larger number of volunteers. Phase III trials are meant to evaluate safety and EFFICACY in large target populations.

New developments

Introduction

Notwithstanding the success of vaccination, several infectious diseases remain against which an effective vaccine is not yet available. New VACCINES against bacterial (e.g., group B meningococci), viral (e.g., human immunodeficiency virus) and parasitic (e.g., malaria) infections are under development. Ideally, these VACCINES should provide lifelong protection in any individual of any age, be absolutely safe, easy to produce in unlimited quantities, stable under varying conditions, easy to administer and cheap. As yet the design of a vaccine with all these ideal characteristics combined remains an important challenge for developers of new and better VACCINES.

Apart from new prophylactic VACCINES, current research is also focused on the development of therapeutic VACCINES, especially for the treatment of chronic diseases such as AIDS and cancer. The rationale of administering VACCINES to patients already suffering from disease is to specifically boost the IMMUNE SYSTEM weakened by the disease.

The number of VACCINES routinely applied is expected to increase, which demands efforts to reduce the number of injections. An obvious way to achieve this is combining separate vaccine components into one vial or syringe. Examples of such COMBINATION VACCINES have been given before. Simply mixing vaccine components, however, may not only pose pharmaceutical problems (e.g., incompatibility of vaccine components and/or excipients), but also bears the risk of immunological interference. For instance, Hib conjugate was reported to be less immunogenic when mixed with DTaP vaccine [7].

Modern technologies

Whereas traditional vaccine development has largely been dependent on empirical methods, a better insight into immune mechanisms and immunogenic structures of infectious organisms has led to a better understanding of what would be the optimal vaccine composition as related to the desired immunological effect. Recent advances in genomics, proteomics and bioinformatics are now being applied to identify putative Ag [8]. Moreover, the advent of (bio)technological advances has enabled scientists to translate the improved immunological knowledge into the rational design of new VACCINES against a variety of life-threatening and chronic diseases. Several classical and modern approaches to the development of a variety of new VACCINES are currently being explored, some of which are schematically shown in Figure 1. Most of these approaches offer the following common advantages over classical VACCINES: (1) relevant EPITOPES of pathogenic organisms or cancer cells are obtained by safer means and (2) in greater quantities, (3) the products are better defined, and (4) EPITOPES of a single or multiple pathogenic agents can be combined easily in one vaccine. Approaches not yet addressed before are briefly discussed below. For more detailed information about modern vaccine technology the reader is referred to specialised textbooks [1, 2, 9, 10].

Recombinant live vaccines

Non-pathogenic or attenuated organisms can be used as carriers for heterologous protein Ag. Such live carriers are called vectors (Fig. 1c). They are obtained by cloning the desired gene and introducing it into an appropriate carrier organism. Both viral (e.g., vaccinia virus, attenuated poliovirus) and bacterial vectors (e.g., Salmonella species, BCG) are being explored as carriers to express a variety of Ag. The properties of recombinant live VACCINES are comparable to those of classical ATTENUATED VACCINES.

Fusion proteins

Fusion proteins are non-toxic proteins containing inserted EPITOPES, larger protein fragments or even entire proteins derived from pathogenic species (Fig. 1e). They are obtained by the insertion of DNA sequences encoding EPITOPES in the gene of the carrier protein, such as HBsAg [11] or a fusion partner with the capability to target the fused Ag to APC (see

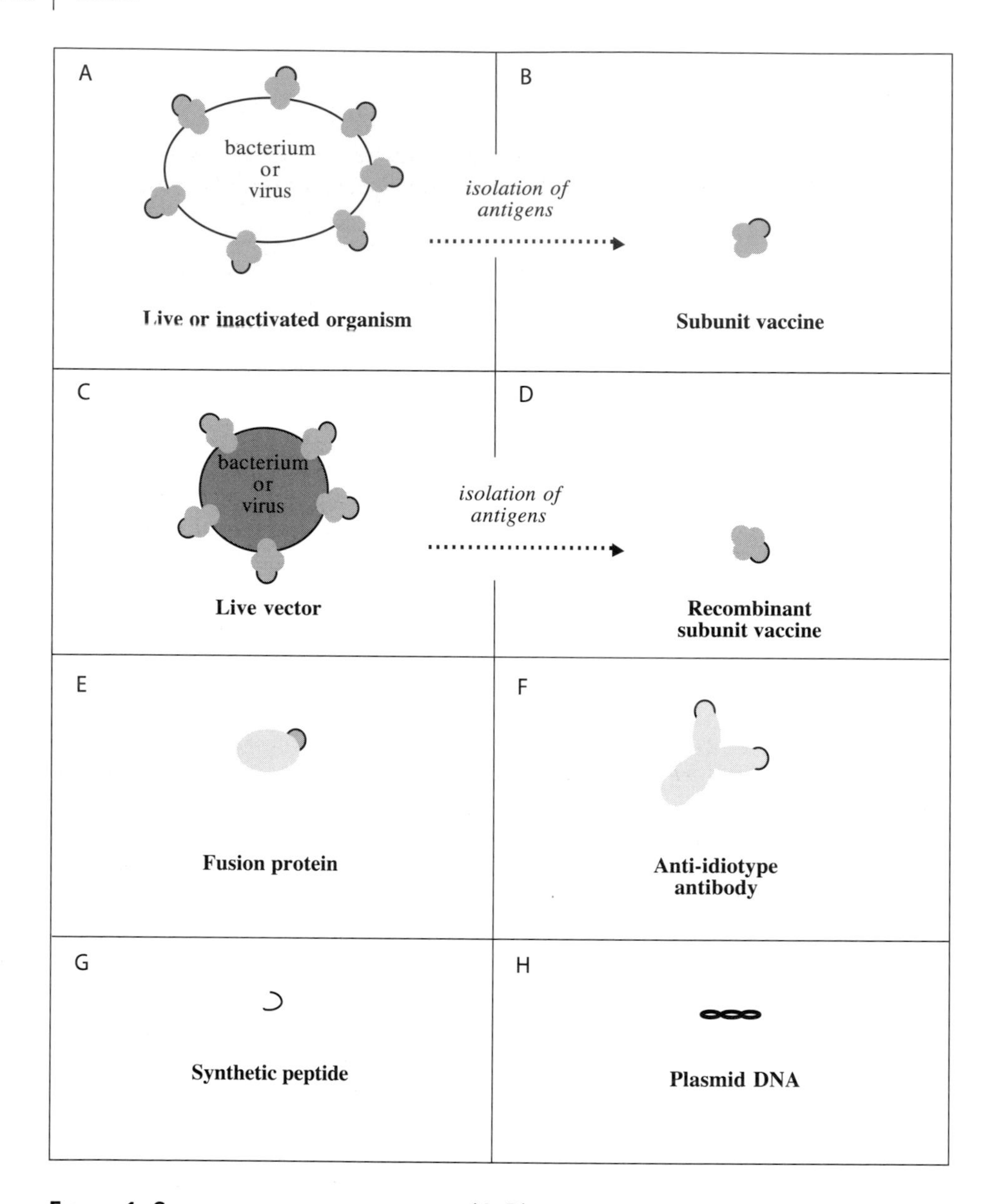

FIGURE 1. SCHEMATIC REPRESENTATION OF (A-B) CLASSICAL VACCINE COMPONENTS AND (C-G) NEW GENERATION VACCINE COMPONENTS

(A) Whole bacterium or virus, live attenuated or killed, with protein antigens (grey objects) containing protective epitope (black semi-circle); (B) subunit vaccine: antigen isolated from pathogenic organism; (C) live vector: antigenic proteins derived from pathogenic organism expressed by live, non-pathogenic bacterium or virus; (D) recombinant subunit vaccine: antigenic protein isolated from heterologous expression system; (E) fusion protein: non-toxic protein containing epitope of protein from pathogen isolated from non-pathogenic organism; (F) anti-idiotype antibody: antibody containing antigen-binding sites mimicking epitope of antigen from pathogen; (G) synthetic peptide with amino acid sequence mimicking epitope of antigen from pathogen; (H) nucleic acid vaccine: plasmid DNA containing the gene encoding antigenic protein or epitope from pathogen.

Box 1. Fusion proteins targeted to APC

Fusion of Ag to Ig-binding proteins can strongly increase the immunogenicity of the fused Ag. Snake toxins, normally weak Ag, have been genetically fused to a carrier protein called ZZ, which is the Ig-binding fragment of staphylococcal protein A (see cartoon).

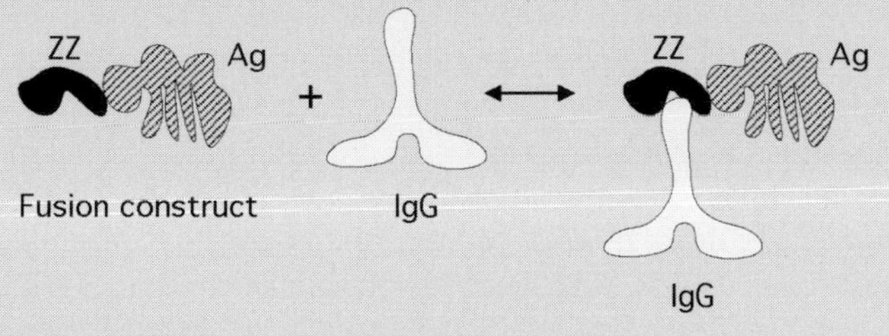

Fusion constructs with ZZ have the capability to target the fused Ag to APC via two pathways: (1) after immunisation the ZZ-fused Ag are targeted to surface Ig of APC by their ZZ moiety; (2) the construct is, before immunisation, complexed with Ig (mAb) directed against APC surface structures such as CD11c; after immunisation, the ZZ-Ag/Ig complex is targeted to APC via the APC-specific Ig. Using this approach, non-adjuvanted ZZ-Ag/Ig complexes were shown to elicit Ab responses in mice ~1000-fold higher than free snake toxin and also enhanced the Ag-specific T CELL response [12].

Box 1) [12]. The recombinant gene is expressed in a suitable organism and the fusion protein is then purified. A drawback of this genetic fusion technology is potential misfolding of the EPITOPE when incorporated in the carrier protein, which would lead to irrelevant immune responses.

Anti-idiotype antibodies

The concept of anti-IDIOTYPE Ab VACCINES involves the generation of a mAb (Ab-1) recognising a relevant EPITOPE of the pathogenic agent, followed by the generation of a second mAb (Ab-2) directed against the IDIOTYPE, i.e., the Ag-binding site of Ab-1. Ab-2 thus mimics the original EPITOPE (Fig. 1f) and can be used for vaccination, thereby eliciting protective Ab (Abs-3). Virtually any desired EPITOPE can be structurally mimicked by anti-IDIOTYPE Ab, whether it be continuous or discontinuous EPITOPES of either protein or polysaccharide nature [1].

Synthetic peptide vaccines

Chemically synthesised peptides belong to the best-defined vaccine components currently under investigation. The synthetic peptide technology allows for the design of VACCINES consisting of selected EPITOPES free from irrelevant or unwanted structures. Large amounts of linear peptides resembling T CELL or B CELL EPITOPES (Fig. 1g) can be prepared by automated methods. The immunogenicity of synthetic peptide Ag is weak, but can be enhanced by conjugation to carrier proteins (analogous to polysaccharide-protein conjugates) or to lipids, or by the construction of multiple different B and T CELL EPITOPES [10, 11]. These options offer the possibility to render synthetic B CELL EPITOPES T CELL-dependent. Furthermore, the conformational freedom of small linear peptides can be restricted by cyclisation, which aims to force them into a conformation reflecting the native structure, which is especially important for peptide analogues of B CELL EPITOPES [13].

Edible vaccines

Since the 1990s attempts have been made to use plants as heterologous expression systems for ANTIGENS by integrating genes encoding the Ag of interest into the plant genome [14, 15]. The goal is to express the Ag preferably in edible parts of plants, such as

Box 2. Presentation forms

Adjuvants comprise a large number of substances of variable chemistry and origin. Examples are colloidal salts, lipid matrices, surface-active compounds and emulsions of mineral, bacterial, vegetable or synthetic nature. Adjuvants have in common that they are not immunogenic per se, but enhance the immunogenicity of co-administered Ag.
Vaccine delivery systems are colloidal carriers with a size which can vary from ~50 nm to the micrometer range, allowing multimeric Ag presentation at their surface. Thereby they mimic the natural presentation of Ag on viral or bacterial surfaces. In general, multimeric presentation of Ag strongly improves their immunogenicity. Colloidal carriers can function as a depot at the administration site, resulting in sustained delivery and a reduction of the number of doses required. Moreover, they can enhance humoral and/or cellular immune reactions, because colloidal particles are taken up more efficiently by APC (in particular dendritic cells) as compared to free Ag. Uptake of colloidal particles by dendritic cells can be further promoted by coupling targeting moieties, specifically recognising dendritic cell receptors, to their surface. As M cells present in mucosal membranes are specialised in the uptake of particulate material and subsequent presentation to immune cells, colloidal particles are also suitable Ag carriers when mucosal immunisation is pursued. In addition, they may protect the Ag from proteolytic attack (e.g., in oral vaccine formulations). Besides Ag, adjuvants are sometimes incorporated into these carrier systems, and many carrier systems have intrinsic adjuvant activity.

leaves or fruits. Among the numerous examples of plants investigated for this purpose (e.g., potato, tobacco, tomato), bananas seem to be most suitable for human applications. This approach offers the possibility of cheap and local production in developing countries, using the standard growing methods of a given region, thereby alleviating logistical and economic problems. Moreover, being edible, the VACCINES would require no syringes and are safe to administer in a patient-friendly way. Examples of ANTIGENS expressed in plants include HBsAg, heat-labile enterotoxin from *E. coli*, and rabies virus glycoprotein. Studies carried out in animals over the past decade, and small tests in people, give hope that edible VACCINES can work. Major obstacles for the use of edible VACCINES are related to quality control.

Nucleic acid vaccines

A development of potential clinical use is genetic immunisation, i.e., direct administration of DNA encoding one or more Ag of interest (Fig. 1h). Upon i.m. immunisation with non-replicating plasmid DNA, the protein encoded is produced and expressed by the host cell. RNA may also be used, but is less suitable because it is rapidly degraded *in vivo* and more expensive to produce. Nucleic acid VACCINES are capable of eliciting both humoral and cellular immunity. A sustained production of Ag after a single immunisation is expected to provide long-term protection [16].

Route of administration

The EFFICACY of vaccination programmes would be enhanced greatly through the availability of needle-free immunisation methods. Several alternative routes of administration have therefore been explored, including mucosal (oral, intranasal, pulmonary) or dermal immunisation.

Mucosal immunisation routes are attractive alternatives to parenteral routes, not only because of the ease of administration, but also because both systemic and mucosal (secretory IgA) responses are induced. The latter is advantageous, because mucosal surfaces are the common port of entrance of many organisms and a strong local immune response may hamper entry into the host by preventing adherence to and colonisation in mucosal surfaces. However, although mucosal immunisation belongs to the oldest means of vaccination (Tab. 2), the number of VACCINES suitable for mucosal immunisation is limited to a few oral VACCINES. Poliovirus can

Box 3. Virosomes

Virosomes are unilamellar vesicles (typical diameter 150 nm) prepared from viral envelopes consisting of phospholipids and viral membrane proteins. Virosomal delivery of membrane proteins generally induces enhanced immune responses against these antigens. In addition, virosomes are useful not only for parenteral delivery but also as Ag carriers for nasal administration. In particular, virosomes derived from influenza virus (also called immunopotentiating reconstituted influenza virosomes, IRIV) have been widely investigated. The presence of viral proteins (haemagglutinin, neuraminidase) has been shown to provide intrinsic APC targeting properties, to promote cell entry and to improve (cytosolic) Ag delivery. This makes IRIV suitable as a delivery system for Ag other than influenza proteins. IRIV-based hepatitis A and influenza vaccines are licensed and are examples of this approach.

be given orally because the virus is relatively resistant to low pH, whereas oral typhoid vaccine is protected from gastric breakdown by formulation in enteric-coated capsules. Other mucosal administration routes, such as nasal or pulmonary delivery, have the advantage that the harsh conditions in the gastrointestinal tract are circumvented. Approaches to augment the immunogenicity of future mucosal VACCINES include the use of vectors (e.g., Salmonella) or VACCINE DELIVERY SYSTEMS and co-administration of ADJUVANTS [17, 18].

TOPICAL administration of Ag to the skin is an appealing delivery route because the skin is a very immune active organ containing a large number of specialised APC, the Langerhans cells [19]. Until recently, the skin was considered almost impermeable for large ANTIGENS. However, chemical penetration enhancers (e.g., surfactants and liposomes) as well as physical approaches (e.g., iontophoresis) to enhance the permeability of the skin for macromolecules have shown some success. Glenn and coworkers have shown that TOPICAL application of Ag-containing patches generate potent immune responses in man [20]. Moreover, targeting to epidermal Langerhans cells is possible by epidermal powder immunization [21].

Vaccine delivery systems and adjuvants

In general it can be stated that the smaller the size of a vaccine component, the weaker its immunogenicity. Therefore, a lot of effort has been and is being put into enhancing the immune response to SUBUNIT VACCINE components by suitable presentation forms, including sophisticated VACCINE DELIVERY SYSTEMS and ADJUVANTS (see Box 2) [9, 10, 22]. Carrier proteins and live vectors are delivery systems already discussed. Other delivery systems include particulate carriers, such as biodegradable microcarriers, nanoparticles, liposomes, immune-stimulating complexes (ISCOMs) and virosomes (see Box 3) [22]. The traditional aluminium phosphate and aluminium hydroxide colloid salts, which are the only ADJUVANTS currently used in licensed products for human use, only stimulate humoral immune responses. Also, they direct the immune response mainly to type 2 (T-helper 2) responses. Often a more balanced Th1/Th2 response is needed for protection. Many novel candidate ADJUVANTS, such as saponines, lipid-A derivatives and bacterial DNA sequences containing CpG-oligodinucleotides, also augment cellular immune responses and mediate their effect through non-specific induction of several CYTOKINES, resulting in balanced Th1/Th2 responses. The B subunits of cholera toxin and of *E. coli* heat-labile enterotoxin are examples of powerful mucosal ADJUVANTS. CYTOKINES such as IL-2, IL-12 and IFN-γ have become of interest as more specific ADJUVANTS, especially in the search for potent VACCINES against AIDS and cancer.

Summary

VACCINES have been very successful in the prevention of infectious diseases. Traditional VACCINES consist of whole (live or killed) bacteria or viruses or components thereof. Several modern approaches, most of which are based on rDNA technologies, are emerging with the aim to generate more effective and safer VACCINES. As a result of the tendency to design smaller, better-defined antigenic components, proper Ag presentation forms are becoming increasingly important. Moreover, to limit the increasing number of injections, needle-free immunisation routes and ways to combine multiple vaccine components in one vial are being explored. It is expected that VACCINES to be marketed in the near future will be based on some of the modern vaccine technologies discussed in this chapter and – like the conventional VACCINES – will make a significant contribution to the improvement of public health.

Selected readings

Plotkin SA, Orenstein WA (eds) (2004) *Vaccines*, fourth edition. WB Saunders Company, Philadelphia

Levine MM, Woodrow GC, Kaper JB, Cobon GS (eds) (1997) *New Generation Vaccines*, second edition. Marcel Dekker Inc., New York

Kaufman S (ed) (2004) *Novel Vaccination Strategies*. Wiley-VCH, Berlin

Recommended websites

NIAID Net News, Infectious Diseases: *http://www.niaid.nih.gov/final/infds/infds.htm* (Accessed March 2005)

The Vaccine Page, Vaccine News and Database: *http://vaccines.org/* (Accessed March 2005)

World Health Organization; Immunization, Vaccines and Biologicals: *http://www.who.int/vaccines/* (Accessed March 2005)

References

1 Plotkin SA, Orenstein WA (eds) (2004) *Vaccines*, second edition. WB Saunders Company, Philadelphia

2 Levine MM, Woodrow GC, Kaper JB, Cobon GS (eds) (1997) *New Generation Vaccines*, second edition. Marcel Dekker Inc., New York

3 Murdin AD, Barreto L, Plotkin S (1996) Inactivated poliovirus vaccine: past and present experience. *Vaccine* 14: 735–746

4 Rappuoli R (1994) Toxin inactivation and antigen stabilization: two different uses of formaldehyde. *Vaccine* 12: 579–581

5 Patel SS, Wagstaff AJ (1996) Acellular pertussis vaccine (Infanrix™-DTPa; SB-3): a review of its immunogenicity, protective efficacy and tolerability in the prevention of *Bordetella pertussis* infection. *Drugs* 52: 254–275

6 Madore DC (1996) Impact of immunization on *Haemophilus influenzae* type b disease. *Infect Agents Dis* 5: 8–20

7 Eskola J, Ölander R-M, Hovi T, Litmanen L, Peltola S, Käyhty H (1996) Randomised trial of the effect of co-administration with acellular pertussis DTP vaccine on immunogenicity of *Haemophilus influenzae* type b conjugate vaccine. *Lancet* 348: 1688–1692

8 Green BA, Baker SM (2002) Recent advances and novel strategies in vaccine development. *Curr Opin Microbiol* 5: 483–488

9 Kaufmann SHE (ed) (1996) *Concepts in Vaccine Development*. Walter de Gruyter, Berlin

10 Powell MF, Newman MJ (eds) (1995) *Vaccine Design: the Subunit and Adjuvant Approach*. Plenum Press, New York

11 BenMohamed L, Wechsler SL, Nesburn AB (2002) Lipopeptide vaccines – yesterday, today, and tomorrow. *Lancet Infect Dis* 2: 425–431

12 Léonetti M, Thai R, Leroy JC, Drevet P, Ducancel F, Boulain JC, Ménez A (1998) Increasing immunogenicity of antigens fused to Ig-binding proteins by cell surface targeting. *J Immunol* 160: 3820–3827

13 Oomen CJ, Hoogerhout P, Bonvin AMJJ, Kuipers B, Brugghe H, Timmermans H, Haseley SR, van Alphen L, Gros P (2003) Immunogenicity of peptide-vaccine candidates predicted by molecular dynamics simulations. *J Mol Biol* 328: 1083–1089

14 Richter L, Kipp PB. (1999). Transgenic plants as edible vaccines. *Curr Top Microbiol Immunol* 240: 159–176

15 Sala F, Rigano MM, Barbante A, Basso B, Walsmsley AM, Castiglione S (2003) Vaccine antigen production in transgenic plants: strategies, gene constructs and perspectives. *Vaccine* 21: 803–808

16 Alpar HO, Bramwell VW (2002) Current status of DNA vaccines and their route of administration. *Crit Rev Ther Drug Syst* 19: 307–383

17 Walker RI (1994) New strategies for using mucosal vaccination to achieve more effective immunization. *Vaccine* 12: 387–400

18 Eriksson K, Holmgren J (2002) Recent advances in mucosal vaccines and adjuvants. *Curr Opin Immunol* 14: 666–672

19 Babiuk S, Baca-Estrada M, Babiuk LA, Ewen C, Foldvari M (2000) Cutaneous vaccination: the skin as an immunologically active tissue and the challenge of antigen delivery. *J Control Rel* 66: 199–214

20 Glenn GM, Taylor DN, Li X, Frankel S, Montemarano A, Alving CR (2000) Transcutaneous immunization: a human vaccine delivery strategy using a patch. *Nature* Med 6: 1403–1406

21 Chen D, Payne LG (2002) Targeting epidermal Langerhans cells by epidermal powder immunization. *Cell Res* 12: 97–104

22 Kersten GFA, Drane D, Pearse M, Jiskoot W, Coulter A (2004) Liposomes and ISCOMs. In: S Kaufman (ed): *Novel Vaccination Strategies*. Wiley-VCH, Berlin 173–196

C2

Sera and immunoglobulins

Gerhard Dickneite, Peter Gronski, Ernst-Jürgen Kanzy and Friedrich R. Seiler

Humoral immunity

Immunoglobulin (Ig) is a family of molecules of essentially similar structure with virtually identical basic components and similar molecule size. The composition of the molecules, however, varies. Their major task lies in the binding and eliminating of substances which are alien to the individual or the species, such as bacteria and viruses. It is well established that the healthy organism is able to produce antibodies (Ab) that act against an astonishingly high number and wide variety of foreign substances, and these Ab are able to attach themselves with high specificity to foreign ANTIGEN (Ag) structures, thereby regulating their further elimination and breakdown.

The organism normally requires a few days to be actively able to provide Ab specific to new or still foreign Ag. This humoral immune response does not result in only one type of Ab which can then be synthesised as a quasi-monoclonal Ab (quasi-mAb) by a single specifically stimulated plasma-cell clone, a situation that rarely arises in healthy organisms (myeloma can be considered as degeneration). Nature normally produces in excess and in variation, and this also applies to the immune response: a considerable number of expanding plasma-cell clones produce – as it were in polyclonal concert – Ab with related and slightly varying specificities and binding properties and, moreover, with activity against various antigenic sections (EPITOPES) of a molecule or pathogen. The interaction of these different types of Ab with Ag usually provides the basis for an effective immunological response (see Chapters A.7–A.8).

In the case of existing or induced immune deficiency, the requisite Ab are either not produced quickly enough or in sufficient amounts, or they are not produced at all. Even a florid infection with a rapidly multiplying pathogen leaves the IMMUNE SYSTEM little time to rally its defences. Particularly for such situations, PASSIVE IMMUNOTHERAPY with Ig preparations has continued to be developed since its discovery in 1890 [1, 2]. Nowadays, Ab of different origin, prepared in different ways and with different specificity are available for intramuscular (i.m.) or intravenous (i.v.) substitution (Tab. 1A).

Homologous preparations for rapid and high-dose i.v. therapy have been available for about 35 years. These products differ with regard to their methods of production, composition, properties (Tab. 2) and MECHANISMS OF ACTION. The functional elements of Ig molecules are dependent on particular structural molecular properties (see also Box 1).

Antibody structure

The structure of Ig reflects its two decisive, major functions. One part is responsible for Ag recognition and varies from molecule to molecule: the variable

Box 1

Ag structures usually comprise a set of antigenic determinants, molecular regions called epitopes, to which epitope-specific Ab molecules are generated by a multitude of B-cell clones in the course of a polyclonal immune response.

Table 1a. Former and present polyclonal Ig-preparations for human use

Heterologous (animal origin)	Homologous (human origin)
immune sera (antitoxins)	*polyspecific Ig* (main component usually IgG) for i.m., s.c. or i.v. application
to scorpion venoms	
to snake venoms	"normal" i.m./s.c. Ig
botulismus antitoxin	"normal" i.v. Ig
diphtheria antitoxin	"normal" IgM-enriched Ig
gas-gangrene antitoxin	
"High-titer" (special) Ig	*"High-titer" (special) Ig*
anti-digitalis Fab	anti-cytomegalo Ig
anti-human lymphocytes (thymocytes) Ig	anti-D Ig
	anti-hepatitis A Ig
	anti-hepatitis B Ig
	anti-rabies Ig
	anti-rubella Ig
	anti-tetanus Ig
	anti-tick-borne encephalitis Ig
	anti-varicella zoster Ig

complementary-determining region (CDR). The other part is responsible for Ag-independent effector functions. This part has a uniform structure: the constant part. The Ab therefore has a bifunctional capacity, particularly effective in Fc-mediated secondary reactions mediated by the crystallisable fragment (Fc) of Ab, that only develops subsequent to binding to the Ag (primary reaction). For physicochemical and structural details see Chapter A.3.

Unlike the variable domains, the constant domains of the heavy (H) chains are responsible for the Fc-mediated effector functions of Ig [3]. Such Fc-mediated reactions are (see Table 3): (1) complement activation in the "classic" manner via the binding of the complement component, Clq; (2) binding to reactive T lymphocytes via Fcγ receptor (FcγR), which enables Ab-dependent cell-mediated cytotoxicity (ADCC) to develop; (3) binding to FcγR on the cell surface of macrophages (MΦ) and neutrophil granulocytes (in addition to binding to receptor [R] for C3b, this mechanism is effective in clearing immune complexes); (4) regulation of both negative and positive feedback in the synthesis of IgG; (5) control of the biological half-life ($t_{1/2}$) and catabolism of the Ig; (6) binding to platelets with ensuing platelet aggregation and activation; and (7) possible binding to other membranes, such as basement membranes of the lung or kidney (see also Box 2).

Box 2

Specific Ag-binding via variable CDR induces Ag-independent effector functions mediated by the constant Fc part.

TABLE 1B. HIGH TITER (SPECIAL) IMMUNOGLOBULIN PREPARATIONS

Anti-virus	Origin	Administration	Manufacturer (selection)
Cytomegalo	human	i.m.	Biotest
		i.v.	Biotest, Bayer
Hepatitis A	human	i.m.	ZLB Behring, Bayer, Biovitrum, Grifols
Hepatitis B	human	i.m.	ZLB Behring, Baxter, Bayer, Biotest, CLB, Grifols, Kedrion
		i.v.	Biotest
Rabies	human	i.m. /intragluteal	ZLB Behring, Bayer, Pasteur MSD
	horse IgG/F(ab')$_2$	i.v.	Thai Red Cross (QSMI)
Tick-borne encephalitis	human	i.m.	Baxter
Varicella zoster	human	i.m.	ZLB Behring, Biotest
		i.v.	Biotest, Cangene
Anti-bacterial toxins			
Botulism	human	i.v.	California DHS, Cangene
	horse IgG/F(ab')$_2$	i.v.	Chiron Behring
Tetanus	human	i.m.	ZLB Behring, Baxter, Bayer, Kedrion
Anti-venoms			
Scorpion venom	horse IgG/F(ab')$_2$	i.v.	Haffkine, Knoll
Snake venoms (various)	horse IgG/F(ab')$_2$	i.v.	CSL, Haffkine, Merck, SAIMR, Thai Red Cross (QSMI)
	sheep IgG/Fab	i.v.	Savage Laboratories
Spider venom	horse IgG	i.m. / i.v.	Merck
Other Ag			
D (Rho)	human	i.m.	ZLB Behring, Baxter, Kedrion, Ortho
		i.v.	Cangene, ZLB Bioplasma
Digitalis	sheep IgG/Fab	i.v.	Roche, Savage Laboratories
Human lymphocytes	horse	i.v.	Sangstat
Human T lymphocytes	rabbit	i.v.	Sangstat

Effector functions of human Ig

Crosslinking of a multivalent Ag by Ab also generates crosslinked Fc parts, a process that often induces multivalent attachment of Ag-Ab complexes to Fc-binding structures that have a $t_{1/2}$ of the binding state sufficiently long to effectively trigger a cascade of reactions either on the surface of a cell or in solution. These reactions define the effector functions of the Abs involved (Tab. 3; Fig. 1 and Box 3).

Immunoglobulin of class M (IgM)

IgM has a molecular weight of about 970 kilodaltons (kDa), and thus the 'M' here stands for macroglobulin.

TABLE 2. VARIOUS I.V.-IMMUNOGLOBULIN PREPARATIONS

Manufacturing process Main process steps						Viral safety Procedures involved in the manufacturing process									IgG constituents			Formulation				Manufacturer (selection)
Ethanol precipitation	PEG precipitation	Adsorption	Ion exchange	Chemical modification	Enzymatic treatment	Partition/Removal	Pasteurization (10 h 60°C)	Low pH	Solvent detergent (TNBP)	Detergents	S-Sulfonation	β–Propiolacton	Enzymatic treatment	Nanofiltration	7S	5S F(ab')$_2$, Fabc	<5S Fab, Fc, peptides	Excipients	Protein	Physical state	pH of solution	Manufacturer
X	X	X	X			X			X						X			sorbitol	5%	liquid	5.5	Alpha
X		X		X		X				X	X				X			glycine	5%	powder	6.8	Aventis Behring
X			X			X			X						X			glycine albumin	5%	powder	6.8	Baxter
X	X		X		X	X							X		X	X	X	glucose	5%	powder	7.0	Baxter
X						X		X	X						X			maltose	5%	liquid	4.25	Bayer
X						X		X	X						X			glycine	10%	liquid	4.25	Bayer
X			X			X		X		X					X			glycine	10%	liquid	4.25	Bayer
X		X		X		X						X			X			glucose	5%	liquid	6.8	Biotest
X						X			X	X					X			glycin	5%	liquid	5.3	Biotest
			X			X	X	X							X			maltose	5%	liquid	4.25	CSL
X	X		X			X	X								X			sorbitol	5%	liquid	5.4	Grifols
X			X			X		X	X						X			maltose	5%	liquid	5.5	Octapharma
X			X			X			X						X			albumin glucose glycin	5%	powder	6.8	Octapharma Nordic
X						X		X					X	X	X	X		sucrose	3%–12%	powder	6.6	ZLB Behring

Examples of preparations produced by different procedures with emphasis on their viral inactivation, main IgG constituents (S means coefficient of sedimentation. 7S corresponds to monomeric IgG, 5S to F(ab')$_2$ or Fab/Fc and <5S mainly to Fab, Fc and/or fragments thereof) and formulation.

Box 3

Crosslinking of Ag by Ab generally crosslinks Fc, thereby inducing effector function.

Box 4

The Ab prototype generated during early immune response is the decavalent IgM.

Box 5

Subclasses IgG_1 and IgG_3 most strongly bind FcγR. FcγR-mediated reactions can cause clinical side-effects.

Its C-binding activity is much stronger than that of IgG: for example, one IgM molecule is capable of lysing one erythrocyte. IgM is the natural Ab in the first phase of the immune response, and phylogenically and ontogenetically it is the earliest Ab molecule.

A cell-membrane-bound, monomeric form of IgM (surface IgM or sIgM) is used by early B LYMPHOCYTES as an ANTIGEN receptor (AgR), sometimes together with surface Ig of class D (sIgD). In viral infections, the appearance of IgM in the serum is considered a sign of primary infection. The decavalent IgM often has specificity for polysaccharides and a particular binding affinity (avidity) for multivalent ANTIGENS (Box 4).

Immunoglobulin of class G (IgG)

IgG accounts for approximately 78–80% of Ig in the human organism and has almost all of the important characteristics of an Ab molecule. Of the four subclasses known in humans, IgG_1 (146 kDa) accounts for 60–70%, IgG_2 (146 kDa) for 14–20%, IgG_3 (165 kDa) for 4–8%, and IgG_4 (146 kDa) for 2–6%. IgG_1 and IgG_3 bind and activate the classic complement pathway via the component C1q in macromolecular C1. The binding takes place in the constant domain number 2 of the H chain (C_H2 domain). IgG is the only Ig that is capable of crossing the placenta and therefore of transmitting passive immunity from the mother to the child.

To a small extent, IgG is also filtered by the kidney and is found in the urine. IgG is also important for self-regulation in the control of IgG synthesis. Above all, IgG_1 and IgG_3 are primarily involved in binding to FcγR on lymphocytes, MONOCYTES, Mϕ and other LEUKOCYTES, which explains the OPSONISATION and PHAGOCYTOSIS of IgG-bound Ag or pathogens and ADCC.

It must be emphasised that, under normal conditions, immune complex-bound Ag is transported to and bound at the site where elimination and breakdown of Ag can occur. In the case of pathophysiological overload, FcγR-mediated reactions can also lead to clinical side-effects (see section on side-effects of Ig therapy) (Box 5).

Immunoglobulin of class A (IgA)

IgA is present in the serum predominantly as a monomer with a molecular weight of 160 kDa for both subclasses IgA_1 and IgA_2, but also as a polymer (dimer to tetramer). Like IgM, the oligomers contain one joining (J) chain per molecule. After IgG, IgA is the second most frequent Ig occurring in amounts of

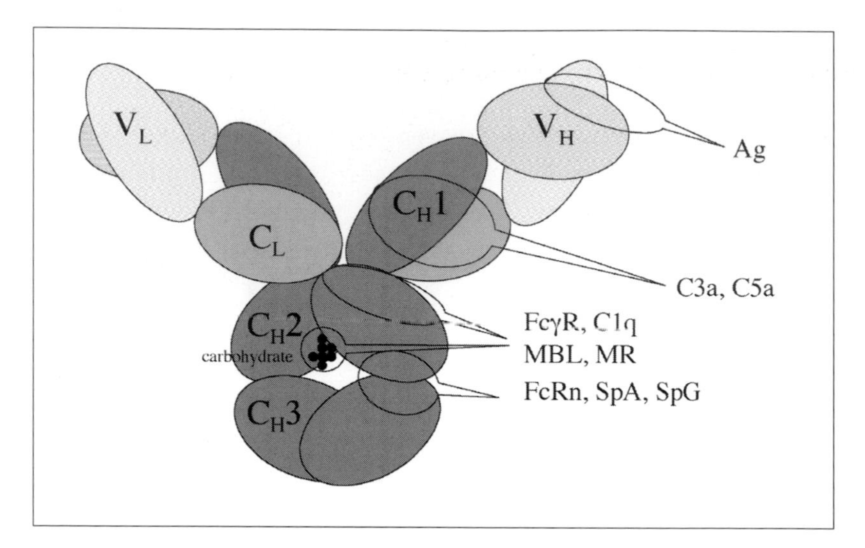

Figure 1. Schematic representation of IgG domain arrangement
Light chain in light grey, heavy chain in dark grey; interchain and intrachain disulfide bonds are not shown. The bubbles indicate areas of the molecule where amino acids that bind important ligands are located (the C3a and C5a binding amino acids have not yet been identified). (C1q, C3a and C5a, complement components; FcγR: FcγRI, FcγRII and FcγRIII; MBL, mannose-binding lectin; MR, mannose receptor; FcRn, neonatal FcR; SpA, staphylococcal protein A; SpG, staphylococcal protein G).

about 15%. The particular clinical significance of secretory IgA lies in its presence in saliva, tear fluid, intestinal secretions, mucous membranes, colostrum and other body fluids. Secretory IgA also contains an additional glycoprotein with a molecular weight of around 60 kDa, the "secretory piece". It is synthesised locally by the epithelial cells of the mucous membranes and is bound covalently to the IgA-dimer, providing it with resistance to acids and enzymes. This secretory piece is also important in the transport of IgA to the surface of mucous membranes, where IgA clearly provides protection against the attachment of microorganisms on the outer surface of the body.

Serum IgA is able to activate C via the alternative pathway; secretory IgA can neither bind nor activate C. Secretory IgA plays an important role at the site of infections, for instance in viral infections of the mucous membranes of the intestine or nose, and it can take part in the neutralisation of bacterial toxins [4] (Box 6).

Immunoglobulin of class D (IgD)

Probably because of its low serum concentration of less than 0.2% of human serum Ig, IgD was not discovered until 1965, as a myeloma protein. It is found in the serum and is membrane-bound as a monomer. It has a molecular weight of 184 kDa and a biological $t_{1/2}$ of only 3 days. IgD has a marked hinge region, which is afforded some protection by carbohydrates, but which can still quite easily be proteolytically cleaved. Most IgD is found together with IgM on the cell-surface membranes of some LYMPHOCYTES, probably functioning as an AgR, much like IgM. It is assumed that IgD may serve as a second R, in addition to IgM, with equal specificity, and can take on an immunoregulatory role in the activation and suppression of LYMPHOCYTES.

Relatively little is known about the functional significance of serum IgD. Specificity has been attributed to it as an autoAb against thyreoglobulin

Box 6

Secretory IgA forms part of the immune barrier of mucous membranes.

TABLE 3. BIOLOGICAL FUNCTIONS OF HUMAN IG CLASSES (ISOTYPES) AND SUBCLASSES

Immunoglobulin	IgG1	IgG2	IgG3	IgG4	IgM	IgA1	IgA2	IgD	IgE
Placental passage	++	+	++	++	–	–	–	–	–
Complement									
Classical pathway of activation	++	(+)	+++	–	+++	–	–	–	–
Alternative pathway of activation	–	–	–	–	–	+	+	–	–
Binding to FcγRI on monocytes, macrophages, neutrophils	++	(+)	+++	+	–	–	–	–	–
Binding to FcγRII on monocytes, neutrophils, eosinophils, basophils, B cells, platelets, MΦ, Langerhans' cells	+	+	+++	(+)	–	–	–	–	–
Binding to FcγRIII on NK-cells, neutrophils, eosinophils, MΦ	+++	(+)	+++	(+)	–	–	–	–	–
Binding to FcεRI on mast cells, basophils	–	–	–	–	–	–	–	–	+++
Binding to FcεRII on B cells, T cells, follicular dendritic cells, inflammatory cells	–	–	–	–	–	–	–	–	+++
Binding to									
Staph. protein A	+	+	–	+	–	–	–	–	–
Strept. protein G	+	+	+	+	–	–	–	–	–
Virus neutralization	+	+	+	+	+	–	+	–	–

NK, natural killer; Staph., Staphylococcus; Strept., Streptococcus

and insulin, and against cell nucleus Ag in autoaggressive illnesses.

Immunoglobulin of class E (IgE)

Like IgG, the IgE molecule is made up of two H and two light (L) chains (ε_2L_2). As in IgM, the H chains have an additional constant domain, C_H4. It has a molecular weight of 188 kDa and a relatively high carbohydrate content like IgM and IgD, and is present in low serum concentrations like IgD. It also has a very short $t_{1/2}$ of 2 days in serum. However, IgE remains fixed for a long time in the skin after binding to specific IgE-FcR (FcεR) on mast cells and basophilic granulocytes. When two molecules of membrane-bound IgE are bound by Ag (allergen) and bridged, vasoactive and chemotactic mediator molecules are released from these cells, and anaphylactic and allergic reactions ensue (hay fever; exogen-mediated, extrinsic asthma; Prausnitz-Küstner skin reaction; Box 7).

The main role of IgE is considered to be the protection of the mucous membranes, for instance after the Ag has penetrated the IgA resistance barrier.

Additional help provided by chemotactically attracted eosinophilic granulocytes and enhanced IgE synthesis as a reaction to parasitic pathogens constitutes a further and important defence mechanism. Circulating IgE cannot activate C via the classic pathway, but it may activate it moderately via the alternative pathway.

Box 7

IgE on surfaces of mast cells and basophils mediates allergic reactions.

Box 8

The mean biological $t_{1/2}$ of IgG is 3–4 weeks. Neonatal FcR (FcRn) regulates IgG serum concentration.

The catabolism of Ig

Ig turnover can be assessed using radioactive-labelled Ig molecules following i.v. application. This shows the biological $t_{1/2}$ of IgG to be around 3–4 weeks. Around 7% of intravascular IgG is catabolised daily, but only around 45% of the total amount of IgG is normally present in the circulation. This means that over half of all IgG must be present extravasally in interstitial fluid. Here it provides for the local immune response to infection after penetration by an Ag.

It is assumed that approximately one-third of Ig is broken down in the liver and a further third in the intestines. IgG catabolism may be enhanced and the biological $t_{1/2}$ shortened during illness. Abnormal increases in IgG concentrations in the serum can inhibit B CELL differentiation to plasma cells as well as IgG synthesis (negative feedback). A diminution of IgG, for instance as a result of plasmapheresis, leads to increased ANTIBODY synthesis (positive feedback).

The so-called neonatal FcR (FcRn) is implicated with the regulation of IgG serum concentration by binding IgG internalised by endothelial cells, for example, and recycling it back into the circulation, thus protecting IgG from lysosomal degradation [5].

IgM and IgA turnover in serum, with a $t_{1/2}$ of around 5 to 6 days, is a good deal faster than that of IgG. About 80% of IgM is intravascular and available for the acute response to infection (Box 8).

Ig preparations for medical use

Preparations containing Ab for "SERUM THERAPY" were introduced by Emil von Behring and Shibasaburo Kitasato over 100 years ago [1] who were recognised with the awarding of the first Nobel Prize for Physiology and Medicine in 1901. These first "antitoxins" were heterologous serum preparations from hyperimmunised animals with Ab-specificity to diphtheria and tetanus toxins. After initial and excellent medical success, particularly with the diphtheria antisera, the side-effects from the alien animal serum proteins – known as serum sickness – became increasingly common. As a result, purer and therefore more compatible preparations were developed.

Early attempts were made at enzymatic treatment (Fig. 2) and purification, particularly with pepsin, trypsin and papain, but only the later introduction of the use of proteases in the 1930s resulted in the "FERMOSERA". Such preparations were at least equivalent to the purified heterologous Ig then available, and they caused far fewer anaphylactic reactions. However, this did not put an end to the risk of serum sickness, which was not resolved until the introduction of homologous human Ig preparations [6]. In the meantime, a number of methods of concentrating and isolating Ig from serum or plasma had been introduced. In most cases, alcohol or neutral salts were used for fractionated precipitation, later additions being polymers such as polyethylene glycol

(PEG) and hydroxyethyl starch, and a number of adsorption substances and ion-exchange resins for chromatographic procedures.

The basic material for the production of common polyclonal and, therefore, polyvalent homologous Ig (or, according to pharmacopoeia terminology, "normal" Ig) for i.m., subcutaneous (s.c.) or i.v. application is a mixture of plasma from at least 1,000 healthy, voluntary donors, in order to guarantee the widest possible spectrum of Ab specificities. Special Ig, however, are gained from the plasma of specifically immunised donors or from the plasma of convalescents. For i.m. application of the smallest possible volumes for a dosage of 0.02–0.66 mL/kg body weight, Ig is prepared as a 16% protein solution (w/v), using a method that should result in antiviral and antibacterial activity ten times higher than in the source plasma. When given i.v., however, normally 5–10% Ig are given at doses of 1–40 mL, with a maximum of 2 g/kg body weight.

Up to the 1960s, Ig therapy was limited to i.m. use, and this was barely changed by the introduction of homologous human products, since these so-called i.m. standard Ig commonly – if not in all cases – led to serious side-effects if given i.v. [7, 8]. The first compatible i.v. Ig (IVIG) was also a homologous "fermo-immunoglobulin" produced by pepsin treatment [9]. For this, the Fc part of the Ab molecule was removed until virtually no more complement activation could be detected.

The wide range of manufacturing methods mentioned above generates differences in the final products and is apparent in the physicochemical and functional profile of IVIG preparations [10–12]. From a theoretical point of view, the attending physician might choose a defined combination of properties within a certain range. The complexity of the *in vivo* situation, which is often poorly understood, sometimes limits adequate therapy.

Historically, the first IgG concentrates contained significant portions of IgG aggregates (>dimers) which could interact with the first component of the complement system, macromolecular C1, upon i.v. application, thereby triggering the classic pathway of activation and causing the *in vivo* generation of anaphylatoxins (C3a, C5a), sometimes with lethal consequences (anaphylactic shock) for patients, who were mainly agamma- or hypogammaglobulinaemic. In addition, we know today that IgG aggregation can also be due to *in vivo* circulating Ag, and that Ig can stimulate further physiological effector systems (Hagemann factor, kallikrein, kinins, PROSTAGLANDINS, CYTOKINES).

This is one example illustrating that some product properties have been determined by aspects of TOLERANCE (e.g., molecular size distribution, anticomplementary activity [aca], content of ISOAGGLUTININS and prekallikrein activator), and these have thus been transformed into special requirements in the pharmacopoeias. Considering the aca test results obtained today for most commercially available products and the relatively low number of side-effects observed, it is more likely that the aca test results are determined by the consistency of the production process.

Nowadays, i.v.-compatible Ig preparations are produced using a range of the least aggressive procedures possible which, on the one hand, lead to bare-

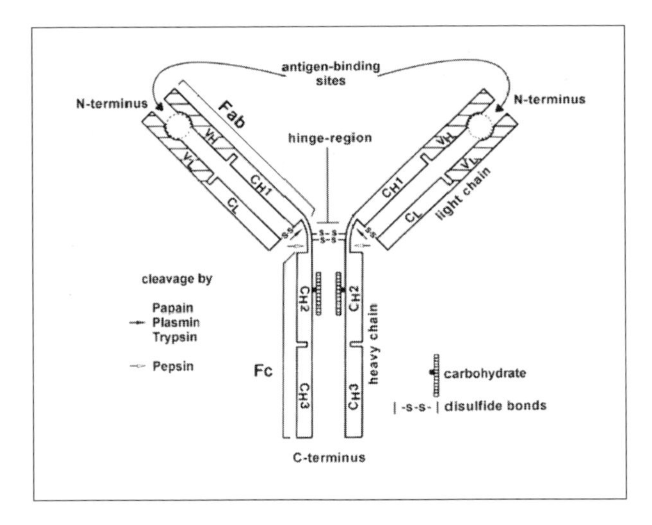

FIGURE 2. IMPORTANT PROTEOLYTIC CLEAVAGE SITES OF AN IMMUNOGLOBULIN MOLECULE

Two identical heavy chains with four H domains (three constant = C, one variable = V) and two identical light chains with two L domains (one constant = C, one variable = V) are combined by interchain disulfide bridges. Intradomain disulfide bridges are not shown.

ly aggregated Ab molecules or, on the other, lead to molecules with structurally modified C_H2 domains and, therefore, altered Clq and FcγR binding properties. Following plasma fractionation by a cold-ethanol method [13, 14], different specific processes are used for further purification and viral inactivation or removal (Tab. 2). The range of products can be divided into three groups (1–3) based on different production methods:

1. *Limited cleavage by proteases.* The main fragmentation product after pepsin treatment is the bivalent part of the Ab molecule, $F(ab')_2$, comprising the 2 monovalent Ag-binding fragments (Fab). Treatment with small amounts of pepsin at pH 4 markedly lowers aca, and the Ab mostly remains intact as a "7S" molecule (S = Svedberg coefficient, a variable that characterises the sedimentation behaviour upon ultracentrifugation). Limited cleavage by immobilised trypsin leads to a preparation containing some portions of a monovalent Ag-binding fragment comprising one Fab and the Fc part (Fabc), Fab and Fc, besides the unsplit IgG.
2. *Chemical modification.* Irreversible chemical modification with β-propiolactone leads to a marked reduction of inherent C binding [15], whereby β-propiolactone selectively modifies the amino acids lysine, cysteine and histidine [16]. Irreversible chemical modification can also be achieved by reduction of the disulfide bridges with dithiothreitol and subsequent alkylation with iodoacetamide. Sulphonation of disulfide bridge-located sulphur (S-sulphonation) provides reversible chemical modification, either by means of sulphite and tetrathionate [17] or with sulphite and copper ions [18]. The interchain disulfide bridges in particular are cleaved and Clq-binding diminished. Repeated reduction and reoxidation enables the SO_3^--groups to be cleaved again *in vitro* and *in vivo*, and the disulfide bridges can be reconstituted.
3. *Special precipitation, absorption and chromatography procedures.* These comprise a series of individual procedures that may also be used in combination. In this context, particular mention should be made of the use of polymers such as PEG for precipitation and stabilisation, as well as the use of adsorbents like Aerosil, bentonite and calcium phosphate. Ion-exchange chromatography is used for polishing the IgG, especially for reduction of the IgA content. All these procedures aim to provide a non-aggressive mode of production to yield, as far as possible, non-cleaved, unaltered and functionally intact 7S-IgG with almost normal IgG-subclass distribution and a physiological $t_{1/2}$.
4. *Virus inactivation and removal.* Single-process steps for the production of IVIG (e.g., enzymatic cleavage, low pH, S-sulphonation and treatment with β-propiolactone) lead to a high level of viral safety because of viral inactivation. Furthermore, additional methods such as heating at 60 °C for 10 h ("pasteurisation"), solvent-detergent (SD) treatment for viral inactivation or nanofiltration for virus removal form part of the production processes.

IVIG is provided in freeze-dried as well as liquid formulations and contains different additives such as sugars, sugar alcohols, amino acids, human serum albumin (HSA) and PEG in varying concentrations. The pH of the solutions is in the range 4.2–7.0.

For special therapeutic purposes, an increasing number of monoclonal IgG preparations with selected specificity and in specific subclasses can be obtained from single-cell, clone-derived cell lines modified by protein engineering based on recombinant deoxyribonucleic acid (DNA) technology. In the humanised versions, only the 6 hypervariable amino acid sequences are of non-human origin and are engineered into the variable domain framework of the IgG molecule. However, recombinant human monoclonal IgG can also be produced. Only the carbohydrate moiety is different from the human plasma-derived IgG species (Box 9).

Indications

From a historical perspective, important indications for the use of Ig are:

- agammaglobulinaemia and hypogammaglobulinaemia.
- deficiencies in certain subclasses of IgG, sometimes combined with deficiency in IgA (e.g., IgA and IgG_2).

Box 9

- Plasma fractionation allows separation and concentration of plasma proteins by means of appropriate procedures such as fractional precipitation, chromatography and filtration, based on different physical principles.
- Aggregated IgG can induce anaphylactic shock *in vivo*.
- Humanised and recombinant human monoclonal IgG of selected specificity and subclass can be produced by modern technology.

- the prophylaxis and treatment of viral infections such as hepatitis A, measles and poliomyelitis.
- the neutralisation of bacterial toxins (e.g., the enterotoxin of Staphylococcus aureus).

Later, the spectrum of indications was supplemented by autoimmune and systemic inflammatory diseases. Consensus panels initiated by the National Institutes of Health (NIH) in May 1990 and by the Australian Society for Blood Transfusion in July 1993 resulted in a nonofficial classification of diseases into categories A, B and C, in which i.v. Ig is recommended for therapy [19]. Its use is suggested as first-line therapy in category A and should be considered in category B; additional clinical data are necessary in category C to make recommendations.

Category A
- idiopathic thrombocytopenic purpura (ITP)
- Kawasaki syndrome

Category B
- Guillain-Barré syndrome
- chronic inflammatory demyelinating polyneuropathy
- autoimmune haemolytic anaemia (AHA)
- autoimmune neutropenia
- acquired haemophilia A
- ITP in pregnancy
- myasthenia gravis (MG)

Category C (selection)
- systemic lupus erythematosus (SLE)
- polymyositis and dermatomyositis
- rheumatoid arthritis (RA)
- insulin-dependent diabetes mellitus
- systemic vasculitis
- antiphospholipid antibodies and recurrent abortions
- multiple sclerosis (MS)

Fc-mediated mechanisms (e.g., ITP: blockade of FcγR) and/or Fab-dependent reactions play – or are thought to play – an important role in the modulation of at least some of the autoimmune disorders listed above. The action of anti-idiotypic Ab, representing a considerable portion of each polyclonal IgG preparation and the main driving force for generation of IgG dimers [20], may be of causative importance in down-regulation at the cellular level [21]. Target-specific application is still impeded by the lack of functional and quantitative characterisation of the specificities necessary, and therefore their content in commercial products is nonstandardisable. The availability of appropriate assays could be of some help in the future.

Moreover, i.m. and IVIG preparations with significantly elevated levels of particular Ab specificities raised to special viruses, bacterial toxins, animal venoms or other antigenic targets of therapeutic interest are available (Tab. 1B). They are obtained from selected and/or immunised human donors or from immunised animals. The heterologous animal Ab preparations are often "fermo-immunoglobulins" (see above) to reduce immunogenicity, but, however, cannot completely prevent it. Therefore, a risk of serum sickness remains in case a repeated donation is required. Sometimes, the manufacturer recommends a skin (urticarial wheal surrounded by a zone of erythema) or conjunctival test (itching of the eye and reddening of the conjunctiva) or even desensitisation when the administration is considered necessary to save life.

Dosage and administration

Originally, the use of Ig preparations in humans was limited to the prophylaxis and therapy of viral and bacterial infections. Additionally, substitution with Ig has proved possible in agamma- and hypogammaglobulinaemic patients [22], where larger amounts of Ig proved to be necessary. In severe infections with a broad spectrum of pathogens, the administration of larger volumes is also necessary. This development finally led to high-dose therapy with IVIG, for instance in ITP, Kawasaki syndrome and other ab-mediated autoimmune disorders in which sometimes up to a maximum of 2 g/kg body weight may be administered per infusion. The general aim here is the immunoregulatory suppression of autoAb synthesis.

The usual recommended average dose for IVIG replacement therapy in agamma- and hypogammaglobulinaemic patients is 400 mg/kg per month [23]. The most suitable form of administration is an infusion with a catheter via the basilic, cephalic or median antebrachial vein. The dose may be adjusted individually since the $t_{1/2}$ of Ig may vary from patient to patient (normal range: 3–4 weeks).

For the treatment of ITP, an initial dose of 0.4 g/kg per day for two consecutive days or a single infusion of 0.8-1.0 g/kg is recommended. If the platelet count remains lower than 30×10^9/L two days after starting therapy, more IVIG may be given up to a maximum dose of 2 g/kg [24].

Kawasaki disease is an acute illness of childhood characterised by INFLAMMATION of the mucous membranes and, as a life-threatening complication, aneurysms caused by vasculitis. The recommended dose is 2 g/kg, which should be administered as a single infusion rather than giving the conventional dose of 0.4 mg/kg per day in multiple administrations.

Side-effects of Ig therapy

Infusion-related adverse effects

Infusion of IVIG is considered safe and is associated with mild side-effects in most cases. While 10–15% of all infusions with early IVIG preparations gave rise to side-effects, the figures have decreased to 2–6% with new products [25].

Mild side-effects include back pain or lower abdominal pain, tachycardia, elevated blood pressure, muscle pain, headache, nausea, flushing, dizziness and chills. It is interesting to note that agammaglobulinaemic patients most frequently have adverse reations. The reason for these reactions is most likely the formation of Ag-Ab (immune) complexes and complement activation when the exogenous Ab interacts with the Ag from possible infectious agents present in the patient. Thus, when IVIG is administered during an active infection, the incidence of side-effects is significantly higher. Adequate antibiotic treatment before Ig infusion is therefore recommended. These moderate inflammatory reactions ("phlogistic reactions") usually disappear when the infusion rate of the IVIG is reduced or the infusion discontinued early.

An initial infusion rate of 60 mL/h is recommended; the flow rate may be increased by 60 mL every 0.5 h, with a maximum of 240 mL/h. During subsequent infusions, flow rate increments should not exceed 30 mL every 30 min with a maximum of 180 mL/h. In some cases, i.v. anti-inflammatory treatment may be considered: either aspirin (5-10 mg/kg) or methylprednisolone (1 mg/kg). The usefulness of ANTI-HISTAMINES in this situation is unclear.

IgA deficiency

Severe anaphylactic reactions may occur in patients with IgA deficiency. Although rare, they may be life-threatening and include hypotension and respiratory or cutaneous reactions such as urticarial skin rash. About 40% of patients with absolute IgA deficiency develop Ab to IgA of either the IgG or IgE type. Binding of IgA-specific IgE in the recipient to the infused IgA-containing IVIG results in activation of mast cells with degranulation and secretion of vasoactive agents such as histamine. Adverse events occur immediately after the start of infusions, and in such instances epinephrine (adrenaline) should be administered immediately.

Aseptic meningitis

Severe headache, vomiting and nuchal rigidity may occur as a complication of high-dose IVIG (up to 2 g/kg) in patients with ITP or chronic neuropathy. This phenomenon, which is called acute aseptic meningitis, is a late event in Ig therapy; it occurs from 10 h up to 1 week after infusion. Analysis of the cerebrospinal fluid is negative for pathogens, although invasion of NEUTROPHILS is usually observed in patients suffering from aseptic meningitis.

The aetiology of this adverse event is not fully understood. It is assumed that IVIG induces the release of serotonin, histamine or prostaglandin and that these substances can affect the meningeal microvascular system and may trigger a migraine headache as well as aseptic meningitis [26].

Viral safety of Ig

As early as the 1940s, the transmission of hepatitis viruses by blood components, especially by plasma, had been recognised as an inherent risk resulting from the practice of pooling plasma from a number of different donors, an occasional one of whom may have been infected. Methods of plasma fractionation were developed which allowed the production of HSA and gamma globulin preparations for i.m. administration. Although the transmission of hepatitis by HSA was not a relevant problem in practice, an efficient virus inactivation procedure was developed for safety reasons, making use of the marked thermal stability of this protein. It was also known that the special procedure of heating an aqueous solution of HSA at 60 °C for 10 h in the presence of acetyltryptophan and sodium caprylate as stabilisers could not be directly applied to other more labile components of human plasma.

There was no reason at that time to develop a similar procedure for gamma globulins, particularly because it was known that these proteins could even be used to prevent infectious hepatitis in humans. Therefore, the presence of protective components in the Ab preparations was reasonable.

Another contribution to safety had its origin in the process of plasma fractionation itself, since certain steps in the manufacturing process can bring about separation and inactivation of viruses. However, since details of the methods used are different and the viruses concerned have distinct physicochemical properties, the virus-reducing capacity of the production procedure cannot be generalised and requires individual evaluation.

Moreover – and again originally introduced as a precautionary measure – manufacturers began to use a separate virus inactivation method (e.g., modified HSA pasteurisation) in the manufacturing procedure of Ig. This proved to be a sensible measure, since reports have occasionally been published on the transmission of HEPATITIS C VIRUS (HCV; previously non-A/non-B hepatitis virus) by certain preparations for i.v. application.

Apart from the presence of specific Ab and the effect of the fractionation and inactivation process, the introduction of serological checks for potential sources of infection, such as the absence of hepatitis B surface ANTIGEN (HBsAg) and for Abs to HCV and human immunodeficiency virus (HIV), have generally been considered to constitute a further improvement in the virus safety of blood and blood-derived products. All single donations to the initial plasma pool are routinely subject to such screening and are eliminated from fractionation if the test results are positive. In addition, plasma for fractionation is quarantined and only released for further processing if the donor has been serologically checked again and found to be negative. This procedure helps to identify early infections at levels below the detection limit.

Despite the rationale underlying plasma screening, the implementation of anti-HCV Ab testing for plasma used for fractionation in the United States caused an outbreak of hepatitis C following administration of a commercial IVIG preparation in early 1994. The basic manufacturing process for this product lacked an additional virus inactivation step. Investigations based on a sensitive validated nucleic acid amplification technique (polymerase chain reaction, PCR) revealed that batches produced before introducing anti-HCV Ab screening were HCV-ribonucleic acid (HCV-RNA) negative, in contrast to almost all lots prepared from single donations without detectable Ab of this specificity. Further experiments suggested a process design appropriate to

Box 10

The capacity to inactivate and remove viruses is an important feature of a particular fractionation procedure.

remove HCV in the presence of anti-HCV Ab and inadequate virus separation at significantly reduced levels of this Ab. Consequently, the integration of an additional efficient validated virus reduction step into the manufacturing process of all i.m. and IVIG preparations has been recommended by the US and European authorities, and is now required.

Plasma pool samples are therefore routinely subjected today to a PCR assay for detection of RNA or DNA of relevant viruses to identify plasma donations with positive test results, which may also help to avoid future HCV transmissions with the final product, for example. From what we know today, HIV and HEPATITIS B VIRUS (HBV) safety has certainly been achieved with the methods commonly used to prepare commercial Ig (Box 10).

Monoclonal Ab (mAb)

The first mAb with defined specificities were introduced more than 25 years ago by Georges Köhler and Cesar Milstein [27], and have now become powerful diagnostic and therapeutic tools. MAb have single specificities and are produced by an immortalised hybridoma resulting from fusion of a mouse spleen B lymphocyte with a defined Ag specificity and a myeloma cell.

The first therapeutic mAb, Orthoclone™ (OKT3), was marketed in 1986 (Tab. 4). It targets the cluster of differentiation number 3 (CD3) molecule on the surface of activated T-cells and is used for IMMUNOSUPPRESSION in transplantation to prevent organ rejection. OKT3 is a murine Ab; all murine Abs share two basic problems: first, they are immunogenic to humans and raise human anti-mouse Ab (HAMA); second, their Fc-mediated R-binding function in ADCC or complement-dependent cytotoxicity (CDC) is impaired because of non-reactivity of the murine mAb with the corresponding human FcR or complement component C1q. In the 1990s, new technologies emerged to replace murine structures in the mAb with human sequences. In the first generation of the genetically engineered Ab, the constant domains of the murine IgG molecule were replaced by human domains, leaving the murine variable domains unchanged. The problem with the Fc-effector function was solved with this so-called chimeric mAb, and although the potential for immunogenicity was considerably reduced, it was not eliminated. ReoPro™, an anti-thrombotic mAb to platelet GPIIb/IIIa (complex of glycoproteins IIb and IIIa, fibrinogen receptor), was the first chimeric Ab which was licensed for the prevention of restenosis after percutaneous transluminal coronary angioplasty (PTCA). Only the CDR sections of the variable domains were conserved in the next generation of humanised Ab, and these mAb molecules contained only 6–8% of the murine sequences, resulting in a less immunogenic product. In the last step in development, the "fully human" Ab made use of the phage display technique [28] and of the transgenic mouse [29].

The production of a therapeutic mAb is a complex process. Different expression systems are used (hybridoma, genetically engineered cell lines, plants and transgenic animals) depending on the molecular form (intact IgG, fragment, conjugate) and origin (murine, humanised, human). Their special properties and influence on important factors in clinical use such as immunogenicity, $t_{1/2}$ and effector functions, which are clearly defined by the expression system (e.g., a difference in glycosylation pattern), also have to be considered [30].

At present, 12 mAb preparations are licensed for the treatment of diseases in humans and can be divided into four therapeutic classes: anti-tumour, anti-inflammatory/transplant rejection, anti-thrombotic and anti-viral treatment (Tab. 4).

TABLE 4. MAB APPROVED FOR TREATMENT OF HUMAN DISEASES

Year FDA Approved	Trade names (Generic)	Target Ag/Indication	Type of Ab	Company
1986	Orthoclone OKT3™ (muromonab)	CD3 Transplantation	Murine	Ortho Biotech
1994	ReoPro™ (abciximab)	GP IIb/IIIa Thrombosis	Chimeric (Fab fragment)	Centocor (J& J)
1997	Rituxan™ (rituximab)	CD20 NHL	Chimeric	IDEC, Genentech
1997	Zenapax™ (daclizumab)	IL-2 receptor Transplantation	Humanized	Hoffmann-La Roche
1998	Simulect™ (basiliximab)	IL-2 receptor Transplantation	Chimeric	Novartis
1998	Synagis™ (palivizumab)	RSV Infection	Humanized	MedImmune
1998	Remicade™ (infliximab)	TNF-α RA, Morbus Crohn	Chimeric	Centocor (J & J)
1998	Herceptin™ (trastuzumab)	HER2 Breast cancer	Humanized	Genentech
2000	Mylotarg™ (gemtuzumab)	CD33 AML	Humanized + calicheamycin	Wyeth
2001	Campath™ (alemtuzumab)	CD52 AML	Humanized	Ilex Oncology
2002	Zevalin™ (ibritumomab)	CD20 NHL	Murine (In-111, Y-90 labeled)	IDEC Pharmaceuticals
2002	Humira™ (adalimumab)	TNF-α RA	Fully human, phage display	Abbott

RA, rheumatoid arthritis; AML, acute myeloic leukemia; NHL, non-Hodgkin's lymphoma

A significant proportion of the mAb-market is associated with the treatment of cancer, an indication with a high medical need. A prerequisite for cancer therapy with an mAb is the detection of an Ag that is, first, highly or selectively expressed on tumour cells, and second, functionally involved in the tumorigenic process. Initial development focused on the treatment of blood-borne cancer. The first anti-cancer mAb, marketed in 1997, was Rituxan, a chimeric mAb targeted against Ag CD20 on malignant LYMPHOCYTES for the treatment of non-Hodgkin's lymphoma (NHL). The anti-cancer activity is mediated via the activation of the phagocytic system (ADCC) and the C system (CDC). The IgG_1 subtype was therefore selected for most of the anti-tumour mAb specificities.

A further product for the treatment of haematological tumours (acute myeloic leukaemia, AML) is Campath™, an IgG_1 directed against CD52 on malignant LYMPHOCYTES. The principle of targeting a cytostatic drug at the tumour site was first introduced by Mylotarg™. This mAb binds to CD33-positive AML cells, and the tumour cells are attacked by the conjugated anti-tumour antibiotic, calicheamycin. The first radioimmunotherapeutic was Zevalin™, a murine mAb, coupled with Indium-111 (In-111) and Yttrium-

Box 10

MAb is usually a monospecific Ab directed against a defined Ag-located epitope and produced by an immortalised B lymphocyte. Therapeutic indications at present are anti-cancer, anti-inflammation and prevention of transplant rejection, anti-thrombosis and anti-infection.

90 (Y-90). Combination cancer treatment with Rituxan™ starts with In-111 Zevalin, and 7–9 days later, Y-90 Zevalin™ is given.

Although it was first assumed that solid tumours might not be suitable for treatment with mAb due to the size of the Ab, an mAb preparation for the treatment of metastatic breast tumours, Herceptin™, has been developed and has been on the market since 1998. It is directed against the human epidermal growth factor R 2 (HER-2). Before treatment with Herceptin™, the HER-2 Ag must be demonstrated in the patient.

The second major class of monoclonal preparations is determined by their immunopharmacological interaction in the treatment of inflammatory and AUTOIMMUNE DISEASES, and the prevention of transplant rejection. The first licensed mAb, Orthoclone OKT-3™, was directed against the CD3 surface R on T-cells. Due to its murine nature, anti-mAb Ig developed in humans, which led to severe unwanted reactions. Two mAb preparations which target the R for INTERLEUKIN 2 (IL-2 R) alpha chain (CD 25) on activated LYMPHOCYTES are Zenapax™ (humanised) and Simulect™ (chimeric). Both are used for the prevention of transplant rejection.

The mAb Remicade™ binds to soluble TUMOR NECROSIS FACTOR α (TNF-α) and neutralises this proinflammatory CYTOKINE. It was initially approved for the treatment of inflammatory bowel disease and Crohn's disease. Later it was also licensed for RA as an adjunct to methotrexate. At the end of 2002, a new anti-TNF-α mAb, Humira™, the first fully human Ab, was licensed for the treatment of RA.

Only one product is licensed as an anti-thrombotic. ReoPro™ is an Ab that targets the fibrinogen surface R on platelets (GpIIb/IIIa). ReoPro™ acts as an anti-thrombotic agent for the treatment of arterial thrombosis after coronary intervention (PTCA). Because a complete IgG molecule would induce the destruction of platelets and result in severe thrombocytopenia, ReoPro™ was designed as a Fab Ab. However, thrombocytopenia and bleeding have been reported from clinical studies after treatment with ReoPro™.

Synagis™ is a humanised mAb belonging to the therapeutic class of anti-infectives. It targets an EPITOPE on the respiratory syncytial virus (RSV). Synagis™ is used for the treatment of lower respiratory tract infections in paediatric patients at high risk of RSV disease (Box 11).

Summary

A century ago, PASSIVE IMMUNOTHERAPY consisted of plasma preparations containing Ab of animal origin. Repeated prophylaxis and treatment with such agents was associated with severe side-effects. This was solved by continuous improvement of the process of protein purification resulting in homologous Ig preparations devoid of aggregates with Fc effector function. The main active component of these products is polyclonal IgG. Nowadays, an increasing number of monoclonal humanised or human IgG preparations with selected specificity and in different subclasses have been developed for special therapeutic purposes. With more than 100 different mAb products in different stages of clinical development, we can expect the number of licensed products to increase considerably over the next few years.

Selected readings

Ravetch JV, Bolland S (2001) IgG Fc receptors. *Annu Rev Immunol* 19: 275–290

Simon HU, Späth PJ (2003) IVIG – mechanisms of action. *Allergy* 58: 543–551

Torphy TJ (2002) Pharmaceutical Biotechnology. Monoclonal antibodies: boundless potential, daunting challenges. *Curr Opin Pharmacol* 13: 589–591

Recommended websites

World Health Organization: *www.who.int* (Accessed December 2004)

European Medicines Agency: *www.emea.eu.int* (Accessed December 2004)

U.S. food and Drug Administration, Center for Biologics Evaluation and Research: *www.fda.gov/cber* (Accessed December 2004)

Paul-Ehrlich-Institut: *www.pei.de* (Accessed December 2004)

Department of Health and Human Services, Centers for Disease Control and Prevention: *www.cdc.gov* (Accessed December 2004)

The National Organization Dedicated to Research, Education and Advocacy for the Primary Immune Deficiency Diseases; Immune Deficiency Foundation: *www.primaryimmune.org* (Accessed December 2004)

References

1 Behring EA, Kitasato S (1890) Über das Zustandekommen der Diphterie-Immunität und der Tetanus-Immunität bei Thieren. *Dtsch Med Wochenschr* 49: 1113–1114

2 Gronski P, Seiler FR, Schwick HG (1991) Discovery of antitoxins and development of antibody preparations for clinical uses from 1890 to 1990. *Molec Immunol* 28: 1321–1332

3 Sedlacek HH, Gronski P, Hofstaetter T, Kanzy EJ, Schorlemmer HU, Seiler FR (1983) The biological properties of immunoglobulin G and its split products (F(ab')$_2$ and Fab). *Klin Wochenschr* 61: 723–736

4 Corthesy B (2003) Recombinant secretory immunoglobulin A in passive immunotherapy: linking immunology and biotechnology. *Curr Pharm Biotechnol* 4: 51–67

5 Ghetie V, Ward ES (1997) FcRn: the MHC class I-related receptor that is more than an IgG transporter. *Immunol Today* 18: 592–598

6 Stokes J Jr, Maris EP, Gellis SS (1944) Chemical, clinical and immunological studies on the products of human plasma fractionation. XI. The use of concentrated normal human serum gamma globulin (human immune serum globulin) in the prophylaxis and treatment of measles. *J Clin Invest* 23: 531–540

7 Janeway CA (1948) The plasma proteins: their functions and uses. *Pediatrics* 2: 489–497

8 Moore GE, Sandberg A. Amos DB (1957) Experimental and clinical adventures with large doses of gamma and other globulins as anticancer agents. *Surgery* 41: 972–983

9 Schultze HE, Schwick G (1962) Über neue Möglichkeiten intravenöser Gammaglobulin-Applikation. *Dtsch Med Wochenschr* 87: 1643–1650

10 Römer J, Späth PJ, Skvaril T, Nydegger UE (1982) Characterization of various immunoglobulin preparations for intravenous application. I. Protein composition and antibody content. *Vox Sang* 42: 62–73

11 Römer J, Morgenthaler JJ, Scherz R, Skvaril F (1982) Characterization of various immunoglobulin preparations for intravenous application. II. Complement activation and binding to Staphylococcus protein A. *Vox Sang* 42: 74–80

12 Suez D (1995) Intravenous immunoglobulin therapy. Indications, potential side-effects, and treatment guidelines. *J Intraven Nurs* 18: 178–190

13 Cohn EJ, Strong LE, Hughes WL, Mulford DJ, Ashworth JN, Melin N, Taylor HJ (1946) Preparation and properties of serum and plasma proteins. IV: A system for the separation into fractions of protein and lipoprotein components of biological tissues and fluids. *J Amer Chem Soc* 68: 459–475

14 Oncley JL, Melin M, Richert DA, Cameron JW, Gross PMJr (1949) Preparation and properties of serum and plasma proteins. XIX. The separation of the antibodies, isoagglutinins, prothrombin, plasminogen, and X-proteins into subfractions of human plasma. *J Amer Chem Soc* 71: 541–550

15 Stephan W (1969) Beseitigung der Komplementfixierung von Gamma-Globulin durch chemische Modifizierung mit Beta-Propiolacton. *Z Klin Chem Klein Biochem* 7: 282–286

16 LoGrippo GA (1960) Investigations of the use of β-propiolactone in virus inactivation. *Ann NY Acad Sci* 83: 587–594

17 Masuho Y, Tomibe K, Matsuzawa K, Ohtsu A (1977)

Development of an intravenous γ-globulin. *Vox Sang* 32: 185–181

18 Gronski P, Hofstaetter T, Kanzy EJ, Lüben G, Seiler FR (1983) S-sulfonation: a reversible chemical modification of human immunoglobulins permitting intravenous application. I. Physiochemical and binding properties of S-sulfonated and reconstituted IgG. *Vox Sang* 45: 144–154

19 Mobini N, Sarela A, Ahmed AR (1995) Intravenous immunoglobulins in the therapy of autoimmune and systemic inflammatory disorders. *Ann Allerg Asth Immunol* 74: 119–128

20 Tankersley DL (1994) Dimer formation in immunoglobulin preparations and speculations on the mechanism of action of intravenous immune globulin in autoimmune diseases. *Immunol Rev* 139: 159–172

21 Ronda N, Hurez V, Kazatchkine MD (1993) Intravenous immunoglobulin therapy of autoimmune and systemic inflammatory diseases. *Vox Sang* 64: 65–72

22 Barandun S, Cottier H, Hässig A, Riva G (eds) (1959) Das Antikörpermangelsyndrom. Benno Schwabe and Co. Verlag, Basel/Stuttgart. Appeared as reprint of Helv Med Acta 26, Fasc. 2–4

23 Cunningham-Rundles C (1988) Intravenous immunoglobulin treatment in the primary immunodeficiency diseases. *Immunol Allergy Clin N Amer* 8: 17–28

24 Mobini N, Sarela A, Ahmed AR (1995) Intravenous immunoglobulins in the therapy of autoimmune and systemic inflammatory disorders. *Ann Allergy Asthma Immunol* 74: 119–128

25 Misbah SA, Chapel HM (1993) Adverse effects of intravenous immunoglobulin. *Drug Safety* 9: 254–262

26 Sekul EA, Cupler EJ, Dalakas MC (1994) Aseptic meningitis associated with high-dose intravenous immunoglobulin therapy: Frequency and risk factors. *Ann Intern Med* 121: 259–262

27 Köhler G, Milstein C (1975) Continuous cultures of fused cells secreting antibodies of predefined specificity. *Nature* 256: 495–497

28 Kretzschmar T, Von Rüden, T (2002) Antibody discovery: phage display. *Biotechnology* 13: 598–602

29 Kellermann S-A, Green LL (2002) Antibody discovery: the use of transgenic mice to generate human monoclonal antibodies for therapeutics. *Biotechnology* 13: 593–597

30 Chadd HE, Chamow SM (2001) Therapeutic antibody expression technology. *Current Opinion in Biotechnology* 12: 188–194

Anti-allergic drugs

Sue McKay and Antoon J.M. van Oosterhout

Introduction

ALLERGY is defined as a disease following a response by the IMMUNE SYSTEM to an otherwise innocuous ANTIGEN. Allergic diseases include allergic rhinitis, atopic dermatitis, systemic anaphylaxis, food ALLERGY, allergic asthma and acute urticaria and are mediated by unwanted type-I hypersensitivity reactions to extrinsic allergen like pollen, house-dust, animal dander, drugs and insect venom. These diseases are characterised by the production of IgE antibodies to the allergen that bind to the high-affinity IgE receptor, FcεRI, on mast cells and BASOPHILS. Binding of allergen to IgE cross-links these receptors and causes the release of chemical mediators from mast cells leading to the development of a type-I hypersensitivity reaction (Fig. 1). This acute response is often followed by a late and more sustained inflammatory response characterised by the recruitment of other effector cells like EOSINOPHILS and T-helper type-2 (Th2) LYMPHOCYTES. This chapter will focus on anti-allergic drugs that target the activation of the mast cell or block the effects of its chemical mediators, in particular histamine.

Disodium cromoglycate and nedocromil sodium (chromones)

Dr. Altounyan first discovered that disodium cromoglycate possessed an anti-asthma action in the 1960s. He induced an asthma attack by inhaling animal dander ANTIGENS, and showed that cromoglycate afforded protection against this bronchial provocation. Disodium cromoglycate was introduced as an anti-allergic drug in 1967. Many companies tried to find improved versions of this compound, using the chemical structure as a starting point, but most of these attempts failed. Nedocromil was discovered and introduced 20 years later by Eady.

The exact MECHANISM OF ACTION of disodium cromoglycate and the related drug nedocromil sodium remains unclear, but their clinical activity probably represents a combination of effects. It was originally suggested that these non-steroidal anti-inflammatory drugs act as mast cell stabilisers [1]. However, although these drugs can prevent histamine release from mast cells, it has been demonstrated that this effect is not the basis of their action in allergic asthma. Other compounds that more potently inhibit mast cell histamine release have not proven to be more effective in the treatment of allergic asthma. Sodium cromoglycate and nedocromil sodium also partly inhibit the IgE-mediated release of other mediators from mast cells, such as prostacyclins and LEUKOTRIENES [2]. In addition, they have been described as exhibiting suppressive effects on inflammatory cells such as MACROPHAGES, MONOCYTES, NEUTROPHILS and EOSINOPHILS, but they do not have any direct effects on smooth muscle and they do not inhibit the actions of smooth muscle contractile agonists [3, 4]. Sodium cromoglycate and nedocromil sodium inhibit the influx of inflammatory cells and the release of inflammatory mediators following provocation with non-specific agents, such as cold air and air pollutants [5–7]. Furthermore, they have been reported to depress the exaggerated neuronal reflexes that are triggered by the stimulation of "irritant" receptors by decreasing neuropeptide release from C fibres and via antagonism of tachykinin receptors [8–10]. A comparison of the activities of sodium cromoglycate and nedocromil sodium on a variety of inflammatory cell types is shown in Table 1. The chemical structures of sodium cromo-

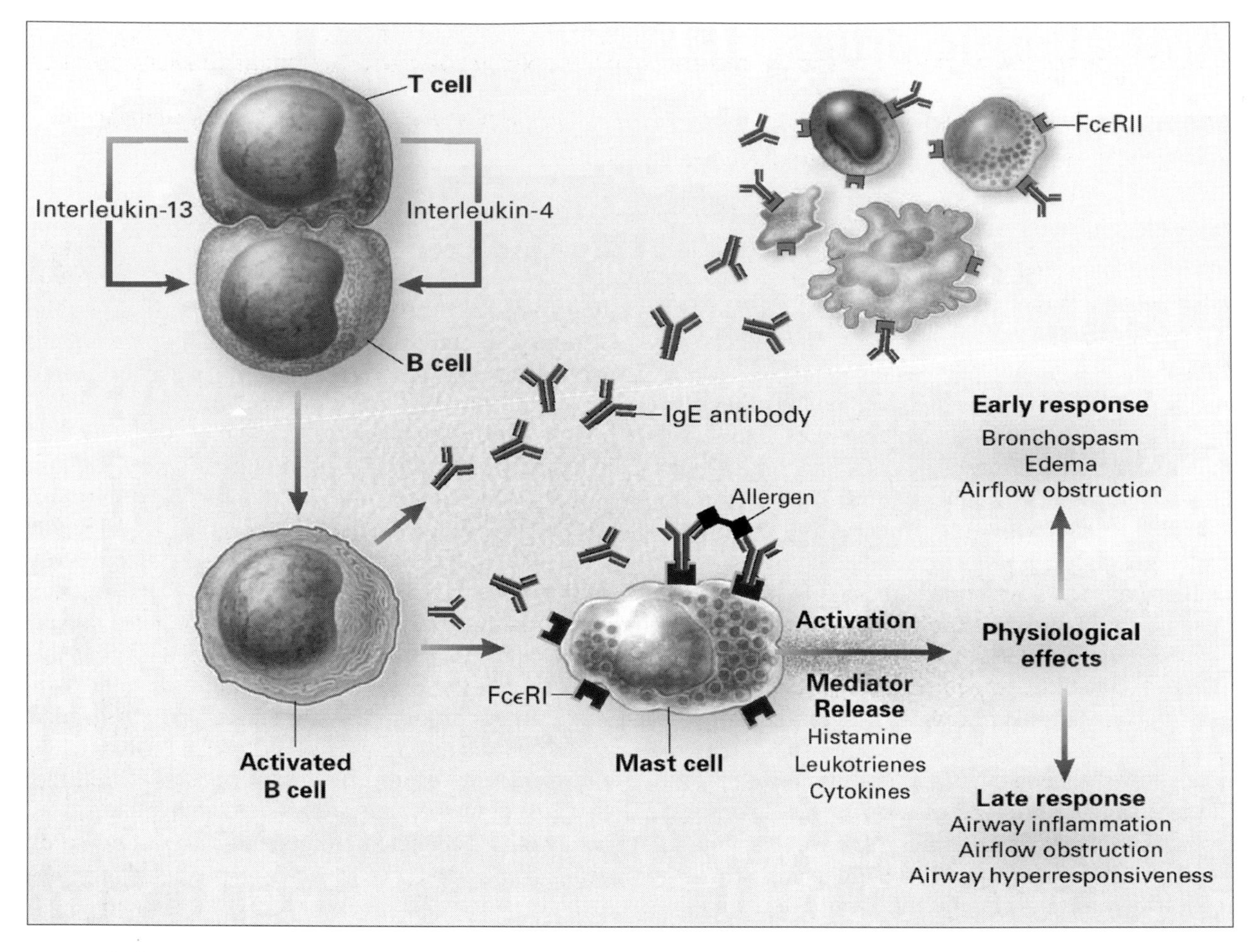

Figure 1. Schematic representation of the induction of IgE synthesis by B-lymphocytes
Once formed, IgE antibody circulates in the blood, eventually binding to high-affinity IgE receptors (FcεRI) on mast cells and low-affinity IgE receptors (FcRII, or CD23) on eosinophils and macrophages. After subsequent encounters with allergen, cross-linking of the high-affinity IgE receptors produces the release of preformed and newly generated mediators. Once present in various tissues, mediators may produce various physiological effects, depending on the target organ [45].

glycate and nedocromil sodium are depicted in Figure 2.

Mechanisms of action

Many studies have been performed to determine the mechanisms by which chromones inhibit the activation of cells. Since these compounds affect a wide variety of cells it has been assumed that a common mechanism must exist. Experiments have been performed to investigate whether this mechanism is regulated through a specific receptor or if it is due to the modulation of a second messenger signal.

Initially Ca^{2+} ions were implicated as the target for sodium cromoglycate [11], but this was discounted after it was shown that sodium cromoglycate could also inhibit mast cell activation in the pres-

TABLE 1. COMPARISON OF EFFECTS OF SODIUM CROMOGLYCATE AND NEDOCROMIL SODIUM ON INFLAMMATORY CELLS

Effect	Nedocromil sodium	Sodium cromoglycate
Mast cells from BAL, lung, conjunctiva, nasal mucosa, gastric mucosa and basophils		
Mediator release following Ascaris Ag or α-IgE Ab-inhibited (histamine, PGD_2, LTC_4)	↓	
Release of cytokines (TNF-α)	↓	↓
Release of histamine	↓	↓
Numbers	↓	↓
Macrophages/monocytes		
Release of cytokines (IL-6)	↓	
Release of lysosomal enzymes and oxygen radicals	↓	
Numbers	↓	
Eosinophils		
Numbers in BAL	↓	↓
Number of activated eos in submucosa	↓	
Release of mediators (preformed and newly generated)	↓	↓
Chemotactic response to PAF and LTB_4	↓	
Chemotactic response to zymosan-activated serum		↓
Activation	↓	
Survival time in presence of IL-5	↓	
Neutrophils		
Activation	↓	↓
Chemotactic response	↓	↓
Release of mediators (TNF-α, IL-6)	↓	↓
Numbers		↓
Platelets		
Release of cytotoxic mediators	↓	
IgE-mediated activation	↓	
Generation of thromboxane B_2 and IP_3	↓	
Epithelial cells		
Release of 15-HETE	↓	
Release of cytokines (TNF-α, IL-8, GM-CSF) and ICAM-1	↓	
Expression of ICAM-1, VCAM-1, E-selectin	↓	↓
B cells		
IgE Ab formation	↓	↓
T cells		
Numbers	↓	↓
Proliferation (allergen- or mitogen-induced)	±	±
Endothelial cells		
Expression of ICAM-1, VCAM-1, E-selectin	↓	↓
Sensory nerve (C fibres) activation		
Release of neuropeptides	↓	↓

↓ reduction; ± no change

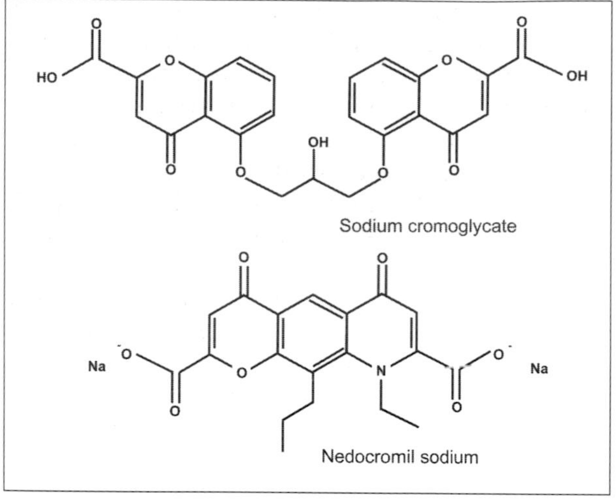

FIGURE 2. CHEMICAL STRUCTURES OF SODIUM CROMOGLYCATE AND NEDOCROMIL SODIUM

ence of Ca^{2+}-chelating agents [12]. Sodium cromoglycate and nedocromil sodium have been shown to reduce Ca^{2+} influx into cells, although it is believed that they do not interfere directly with Ca^{2+} channels [11]. These compounds have also been shown to phosphorylate intracellular proteins preceding mediator release from rat peritoneal mast cells; however, this does not hold true for mast cells isolated from the macaque [1]. Additionally, activation of protein kinase C (PKC) has been suggested as the molecular target for sodium cromoglycate [11, 13, 14], although other researchers report an inhibition of PKC activity [15]. The ability of sodium cromoglycate and nedocromil sodium to affect intracellular targets directly is unlikely if we consider the physical properties of these compounds. Both compounds are extremely polar and hydrophilic at pH 7.4, and therefore unlikely to penetrate the cell membrane. Consequently, it was assumed that they must function via a cell membrane component, possibly a receptor. Mazurek and colleagues have described a "sodium cromoglycate-binding protein", in rat basophil leukaemia (RBL) cells, that may be involved in Ca^{2+} mobilisation [16–18]. Eady and co-workers also identified proteins on rat peritoneal mast cells and Chinese hamster ovary cells that may act as receptors. Unfortunately, the identification of a specific receptor has not yet been established.

There is an increasing amount of evidence suggesting that modulation of chloride channel activity is possibly the common mechanism to explain the effects of sodium cromoglycate and nedocromil sodium. An accumulation of intracellular Ca^{2+} (Ca^{2+}_i) often precedes cell activation and mediator secretion in many cells. This accumulation of Ca^{2+}_i can result from a Ca^{2+} influx due to a negative membrane potential, which in turn is the result of an inward flow of Cl^- ions through chloride channels. Degranulation is dependent on a sustained elevation of intracellular Ca^{2+} – due to release of Ca^{2+} from intracellular stores and influx of Ca^{2+} ions. A small-conductance chloride channel (0.5-1 pS), identified in rat peritoneal mast cells, can achieve this by providing the negative membrane potential necessary for maintaining Ca^{2+} influx and its sustained elevation. The Ca^{2+} current activated by this mechanism is described as I_{CRAC} (Ca^{2+} release-activated Ca^{2+} current). By replacing extracellular Cl^- ions with non-permeant isethionate or gluconate anions, Friis and colleagues were able to inhibit ANTIGEN-stimulated histamine secretion from rat peritoneal mast cells, although some histamine secretion still occurred [19]. Sodium cromoglycate can also block intermediate conductance chloride channels on RBL cell membranes [11].

Studies on epithelial cells provide more evidence that sodium cromoglycate and nedocromil sodium affect chloride transport. Alton and colleagues [20] showed that these compounds are able to block the activity of a chloride channel present on the mucosal surface of airway epithelial cells. Moreover, epithelial cells are sensitive to the concentration of solutes in their environment, and chloride currents are believed to be involved in the regulation of cell volume. Nedocromil and cromoglycate can inhibit the chloride current induced in epithelial cells in response to osmotic changes, thereby inhibiting cell swelling [21]. Furthermore, Paulmichl and colleagues show that sodium cromoglycate and nedocromil sodium inhibit hypotonic saline-induced activation of a chloride channel in mouse 3T3 fibroblasts [22].

Further, evidence from *in vitro* studies suggests that these compounds affect neuronal chloride transport – chloride efflux from sensory nerves leads to depolarisation and the generation of action potentials. Nedocromil sodium prevents the contraction of guinea pig bronchus that is induced by electric field stimulation in the presence of atropine [10]. Bronchoconstriction is probably mediated by the release of neuropeptides from C fibre terminals. This is supported by other studies that show nedocromil inhibition of substance P-induced potentiation of the cholinergic neural responses in rabbit trachea [23] as well as inhibition of tachykinin release [24].

Chloride channel activation is a mechanism that occurs when cells are activated. By preventing chloride channel activation, sodium cromoglycate and nedocromil sodium would be expected to maintain cells in a normal resting physiological state, and this is associated with the relative lack of toxicity of these compounds.

Biochemical and pharmacological effects

The anti-inflammatory effects of sodium cromoglycate and nedocromil can result in a number of biological effects:

- Inhibition of mediator release from human mast cells isolated from bronchoalveolar lavage (BAL) fluid and from mast cells derived from lung, conjunctiva and nasal mucosa. Human skin-derived mast cells, however, do not respond to cromoglycate or nedocromil. Histamine secretion by enterochromaffin-like cells from the gastric mucosa can also be inhibited by sodium cromoglycate. Mediator release from EOSINOPHILS, NEUTROPHILS and platelets is also inhibited.
- The numbers of inflammatory cells such as EOSINOPHILS, NEUTROPHILS, BASOPHILS (though not unequivocally confirmed for nedocromil sodium), mast cells, MACROPHAGES and T LYMPHOCYTES are reduced, in both tissues and blood.
- Inflammatory cell infiltration depends on the activating effects of chemotactic factors that are often released by infiltrating inflammatory cells. Cromoglycate and nedocromil can completely suppress the activating effects of chemo-attractant peptides on human EOSINOPHILS, NEUTROPHILS and MONOCYTES.
- Inflammatory cell infiltration also depends on the expression of ADHESION MOLECULES. These compounds can inhibit the expression of various ADHESION MOLECULES, such as ICAM-1, VCAM-1 and E-SELECTIN, which are crucial for the passage of inflammatory cells from the blood to peripheral tissues.
- Cell activation and cytokine release from inflammatory cells such as T LYMPHOCYTES, MACROPHAGES and mast cells is inhibited.
- Inhibition of IL-4-induced IgE ISOTYPE switching and suppression of IgG_4 production, without further effects on B CELLS that have already undergone switching.
- Sensory nerve (C-fibres) activation is inhibited resulting in reduced release of neuropeptides such as substance P and tachykinins.
- Survival of platelets can be increased and these compounds inhibit IgE activation of platelets.
- MICROVASCULAR LEAKAGE is reduced, presumably through functional antagonism of tachykinin receptors.

Pharmacokinetics

Disodium cromoglycate and nedocromil are poorly absorbed from the gastrointestinal tract, and are therefore given locally per inhalation; as either an aerosol, a nebulised solution or in powder form; or they are given as eye drops. Nedocromil and disodium cromoglygate are not metabolised and are excreted unchanged. Their plasma half-life is approximately 90 minutes.

Clinical indications

Therapeutic studies have revealed and confirmed the clinical EFFICACY, protective effects and high safety/low side-effect profile of these drugs. The diverse clinical effects, including the anti-inflammatory character, of these drugs have been described in detail.

Allergic bronchial asthma

Sodium cromoglycate and nedocromil sodium have been reported to demonstrate protective effects on the immediate ASTHMATIC RESPONSE (IAR) as well as the late ASTHMATIC RESPONSE (LAR) induced by bronchial challenge with allergen. The delayed ASTHMATIC RESPONSE, however, was altered by nedocromil but not by cromoglycate. These compounds not only reduce the numbers of inflammatory cells in the BAL fluid, but they also decrease the activation and/or stimulation state of these cells. They also reduce the number of circulating LEUKOCYTES (EOSINOPHILS, NEUTROPHILS and BASOPHILS) during the IAR and the LAR following allergen challenge as well as decrease the activation of circulating T LYMPHOCYTES. Cromoglycate and nedocromil can also inhibit the IAR that occurs during exercise-induced asthma and reduce bronchoconstriction due to non-specific hyperreactivity mechanisms. However, not all asthma subjects respond to these drugs and children respond more often than adults do [1, 25, 26].

Allergic rhinitis

Similar to the treatment of allergic asthma, sodium cromoglycate and nedocromil sodium have demonstrated protective effects on the immediate nasal response (INR) as well as the late nasal response (LNR) induced by nasal challenge with allergen. The delayed nasal response, however, was not altered by cromoglycate and only partially prevented by nedocromil. Prophylactic treatment with these compounds significantly reduces inflammatory cell infiltration and epithelial cell numbers as determined by cytological analysis of nasal secretions following allergen challenge in patients with allergic rhinitis [27]. Degranulation of mast cells, BASOPHILS and EOSINOPHILS is inhibited and the expression of ICAM-1 on epithelial cells is downregulated [28, 29].

Allergic conjunctivitis

The immediate and late responses to allergen challenge are prevented in the eye. Sodium cromoglycate and nedocromil sodium inhibit the emergence of conjunctival oedema and erythema and reduce mast cell degranulation as well as vascular leakage [30–32].

Food allergy

Adverse reactions to food, including ALLERGY, can lead to unwanted organ responses. The various organ responses to food ingestion challenge can be inhibited by sodium cromoglycate treatment. This compound significantly inhibits immediate and late types of asthmatic, nasal, paranasal sinus, middle ear, conjunctival, migraine, atopic eczema, urticarial and Quincke's oedema responses to food ingestion ALLERGY [1, 27, 33].

It can be concluded that sodium cromoglycate and nedocromil sodium are effective drugs in the prophylaxis of allergic bronchial asthma, allergic rhinitis, allergic conjunctivitis and related allergic disorders. However, some authors suggest that it is not justified to recommend sodium cromoglycate as a first line prophylactic agent in childhood asthma [34].

Unwanted effects

Unwanted effects are infrequent and consist predominantly of the effects of irritation in the upper airway. Hypersensitivity reactions have been reported and include urticaria, bronchospasm, angio-oedema and anaphylaxis, but these are uncommon.

Histamine receptor antagonists

Histamine was first identified in 1910 and the first histamine receptor antagonists were synthesised over 20 years later. Early anti-histamine studies were qualitative, for example, the demonstration of their effectiveness in protecting against histamine-induced bronchospasm. Nevertheless, these studies introduced compounds, such as mepyramine, that remain major ligands to define histamine receptors. It became apparent in the 1950s that there were multiple histamine receptors, and research still continues to identify novel histamine receptors.

Histamine

Histamine (2-(4-imidazolyl)ethylamine or 5-amino-ethylimidazole) plays a significant role in the regulation of physiological processes and it is an important mediator during allergic reactions. It is synthesised from L-histidine by histidine decarboxylation and stored in various cells including mast cells, BASOPHILS, neurones and enterochromaffin-like cells. Other cells, predominantly from the hematopoietic lineage, can also synthesise and secrete histamine, although these cells lack specific storage granules. Histamine can closely mimic the anaphylactic response that usually results from an ANTIGEN-ANTIBODY reaction in sensitised tissue. Once released, histamine can be metabolised by diamine oxidase (DAO) and histamine N-methyltransferase (HMT). The effect of histamine is produced by its action on specific receptors, which are subdivided into several groups – H_1, H_2, H_3, and H_4-receptors. All subtypes are members of the seven membrane-spanning G protein-coupled receptor (GPCR) family.

Characterisation of H-receptors

The receptor subtype determines the biological effects of histamine. H_1-receptors are expressed on most smooth muscle cells, endothelial cells, adrenal medulla, heart and central nervous system (CNS). They have also been reported to be expressed on bronchial epithelial cells, fibroblasts, T CELLS, MONOCYTES, MACROPHAGES, DENDRITIC CELLS and B CELLS. Their stimulation leads to smooth muscle contraction, stimulation of NO formation, endothelial cell contraction, stimulation of hormone release, negative ionotropism, depolarisation and increased neuronal firing as well as increased vascular permeability. Stimulation of H_1-receptors can also lead to pro-inflammatory reactions such as induction of the expression of ADHESION MOLECULES on endothelial cells and the production of CYTOKINES by these cells as well as the induction of co-stimulatory molecules on DENDRITIC CELLS. H_1-receptor expression can be modified during inflammatory reactions. H_2-receptors are expressed on parietal cells in the gut, vascular smooth muscle cells, heart, suppressor T CELLS, NEUTROPHILS, CNS and BASOPHILS. Their stimulation triggers gastric acid secretion, vascular smooth muscle relaxation, positive chronotropic and ionotropic effects on cardiac muscle, inhibition of lymphocyte function, basophil chemotaxis and other immune responses. H_3-receptors are found mainly on cells in the CNS and peripheral nervous system as pre-synaptic receptors. They have also been identified on endothelium and enterochromaffin cells. These receptors control release of histamine and other neurotransmitters, such as acetylcholine and dopamine, from neurones. H_4-receptors have only recently been described and appear to be expressed on cells of the haematopoietic lineage and on immunocompetent cells such as mast cells, BASOPHILS, T CELLS, DENDRITIC CELLS, NEUTROPHILS and EOSINOPHILS. Stimulation of H4-receptors mediates chemotaxis of mast cells, NEUTROPHILS and regulates CYTOKINE release from $CD8^+$ T CELLS [35–40].

H_1-receptor antagonists are clinically effective when used to treat inflammatory and allergic reactions. The main clinical effect of H_2-receptor antagonists is on gastric secretion and the clinical relevance of H_3-receptor antagonists is still being explored, although they appear to be effective in the treatment of CNS disorders. Neither H_2-receptor antagonists nor H_3-receptor antagonists are considered to be clinically effective in anti-allergic therapies. H_4-receptor antagonists are being developed and it is thought that may be useful in the treatment of allergic diseases such as allergic rhinitis and asthma in the future [37].

Mechanisms of action of H_1-receptor antagonists

The term ANTI-HISTAMINE conventionally refers to H_1-receptor antagonists and these drugs are discussed in this section. The EFFICACY of these drugs is attributed principally to down-regulation of H_1-receptor activity. In Table 2 a number of first, second and third generation H_1-receptor antagonists are shown.

Signal transduction by H_1-receptors (and probably also for H_4-receptors) occurs through the hydrolysis of phosphatidylinositols. Histamine binds to the receptor, which in turn activates the $G_{\alpha q}$ protein ($G_{i/o}$

TABLE 2. FIRST, SECOND AND THIRD GENERATION H_1-RECEPTOR ANTAGONISTS

Antagonist	Disorder
First generation	
Clemastine	Anaphylactic reactions to insect bites or food allergy
Dexchlorofeniramine	Allergic conditions
Dimethindene	Allergic conditions
Diphenhydramine	Rhinitis/urticaria
Emedastine	Conjunctivitis/rhinitis
Hydroxyzine	Pruritis/chronic urticaria
Mebhydroline	Allergic conditions
Oxatomide	Allergic conditions
Promethazine	Allergic conditions/anaphylactic shock
Second generation	
Acrivastine	Allergic rhinitis/hayfever
Astemizole	Urticaria
Cetirizine	Allergic rhinitis/conjunctivitis/ urticaria
Fexofenadine	Allergic rhinitis/chronic urticaria
Ketotifen	Allergic rhinitis/allergic skin conditions/prophylactic for asthma
Levocabastine	Allergic rhinitis/conjunctivitis
Levocetirizine	Allergic rhinitis/chronic urticaria
Terfenadine	Allergic rhinitis/conjunctivitis/allergic skin disorders
Third generation	
Azelastine	Allergic rhinitis/conjunctivitis
Desloratidine	Allergic rhinitis
Ebastine	Allergic rhinitis/conjunctivitis
Loratadine	Allergic rhinitis/conjunctivitis/ chronic urticaria/pruritis
Mizolastine	Allergic rhinitis/conjunctivitis/urticaria

protein of H_4-receptors). Activation of these G proteins precedes activation of phospholipase C (PLC) which cleaves phosphatidylinositol bisphosphate (PIP_2) to form inositol tri-phosphate (IP_3) and diacylglycerol (DAG). IP_3 activates IP_3 receptors on the endoplasmic reticulum, causing the release of intracellular Ca^{2+}, as depicted in Figure 3. The various biological effects follow the rise in Ca^{2+}.

Pharmacological effects of H_1-receptor antagonists

H_1-receptors modulate inflammatory and allergic responses by controlling NO formation and smooth muscle and endothelial cell contraction, which subsequently result in increased vascular permeability. Histamine can also stimulate sensory nerve endings,

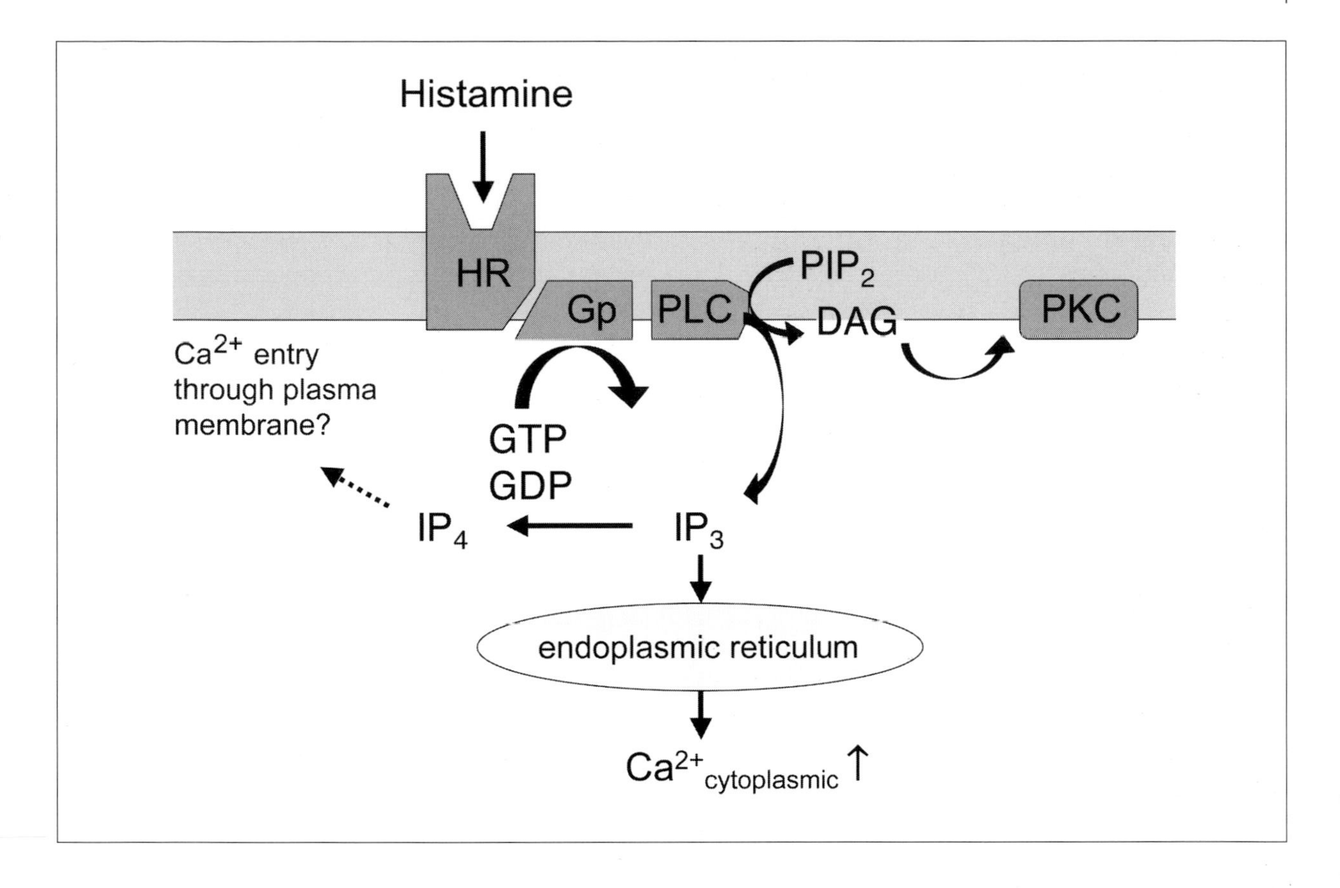

FIGURE 3. SIGNALLING PATHWAY OF H_1 RECEPTOR (AND POSSIBLY H_4 RECEPTOR)
Histamine binds to the receptor, which in turn activates the $G_{\alpha q}$ protein ($G_{i/o}$ protein of H_4-receptors). Activation of these G proteins precedes activation of PLC which hydrolyses IP_3. IP_3 activates IP_3 receptors on the endoplasmic reticulum, causing the release of intracellular Ca^{2+}. The rise in Ca^{2+} is followed by the biological effect. Abbreviations: HR, histamine receptor; Gp, G protein; PLC, phospholipase C; PIP_2 phosphatidylinositol bisphosphate; DAG, diacyglycerol; PKC, protein kinase C; GTP, guanosine triphosphate; GDP, guanosine diphosphate; IP_3, inositol triphosphate; Ca^{2+}, calcium.

thereby causing itching of the mucosa and skin through stimulation of C fibres. Regulation of the transcription factor NF-κB and subsequent generation of ADHESION MOLECULES (ICAM-1 and P-selectin) and CYTOKINES (IL-6, IL-8, GM-CSF, RANTES) are also H_1-receptor dependent. The pharmacological actions of H_1-receptor antagonists are therefore useful for the inhibition of contraction of the smooth muscle and the inhibition of histamine-induced vascular permeability. Additionally, H_1-receptor antagonists inhibit the constitutive activation of NF-κB which results in an inhibition of CYTOKINE production. H_1-receptor antagonists inhibit histamine-induced bronchospasm in the guinea pig, but they are not effective in decreasing allergen-induced bronchospasm in human airways. Furthermore, first generation H_1-receptor antagonists can exhibit anti-serotoninergic, anti-emetic and/or anti-cholinergic characteristics, depending on the particular H_1-receptor antagonist used. Second generation H1-receptor antagonists, however, are more selective, have minimal sedative effects and have little affinity for muscarinic cholinergic, α-adrenergic or serotoninergic receptors, although they still have dose-related adverse effects at high doses. Third generation H_1-receptor antagonists are either active metabolites or enantiomers of second generation compounds and show reduced adverse effects.

Pharmacokinetics

Most H_1-receptor antagonists are given orally, some are given as nose sprays or eye drops, they are well absorbed and reach their peak effect in 1–2 hours. The duration of activity depends largely on whether first, second or third generation drugs are administered, and can vary between 2 hours and a few days. Most of these drugs are widely distributed throughout the body, but the third generation drugs do not pass the blood-brain barrier. They are largely metabolised in the liver and excreted in the urine.

Clinical indications

H_1-receptor antagonists can effectively control allergic disorders with mild symptoms, especially of the upper airways and skin. Allergic disorders with more severe symptoms and a complicated clinical picture, such as severe asthma, require other therapies but H_1-receptor antagonists may be supplemental.

Allergic rhinitis

H_1-receptor antagonists, administered either orally or topically to mucosal surfaces, are the most frequently used first-line medication for intermittent (seasonal) and persistent (perennial) allergic rhinitis. Non-sedating second-generation H_1-receptor antagonists such as cetirizine, fexofenadine and loratadine have been proven effective in short and long-term studies. Also desloratadine, levocetirizine and tecastemizole have been found to be effective in allergic rhinitis. H_1-receptor antagonists reduce sneezing and rhinorrhea as well as itchy, watery, red eyes. They reduce itchy nose, palate, or throat and sometimes reduce nasal congestion. There are very few clinical differences between the H_1-receptor antagonists. Some investigators have reported that H_1-receptor antagonists reduce the influx of inflammatory cells into nasal secretion whereas other investigators report a lack of inhibitory effects.

Mild atopic asthma

Cetirizine, desloratidine and loratidine have been reported to improve mild "seasonal" asthma symptoms and to reduce the amount of β_2-agonist usage, as well as improve pulmonary function. These compounds prevent and relieve allergic INFLAMMATION in both the upper and lower airways. H_1-receptor antagonists have not been reported to be clinically effective in the treatment of bronchial asthma and are therefore not a first choice for the treatment of this disorder. They do not affect histamine- or methacholine-induced bronchospasm but they may be useful as additional or supplemental therapy.

Allergic conjunctivitis

This disorder usually occurs as part of an allergic syndrome, i.e., together with allergic rhinitis. If acute symptoms arise, the eyes can be treated locally with azelastine, emadine, ketotifen, levocabastine or olopatadine. The main symptoms are red, itchy, watery eyes.

Acute and chronic urticaria

This disorder can be treated with oral, H1-receptor antagonists. These compounds reduce itching as well as the number, size and duration of urticarial lesions. Erythema may not be completely inhibited because the vascular effects of histamine are also mediated via H_2-receptors as well as by other vasoactive substances such as proteases, eicosanoids (LEUKOTRIENES, prostaglandin E_1) and neuropeptides (substance P). First and second generation H_1-receptor antagonists have been shown to be equally effective. Urticarial vasculitis cannot be satisfactorily treated with H_1-receptor antagonists.

Treatment of anaphylactic shock

The initial treatment of choice is epinephrine, but treatment of anaphylactic shock can be accomplished with intramuscular or subcutaneous injections of epinephrine. The H_1-receptor antagonist clemastine may be given intravenously (2 mg) as an ADJUVANT. H_1-receptor antagonists may also be useful in the ancillary treatment of pruritis, urticaria and angio-oedema.

Atopic dermatitis (= eczema)

is often treated with oral H_1-receptor antagonists, that also exhibit a sedative action, in conjunction with TOPICAL glucocorticoids to relieve itching. Second generation H_1-receptor antagonists are generally less effective than first generation drugs.

Unwanted effects

Most H_1-receptor antagonists have few unwanted effects when used at the recommended doses, although adverse CNS effects have been observed. The first generation H_1-receptor antagonists have marked sedative effects due to the fact that they can cross the blood-brain barrier. The second generation H_1-receptor antagonists are more specific for the H_1-receptor than first generation drugs, and therefore have little affinity for muscarinic cholinergic, α-adrenergic or serotoninergic receptors, these compounds have less sedative effects. Third generation H_1-receptor antagonists have been recently developed to further reduce adverse effects. Dry mouth, urinary dysfunction, constipation, tachycardia and other unwanted effects are consequently not often observed. Allergic dermatitis following TOPICAL application has been reported.

Suppression of mediator release from mast cells and BASOPHILS are thought to occur independently of the H_1-receptor. Neither mast cells nor BASOPHILS express H_1-receptors on their surface. However, it was recently shown that mast cells, BASOPHILS and EOSINOPHILS express the H_4-receptor on their surface.

H_4-receptor antagonists

Increased numbers of mast cells are found in the airways of allergic asthma and allergic rhinitis patients. Hofstra and colleagues suggest that migration of these mast cells towards inflamed tissues may be mediated through histamine. To support their hypothesis they have recently shown that stimulation of H_4-receptors with histamine mediates cell signalling and mast cell chemotaxis in a dose-dependent manner [37]. Redistribution of mast cells during allergic episodes may be mediated by this mechanism, indicating that specific H_4-receptor antagonists may prove to be useful in the treatment of allergic diseases in the future. Since EOSINOPHILS have also been demonstrated to express H_4-receptors, we can speculate that specific H_4-receptor antagonists may also inhibit their migration and infiltration of tissues during allergic reactions. Furthermore, H_4-receptors may play a role in the control of CYTOKINE release from $CD8^+$ T CELLS, and possibly other cells that express the H_4-receptor, during inflammatory disorders such as asthma [35].

Some of the currently available H_3-receptor agonists and antagonists are also recognised by the H_4-receptor, although they are much less potent for the H_4-receptors. Using cells transfected with the H_4-receptor, it has been shown that specific H_1 and H_2 receptor agonists and antagonists do not bind to the H_4-receptor. Specific antagonists for the H_4-receptor need to be developed and tested.

Anti-IgE

Ever since the discovery of the function of IgE, more than 3 decades ago [41], research focussed on the selective inhibition of either the activity or the production of IgE. Omalizumab, marketed as Xolair, is a monoclonal ANTIBODY that targets IgE. Xolair is a recombinant DNA-derived humanised $IgG_{1\kappa}$ monoclonal ANTIBODY that selectively binds to human IgE. It is a humanised mouse ANTIBODY that contains only 5% amino acid sequence derived from the mouse. The ANTIBODY has a molecular weight of approximately 149 kilo Daltons and is produced by Chinese hamster ovary cells. Xolair has been approved by the FDA for the treatment of adults and adolescents (12 years of age and older) with moderate to severe persistent asthma who have a positive skin test or *in vitro* reactivity to a perennial aeroallergen and whose

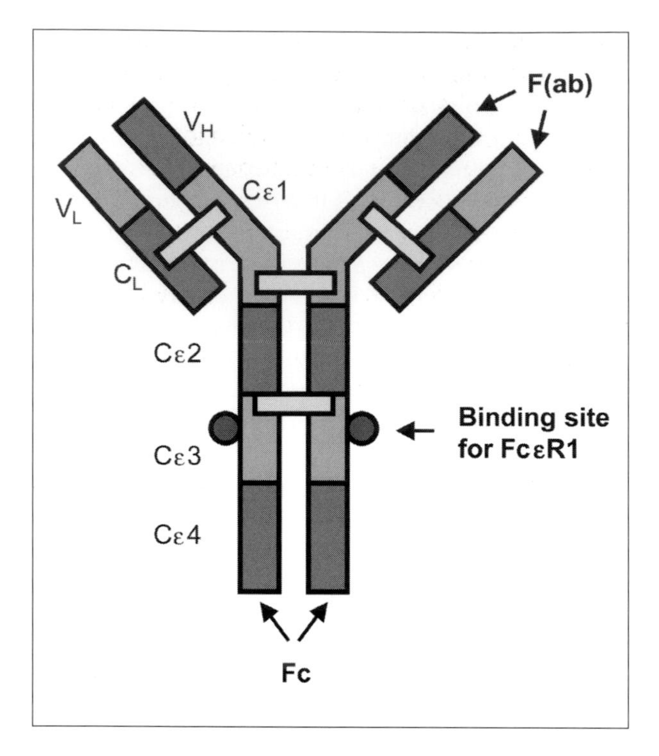

FIGURE 4. STRUCTURE OF IGE AND POSITION OF BINDING SITE FOR FCεRI [43]

symptoms are inadequately controlled with inhaled corticosteroids. Xolair is given by subcutaneous injection at a dose of 150 mg to 375 mg every 2 to 4 weeks based upon pre-treatment serum IgE levels and total body weight.

Biochemical and pharmacological effects

Xolair binds to free IgE at the FcεRI binding site on the Cε3 domain of the IgE ANTIBODY, thereby inhibiting the binding to FcεRI on the surface of mast cells, BASOPHILS and other $Fc\varepsilon RI^+$ cells (Fig. 4). The reduction of surface-bound IgE on FcεRI-bearing cells prevents or limits the release of chemical mediators of the allergic response. Levels of serum-free IgE decrease by >90% from pre-treatment values within 24 hours of subcutaneous administration of Xolair. In addition, the expression of FcεRI appears to be down-regulated as demonstrated on BASOPHILS after continued (3 months) Xolair treatment [42].

Clinical trials

Several clinical studies using recombinant humanised monoclonal ANTIBODY to human IgE (E25, Xolair, Omalizumab) have been conducted and published (reviewed in [43]).

The approval of Xolair was based upon EFFICACY data from several multicenter placebo-controlled phase-III clinical trials with symptomatic patients with moderate to severe persistent asthma who had a positive skin test reaction to a perennial aeroallergen [44–46]. All patients were also being treated with inhaled corticosteroids and short-acting β-agonists. EFFICACY in these trials was based upon the number of asthma exacerbations per patient, defined as a worsening of asthma that required treatment with systemic corticosteroids or a doubling of the baseline dose of inhaled corticosteroid. Study results showed that the number of exacerbations per patient was reduced in patients receiving Xolair compared with placebo. Reduction of asthma exacerbations was not observed in Xolair-treated patients who had baseline $FEV_1 > 80\%$ or in patients who required oral steroids as maintenance therapy. Comparative clinical studies between Xolair and other agents used to treat asthma are currently not available. Xolair has not been shown to alleviate asthma exacerbations acutely and it is not indicated for the treatment of bronchospasm or status asthmaticus.

The following effects of Xolair have been observed in clinical trials with allergic asthma patients:

- >90% drop in serum IgE levels within hours after injection [47, 48].
- Partial reduction of drop in FEV_1 during early- and LATE-PHASE ASTHMATIC RESPONSE after allergen challenge [49, 50].
- Reduction of asthma symptom scores and increase of quality-of-life scores; decreased use of β-agonists after high dose of Xolair; reduction of oral or inhaled steroid use [45, 47].
- Reduction of asthma exacerbations during stable- and steroid-reduction phase; steroid sparing effect [45, 46].

- Steroid sparing effect in asthmatic children; no change in asthma symptom scores [51].

The following effects of Xolair have been observed in clinical trials with seasonal allergic rhinitis patients:
- No significant difference in daily symptom scores; no significant difference in the use of rescue medication [48].
- Reduction of rescue ANTI-HISTAMINES; decreased daily nasal symptom severity scores [52].
- Dose-dependent reduction of nasal and ocular symptom severity and duration scores; reduction of rescue ANTI-HISTAMINES [53].

Unwanted effects

Single- and multiple-dose trials in adults with and without allergic diseases have demonstrated that Xolair is well tolerated. Immune-complex formation of Xolair with IgE leads to relatively small complexes (<1000 kDa) that are not complement-fixing and do not accumulate within any organ system [54]. The immune complexes have a serum half-life of approximately 20 days and are cleared via Fcγ receptors of the reticuloendothelial system [54]. The most commonly reported adverse effect is an urticarial rash, with an incidence of 0.5 to 0.7%. The rash develops within one hour of receipt of the first dose and responds to ANTI-HISTAMINE therapy. Other adverse effects associated with Xolair are headache, fatigue and vertigo. There are no reports of systemic anaphylactic reactions, and no antibodies to the anti-IgE ANTIBODY have been detected.

Conclusions

The IgE-mast-cell pathway plays a central role in the pathogenesis of allergic diseases. It is therefore not surprising that many of the anti-allergic drugs are targetted at this pathway. These drug are either preventing or reversing allergen-induced mast-cell activation by blocking IgE (Xolair) or stabilizing mast-cells (chromones) or block the biological effects of histamine released upon allergen-induced mast-cell activation (ANTI-HISTAMINES). ANTI-HISTAMINES have proven to be safe and successful, although they may not block all effects of mast cells. Chromones can be used as prophylactic drugs for allergic diseases but their clinical EFFICACY is not undisputed. Xolair is very effective in reducing serum IgE levels and the prevention of mast-cell activation. However, this protein drug needs to be administered parenterally and the costs of treatment is estimated to be very high compared to other anti-allergic treatments. Thus, Xolair will probably not become the first line of prophylactic treatment for allergic diseases.

Summary

Allergy is defined as a disease following a response by the IMMUNE SYSTEM to an otherwise innocuous agent. This chapter describes the actions of anti-allergic drugs and therapeutics on diseases such as allergic rhinitis, atopic dermatitis, allergic conjunctivitis, systemic anaphylaxis, food ALLERGY, allergic asthma and acute urticaria. The putative MECHANISMS OF ACTION of disodium cromoglycate and nedocromil sodium (chromones) are discussed in detail as well as the biochemical and pharmacological effects, clinical applications and unwanted effects of these drugs. The characterisation of histamine receptors is detailed and the MECHANISMS OF ACTION as well as the observed biological effects of H_1 and the relatively new H_4 receptors are explained. Further, the biochemical and pharmacological effects of anti-IgE therapy are described and recent clinical trials with these drugs are briefly reviewed.

Selected readings

Hill SJ, Ganellin CR, Timmerman H, Schwartz JC, Shankley NP, Young JM, Schunack W, Levi R, Haas HL (1997) International Union of Pharmacology. XIII. Classification of Histamine Receptors. *Pharm Rev* 49: 253–278 (see also: *http://pharmrev.aspetjournals.org/cgi/reprint/49/3/253*, accessed December 2004)

References

1 Eady RP, Norris AA (1997) Nedocromil sodium and sodium cromoglycate: Pharmacology and putative modes of action. In: AB Kay (ed): *Allergy and Allergic Diseases*. Blackwell, Oxford, 584–595
2 Eady RP (1986) The pharmacology of nedocromil sodium. *Eur J Resp Dis* 147: 112–119
3 Janssen LJ, Wattie J, Betti PA (1998) Effects of cromolyn and nedocromil on ion currents in canine tracheal smooth muscle. *Eur Resp J* 12: 50–56
4 Rang HP, Dale MM, Ritter JM (1999) The respiratory system. In: L Hunter (ed): *Pharmacology*. Churchill Livingston, London, 347
5 Dixon CM, Fuller RW, Barnes PJ (1987) Effect of nedocromil sodium on sulphur dioxide induced bronchoconstriction. *Thorax* 42: 462–465
6 Dixon CMS, Ind PW (1990) Inhaled sodium metabisulphate induced bronchoconstriction: inhibition by nedocromil sodium and sodium cromoglycate. *Br J Clin Pharmacol* 30: 371–376
7 Bigby B, Boushey H (1993) Effects of nedocromil sodium on the bronchomotor response to sulfur dioxide in asthmatic patients. *J Allergy Clin Immunol* 92: 195–197
8 Yamawaki I, Tamaoki J, Takeda Y, Nagai A (1997) Inhaled cromoglycate reduces airway neurogenic inflammation via tachykinin antagonism. *Res Com Mol Pathol Pharmacol* 98: 265–272
9 Dixon M, Jackson DM, Richards IM (1980) The action of sodium cromoglycate on 'C' fibre endings in the dog lung. *Br J Pharmacol* 70: 11–13
10 Verleden GM, Belvisi MG, Stretton CD, Barnes PJ (1991) Nedocromil sodium modulates nonadrenergic, noncholinergic bronchoconstrictor nerves in guinea pig airways *in vitro*. *Am Rev Respir Dis* 143: 114–118
11 Foreman JC, Hallett MB, Mongar JL (1977) Site of action of the antiallergic drugs cromoglycate and doxantrazole. *Br J Pharmacol* 59: 473P–474P
12 Ennis M, Atkinson S, Pearce FL (1980) Inhibition of histamine release induced by Compound F48/80 and peptide 401 in the presence and absence of calcium: implications for the mode of action of anti-allergic compounds. *Agents Actions* 10: 222–228
13 Sagi-Eisenberg R (1987) The role of protein kinase C in histamine secretion: implications for the mode of action of the antiasthmatic drug cromoglycate. *Curr Topics Pulm Pharmacol Toxicol* 2: 24–42
14 Sagi-Eisenberg R, Mazurek N, Pecht I (1984) Calcium fluxes and protein phosphorylation in stimulus-secretion coupling of basophils. *Mol Immunol* 21: 175–181
15 Lucas AM, Shuster S (1987) Cromolyn inhibition of protein kinase C activity. *Biochem Pharmacol* 36: 562–565
16 Mazurek N, Bashkin P, Loyter A, Pecht I (1983) Restoration of Ca2+ influx and degranulation capacity of variant RBL-2H3 cells upon implantation of isolated cromolyn binding protein. *Proc Natl Acad Sci USA* 80: 6014–6018
17 Mazurek N, Geller-Bernstein C, Pecht I (1980) Affinity of calcium ions to the anti-allergic drug, dicromoglycate. *FEBS Lett* 111: 194–196
18 Mazurek N, Schindler H, Schurholz T, Pecht I (1984) The cromolyn binding protein constitutes the Ca2+ channel of basophils opening upon immunological stimulus. *Proc Natl Acad Sci USA* 81: 6841–6845
19 Friis UG, Johansen T, Hayes NA, Foreman JC (1994) IgE-receptor activated chloride uptake in relation to histamine secretion from rat mast cells. *Br J Pharmacol* 111: 1179–1183
20 Alton EWFW, Norris AA (1996) Chloride transport and the actions of nedocromil sodium and cromolyn sodium in asthma. *J Allergy Clin Immunol* 98: S102–S106
21 Anderson SD, Rodwell LT, Daviskas E, Spring JF, du Toit J (1996) The protective effect of nedocromil sodium and other drugs on airway narrowing provoked by hyperosmolar stimuli: A role for the airway epithelium? *J Allergy Clin Immunol* 98: S124–S134
22 Paulmilchl M, Norris AA, Rainey DK (1995) Role of chloride channel modulation in the mechanism of action of nedocromil sodium. *Int Arch Allergy Appl Immunol* 107: 416
23 Armour CL, Johnson PRA, Black JL (1991) Nedocromil sodium inhibits substance P-induced potentiation of the cholinergic neural responses in the isolated innervated rabit trachea. *J Auton Pharmacol* 11: 167–172
24 Javdan P, Figini M, Emanueli C, Geppetti P (1995) Nedocromil sodium reduces allergen-induced plasma extavasation in the guinea-pig nasal mucosa by inhibition of tachykinin release. *Allergy* 50: 825–829
25 Pelikan Z, Pelikan-Filipek M, Remeijer L (1988) Effects of disodium cromoglycate and beclomethasone dipropionate on the asthmatic response to allergen

challenge II. Late response (LAR). *Ann Allergy* 60: 217–225

26 Pelikan Z, Pelikan-Filipek M, Schoemaker MC, Berger MP (1988) Effects of disodium cromoglycate and beclomethasone dipropionate on the asthmatic response to allergen challenge I. Immediate response (IAR). *Ann Allergy* 60: 211–216

27 Pelikan Z (2001) Late type of the nasal allergic response review. *Scripta Medica (BRNO)* 74: 303–344

28 Larsson K, Larsson BM, Sandstrom T, Sundblad BM, Palmberg L (2001) Sodium cromoglycate attenuates pulmonary inflammation without influencing bronchial responsiveness in healthy subjects exposed to organic dust. *Clin Exp Allergy* 31: 1356–1368

29 Hoshino M, Nakamura Y (1997) The effect of inhaled sodium cromoglycate on cellular infiltration into the bronchial mucosa and the expression of adhesion molecules in asthmatics. *Eur Respir J* 10: 858–865

30 James IG, Campbell LM, Harrison JM, Fell PJ, Ellers-Lenz B, Petzold U (2003) Comparison of the efficacy and tolerability of topically administered azelastine, sodium cromoglycate and placebo in the treatment of seasonal allergic conjunctivitis and rhino-conjunctivitis. *Curr Med Res Opinions* 19: 313–320

31 Katelaris CH, Ciprandi G, Missotten L, Turner FD, Bertin D, Berdeaux G (2002) A comparison of the efficacy and tolerability of olopatadine hydrochloride 0.1% ophthalmic solution and cromolyn sodium 2% ophthalmic solution in seasonal allergic conjunctivitis. *Clin Ther* 24: 1561–1575

32 Tauber J (2002) Nedocromil sodium ophthalmic solution 2% twice daily in patients with allergic conjunctivitis. *Adv Ther* 19: 73–84

33 Pelikan Z, Pelikan-Filipek M (1989) Effects of oral cro molyn on the nasal response due to foods. *Arch Otolaryngol Head Neck Surg* 115: 1238–1243

34 Tasche MJ, Uijen JH, Bernsen RM, de Jongste JC, van der Wouden JC (2000) Inhaled disodium cromoglycate (DSCG) as maintenance therapy in children with asthma: a systematic review. *Thorax* 55: 913–920

35 Gantner F, Sakai K, Tusche MW, Cruikshank WW, Center DM, Bacon KB (2002) Histamine h(4) and h(2) receptors control histamine-induced interleukin-16 release from human CD8(+) T cells. *J Pharmacol Exp Ther* 303: 300–307

36 Takeshita K, Sakai K, Bacon KB, Gantner F (2003) Critical role of histamine H_4 receptor in LTB_4 production and mast cell-dependent neutrophil recruitment induced by zymosan *in vivo*. *J Pharmacol Exp Ther* 9: 9

37 Hofstra CL, Desai PJ, Thurmond RL, Fung-Leung WP (2003) Histamine H_4 receptor mediates chemotaxis and calcium mobilization of mast cells. *J Pharmacol Exp Ther* 305: 1212–1221

38 Hill SJ, Ganellin CR, Timmerman H, Schwartz JC, Shankley NP, Young JM, Schunack W, Levi R, Haas HL (1997) International Union of Pharmacology. XIII. Classification of histamine receptors. *Pharmacol Rev* 49: 253–278

39 Schneider E, Rolli-Derkinderen M, Arock M, Dy M (2002) Trends in histamine research: new functions during immune responses and hematopoiesis. *Trends Immunol* 23: 255–263

40 Lovenberg TW, Pyati J, Chang H, Wilson SJ, Erlander MG (2000) Cloning of rat histamine H(3) receptor reveals distinct species pharmacological profiles. *J Pharmacol Exp Ther* 293: 771–778

41 Ishizaka K, Ishizaka T (1970) Biological function of gamma E antibodies and mechanisms of reaginic hypersensitivity. *Clin Exp Immunol* 6: 25-42.

42 Macglashan DW, Bochner BS, Adelman DC, Jardieu PM, Togias A, McKenziewhite J, Sterbinsky SA, Hamilton RG, Lichtenstein LM (1997) Down-regulation of Fc epsilon RI expression on human basophils during *in vivo* treatment of atopic patients with anti-IgE antibody. *J Immunol* 158: 1438–1445

43 Hamelmann E, Rolinck-Werninghaus C, Wahn U (2002) From IgE to anti-IgE: where do we stand? *Allergy* 57: 983–994

44 Finn A, Gross G, van Bavel J, Lee T, Windom H, Everhard F, Fowler-Taylor A, Liu J, Gupta N (2003) Omalizumab improves asthma-related quality of life in patients with severe allergic asthma. *J Allergy Clin Immunol* 111: 278–284

45 Busse W, Corren J, Lanier BQ, McAlary M, Fowler-Taylor A, Cioppa GD, van As A, Gupta N (2001) Omalizumab, anti-IgE recombinant humanized monoclonal antibody, for the treatment of severe allergic asthma. *J Allergy Clin Immunol* 108: 184–190

46 Soler M, Matz J, Townley R, Buhl R, O'Brien J, Fox H, Thirlwell J, Gupta N, Della Cioppa G (2001) The anti-IgE antibody omalizumab reduces exacerbations and steroid requirement in allergic asthmatics. *Eur Respir J* 18: 254–261

47 Milgrom H, Fick RB, Jr., Su JQ, Reimann JD, Bush RK, Watrous ML, Metzger WJ (1999) Treatment of allergic asthma with monoclonal anti-IgE antibody. rhuMAb-E25 Study Group. *N Engl J Med* 341: 1966–1973
48 Casale TB, Bernstein IL, Busse WW, LaForce CF, Tinkelman DG, Stoltz RR, Dockhorn RJ, Reimann J, Su JQ, Fick RB, Jr. et al (1997) Use of an anti-IgE humanized monoclonal antibody in ragweed-induced allergic rhinitis. *J Allergy Clin Immunol* 100: 110–121
49 Boulet LP, Chapman KR, Cote J, Kalra S, Bhagat R, Swystun VA, Laviolette M, Cleland LD, Deschesnes F, Su JQ, et al (1997) Inhibitory effects of an anti-IgE antibody E25 on allergen-induced early asthmatic response. *Am J Respir Crit Care Med* 155: 1835–1840
50 Fahy JV, Fleming HE, Wong HH, Liu JT, Su JQ, Reimann J, Fick RB, Boushey HA (1997) The effect of an anti-IgE monoclonal antibody on the early-and late-phase responses to allergen inhalation in asthmatic subjects. *Am J Respir Crit Care Med* 155: 1828–1834
51 Milgrom H, Berger W, Nayak A, Gupta N, Pollard S, McAlary M, Taylor AF, Rohane P (2001) Treatment of childhood asthma with anti-immunoglobulin E antibody (omalizumab). *Pediatrics* 108: E36
52 Adelroth E, Rak S, Haahtela T, Aasand G, Rosenhall L, Zetterstrom O, Byrne A, Champain K, Thirlwell J, Cioppa GD et al (2000) Recombinant humanized mAb-E25, an anti-IgE mAb, in birch pollen-induced seasonal allergic rhinitis. *J Allergy Clin Immunol* 106: 253–259
53 Casale TB, Condemi J, LaForce C, Nayak A, Rowe M, Watrous M, McAlary M, Fowler-Taylor A, Racine A, Gupta N, et al (2001) Effect of omalizumab on symptoms of seasonal allergic rhinitis: a randomized controlled trial. *JAMA* 286: 2956–2967
54 Fox JA, Hotaling TE, Struble C, Ruppel J, Bates DJ, Schoenhoff MB (1996) Tissue distribution and complex formation with IgE of an anti-IgE antibody after intravenous administration in cynomolgus monkeys. *J Pharmacol Exp Ther* 279: 1000–1008

Drugs for the treatment of asthma and COPD

Peter J. Barnes

Introduction

Both ASTHMA and chronic obstructive pulmonary disease (COPD) are characterized by airflow obstruction and chronic INFLAMMATION of the airways, but there are important differences in inflammatory mechanisms and response to therapy between these diseases [1]. This chapter discusses the pharmacology of the drugs used in the treatment of obstructive airways diseases. These drugs include bronchodilators, which act mainly by reversing airway smooth muscle contraction, and anti-inflammatory drugs which in ASTHMA suppress the inflammatory response in the airways. In COPD no effective anti-inflammatory drugs are available, but several new classes of drug are now in development.

Bronchodilators

Bronchodilator drugs have an "anti-bronchoconstrictor" effect, which may be demonstrated directly *in vitro* by a relaxant effect on precontracted airways. Bronchodilators cause immediate reversal of airway obstruction in ASTHMA *in vivo*, and this is believed to be due to an effect on airway smooth muscle, although additional pharmacological effects on other airway cells (such as reduced MICROVASCULAR LEAKAGE and reduced release of bronchoconstrictor mediators from inflammatory cells) may contribute to the reduction in airway narrowing. Three main classes of bronchodilator are in current clinical use:

- β-Adrenergic agonists (sympathomimetics)
- Theophylline (methylxanthines)
- Anticholinergics (muscarinic receptor antagonists)

Drugs such as sodium cromoglycate, which prevent bronchoconstriction, have no direct bronchodilator action and are ineffective once bronchoconstriction has occurred. Anti-LEUKOTRIENES (leukotriene receptor antagonists and 5'-lipoxygenase inhibitors) have a small bronchodilator effect in some patients and appear to act more to prevent bronchoconstriction. Corticosteroids, while gradually improving airway obstruction in asthma, have no direct effect on contraction of airway smooth muscle and are not therefore considered to be bronchodilators.

β_2-Adrenergic agonists

Inhaled β_2-agonists are the bronchodilator treatment of choice in asthma, as they are the most effective bronchodilators and have minimal side-effects when used correctly [2]. There is no place for short-acting and non-selective β-agonists, such as isoprenaline or metaproterenol.

Chemistry

The development of β_2-agonists was a logical development of substitutions in the catecholamine structure of noradrenaline. The catechol ring consists of hydroxyl groups in the 3 and 4 positions of the benzene ring (Fig. 1). Noradrenaline differs from adrenaline (epinephrine) only in the terminal amine group, which therefore indicates that modification at this site confers β-receptor selectivity. Further substitution of the terminal amine resulted in β_2-receptor selectivity, as in albuterol and terbutaline. Catecholamines are rapidly metabolised by the enzyme catechol-o-methyl transferase (COMT), which methylates in the 3-hydroxyl position, and accounts for the short duration of action of catecholamines. Modification of the catechol ring, as in salbutamol and terbutaline, prevents this degradation and therefore prolongs their

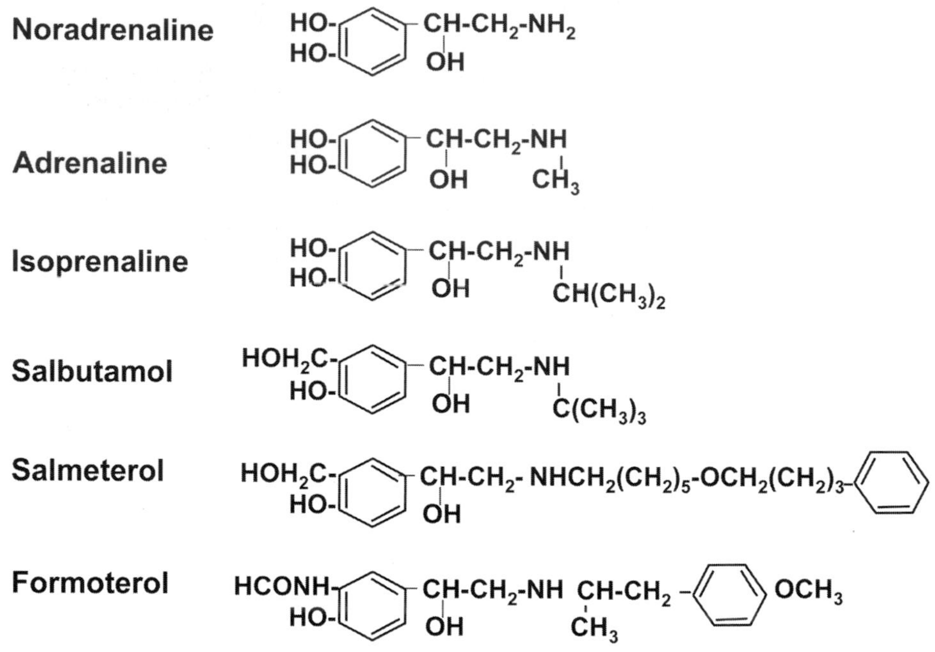

FIGURE 1. CHEMICAL STRUCTURE OF SOME ADRENERGIC AGONISTS SHOWING DEVELOPMENT FROM CATECHOLAMINES

effect. Catecholamines are also broken down by monoamine oxidase (MAO) in sympathetic nerve terminals and in the gastrointestinal tract which cleaves the side chain. Isoprenaline, which is a substrate for MAO, is therefore metabolised in the gut, making absorption variable. Substitution in the amine group confers resistance to MAO and ensures reliable absorption. Many other β_2-selective agonists have now been introduced and, while there may be differences in potency, there are no clinically significant differences in selectivity. Inhaled β_2-selective drugs in current clinical use (apart from rimiterol which is broken down by COMT) have a similar duration of action of 3–6 h. The inhaled long-acting inhaled β_2-agonists salmeterol and formoterol have a much longer duration of effect, providing bronchodilation and bronchoprotection for over 12 h [3]. Formoterol has a bulky substitution in the aliphatic chain and has a high lipophilicity which keeps the drug within the membrane close to the receptor. Salmeterol has a long aliphatic chain and its long duration may be due to binding within the receptor binding cleft ("exosite") that anchors the drug in the binding cleft [4].

Mode of action

β-Agonists produce bronchodilation by directly stimulating β_2-receptors in airway smooth muscle, which leads to relaxation [5]. This can be demonstrated *in vitro* by the relaxant effect of isoprenaline on human bronchi and lung strips (indicating an effect on peripheral airways) and *in vivo* by a rapid decrease in airway resistance. β-Receptors have been demonstrated in airway smooth muscle by direct receptor binding techniques and autoradiographic studies

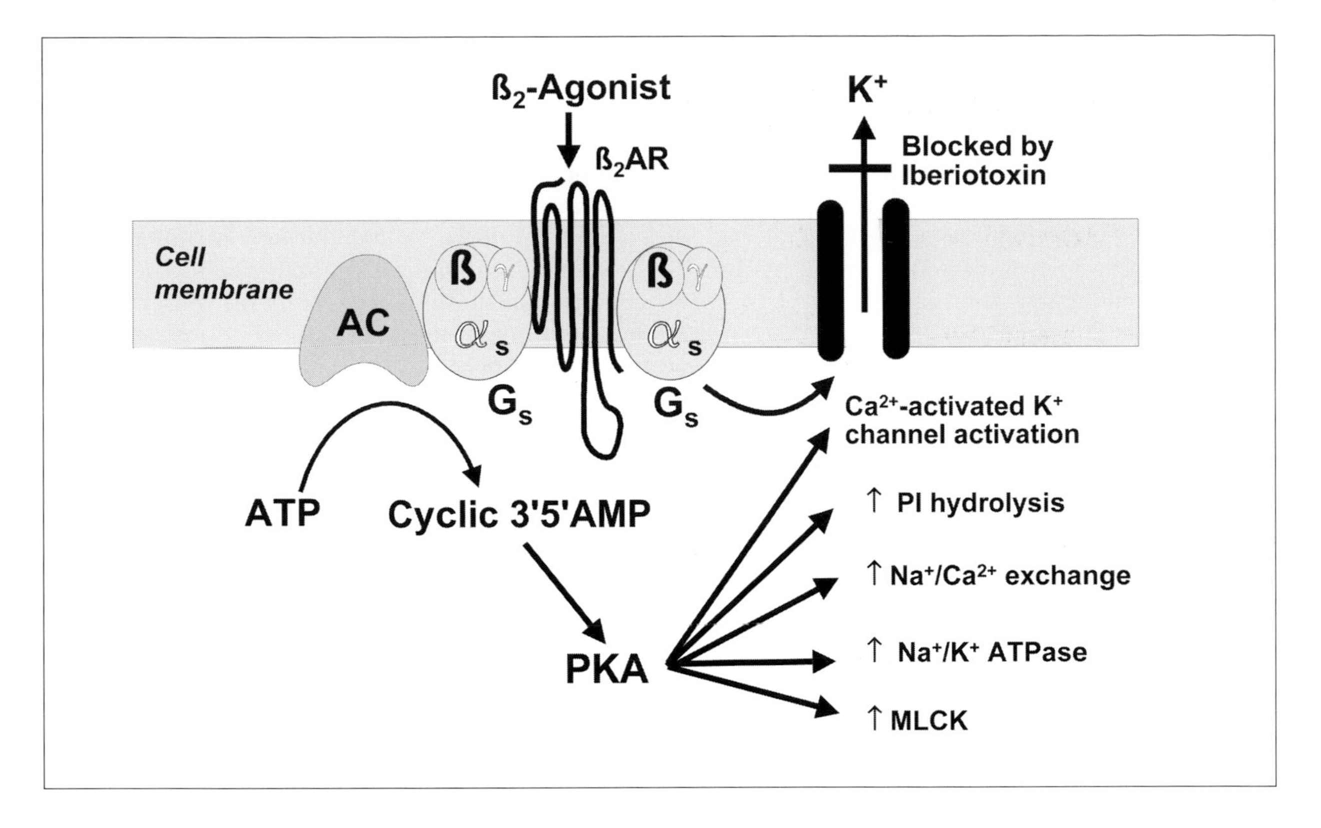

Figure 2. Molecular mechanism of action of β_2-agonists on airway smooth muscle cells
Activation of β_2-receptors (β_2AR) results in activation of adenylyl cyclase (AC) via a stimulatory G-protein (G_s) and increase in cyclic 3'5' adenosine monophosphate (AMP). This activates protein kinase A which then phosphorylates several target proteins which result in opening of calcium-activated potassium channels (K_{Ca}) or maxi-K channels, decreased phosphoinositide (PI) hydrolysis, increased sodium/calcium ion (Na^+/Ca^{2+}) exchange, increased Na^+/K^+ ATPase and decreased myosin light chain kinase (MLCK) activity. In addition, β_2-receptors may be coupled directly via G_s to K_{Ca}. ATP, adenosine triphosphate.

indicate that β-receptors are localized to smooth muscle of all airways from trachea to terminal bronchioles. The molecular mechanisms by which β-agonists induce relaxation of airway smooth muscle have been extensively investigated.

Occupation of β_2-receptors by agonists results in the activation of adenylyl cyclase via the stimulatory G-protein (G_s). This increases intracellular cyclic adenosine 3',5'-monophosphate (cAMP), leading to activation of a specific kinase (protein kinase A) which phosphorylates several target proteins within the cell, leading to relaxation (Fig. 2). These processes include:

- Lowering of intracellular calcium ion (Ca^{2+}) concentration by active removal of Ca^{2+} from the cell and into intracellular stores
- An inhibitory effect on phosphoinositide hydrolysis
- Inhibition of myosin light chain kinase
- Activation of myosin light chain phosphatase
- Opening of a large conductance calcium-activated potassium channel (K_{Ca}) which repolarizes the smooth muscle cell and may stimulate the sequestration of Ca^{2+} into intracellular stores [6]. β_2-Receptors are also directly coupled to K_{Ca} via G_s so that relaxation of airway smooth muscle may occur independently of an increase in cAMP.

TABLE 1. EFFECTS OF β-ADRENERGIC AGONISTS ON AIRWAYS

- Relaxation of airway smooth muscle (proximal and distal airways)
- Inhibition of mast-cell mediator release
- Inhibition of plasma exudation and airway oedema
- Increased mucociliary clearance
- Increased mucus secretion
- Decreased cholinergic neurotransmission
- Decreased cough
- No effect on chronic inflammation

Recently it has been recognized that several actions of β_2-agonists are not mediated via PKA and that there are other cAMP-regulated proteins [7].

β_2-Agonists act as functional antagonists and reverse bronchoconstriction irrespective of the contractile agent. This is an important property in asthma, since many bronchoconstrictor mechanisms (neurotransmitters and mediators) are likely to be contributory in asthma. In COPD the major MECHANISM OF ACTION is likely to be reduction of cholinergic neural bronchoconstriction.

β_2-Agonists may have additional effects on airways, and β-receptors are localized to several different airway cells [8] (Tab. 1). β_2-Agonists may therefore cause bronchodilation by a direct action on airways smooth muscle, but also indirectly by inhibiting the release of bronchoconstrictor mediators from inflammatory cells and of bronchoconstrictor neurotransmitters from airways nerves (Fig. 3). These additional effects include:

- Prevention of mediator release from isolated human lung mast cells (via β_2-receptors) [9].
- Prevention of MICROVASCULAR LEAKAGE and thus the development of bronchial mucosal oedema after exposure to mediators such as histamine and leukotriene D_4.
- Increase in mucus secretion from submucosal glands and ion transport across airway epithelium; these effects may enhance mucociliary CLEARANCE, and therefore reverse the defect in CLEARANCE found in asthma. β_2-Agonists appear to selectively stimulate mucous rather than serous cells, which may result in a more viscous mucus secretion, although the clinical significance of this is uncertain.
- Reduction in neurotransmission in human airway cholinergic nerves by an action at pre-junctional β_2-receptors to inhibit acetylcholine release. This may contribute to their bronchodilator effect by reducing cholinergic reflex bronchoconstriction. In animal studies β_2-receptors on sensory nerves inhibit the release of bronchoconstrictor and inflammatory peptides, such as substance P.

Although these additional effects of β_2-agonists may be relevant to the prophylactic use of these drugs against various challenges, their rapid bronchodilator action can probably be attributed to a direct effect on airway smooth muscle.

Anti-inflammatory effects

Whether β_2-agonists have anti-inflammatory effects in ASTHMA is an important issue, in view of their increasing use and the introduction of long-acting inhaled β_2-agonists. The inhibitory effects of β_2-agonists on mast-cell mediator release and MICROVASCULAR LEAKAGE are clearly anti-inflammatory, suggesting that β_2-agonists may modify acute INFLAMMATION. However β_2-agonists do not appear to have a significant inhibitory effect on the chronic INFLAMMATION of asthmatic airways, which is suppressed by corticosteroids. This has now been confirmed by several biopsy and bronchoalveolar lavage studies in patients with ASTHMA who are taking regular β_2-agonists (including long-acting inhaled β_2-agonists), that demonstrate no significant reduction in the number or activation in inflammatory cells in the airways, in contrast to resolution of the INFLAMMATION which occurs with inhaled corticosteroids [10]. This is likely to be related to the fact that β_2-agonists do not have a prolonged inhibitory effect on MACROPHAGES, EOSINOPHILS or LYMPHOCYTES and the low density of β_2-receptors on these cells is rapidly down-regulated.

Clinical use

Inhaled β_2-agonists are the most widely used and effective bronchodilators in the treatment of ASTHMA.

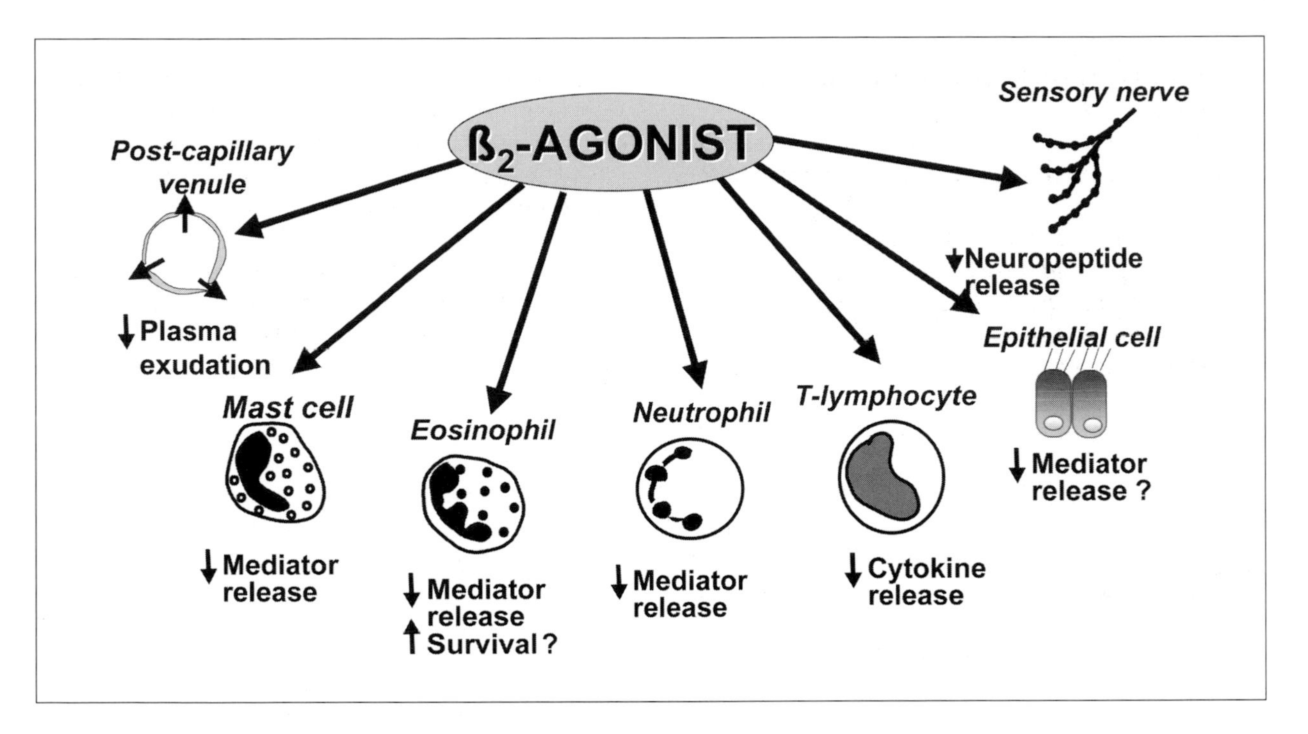

FIGURE 3. ANTI-INFLAMMATORY EFFECTS OF β_2-AGONISTS
β_2-Agonists relax airway smooth muscle cells directly, but may have several non-bronchodilator actions.

When inhaled from metered dose aerosols they are convenient, easy to use, rapid in onset and without significant side-effects. In addition to an acute bronchodilator effect, they are effective in protecting against various challenges, such as exercise, cold air and allergen. They are the bronchodilators of choice in treating acute severe ASTHMA, when the nebulized route of administration is as effective as and easier and safer than intravenous use [11]. The inhaled route of administration is preferable to the oral route because side-effects are less, and also because it may be more effective. Short-acting inhaled β_2-agonists, such as salbutamol and terbutaline, should be used "as required" by symptoms and not on a regular basis in the treatment of mild ASTHMA, as increased usage serves as an indicator for the need for more anti-inflammatory therapy.

Oral β_2-agonists are indicated as an additional bronchodilator. Slow-release preparations (such as slow-release salbutamol and bambuterol) may be indicated in nocturnal ASTHMA, but are less useful than inhaled β-agonists because of an increased risk of side-effects. The once daily β_2-agonist bambuterol (a prodrug that slowly releases terbutaline) is as effective as inhaled salmeterol as an add-on therapy, although systemic side-effects are more frequent [12].

Therapeutic choices

Several β_2-selective agonists are now available. These drugs are as effective as non-selective agonists in their bronchodilator action, since beneficial airway effects are mediated only by β_2-receptors. However, they are less likely to produce cardiac stimulation than isoprenaline because β_1-receptors are stimulated relatively less. With the exception of rimiterol (which retains the catechol ring structure and is therefore susceptible to rapid enzymatic degradation), they have a longer duration of action because they are resistant to uptake and enzymatic degradation by COMT and MAO. There is little to choose between the various short-acting β_2-agonists currently available; all are usable by inhalation and orally, have a similar duration of action (usually 3–4 h but less in severe ASTHMA) and similar side-effects. Differences in β_2-selectivity have been claimed but are not clinically important. Drugs in clinical use include salbutamol, terbutaline, fenoterol, tulobuterol, rimiterol and pirbuterol. It was claimed that fenoterol is less β_2-selective than salbutamol and terbutaline, resulting in increased cardiovascular side-effects, but this evidence is controversial, since all of these effects are mediated via β_2-receptors. The increased incidence of cardiovascular effects is more likely to be related to the greater effective dose of fenoterol which is used and perhaps to more rapid absorption into the circulation.

Salbutamol is a racemic mixture of active *R*- and inactive *S*-isomers. Animal studies have suggested that the *S*-isomer may increase airway responsiveness, providing a rationale for the development of *R*-salbutamol (levasalbutamol) [13]. Although the *R*-isomer is more potent than racemic *RS*-salbutamol in some studies, careful dose-responses show no advantage in terms of EFFICACY and no evidence that the *S*-salbutamol is detrimental in asthmatic patients [14]. As levasalbutamol is considerably more expensive than normally used racemic salbutamol, this therapy cannot be recommended.

Long-acting inhaled β_2-agonists

The long-acting inhaled β_2-agonists (LABA) salmeterol and formoterol have proved to be a major advance in ASTHMA therapy. Both drugs have a bronchodilator action of >12 h and also protect against bronchoconstriction for a similar period [3]. They are useful in treating nocturnal ASTHMA. Both improve ASTHMA control (when given twice daily) compared with regular treatment with short-acting β_2-agonists four times daily [15]. Both drugs are well tolerated. TOLERANCE to the bronchodilator effect of formoterol and the bronchoprotective effects of formoterol and salmeterol have been demonstrated, but this is not a loss of protection, does not appear to be progressive and is of doubtful clinical significance [3]. While both drugs have a similar duration of effect in clinical studies, there are some differences. Formoterol has a more rapid onset of action and is a full agonist, whereas salmeterol is a partial agonist. This might confer a theoretical advantage in more severe ASTHMA, whereas it may also make it more likely to induce TOLERANCE. However no difference between salmeterol and formoterol was found in the treatment of patients with severe ASTHMA [16].

Recent studies have suggested that inhaled LABA might be introduced earlier in therapy. In asthmatic patients not controlled on 400 or 800 μg inhaled corticosteroids, addition of salmeterol gives better control of ASTHMA than increasing the dose of inhaled steroid and also reduces exacerbations of ASTHMA [17–19]. This suggests that LABA may be added to low-dose inhaled steroids if ASTHMA is not controlled in preference to the previous recommendation of increasing the dose of inhaled corticosteroids. LABA have also been shown to be beneficial in patients with COPD, improving symptoms and health status [20].

Formoterol, but not salmeterol, can also be used as required for symptom control, taking advantage of a more prolonged action than salbutamol and the flexibility of dosing that is not possible with salmeterol. This may improve ASTHMA control, but it is more expensive than using short-acting inhaled β_2-agonists as needed [21]. At present it is recommended that LABA should only be used in ASTHMA patients who are also prescribed inhaled steroids. In the future long-acting inhaled β_2-agonists may be used in fixed combination inhalers (salmeterol + fluticasone, formoterol + budesonide) in order to improve compliance and reduce the risk of patients using these drugs as sole long-term treatment.

Combination inhalers

Combination inhalers that contain a LABA and a corticosteroid have now been introduced and appear to be the most effective therapies currently available for controlling ASTHMA [22]. There is a strong scientific rationale for combining a LABA with a corticosteroid in ASTHMA as these treatments have complementary actions and may also interact positively with the corticosteroids, enhancing the effect of the LABA and the LABA potentiating the effect of the corticosteroid [23]. The combination inhaler is more convenient for the patients, simplifies therapy and improves compliance with inhaled corticosteroids, but there may be an additional advantage as delivering the two drugs

TABLE 2. SIDE-EFFECTS OF β_2-AGONISTS

- Muscle tremor (direct effect on skeletal muscle β_2-receptors)
- Tachycardia (direct effect on atrial β_2-receptors, reflex effect from increased peripheral vasodilation via β_2-receptors)
- Hypokalaemia (direct effect on skeletal muscle uptake of K^+ via β_2-receptors)
- Restlessness
- Hypoxemia (increased V/Q mismatch due to pulmonary vasodilation)

in the same inhaler ensures that they are delivered to the same cells in the airways, allowing the beneficial molecular interactions between LABA and corticosteroids to occur [24]. It is likely that these inhalers will become the preferred therapy for all patients with persistent ASTHMA. These combination inhalers are also more effective in COPD patients than LABA and inhaled corticosteroids alone, but the mechanisms accounting for this beneficial interaction are less well understood than in patients with ASTHMA [25, 26].

Side-effects

Unwanted effects are dose-related and are due to stimulation of extra-pulmonary β-receptors (Tab. 2). Side-effects are not common with inhaled therapy, but more common with oral or intravenous administration.

- *Muscle tremor* is due to stimulation of β_2-receptors in skeletal muscle, and is the commonest side-effect. It may be more troublesome with elderly patients so is a greater problem in COPD patients.
- *Tachycardia and palpitations* are due to reflex cardiac stimulation secondary to peripheral vasodilation, from direct stimulation of atrial β_2-receptors (human heart is unusual in having a relatively high proportion of β_2-receptors), and possibly also from stimulation of myocardial β_1-receptors as the doses of β_2-agonist are increased. These side-effects tend to disappear with continued use of the drug, reflecting the development of TOLERANCE.
- *Metabolic* effects (increase in free fatty acid, insulin, glucose, pyruvate and lactate) are usually seen only after large systemic doses.
- *Hypokalaemia* is a potentially more serious side-effect. This is due to β_2-receptor stimulation of potassium entry into skeletal muscle, which may be secondary to a rise in insulin secretion. Hypokalaemia might be serious in the presence of hypoxia, as in acute ASTHMA, when there may be a predisposition to cardiac dysrrhythmias. In practice, significant arrhythmias after nebulized β_2-agonist have not been reported in acute ASTHMA, however.
- *Ventilation-perfusion(V/Q) mismatching* by causing pulmonary vasodilation in blood vessels previously constricted by hypoxia, resulting in the shunting of blood to poorly ventilated areas and a fall in arterial oxygen tension. Although in practice the effect of β_2-agonists on P_{AO_2} is usually very small (<5 mm Hg fall), occasionally in severe COPD it is large, although it may be prevented by giving additional inspired oxygen.

Tolerance

Continuous treatment with an agonist often leads to TOLERANCE (desensitisation, subsensitivity), which may be due to down-regulation of the receptor. For this reason there have been many studies of bronchial β-receptor function after prolonged therapy with β-agonists [27]. TOLERANCE of non-airway β_2-receptor mediated responses, such as tremor and cardiovascular and metabolic responses, is readily induced in normal and asthmatic subjects. TOLERANCE of human airway smooth muscle to β_2-agonists *in vitro* has been demonstrated, although the concentration of agonist necessary is high and the degree of desensitization is variable. Animal studies suggest that airway smooth muscle β_2-receptors may be more resistant to desensitization than β_2-receptors elsewhere due to a high receptor reserve. In normal subjects bronchodilator TOLERANCE has been demonstrated in some studies after high-dose inhaled salbutamol, but not in others. In asthmatic patients TOLERANCE to the bronchodilator effects of β_2-agonists has not usually been found. However, TOLERANCE develops to the bronchoprotective effects of β_2-agonists and this is more marked with indirect constrictors such as adenosine, allergen and exercise (that activate mast cells) than with direct constrictors such as histamine and methacholine [28, 29]. The reason for the relative resistance of airway smooth muscle β_2-receptors to desensitization remains uncertain, but may reflect the fact that there is a large receptor reserve, so that >90% of β_2-receptors may be lost without any reduction in the relaxation response. The high level of β_2-receptor gene expression in airway smooth muscle compared to peripheral lung [30] may also contribute to the resistance to TOLERANCE since there is likely to be a high rate of β-receptor synthesis. Another possibility is that the expression of the enzyme β-adrenergic receptor kinase (βARK) that phosphorylates and inactivates the occupied β_2-receptor is very low in airway smooth muscle [31]. By contrast there is no receptor reserve in inflammatory cells and βARK expression is high, so that indirect effects of β_2-agonists are more readily lost. TOLERANCE to the bronchodilator effects of formoterol has been reported, however, possibly reflecting the fact that it is a full agonist [32].

Experimental studies have shown that corticosteroids prevent the development of TOLERANCE in airway smooth muscle, and prevent and reverse the fall in pulmonary β-receptor density [33]. However, inhaled corticosteroids do not appear to prevent the TOLERANCE to the bronchoprotective effect of inhaled β_2-agonists, possibly because they do not reach airway smooth muscle in a high enough concentration [34, 35].

Long-term safety

Because of a possible relationship between adrenergic drug therapy and the rise in asthma deaths in several countries during the early 1960s, doubts were cast on the safety of β-agonists. A causal relationship between β-agonist use and mortality has never been firmly established, although in retrospective studies this would not be possible. A particular β_2-agonist, fenoterol, was linked to the recent rise in asthma deaths in New Zealand since significantly more of the fatal cases were prescribed fenoterol than the case-matched control patients. This association was strengthened by two subsequent studies and since fenoterol has not been available the asthma mortality has fallen dramatically [36]. An epidemiological study based in Saskatchewan, Canada, examined the links between drugs prescribed for ASTHMA and death or near death from asthma attacks, based on computerized records of prescriptions. There was a marked increase in the risk of death with high doses of all inhaled β_2-agonists [37]. The risk was greater with fenoterol, but when the dose is adjusted to the equivalent dose for salbutamol there is no significant difference in the risk for these two drugs. The link between high β_2-agonist usage and increased asthma mortality does not prove a causal association, since patients with more severe and poorly controlled ASTHMA, and who are therefore more likely to have an increased risk of fatal attacks, are more likely to be using higher doses of β_2-agonist inhalers and less likely to be using effective anti-inflammatory treatment. Indeed in the patients who used regular inhaled steroids there was a significant reduction in risk of death [38].

Regular use of inhaled β_2-agonists has also been linked to increased asthma morbidity. In a controversial study from New Zealand the regular use of fenoterol was associated with poorer control and a small increase in airway hyperresponsiveness compared with patients using fenoterol 'on demand' for symptom control over a 6-month period [39]. However this was not found in studies with salbutamol [40, 41]. There is some evidence that regular inhaled β_2-agonists may increase allergen-induced ASTHMA and sputum eosinophilia [42, 43]. One possible mechanism is that β-agonists may inhibit the anti-inflammatory action of glucocorticoids [44]. Another mechanism is that β_2-agonists activate phospholipase C via coupling through G_q, resulting in augmentation of the bronchoconstrictor responses to cholinergic agonists and mediators [45].

While it is unlikely that normally recommended doses of β_2-agonists worsen ASTHMA, it is possible that this could occur with larger doses. Furthermore, some patients may be more susceptible if they have polymorphic forms of the β_2-receptor that more rapidly down-regulate [46]. Short-acting inhaled β_2-agonists should only be used 'on demand' for symptom control and if they are required frequently (more than three times weekly) then an inhaled corticosteroid is needed. There is an association between increased risk of death from ASTHMA and the use of high doses of inhaled β_2-agonists; while this may reflect severity, it is also possible that high doses of β_2-agonists may have a deleterious effect on ASTHMA. High concentrations of β_2-agonists interfere with the anti-inflammatory action of steroids [47]. Patients on high doses of β_2-agonists (>1 canister per month) should be treated with inhaled corticosteroids and attempts should be made to reduce the daily dose of inhaled β_2-agonist. All patients with given LABA should also have corticosteroids and are best treated with a combination inhaler.

Future trends

β_2-Agonists will continue to be the bronchodilators of choice for the foreseeable future, as they are effective in all patients and have few or no side-effects when used in low doses. It would be very difficult to find a bronchodilator that improves on the EFFICACY and safety of inhaled β_2-agonists. Although some concerns have been expressed about the long-term effects of inhaled β_2-agonists, the evidence suggests that when used as required for symptom control, inhaled β_2-agonists are safe. In patients who are using large doses, their ASTHMA should be assessed and appropriate anti-inflammatory treatment given and attempts should be made to reduce the dose. LABA are very useful for long-term control in ASTHMA and COPD. In the future formoterol may also be useful for treatment of acute exacerbations. There is little advantage to be gained by improving β_2-receptor selectivity, since most of the side-effects of β-agonists are due to β_2-receptor stimulation (muscle tremor, tachycardia, hypokalaemia). Several once-daily inhaled β_2-agonists are now in development and are likely to replace salmeterol and formoterol in the future. There is now increasing use of combination inhalers and it is likely that these will become standard therapy for all patients with persistent ASTHMA. A combination of once-daily LABA and corticosteroid is likely to be developed.

Theophylline

Methylxanthines such as theophylline, which are related to caffeine, have been used in the treatment of ASTHMA since 1930. Indeed, theophylline is still the most widely used anti-asthma therapy world-wide because it is inexpensive. Theophylline became more useful with the availability of rapid plasma assays and the introduction of reliable slow-release preparations [48]. However, the frequency of side-effects and the relative low EFFICACY of theophylline have recently led to reduced usage, since inhaled β_2-agonists are far more effective as bronchodilators and inhaled steroids have a greater anti-inflammatory effect. In patients with severe ASTHMA and in COPD it still remains a very useful drug, however. There is increasing evidence that theophylline has an anti-inflammatory or IMMUNOMODULATORY effect [49].

Chemistry

Theophylline is a methylxanthine similar in structure to the common dietary xanthines caffeine and theobromine. Several substituted derivatives have been synthesised but none has any advantage over theophylline, apart from the 3-propyl derivative, enprofylline, which is more potent as a bronchodilator and may have fewer toxic effects. Many salts of theophylline have also been marketed, the most common being aminophylline, which is the ethylenediamine salt used to increase solubility at neutral pH. Other salts, such as choline theophyllinate, do not have any advantage and others, such as acepifylline, are virtually inactive, so that theophylline remains the only methylxanthine in clinical use.

Mode of action

Although theophylline has been in clinical use for more than 70 years, its MECHANISM OF ACTION is still uncertain. In addition to its bronchodilator action, theophylline has many other actions that may be rel-

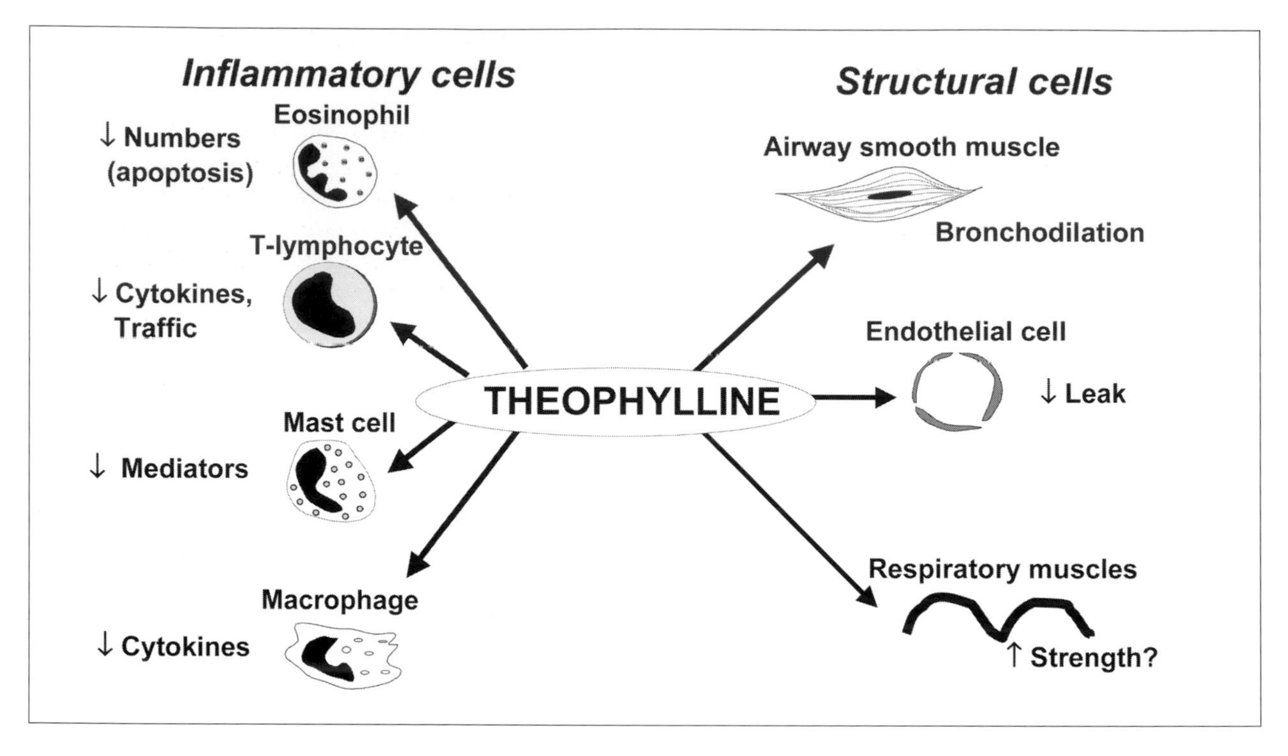

FIGURE 4. THEOPHYLLINE HAS EFFECTS ON SEVERAL OTHER CELLS IN ADDITION TO AIRWAYS SMOOTH MUSCLE
Some of these effects are mediated via inhibition of phosphodiesterases (PDE).

evant to its anti-asthma effect (Fig. 4). Many of these effects are seen only at high concentrations that far exceed the therapeutic range.

Non-bronchodilator effects. Theophylline proved to have clinical benefit in ASTHMA and COPD at doses that give plasma concentrations <10mg/L, so these are unlikely to be explained by a bronchodilator effect. There is increasing evidence that theophylline has anti-inflammatory effects in ASTHMA [50]. Chronic oral treatment with theophylline inhibits the late response to inhaled allergen [51] and a reduced infiltration of EOSINOPHILS and $CD4^+$ LYMPHOCYTES into the airways after allergen challenge [52, 53]. In patients with mild ASTHMA, low doses of theophylline (mean plasma concentration ~5 mg/L) reduce the numbers of EOSINOPHILS in bronchial biopsies, bronchoalveolar lavage and induced sputum [54], whereas in severe ASTHMA, withdrawal of theophylline results in increased numbers of activated $CD4^+$ cells and EOSINOPHILS in bronchial biopsies [55]. In patients with COPD, theophylline reduces the total number and proportion of NEUTROPHILS in induced sputum, the concentration of IL-8 and NEUTROPHIL chemotactic responses, suggesting that it may have an anti-inflammatory effect [56].

Several MECHANISMS OF ACTION have been proposed for theophylline (Tab. 3).

Inhibition of phosphodiesterases. Phosphodiesterases (PDE) break down cyclic nucleotides in the cell, thereby leading to an increase in intracellular cAMP and cyclic guanosine 3'5 monophosphate (cGMP) concentrations (Fig. 5). Theophylline is a non-selective PDE inhibitor, but the degree of inhibition is minor at concentrations of theophylline which are within the "therapeutic range". PDE inhibition almost certainly accounts for the bronchodilator action of theophylline [57], but this is unlikely to account for

TABLE 3. MECHANISMS OF ACTION OF THEOPHYLLINE

- Phosphodiesterase inhibition (non-selective)
- Adenosine receptor antagonism (A_1, A_{2A}, A_{2B}-receptors)
- Increased interleukin-10 release
- Stimulation of catecholamine (adrenaline) release
- Mediator inhibition (prostaglandins, tumour necrosis factor-α)
- Inhibition of intracellular calcium release
- Inhibition of nuclear factor-κB (↓ nuclear translocation)
- Increased apoptosis
- ↑ Histone deacetylase activity (↑ efficacy of corticosteroids)

the non-bronchodilator effects of theophylline that are seen at sub-bronchodilator doses. Inhibition of PDE should lead to synergistic interaction with β-agonists, but this has not been convincingly demonstrated *in vivo*. Several isoenzyme families of PDE have now been recognized and some are more important in smooth muscle relaxation, including PDE3, PDE4 and PDE5 [58].

Adenosine receptor antagonism. Theophylline inhibits adenosine receptors at therapeutic concentrations. Of particular importance may be the adenosine A_{2B} receptor on mast cells which is activated by adenosine in asthmatic patients [59]. *In vitro* adenosine has little direct effect on human airway smooth muscle, but causes bronchoconstriction in airways from asthmatic patients by releasing histamine and LEUKOTRIENES [60]. Adenosine antagonism is unlikely to account for the anti-inflammatory effects of theophylline but may be responsible for serious side-effects, including cardiac arrhythmias and seizures.

Interleukin-10 release. IL-10 has a broad spectrum of anti-inflammatory effects and there is evidence that its secretion is reduced in ASTHMA [61]. IL-10 release is increased by theophylline and this effect may be mediated via PDE inhibition [62], although this has not been seen at the low doses that are effective in ASTHMA [63].

Effects on gene transcription. Theophylline prevents the translocation of the proinflammatory transcription factor NUCLEAR FACTOR κB (NF-κB) into the nucleus, thus potentially reducing the expression of inflammatory genes in ASTHMA and COPD [64]. Inhibition of NF-κB appears to be due to a protective effect against the degradation of the inhibitory protein I-κBα, so that nuclear translocation of activated NF-κB is prevented [65]. However, these effects are seen at high concentrations and may be mediated by inhibition of PDE.

Effects on apoptosis. Prolonged survival of granulocytes due to a reduction in APOPTOSIS may be important in perpetuating chronic INFLAMMATION in ASTHMA (EOSINOPHILS) and COPD (NEUTROPHILS). Theophylline promotes inhibits APOPTOSIS in EOSINOPHILS and NEUTROPHILS *in vitro* [66]. This is associated with a reduction in the anti-apoptotic protein Bcl-2 [67]. This effect is not mediated via PDE inhibition, but in NEUTROPHILS may be mediated by antagonism of adenosine A_{2A}-receptors [68]. Theophylline also induces APOPTOSIS of T LYMPHOCYTES, thus reducing their survival and this effect appears to be mediated via PDE inhibition [69].

Other effects. Several other effects of theophylline have been described, including an increase in circulating catecholamines, inhibition of calcium influx into inflammatory cells, inhibition of prostaglandin effects, and antagonism of TUMOR NECROSIS FACTOR-α. These effects are generally seen only at high concentrations of theophylline that are above the therapeutic range in ASTHMA and are therefore unlikely to contribute to the anti-inflammatory actions of theophylline.

Histone deacetylase activation. Recruitment of histone deacetylase-2 (HDAC2) by glucocorticoid receptors switches off inflammatory genes (see below). Theophylline is an activator of HDAC at therapeutic concentrations, thus enhancing the anti-inflammatory effects of corticosteroids (Fig. 6) [70]. This mechanism is independent of PDE inhibition or

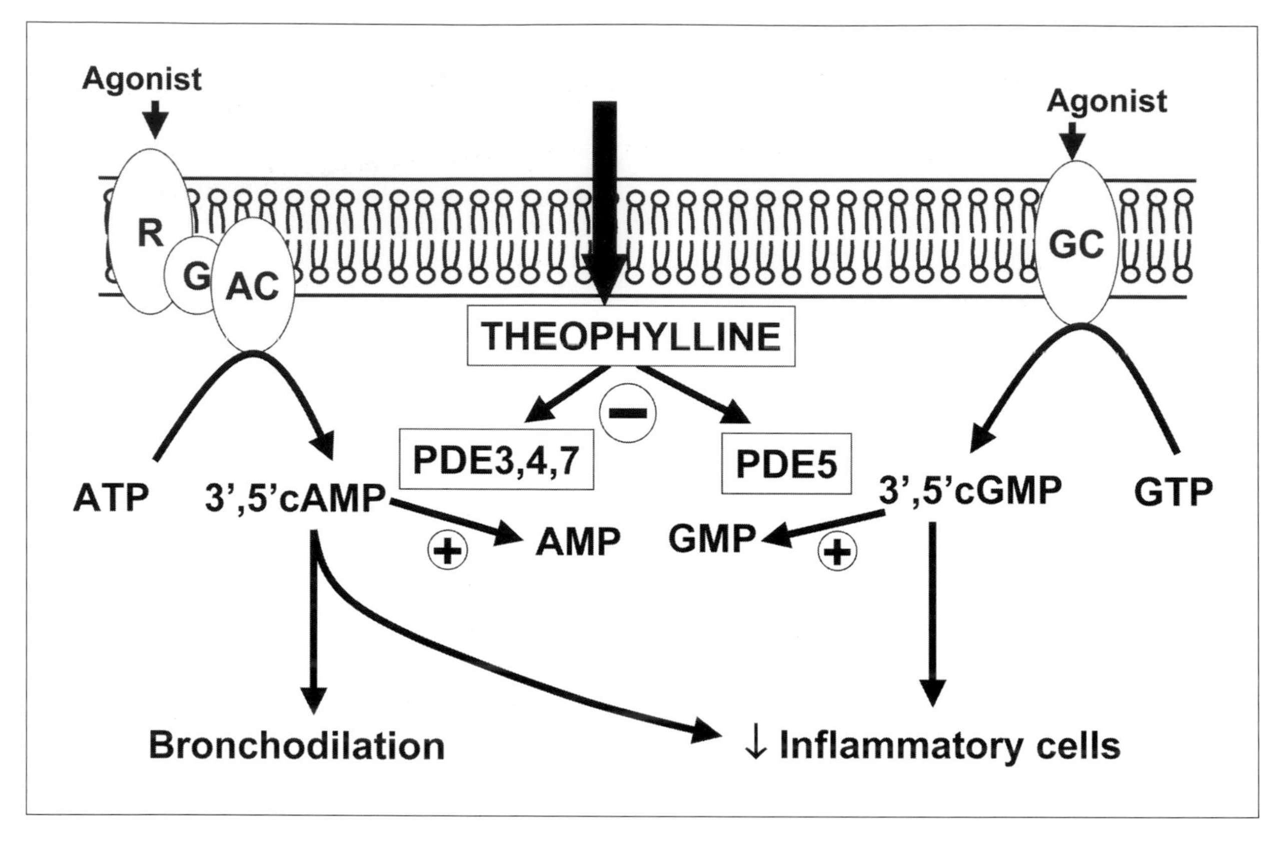

FIGURE 5
The inhibitory effect of theophylline on phosphodiesterases (PDE) may result in bronchodilation and inhibition of inflammatory cells. ATP, adenosine triphosphate; AMP, adenosine monophosphate; PKA, protein kinase A; GTP, guanosine triphosphate; GMP, guanosine monophosphate; PKG, protein kinase G.

adenosine antagonism. The anti-inflammatory effects of theophylline are inhibited by a HDAC inhibitor trichostatin A. Low doses of theophylline increase HDAC activity in bronchial biopsies of asthmatic patients and correlate with the reduction in eosinophil numbers in the biopsy.

Pharmacokinetics

There is a close relationship between improvement in airway function and serum theophylline concentration. Below 10 mg/L therapeutic effects (at least in terms of bronchodilation) are small and above 25 mg/L additional benefits are outweighed by side-effects, so that the therapeutic range is usually taken as 10-20 mg/L. It is now clear that theophylline has anti-asthma effects other than bronchodilation and that these may be seen below 10mg/L. A more useful therapeutic range is 5-15 mg/L. The dose of theophylline required to give these therapeutic concentrations varies between subjects, largely because of differences in CLEARANCE. In addition, there may be differences in bronchodilator response to theophylline and, with acute bronchoconstriction, higher concentrations may be required to produce bronchodilation. Theophylline is rapidly and completely absorbed, but there are large inter-individual variations in CLEARANCE, due to differences in hepatic metabolism (Tab. 4). Theophylline is metabolized in the liver by the cytochrome P450 microsomal

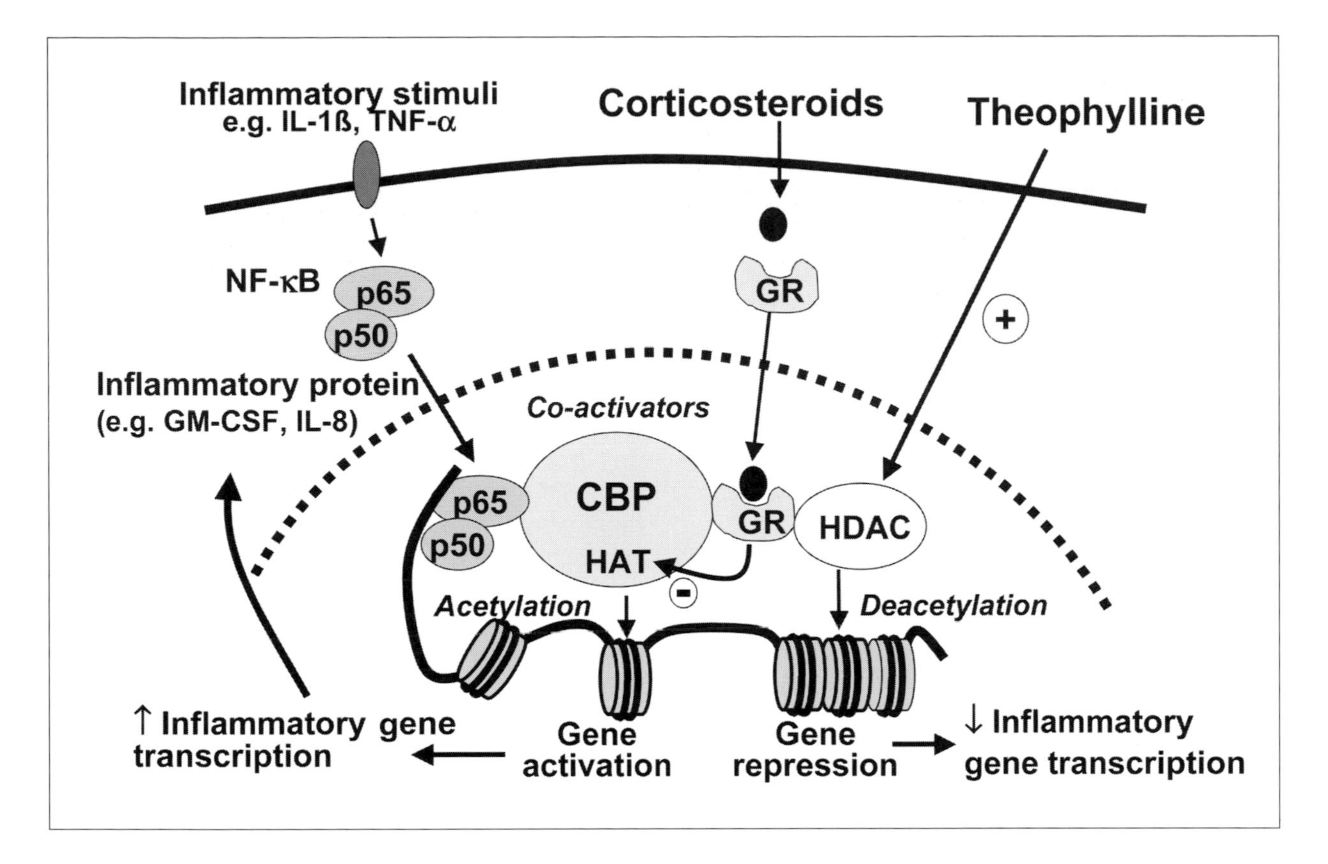

FIGURE 6

Theophylline directly activates histone deacetylases (HDACs) which deacetylate core histones that have been acetylated by the histone acetyltransferase (HAT) activity of co-activators, such as CREB-binding protein (CBP). This results in suppression of inflammatory genes and proteins, such as granulocyte-macrophage colony-stimulating factor (GM-CSF) and interleukin-8 (IL-8) that have been switched on by proinflammatory transcription factors, such as nuclear factor-κB (NF-κB). Corticosteroids also activate HDACs, but through a different mechanism resulting in the recruitment of HDACs to the activated transcriptional complex via activation of the glucocorticoid receptors (GR) which function as a molecular bridge. This predicts that theophylline and corticosteroids may have a synergistic effect in repressing inflammatory gene expression.

enzyme system (mainly CYP1A2), and a large number of factors may influence hepatic metabolism [71].

Increased clearance is seen in children (1–16 yrs), and in cigarette and marijuana smokers. Concurrent administration of phenytoin and phenobarbitone increases activity of P450, resulting in increased metabolic breakdown, so that higher doses may be required.

Reduced clearance is found in liver disease, pneumonia and heart failure and doses need to be reduced to half and plasma levels monitored carefully. Increased CLEARANCE is also seen with certain drugs, including erythromycin, certain quinolone antibiotics (ciprofloxacin, but not ofloxacin), allopurinol, cimetidine (but not ranitidine), fluoxamine and zafirlukast, which interfere with cytochrome P450 function. Thus, if a patient on maintenance

Table 4. Factors affecting clearance of theophylline

Increased clearance
- Enzyme induction (rifampicin, phenobarbitone, ethanol)
- Smoking (tobacco, marijuana)
- High-protein, low-carbohydrate diet
- Barbecued meat
- Childhood

Decreased clearance
- Enzyme inhibition (cimetidine, erythromycin, ciprofloxacin, allopurinol, zileuton, zafirlukast)
- Congestive heart failure
- Liver disease
- Pneumonia
- Viral infection and vaccination
- High-carbohydrate diet
- Old age

theophylline requires a course of erythromycin, the dose of theophylline should be halved. Viral infections and vaccination may also reduce CLEARANCE, and this may be particularly important in children. Because of these variations in CLEARANCE, individualization of theophylline dosage is required and plasma concentrations should be measured 4 h after the last dose with slow-release preparations when steady state has usually been achieved. There is no significant circadian variation in theophylline metabolism [72], although there may be delayed absorption at night, which may relate to the supine posture.

Routes of administration

Intravenous aminophylline has been used for many years in the treatment of acute severe ASTHMA. The recommended dose is now 6 mg/kg given intravenously over 20–30 min, followed by a maintenance dose of 0.5 mg/kg/h. If the patient is already taking theophylline, or there are any factors which decrease CLEARANCE, these doses should be halved and the plasma level checked more frequently.

Oral plain theophylline tablets or elixir, which are rapidly absorbed, give wide fluctuations in plasma levels and are not recommended. Several sustained-release preparations are now available which are absorbed at a constant rate and provide steady plasma concentrations over a 12–24-h period. Although there are differences between preparations, these are relatively minor and of no clinical significance. Both slow-release aminophylline and theophylline are available and are equally effective (although the ethylene diamine component of aminophylline has very occasionally been implicated in allergic reactions). For continuous treatment, twice-daily therapy (approximately 8 mg/kg twice daily) is needed, although some preparations are designed for once-daily administration. For nocturnal ASTHMA, a single dose of slow-release theophylline at night is often effective. Once optimal doses have been determined, plasma concentrations usually remain stable, providing no factors which alter CLEARANCE change.

Aminophylline may be given as a suppository, but rectal absorption is unreliable and proctitis may occur, so is best avoided. Inhalation of theophylline is irritant and ineffective. Intramuscular injections of theophylline are very painful and should never be given. Inhaled administration of theophylline is ineffective.

Clinical use

In patients with acute ASTHMA, intravenous aminophylline is less effective than nebulized β_2-agonists, and should therefore be reserved for those patients who fail to respond to β-agonists. Theophylline should not be added routinely to nebulized β_2-agonists since it does not increase the bronchodilator response and may only increase their side-effects [73].

Theophylline has little or no effect on bronchomotor tone in normal airways, but reverses bronchoconstriction in asthmatic patients, although it is less effective than inhaled β_2-agonists and is more likely to have unwanted effects. Indeed, any role of theophylline in the management of asthma has been questioned. There is good evidence that theophylline and β_2-agonists have additive effects, even if true synergy is not seen, and there is evidence that theophylline may provide an additional bronchodilator effect even when maximally effective doses of β_2-agonist have been given [74]. This means that, if adequate bronchodilation is not achieved by a β-agonist alone, theophylline may be added to the maintenance therapy with benefit. Addition of low-dose theophylline to either high- or low-dose inhaled corticosteroids in patients who are not adequately controlled provides better symptom control and lung function that doubling the dose of inhaled steroid [75–77], although it is less effective as an add-on therapy than a LABA [78]. Theophylline may be useful in patients with nocturnal ASTHMA, since slow-release preparations are able to provide therapeutic concentrations overnight although it is less effective than a LABA [79]. Studies have also documented steroid-sparing effects of theophylline [80]. Although theophylline is less effective than a β_2-agonist and corticosteroids, there are a minority of asthmatic patients who appear to derive unexpected benefit, and even patients on oral steroids may show a deterioration in lung function when theophylline is withdrawn [55, 81]. Theophylline has been used as a controller in the management of mild persistent ASTHMA [82], although it is usually found to be less effective than low doses of inhaled corticosteroids [83, 84]. Theophylline is currently a less preferred option than inhaled corticosteroids and is recommended as a second-line choice of controller in management of asthmatic patients at Step 2 [85]. Although LABA are more effective as an add-on therapy at Steps 3 and 4 of the GINA 2002 guidelines, theophylline is considerably less expensive and may be the only affordable add-on treatment when the costs of medication are limiting.

Theophylline is still used as a bronchodilator in COPD, but inhaled anticholinergics and β_2-agonists are preferred [86]. Theophylline tends to be added to these inhaled bronchodilators in more severe patients and has been shown to give additional clinical improvement when added to a long-acting β_2-agonist [87]. As in ASTHMA, patients with severe COPD deteriorate when theophylline is withdrawn from their treatment regime [88]. A theoretical advantage of theophylline is that its systemic administration may have effects on small airways, resulting in reduction of hyperinflation and thus a reduction in dyspnea [89].

TABLE 5. SIDE-EFFECTS OF THEOPHYLLINE

- Nausea and vomiting
- Headaches
- Gastric discomfort
- Diuresis
- Behavioural disturbance (?)
- Cardiac arrhythmias
- Epileptic seizures

Side-effects

Unwanted effects of theophylline are usually related to plasma concentration and tend to occur when plasma levels exceed 20 mg/L. However, some patients develop side-effects even at low plasma concentrations. To some extent side-effects may be reduced by gradually increasing the dose until therapeutic concentrations are achieved.

The commonest side-effects are headache, nausea and vomiting (due to inhibition of PDE4), abdominal discomfort and restlessness (Tab. 5). There may also be increased acid secretion and diuresis (due to inhibition of adenosine A_1-receptors). There was concern that theophylline, even at therapeutic concentrations, may lead to behavioural disturbance and learning difficulties in school children, although it is difficult to design adequate controls for such studies.

At high concentrations cardiac arrhythmias may occur as a consequence of PDE3 inhibition and adenosine A_1-receptor antagonism and at very high concentrations seizures may occur (due to central A_1-receptor antagonism). Use of low doses of theophylline which give plasma concentrations of 5–10 mg/L largely avoids side-effects and drug interactions and makes it unnecessary to monitor plasma concentrations (unless checking for compliance).

Future developments

Theophylline use has been declining, partly because of the problems with side-effects, but mainly because more effective therapy with inhaled corticosteroids has been introduced. Oral theophylline is still a very useful treatment in some patients with difficult ASTHMA and appears to have effects beyond those provided by steroids. Rapid-release theophylline preparations are cheap and are the only affordable anti-asthma medication in some developing countries. There is increasing evidence that theophylline has some anti-asthma effect at doses that are lower than those needed for bronchodilation and plasma levels of 5-10 mg/L are recommended, instead of the previously recommended 10-20 mg/L. Adding a low dose of theophylline gives better control of ASTHMA than doubling the dose of inhaled steroids in patients who are not controlled, and is a less expensive alternative add-on therapy than a LABA or anti-leukotriene.

Now that the molecular mechanisms for the anti-inflammatory effects of theophylline are better understood, there is a strong scientific rationale for combining low-dose theophylline with inhaled corticosteroids, particularly in patients with more severe ASTHMA. The synergistic effect of low-dose theophylline and corticosteroids on inflammatory gene expression may account for the add-on benefits of theophylline in ASTHMA [49, 90]. The potentiation of the anti-inflammatory actions of corticosteroids in ASTHMA may result in the use of lower doses of inhaled corticosteroids or even combined therapy with low-dose theophylline and a low dose of oral corticosteroids that does not have significant systemic side-effects. In COPD, low-dose theophylline is the first drug to demonstrate clear anti-inflammatory effects and thus it may even have a role in preventing progression of the disease. Furthermore, the reversal of the steroid resistance induced by oxidative stress suggests that theophylline may increase responsiveness to corticosteroids. This may mean that theophylline could "unlock" steroid resistance that is characteristic of COPD and allow corticosteroids to suppress the chronic INFLAMMATION.

Anticholinergics

Datura plants, which contain the muscarinic antagonist strammonium, were smoked for relief of ASTHMA two centuries ago. Atropine, a related naturally occurring compound, was also introduced for treating ASTHMA, but these compounds gave side-effects, particularly drying of secretions, so less soluble quaternary compounds, such as atropine methylnitrate and ipratropium bromide, were developed. These compounds are topically active and are not significantly absorbed from the respiratory tract or from the gastrointestinal tract.

Mode of action

Anticholinergics are specific antagonists of muscarinic receptors and, in therapeutic use, have no other significant pharmacological effects. In animals and man there is a small degree of resting bronchomotor tone which is probably due to tonic vagal nerve impulses which release acetylcholine in the vicinity of airway smooth muscle, since it can be blocked by anticholinergic drugs. Acetylcholine may also be released from other airway cells, including epithelial cells [91]. The synthesis of acetylcholine in epithelial cells is increased by inflammatory stimuli (such as TNF-α) which increases the expression of choline acetyltransferase and this could therefore contribute to cholinergic effects in airway diseases. Since muscarinic receptors are expressed in airway smooth muscle of small airways which do not appear to be innervated by cholinergic nerves, this might be important as a mechanism of cholinergic narrowing in peripheral airways that could be relevant in COPD (Fig. 7).

Cholinergic pathways may play an important role in regulating acute bronchomotor responses in animals and there are a wide variety of mechanical, chemical and immunological stimuli which elicit reflex bronchoconstriction via vagal pathways. This suggested that cholinergic mechanisms might underlie airway hyperresponsiveness and acute bronchoconstrictor responses in ASTHMA, with the implication that anticholinergic drugs would be effective bronchodilators. While these drugs may afford protection against acute challenge by sulphur dioxide, inert dusts, cold air and emotional factors, they are less effective against ANTIGEN challenge, exercise and fog. This is not surprising, as anticholinergic drugs will only inhibit reflex cholinergic bronchoconstriction and could have no significant blocking effect on

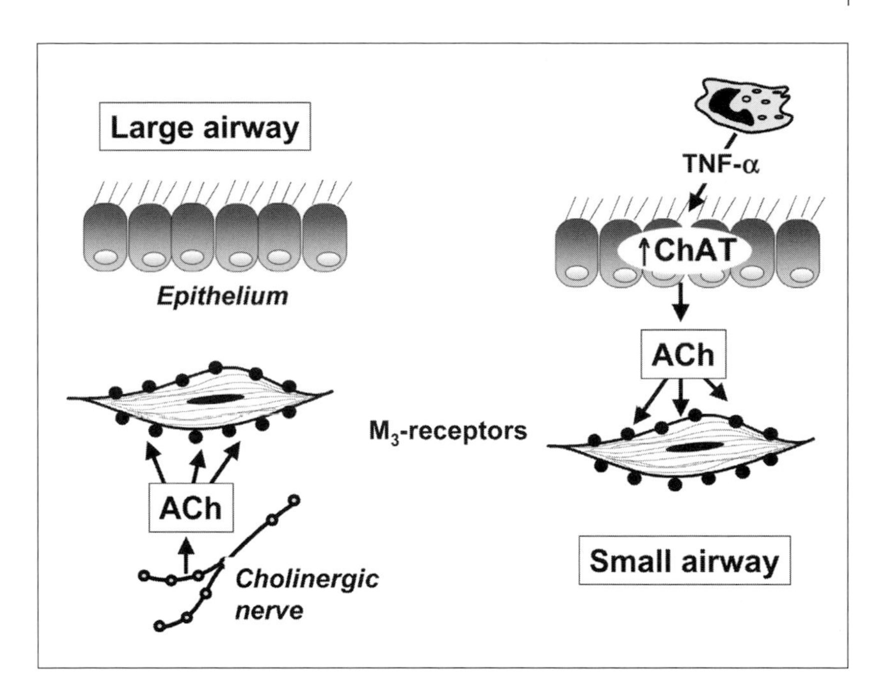

FIGURE 7
In proximal airways acetylcholine (ACh) is released from vagal parasympathetic nerves to activate M_3-receptors on airway smooth muscle cells. In peripheral airways M_3-receptors are expressed but there is no cholinergic innervation, but these may be activated by ACh released from epithelial cells that may express choline acetyltransferase (ChAT) in response to inflammatory stimuli, such as tumor necrosis factor-α (TNF-α).

the direct effects of inflammatory mediators such as histamine and LEUKOTRIENES on bronchial smooth muscle. Furthermore, cholinergic antagonists probably have little or no effect on mast cells, microvascular leak or the chronic inflammatory response.

Theoretically, anticholinergics may reduce airway mucus secretion and reduce mucus CLEARANCE, but this does not appear to happen in clinical studies. Oxitropium bromide in high doses has been shown to reduce mucus secretion in patients with COPD [92].

Clinical use

In asthmatic subjects anticholinergic drugs are less effective as bronchodilators than β-agonists and offer less efficient protection against various bronchial challenges. These drugs may be more effective in older patients with ASTHMA in whom there is an element of fixed airway obstruction. Anticholinergics are currently used as an additional bronchodilator in patients not controlled on a LABA. Nebulized anticholinergic drugs are effective in acute severe ASTHMA, although they are less effective than β_2-agonists in this situation [93]. Nevertheless, in the acute and chronic treatment of ASTHMA, anticholinergic drugs may have an additive effect with β_2-agonists and should therefore be considered when control of ASTHMA is not adequate with β_2-agonists, particularly if there are problems with theophylline, or inhaled β_2-agonists give troublesome tremor in elderly patients.

In COPD, anticholinergic drugs may be as effective as, or even superior to, β-agonists [94]. Their relatively greater effect in COPD than in ASTHMA may be explained by an inhibitory effect on vagal tone which, while not necessarily being increased in COPD, may be the only reversible element of airway obstruction which is exaggerated by geometric factors in a narrowed airway (Fig. 8).

Therapeutic choices

Ipratropium bromide is the most widely used anticholinergic inhaler and is available as a MDI and nebulized preparation. The onset of bronchodilation is relatively slow and is usually maximal 30-60 min after inhalation, but may persist for 6-8h. It is usually given by MDI four times daily on a regular basis,

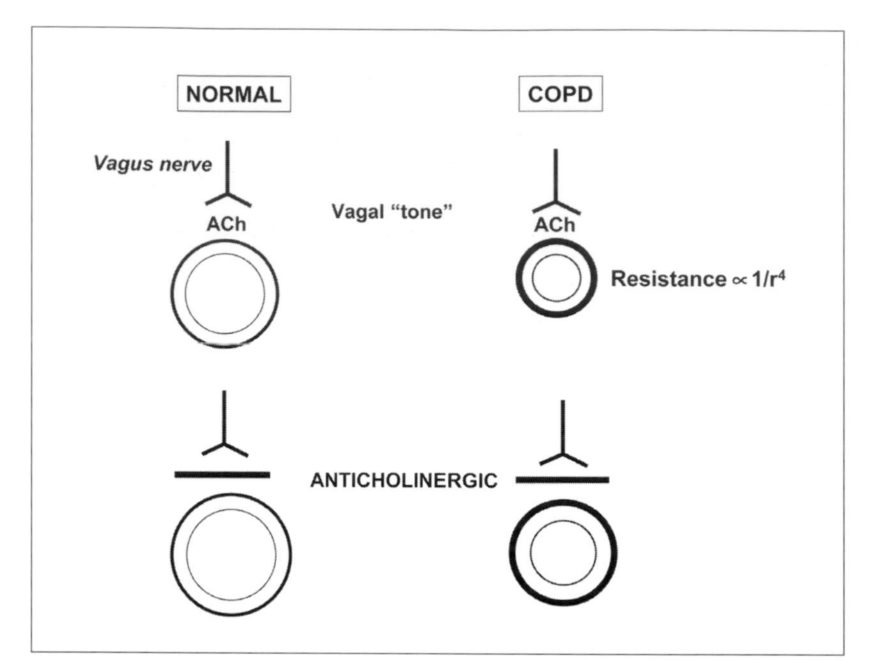

FIGURE 8
Anticholinergic drugs inhibit vagally-mediated airway tone leading to bronchodilation. This effect is small in normal airways but is greater in airways of patients with chronic obstructive pulmonary disease (COPD) which are structurally narrowed as airway resistance (R) is inversely related to the fourth power of the radius (r).

rather than intermittently for symptom relief, in view of its slow onset of action.

Oxitropium bromide is a quaternary anticholinergic bronchodilator which is similar to ipratropium bromide in terms of receptor blockade [95]. It is available in higher doses by inhalation and may therefore have a more prolonged effect. Thus, it may be useful in some patients with nocturnal ASTHMA [96].

Combination inhalers of an anticholinergic and β_2-agonist are popular, particularly in patients with COPD. Several studies have demonstrated additive effects of these two drugs, thus providing an advantage over increasing the dose of β_2-agonist in patients who have side-effects [97].

Tiotropium bromide is a new long-acting anticholinergic drug that is suitable for once-daily dosing [98]. It binds with equal affinity to all muscarinic receptor subtypes but dissociates very slowly from M_3 and M_1 receptors, giving it a degree of receptor selectivity [99]. It is an effective bronchodilator in patients with COPD and is more effective than ipratropium four times daily without any loss of EFFICACY over a one-year treatment period [100, 101]. Ipratropium is given as a dry powder inhaler once daily and is well tolerated, the only side-effect of note being transient dryness of the mouth in about 10% of patients. It is now the bronchodilator of choice in COPD patients.

Side-effects

Inhaled anticholinergic drugs are usually well tolerated and there is no evidence for any decline in responsiveness with continued use. On stopping inhaled anticholinergics a small rebound increase in responsiveness has been described [102], but the clinical relevance of this is uncertain. Atropine has side-effects which are dose-related and are due to cholinergic antagonism in other systems which may lead to dryness of the mouth, blurred vision and urinary retention. Systemic side-effects after ipratropium bromide and tiotropium bromide are very uncommon because there is virtually no systemic absorption. Because cholinergic agonists stimulate mucus secretion, there have been several studies of mucus secretion with anticholinergic drugs as there has been concern that they may reduce secretion and lead to more viscous mucus. Atropine reduces mucociliary CLEARANCE in normal subjects and in patients with ASTHMA and chronic bronchitis, but ipratropium bromide, even in high doses, has no detectable effect in either normal subjects or in patients with airway disease. A significant unwanted effect is the unpleasant bitter taste of inhaled ipratropium, which may contribute to poor compliance with this drug. Nebulized ipratropium bromide may precipitate glaucoma in elderly patients due to a direct effect of the nebulized drug on the eye. This may be prevented by nebulization with a mouthpiece rather than a face mask. Reports of paradoxical bronchoconstriction with ipratropium bromide, particularly when given by nebulizer, were largely explained by the hypotonicity of the nebulizer solution and by antibacterial additives, such as benzalkonium chloride and EDTA. This has not been described with tiotropium. Nebulizer solutions free of these problems are less likely to cause bronchoconstriction. Occasionally, bronchoconstriction may occur with ipratropium bromide given by MDI. It is possible that this is due to blockade of pre-junctional M_2-receptors on airway cholinergic nerves which normally inhibit acetylcholine release.

Future developments

Anticholinergics are the bronchodilators of choice in COPD and therefore have a large market. There has been a search for selective muscarinic antagonists. Ipratropium bromide and oxitropium bromide are non-selective and therefore, as well as blocking M_3-receptors in airway smooth muscle and M_1 receptors in ganglia, they also block prejunctional M_2-receptors and may therefore increase the release of acetylcholine from airway cholinergic nerves [103]. This suggests that selective M_3 or mixed M_1/M_3 antagonists may have an advantage over non-selective antagonists. However, it has been difficult to develop such drugs and antagonists such as darifenacin (M_3-selective) and revapropate do not have a high degree of selectivity [104]. Tiotropium bromide has a kinetic selectivity for M_3 and M_1 receptors, although it is not certain whether this contributes to its clinical EFFICACY. Several other long-acting anticholinergics are now in clinical development, including LAS

TABLE 6. NEW BRONCHODILATORS

- Long-acting inhaled β_2-agonists (once daily)
- Selective/long-acting muscarinic antagonists (e.g., tiotropium)
- Potassium channel activators (e.g., levcromakalim)
- Magnesium sulphate
- Nitrovasodilators
- Vasoactive intestinal peptide analogues
- Atrial natriuretic peptide and analogues (urodilatin)

34273 [105]. Combination inhalers of long-acting anticholinergic and long-acting β_2-agonist are also in development [106].

New bronchodilators

Several new classes of bronchodilator are under development (Tab. 6), but it is difficult to envisage a more effective bronchodilator than inhaled β_2-agonists for ASTHMA and the long-acting anticholinergic tiotropium bromide for COPD. It has been difficult to find new classes of bronchodilator and several new potential drugs have often had problems with vasodilator side-effects.

Magnesium sulphate

There is increasing evidence that magnesium sulphate is useful as an additional bronchodilator in patients with acute severe ASTHMA. Intravenous magnesium sulphate appears to benefit adults and children with severe exacerbations (FEV_1 <30% predicted), giving an improvement in lung function when added to nebulized β_2-agonist and a reduction in hospital admissions [107, 108]. The treatment is well tolerated and the side-effects include flushing and nausea but are usually minor. It appears to act as a bronchodilator and may reduce cytosolic calcium ion concentrations in airway smooth muscle cells. The concentration of magnesium is lower in serum and ERYTHROCYTES in asthmatic patients compared to normal controls and correlates with airway hyperresponsiveness [109], although the improvement in acute severe ASTHMA after magnesium does not correlate with plasma concentrations. Nebulized isotonic magnesium sulphate has also been shown to be effective in acute severe ASTHMA when added to nebulized salbutamol [110].

K^+ channel openers

Potassium (K^+) channels are involved in recovery of excitable cells after depolarization and therefore are important in stabilisation of cells. K^+ channel openers such as cromakalim or levcromakalim (the levo-isomer of cromakalim) open ATP-dependent K^+ channels in smooth muscle and therefore relax airway smooth muscle [111]. This suggests that K^+ channel activators may be useful bronchodilators [112]. Clinical studies in ASTHMA have been disappointing with no bronchodilation or protection against bronchoconstrictor challenges [113]. The cardiovascular side-effects of these drugs (postural hypotension, flushing) limit the oral dose, however. Inhaled K^+ channel openers also have problems, but new developments include K^+ channel openers that open calcium-activated large conductance K^+ channels (maxi-K channels) that are also opened by β_2-agonists and these drugs may be better tolerated . Maxi-K channel openers also inhibit mucus secretion and cough, so may be of particular value in the treatment of COPD [114, 115].

Atrial natriuretic peptides

Atrial natriuretic peptide (ANP) activates guanylyl cyclase and increases cyclic GMP, leading to bronchodilation. ANP and the related peptide urodilatin are bronchodilators in ASTHMA and give comparable effects to β_2-agonists [116, 117]. Since they work via a different mechanism from β_2-agonists, they may give additional bronchodilation that may be useful in acute severe ASTHMA when β_2-receptor function might be impaired.

VIP analogues

Vasoactive intestinal polypeptide (VIP) is a potent bronchodilator of human airways *in vitro*, but is not effective in patients as it is metabolized and also causes vasodilator side-effects. More stable analogues of VIP, such as Ro 25-1533, which selectively stimulates the VIP receptor in airway smooth muscle (VPAC2 receptor) have been synthesised. Inhaled Ro 25-1533 has a rapid bronchodilator effect in asthmatic patients but is not as prolonged as formoterol [118].

Controller drugs

INFLAMMATION underlies several airway diseases, although the type of inflammatory responses differs between diseases. Anti-inflammatory drugs suppress the inflammatory response by inhibiting components of the INFLAMMATION, such as inflammatory cell infiltration and activation or release, synthesis and effects of inflammatory mediators. Corticosteroids have an anti-inflammatory effect in ASTHMA, whereas other drugs may improve control of ASTHMA without acute bronchodilator effects. These drugs are now classified as controllers.

Corticosteroids

Corticosteroids are used in the treatment of several lung diseases. They were introduced for the treatment of ASTHMA shortly after their discovery in the 1950s and remain the most effective therapy available for ASTHMA. However, side-effects and fear of adverse effects have limited and delayed their use and there has therefore been considerable research into discovering new or related agents which retain the beneficial action on airways without unwanted effects. The introduction of inhaled corticosteroids, initially as a way of reducing the requirement for oral steroids, has revolutionized the treatment of chronic ASTHMA [119]. Now that ASTHMA is viewed as a chronic inflammatory disease, inhaled steroids are considered as first-line therapy in all but the mildest of patients. By contrast, inhaled corticosteroids are much less effective in COPD and should only be used in patients with severe disease. Oral steroids are indicated in the treatment of several other pulmonary diseases, such as sarcoidosis, interstitial lung diseases and pulmonary eosinophilic syndromes.

Chemistry

The adrenal cortex secretes cortisol (hydrocortisone) and, by modification of its structure, it was possible to develop derivatives, such as prednisolone and dexamethasone, with enhanced corticosteroid effects but with reduced minerallocorticoid activity. These derivatives with potent glucocorticoid actions were effective in ASTHMA when given systemically but had no anti-asthmatic activity when given by inhalation. Further substitution in the 17α ester position resulted in steroids with high TOPICAL activity, such as beclomethasone dipropionate (BDP), triamcinolone, flunisolide, budesonide and fluticasone propionate, which are potent in the skin (dermal blanching test) and were later found to have significant anti-asthma effects when given by inhalation.

Mode of action

Corticosteroids enter target cells and bind to glucocorticoid receptors in the cytoplasm. There is only one type of glucocorticoid receptor that binds corticosteroids and no evidence for different subtypes which might mediate different aspects of corticosteroid action [120, 121]. The steroid-receptor complex is transported to the nucleus where it binds to specific sequences on the upstream regulatory element of certain target genes, resulting in increased (or rarely decreased) transcription of the gene which leads to increased (or decreased) protein synthesis. Glucocorticoid receptors may also interact with protein transcription factors and coactivator molecules in the nucleus and thereby influence the synthesis of certain proteins independently of any direct interaction with DNA. The repression of transcription factors, such as ACTIVATOR PROTEIN-1 (AP-1) and NF-κB, is likely to account for many of the anti-inflammatory effects of steroids in ASTHMA. In particular, corticosteroids reverse the activating effect of these proinflammatory transcription factors on histone acetylation by recruiting HDAC2.

The MECHANISMS OF ACTION of corticosteroids in ASTHMA are still poorly understood, but are most likely to be related to their anti-inflammatory properties. Corticosteroids have widespread effects on gene transcription, increasing the transcription of anti-inflammatory genes and suppressing transcription of inflammatory genes (Tab. 7). Steroids have inhibitory effects on many inflammatory and structural cells that are activated in ASTHMA (Fig. 9). There is compelling evidence that ASTHMA and airway hyperresponsiveness are due to an inflammatory process in the airways and there are several components of this inflammatory response which might be inhibited by steroids. Many studies of bronchial biopsies in ASTHMA have demonstrated a reduction in the number and activation of inflammatory cells in the epithelium and submucosa after regular inhaled steroids, together with a healing of the damaged epithelium. Indeed in mild asthmatics the INFLAMMATION may be completely resolved after inhaled steroids. Steroids potently inhibit the formation of CYTOKINES, such as IL-1, IL-2, IL-3, IL-4, IL-5, IL-9, IL-13, TNF-α and GM-CSF by T LYMPHOCYTES and MACROPHAGES. Corticosteroids also decrease eosinophil survival by inducing apop-

TABLE 7. EFFECTS OF CORTICOSTEROIDS ON GENE TRANSCRIPTION

Increased transcription
- Lipocortin-1
- β_2-Adrenoceptor
- Secretory leukocyte inhibitory protein
- IκB-α (inhibitor of NF-κB)
- Anti-inflammatory or inhibitory cytokines
  IL-10, IL-12, IL-1 receptor antagonist

Decreased transcription
- Inflammatory cytokines
  IL-2, IL-3, IL-4, IL-5, IL-6, IL-11, IL-13, IL-15, TNF-α, GM-CSF, SCF
- Chemokines
  IL-8, RANTES, MIP-1α, eotaxin
- Inducible nitric oxide synthase (iNOS)
- Inducible cyclo-oxygenase (COX-2)
- Inducible phospholipase A_2 ($cPLA_2$)
- Endothelin-1
- NK_1-receptors
- Adhesion molecules (ICAM-1, VCAM-1)

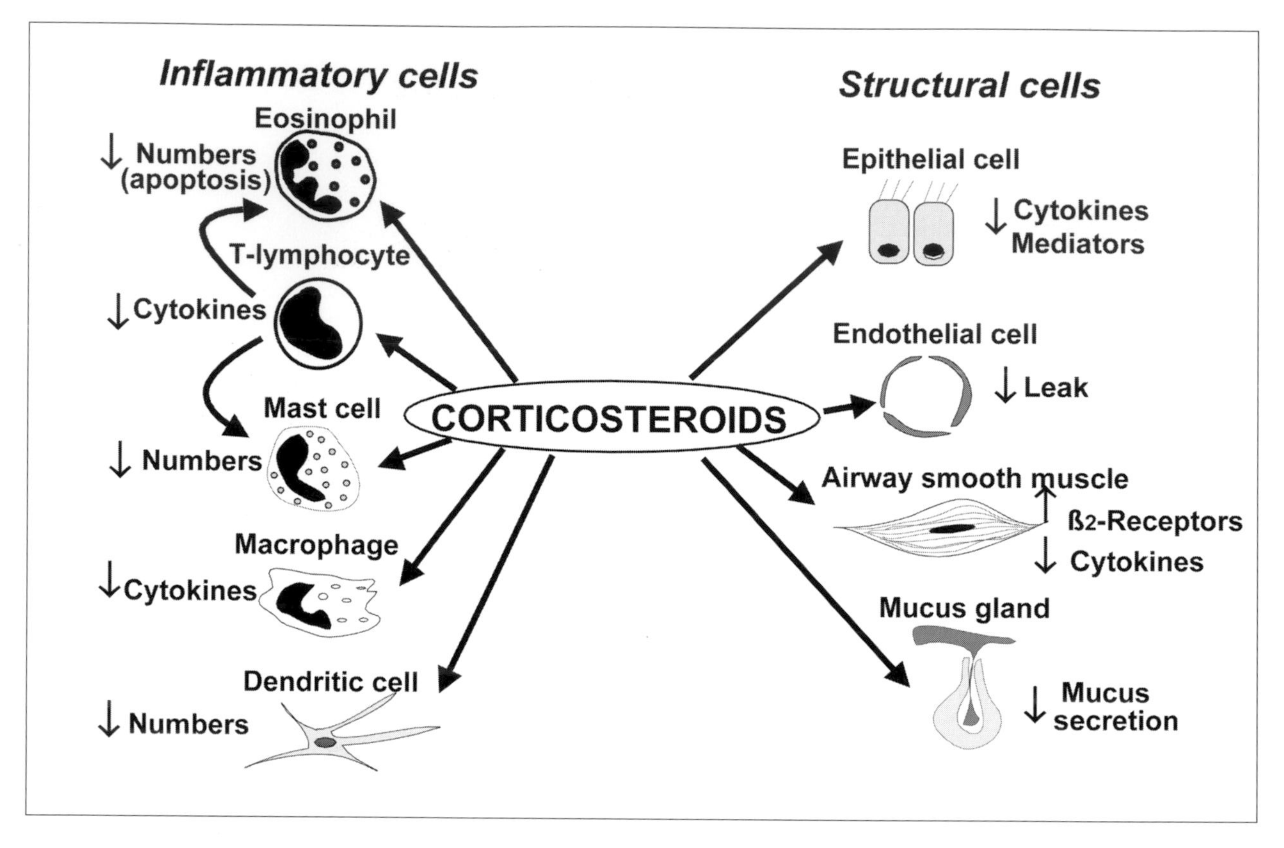

FIGURE 9. EFFECT OF CORTICOSTEROIDS ON INFLAMMATORY AND STRUCTURAL CELLS IN THE AIRWAYS

tosis. They also inhibit the expression of multiple inflammatory genes in airway epithelial cells. Indeed, this may be the most important action of inhaled corticosteroids in suppressing asthmatic INFLAMMATION (Fig. 10). Steroids prevent and reverse the increase in vascular permeability due to inflammatory mediators in animal studies and may therefore lead to resolution of airway oedema. Steroids also have a direct inhibitory effect on mucus glycoprotein secretion from airway submucosal glands, as well as indirect inhibitory effects by down-regulation of inflammatory stimuli that stimulate mucus secretion.

Steroids have no direct effect on contractile responses of airway smooth muscle and improvement in lung function is presumably due to an effect on the chronic airway INFLAMMATION and airway hyperresponsiveness. After a single dose, inhaled steroids have no effect on the early response to allergen (reflecting their lack of effect on mast-cell mediator release), but inhibit the late response (which may be due to an effect on MACROPHAGES and EOSINOPHILS) and also inhibit the increase in airway hyperresponsiveness. Inhaled steroids also reduce airway hyperresponsiveness but this effect may take several weeks or months and presumably reflects the slow healing of the damaged inflamed airway.

It is important to recognize that steroids suppress INFLAMMATION in the airways but do not cure the underlying disease. When steroids are withdrawn there is a recurrence of the same degree of airway hyperresponsiveness, although in patients with mild ASTHMA it may take several months to return [122].

Steroids increase β-adrenergic responsiveness, but whether this is relevant to their effect in ASTHMA

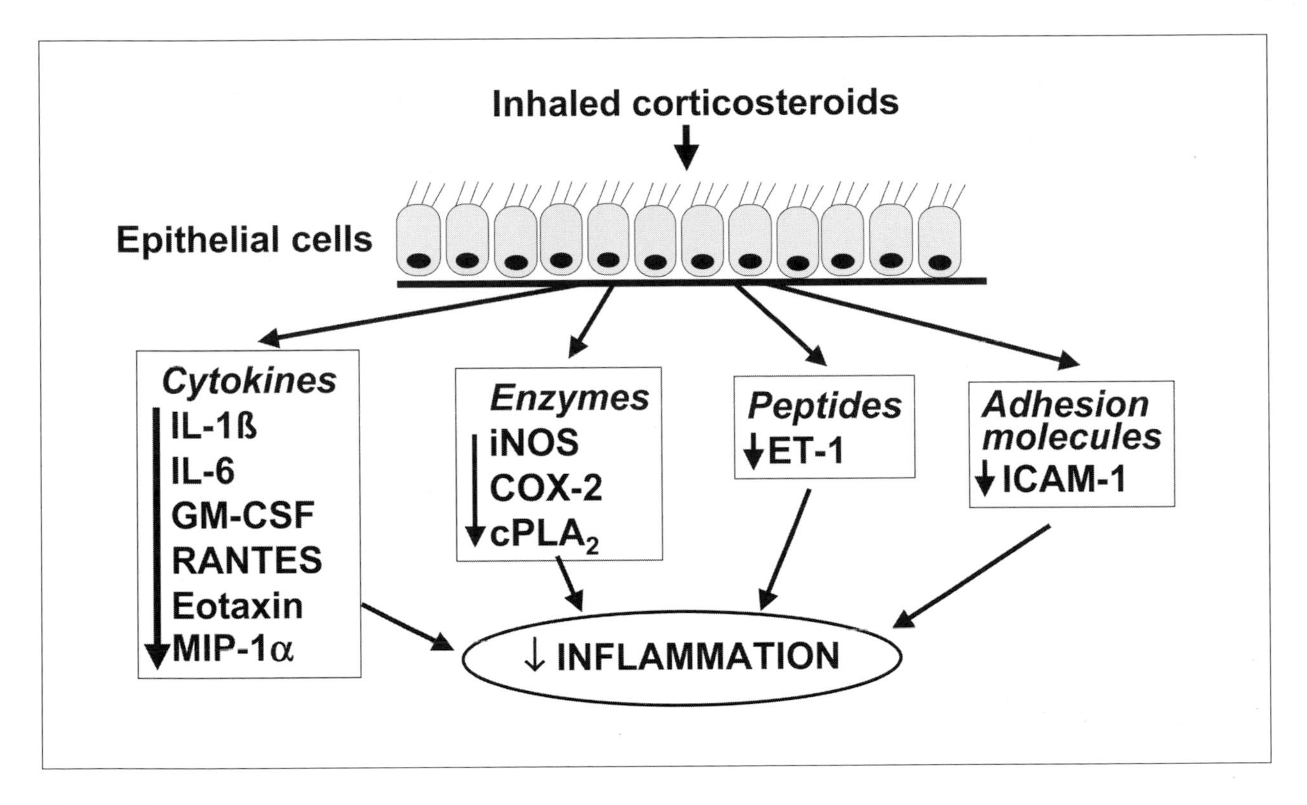

FIGURE 10
Inhaled corticosteroids may inhibit the transcription of several "inflammatory" genes in airway epithelial cells and thus reduce inflammation in the airway wall. NF-κB, nuclear factor κB; AP-1, activator protein-1; GM-CSF, granulocyte-macrophage colony-stimulating factor; IL-1, interleukin-1; iNOS, inducible nitric oxide synthase; NO, nitric oxide; COX-2, inducible cyclooxygenase; cPLA$_2$, cytoplasmic phospholipase A$_2$; PG, prostaglandin; ET, endothelin; ICAM, intercellular adhesion molecule.

is uncertain. Steroids potentiate the effects of β-agonists on bronchial smooth muscle and prevent and reverse β-receptor desensitisation in airways *in vitro* and *in vivo*. At a molecular level, steroids increase the gene transcription of β_2-receptors in human lung *in vitro* [123] and in the respiratory mucosa *in vivo* [124]. Systemic glucocorticoids prevent down-regulation of β_2-receptors in animal lungs [33]. Inhaled steroids do not appear to prevent the development of TOLERANCE to the bronchodilator action of inhaled β_2-agonists in asthmatic patients [34], but it is probable that they prevent the loss of non-bronchodilator responses to β_2-agonists, such as mast-cell stabilisation [125].

Pharmacokinetics

Prednisolone is readily and consistently absorbed after oral administration with little inter-individual variation. Enteric coatings to reduce the incidence of dyspepsia delay absorption but not the total amount of drug absorbed. Prednisolone is metabolized in the liver and drugs such as rifampicin, phenobarbitone or phenytoin, which induce P450 enzymes, lower the plasma half-life of prednisolone [126]. The plasma half-life is 2–3 h, although its biological half-life is approximately 24h, so that it is suitable for daily dosing. There is no evidence that previous exposure to steroids changes their subsequent metabolism. Pred-

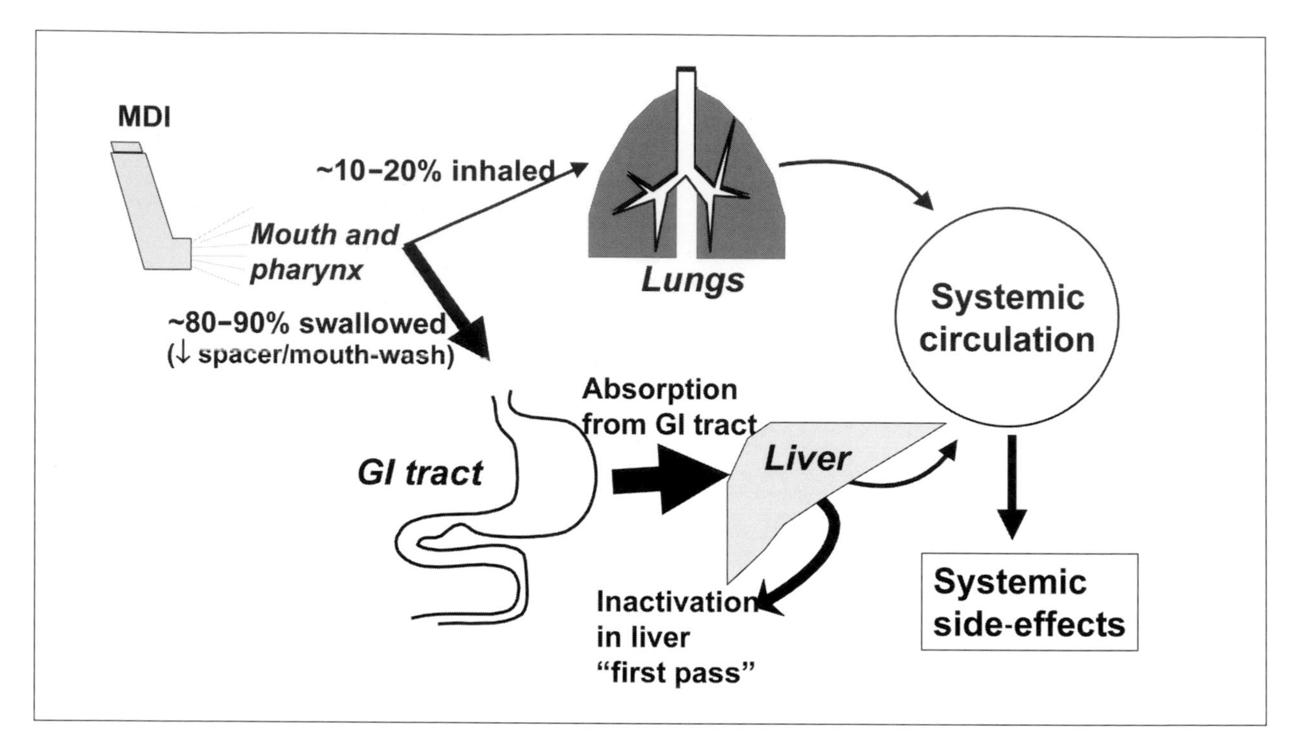

FIGURE 11. PHARMACOKINETICS OF INHALED CORTICOSTEROIDS
GI, gastrointestinal

nisolone is approximately 92% protein-bound, the majority to a specific protein transcortin and the remainder to albumin; it is the unbound fraction which is biologically active.

Some patients, usually with severe ASTHMA, apparently fail to respond to corticosteroids [127]. "Steroid-resistant" ASTHMA is not due to impaired absorption or metabolism of steroids, but is due to reduced anti-inflammatory actions of corticosteroids.

The pharmacokinetics of inhaled corticosteroids is important in relation to systemic effects [128]. The fraction of steroid which is inhaled into the lungs acts locally on the airway mucosa; this may be absorbed from the airway and alveolar surface and therefore reach the systemic circulation. The fraction of inhaled steroid which is deposited in the oropharynx is swallowed and absorbed from the gut. The absorbed fraction may be metabolized in the liver before reaching the systemic circulation (first-pass metabolism) (Fig. 11). Budesonide and fluticasone propionate have a greater first-pass metabolism than BDP and are therefore less likely to produce systemic effects at high inhaled doses. The use of a large volume spacer chamber reduces oropharyngeal deposition and therefore reduces systemic absorption of corticosteroids, although this effect is minimal in corticosteroids with a high first-pass metabolism. Mouth rinsing and discarding the rinse has a similar effect and this procedure should be used with high-dose dry powder steroid inhalers, since spacer chambers cannot be used with these devices.

Routes of administration

Oral. Prednisolone or prednisone are the most commonly used oral steroids. Prednisone is converted to prednisolone in the liver. Deflazacort is an oral steroid that was claimed to have fewer systemic effects than prednisolone, but this has not been confirmed and it is considerably more expensive [129]. Clinical improvement with oral steroids may take several days and the maximal beneficial effect is usually achieved with 30–40 mg prednisone daily, although a few patients may need 60–80 mg daily to achieve control of symptoms. The usual maintenance dose is in the order of 10–15 mg/day. Oral steroids are usually given as a single dose in the morning, since this coincides with the normal diurnal increase in plasma cortisol and there is therefore less adrenal suppression than if given in divided doses or at night. Alternate day treatment has the advantage of less adrenal suppression, although in many patients control of ASTHMA is not optimal on this regime.

Intravenous. Parenteral steroids are indicated in acute severe exacerbations of asthma. Hydrocortisone is the steroid of choice as it has then most rapid onset (5–6 h after administration), being more rapid than prednisolone (8 h). The dose required is still uncertain, but it is common to give hydrocortisone 4 mg/kg initially followed by a maintenance dose of 3 mg/kg/6 h. These doses are based on the argument that it is necessary to maintain "stress" levels of plasma cortisol. Although intravenous hydrocortisone is traditionally used for the treatment of acute severe ASTHMA, oral prednisolone is just as effective and is easier to administer [130].

Inhaled. Inhaled TOPICAL steroids have been a great advance in the management of chronic ASTHMA as it is possible to control symptoms without adrenal suppression or side-effects, and allows a reduction in the dose of oral maintenance steroids. The high TOPICAL activity of inhaled steroids means that only small doses are required and any swallowed drug is metabolised by the liver. Only when much larger doses are inhaled is sufficient steroid absorbed (from the gastrointestinal tract and across the alveolar surface) to cause adrenal suppression.

Many patients get a maximal response at a dose of 400–500 μg per day of BDP or equivalent, but some patients (with a relative steroid resistance) may benefit from higher doses (up to 2000 μg/day). Traditionally steroid inhalers are used twice daily, but four-times-daily administration may be more effective in severe disease [131]. Once-daily administration of some steroids, such as budesonide and mometasone, is effective when doses of 400 μg or less are needed [132].

Nebulized corticosteroids (budesonide) are useful in the treatment of small children who are not able to use other inhaler devices [133]. Nebulized budesonide has been used as a means of delivering high doses of inhaled corticosteroid in patients with severe ASTHMA, but there is little evidence for any advantage of this expensive therapy.

Clinical use

Hydrocortisone is given intravenously in acute severe ASTHMA. While the value of corticosteroids in acute severe ASTHMA has been questioned, others have found that they speed the resolution of attacks. There is no apparent advantage in giving very high doses of intravenous steroids (such as methylprednisolone 1 g). Intravenous steroids are indicated in acute ASTHMA if lung function is <30% predicted and in whom there is no significant improvement with nebulized β_2-agonist. Intravenous therapy is usually given until a satisfactory response is obtained and then oral prednisolone may be substituted. Oral prednisolone (40–60 mg) has a similar effect to intravenous hydrocortisone and is easier to administer [130]. Inhaled steroids have no proven effect in the management of acute ASTHMA [134], but trials with nebulized steroids are underway.

Inhaled steroids are recommended as first-line therapy for all patients with persistent ASTHMA [85]. Inhaled steroids should be started in any patient who needs to use a β_2-agonist inhaler for symptom control more than three times a week. They are effective in mild, moderate and severe ASTHMA and in children as well as adults [128, 135]. Although it was recommended that inhaled corticosteroids should be initiated at a relatively high dose and then the dose reduced once control was achieved, there is no evidence that this is more effective than starting with the maintenance dose. Dose-response studies for inhaled corticosteroids are relatively flat, with most of the benefit derived from doses < 400 μg BDP or equivalent [136–138].

Oral steroids are reserved for patients who cannot be controlled on other therapy, the dose being titrated to the lowest which provides acceptable control of symptoms. For any patient taking regular oral steroids, objective evidence of steroid responsiveness should be obtained before maintenance therapy is instituted. Short courses of oral steroids (30–40 mg prednisolone daily for 1-2 weeks) are indicated for exacerbations of ASTHMA, and the dose may be tailed off over 1 week once the exacerbation is resolved (although the tail-off period is not strictly necessary, patients find it reassuring). For most patients inhaled steroids should be used twice daily, which improves compliance once control of ASTHMA has been achieved (which may require four-times-daily dosing initially or a course of oral steroids if symptoms are severe). If a dose of more than 800 μg via MDI daily is used, a spacer device should be used as this reduces the risk of orpharyngeal side-effects. Inhaled steroids may be used in children in the same way as adults and at doses of 400 μg daily or less there is no evidence of growth suppression. The dose of inhaled corticosteroid should be the minimal dose that controls ASTHMA and once control is achieved the dose should be slowly reduced [139].

COPD patients occasionally respond to steroids and these patients are likely to also have ASTHMA. Steroids have no objective short-term benefit of airway function in patients with true COPD, although they may often produce subjective benefit because of their euphoric effect. Corticosteroids do not appear to have any significant anti-inflammatory effect in COPD and there appears to be an active resistance mechanism which may be explained by impaired activity of histone deacetylases [140]. Inhaled corticosteroids have no effect on the progression of COPD, even when given to patients with presymptomatic disease [141]. Inhaled corticosteroids reduce the number of exacerbations in patients with severe COPD (FEV_1<50% predicted) who have frequent exacerbations and are recommended in these patients. Oral corticosteroids are used to treat acute exacerbations of COPD and reduce the rate of recovery, although the effect is small [142, 143].

Side-effects

Corticosteroids inhibit ACTH and cortisol secretion by a negative feedback effect on the pituitary gland. Hypothalamo-pituitary-adrenal (HPA) axis suppression is dependent on dose, and usually only occurs when a dose of prednisolone >7.5–10 mg daily is used. Significant suppression after short courses of steroid therapy is not usually a problem, but prolonged suppression may occur after several months or years. Steroid doses after prolonged oral therapy must therefore be reduced slowly. Symptoms of "steroid withdrawal syndrome" include lassitude, musculoskeletal pains and occasionally fever. HPA suppression with inhaled steroids is seen only when the daily inhaled dose exceeds 2000 μg BDP or equivalent daily.

Side-effects of long-term oral corticosteroid therapy are well described and include fluid retention, increased appetite, weight gain, OSTEOPOROSIS, capillary fragility, hypertension, peptic ulceration, diabetes, cataracts and psychosis. Their frequency tends to increase with age. Very occasionally adverse reactions (such as anaphylaxis) to intravenous hydrocortisone have been described, particularly in aspirin-sensitive asthmatics.

The incidence of systemic side-effects after inhaled steroids is an important consideration [128] (Tab. 8). Initial studies suggested that adrenal suppression only occurred when inhaled doses of over 1500–2000 μg daily were used. More sensitive measurements of systemic effects include indices of bone metabolism, such as serum osteocalcin and urinary pyridinium cross-links, and in children knemometry, which may be increased with inhaled doses as low as 400 μg in some patients. The clinical relevance of these measurements is not yet clear, however. Nevertheless it is important to reduce the likelihood of systemic effects by using the lowest dose of inhaled steroid needed to control the ASTHMA, by the use of a large volume spacer to reduce oropharyngeal deposition.

Several systemic effects of inhaled steroids have been described and include dermal thinning and skin capillary fragility, which is relatively common in elderly patients after high-dose inhaled steroids [144]. Other side-effects such as cataract formation and OSTEOPOROSIS are reported, but often in patients who are also receiving courses of oral steroids. There has been particular concern about the use of inhaled steroids in children because of growth sup-

Table 8. Side-effects of inhaled corticosteroids

Local side-effects
- Dysphonia
- Oropharyngeal candidiasis
- Cough

Systemic side-effects
- Adrenal suppression and insufficiency
- Growth suppression
- Bruising
- Osteoporosis
- Cataracts
- Glaucoma
- Metabolic abnormalities (glucose, insulin, triglycerides)
- Psychiatric disturbances (euphoria, depression)

pression [145]. Most studies have been reassuring in that doses of 400 µg or less have not been associated with impaired growth, and there may even be a growth spurt as ASTHMA is better controlled [146]. There is some evidence that use of high-dose inhaled corticosteroids is associated with cataract [147] and glaucoma [148], but it is difficult to dissociate the effects of inhaled corticosteroids from the effects of courses of oral steroids which these patients usually require.

Inhaled steroids may have local side-effects due to the deposition of inhaled steroid in the oropharynx. The most common problem is hoarseness and weakness of the voice (dysphonia) which is due to laryngeal deposition. It may occur in up to 40% of patients and is noticed particularly by patients who need to use their voices during their work (lecturers, teachers and singers) [149]. It may be due to atrophy of the vocal cords. Throat irritation and coughing after inhalation are common with MDIs and appear to be due to the additives, since these problems are not usually seen if the patient switches to the dry powder inhalers. Oropharyngeal candidiasis occurs in 5% of patients. The incidence of local side-effects may be related to the local concentrations of steroid deposited and may be reduced by the use of large volume spacers, which markedly reduce oropharyngeal deposition. Local side-effects are also less likely when inhaled steroids are used twice daily rather than four times daily. There is no evidence for atrophy of the lining of the airway, or of an increase in lung infections (including tuberculosis) after inhaled steroids.

Recently it has become clear that it is difficult to extrapolate systemic side-effects of corticosteroids from studies in normal subjects. In asthmatic patients, systemic absorption from the lung is reduced, presumably because of reduced and more central deposit of the inhaled drug, particularly in more severe patients [150, 151]. Another important issue is the distribution of inhaled corticosteroids in patients with ASTHMA, since most of the drug deposits in larger airways. As INFLAMMATION is also found in small airways, particularly in severe patients, the inhaled corticosteroids may not adequately suppress INFLAMMATION in these airways [152]. Corticosteroid metered-dose inhalers with hydrofluroalkane (HFA) propellants produce smaller aerosol particles and may have a more peripheral deposition, making them useful in treating patients with more severe ASTHMA [153, 154].

Therapeutic choices

Several inhaled corticosteroids are now available, including beclomethasone dipropionate (BDP), triamcinolone, flunisolide, budesonide, fluticasone propionate and mometasone. All are equally effective as anti-asthma drugs, but there are differences in pharmacokinetics, in that budesonide, fluticasone and mometasone have a lower oral BIOAVAILABILITY than BDP, as there is a greater first-pass hepatic metabolism, resulting in reduced systemic absorption from the fraction of the inhaled drug that is swallowed [128, 155]. At high doses (>1000 µg), budesonide and fluticasone have fewer systemic effects than BDP and triamcinolone, and are preferred in patients who need high doses of inhaled corticosteroids and in children. The type of delivery system is important in the comparison of inhaled steroids. When doses of inhaled steroid exceed 800 µg daily, a large volume spacer is recommended as this reduces oropharyngeal deposition and systemic absorption in the case of BDP.All currently available inhaled corticosteroids are absorbed from the lung into the systemic circulation, so that some systemic absorption is inevitable. However, the amount of drug absorbed does not appear to have clinical effects in doses of <800 μg BDP equivalent. Although there are potency differences between corticosteroids, there are relatively few comparative studies, partly because dose comparison of corticosteroids is difficult due to their long-time course of action and the relative flatness of their dose response. Triamcinolone and flunisolide appear to be the least potent, with BDP and budesonide approximately of equal potency, whereas FP is approximately twice as potent as BDP

Future developments

There has been a dramatic increase in the use of inhaled corticosteroids in asthma treatment. This is due to the recognition that ASTHMA is an inflammatory condition and to the introduction of treatment guidelines that emphasize the early use of inhaled corticosteroids. There is also increasing recognition that at the dose of inhaled steroids needed to control ASTHMA in most patients there are no systemic effects. However, there is an increasing recognition that there may be systemic effects when inhaled corticosteroids are used long-term in high doses. It is also apparent that the dose-response curve for inhaled corticosteroids is relatively flat, so that relatively little clinical improvement is obtained, whereas the risk of systemic effects is increased. This has led to the use of alternative add-on therapies (LABA, theophylline, anti-leukotrienes) in patients not controlled on 400–800 μg daily. In particular combination inhalers of LABA and corticosteroids have been introduced which simplify asthma management and increase compliance with regular therapy.

Early treatment with inhaled steroids in both adults and children gives a greater improvement in lung function than if treatment with inhaled steroids is delayed (and other treatments such as bronchodilators are used) [156–158]. This may reflect the fact that steroids are able to modify the underlying inflammatory process and prevent any structural changes (fibrosis, smooth muscle hyperplasia, etc.) in the airway as a result of chronic INFLAMMATION. Inhaled corticosteroids are currently recommended for patients with persistent asthma symptoms, but have been advocated even in patients with episodic ASTHMA. In a large study of once-daily low-dose inhaled budesonide in patients with mild persistent ASTHMA, there was a reduction in exacerbations and improved asthma control, although there were no differences in lung function [135]. There is evidence for airway INFLAMMATION in patients with episodic ASTHMA, but at present it is recommended that inhaled steroids are introduced only when there are chronic symptoms (e.g., use of an inhaled β_2-agonist on a daily basis).

There has been a search for new corticosteroids that may have fewer systemic effects. Corticosteroids that are metabolised in the airways, such as butixocort and tipredane, proved disappointing in clinical trials as they are probably broken down too rapidly before they are able to exert any anti-inflammatory effect. There is a search for corticosteroids that are metabolized rapidly in the circulation after absorption from the lungs. Ciclesonide is a new steroid that is an inactive prodrug. The active metabolite is released by esterases in the lung and this reduces local side-effects. A high degree of protein binding ensures low systemic effects, giving this steroid a very favourable therapeutic ratio [159]. The anti-inflammatory actions of corticosteroids may be mediated via different molecular mechanisms from side-effects (which are endocrine and metabolic actions of corticosteroids). It has been possible to develop corticosteroids that dissociate the DNA-binding effect of corticosteroids (which mediates most of the adverse effects) from the transcription factor binding action (which mediates much of the anti-inflammatory effect) [160]. These "dissociated steroids" should therefore retain anti-inflammatory activity but have a reduced risk of adverse effects, although achieving this separation is difficult *in vivo* [161, 162].

Cromones

Sodium cromoglycate is a derivative of khellin, an Egyptian herbal remedy which was found to protect against allergen challenge without bronchodilator effect. A structurally related drug, nedocromil sodium, which has an identical mode of action, was subsequently developed. Although sodium cromoglycate was popular in the past, its use has sharply declined with the more widespread use of the more effective inhaled corticosteroids, particularly in children.

Mode of action

Initial investigations indicated that cromoglycate inhibited the release of mediators by allergen in passively sensitised human and animal lung, and inhibited passive cutaneous anaphylaxis in rat, although it was without effect in guinea pig. This activity was attributed to stabilisation of the mast-cell membrane and thus sodium cromoglycate was classified as a mast-cell stabilizer. However, cromoglycate has a rather low potency in stabilising human lung mast cells, and other drugs which are more potent in this respect have little or no effect in clinical ASTHMA. This has raised doubts about mast-cell stabilization as the mode of action of cromoglycate.

Cromoglycate and nedocromil potently inhibit bronchoconstriction induced by sulphur dioxide, metabisulphite and bradykinin, which are believed to act through activation of sensory nerves in the airways. In dogs, cromones suppress firing of unmyelinated C-fibre nerve endings, reinforcing the view that it might be acting to suppress sensory nerve activation and thus neurogenic INFLAMMATION. Cromones have variable inhibitory actions on other inflammatory cells which may participate in allergic INFLAMMATION, including MACROPHAGES and EOSINOPHILS. *in vivo* cromoglycate is capable of blocking the early response to allergen (which may be mediated by mast cells) but also the late response and airway hyperresponsiveness, which are more likely to be mediated by MACROPHAGE and eosinophil interaction. There is no convincing evidence that cromones reduce INFLAMMATION in asthmatic airways [163]. The molecular MECHANISM OF ACTION of cromones is not understood, but some evidence suggests that they may block a particular type of chloride channel that may be expressed in sensory nerves, mast cells and other inflammatory cells [164].

Current use

Cromoglycate is a prophylactic treatment and needs to be given regularly. Cromoglycate protects against various indirect bronchoconstrictor stimuli, such as exercise and fog. It is only effective in mild ASTHMA, but is not effective in all patients and there is no way to predict which patients will respond. Cromoglycate was often used previously as the controller drug of choice in children because it lacks side-effects. However, there is an increasing tendency to use low-dose inhaled corticosteroids instead as they are safe and far more effective. In adults, inhaled corticosteroids are preferable to cromones, as they are effective in all patients. Cromoglycate has a short duration of action and has to be given four times daily to provide good protection, which makes it much less useful than inhaled steroids which may be given once or twice daily. It may also be taken prior to exercise in children with exercise-induced ASTHMA that is not blocked by an inhaled β_2-agonist. In clinical practice, nedocromil has a similar EFFICACY to cromoglycate, but has the disadvantage of an unpleasant taste [165]. Systematic reviews indicate that cromoglycate is largely ineffective in long-term control of ASTHMA in children and its use has now markedly declined [166, 167]. The introduction of anti-leukotrienes has further eroded the market for cromones, as these drugs are of comparable or greater clinical EFFICACY and are more conveniently taken by mouth.

Side-effects

Cromoglycate is one of the safest drugs available and side-effects are extremely rare. The dry powder inhaler may cause throat irritation, coughing and, occasionally, wheezing, but this is usually prevented by prior administration of a β_2-agonist inhaler. Very rarely a transient rash and urticaria are seen and a few cases of pulmonary eosinophilia have been reported, all of which are due to hypersensitivity.

Mediator antagonists

Many inflammatory mediators have been implicated in ASTHMA, suggesting that inhibition of synthesis of effects of these mediators may be beneficial. However, because these mediators have similar effects, specific inhibitors have usually been disappointing in ASTHMA treatment.

Anti-histamines

Histamine mimics many of the features of ASTHMA and is released from mast cells in acute ASTHMATIC RESPONSES, suggesting that ANTI-HISTAMINES may be useful in ASTHMA therapy. Many trials of ANTI-HISTAMINES have been conducted, but there is little evidence of any useful clinical benefit, as demonstrated by a meta-analysis [168]. New ANTI-HISTAMINES, including cetirizine and azelastine, have some beneficial effects, but this may be unrelated to their H_1-receptor antagonism. In addition, ANTI-HISTAMINES are effective in controlling rhinitis and this may help to improve overall asthma control [169].

Ketotifen is described as a prophylactic anti-asthma compound. Its predominant effect is H_1-receptor antagonism and it is this antihistaminic effect that accounts for its sedative effect. Ketotifen has little effect in placebo-controlled trials in clinical ASTHMA, either in acute challenge, on airway hyperresponsiveness or on symptoms [170]. Ketotifen does not have a steroid-sparing effect in children maintained on inhaled corticosteroids [171]. It is claimed that ketotifen has disease-modifying effects if started early in ASTHMA in children and may even prevent the development of ASTHMA in atopic children [172]. More carefully controlled studies are needed to assess the validity of these claims.

Anti-leukotrienes

There is considerable evidence that cysteinyl-leukotrienes (LT) are produced in ASTHMA and that they have potent effects on airway function, inducing bronchoconstriction, airway hyperresponsiveness, plasma exudation and mucus secretion and possibly on eosinophilic INFLAMMATION [173] (Fig. 12). This suggested that blocking the leukotriene pathways may be useful in the treatment of ASTHMA, leading to the development of 5'-lipoxygenase (5-LO) enzyme inhibitors (of which zileuton is the only drug marketed) and several antagonists of the cys-LT_1-receptor, including montelukast, zafirlukast and pranlukast.

Clinical studies

Leukotriene antagonists inhibit the bronchoconstrictor effect of inhaled LTD_4 in normal and asthmatic volunteers. They also inhibit bronchoconstriction induced by a variety of challenges, including allergen, exercise and cold air, with approximately 50% inhibition. This suggests that LEUKOTRIENES account for approximately half of these responses. With aspirin challenge, in aspirin-sensitive asthmatic patients, there is almost complete inhibition of the response [174]. Similar results have been obtained with the 5-LO inhibitor zileuton. This suggests that there may be no additional advantage in blocking LTB_4 in addition to cysteinyl-LTs, although this may not be the case in other inflammatory diseases, such as rheumatoid arthritis and inflammatory bowel disease. These drugs are active by oral administration and this may confer an important advantage in chronic treatment, particularly in relation to compliance.

Anti-leukotrienes have been intensively investigated in clinical studies [175]. In patients with mild to moderate ASTHMA there is a significant improvement in lung function (clinic FEV_1 and home peak expiratory flow measurements) and asthma symptoms, with a reduction in the use of rescue inhaled β_2-agonists. In several studies there is evidence for a bronchodilator effect, with an improvement in baseline lung function, suggesting thatLEUKOTRIENES are contributing to the baseline bronchoconstriction in ASTHMA, although this varies between patients. However, anti-leukotrienes are considerably less effective than inhaled corticosteroids in the treatment of mild ASTHMA and cannot be considered as the treatment of first choice [176]. Anti-leukotrienes are therefore indicated more as an add-on therapy in patients not well controlled on inhaled corticosteroids. While they have add-on benefits, this effect is small and equivalent to doubling the dose of inhaled corticosteroid [177] and less effective than adding a LABA [178, 179]. In patients with severe ASTHMA who are not controlled on high doses of inhaled corticosteroids and LABA, anti-leukotrienes do not appear to pro-

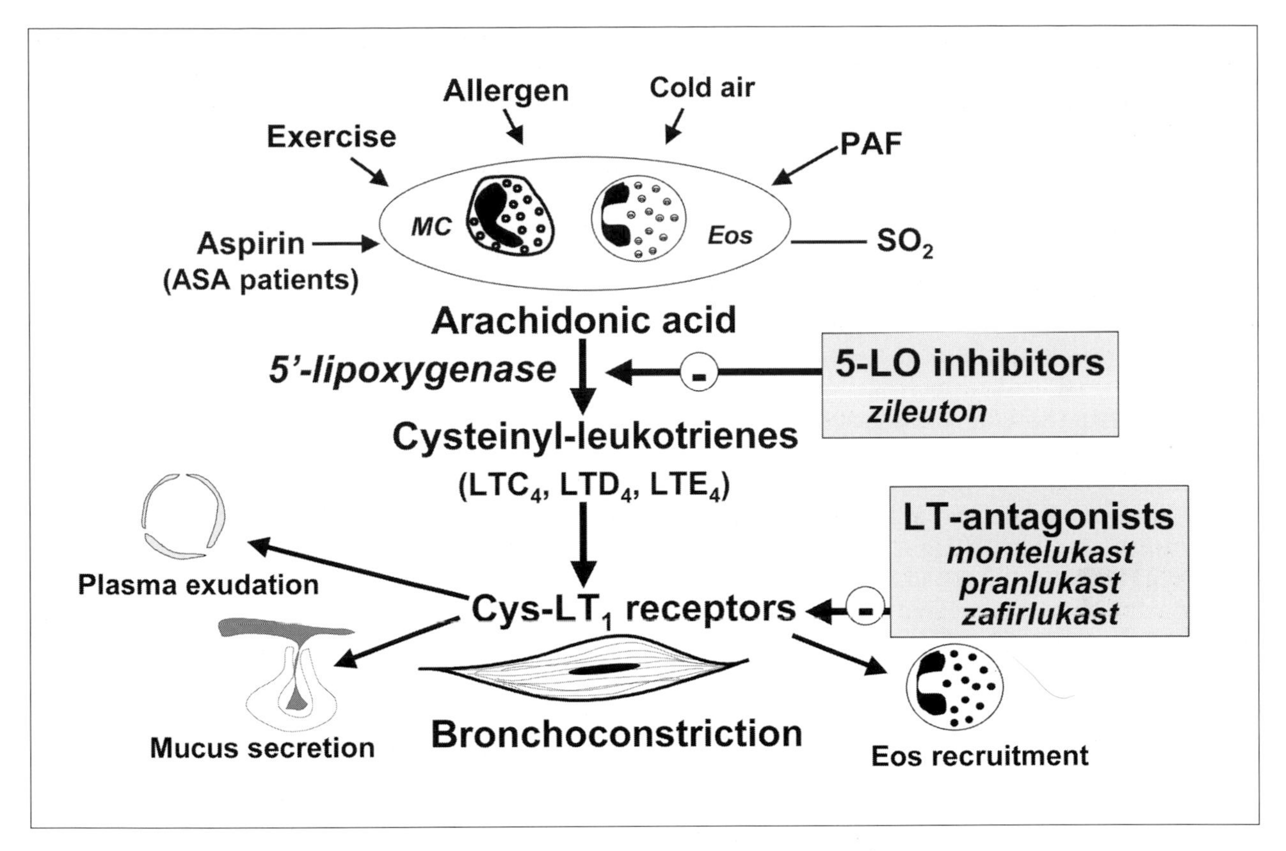

FIGURE 12. EFFECTS OF CYSTEINYL-LEUKOTRIENES ON THE AIRWAYS AND THEIR INHIBITION BY ANTI-LEUKOTRIENES
PAF, platelet-activating factor; LT, leukotriene; 5-LO, 5'-lipoxygenase; ASA, aspirin-sensitive asthmatic.

vide any additional benefit [180]. Theoretically anti-leukotrienes should be of particular value in patients with aspirin-sensitive ASTHMA as they block the airway response to aspirin challenge, but although anti-leukotrienes have some benefit in these patients, it is no greater than in other types of ASTHMA [181]. Anti-leukotrienes are also effective in preventing exercise-induced ASTHMA, although they are similar in EFFICACY to LABA in this respect [182]. Anti-leukotrienes also have a weak effect in rhinitis that may be additive to the effects of an anti-histamine [183].

An important issue is whether anti-leukotrienes are anti-inflammatory. By definition, anti-leukotrienes (like anti-histamines) must be anti-inflammatory, asLEUKOTRIENES themselves (like histamine) are inflammatory mediators. Studies have demonstrated weak anti-inflammatory effects of anti-leukotrienes in reducing EOSINOPHILS in sputum or in biopsies [184, 185], but this is much less marked than with an inhaled corticosteroid and there is no additional anti-inflammatory effect when added to an inhaled corticosteroid [186]. Anti-leukotrienes therefore appear to act mainly as anti-bronchoconstrictor drugs, although they are clearly less effective in this respect than β_2-agonists.

Cys-LT have no role in COPD and are not elevated in exhaled breath condensate as in ASTHMA [187] and cys-LT1 receptor antagonists therefore have no role. By contrast LTB_4, a potent NEUTROPHIL chemoattractant, is elevated in COPD, indicating that 5-LO inhibitors may have some potential benefit by reducing NEUTROPHIL INFLAMMATION. A pilot study did not indicate any clear benefit of a 5-LO inhibitor in COPD patients, however [188].

Side-effects

To date, anti-leukotrienes have been remarkably free of class-specific side-effects. Zileuton causes an increased level of liver enzymes, so that monitoring of liver enzymes necessary with this drug, and high doses of zafirlukast may be associated with abnormal liver function. Montelukast is well tolerated in adults and children, with no significant adverse effects. The lack of side-effects implies thatLEUKOTRIENES do not appear to be important in any physiological functions. Several cases of Churg-Strauss syndrome have been associated with the use of zafirlukast and montelukast [189]. Churg-Strauss syndrome is a very rare vasculitis that may affect the heart, peripheral nerves and kidney and is associated with increased circulating EOSINOPHILS and ASTHMA. It is likely that the cases so far reported are due to a reduction in oral or inhaled corticosteroid dose, rather than as a direct effect of the drug [190], although a few cases of Churg-Strauss syndrome have been described in patients on anti-leukotrienes who were not on concomitant corticosteroids therapy.

Future use

One of the major advantages of anti-leukotrienes is that they are active in tablet form. This may increase the compliance with chronic therapy and will make treatment of children easier. Montelukast is effective as a once-daily preparation (10 mg in adults and 5 mg in children) and is therefore easy for patients to use. In addition, oral administration may treat concomitant allergic rhinitis. However, the currently available clinical studies indicate a relatively modest effect on lung function and symptom control, which is less for every clinical parameter measured than with inhaled corticosteroids. This is not surprising, as there are many more mediators than cys-LTs involved in the pathophysiology of ASTHMA, and it is unlikely that antagonism of a single mediator could ever be as effective as steroids, which inhibit all aspects of the inflammatory process in ASTHMA. Similarly, if anti-leukotrienes are functioning in ASTHMA as bronchodilators and anti-constrictors, it is unlikely that they will be as effective as a β_2-agonist, which will counteract bronchoconstriction, irrespective of the spasmogen. It is likely that anti-leukotrienes will be used less in the future as combination inhalers are used as the mainstay of ASTHMA therapy [191].

An interesting feature of the clinical studies is that some patients appear to show better responses than others, suggesting thatLEUKOTRIENES may play a more important role in some patients. The variability in response to anti-leukotrienes may reflect differences in production of or responses toLEUKOTRIENES in different patients, and this in turn may be related to POLYMORPHISMS of 5-LO, LTC_4 synthase or cys-LT_1-receptors that are involved in the synthesis ofLEUKOTRIENES [192].

It is unlikely that further advances can be made in cys-LT_1-receptor antagonists as montelukast is a once-daily drug that probably gives maximal receptor blockade. Cys-LT_2-receptors may be important in vascular and airway smooth muscle proliferative effects of cys-LT and are not inhibited by current cys-LT_1-receptor antagonists [193]. The role of this receptor in ASTHMA is unknown so the value of also blocking cys-LT_2 is uncertain.

Steroid-sparing therapies

Immunosuppressive therapy has been considered in ASTHMA when other treatments have been unsuccessful or to reduce the dose of oral steroids required [194]. They are therefore only indicated in a very small proportion of asthmatic patients at present (probably ~1% of all patients). Most immunosuppressive treatments have a greater propensity to side-effects than oral corticosteroids and therefore cannot be routinely recommended.

Methotrexate

Low-dose methotrexate (15 mg weekly) has a steroid-sparing effect in ASTHMA [195] and may be indicated when oral steroids are contraindicated because of unacceptable side-effects (e.g., in post-menopausal women when OSTEOPOROSIS is a problem). Some patients show better responses than others, but whether patients will have a useful steroid-sparing effect is unpredictable. Overall, methotrexate has a small steroid-sparing effect that is insufficient to significantly reduce side-effects of systemic steroids and this needs to be offset against the relatively high risk of side-effects [196]. Side-effects of methotrexate are relatively common and include nausea (reduced if methotrexate is given as a weekly injection), blood dyscrasias and hepatic damage. Careful monitoring of such patients (monthly blood counts and liver enzymes) is essential. Pulmonary infections, pulmonary fibrosis may rarely occur and even death. Methotrexate is disappointing in clinical practice and is now little used.

Gold

Gold has long been used in the treatment of chronic arthritis. A controlled trial of an oral gold preparation (Auranofin) demonstrated some steroid-sparing effect in chronic asthmatic patients maintained on oral steroids [197]. Side-effects such as skin rashes and nephropathy are a limiting factor. Overall gold provides little benefit in view of its small therapeutic ration and is not routinely used [198].

Cyclosporin A

Cyclosporin A is active against $CD4^+$ LYMPHOCYTES and therefore should be useful in ASTHMA, in which these cells are implicated. A trial of low-dose oral cyclosporin A in patients with steroid-dependent ASTHMA indicate that it can improve control of symptoms in patients with severe ASTHMA on oral steroids [199], but other trials have been unimpressive and overall its poor EFFICACY is outweighed by its side-effects [200]. Side-effects, such as nephrotoxicity and hypertension, are common and there are concerns about long-term IMMUNOSUPPRESSION. In clinical practice it has proved to be very disappointing as a steroid-sparing agent.

Intravenous immunoglobulin

Intravenous immunoglobulin (IVIG) has been reported to have steroid-sparing effects in severe steroid-dependent ASTHMA when high doses are used (2 g/kg) [201], although a placebo-controlled double-blind trial in children shows that it is ineffective [202]. IVIG reduces the production of IgE from B LYMPHOCYTES and this may be the rationale for its use in severe asthmas [203]. The relatively poor effectiveness and very high cost of this treatment mean that it is not generally recommended.

Immunotherapy

Theoretically, specific IMMUNOTHERAPY with common ALLERGENS should be effective in preventing ASTHMA. While this treatment is effective in allergic rhinitis, there is little evidence that desensitizing injections to common ALLERGENS are very effective in controlling chronic ASTHMA [204]. Double-blind placebo-controlled studies have demonstrated poor effect in chronic ASTHMA in adults and children [205, 206]. Overall the benefits of specific IMMUNOTHERAPY are small in ASTHMA and there are no well-designed studies comparing this treatment to effective treatments such as inhaled corticosteroids [207]. Because there is a risk of anaphylactic and local reactions, and because the course of treatment is time-consuming, this form of therapy is controversial. The cellular mechanisms of specific IMMUNOTHERAPY are of interest as this might lead to safer and more effective approaches in the future (see chapter C.14). Specific IMMUNOTHERAPY induces the secretion of the anti-inflammatory CYTOKINE interleukin-10 (IL-10) from regulatory helper T LYMPHOCYTES (Tr1) and this blocks co-stimulatory signal transduction in T LYMPHOCYTES (via CD28) so that they are unable to react to ALLERGENS presented by ANTIGEN-PRESENTING CELLS [208, 209]. In the future more specific immunotherapies may be developed with cloned allergen EPITOPES or T-cell reactive peptide fragments of ALLERGENS [210, 211].

Anti-IgE

Increased specific IgE is a fundamental feature of allergic ASTHMA. Omalizumab (E25) is a humanized monoclonal ANTIBODY that blocks the binding of IgE

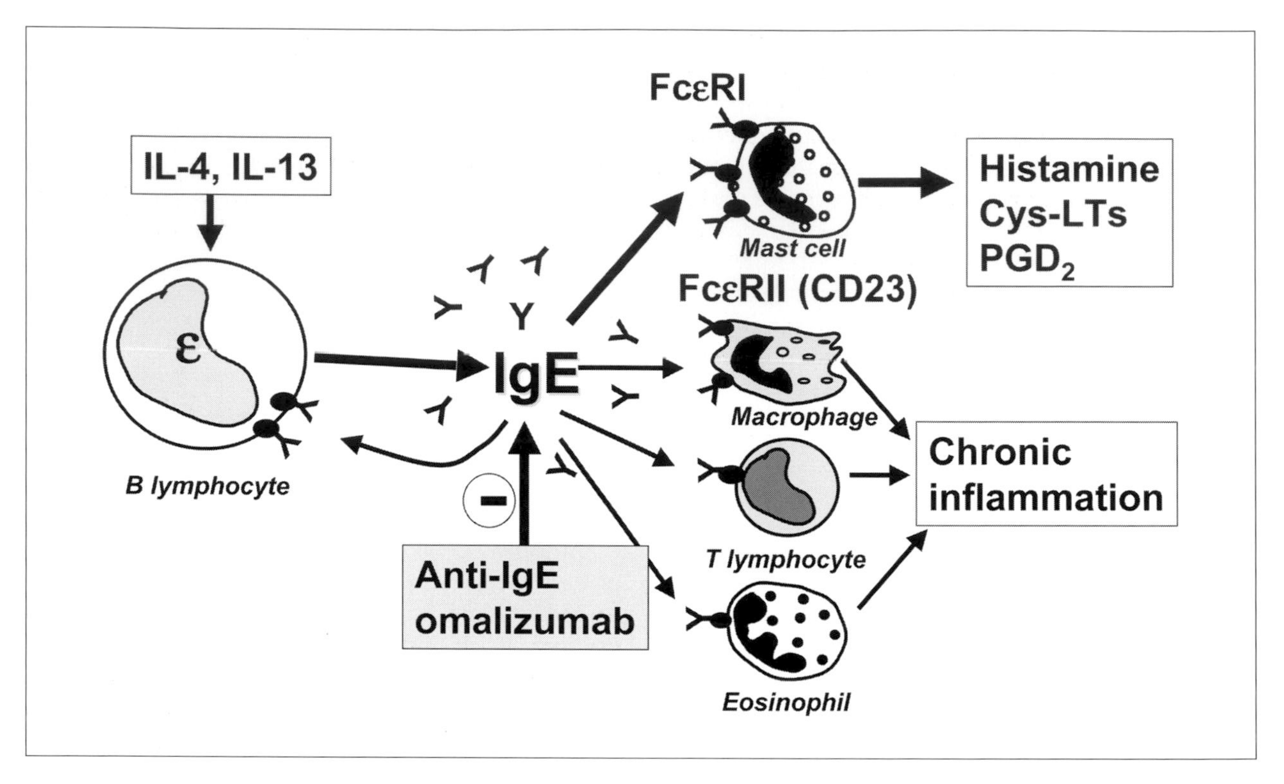

FIGURE 13

IgE play a central role in allergic diseases and blocking IgE using an antibody, such as omalizumab, is a logical approach. IgE may activate high-affinity receptors (FcεRI) on mast cells as well as low-affinity receptors (FcεRII, CD23) on other inflammatory cells. IL, interleukin; cys-LT, cysteinyl-leukotriene; PG, prostaglandin.

to high-affinity IgE receptors (FcεR1) on mast cells and thus prevents their activation by ALLERGENS [212, 213] (Fig. 13). It also blocks binding on IgE to low-affinity IgE receptors (FcεRII, CD23) on other inflammatory cells, including T and B LYMPHOCYTES, MACROPHAGES and possibly EOSINOPHILS to inhibit chronic INFLAMMATION. It also results in a reduction in circulating IgE levels. The ANTIBODY has a high affinity and blocks IgE receptors by over 99% which is necessary because of the amplification of these receptors.

Omalizumab has recently been introduced for the treatment of patients with severe ASTHMA. The ANTIBODY is administered by subcutaneous injection every 2-4 weeks and the dose is determined by the titre of circulating IgE. Omalizumab reduces the requirement for oral and inhaled corticosteroids and markedly reduces ASTHMA exacerbations [214–216]. It is also beneficial in treating allergic rhinitis [217]. Because of its very high cost this treatment is likely to be used only in patients with very severe ASTHMA who are poorly controlled even on oral corticosteroids and in patients with very severe concomitant allergic rhinitis [218]. It may not be useful in concomitant atopic dermatitis due to the high levels of circulating IgE which cannot be neutralised. It may also be of value in protecting against anaphylaxis during specific IMMUNOTHERAPY.

New drugs for asthma and COPD

Several new classes of drug are now in development for ASTHMA and COPD that are directed at the under-

Table 9. New anti-inflammatory drugs for asthma and COPD

- New glucocorticoids (ciclesonide, dissociated steroids)
- Immunomodulators (inhaled cyclosporin, tacrolimus, rapamycin, mycophenolate mofetil)
- Phosphodiesterase-4 inhibitors (cilomilast, roflumilast)
- p38 MAP kinase inhibitors
- NF-κB inhibitors (IKK-2 inhibitors)
- Adhesion molecule blockers (VLA4 antibody, selectin inhibitors)
- Cytokine inhibitors (anti-IL-4, anti-IL-5, anti-IL-13, anti-TNF antibodies)
- Anti-inflammatory cytokines (IL-1ra, IFN-γ, IL-10, IL-12)
- Chemokine receptor antagonists (CCR3, CCR2, CXCR2 antagonists)
- Peptides for immunotherapy
- Vaccines

lying chronic inflammatory process [219–222] (Tab. 9). The INFLAMMATION in ASTHMA and COPD are different, so that different approaches are needed, but there are some common inflammatory mechanisms. Indeed, patients with severe ASTHMA have an inflammatory process that becomes more similar to that in COPD, suggesting that drugs that are effective in COPD may also be useful in patients with severe ASTHMA that is not well controlled with corticosteroids. In ASTHMA many new therapies have targeted eosinophilic INFLAMMATION (Fig. 14). In COPD a better understanding of the inflammatory process has highlighted several targets for inhibition (Fig. 15).

The need for new treatments

Asthma

Current ASTHMA therapy is highly effective and the majority of patients can be well controlled with inhaled corticosteroids and short- and long-acting β_2-agonists, particularly combination inhalers. These treatments are not only effective, but safe and relatively inexpensive. This poses a challenge for the development of new treatments, since they will need to be safer or more effective than existing treatments, or offer some other advantage in long-term disease management. However, there are several problems with existing therapies:

- Existing therapies have side-effects as they are non-specific. Inhaled β_2-agonists may have side-effects and there is some evidence for the development of TOLERANCE, especially to their bronchoprotective effects. Inhaled corticosteroids also may have local and systemic side-effects at high doses and there is still a fear of using long-term steroid treatment in many patients. Other treatments, such as theophylline, anticholinergics and anti-leukotrienes are less effective and are largely used as add-on therapies.
- There is still a major problem with poor compliance (adherence) in the long-term management of ASTHMA, particularly as symptoms come under control with effective therapies [223]. It is likely that a once-daily tablet or even an infrequent injection may give improved compliance. However, oral therapy is associated with a much greater risk of systemic side-effects and therefore needs to be specific for the abnormality in ASTHMA.
- Patients with severe ASTHMA are often not controlled on maximal doses of inhaled therapies or may have serious side-effects from therapy, especially oral corticosteroids. These patients are relatively resistant to the anti-inflammatory actions of corticosteroids and require some other class of therapy to control the asthmatic process.

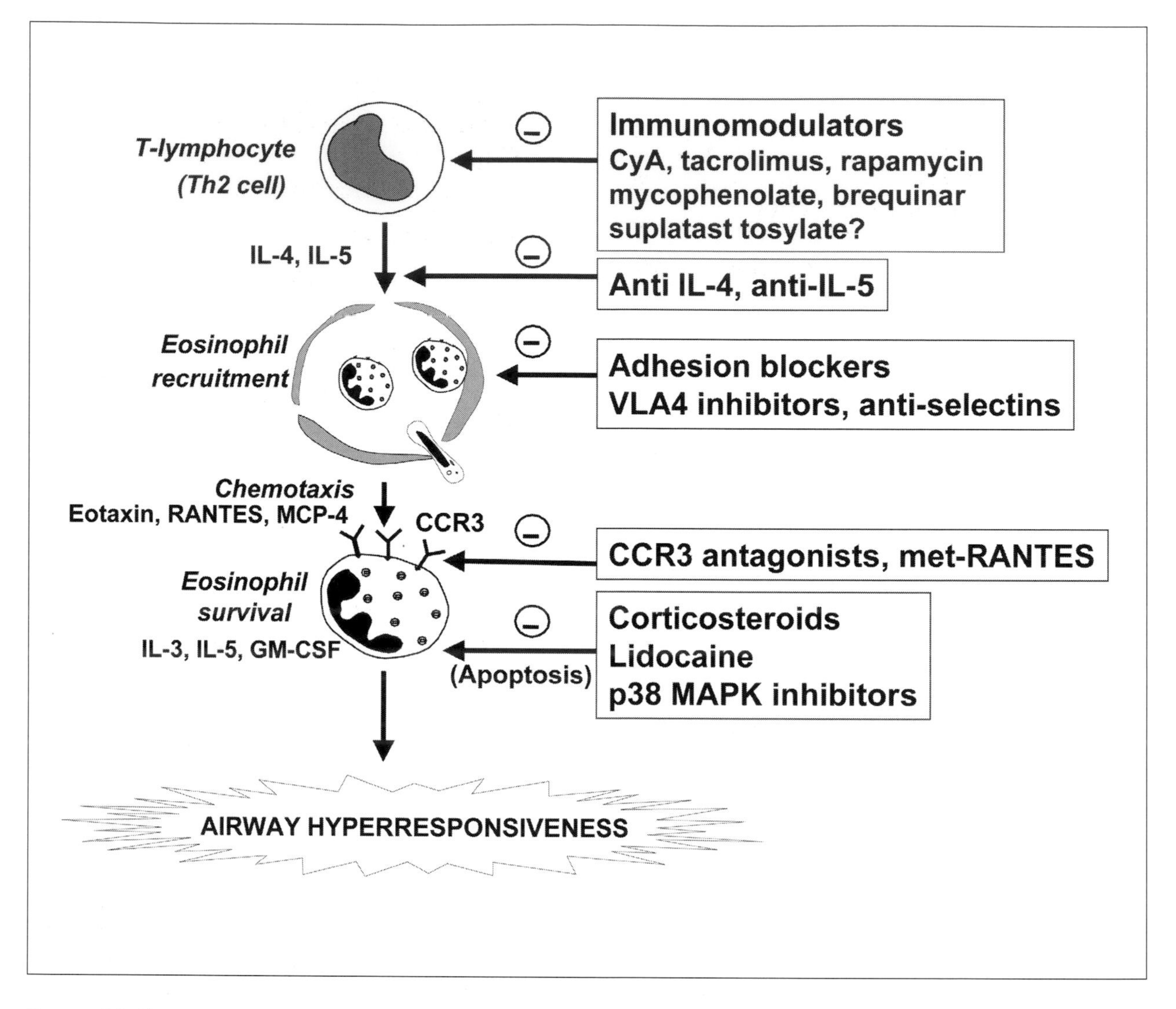

FIGURE 14. SEVERAL APPROACHES TO BLOCKING EOSINOPHILIC INFLAMMATION
IL, interleukin; Th2, T helper 2; VLA, vascular leukocyte adhesion molecule; CCR, CC chemokine receptor; MCP, macrophage chemotactic protein; GM-CSF, granulocyte-macrophage colony-stimulating factor (GM-CSF).

- None of the existing treatments for ASTHMA is disease-modifying, which means that the disease recurs as soon as treatment is discontinued.
- None of the existing treatments is curative, although it is possible that therapies which prevent the immune aberration of ALLERGY may have the prospects for a cure in the future.

COPD

In sharp contrast to ASTHMA there are few effective therapies in COPD, despite the fact that it is a common disease that is increasing world-wide [221,222].

The neglect of COPD is probably as a result of several factors:

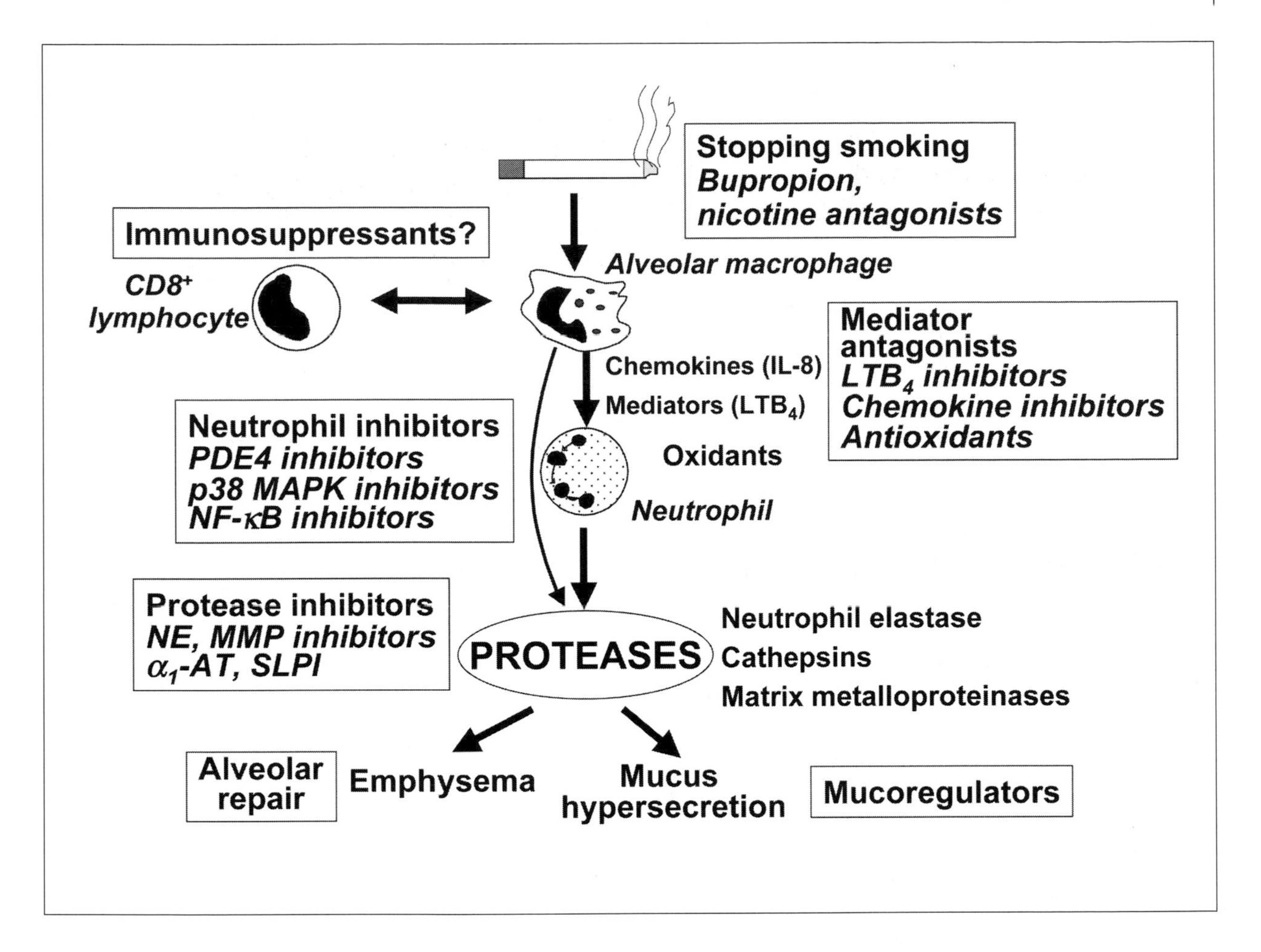

FIGURE 15. TARGETS FOR COPD THERAPY BASED ON CURRENT UNDERSTANDING OF THE INFLAMMATORY MECHANISMS
Cigarette smoke (and other irritants) activate macrophages in the respiratory tract that release neutrophil chemotactic factors, including interleukin-8 (IL-8) and leukotriene B_4 (LTB_4). These cells then release proteases that break down connective tissue in the lung parenchyma, resulting in emphysema, and also stimulate mucus hypersecretion. These enzymes are normally counteracted by protease inhibitors, including α_1-antitrypsin, secretory leukoprotease inhibitor (SLPI) and tissue inhibitor of matrix metalloproteinases (TIMP). Cytotoxic T cells ($CD8^+$) may also be involved in the inflammatory cascade.

- COPD is regarded as largely irreversible and is treated as poorly responsive ASTHMA.
- COPD is self-inflicted and therefore does not deserve investment.
- There are few satisfactory animal models.
- Relatively little is understood about the cell and molecular biology of this disease or even about the relative roles of small airways disease and parenchymal destruction.

None of the treatments currently available prevent the progression of the disease, and yet the disease is associated with an active inflammatory process that results in progressive obstruction of small airways and destruction of lung parenchyma. Increased understanding of COPD will identify novel targets for future therapy [224].

Development of new therapies

Several strategies have been adopted in the search for new therapies:

- Improvement of an existing class of drug. This is well exemplified by the increased duration of β_2-agonists with salmeterol and formoterol and of anticholinergics with tiotropium bromide, and with the improved pharmacokinetics of the inhaled corticosteroids fluticasone propionate, mometasone and budesonide, with increased first-pass metabolism and therefore reduced systemic absorption.
- Development of novel therapies through better understanding of the disease process. Examples are anti-IL-5 as a potential treatment of ASTHMA and PDE-4 inhibitors as an anti-inflammatory therapy for COPD.
- Serendipitous observations, often made in other therapeutic areas. Examples are TNF-α antagonists for airway diseases, derived from observations in other chronic inflammatory diseases.
- Identification of novel targets through gene and protein profiling. This approach will be increasingly used to identify the abnormal expression of genes (molecular genomics) and proteins (proteomics) from diseased cells, and through identification of single nucleotide POLYMORPHISMS (SNPs) that contribute to the disease process [225].

Mediator antagonists

Blocking the receptors or synthesis of inflammatory mediators is a logical approach to the development of new treatments for ASTHMA and COPD (Tab. 10). However, in both diseases many different mediators are involved and therefore blocking a single mediator may not be very effective, unless it plays a key role in the disease process [226]. Several specific mediator antagonists have been found to be ineffective in ASTHMA, including antagonists/inhibitors of histamine, thromboxane, platelet-activating factor, bradykinin and tachykinins. However, these blockers have often not been tested in COPD, in which different mediators are involved.

Inhibitors of leukotriene pathways

Cys-LT_1 receptor antagonists are currently used to treat patients with ASTHMA but are less effective than inhaled corticosteroids as discussed above. Cys-LTs play no clear role in COPD, but LTB_4 is elevated and may contribute to neutrophilic INFLAMMATION through the activation of BLT_1-receptors [227]. Several BLT_1-receptor antagonists are in development but may not be effective as there are several other chemotactic CHEMOKINES involved in COPD in addition to LTB_4. 5'-Lipoxygenase inhibitors are potentially useful in ASTHMA and COPD by blocking synthesis of cys-LT and LTB_4, but suffer the same limitations of receptor antagonists. More potent 5'-LO inhibitors than zileuton have proved difficult to develop. Phospholipase A_2 inhibitors which should prevent the availability of arachidonic acid, the precursor molecule for LEUKOTRIENES, have also proved very difficult to develop.

Endothelin antagonists

Endothelin is a peptide mediator which has potent bronchoconstrictor and vasoconstrictor effects [228]. There is evidence for increased expression in asthmatic airways and in pulmonary vessels of COPD patients with pulmonary hypertension. Endothelin also stimulates fibrosis and may be involved in the structural remodelling in both ASTHMA and COPD. Several endothelin receptor antagonists are in clinical development [229], but it may be difficult to assess their effect in airway disease as it may be necessary to measure their effects over very prolonged periods.

Anti-oxidants

Oxidative stress is important in ASTHMA and COPD, particularly during exacerbations [230]. Existing antioxidants include vitamins C and E and *N*-acetylcysteine and these drugs have weak effects, but more

Table 10. Inflammatory mediator inhibitors

Mediator	Inhibitor
Histamine	terfenadine, loratadine, cetirizine
Leukotriene D_4	zafirlukast, montelukast, pranlukast, zileuton, Bay-x1005
Leukotriene B_4	LY 293111
PAF	apafant, modipafant, bepafant
Thromboxane	ozagrel
Bradykinin	icatibant, WIN 64338
Adenosine	theophylline
Reactive oxygen species	N-acetyl-cysteine, ascorbic acid, glutathione analogs
Nitric oxide	aminoguanidine, L-NIL
Endothelin	bosentan
IL-1β	recombinant IL-1 receptor antagonist
TNF-α	TNF-antibody (infliximab), TNF soluble receptors (etanercept)
IL-4	IL-4 antibody
IL-5	IL-5 antibody (mepolizumab)
IL-8	CXCR2 antagonist
Eotaxin	CCR3 antagonist
Mast cell tryptase	APC366
Eosinophil basic proteins	Heparin

potent anti-oxidants are in development [231]. It is likely that these drugs would be useful in COPD and COPD, as oxidative stress is likely to be an important component of the disease, particularly during exacerbations, and it may contribute to corticosteroid resistance [140].

Inducible nitric oxide synthase inhibitors

NO production is increased in ASTHMA and COPD as a result of increased inducible NO synthase (iNOS) expression in the airways. NO and oxidative stress generate peroxynitrite which may nitrate proteins, leading to altered cell function. Several selective iNOS inhibitors are now in development [232] and one of these, L-N^6-(1-imminoethyl)lysine (L-NIL), gives a profound and long-lasting reduction in the concentrations of nitric oxide in exhaled breath [233].

Cytokine modifiers

CYTOKINES play a critical role in perpetuating and amplifying the INFLAMMATION in ASTHMA and COPD, suggesting that anti-CYTOKINES may be beneficial as therapies [234, 235]. Although most attention has focused on inhibition of CYTOKINES, some CYTOKINES are anti-inflammatory and might have therapeutic potential [236]. There are several approaches to inhibition of CYTOKINES, including inhibiting their synthesis, blocking with antibodies or soluble receptors and antagonism of their receptors and signal transduction pathways. Some CYTOKINES may play a critical role in the allergic inflammatory process, whereas others play a proinflammatory role in both diseases.

Anti-IL-5. IL-5 plays a pivotal role in eosinophilic INFLAMMATION and is also involved in eosinophil survival and priming. It is an attractive target in ASTHMA, as it is essential for eosinophilic INFLAMMATION, there do not appear to be any other CYTOKINES with a similar role, and lack of IL-5 in gene knock-out mice does not have any deleterious effect [237]. The major strategy for inhibiting IL-5 has been the development of blocking antibodies which inhibit eosinophilic INFLAMMATION and airway hyperresponsiveness in animal models of ASTHMA. This blocking effect may last for over 3 months after a single injection, making treatment of chronic ASTHMA with such a therapy a feasible proposition. Humanized monoclonal antibodies to IL-5 have been developed and a single injection reduces blood EOSINOPHILS for over 3 months and prevents eosinophil recruitment into the airways after allergen challenge [238]. However, this treatment has no effect on the early or late response to allergen challenge or on AHR, suggesting that EOSINOPHILS may be less important for these responses than previously believed. In a clinical study, anti-IL-5 similarly markedly reduced circulating EOSINOPHILS but had no effect on clinical parameters in asthmatic patients not controlled on inhaled corticosteroids [239]. This has suggested that EOSINOPHILS may be less critical in ASTHMA than previously believed, although EOSINOPHILS are not completely suppressed in the airways after anti-IL-5 therapy [239, 240].

Anti-IL-4. IL-4 is critical for the synthesis of IgE by B LYMPHOCYTES and to the development of Th2 cells. IL-4 receptor-blocking antibodies inhibit allergen-induced AHR, goblet cell metaplasia and pulmonary eosinophilia in a murine model [241]. Inhibition of IL-4 may therefore be effective in inhibiting allergic diseases. Nebulized soluble IL-4 receptors have shown some clinical EFFICACY in patients with moderate ASTHMA, and were effective when given as a once-weekly nebulization in preventing the deterioration in ASTHMA following reduction in inhaled corticosteroids [242, 243], but were not very effective in larger trials and have been discontinued.

Anti-IL-13. There is increasing evidence that IL-13 in mice mimics many of the features of ASTHMA, including AHR, increased IgE, mucus hypersecretion, fibrosis and release of eotaxin [244]. IL-13 signals through the IL-4 receptor α-chain, but may also activate different intracellular pathways, so that it may be an important target for the development of new therapies. A soluble IL-13Rα2-Fc fusion protein, which blocks the effects of IL-13 but not IL-4, has been used successfully to neutralise IL-13 both in mice *in vivo* [245]. The IL-13Rα2-Fc fusion protein markedly inhibits the eosinophilic INFLAMMATION, AHR and mucus secretion induced by allergen exposure. IL-13 is expressed in ASTHMA to a much greater extent than IL-4, indicating that it may be a more important target. This suggests that development of IL-13 blockers, such as a humanised IL-13 ANTIBODY or the IL-13Rα2, may be a useful approach to the treatment of established allergic diseases.

Anti-IL-9. IL-9 is produced by Th2 cells and appears to have an amplifying effect on the expression of IL-4 and IL-5 and on mucus secretion and mast-cell survival. This suggests that IL-9 may be a useful upstream target in ASTHMA and humanised monoclonal IL-9 antibodies are now in clinical development [246].

Anti-TNF. TNF-α may play a key role in amplifying atopic INFLAMMATION, through the activation of NF-κB, AP-1 and other transcription factors. TNF-α production is increased in ASTHMA and COPD, and in the latter may be associated with the cachexia and weight loss that occurs in some patients with severe COPD [247, 248]. In rheumatoid arthritis and inflammatory bowel disease a blocking ANTIBODY to TNF-α (infliximab) has produced remarkable clinical responses, even in patients who are relatively unresponsive to steroids. TNF antibodies or soluble TNF-receptors (entanercept) are a logical approach to ASTHMA therapy, particularly in patients with severe disease. There is also a search for small molecule inhibitors of TNF-α, of which the most promising are inhibitors of TNF-α converting enzyme (TACE) as these may be given orally. Other new anti-inflammatory treatments, including phosphodiesterase-4 inhibitors and p38 mitogen-activated protein (MAP) kinase inhibitors are also effective in inhibiting TNF-α release from inflammatory cells.

Chemokine inhibitors

Many CHEMOKINES are involved in COPD and ASTHMA and play a key role in recruitment of inflammatory cells, such as EOSINOPHILS, NEUTROPHILS, MACROPHAGES and LYMPHOCYTES [249]. Chemokine receptors are attractive targets, as they are 7 transmembrane spanning proteins, like adrenergic receptors, and small molecule inhibitors are now in development [250].

CCR2 inhibitors. MCP-1 recruits and activates CCR2 on MONOCYTES and T LYMPHOCYTES and blocking MCP-1 with neutralising antibodies reduces recruitment of both T CELLS and EOSINOPHILS in a murine model of ovalbumin-induced airway INFLAMMATION, with a marked reduction in AHR [251]. CCR2 may also play an important role in COPD, as MCP-1 levels are increased in sputum and lungs of patients with COPD [252]. MCP-1 is a potent chemoattractant of MONOCYTES and may therefore be involved in the recruitment of MACROPHAGES in COPD. Indeed the chemoattractant effect of induced sputum from patients with COPD is abrogated by an ANTIBODY to CCR2.

CCR3 antagonists. In ASTHMA most attention has focused on blockade of CCR3 receptors that are predominantly expressed on EOSINOPHILS and mediate the chemotactic effect of eotaxin, RANTES and MCP-4. Several small molecule inhibitors of CCR3, including UCB35625, SB-297006 and SB-328437, are effective in inhibiting eosinophil recruitment in allergen models of ASTHMA [253, 254] and drugs in this class are currently undergoing clinical trials in ASTHMA. It was originally thought that CCR3 were restricted to EOSINOPHILS, but there is evidence for their expression on Th2 cells and mast cells, so that these inhibitors may have a more widespread effect than on EOSINOPHILS alone, making them potentially more valuable in asthma treatment.

CCR4 antagonists. CCR4 are expressed on Th2 cells and may be important in the recruitment of Th2 cells to the asthmatic airways in response to macrophage-derived CYTOKINE (MDC) [255].

CXCR antagonists. In COPD the focus of attention has been the blockade of IL-8, which attracts and activates NEUTROPHILS via two receptors, the specific low-affinity CXCR1 and the high-affinity CXCR2, which is shared by other CXC CHEMOKINES. A small molecule inhibitor of CXCR2 has now been developed and may lead to clinical trials in COPD [256].

Inhibitory cytokines

Although most CYTOKINES have anti-inflammatory effects, others have the potential to inhibit INFLAMMATION in ASTHMA and COPD [236]. While it may not be feasible or cost-effective to administer these proteins as long-term therapy, it may be possible to develop drugs that increase the release of these endogenous CYTOKINES or activate their receptors and specific signal transduction pathways.

IL-1 receptor antagonist. IL-1 receptor antagonist (anakinra) binds to IL-1 receptors and blocks the action of IL-1β. In experimental animals it reduces AHR and clinical studies are in progress [257].

IL-10. IL-10 is a potent anti-inflammatory CYTOKINE that inhibits the synthesis of many inflammatory proteins, including CYTOKINES (TNF-α, GM-CSF, IL-5, CHEMOKINES) and inflammatory enzymes (iNOS) that are over-expressed in ASTHMA [61]. In sensitized animals IL-10 suppresses the inflammatory response to allergen, suggesting that IL-10 might be effective as a treatment in ASTHMA. Indeed, there may be a defect in IL-10 transcription and secretion from MACROPHAGES in ASTHMA [258, 259] and reduced levels of IL-10 are found in sputum of asthmatic and COPD patients [260].

IL-10 might also be a therapeutic strategy in COPD, since it inhibits TNF-α, IL-8 and matrix metalloproteinase secretion, and increases tissue inhibitors of metalloproteinases. Recombinant human IL-10 has proved to be effective in controlling inflammatory bowel disease, where similar CYTOKINES are expressed and may be given as a weekly injection. In the future, drugs which activate the unique signal transduction pathways activated by the IL-10 receptor or drugs that increase endogenous production of IL-10 may be developed.

Interferon-γ. IFN-γ inhibits Th2 cells and should therefore reduce atopic INFLAMMATION. In sensitized animals, nebulised IFN-γ inhibits eosinophilic INFLAMMATION induced by allergen exposure. Administration of IFN-γ by nebulization to asthmatic patients did not significantly reduce eosinophilic INFLAMMATION, however, possibly due to the difficulty in obtaining a high enough concentration locally in the airways [261].

IL-12. IL-12 is the endogenous regulator of Th1 cell development and determines the balance between Th1 and Th2 cells. IL-12 administration to rats inhibits allergen-induced INFLAMMATION and inhibits sensitisation to ALLERGENS. Recombinant human IL-12 has been administered to humans and has several toxic effects which are diminished by slow escalation of the dose. Infusion of human recombinant IL-12 has an inhibitory effect on EOSINOPHILS in asthmatic patients but has significant systemic side-effects that preclude its clinical development [262].

Neuromodulators

Neural mechanisms are important in symptoms of ASTHMA and COPD and cough is a prominent symptom that may require specific therapy. Airway sensory nerves become sensitized in ASTHMA and COPD and may be a target for inhibition. Many prejunctional receptors are localized to sensory nerve endings, including opioid and cannabinoid receptors [263, 264]. These nerve endings may be activated via ion channels, such as vanilloid receptors, for which selective inhibitors have now been developed. These drugs may be of particular value in treating cough (see below). There is some evidence that neurogenic INFLAMMATION may be implicated in ASTHMA and COPD, and there are various strategies to inhibit neurogenic INFLAMMATION, which is due to the release of neuropeptides such as substance P from unmyelinated sensory nerves in the airways. These include tachykinin receptor antagonists, such as the non-peptide antagonist CP99,994 and sendide, or strategies to inhibit sensory nerve activation.

Enzyme inhibitors

Several enzymes are involved in chronic INFLAMMATION and inhibitors of several enzymes are in development for the treatment of airway diseases. Enzymes may result in the formation of inflammatory mediators, such as eicosanoids, or may have direct inflammatory effects, such as tryptase. Enzymes are also involved in the tissue remodelling that occurs in ASTHMA and COPD and a range of proteinases is implicated in the tissue destruction of emphysema.

Tryptase inhibitors

Mast-cell tryptase has several effects on airways, including increasing responsiveness of airway smooth muscle to constrictors, increasing plasma exudation, potentiating eosinophil recruitment and stimulating fibroblast proliferation. Some of these effects are mediated by activation of the proteinase-activated receptor PAR2. A tryptase inhibitor, APC366, is effective in a sheep model of allergen-induced ASTHMA [265], but was only poorly effective in asthmatic patients [266]. More potent tryptase inhibitors and PAR2 antagonists are now in development. It is unlikely that tryptase inhibitors will have a role in COPD as mast-cell tryptase is not involved.

Neutrophil elastase inhibitors

NEUTROPHIL elastase (NE), a neutral serine protease, is a major constituent of lung elastolytic activity. In addition it potently stimulates mucus secretion and induces IL-8 release from epithelial cells and may therefore perpetuate the inflammatory state. This has led to a search for NEUTROPHIL elastase inhibitors. Peptide and non-peptide inhibitors have been developed, but there are few clinical studies with NE inhibitors in COPD. The NE inhibitor MR889 administered for 4 weeks showed no overall effect on plasma elastin-derived peptides or urinary desmosine (markers of elastolytic activity) [267].

Cathepsin inhibitors

NE is not the only proteolytic enzyme secreted by NEUTROPHILS. Cathepsin G and proteinase 3 have elastolytic activity and may need to be inhibited together with NEUTROPHIL elastase. Cathepsins (cathepsins B, L and S) are cysteine proteases that are also released from MACROPHAGES. Cathepsin inhibitors are now in development.

Matrix metalloproteinase inhibitors

Matrix metalloproteinases (MMP) are a group of over 25 closely related endopeptidases that are capable of degrading all of the components of the extracellular matrix of lung parenchyma, including elastin, collagen, proteoglycans, laminin and fibronectin. MMPs are produced by NEUTROPHILS, alveolar MACROPHAGES and airway epithelial cells. Increased levels of collagenase (MMP-1) and gelatinase B (MMP-9) have been detected in bronchoalveolar lavage fluid of patients with emphysema. Lavaged MACROPHAGES from patients with emphysema express more MMP-9 and MMP-1 than cells from control subjects, suggesting that these cells, rather than NEUTROPHILS, may be the major cellular source [268]. Alveolar MACROPHAGES also express a unique MMP, MACROPHAGE metalloelastase (MMP-12). MMP-12 knock-out mice do not develop emphysema and do not show the expected increases in lung MACROPHAGES after long-term exposure to cigarette smoke [269]. Tissue inhibitors of metalloproteinases (TIMP) are endogenous inhibitors of these potent enzymes. There are several approaches to inhibiting MMPs [270]. One approach is to enhance the secretion of TIMPs and another is to inhibit the induction of MMPs in COPD. MMPs may show increased expression with cigarette smoking through induction in response to inflammatory CYTOKINES, oxidants and other enzymes, such as NE. Non-selective MMP inhibitors, such as marimastat (BB-2516), have been developed. Side-effects of such drugs may be a problem in long-term use, however. More selective inhibitors of individual MMPs, such as MMP-9 and MMP-12, are in development and are likely to be better tolerated in chronic therapy. However, it is still not clear whether there is one predominant MMP in COPD or whether a broad spectrum inhibitor will be necessary.

α_1-Antitrypsin

The association of α_1-AT deficiency with early-onset emphysema suggested that this endogenous inhibitor of NE may be of therapeutic benefit in COPD. Cigarette smoking inactivates α_1-AT, resulting in unopposed activity of NE and cathepsins. Extraction of α_1-AT from human plasma is very expensive and extracted α_1-antitrypsin is only available in a few countries. This treatment has to be given intravenously and has a half-life of only 5 days. This has led to the development of inhaled formulations, but these are inefficient and expensive. Recombinant α_1-AT with amino acid substitutions to increase stability may result in a more useful product [271]. GENE THERAPY is another possibility using an adenovirus vector or liposomes, but there have been major problems in developing efficient delivery systems. There is a particular problem with gene transfer in α_1-AT deficiency in that large amounts of protein (1–2 g) need to be synthesized each day. There is no evidence that α_1-AT treatment would halt the progression of COPD and emphysema in patients who have normal plasma concentrations.

Serpins

Other serum protease inhibitors (serpins), such as elafin, may also be important in counteracting elastolytic activity in the lung. Elafin, an elastase-specific inhibitor, is found in bronchoalveolar lavage and is synthesized by epithelial cells in response to inflammatory stimuli [272]. Serpins may not be able to inhibit NE at the sites of elastin destruction, due to tight adherence of the inflammatory cell to connective tissue. Furthermore, these proteins may become inactivated by the inflammatory process and the action of oxidants, so that they may not be able to adequately counteract elastolytic activity in the lung unless used in conjunction with other therapies.

Secretory leukoprotease inhibitor (SLPI) is a 12 kDa serpin that appears to be a major inhibitor of elastase activity in the airways and is secreted by airway epithelial cells [273]. Recombinant human SLPI given by aerosolization increases anti-neutrophil elastase activity in epithelial lining fluid for over 12 h, indicating potential therapeutic use [274].

New anti-inflammatory drugs

Inhaled corticosteroids are by far the most effective therapy for ASTHMA, yet are ineffective in COPD. Thus for ASTHMA, one strategy has been to develop safer

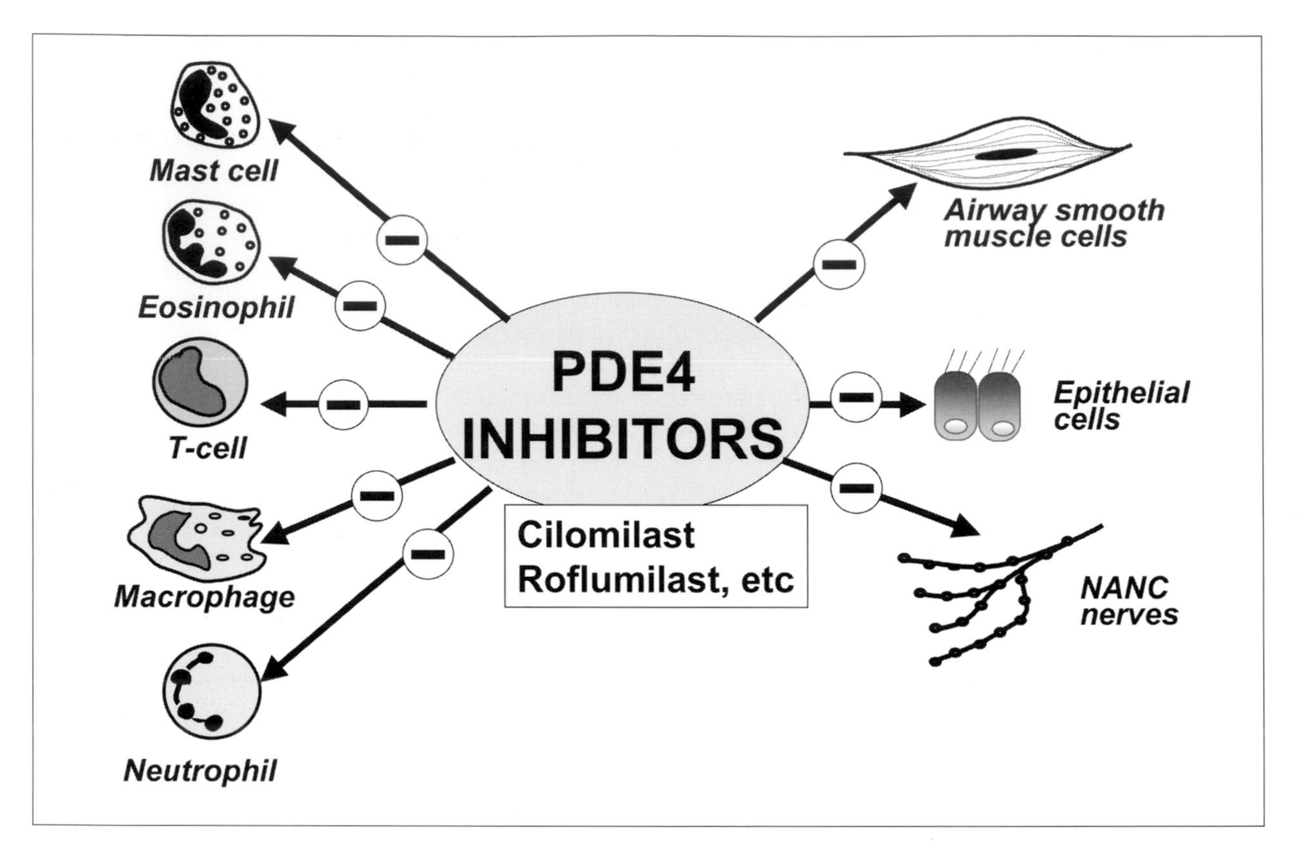

FIGURE 16
Inhibitors of phosphodiesterase (PDE)-4 may be useful anti-inflammatory treatments in COPD and asthma as they inhibit several aspects of the inflammatory process.

inhaled corticosteroids or drugs that mimic their effects, whereas in COPD non-steroidal anti-inflammatory treatments are needed.

Novel corticosteroids

Systemic side-effects of corticosteroids are largely mediated via binding of glucocorticoid receptors (GR) to DNA and increased gene transcription, whereas their anti-inflammatory effects are due to interaction of GR with coactivator molecules and repression of inflammatory genes. It may be possible to dissociate DNA binding, which requires a GR dimer, from coactivator molecule binding that requires only a monomer [121]. Several dissociated corticosteroids have now been synthesized and a separation between *trans*-activation (DNA binding) and *trans*-repression (coactivator interaction) has been demonstrated in gene reporter systems and in intact cells *in vitro* [160], although this is less clear in animal models *in vivo* [275]. Determination of the crystal structure of the glucocorticoid receptor may facilitate the development of novel corticosteroids [276].

Phosphodiesterase inhibitors

PDEs break down cyclic nucleotides that inhibit cell activation and at least 10 families of enzymes have now been discovered. Theophylline, long used as an ASTHMA treatment, is a weak but non-selective PDE inhibitor. PDE4 is the predominant family of PDEs in inflammatory cells, including mast cells, EOSINOPHILS, NEUTROPHILS, T LYMPHOCYTES, MACROPHAGES and structural cells such as sensory nerves and epithelial cells [58, 277] (Fig. 16). This has suggested that PDE4 inhibitors would be useful as an anti-inflammatory treatment in both ASTHMA and COPD. In animal models of ASTHMA, PDE4 inhibitors reduce eosinophil infiltration and AHR responses to allergen [58, 278]. Several PDE4 inhibitors have been tested in ASTHMA, but with disappointing results, as the dose has been limited by side-effects, particularly nausea and vomiting. However, in COPD a PDE4 inhibitor has shown promising results in terms of improvement in lung function, reduced symptoms and improved quality of life [279]. This is associated with an anti-inflammatory effect, with reduced numbers of MACROPHAGES and LYMPHOCYTES [280]. Side-effects, particularly nausea and emesis, headaches and gastrointestinal disturbance are the major side-effects that limit the dose and appear to be mechanism-related. There are 4 subfamilies of PDE4 and it now seems that PDE4D is the major enzyme mediating vomiting, whereas PDE4B is important in anti-inflammatory effects [281]. This suggests that more selective drugs might have a greater therapeutic ration.

PDE7 is a novel subtype of PDE that is expressed in a number of cell types, including T LYMPHOCYTES, but PDE7 inhibition does not have consistent anti-inflammatory effects [282].

Transcription factor inhibitors

Transcription factors, such as NF-κB and AP-1, play an important role in the orchestration of chronic INFLAMMATION and many of the inflammatory genes that are expressed in ASTHMA and COPD are regulated by these transcription factors [283]. This has prompted a search for specific blockers of these transcription factors [284]. NF-κB is naturally inhibited by the inhibitory protein IκB, which is degraded after activation by specific kinases. Small molecule inhibitors of the IκB kinase IKK-2 are now in clinical development [285]. These drugs may be of particular value in COPD where corticosteroids are largely ineffective. However, there are concerns that inhibition of NF-κB may cause side-effects, such as increased susceptibility to infections, which has been observed in gene disruption studies when components of NF-κB are inhibited.

MAP kinase inhibitors

There are three major mitogen-activated protein (MAP) kinase pathways and there is increasing recognition that these pathways are involved in chronic INFLAMMATION [286]. There has been particular interest in the p38 MAP kinase pathway that is blocked by a novel class of drugs, the CYTOKINE suppressant anti-inflammatory drugs (CSAIDs), such as SB203580 and RWJ67657. These drugs inhibit the synthesis of many inflammatory CYTOKINES, CHEMOKINES and inflammatory enzymes [287]. Interestingly, they appear to have a preferential inhibitory effect on synthesis of Th2 compared to Th1 CYTOKINES, indicating their potential application in the treatment of atopic diseases [288]. Furthermore, p38 MAPK inhibitors have also been shown to decrease eosinophil survival by activating apoptotic pathways [289]. p38 MAPK inhibitors are also indicated in COPD as inhibition of this enzyme inhibits the expression of TNF-α and IL-8, as well as MMPs [290]. Whether this new class of anti-inflammatory drugs will be safe in long-term studies remains to be established.

Tyrosine kinase inhibitors

Syk kinase is a protein tyrosine kinase that plays a pivotal role in signalling of the high affinity IgE receptor (FcεRI) in mast cells and in *syk*-deficient mice mast-cell degranulation is inhibited, suggesting that this might be an important potential target for the development of mast-cell stabilising drugs [291]. *Syk* is also involved in ANTIGEN receptor signalling of B and T LYMPHOCYTES and in eosinophil survival in response to IL-5 and GM-CSF [292], so that *syk* inhibitors might have several useful beneficial effects in atopic diseases. Another tyrosine kinase, *lyn*, is upstream of syk and an inhibitor of lyn kinase, PP1, has an inhibitory effect on inflammatory and mast-cell activation [293]. Since *lyn* and *syk* are widely distributed in the IMMUNE SYSTEM, there are doubts about the long-term safety of selective inhibitors, however.

Epidermal growth factor receptor kinases play a critical role in mucus hypersecretion in response to irritants [294] and therefore inhibition of this enzyme by small molecule inhibitors, such as gefitinib, may be a useful strategy in treating chronic bronchitis [295].

Immunosuppressants

T LYMPHOCYTES may play a critical role in initiating and maintaining the inflammatory process in ASTHMA via the release of CYTOKINES that result in eosinophilic INFLAMMATION, suggesting that T-cell inhibitors may be useful in controlling asthmatic INFLAMMATION. The non-specific immunomodulator cyclosporin A has limited EFFICACY in ASTHMA, and side-effects, particularly nephrotoxicity, limit its clinical use. The possibility of using inhaled cyclosporin A is now being explored, since in animal studies the inhaled drug is effective in inhibiting the inflammatory response in experimental ASTHMA [296]. IMMUNOMODULATORS, such as tacrolimus (FK506) and rapamycin, appear to be more potent but are also toxic and may offer no real advantage. Novel IMMUNOMODULATORS that inhibit purine or pyrimidine pathways, such as mycophenolate mofetil, leflunomide and brequinar sodium, may be less toxic and therefore of greater potential value in ASTHMA therapy [297]. One problem with these non-specific IMMUNOMODULATORS is that they inhibit both Th1 and Th2 cells, and therefore do not restore the imbalance between these Th1 and Th2 cells in atopy. They also inhibit suppresser T CELLS (Tc cells) that may modulate the inflammatory response. Selective inhibition of Th2 cells may be more effective and better tolerated and there is now a search for such drugs.

The role of IMMUNOMODULATORS in COPD is even less certain. There is an increase in Tc1 cells in patients with COPD and these may play a role in emphysema [298], but the usefulness of IMMUNOMODULATORS in COPD has not yet been assessed.

Cell adhesion blockers

Infiltration of inflammatory cells into tissues is dependent on adhesion of blood-borne inflammatory cells to endothelial cells prior to migration to the inflammatory site. This depends upon specific glycoprotein ADHESION MOLECULES, including integrins and selectins, on both LEUKOCYTES and on endothelial cells, which may be up-regulated or show increased binding affinity in response to various inflammatory stimuli, such as CYTOKINES or lipid mediators. Drugs which inhibit these ADHESION MOLECULES therefore may prevent inflammatory cell infiltration and may be useful in airway disease. The interaction between VLA-4 and VCAM-1 is important for eosinophil INFLAMMATION and humanized antibodies to VLA-4 ($\alpha_4\beta_1$) have been developed [299]. Small molecule peptide inhibitors of VLA-4 have subsequently been developed which are effective in inhibiting allergen-induced responses in sensitised sheep [300].

Inhibitors of selectins, particularly L-selectin and E-SELECTIN, based on the structure of sialyl-Lewis^x inhibit the influx of inflammatory cells in response to inhaled allergen [301]. These glycoprotein inhibitors, which may inhibit neutrophilic and eosinophilic INFLAMMATION, are now in trial in ASTHMA and COPD.

Anti-allergic drugs

Atopy underlies most ASTHMA and this has prompted a search for anti-inflammatory agents that would selectively target the atopic disease process. Such treatments then may be effective in controlling concomitant allergic diseases.

Co-stimulation inhibitors

Co-stimulatory molecules may play a critical role in augmenting the interaction between ANTIGEN-PRESENTING CELLS and $CD4^+$ T LYMPHOCYTES. The interaction between B7 and CD28 may determine whether a Th2 type cell response develops, and there is some evidence that B7-2 (CD86) skews towards a Th2 response. Blocking antibodies to B7-2 inhibit the development of specific IgE, pulmonary eosinophilia and AHR in mice, whereas antibodies to B7-1 (CD80) are ineffective [302]. A molecule on activated T CELLS, CTL4, appears to act as an endogenous inhibitor of T-cell activation and a soluble fusion protein construct, CTLA4-Ig, is also effective in blocking AHR in a murine model of ASTHMA [303]. Anti-CD28, anti-B7-2 and CTLA4-Ig also block the proliferative response of T CELLS to allergen [304], indicating that these are potential targets for novel therapies that should be effective in all atopic diseases.

Th2 cell inhibitors

Non-selective T CELL suppressants, such as cyclosporin A and tacrolimus, may be relatively ineffective in ASTHMA as they inhibit all types of T CELL. $CD4^+$ T CELLS have been implicated in ASTHMA and a chimeric ANTIBODY directed against $CD4^+$ (keliximab) which reduces circulating $CD4^+$ cells appears to have some beneficial effect in ASTHMA [305], although long-term safety of such a treatment is a problem. There has been a search for selective inhibitors of Th2 cells by identifying features that differentiate Th1 and Th2 cells. The transcription factor GATA-3 appears to be of particular importance in murine and human Th2 cells [306] and may be a target for selective IMMUNOMODULATORY drugs.

Specific immunotherapy

Subcutaneous injection of small amounts of purified allergen has been used for many years in the treatment of ALLERGY, but it is not very effective in ASTHMA and has a risk of severe, sometimes fatal, anaphylactic responses. Cloning of several common allergen genes has made it possible to prepare recombinant ALLERGENS for injection, although this purity may detract from their allergenicity as most natural ALLERGENS contain several proteins. Intramuscular injection of rats with plasmid DNA expressing house-dust-mite allergen results in its long-term expression and prevents the development of IgE responses to inhaled allergen [307]. This suggests that allergen gene immunisation might be a useful therapeutic strategy in the future.

Small peptide fragments of allergen (EPITOPES) are able to block allergen-induced T-cell responses without inducing anaphylaxis. T-cell-derived peptides from cat allergen (*fel* d1) appear to be effective in blocking allergen responses to cat dander, although they may provoke an isolated later reaction [211].

Vaccination

A relative lack of infections may be a factor predisposing to the development of atopy in genetically predisposed individuals, leading to the concept that vaccination to induce protective Th1 responses prevent sensitisation and thus prevent the development of atopic diseases. BCG inoculation in mice 14 days before allergen sensitization reduces the formation of specific IgE in response to allergen and the eosinophilic response and AHR responses to allergen, with an increase in production of IFN-γ [308]. This has prompted several clinical trials of BCG and heat-killed *Mycobacterium vaccae* to prevent the development of atopy, although so far results are not impressive [309, 310]. Immunostimulatory DNA sequences, such as unmethylated cytosine-guanosine dinucleotide-containing oligonucleotides (CpG ODN), are also potent inducers of Th1 CYTOKINES, and in mice, administration of CpG ODN increases the ratio of Th1 to Th2 cells, decreases formation of specific IgE and reduces the eosinophilic response to allergen, an effect which lasts for over 6 weeks [311, 312]. These promising animal studies encourage the possibility that vaccination might prevent or even cure atopic ASTHMA in the future.

Gene therapy

Since atopic diseases are polygenic, it is unlikely that GENE THERAPY will be of value in long-term therapy. However, understanding the genes involved in atopic diseases and in disease severity may identify new molecular targets [313, 314] and may also predict the response to different forms of therapy (pharmacogenetics) [315].

Gene transfer

Transfer of anti-inflammatory genes may provide specific anti-inflammatory or inhibitory proteins in a convenient manner and gene transfer has been shown to be feasible in animals using viral vectors [316]. Anti-inflammatory proteins relevant to ASTHMA include IL-10, IL-12, IL-18 and IκB. Antiproteases, such as α_1-antitrypsin, TIMPs and SLPI, may also be useful in treating COPD, although large amounts of protein are needed to neutralize proteases [317].

Antisense oligonucleotides

An inhaled antisense oligonucleotide directed against the adenosine A_1-receptor reduces AHR in a rabbit model of ASTHMA, demonstrating the potential of this approach in treating ASTHMA [318]. Respirable antisense oligonucleotides (RASONS) are a novel approach to ASTHMA therapy and clinical trials with the A_1-receptor oligonucleotide (EPI-2010) have shown that this therapy is well tolerated [319]. Suitable target genes may be IL-13 or CCR3, as well as novel genes discovered through the human genome project.

Impact of molecular genetics

There is much interest in the genetics of ASTHMA as a means of identifying novel and more specific drug targets. ASTHMA clearly involves many genes and environmental stimuli are very important in interacting with genetic factors, so it will be impossible to identify a single target. Genetic POLYMORPHISMS (variations in gene structure that result in differences in protein and often function) may account for differences in severity of ASTHMA between patients, but also differences in response to therapy. In the future it may be possible to predict which patients are most likely to show response to particular drugs (pharmacogenomics).

Other drugs

Several other types of drug are used in the treatment of airway diseases and these often have poor EFFICACY.

Mucoregulators

Many pharmacological agents may influence the secretion of mucus in the airways, but there are few drugs which have useful clinical effects [320]. Mucus hypersecretion occurs in chronic bronchitis and ASTHMA. In chronic bronchitis mucus hypersecretion is related to chronic irritation by cigarette smoke and may involve neural mechanisms and the activation of NEUTROPHILS to release enzymes such as NEUTROPHIL elastase and proteinase-3, which have powerful effects on mucus secretion [321]. Mast-cell-derived chymase is also a potent mucus secretagogue. This suggests that several classes of drug may be developed to control mucus hypersecretion.

Sensory nerves and neuropeptides are important in submucosal gland secretion (which predominates in proximal airways) and goblet cell secretion (more peripheral airways) [322]. Opioids and K^+ channel openers inhibit mucus secretion mediated via sensory neuropeptide release, and peripherally acting opioids may be developed to control mucus hypersecretion due to irritants in the future [323].

Several drugs reduce the viscosity of sputum *in vitro*. These drugs are usually derivatives of cysteine and reduce the disulfide bridges that bind glycoproteins to other proteins such as albumin and secretory IgA. In addition these drugs act as antioxidants and may therefore reduce airway INFLAMMATION. Orally administered *N*-acetylcysteine, carbocysteine, methylcysteine and bromhexine are well tolerated, but clinical studies in chronic bronchitis, ASTHMA and bronchiectasis have been disappointing. However, a recent systematic review of many studies has demonstrated a small benefit in terms of reducing exacerbations [324]. Most of the benefit is derived from N-acetylcysteine [325], but it is uncertain whether this relates to its mucolytic activity or to its action as an antioxidant.

Epidermal growth factor plays a critical role in airway mucus secretion from goblet cells and submucosal glands and appears to mediate the mucus secretory response to several secretagogues, including oxidative stress, cigarette smoke and inflammatory CYTOKINES [294, 326]. Small molecule inhibitors of EGF receptor kinase, such as gefitinib, have now been developed for clinical use. There has been some concern about interstitial lung disease in some patients with small-cell lung cancer treated with gefitinib, but it is not yet certain if this is related to EGF inhibition [327].

Another novel approach involves inhibition of calcium-activated chloride channels (CACC), which are important in mucus secretion from goblet cells. Activation of human hCLCA1 induces mucus secretion and mucus gene expression and may therefore be a target for inhibition. Small molecule inhibitors of CACC, such as niflumic acid and MSI 1956, have been developed [328].

Expectorants are drugs which taken orally enhance the CLEARANCE of mucus. Although they are commonly prescribed, there is little or no objective evidence for their EFFICACY. Such drugs are often emetics (guafenesin, ipecac, ammonium chloride) that are given in sub-emetic doses on the basis that gastric irritation may stimulate an increase in mucus production via a reflex mechanism. However there is no good evidence for this assumption. In patients who find it difficult to clear mucus, adequate hydration and inhalation of steam may be of some benefit.

DNAse (dornase alfa) reduces mucus viscosity in sputum of patients with cystic fibrosis and is indicated if there is significant symptomatic and lung function improvement after a trial of therapy [329]. There is no evidence that dornase alfa is effective in COPD or ASTHMA.

Antitussives

Despite the fact that cough is a common symptom of airway disease, its mechanisms are poorly understood and current treatment in unsatisfactory [330]. Because cough is a defensive reflex, its suppression may be inappropriate in bacterial lung infection. Before treatment with antitussives it is important to identify underlying causal mechanisms which may require therapy. Treatments, such as opioids, may act centrally on the "cough centre", whereas other treatments may act on airway sensory nerves.

Opiates have a central MECHANISM OF ACTION on the medullary cough centre, but there is some evidence that they may have additional peripheral action on cough receptors in the proximal airways. Codeine and pholcodeine are commonly used, but there is little evidence that they are clinically effective [331]. Morphine and methadone are effective but only indicated in intractable cough associated with bronchial carcinoma. A peripherally acting opioid agonist 443C81 does not appear to be effective [332].

ASTHMA commonly presents as cough and the cough will usually respond to bronchodilators, to inhaled steroids. A syndrome characterized by cough in association with sputum eosinophilia but no airway hyperresponsiveness and termed 'eosinophilic bronchitis' responds to steroids and may be regarded as pre-asthma [333]. Non-asthmatic cough does not respond to inhaled steroids, but sometimes responds to cromones or anticholinergic therapy. The cough associated with post-nasal drip of sinusitis responds to antibiotics, nasal decongestants and intranasal steroids. The cough associated with ACE inhibitors responds to withdrawal of the drug and to cromones. In some patients there may be underlying gastroesophageal reflux, which leads to cough by a reflex mechanism and occasionally by acid aspiration. This cough responds to effective suppression of gastric acid with an H_2-receptor antagonist or more effectively to a proton pump inhibitor omeprazole [334].

Some patients have an intractable cough which often starts following a severe respiratory tract infection. When no other causes for this cough are found it is termed idiopathic and may be due to hyperesthesia of airway sensory nerves. This is supported by the fact that these patients have an increased responsiveness to tussive stimuli such as capsaicin. This form of cough is difficult to manage. It may respond to nebulized lidocaine, but this is not practical for long-term management and novel therapies are needed [335].

There is clearly a need to develop new more effective therapies for cough, particularly drugs which act peripherally. There are close analogies between chronic cough and sensory hyperesthesia, so it is likely that new therapies are likely arise from pain research [336].

Summary

Pharmacological therapies are the mainstay of management of ASTHMA and COPD. Current bronchodilators relax airway smooth muscle and include β_2-adrenergic agonists, muscarinic receptor antagonists and theophylline. Long-acting inhaled β_2-agonists are an important advance and these are a very effective add-on therapy to controller drugs. Although new classes of bronchodilator, such as potassium channel openers and vasoactive intestinal peptide analogues, have been explored, they are unlikely to be as effective as β_2-agonists which act as functional antagonists to counteract all bronchoconstrictor mechanisms. Controller drugs act on the underlying disease by suppressing the inflammatory process. Corticosteroids are by far the most effective anti-inflammatory treatment in ASTHMA and are effective in almost all patients, but are poorly effective in COPD in which the INFLAMMATION is steroid-resistant. Cromones are much less effective in ASTHMA and are now little used. Theophylline may also have anti-inflammatory effects in low doses and is a useful add-on therapy in more severe ASTHMA. Many mediators are involved in ASTHMA and COPD, so blocking a single mediator is unlikely to have a major beneficial effect. Anti-leukotrienes have relatively weak effects in ASTHMA compared to corticosteroids. Steroid-sparing therapies, such as methotrexate and cyclosporin A, are only indicated in patients who have side-effects from maintenance oral steroids and these drugs often have worse side-effects, so are little used. Anti-IgE has recently been introduced for the treatment of very severe ASTHMA. There are several new classes of controller drug now in development for ASTHMA and COPD, including phosphodiesterase-4 inhibitors and chemokine receptor antagonists.

Selected readings

Barnes PJ (2004) New drugs for asthma. *Nat Rev Drug Discov* 3: 831–844

Barnes PJ, Adcock IM (2003) How do corticosteroids work in asthma? *Ann Intern Med* 139: 359–370

Barnes PJ, Hansel TT (2004) Prospects for new drugs for chronic obstructive pulmonary disease. *Lancet* 364: 985–996

Rennard SI (2004) Treatment of stable chronic obstructive pulmonary disease. *Lancet* 364: 791–802

Sin DD, McAlister FA, Man SF, Anthonisen NR (2003) Contemporary management of chronic obstructive pulmonary disease: scientific review. *JAMA* 290: 2301–2312

Tattersfield AE, Knox AJ, Britton JR, Hall IP (2002) Asthma. *Lancet* 360: 1313–1322

Recommended websites

Global Initiative for Asthma (GINA) (2002) Global strategy for asthma management and prevention. NHLBI/WHO Workshop Report. NIH Publication 02-3659: *www.ginasthma.com* (Accessed March 2005)

Global Initiative for Chronic Obstructive Lung Disease (GOLD) (2003) Global strategy for the diagnosis, management of chronic obstructive pulmonary disease. NHLBI/WHO Workshop Report: *www.goldcopd.com/workshop/index.html* (Accessed March 2005)

References

1 Barnes PJ (2000) Mechanisms in COPD: differences from asthma. *Chest* 117: 10S–14S

2 Nelson HS (1995) Beta-adrenergic bronchodilators. *N Engl J Med* 333: 499–506

3 Kips JC, Pauwels RA (2001) Long-acting inhaled β_2-agonist therapy in asthma. *Am J Respir Crit Care Med* 164: 923–932

4 Green SA, Spasoff AP, Coleman RA, Johnson M, Liggett SB (1996) Sustained activation of a G protein-coupled receptor via "anchored" agonist binding. Molecular localization of the salmeterol exosite within the 2-adrenergic receptor. *J Biol Chem* 271: 24029–24035

5 Barnes PJ (1995) Beta-adrenergic receptors and their regulation. *Am J Respir Crit Care Med* 152: 838–860

6 Kotlikoff MI, Kamm KE (1996) Molecular mechanisms of β-adrenergic relaxation of airway smooth muscle. *Annu Rev Physiol* 58: 115–141

7 Staples KJ, Bergmann M, Tomita K, Houslay MD, McPhee I, Barnes PJ, Giembycz MA, Newton R (2001) Adenosine 3',5'-cyclic monophosphate (cAMP)-dependent inhibition of IL-5 from human T lymphocytes is not mediated by the cAMP-dependent protein kinase A. *J Immunol* 167: 2074–2080

8 Barnes PJ (1999) Effect of beta agonists on inflammatory cells. *J Allergy Clin Immunol* 104: 10–17

9 Weston MC, Peachell PT (1998) Regulation of human

mast cell and basophil function by cAMP. *Gen Pharmacol* 31: 715–719

10 Howarth PH, Beckett P, Dahl R (2000) The effect of long-acting β_2-agonists on airway inflammation in asthmatic patients. *Respir Med* 94 (Suppl F): S22–S25

11 Travers AH, Rowe BH, Barker S, Jones A, Camargo CA, Jr. (2002) The effectiveness of IV β-agonists in treating patients with acute asthma in the emergency department: a meta-analysis. *Chest* 122: 1200–1207

12 Crompton GK, Ayres JG, Basran G, Schiraldi G, Brusasco V, Eivindson A, Jamieson AH, Olsson H (1999) Comparison of oral bambuterol and inhaled salmeterol in patients with symptomatic asthma and using inhaled corticosteroids. *Am J Respir Crit Care Med* 159: 824–828

13 Handley DA (2001) Single-isomer beta-agonists. *Pharmacotherapy* 21: 21S–27S

14 Lotvall J, Palmqvist M, Arvidsson P, Maloney A, Ventresca GP, Ward J (2001) The therapeutic ratio of R-albuterol is comparable with that of *RS*-albuterol in asthmatic patients. *J Allergy Clin Immunol* 108: 726–731

15 Walters EH, Walters JA, Gibson PW (2002) Regular treatment with long acting beta agonists *versus* daily regular treatment with short acting beta agonists in adults and children with stable asthma. *Cochrane Database Syst Rev*, CD003901

16 Nightingale JA, Rogers DF, Barnes PJ (2002) Comparison of the effects of salmeterol and formoterol in patients with severe asthma. *Chest* 121: 1401–1406

17 Shrewsbury S, Pyke S, Britton M (2000) Meta-analysis of increased dose of inhaled steroid or addition of salmeterol in symptomatic asthma (MIASMA). *BMJ* 320: 1368–1373

18 Pauwels RA, Lofdahl CG, Postma DS, Tattersfield AE, O'Byrne P, Barnes PJ, Ullman A (1997) Effect of inhaled formoterol and budesonide on exacerbations of asthma. *N Engl J Med* 337: 1405–1411

19 O'Byrne PM, Barnes PJ, Rodriguez-Roisin R, Runnerstrom E, Sandstrom T, Svensson K, Tattersfield A (2001) Low dose inhaled budesonide and formoterol in mild persistent asthma: the OPTIMA randomized trial. *Am J Respir Crit Care Med* 164: 1392–1397

20 Appleton S, Smith B, Veale A, Bara A (2000) Long-acting β_2-agonists for chronic obstructive pulmonary disease. *Cochrane Database Syst Rev* 2: CD001104: CD001104

21 Tattersfield AE, Lofdahl CG, Postma DS, Eivindson A, Schreurs AG, Rasidakis A, Ekstrom T (2001) Comparison of formoterol and terbutaline for as-needed treatment of asthma: a randomised trial. *Lancet* 357: 257–261

22 Nelson HS (2001) Advair: combination treatment with fluticasone propionate/salmeterol in the treatment of asthma. *J Allergy Clin Immunol* 107: 398–416

23 Barnes PJ (2002) Scientific rationale for combination inhalers with a long-acting β_2-agonists and corticosteroids. *Eur Respir J* 19: 182–191

24 Nelson HS, Chapman KR, Pyke SD, Johnson M, Pritchard JN (2003) Enhanced synergy between fluticasone propionate and salmeterol inhaled from a single inhaler *versus* separate inhalers. *J Allergy Clin Immunol* 112: 29–36

25 Szafranski W, Ramirez A, Petersen S (2002) Budesonide/formoterol in a single inhaler provides sustained improvements in lung function in patients with moderate to severe COPD. *Eur Resp J* 20: 397S

26 Calverley P, Pauwels R, Vestbo J, Jones P, Pride N, Gulsvik A, Anderson J, Maden C; Trial of Inhaled Steroids and long-acting beta2 agonists study group (2003) Combined salmeterol and fluticasone in the treatment of chronic obstructive pulmonary disease: a randomised controlled trial. *Lancet* 361: 449–456

27 Grove A, Lipworth BJ (1995) Tolerance with β_2-adrenoceptor agonists: time for reappraisal. *Br J Clin Pharmacol* 39: 109–118

28 O'Connor BJ, Aikman SL, Barnes PJ (1992) Tolerance to the non-bronchodilator effects of inhaled β_2-agonists. *N Engl J Med* 327: 1204–1208

29 Cockcroft DW, McParland CP, Britto SA, Swystun VA, Rutherford BC (1993) Regular inhaled salbutamol and airway responsiveness to allergen. *Lancet* 342: 833–837

30 Hamid QA, Mak JC, Sheppard MN, Corrin B, Venter JC, Barnes PJ (1991) Localization of β_2-adrenoceptor messenger RNA in human and rat lung using *in situ* hybridization: correlation with receptor autoradiography. *Eur J Pharmacol* 206: 133–138

31 McGraw DW, Liggett SB (1997) Heterogeneity in beta-adrenergic receptor kinase expression in the lung accounts for cell-specific desensitization of the beta2-adrenergic receptor. *J Biol Chem* 272: 7338–7344

32 Yates DH, Sussman H, Shaw MJ, Barnes PJ, Chung KF (1995) Regular formoterol treatment in mild asthma: effect on bronchial responsiveness during and after treatment. *Am J Resp Crit Care Med* 152: 1170–1174

33 Mak JCW, Nishikawa M, Shirasaki H, Miyayasu K, Barnes

PJ (1995) Protective effects of a glucocorticoid on down-regulation of pulmonary β_2-adrenergic receptors *in vivo*. *J Clin Invest* 96: 99–106

34 Yates DH, Kharitonov SA, Barnes PJ (1996) An inhaled glucocorticoid does not prevent tolerance to salmeterol in mild asthma. *Am J Respir Crit Care Med* 154: 1603–1607

35 Kalra S, Swystun VA, Bhagat R, Cockcroft DW (1996) Inhaled corticosteroids do not prevevt the development of subsensitivity to salbutamol after twice daily salmeterol. *Chest* 109: 953–956

36 Beasley R, Pearce N, Crane J, Burgess C (1999) Beta-agonists: what is the evidence that their use increases the risk of asthma morbidity and mortality? *J Allergy Clin Immunol* 104: S18–S30

37 Spitzer WO, Suissa S, Ernst P, Horwitz RI, Habbick B, Cockcroft D, Boivin JF, McNutt M, Buist AS, Rebuck AS (1992) The use of β-agonists and the rate of death and near-death from asthma. *N Engl J Med* 326: 501–506

38 Suissa S, Ernst P, Benayoun S, Baltzan M, Cai B (2000) Low-dose inhaled corticosteroids and the prevention of death from asthma. *N Engl J Med* 343: 332–336

39 Sears MR, Taylor DR, Print CG, Lake DC, Li QQ, Flannery EM, Yates DM, Lucas MK, Herbison GP (1990) Regular inhaled beta-agonist treatment in bronchial asthma. *Lancet* 336: 1391–1396

40 Drazen JM, Israel E, Boushey HA, Chinchilli VM, Fahy JV, Fish JE, Lazarus SC, Lemanske RF, Martin RJ, Peters SP et al (1996) Comparison of regularly scheduled with as needed use of albuterol in mild asthma. *N Engl J Med* 335: 841–847

41 Dennis SM, Sharp SJ, Vickers MR, Frost CD, Crompton GK, Barnes PJ, Lee TH (2000) Regular inhaled salbutamol and asthma control: the TRUST randomised trial. *Lancet* 355: 1675–1679

42 Gauvreau GM, Jordana M, Watson RM, Cockroft DW, O'Byrne PM (1997) Effect of regular inhaled albuterol on allergen-induced late responses and sputum eosinophils in asthmatic subjects. *Am J Respir Crit Care Med* 156: 1738–1745

43 Mcivor RA, Pizzichini E, Turner MO, Hussack P, Hargreave FE, Sears MR (1998) Potential masking effects of salmeterol on airway inflammation in asthma. *Am J Respir Crit Care Med* 158: 924–930

44 Adcock IM, Stevens DA, Barnes PJ (1996) Interactions between steroids and β_2-agonists. *Eur Respir J* 9: 160–168

45 McGraw DW, Almoosa KF, Paul RJ, Kobilka BK, Liggett SB (2003) Antithetic regulation by β-adrenergic receptors of Gq receptor signaling via phospholipase C underlies the airway β-agonist paradox. *J Clin Invest* 112: 619–626

46 Liggett SB (2002) Polymorphisms of the beta2-adrenergic receptor. *N Engl J Med* 346: 536–538

47 Peters MJ, Adcock IM, Brown CR, Barnes PJ (1995) β-Adrenoceptor agonists interfere with glucocorticoid receptor DNA binding in rat lung. *Eur J Pharmacol* (Molec Pharmacol Section) 289: 275–281

48 Weinberger M, Hendeles L (1996) Theophylline in asthma. *N Engl J Med* 334: 1380–1388

49 Barnes PJ (2003) Theophylline: new perspectives on an old drug. *Am J Respir Crit Care Med* 167: 813–818

50 Barnes PJ, Pauwels RA (1994) Theophylline in asthma: time for reappraisal? *Eur Resp J* 7: 579–591

51 Ward AJM, McKenniff M, Evans JM, Page CP, Costello JF (1993) Theophylline – an immunomodulatory role in asthma? *Am Rev Respir Dis* 147: 518–523

52 Sullivan P, Bekir S, Jaffar Z, Page C, Jeffery P, Costello J (1994) Anti-inflammatory effects of low-dose oral theophylline in atopic asthma. *Lancet* 343: 1006–1008

53 Jaffar ZH, Sullivan P, Page C, Costello J (1996) Low-dose theophylline modulates T-lymphocyte activation in allergen-challenged asthmatics. *Eur Respir J* 9: 456–462

54 Lim S, Tomita K, Caramori G, Jatakanon A, Oliver B, Keller A, Adcock I, Chung KF, Barnes PJ (2001) Low-dose theophylline reduces eosinophilic inflammation but not exhaled nitric oxide in mild asthma. *Am J Respir Crit Care Med* 164: 273–276

55 Kidney J, Dominguez M, Taylor PM, Rose M, Chung KF, Barnes PJ (1995) Immunomodulation by theophylline in asthma: demonstration by withdrawal of therapy. *Am J Resp Crit Care Med* 151: 1907–1914

56 Culpitt SV, de Matos C, Russell RE, Donnelly LE, Rogers DF, Barnes PJ (2002) Effect of theophylline on induced sputum inflammatory indices and neutrophil chemotaxis in COPD. *Am J Respir Crit Care Med* 165: 1371–1376

57 Rabe KF, Magnussen H, Dent G (1995) Theophylline and selective PDE inhibitors as bronchodilators and smooth muscle relaxants. *Eur Respir J* 8: 637–642

58 Torphy TJ (1998) Phosphodiesterase isoenzymes. *Am J Respir Crit Care Med* 157: 351–370

59 Feoktistov I, Biaggioni I (1997) Adenosine A2B receptors. *Pharmacol Rev* 49: 381–402

60 Björk T, Gustafsson LE, Dahlén S-E (1992) Isolated bronchi from asthmatics are hyperresponsive to adenosine, which apparently acts indirectly by liberation of leukotrienes and histamine. *Am Rev Respir Dis* 145: 1087–1091
61 Barnes PJ (2001) IL-10: a key regulator of allergic disease. *Clin Exp Allergy*;31: 667–669
62 Mascali JJ, Cvietusa P, Negri J, Borish L (1996) Anti-inflammatory effects of theophylline: modulation of cytokine production. *Ann Allergy Asthma Immunol* 77: 34–38
63 Oliver B, Tomita K, Keller A, Caramori G, Adcock I, Chung KF, Barnes PJ, Lim S (2001) Low-dose theophylline does not exert its anti-inflammatory effects in mild asthma through upregulation of interleukin-10 in alveolar macrophages. *Allergy* 56: 1087–1090
64 Tomita K, Chikumi H, Tokuyasu H, Yajima H, Hitsuda Y, Matsumoto Y, Sasaki T (1999) Functional assay of NF-kappaB translocation into nuclei by laser scanning cytometry: inhibitory effect by dexamethasone or theophylline. *Naunyn Schmiedebergs Arch Pharmacol* 359: 249–255
65 Ichiyama T, Hasegawa S, Matsubara T, Hayashi T, Furukawa S (2001) Theophylline inhibits NF-κB activation and IκBα degradation in human pulmonary epithelial cells. *Naunyn Schmiedebergs Arch Pharmacol* 364: 558–561
66 Yasui K, Hu B, Nakazawa T, Agematsu K, Komiyama A (1997) Theophylline accelerates human granulocyte apoptosis not via phosphodiesterase inhibition. *J Clin Invest* 100: 1677–1684
67 Chung IY, Nam-Kung EK, Lee NM, Chang HS, Kim DJ, Kim YH, Park CS (2000) The downregulation of bcl-2 expression is necessary for theophylline-induced apoptosis of eosinophil. *Cell Immunol* 203: 95–102
68 Yasui K, Agematsu K, Shinozaki K, Hokibara S, Nagumo H, Nakazawa T, Komiyama A (2000) Theophylline induces neutrophil apoptosis through adenosine A_{2A} receptor antagonism. *J Leukoc Biol* 67: 529–535
69 Ohta K, Yamashita N (1999) Apoptosis of eosinophils and lymphocytes in allergic inflammation. *J Allergy Clin Immunol* 104: 14–21
70 Ito K, Lim S, Caramori G, Cosio B, Chung KF, Adcock IM, Barnes PJ (2002) A molecular mechanism of action of theophylline: Induction of histone deacetylase activity to decrease inflammatory gene expression. *Proc Natl Acad Sci USA* 99: 8921–8926
71 Zhang ZY, Kaminsky LS (1995) Characterization of human cytochromes P450 involved in theophylline 8-hydroxylation. *Biochem Pharmacol*; 50: 205–211
72 Taylor DR, Ruffin D, Kinney CD, McDevitt DG (1983) Investigation of diurnal changes in the disposition of theophylline. *Br J Clin Pharmac* 16: 413–416
73 Parameswaran K, Belda J, Rowe BH (2000) Addition of intravenous aminophylline to beta2-agonists in adults with acute asthma. *Cochrane Database Syst Rev*, CD002742
74 Rivington RN, Boulet LP, Cote J, Kreisman H, Small DI, Alexander M, Day A, Harsanyi Z, Darke AC (1995) Efficacy of Uniphyl, salbutamol, and their combination in asthmatic patients on high-dose inhaled steroids. *Am J Respir Crit Care Med* 151: 325–332
75 Evans DJ, Taylor DA, Zetterstrom O, Chung KF, O'Connor BJ, Barnes PJ (1997) A comparison of low-dose inhaled budesonide plus theophylline and high-dose inhaled budesonide for moderate asthma. *N Engl J Med* 337: 1412–1418
76 Ukena D, Harnest U, Sakalauskas R, Magyar P, Vetter N, Steffen H, Leichtl S, Rathgeb F, Keller A, Steinijans VW (1997) Comparison of addition of theophylline to inhaled steroid with doubling of the dose of inhaled steroid in asthma. *Eur Respir J* 10: 2754–2760
77 Lim S, Jatakanon A, Gordon D, Macdonald C, Chung KF, Barnes PJ (2000) Comparison of high dose inhaled steroids, low dose inhaled steroids plus low dose theophylline, and low dose inhaled steroids alone in chronic asthma in general practice. *Thorax* 55: 837–841
78 Wilson AJ, Gibson PG, Coughlan J (2000) Long acting beta-agonists *versus* theophylline for maintenance treatment of asthma. *Cochrane Database Syst Rev*, CD001281
79 Shah L, Wilson AJ, Gibson PG, Coughlan J (2003) Long acting beta-agonists *versus* theophylline for maintenance treatment of asthma. *Cochrane Database Syst Rev*, CD001281
80 Markham A, Faulds D (1998) Theophylline. A review of its potential steroid sparing effects in asthma. *Drugs* 56: 1081–1091
81 Brenner MR, Berkowitz R, Marshall N, Strunk RC (1988) Need for theophylline in severe steroid-requiring asthmatics. *Clin Allergy* 18: 143–150
82 Tinkelman DG, Reed CE, Nelson HS, Offord KP (1993) Aerosol beclomethasone diprionate compared with

theophylline as primary treatment of chronic, mild to moderately severe asthma in children. *Pediatr* 92: 64–77

83 Reed CE, Offord KP, Nelson HS, Li JT, Tinkelman DG (1998) Aerosol beclomethasone dipropionate spray compared with theophylline as primary treatment for chronic mild-to-moderate asthma. The American Academy of Allergy, Asthma and Immunology Beclomethasone Dipropionate-Theophylline Study Group. *J Allergy Clin Immunol* 101: 14–23

84 Dahl R, Larsen BB, Venge P (2002) Effect of long-term treatment with inhaled budesonide or theophylline on lung function, airway reactivity and asthma symptoms. *Respir Med* 96: 432–438

85 Global Initiative for Asthma (2002) Global strategy for asthma management and prevention. NHLBI/WHO Workshop Report. *NIH Publication* 02–3659

86 Pauwels RA, Buist AS, Calverley PM, Jenkins CR, Hurd SS (2001) Global strategy for the diagnosis, management, and prevention of chronic obstructive pulmonary disease. NHLBI/WHO Global Initiative for Chronic Obstructive Lung Disease (GOLD) Workshop summary. *Am J Respir Crit Care Med* 163: 1256–1276

87 ZuWallack RL, Mahler DA, Reilly D, Church N, Emmett A, Rickard K, Knobil K (2001) Salmeterol plus theophylline combination therapy in the treatment of COPD. *Chest* 119: 1661–1670

88 Kirsten DK, Wegner RE, Jorres RA, Magnussen H (1993) Effects of theophylline withdrawal in severe chronic obstructive pulmonary disease. *Chest* 104: 1101–1107

89 Chrystyn H, Mulley BA, Peake MD (1988) Dose response relation to oral theophylline in severe chronic obstructive airway disease. *Br Med J* 297: 1506–1510

90 Ito K, Caramori G, Lim S, Oates T, Chung KF, Barnes PJ, Adcock IM (2002) Expression and activity of histone deacetylases (HDACs) in human asthmatic airways. *Am J Respir Crit Care Med* 166: 392–396

91 Wessler IK, Kirkpatrick CJ (2001) The non-neuronal cholinergic system: an emerging drug target in the airways. *Pulm Pharmacol Ther* 14: 423–434

92 Tamaoki J, Chiyotani A, Tagaya E, Sakai N, Konno K (1994) Effect of long term treatment with oxitropium bromide on airway secretion in chronic bronchitis and diffuse panbronchiolitis. *Thorax* 49: 545–548

93 Stoodley RG, Aaron SD, Dales RE (1999) The role of ipratropium bromide in the emergency management of acute asthma exacerbation: a metaanalysis of randomized clinical trials. *Ann Emerg Med* 34: 8–18

94 Rennard SI, Serby CW, Ghafouri M, Johnson PA, Friedman M (1996) Extended therapy with ipratropium is associated with improved lung function in patients with COPD. A retrospective analysis of data from seven clinical trials. *Chest* 110: 62–70

95 Frith PA, Jenner B, Dangerfield R, Atkinson J, Drennan C (1986) Oxitropium bromide. Dose response and time-response study of a new anticholinergic bronchodilator drug. *Chest* 89: 249–253

96 Coe CI, Barnes PJ (1986) Reduction of nocturnal asthma by an inhaled anticholinergic drug. *Chest* 90: 485–488

97 Combivent inhlation study group (1997) Routine nebulized ipratropium and albuterol together are better than either alone in COPD. *Chest* 112: 1514–1521

98 Hansel TT, Barnes PJ (2002) Tiotropium bromide: a novel once-daily anticholinergic bronchodilator for the treatment of COPD. *Drugs Today* 38: 585–600

99 Disse B, Speck GA, Rominger KL, Witek TJ, Hammer R (1999) Tiotropium (Spiriva): mechanistical considerations and clinical profile in obstructive lung disease. *Life Sci* 64: 457–464

100 Vincken W, van Noord JA, Greefhorst AP, Bantje TA, Kesten S, Korducki L, Cornelissen PJ; Dutch/Belgian Tiotropium Study Group (2002) Improved health outcomes in patients with COPD during 1 yr's treatment with tiotropium. *Eur Respir J* 19: 209–216

101 Casaburi R, Mahler DA, Jones PW, Wanner A, San PG, ZuWallack RL, Menjoge SS, Serby CW, Witek T Jr (2002) A long-term evaluation of once-daily inhaled tiotropium in chronic obstructive pulmonary disease. *Eur Respir J* 19: 217–224

102 Newcomb R, Tashkin DP, Hui KK, Connolly ME, Lee E, Dauphinee B (1985) Rebound hyperresponsiveness to muscarinic stimulation after chronic therapy with an inhaled muscarinic antagonist. *Am Rev Respir Dis* 132: 12–15

103 Patel HJ, Barnes PJ, Takahashi T, Tadjkarimi S, Yacoub MH, Belvisi MG (1995) Characterization of prejunctional muscarinic autoreceptors in human and guinea-pig trachea *in vitro*. *Am J Resp Crit Care Med* 152: 872–878

104 Alabaster VA (1997) Discovery and development of selective M3 antagonists for clinical use. *Life Sci* 60: 1053–1060

105 Schelfhout VJ, Joos GF, Ferrer P, Luria X, Pauwels RA (2003) Activity of LAS 34273, a new long-acting anticholinergic antagonist. *Am J Resp Crit Care Med* 167: A93

106 Tennant RC, Erin EM, Barnes PJ, Hansel TT (2003) Long-acting β_2-adrenoceptor agonists or tiotropium bromide for patients with COPD: is combination therapy justified? *Curr Opin Pharmacol* 3: 270–276

107 Rowe BH, Bretzlaff JA, Bourdon C, Bota GW, Camargo CA, Jr. (2000) Intravenous magnesium sulfate treatment for acute asthma in the emergency department: a systematic review of the literature. Ann Emerg Med 36: 181–190

108 Silverman RA, Osborn H, Runge J, Gallagher EJ, Chiang W, Feldman J, Gaeta T, Freeman K, Levin B, Mancherje N et al; Acute Asthma/Magnesium Study Group (2002) IV magnesium sulfate in the treatment of acute severe asthma: a multicenter randomized controlled trial. *Chest* 122: 489–497

109 Emelyanov A, Fedoseev G, Barnes PJ (1999) Reduced intracellular magnesium concentrations in asthmatic subjects. *Eur Respir J* 13: 38–40

110 Hughes R, Goldkorn A, Masoli M, Weatherall M, Burgess C, Beasley R (2003) Use of isotonic nebulised magnesium sulphate as an adjuvant to salbutamol in treatment of severe asthma in adults: randomised placebo-controlled trial. *Lancet* 361: 2114–2117

111 Black JL, Armour CL, Johnson PRA, Alouan LA, Barnes PJ (1990) The action of a potassium channel activator BRL 38227 (lemakalim) on human airway smooth muscle. *Am Rev Respir Dis* 142: 1384–1389

112 Pelaia G, Gallelli L, Vatrella A, Grembiale RD, Maselli R, De Sarro GB, Marsico SA (2002) Potential role of potassium channel openers in the treatment of asthma and chronic obstructive pulmonary disease. *Life Sci* 70: 977–990

113 Kidney JC, Fuller RW, Worsdell Y-M, Lavender EA, Chung KF, Barnes PJ (1993) Effect of an oral potassium channel activator BRL 38227 on airway function and responsiveness in asthmatic patients: comparison with oral salbutamol. *Thorax* 48: 130–134

114 Ramnarine SI, Liu YC, Rogers DF (1998) Neuroregulation of mucus secretion by opioid receptors and K(ATP) and BK(Ca) channels in ferret trachea *in vitro*. *Br J Pharmacol* 123: 1631–1638

115 Fox AJ, Barnes PJ, Venkatesan P, Belvisi MG (1997) Activation of large conductance potassium channels inhibits the afferent and efferent function of airway sensory nerves. *J Clin Invest* 99: 513–519

116 Angus RM, Mecallaum MJA, Hulks G, Thomson NC (1993) Bronchodilator, cardiovascular and cyclic guanylyl monophosphate response to high dose infused atrial natriuretic peptide in asthma. *Am Rev Respir Dis* 147: 1122–1125

117 Fluge T, Forssmann WG, Kunkel G, Schneider B, Mentz P, Forssmann K, Barnes PJ, Meyer M (1999) Bronchodilation using combined urodilatin – albuterol administration in asthma: a randomized, double-blind, placebo-controlled trial. *Eur J Med Res* 4: 411–415

118 Linden A, Hansson L, Andersson A, Palmqvist M, Arvidsson P, Lofdahl CG, Larsson P, Lotvall J (2003) Bronchodilation by an inhaled VPAC(2) receptor agonist in patients with stable asthma. *Thorax* 58: 217–221

119 Barnes PJ (1995) Inhaled glucocorticoids for asthma. *N Engl J Med* 332: 868–875

120 Barnes PJ (1998) Antiinflammatory actions of glucocorticoids: molecular mechanisms. *Clin Sci* 94: 557–572

121 Barnes PJ, Adcock IM (2003) How do corticosteroids work in asthma? *Ann Intern Med* 139: 359–370

122 Juniper EF, Kline PA, Yan Zieleshem MA, Ramsdale EH, O'Byrne PM, Hargreave FE (1990) Long-term effects of budesonide on airway responsiveness and clinical asthma severity in inhaled steroid-dependent asthmatics. *Eur Respir J* 3: 1122–1127

123 Mak JCW, Nishikawa M, Barnes PJ (1995) Glucocorticosteroids increase β_2-adrenergic receptor transcription in human lung. *Am J Physiol* 12: L41–L46

124 Baraniuk JN, Ali M, Brody D, Maniscalco J, Gaumond E, Fitzgerald T, Wong G, Yuta A, Mak JC, Barnes PJ et al (1997) Glucocorticoids induce β_2-adrenergic receptor function in human nasal mucosa. *Am J Respir Crit Care Med* 155: 704–710

125 Chong LK, Drury DE, Dummer JF, Ghahramani P, Schleimer RP, Peachell PT (1997) Protection by dexamethasone of the functional desensitization to β_2-adrenoceptor-mediated responses in human lung mast cells. *Br J Pharmacol* 121: 717–722

126 Gambertoglio JG, Amend WJC, Benet LZ (1980) Pharmacokinetics and bioavailability of prednisone and prednisolone in healthy volunteers and patients: a review. *J Pharmacokin Biopharm* 8: 1–52

127 Barnes PJ, Greening AP, Crompton GK (1995) Gluco-

corticoid resistance in asthma. *Am J Respir Crit Care Med* 152: 125S–140S

128 Barnes PJ, Pedersen S, Busse WW (1998) Efficacy and safety of inhaled corticosteroids: an update. *Am J Respir Crit Care Med* 157: S1–S53

129 Markham A, Bryson HM (1995) Deflazacort. A review of its pharmacological properties and therapeutic efficacy. *Drugs* 50: 317–333

130 Harrison BDN, Stokes TC, Hart GJ, Vaughan DA, Ali NJ, Robinson AA (1986) Need for intravenous hydrocortisone in addition to oral prednisolone in patients admitted to hospital with severe asthma without ventilatory failure. *Lancet* 1: 181–184

131 Malo JL, Cartier A, Merland N, Ghezzo H, Burek A, Morris J, Jennings BH (1989) Four-times-a-day dosing frequency is better than twice-a-day regimen in subjects requiring a high-dose inhaled steroid, budesonide, to control moderate to severe asthma. *Am Rev Respir Dis* 140: 624–628

132 Metzger WJ, Hampel FC Jr, Sugar M (2002) Once-daily budesonide inhalation powder (Pulmicort Turbuhaler) is effective and safe in adults previously treated with inhaled corticosteroids. *J Asthma* 39: 65–75

133 Szefler SJ, Eigen H (2002) Budesonide inhalation suspension: a nebulized corticosteroid for persistent asthma. *J Allergy Clin Immunol* 109: 730–742

134 Edmonds ML, Camargo CA, Jr.., Pollack CV, Jr., Rowe BH (2003) Early use of inhaled corticosteroids in the emergency department treatment of acute asthma. *Cochrane Database Syst Rev*, CD002308

135 Pauwels RA, Pedersen S, Busse WW, Tan WC, Chen YZ, Ohlsson SV, Ullman A, Lamm CJ, O'Byrne PM; START Investigators Group (2003) Early intervention with budesonide in mild persistent asthma: a randomised, double-blind trial. *Lancet* 361: 1071–1076

136 Adams N, Bestall J, Jones P (2001) Inhaled beclomethasone at different doses for long-term asthma. *Cochrane Database Syst Rev*, CD002879

137 Adams N, Bestall J, Jones PW (2001) Budesonide at different doses for chronic asthma (Cochrane Review). *Cochrane Database Syst Rev* 4: CD003271

138 Adams N, Bestall JM, Jones PW (2002) Inhaled fluticasone at different doses for chronic asthma. *Cochrane Database Syst Rev*, CD003534

139 Hawkins G, McMahon AD, Twaddle S, Wood SF, Ford I, Thomson NC (2003) Stepping down inhaled corticosteroids in asthma: randomised controlled trial. *Br Med J* 326: 1115

140 Barnes PJ, Ito K, Adcock IM (2004) A mechanism of corticosteroid resistance in COPD: inactivation of histone deacetylase. *Lancet* 363: 731–733

141 Alsaeedi A, Sin DD, McAlister FA (2002) The effects of inhaled corticosteroids in chronic obstructive pulmonary disease: a systematic review of randomized placebo-controlled trials. *Am J Med* 113: 59–65

142 Davies L, Angus RM, Calverley PM (1999) Oral corticosteroids in patients admitted to hospital with exacerbations of chronic obstructive pulmonary disease: a prospective randomised controlled trial. *Lancet* 354: 456–460

143 Niewoehner DE, Erbland ML, Deupree RH, Collins D, Gross NJ, Light RW, Anderson P, Morgan NA (1999) Effect of systemic glucocorticoids on exacerbations of chronic obstructive pulmonary disease. *N Engl J Med* 340: 1941–1947

144 Lipworth BJ (1999) Systemic adverse effects of inhaled corticosteroid therapy: A systematic review and meta-analysis. *Arch Intern Med* 159: 941–955

145 Pedersen S (2001) Do inhaled corticosteroids inhibit growth in children? *Am J Respir Crit Care Med* 164: 521–535

146 Agertoft L, Pedersen S (2000) Effect of long-term treatment with inhaled budesonide on adult height in children with asthma. *N Engl J Med* 343: 1064–1069

147 Cumming RG, Mitchell P, Leeder SR (1997) Use of inhaled corticosteroids and the risk of cataracts. *N Engl J Med* 337: 8–14

148 Garbe E, LeLorier J, Boivin J-F, Suissa S (1997) Inhaled and nasal glucocorticoids and the risks of ocular hypertension or open-angel glaucoma. *JAMA* 227: 722–727

149 Williamson IJ, Matusiewicz SP, Brown PH, Greening AP, Crompton GK (1995) Frequency of voice problems and cough in patients using pressurised aerosol inhaled steroid preparations. *Eur Resp J* 8: 590–592

150 Harrison TW, Wisniewski A, Honour J, Tattersfield AE (2001) Comparison of the systemic effects of fluticasone propionate and budesonide given by dry powder inhaler in healthy and asthmatic subjects. *Thorax* 56: 186–191

151 Brutsche MH, Brutsche IC, Munawar M, Langley SJ, Masterson CM, Daley-Yates PT, Brown R, Custovic A, Woodcock A (2000) Comparison of pharmacokinetics and

systemic effects of inhaled fluticasone propionate in patients with asthma and healthy volunteers: a randomised crossover study. *Lancet* 356: 556–561

152 Martin RJ (2002) Therapeutic significance of distal airway inflammation in asthma. *J Allergy Clin Immunol* 109: S447–S460

153 Leach CL, Davidson PJ, Boudreau RJ (1998) Improved airway targeting with the CFC-free HFA-beclomethasone metered-dose inhaler compared with CFC-beclomethasone. *Eur Respir J* 12: 1346–1353

154 Hauber HP, Gotfried M, Newman K, Danda R, Servi RJ, Christodoulopoulos P, Hamid Q (2003) Effect of HFA-flunisolide on peripheral lung inflammation in asthma. *J Allergy Clin Immunol* 112: 58–63

155 Martin RJ, Szefler SJ, Chinchilli VM, Kraft M, Dolovich M, Boushey HA, Cherniack RM, Craig TJ, Drazen JM, Fagan JK et al (2002) Systemic effect comparisons of six inhaled corticosteroid preparations. *Am J Respir Crit Care Med* 165: 1377–1383

156 Haahtela T, Jarvinen M, Kava T, Kiviranta K, Koskinen S, Lehtonen K, Nikander K, Persson T, Selroos O, Sovijarvi A et al (1994) Effects of reducing or discontinuing inhaled budesonide in patients with mild asthma. *N Engl J Med* 331: 700–705

157 Agertoft L, Pedersen S (1994) Effects of long-term treatment with an inhaled corticosteroid on growth and pulmonary function in asthmatic children. *Resp Med* 5: 369–372

158 Selroos O, Pietinalcho A, Lofroos A-B, Riska A (1995) Effect of early and late intervention with inhaled corticosteroids in asthma. *Chest* 108: 1228–1234

159 Dent G (2002) Ciclesonide. *Curr Opin Investig Drugs* 3: 78–83

160 Vayssiere BM, Dupont S, Choquart A, Petit F, Garcia T, Marchandeau C, Gronemeyer H, Resche-Rigon M (1997) Synthetic glucocorticoids that dissociate transactivation and AP-1 transrepression exhibit antiinflammatory activity *in vivo*. *Mol Endocrinol* 11:1245–1255

161 Belvisi MG, Brown TJ, Wicks S, Foster ML (2001) New Glucocorticosteroids with an improved therapeutic ratio? *Pulm Pharmacol Ther* 14: 221–227

162 Schacke H, Docke WD, Asadullah K (2002) Mechanisms involved in the side-effects of glucocorticoids. *Pharmacol Ther* 96: 23–43

163 Manolitsas ND, Wang J, Devalia J, Trigg CJ, McAuley AE, Davies RJ (1995) Regular albuterol, nedocromil sodium and bronchial inflammation in asthma. *Am J Respir Crit Care Med* 152: 1925–1930

164 Norris AA (1996) Pharmacology of sodium cromoglycate. *Clin Exp Allergy* 26 (Suppl 4): 5–7

165 Thomson NC (1989) Nedocromil sodium: an overview. *Resp Med* 83: 269–276

166 Tasche MJ, Uijen JH, Bernsen RM, de Jongste JC, van Der W (2000) Inhaled disodium cromoglycate (DSCG) as maintenance therapy in children with asthma: a systematic review. *Thorax* 55: 913–920

167 Wouden JC, Tasche MJ, Bernsen RM, Uijen JH, Jongste JC, Ducharme FM (2003) Inhaled sodium cromoglycate for asthma in children. *Cochrane Database Syst Rev*, CD002173

168 van Ganse E, Kaufman L, Derde MP, Yernault JC, Delaunois L (1997) Effects of antihistamines in adult asthma: a meta-analysis of clinical trials. *Eur Respir J* 10: 2216–2224

169 Lordan JL, Holgate ST (2002) H_1-antihistamines in asthma. *Clin Allergy Immunol* 17: 221–248

170 Grant SM, Goa KL, Fitton A, Sorkin EM (1990) Ketotifen. A review of its pharmacodynamic and pharmacokinetic properties, and therapeutic use in asthma and allergic disorders. *Drugs* 40: 412–448

171 Canny GJ, Reisman J, Levison H (1997) Does ketotifen have a steroid-sparing effect in childhood asthma? *Eur Respir J* 10: 65–70

172 Bustos GJ, Bustos D, Romero O (1995) Prevention of asthma with ketotifen in preasthmatic children: a three-year follow-up study. *Clin Exp Allergy* 25: 568–573

173 Leff AR (2001) Regulation of leukotrienes in the management of asthma: biology and clinical therapy. *Annu Rev Med* 52: 1–14

174 Szczeklik A, Stevenson DD (2003) Aspirin-induced asthma: advances in pathogenesis, diagnosis, and management. *J Allergy Clin Immunol* 111: 913–921

175 Calhoun WJ (2001) Anti-leukotrienes for asthma. *Curr Opin Pharmacol* 1: 230–234

176 Ducharme FM (2003) Inhaled glucocorticoids *versus* leukotriene receptor antagonists as single agent asthma treatment: systematic review of current evidence. *Br Med J* 326: 621–624

177 Ducharme F (2002) Anti-leukotrienes as add-on therapy to inhlaed glucocorticoids in patients with asthma: systematic review of current evidence. *Br Med J* 324: 1545–1548

178 Nelson HS, Busse WW, Kerwin E, Church N, Emmett A,

Rickard K, Knobil K (2000) Fluticasone propionate/ salmeterol combination provides more effective asthma control than low-dose inhaled corticosteroid plus montelukast. *J Allergy Clin Immunol* 106: 1088–1095
179 Ringdal N, Eliraz A, Pruzinec R, Weber HH, Mulder PG, Akveld M, Bateman ED; International Study Group (2003) The salmeterol/fluticasone combination is more effective than fluticasone plus oral montelukast in asthma. *Respir Med* 97: 234–241
180 Robinson DS, Campbell DA, Barnes PJ (2001) Addition of an anti-leukotriene to therapy in chronic severe asthma in a clinic setting: a double-blind, randomised, placebo-controlled study. *Lancet* 357: 2007–2011
181 Dahlen SE, Malmstrom K, Nizankowska E, Dahlen B, Kuna P, Kowalski M, Lumry WR, Picado C, Stevenson DD, Bousquet J et al (2002) Improvement of aspirin-intolerant asthma by montelukast, a leukotriene antagonist: a randomized, double-blind, placebo-controlled trial. *Am J Respir Crit Care Med* 165: 9–14
182 Coreno A, Skowronski M, Kotaru C, McFadden ER, Jr. (2000) Comparative effects of long-acting beta2-agonists, leukotriene receptor antagonists, and a 5-lipoxygenase inhibitor on exercise-induced asthma. *J Allergy Clin Immunol* 106: 500–506
183 Nathan RA (2003) Pharmacotherapy for allergic rhinitis: a critical review of leukotriene receptor antagonists compared with other treatments. *Ann Allergy Asthma Immunol* 90: 182–190
184 Diamant Z, Hiltermann JT, van Rensen EL, Callenbach PM, Veselic-Charvat M, van der Veen H, Sont JK, Sterk PJ (1997) Te effect of inhaled leukotriene D4 and methacholine on sputum cell differentials in asthma. *Am J Respir Crit Care Med* 155: 1247–1253
185 Minoguchi K, Kohno Y, Minoguchi H, Kihara N, Sano Y, Yasuhara H, Adachi M (2002) Reduction of eosinophilic inflammation in the airways of patients with asthma using montelukast. *Chest* 121: 732–738
186 O'Sullivan S, Akveld M, Burke CM, Poulter LW (2003) Effect of the addition of montelukast to inhaled fluticasone propionate on airway inflammation. *Am J Respir Crit Care Med* 167: 745–750
187 Montuschi P, Kharitonov SA, Ciabattoni G, Barnes PJ (2003) Exhaled leukotrienes and prostaglandins in COPD. *Thorax* 58: 585–588
188 Gompertz S, Stockley RA (2002) A randomized, placebo-controlled trial of a leukotriene synthesis inhibitor in patients with COPD. *Chest* 122: 289–294
189 Keogh KA, Specks U (2003) Churg-Strauss syndrome. clinical presentation, antineutrophil cytoplasmic antibodies, and leukotriene receptor antagonists. *Am J Med* 115: 284–290
190 Lilly CM, Churg A, Lazarovich M, Pauwels R, Hendeles L, Rosenwasser LJ, Ledford D, Wechsler ME (2002) Asthma therapies and Churg-Strauss syndrome. *J Allergy Clin Immunol* 109: S1–19
191 Barnes PJ (2003) Anti-leukotrienes: here to stay? *Curr Opin Pharmacol* 3: 257–263
192 Palmer LJ, Silverman ES, Weiss ST, Drazen JM (2002) Pharmacogenetics of asthma. *Am J Respir Crit Care Med* 165: 861–866
193 Back M (2002) Functional characteristics of cysteinyl-leukotriene receptor subtypes. *Life Sci* 71: 611–622
194 Hill SJ, Tattersfield AE (1995) Corticosteroid sparing agents in asthma. *Thorax* 50: 577–582
195 Shiner RJ, Nunn AJ, Chung KF, Geddes DM (1990) Randomized, double-blind, placebo-controlled trial of methotrexate in steroid-dependent asthma. *Lancet* 336: 137–140
196 Davies H, Olson L, Gibson P (2000) Methotrexate as a steroid sparing agent for asthma in adults. *Cochrane Database Syst Rev* 2: CD000391:CD000391
197 Nierop G, Gijzel WP, Bel EH, Zwinderman AH, Dijkman JH (1992) Auranofin in the treatment of steroid dependent asthma : a double blind study. *Thorax* 47: 349–354
198 Evans DJ, Cullinan P, Geddes DM (2001) Gold as an oral corticosteroid sparing agent in stable asthma. *Cochrane Database Syst Rev*, CD002985
199 Lock SH, Kay AB, Barnes NC (1996) Double-blind, placebo-controlled study of cyclosporin A as a corticosteroid-sparing agent in corticosteroid-dependent asthma. *Am J Respir Crit Care Med* 153: 509–514
200 Evans DJ, Cullinan P, Geddes DM (2001) Cyclosporin as an oral corticosteroid sparing agent in stable asthma (Cochrane Review). *Cochrane Database Syst Rev* 2: CD002993
201 Salmun LM, Barlan I, Wolf HM, Eibl M, Twarog FJ, Geha RS, Schneider LC (1999) Effect of intravenous immunoglobulin on steroid consumption in patients with severe asthma: a double-blind, placebo-controlled, randomized trial. *J Allergy Clin Immunol* 103: 810–815
202 Niggemann B, Leupold W, Schuster A, Schuster R, v Berg A, Grubl A, v d Hardt H, Eibl MM, Wahn U (1998)

Prospective, double-blind, placebo-controlled, multicentre study on the effect of high-dose, intravenous immunoglobulin in children and adolescents with severe bronchial asthma. *Clin Exp Allergy* 28: 205–210
203 Sigman K, Ghibu F, Sommerville W, Toledano BJ, Bastein Y, Cameron L, Hamid QA, Mazer B (1998) Intravenous immunoglobulin inhibits IgE production in human B lymphocytes. *J Allergy Clin Immunol* 102: 421–427
204 Barnes PJ (1996) Is there a role for immunotherapy in the treatment of asthma? No. *Am J Respir Crit Care Med* 154: 1227–1228
205 Creticos PS, Reed CE, Norman PS, Khoury J, Adkinson NF, Jr., Buncher CR, Busse WW, Bush RK, Gadde J, Li JT et al (1996) Ragweed immunotherapy in adult asthma. *N Engl J Med* 334: 501–506
206 Adkinson NF, Jr., Eggleston PA, Eney D, Goldstein EO, Schuberth KC, Bacon JR, Hamilton RG, Weiss ME, Arshad H, Meinert CL et al (1997) A controlled trial of immunotherapy for asthma in allergic children. *N Engl J Med* 336: 324–331
207 Abramson MJ, Puy RM, Weiner JM (2000) Allergen immunotherapy for asthma. *Cochrane Database Syst Rev* 2: CD001186:CD001186
208 Akdis CA, Blesken T, Akdis M, Wuthrich B, Blaser K (1998) Role of interleukin 10 in specific immunotherapy. *J Clin Invest* 102: 98–106
209 Jutel M, Akdis M, Budak F, Aebischer-Casaulta C, Wrzyszcz M, Blaser K, Akdis CA (2003) IL-10 and TGF-β cooperate in the regulatory T cell response to mucosal allergens in normal immunity and specific immunotherapy. *Eur J Immunol* 33: 1205–1214
210 Haselden BM, Kay AB, Larche M (2000) Peptide-mediated immune responses in specific immunotherapy. *Int Arch Allergy Immunol* 122: 229–237
211 Oldfield WL, Larche M, Kay AB (2002) Effect of T-cell peptides derived from Fel d 1 on allergic reactions and cytokine production in patients sensitive to cats: a randomised controlled trial. *Lancet* 360: 47–53
212 Fahy JV (2000) Reducing IgE levels as a strategy for the treatment of asthma. *Clin Exp Allergy* 30 (Suppl 1): 16–21
213 Barnes PJ (2000) Anti-IgE therapy in asthma: rationale and therapeutic potential. *Int Arch Allergy Immunol* 123: 196–204
214 Milgrom H, Fick RB Jr, Su JQ, Reimann JD, Bush RK, Watrous ML, Metzger WJ (1999) Treatment of allergic asthma with monoclonal anti-IgE antibody. *N Engl J Med* 341: 1966–1973
215 Soler M, Matz J, Townley R, Buhl R, O'Brien J, Fox H, Thirlwell J, Gupta N, Della Cioppa G (2001) The anti-IgE antibody omalizumab reduces exacerbations and steroid requirement in allergic asthmatics. *Eur Respir J* 18: 254–261
216 Corren J, Casale T, Deniz Y, Ashby M (2003) Omalizumab, a recombinant humanized anti-IgE antibody, reduces asthma-related emergency room visits and hospitalizations in patients with allergic asthma. *J Allergy Clin Immunol* 111: 87–90
217 Chervinsky P, Casale T, Townley R, Tripathy I, Hedgecock S, Fowler-Taylor A, Shen H, Fox H (2003) Omalizumab, an anti-IgE antibody, in the treatment of adults and adolescents with perennial allergic rhinitis. *Ann Allergy Asthma Immunol* 91: 160–167
218 Walker S, Monteil M, Phelan K, Lasserson TJ, Walters EH (2003) Anti-IgE for chronic asthma. *Cochrane Database Syst Rev*, CD003559
219 Barnes PJ (2000) New treatments for asthma. *Eur J Int Med* 11: 9–20
220 Barnes PJ (1999) Therapeutic strategies for allergic diseases. *Nature* 402: B31–B38
221 Barnes PJ (2002) New treatments for COPD. *Nature Rev Drug Disc* 1: 437–445
222 Barnes PJ (2003) New treatments for COPD. *Thorax* 58: 803–808
223 Cochrane GM, Horne R, Chanez P (1999) Compliance in asthma. *Respir Med* 93: 763–769
224 Barnes PJ (2003) New concepts in COPD. *Ann Rev Med* 54: 113–129
225 Roses AD (2000) Pharmacogenetics and future drug development and delivery. *Lancet* 355: 1358–1361
226 Barnes PJ, Chung KF, Page CP (1998) Inflammatory mediators of asthma: an update. *Pharmacol Rev* 50: 515–596
227 Beeh KM, Kornmann O, Buhl R, Culpitt SV, Giembycz MA, Barnes PJ (2003) Neutrophil chemotactic activity of sputum from patients with COPD: role of interleukin 8 and leukotriene B4. *Chest* 123: 1240–1247
228 Goldie RG, Henry PJ (1999) Endothelins and asthma. *Life Sci* 65: 1–15
229 Benigni A, Remuzzi G (1999) Endothelin antagonists. *Lancet* 353: 133–138
230 Macnee W (2001) Oxidative stress and lung inflammation in airways disease. *Eur J Pharmacol* 429: 195–207

231 Cuzzocrea S, Riley DP, Caputi AP, Salvemini D (2001) Antioxidant therapy: a new pharmacological approach in shock, inflammation, and ischemia/reperfusion injury. *Pharmacol Rev* 53: 135–159

232 Hobbs AJ, Higgs A, Moncada S (1999) Inhibition of nitric oxide synthase as a potential therapeutic target. *Annu Rev Pharmacol Toxicol* 39: 191–220

233 Hansel TT, Kharitonov SA, Donnelly LE, Erin EM, Currie MG, Moore WM, Manning PT, Recker DP, Barnes PJ (2003) A selective inhibitor of inducible nitric oxide synthase inhibits exhaled breath nitric oxide in healthy volunteers and asthmatics. *FASEB J* 17: 1298–1300

234 Barnes PJ (2001) Cytokine modulators as novel therapies for airway disease. *Eur Respir J* 34 (Suppl): 67s–77s

235 Barnes PJ (2002) Cytokine modulators as novel therapies for asthma. *Ann Rev Pharmacol Toxicol* 42: 81–98

236 Barnes PJ, Lim S (1998) Inhibitory cytokines in asthma. *Mol Medicine Today* 4: 452–458

237 Egan RW, Umland SP, Cuss FM, Chapman RW (1996) Biology of interleukin-5 and its relevance to allergic disease. *Allergy* 51:71–81

238 Leckie MJ, ten Brinke A, Khan J, Diamant Z, O'Connor BJ, Walls CM, Mathur AK, Cowley HC, Chung KF, Djukanovic R et al (2000) Effects of an interleukin-5 blocking monoclonal antibody on eosinophils, airway hyperresponsiveness and the late asthmatic response. *Lancet* 356: 2144–2148

239 Kips JC, O'Connor BJ, Langley SJ, Woodcock A, Kerstjens HA, Postma DS, Danzig M, Cuss F, Pauwels RA (2003) Effect of SCH55700, a humanized anti-human interleukin-5 antibody, in severe persistent asthma: a pilot study. *Am J Respir Crit Care Med* 167: 1655–1659

240 Flood-Page PT, Menzies-Gow AN, Kay AB, Robinson DS (2003) Eosinophil's role remains uncertain as anti-interleukin-5 only partially depletes numbers in asthmatic airways. *Am J Respir Crit Care Med* 167: 199–204

241 Gavett SH, O'Hearn DJ, Karp CL, Patel EA, Schofield BH, Finkelman FD, Wills-Karp M (1997) Interleukin-4 receptor blockade prevents airway responses induced by antigen challenge in mice. *Am J Physiol* 272: L253–L261

242 Borish LC, Nelson HS, Lanz MJ, Claussen L, Whitmore JB, Agosti JM, Garrison L (1999) Interleukin-4 Receptor in moderate atopic asthma. A phase I/II randomized, placebo-controlled trial. *Am J Respir Crit Care Med* 160: 1816–1823

243 Borish LC, Nelson HS, Corren J, Bensch G, Busse WW, Whitmore JB, Agosti JM; IL-4R Asthma Study Group (2001) Efficacy of soluble IL-4 receptor for the treatment of adults with asthma. *J Allergy Clin Immunol* 107: 963–970

244 Wills-Karp M, Chiaramonte M (2003) Interleukin-13 in asthma. *Curr Opin Pulm Med* 9: 21–27

245 Wills-Karp M, Luyimbazi J, Xu X, Schofield B, Neben TY, Karp CL, Donaldson DD (1998) Interleukin-13: central mediator of allergic asthma. *Science* 282: 2258–2261

246 Zhou Y, McLane M, Levitt RC (2001) Interleukin-9 as a therapeutic target for asthma. *Respir Res* 2: 80–84

247 Shah A, Church MK, Holgate ST (1995) Tumour necrosis factor alpha: a potential mediator of asthma. *Clin Exp Allergy* 25: 1038–1044

248 Keatings VM, Collins PD, Scott DM, Barnes PJ (1996) Differences in interleukin-8 and tumor necrosis factor-α in induced sputum from patients with chronic obstructive pulmonary disease or asthma. *Am J Respir Crit Care Med* 153: 530–534

249 Luster AD (1998) Chemokines – chemotactic cytokines that mediate inflammation. *N Engl J Med* 338: 436–445

250 Proudfoot AE (2002) Chemokine receptors: multifaceted therapeutic targets. *Nat Rev Immunol* 2: 106–115

251 Campbell EM, Charo IF, Kunkel SL, Strieter RM, Boring L, Gosling J, Lukacs NW (1999) Monocyte chemoattractant protein-1 mediates cockroach allergen-induced bronchial hyperreactivity in normal but not CCR2-/- mice: the role of mast cells. *J Immunol* 163: 2160–2167

252 Traves SL, Culpitt S, Russell REK, Barnes PJ, Donnelly LE (2002) Elevated levels of the chemokines GRO-α and MCP-1 in sputum samples from COPD patients. *Thorax* 57: 590–595

253 Sabroe I, Peck MJ, Van Keulen BJ, Jorritsma A, Simmons G, Clapham PR, Williams TJ, Pease JE (2000) A small molecule antagonist of chemokine receptors CCR1 and CCR3. Potent inhibition of eosinophil function and CCR3-mediated HIV-1 entry. *J Biol Chem* 275: 25985–25992

254 White JR, Lee JM, Dede K, Imburgia CS, Jurewicz AJ, Chan G, Fornwald JA, Dhanak D, Christmann LT, Darcy MG et al (2000) Identification of potent, selective non-peptide CC chemokine receptor-3 antagonist that

inhibits eotaxin-, eotaxin-2-, and monocyte chemotactic protein-4-induced eosinophil migration. *J Biol Chem* 275: 36626–36631
255 Andrew DP, Chang MS, McNinch J, Wathen ST, Rihanek M, Tseng J, Spellberg JP, Elias CG 3rd (1998) STCP-1 (MDC) CC chemokine acts specifically on chronically activated Th2 lymphocytes and is produced by monocytes on stimulation with Th2 cytokines IL-4 and IL-13. *J Immunol* 161: 5027–5038
256 White JR, Lee JM, Young PR, Hertzberg RP, Jurewicz AJ, Chaikin MA, Widdowson K, Foley JJ, Martin LD, Griswold DE, Sarau HM (1998) Identification of a potent, selective non-peptide CXCR2 antagonist that inhibits interleukin-8-induced neutrophil migration. *J Biol Chem* 273: 10095–10098
257 Rosenwasser LJ (1998) Biologic activities of IL-1 and its role in human disease. *J Allergy Clin Immunol* 102: 344–350
258 Borish L, Aarons A, Rumbyrt J, Cvietusa P, Negri J, Wenzel S (1996) Interleukin-10 regulation in normal subjects and patients with asthma. *J Allergy Clin Immunol* 97: 1288–1296
259 John M, Lim S, Seybold J, Jose P, Robichaud A, O'Connor B, Barnes PJ, Chung KF (1998) Inhaled corticosteroids increase IL-10 but reduce MIP-1α, GM-CSF and IFN-γ release from alveolar macrophages in asthma. *Am J Respir Crit Care Med* 157: 256–262
260 Takanashi S, Hasegawa Y, Kanehira Y, Yamamoto K, Fujimoto K, Satoh K, Okamura K (1999) Interleukin-10 level in sputum is reduced in bronchial asthma, COPD and in smokers. *Eur Respir J* 14: 309–314
261 Boguniewicz M, Martin RJ, Martin D, Gibson U, Celniker A (1995) The effects of nebulized recombinant interferon-γ in asthmatic airways. *J Allergy Clin Immunol* 95: 133–135
262 Bryan SA, O'Connor BJ, Matti S, Leckie MJ, Kanabar V, Khan J, Warrington SJ, Renzetti L, Rames A, Bock JA et al (2000) Effects of recombinant human interleukin-12 on eosinophils, airway hyperreactivity and the late asthmatic response. *Lancet* 356: 2149–2153
263 Barnes PJ (1992) Modulation of neurotransmission in airways. *Physiol Rev* 72: 699–729
264 Undem BJ, Carr MJ (2002) The role of nerves in asthma. *Curr Allergy Asthma Rep* 2: 159–165
265 Clark JM, Abraham WM, Fishman CE, Forteza R, Ahmed A, Cortes A, Warne RL, Moore WR, Tanaka RD (1995) Tryptase inhibitors block allergen-induced airway and inflammatory responses in allergic sheep. *Am J Respir Crit Care Med* 152: 2076–2083
266 Krishna MT, Chauhan A, Little L, Sampson K, Hawksworth R, Mant T, Djukanovic R, Lee T, Holgate S (2001) Inhibition of mast cell tryptase by inhaled APC 366 attenuates allergen-induced late-phase airway obstruction in asthma. *J Allergy Clin Immunol* 107: 1039–1045
267 Luisetti M, Sturani C, Sella D, Madonini E, Galavotti V, Bruno G, Peona V, Kucich U, Dagnino G, Rosenbloom J et al (1996) MR889, a neutrophil elastase inhibitor, in patients with chronic obstructive pulmonary disease: a double-blind, randomized, placebo-controlled clinical trial. *Eur Respir J* 9: 1482–1486
268 Russell RE, Culpitt SV, DeMatos C, Donnelly L, Smith M, Wiggins J, Barnes PJ (2002) Release and activity of matrix metalloproteinase-9 and tissue inhibitor of metalloproteinase-1 by alveolar macrophages from patients with chronic obstructive pulmonary disease. *Am J Respir Cell Mol Biol* 26: 602–609
269 Hautamaki RD, Kobayashi DK, Senior RM, Shapiro SD (1997) Requirement for macrophage metalloelastase for cigarette smoke-induced emphysema in mice. *Science* 277: 2002–2004
270 Cawston TE (1996) Metalloproteinase inhibitors and the prevention of connective tisue breakdown. *Pharmaol Ther* 70: 163–182
271 Carrell RW, Lomas DA (2002) Alpha1-antitrypsin deficiency–a model for conformational diseases. *N Engl J Med* 346: 45–53
272 Sallenave JM, Shulmann J, Crossley J, Jordana M, Gauldie J (1994) Regulation of secretory leukocyte proteinase inhibitor (SLPI) and elastase-specific inhibitor (ESI/elafin) in human airway epithelial cells by cytokines and neutrophilic enzymes. *Am J Respir Cell Mol Biol* 11: 733–741
273 Sallenave JM, Si Ta har M, Cox G, Chignard M, Gauldie J (1997) Secretory leukocyte proteinase inhibitor is a major leukocyte elastase inhibitor in human NEUTROPHILS. *J Leukoc Biol* 61: 695–702
274 McElvaney NG, Doujaiji B, Moan MJ, Burnham MR, Wu MC, Crystal RG (1993) Pharmacokinetics of recombinant secretory leukoprotease inhibitor aerosolized to normals and individuals with cystic fibrosis. *Am Rev Respir Dis* 148: 1056–1060
275 Belvisi MG, Wicks SL, Battram CH, Bottoms SE, Redford JE, Woodman P, Brown TJ, Webber SE, Foster ML (2001)

Therapeutic benefit of a dissociated glucocorticoid and the relevance of *in vitro* separation of transrepression from transactivation activity. *J Immunol* 166: 1975–1982
276 Bledsoe RK, Montana VG, Stanley TB, Delves CJ, Apolito CJ, McKee DD, Consler TG, Parks DJ, Stewart EL, Willson TM et al (2002) Crystal structure of the glucocorticoid receptor ligand binding domain reveals a novel mode of receptor dimerization and coactivator recognition. *Cell* 110: 93–105
277 Soderling SH, Beavo JA (2000) Regulation of cAMP and cGMP signaling: new phosphodiesterases and new functions. *Curr Opin Cell Biol* 12: 174–179
278 Essayan DM (2001) Cyclic nucleotide phosphodiesterases. *J Allergy Clin Immunol* 108: 671–680
279 Compton CH, Gubb J, Nieman R, Edelson J, Amit O, Bakst A, Ayres JG, Creemers JP, Schultze-Werninghaus G, Brambilla C et al (2001) Cilomilast, a selective phosphodiesterase-4 inhibitor for treatment of patients with chronic obstructive pulmonary disease: a randomised, dose-ranging study. *Lancet* 358: 265–270
280 Gamble E, Grootendorst DC, Brightling CE, Troy S, Qiu Y, Zhu J, Parker D, Matin D, Majumdar S, Vignola AM et al (2003) Anti-inflammatory effects of the phosphodiesterase 4 inhibitor cilomilast (Ariflo) in COPD. *Am J Respir Crit Care Med* 168: 976–982
281 Jin SL, Conti M (2002) Induction of the cyclic nucleotide phosphodiesterase PDE4B is essential for LPS-activated TNF-alpha responses. *Proc Natl Acad Sci USA* 99: 7628–7633
282 Smith SJ, Brookes-Fazakerley S, Donnelly LE, Barnes PJ, Barnette MS, Giembycz MA (2003) Ubiquitous expression of phosphodiesterase 7A in human proinflammatory and immune cells. *Am J Physiol Lung Cell Mol Physiol* 284: L279–L289
283 Barnes PJ, Adcock IM (1998) Transcription factors and asthma. *Eur Respir J* 12: 221–234
284 Manning AM (1996) Transcription factors: a new frontier in drug discovery. *Drug Disc Today* 1: 151–160
285 Kishore N, Sommers C, Mathialagan S, Guzova J, Yao M, Hauser S, Huynh K, Bonar S, Mielke C, Albee L et al (2003) A selective IKK-2 inhibitor blocks NF-kB-dependent gene expression in IL-1β stimulated synovial fibroblasts. *J Biol Chem* 278: 32861–32871
286 Johnson GL, Lapadat R (2002) Mitogen-activated protein kinase pathways mediated by ERK, JNK, and p38 protein kinases. *Science* 298: 1911–1912
287 Lee JC, Kumar S, Griswold DE, Underwood DC, Votta BJ, Adams JL (2000) Inhibition of p38 MAP kinase as a therapeutic strategy. *Immunopharmacology* 47: 185–201
288 Schafer PH, Wadsworth SA, Wang L, Siekierka JJ (1999) p38α Mitogen-activated protein kinase is activated by CD28-mediated signaling and is required for IL-4 production by human $CD4^+CD45RO^+$ T cells and Th2 effector cells. *J Immunol* 162: 7110–7119
289 Kankaanranta H, Giembycz MA, Barnes PJ, Lindsay DA (1999) SB203580, an inhibitor of p38 mitogen-activated protein kinase, enhances constitutive apoptosis of cytokine-deprived human eosinophils. *J Pharmacol Exp Ther* 290: 621–628
290 Underwood DC, Osborn RR, Bochnowicz S, Webb EF, Rieman DJ, Lee JC, Romanic AM, Adams JL, Hay DW, Griswold DE (2000) SB 239063, a p38 MAPK inhibitor, reduces neutrophilia, inflammatory cytokines, MMP-9, and fibrosis in lung. *Am J Physiol Lung Cell Mol Physiol* 279: L895–L902
291 Costello PS, Turner M, Walters AE, Cunningham CN, Bauer PH, Downward J, Tybulewicz VL (1996) Critical role for the tyrosine kinase Syk in signalling through the high affinity IgE receptor of mast cells. *Oncogene* 13: 2595–2605
292 Yousefi S, Hoessli DC, Blaser K, Mills GB, Simon HU (1996) Requirement of Lyn and Syk tyrosine kinases for the prevention of apoptosis by cytokines in human eosinophils. *J Exp Med* 183: 1407–1414
293 Amoui M, Draber P, Draberova L (1997) Src family-selective tyrosine kinase inhibitor, PP1, inhibits both Fc epsilonRI- and Thy-1-mediated activation of rat basophilic leukemia cells. *Eur J Immunol* 27: 1881–1886
294 Takeyama K, Jung B, Shim JJ, Burgel PR, Dao-Pick T, Ueki IF, Protin U, Kroschel P, Nadel JA (2001) Activation of epidermal growth factor receptors is responsible for mucin synthesis induced by cigarette smoke. *Am J Physiol Lung Cell Mol Physiol* 280: L165–L172
295 Wakeling AE (2002) Epidermal growth factor receptor tyrosine kinase inhibitors. *Curr Opin Pharmacol* 2: 382–387
296 Morley J (1992) Cyclosporin A in asthma therapy: a pharmacological rationale. *J Autoimmunity* 5 (Suppl A): 265–269
297 Thompson AG, Starzl TC (1993) New immunosuppres-

sive drugs: mechanistic insights and potential therapeutic advances. *Immunol Rev* 136: 71–98

298 Cosio MG, Majo J, Cosio MG (2002) Inflammation of the airways and lung parenchyma in COPD: role of T cells. *Chest* 121: 160S–165S

299 Yuan Q, Strauch KL, Lobb RR, Hemler ME (1996) Intracellular single-chain antibody inhibits integrin VLA-4 maturation and function. *Biochem J* 318: 591–596

300 Lin K, Ateeq HS, Hsiung SH, Chong LT, Zimmerman CN, Castro A, Lee WC, Hammond CE, Kalkunte S, Chen LL et al (1999) Selective, tight-binding inhibitors of integrin alpha4beta1 that inhibit allergic airway responses. *J Med Chem* 42: 920–934

301 Romano SJ, Slee DH (2001) Targeting selectins for the treatment of respiratory diseases. *Curr Opin Investig Drugs* 2: 907–913

302 Haczku A, Takeda K, Redai I, Hamelmann E, Cieslewicz G, Joetham A, Loader J, Lee JJ, Irvin C, Gelfand EW (1999) Anti-CD86 (B7.2) treatment abolishes allergic airway hyperresponsiveness in mice. *Am J Respir Crit Care Med* 159: 1638–1643

303 Van Oosterhout AJ, Hofstra CL, Shields R, Chan B, Van Ark I, Jardieu PM, Nijkamp FP (1997) Murine CTLA4-IgG treatment inhibits airway eosinophilia and hyperresponsiveness and attenuates IgE upregulation in a murine model of allergic asthma. *Am J Respir Cell Mol Biol* 17: 386–392

304 van Neerven RJ, Van de Pol MM, van der Zee JS, Stiekema FE, De Boer M, Kapsenberg ML (1998) Requirement of CD28-CD86 costimulation for allergen-specific T cell proliferation and cytokine expression. *Clin Exp Allergy* 28: 808–816

305 Kon OM, Compton CH, Kay AB, Barnes NC (1997) A dopuble-blind placebo-controlled trial of an anti-CD4 monoclonal antibody SB210396. *Am J Respir Crit Care Med* 155: A203

306 Zhang DH, Yang L, Cohn L, Parkyn L, Homer R, Ray P, Ray A (1999) Inhibition of allergic inflammation in a murine model of asthma by expression of a dominant-negative mutant of GATA-3. *Immunity* 11: 473–482

307 Hsu CH, Chua KY, Tao MH, Lai YL, Wu HD, Huang SK, Hsieh KH (1996) Immunoprophylaxis of allergen-induced immunoglobulin E synthesis and airway hyperresponsiveness *in vivo* by genetic immunization. *Nat Med* 2: 540–544

308 Herz U, Gerhold K, Gruber C, Braun A, Wahn U, Renz H, Paul K (1998) BCG infection suppresses allergic sensitization and development of increased airway reactivity in an animal model. *J Allergy Clin Immunol* 102: 867–874

309 Choi IS, Koh YI (2002) Therapeutic effects of BCG vaccination in adult asthmatic patients: a randomized, controlled trial. *Ann Allergy Asthma Immunol* 88: 584–591

310 Shirtcliffe PM, Easthope SE, Cheng S, Weatherall M, Tan PL, Le Gros G, Beasley R (2001) The effect of delipidated deglycolipidated (DDMV) and heat-killed *Mycobacterium vaccae* in asthma. *Am J Respir Crit Care Med* 163: 1410–1414

311 Sur S, Wild JS, Choudhury BK, Sur N, Alam R, Klinman DM (1999) Long term prevention of allergic lung inflammation in a mouse model of asthma by CpG oligodeoxynucleotides. *J Immunol* 162: 6284–6293

312 Horner AA, Van Uden JH, Zubeldia JM, Broide D, Raz E (2001) DNA-based immunotherapeutics for the treatment of allergic disease. *Immunol Rev* 179: 102–118

313 Cookson WO (2002) Asthma genetics. *Chest* 121: 7S–13S

314 Sandford AJ, Silverman EK (2002) Chronic obstructive pulmonary disease. 1: Susceptibility factors for COPD the genotype-environment interaction. *Thorax* 57: 736–741

315 Hall IP (2000) Pharmacogenetics of asthma. *Eur Respir J* 15: 449–451

316 Xing Z, Ohkawara Y, Jordana M, Graham FL, Gauldie J (1996) Transfer of granulocyte-macrophage colony-stimulating factor gene to rat induces eosinophilia, monocytosis and fibrotic lesions. *J Clin Invest* 97: 1102–1110

317 Stecenko AA, Brigham KL (2003) Gene therapy progress and prospects: alpha-1 antitrypsin. *Gene Ther* 10: 95–99

318 Nyce JW, Metzger WJ (1997) DNA antisense therapy for asthma in an animal model. *Nature* 385: 721–725

319 Sandrasagra A, Leonard SA, Tang L, Teng K, Li Y, Ball HA, Mannion JC, Nyce JW (2002) Discovery and development of respirable antisense therapeutics for asthma. *Antisense Nucleic Acid Drug Dev* 12: 177–181

320 Barnes PJ (2002) Current and future therapies for airway mucus hypersecretion. *Novartis Found Symp* 248: 237–249

321 Sommerhoff CP, Krell RD, Williams JL, Gomes BC, Strimpler AM, Nadel JA (1991) Inhibition of human neu-

trophil elastase by ICI 200,355. *Eur J Pharmacol* 193: 153–158

322 Kuo H-P, Barnes PJ, Rogers DF (1992) Cigarette smoke-induced airway goblet cell secretion: dose dependent differential nerve activation. *Am J Physiol* 7: L161–L167

323 Rogers DF (2002) Pharmacological regulation of the neuronal control of airway mucus secretion. *Curr Opin Pharmacol* 2: 249–255

324 Poole PJ, Black PN (2001) Oral mucolytic drugs for exacerbations of chronic obstructive pulmonary disease: systematic review. *Br Med J* 322: 1271–1274

325 Grandjean EM, Berthet P, Ruffmann R, Leuenberger P (2000) Efficacy of oral long-term N-acetylcysteine in chronic bronchopulmonary disease: a meta-analysis of published double-blind, placebo-controlled clinical trials. *Clin Ther* 22: 209–221

326 Nadel JA, Burgel PR (2001) The role of epidermal growth factor in mucus production. *Curr Opin Pharmacol* 1: 254–258

327 Inoue A, Saijo Y, Maemondo M, Gomi K, Tokue Y, Kimura Y, Ebina M, Kikuchi T, Moriya T, Nukiwa T (2003) Severe acute interstitial pneumonia and gefitinib. *Lancet* 361: 137–139

328 Zhou Y, Shapiro M, Dong Q, Louahed J, Weiss C, Wan S, Chen Q, Dragwa C, Savio D, Huang M (2002) A calcium-activated chloride channel blocker inhibits goblet cell metaplasia and mucus overproduction. *Novartis Found Symp* 248: 150–165

329 Bush A (1998) Early treatment with dornase alfa in cystic fibrosis: what are the issues? *Pediatr Pulmonol* 25: 79–82

330 Madison JM, Irwin RS (2003) Pharmacotherapy of chronic cough in adults. *Expert Opin Pharmacother* 4: 1039–1048

331 Fuller RW, Jackson DM (1990) Physiology and treatment of cough. *Thorax* 45: 425–430

332 Choudry NB, Gray SJ, Posner J, Fuller RW (1991) The effect of 443C81, a μ-opioid receptor agonist, on the response to inhaled capsaicin in healthy volunteers. *Br J Clin Pharmacol* 32: 683–686

333 Ayik SO, Basoglu OK, Erdinc M, Bor S, Veral A, Bilgen C (2003) Eosinophilic bronchitis as a cause of chronic cough. *Respir Med* 97: 695–701

334 Waring JP, Lacayo L, Hunter J, Katz E, Suwak B (1995) Chronic cough and hoarseness in patients with severe gastroesophageal reflux disease. Diagnosis and response to therapy. *Dig Dis Sci* 40: 1093–1097

335 Udezue E (2001) Lidocaine inhalation for cough suppression. *Am J Emerg Med* 19: 206–207

336 Chung KF (2002) Cough: potential pharmacological developments. *Expert Opin Investig Drugs* 11: 955–963

Immunostimulants in cancer therapy

James E. Talmadge

Introduction

Immunoaugmenting agents have been used to treat disease since William B. Coley treated cancer patients with mixed bacterial toxins early in the 20th century [1]. These early studies spawned the clinical use of such microbially derived substances such as Bacille Calmette-Guérin (BCG) (bladder cancer, USA), Krestin, Picibanil and Lentinan (gastric & other cancers, Japan), and Biostim and Broncho-Vaxom (recurrent infections, Europe). While these "crude" drugs induce numerous immunopharmacological activities, they pose considerable regulatory obstacles due to impurity, lot-to-lot variability, unreliability and adverse side-effects. Similarly, traditional herbal medicines (Asia) also provide a source of active substances for IMMUNOTHERAPY. The purification, characterization and synthetic production of the active moieties from natural products (Bestatin, Taxol) and culture supernatants (FK-506, Rapamycin, Deoxyspergualin and Cyclosporin) have also provided valuable drugs. The current focus within IMMUNOTHERAPY is on the use of RECOMBINANT proteins (CYTOKINES), although the utility of these drugs can be limited due to BIOACTIVITY and pharmacological deficiencies. Thus, there remains a potential role for classical BIOLOGICAL RESPONSE MODIFIERS (BRMs) due to their potential for oral BIOAVAILABILITY and ability to induce multiple CYTOKINES for immune augmentation and hematopoietic restoration.

In 2002, the 20th anniversary of the first approved biopharmaceutical-RECOMBINANT insulin (Humulin: Genotech, South San Francisco, CA, USA) was observed. Today, biotechnological drugs incorporate not only immunoregulatory proteins, enzymes and biologicals derived from natural sources, but also engineered (manipulated) MONOCLONAL ANTIBODIES (Abs) and CYTOKINES in addition to GENE THERAPY and TISSUE ENGINEERING strategies. As of the year 2000, there were 84 BIOPHARMACEUTICALS approved in the United States and/or the European Union for use in humans [2]. Since then, 60 additional BIOPHARMACEUTICALS have been approved for use in humans and some 400–600 are currently undergoing clinical evaluation [3].

The overall approach in this chapter is to limit the discussion of BIOTHERAPEUTICS to the RECOMBINANT, natural and synthetic drugs that are currently approved for clinical use against cancer (Box 1). We have focused on RECOMBINANT proteins (except antithrombotics, VACCINES and monoclonal Abs) and will not discuss nucleic acid-based or tissue-engineered gene therapeutic products. A brief discussion follows regarding COMBINATION THERAPY and CELLULAR THERAPY as future prospects.

Recombinant proteins

RECOMBINANT proteins have emerged as an important class of drugs for the treatment of cancer, IMMUNOSUPPRESSION, myeloid dysplasia and infectious diseases; however, our limited understanding of their pharmacology and MECHANISM OF ACTION (MOA) has hindered their development (Tab. 1). To facilitate advancement, information is needed on their pharmacology, MOA and toxicology [4, 5]. One approach to the development of BIOTHERAPEUTICS is to identify a CLINICAL HYPOTHESIS based upon therapeutic SURROGATE(s) identified during preclinical pharmacological studies [6, 7]. A SURROGATE for clinical EFFICACY may be a phenotypic, biochemical, enzymatic, functional (immunological, molecular or hematological) or quality-of-life measurement that is believed to be associated with therapeutic activity. Phase I clinical

Box 1. Types of cancer amenable to immunotherapy

- Melanoma and renal carcinoma. These tumors are highly antigenic and frequently have a significant histocytic and lymphocytic cellular infiltration. As such, they have demonstrated significant responsiveness to intervention with biotherapeutics.
- Bladder cell carcinoma. Superficial transitional cell carcinoma has proven highly responsive to therapy with BCG as well as cytokines such as interferon-α (IFN-α).
- Head and neck cancer. Head and neck cancer is responsive to cytokine therapy, due in part to its availability for therapeutic intervention and has been shown to be highly responsive to low-dose paralymphatic administration of interleukin-2 (IL-2).
- Solid tumors. A number of solid tumors have shown a response to immune-augmenting agents, but in general, Phase III trials have not been undertaken to demonstrate responsiveness.
- Leukemias and lymphomas. In general, these 'liquid' tumors have shown responsiveness to several immune-augmenting agents, including intervention with IFNs.

trials can then be designed to identify the OPTIMAL IMMUNOMODULATORY DOSE (OID) and treatment schedule that maximizes the augmentation of SURROGATE end point(s). Subsequent phase II/III trials can be established to determine if the changes in the SURROGATE levels correlate with therapeutic activity. Table 1 lists the immunologically and hematologically active CYTOKINES that are approved for use in the United States. This information is expanded in Appendix 1.

In contrast to strategies based on the identification of SURROGATES for therapeutic EFFICACY, protocols for RECOMBINANT proteins are often identified based on practices developed for conventional drugs that may not be advantageous for the identification of therapeutic efficiency by CYTOKINES (Box 2).

Interferon-α (IFN-α)

Clinical activity against cancer

The initial, nonrandomized, clinical studies with IFN-α suggested that it had therapeutic activity for malignant melanoma, osteosarcoma and various lymphomas [13]. Subsequent randomized trials, however, demonstrated significant therapeutic activity against only less common tumor histiotypes, including hairy cell (HCL) and chronic myelogenous (CML) leukemia [13–15] and a few types of lymphoma [15], including low-grade non-Hodgkin's lymphoma [16] and cutaneous T CELL lymphoma [17]. Currently, the list of responding indications has expanded to include malignant melanoma [18], acquired immune deficiency syndrome (AIDS) and Kaposi's sarcoma [19], genital warts, and hepatitis B and C.

Pharmacological actions – dose response

It has taken almost three decades to translate the concept of IFN-α as an anti-viral to its routine utility in clinical oncology and infectious diseases. Despite extensive study, the development of IFN-α is still in its early stages, and basic parameters such as optimal dose and therapeutic schedule remain to be determined [14, 15, 20]. The MOA is also controversial since IFN-α has been shown to have dose-dependent antitumor activities *in vitro*, yet to be active at low doses for HCL [14, 15]. Immunomodulation as the mechanism of therapeutic activity with IFN-α is perhaps best supported by its action against HCL. Treatment with IFN-α is associated with a 90–95% response rate; however, this is not fully achieved until the patients have been on the protocol for a year, and it appears that low doses of IFN-α are as active as higher doses [21]. It should be noted that the clinical

TABLE 1. MULTIPLE-TARGETED DRUGS

Drug	Target	Approval date
Interferon-γ (IFN-γ)	Management of chronic granulomatous disease and osteopetrosis.	December 90
Interferon-α (IFN-α)	Treatment of relapsing multiple sclerosis, chronic hepatitis C viral infection, AIDS-related Kaposi's sarcoma, follicular lymphoma, genital warts, hairy cell leukemia, hepatitis B and C, malignant melanoma.	May 96
Bone morphology protein-2	Treatment of spinal degenerative disc disease.	July 02
TNFR:Fc	Moderate to severe active rheumatoid arthritis and juvenile rheumatoid arthritis, active ankylosing spondylitis.	November 98
rEPO	Anemia caused by chemotherapy, anemia, chronic renal failure, anemia in Retrovir® treated HIV-infected, chronic renal failure, dialysis. Surgical blood loss.	June 89
rHuGM-CSF	Allogeneic and autologous bone marrow transplantation, neutropenia resulting from chemotherapy, peripheral blood progenitor cell mobilization.	March 91
LFA-1/IgG1	Moderate to severe chronic plaque psoriasis.	January 03
rHu IL-11	Chemotherapy-induced thrombocytopenia.	November 97
rG-CSF	Acute myelogenous leukemia, autologous or allogeneic bone marrow transplantation, chemotherapy-induced neutropenia, chronic severe neutropenia, peripheral blood progenitor cell transplantation.	February 91
IL-2	CTCL, metastatic melanoma, renal cell carcinoma.	May 92
rHPDGF-BB	Diabetic neuropathy, foot ulcers.	1997
Stem cell factor	Mobilization.	1997
Interferon-n	Treatment of hepatitis C.	March 99

use of IFN-α has been supplanted by other even more effective drugs; however, approval in HCL precipitated expanded studies of IFN-α for other diseases.

The initial dose-finding studies determined that a dose of 12×10^6 U/M^2 of IFN-α was not tolerable in patients with HCL [15]. Subsequently, it was demonstrated that highly purified natural IFN-α (2×10^6 U/M^2) was both well tolerated and effective when administered three times per week for 28 days [22]; however, it retained some toxicity, including myelosuppression as well as neurotoxicity and cardiotoxicity. In these studies, a lower dose (2×10^5 U/M^2) was also administered for 28 days and found to be better tolerated than the standard dose, while inducing equivalent improvements in NEUTROPHIL and platelet counts. Furthermore, substantial clinical improvement (increased platelet and NEUTROPHIL counts) was observed within the first 4 to 8 weeks of treatment as well as an improved quality of life (decreased cardiac and neurological toxicity, myelosuppression, flu-like syndrome, platelet transfusions, and bacterial infection incidences). However, IFN-α also generates a therapeutic dose-response effect, whereby higher doses of IFN-α will induce a quantitatively greater anti-leukemic response than that observed with low doses of IFN-α. Thus, therapy with IFN-α initially is at 2×10^5 U/M^2, allowing patients to

Box 2. Pharmacological and dose-relationship considerations with cytokines

Cytokines have in several instances shown increased bioactivity following delivery by slow release [8–10]. The covalent attachment of polyethylene glycol (PEG) to cytokines (pegylation) (Tab. 2), including IFN-α and granulocyte colony-stimulating factor (G-CSF), has demonstrated significant biological activity due, in part, to their improved pharmacological profile [9, 10]. Thus, strategies to limit the pharmacological deficiencies are critical to the development of recombinant biotherapeutics. The pharmacological attributes of recombinant biotherapeutics are improved with targeted delivery, which prolongs their short half-life, limiting biological activity [11]. Furthermore, there can be unexpected relationships between the dose administered and the biological effect in recombinant biotherapeutics, including a nonlinear dose-relationship, described as "bell-shaped" [12]. This lack of a linear dose-response relationship may be due to a nonlinear dispersal throughout the body, a poor ability to enter into a saturatable receptor-mediated transport process, chemical instability, sequence of administration with other agents, an incorrect time of administration, an inappropriate location, and/or response of the target cells. Furthermore, a "bell-shaped" dose-response curve may be associated with receptor tachyphylaxis expression or a signal transduction mechanism whereby the cells become refractory to subsequent receptor-mediated augmentation.

become tolerant to the acute toxicity of IFN-α with a subsequent dose increase to 2×10^6 U/M^2 to obtain a greater anti-leukemic effect. In CML, sustained therapeutic responses are found in more than 75% of patients [23, 24]. In addition to reducing leukemic-cell mass, there is also a gradual reduction in the frequency of cells bearing a 9–22 chromosomal translocation [25] (Box 3).

The unique cellular and molecular activities of IFN-α can potentially complement the MOAs by other therapies [30]. At present, therapeutic applications for IFN-α are focused on synergistic or additive effects with IFN-γ, granulocyte-monocyte colony-stimulating factor (GM-CSF) and interleukin- (IL-) 2. The MOA by IFN-α, as well as the optimal dose, remains controversial. A recent meta-analysis of twelve clinical studies for high-risk melanoma showed a significant recurrence free survival (RFS) following treatment with IFN-α. However, the benefits of IFN-α therapy on overall survival (OS) are less clear [31]. There is a significant trend between increasing dose and RFS, but not OS. Similarly, studies on immune augmentation in melanoma patients receiving high- or low-dose IFN-α revealed no association between immune response and baseline phenotypic and functional immunity. However, numerous immune surrogates are augmented by IFN-α treatment and are associated with dosage. Administration of IFN-α has been shown to significantly upregulate class II major histocompatibility complex (MHC) and intercellular adhesion molecule- (ICAM-)1 expression on tumor cells in a dose dependent manner. In addition, natural killer (NK) cell and T cell functions are augmented in a dose-dependent manner by IFN-α, as are changes in T cell phenotypes. Furthermore, high-dose IFN-α regulates immune parameters more rapidly than low-dose IFN-α. However, in this study of 51 high-dose, 54 low-dose and 43 no IFN-α patients, there was no relationship between immune augmentation and RFS [32].

Interferon-γ (IFN-γ)

Effect in animal models and MOA

IFN-γ has multiple potential mechanisms that may be involved in tumor protection and therapy. These include, but are not limited to: 1) anti-proliferative activity for tumor growth/survival, 2) induction of angiostasis, and 3) augmentation of both innate and adoptive immunity (Fig. 1). However, it is unclear which, if any, of these potential mechanisms are critical for the activity of this pleiotropic cytokine.

Box 3. Gleevec

Until the induction of Imatinib mesylate, IFN-α was the standard treatment for patients in chronic-phase CML. Imatinib mesylate (STI571, Gleevec), a signal transduction inhibitor, has demonstrated significant activity as a single agent for the treatment of CML. Patients who failed IFN-α therapy are responsive to Imatinib; therefore, at present, it is used either as first-line monotherapy or to rescue patients who have failed IFN-α therapy [26, 27]. In addition to its activity in chronic phase CML, Gleevec also has therapeutic activity for patients with accelerated and blast phase CML [27]. Long-term survival and apparent cure are achieved with allogeneic stem cell transplantation (SCT) for CML patients who have a donor and who can tolerate the transplant and remains the therapy of choice [28, 29]. Because of the limited experience with Imatinib, no "curative" information exists for this drug and thus, allogeneic SCT remains the therapy of choice when a donor is available, although not necessarily the primary therapy of choice [27]. Furthermore, resistance to Imatinib can develop, suggesting that Imatinib may not be sufficient to "cure" a significant number of patients, supporting a need for studies into sequential/concomitant therapies, which are likely to be with IFN-α despite its toxicities.

1) Anti-proliferative effects on tumor growth/survival: IFN-γ has direct anti-proliferative and anti-metabolic effects on tumor cells. It can also induce the APOPTOSIS of tumor cells via conventional signaling mechanisms, resulting in the induction of genes that promote cellular apoptosis, including caspase-1 (IL-1α converting enzyme or ICE) [33] and Fas ligand (FasL) [34].

2) Induction of angiostasis: The anti-tumor activity of IFN-γ can also be mediated by an inhibition of NEOANGIOGENESIS. The growth of solid tumors requires new blood vessel formation, which is a result of tumor-induced angiogenesis [35]. Pro-angiogenic molecules are secreted by tumors, including vascular endothelial cell growth factor (VEGF) and basic fibroblast growth factor (FGF). However, IFN-γ induces CHEMOKINES with potent angiostatic actions, including inducible protein- (IP-) 10 [36]. In addition, IP-10 belongs to a family of CXC non-ELR CHEMOKINES that all have angiostatic activity and whose expression is regulated by IFN-γ.

3) Augmentation of both innate and adoptive immune responses. IFN-γ is a potent macrophage-activating CYTOKINE capable of augmenting macrophage-mediated tumoricidal activity *in vitro* and *in vivo* [37, 38]. IFN-γ activated MACROPHAGES express multiple tumoricidal mechanisms, including the production of reactive oxygen and/or nitrogen intermediates and up-regulated expression of cytotoxic ligands, such as FasL [39], TUMOR NECROSIS FACTOR- (TNF-)α, and TNF-related apoptosis-inducing ligand (TRAIL) [40]. In addition, IFN-γ significantly enhances IL-12 secretion by MACROPHAGES and DENDRITIC CELLS (DCs) [41]. NK cells also have a potential role in promoting anti-tumor responses, via at least two mechanisms. NK cells are important sources of IFN-γ and also exert direct cytocidal activity against tumors through mechanisms involving perforin [42] and TRAIL [43]. In addition, IFN-γ can markedly enhance adaptive T CELL responses to tumors. IFN-γ has also an important role in regulating the T helper cell 1/T helper cell 2 (Th1/Th2) balance during the host response to a tumor [43].

Dose-response relationship

Preclinical studies have suggested that IFN-γ has significant therapeutic activity with a "BELL-SHAPED" dose-response curve [12]. Studies of immune response in normal animals have revealed the same "BELL-SHAPED" dose-response curve for the augmentation of MACROPHAGE tumoricidal activity [12, 22]. Thus, optimal therapeutic activity is observed with the same dose and protocol of IFN-γ, but with significantly less therapeutic activity at lower and higher doses. A significant correlation between MACRO-

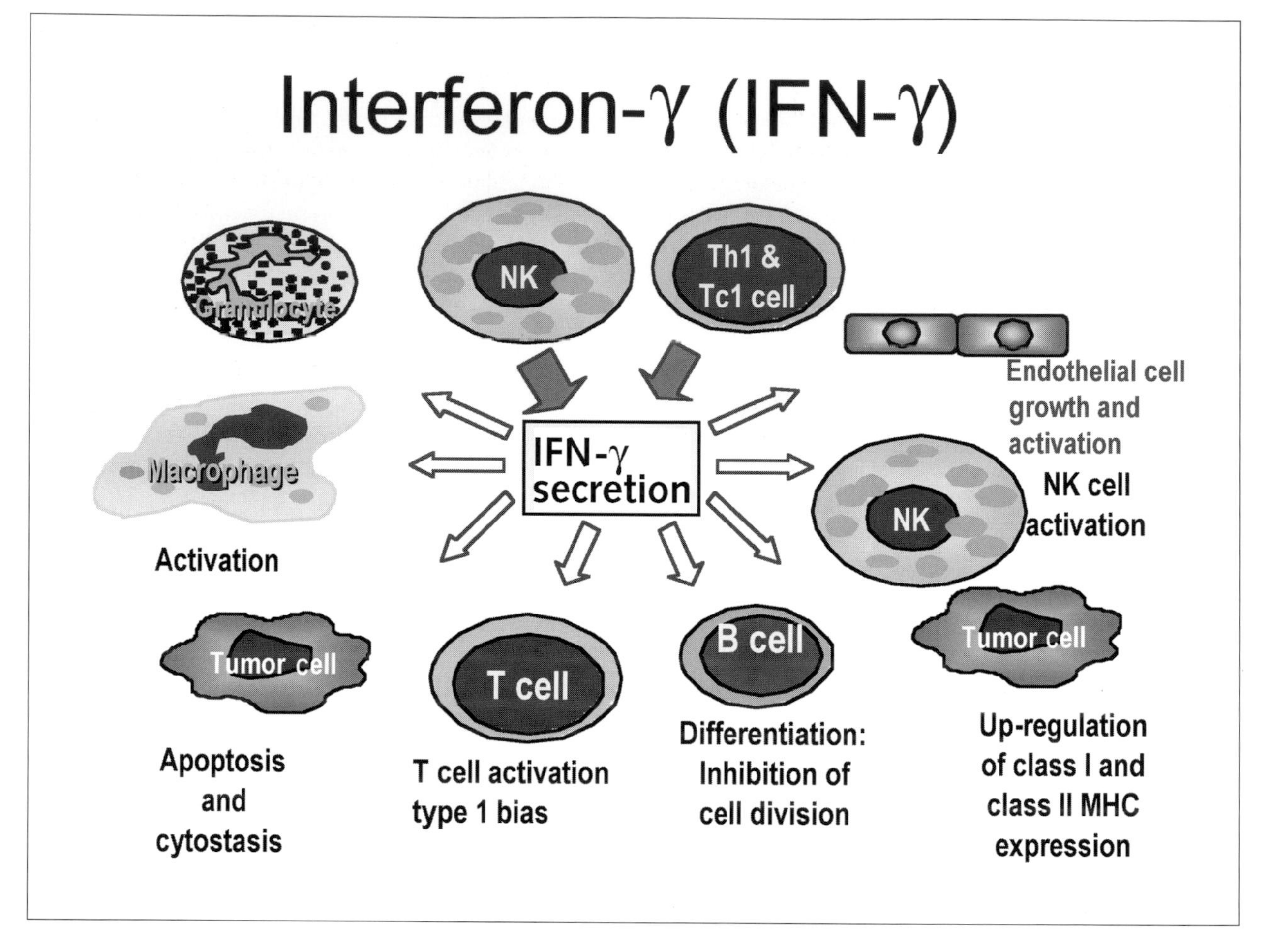

FIGURE 1. REGULATION OF IMMUNE RESPONSES BY INTERFERON-γ (IFN-γ)
IFN-γ is produced predominately by Th1, Tc1 and NK cells, resulting in the activation of T cells, NK cells, macrophages, and granulocytes. In addition, it up-regulates class I and class II MHC expression on tumor cells as well as the expression of a wide variety of receptors on both tumor cells and epithelial cells. IFN-γ, interferon-γ, NK, natural killer cell; Th1, T helper cell type 1; Tc1, T cytotoxic cell type 1; MHC, major histocompatibility complex.

PHAGE augmentation and therapeutic EFFICACY has been reported [22], suggesting that immunological augmentation provides one mechanism for the therapeutic activity of IFN-γ and supports the hypothesis that treatment with the maximum tolerated dose (MTD) of IFN-γ may not be optimal in an adjuvant setting. The PRECLINICAL HYPOTHESIS of a "BELL-SHAPED" dose-response curve for IFN-γ has been confirmed in clinical studies on the immunoregulatory effects of IFN-γ, which defined an OID [44, 45]. In general, the OID for IFN-γ has been found to be between 0.1 and 0.3 mg/M^2 [44]. In contrast, the MTD for IFN-γ may range from 3 to 10 mg/M^2, depending upon the source of IFN-γ and/or the clinical center. The identification of an OID for IFN-γ in patients with minimal tumor burden has resulted in the development of clinical trials to test the hypothesis that the immunological enhancement induced by IFN-γ will result in

Box 4. IFN-γ in CGD

Studies in patients with CGD suggest that the mechanism of therapeutic activity by IFN-γ is associated with enhanced phagocytic oxidase and superoxide activity by polymorphic nuclear (PMN) cells. However, more recent data suggest that the majority of CGD patients obtain clinical benefit by prolonging IFN-γ therapy and that the MOA may not be due to enhanced PMN oxidase activity, but rather the correction of a respiratory-burst deficiency in monocytes [47]. Furthermore, IFN-γ administration could induce nitric oxide (NO) synthetase activity by PMNs in patients with CGD [48]. Following two days of IFN-γ administration, a significant increase in PMN-produced NO activity is observed in association with an increase in bactericidal capacity of PMN [48]. As these PMN lack the capacity to produce superoxide anions, at least in patients with CGD, it is possible that the increased NO release and *in vitro* bactericidal activity is instrumental in augmenting host defenses and reducing the morbidity of CGD [48]. Similarly, IFN-γ increases PMN expression of Fc γ receptor I (FcγRI) and improves FcγR-mediated phagocytosis, at least in normal subjects [49]. Therefore, the MOA critical to reducing the frequency of infections in patients with CGD by IFN-γ may not be associated with increased oxidase and superoxide production, but rather with other mechanisms of granulocyte activity, including NO production.

prolongation of the disease-free period and OS of patients in an adjuvant setting [45].

Clinical therapeutic activity – MOA

IFN-γ was found on an empirical basis to have therapeutic activity in chronic granulomatous disease (CGD) [46] and it was for this indication that the United States Food And Drug Administration (FDA) first approved IFN-γ (Box 4).

In addition to its approval for CGD, IFN-γ has also been approved for the treatment of rheumatoid arthritis (RA) in Germany and most recently, severe malignant osteopetrosis in the United States. In a randomized Phase III trial, IFN-γ was also reported to have activity in women receiving first-line platinum-based chemotherapy against ovarian cancer [50]. In this study, there was a significantly higher response rate and longer progression-free survival (PFS) in woman receiving IFN-γ plus chemotherapy when compared to chemotherapy alone. However, there was no statistically significant improvement in OS. The IFN-γ MOA is unknown for this study, although the authors speculated that it may be associated with the inhibition of HER-2/neu expression. Currently, IFN-γ is being studied for the treatment of idiopathic pulmonary fibrosis [51].

Interleukin-2 (IL-2)

Pharmalogical actions relevant to cancer

IL-2, a T cell growth factor, has a significant role in regulating immunity to infectious and neoplastic diseases (Fig. 2). IL-2 is produced primarily by activated DCs, NK and T cells and induces pleiotrophic biological responses after binding to one of three receptors.

IL-2 stimulates the growth of naïve T cells following Ag (Ag) activation, and later induces activation-induced cell death (AICD) [52, 53]. IL-2 also has effects on several other immune cells, including NK cells [54], B cells [55], monocyte/macrophages [56], and PMNs [57]. The ability of IL-2 to stimulate NK and T cell lysis of tumor cells stimulated clinical interest in IL-2 [58]. NK cells are part of the innate immune system and comprise 10–15% of peripheral blood lymphocytes (PBL). Functionally, NK cells are characterized by NK and lymphokine-activated killing (LAK), antibody-dependent cellular cytotoxicity (ADCC), and immunoregulatory cytokine production. NK cells also express NK receptors (NKRs) that recognize MHC class I ligands and regulate, inhibit, and activate a response to target cells [59].

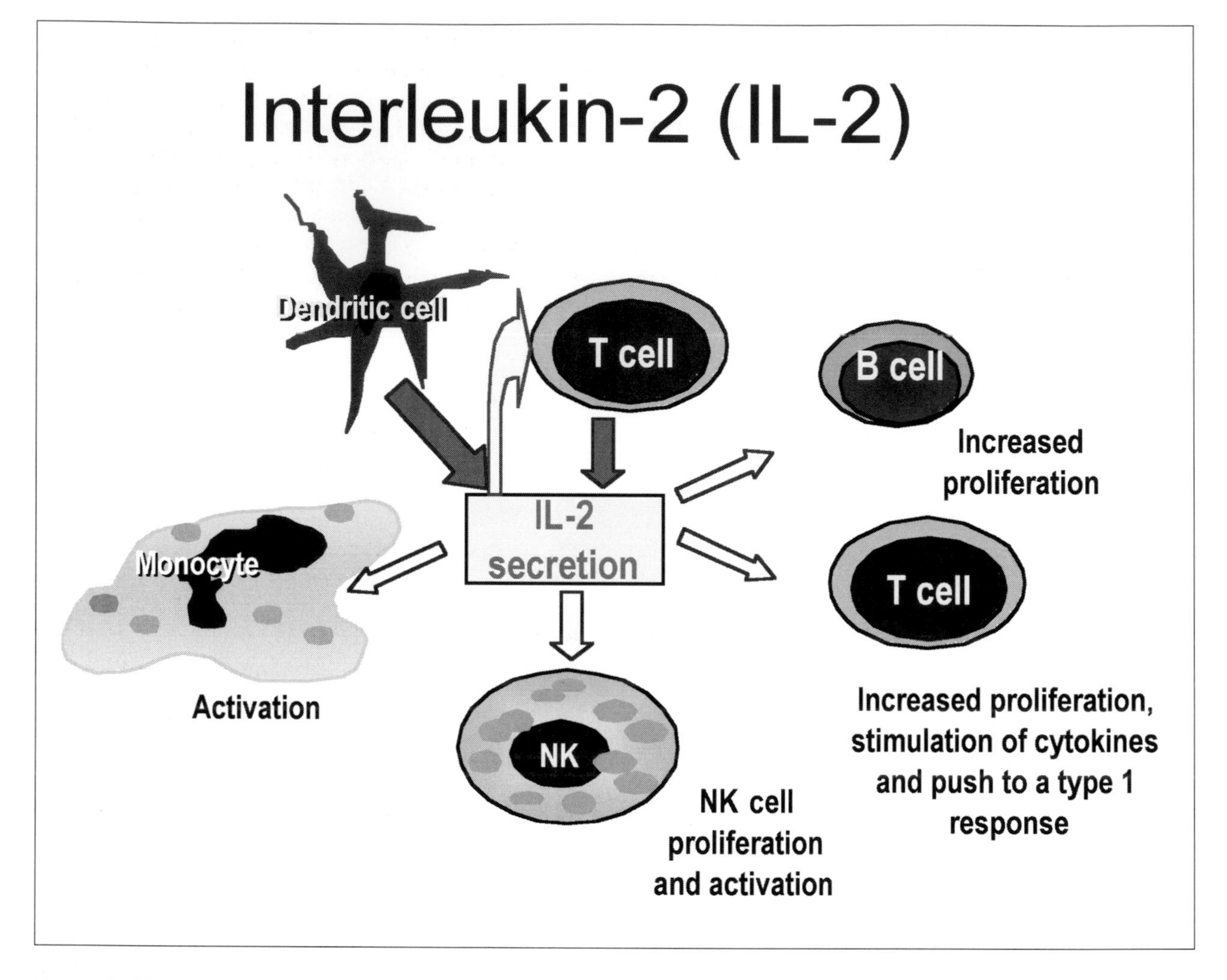

Figure 2. Regulation of immune responses by interleukin-2 (IL-2).
IL-2 production by T cells and DCs supports the proliferation of T cells, B cells, and NK cells, in addition to establishing a bias towards a type 1 T cell response. IL-2 can also activate monocytes and NK cells, resulting in increased cytotoxicity. DC, dendritic cell; IL-2, interleukin-2; NK, natural killer cell.

One important *in vivo* role of IL-2 for T cell responses is the promotion of thymic development and peripheral expansion of $CD4^{+}CD25^{+}$ T cells known as T regulatory (Treg) cells. Loss of Treg activity in IL-2- or IL-2R-deficient mice results in lymphadenopathy and autoimmune disease. It appears that IL-2-dependent Treg cells regulate homeostatic and Ag-induced T cell proliferative responses, resulting in pathological autoreactivity [60]. IL-2 and its high-affinity receptor are rapidly up-regulated following binding with a cognate Ag and the absence of either IL-2 or IL-2R can result in a loss of immune reactivity [61]. Furthermore, the induction of $CD8^{+}$ T cell responses is dependent upon the presence of $CD4^{+}$ T cell help and their secretion of IL-2 [62]. Recently, IL-2 production by DCs has also been shown to be essential for the initiation of both $CD4^+$ and $CD8^+$ T CELL responses [63].

IL-15 was identified based upon its ability to stimulate proliferation of IL-2-dependent T CELLS in the presence of neutralizing IL-2 Abs [64]. IL-15 shares two of the three IL-2Rs, including IL-2/15R beta (β) and the γc receptor subunits [64]. Similar to IL-2, resting T CELLS do not respond to IL-15, and T CELL Ag receptor (TCR) ligation induces IL-15Rα expression, allowing a response to IL-15. Both IL-2 and IL-15 can induce the proliferation of activated T CELLS and the differentiation of cytotoxic T LYMPHOCYTES (CTL) and LAK cells expressing IL-2R or IL-15R *in vitro*. However, despite their shared receptor usage, distinct roles for IL-2 and IL-15 are observed for the proliferation and survival of $CD4^+$ and $CD8^+$ T CELLS [53]. Both IL-2 and IL-15 promote the proliferation of $CD4^+$ T CELLS, but continued stimulation with IL-2 promotes AICD, IL-15 has the opposite effect and can even inhibit IL-2-induced AICD of $CD4^+$ T CELLS [65]. IL-15 can also selectively stimulate the proliferation of memory $CD8^+$ T CELLS, in contrast to IL-2, which inhibits $CD8^+$ memory T CELL proliferation [66]. Thus, IL-15 and IL-2 have similar biological activity *in vitro*, while their critical and non-redundant functions *in vivo* are distinct [67, 68]. Furthermore, IL-15 is better than IL-2 in mediating NK cell differentiation and promoting the memory $CD8^+$ T CELLS' survival. Given IL-15's role in the maintenance and proliferation of T CELLS, including memory $CD8^+$ T CELLS, a clinical investigation is likely [69].

Clinical effects against cancer

IL-2 has been approved for use as a single agent in the treatment of renal cell carcinoma, metastatic melanoma and hepatitis C. It is also administered in conjunction with LAK or T CELL infiltrating LYMPHOCYTES (TILs) in ADOPTIVE CELLULAR THERAPY protocols. TILs are T CELLS obtained from a tumor and expanded *in vitro* with low levels of IL-2 and, in the presence of tumor Ag, this results in a population of tumor-specific cytotoxic T CELLS. However, it has been questioned whether the ADOPTIVE transfer of LAK or TIL cells is necessary or adds to the clinical EFFICACY of IL-2. Indeed, there has been little indication of an improved therapeutic effect of IL-2 plus LAK cells *versus* IL-2 alone [70, 71]. When the clinical trials with IL-2 are rigorously examined, neither strategy has impressive (as opposed to significant) therapeutic activity [71, 72]. The overall response rate with IL-2 is 7-14% and is associated with considerable toxicity [73]; however, it should be remarked that most of these responses are durable. In one of the first clinical studies [74], partial responses were observed in 4 out of 31 patients. Interestingly, these partial responders did not correspond to patients with increased LAK or NK cell activity. The antitumor effect of both TIL and LAK cells could be due to either a direct effect or secondary to the induction of other CYTOKINE mediators. This is suggested by the observation that IL-2-stimulated LYMPHOCYTES produce IFN-γ and TNF as well as other CYTOKINES and that the therapeutic activity of IL-2 may be synergistic with these CYTOKINES [74]. Recently, IL-2 has also been examined as an adjuvant to augment the tumor host response to human immunodeficiency virus (HIV) [75] and anticancer VACCINES [76].

Many of the IL-2 clinical trials in metastatic renal cell carcinoma, with or without LAK cells, have used an MTD of IL-2. A study by Fefer et al. [77] compared maintenance IL-2 therapy at the MTD or a dose 60% lower. They found that it was possible to maintain patients on therapy for a median of 4 days at the IL-2 MTD, but in the presence of severe hypertension and capillary-leak syndrome. In the lower-dose protocol, none of the patients experienced severe hypertension or capillary-leak syndrome and the median duration of maintenance IL-2 therapy was 9 days. Furthermore, there was a total response rate of 41% in the lower-dose protocol compared to a 22% response rate for the higher-dose protocol and a shorter duration of administration. These investigators suggest that there may be an improved therapeutic activity associated with a longer IL-2 maintenance protocol that can be achieved at lower doses.

Dose-response, toxicity and pharmacological studies

Several IL-2 dose-response studies have examined the effect of IL-2 administration on CYTOKINE messenger ribonucleic acid (mRNA) levels in the PBL of cancer patients. The results from one IL-2 dose-response study suggested that: 1) doses of IL-2 as low as 3×10^4 U per day administration by continuous infusion could augment T CELL function and 2) doses of IL-2 $\geq 1 \times 10^5$ U/day increases not only T CELL, but also MACROPHAGE function [78]. The latter was measured as an upregulation of TNF mRNA, which was observed at the higher dose of IL-2. The increased TNF mRNA levels combined with the increased levels of IFN-α observed at the lower dose of IL-2 may be responsible for IL-2 toxicity [79]. The effect of low-dose, subcutaneous (s.c.) IL-2 administration was studied in healthy males at 1,000 or 10,000 international units (IU)/kg [80]. No consistent changes were observed with 1,000 IU/kg; however, phenotypic and immunoregulatory changes were observed following administration of 10,000 IU/kg of IL-2. This dose significantly depressed the number of circulating LYMPHOCYTES, including $CD4^+$, $CD8^+$, and activated T, B, and NK cells. In contrast, the number of NEUTROPHILS and MONOCYTES were increased. There was also a significant increase in IL-4 serum levels, while the levels of IFN-γ and IL-2 receptors were significantly depressed. Kinetic studies revealed that these effects varied with time, but occurred at IL-2 serum levels sufficient to saturate the high-affinity receptor and, by three hours following injection of 10,000 IU/kg of IL-2, remained elevated for approximately 12 hours. Together, these studies suggest that IL-2 mediated immune augmentation can occur with low doses of IL-2 administered s.c. or by continuous infusion.

The potential for IL-2 to be active at low doses was also shown in patients with squamous cell carcinoma of the oral cavity and oropharynx [81]. In this study, 202 patients were randomly assigned to receive either surgery and radiotherapy or surgery, radiotherapy, and daily IL-2. The IL-2 was injected perilymphatically at 5,000 units of IL-2, daily for 10 days prior to and following surgery. IL-2 was also given post-surgery for 12 monthly cycles, each consisting of 5 daily injections. This study revealed a significant increase in disease-free survival (DFS) and OS in the IL-2 treated patients.

In a study of renal cell cancer patients, cohorts were randomized to receive either a high-dose intravenous (i.v.) IL-2 regimen or one using one-tenth the dose (72,000 IU/kg/8 h) administered by the same schedule [82]. A third arm using the low dose of IL-2, where it was given daily by s.c. administration, was added later. In the most recent interim report [72], 156 patients were assigned to the high-dose arm, 150 patients to the low-dose i.v. arm and 94 patients to the low-dose s.c. arm. Toxicities were less frequent when IL-2 was given at the low doses, especially hypotension, but there were no IL-2 associated deaths in any arm. A higher response rate was observed in the patients given the high dose of IL-2 (21%) *versus* low-dose i.v. IL-2 (13%), but no OS differences were observed. The response rate of s.c. IL-2 (10%) was similar to the low-dose i.v. IL-2, but was significantly different from high-dose IL-2 therapy ($P = 0.033$). The response duration and survival in the complete responders were significantly better in patients receiving high-dose IL-2, as compared to low-dose IL-2 i.v. therapy ($P = 0.04$). Thus, tumor regressions, as well as complete responses, were seen with all IL-2 regimes. Furthermore, it was suggested that IL-2 is more clinically active at the higher dose given i.v., although this did not provide an OS benefit and only a small percentage of patients achieved a durable clinical response.

Granulocyte-monocyte colony-stimulating factor (GM-CSF)

Pharmacological activity related to cancer therapy

GM-CSF was initially defined by its ability to support the growth of both granulocyte and MACROPHAGE colonies from hematopoietic precursor cells [83]. However, GM-CSF-deficient mice have no obvious deficiency in myeloid cells [84], suggesting the presence of redundant growth factor(s). GM-CSF can also potentiate the functions of mature granulocytes and MACROPHAGES [85], in addition to its role as a hemopoietic regulator (Fig. 3) [86].

Similar to other proinflammatory CYTOKINES, the production and activity of GM-CSF occurs at the site of INFLAMMATION and increased levels of GM-CSF mRNA are observed in skin biopsies from allergic patients with cutaneous reactions [87] and in arthritic joints [88]. At present, GM-CSF is considered an

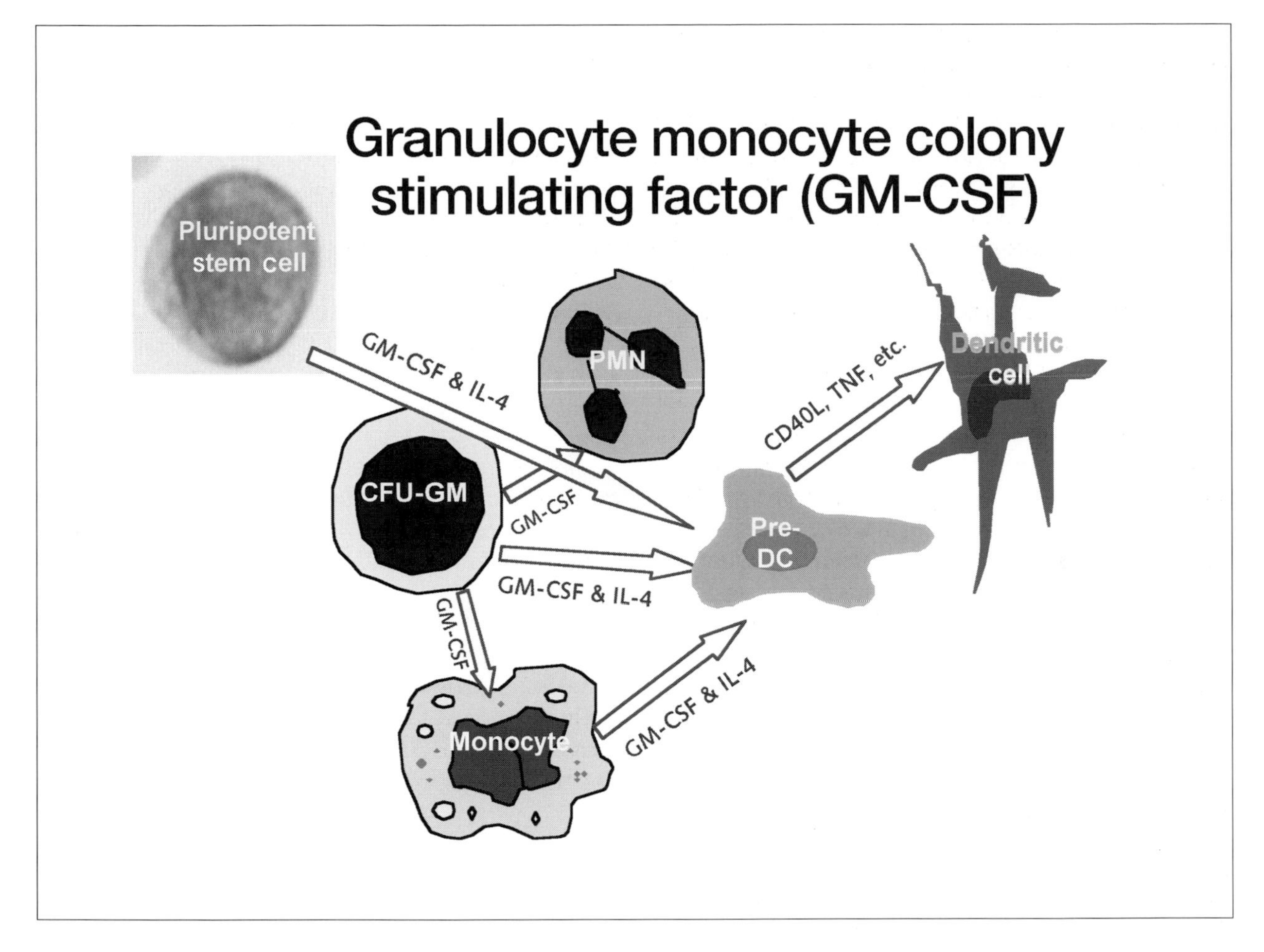

Figure 3. Maturation of precursors into dendritic cells (DCs) by granulocyte-monocyte colony-stimulating factor (GM-CSF)

GM-CSF and IL-4 can drive pluripotent stem cells, PMNs, monocytes, and CFU-GM genetic precursors into pre-DCs, which following activation can become mature activated DCs. DC, dendritic cell; GM-SCF, granulocyte-monocyte colony-stimulating factor; IL-4, interleukin-4; PMN, polymorphonuclear neutrophil; CFU-GM, colony-forming unit-granulocyte monocyte; CD40L, CD40 ligand; TNF, tumor necrosis factor.

important regulator (proliferation, maturation and activation) of granulocyte and MACROPHAGE lineage populations at all stages of maturation. GM-CSF in combination with TNF-α or IL-4 can differentiate MONOCYTES into DCs *in vitro* [89], which are then used as VACCINES. Treatment of mice with PEGYLATED GM-CSF has also been shown to expand DCs in the spleen, [90] and clinically, the number of circulating DCs are expanded in the peripheral blood by GM-CSF and IL-4 administration [91].

Clinical use in cancer patients

GM-CSF was approved in 1991 by the FDA to support transplant-associated neutropenia and to mobilize STEM CELLS. In Europe, it is also approved for prophy-

TABLE 2. SLOW-RELEASE VARIANTS

Drug	Target	Stage
Interferon α-2a, PEG	First-line treatment of chronic hepatitis C	October 02
Pegfilgrastim, Neulasta, G-CSF-PEG	Febrile neutropenia in patients receiving chemotherapy	January 03
Peginterferon α-2b (PEG-Intron™)	Treatment of chronic hepatitis C in patients not previously treated with IFN-α	January 01
Aranesp (novel erythropoiesis-stimulating protein or NESP)	Anemia associated with chronic kidney disease (Europe)	June 01

lactic treatment following dose-intensive chemotherapy. However, the rate of absolute NEUTROPHIL count (ANC) recovery in response to treatment with GM-CSF in patients receiving myelosuppressive chemotherapy or in the mobilization of STEM CELLS into the peripheral blood (PB) with GM-CSF is one day slower than that observed with G-CSF. It also has a toxicity profile, including low-grade fever, myalgias and bone pain that may be slightly greater than that observed with G-CSF. Because of the real and perceived problems, GM-CSF has been used to a lesser extent then G-CSF. However, patients receiving chronic GM-CSF therapy post-marrow-graft failure have been shown to have a significantly improved survival as compared to historically matched controls [92]. Similarly, significantly reduced hospital and antibiotic therapy duration have been reported for several phase II and III clinical trials [93]. Thus, patients undergoing autologous BONE MARROW transplantation (BMT) or SCT and receiving prophylactic GM-CSF administration have demonstrated significant improvements in ANC recovery, fewer cases of infection and fewer days spent in the hospital [94].

GM-CSF effects on histiocytes

In addition to an effect on HEMATOPOIESIS, administration of GM-CSF can also affect MACROPHAGES and DCs. MACROPHAGES are functionally activated following stimulation with GM-CSF and preclinical and clinical data suggest that GM-CSF can act as an adjuvant for VACCINES [95, 96]. DCs are Ag-presenting cells that play a major role in the induction of primary and secondary T CELL immune responses against cancer. Because GM-CSF is a mediator of proliferation, maturation and migration of DCs [97], it has been used to enhance the induction of Ag-specific cytotoxic T CELLS. Due to the effect of GM-CSF alone or in combination with IL-4 on DCs, it has been used both *ex vivo* and *in vivo* to expand DCs for use as VACCINES, [91, 98, 99] following transfection or priming with peptide Ags [91]. Furthermore, preclinical studies have suggested that GM-CSF transfected tumor cells can provide prophylactic and, in some instances, therapeutic anti-cancer activity [98–100]. Thus, GM-CSF has potential utility not only to address neutropenia, but also as a direct therapeutic agent within an adjuvant protocol or as a VACCINE adjuvant.

Cancer patients have been shown to have defective MACROPHAGE and DC function [101, 102]. Preclinical studies have shown that MONOCYTES and tumor-associated histocytes can be stimulated by GM-CSF to become cytotoxic against tumor cells [103]. Because of the functional attributes of GM-CSF, a few clinical trials have studied GM-CSF either alone or in combination with IL-2 for the potential to improve anticancer immunity. In a recent study, 48 cancer patients with surgically resected stage III or stage IV melanomas received GM-CSF s.c. for 14 days followed by treatment monthly for one year or until disease recurrence and the outcomes were compared to matched historic controls. A median survival time of 37.5 months was observed with GM-CSF therapy *versus* 12.2 months for historical controls [104].

Box 5. Pegylation

- Prolongs half-life and bioavailability.
- Protects against binding to monocyte and PMNs.
- Protects against enzyme degradation and Ab induction.
- Reduces number of injections required.
- Improves patient compliance.
- May result in increased bioactivity or a new profile of activity.

Engineered recombinant proteins

Several new RECOMBINANT BIOPHARMACEUTICALS have been engineered to improve their pharmacological properties (Tab. 2).

The primary strategy has been to improve the pharmacological half-life by PEGYLATION.

PEGYLATION improves the BIOACTIVITY of a CYTOKINE by decreasing the rate of systemic CLEARANCE, which reduces the frequency of dosing, improves PATIENT COMPLIANCE, and lowers cost. (Box 5)

Similarly, genetic manipulation to increase the glycosylation of erythropoietin (EPO) has been shown to prolong the pharmacological half-life as well. Several PEGYLATED IFNs have been studied, including native IFNs that have a relatively short half-life, typically one to four hours. In contrast, PEGYLATION increases this to twenty-four hours or longer. As an example, IntronA (IFN-α2b) is typically administered three times a week, whereas PEGYLATED IFN is administered as a single weekly injection. Similarly, thrice-weekly injections of Roferon (IFN-α2a) have been replaced by a single weekly injection of Pegasys, resulting in an increased EFFICACY for the treatment of hepatitis C. Most recently, PEGYLATED G-CSF has allowed a single injection of PEGYLATED G-CSF, replacing the requirement for daily or twice-daily injections of G-CSF for five or more days.

Similar to PEGYLATION, the sequence of EPO was mutated to alter oligosaccharide sequences, resulting in improved pharmacokinetics. The sialic acid content of the carbohydrate component of glycoproteins has a significant effect on a protein's half-life, which was exploited to create this novel EPO sequence (Aranesp). This resulted in increased sialic acid content from the addition of two extra glycosylation sites in the EPO backbone, and a longer half-life. Additional strategies are being used to prolong the half-life of proteins, including nanoparticles, liposomes and poloxamer matrices, which allow the slow release of a protein. These formulations provide not only a slow release, but also the potential for specific targeting to organs or tumors via modification of the formulation.

Natural and synthetic BRMS

Natural BRMs

The use of BRMs to treat human disease has its origins in the use of bacterial toxins to treat cancer by William B. Coley [105]. This form of therapy can occur by multiple mechanisms as shown in Box 6.

These early studies resulted in the clinical use of microbially derived substances such as BCG VACCINE or Picibanil, carbohydrates from plants or fungi such as Krestin and Lentinan, other products such as Biostim and Broncho-Vaxom, as well as thymic extracts (Tab. 3).

However, there is considerable lot-to-lot variation in the purity of these compounds. In addition, due to the particulate nature of some of the BRMs, i.v. injections can result in pulmonary thrombosis and respiratory distress as well as the development of granulomatous disease following dermal administration and scarification.

Box 6. Biological response modifiers

- Stop, control, or suppress processes that permit cancer growth, inflammatory or autoimmune process, including overcoming immunosuppressive processes
- Introduce processes that result in cancer cells being more recognizable and susceptible to destruction by the immune system
- Boost the cytotoxic activity of effector cells, such as T cells, NK cells, and macrophages
- Alter the growth patterns of cancer cells to promote a cellular behavior like that of healthy cells
- Block or reverse the process that changes a normal cell or a precancerous cell into a cancerous cell
- Enhance the body's ability to restore normal cells damaged or destroyed by treatment, such as chemotherapy or radiation
- Prevent cancer cells from spreading (metastasizing) to other parts of the body

BCG

The most commonly used microorganism for cancer therapy in the United States is BCG. It has been used systemically for the treatment of metastatic disease or adjuvant therapy, intralesionally (especially for cutaneous malignant melanoma), topically for superficial bladder cancer, and in combination with other immune modulators, tumor VACCINES and chemotherapy. When given intravesically, it can treat superficial bladder cancer in residual disease and in the adjuvant setting [106]. A well-controlled, randomized study has shown a prolonged disease-free interval and time-to-progression in patients treated with intradermal and intravesical BCG as compared to controls [107]. The mechanism by which BCG mediates its antitumor response is not known, but BCG treatment induces granulomatous INFLAMMATION in the bladder [108] and elevates IL-2 levels in the urine of treated patients [109], suggesting that an augmented local immune response may be important.

Clinical studies have shown that intravesical installation of BCG in patients with superficial bladder cancer results in a significant increase in IL-1β, IL-2, IL-6, TNF-α, IFN-γ and MACROPHAGE COLONY-

Table 3 – Natural BRMs

Agent	Chemical nature	Action	Clinical use
BCG (USA & EUR)	Live mycobacteria	Macrophage activator	Bladder cancer
Picibanil (OK432)(Jap)	Extract *Strep. Pyogenes*	Macrophage activator	Gastric/other cancers
Krestin (PSK)(Jap)	Fungal polysaccharide	Macrophage activator	Gastric/other cancers
Lentinan (Jap)	Fungal polysaccharide	Macrophage activator	Gastric/other cancers
Biostim (Eur)	Extract *Klebsiella penum.*	Macrophage activator	Chronic or recurrent infections
Thymostimulin (Eur)	Thymic peptide extract	T cell stimulator	Cancer & infection
T-activin (Russia)	Thymic peptide extract	T cell stimulator	Cancer & infection
Thym-Uvocal (FRG)	Thymic peptide extract	T cell stimulator	Cancer & infection

TABLE 4. CHEMICALLY DEFINED BRMS

Agent	Chemical nature	Action	Clinical use
Azathiopine Imuran	Purine antimetabolite	Immunosuppressant	Graft *versus* host disease, allograft
Bestatin (Jap)	Dipeptide	Macrophage and T cell stimulant	Acute myelosis leukemia
Cytosine-phosphate-guanosine	Nucleotide	Binding to TLR9 and DC activity	Vaccine adjuvant
Cyclosporine Sandimmune®	Cyclic undecapeptide which is a metabolite of soil fungus	Immunosuppressant	Graft *versus* host disease, allografts, rheumatoid arthritis, psoriasis
Deoxyspergualin	Peptide fermentation product of bacillus laterosporus	Immunosuppressant	Acute renal rejection
Everolimus (Certican)	TOR kinase inhibitor	Immunosuppressant	Cardiac transplant
Isoprinosine (Eur)	Inosine: salt complex	T cell stimulant	Infection
Levamisole (USA)	Phenylimidothiazole	T cell stimulant	Cancer
Mycophenolate moeftil (MMF) Celleept®	2-morpholine ethylester of mycophenolic acid	Immunosuppressant	
Rapamycin Rapamune® (sirolimus)	Metabolite of *Streptomyces hygroscopicus*	Immunosuppressant	Solid organ transplantation, graft *versus* host disease
Romurtide (Jap)	18 lys MDP	Macrophage stimulant	Bone marrow recovery
Tacrolimus, FK505, Prograf®	Macrolide lactone which is a fermentation product of *Streptomyces*	Immunosuppressant	Eczema, solid organ transplantation, graft *versus* host disease
Thalidomide (USA)	α-(N-phthalimido) glutarimide	Suppresses TNF and adhesion molecule expression	Erythema nodosum leprosum
Thymopentin TP-5 (Italy & FRG)	Pentapeptide	T cell stimulant	Rheumatoid arthritis infection and cancer

STIMULATING FACTOR (M-CSF) with a concomitant and significant increase in serum levels of IL-2 and IFN-α [110]. There appears to be a relationship between CYTOKINE production and therapeutic EFFICACY as a multivariant logistical analysis demonstrated that IL-2 induction was a discriminating parameter for remission in patients receiving BCG treatment for superficial bladder carcinoma [111]. This association of increased IL-2 levels was recently confirmed in a study of 20 bladder cancer *in situ* patients [112].

Chemically defined BRMs

Synthetic BRMs

The use of nonspecific immunostimulants has also been extensively studied (Tab. 4). The microbially derived agents have in common widespread effects on the IMMUNE SYSTEM and side-effects akin to infection (e.g., fever, malaise, myalgia, etc.). These agents can enhance nonspecific resistance to microbial or neoplastic challenge when administered prior to challenge (immunoprophylactic), but rarely when administered following challenge (immunotherapeutic). This is an important distinction in that the primary objective for the oncologist is the treatment of preexistent metastatic disease.

Levamisole

Following a long history of experimental use in many different cancers and diseases, LEVAMISOLE, a chemically defined, orally active immunostimulant, demonstrated significant therapeutic activity (meta-analysis) [113, 114]. It was approved for the treatment of Duke's C colon cancer in combination with 5-fluorouracil (5-FU). LEVAMISOLE promotes T lymphocyte, MACROPHAGE and NEUTROPHIL function, suggesting multiple MOA. LEVAMISOLE stimulates T CELL function *in vivo*, particularly in immunodeficient individuals, presumably through the action of its sulphur moiety. One study with LEVAMISOLE demonstrated a significant increase in the frequency of PB mononuclear cells expressing the NK cell Ag CD16 at all dose levels, although lower toxicity was observed at the lower doses of LEVAMISOLE [115]. The authors suggested that short-term LEVAMISOLE administration was only minimally IMMUNOMODULATORY and that chronic administration at low doses may be better tolerated and provide similar levels of immune modulation to that observed with higher doses [115]. It is relatively nontoxic (flu-like symptoms, gastrointestinal upset, metallic taste, skin rash and Antabuse reaction), but can produce an agranulocytosis, particularly in human leukocyte Ag (HLA) B-27$^+$ patients with RA where its use has been discontinued. It is noted that the adjuvant therapeutic activity of LEVAMISOLE has been questioned in recent years. In one phase III trial comparing 5-FU with leucovorin to 5-FU with LEVAMISOLE, it was found that the 5-FU and LEVAMISOLE significantly prolonged DFS and OS in patients with type III colon cancer who had undergone curative resection relative to adjuvant therapy with LEVAMISOLE [116].

Muramyl dipeptides (MDP) – synthetic natural BRMs

One of the largest and best-studied classes of synthetic agents are the MURAMYL DIPEPTIDES (MDP). MDP was discovered based on the isolation of the minimally active substitute for intact BCG in Freund's adjuvant [117]. Unfortunately, as with many of the polypeptides that have low molecular weights, MDP has a short serum half-life and requires frequent administration of high doses to be active. In addition, agents such as MDP are strongly pyrogenic, presumably in association with their ability to induce IL-1. MDP has also been incorporated into multilamellar vesicles (MLV) for higher stability and to facilitate monocytic PHAGOCYTOSIS.

The first MDP that was approved for clinical use was Romurtide (Japan), which induces BM recovery following cancer chemotherapy [118] via growth factor induction. Its MOA is the activation of MACROPHAGES to secrete COLONY-STIMULATING FACTORS (CSF), IL-1 and TNF, resulting in the expansion of marrow precursors and subsequent commitment and differentiation into mature granulocytes and MONOCYTES. Therefore, the period of granulocytopenia and the risk of secondary infections are reduced, allowing more frequent and/or intense chemotherapy. MURABUTIDE (Tab. 4), an orally active form of MDP that does not induce fevers, is currently in clinical trials with cancer and infectious disease (France). In order to further stabilize the incorporation of MDP into MLV, lipophilic analogues of MDP such as muramyl tripeptide phosphatidylethanolamide [MTP-PE (ImmTher)] have been developed. MTP-PE has shown significant therapeutic benefit for pediatric patients with osteosarcoma [119] and is also being studied in patients with resectable melanoma [120] (USA & Europe). Preclinically, MTP-PE has also shown protection of the mucosal epithelium from cytoreduction therapy [121]. The MDPs are also potent adjuvants, either alone or an as oil emulsion and are under consideration for use as an adjuvant with HIV peptide VACCINES.

Bestatin – engineered synthetic natural BRMs

Bestatin (ubenimex) is a potent inhibitor of aminopeptidase N and aminopeptidase B [122], which was isolated from a culture filtrate of *Streptomyces olivoreticuli* during the search for specific inhibitors of enzymes present on the membrane of eukaryotic cells [123]. Inhibitors of aminopeptidase activity are associated with MACROPHAGE activation and differentiation, and Bestatin has shown significant therapeutic effects in several clinical trials [124]. In one multi-institutional study, patients with acute non-lymphocytic leukemia (ANLL) were randomized to receive either Bestatin or control [125] orally after completion of induction and consolidation therapy, and concomitant with maintenance chemotherapy. Remission duration was prolonged in the Bestatin group, although this difference did not reach statistical significance; however, OS was prolonged in the Bestatin group. Recently, a confirmatory phase III trial in ANLL was reported which extended the observation to a significant prolongation of remission [126]. Bestatin has also shown adjuvant activity when administered to acute leukemia and chronic myelogenous leukemia patients who did not develop graft-*versus*-host disease (GVHD) within 30 days following BMT [127]. Bestatin-treated acute leukemia patients had an increased incidence of chronic low-grade GVHD compared with the control arm and a lower relapse rate. Recently, a phase III study of resected stage 1 squamous cell lung cancer patients treated with either Bestatin or placebo daily per OS for two years revealed that five-year cancer-free survival was significantly greater in the Bestatin group as compared to the placebo group. In this study, the five-year cancer-free survival was 71% for the Bestatin cohort and 62% for the placebo group. OS was also significantly improved as was cancer-free survival [128].

Oligonucleotides (ODNs) as natural/synthetic BRMs

BACTERIAL EXTRACTS can activate both INNATE and ADAPTIVE IMMUNITY. It was recently revealed how the INNATE immune system detects infectious agents and distinguishes different classes of pathogens. The immune system uses 'PATTERN-RECOGNITION RECEPTORS' that are expressed on certain INNATE immune cells, to trigger cellular activation when they recognize conserved microbial-specific molecules [129, 130]. These molecules, originally thought of as nonspecific immune activators as discussed above, are now known to be specifically recognized by receptors that are expressed in a cell-specific and compartmentalized manner. The best-characterized family of PATTERN-RECOGNITION RECEPTORS is the TOLL-LIKE RECEPTOR (TLR) family. One of these, TLR9, is expressed in the endosomal compartment of plasmacytoid DCs and B CELLS [131], and is essential for the recognition of viral and intracellular bacterial DNA [132]. Now that specific ligands have been identified, IMMUNOTHERAPY has begun to grow beyond the nonspecific effects of whole BACTERIAL EXTRACTS, and to develop synthetic TOLL-LIKE RECEPTOR ligands (TLRLs). One example of such synthetic C IMMUNOMODULATORY molecules is the short ODNs that mimic the INNATE immune response to microbial DNA and which contain one or more cytosine-phosphate-guanosine (CpG) dinucleotide-containing motifs with unmethylated cytosine residues and are recognized by TLR-9. The immune effects of CpG ODNs occur in two stages: an early stage of INNATE immune activation and a later stage of enhanced ADAPTIVE IMMUNITY (Fig. 4).

Within minutes of the exposure of B CELLS or plasmacytoid DCs to CpG, the expression of costimulatory molecules, resistance to apoptosis, upregulation of the chemokine receptor CCR7 (associated with trafficking to the T-cell zone of the lymph nodes) and secretion of Th1-promoting CHEMOKINES and CYTOKINES, including as macrophage inflammatory protein-1 (IP10) and other IFN-inducible genes [133], are observed. pDCs secrete IFN-α and mature into highly effective Ag-presenting cells (APCs) [134]. The CpG-induced secretion of IFN-α, TNF-α, and other CYTOKINES and CHEMOKINES induce, within hours, secondary effects, including NK cell activation and enhanced expression of Fc receptors, resulting in increased ADCC. This INNATE immune activation and pDC maturation into myeloid DCs is followed by the induction of adaptive immune responses. B CELLS are strongly costimulated if they bind specific Ag at the same time as CpG, which selectively enhances the development of Ag-specific Abs [135]. CpG binding also activates B CELLS to proliferate, secrete IL–6, and differentiate into plasma cells [136]. The CpG-enhanced APC function occurs via the upregulated expression of costimulatory molecules, including CD40, CD80, and CD86 [137]. In mixed cell populations, a cascade of secondary responses, including activation of MACROPHAGES and NK cells and the induction of IFN-γ by Th1 cells also occurs

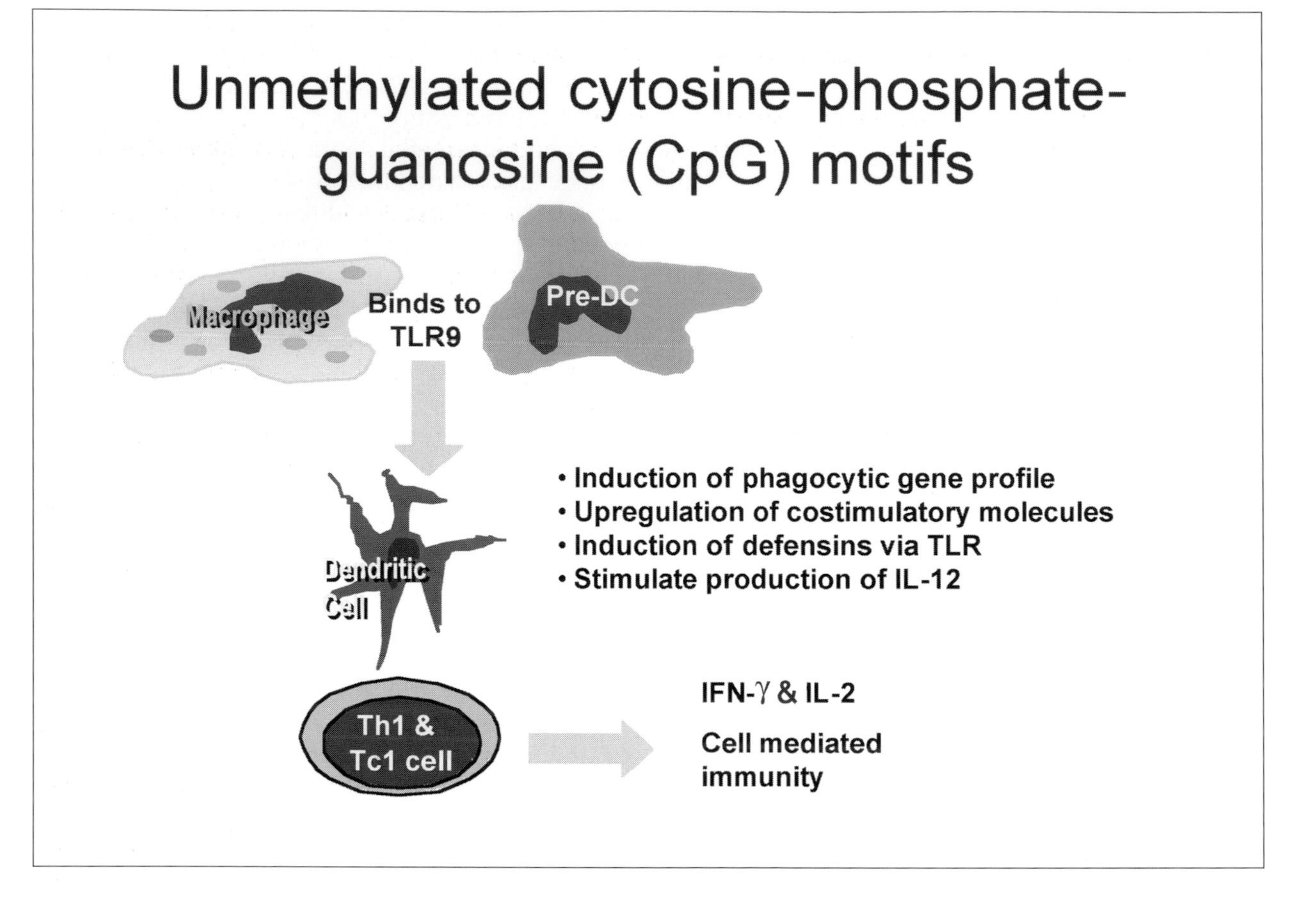

FIGURE 4. REGULATION OF INNATE AND ADAPTIVE IMMUNITY BY CYTOSINE-PHOSPHATE-GUANOSINE (CPG) MOTIFS
Co-incubation in the presence of or injection of unmethylated CpG motifs results in the binding to TRL9, ensuing in the activation and maturation of DCs, which can interact with Th1 or Tc1 cells and up-regulate cell-mediated immunity. In addition, CpG can induce phagocytosis, upregulation of co-stimulatory molecules, defensins, and IL-12. CpG, cytosine-phosphate-guanosine; TLR9, toll-like receptor 9; DC, dendritic cell; Th1, T helper cell type 1; Tc1, T cytotoxic cell type 1; IFN-γ, interferon-γ; IL-2, interleukin-2; IL-12, interleukin-12.

[137–139]. The efficient activation of APCs and induction of IL-12, IL-18, IFN-α, and IFN-γ is directly associated with their ability to induce T-helper cell 1 (Th)1-polarization, adjuvant activity [140] inhibition of Th2 responses [141] and stimulate CD8 T-cell responses [138].

CpG ODNs are also very effective as VACCINE adjuvants to enhance adaptive Th1 cellular immune responses [142–144]. In mice, CpG ODNs can trigger strong TH1 responses [143, 144], enhancing the number and function of tumor-specific CTLs and IFN-g secreting T CELLS [145, 146]. CpG ODNs also enhance therapeutic responses to other immune therapies, such as to VACCINES, including DC VACCINES [147], proteins, irradiated cells transduced with GM-CSF [148, 149], and long peptide VACCINES [145]. This has resulted in therapeutic VACCINES in mouse tumor models where no other approach has shown comparable EFFICACY, even with 1-cm established tumors [147, 150]. Even without a VACCINE, CpG ODNs can induce $CD8^+$ T-cell-mediated regression of established tumors with durable memory responses [151].

BOX 7. COMBINATION CHEMO-IMMUNOTHERAPY (BIOCHEMOTHERAPY)

- Immunotherapy is most active against minimal residual disease.
- Immunotherapy utilizes mechanisms of action that differ from chemotherapy and as such has a different resistance profile.
- Chemotherapy generally results in a reduction in the extent of neoplastic disease but frequently does not remove all residual tumor cells.
- Tumor resistance to chemotherapy frequently occurs, such that dose intensity is often needed to assure maximal effect.
- In part because immunotherapy utilizes different mechanisms of action, additional efficacy can be expected.
- Chemotherapy and other cytoreductive therapies are generally toxic, while immunotherapy as a generality is minimally toxic.
- Chemotherapy can reduce the number of host effector cells capable of responding to a tumor and limiting the extent of immunotherapy.
- In contrast, chemotherapy in a dose-dependent manner often can be immune-augmenting via a reduction in immune suppressor cells.
- Based on the above considerations, significant insight into the pharmacology and toxicology of the therapeutic agents are needed in order to successfully combine chemotherapy and immunotherapy.

Combination immunotherapy and cellular therapy

Stem cell transplantation (SCT): demonstration of T cell therapeutic activity

Because CYTOKINES have unique MECHANISMS OF ACTION, they are ideal candidates for COMBINATION THERAPY with chemotherapeutic agents (Box 7).

However, increased knowledge and consideration of the potential interactions are necessary for successful clinical use. The use of HIGH-DOSE CHEMOTHERAPY (HDT), which is myeloablative and requires stem cell rescue, provides the ultimate in cytoreductive therapy. Further, SCT provides one of the few statistically supported demonstrations of clinical therapeutic EFFICACY by T CELLS based on the survival of patients receiving an allogeneic *versus* an autologous transplant [152]. Thus, strategies to upregulate T CELL function post-autologous SCT provides one focus for CYTOKINE therapy. This is important as the return of immunological function in transplant patients is slow and accompanied by depressed numbers of $CD4^+$ T CELLS, a low CD4/CD8 T CELL ratio, and suppressed T CELL responses [153]. The role of T CELLS in controlling neoplastic disease is described as a GRAFT-*versus*-TUMOR (GVT) response. This role of T CELLS in treating neoplasia is supported by the significantly higher risk of relapse in patients receiving an allogeneic SCT, which has been T CELL-DEPLETED (TCD), or who receive cyclosporin A (CSA) to prevent GVHD [154]. However, GVHD also has unfavorable effects on transplant-related morbidity and mortality. In first remission, the decreased relapse rate with acute and/or chronic GVHD is more than offset by the increased risk of death from other causes. Consequently, patients with GVHD have a lower risk of treatment failure, but an increased risk of morbidity due to GVHD.

TCD markedly reduces the incidence of severe GVHD; however, as discussed above, TCD is also associated with an increased rate of severe, and often, fatal infections, a higher incidence of graft rejection, and an increased risk of disease recurrence. The increased risk of infectious complications is associated with the slow recovery of $CD4^+$ and $CD8^+$ T CELLS that occurs following SCT, as the initial T CELL recovery that occurs with an unmanipulated stem cell product is associated with the T CELLS transplanted with the STEM CELLS [152]. Similarly, the increased graft failure that is observed following transplantation with a TCD product is likely to reflect the contribution that infused T CELLS make towards the eradication of residual host T CELLS following the transplant preparatory regimen. Due to the increased incidence of infections and relapse, donor leukocyte infusions (DLI) may be used to reduce the incidence of graft loss, disease relapse and secondary infections [155]. However, DLI is also associated with an increased risk of GVHD, and thus, alternatives to TCD and DLI, such as strategies that can induce Ag-specific TOLERANCE shortly after allogeneic SCT, are appealing as they might prevent GVHD without resulting in a requirement for post-graft IMMUNOSUPPRESSION.

SCT and immunotherapy

Adjuvant studies in patients receiving HDT and SCT include a focus on IMMUNOTHERAPY. The dose-intense preparatory regimens, commonly referred to as HDT, are administered before transplantation and a number of CYTOKINE- and/or VACCINE-associated protocols are given following transplant with a TCD product or intact product in an attempt to improve immunological function, particularly ones directed against tumor cells. One THERAPEUTIC STRATEGY is the use of VACCINES capable of inducing Ag-specific effector T CELLS. In addition, T CELLS from the donor may be stimulated *ex vivo*, expanded and then reinfused. Strategies have also focused on the initiation of CTL response to viruses, which can reduce the incidence of treatment-related Epstein-Barr virus (EBV)-associated lymphomas or infections such as cytomegalovirus (CMV) [156, 157].

SCT combination therapy with IL-2

One approach to improving survival of cancer patients has been the use of IL-2 IMMUNOTHERAPY following HDT and SCT to induce an autologous GVT response. Based on this strategy, studies using IL-2 alone following SCT have shown an increase in NK cell phenotype and function [158–160]. In one study [161] with 18 patients suitable for evaluation, three responses were observed. In another study, IL-2 was infused for a median of 85 days following both autologous and allogeneic SCT [160]. Toxicity was minimal and the treatment was undertaken in the outpatient setting via a Hickman catheter. In this study, no patient developed any signs of GVHD, hypertension or pulmonary capillary-leak syndrome. Despite the administration of low-dose IL-2, significant immunological changes were noted with a five- to 40-fold increase in NK cell number. In addition, there was a significant augmentation of *ex vivo* cytotoxicity against K-562 and colon tumor targets. In a similar study, it was shown that following continuous infusion of IL-2 in patients receiving autologous SCT, the $CD3^+$ and $CD16^+$ cells secreted increased levels of IFN-γ and TNF following *in vitro* culture and there was a significant increase in serum levels of IFN-γ, but not TNF [160]. Recently, post-transplantation IL-2 administration has been extended to include the use of IL-2 or IL-2 and G-CSF for the mobilization of STEM CELLS [162, 163]. The objective of using IL-2 as a mobilizing agent is to mobilize T CELLS or change the population of T CELLS to ones that may have improved antitumor activity as well as the potential to reduce secondary infections.

SCT combination therapy with IFN-α

Similar post-transplantation strategies with IFN-α have been undertaken with the suggestion of a reduced risk of relapse and an increase in myelosuppression [164, 165]. In an early study of the prophylactic use of IFN-α following allogeneic BMT, the Seattle group [165] found that adjuvant treatment with IFN-α had no effect on the probability or severity of CMV infections or GVHD in acute lymphocytic leukemia (ALL) patients who were in remission at the time of transplantation. In this large study, there was a significant reduction in the probability of relapse in the IFN-α recipients ($p = 0.004$) as compared to transplant patients who did not receive IFN-α, although survival rates did not differ between the two groups. It was suggested that the administration of IFN-α following transplantation reduced the risks of relapse, but did not affect CMV infection, perhaps because IFN-α was not initiated until a median of 18 days following transplantation and was not administered chronically.

SCT combination therapy with CSA

Recently, Ratanatharathorn et al. [166] undertook studies to induce a GVT reaction by administration of both CSA and IFN-α to augment GVHD in autologous transplant patients. Twenty-two patients were enrolled of which 17 were considered suitable for evaluation. Thirteen of the patients who received Hu-IFN-α2a developed GVHD regardless of whether they received CSA, whereas only 2 of the 4 patients who received CSA alone developed detectable GVHD. Patients receiving 1×10^6 U/day of Hu-IFN-α2a concomitant with CSA showed a trend towards increased severity of clinical GVHD as compared to patients receiving CSA alone ($p = 0.06$). They concluded that IFN-α administration can be safely started on day 0 of autologous BMT and can induce autologous GVHD as a single agent with the potential to improve therapy.

In similar studies, Kennedy et al. [167] treated women with advanced breast cancer with combined therapy of CSA for 28 days using a dose of 0.025 mg/m^2 of s.c. IFN-γ every other day on days 7–28 after HDT and autologous BMT. They observed that autologous GVHD developed in 56% of the patients, an incidence comparable to that previously observed with CSA alone. The severity of GVHD was greater with CSA plus IFN-γ than with CSA alone, as 16 patients required corticosteroid therapy for dermatological GVHD. Recently, IFN-γ therapy was administered following DLI on a patient who received a matched sibling graft to treat first chronic phase CML [168].

SCT combination therapy with IL-12

Furthermore, adjuvant IMMUNOTHERAPY studies following autologous peripheral SCT (PSCT), have used patients with hematological malignancies and dose escalation IL-12 studies following myeloid recovery. IL-12 is a heterodimeric CYTOKINE that can stimulate both INNATE and ADAPTIVE IMMUNITY [169]. It is produced predominately by DCs, supports the proliferation of activated T CELLS and promotes the differentiation of T CELLS into type 1, helper and effector cells [170, 171]. Furthermore, IL-12 activates and augments the cytolytic activity of NK cells [172] and stimulates IFN-γ production by both NK and T CELLS [173]. In preclinical tumor models, IL-12 therapy has been shown to induce regression of established primary tumors, inhibit the formation of tumor metastasis and to prolong the survival of tumor-bearing animals [174, 175]. Further, objective, complete and partial responses have been seen in patients with advanced solid tumors who have received IL-12 in phase I clinical trials [176]. Based on these results, a dose-finding study was established [177] in which IL-12 was administered as a single bolus i.v. injection, followed two weeks later with daily injections for five consecutive days with cycles repeated every three weeks. A transient neutropenia and thrombocyopenia were seen at all doses. Biological activities included an increase in serum IFN-γ levels and effects on T CELLS, B CELLS and NK cell numbers were observed at all three dose levels. A slight neutropenia occurred with IL-12 administration and a significant decrease in total CD4 cells. In contrast, there was an increase in CD8 and NK cells, but no effect on B CELLS following the administration of IL-12. Serum INF-γ levels peaked from 12 to 24 hours following administration at 100 and 250 ng/kg. Based on these studies, it appears that an appropriate dose of IL-12 has been identified for future studies to examine EFFICACY following PSCT.

Conclusion

The goal of regulating the host's immune responses, as a THERAPEUTIC STRATEGY for neoplastic, infectious, autoimmune and inflammatory diseases, has been achieved for some indications. Optimism for this approach has fluctuated, but at present, numerous immunoregulatory drugs have been approved and currently IMMUNOTHERAPEUTICS represent a quarter of all drug approvals in the United States. During the last decade, we have observed an explosion in the cloning of immunoregulatory genes and their receptors as well as the development of novel therapeutic approaches. These critical advances represent the culmination of efforts with crude and fractionated natural products, supernatants and cell products. Thus, immunotherapeutics can be subdivided into RECOMBINANT proteins and natural or synthetic products. The latter do not currently share the enthusiasm directed towards RECOMBINANT proteins, yet have given us therapeutically important drugs such as Bestatin, FK-506, LEVAMISOLE and natural products such as BCG, currently used to treat bladder cancer.

In the last 20 years, NON-SPECIFIC IMMUNOSTIMULATION has progressed from initial trials with crude microbial mixtures and extracts to more sophisticated uses with a large collection of targeted immunopharmacologically active compounds (only a few of which are discussed here) having diverse actions on the immune system. Furthermore, a body of immunopharmacological knowledge has evolved with these BRMs, which show substantial divergence from conventional pharmacology, particularly in terms of the relationship of dosing schedules to immunopharmacodynamics. This knowledge is important in evaluating agents and predicting appropriate use and EFFICACY. While much remains to be learned and new compounds remain to be extracted and/or cloned, the future of IMMUNOTHERAPY seems bright. A number of the CYTOKINES have been approved as well as numerous supplemental indications [178] in the United States, Europe and Asia. However, it is apparent that the combinations of CYTOKINES and BRMs will have optimal activity when used as adjuvants with more traditional therapeutic modalities. Please see the chapter entitled "Cytokines" by Resch et al. in this book for an extension of studies on CYTOKINES.

Summary

IMMUNOTHERAPY as a treatment modality is attractive because of the exquisite specificity of the immune response and the associated lack of toxicity. In the last one hundred years, IMMUNOTHERAPY has progressed from the use of ill-defined therapeutic modalities, using plant or BACTERIAL EXTRACTS, to well-defined, molecularly targeted drugs. The identification and development of synthetic and recombinant therapeutics, and in a few instances natural products, now provide the standard of care for several diseases. Perhaps the most impressive form of IMMUNOTHERAPY is the use of BCG in the treatment of transitional cell carcinoma of the bladder that demonstrates a consistent, complete response rate of greater than sixty percent. Furthermore, immune augmentation also encompases cellular and gene therapeutic strategies, which were not considered when immune intervention was first undertaken.

Activation of a tumor-specific response can be accomplished by several different strategies including targeting tumor-associated self-antigens, viral ANTIGENS, or uncharacterized and mutated ANTIGENS with the use of whole tumor cell or tumor lysate-based strategies or with vectors with genes associated with tumorigenicity. This approach allows the immune system to naturally induce T and B CELL responses to immunogenic target ANTIGENS. Other, untargeted strategies currently being developed include increasing the immunogenicity of a tumor by genetic manipulation of tumor cells with transgenes for CYTOKINES or costimulatory molecules or immunization with tumor cells fused with ANTIGEN-PRESENTING CELLS. Similarly, CELLULAR THERAPY is being developed with the adoptive transfer of ANTIGEN-specific T CELLS, or the *ex vivo* manipulation, expansion or augmentation of DCs incorporating ANTIGENS delivered via DNA, adenovirus or Pox vectors, or protein or peptide pulsing. Such strategies include the adoptive transfer of LYMPHOCYTES, educated *ex vivo* against a specific tumor or viral ANTIGEN or DCs presenting autologous tumor ANTIGENS or recombinant ANTIGENS, as well as peptides pulsed onto autologous DC. These clinical studies, testing novel immune-based interventions, have resulted in immunological responses and in some instances affected recurrence and survival of cancer patients.

Therapeutics based on the use of immunostimulants have progressed not only in terms of EFFICACY and pharmacological attributes, but have also expanded beyond the initial therapeutics. The initial recombinant and synthetic immune augmenting agents have been joined by contact pharmacophores or peptide mimetics and new formulations which are all being currently studied clinically. These second-generation immune-augmenting drugs are joined by second-generation immune-augmenting agents that have been manipulated to expand their pharmacological profile, based on pharmacokinetics, biodistribution, or tissue-targeting approaches. Thus, pegylation or fusion molecules are being developed to prolong the half-life and BIOAVAILABILITY. Similarly, nanoparticles or slow-release formulations are used to protect against metabolism and/or to target the biological to the tumor.

Our expanded understanding of immunopharmacology has evolved with these drugs and allows us to improve not only clinical protocols, but also delivery of bioactive drugs based on their interactions with other therapeutics such as hormonal and cytotoxic therapies. This has proven critical, as some immunostimulants have pharmacological profiles substantially divergent from conventional therapeutics. This can have a significant impact in terms of the immunopharmcological dynamics and dosing schedules based on response rates and therapeutic impact. While we stand on the threshold of understanding how best to use immunostimulants in oncology, infectious diseases, congenital diseases, as well as inflammatory diseases, this basic understanding is critical to the evaluation and development of these novel therapeutic agents.

The number of therapeutically active immunostimulants approved by the FDA remains low and the number of patients with disease approved for targeted therapies and capable of responding is also low in number. However, our improved understanding will allow us to broaden the indications targeted and the number of patients who will benefit from these therapies. Furthermore, the number of new genes, including INTERLEUKINS and CHEMOKINES and their receptors, being identified with potential therapeutic application are rapidly expanding, suggesting that the theme of immune intervention is just beginning. However, for this to be successful, we need to improve our understanding of the pharmacology and toxicology of these drugs and how best to apply these novel therapeutics, not only for specific indications, but also in combination with the therapeutics that represent our current standard of care to optimally obtain adjuvant therapeutic activity.

The future is bright for immunostimulating agents. While their use is less controversial than a decade ago, their development is still in the beginning stages and further development is needed even for those indications for which they are currently approved. In addition, we are identifying new drug candidates, based on the information and data newly available from genomics and proteomics. Thus, new targets, as well as new drugs, are needed to take advantage of the promise provided by these exciting new therapeutics. Furthermore, new strategies are needed to overcome the IMMUNOSUPPRESSION and TOLERANCE associated with tumor growth and conventional therapy. Currently several approaches are being considered/utilized; these include the use of molecular therapeutics such as COX-2 inhibitors, or tyrosine kinase inhibitors for VEGF to overcome the high levels of such immunosuppressive CYTOKINES/enzymes. In addition, antibodies to TGF-β and IL-25 (found on T-REGULATORY CELLS) and vitamin D3 or all trans-retinoic acid are being used or have been proposed to regulate immature myeloid suppressor cells to overcome tumor-associated IMMUNOSUPPRESSION and improve host immunity. These approaches are being used either alone or in combination with other modalities of immune intervention that may be limited by these cells, enzymes and CYTOKINES. Therefore, future successes with immunostimulants are dependent not only on new or already identified immune augmenting drugs, but also the identification of appropriate clinical targets and the concomitant delivery of drugs to overcome tumor and iratrogenic associated IMMUNOSUPPRESSION. In summary, while the future is bright and proof of principle achieved with some therapeutics, the future is still assured and significant work remains.

Selected readings

Thomson A, Lotze M (eds) (2003) *The Cytokine Handbook Two Volume Set*, Fourth Edition. Academic Press, New York

Giaccone G, Schilsky R, Sondel (eds) (2001) *Cancer Chemotherapy and Biological Response Modifiers*. Annual 19, Vol. 19. Elsevier Health Sciences, New York

Recommended websites:

The Cytokine Web: *http://cmbi.bjmu.edu.cn/cmbidata/cgf/CGF_Database/cytweb/* (Accessed December 2004)

Cytokines Online Pathfinder Encyclopedia: *http://www.copewithcytokines.de/cope.cgi* (Accessed December 2004)

References

1 Smyth MJ, Godfrey DI, Trapani JA (2001) A fresh look at tumor immunosurveillance and immunotherapy. *Nat Immunol* 2: 293–299
2 Walsh G (2000) Biopharmaceutical benchmarks. *Nat Biotechnol* 18: 831–833
3 Walsh G (2003) Biopharmaceutical benchmarks–2003. *Nat Biotechnol* 21: 865–870
4 Talmadge JE, Herberman RB (1986) The preclinical screening laboratory: evaluation of immunomodulatory and therapeutic properties of biological response modifiers. *Cancer Treat Rep* 70: 171–182
5 Mihich E (1986) Future perspectives for biological response modifiers: a viewpoint. *Semin Oncol* 13: 234–254
6 Ellenberg SS (1993) Surrogate endpoints. *Br J Cancer* 68: 457–459
7 Holden C (1993) Okays surrogate markers. *Science* 259: 32
8 Talmadge JE, Phillips H, Schindler J, Tribble H, Pennington R (1987) Systematic preclinical study on the therapeutic properties of recombinant human interleukin 2 for the treatment of metastatic disease. *Cancer Res* 47: 5725–5732
9 van Der Auwera P, Platzer E, Xu ZX, Schulz R, Feugeas O, Capdeville R, Edwards DJ (2001) Pharmacodynamics and pharmacokinetics of single doses of subcutaneous pegylated human G-CSF mutant (Ro 25-8315) in healthy volunteers: comparison with single and multiple daily doses of filgrastim. *Am J Hematol* 66: 245–251
10 Jen JF, Glue P, Ezzet F, Chung C, Gupta SK, Jacobs S, Hajian G (2001) Population pharmacokinetic analysis of pegylated interferon alfa-2b and interferon alfa-2b in patients with chronic hepatitis C. *Clin Pharmacol Ther* 69: 407–421
11 Tomlinson E (1991) Site-specific proteins. In: RC Hider, D Barlow (eds): *Polypeptide and Protein Drugs: Production, Characterization and Formulation*. Ellis Horwood Ltd., Chichester: 251–364
12 Talmadge JE, Tribble HR, Pennington RW, Phillips H, Wiltrout RH (1987) Immunomodulatory and immunotherapeutic properties of recombinant gamma-interferon and recombinant tumor necrosis factor in mice. *Cancer Res* 47: 2563–2570
13 Misset JL, Mathe G, Gastiaburu J, Goutner A, Dorval T, Gouveia J, Schwarzenberg L, Machover D, Ribaud P, de Vassal F (1982) Treatment of leukemias and lymphomas by interferons: II. phase II of the trial treatment of chronic lymphoid leukemia by human interferon alpha+. *Biomedicine and Pharmacotherapy* 36: 112–116
14 Golomb HM, Fefer A, Golde DW, Ozer H, Portlock C, Silber R, Rappeport J, Ratain MJ, Thompson J, Bonnem E et al (1988) Report of a multi-institutional study of 193 patients with hairy cell leukemia treated with interferon-alfa2b. *Semin Oncol* 15: 7–9
15 Quesada JR, Reuben J, Manning JT, Hersh EM, Gutterman JU (1984) Alpha interferon for induction of remission in hairy-cell leukemia. *N Engl J Med* 310: 15–18
16 O'Connell MJ, Colgan JP, Oken MM, Ritts RE, Jr., Kay NE, Itri LM (1986) Clinical trial of recombinant leukocyte a interferon as initial therapy for favorable histology non-Hodgkin's lymphomas and chronic lymphocytic leukemia. An Eastern Cooperative Oncology Group Pilot Study. *J Clin Oncol* 4: 128–136
17 Bunn PA, Jr., Foon KA, Ihde DC, Longo DL, Eddy J, Winkler CF, Veach SR, Zeffren J, Sherwin S, Oldham R (1984) Recombinant leukocyte A interferon: an active agent in advanced cutaneous T-cell lymphomas. *Ann Intern Med* 101: 484–487
18 Kirkwood JM, Strawderman MH, Ernstoff MS, Smith TJ, Borden EC, Blum RH (1996) Interferon alfa-2b adjuvant therapy of high-risk resected cutaneous melanoma: the Eastern Cooperative Oncology Group Trial EST 1684. *J Clin Oncol* 14: 7–17
19 Lane HC, Kovacs JA, Feinberg J, Herpin B, Davey V, Walker R, Deyton L, Metcalf JA, Baseler M, Salzman N et al (1988) Anti-retroviral effects of interferon-alpha in AIDS-associated Kaposi's sarcoma. *Lancet* 2: 1218–1222
20 Pfeffer LM, Dinarello CA, Herberman RB, Williams BR, Borden EC, Bordens R, Walter MR, Nagabhushan TL, Trotta PP, Pestka S (1998) Biological properties of recombinant alpha-interferons: 40th anniversary of the discovery of interferons. *Cancer Res* 58: 2489–2499
21 Teichmann JV, Sieber G, Ludwig WD, Ruehl H (1988) Modulation of immune functions by long-term treatment with recombinant interferon-alpha 2 in a patient with hairy-cell leukemia. *J Interferon Res* 8: 15–24
22 Black PL, Phillips H, Tribble HR, Pennington R, Schneider M, Talmadge JE (1993) Antitumor response to recombinant murine interferon gamma correlates

with enhanced immune function of organ-associated, but not recirculating cytolytic T lymphocytes and macrophages. *Cancer Immunol Immunother* 37: 299–306

23 The Italian Cooperative Study Group on Chronic Myeloid Leukemia (1994) Interferon alpha-2a as compared with conventional chemotherapy for the treatment of chronic myeloid leukemia. *N Engl J Med* 330: 820–825

24 Alimena G, Morra E, Lazzarino M, Liberati AM, Montefusco E, Inverardi D, Bernasconi P, Mancini M, Donti E, Grignani F et al (1988) Interferon alpha-2b as therapy for Ph'-positive chronic myelogenous leukemia: a study of 82 patients treated with intermittent or daily administration. *Blood* 72: 642–647

25 Guilhot F, Chastang C, Michallet M, Guerci A, Harousseau JL, Maloisel F, Bouabdallah R, Guyotat D, Cheron N, Nicolini F et al (1997) Interferon alfa-2b combined with cytarabine *versus* interferon alone in chronic myelogenous leukemia. French chronic myeloid leukemia study group. *N Engl J Med* 337: 223–229

26 Zonder JA, Schiffer CA (2003) Practical aspects of the treatment of chronic myelogenous leukemia with imatinib mesylate. *Curr Hematol Rep* 2: 57–64

27 Hehlmann R (2003) Current CML therapy: progress and dilemma. Leukemia 17: 1010–1012

28 Hansen JA, Gooley TA, Martin PJ, Appelbaum F, Chauncey TR, Clift RA, Petersdorf EW, Radich J, Sanders JE, Storb RF et al (1998) Bone marrow transplants from unrelated donors for patients with chronic myeloid leukemia. *N Engl J Med* 338: 962–968

29 Gratwohl A, Hermans J, Goldman JM, Arcese W, Carreras E, Devergie A, Frassoni F, Gahrton G, Kolb HJ, Niederwieser D et al (1998) Risk assessment for patients with chronic myeloid leukaemia before allogeneic blood or marrow transplantation. Chronic Leukemia Working Party of the European Group for Blood and Marrow Transplantation. *Lancet* 352: 1087–1092

30 Wadler S, Schwartz EL (1990) Antineoplastic activity of the combination of interferon and cytotoxic agents against experimental and human malignancies: a review. *Cancer Res* 50: 3473–3486

31 Wheatley K, Ives N, Hancock B, Gore M, Eggermont A, Suciu S (2003) Does adjuvant interferon-alpha for high-risk melanoma provide a worthwhile benefit? A meta-analysis of the randomised trials. *Cancer Treat Rev* 29: 241–252

32 Kirkwood JM, Richards T, Zarour HM, Sosman J, Ernstoff M, Whiteside TL, Ibrahim J, Blum R, Wieand S, Mascari R (2002) Immunomodulatory effects of high-dose and low-dose interferon alpha2b in patients with high-risk resected melanoma: the E2690 Laboratory Corollary of Intergroup Adjuvant Trial E1690. *Cancer* 95: 1101–1112

33 Detjen KM, Farwig K, Welzel M, Wiedenmann B, Rosewicz S (2001) Interferon gamma inhibits growth of human pancreatic carcinoma cells via Caspase-1 dependent induction of apoptosis. *Gut* 49: 251–262

34 Xu X, Fu XY, Plate J, and Chong AS (1998) IFN-gamma induces cell growth inhibition by Fas-mediated apoptosis: requirement of STAT1 protein for up-regulation of Fas and FasL expression. *Cancer Res* 58: 2832–2837

35 Folkman J (1999) Angiogenic zip code. *Nat Biotechnol* 17: 749

36 Luster AD, Ravetch JV (1987) Biochemical characterization of a gamma interferon-inducible cytokine (IP-10). *J Exp Med* 166: 1084–1097

37 Key ME, Talmadge JE, Fogler WE, Bucana C, Fidler IJ (1982) Isolation of tumoricidal macrophages from lung melanoma metastases of mice treated systemically with liposomes containing a lipophilic derivative of muramyl dipeptide. *J Natl Cancer Inst* 69: 1198

38 Fogler WE, Talmadge JE, Fidler IJ (1983) The activation of tumoricidal properties in macrophages of endotoxin responder and nonresponder mice by liposome-encapsulated immunomodulators. *J Reticuloendothel Soc* 33: 165–174

39 Singh RK, Varney ML, Buyukberber S, Ino K, Ageitos AG, Reed E, Tarantolo S, Talmadge JE (1999) Fas-FasL-mediated CD4^{+} T-cell apoptosis following stem cell transplantation. *Cancer Res* 59: 3107–3111

40 Griffith TS, Wiley SR, Kubin MZ, Sedger LM, Maliszewski CR, Fanger NA (1999) Monocyte-mediated tumoricidal activity via the tumor necrosis factor-related cytokine, TRAIL. *J Exp Med* 189: 1343–1354

41 Jackson JD, Yan Y, Brunda MJ, Kelsey LS, Talmadge JE (1995) Interleukin-12 enhances peripheral hematopoiesis *in vivo*. *Blood* 85: 2371–2376

42 Smyth MJ, Thia KY, Street SE, MacGregor D, Godfrey DI, Trapani JA (2000) Perforin-mediated cytotoxicity is critical for surveillance of spontaneous lymphoma. *J Exp Med* 192: 755–760

43 Takeda K, Hayakawa Y, Smyth MJ, Kayagaki N, Yamaguchi N, Kakuta S, Iwakura Y, Yagita H, Okumura K (2001) Involvement of tumor necrosis factor-related apoptosis-inducing ligand in surveillance of tumor metastasis by liver natural killer cells. *Nat Med* 7: 94–100

44 Maluish AE, Urba WJ, Longo DL, Overton WR, Coggin D, Crisp ER, Williams R, Sherwin SA, Gordon K, Steis RG (1988) The determination of an immunologically active dose of interferon-gamma in patients with melanoma. *J Clin Oncol* 6: 434–445

45 Jaffe HS, Herberman RB (1988) Rationale for recombinant human interferon-gamma adjuvant immunotherapy for cancer. *J Natl Cancer Inst* 80: 616–618

46 The International Chronic Granulomatous Disease Cooperative Study Group (1991) A controlled trial of interferon gamma to prevent infection in chronic granulomatous disease. *N Engl J Med* 324: 509–516

47 Woodman RC, Erickson RW, Rae J, Jaffe HS, Curnutte JT (1992) Prolonged recombinant interferon-gamma therapy in chronic granulomatous disease: evidence against enhanced neutrophil oxidase activity. *Blood* 79: 1558–1562

48 Ahlin A, Larfars G, Elinder G, Palmblad J, Gyllenhammar H (1999) Gamma interferon treatment of patients with chronic granulomatous disease is associated with augmented production of nitric oxide by polymorphonuclear neutrophils. *Clin Diagn Lab Immunol* 6: 420–424

49 Schiff DE, Rae J, Martin TR, Davis BH, Curnutte JT (1997) Increased phagocyte Fc gammaRI expression and improved Fc gamma-receptor-mediated phagocytosis after *in vivo* recombinant human interferon-gamma treatment of normal human subjects. *Blood* 90: 3187–3194

50 Windbichler GH, Hausmaninger H, Stummvoll W, Graf AH, Kainz C, Lahodny J, Denison U, Muller-Holzner E, Marth C (2000) Interferon-gamma in the first-line therapy of ovarian cancer: a randomized phase III trial. *Br J Cancer* 82: 1138–1144

51 Raghu G, Brown KK, Bradford WZ, Starko K, Noble PW, Schwartz DA, King TE Jr (2004) A placebo-controlled trial of interferon gamma-1b in patients with idiopathic pulmonary fibrosis. *N Engl J Med* 350: 125–133

52 Smith KA (1988) Interleukin-2: Inception, impact, and implications. *Science* 240: 1169–1176

53 Waldmann TA, Dubois S, Tagaya Y (2001) Contrasting roles of IL-2 and IL-15 in the life and death of lymphocytes: implications for immunotherapy. *Immunity* 14: 105–110

54 Robertson MJ, Ritz J (1990) Biology and clinical relevance of human natural killer cells. *Blood* 76: 2421–2438

55 Mingari MC, Gerosa F, Carra G, Accolla RS, Moretta A, Zubler RH, Waldmann TA, Moretta L (1984) Human interleukin-2 promotes proliferation of activated B cells via surface receptors similar to those of activated T cells. *Nature* 312: 641–643

56 Espinoza-Delgado I, Bosco MC, Musso T, Gusella GL, Longo DL, Varesio L (1995) Interleukin-2 and human monocyte activation. *J Leukoc Biol* 57: 13–19

57 Ferrante A (1992) Activation of neutrophils by interleukins-1 and -2 and tumor necrosis factors. *Immunol Ser* 57: 417–436

58 Rosenberg SA (2001) Progress in human tumour immunology and immunotherapy. *Nature* 411: 380–384

59 Moretta A, Bottino C, Vitale M, Pende D, Cantoni C, Mingari MC, Biassoni R, Moretta L (2001) Activating receptors and coreceptors involved in human natural killer cell-mediated cytolysis. *Annu Rev Immunol* 19: 197–223

60 Nelson BH (2004) IL-2, regulatory T cells, and tolerance. *J Immunol* 172: 3983–3988

61 Gillis S, Smith KA (1977) Long term culture of tumour-specific cytotoxic T cells. *Nature* 268: 154–156

62 Keene JA, Forman J (1982) Helper activity is required for the *in vivo* generation of cytotoxic T lymphocytes. *J Exp Med* 155: 768–782

63 Andrews DM, Andoniou CE, Granucci F, Ricciardi-Castagnoli P, Degli-Esposti MA (2001) Infection of dendritic cells by murine cytomegalovirus induces functional paralysis. *Nat Immunol* 2: 1077–1084

64 Grabstein KH, Eisenman J, Shanebeck K, Rauch C, Srinivasan S, Fung V, Beers C, Richardson J, Schoenborn MA, Ahdieh M et al (1994) Cloning of a T cell growth factor that interacts with the beta chain of the interleukin-2 receptor. *Science* 264: 965–968

65 Marks-Konczalik J, Dubois S, Losi JM, Sabzevari H, Yamada N, Feigenbaum L, Waldmann TA, Tagaya Y (2000) IL-2-induced activation-induced cell death is inhibited in IL-15 transgenic mice. *Proc Natl Acad Sci USA* 97: 11445–11450

66 Zhang X, Sun S, Hwang I, Tough DF, Sprent J (1998) Potent and selective stimulation of memory-phenotype $CD8^+$ T cells *in vivo* by IL-15. *Immunity* 8: 591–599

67 Kennedy MK, Glaccum M, Brown SN, Butz EA, Viney JL, Embers M, Matsuki N, Charrier K, Sedger L, Willis CR et al (2000) Reversible defects in natural killer and memory CD8 T cell lineages in interleukin 15-deficient mice. *J Exp Med* 191: 771–780

68 Lodolce JP, Boone DL, Chai S, Swain RE, Dassopoulos T, Trettin S, and Ma A (1998) IL-15 receptor maintains lymphoid homeostasis by supporting lymphocyte homing and proliferation. *Immunity* 9: 669–676

69 Fehniger TA, Cooper MA, Caligiuri MA (2002) Interleukin-2 and interleukin-15: immunotherapy for cancer. *Cytokine Growth Factor Rev* 13: 169–183

70 Rosenberg SA, Lotze MT, Yang JC, Topalian SL, Chang AE, Schwartzentruber DJ, Aebersold P, Leitman S, Linehan WM, Seipp CA et al (1993) Prospective randomized trial of high-dose interleukin-2 alone or in conjunction with lymphokine-activated killer cells for the treatment of patients with advanced cancer. *J Natl Cancer Inst* 85: 622–632

71 West WH, Tauer KW, Yannelli JR, Marshall GD, Orr DW, Thurman GB, Oldham RK (1987) Constant-infusion recombinant interleukin-2 in adoptive immunotherapy of advanced cancer. *N Engl J Med* 316: 898–905

72 Yang JC, Sherry RM, Steinberg SM, Topalian SL, Schwartzentruber DJ, Hwu P, Seipp CA, Rogers-Freezer L, Morton KE, White DE et al (2003) Randomized study of high-dose and low-dose interleukin-2 in patients with metastatic renal cancer. *J Clin Oncol* 21: 3127–3132

73 Lotze MT, Chang AE, Seipp CA, Simpson C, Vetto JT, Rosenberg SA (1986) High-dose recombinant interleukin 2 in the treatment of patients with disseminated cancer. Responses, treatment-related morbidity, and histologic findings. *JAMA* 256: 3117–3124

74 Heslop HE, Gottlieb DJ, Bianchi AC, Meager A, Prentice HG, Mehta AB, Hoffbrand AV, Brenner MK (1989) *In vivo* induction of gamma interferon and tumor necrosis factor by interleukin-2 infusion following intensive chemotherapy or autologous marrow transplantation. Blood 74: 1374–1380

75 Barouch DH, Santra S, Steenbeke TD, Zheng XX, Perry HC, Davies ME, Freed DC, Craiu A, Strom TB, Shiver JW et al (1998) Augmentation and suppression of immune responses to an HIV-1 DNA vaccine by plasmid cytokine/Ig administration. *J Immunol* 161: 1875–1882

76 Oosterwijk-Wakka JC, Tiemessen DM, Bleumer I, De Vries IJ, Jongmans W, Adema GJ, Debruyne FM, de Mulder PH, Oosterwijk E, Mulders PF (2002) Vaccination of patients with metastatic renal cell carcinoma with autologous dendritic cells pulsed with autologous tumor Ags in combination with interleukin-2: a phase 1 study. J Immunother 25: 500–508

77 Thompson JA, Shulman KL, Kenyunes MC, Lindgren CG, Collins C, Lange PH, Bush WH, Jr., Benz LA, and Fefer A (1992) Prolonged continuous intravenous infusion interleukin-2 and lymphokine-activated killer-cell therapy for metastatic renal cell carcinoma. *J Clin Oncol* 10: 960–968

78 Hladik F, Tratkiewicz JA, Tilg H, Vogel W, Schwulera U, Kronke M, Aulitzky WE, Huber C (1994) Biologic activity of low dosage IL-2 treatment *in vivo*. Molecular assessment of cytokine network interaction. *J Immunol* 153: 1449–1454

79 Mier JW, Vachino G, van der Meer JW, Numerof RP, Adams S, Cannon JG, Bernheim HA, Atkins MB, Parkinson DR, Dinarello CA (1988) Induction of circulating tumor necrosis factor (TNF-alpha) as the mechanism for the febrile response to interleukin-2 (IL-2) in cancer patients. *J Clin Immunol* 8: 426–432

80 Lange T, Marshall L, Spath-Schwalbe E, Fehm HL, Born J (2002) Systemic immune parameters and sleep after ultra-low dose administration of IL-2 in healthy men. *Brain Behav Immun* 16: 663–674

81 De Stefani A, Forni G, Ragona R, Cavallo G, Bussi M, Usai A, Badellino F, and Cortesina G (2002) Improved survival with perilymphatic interleukin 2 in patients with resectable squamous cell carcinoma of the oral cavity and oropharynx. *Cancer* 95: 90–97

82 Yang JC, Topalian SL, Parkinson D, Schwartzentruber DJ, Weber JS, Ettinghausen SE, White DE, Steinberg SM, Cole DJ, Kim HI (1994) Randomized comparison of high-dose and low-dose intravenous interleukin-2 for the therapy of metastatic renal cell carcinoma: an interim report. *J Clin Oncol* 12: 1572–1576

83 Burgess AW, Metcalf D (1980) The nature and action of granulocyte-macrophage colony stimulating factors. *Blood* 56: 947–958

84 Stanley E, Lieschke GJ, Grail D, Metcalf D, Hodgson G, Gall JA, Maher DW, Cebon J, Sinickas V, Dunn AR (1994) Granulocyte/macrophage colony-stimulating factor-deficient mice show no major perturbation of hematopoiesis but develop a characteristic pul-

monary pathology. *Proc Natl Acad Sci USA* 91: 5592–5596

85 Hamilton JA, Stanley ER, Burgess AW, Shadduck RK (1980) Stimulation of macrophage plasminogen activator activity by colony-stimulating factors. *J Cell Physiol* 103: 435–445

86 Hamilton JA, Stanley ER, Burgess AW, Shadduck RK (1980) Stimulation of macrophage plasminogen activator activity by colony-stimulating factors. *J Cell Physiol* 103: 435–445

87 Kay AB, Ying S, Varney V, Gaga M, Durham SR, Moqbel R, Wardlaw AJ, Hamid Q (1991) Messenger RNA expression of the cytokine gene cluster, interleukin 3 (IL-3), IL-4, IL-5, and granulocyte/macrophage colony-stimulating factor, in allergen-induced late-phase cutaneous reactions in atopic subjects. *J Exp Med* 173: 775–778

88 Williamson DJ, Begley CG, Vadas MA, Metcalf D (1988) The detection and initial characterization of colony-stimulating factors in synovial fluid. *Clin Exp Immunol* 72: 67–73

89 Inaba K, Inaba M, Romani N, Aya H, Deguchi M, Ikehara S, Muramatsu S, Steinman RM (1992) Generation of large numbers of dendritic cells from mouse bone marrow cultures supplemented with granulocyte/macrophage colony-stimulating factor. *J Exp Med* 176: 1693–1702

90 Daro E, Pulendran B, Brasel K, Teepe M, Pettit D, Lynch DH, Vremec D, Robb L, Shortman K, McKenna HJ et al (2000) Polyethylene glycol-modified GM-CSF expands CD11b(High)CD11c(High) but not CD11b(Low) CD11c(High) murine dendritic cells *in vivo*: a comparative analysis with Flt3 ligand. *J Immunol* 165: 49–58

91 Kiertscher SM, Gitlitz BJ, Figlin RA, Roth MD (2003) Granulocyte/macrophage-colony stimulating factor and interleukin-4 expand and activate type-1 dendritic cells (DC1) when administered *in vivo* to cancer patients. *Int J Cancer* 107: 256–261

92 Nemunaitis J, Singer JW, Buckner CD, Durnam D, Epstein C, Hill R, Storb R, Thomas ED, Appelbaum FR (1990) Use of recombinant human granulocyte-macrophage colony-stimulating factor in graft failure after bone marrow transplantation. *Blood* 76: 245–253

93 Brandt SJ, Peters WP, Atwater SK, Kurtzberg J, Borowitz MJ, Jones RB, Shpall EJ, Bast RC Jr, Gilbert CJ, Oette DH (1988) Effect of recombinant human granulocyte-macrophage colony-stimulating factor on hematopoietic reconstitution after high-dose chemotherapy and autologous bone marrow transplantation. *N Engl J Med* 318: 869–876

94 Rowe JM, Andersen JW, Mazza JJ, Bennett JM, Paietta E, Hayes FA, Oette D, Cassileth PA, Stadtmauer EA, Wiernik PH (1995) A randomized placebo-controlled phase III study of granulocyte-macrophage colony-stimulating factor in adult patients (> 55 to 70 years of age) with acute myelogenous leukemia: a study of the eastern cooperative oncology group (E1490). *Blood* 86: 457–462

95 Ou-Yang P, Hwang LH, Tao MH, Chiang BL, Chen DS (2002) Co-delivery of GM-CSF gene enhances the immune responses of hepatitis C viral core protein-expressing DNA vaccine: role of dendritic cells. *J Med Virol* 66: 320–328

96 Levitsky HI, Montgomery J, Ahmadzadeh M, Staveley-O'Carroll K, Guarnieri F, Longo DL, Kwak LW (1996) Immunization with granulocyte-macrophage colony-stimulating factor – transduced, but not B7-1-transduced, lymphoma cells primes idiotype – specific T cells and generates potent systemic antitumor immunity. *J Immunol* 156: 3858–3865

97 Beyer J, Schwella N, Zingsem J, Strohscheer I, Schwaner I, Oettle H, Serke S, Huhn D, Stieger W (1995) Hematopoietic rescue after high-dose chemotherapy using autologous peripheral-blood progenitor cells or bone marrow: a randomized comparison. *J Clin Oncol* 13: 1328–1335

98 Soiffer R, Hodi FS, Haluska F, Jung K, Gillessen S, Singer S, Tanabe K, Duda R, Mentzer S, Jaklitsch M et al (2003) Vaccination with irradiated, autologous melanoma cells engineered to secrete granulocyte-macrophage colony-stimulating factor by adenoviral-mediated gene transfer augments antitumor immunity in patients with metastatic melanoma. *J Clin Oncol* 21: 3343–3350

99 Dranoff G (2002) GM-CSF-based cancer vaccines. *Immunology Review* 188: 147–154

100 Colombo MP, Ferrari G, Stoppacciaro A, Parenza M, Rodolfo M, Mavilio F, Parmiani G (1991) Granulocyte colony-stimulating factor gene transfer suppresses tumorigenicity of a murine adenocarcinoma *in vivo*. *J Exp Med* 173: 889–897

101 Marrogi AJ, Munshi A, Merogi AJ, Ohadike Y, El Habashi A, Marrogi OL, Freeman SM (1997) Study of tumor infiltrating lymphocytes and transforming growth factor-

beta as prognostic factors in breast carcinoma. *Int J Cancer* 74: 492–501

102 Ohm JE, Shurin MR, Esche C, Lotze MT, Carbone DP, and Gabrilovich DI (1999) Effect of vascular endothelial growth factor and FLT3 ligand on dendritic cell generation *in vivo*. *J Immunol* 163: 3260–3268

103 Wing EJ, Magee DM, Whiteside TL, Kaplan SS, Shadduck RK (1989) Recombinant human granulocyte/macrophage colony-stimulating factor enhances monocyte cytotoxicity and secretion of tumor necrosis factor alpha and interferon in cancer patients. *Blood* 73: 643–646

104 Spitler LE, Grossbard ML, Ernstoff MS, Silver G, Jacobs M, Hayes FA, Soong SJ (2000) Adjuvant therapy of stage III and IV malignant melanoma using granulocyte-macrophage colony-stimulating factor. *J Clin Oncol* 18: 1614–1621

105 Nauts HC (1975) *The bibliography of reports concerning the experimental clinical use of coley toxins, New York*. Cancer Research Institute Publication

106 Haaff EO, Dresner SM, Ratliff TL, Catalona WJ (1986) Two courses of intravesical bacillus Calmette-Guérin for transitional cell carcinoma of the bladder. *J Urol* 136: 820

107 Pinsky CM, Camacho FJ, Kerr D, Geller NL, Klein FA, Herr HA, Whitmore WF, Oettgen HF (1985) Intravesical administration of bacillus Calmette-Guérin in patients with recurrent superficial carcinoma of the urinary bladder: report of a prospective, randomized trial. *Cancer Treat Rep* 69: 47

108 Lage JM, Bauer WC, Kelley DR, Ratliff TL, Catalona WJ (1986) Histological parameters and pitfalls in the interpretation of bladder biopsies in bacillus Calmette-Guérin treatment of superficial bladder cancer. *J Urol* 135: 916

109 Haaff EO, Caralona WJ, Ratliff TL (1986) Detection of interleukin-2 in the urine of patients with superficial bladder tumors after treatment with intravesical BCG. *J Urol* 136: 970

110 Taniguchi K, Koga S, Nishikido M, Yamashita S, Sakuragi T, Kanetake H, Saito Y (1999) Systemic immune response after intravesical instillation of bacille Calmette-Guérin (BCG) for superficial bladder cancer. *Clin Exp Immunol* 115: 131–135

111 Kaempfer R, Gerez L, Farbstein H, Madar L, Hirschman O, Nussinovich R, Shapiro A (1996) Prediction of response to treatment in superficial bladder carcinoma through pattern of interleukin-2 ene expression. *J Clin Oncol* 14: 1778–1786

112 Watanabe E, Matsuyama H, Matsuda K, Ohmi C, Tei Y, Yoshihiro S, Ohmoto Y, Naito K (2003) Urinary interleukin-2 may predict clinical outcome of intravesical bacillus Calmette-Guérin immunotherapy for carcinoma *in situ* of the bladder. *Cancer Immunol Immunother* 52: 481–486

113 Amery WK, Bruynseels JP (1992) Levamisole, the story and the lessons. *Int J Immunopharmacol* 14: 481–486

114 Mutch RS, Hutson PR (1991) Levamisole in the adjuvant treatment of colon cancer. *Clin Pharm* 10: 95–109

115 Holcombe RF, Milovanovic T, Stewart RM, Brodhag TM (2001) Investigating the role of immunomodulation for colon cancer prevention: results of an *in vivo* dose escalation trial of levamisole with immunologic endpoints. *Cancer Detect Prev* 25: 183–191

116 Porschen R, Bermann A, Loffler T, Haack G, Rettig K, Anger Y, Strohmeyer G (2001) Fluorouracil plus leucovorin as effective adjuvant chemotherapy in curatively resected stage III colon cancer: results of the trial AdjCCA-01. *J Clin Oncol* 19: 1787–1794

117 Wood DD, Staruch MJ, Durette PL, Melvin WV, Graham BK (1983) Role of interleukin-1 in the adjuvanticity of muramyl dipeptide *in vivo*. In: JJ Oppenheim, S Cohen: *Interleukins, Lymphokines and Cytokines*. Raven Press, New York, 691

118 Ellouz F, Adam A, Ciorbaru R, Lederer E (1974) Minimal structural requirements for adjuvant activity of bacterial peptidoglycan derivatives. *Biochem Biophys Res Commun* 59: 1317–1325

119 Fedorocko P, Hoferova Z, Hofer M, Brezani P (2003) Administration of liposomal muramyl tripeptide phosphatidylethanolamine (MTP-PE) and diclofenac in the combination attenuates their anti-tumor activities. *Neoplasma* 50: 176–184

120 Gianan MA, Kleinerman ES (1998) Liposomal muramyl tripeptide (CGP 19835A Lipid) therapy for resectable melanoma in patients who were at high risk for relapse: an update. *Cancer Biother Radiopharm* 13: 363–368

121 Killion JJ, Bucana CD, Radinsky R, Dong Z, O'Reilly T, Bilbe G, Tarcsay L, Fidler IJ (1996) Maintenance of intestinal epithelium structural integrity and mucosal leukocytes during chemotherapy by oral administration of muramyl tripeptide phosphatidylethanolamine. *Cancer Biother Radiopharm* 11: 363–371

122 Aoyagi T, Suda H, Nagai M, Ogawa K, Suzuki J (1976) Aminopeptidase activities on the surface of mammalian cells. *Biochimica Et Biophysica Acta* 452: 131–143

123 Morahan PS, Edelson PJ, Gass K (1980) Changes in macrophage ectoenzymes associated with anti-tumor activity. *J Immunol* 125: 1312–1317

124 Urabe A, Mutoh Y, Mizoguchi H, Takaku F, Ogawa N (1993) Ubenimex in the treatment of acute nonlymphocytic leukemia in adults. *Ann Hematol* 67: 63–66

125 Yasumitsu T, Ohshima S, Nakano N, Kotake Y, Tominaga S (1990) Bestatin in resected lung cancer. A randomized clinical trial. Acta Oncologica 29: 827

126 Hiraoka A, Shibata H, Masaoka T (1992) Immunopotentiation with Ubenimex for prevention of leukemia relapse after allogeneic BMT. The Study Group of Ubenimex for BMT. *Transplant Proc* 24: 3047–3048

127 Goldstein AL (1984) *Thymic Hormones and Lymphokines*. Plenum Press ((A: city?))

128 Ichinose Y, Genka K, Koike T, Kato H, Watanabe Y, Mori T, Iioka S, Sakuma A, Ohta M (2003) Randomized double-blind placebo-controlled trial of bestatin in patients with resected stage I squamous-cell lung carcinoma. *J Natl Cancer Inst* 95: 605–610

129 Janeway CA, Jr., Medzhitov R (2002) Innate immune recognition. *Annu Rev Immunol* 20: 197–216

130 Kadowaki N, Ho S, Antonenko S, Malefyt RW, Kastelein RA, Bazan F, Liu YJ (2001) Subsets of human dendritic cell precursors express different Toll-like receptors and respond to different microbial ags. *J Exp Med* 194: 863–869

131 Ahmad-Nejad P, Hacker H, Rutz M, Bauer S, Vabulas RM, Wagner H (2002) Bacterial CpG-DNA and lipopolysaccharides activate Toll-like receptors at distinct cellular compartments. *Eur J Immunol* 32: 1958–1968

132 Hemmi H, Takeuchi O, Kawai T, Kaisho T, Sato S, Sanjo H, Matsumoto M, Hoshino K, Wagner H, Takeda K et al (2000) A Toll-like receptor recognizes bacterial DNA. *Nature* 408: 740–745

133 Krieg AM (2002) CpG motifs in bacterial DNA and their immune effects. *Annu Rev Immunol* 20: 709–760

134 Krug A, Rothenfusser S, Hornung V, Jahrsdorfer B, Blackwell S, Ballas ZK, Endres S, Krieg AM, Hartmann G (2001) Identification of CpG oligonucleotide sequences with high induction of IFN-alpha/beta in plasmacytoid dendritic cells. *Eur J Immunol* 31: 2154–2163

135 Krieg AM, Yi AK, Matson S, Waldschmidt TJ, Bishop GA, Teasdale R, Koretzky GA, Klinman DM (1995) CpG motifs in bacterial DNA trigger direct B-cell activation. *Nature* 374: 546–549

136 Hartmann G, Krieg AM (2000) Mechanism and function of a newly identified CpG DNA motif in human primary B cells. *J Immunol* 164: 944–953

137 Marshall JD, Fearon K, Abbate C, Subramanian S, Yee P, Gregorio J, Coffman RL, Van Nest G (2003) Identification of a novel CpG DNA class and motif that optimally stimulate B cell and plasmacytoid dendritic cell functions. *J Leukoc Biol* 73: 781–92

138 Cho HJ, Takabayashi K, Cheng PM, Nguyen MD, Corr M, Tuck S, and Raz E (2000) Immunostimulatory DNA-based vaccines induce cytotoxic lymphocyte activity by a T-helper cell-independent mechanism. *Nat Biotechnol* 18: 509–514

139 Ballas ZK, Rasmussen WL, Krieg AM (1996) Induction of NK activity in murine and human cells by CpG motifs in oligodeoxynucleotides and bacterial DNA. *J Immunol* 157: 1840–1845

140 Roman M, Martin-Orozco E, Goodman JS, Nguyen MD, Sato Y, Ronaghy A, Kornbluth RS, Richman DD, Carson DA, Raz E (1997) Immunostimulatory DNA sequences function as T helper-1-promoting adjuvants. *Nat Med* 3: 849–854

141 Shirota H, Sano K, Kikuchi T, Tamura G, Shirato K (2000) Regulation of murine airway eosinophilia and Th2 cells by Ag-conjugated CpG oligodeoxynucleotides as a novel Ag-specific immunomodulator. *J Immunol* 164: 5575–5582

142 Kim SK, Ragupathi G, Musselli C, Choi SJ, Park YS, Livingston PO (1999) Comparison of the effect of different immunological adjuvants on the antibody and T-cell response to immunization with MUC1-KLH and GD3-KLH conjugate cancer vaccines. *Vaccine* 18: 597–603

143 Chu RS, Targoni OS, Krieg AM, Lehmann PV, Harding CV (1997) CpG Oligodeoxynucleotides act as adjuvants that switch on T helper 1 (Th1) immunity. *J Exp Med* 186: 1623–1631

144 Davis HL (2000) Use of CpG DNA for enhancing specific immune responses. *Curr Top Microbiol Immunol* 247: 171–183

145 Zwaveling S, Ferreira Mota SC, Nouta J, Johnson M, Lipford GB, Offringa R, van der Burg SH, Melief CJ (2002) Established human papillomavirus type 16-expressing

tumors are effectively eradicated following vaccination with long peptides. *J Immunol* 169: 350–358

146 Stern BV, Boehm BO, Tary-Lehmann M (2002) Vaccination with tumor peptide in CpG adjuvant protects via IFN-gamma-dependent CD4 cell immunity. *J Immunol* 168: 6099–6105

147 Heckelsmiller K, Beck S, Rall K, Sipos B, Schlamp A, Tuma E, Rothenfusser S, Endres S, Hartmann G (2002) Combined dendritic cell- and CpG oligonucleotide-based immune therapy cures large murine tumors that resist chemotherapy. *Eur J Immunol* 32: 3235–3245

148 Liu HM, Newbrough SE, Bhatia SK, Dahle CE, Krieg AM, Weiner GJ (1998) Immunostimulatory CpG oligodeoxynucleotides enhance the immune response to vaccine strategies involving granulocyte-macrophage colony-stimulating factor. *Blood* 92: 3730–3736

149 Sandler AD, Chihara H, Kobayashi G, Zhu X, Miller MA, Scott DL, Krieg AM (2003) CpG oligonucleotides enhance the tumor Ag-specific immune response of a granulocyte macrophage colony-stimulating factor-based vaccine strategy in neuroblastoma. *Cancer Res* 63: 394–399

150 Heckelsmiller K, Rall K, Beck S, Schlamp A, Seiderer J, Jahrsdorfer B, Krug A, Rothenfusser S, Endres S, Hartmann G (2002) Peritumoral CpG DNA elicits a coordinated response of CD8 T cells and innate effectors to cure established tumors in a murine colon carcinoma model. *J Immunol* 169: 3892–3899

151 Ballas ZK, Krieg AM, Warren T, Rasmussen W, Davis HL, Waldschmidt M, Weiner GJ (2001) Divergent therapeutic and immunologic effects of oligodeoxynucleotides with distinct CpG motifs. *J Immunol* 167: 4878–4886

152 Storek J, Storb R (2000) T-cell reconstitution after stem-cell transplantation–by which organ? *Lancet* 355: 1843–1844

153 Talmadge JE, Reed E, Ino K, Kessinger A, Kuszynski C, Heimann D, Varney M, Jackson J, Vose JM, Bierman PJ (1997) Rapid immunologic reconstitution following transplantation with mobilized peripheral blood stem cells as compared to bone marrow. *Bone Marrow Transplant* 19: 161–172

154 Horowitz MM, Gale RP, Sondel PM, Goldman JM, Kersey J, Kolb HJ, Rimm AA, Ringden O, Rozman C, Speck B et al (1990) Graft-*versus*-leukemia reactions after bone marrow transplantation. *Blood* 75: 555–562

155 Champlin R, Ho W, Gajewski J, Feig S, Burnison M, Holley G, Greenberg P, Lee K, Schmid I, Giorgi J et al (1990) Selective depletion of CD8⁺ T lymphocytes for prevention of graft-*versus*-host disease after allogeneic bone marrow transplantation. *Blood* 76: 418–423

156 Riddell SR, Watanabe KS, Goodrich JM, Li CR, Agha ME, Greenberg PD (1992) Restoration of viral immunity in immunodeficient humans by the adoptive transfer of T cell clones. *Science* 257: 238–241

157 Walter EA, Greenberg PD, Gilbert MJ, Finch RJ, Watanabe KS, Thomas ED, Riddell SR (1995) Reconstitution of cellular immunity against cytomegalovirus in recipients of allogeneic bone marrow by transfer of T-cell clones from the donor. *N Engl J Med* 333: 1038–1044

158 Higuchi CM, Thompson JA, Petersen FB, Buckner CD, Fefer A (1991) Toxicity and immunomodulatory effects of interleukin-2 after autologous bone marrow transplantation for hematologic malignancies. *Blood* 77: 2561–2568

159 Blaise D, Olive D, Stoppa AM, Viens P, Pourreau C, Lopez M, Attal M, Jasmin C, Monges G, Mawas C et al (1990) Hematologic and immunologic effects of the systemic administration of recombinant interleukin-2 after autologous bone marrow transplantation. *Blood* 76: 1092–1097

160 Soiffer RJ, Murray C, Cochran K, Cameron C, Wang E, Schow PW, Daley JF, Ritz J (1992) Clinical and immunologic effects of prolonged infusion of low-dose recombinant interleukin-2 after autologous and T-cell-depleted allogeneic bone marrow transplantation. *Blood* 79: 517–526

161 Negrier S, Ranchere JY, Philip I, Merrouche Y, Biron P, Blaise D, Attal M, Rebattu P, Clavel M, Pourreau C et al (1991) Intravenous interleukin-2 just after high dose BCNU and autologous bone marrow transplantation. Report of a multicentric French pilot study. *Bone Marrow Transplant* 8: 259–264

162 Sosman JA, Stiff P, Moss SM, Sorokin P, Martone B, Bayer R, van Besien K, Devine S, Stock W, Peace D et al (2001) Pilot trial of interleukin-2 with granulocyte colony-stimulating factor for the mobilization of progenitor cells in advanced breast cancer patients undergoing high-dose chemotherapy: expansion of immune effectors within the stem-cell graft and post-stem-cell infusion. *J Clin Oncol* 19: 634–644

163 Toh HC, McAfee SL, Sackstein R, Multani P, Cox BF, Garcia-Carbonero R, Colby C, Spitzer TR (2000) High-dose cyclophosphamide + carboplatin and interleukin-2 (IL-2) activated autologous stem cell transplantation

followed by maintenance IL-2 therapy in metastatic breast carcinoma – a Phase II study. *Bone Marrow Transplant* 25: 19–24

164 Klingemann HG, Grigg AP, Wilkie-Boyd K, Barnett MJ, Eaves AC, Reece DE, Shepherd JD, Phillips GL (1991) Treatment with recombinant interferon (alpha-2b) early after bone marrow transplantation in patients at high risk for relapse. *Blood* 78: 3306–3311

165 Meyers JD, Flournoy N, Sanders JE, McGuffin RW, Newton BA, Fisher LD, Lum LG, Appelbaum FR, Doney K, Sullivan KM et al (1987) Prophylactic use of human leukocyte interferon after allogeneic marrow transplantation. *Ann Intern Med* 107: 809–816

166 Ratanatharathorn V, Uberti J, Karanes C, Lum LG, Abella E, Dan ME, Hussein M, Sensenbrenner LL (1994) Phase I study of alpha-interferon augmentation of cyclosporine-induced graft *versus* host disease in recipients of autologous bone marrow transplantation. *Bone Marrow Transplant* 13: 625–630

167 Kennedy MJ, Vogelsang GB, Jones RJ, Farmer ER, Hess AD, Altomonte V, Huelskamp AM, Davidson NE (1994) Phase I trial of interferon gamma to potentiate cyclosporine-induced graft-*versus*-host disease in women undergoing autologous bone marrow transplantation for breast cancer. *J Clin Oncol* 12: 249–257

168 Leda M, Ladon D, Pieczonka A, Boruczkowski D, Jolkowska J, Witt M, Wachowiak J (2001) Donor lymphocyte infusion followed by interferon-alpha plus low dose cyclosporine A for modulation of donor CD3 cells activity with monitoring of minimal residual disease and cellular chimerism in a patient with first hematologic relapse of chronic myelogenous leukemia after allogeneic bone marrow transplantation. *Leuk Res* 25: 353–357

169 Trinchieri G (1995) Interleukin-12: a proinflammatory cytokine with immunoregulatory functions that bridge innate resistance and Ag-specific adaptive immunity. *Annu Rev Immunol* 13: 251–276

170 Gately MK, Wolitzky AG, Quinn PM, Chizzonite R (1992) Regulation of human cytolytic lymphocyte responses by interleukin-12. *Cell Immunol* 143: 127–142

171 Trinchieri G (1993) Interleukin-12 and its role in the generation of TH1 cells. *Immunol Today* 14: 335–338

172 Robertson MJ, Soiffer RJ, Wolf SF, Manley TJ, Donahue C, Young D, Herrmann SH, Ritz J (1992) Response of human natural killer (NK) cells to NK cell stimulatory factor (NKSF): cytolytic activity and proliferation of NK cells are differentially regulated by NKSF. *J Exp Med* 175: 779–788

173 Chan SH, Perussia B, Gupta JW, Kobayashi M, Pospisil M, Young HA, Wolf SF, Young D, Clark SC, Trinchieri G (1991) Induction of interferon gamma production by natural killer cell stimulatory factor: characterization of the responder cells and synergy with other inducers. *J Exp Med* 173: 869–879

174 Brunda MJ, Luistro L, Warrier RR, Wright RB, Hubbard BR, Murphy M, Wolf SF, Gately MK (1993) Antitumor and antimetastatic activity of interleukin 12 against murine tumors. *J Exp Med* 178: 1223–1230

175 Mu J, Zou JP, Yamamoto N, Tsutsui T, Tai XG, Kobayashi M, Herrmann S, Fujiwara H, Hamaoka T (1995) Administration of recombinant interleukin 12 prevents outgrowth of tumor cells metastasizing spontaneously to lung and lymph nodes. *Cancer Res* 55: 4404–4408

176 Atkins MB, Robertson MJ, Gordon M, Lotze MT, DeCoste M, Dubois JS, Ritz J, Sandler AB, Edington HD, Garzone PD et al (1997) Phase I evaluation of intravenous recombinant human interleukin 12 in patients with advanced malignancies. *Clin Cancer Res* 3: 409–417

177 Robertson MJ, Pelloso D, Abonour R, Hromas RA, Nelson RP, Jr., Wood L, Cornetta K (2002) Interleukin 12 immunotherapy after autologous stem cell transplantation for hematological malignancies. *Clin Cancer Res* 8: 3383–3393

178 Gosse ME, Nelson TF (1977) Approval times for supplemental indications for recombinant proteins. *Nat Biotechnol* 15: 130–134

Anti-infective activity of immunomodulators

K. Noel Masihi

Introduction

Infectious diseases continue to cause suffering and death in this teeming world of almost 6 billion people. Every individual is vulnerable to microbial infections regardless of socioeconomic status, gender, age group or ethnic background. Increased international air travel, mass tourism and the expansion of globalization have meant that disease-producing pathogens can be transported and spread rapidly from one geographic region to another. This has been dramatically demonstrated by the sudden and completely unexpected multi-country outbreak of severe acute respiratory syndrome (SARS) in the beginning of 2003 [1]. This atypical pneumonia, caused by a human coronavirus, led the World Health Organization to issue an exceptional global alert and caution against travel to certain countries. The unfolding tragedy of AIDS, particularly in many developing nations, continues to be vividly highlighted by the scientific and public media. According to current estimates, over 30 million people are virus carriers and around 15000 new cases of HIV occur daily. Tuberculosis and other infections which caused ravages in the nineteenth century are once again resurgent. Several infectious agents, such as neurological variants of Creutzfeldt-Jakob disease in Europe, vector-borne West Nile virus in the USA, vancomycin-resistant forms of bacteria in Japan, and virulent avian influenza in humans in Hong Kong, have emerged as public health concerns over the past few years. The sheer adaptability and the destructiveness of infectious microbes pose threats of unforeseen diseases in the 21st century. According to the World Health Organization, at least 30 new diseases have emerged over the past two decades and underscore the need to be vigilant.

Antimicrobial drugs have been instrumental in saving the lives of millions of people worldwide. The antibiotic magic cure of the previous 50 years is, however, being steadily eroded by the emergence of DRUG-RESISTANT microorganisms [2]. This is shown by the adverse effects on the control and treatment of deadly diseases caused by *Mycobacterium tuberculosis* and *Plasmodium falciparum*. The acute respiratory infections in children, mostly caused by *Pneumococci* and *Haemophilus influenzae*, are becoming more drug-resistant. Over 90% of *Staphylococcus aureus* strains and about 40% of *Pneumococci* strains are resistant to penicillin which was introduced in the 1940s. *Salmonella typhi*, *Streptococcus pneumoniae*, *Enterococcus faecium*, and *Shigella dysenteriae* have been reported as multidrug-resistant. Crucial drug choices for the treatment of common infections by bacteria, viruses, parasites, and fungi are becoming limited and even nonexistent in some cases. This development has not been parallelled by an effective increase in the discovery of new medicines for most microbial pathogens [3]. The struggle to control infectious diseases, far from being over, has acquired a new poignancy. Novel concepts acting as adjuncts to established therapies are urgently needed.

The immune system can be manipulated specifically by vaccination or nonspecifically by immunomodulation. IMMUNOMODULATORS include both immunostimulatory and immunosuppressive agents. This chapter will concentrate on immunostimulatory agents capable of enhancing host defence mechanisms to provide protection against infections. Synonymous terms for IMMUNOMODULATORS include biological response modifiers, immunoaugmentors, immunostimulants or immunorestoratives. Their mode of action includes augmentation of the anti-infectious immunity by the cells of the immune sys-

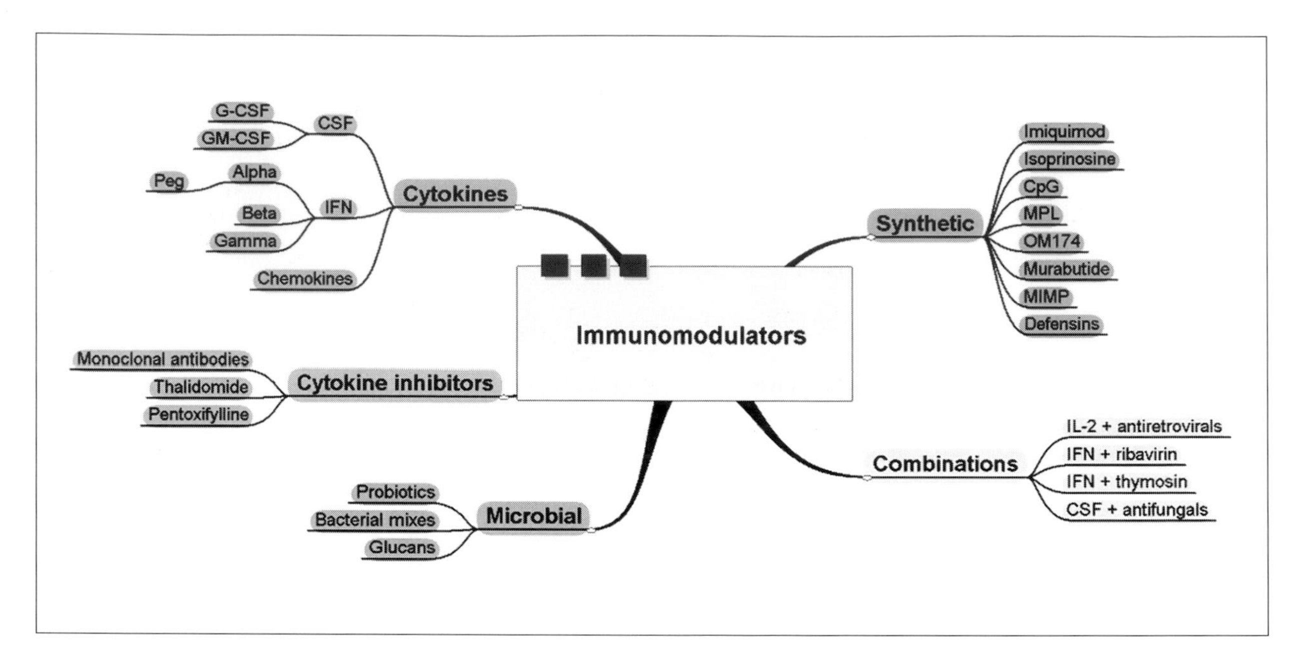

FIGURE 1. IMMUNOMODULATORS

tem encompassing lymphocyte subsets, MACROPHAGES, dendritic and natural killer (NK) cells. Further mechanisms can involve induction or restoration of immune effector functions and tilting the balance towards CYTOKINE pathways germane to protection. A diverse array of recombinant, synthetic, and natural IMMUNOMODULATORY preparations for prophylaxis and treatment of various infections are available today [4–6]. A concise mind-map of the IMMUNOMODULATORS discussed in this chapter is summarised in Figure 1.

Cytokine immunomodulators

The interactions between a host and an infectious pathogen are complex, diverse and intricately regulated. CYTOKINES, hormone-like polypeptides possessing pleiotropic properties, are crucial in orchestrating the appropriate immune responses critical for the outcome of an infection. Certain CYTOKINES stimulate the production of other CYTOKINES and interact in synergistic or antagonistic networks. CYTOKINES exhibit specific IMMUNOMODULATORY properties that can enable manipulation of the host response to enhance overall immunogenicity and direct the nature of the response either toward a type 1 or type 2 pathway. In the type 1 response, Th1 cells produce interferon-γ (IFN-γ), tumor necrosis factor-β (TNF-β) and INTERLEUKIN- (IL-)2 that are required for effective development of cell-mediated immune responses to intracellular microbes. In the type 2 response, Th2 cells produce IL-4 and IL-5 that enhance humoral immunity to T-dependent ANTIGENS and are necessary for immunity to helminth infections. Recent studies have shown that lymphoid progenitor-derived DENDRITIC CELLS (DC1) can also express IL-12 and preferentially induce type 1 T CELL responses, whereas myeloid progenitor-derived DENDRITIC CELLS (DC2) express IL-10 and induce type 2 T CELL responses. Local and systemic effects of CYTOKINES are, thus, intimately involved in the host control of infections (Fig. 2) (also see chapters A4, A7 and C5). Several recombinant and natural CYTOKINE preparations such as INTERFERONS, granulocyte COLONY-STIMULATING FACTOR, and IL-2 are already licensed for use in patients.

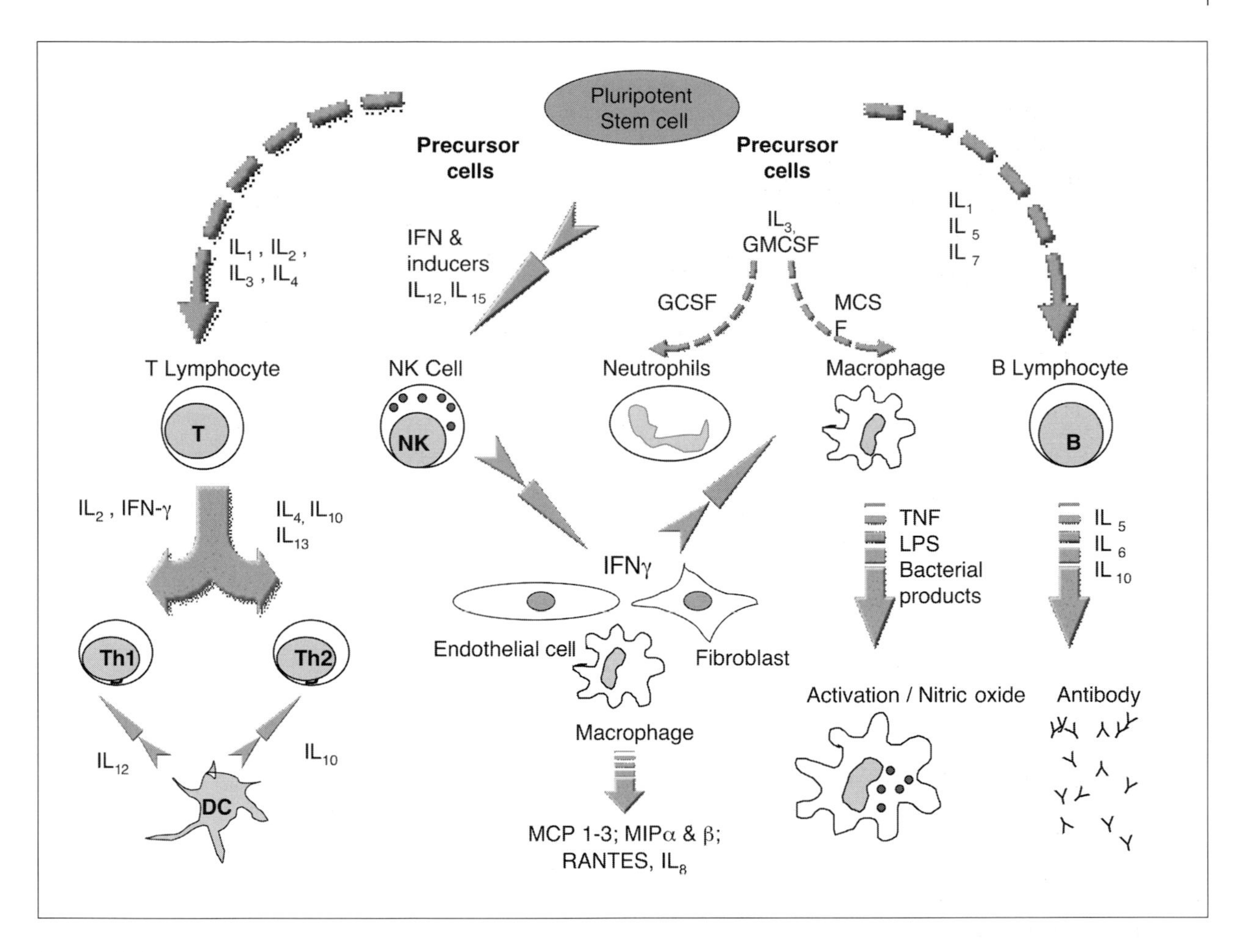

FIGURE 2. INTERACTION OF IMMUNE SYSTEM CELLS AND CYTOKINES

Interferons and combinations

INTERFERONS play an important role in host defense against infectious pathogens and in the regulation of immune responses. INTERFERON-α (IFN-α) is a clinically effective therapy used in a wide range of viral infections besides its application in malignant melanoma, basal cell carcinoma and warts. Natural IFN-α obtained from human serum and LEUKOCYTES is currently licensed for the treatment of a rare form of cancer, hairy cell leukemia. Recombinant IFN-α2a is licensed for treatment of chronic active hepatitis B and for HEPATITIS C VIRUS infections. However, only a small subset of patients with hepatitis B and around 40% of cases with hepatitis C are generally responsive to interferon therapy. IFN-α is also approved for treating condyloma acuminata caused by human papilloma virus and for Kaposi sarcoma in patients with HIV infection.

HEPATITIS B VIRUS (HBV) infects around 400 million people and kills between 1 and 2.5 million people a year. Overall, 15 to 25% of HBV carriers die from chronic hepatitis, cirrhosis or hepatocellular carcinoma. At present, only two licensed therapies for treating liver disease caused by infection with the HBV are available. These are IFN-α and lamivudine.

Several promising studies have shown the effectiveness of lamivudine-interferon (LAM/IFN) COMBINATION THERAPY and it is gaining increasing favour in the treatment of chronic hepatitis B [7,8].

Thymosin-α1 (T-α1) influences T-cell maturation, production of Th1-type CYTOKINES, and activity of NK cell-mediated cytotoxicity [9]. Patients with chronic hepatitis B treated with T-α1 had augmented NKT CELLS and $CD8^+$ cytotoxic T LYMPHOCYTES in the liver. Forty-eight weeks after T-α1 treatment some patients showed normalized alanine aminotransferase (ALT) and decreased HBV-DNA to undetectable level from serum [10]. The combination of T-α1, ribavirin and either PEGYLATED IFN-α, -α2a, -β and -γ are stated to simultaneously and substantially reduce or eliminate the side-effects normally associated with the administration of IFN alone.

HEPATITIS C VIRUS (HCV) infection is a common cause of chronic viral liver disease and is a leading indication for liver transplantation. Almost all such patients show recurrent hepatitis C viremia. Progressive fibrosis and cirrhosis after liver transplantation have been observed with some of the patients progressing to cirrhosis within 5 years of transplantation. Genotype 1 is considered to be the most resistant to therapy, whereas genotypes 2 and 3 are more responsive to therapy; genotype 4 seems to be similar to genotype 1 in this respect (Box 1). IFN-α2b plus ribavirin COMBINATION THERAPY has been found to be effective in chronic HCV patients with genotypes 2 and 3 who are virological nonresponders to IFN monotherapy [11]. The treatment duration is 48 weeks for genotype 1 and 24 weeks for other genotypes.

Standard IFN-α has the drawbacks of a short serum half-life and rapid clearance. To overcome this problem, PEGYLATED forms of IFN have been developed and tested clinically. PEGYLATED IFN-α2b is formed by covalent conjugation of a 12-kd monomethoxy polyethylene glycol (PEG) molecule to IFN-α2b, and PEGYLATED IFN-α2a by covalent conjugation of a 40-kd branched monomethoxy PEG molecule to IFN-α2a. PEGYLATED IFN-α2b has a prolonged serum half-life (40 hours) relative to standard IFN-α2b (7–9 hours). The greater polymer size of PEGYLATED IFN-α2a acts to reduce glomerular filtration, markedly prolonging its serum half-life (72–96 hours) compared with standard IFN-α2a (6–9 hours). Peginterferon α2a (40 KD) has superior virological EFFICACY to IFN-α2a, and elicits histological improvements in chronic HCV genotype 1 patients with and without sustained virological response. Peginterferon α2a (40 KD) is effective in patients infected with viral genotype 1 and those with liver cirrhosis. The addition of ribavirin to peginterferon α-2a (40 KD) further enhances the therapeutic benefit for patients with HCV [12]. Once-weekly dosing with either PEGYLATED IFN-α2a or PEGYLATED IFN-α2b has been shown to produce significantly higher rates of viral eradication than standard thrice-weekly IFN-α therapy. With respect to the treatment of chronic hepatitis C (CHC), the greatest anti-HCV efficacy has been achieved with the combination of once-weekly PEGYLATED IFN and ribavirin [13]. The use of ribavirin and PEGYLATED IFN-α, particularly PEGYLATED IFN-α2b, for the treatment of CHC infections, which involves a first and a second treatment time period, each of at least 20 to 30 weeks, has been reported. In clinical studies to assess sustained loss of serum HCV-RNA 12 weeks post-treatment, a regimen employing ribavirin plus PEGYLATED IFN-α2b for 48 weeks gave a successful response in 54% of patients, with a 41% response in those with HCV genotype 1.

INTERFERON-β (IFN-β) obtained from human FS-4 fibroblast cell lines is licensed for use in severe uncontrolled virus-mediated diseases in cases of viral encephalitis, herpes zoster and varicella in immunosuppressed patients. A further indication is viral infection of the inner ear with loss of hearing. The standard treatment for multiple sclerosis, a disease without a definitively elucidated etiology, is currently with IFN-β.

INTERFERON-γ (IFN-γ) is the major mediator of host resistance during the acute and chronic phases of infection and is pivotal in protection against a variety of intracellular pathogens. IFN-γ is produced by both $CD4^+$ and $CD8^+$ T CELLS and can induce class I and class II MHC products. Patients with chronic granulomatous disease are unable to generate an oxidative respiratory burst. As a consequence, they develop recurring catalase-positive bacterial infections such as *S. aureus*, *Pseudomonas cepacia*, and *Chromobacterium violaceum*. Multicenter clinical trials have shown that sustained administration of IFN-γ to

Box 1. Hepatitis C virus genotypes

Genotype	Sensitivity to interferon therapy
1	resistant
2	responsive
3	responsive
4	resistant

chronic granulomatous disease patients markedly reduced the relative risk of serious infection. IFN-γ is licensed as a therapeutic adjunct for use in patients with chronic (septic) granulomatosis for reduction of the frequency of serious infections. Actimmune™, a IFN-γ preparation, is marketed for chronic granulomatous disease, mycobacterial and fungal infections.

The major side-effects of all interferon therapies include flu-like syndromes, fever, myalgia, headache and fatigue. Hypotension, granulocytopenia, and thrombocytopenia can also occur. Deleterious effects on the central nervous system (CNS), particularly at high doses, have been observed.

Colony-stimulating factors

A high incidence of HIV-infected patients have neutropenia, which considerably increases risk for bacterial and fungal infections. G-CSF preparations such as Filgrastim (r-metHuG-CSF) can significantly enhance NEUTROPHIL functions in patients with AIDS [14] and reverse neutropenia associated with HIV and cytomegalovirus (CMV) infections. Filgrastim has been granted a license extension to cover the treatment of persistent neutropenia at an advanced stage of HIV infection. In one study, Filgrastim-treated patients have been shown to have 54% fewer severe bacterial infections and 45% fewer days in hospital for any bacterial infections [15]. Determination of absolute numbers of CD34+ progenitor cells and progenitor cell function in HIV-infected patients showed that G-CSF mainly increases the number and differentiation of myeloid progenitors. G-CSF and GM-CSF are used to reverse leukopenia as adjunctive therapy for HIV-associated infections. In a study conducted at 27 European centers on AIDS patients with CMV infection, G-CSF (lenograstim) was found to be suitable for the treatment of ganciclovir-induced neutropenia [16]. Multiple deficiencies are involved in the progression of fungal infections in neutropenic and nonneutropenic cancer patients. Although clinical experience is still limited, G-CSF, GM-CSF, and macrophage CSF show promise as adjuvant therapy for fungal infections [17].

Emerging cytokine therapies

IL-2 combinations

Highly active antiretroviral therapy (HAART), a combination of nucleoside and non-nucleoside reverse transcriptase, protease and fusion drug inhibitors, has improved the morbidity and survival of patients with HIV infection. Currently available antiretroviral agents can suppress viral replication and partially reverse cellular immunity defects in HIV patients. Combination antiretroviral drug therapy has been pivotal in the beneficial outcome in many HIV-1 infections and has rekindled a renewed interest in the use of IMMUNOTHERAPY as an adjunct treatment [18, 19]. The ability of IL-2 to induce expansion of the $CD4^+$ T lymphocyte pool has made it the most studied CYTOKINE in the treatment of HIV infection. Initial studies of high-dose IL-2 therapy were reported over a decade ago but were severely limited by toxicity problems. The current approach of decreasing the IL-2 amounts and using intermittent dosing has dramatically increased $CD4^+$ cell counts in HIV-infected persons. A majority of recent combinatory studies in HIV have, thus, been been performed with IL-2 and HAART. Immune-based therapy with IL-2 when used as adjunctive therapy to antiretroviral therapy may further improve immune responses, as demonstrated by an increase in $CD4^+$ T-lymphocyte counts in recent clinical trials [19]. An IL-2 regimen consisting of 5-day cycles administered every 4-8 weeks aims at expanding the $CD4^+$ T lymphocyte pool.

Combined antiretroviral treatment in some HIV-infected persons does not lead to a rapid increase in CD4 cell counts. Patients with CD4 cell counts <200 cells after 9 or more months of combined antiretroviral treatment received IL-2 IMMUNOTHERAPY (4.5×10^6 IU twice daily for 5 days every 6 weeks). After only 3 cycles, their CD4 cell counts increased significantly (mean cell count from 123 cells to 229 cells/mm^3). An increase in the naive $CD45RA^+$ T CELL subset was noted. IL-2 IMMUNOTHERAPY appears to accelerate the recovery of $CD4^+$ LYMPHOCYTES in persons whose CD^4+ cell counts fail to increase rapidly in response to combined antiretroviral treatment [20].

Chemokines and chemokine modulators

CHEMOKINES have been historically regarded as leukocyte chemoattractants capable of regulating cellular trafficking into inflammatory sites. The term chemokine is abbreviated from chemotactic CYTOKINES. The major classes of CHEMOKINES are constituted by the CXC or α CHEMOKINES, the CC or β CHEMOKINES, the C or γ CHEMOKINES, and the CX3C CHEMOKINES. Accumulating evidence suggests that CHEMOKINES have a broad range of functions, including macrophage activation, NEUTROPHIL degranulation, dendritic cell maturation, T CELL activation and B CELL ANTIBODY class switching. Thus, CHEMOKINES can influence both the innate and acquired phases of an immune response. CHEMOKINES can be mediators of angiogenesis and also play important roles in the development of the immune, circulatory and central nervous system. Currently, the chemokine family has expanded to nearly 50 ligands; 16 CXCL, 28 CCL, 2 XCl and 1 CX3CL CHEMOKINES have been identified. Furthermore, around 18 chemokine receptors, including 6 CXCR, 10 CCR, 1 XCR and 1 CX3CR, have been described (also see Chapter A.4).

The human immunodeficiency virus (HIV) can infect a wide range of human cells but has a particular tropism for $CD4^+$ T CELLS and the MONOCYTE-MACROPHAGE cell lineage [21]. HIV induces T-cell dysfunction and $CD8^+$ T CELL apoptosis, decreasing the number of T-cells that ultimately leads to immunodeficiency. B-cell dysfunction can be caused by HIV and is characterized by hypergammaglobulinemia, polyclonal activation, and absence of specific ANTIBODY responses. HIV entry into the host cell is mediated through the CD4 receptor and a variety of coreceptors. Chemokine receptors–mainly CCR5 and CXCR4–have been discovered to be necessary for HIV entry [22–25]. Binding of the HIV-1 envelope glycoprotein gp120 to CD4 and an appropriate chemokine receptor triggers conformational changes facilitating the fusion of the viral and host cell membranes. HIV-1 found in the vagina, rectal mucosa, CNS and in resident MACROPHAGES of many other tissues is mainly CCR-5-dependent. Macrophage-tropic (R5) HIV-1 variants predominantly make use of the CCR5 coreceptors [26,27]. The T-cell-tropic (X4) and dualtropic (R5X4) HIV-1 strains, generally associated with the clinical manifestations of AIDS, emerge after a latency of several years, although pathogenesis of the CNS and related symptoms are normally associated to M-tropic (R5) HIV-1 strains. X4 HIV-1 infection augments the expression of CHEMOKINES such as MIP-1α (macrophage inflammatory protein) and RANTES (regulated upon activation normal T CELL expressed and secreted).

HIV infection can be inhibited by CHEMOKINES and chemokine-related molecules that are ligands for those chemokine receptors that function as coreceptors. Chemokine receptors, thus, represent important targets for intervention in HIV and the search for molecules that have a therapeutic potential as inhibitors of these receptors has been intense [28–30]. Chemokine receptors belong to the rhodopsin family of the G-protein-coupled receptor family. Intervention strategies based on chemokine antagonists that could be useful for the therapeutics of HIV include receptor-ligand interaction, prevention of the chemokine-glycosaminoglycan interaction, interfering with the signalling pathways that are induced upon receptor activation, and modification of receptor pathways [31,32].

Azaheterocyclylcarboxamide, piperidinyl, and pyrrolidine compounds and their derivatives targeting CCR-5 have been advocated recently for the treatment of AIDS. β-CHEMOKINES such as RANTES exhibit clear-cut suppressive effects on HIV replication in peripheral blood mononuclear cells of HIV-infected individuals. CCR5 can be blocked and/or downmodulated by RANTES and by antibodies to three extracellular domains of CCR5. Recently, modified CHEMOKINES and small derivative molecules maintaining the HIV-inhibitory function, but devoid of receptor-activating capability, have been generated [33].

Cytokine inhibitors

Many infectious diseases, including HIV and malaria, and chronic inflammatory conditions can induce deleterious overproduction of proinflammatory CYTOKINES such as TNF-α and IL-1. This has generated intense interest in developing agents that can block the activity of such CYTOKINES. Inhibiting TNF activity has been singularly successful in the treatment of AUTOIMMUNE DISEASES. Monoclonal antibodies including adalimumab (Humira; Abbott), etanercept (Enbrel; Amgen/Wyeth), and infliximab (Remicade; Centocor) are being used in the treatment of rheumatoid arthritis. An INTERLEUKIN-1 (IL-1) receptor antagonist, Kineret, has also been recently licensed. Another potent TNF inhibitor, thalidomide, has been used in trials in HIV patients. Structural analogues of thalidomide with improved TNF-α inhibitory activity are currently being developed. Pentoxifylline, a methylxanthine usually used in the treatment of peripheral arterial circulatory disorders, has been shown to inhibit TNF synthesis. Currently, clinical trials are ongoing with phosphodiesterase inhibitors and small-molecule TNF-converting enzyme that specifically interrupt the signalling pathways of TNF.

Synthetic immunomodulators

Imiquimod derivatives

TOLL-LIKE RECEPTORS (TLRs) are type I transmembrane proteins that recognize microbial conserved structures involved in INNATE IMMUNITY (Box 2). Recent studies have shown that TLR3 recognizes dsRNA, a viral product, whereas TLR9 recognizes unmethylated CpG motifs frequently found in the genome of bacteria and viruses, but not vertebrates. In humans, TLR7 and TLR9 are expressed on the plasmacytoid DENDRITIC CELLS (PDC) which can rapidly synthesize large amounts of IFN-α and IFN-β in response to viral infection. These observations indicate that TLR3, TLR7 and TLR9 may play an important role in combating viral infections. Interestingly, TLR7 recognizes synthetic IMMUNOMODULATORS such as imidazoquinolone compounds that are used against viral infections.

BOX 2. TOLL-LIKE RECEPTORS

TLR	Ligands
TLR1	Lipoprotein
TLR2	Lipoprotein, peptidoglycan
TLR3	dsRNA
TLR4	Lipopolysaccharide, lipoteichoic acid
TLR5	Flagellin
TLR6	Lipoprotein
TLR7	Imiquimod
TLR8	Imiquimod
TLR9	Unmethylated CpG
TLR10	?

IMIQUIMOD is a fully synthetic immune response enhancing imidazoquinoline aminc, (S-26308, R-837) (1-(2-methylpropyl)-1H-imidazo[4,5-c]quinolin-4 amine) [34]. IMIQUIMOD is marketed as Aldara™ for genital warts, but is widely used for basal cell carcinoma, actinic keratosis and molluscum contagiosum. It has shown promise in lentigo maligna and cutaneous metastases of malignant melanoma. IMIQUIMOD and its derivatives such as RESIMIQUIMOD (R-848) improve antigen-presentation by DENDRITIC CELLS and also act on B CELLS leading to the synthesis of antibodies such as IgG2a. Furthermore, these agents activate MACROPHAGES and other cells via binding to cell surface receptors, such as TLR7, inducing secretion of proinflammatory CYTOKINES such as IFN-α, TNF-α and IL-12. The presence of this CYTOKINE milieu biases towards a Th1-type immune response and has been exploited clinically in the treatment of viral infections (human papillomavirus, herpes simplex virus, molluscum contagiosum). In randomized, double-blind, placebo-controlled stud-

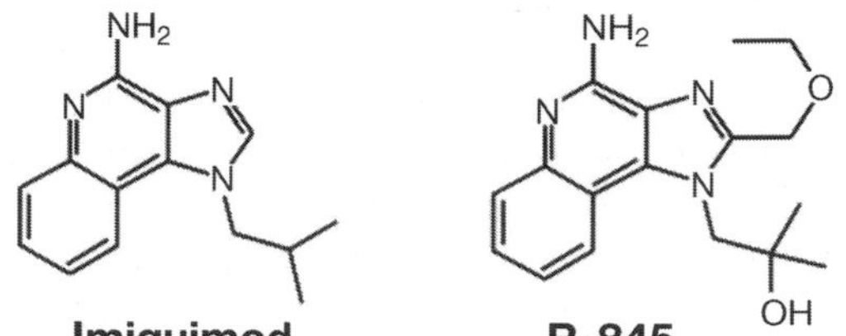

FIGURE 3. CHEMICAL STRUCTURE OF IMIQUIMOD AND RESIMIQUIMOD (R-848)

ies, IMIQUIMOD cream has been shown to be significantly effective in eliminating genital warts in patients with clinical, histopathological and polymerase chain reaction confirmed diagnosis of human papilloma virus infection and in treatment of external genital and perianal warts. Topically applied IMIQUIMOD cream reduced wart area in HIV-infected patients (Fig. 3).

Isoprinosine

ISOPRINOSINE (Inosiplex) is a complex of the *p*-acetamidobenzoate salt of *N,N*-dimethylamino-2-propanol: inosine in a 3:1 molar ratio. It is a white cystalline powder soluble in water. The inosine portion of ISOPRINOSINE is metabolically labile and half-life in rhesus monkeys has been found to be 3 minutes after intravenous and 50 minutes after oral administration. ISOPRINOSINE has been shown to augment production of CYTOKINES such as IL-1, IL-2 and IFN-γ. It increases proliferation of LYMPHOCYTES in response to mitogenic or antigenic stimuli, increases active T-cell rosettes and induces T-cell surface markers on prothymocytes. ISOPRINOSINE is currently licensed in Europe for treatment of herpes simplex infections, subacute sclerosing panencephalitis, acute viral encephalitis caused by herpes simplex, Epstein-Barr and measles viruses and for treatment of these viral infections in immunosuppressed patients. ISOPRINOSINE has been reported to have minor CNS depressant but no neuromuscular, sedative, or antipyretic activities in pharmacological studies in animals. In humans ISOPRINOSINE may cause transient nausea and rise of uric acid in serum and urine at high doses.

Emerging therapies with synthetic immunomodulators

CpG oligodeoxynucleotides

Over the last few years, nucleotide sequences containing non-methylated cytosine-phosphate-guanosine (CpG) dinucleotides, with flanking regions of two 5' purines and two 3' pyrimidines, have been discovered to play an important role in INNATE IMMUNITY. Effector cells such as MACROPHAGES, DENDRITIC CELLS, B CELLS and NK cells possess pattern recognition receptors. These can bind to the microbe-restricted structure of CpG motifs and trigger protective immune responses. Although DNA containing CpG motifs (CpG DNA) evolved as a defense system in eukaryotes to probably protect against infection, it is possible to use CpG DNA as an immunomodulator for therapeutic applications. Most interesting for the induction of antimicrobial effects, CpG DNA triggers a predominantly T helper cell type-1 (Th1) pattern of immune activation. CpG DNA triggers humoral B CELL responses and also activates MONOCYTES and MACROPHAGES to secrete CYTOKINES, especially IL-12, TNF, and IFN. The exact mechanisms by which CpG DNA induces DENDRITIC CELLS, MACROPHAGES and NK cells to produce immunomodulating CYTOKINES are being elucidated. Recent reports describe molecular mechanisms involved in CpG DNA-mediated activation of immune cells, including cellular recognition of CpG motifs by a member of the TLR family, TLR-9. The finding that synthetic oligodeoxynucleotides can mimic the action of bacterial DNA has galvanized research in this area. Administration of synthetic oligodeoxynucleotides (ODNs) containing CpG motifs [35] has been shown to confer protection or act as an adjuvant in experimental infections by influenza virus, hepatitis virus, *Listeria monocytogenes*, *Francisella tularensis*, *Trypanosoma cruzi*,

Leishmania and markedly increase resistance against acute poly-microbial sepsis. CpG ODNs have been administered to over 300 humans and found to be well tolerated.

Defensins

DEFENSINS are a family of structurally related cysteine-rich cationic polypeptides of the innate immune system produced in response to microbial infection in humans, animals, insects and plants. Six human α-DEFENSINS have been described; HD1, 2, 3, and 4 are secreted primarily by granulocytes and certain LYMPHOCYTES, whereas HD5 and 6 are expressed by intestinal Paneth cells. β-DEFENSINS are produced in response to microbial infection of mucosal tissue and skin. The unusual circular minidefensins were identified recently in rhesus monkeys. DEFENSINS of vertebrate animals have been reviewed elsewhere [36, 37].

Recent reports have highlighted the anti-HIV-1 activities of DEFENSINS, whose structure and charge resemble portions of the HIV-1 transmembrane envelope glycoprotein gp41. CD8 T LYMPHOCYTES from certain immunologically stable HIV-1-infected individuals secrete soluble factors that suppress HIV-1 replication. CD8 T CELLS from long-term nonprogressors with HIV-1 infection were found to secrete a cluster of proteins identified as α-DEFENSIN 1, 2, and 3 on the basis of specific ANTIBODY recognition and amino acid sequencing [38]. It is of interest to note that α-DEFENSINS have been shown to effectively suppress HIV replication *in vitro* about 10 years ago. In earlier studies, synthetic peptides derived from a region of HIV-1 gp41 exhibited a sequence similar to DEFENSINS. Like DEFENSINS, which can insert themselves into lipid bilayers, gp41 peptides have the ability to associate with liposomes and lymphocyte membranes. An antimicrobial peptide homologous to rhesus monkey circular minidefensins (δ-DEFENSINS) has been prepared by solid-phase peptide synthesis and named "retrocyclin" [39]. Retrocyclin has a remarkable ability to inhibit proviral DNA formation and to protect $CD4^+$ LYMPHOCYTES from *in vitro* infection by both T-tropic and M-tropic strains of HIV-1. Retrocyclin interferes with an early stage of HIV-1 infection and retrocyclin-like agents might be useful topical agents to prevent sexually acquired HIV-1 infections.

DNA encoding gp120 fused with proinflammatory chemoattractants of immature DENDRITIC CELLS, such as β-DEFENSIN 2, elicited anti-gp120 antibodies with high titers of virus-neutralizing activity. The use of chemokine fusion constructs with gp140, gp120 linked to the extracellular domain of gp41 via a 14-amino acid spacer peptide sequence further augmented immunogenicity. This construct elicited antibodies with more effective neutralizing activity than corresponding constructs expressing gp120 [40]. A fusion polypeptide comprising a DEFENSIN and a viral ANTIGEN and its use for producing an immune response in a subject and for treating viral infections, including HIV infection, have been detailed by the US Department of Health & Human Services, National Institute of Health, Office of Technology Transfer. The viral ANTIGEN may be an HIV ANTIGEN, and the DEFENSIN may be an α-DEFENSIN or a β-DEFENSIN. It is interesting to note that the peptides derived from HIV-1 envelope glycoproteins corresponding predominantly to membrane-active domains of gp120 and gp41can exert antimicrobial activity against laboratory strains of *Escherichia coli* and *Listeria monocytogenes*. Human β-DEFENSIN-3 (hβD-3) and fragments of hβD-3 for the treatment of various infections, particularly respiratory infections, have been described. The peptides are claimed to be of use in the treatment of infection caused by *Pseudomonas aeruginosa* and *Burkholderia cepacia*, other gram-negative and gram-positive bacterial infections and yeast infections. The peptides were synthesized and tested with a range of bacteria such as *Enterococcus faecalis*, *Escherichia coli*, *Klebsiella pneumoniae*, *Staphylococcus aureus*, *Streptococcus pneumoniae*, *P. aeruginosa* and *B. cepacia*.

Methyl inosine monophosphate (MIMP)

MIMP is a newer generation thymomimetic purine immunomodulator capable of enhancing a wide variety of immune responses. It can induce in human prothymocytes the expression of T lymphocyte differentiation markers (CD3, CD4, CD8) and IL-2 receptors

(CD25), enhance mitogen-induced proliferation of LYMPHOCYTES, augment IgM plaque-forming cells, induce delayed-type hypersensitivity and normalize an impaired response to IL-2 and suggest MIMP as a potential IMMUNOTHERAPY for early HIV infection and for inducing protective effects against human viruses. Intranasal administration of MIMP improves survival rates and incorporation of MIMP in squalane-saline emulsion confers complete protection against influenza infection. Complete survival and reduction of viral load after influenza challenge infection suggests effective stimulation by MIMP of protective responses against influenza virus [41].

Microbial-derived synthetic immunomodulators

Cell walls of gram-negative bacteria contain endotoxic lipopolysaccharide which is a potent stimulator of the immune system, even in nanogram quantities. Clinical application of endotoxin has, however, been hampered as it plays a major role in the pathophysiology of gram-negative bacterial sepsis. Concerted efforts have been made to modify endotoxin for possible therapeutic use in humans. Investigations on endotoxic lipopolysaccharide have led to the identification of lipid A as an important constituent of the endotoxin molecule capable of exhibiting various immunopharmacological activities of bacterial lipopolysaccharide. During the last ten years, selective reduction of the toxicity of endotoxin whilst retaining its beneficial adjuvant property has been achieved by effective chemical treatment and by synthesis of nontoxic lipid A analogues.

Monophosphoryl lipid A

Prophylactic administration of monophosphoryl lipid A (MPL), a nontoxic derivative of endotoxin, has been shown to mitigate the sepsis syndrome by reducing chemokine production of pulmonary and hepatic MIP-1α, MIP-1β, MIP-2, and MCP-1 mRNA and attenuating the production of proinflammatory CYTOKINES. MPL has been used as an adjuvant with leishmanial, malarial, tuberculosis and cancer ANTIGENS. It has been administered to over thirty thousand humans and is a component of a melanoma Melacine vaccine which is approved for use in Canada.

OM-174

IMMUNOMODULATOR OM-174 is a chemically defined new generation lipid A analogue. Animal experiments using aerosol influenza virus showed that the survival in the group receiving H1N1 SUBUNIT VACCINE alone could be improved after the booster immunization. In contrast, the vaccine adjuvanted with OM-174 could confer complete protection already after the primary immunization. OM-174 has shown adjuvant activity with several microbial ANTIGENS and has already been administered intramuscularly to humans in a limited number of trials.

Murabutide

Muramyl peptides (MDP) represent structures that are essential for mycobacterial adjuvanticity. MURABUTIDE, an MDP analogue, significantly reduced reduced CD4 and CCR5 receptors, secreted high levels of β-CHEMOKINES and suppressed HIV-1viral replication in acutely infected MONOCYTE-derived MACROPHAGES and DENDRITIC CELLS. Clinical TOLERANCE and biological effects of IMMUNOTHERAPY with the synthetic immunomodulator MURABUTIDE have been evaluated in HIV-1 patients [42]. MURABUTIDE recipients showed a significant increase in CD4 cells and platelet counts in HIV-1 patients with weak immune reconstitution and ineffective virus suppression following long-term HAART therapy.

Microbial immunomodulators

Probiotics

Ingestion of living microorganisms for attaining beneficial effects on health has been termed as a probiotic approach which is gaining increasing popularity. Several fortified products are available to consumers. PROBIOTICS have been used for conditions such as diarrhea, candidal vaginitis and urinary tract infections [43]. Whole microbes such as *Lactobacil-*

lus casei, *Bifidobacterium* species and *Saccharomyces boulardii* have been used successfully to prevent antibiotic-associated diarrhea, and to treat other diarrheal illnesses caused by bacteria [44]. *Bifidobacterium breve* has been shown to augment ANTIBODY production and induce significant reduction in the frequency of rotavirus shedding in stool samples of infants [45]. The major *Lactobacillus* species present in the human gastrointestinal mucosa, *L. plantarum*, *L. rhamnosus* and *L. paracasei*, have been found to be potent stimulators of INTERLEUKIN-12 [46].

Bacterial extracts

BACTERIAL EXTRACTS have been widely used as IMMUNOMODULATORS to prevent recurrent infections of the respiratory tract. Several preparations containing bacterial lysates (e.g., Broncho-Munal®, Luivac®) are licensed for use in Europe. There exists a large body of primary literature on placebo-controlled and double-blind clinical trials which have been conducted with these agents. The current concept of the mucosal immune system postulates that stimulation of the gut-associated lymphoid tissue can lead to the induction of a generalized response by the whole mucosal-associated lymphoid tissue. Patients, most often children, having recurrent episodes of infections of the respiratory tract and of the ear, nose and throat have been treated with oral bacterial lysates. In most cases the frequency and the severity of infections was reported to be reduced and both the physician and the patient considered the treatment to be beneficial. It is of interest that many studies have reported a decrease in antibiotic consumption in bacterial lysate-treated patients. A multicenter study with oral bacterial lysate immunomodulator LW 50020 involving 4965 recurrent respiratory tract infection patients in 14 countries in Europe, Latin America, and Asia has recently been conducted. An overall reduction of at least 50% in the number, severity, and duration of respiratory tract infections, the number of antibiotic and symptomatic treatments, and the number of days absent from school or work was reported [47]. Hemodialysis patients suffer from several immune defects which increase their risk of developing bacterial infections, particularly of the respiratory tract. In a double-blind placebo-controlled prospective study, oral IMMUNOTHERAPY with an immunomodulating bacterial extract significantly reduced the number of patients with respiratory tract infections and consequently the number of antibiotic treatment courses [48].

Glucans

Immunomodulating polysaccharides, poly-(1-6)-β-glucotriosyl-(1-3)-β-glucopyranose, PGG-GLUCAN, marketed as Betafectin, have been found to be useful in prevention of mortality and abscess formation associated with experimental intraabdominal sepsis in the absence of antimicrobial therapy. The safety and EFFICACY of PGG-GLUCAN in surgical patients at high risk for postoperative infection who underwent major thoracic or abdominal surgery showed that PGG-GLUCAN had significantly fewer infectious complications, decreased intravenous antibiotic requirement and shorter intensive care unit length of stay. Prophylaxis to prevent staphylococcal wound infection could be synergistically enhanced by adding PGG-GLUCAN to the antibiotic cefazolin. Perioperative administration of PGG-GLUCAN reduced serious postoperative infections or death by 39% after high-risk noncolorectal operations.

LENTINAN is chemically well defined as 1-3-β-D-glucan with 1-6-β-D-glucopyranoside branches and is isolated from an edible Japanese mushroom *Lentinus edodes* (Shiitake mushroom) and has immunomodulating properties. It is licensed as an adjunctive for antitumor therapy in Japan. Experimental studies have demonstrated that LENTINAN can confer protection against influenza virus, *Listeria* and prevent relapse of *Mycobacterium tuberculosis*. Yeast glucan could enhance resistance against herpes simplex virus types I and II, and murine hepatitis virus, and polysaccharide schizophyllan could confer protection on mice against lethal Sendai virus infection. *S. cervisiae* glucan induced nonspecific resistance against *K. pneumoniae* infection and yeast glucan can protect patients from sepsis, bacteremia and peritonitis resulting from *Escherichia coli*, *Staphylococcus aureus* and *Pseudomonas aeruginosa* infections. Yeast glucan

can prolong survival against parasitic infections by *Plasmodium berghei* and *Leishmania donovani* and exert antifungal activity against *Candida*, *Cryptococcus* and *Sporotichum*. Phase I/II placebo-controlled trials with LENTINAN or LENTINAN in combination with didanosine in HIV-positive patients showed a trend toward increases in CD4 cells and in NEUTROPHIL activity.

Perspective

Preventing existing and emerging infectious diseases is a multidisciplinary and multifaceted endeavor. The traditional treatment of infections has focused on antimicrobial agents and on the induction of specific immune defenses. The forefront of immunotherapeutics in the coming decades will be on the modulation of the host response against infectious diseases. Selective stimulation of discrete lymphocyte subpopulations important in protective effector mechanisms against a given infection by suitable IMMUNOMODULATORS will play an increasingly important role. Polarized type-1 and type-2 immune responses can be readily achieved, irrespective of the genetic bias of the host and of the nature of the protein ANTIGEN, through the choice of appropriate IMMUNOMODULATORS. The major promise of new IMMUNOMODULATORS lies in their broad activity against diverse microbial pathogens and in a mode of action that is distinctly different from the direct microbicidal action of established antibiotics, antivirals and antifungals. Adjuvant CYTOKINES such as IL-2, IFN-γ, and GM-CSF, can be useful for shortening the duration of treatment. It is now also possible to control the deleterious overproduction of inflammatory CYTOKINES observed in certain disease states with several IMMUNOMODULATORY agents. Activation of innate non-antigen-specific host defenses by recently recognized IMMUNOMODULATORY agents such as CpG DNA will increasingly find a beneficial role in the prevention of infections. It is noteworthy that around 60 biotechnology protein drugs, including recombinant proteins and MONOCLONAL ANTIBODIES, that are currently available represent over a quarter of the drugs introduced worldwide recently. New methodologies currently available have the potential to identify novel targets and foster the development of individually tailored IMMUNOMODULATORY drug treatments. The human genome project has indicated that the human genome contains 30,000–40,000 protein-encoding genes with an estimated 90,000 or more distinct proteins. It would require only a fraction, 1 or 2%, of these proteins to be developed as drugs to be able provide hundreds of novel therapeutic agents. IMMUNOMODULATORS are becoming a viable adjunct to established modalities offering a novel approach for the treatment of infectious diseases in the coming decades of the 21st century.

Summary

The availability of a vast array of recombinant and synthetic IMMUNOMODULATORS is a significant milestone toward the development of effective therapies for infectious diseases. This is evinced by licensing of several recombinant human CYTOKINES, including COLONY-STIMULATING FACTORS, INTERLEUKINS, INTERFERONS and erythropoietin, for clinical use in patients. Diverse combinations with INTERFERONS and other CYTOKINES for the treatment of various infections have been proposed. Others, including various CHEMOKINES, synthetic CpG oligodeoxynucleotides and glucans, are extensively being investigated in clinical and preclinical studies. Considerable advances have been made on compounds exhibiting CYTOKINE inhibitory properties useful for new treatments of infectious and inflammatory diseases. Many of the major developments and current trends are highlighted in this review. Novel strategies based on the engineering of CYTOKINES and inhibitors are poised to revolutionize therapeutic options contingent upon scientific evidence rather than dictates of discursive empiricism in the coming decades.

Recommended websites

The Cleveland Clinic (2002) Biologictherapy.org: *http://www.biologictherapy.org* (Accessed March 2005)

Horst Ibelgaufts' COPE (2003) Cytokines Online Pathfinder Encyclopaedia: *http://www.copewithcytokines.de* (Accessed March 2005)

Beta Glucan Research Organization (2003) Non-Commercial Beta Glucan Research Site Sorted by Health Condition: *http://www.betaglucan.org* (Accessed March 2005)

References

1 Ksiazek TG, Erdman D, Goldsmith C, Zaki SR, Peret T, Emery S, Tong S, Urbani C, Comer JA, Lim W et al (2003) A novel coronavirus associated with severe acute respiratory syndrome. *N Engl J Med* 348: 1953–1966
2 Taylor PW, Stapleton PD, Paul LJ (2002) New ways to treat bacterial infections. *Drug Discov Today* 7: 1086–1091
3 Amyes SGB (2000) The rise in bacterial resistance is partly because there have been no new classes of antibiotics since the 1960s. *Br Med J* 320: 199–200
4 Hengel H, Masihi KN (2003) Combinatorial immunotherapies for infectious diseases. *Int Immunopharmacol* 3: 1159–1167
5 Masihi KN (2003) Concepts of immunostimulation to increase antiparasitic drug action. *Parasitol Res* 90 (Suppl 2): S97–S104
6 Masihi KN (2001) Fighting infection using immunomodulatory agents. *Expert Opin Biol Ther* 1: 641–653
7 Koliouskas D, Sidiropoulos I, Masmanidou M, Dokas S, Ziakas A (2002) Comparative analysis of Peg-interferon alpha-2b and lamivudine in the treatment of chronic hepatitis B patients: preliminary results. *J Hepatol* 36: 237–238
8 Santantonio T, Anna N, Sinisi E, Leandro G, Insalata M, Guastadisegni A, Facciorusso D, Gravinese E, Andriulli A, Pastore G (2002) Lamivudine/interferon combination therapy in anti-HBe positive chronic hepatitis B patients: a controlled pilot study. *J Hepatol* 36: 799–804
9 Billich A (2002) Thymosin alpha1. SciClone Pharmaceuticals. *Curr Opin Investig Drugs* 3: 698–707
10 Sugahara S, Ichida T, Yamagiwa S, Ishikawa T, Uehara K, Yoshida Y, Yang XH, Nomoto M, Watanabe H, Abo T et al (2002) Thymosin-[alpha]1 increases intrahepatic NKT cells and CTLs in patients with chronic hepatitis B. *Hepatol Res* 24: 346–354
11 Pimstone NR, Canio JB, Chiang MH (2001) Ribavirin/interferon-2b therapy is very effective in the treatment of chronic hepatitis C genotype 2 and 3 patients who have failed to respond virologically to IFN monotherapy. *Gastroenterology* 120: A-382
12 Luxon BA, Grace M, Brassard D, Bordens R (2002) Pegylated interferons for the treatment of chronic hepatitis C infection. *Clin Ther* 24: 1363–1383
13 Rajender R, Modi MW, Pedder S (2002) Use of peginterferon alfa-2a (40 KD) (Pegasys(R)) for the treatment of hepatitis C. *Adv Drug Deliv Rev* 54: 571–586
14 Welch W, Foote M (1999) The use of Filgrastim in AIDS-related neutropenia. *J Hematother Stem Cell Res* 8 (Suppl 1): S9–S16
15 Kuritzkes DR (1999) Clinical experience with Filgrastim in AIDS. *J Hematother Stem Cell Res* 8 (Suppl 1): S17–S19
16 Dubreuil-Lemaire ML, Gori A, Vittecoq D, Panelatti G, Tharaux F, Palisses R, Gharakhanian S, Rozenbaum W; GCS 309 European Study Group (2000): Lenograstim for the treatment of neutropenia in patients receiving ganciclovir for cytomegalovirus infection: a randomised, placebo-controlled trial in AIDS patients. *Eur J Haematol* 65: 337–343
17 Rodriguez-Adrian LJ, Grazziutti ML, Rex JH, Anaissie EJ (1998) The potential role of cytokine therapy for fungal infections in patients with cancer: is recovery from neutropenia all that is needed? *Clin Infect Dis* 26: 1270–1278
18 Masihi KN (2003) Progress on novel immunomodulatory agents for HIV 1 infection and other infectious diseases. *Expert Opin Ther Patents* 13: 867–882
19 Pett SL, Emery S (2001) Immunomodulators as adjunctive therapy for HIV-1 infection. *J Clin Virol* 22: 289–295
20 David D, Nait-Ighil L, Dupont B, Maral J, Gachot B, Theze J (2001) Rapid effect of interleukin-2 therapy in human immunodeficiency virus-infected patients whose CD4 cell counts increase only slightly in response to combined antiretroviral treatment. *J Infect Dis* 183: 730–735
21 Aquaro S, Calio R, Balzarini J, Bellocchi MC, Garaci E, Perno CF (2002) Macrophages and HIV infection: therapeutical approaches toward this strategic virus reservoir. *Antiviral Res* 55: 209–225
22 Deng H, Liu R, Ellmeier W, Choe S, Unutmaz D, Burkhart M, Di Marzio P, Marmon S, Sutton RE, Hill CM et al (1996) Identification of a major co-receptor for primary isolates of HIV-1. *Nature* 381: 661–666
23 Dragic T (2001) An overview of the determinants of

CCR5 and CXCR4 co-receptor function. *J Gen Virol* 82: 1807–1814

24 Lehner T (2002) The role of CCR5 chemokine ligands and antibodies to CCR5 coreceptors in preventing HIV infection. *Trends Immunol* 23: 347–351

25 Nansen A, Christensen JP, Andreasen SO, Bartholdy C, Christensen JE, Thomsen AR (2002) The role of CC chemokine receptor 5 in antiviral immunity. *Blood* 99: 1237–1245

26 Alkhatib G, Combadiere C, Broder CC, Feng Y, Kennedy PE, Murphy PM, Berger EA (1996) CC CKR5: a RANTES, MIP-1alpha, MIP-1beta receptor as a fusion cofactor for macrophage-tropic HIV-1. *Science* 272: 1955–1958

27 Blanpain C, Libert F, Vassart G, Parmentier M (2002) CCR5 and HIV infection. *Receptors Channels* 8: 19–31

28 Lusso P (2002) HIV and chemokines: implications for therapy and vaccine. *Vaccine* 20: 1964–1967

29 Scozzafava A, Mastrolorenzo A, Supuran CT (2002) Non-peptidic chemokine receptors antagonists as emerging anti-HIV agents. *J Enzyme Inhib Med Chem* 17: 69–76

30 Verani A, Lusso P (2002) Chemokines as natural HIV antagonists. *Curr Mol Med* 2: 691–702

31 Onuffer JJ, Horuk R (2002) Chemokines, chemokine receptors and small-molecule antagonists: recent developments. *Trends Pharmacol Sci* 23: 459–467

32 Regoes RR, Bonhoeffer SEBA (2002) HIV coreceptor usage and drug treatment. *J Theor Biol* 217: 443–457

33 Polo S, Nardese V, De Santis C, Arcelloni C, Paroni R, Sironi F, Verani A, Rizzi M, Bolognesi M, Lusso P (2000) Enhancement of the HIV-1 inhibitory activity of RANTES by modification of the N-terminal region: dissociation from CCR5 activation. *Eur J Immunol* 30: 3190–3198

34 Stanley MA (2002): Imiquimod and the imidazoquinolones: mechanism of action and therapeutic potential. *Clin Exp Dermatol* 27: 571–577

35 Krieg AM, Davis HL (2001) Enhancing vaccines with immune stimulatory CpG DNA. *Curr Opin Mol Ther* 3: 15–24

36 Lehrer RI, Ganz T (2002) Defensins of vertebrate animals. *Curr Opin Immunol* 14: 96–102

37 Raj PA, Dentino AR (2002) Current status of defensins and their role in innate and adaptive immunity. *FEMS Microbiol Lett* 206: 9–18

38 Zhang L, Yu W, He T, Yu J, Caffrey RE, Dalmasso EA, Fu S, Pham T, Mei J, Ho JJ et al (2002) Contribution of human alpha-defensin 1, 2, and 3 to the anti-HIV-1 activity of CD8 antiviral factor. *Science* 298: 995–1000

39 Cole AM, Hong T, Boo LM, Nguyen T, Zhao C, Bristol G, Zack JA, Waring AJ, Yang OO, Lehrer RI (2002) Retrocyclin: a primate peptide that protects cells from infection by T- and M-tropic strains of HIV-1. *Proc Natl Acad Sci USA* 99: 1813–1818

40 Biragyn A, Belyakov IM, Chow YH, Dimitrov DS, Berzofsky JA, Kwak LW (2002) DNA vaccines encoding human immunodeficiency virus-1 glycoprotein 120 fusions with proinflammatory chemoattractants induce systemic and mucosal immune responses. *Blood* 100: 1153–1159

41 Masihi KN, Hadden JW (2002) Protection by methyl inosine monophosphate (MIMP) against aerosol influenza virus infection in mice. *Int Immunopharmacol* 2: 835–841

42 Bahr GM (2003) Non-specific immunotherapy of HIV-1 infection: potential use of the synthetic immunodulator murabutide. *J Antimicrob Chemother* 51: 5–8

43 Mombelli B, Gismondo M (2000) The use of probiotics in medical practice. *Int J Antimicrob Agents* 16: 531–536

44 Cross ML (2002) Microbes *versus* microbes: immune signals generated by probiotic lactobacilli and their role in protection against microbial pathogens. *FEMS Immunol Med Microbiol* 34: 245–253

45 Yasui H, Shida K, Matsuzaki T, Yokokura T (1999) Immunomodulatory function of lactic acid bacteria. *Antonie Van Leeuwenhoek* 76: 383–389

46 Hessle C, Hanson LÅ, Wold AE (1999) Lactobacilli from human gastrointestinal mucosa are strong stimulators of IL-12 production. *Clin Exp Immunol* 116: 276–282

47 Grevers G, Palacios OA, Rodriguez B, Abel S, van Aubel A (2000) Treatment of recurrent respiratory tract infections with a polyvalent bacterial lysate: results of an open, prospective, multinational study. *Adv Ther* 17: 103–116

48 Tielemans C, Gastaldello K, Husson C, Marchant A, Delville JP, Vanherweghem JL, Goldman M (1999) Efficacy of oral immunotherapy on respiratory infections in hemodialysis patients: a double-blind, placebo-controlled study. *Clin Nephrol* 51: 153–160

Mild plant and dietary immunomodulators

Michael J. Parnham and Donatella Verbanac

Introduction

Plants and minerals have been used since ancient times for the treatment of many ailments and diseases. Most were used for mystical reasons and others relied on the "doctrine of signatures", which stated that the shape of the plant reflected its potential medicinal use. The root of the mandrake or ginseng, for instance, is shaped like that of the human body and has been used as a general tonic for a variety of illnesses [1]. It is claimed by herbalists to have immunostimulant properties. Siberian ginseng or Taiga root (*Eleutherococcus senticosus*) is also used as a tonic and has been reported to exhibit immunostimulatory properties. The pharmacological bases of these actions are unclear, so that these plant medicines cannot be considered unequivocally as immunostimulants.

In recent years, many folklore remedies have been subjected to pharmacological study and some have been shown to exhibit therapeutic immunostimulant properties.

Antioxidant dietary constituents also have been shown to exert immunoprotective or immunostimulant properties and are widely sold as prophylactic nutritional supplements. Some of the compounds for which clear IMMUNOMODULATORY actions have been described are discussed in this chapter. Combination products are not considered, since little scientific basis is available for their EFFICACY.

Plant immunostimulants

Purple coneflower (Echinacea)

History

The PURPLE CONEFLOWER (Fig. 1) is indigenous to North America and was used by the American Indians of the Great Plains as a universal remedy, particularly for colds, sore throats and pain [2]. Extracts of *E. angustifolia* (narrow-leaved PURPLE CONEFLOWER) were introduced into medical practice in the United States at the end of the nineteenth century, becoming the most widely used medicinal plants by the 1930s. With the introduction of antibiotics, Echinacea fell into disuse. In Europe, *E. angustifolia* was introduced into homeopathic practice in response to publications from the United States. In 1937, a general lack of supplies subsequently led to the introduction of the common PURPLE CONEFLOWER (*E. purpurea*) to Germany, where the squeezed sap was marketed. Many of the pharmacological studies on Echinacea have been performed in Germany on this preparation.

Chemical constituents

Compounds isolated from Echinacea species include caffeic derivatives, FLAVONOIDS, ethereal oils, polyacetylenes, alkylamides, alkaloids and polysaccharides [3]. Ingredients thought to contribute to the immunostimulatory properties of Echinacea include cichoric acid, polysaccharides and alkylamides, the latter being the most potent stimulators of INNATE IMMUNITY [4]. Commercial preparations contain 60-80g squeezed sap per 100 g, but the relative proportions of the various constituents vary markedly

Figure 1. Purple coneflower (*Echinacea purpurea*)

between different products. Standardisation on the basis of the active ingredients is clearly needed.

Modes of action and pharmacological effects

The squeezed sap of *E. purpurea* stimulates the phagocytic activity of NEUTROPHILS and MACROPHAGES *in vitro* and *in vivo*. The response is moderate, but a significant increase in NEUTROPHIL phagocytosis has been observed following repeated oral administration to healthy volunteers [3]. With many phytopharmaceuticals, pharmacological effects are thought to be due to a combination of constituents. Stimulation of macrophage phagocytosis and release of TNF-α and nitric oxide, however, appear to be most pronounced with the alkylamides (dodeca-2E, 4E, 8Z, 10E/Z-teraenoic acid isobutylamides) present in the *E. purpurea* preparation [4]. Other actions, such as stimulation of leukocyte adherence, also probably contribute to the oral activity of the preparation. Administered topically to the skin, the squeezed sap of *E. purpurea* enhances wound healing, probably by inhibiting hyaluronidase leading to increased hyaluronic acid secretion.

Cellular pharmacokinetics

Recently, dodeca-2E, 4E, 8Z, 10E/Z-teraenoic acid isobutylamides have been detected in human blood after oral administration of *E. purpurea* extract [5]. These findings have been confirmed *in vitro* on studies with Caco-2 colonic cell lines, showing that the alkylamides are transported across the cell monolayers within 4h without significant metabolism [6]. These constituents are therefore likely to be major contributors to immunostimulatory activity *in vivo*.

Clinical indications

Non-homeopathic preparations of *E. purpurea* are used mainly for the oral adjuvant treatment of respiratory and urinary tract infections and also topically for wound healing. Double-blind, controlled clinical trials have confirmed moderate EFFICACY in the treatment of mild respiratory infections [7]. When administered within a few hours of symptoms arising, *E. purpurea* has been reported to shorten the course of the common cold, but without reducing symptom intensity [8]. EFFICACY in the treatment of vaginal candidiasis has only been reported in open studies.

Side-effects

Common adverse reactions reported with Echinacea include headache, dizziness, nausea, constipation and mild epigastric pain. No adverse effects that are specific to *E. purpurea* have been observed, but as with all plant extracts, hypersensitivity responses (dermatitis, anaphylaxis) have been reported which in rare cases can be severe.

Mistletoe (*Viscum album*)

History

Extract of the leaves (not berries) of mistletoe (Fig. 2) has been used for centuries in Europe as a traditional herbal treatment for infections. In the last century, Rudolf Steiner, the originator of ANTHROPOSOPHY, suggested its use as a remedy for cancer. Biochemical analysis of mistletoe constituents led in the 1980s to the isolation and characterization of specific cytotoxic LECTINS that are responsible for the proposed anti-tumor activity of the extract [9]. The herb is not approved for use in the United States.

FIGURE 2. MISTLETOE (*VISCUM ALBUM*)
Photo can be downloaded from: http://www.cancer-plants.com/medicinal_plants/viscum_album.html (last accessed September 2005)

Chemical constituents

The main immunostimulatory constituents of mistletoe are the glycosylated LECTINS, ML-I, ML-II and ML-III. The major component is ML-I (viscumin), a member of the type II ribosome-inactivating proteins, which is used to standardize mistletoe extracts. It consists of two polypeptide chains linked by a disulphide bridge. The A-chain has enzymatic rRNA-cleaving activity and the B-chain binds to the target cell. Other constituents of mistletoe include FLAVONOIDS, viscotoxins, terpenoids and polysaccharides. Amines, such as acetylcholine, histamine and tyramine, are also present and may contribute to hypotensive effects of mistletoe preparations.

Modes of action and pharmacological effects

ML-I has a broad range of affinities for α/β-linked galactopyranosyl residues. High nanogram concentrations of all three mistletoe LECTINS are cytotoxic. This action is due to ribosome inactivation by the rRNA N-glycosidase A-chain [10], leading to induction of apoptosis, possibly through activation of cation channels. At lower concentrations, ML-I and ML-I-standardized mistletoe extracts stimulate release of IL-1, IL-6 and TNF-α from peripheral blood mononuclear and skin cells [11]. Repeated doses of mistletoe extract at an ML-I-equivalent of 1 ng/kg s.c. in cancer patients, cause an increase in body temperature, increases in circulating Th cells and NK cells and enhanced expression of IL-2 receptors (CD25) on LYMPHOCYTES [12]. A direct stimulatory action on T CELLS is likely. In mice, mistletoe extract reduces formation of melanoma metastases. Mistletoe extract is inactive on oral administration and only exhibits immunostimulatory activity on parenteral injection.

Cellular pharmacokinetics

Using gold-labelled ML-I, the lectin has been shown to be taken up into L1210 leukaemia cells *in vitro* via coated pits and via plasma membrane endocytosis. These kinetics correlate – both with regard to time and concentration – with the cytotoxicity of ML-I [13].

Table 1. Daily recommended intakes for zinc (in milligrams, mg)

Persons	U.S. (mg)	Canada (mg)
Infants and children		
Birth to 3 years of age	5–10	2–4
4 to 6 years of age	10	5
7 to 10 years of age	10	7–9
Adolescent and adult males	15	9–12
Adolescent and adult females	12	9
Pregnant females	15	15
Breast-feeding females	16–19	15

Taken from [16]

Clinical indication

In German-speaking countries, ML-I-standardized mistletoe extract is administered intracutaneously at 0.5–1.0 ng/kg twice weekly for at least 3 months in cancer patients. Quality of life was reported to be improved in some recent clinical trials, but these were predominantly uncontrolled. A clear improvement in survival time has not been shown unequivocally. Case reports of improved quality of life in patients with hepatitis C have also appeared, but further studies are needed to confirm this effect. Non-glycosylated recombinant ML-I (rViscumin) has been shown to exhibit cytotoxicity in animal tumour models and is under investigation in cancer patients by intravenous infusion [14].

Side-effects

Mistletoe extract can cause fever, headache, leukocytosis, orthostatic hypotension, bradycardia, diarrhea and hypersensitivity reactions. Toxic doses may cause coma, seizures and death.

Zinc

History

Zinc is an essential mineral that is found in almost every cell. Its importance for human health was first documented in 1963. Zinc is required for many biological functions [15]. It stimulates the activity of approximately 100 enzymes, promoting various biochemical reactions in the body. Zinc is needed for DNA synthesis, proper immune response, and wound healing, and helps maintain sense of taste and smell. Furthermore, zinc supports normal growth and development during pregnancy, childhood, and adolescence. The element is found in almost all food, but the majority of zinc in the diet is provided by sea food, red meat and poultry. Other good food sources are beans, nuts, whole grains, fortified breakfast cereals, and dairy products [16]. Zinc absorption is greater from a diet high in animal protein than a diet rich in plant proteins, because of the presence of PHYTATES, which are found in whole grain cereals and legumes and can interfere with zinc absorption.

Recommendations for adequate dietary zinc intake by humans in the US and Canada have been revised recently, and a summary is given in Table 1.

Pharmacology

Deficiency of zinc in humans due to nutritional factors and several disease states has been gradually recognized over the past 25 years. Alcoholism, malabsorption, sickle cell anemia, chronic renal disease, and chronically debilitating diseases are known to be predisposing factors for zinc deficiency. Vegetarians may need as much as 50% more zinc than non-vegetarians because of the lower absorption of zinc from plant foods [16]. Individuals who have had gastrointestinal surgery or who have digestive disorders that result in malabsorption, including sprue, Crohn's disease and short bowel syndrome, are also at greater risk of a zinc deficiency. The immune system is adversely affected by even moderate degrees of zinc deficiency, while severe zinc deficiency depresses immune function [17]. Zinc is unequivocally important for the INNATE, as well as for the ADAPTIVE IMMUNE RESPONSE. Decreased zinc concentrations impair NK (NATURAL KILLER) cell activity, NEUTROPHIL and MACROPHAGE PHAGOCYTOSIS, CHEMOTAXIS and oxidative burst generation [18]. It is required for recognition by killer cell inhibitory receptors (KIR) on NK CELLS of MHC class I molecules, predominantly HLA-C, on target cells. In this way zinc can influence NK CELL-mediated killing of virus-infected and tumor cells and modulate the action of cytolytic T LYMPHOCYTES. In addition, zinc is required for the development and activation of T LYMPHOCYTES [16]. The element is an essential cofactor in stabilizing thymulin, the thymic hormone that is a key factor for differentiation and maturation of immature T LYMPHOCYTES in the thymus and the periphery. Thymulin acts on CYTOKINE secretion by peripheral blood mononuclear (PMN) cells and together with IL-2 induces proliferation of CD8 positive T CELLS. Zinc also influences mature T CELLS, inducing the expression of the high-affinity receptor (CD25) for IL-2. This could be the reason why decreased T-CELL proliferation and anergy is observed after mitogen stimulation during zinc deficiency. Another possible reason could be the fact that zinc is essential for binding of the protein tyrosine kinase p56 (Lck) to T-CELL co-receptors CD4 and CD8 alpha chain, a signalling step necessary for T-LYMPHOCYTE development and activation. Association of p56 (Lck) with CD4 requires two conserved cysteine residues in the cytosolic domain of CD4 and two in the amino terminus of p56(lck), and zinc is essential for this complex formation [19].

It has also been known for more than 25 years, that zinc ions can induce blast transformation in human LYMPHOCYTES. Therefore zinc can be considered the simplest mitogen known. When added to peripheral blood mononuclear (PMN) cells at stimulatory concentrations, it induces the release of IL-1, IL-6, TNF-α and IFN-γ, an effect independent of the presence of LYMPHOCYTES. In contrast to this direct stimulation of MONOCYTES, the stimulatory effect on T CELLS is indirect and strictly dependent on the presence of MONOCYTES in the cell culture.

When zinc supplements are given to individuals with low zinc levels, the numbers of circulating T LYMPHOCYTES increase and the ability of LYMPHOCYTES to fight viral and bacterial infection improves. For instance, malnourished children in India, Africa, South America, and Southeast Asia experience shorter courses of infectious diarrhea after taking zinc supplements [20]. The importance of zinc supplementation during aging has also been recognised.

Cellular pharmacokinetics

Zinc supplements are available in oral and parenteral formulations. Oral formulations available include zinc sulphate, zinc gluconate, zinc picolinate and the newest form of supplementary zinc – zinc monomethionine – while zinc chloride and zinc sulphate are available as injections. Zinc monomethionine is the most bioavailable form of zinc, because the molecule is transported through small intestine epithelial cells using the endogenous transport system for methionine. Foods containing high amounts of phosphorus, calcium, or PHYTATES (found in bran, brown bread) can reduce the amount of zinc absorbed. The same effect has been observed with caffeine-rich beverages and food. Commonly used sweeteners such as sorbitol, mannitol, and citric acid make zinc lozenges ineffective. Drug-zinc interactions have been observed with quinolone and tetracycline antibiotics and penicillamine. It is recommended, therefore, to take zinc 6 h before or 2 h after antibiotics.

Clinical indications

Zinc is an essential trace element used to treat zinc deficiency and delayed wound healing associated with zinc deficiency. It is also used for herpes simplex, Hansen's disease, diabetes, dental plaque, Alzheimer's disease, Wilson's disease, colds, acne and other skin problems, and to stimulate the immune system to fight infection.

Box 1: New technology to study the influence of nutraceuticals on human health – massive screening of gene expression profile

1. Zinc

Recently, cDNA microarray and quantitative polymerase chain reaction (PCR) technologies have been used to investigate the responsiveness of genes known to influence zinc homeostasis and to identify genes related to phenotypic outcomes of altered dietary zinc intake [21]. Microarrays composed of about 22,000 cDNA elements were used to identify genes responsive either to depletion, zinc supplementation, or both. THP-1 cells were either acutely zinc-depleted, using TPEN (N,N,N,N'-teterakis (2-pyridilmethyl)ethylenediamine – a cell-permeable zinc-specific chelator, or were supplemented with zinc sulphate to alter intracellular zinc concentrations. Among 22,000 elements analysed, 104 genes showed a positive linear correlation with zinc, while 86 genes showed a negative correlation responsiveness. Categorisation by function revealed numerous zinc-responsive genes to be needed for host defence, including cytokine receptors and genes associated with amplification of the Th1 immune response.

2. Polyphenols

cDNA microarray has been used to elucidate how the flavonoid EGCG from green tea alters the profile of gene expression in prostate carcinoma LNCaP cells. EGCG induced a subset of genes mediating inhibitory effects on cell growth. These genes mostly belong to the G-protein signalling network. Only protein kinase C-α (PKC-α) was selectively repressed by EGCG, while the expression of six other PKC isoforms ($\beta, \delta, \varepsilon, \mu, \eta$, and ζ) was unaffected [22].

Side-effects

Rare side-effects with large (< 150 mg/day) doses of zinc include chills, sustained ulcers or sores in the mouth or throat, fever, heartburn, indigestion, nausea, sore throat, unusual tiredness or weakness. The symptoms of overdose (>150 mg/day) include chest pain, dizziness, fainting, shortness of breath, vomiting, yellow eyes or skin. Cases of zinc toxicity have been seen in both acute and chronic forms. Intakes of 150 to 450 mg of zinc per day may be accompanied by low copper status, altered iron function, reduced immune function, and reduced levels of high-density lipoproteins [23] (see Box 1).

Dietary antioxidants

Several constituents of the normal diet help to protect against the damaging effects on lipid membranes of reactive oxygen species, such as the SUPEROXIDE ANION, H_2O_2, and the hydroxyl radical, which are formed during a variety of physiological oxidation processes. Vitamin E is a lipid-soluble antioxidant which breaks the chain reaction of lipid peroxidation by scavenging peroxy radicals. The resulting vitamin E radical is transferred to water soluble antioxidants, such as vitamin C (ascorbate), for excretion in the urine. Selenium is an essential dietary trace element which is incorporated into the active site of the enzyme GLUTATHIONE PEROXIDASE (GSH-Px). GSH-Px catalyses the breakdown of hydroperoxides to hydroxy acid (oxidizing GSH to GSSG), thereby complementing the action of vitamin E (Fig. 3). Among these nutrients, vitamins C and E and selenium have been shown to have clear immunostimulant/immunoprotective properties and play a role in disease prophylaxis.

Selenium

History

The element selenium (Se) was discovered in 1818 by the Swedish chemist Berzelius, who named it after Selene, the Greek goddess of the moon. A biological role for Se was first demonstrated in 1957 by Klaus

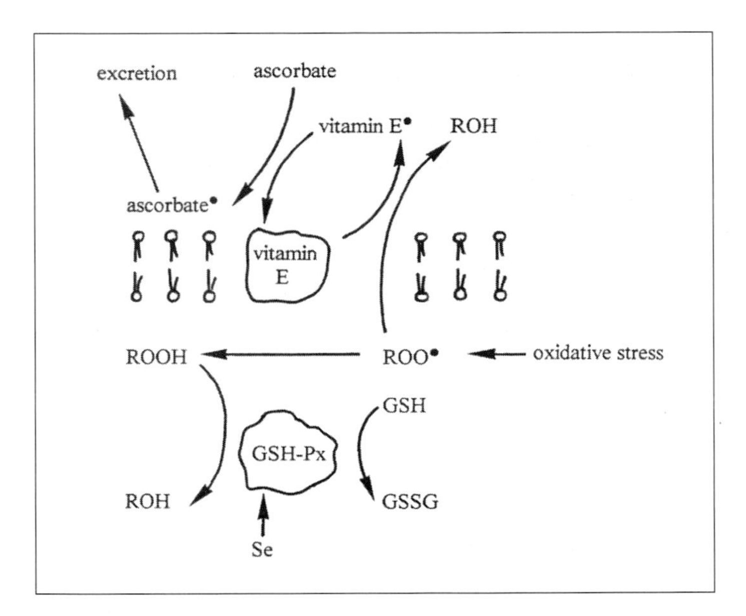

FIGURE 3. INTERPLAY BETWEEN SELENIUM, VITAMIN E AND VITAMIN C IN PROTECTION OF MEMBRANES FROM OXIDATIVE DAMAGE *(ROOH, lipid hydroperoxide; ROO˙, hydroperoxy radical; ascorbate˙, semidehydroascorbate radical; vitamin E, vitamin E radical)*

Schwarz who found that Se protected against dietary liver degeneration in rats. Subsequently, in 1973, Flohé in Germany and Rotruck in the United States showed that Se is present at the active site of the enzyme GSH-Px, incorporated as selenocysteine. Since then 30 new SELENOPROTEINS have been identified, of which 15 have been characterized.

Pharmacology

It is well recognized that dietary Se is important for a healthy immune response. There is also evidence that Se has a protective effect against some forms of cancer, that it may enhance male fertility, decrease cardiovascular disease mortality, and regulate inflammatory mediators in ASTHMA.

Incorporated as selenoprotein, the pharmacological activity of Se is mainly expressed through physiologically important enzymes. GLUTATHIONE PEROXIDASE enzymes (GPX-1, gastrointestinal GPX-2, plasma GPX-3, phospholipid hydroperoxide GPX-4) represent a major class of functionally important SELENOPROTEINS. Thioredoxin reductase (TR) is a recently identified selenocysteine-containing enzyme. It catalyzes the NADPH-dependent reduction of thioredoxin. About 60% of Se in plasma is incorporated into selenoprotein-P which probably serves as a transport protein for Se. However, since it is expressed in many tissues, other functions than just transport are possible. A second major class of SELENOPROTEINS is represented by the iodothyronine deiodinase enzymes which catalyse the 5'5-monodeiodination of the pro-hormone thyroxine (T4) to the active thyroid hormone 3,3'5-triiodothyronine (T3). In addition, sperm capsule selenoprotein, localized in the mid-piece portion of spermatozoa, stabilizes the integrity of the sperm flagella. Selenium intake also affects tissue concentrations of selenoprotein W which is reported to be necessary for muscle metabolism [24].

Diet-induced Se deficiency is associated with a variety of defects in NEUTROPHIL and LYMPHOCYTE functions in experimental and domestic animals that are reversed by Se supplementation [25]. These defects are considered to be due to a reduction in the activity of protective GSH-Px in association with increased production of reactive oxygen species, such as that occurring during the oxidative burst of PHAGOCYTES. As a result, cells in the vicinity of actively phagocytosing cells are damaged. This process also occurs to some extent in Se-adequate animals, in which GSH-Px activity decreases in cells at local sites of INFLAMMATION. In NEUTROPHILS from humans with a low Se status, sodium selenite added *in vitro* is able to enhance the phagocytic and bactericidal activities of the cells [26], probably by protecting them from autolytic damage.

In addition to protecting PHAGOCYTES from damage, inorganic Se administered to animals in nutritional excess has been shown to enhance ANTIBODY titres in response to VACCINES or sensitization to ERYTHROCYTES. Studies on human LYMPHOCYTES *in vitro* suggest that sodium selenite selectively enhances the synthesis of IgG antibodies [27]. However, the effective dose range for Se supplementation above nutritional requirements is relatively narrow, since increasing the dose leads to IMMUNOSUPPRESSION. Enhancement of cytotoxic lymphocyte activity is a consistent response to Se supplementation of animals and humans in nutritional excess, with increased expression of IL-2 receptors on peripheral T CELLS [25]. Administered to patients on hemodialysis, Se supplementation (200–500 µg, 3 times weekly) enhanced T CELL responses to mitogens and delayed hypersensitivity responses [28].

Cellular pharmacokinetics

Cellular uptake of Se, GPx activities, and cytoprotection has been compared in human hepatoma cells (HepG2). Selenite and selenocysteine serve as Se donors with high BIOAVAILABILITY. In contrast, Se from selenomethionine is usually incorporated into cellular proteins but has no effect on GPx activities or cytoprotection. Consequently, not all donor forms of Se provide Se in a bioactive form to act as an antioxidant. Cellular Se content, in general, does not correlate with the cytoprotective activity of this trace element, in contrast to cellular GPx activities, which always correlate, irrespective of the Se donor, with protection against lipid hydroperoxides. Thus, cellular GPx represents a more reliable marker of adequate Se supply [29].

After injection of radiolabelled Se, the metabolic turnover of selenoprotein P (Sel P) in plasma peaks at 6–9 h, whereas that of extracellular GLUTATHIONE PEROXIDASE (eGPx) is sustained for at least 24 h. Selenium is rapidly incorporated into hepatic Sel P in the liver, followed by slow and steady incorporation into renal eGPx [30].

Clinical indications

Sodium selenite or seleno-yeast is widely available as a nutritional supplement, providing 50–100 µg Se/day. This is of benefit immunologically in subjects with inadequate Se intake, including patients on total parenteral nutrition. Although serum Se status is low in various inflammatory skin diseases and RHEUMATOID ARTHRITIS, clear therapeutic benefit of nutritional supplementation with Se has yet to be demonstrated. Like vitamin E, Se supplementation enhances lymphocyte proliferation responses in the elderly [31]. There is growing evidence that prolonged Se intake in nutritional excess is associated with a reduced incidence of a variety of cancers. Because many geographical areas – particularly Finland, parts of China, New Zealand and the UK – have low soil Se content, nutritional supplementation with sodium selenite or Se-enriched yeast is widespread.

Side-effects

Selenium as sodium selenite or selenomethionine is considered to be non-toxic on repeated ingestion up to approximately 1000 µg Se per day. Above this dose, hair and nail loss and skin lesions can arise. At higher intakes, nervous system abnormalities, including numbness, convulsions and paralysis occur.

Vitamin C (ascorbic acid)

History

The great seafaring voyages of the Middle Ages meant that sailors were at sea for many months on very poor food rations. Many suffered exhaustion and depression, bleeding gums, hemorrhaging and bruising with fatal diarrhea, lung and kidney damage – the symptoms of SCURVY. In 1747, the British Physician, J. Lind, found that 2 oranges and 1 lemon a day could relieve the symptoms of SCURVY, but it wasn't until 1795 that the Royal Navy decreed that all sailors should be given regular lime juice. The “scorbutic principle” was only identified after 1928, the year in which Albert Szent-Gyorgi isolated hexuronic acid as the factor that prevented browning of decaying fruit.

The name was changed to vitamin C following structural identification and to ascorbic acid in recognition of its ability to prevent SCURVY. Szent-Gyorgi received the Nobel Prize for Physiology and Medicine in 1937.

Pharmacology

Vitamin C is present at high concentrations in NEUTROPHILS and is required for optimal PHAGOCYTOSIS [32]. During vitamin C deficiency (SCURVY), almost every component of the immune system is compromised [33]. This is mainly due to the fact that vitamin C, being a water-soluble antioxidant, is able to scavenge free radicals in the extracellular compartments, representing the prime antioxidant defense in plasma. Lack of vitamin C opens circulating white blood cells to radical attack with subsequent membrane damage. Vitamin C *in vitro* inhibits activation of NF-κB, the transcription factor for CYTOKINE expression and inhibits T-CELL APOPTOSIS [34,35]. The extent to which vitamin C, at doses above the dietary requirement, is able to further enhance immune responses is still unclear, as many studies have been confounded by the administration of additional antioxidants.

Cellular pharmacokinetics

Human LEUKOCYTES actively take up the oxidized form of ascorbic acid, dehydroascorbic acid, via glucose transporters. This uptake is facilitated by stimulation of the cells. Oxidation of ascorbic acid by superoxide anion, generated by HL-60 NEUTROPHILS undergoing an oxidative burst, leads to enhanced dehydroascorbic acid uptake by all cells in the vicinity and its immediate reconversion to ascorbic acid intracellularly [36]. This provides a feedback mechanism to enhance intracellular levels of protective vitamin C in activated LEUKOCYTES that are generating large amounts of oxygen radicals.

Clinical indications

Supplementation of subjects deficient in vitamin C (e.g., some poorly nourished elderly persons) clearly restores deficient immune responses. However, despite data from a number of clinical trials, there is no unequivocal evidence that mega-doses (<1 g/day) of vitamin C alone are able to stimulate immune responses in healthy individuals.

Side-effects

In allergic persons even small amounts of vitamin C (50 mg) may cause breathing problems, tightness in the throat or chest, chest pain, skin hives, rash, itchy or swollen skin. Taking large amounts (in grams) may cause diarrhea (loose stools).

Vitamin E

History

Vitamin E was discovered in 1922 by H. Evans and K. Bishop as a dietary factor required for normal rat reproduction. It was officially recognized only in 1968. Vitamin E is a generic description for all tocol and tocotrienol derivatives exhibiting the biological activity of α-tocopherol.

Pharmacology

LYMPHOCYTES and mononuclear cells have the highest vitamin E content of any cells in the body. Exposure of these cells to oxidative stress, such as that which occurs during INFLAMMATION or infection, leads to a loss of vitamin E, damage to cell membranes and cellular dysfunction. Addition of vitamin E *in vitro* to LYMPHOCYTES which have been subjected to lipid peroxidation reverses IMMUNOSUPPRESSION, measured in terms of cell proliferation and ANTIBODY formation. This protective action of vitamin E is seen most clearly in experimental vitamin E deficiency in animals. Under these conditions, ANTIBODY titres and ANTIBODY-forming cells are severely depressed, T-CELL responses, including proliferation and IL-2 production, are decreased and mortality to various infections is enhanced [15]. In all cases, supplementation with vitamin E reverses the IMMUNOSUPPRESSION. Prolonged vitamin E supplementation of mice also partially reverses IMMUNOSUPPRESSION caused by retrovirus infection [37].

Cellular pharmacokinetics

Vitamin E is highly lipid-soluble and therefore rapidly absorbed after oral ingestion and incorporated into cell membranes.

Clinical indications

Vitamin E supplements are available for the treatment of deficiency symptoms and for the protection of muscles, blood vessels and the immune system from the effects of oxidation. Vitamin E deficiency in humans is rare, but can arise in preterm infants, in association with impaired NEUTROPHIL phagocytic capacity. Phagocytic activity can be restored by administration of vitamin E to newborn children, including those with glutathione deficiency [38].

The most convincing indication for clinical supplementation with vitamin E to achieve immunostimulation is the aging subject. The activities of antioxidant enzymes decrease with age, leading to a general increase in lipid peroxide tone in the body. In elderly people, nutrition can be sub-optimal and in such subjects supplementation with vitamin E has been shown to increase DTH skin test responses, enhance ANTIBODY responses to vaccines as well as mitogen-induced lymphocyte proliferation and IL-2 production [39, 40]. Whether this enhancement of immune responses by prophylactic vitamin E supplementation leads to an increase in resistance to infectious diseases in humans remains to be demonstrated. Epidemiological studies that suggest a protective effect of vitamin E against cancer cannot be interpreted solely on the basis of possible effects on the immune system, because protection against cell damage in general is also involved in the response to vitamin E. However, recent studies show that vitamin E can inhibit angiogenesis, a crucial requirement for tumor growth, via suppression of INTERLEUKIN 8 and modulation of ADHESION MOLECULES [40].

Side-effects

The evidence is compelling that intake of vitamin E above the recommended daily allowance (RDA) is of benefit to health. Vitamin E, given orally, has a very low toxicity. The RDA in the USA is 10mg. At daily doses up to 3000mg vitamin E is without any overt side-effects. However, there may be an upper limit for immunostimulation, since 300 mg/day of vitamin E depressed bactericidal activity and proliferation of peripheral leucocytes in humans [41].

Phenolic compounds as immunomodulators

Many epidemiological studies have shown that a diet rich in fruits and vegetables can protect against the development of cardiovascular disease [42].

Researchers have examined the composition of these foods, and identified the physiologically active components as phytochemicals. Plant phytochemicals can be divided into plant sterols, FLAVONOIDS, and plant sulphur compounds. FLAVONOIDS are a group of naturally occurring compounds that are widely distributed in nature and are ubiquitous in vegetables, berries, fruits and other foods, and provide much of the flavour and colour to fruits and vegetables. At present more than 500 FLAVONOIDS are known and described, while probably more than 4,000 are present in various plants and extracts. A large number of studies have demonstrated the beneficial effects of flavonoid consumption against the development of cancer and cardiovascular disease. The general opinion, however, is that these compounds can serve to help prevent diseases, but are not effective enough to be used as specific therapies.

History

The FLAVONOIDS were first isolated in the 1930s by Albert Szent-Gyorgyi, who also discovered vitamin C. Szent-Gyorgyi found that FLAVONOIDS strengthened capillary walls in ways in which vitamin C could not and, at first, they were referred to as vitamin P. But the chemical diversity of FLAVONOIDS precludes their classification as a single vitamin.

Chemical constituents

The FLAVONOIDS have been divided into six major groups:

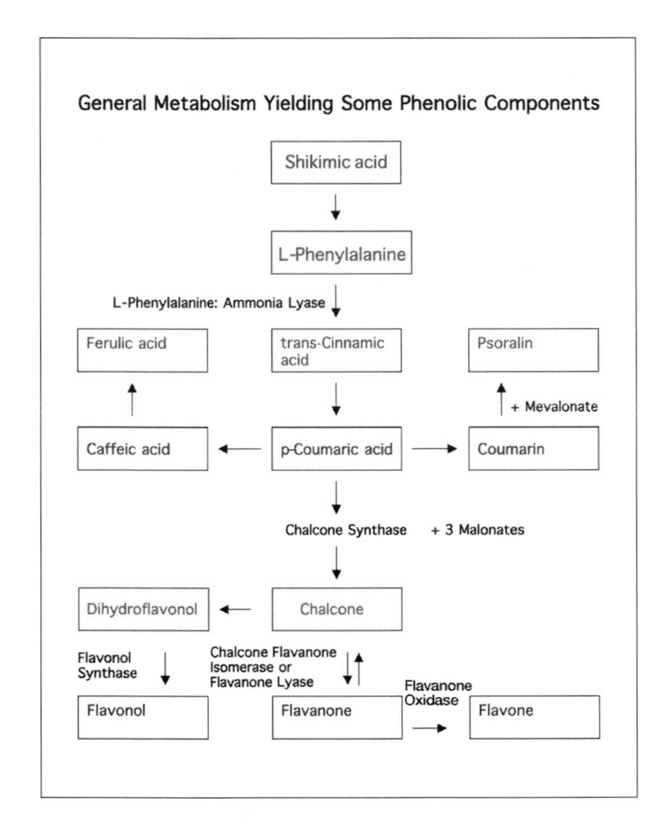

Figure 4. Metabolism of shikimic acid in plants to produce various flavonoids

Figure 5. Tea (*Camellia sinensis*)
Photo can be downloaded from: http://www.illustratedgarden.org/mobot/rarebooks/page.asp?relation=QK99A1K6318831914B2&identifier=0616 (last accessed September 2005)

- Anthocyanidins
- Catechins or flavanols
- Flavones
- Flavonols
- Flavanones
- Isoflavones

The biosynthesis of FLAVONOIDS in plant tissues has been extensively studied in many plants, and several biosynthetic steps have been elucidated. The general metabolism includes shikimic acid, L-phenylalanine and p-coumaric acid. The scheme is given in Figure 4. FLAVONOIDS include flavanones, flavones, chalcones, dihydroflavonols, flavonols, and any glycosides with an aglycone based on a C6-C3-C6 structure. Dihydroflavonols are included in the flavone and flavonol class. Here, for reasons of space, we shall only discuss FLAVONOIDS from tea, as these compounds have been most extensively studied.

Tea (*Camellia sinensis*) (Fig. 5) has been consumed as an infusion for millennia. The first documented use of this particular drink is dated 2700 BC, while the first report of its beneficial effects on human health was written in 1211 by the Japanese monk Eisai [43]. In the 16th century, European explorers used tea extracts to fight fever, headache, stomach ache, and joint pain. Today tea is for the most part simply considered a tasteful drink, but the scientific community has recently re-discovered the therapeutic potential of this beverage.

Tea is known as a rich source of antioxidant polyphenols (catechins, flavones, theaflavins, and thearubigins) and many confer a cardioprotective effect by decreasing LDL oxidative susceptibility, inhibiting LDL lipid peroxidation [44]. Common green and black tea leaves consist of about 25–30 percent FLAVONOIDS, while the primary sources of polyphenols in green tea are cathechins and flavones. Flavones include epicatechin, epicatechin-3-gallate, epigallocatechin, and epigallocatechin-3-gallate (EGCG) [45].

Pharmacology

These compounds from tea have been tested *in vitro*, and EGCG especially shows inhibitory activity on metallo- and serine proteases that are involved in matrix degradation and act as crucial factors in tumor invasion. In animals, green tea significantly increases the activity of antioxidants and detoxifying enzymes, such as glutathione S-transferase, catalase, and quinone reductase, in the lungs, liver, and small intestine [45].

EGCG and other selected polyphenols undergo structural rearrangements under physiological conditions that result in remarkably increased inhibition of telomerase, an enzyme involved in cancer. In nude mice models bearing both telomerase-dependent and -independent xenograft tumors cloned from a single human cancer progeny, only the telomerase-dependent tumors responded to prolonged oral administration of EGCG [46].

IMMUNOMODULATORY effects also have been observed with aqueous extracts of the tea, *Camellia sinensis*, as they slightly enhance neopterin production (a sensitive marker of CELL-MEDIATED IMMUNITY) in unstimulated human peripheral mononuclear cells *in vitro*, whereas a reduction of neopterin formation is seen in cells stimulated with MITOGENS [47].

Several studies have shown that polyphenolic compounds isolated from green tea afford protection against UVB-induced inflammatory responses and photocarcinogenesis in murine models. TOPICAL application of (–)-epigallocatechin-3-gallate (EGCG) before UVB exposure demonstrated that EGCG protects against UVB-induced IMMUNOSUPPRESSION and TOLERANCE induction by: (1) blocking UVB-induced infiltration of $CD11b^+$ cells into the skin; (2) reducing IL-10 production in skin as well as in draining lymph nodes (DLN); (3) markedly increasing IL-12 production in DLN [48]. Protection against UVB-induced IMMUNOSUPPRESSION by EGCG may be associated with protection against UVB-induced photocarcinogenesis.

However, clinical trials of green tea in cancer have not demonstrated clear beneficial effects. It seems that many of green tea's alleged antitumor actions require prolonged exposure to the agent. Inhibition of proteolytic enzymes to prevent metastases, alteration in cell communication, and antiangiogenesis have been touted as explanations for antineoplastic effects of green tea *in vitro* and on animals, but many of these mechanisms lead to tumor regression only after prolonged exposure to an antineoplastic agent [49].

Black tea also contains smaller concentrations of quercetin and thea flavins that inhibit growth of virally-transformed human lung cells and cell proliferation of a colon-cancer cell line [50]. This inhibitory effect appears to be the result of induction of APOPTOSIS and inhibition of the expression of the COX-2 gene.

Cellular pharmacokinetics

No studies on cellular pharmacokinetics of tea constituents have been performed, but EGCG is well absorbed into the plasma after repeated oral administration of EGCG or green tea extract to human volunteers [51].

Clinical indications

Green tea is an herbal medicine used in the adjuvant treatment of cancer, dental plaque, and heart disease. Green tea extracts are recommended for general use at 300 to 400 milligrams daily, which "translated" into the amount of tea as a beverage, means about 3 cups daily. The FLAVONOIDS are an important reminder that the nutritional benefits of wholesome foods go beyond familiar vitamins and minerals. Although it may be convenient to reach for a high-potency flavonoid tablet, the best way to obtain a broad selection of FLAVONOIDS is by eating fruit, vegetables, tea, and soy.

Side-effects

Taking green tea as capsules or beverages can provoke allergic reactions. Other more severe side-effects include: muscle spasms or twitches, nervousness, insomnia (sleeplessness), rapid heart rate, high levels of stomach acid, and heartburn (due to the caffeine content in the tea). Green tea may change the way iron is used in the body, and there are some cases of anemia in children drinking an average of 250 ml of green tea per day.

Emerging therapies and summary

Throughout history people have used plants and natural derived products to cure and prevent diseases. Although the healing properties of plants have been known for a long time, the ability to better exploit the uniqueness of plant therapeutics has been acquired only recently as a result of the dramatic developments in biochemical engineering, molecular genomics, analytical chemistry separation techniques, molecular characterizations and screening for new pharmaceuticals. It seems that the discovery, development and manufacturing of botanical therapeutics, either isolated from plants and different organisms or delivered as food constituents, is likely to be a major area of expansion in plant biotechnology in the twenty-first century.

In addition to the search for the active immunostimulatory agents in extracts of *Echinacea purpurea*, *Viscum album* and *Eleutherococcus senticosus*, a wide variety of plants are under investigation worldwide for immunostimulants, antibacterial and anticancer constituents. Plants are a rich source of novel compounds which can act as prototypes for synthetic substances.

Based upon the role of Se in GSH-Px, a benzisoselenazolone, ebselen, which has GSH-Px-like activity, was until recently under clinical development as an anti-inflammatory agent, specifically for cerebral ischemia; the discovery of this compound has stimulated a search for other anti-inflammatory or IMMUNOMODULATORY agents [52]. Ebselen and a variety of other seleno-organic compounds have been found to be CYTOKINE inducers *in vitro* and *in vivo* and have been proposed as potential anti-viral agents [53].

The immune system is subjected to a wide variety of stress factors in western society. These include overwork, lack of exercise, air pollution and processed foods. Although it is often financially impracticable to perform extensive clinical studies on mild immunostimulants, it is widely agreed that in view of these stress factors, the benefit of dietary antioxidants in nutritional excess is probably greater than has been demonstrable so far. It is likely that with increasing emphasis on self-medication to reduce health budgets, the commercial importance of plant and dietary immunostimulants for the therapy and prophylaxis of mild infectious disorders will increase. As a result, further scientific data on their pharmacological and clinical effects are expected in the future.

Selected readings

Barrett B (2003) Medicinal properties of Echinacea: a critical review. *Phytomedicine* 10: 66–86

Field CJ, Johnson IR, Schley PD (2002) Nutrients and their role in host resistance to infection. *J Leuk Biol* 71: 16–32

Recommended websites

Memorial Sloan-Kettering Cancer Center – About Herbs, Botanicals and Other Products: *www.mskcc.org/aboutherbs* (Accessed November 2004)

Positive Health. Complementary Medicine Magazine: *www.positivehealth.com/permit/Articles/Nutrition* (Accessed November 2004)

References

1 Leake CD (1975) *An histological account of pharmacology to the twentieth century*. C.C. Thomas, Springfield

2 Foster S (1991) *Echinacea. Nature's immune enhancer*. Healing Arts Press, Rochester

3 Bauer R, Wagner H (1990) *Echinacea. Handbuch für Ärzte, Apotheker und andere Naturwissenschaftler.* Wissenschaftliche Verlag, Stuttgart

4 Goel V, Chang C, Slama JV, Barton R, Bauer R, Gahler R, Basu TK (2002) Alkylamides of *Echinacea purpurea*

stimulate alveolar macrophage function in normal rats. *Int Immunopharmacol* 2: 381–387
5 Dietz B, Heilmann J, Bauer R (2001) Absorption of dodeca-2E, 4E, 8Z, 10E/Z-tetraenoic acid isobutylamides after oral application of *Echinacea purpurea* tincture. *Planta Med* 67: 863–864
6 Jager H, Meinel L, Dietz B, Lapke C, Bauer R, Merckle HP, Heilmann J (2002) Transport of alkylamides from Echinacea species through Caco-2 monolayers. *Planta Med* 68: 469–471
7 Parnham MJ (1996) Benefit-risk assessment of the squeezed sap of the purple coneflower (*Echinacea purpurea*) for long-term oral immunostimulation. *Phytomedicine* 3: 95–102
8 Hoheisel O, Sandberg M, Bertram S, Bulitta M, Schaefer M (1997) Echinagard® treatment shortens the course of the common cold: a double-blind, placebo-controlled clinical trial. *Eur J Clin Res* 9: 261–268
9 Holtskog R, Sandvig K, Olsnes S (1988) Characterization of a toxic lectin in Iscador, a mistletoe preparation with alleged cancerostatic properties. *Oncology* 45: 171–179
10 Eck J, Langer M, Mockel B, Witthohn K, Zinke H, Lentzen H (1999) Characterization of recombinant and plant-derived mistletoe lectin and their B-chains. *Eur J Biochem* 265: 788–797
11 Joller PW, Menrad JM, Schwarz T, Pfüller U, Parnham MJ, Weyhenmeyer R, Lentzen H (1996) Stimulation of cytokine production via a special standardized mistletoe preparation in an *in vitro* human skin bioassay. *Arzneim-Forsch/Drug Res* 46: 649–653
12 Beuth J, Ko HL, Gabius H-J, Burrichter H, Oette K, Pulverer (1992) Behavior of lymphocyte subsets and expression of activation markers in response to immunotherapy with galactoside-specific lectin from mistletoe in breast cancer patients. *Clin Invest* 70: 658–661
13 Walzel H, Jonas L, Rosin T, Brock J (1990) Relationship between internalization kinetics and cytotoxicity of mistletoe lectin I to L1210 leukaemia cells. *Folia Biol (Praha)* 36: 181–188
14 Habeck M (2003) Mistletoe compound enters clinical trials. *Drug Discovery Today* 8: 52–53
15 Sandstead HH (1994) Understanding zinc: Recent observations and interpretations. *J Lab Clin Med* 124: 322–327
16 Institute of Medicine. Food and Nutrition Board. (2001) *Dietary Reference Intakes for Vitamin A, Vitamin K, Arsenic, Boron, Chromium, Copper, Iodine, Iron, Manganese, Molybdenum, Nickel, Silicon, Vanadium, and Zinc.* National Academy Press, Washington, DC
17 Shankar AH, Prasad AS (1998) Zinc and immune function: the biological basis of altered resistance to infection. *Am J Clin Nutr* 68: 447S–463S
18 Reinhold D, Ansorge S, Grungreiff K (1999) Immunobiology of zinc and zinc therapy. *Immunol Today* 20: 102–103
19 Lin RS, Rodriguez C, Veillette A, Lodish HF (1998) Zinc is essential for binding of p56(lck) to CD4 and CD8alpha. *J Biol Chem* 273: 32878–3282
20 Black RE (1998) Therapeutic and preventive effects of zinc on serious childhood infectious diseases in developing countries. *Am J Clin Nutr* 68: 476S–479S
21 Cousins RJ, Blanchard RK, Popp MP, Liu L, Cao J, Moore JB, Green CL (2003) A global view of the selectivity of zinc deprivation and excess on genes expressed in human THP-1 mononuclear cells. *Proc Natl Acad Sci USA* 100: 6952–6957
22 Wang SI, Mukhtar H (2002) Gene expression profile in human prostate LNCaP cancer cells by (–)-epigallocatechin-3-gallate. *Cancer Lett* 182: 43–51
23 Hooper PL, Visconti L, Garry PJ, Johnson GE (1980) Zinc lowers high-density lipoprotein-cholesterol levels. *J Am Med Assoc* 244:1960–1961
24 Brown KM, Arthur JR (2001) Selenium, selenoproteins and human health: a review. *Public Health Nutr* 4: 593–599
25 Finch JM, Turner RJ (1996) Effect of selenium and vitamin E on the immune response of domestic animals. *Res Vet Sci* 60: 97–106
26 Urban T, Jarstrand C (1986) Selenium effects on human neutrophilic granulocyte function *in vitro*. *Immunopharmacology* 12: 167–172
27 Reinhold U, Pawelec G, Enczmann J, Werner P (1989) Class-specific effects of selenium on PWM-driven human antibody synthesis *in vitro*. *Biol Trace Element* Res 20: 45–58
28 Bonomini M, Forster S, De Risio F, Rychly J, Nebe B, Manfrini V, Klinkmann H, Albertazzi A (1995) Effects of selenium supplementation on immune parameters in chronic uraemic patients on hemodialysis. *Nephrol Dial Transplant* 10: 1654–1661
29 Leist M, Maurer S, Schultz M, Elsner A, Gawlik D, Brigelius-Flohe R (1999) Cytoprotection against lipid

hydroperoxides correlates with increased glutathione peroxidase activities, but not selenium uptake from different selenocompounds. *Biol Trace Elem Res* 68: 159–174

30 Suzuki KT, Ishiwata K, Ogra Y (1999) Incorporation of selenium into selenoprotein P and extracellular glutathione peroxidase: HPLC-ICPMS data with enriched selenite. *Analyst* 124: 1749–1753

31 Peretz A, Nève J, Desmedt J, Duchateau J, Dramaix M, Famaey JP (1991) Lymphocyte response is enhanced by supplementation of elderly subjects with selenium-enriched yeast. *Am J Clin Nutr* 53: 1323–1328

32 Bendich A (1990) Antioxidant vitamins and their functions in immune responses. *Adv Exp Med Biol* 262: 35–55

33 Basu TK (1996) *Vitamins in Human Health and Disease.* CAB International, Wallingford

34 Schwager J, Schulze J (1998) Modulation of interleukin production by ascorbic acid. *Vet Immunol Immunopathol* 64: 45–57

35 Campbell JD, Cole M, Bunditrutavorn B, Vella AT (1999) Ascorbic acid is a potent inhibitor of various forms of T cell apoptosis. *Cell Immunol* 194: 1–5

36 Nualart FJ, Rivas CI, Montecinos VP, Godoy AS, Guaiquil VH, Golde DW, Vera JC (2003) Recycling of vitamin C by a bystander effect. *J Biol Chem* 278: 10128–10133

37 Wang YJ, Huang DS, Eskelson CD, Watson RR (1994) Long-term dietary vitamin-E retards development of retrovirus-induced disregulation in cytokine production. *Clin Immunol Immunopathol* 72: 70–75

38 Boxer LA, Oliver JM, Spielberg SP, Allen JM, Schulman JD (1979) Protection of granulocytes by vitamin E in glutathione synthesis deficiency. *New Engl J Med* 301: 901–905

39 Meydani SN, Meydani M, Blumberg JB, Leka LS, Siber G, Loszewski R, Thompson C, Pedrosa MC, Diamond RD, Stollar BD (1997) Vitamin E supplementation and *in vivo* immune response in healthy elderly subjects. A randomzed controlled trial. *Am J Med Assoc* 277: 1380–1386

40 Meydani M (2002) Nutrition interventions in aging and age-associated disease. *Proc Nutr Soc* 61: 165–171

41 Prasad JS (1980) Effect of vitamin E supplementation on leukocyte function. *Am J Clin Nutr* 33: 606–608

42 Bazzano LA, Serdula MK, Liu S (2003) Dietary intake of fruits and vegetables and risk of cardiovascular disease. *Curr Atheroscler Rep* 5: 492–499

43 Benelli R, Vene R, Bisacchi D, Garbisa S, Albini A (2002) Anti-invasive effects of green tea polyphenol epigallocatechin-3-gallate (EGCG), a natural inhibitor of metallo and serine proteases. *Biol Chem* 383: 101–105

44 Weisburger JH (1999) Tea and health: the underlying mechanisms. *Proc Soc Exp Biol Med* 220: 271–275

45 Bushman JL (1998) Green tea and cancer in humans: a review in literature. *Nutr Cancer* 31: 151–158

46 Naasani I, Oh-Hashi F, Oh-Hara T, Feng WY, Johnston J, Chan K, Tsuruo T (2003) Blocking telomerase by dietary polyphenols is a major mechanism for limiting the growth of human cancer cells *in vitro* and *in vivo*. *Cancer Res* 63: 824–830

47 Zvetkova E, Wirleitner B, Tram NT, Schennach H, Fuchs D (2001) Aqueous extracts of *Crinum latifolium* (L.) and *Camellia sinensis* show immunomodulatory properties in human peripheral blood mononuclear cells. *Int Immunopharmacol* 1: 2143–2150

48 Lu YP, Lou YR, Xie JG, Peng QY, Liao J, Yang CS, Huang MT, Conney AH (2002) Topical applications of caffeine or (−)-epigallocatechin gallate (EGCG) inhibit carcinogenesis and selectively increase apoptosis in UVB-induced skin tumors in mice. *Proc Natl Acad Sci USA* 99 : 12455–12460

49 Jatoi A, Ellison N, Burch PA, Sloan JA, Dakhil SR, Novotny P, Tan W, Fitch TR, Rowland KM, Young CY et al (2003) A phase II trial of green tea in the treatment of patients with androgen independent metastatic prostate carcinoma. *Cancer* 97: 1442–1446

50 Senior K (2001) Tea: a rich brew of anti-cancer magic bullets? *Drug Discovery Today* 6: 1079–1080

51 Lamy S, Gingras D, Beliveau R (2002) Green tea catechins inhibit vascular endothelial growth factor receptor phosphorylation. *Cancer Res* 62 : 381–385

52 Parnham MJ (1996) The pharmaceutical potential of seleno-organic compounds. *Exp Opin Invest Drugs* 5: 861–870

53 Inglot AD, Mlochowski J, Zielínska-Jenczylik J, Piasecki E, Ledwon TK, Kloc K (1996) Seleno-organic compounds as immunostimulants: an approach to the structure-activity relationship. *Arch Immunol Ther Exp* 44: 67–75

Influence of antibacterial drugs on the immune system

Marie-Thérèse Labro

Introduction

Interference of ANTIBACTERIAL AGENTS with the immune system, and its possible clinical implications, have long been a focus of attention worldwide. Toxic effects with immunological implications (neutropenia, ALLERGY, etc.) influenced the therapeutic choice early in the antimicrobial era, but attention has gradually shifted to beneficial "IMMUNOMODULATORY" properties. Many papers in this field have been published and periodically reviewed over the last three decades [1–9]. Interest in "immunostimulation" peaked in the 1970s. Only recently have the potential benefits of down-regulating immune effectors come into the limelight, with the understanding that immune hyperactivation (in sepsis and inflammatory/AUTOIMMUNE DISEASES, for example) can also have disastrous consequences. Incidental observations that some non-infectious diseases, including inflammatory disorders, are improved by antibacterials have bolstered interest in the IMMUNOMODULATORY activity of antibacterial drugs. A better knowledge of the IMMUNE SYSTEM and its regulation, as well as technological advances, have facilitated such investigations.

Antibacterial agents (ABA) and the immune system

Generalities

There are many possibilities for host-(microbe)-ABA interplay. Two clinically relevant effects of antibacterials are currently recognized, namely toxic and immunotoxic effects, and intracellular BIOACTIVITY. Other possibilities are modulation of bacterial virulence (leading to antibacterial synergy or a proinflammatory effect); ABA activation/inactivation by host cell activities; and modulation of cell functions or products by ABA (resulting in immunostimulation, immunodepression or antiinflammatory activity). These general aspects of ABA IMMUNOMODULATORY effects are reviewed in [4, 6, 9] and summarized here in Table 1 [9–16].

Modulation (stimulation/inhibition) of host cell activities is the most widely investigated and most controversial area. Some ABA can either increase or decrease a given function *in vitro*, depending on the cell type, the technique used, and other variables such as the drug concentration and cell activation status. Effects observed *in vitro* may differ from those occurring *ex vivo*/*in vivo*, which result from the combined activities of the different players: the host and its cellular and humoral (redundant) effectors, the pathogen and its virulence mechanisms, and the physicochemical, pharmacokinetic (concentration, tissue distribution, metabolism) and antibacterial properties of the ABA in question. In vivo studies are the gold standard, but are subject to multiple pitfalls such as species differences in the composition and functions of the IMMUNE SYSTEM [17], as well as ethical problems and interindividual variability. This complexity of potential ABA interference with the IMMUNE SYSTEM helps to explain why the search for new therapeutic applications is so arduous. Nonetheless, major progress has been made in developing IMMUNOMODULATORY ABA, as described below (Part III). A simplified representation of the IMMUNOMODULATORY profile of ABA is given in Figure 1 (modified from [6], [8] and [9]).

Table 1. Possible interferences of ABA with the immune system

Effects	Consequences	Main ABA
ABA effects on host		
1 Toxic/immunotoxic side-effects [9]	Agranulocytosis, allergy, auto-immunity	β-Lactams, chloramphenicol, sulfonamides
2 Cellular uptake [9–14]	IC Bioactivity*, targeted delivery	Macrolides, quinolones, ansamycins
3 Alterations of immune cell functions [1–9]	↑/↓ cell activity (↑ bacterial killing, ↓ inflammation)	(See "Class-specific immuno-modulatory effects", next page)
4 Scavenging/inhibition oxidants/ enzymes [9]	Cell/tissue protection	β-Lactams, cyclines, INH, clofazimine, rifampicin
5 Alteration of non-immune cell functions, gene expression	↑/↓ Immune system activity	Macrolides? cyclines? ansamycins?
6 Disturbances of host microflora	Superinfections/toxin-mediated diseases Defective immune response?	
ABA effects on pathogens		
AB effect (cidal/static)	↓ bacterial load (↓ specific response?)	all ABA (bactericidal ABA?)
morphology**	PMN apoptosis/necrosis/stimulation	some β-Lactams
8 Release of pro-inflammatory mediators [15]	cell stimulation (shock)	β-Lactams, quinolones? amino-glycosides
9 ↓ production of virulence factors [16]	↑ susceptibility (PALE)	all ABA (sub/supra MICs)
10 Alteration of Ag structure	↑/↓ specific response (relapses, chronic inflammation)	?
11 Binding to LPS	↓ inflammation	Polymyxin B
Host effects on ABA		
12 Alteration of ABA (Metabolism, oxidation...)	↑/↓ AB activity, toxicity	Chloramphenicol, INH
Host effects on pathogens		
13 IC sequestration	↓ AB activity	non-cell-penetrating AB
14 Alteration of structure, metabolism	↑/↓ AB activity, synergy	macrolides, quinolones, β-lactams

AB, antibacterial; IC, intracellular; INH, Isoniazid; PALE. post-antibiotic leukocyte enhancement
**, intracellular bioactivity results from many factors, including cellular uptake*
***, (filament, spheroplast)*

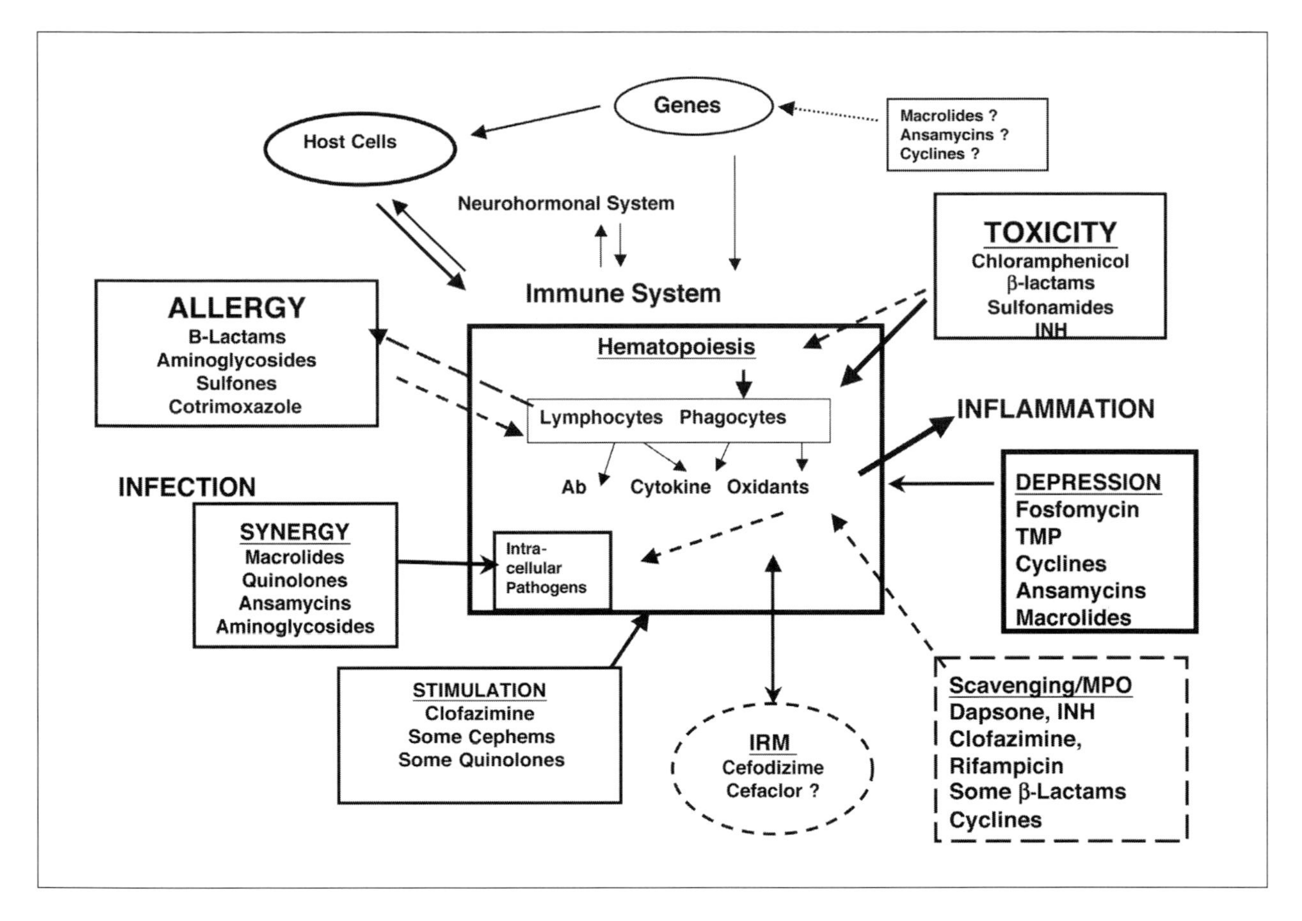

FIGURE 1. SCHEMATIC CLASSIFICATION OF ANTIBACTERIAL AGENTS AS IMMUNOMODULATORS
Ab, antibodies; INH, Isoniazid; IRM, Immune response modifiers; MPO, myeloperoxidase

Class-specific immunomodulatory effects

A very early study suggested that the IMMUNOMODULATORY properties of antimicrobial drugs could be predicted from their mode of action on microbial cells [1]. However, this hypothesis has not been confirmed 20 years later. Even within a given class of drugs, variable effects are obtained, owing to particular chemical structures and physicochemical properties. This chapter will examine the main effects of ABA (presented in alphabetical order) on immune functions. The chemical structure of lead compounds in the most investigated ABA classes (e.g., ansamycins, β-lactams, cyclines, macrolides and quinolones) is given in Figure 2. A synthesis of *in vitro*, *ex vivo*, *in vivo* effects is given in Tables 2–4 for three ABA families which have demonstrated class-dependent effects (e.g., ansamycins, cyclines and erythromycin A-derived macrolides).

Aminoglycosides

Aminoglycosides interfere with bacterial protein synthesis by acting on the 30S ribosomal subunit. Although they are considered as extracellular ANTI-

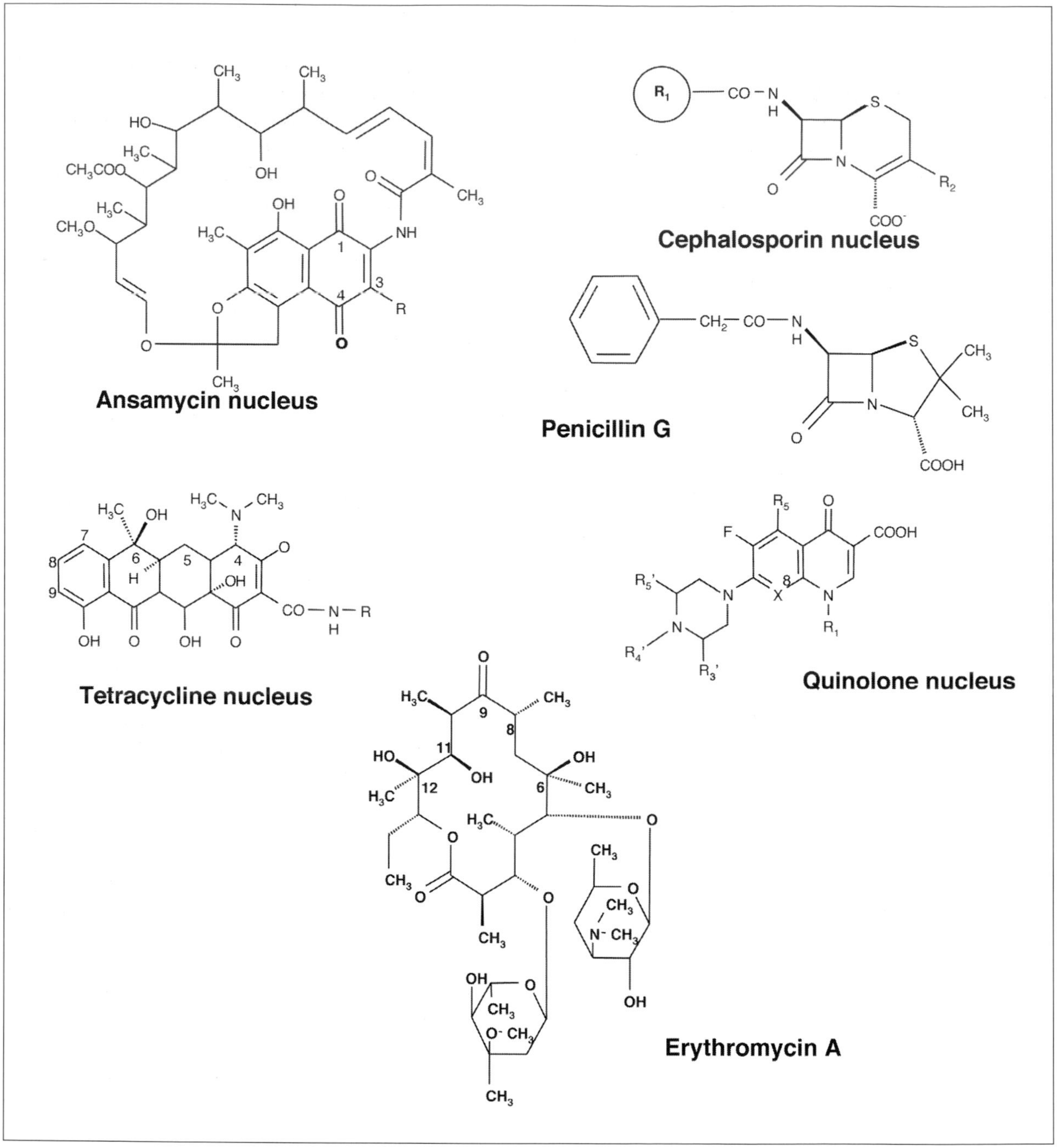

Ansamycin nucleus

Cephalosporin nucleus

Penicillin G

Tetracycline nucleus

Quinolone nucleus

Erythromycin A

FIGURE 2. CHEMICAL STRUCTURES OF IMPORTANT LEAD COMPOUNDS

Ansamycin nucleus, cephalosporin nucleus, penicillin G, tetracycline nucleus (tetracycline, chlortetracycline, oxytetracycline, doxycline, minocycline), quinolone nucleus (7-piperazinyl derivatives, e.g., ciprofloxacin, pefloxacin, grepafloxacin, sparfloxacin, fleroxacin, lomefloxacin, gatifloxacin, enoxacin, etc.) and erythromycin A.

BACTERIAL AGENTS, they accumulate slowly in host cells (over days in MACROPHAGES), by fluid-phase pinocytosis and remain a "gold standard" in several intracellular infections [13]. Streptomycin BIOACTIVITY on intracellular *Escherichia coli* is thought to rely on stimulation of (synergy with?) macrophagic oxygen-dependent bactericidal mechanisms. The use of gentamicin as an extracellular bactericidal agent in many cell systems has recently been questioned [18]. Conflicting data have been published on the *in-vitro* inhibitory effect of aminoglycosides (at therapeutic concentrations) on POLYMORPHONUCLEAR NEUTROPHIL (PMN) chemotaxis, oxidative metabolism and yeast killing. The underlying mechanisms include binding to negatively charged membrane phospholipids (leading to membrane disturbances), specific binding to inositol biphosphate (resulting in phospholipase C inhibition), and protein kinase C (PKC) inhibition. Interestingly, amikacin (contrary to other aminoglycosides) enhances the PMN oxidative burst at low concentrations *in vitro*, whereas concentrations higher than 1 g/l inhibit this phenomenon, which is likely to be as a result of oxidant scavenging [19]. Gentamicin, but not amikacin, significantly decreases lymphokine-activated killer cell activity *in vitro*, at clinically relevant concentrations. No clinical relevance of these observations has been demonstrated. Cystic fibrosis (CF), a systemic autosomal recessive inherited disorder, results from mutations in the cystic fibrosis transmembrane conductance regulator (CFTR) gene. In Class I defects the CFTR mutation results in a premature stop codon which interferes with translation of CFTR. Distinct aminoglycosides (specifically gentamicin but not tobramycin) have been shown to bind to the rRNA, causing misreading of the abnormal stop codon and insertion of alternate amino acids that then restore full-length translation. In cell lines, gentamicin treatment restores full-length CFTR to 10–20% of normal levels. Two groups have conducted Phase I trials of intravenous and inhaled gentamicin in CF patients with class I mutations. Preliminary results have shown an increase in CFTR function, as measured by nasal potential difference after short exposures. Long-term effectiveness and safety are under investigation [20].

Ansamycins (Tab. 2)

Antibacterial ansamycins are a group of macrocyclic antibiotics containing a chromophoric naphthohydroquinone system spanned by a long aliphatic bridge (Fig. 2). They are mainly effective against mycobacteria, altering RNA biosynthesis by interfering with RNA polymerase activity.

In vitro effects

Uptake of rifampicin, the most important representative of this group, is only moderate, but rifapentine, a new cyclopentyl rifamycin derivative, is avidly concentrated (62- to 88-fold) by PHAGOCYTES, possibly through membrane binding. Ansamycins are active against intracellular pathogens in both normal and deficient PHAGOCYTES. Ansamycins exert depressive effects on phagocyte and lymphocyte activities. Ansamycins impair various PMN functions such as chemotaxis (through competition with PMN receptors for small chemoattractant peptides) and the oxidative burst (although light-absorbing activity and superoxide anion scavenging can interfere with the detection method). The most active new compounds are derivatives carrying an acidic substituent at C3 [21, 22]. PMN from patients with rheumatoid arthritis (RA) are more susceptible to the depressive effect of rifamycin SV than are cells from healthy subjects. Rifampicin decreases phytohaemagglutinin (PHA)- and concanavalin (ConA)-induced proliferation of peripheral blood cells from healthy volunteers, and strongly inhibits chemiluminescence and leukocyte-induced angiogenesis [23]. Suppression of IFN-γ production by mouse T LYMPHOCYTES cultured for 4 days with *Listeria monocytogenes* has been observed with rifampicin [24]. Rifamycin B, rifapentine, rifamycin SV, rifabutin and rifampicin inhibit both TNF-α and PMA-induced NF-κB activation in Jurkat T CELLS [25]. Recently, two new ansamycin derivatives that induce a heat-shock response (HSR) *in vitro* and inhibit tyrosine kinase activity also increase mRNA levels of the inhibitory protein Iκ-Bα, suggesting that inhibition of NF-κB activation could help explain their immunosuppressive effects [26]. Among these derivatives, geldanamycin, a benzoquinoid ansamycin, induces an HSR in rat astrocytes and glioma cells and concen-

TABLE 2. EFFECT OF ANSAMYCINS ON THE IMMUNE SYSTEM

In vitro	*Ex vivo / In vivo*
1 *Cellular uptake*	
C/E: 2–14 (rifampicin)	Antibacterial activity
61–88 (rifapentine)	(*Mycobacteria*, intracellular pathogens)
2 Phagocyte functions	Antiinflammatory?
(↓ oxidants, chemotaxis)	↓ oxidant? (meningitis model)
↓ leukocyte-induced angiogenesis	
↓ NO	
3 *Scavenging* $O_2^{-\cdot}$	
4 *Immunodepression*	
↓ PHA/ConA-induced proliferation	↓ cellular immunity
↓ IFN-γ production	↑ Ab against SRBC
5 ↑CD1b expression	↑ mycobacterial Ag presentation ?
(macrophage membrane)	
6 *Others*	
↓ iNOS mRNA (geldanamycin)	↓ disease onset (EAE model), adjuvant-induced arthritis
↓ collagenase activity (rifamycins)	antiinflammatory?
↓ cholesterol	atherosclerosis model (CGP43371)
	↑ P-gP expression (human gut)
	↑ MDR1 mRNA (lymphocytes) (resistance of tumour cells?)
MECHANISMS?	*THERAPEUTIC APPLICATIONS*
Antibacterial activity	
Immunomodulatory effect	
ligand for fMLP- R (PMN)	Rheumatoid arthritis (RA)?
	Ankylosing spondylitis, Juvenile pauci/poly RA
↓ NF-κB activation	(rifamycin SV)
activation of glucocorticoid receptor	Crohn's disease (rifabutin + macrolides)
	Future prospects: atherosclerosis,
Geldanamycin	neuro-inflammatory diseases, cancer
↓ Tyr kinase activity	
induces HSR, ↑ Iκ-B mRNA	
	Paradoxical effects: pro-inflammatory side-effects (uveitis, arthritis)
	(high doses or combination with macrolides)

Ab, antibody; C/E, cellular/extracellular concentration ratio; EAE, experimental auto-immune encephalitis; fMLP-R, formyl-methionyl-leucyl-phenylalanine receptor; HSR, heat shock response; MDR, multiple drug resistance; NO, nitric oxide; P-gP, P-glycoprotein; iNOS, inducible NO synthase; PMN, polymorphonuclear neutrophils; RA, rheumatoid arthritis; Tyr, tyrosine tration-dependently reduces nitric oxide synthase mRNA levels. The immunosuppressive potential of rifampicin may also be due to binding and activation of glucocorticoid receptors [27, 28]: this drug inhibits anti-CD95-mediated APOPTOSIS of Jurkat T CELLS and peripheral blood LYMPHOCYTES, at least partly via glucocorticoid receptor activation and the NF-κB signaling pathway. The effects of rifampicin have also been studied with respect to ANTIGEN presentation by MACROPHAGES and double-negative (DN) T CELL cytolytic function [29, 30]: a non-classical MAJOR HISTOCOMPATIBILITY COMPLEX-independent system (CD1-restricted T-cell responses) is involved in T-cell immunity against nonpeptide ANTIGENS, including several known constituents of mycobacterial cell walls. Among the four known human CD1 isoforms, CD1b protein is the best characterized with regard to its ANTIGEN-presenting function. Non-classical ANTIGEN presentation by CD1 molecules expressed on cytokine-activated MONOCYTES, and cell-mediated responses supported by DN $CD8^+$ responder αβ T CELLS, are involved in host resistance against mycobacterial infection. Clinically achievable concentrations of rifampicin produced a marked increase in CD1b expression on the plasma membrane, resulting from increased CD1b gene expression. In addition, treatment of effector cells with rifampicin did not reduce DN T CELL-mediated cytolysis of lymphoblastoid cells transfected with CD1b cDNA and pulsed with *Mycobacterium tuberculosis*. These results suggest that rifampicin could be useful for improving mycobacterial ANTIGEN presentation, without impairing responder T CELL function. Other potentially IMMUNOMODULATORY properties of ansamycins have been described. Matrix metalloproteinases (MMPs) play an important role in articular cartilage breakdown in several diseases, including osteoarthritis and RA. Rifamycins inhibit collagenase type XI with an $IC_{50\%}$ ranging from 13 to 20.7 μM [31].

In vivo, ex vivo effects

Recently, in a rabbit model of *Streptococcus pneumoniae* meningitis, it was shown that LEUKOCYTES (NEUTROPHILS and MONOCYTES) from rifampicin-treated rabbits produced smaller amounts of reactive oxygen species than LEUKOCYTES from ceftriaxone-treated animals [32]; although rifampicin triggers less proinflammatory mediator release from bacteria than do β-lactams, a direct effect on LEUKOCYTES present in cerebrospinal fluid cannot be excluded. *In vivo*, rifampicin slightly stimulates production of antibodies against sheep red blood cells (SRBC) and suppresses cellular responses in B6AF1 mice [23]. *Ex vivo* modulation of CYTOKINE production has rarely been investigated. In a model of experimental chronic osteomyelitis due to *Staphylococcus aureus*, the combination of azithromycin and rifampicin induced a marked reduction in bacterial counts in bone, but levels of TNF-α (both activity and mRNA) remained elevated throughout the observation period [33]. A single injection of geldanamycin reduces disease onset in mice actively immunized to develop experimental auto-immune encephalitis (EAE) [26]. This drug also confers protection against INFLAMMATION-associated acute lung injury and suppresses the progression of adjuvant-induced arthritis. Ansamycins are hypolipidemic compounds which, when administered to various animal species, markedly lower high-density lipoprotein (HDL) cholesterol levels. CGP 43371, a new rifamycin derivative, has been tested for its hypolipidemic and anti-atherosclerotic activity in cholesterol-fed rabbits [34]. Compared with control rabbits, CGP 43371 treatment lowered total cholesterol levels and lipoprotein cholesterol levels and inhibited aortic atherosclerosis. CGP 43371-treated rabbits developed striking splenomegaly, with massive accumulation of macrophagic foam cells in the splenic red pulp. The authors speculated that CGP 43371 inhibited the development of atherosclerotic lesions in rabbits by both a hypolipidemic mechanism and by accumulation of macrophage foam cells in the spleen. Lastly, despite significant interindividual variability, a significant increase in the expression of multidrug resistance-1 (MDR1) mRNA in human LYMPHOCYTES has been observed following rifampicin administration [35]. P-glycoprotein (P-gp) is also inducible by rifampicin in the human gut, as shown by intestinal biopsy. The therapeutic antiinflammatory effect of various ansamycins has received attention (see below, "Non-antibacterial effects of antibacterial agents. Therapeutic implications").

Benzylpyrimidines. Trimethoprim and analogues

Benzylpyrimidines [trimethoprim (TMP), tetroxoprim, epiroprim and brodimoprim] inhibit dihydrofolate reductase. TMP is generally used in combination with another antifolate (sulfamethoxazole). TMP accumulates 6- to 20-fold in host cells. In most studies, TMP, alone or in combination, has an inhibitory effect on PMN functions. Interestingly, TMP-SMX (sulfamethoxazole) increases nitric oxide (NO) production by PMN from patients with chronic granulomatous disease [36]. Brodimoprim, in which the methoxy group in position 4 of the TMP benzyl ring is replaced by a bromine atom, displays greater lipophilicity and cellular uptake (74-fold) than TMP, and no inhibitory effects on PMN functions. At high (therapeutically irrelevant) concentrations, TMP impairs the activity of phosphatidate phosphohydrolase, an enzyme involved in the transduction pathway of NADPH oxidase activation.

Betalactams

β-Lactam antibiotics comprise five groups of compounds, namely penams (penicillins (Fig. 2) and β-lactamase inhibitors), penems (faropenem), carbapenems (imipenem, meropenem), cephems (cephalosporins (Fig. 2), cephamycins, oxa-and carbacephems) and monobactams (aztreonam), all of which inhibit various enzymes (PBP, penicillin-binding protein) involved in peptidoglycan synthesis.

In vitro effects

Cellular uptake of β-lactams is poor, but intracellular activity has been observed in some models. Passive diffusion seems to be the predominant uptake mechanism with most penicillin derivatives, at least in their ionized forms (weak organic acids). A proton-dependent dipeptide transport carrier has been described for some cephalosporins, and a probenecid/gemfibrozil-inhibitable organic anion transporter may be involved in penicillin G efflux. Alterations of cefazolin targets by PMN have been proposed to explain the decrease in the antibacterial activity of this cephalosporin [37]. By contrast, bactericidal synergy between β-lactams and the specific and non-specific IMMUNE SYSTEMS has been reported [38]. An indirect impact of β-lactams on PMN death after *E. coli* killing has been observed: β-lactams which induce filament formation promoted PMN necrosis, while imipenem, which induces spheroplast formation, promoted PMN APOPTOSIS [39]. Many data are available on the direct *in vitro* effects of these drugs on immune cells and effectors, but no class or subgroup effect has been demonstrated [40]. In contrast, particular behaviors have been linked to certain chemical features. Some β-lactams (e.g., cefdinir, aminothiazolyl cephalosporins, penicillin, ampicillin) can decrease the PMN oxidative burst by scavenging oxidants or inhibiting myeloperoxidase (MPO). A decrease in bacterial killing has been reported with carbenicillin and cephalothin, whereas cefotaxime has been shown to potentiate it, which is likely to be owing to an enhancement of the oxidative response of PMN stimulated with complement-opsonized particles; the presence of an acetoxy at position 3 of the cephem ring is crucial for this effect. High concentrations of meropenem decrease superoxide anion production by PMN. However, meropenem potentiates phagocytosis and bacterial killing by human MACROPHAGES. Three chemically unrelated β-lactams (cefmetazole, imipenem and cefoxitin) have similar stimulatory effects on various PMN functions (phagocytosis, oxidative burst and ANTIBODY-dependent toxicity) and display chemoattractant activity. These antibiotics also significantly stimulate protein carboxy methylation, increase intracellular cyclic GMP levels, and decrease ascorbate content; cefaclor and cefetamet have been reported to increase phagocytosis and bactericidal activity, and to decrease leukotriene B_4 (LTB_4) production by PMN. Faropenem enhances superoxide anion production by PMN, possibly due to an interference at a site where Ca^{2+} regulates NADPH oxidase activation. Few and often controversial data have been reported on the effect of β-lactams on CYTOKINE release. Recently, the amoxicillin-clavulanic acid combination, which increases the phagocytic and microbicidal activity of PMN, was also shown to elicit the production of IL-1β and IL-8 by LPS- and *Klebsiella*-stimulated PMN. Benzylpenicillin and serum from patients with sickle cell disease have a combined suppressive effect on normal lymphocyte production of IFN-γ and TNF-α, while increasing that of IL-2 [41]. Benzylpenicillin conjugates to IFN-γ and reduces its activity; it also conjugates to IL-1β, -2, -5, -13 and TNF-α, but without altering their biological activity [42]. Inhibition of PMN elastase has been obtained with some cephem compounds. Special mention must be made on the 2-amino-5-thiazolyl cephalosporin cefodizime which has been the subject of worldwide interest in the nineties and was coined as an "immune response modifier" (IRM) antibiotic. Cefodizime stimulates the proliferative response of LYMPHOCYTES, increases the phagocytotic and bactericidal activity of PMN, and down-regulates the production of pro-inflammatory CYTOKINES by stimulated MONOCYTES. Contrary to all other β-lactams, cefodizime significantly increases colony formation by granulocyte-MONOCYTE progenitors. The chemical structure responsible for these IMMUNOMODULATORY properties has been identified as the thio-thiazolyl moiety at position 3 of the cephem ring, but the cellular mechanism remains to be elucidated. Other novel β-lactams have potential as anticancer agents [43].

In vivo/ex vivo effects

Modulation of CYTOKINE production by β-lactams has been observed *ex vivo*: IL-6 concentrations were significantly elevated after ceftazidime administration to both septic rats (independently of LPS levels) and non-septic rats, and TNF-α levels were also increased in non-septic rats, suggesting a direct effect of ceftazidime on CYTOKINE production [44]. Other authors have suggested that the myelosuppressive effect of this drug is mediated by TNF release. However, in a randomized trial comparing imipenem and ceftazidime in the treatment of urosepsis, no differences in plasma endotoxin, IL-6 or TNF levels, or urinary endotoxin, IL-6 or IL-8 levels, were found between the two treatment groups [45]. The IMMUNOMODULATORY activity of cefodizime has been investigated *ex vivo* and *in vivo*, in both humans and animals (healthy and immunocompromised) [46]. Animal studies have demonstrated the benefit of cefodizime in the prevention and/or treatment of various infections caused by resistant pathogens such as *Plasmodium berghei*, *Candida albicans* and *Toxoplasma gondii*, and its better EFFICACY compared to other cephalosporins in immunocompromised animals. *Ex vivo* studies have shown strain- and cefodizime concentration-dependent immune responsiveness with regard to delayed-type hypersensitivity, ANTIBODY production and lymphocyte proliferation. Different results were obtained between healthy subjects and immunocompromised patients: immune parameters in healthy individuals given cefodizime were little if at all affected, whereas in immunocompromised individuals (cancer, hemodialysis, old age, surgical stress, etc.), cefodizime administration restored depressed phagocytic functions. The effects of cefodizime and cefuroxime on NEUTROPHIL phagocytosis and oxidant production have been studied in 54 patients undergoing elective coronary artery bypass grafting. In the cefodizime group a normalization of *E. coli* and *S. aureus* phagocytosis was seen on day 5, whereas in cefuroxime-treated patients phagocytic ability remained depressed. The depressed oxidant production was not modified by the two cephalosporins. Postoperative serum levels of the C-reactive protein on days 2 and 7 were lower in cefodizime-treated patients. In addition to cefodizime's antimicrobial activity during perioperative prophylaxis, its use in coronary artery bypass grafting could prevent procedure-related prolonged postoperative NEUTROPHIL phagocytosis impairment [47]. In another study by the same group [48], phagocyte function recovered significantly earlier in a group of 15 infected patients receiving cefodizime than in a comparable group treated with ceftriaxone, but the only apparent clinical advantage was earlier defervescence in the cefodizime group. Recently, cefaclor has been presented as another candidate IRM [49].

Chloramphenicol

Chloramphenicol impairs bacterial protein biosynthesis by acting on the 50S ribosomal subunit. Owing to its lipid solubility, chloramphenicol accumulates moderately (2- to 9- fold) in host cells and is active intracellularly. Chloramphenicol is a leading cause of neutropenia by cytotoxic mechanisms involving its

Table 3. Effect of cyclines on the immune system

In vitro	*Ex vivo / In vivo*
1 Cellular uptake	
C/E = 2-4	
2 Phagocyte functions	Antiinflammatory?
↓ oxidants, phagocytosis, CT	↓ oxidants (myocardial infarction, acne inflammation)
	↓ inflammatory damages (*H.pylori*-associated gastritis)
↑ IL-1β (LPS-stimulated monocytes)	↓ TNF-α, IL-1β (LPS challenge)
(minocycline > tetracycline)	prevention of endotoxic shock / tetracycline)
↓ NO production (minocycline only)	↓ iNOS, caspase (minocycline: mouse model of Parkinson's disease)
	Animal models (ischemia/reperfusion injury, stroke, Huntington's disease,human amyotrophic lateral sclerosis) (minocycline)
↓ angiogenesis (minocycline, doxycycline)	
3 ↓ IgE production	
(PMBC from asthmatic patients)	
4 Scavenging HOCl	
5 ↓ MMP activity	Animal models of multiple sclerosis
↓ MMP-9 (protein, mRNA)	↓ MMP-9 (protein, mRNA) patients with AAA treated with
(THP-1 monocytes, doxycycline)	doxycycline)
↓ gelatinase, collagenase (PMN, tumour cells)	
↓ T leukemia cell growth	
(apoptosis/↓ MMP-2 and caspase-3 activation)	
6 MDR1 overexpression	↑ Resistance of cancer cells ?
(breast carcinoma cells)	
MECHANISMS?	*THERAPEUTIC APPLICATIONS*
Antibacterial activity	Acne vulgaris and rosacea
(*C. pneumoniae, Mycoplasma Propionibacterium*, others?)	RA (minocycline)
	Reactive arthritis (uroarthritis)
Immunomodulatory effect	Scleroderma (minocycline)
Ca^{2+} chelation	control of calcinosis in systemic sclerosis
	Skin diseases
Photodamage	Periodontal diseases
PKC inhibition)	
	Future prospects
	ischemia/reperfusion injury; endotoxic shock; acute lung injury, abdominal aortic aneurysm, acute myocardial infarction, Parkinson's disease, asthma, cancer, multiple sclerosis? Glomerulonephritis?

AAA, abdominal aortic aneurysm; C/E, cellular/extracellular concentration ratio; CT, chemotaxis; MMP, matrix metalloproteinase; MDR, multiple drug resistance; NO, nitric oxide; iNOS, inducible nitric oxide synthase; PMN, polymorphonuclear neutrophils; PKC, protein kinase C; RA, rheumatoid arthritis nitroso- and dehydro-derivatives. Reports conflict on a potential chloramphenicol-induced reduction in host cell functions. Chloramphenicol decreases lymphocyte functions *in vitro* and impairs ANTIBODY production *in vivo*. Thiamphenicol synergizes with PMN to increase their bactericidal activity.

Cyclines (Tab. 3)

Cyclines (Fig. 2) interfere with bacterial protein synthesis, by acting on the 30S ribosomal subunit.

In vitro effects

Cyclines are passively accumulated (about 2- to 4-fold) by PHAGOCYTES and exert intracellular activity. Many reports confirm their inhibitory action on various phagocyte functions at therapeutic concentrations. The underlying mechanisms include Ca^{2+} chelation, binding of intracellular Mg^{2+}, photodamage of PMN, and scavenging of hypochlorous acid. Structure-activity studies indicate a parallel increase in lipid solubility and inhibitory properties, although others stress the different chemical reactivity of the various molecules to UV light. Few studies have investigated the effect of cyclines on CYTOKINE production: minocycline and, to a lesser extent, tetracycline, increase IL-1β secretion by LPS-stimulated human MONOCYTES. IgE production by PMBC (peripheral mononuclear blood cells) from asthmatic patients was reduced when the cells were co-cultured with IL-4 and CD-40 in the presence of minocycline or doxycycline [50]. A potential new mechanism by which minocycline could exert its beneficial effects in joint diseases was recently described [51]: minocycline (but not doxycycline, tetracycline or meloxicam) led to a concentration-dependent decrease in IL-1-stimulated NO production by chondrocytes, by inhibiting mRNA and protein expression of inducible NO synthase. Doxycycline inhibits the growth of a T lymphoblastic leukemia cell line, by inducing APOPTOSIS via caspase-3 activation and MMP-2 inhibition [52]. Similar findings have been made with chemically modified tetracyclines [53]. Cyclines impair collagenase/gelatinase activity, an effect that appears to be specific for NEUTROPHIL and tumor cell-derived enzymes and suggest interesting possibilities as antitumor agents, particularly for doxycycline. Doxycycline inhibits aortic smooth muscle cell MMP-2 production in part by reducing MMP-2 mRNA stability. As MMP activity contributes to extracellular matrix degradation in atherosclerotic plaque and abdominal aortic aneurysms, doxycycline may have potential value in treating these diseases. Minocycline and doxycycline inhibit angiogenesis *in vitro* by a metalloproteinase-independent mechanism, with implications for the treatment of disorders linked to extracellular matrix degradation, including periodontal disease, arthritis and tumor angiogenesis. A new host-cycline interaction has just been observed *in vitro* [54]: MCF-7 breast carcinoma cells treated with doxycycline overexpressed P-glycoprotein and MDR1 mRNA and accumulated the antineoplastic drug doxorubicin less strongly relative to control cells. The authors suggested that the use of doxycycline in cancer patients could induce P-gP expression by cancer cells, resulting in multidrug resistance.

In vivo/ex vivo effects

Decreased oxidant production by PMN from patients with acute myocardial infarction and treated with doxycycline has been observed *ex vivo* [55]. Patients with acne INFLAMMATION show a significantly increased level of hydrogen peroxide produced by NEUTROPHILS compared to patients with acne comedones and healthy controls. Oral administration of minocycline hydrochloride to these patients significantly decreased the ability of their NEUTROPHILS to produce hydrogen peroxide in accordance with a decrease in the inflammatory activity of acne lesions. In animal models, tetracycline prevents endotoxic shock by inhibiting LPS-induced TNF-α and IL-1β secretion. Recently, Du et al. [56] reported that minocycline had neuroprotective effects in animal models of stroke/ischemic injury and Huntington's disease, and that it prevented nigrostriatal dopaminergic neurodegeneration in the 1-methyl-4-phenyl-1,2,3,6-tetrahydropyridine (MPTP) mouse model of Parkinson's disease. The neuroprotective effect of minocycline was associated with marked reductions in inducible NO synthase (iNOS) and caspase 1 expression. Neuroprotective tetracyclines may prevent or slow the progression of Parkinson's and other neurodegenerative diseases. Because cyclines inhibit several phagocyte functions, their potential benefit has been studied in ischemia-reperfusion injury [57, 58]. A benefit of minocycline in acute ischemic stroke has been suggested in a rat model of transient middle cerebral artery occlusion [59]. The effects of minocycline has been tested in mice expressing a mutant superoxide dismutase linked to human amyotrophic lateral sclerosis (ALS) [60]. Administration of minocycline into the diet, beginning at the late presymptomatic stage, delayed the onset of motor neuron degeneration, muscle strength decline, and it increased the longevity of mice by approximately 5 weeks for approximately 70% of tested mice. Moreover, less activation of microglia was detected at the early symptomatic stage and at the end stage of disease in the spinal cord of mice treated with minocycline. The benefit of minocycline was confirmed in another mouse model for familial ALS [61]: this drug delayed disease onset and dose-dependently extended the survival of mice; minocycline protected mice from loss of motor neurons and from vacuolization. Minocycline, which is clinically well tolerated, may represent a novel and effective drug for treatment of ALS. Multiple sclerosis is characterized by the infiltration of LEUKOCYTES into the central nervous system (CNS). As MMPs facilitate the passage of LEUKOCYTES across matrix barriers, the hypothesis that targeting MMPs could attenuate neuro-INFLAMMATION has been assessed in experimental autoimmune encephalomyelitis (EAE), an animal model of multiple sclerosis [62]. Minocycline inhibited MMP activity, reduced production of MMP-9 and decreased the transmigration of T LYMPHOCYTES across a fibronectin matrix barrier. In addition, minocycline was efficacious against both mild and severe EAE in mice. Minocycline pre-treatment delayed the course of the disease, and when maximal disease activity occurred in vehicle-treated EAE mice, minocycline-treated animals were relatively normal and had minimal signs of INFLAMMATION and demyelination in the CNS. In mice afflicted with mild EAE, minocycline attenuated the clinical severity of disease throughout the course of treatment. Therapeutic treatment with minocycline dramatically suppressed ongoing disease activity and limited disease progression [63]. Proteolytic degradation of the aortic wall by MMPs is considered important in the pathogenesis of abdominal aortic aneurysms (AAA), and tetracyclines, by virtue of their MMP-inhibiting effect, have also been proposed as a novel therapeutic strategy. An *ex-vivo* study of AAA patients receiving doxycycline before surgery showed a 2.5-fold reduction in MMP-9 and a 5.5-fold reduction in MMP-9 mRNA in aneurysm tissue compared to untreated patients; in addition, exposure to doxycycline also reduced MMP-9 protein and mRNA expression in THP-1 MONOCYTES [64]. However, MMPs play a pivotal role in protecting against pulmonary remodeling, and doxycycline-treated rats in chronic hypoxic conditions (10% O2 for 15 days) had higher pulmonary artery pressure and more severe right ventricular failure than controls [65]. Among other therapeutic prospects for cyclines, a recent trial has assessed the ability of tetracycline to decrease the inflammatory process involving *Helicobacter pylori*-associated gastritis which is thought to lead to epithelial damage and contribute to the development of gastric cancer [66]. In a 16-week placebo-controlled clinical trial involving 374 H. pylori-associated gastritis patients, tetracycline-treated patients showed a reduction in INFLAMMATION and epithelial damage independent of a change in H. pylori density, which supports the hypothesis that tetracycline can decrease INFLAMMATION independent of a reduction in the bacterial load. The therapeutic antiinflammatory potential of cyclines is widely acknowledged (see below, "Non-antibacterial effects of antibacterial agents. Therapeutic implications").

Fosfomycin

Fosfomycin (1-cis-1,2-epoxypropylphosphoric acid) is a broad-spectrum bactericidal antibiotic which interferes with bacterial cell wall biosynthesis by inhibiting pyruvate-uridine-diphosphate-*N*-acetylglucosamine transferase. It is passively (2-fold) accumulated by PMN. *In vitro*, fosfomycin increases PMN bactericidal activity, extracellular oxidant production and intracellular Ca^{2+} concentrations [67]. By contrast, other authors have noted an inhibitory effect of fosfomycin on PMA-stimulated oxidant production by PMN, suggesting an effect on PKC-dependent activation pathways [68]. Fosfomycin decreases TNF-α and IL-1 synthesis but increases that of IL-6. It suppresses IL-8 production and mRNA expression by MONOCYTES, by inhibiting PMN functions [69]. The inhibitory effect of fosfomycin on proinflammatory CYTOKINE production seems to be related to inhibition of NF-κB activation [70]. Fosfomycin has IMMUNOMODULATORY activity on B and T lymphocyte functions, increases sensitivity to TGF-β [71] and inhibits histamine release from BASOPHILS. The IMMUNOMODULATORY activity of fosfomycin (and of its enantiomer, which lacks antimicrobial activity) has been demonstrated in various animal models. Fosfomycin and its enantiomer significantly increased the survival rate in a model of gut-derived P. aeruginosa sepsis and reduced serum levels of TNF-α, IL-1 and IL-6. In mice injected with LPS, fosfomycin significantly lowered peak serum levels of TNF-α and IL-1β. In a rat air-pouch model, after carrageenan challenge, fosfomycin decreased the amount of prostaglandin E_2 (PGE_2), TNF-α and mRNA encoding cyclooxygenase-2 [72]. The therapeutic benefit of fosfomycin has been assessed in an animal model of Sjögren's syndrome [73].

Fusidic acid

Fusidic acid, a tetracyclic triterpenoic structure used mainly as an antistaphyloccocal agent interferes with protein biosynthesis. It decreases PMN functions *in vitro*, without markedly altering MONOCYTE functions. Fusidic acid protects mice from LPS- and staphylococcal enterotoxin B-induced lethality, and suppresses TNF-α and IFN-γ release *in vivo*. Prophylactic administration of fusidin significantly increases the survival of neonatal mice challenged with *Salmonella enteritidis* LPS, and also reduces peak values of TNF-α. The potential IMMUNOMODULATORY effect of fusidic acid has also been demonstrated in a model of Con A-induced liver damage: prophylactic administration of fusidic acid protected mice from ConA-induced hepatitis, and this was accompanied by markedly diminished plasma levels of IL-2, IFN-γ and TNF-α, along with increased levels of IL-6. Fusidic acid has proved beneficial in experimental autoimmune neuritis in rats (a model of Guillain-Barré syndrome) and decreased proinflammatory CYTOKINE release [74].

Gyrase B inhibitors

Novobiocin and coumermycin impair bacterial DNA replication by inhibiting gyrase B activity. Few studies have been done with these compounds. At therapeutic concentrations, coumermycin impairs PMN chemotaxis, superoxide anion production and intracellular killing. Novobiocin interferes with metabolic processes in eukaryotic cells and is a potent inhibitor of ADP ribosylation. Novobiocin effectively suppresses the production of proinflammatory CYTOKINES (IL-1, IL-6) and the antiinflammatory CYTOKINE IL-10 by LPS-stimulated MONOCYTES. It also induces CD14 shedding and modulates the expression of other surface ANTIGENS. No immune modulating effects have been demonstrated in animal models.

Isoniazide

Isoniazide, an isonicotinic acid hydrazide, is an antituberculous agent. Its antimycobacterial activity has been attributed to its oxidative metabolism by mycobacterial peroxidases. This chemical reactivity explains its inhibition of the MPO-H_2O_2-halide system and also its potential toxicity after oxidization by activated LEUKOCYTES.

Lincosamides

Lincomycin and clindamycin interact with bacterial protein synthesis at the level of the 50S ribosomal subunit. The nucleoside transport system has been suggested to explain the cellular accumulation of clindamycin (12- to 20-fold). Clindamycin was presented as a possible immunomodulator in infection in the early 1980s. However, controversial effects on PHAGOCYTE functions (enhancement, decrease or no action) have been reported with various techniques and drug concentrations. Interest in this drug was stimulated by its potential prophylactic effect in LPS-induced septic shock, through inhibition of proinflammatory CYTOKINE release *in vitro* and *in vivo* [75, 76]. Interestingly, modulation of CYTOKINE release *in vitro* is not accompanied by a parallel change in mRNA expression [77]. In a canine model of *Babesia gibsoni* infection, clindamycin damages (but does not eliminate) the parasite, stimulating efficient humoral and cellular immune responses and improving clinical outcome [78].

Macrolides (Tab. 4)

Macrolide antibiotics have a 12- to 16-membered macrocyclic lactone nucleus, with few double bonds, substituted by several amino and/or neutral sugars. Modern therapeutic agents, and particularly semi-synthetic derivatives of erythromycin A (Fig. 2) (roxithromycin, clarithromycin, azithromycin), have been obtained by adding new substituents, or by introducing a nitrogen atom into the lactone ring (azalides), or, more recently, by removing the L-cladinose at position 3 of the lactone ring and oxidization into a 3-keto function (ketolides). Macrolides impair bacterial protein synthesis by acting on the 50S bacterial ribosomal subunit. Extensive analyses of macrolides have been done in the context of INFLAMMATION and cancer.

In vitro effects

All macrolide antibiotics are able to concentrate (10- up to 300-fold) within host cells (particularly PHAGOCYTES) [11]. This property contributes to BIOACTIVITY against facultative and obligate intracellular pathogens. Many data point to the existence of an active transport system in PMN and other mature phagocytic cells. There is some evidence of a link between the P-gP family and the macrolide carrier. Cellular accumulation of macrolides is likely to be responsible for their impact on host cell functions [79–83]. Structure-activity studies have shown that only erythromycin A derivatives, including the azalide azithromycin, impair the PHAGOCYTE oxidative burst in a time- and concentration-dependent manner and directly stimulate exocytosis by human NEUTROPHILS. The chemical entity responsible for these effects is the L cladinose at position 3 of the lactone ring, but other structures may also interfere with phagocytic transduction targets [84]. The transduction pathway by which erythromycin A derivatives interfere with NEUTROPHILS seems to be the phospholipase D-phosphatidate phosphohydrolase (PLD-PPH) pathway, which is crucial for the activation of exocytosis and oxidant production. Macrolides (including ketolides [85]) decrease proinflammatory CYTOKINE production by stimulated PHAGOCYTES, while increasing that of the antiinflammatory CYTOKINE IL-10. Individual susceptibility to the IMMUNOMODULATORY activity of macrolides has been shown for both CYTOKINE production and oxidant production. The underlying mechanism (antibiotic uptake, cellular target, etc.) is not known. Roxithromycin has been shown to correct the dysregulated immunological response observed in cells from patients with mite-antigen-sensitive bronchial ASTHMA, suppressing mite-antigen-specific upregulation of IL-4 and IL-5 production, and upregulating the subnormal production of IFN-γ [86]. Modulation of proinflammatory CYTOKINE production has also been observed in eosinophils and non-phagocytic cells, including normal and transformed human bronchial cells, nasal epithelial cells from polyps of patients with chronic sinusitis, and a lung fibroblast cell line. In general, the suppression of CYTOKINE release is accompanied by a parallel decrease in mRNA expression. Depending on the cell type, inhibition of various transcription factors (AP-1, NF-κB and NF-AT) has been described [87-90]. Recently, Abeyama and coll. [91] have provided a link between macrolide-induced impairment of the oxidative response and modulation of CYTOKINE production: roxithromycin, erythromycin A and clarithromycin decreased oxidants produced by LPS-stimulated LEUKOCYTES and THP-1 MONOCYTES, and preferentially inhibited ROS-mediated "pro-inflammatory events" including NF-κB activation and subsequent proinflammatory CYTOKINE production. Other macrolide-induced modifications of mammalian cell functions or metabolism include accelerated NEUTROPHIL APOPTOSIS [92], decreased glycoconjugate secretion by cultured human airway cells, and decreased NO synthase and cyclooxygenase expression in rat MACROPHAGES. This raises the possible existence of

TABLE 4. EFFECT OF ERYTHROMYCIN A-DERIVED MACROLIDES ON THE IMMUNE SYSTEM

In vitro	*Ex vivo / In vivo*
1 Cellular uptake	Tissue accumulation, intracellular bioactivity
C/E ≥ 10 – ≥ 200–300	
(active mechanism, relation with P-gP?)	
2 Phagocyte functions	Antiinflammatory?
(↓ oxidants, CT? ↑ degranulation	Short-term treatment: ↑ phagocyte responses
cytokines: ↓ TNF, IL-1, -8, -6 ; ↑ IL-10)	Long-term treatment : ↓ " "
↓ LTB4, ↓ NO, ↑ apoptosis, ↑ APC activity	
3 ↓ Angiogenesis	Various animal models: ↓ PMN migration,
4 ↓ Bronchial responsiveness	↓ adhesion molecules
↓ Mucus production	Patients: DPB, asthma, sinusitis, CAD
5 ↓ Proliferation (fibroblasts)	(↓ inflammatory parameters: cytokines, PMN infiltration,
	adhesion molecules, NO levels, etc.)
6 Cancer cells: ↓ proliferation?	Animal models: ↓ metastasis, tumor growth, ↑ survival
↓ production:TGF-β, TNF-α,	Lung cancer (humans): ↑ survival
7 ↓ Metalloprotease	
8 Reverse MDR	↑ P-gP (protein and mRNA)
MECHANISMS	*THERAPEUTIC APPLICATIONS*
Antibacterial activity	Asthma?
(*C. pneumoniae*, *Mycoplasma*,	DPB,
H. pylori, *P. aeruginosa*, ?)	CF?
Bacterial virulence	Bronchiectasis, sinusitis, polyposis
Steroid-sparing effect	Psoriasis
Immunomodulatory effect	Crohn's disease
modulation of the PLD-PPH pathway	
(PMN);	
↓ Ca^{2+} influx	
modulation of genes:	Future prospects
induction of protein expression	Atherosclerosis?
(P-gP, others),	Cancer? (lung cancer)
↓ NF-κB (T cells), AP-1 (bronchial cells);	
↓↑ gene expression	

APC, antigen-presenting cells; CAD, coronary artery disease; C/E, cellular/extracellular concentration ratio; CF, cystic fibrosis; CT, chemotaxis; DPB, diffuse panbronchiolitis; LTB4, leukotriene B4; MDR, multiple drug resistance; NO, nitric oxide; P-gP, P-glycoprotein; PLD-PPH, phospholipase D-phosphatidate phosphohydrolase; PMN, polymorphonuclear neutrophils one or more common cellular targets of macrolide action. In addition, erythromycin suppresses the gelatinolytic activity of cell (U937 and spleen macrophage)-derived MMP-9 and down-regulates the expressions of MMP-9 protein and MMP-9 mRNA in a concentration-dependent manner. Macrolides interfere with calcium influx: in particular, erythromycin suppresses ATP-induced Ca^{2+} influx in A549 cells [93]. Modifications of cell functions may also result from modulation of gene expression: for instance, although clarithromycin is not directly cytotoxic for tumor cells, it does inhibit the expression of the TGF-β, TNF-α, and MMP-9 genes. Yamanaka and colleagues have observed that erythromycin, clarithromycin but not josamycin alter gene expression profiles in long-term-cultured small airway epithelial cells [94]. Among the 9 genes whose expression was altered by erythromycin A, out the roughly 20 000 genes examined, two genes (filamin Aα and HSP70/HSP90-organizing protein [HOP]), repressed by erythromycin A, play a role in the IMMUNE SYSTEM. Seven genes were modulated by clarithromycin. A direct inhibitory effect on tumor cell growth has been shown with roxithromycin and the human myeloid leukemia cell line HL-60, and also with a 17-membered-ring azalide and five different human cancer cell lines. Finally, several lines of evidence suggest that macrolides can interfere with the functioning of membrane-associated proteins belonging to the ATP-binding cassette (ABC) transporter proteins, that include P-glycoproteins and multidrug resistance-associated protein, reversing the Multiple Drug Resistance phenotype. This property could be useful to bolster the anticancer activity of classical chemotherapeutic agents.

In vivo/ex vivo effects

Animal models. Various animal models have been used to explore immune mechanisms underlying macrolide actions. Aseptic models (surgical trauma, carrageenin-induced pleurisy, extrinsic allergic alveolitis in mice after *Trichosporon mucoides* challenge, LPS inhalation, etc.) have emphasized the antiinflammatory action of macrolides. In a model of bleomycin-induced acute lung injury with subsequent fibrosis, 14-membered macrolides attenuated the migration of inflammatory cells (especially PMN) to the lung, decreased lung injury and fibrosis, inhibited mRNA expression of the cell ADHESION MOLECULE VCAM-1, and attenuated the mRNA expression of ICAM-1 but not that of E- and P-selectins [95]. In a model of *S. aureus*-induced osteomyelitis, roxithromycin suppressed local expression of IL-1β and TNF-α, without modifying bacterial counts [96]. A prolongation of survival in animals with experimental sepsis caused by multidrug-resistant *P. aeruginosa* was achieved after coadministration of clarithromycin and amikacin and the increased survival was probably attributable to the IMMUNOMODULATORY properties of clarithromycin [97]. Erythromycin has also been shown to enhance Th1 responses to plasmid DNA in immunized mice, through an increase in the accessory cell activity of ANTIGEN-presenting cells. Administration of roxithromycin for 4 weeks suppresses the expression of the costimulatory molecules CD 40 and CD 86 in splenic B LYMPHOCYTES; CD 80 expression was also inhibited after 8 weeks of treatment [98]. The possible benefits of macrolides in cancer have been studied in animals. Erythromycin A derivatives increase the survival of tumor-bearing mice (in both allogeneic and syngeneic systems), retard tumor growth, and inhibit metastasis. In a rat model of transplanted mammary adenocarcinoma, clarithromycin displays a synergistic effect with cyclophosphamide or surgery to reduce mortality. Spleen cells from clarithromycin-treated tumor-bearing rats exhibit greater tumor-neutralizing activity and lower IL-6 and TGF-β gene expression than do cells from control rats. Roxithromycin and clarithromycin potentiate the antitumoral effect of cyclophosphamide, adriamycin and vindesine *in vivo*, without altering their direct cytotoxicity *in vitro*. Finally, the inhibitory effect of roxithromycin and clarithromycin on tumor growth and lung metastasis, and their antiangiogenic activity, has been demonstrated in an animal model of tumor-induced angiogenesis, whereas azithromycin and josamycin were not effective in these systems. Interestingly, clarithromycin does not alter metastatic development of two different human non-small-cell lung cancers in severe combined immunodeficient (SCID) mice. Induction of various proteins has been noted in some models: for instance, in old monkeys erythromycin induced the expression of P-gP and PgP mRNA in the liver, while in rats, clarithromycin increased the serum level of α1-acid glycoprotein (AGP) and the steady-state level of AGP mRNA, both in the liver and in primary cultured hepatocytes, by a glucocorticoid-mediated mechanism.

Human studies. In general, short-term macrolide treatment enhances immune responses, but long-term exposure results in gradual inhibition [99]. When volunteers received a 30-min infusion of erythromycin, the production of two C-X-C CHEMOKINES, IL-8 and ENA-78, by whole blood cells stimulated *ex vivo* with heat-killed *S. pneumoniae*, was decreased 0.5 to 1 h post-infusion, whereas the release of various granular components (the cationic protein BPI, NEUTROPHIL elastase, and lactoferrin) was increased at 4 h [100]. Various erythromycin A derivatives have proven beneficial in inflammatory diseases such as diffuse panbronchiolitis (DPB) and cystic fibrosis (see below, "Non-antibacterial effects of antibacterial agents. Therapeutic implications"), and modulation of various inflammatory parameters has been observed in patients receiving macrolide therapy. In DPB patients, NEUTROPHIL infiltration and IL-8, LTB4 and elastase levels in bronchoalveolar lavage (BAL) fluid fell in parallel to the the clinical improvement during erythromycin A therapy. A significant decrease in human β-DEFENSIN (HBD)-2 levels in BAL fluid [101] and an increase in Cu-Zn superoxide dismutase activity in alveolar MACROPHAGES from DPB patients have also been observed [102]. In ASTHMA, inhibition of eosinophil or NEUTROPHIL activation has been proposed to explain the reduction in airway hyperresponsiveness during therapy, and lymphocyte APOPTOSIS has also been reported [103]. In an open-label pilot study of patients with mild to moderate ASTHMA, clarithromycin showed significant synergy with dexamethasone in suppressing lymphocyte activation *ex vivo* [104]. In patients with sinusitis, clarithromycin and roxithromycin modulated PMN expression of L-selectin and Mac-1, thereby attenuating PMN adhesiveness [105]. In addition, when MACROPHAGES isolated from nasal polyps of macrolide-treated patients were studied immunohistochemically, the percentage of cells positive for CD80 (a costimulatory molecule) was higher than in controls, and this correlated negatively with the number of eosinophils infiltrating the polyps. HLA-DR expression was not significantly modified [106]. Elevated levels of NO in nasal fluid and serum, and increased spontaneous or antigen-stimulated NO production by mononuclear cells from patients with allergic rhinitis, sinusitis and ASTHMA have been observed. Macrolides can decrease NO levels in these patients. In patients with coronary artery disease (CAD), azithromycin therapy is associated with decreased CYTOKINE levels and an overall attenuation of INFLAMMATION [107].

Peptides

Peptide antibiotics comprise polypeptides (tyrocidins, gramicidins and bacitracin), polymyxins, streptogramins, glycopeptides (vancomycin, oritavancin), the lipopeptide daptomycin and the lipoglycopeptide teicoplanin. The mechanisms underlying the antibacterial activity of these drugs differ: polymyxins act by increasing the permeability of the cytoplasmic membrane, while glycopeptides interfere with cell-wall biosynthesis and streptogramins impair bacterial protein biosynthesis by acting on the 50S ribosomal subunit. Cellular uptake of these drugs also varies: daptomycin accumulates little (0.6-fold), teicoplanin (mainly membrane-bound) by about 50-fold (like streptogramins), and oritavancin reaches more than 300-fold [108]. In general, peptide antibiotics do not significantly alter immune functions at therapeutic concentrations. The drug most extensively studied in this respect is polymyxin B, one of the first recognized inhibitors of protein kinase C. Polymyxin B has a stimulatory effect on MONOCYTE function, stimulating the production of IL-1, IL-6, GM-CSF and complement components. The capacity of polymyxin B to bind the lipid A portion of LPS is unfortunately associated with toxicity, ruling out its general use in septic shock. Bacitracin binds Ca^{2+} and Mg^{2+}, a property that has been held responsible for the inhibitory effect of this drug on phagocytosis. Colistin increases the activity of human NEUTROPHIL elastase and P. aeruginosa elastase, two proteases that contribute to the pathogenesis of cystic fibrosis. Vancomycin and teicoplanin depress some PMN functions, but only at very high, clinically irrelevant concentrations. Vancomycin can induce neutropenia and anaphylactoid reactions and promotes histamine release from rat peritoneal mast cells *in vitro* [109]. At a concentration of 50 mg/l, teicoplanin also increases the production of TNF-α, IL-1 and IL-6 by concanavalin A-stimulated human MONOCYTES.

Quinolones

Quinolones are synthetic antibacterial compounds whose first representative (nalidixic acid) was synthesized in 1962. Thousands of compounds have since been made, of which the 6-fluorinated molecules (fluoroquinolones) represent a breakthrough in quinolone research (Fig. 2). The antibacterial activity of quinolones stems from their inhibitory effect on bacterial DNA gyrase (topoisomerase II), and thus on DNA replication. Quinolones might also affect mammalian DNA metabolism, as mammalian cells also contain an essential type II DNA topoisomerase.

In vitro effects

Most quinolones seem to enter and exit from loaded PHAGOCYTES by passive diffusion, although an unidentified active transporter has been proposed for pefloxacin and ciprofloxacin. Some quinolones, such as difloxacin, ciprofloxacin and ofloxacin, may use a P-glycoprotein-like receptor as efflux carrier. Cellular (mainly cytosolic) concentrations are about 4 to 10 times higher than extracellular concentrations, except for grepafloxacin (about 66-fold). Significant intracellular bactericidal synergy with oxidants has been reported with some molecules, and a synergistic effect was recently observed between G-CSF and ofloxacin on PMN bactericidal activity. The underlying mechanisms (G-CSF-induced increase in ofloxacin uptake or in superoxide anion production) were not investigated. At therapeutic concentrations, quinolones affect phagocytosis, adhesion, and oxidant production by rat peritoneal MACROPHAGES and human PMN differently; their effect (increase, decrease, no effect) on oxidant production appears to depend on the animal species and the quinolone structure. Ofloxacin induces an increase in the PMN oxidative response by enhancement of PKC activity, whereas norfloxacin increases oxidant production by mouse MACROPHAGES through enhanced mobilization of NADPH oxidase subunits [110]. A similar transient potentiation of the oxidative burst of rat MACROPHAGES has been reported with ofloxacin, fleroxacin, sparfloxacin and levofloxacin. Lower concentrations were more effective than higher concentrations. All quinolones modestly but significantly impair rat macrophage chemotaxis, in a concentration-dependent manner. The effects of quinolones on CYTOKINE production by MONOCYTES are widely documented (reviewed in [111, 112]). At high concentrations, pefloxacin and ciprofloxacin decrease IL-1 production by LPS-stimulated human MONOCYTES, and ciprofloxacin and ofloxacin (>25 mg/l) decrease TNF-α, production. These depressive effects may be linked to cyclic AMP accumulation. A suppressive effect of therapeutically achievable trovafloxacin concentrations on the synthesis of IL-1α, and β, IL-6, IL-10, GM-CSF and TNF-α by LPS-stimulated human MONOCYTES has also been reported. Similar results have been obtained with moxifloxacin. Grepafloxacin inhibits the production of IL-1α and β, and the expression of IL-1α and β, IL-6 and IL-8 mRNA, suggesting an effect at the gene transcription level. The effects of grepafloxacin [113] and moxifloxacin [114] on the release of CYTOKINES, chemical mediators, hydrolytic enzyme activities, and lipoxygenation in zymogen A- or *S. aureus*-stimulated human THP-1 MONOCYTES have been evaluated: within the first hour, the release of CYTOKINES (TNF-α, IL-1, IL-6, and IL-8), chemical oxidants (nitric oxide and H_2O_2) and hydrolytic enzymes was increased. A second phase between 2–4 h was characterized by the suppression of mediators involved in INFLAMMATION. The third response, an apparent bacteriostatic inhibition of DNA synthesis, caused bacterial death. The quinolones appear initially to activate MONOCYTES to kill bacteria through the innate immune process by releasing oxidants and lysosomal hydrolytic enzymes; at a later time, the bacteria are killed and there is a reversal of the effects of quinolones on CYTOKINE release, free radical generation and hydrolytic enzymes so that lipid peroxidation and tissue destruction by the infection process is suppressed. Several authors have observed that quinolones alter T and B lymphocyte functions and delay or suppress the proliferative response of human mononuclear cells. Most derivatives superinduce IL-2 synthesis and, to a lesser extent, IFN-γ synthesis, by mitogen-activated LYMPHOCYTES; lymphokine mRNA profiles are also upregulated, suggesting a mammalian stress and/or DNA damage response. The activity of transcriptional regulatory factors (NFAT-1 and AP-1) is increased. IL-8 production and E-SELECTIN expression by human endothelial cells are increased by trovafloxacin.

In vivo/ex vivo effects

The potential value of quinolones as antiinflammatory agents was recently proposed, based on the modulation of CYTOKINE responses. Accordingly, an *in vivo* model was developed to determine the protective effect of three quinolones in mice injected with a lethal dose of LPS. Trovafloxacin, ciprofloxacin and tosufloxacin significantly protected mice against death and diminished serum levels of IL-6. A beneficial effect of various quinolones (ciprofloxacin, rufloxacin, difloxacin, trovafloxacin and temafloxacin) has also been observed in a model of *Bacteroides fragilis* (a resistant pathogen)-induced intraabdominal abscesses; the protective effect was related to a decrease in TNF-α production. The effect of oral ofloxacin prophylaxis on endotoxin/CYTOKINE release in aortic aneurysm repair has been evaluated in 25 patients with infrarenal aortic aneurysm [115]: outcome parameters included complications after operation, endotoxin and endotoxin neutralizing capacity (ENC), IL-6, procalcitonin and neopterin. Ofloxacin had no effect on the occurrence of complications or on the peripheral endotoxin levels but ofloxacin-treated patients showed increased ENC and increased IL-6 levels preoperatively and 30 min after clamping. To date, quinolones have not proven beneficial in inflammatory diseases. The combination of metronidazole and ciprofloxacin could be useful for the treatment of patients with active Crohn's disease [116]. Interest in the potential "immunostimulating" properties of some fluoroquinolones is growing. Contrary to pefloxacin, ciprofloxacin and moxifloxacin demonstrate a beneficial effect on HEMATOPOIESIS, as observed in various animal models. Accelerated NEUTROPHIL recovery has also been obtained after ciprofloxacin administration to BONE MARROW-transplanted patients and patients suffering from breast cancer under chemotherapy.

Riminophenazines

In structural terms, riminophenazines are phenazine compounds in which a substituent (R) is included in the "imino" part of the molecule. The first compound developed for clinical use was clofazimine (Lamprene), in mycobacterial diseases. The antimycobacterial mechanism of these drugs has not yet been found. Intracellular (phagocytic) accumulation of riminophenazines is a key factor in their BIOACTIVITY against mycobacteria, which are obligate intracellular pathogens. This intracellular activity is potentiated by PHAGOCYTE treatment with IFN-γ or TNF-α. Clofazimine increases superoxide anion production and degranulation by stimulated NEUTROPHILS, and TNF-α potentiates this enhancement. The mechanism underlying this pro-oxidative effect seems to involve stimulation of phospholipase A_2 (PLA_2) activity, with subsequent accumulation of arachidonic acid and lysophospholipids, which act as second messengers to activate the oxidase. In addition, PLA_2 activation and lysophosphotidylcholine accumulation have been held responsible for inhibition of the membrane Na^+, K^+-ATPase, a key enzyme in various lymphocyte functions. Cyclosporin A potentiates the immunosuppressive activity of clofazimine through a PLA_2-Na^+, K^+-ATPase-dependent mechanism. Other immunosuppressive/antiinflammatory effects of clofazimine could be related to its capacity to scavenge chlorinating oxidants and to stimulate PGE_2 production by NEUTROPHILS.

Sulfones/sulfonamides

Dapsone (4,4' diaminophenyl sulfone) was initially developed as an antitubercular drug; it was tested in leprosy in the early 1950s and is still a part of drug combinations used in this disease. It was later tested in malaria and some inflammatory diseases. Its antibacterial activity is due to inhibition of dihydropteroate synthase. The antiinflammatory activity of dapsone is less well understood. Dapsone inhibits NEUTROPHIL functions such as chemotaxis and oxidant production. It also impairs NEUTROPHIL adherence to antibodies bound to the basement membrane (this is likely to be by direct interference with the antibodies). In addition, it irreversibly inhibits MPO, by converting the enzyme into its inactive (ferryl) form. The hematological toxicity of dapsone is linked to its oxidative metabolism. Dapsone impairs the production of PGE_2 by NEUTROPHILS, possibly explaining dapsone-induced potentiation of cell-mediated immunity.

Sulfonamides also inhibit the dihydropteroate synthase. Synergy with dihydrofolate reductase inhibitors led to COMBINATION THERAPY with agents from the two classes. The most frequently used antibacterial sulfamide is sulfamethoxazole, in combination with trimethoprim (cotrimoxazole). In general, sulfonamides inhibit PHAGOCYTE functions, and many agents in this class have been switched from infections to antiinflammatory indications. The mechanisms underlying these IMMUNOMODULATORY effects are unclear. Inhibition of the elevation of intracellular Ca^{2+} after stimulation has been reported with sulfasalazine and sulfapyridine. Sulfapyridine, but not sulfamethoxazole, has been reported to scavenge HOCl.

Other antibiotics

Some authors have studied the *in vitro* effects of ethambutol, nitrofurans and minimally substituted imidazoles (metronidazole and tinidazole) on immune cell functions, but no significant alterations have been found. Except for cellular uptake, no data are available on the recently introduced oxazolidinone class.

Non-antibacterial effects of antibacterial agents. Therapeutic implications

The therapeutic relevance of the IMMUNOMODULATORY actions of ANTIBACTERIAL AGENTS is controversial, and there is no general agreement on whether these effects must be taken into account when choosing an antibacterial treatment. The clinical benefit of the immunostimulating/restoring effects of ANTIBACTERIAL AGENTS is considered minimal compared to their direct antibacterial activity. By contrast, some antibacterials with immunodepressive potential are showing promise in inflammatory diseases [117].

Immunostimulating antibacterial agents

In the 1990s, cefodizime was presented as a forerunner IRM antibiotic with both classical antibacterial activity and innovative IMMUNOMODULATORY potential. Despite the abundance of published data, the development of cefodizime as an IMMUNOMODULATORY antibiotic has been unsuccessful. No reports are available on the consequences of prophylactic administration of cefodizime in patients at risk of infections.

Immunodepressive antibacterial agents

The use of ANTIBACTERIAL AGENTS in inflammatory diseases has been supported by two hypotheses: either microorganisms can initiate an excessive and/or chronic inflammatory reaction, and ABA downregulate INFLAMMATION by suppressing its bacterial origin, or ABA directly affect the IMMUNE SYSTEM and modulate the inflammatory response (or correct an immunological dysfunction). Various ABA are used in inflammatory diseases (Tab. 5). Three classes have stimulated widespread interest in the context of inflammatory diseases, namely macrolides, cyclines and ansamycins.

Macrolides

Macrolides display IMMUNOMODULATORY properties that may confer beneficial effects on patients with respiratory diseases associated with chronic INFLAMMATION [117–127]: they attenuate inflammatory responses in the lung, regulate mucus production, and decrease bronchial responsiveness. They also

Table 5. Therapeutic indications of ABA in the context of inflammatory diseases

ABA	Therapeutic uses
Dapsone	Neutrophilic dermatoses, dermatitis herpetiformis, leukocytoclastic vasculitis, bullous lupus erythematosus, pustular psoriasis, erythema elevatum diutinum Crohn's disease
Clofazimine	Vitiligo, discoid lupus erythematosus, pyoderma gangrenosum, pustular psoriasis
Sulfonamides	Wegener's granulomatosis
Cyclines	Acne vulgaris, rosacea (doxycycline) [139]
	Skin diseases: pemphigus vulgaris, foliaceous and bullous pemphigoid
	Periodontal diseases
	Reactive arthritis? (uroarthritis)
	Rheumatoid arthritis (minocycline) [140, 141]
	Scleroderma (minocycline) [142]
	Control of calcinosis in systemic sclerosis (minocycline) [143]
	Abdominal aortic aneurysms (doxycycline) [144]
Ansamycins	Rheumatoid arthritis (rifampicin; controversial)
	Juvenile pauci/polyarticular rheumatoid arthritis, ankylosing spondylitis (intraarticular rifamycin SV) [147]
	Crohn's disease? (+ macrolides [150], rifaximin [151])
Macrolides	Erythromycin A and derivatives (roxithromycin, clarithromycin, azithromycin)
Immunomodulatory effects	Diffuse panbronchiolitis, cystic fibrosis, chronic bronchitis, bronchiectasis, chronic obstructive pulmonary disease, rhinosinusitis, nasal polyposis [119–125]
	Skin diseases: prurigo pigmentosa, confluent and reticulated papillomatosis, psoriasis [128]
	Chronic recurrent multifocal osteomyelitis (azithromycin)[129]
	Cancer: non-small-cell lung cancer (clarithromycin) [136]
	Waldenström's macroglobulinemia
	(Clarithromycin + low dose thalidomide + dexamethasone) [137]
Antibacterial effects?	Hodgkin's sarcoma [138]
	Asthma [126, 127]
	Crohn's disease [135]
	Coronary artery diseases (azithromycin, roxithromycin) [107, 130–133]
	Small abdominal aortic aneurysms [134]
Other antibacterial agents (New prospects)	
Gentamicin	Cystic fibrosis (class 1 mutation) [20]
Some cephalosporins	Cancer? Inflammation?
Fosfomycin	Inflammation?
Fusidic acid	Inflammation?
Quinolones	Neutrophil recovery after chemotherapy?

increase mucociliary CLEARANCE, improve sinusitis symptoms, and decrease nasal secretions and polyp size in patients with sinusitis. In patients with ASTHMA, macrolides reduce airway hyperresponsiveness and improve pulmonary function, and have historically been selected for their "steroid-sparing" effect. Preliminary data from studies of patients with chronic obstructive pulmonary disease (COPD) have shown improvements in symptom scores and respiratory function after macrolide treatment. Macrolides have the potential to improve the outcomes of patients with inflammatory airway diseases: DPB and cystic fibrosis are two main targets for macrolide action. The use of macrolides in other inflammatory diseases is increasing. Macrolides show promise in various skin diseases such as prurigo pigmentosa, confluent and reticulated papillomatosis and, possibly, linear IgA disease and the related chronic immunobullous disease of childhood. Macrolides have been reported to improve the clinical outcome of psoriasis [128]. Long-term treatment with azithromycin has been used successfully in the SAPHO (synovitis, acne, pustulosis, hyperostosis, osteitis) syndrome. Recently, this macrolide yielded a rapid clinical improvement in 7 out of 13 patients with chronic recurrent multifocal osteomyelitis (CRMO), an entity covered by the SAPHO syndrome [129]. The origin of macrolide EFFICACY (antibacterial activity or IRM property) still remains controversial, particularly in the field of coronary artery diseases. Atherosclerosis may result from an excessive immune response to various inflammatory (including microbial) stimuli, leading to vascular endothelial injury and a link between C. pneumoniae infection and atherogenesis has been proposed. Controversial results (benefit or no effects) have been obtained in various trials [107, 130, 131]. The WIZARD trial (with zithromax against atherosclerotic-related disorders, weekly intervention with zithromax for atherosclerosis and its related disorders, the largest trial of antibiotic therapy for coronary artery disease involving 7747 adults enrolled by 271 centers in 9 countries) is adequately powered to detect a clinically significant effect on cardiovascular events [132, 133]. The results demonstrate no significant risk reduction in the likelihood of a primary event with azithromycin vs placebo after a median of 14 months of follow-up. However, early benefits of azithromycin on the primary event and on death or reinfarction were suggested, but these decreased over time. Other trials are ongoing. In another context, roxithromycin proved beneficial, compared to placebo, on the expansion rate of small abdominal aortic aneurysms (which some investigators have linked to *C. pneumoniae* infection) [134]. Whereas the use of antibiotics in CAD was based on the hypothesis of an infection-linked etiology, the possibility of a direct antiinflammatory action of macrolides in such settings has been advocated by several investigators. The antibacterial activity of macrolides has also been suspected to sustain their therapeutic use in ASTHMA [126, 127] and Crohn's disease: an open-label study showed an impressive response to clarithromycin in a group of patients with active Crohn's disease, many of whom had been resistant to other therapies [135]. Large-scale, placebo-controlled clinical trials designed to assess long-term EFFICACY and safety in these diseases are warranted. Other interesting prospects for erythromycin A derivatives concern their potential benefit in cancer. Most studies have involved Japanese patients with unresectable non-small-cell lung cancer: clarithromycin (400 mg/day, as long as the patients could tolerate it) significantly increased the survival and slowed the progression of cancer-associated wasting [136]. The combination of clarithromycin (Biaxin) with low-dose thalidomide and dexamethasone (BLT-D) has been proposed as a salvage regimen in Waldenström's macroglobulinemia [137]. A regression of pulmonary Hodgkin's disease has been observed after prolonged treatment with ciprofloxacin and clarithromycin which lead the authors to hypothesize a bacterial origin for this disease [138].

Cyclines

Acne is a therapeutic target of cyclines. One mechanism by which this drug exerts its effect is by inhibiting the proliferation of *Propionibacterium acnes*. However, the lack of correlation between the drug dose regimen and cutaneous bacterial counts has led to speculation that this drug also interferes with the inflammatory reaction. Acne vulgaris and rosacea present therapeutic challenges due to their chronicity, potential for disfigurement, and psychosocial impact. Although pathophysiologically distinct, both conditions have major inflammatory components. Subantimicrobial dose doxycycline has proved beneficial in a double-blind, placebo-controlled trial in the treatment of moderate facial acne as well as in an-open label study in the treatment of rosacea [139]. Another target of cyclines concerns RA. An infectious etiology such as persistent *Mycoplasma* infections has been proposed to explain the benefit of lengthy courses of tetracyclines in this disease. The EFFICACY of minocycline in RA has been reported in two open trials and in three double-blind controlled studies [140]. In a 2-year, double-blind protocol, the EFFICACY of minocycline has been compared with that of hydroxychloroquine in patients with early seropositive rheumatoid arthritis [141]: patients treated with minocycline achieved a better improvement of disease and received less prednisone than those receiving hydroxychloroquine and they were more likely to have discontinued treatment with prednisone at 2 years. Tetracyclines have been also used in reactive arthritis, i.e., non-purulent INFLAMMATION of a joint following urogenital, gastrointestinal or lower respiratory tract infections, but they seem to be effective in the case of uroarthritis only. Minocycline has also been used in early diffuse scleroderma [142] and may be effective in the control of calcinosis in systemic sclerosis [143]. Inhibition of MMPs, anti-inflammatory effects, and calcium-binding properties may play a role in this setting. An unexpected effect was the darkening of the calcinosis deposits to a blue/black colour. Cyclines are also effective in various skin diseases, including immunobullous disorders (pemphigous vulgaris, foliaceous and bullous pemphigoid), with fewer side-effects than immunosuppressive drugs, although frequent hyperpigmentation has been observed with minocycline. Eradication of *C. pneumoniae* infection and inhibition of elastolytic MMPS by doxycycline have been suggested to reduce the growth rates of small abdominal aortic aneurysms (AAA) [144]. Doxycycline treatment had no clear effect on antichlamydial ANTIBODY titers. Among future prospects for tetracyclines, their broad-spectrum anti-inflammatory properties make them attractive candidates for use in the prevention of acute lung injury [145]. The first report of possible benefits of a metalloproteinase inhibitor (doxycycline) in glomerulonephritis in humans has been published recently: a young man presenting crescentic glomerulonephritis treated with cyclophosphamide and prednisone developed steroid-induced acne that was treated with long-term oral doxycycline therapy. During the period the patient was administered doxycycline, proteinuria decreased by 70% and recurred when doxycycline was stopped, suggesting a favourable effect of this antibiotic [146].

Ansamycins

Some anecdotal reports of RA improving in patients with coexisting tuberculosis treated with rifampicin have suggested a potential application of this drug in this disease. However, various studies in larger groups have failed to confirm the usefulness of rifampicin in the treatment of RA in early stages of disease. Over the past 20 years Caruso and his group have developed a novel therapeutic approach based on the intra-articular administration of rifamycin [147]. Intra-articular rifamycin is effective against active synovitis and can profitably be combined with any basic therapy with slow-acting antirheumatic drugs. In an open study, this treatment modality was evaluated in 22 patients with active ankylosing spondylitis and compared with oral treatment [148]. Clinical improvements observed at the end of the 10-week treatment cycle persisted for 12 months. The therapeutic activity of rifamycin SV administered by the intra-articular route has also been evaluated in children with juvenile rheumatoid arthritis (oligopolyarthritis) [149]. At 24 months, 62% of patients with oligoarthritis and 24% with polyarthritis showed complete remission in all affected joints and recovered movement in all those joints which had shown limitations at baseline. There was also a normalization of inflammatory indexes (ESR, C-reactive protein) and regression of general features of disease. However, rifamycin SV is less useful than triamcinolone acetonide in the local treatment of rheumatoid synovitis. Other targets for ansamycins concern Crohn's disease [150]: treatment with rifabutin and clarithromycin or azithromycin may result in a substantial clinical improvement in this disease and justify the conduct of a randomized controlled trial. Rifaximin, a non-absorbable broad-spectrum antibiotic, may be useful in treatment of ulcerative colitis and pouchitis, since its absorption through inflamed mucosa is negligible; it maintains a TOPICAL action without systemic effects and the lack of resistant bacterial strains may allow prolonged and repeated treatments [151].

Conclusions

Interest in the IMMUNOMODULATORY potential of ANTIBACTERIAL AGENTS is growing, and it is difficult to provide an up-to-date panorama of recent and ongoing studies, including many clinical trials. A number of conclusions can be drawn from the abundant literature. In addition to their antibacterial activity, many ANTIBACTERIAL AGENTS have IMMUNOMODULATORY properties with potential therapeutic importance. Modulation of immune functions is now a major focus of attention, particularly in inflammatory diseases. The etiology of inflammatory disorders involves many cellular, plasmatic and humoral pathways of signalling, culminating in the production of enzymatic and free radical-mediated tissue damage. Multiple redundant pathways of initiation and elusive temporal expression of initiators pose formidable barriers to effective treatment. In addition to their anti-microbial properties, tetracyclines, macrolides, and, to a lesser extent, ansamycins, exhibit inhibitory activity toward several initiators of the inflammatory cascade and mediators of tissue damage. To successfully diminish INFLAMMATION, a drug must inhibit the inflammatory process at more than one step. It is likely that these multi-faceted antiinflammatory (at the same time antibacterial) drugs will have more opportunity to jugulate the pathological process. However, lengthy administration and absence of selectivity of these antimicrobial IMMUNOMODULATORS can lead to the induction of microbial resistance. Intensive research is ongoing to identify IMMUNOMODULATORY antibiotic derivatives devoid of antibacterial activity, most notably with tetracycline and sulfamide derivatives.

Summary

The IMMUNE SYSTEM (IS) is a complex network involving a pleiad of cellular effectors (particularly PHAGOCYTES, T and B LYMPHOCYTES) and humoral mediators with redundant and pleiotropic effects. Although this system plays a critical and beneficial role in host defence, and, more generally, in host homeostasis, its inappropriate or excessive activation can lead to detrimental and disease-generating effects. With a better understanding of the correct functioning of the IS, immunomodulation (i.e., the way to influence the IS by either decreasing a damaging hyperimmune response or potentially increasing an inadequate response) has become a therapeutic challenge. Among the possible agents that can interact with the IS, ANTIBACTERIAL AGENTS (ABA) have received much attention from the early beginning of their therapeutic use. ABA can interfere with the IS in different ways, some of which are widely acknowledged and have clinical significance, such as toxic and immunotoxic effects (neutropenia, ALLERGY, auto-immunity, etc.) and intracellular BIOACTIVITY. Modification of bacterial virulence is also recognized *in vitro*, but its *in vivo* consequences are still doubtfull. Modulation (stimulation/inhibition) of host cell activities is the most widely investigated and most controversial area. A very early study suggested that the IMMUNOMODULATORY properties of antimicrobial drugs could be predicted from their mode of action on microbial cells. However, this hypothesis has not been confirmed 20 years later. Even within a given class of drugs, variable effects are obtained, owing to particular chemical structures and physicochemical properties. This chapter summarizes the literature related to the main characteristics of the IMMUNOMODULATORY effects of ABA and reviews some potential therapeutic applications.

Ansamycins exert depressive effects on PHAGOCYTE and lymphocyte activities. Rifamycin B, rifapentine, rifamycin SV, rifabutin and rifampicin inhibit both TNF-α and PMA-induced NF-κB activation in Jurkat T CELLS. The immunosuppressive potential of rifampicin may also be due to binding and activation of glucocorticoid receptors. Many data are available on the direct *in vitro* effects of β-lactams on immune cells and effectors, but no class or subgroup effect has been demonstrated. In contrast, particular behaviors have been linked to certain chemical features. Special mention must be made of the 2-amino-5-thiazolyl cephalosporin, cefodizime, which has been the subject of worldwide interest in the nineties and was coined as an "immune response modifier" (IRM) antibiotic. Cefodizime stimulates the proliferative response of LYMPHOCYTES, increases phagocytosis and bacterial killing, and down-regulates the production of pro-inflammatory CYTOKINES by stimulated MONOCYTES. Contrary to all other β-lactams, cefodizime significantly increases colony formation by granulocyte-MONOCYTE progenitors. The chemical structure responsible for these IMMUNOMODULATORY properties has been identified as the thio-thiazolyl moiety at position 3 of the cephem ring, but the cellular mechanism remains to be elucidated. The IMMUNOMODULATORY activity of cefodizime has been demonstrated *ex vivo* and *in vivo*, in both humans and animals (healthy and immunocompromised). Cyclines display an inhibitory action on various PHAGOCYTE functions at therapeutic concentrations. The underlying mechanisms include Ca^{2+} chelation, binding of intracellular Mg^{2+}, photodamage of cells, and scavenging of hypochlorous acid. Cyclines impair collagenase/gelatinase activity, an effect that appears to be specific for NEUTROPHIL and tumor cell-derived enzymes and suggest interesting possibilities as anti-tumor agents. Minocycline and doxycycline inhibit angiogenesis *in vitro* by a metalloproteinase-independent mechanism, with implications for the treatment of disorders linked to extracellular matrix degradation, including periodontal disease, arthritis and tumor angiogenesis. In animal models, tetracycline prevents endotoxic shock by inhibiting LPS-induced TNF-α and IL-1β secretion. Minocycline has neuroprotective effects in animal models of stroke/ischemic injury, Huntington's disease, and Parkinson's disease. Minocycline may represent a novel and effective drug for treatment of amyotrophic lateral sclerosis and multiple sclerosis. Tetracyclines, by virtue of their metalloprotease-inhibiting effect, have also been proposed as a novel therapeutic strategy in abdominal aortic aneurysms. Clindamycin was presented as a possible immunomodulator in infection in the early 1980s. However, controversial effects on PHAGOCYTE functions (enhancement, decrease or no action) have been reported with various techniques and drug concentrations. Interest in this drug was stimulated by its potential prophylactic effect in LPS-induced septic shock, through inhibition of proinflammatory CYTOKINE release *in vitro* and *in vivo*. All macrolide antibiotics are able to concentrate within host cells (particularly PHAGOCYTES). This property contributes to their intracellular BIOACTIVITY and is likely to be responsible for their impact on host cell functions. Only erythromycin A derivatives, including the azalide azithromycin, impair the PHAGOCYTE oxidative burst and directly stimulate exocytosis by human NEUTROPHILS. The chemical entity responsible for these effects is the L cladinose at position 3 of the lactone ring, but other structures may also interfere with phagocytic transduction targets. The transduction pathway by which erythromycin A derivatives interfere with NEUTROPHILS seems to be the phospholipase D-phosphatidate phosphohydrolase pathway. Macrolides decrease proinflammatory CYTOKINE production by stimulated PHAGOCYTES, while increasing that of the antiinflammatory CYTOKINE IL-10. Modulation of proinflammatory CYTOKINE production has also been observed in eosinophils and non-phagocytic cells. Depending on the cell type, inhibition of various transcription factors (AP-1, NF-κB and NF-AT) has been described. Modifications of cell functions may also result from modulation of gene expression. Finally, several lines of evidence suggest that macrolides can interfere with the functioning of membrane-associated proteins belonging to the ATP-binding cassette (ABC) transporter proteins, that include P-glycoproteins and multidrug resistance-associated protein. Various animal models have emphasized the anti-inflammatory action of macrolides. The possible benefits of macrolides in cancer have also been studied in animals. In humans, short-term macrolide treatment enhances immune responses, but long-term exposure results in gradual inhibition. At therapeutic concentrations, quinolones affect phagocytosis, adhesion, and oxidant production by PHAGOCYTES differently; their effect on oxidant production appears to depend on the animal species and the quinolone structure. The effects of quinolones on CYTOKINE production by

MONOCYTES are widely documented. The potential value of quinolones as antiinflammatory agents was recently proposed, based on the modulation of CYTOKINE responses. Interest in the potential "immunostimulating" properties of some fluoroquinolones is growing. Contrary to pefloxacin, ciprofloxacin and moxifloxacin demonstrate a beneficial effect on HEMATOPOIESIS, as observed in various animal models. Accelerated NEUTROPHIL recovery has also been obtained after ciprofloxacin administration to BONE MARROW-transplanted patients and patients suffering from breast cancer under chemotherapy.

The therapeutic relevance of the IMMUNOMODULATORY actions of ABA is controversial, and there is no general agreement on whether these effects must be taken into account when choosing an antibacterial treatment. The clinical benefit of the immunostimulating/restoring effects of ANTIBACTERIAL AGENTS is considered minimal compared to their direct antibacterial activity. By contrast, some antibacterials with immunodepressive potential (erythromycin A-derived macrolides, cyclines and ansamycins) are showing promise in inflammatory diseases. Macrolides display IMMUNOMODULATORY properties that may confer beneficial effects on patients with respiratory diseases associated with chronic INFLAMMATION. Diffuse panbronchiolitis and cystic fibrosis are two main targets for macrolide action. The use of macrolides in other inflammatory diseases is increasing. Macrolides show promise in various skin diseases and improve the clinical outcome of psoriasis, of SAPHO (Synovitis, Acne, Pustulosis, Hyperostosis, Osteitis) syndrome and chronic recurrent multifocal osteomyelitis. The origin of macrolide EFFICACY (antibacterial activity or IMMUNOMODULATORY property) still remains controversial, particularly in the field of coronary artery diseases (CAD). The antibacterial or antiinflammatory activity of macrolides has also been questioned in ASTHMA and Crohn's disease. Other interesting prospects for erythromycin A derivatives concern their potential benefit in cancer (unresectable non small-cell lung cancer). Acne is a therapeutic target of the antibacterial and antiinflammatory action of cyclines. The EFFICACY of minocycline in rheumatoid arthritis has been acknowledged and tetracyclines have been also used in reactive arthritis. Minocycline has also been used in early diffuse scleroderma and may be effective in the control of calcinosis in systemic sclerosis. Cyclines are also effective in various skin diseases, including immunobullous disorders. Among future prospects for tetracyclines, their broad spectrum antiinflammatory properties make them attractive candidates for use in the prevention of acute lung injury. Intra-articular rifamycin SV is effective against active synovitis, active ankylosing spondylitis and in children with juvenile rheumatoid arthritis (oligopolyarthritis). Other targets for ansamycins concern Crohn's disease in association with clarithromycin or azithromycin.

Interest in the IMMUNOMODULATORY potential of ABA is growing. However, lengthy administration of these antimicrobial IMMUNOMODULATORS can lead to the induction of microbial resistance. Intensive research is ongoing to identify IMMUNOMODULATORY antibiotic derivatives devoid of antibacterial activity.

Selected readings

(1995) Phagocytosis. *Trends Cell Biol* (special issue) 5: 85–141

Hellewell PG, Williams TJ (eds) (1994) *Immunopharmacology of Neutrophils*. Academic Press, London, San Diego, New York

Leffell MS, Donnenberg AD, Rose NR (eds) (1997) *Handbook of Human Immunology*. CRC Press, Boca Raton, New York

Shearer WT, Li JT (eds) (2003) 5th Primer on allergic and immunologic diseases. *J Allergy Clin Immunol* 111 (2 Suppl): 441–778

Rankin JA (2004) Biological mediators of acute inflammation. *AACN Clin Issues* 15: 3–17

Zychlinsky A, Weinrauch Y, Weiss J (eds) (2003) Microbes and Infection. *Forum in Immunology on Neutrophils*, Vol. 5, Issue 14, 1289–1344

Kresina TF (ed) (1998) *Immune Modulating Agents*. Marcel Dekker, New York, Basel, Hong-Kong

Schönfeld W, Kirst H (eds) (2002) *Macrolide Antibiotics*. Birkhäuser Verlag AG, Basel

Gotfried MH (2004) Macrolides for the Treatment of Chronic Sinusitis, Asthma, and COPD. *Chest* 125 (2 Suppl): 52S–S61

Recommended website

Pub med: *http://www.ncbi.nlm.nih.gov* (Accessed November 2004)

References

1 Thong YH (1982) Immunomodulation by antimicrobial drugs. *Med Hypotheses* 8: 361–370
2 Mandell LA (1982) Effects of antimicrobial and antineoplastic drugs on the phagocytic and microbicidal function of polymorphonuclear leukocyte. *Rev Infect Dis* 4: 683–697
3 Anderson RA (1985) The effect of antibiotics and of drug associations including antibiotics on the immunodefense system. In: M Neumann (ed): *Useful and Harmful Interactions of Antibiotics*. CRC Press, Boca Raton, 185–203
4 Van den Broek PJ (1989) Antimicrobial drugs, microorganisms, and phagocytes. *Rev Infect Dis* 11: 213–245
5 Ritts RE (1990) Antibiotics as biological response modifiers. *J Antimicrob Chemother* 26 (Suppl C): 31–36
6 Labro MT (1993) Immunomodulation by antibacterial agents. Is it clinically relevant? *Drugs* 45: 319–328
7 Barrett JF (1995) The immunomodulatory activities of antibacterials. *Exp Op Invest Drugs* 4: 551–557
8 Labro MT (1996) Immunomodulatory actions of antibacterial agents. *Clin Immunother* 6: 454–464
9 Labro MT (2000) Interference of antibacterial agents with phagocyte functions: immunomodulation or "immuno-fairy" tales. *Clin Microbiol Rev* 13: 615–650
10 Butts JD (1994) Intracellular concentrations of antibacterial agents and related clinical implications. *Clin Pharmacokinet* 27: 63–84
11 Labro MT (2002) Cellular accumulation of macrolide antibiotics. Intracellular bioactivity. In: W Schönfeld, H Kirst (eds): *Macrolide Antibiotics*. Birkhäuser Verlag AG, Basel, 37–52
12 Silverstein SC, Kabbash C (1994) Penetration, retention, intracellular localization, and antimicrobial activity of antibiotics within phagocytes. *Curr Op Hematol* 1: 85–91
13 Maurin M, Raoult D (2001) Use of aminoglycosides in treatment of infections due to intracellular bacteria. *Antimicrob Agents Chemother* 45: 2977–2986
14 Mandell GL, Coleman E (2001) Uptake, transport, and delivery of antimicrobial agents by human polymorphonuclear neutrophils. *Antimicrob Agents Chemother* 45: 1794–1798
15 Prins JM, van Deventer SHJ, Kuijper EL, Speelman P (1994) Clinical relevance of antibiotic-induced endotoxin release. *Antimicrob Agents Chemother* 38: 1211–1218
16 Gemmell CG (1995) Antibiotics and the expression of staphylococcal virulence. *J Antimicrob Chemother* 36: 283–291
17 Haley PJ (2003) Species differences in the structure and function of the immune system. *Toxicology* 3: 49–71
18 Hamrick TS, Diaz AH, Havell EA, Horton JR, Orndorff PE (2003) Influence of extracellular bactericidal agents on bacteria within macrophages. *Infect Immun* 71: 1016–1019
19 Gressier B, Brunet C, Dine T, Luycks M, Ballester L, Cazin M, Cazin JC (1998) *In vitro* activity of aminoglycosides on the respiratory burst response in human polymorphonuclear neutrophils. *Methods Find Exp Clin Pharmacol* 20: 819–824
20 Lim M, Zeitlin PL (2001) Therapeutic strategies to correct malfunction of CFTR. *Paediatr Respir Rev* 2: 159–164
21 Spisani S, Traniello S, Martuccio C, Rizzuti O, Cellai L (1997) Rifamycins inhibit human neutrophil functions: new derivatives with potential antiinflammatory activity. *Inflammation* 21: 391–400
22 Spisani S, Traniello S, Onori AM, Rizzuti O, Martuccio C, Cellai L (1998) 3-(Carboxyalkylthio) rifamycin S and SV derivatives inhibit human neutrophil functions. *Inflammation* 22: 459–469
23 Demkow U, Radomska D, Chorostowska-Wynimko J, Skopinska-Rozewska E (1998) The influence of rifampicin on selected parameters of immunologic response. *Pneumonol Alergol Pol* 66: 45–53
24 Sacha PT, Zaremba ML, Jakoniuk P (1999) The effect of selected antibacterial antibiotics on production of interferon gamma by mouse T lymphocytes stimulated by Listeria monocytogenes. *Med Dosw Mikrobiol* 51: 413–419
25 Pahlevan AA, Wright DJ, Bradley L, Smith C, Foxwell BM (2002) Potential of rifamides to inhibit TNF-induced NF-kappaB activation. *J Antimicrob Chemother* 49: 531–534

26 Murphy P, Sharp A, Shin J, Gavrilyuk V, Dello Russo C, Weinberg G, Sharp FR, Lu A, Heneka MT, Feinstein DL (2002) Suppressive effects of ansamycins on inducible nitric oxide synthase expression and the development of experimental autoimmune encephalomyelitis. *J Neurosci Res* 67: 461–470
27 Yerramasetti R, Gollapudi S, Gupta S (2002) Rifampicin inhibits CD95-mediated apoptosis of Jurkat T cells via glucocorticoid receptors by modifying the expression of molecules regulating apoptosis. *J Clin Immunol* 22: 37–47
28 Gollapudi S, Jaidka S, Gupta S (2003) Molecular basis of rifampicin-induced inhibition of anti CD95-induced apoptosis of peripheral blood T lymphocytes: the role of CD95 ligand and FLIPs. *J Clin Immunol* 23: 11–22
29 Giuliani A, Porcelli SA, Tentori L, Graziani G, Testorelli C, Prete SP, Bussini S, Cappelletti D, Brenner MB, Bonmassar E et al (1998) Effect of rifampicin on CD1b expression and double-negative T cell responses against mycobacteria-derived glycolipid antigen. *Life Sci* 63: 985–994
30 Tentori L, Graziani G, Porcelli SA, Sugita M, Brenner MB, Madaio R, Bonmassar E, Giuliani A, Aquino A (1998) Rifampicin increases cytokine-induced expression of the CD1b molecule in human peripheral blood monocytes. *Antimicrob Agents Chemother* 42: 550–554
31 Barracchini A, Franceschini N, Di Giulio A, Amicosante G, Oratore A, Minisola G, Pantaleoni GC (1999) Metalloproteinase inhibition: therapeutic application in rheumatic diseases. *Clin Ter* 150: 295–299
32 Bottcher T, Gerber J, Wellmer A, Smirnov AV, Fakhrjanali F, Mix E, Pilz J, Zettl UK, Nau R (2000) Rifampin reduces production of reactive oxygen species of cerebrospinal fluid phagocytes and hippocampal neuronal apoptosis in experimental *Streptococcus pneumoniae* meningitis. *J Infect Dis* 181: 2095–2098
33 Littlewood-Evans AJ, Hattenberger M, Zak O, O'Reilly T (1997) Effect of combination therapy of rifampicin and azithromycin on TNF levels during a rat model of chronic osteomyelitis. *J Antimicrob Chemother* 39: 493–498
34 Feldman DL, Sawyer WK, Jeune MR, Mogelesky TC, Von Linden-Reed J, Forney Prescott M (2001) CGP 43371 paradoxically inhibits development of rabbit atherosclerotic lesions while inducing extra-arterial foam cell formation. *Atherosclerosis* 154: 317–328
35 Asghar A, Gorski JC, Haehner-Daniels B, Hall SD (2002) Induction of multidrug resistance-1 and cytochrome P450 mRNAs in human mononuclear cells by rifampicin. *Drug Metab Dispos* 30: 20–26
36 Tsuji S, Taniuchi S, Hasui M, Yamamoto A, Kobayashi Y (2002) Increased nitric oxide production by neutrophils from patients with chronic granulomatous disease on trimethoprim-sulfamethoxazole. *Nitric Oxide* 7: 283–288
37 Bamberger DM, Heradon BL, Fitch J, Florkowski A, Parkhurst V (2002) Effects of neutrophils on cefazolin activity and penicillin-binding proteins in *Staphylococcus aureus* abscesses. *Antimicrob Agents Chemother* 46: 2878–2884
38 Casal J, Gimenez MJ, Aguilar L, Yuste J, Jado I, Tarrago D, Fenoll A (2002) Beta-lactam activity against resistant pneumococcal strains is enhanced by the immune system. *J Antimicrob Chemother* 50 (Suppl S2): 83–86
39 Matsuda T, Saito H, Fukatsu K, Han I, Inoue T, Furukawa S, Ikeda S, Hidemura A, Kang W (2002) Differences in neutrophil death among beta-lactams antibiotics after *in vitro* killing of bacteria. *Shock* 18: 69–74
40 Labro MT (1995) Resistance to and immunomodulation effect of cephalosporin antibiotics. *Clin Drug Invest* 9 (Suppl 3): 31–44
41 Taylor SC, Shacks SJ, Qu Z, Bryant P (2002) Combined effects of *in vitro* penicillin and sickle cell disease sera on normal lymphocyte functions. *J Natl Med Assoc* 94: 678–685
42 Brooks BM, Thomas AL, Coleman JW (2003) Benzylpenicillin differentially conjugates to IFN-γ, TNF-α, IL-1β, IL-4 and IL-13 but selectively reduces IFN-γ activity. *Clin Exp Immunol* 131: 268–274
43 Smith DM, Kazi A, Smith L, Long TE, Heldreth B, Turos E, Dou PQ (2002) A novel β-lactam antibiotic activates tumor cell apoptotic program by inducing DNA damage. *Mol Pharmacol* 61: 1348–1358
44 Alkharfy K, Kellum JA, Frye RF, Matzke GR (2000) Effect of ceftazidime on systemic cytokine concentration in rats. *Antimicrob Agents Chemother* 44: 3217–3219
45 Luchi M, Morrison DC, Opal S, Yoneda K, Slotman G, Chambers H, Wiesenfeld H, Lemke J, Ryan JL, Horn D (2000) A comparative trial of imipenem *versus* ceftazidime in the release of endotoxin and cytokine generation in patients with gram-negative sepsis. *J Endotoxin Res* 6: 25–31
46 Labro MT (1990) Cefodizime as a biological response modifier: a review of its *in-vivo*, *ex-vivo* and *in-vitro*

immunomodulatory properties. *J Antimicrob Chemother* 26 (Suppl C): 37–47

47 Wenisch C, Bartunek A, Zedtwitz-Liebenstein K, Hiesmayr M, Parschalk B, Pernerstorfer T, Graninger W (1997) Prospective randomized comparison of cefodizime *versus* cefuroxime for perioperative prophylaxis in patients undergoing coronary artery bypass grafting. *Antimicrob Agents Chemother* 41: 1584–1588

48 Wenisch C, Parshalk B, Hasenhundl M, Wiesinger E, Graninger W (1995) Effects of cefodizime and ceftriaxone on phagocytic functions in patients with severe infections. *Antimicrob Agents Chemother* 39: 672–676

49 Dabrowski MP, Stankiewicz W (2001) Desirable and undesirable immunotropic effects of antibiotics: immunomodulating properties of cefaclor. *J Chemother* 13: 615–620

50 Smith-Norowitz TA, Bluth MH, Drew H, Norowitz KB, Chice S, Shah VN, Nowakowski M, Josephson AS, Durkin HG, Joks R (2002) Effect of minocycline and doxycycline on IgE responses. *Ann Allergy Asthma Immunol* 89: 172–179

51 Sadowski T, Steinmeyer J (2001). Minocycline inhibits the production of inducible nitric oxide synthase in articular chondrocytes. *J Rheumatol* 28: 336–340

52 Iwasaki H, Inoue H, Mitsuke Y, Badran A, Ikegaya S, Ueda T (2002) Doxycycline induces apoptosis by way of caspase-3 activation with inhibition of matrix metalloproteinase in human T-lymphoblastic leukemia CCRF-CEM cells. *J Lab Clin Med* 140: 382–386

53 D'Agostino P, Ferlazzo V, Milano S, La Rosa M, Di Bella C, Caruso R, Barbera C, Grimaudo S, Tolomeo M, Feo S et al (2003) Chemically modified tetracyclines induce cytotoxic effects against J774 tumour cell line by activating the apoptotic pathway. *Int Immunopharmacol* 3: 63–73

54 Mealey KL, Barhoumi R, Burghardt RC, Safe S, Kochevar DT (2002) Doxycycline induces expression of P-glycoprotein in MCF-7 breast carcinoma cells. *Antimicrob Agents Chemother* 46: 755–761

55 Takeshita S, Ono Y, Kozuma K, Suzuki M, Kawamura Y, Yokoyama N, Furukawa S, Isshiki T (2002) Modulation of oxidative burst of neutrophils by doxycycline in patients with acute myocardial infarction. *J Antimicrob Chemother* 49: 411–413

56 Du Y, Ma Z, Lin S, Dodel RC, Gao F, Bales KR, Triarhou LC, Chernet E, Perry KW, Nelson DL et al (2001) Minocycline prevents nigrostriatal dopaminergic neurodegeneration in the MPTP model of Parkinson's disease. *Proc Natl Acad Sci USA* 98: 14669–14674

57 Reasoner DK, Hindman BJ, Dexter F, Subieta A, Cutkomp J, Smith T (1997) Doxycycline reduces early neurologic impairment after cerebral arterial air embolism in the rabbit. *Anesthesiol* 87: 569–576

58 Smith JR, Gabler WL (1995) Protective effects of doxycycline in mesenteric ischemia and reperfusion. *Res Commun Mol Pathol Pharmacol* 88: 303–315

59 Clark WM, Lessov N, Lauten JD, Hazel K (1997) Doxycycline treatment reduces ischemic brain damage in transient middle cerebral artery occlusion in the rat. *J Mol Neurosci* 9: 103–108

60 Kriz J, Nguyen MD, Julien JP (2002) Minocycline slows disease progression in a mouse model of amyotrophic lateral sclerosis. *Neurobiol Dis* 10: 268–278

61 Van Den Bosch L, Tilkin P, Lemmens G, Robberecht W (2002) Minocycline delays disease onset and mortality in a transgenic model of ALS. *Neuroreport* 13: 1067–1070

62 Brundula V, Rewcastle NB, Metz LM, Bernard CC, Yong VW (2002) Targeting leukocyte MMPs and transmigration: minocycline as a potential therapy for multiple sclerosis. *Brain* 125: 1297–1308

63 Popovic N, Schubart A, Goetz BD, Zhang SC, Linington C, Duncan ID (2002) Inhibition of autoimmune encephalomyelitis by a tetracycline. *Ann Neurol* 51: 215–223

64 Curci JA, Mao D, Bohner DG, Allen BT, Rubin BG, Reilly JM, Sicard GA, Thompson RW (2000) Preoperative treatment with doxycycline reduces aortic wall expression and activation of matrix metalloproteinases in patients with abdominal aortic aneurysms. *J Vasc Surg* 31: 325–342

65 Vieillard-Baron A, Frisdal E, Eddahibi S, Deprez I, Baker AH, Newby AC, Berger P, Levame M, Raffestin B, Adnot S et al (2000) Inhibition of matrix metalloproteinases by lung TIMP-1 gene transfer or doxycycline aggravates pulmonary hypertension in rats. *Circ Res* 87: 418–425

66 Fischbach LA, Correa P, Ramirez H, Realpe JL, Collazos T, Ruiz B, Bravo LE, Bravo JC, Casabon AL, Schmidt BA (2001) Anti-inflammatory and tissue-protectant drug effects: results from a randomized placebo-controlled trial of gastritis patients at high risk for gastric cancer. *Aliment Pharmacol Ther* 15: 831–841

67 Krause R, Patruta S, Daxböck F, Fladerer P, Wenisch C

(2001) The effect of fosfomycin on neutrophil function. *J Antimicrob Chemother* 47: 141–146

68 Hamada M, Honda J, Yoshimoto T, Fumimori T, Okamoto M, Aizawa H (2002) Fosfomycin inhibits neutrophil function via a protein kinase C-dependent signaling pathway. *Int Immunopharmacol* 2: 511–518

69 Honda J, Okubo Y, Kusaba M, Kumagai M, Saruwatari N, Oizumi K (1998) Fosfomycin (FOM: 1R-2S-epoxypropylphosphonic acid) suppresses the production of IL-8 from monocytes via the suppression of neutrophil function. *Immunopharmacol* 39: 149–155

70 Yoneshima Y, Ichiyama T, Ayukawa H, Matsubara T, Furukawa S (2003) Fosfomycin inhibits NF-κB activation in U-937 and Jurkat cells. Int J Antimicrob Agents 21: 589–592

71 Ishizaka S, Takeuchi H, Kimoto M, Kanda S, Saito S (1998) Fosfomycin, an antibiotic, possessed TGF-beta-like immunoregulatory activities. *Int J Immunopharmacol* 20: 765–779

72 Morikawa K, Nonaka M, Torii I, Morikawa S (2003) Modulatory effect of fosfomycin on acute inflammation in the rat air pouch model. *Int J Antimicrob Agents* 21: 334–339

73 Ishimaru N, Haneji N, Yanagi K, Hayashi Y (2000) Therapeutic effet of fosfomycin in animal model of Sjögren's syndrome. *Chemother (Tokyo)* 48: 775–779

74 Di Marco R, Khademi M, Wallstrom E, Muhallab S, Nicoletti F, Olsson T (1999) Amelioration of experimental allergic neuritis by sodium fusidate (fusidin): suppression of IFN-gamma and TNF-alpha and enhancement of IL-10. *J Autoimmun* 13: 187–195

75 Kishi K, Hirai K, Hiramatsu K, Yamasaki T, Nasu M (1999) Clindamycin suppresses endotoxin released by ceftazidime-treated *Escherichia coli* O55:B5 and subsequent production of tumor necrosis factor alpha and interleukin-1β. *Antimicrob Agents Chemother* 43: 616–622

76 Hirata N, Hiramatsu K, Kishi K, Yamasaki T, Ichimaya T, Nasu M (2001) Pretreatment of mice with clindamycin improves survival of endotoxic shock by modulating the release of inflammatory cytokines. *Antimicrob Agents Chemother* 45: 2638–2642

77 Nakano T, Hiramatsu K, Kishi K, Hirata N, Kadota J, Nasu M (2003) Clindamycin modulates inflammatory cytokine production in lipopolysaccharide-stimulated mouse peritoneal macrophages. *Antimicrob Agents Chemother* 47: 363–367

78 Wulansari R, Wijaya A, Ano H, Horii Y, Makimura S (2003) Lymphocyte subsets and specific IgG antibody levels in clindamycin-treated and untreated dogs experimentally infected with *Babesia gibsoni*. *J Vet Med Sci* 65: 579–584

79 Labro MT (1998) Antiinflammatory activity of macrolides: a new therapeutic potential? *J Antimicrob Chemother* 41 (Suppl B): 37–46

80 Labro MT (1998) Immunological effects of macrolides. *Curr Op Infect Dis* 11: 681–688

81 Culic O, Erakovic V, Parnham MJ (2001) Anti-inflammatory effects of macrolide antibiotics. *Eur J Pharmacol* 429: 209–229

82 Labro MT, Abdelghaffar H (2001) Immunomodulation by macrolide antibiotics. *J Chemother* 13: 3–8

83 Zalewska-Kaszubska J, Gorska D (2001) Anti-inflammatory capabilities of macrolides. *Pharmacol Res* 44: 451–454

84 Abdelghaffar H, Kirst H, Soukri A, Babin-Chevaye C, Labro MT (2002) Structure-activity relationships among 9-N-alkyl derivatives of erythromycylamine and their effect on the oxidative burst of human neutrophils *in vitro*. *J Chemother* 14: 132–139

85 Araujo FG, Slifer TL, Remington JS (2002) Inhibition of secretion of interleukin-1β and tumor necrosis factor alpha by the ketolide antibiotic telithromycin. *Antimicrob Agents Chemother* 46: 3327–3330

86 Noma T, Aoki K, Hayashi M, Yoshizawa I, Kawano Y (2001) Effect of roxithromycin on T lymphocyte proliferation and cytokine production elicited by mite antigen. *Internat Immunopharmacol* 1: 201–210

87 Abe S, Nakamura H, Inoue S, Takeda H, Saito H, Kato S, Mukaida N, Matsushima K, Tomoike H (2000) Interleukin-8 gene repression by clarithromycin is mediated by the activator protein-1 binding site in human bronchial cells. *Am J Respir Cell Mol Biol* 22: 51–60

88 Aoki Y, Kao PN (1999) Erythromycin inhibits transcriptional activation of NF-κB, but not NF-AT, through calcineurin-independent signaling in T cells. *Antimicrob Agents Chemother* 43: 2678–2684

89 Ichiyama T, Nishikawa M, Yoshitomi T, Hasegawa S, Matsubara T, Hayashi T, Furukawa S (2001) Clarithromycin inhibits NF-kB activation in human peripheral blood mononuclear cells and pulmonary epithelial cells. *Antimicrob Agents Chemother* 45: 44–47

90 Kikuchi T, Hagiwara K, Honda Y, Gomi K, Kobayashi T, Takahashi H, Tokue Y, Watanabe A, Nukiwa T (2002)

Clarithromycin suppresses lipopolysaccharide-induced interleukin-8 production by human monocytes through AP-1 and NF-kappa B transcription factors. *J Antimicrob Chemother* 49: 745–755

91 Abeyama K, Kawahara K-I, Iino S, Hamada T, Arimura S-I, Matsushita T, Nakajima T, Maruyama I (2003) Antibiotic cyclic AMP signaling by "primed" leukocytes confers anti-inflammatory cytoprotection. *J Leukoc Biol* 74: 908–915

92 Yamaryo T, Oishi K, Yoshimine H, Tsuchihashi Y, Matsushima K, Nagatake T (2003) Fourteen-member macrolides promote the phosphatidylserine receptor-dependent phagocytosis of apoptotic neutrophils by alveolar macrophages. *Antimicrob Agents Chemother* 47: 48–53

93 Zhao DM, Xue HH, Chida K, Suda T, Oki Y, Kanai M, Uchida C, Ichiyama A, Nakamura H (2000) Effect of erythromycin on ATP-induced intracellular calcium response in A-549 cells. *Am J Physiol Lung Cell Mol Physiol* 278: 726–736

94 Yamanaka Y, Tamari M, Nakahata T, Nakamura Y (2001) Gene expression profiles of human small airway epithelial cells treated with low doses of 14- and 16-membered macrolides. *Biochem Biophys Res Comm* 287: 198–203

95 Li Y, Azuma A, Takahashi S, Usuki J, Matsuda K, Aoyama A, Kudoh S (2002) Fourteen-membered ring macrolides inhibit vascular cell adhesion molecule 1 messenger RNA induction and leukocyte migration: role in preventing lung injury and fibrosis in bleomycin-challenged mice. *Chest* 122: 2137–2145

96 Yoshii T, Magara S, Miyai D, Kuroki E, Nishimura H, Furudoi S, Komori T (2002) Inhibitory effect of roxithromycin on the local levels of bone-resorbing cytokines in an experimental model of murine osteomyelitis. *J Antimicrob Chemother* 50: 289–292

97 Giamarellos-Bourboulis EJ, Adamis T, Laoutaris G, Sabracos L, Koussoulas V, Mouktaroudi M, Perrea D, Karayannacos PE, Giamarellou H (2004) Immunomodulatory clarithromycin treatment of experimental sepsis and acute pyelonephritis caused by multidrug-resistant *Pseudomonas aeruginosa*. *Antimicrob Agents Chemother* 48: 93–99

98 Suzuki M, Asano K, Yu M, Hisamitsu T, Suzuki H (2002) Inhibitory action of a macrolide antibiotic, roxithromycin, on co-stimulatory molecule expression *in vitro* and *in vivo*. *Mediators Inflamm* 11: 235–244

99 Culic O, Erakovic V, Cepelak I, Barisic K, Brajsa K, Ferencic Z, Galovic R, Glojnaric I, Manojlovic Z, Munic V et al (2002) Azithromycin modulates neutrophil function and circulating inflammatory mediators in healthy human subjects. *Eur J Pharmacol* 450: 277–289

100 Schultz MJ, Speelman P, Hack CE, Buurman WA, van Deventer SJH, van der Poll T (2000) Intravenous infusion of erythromycin inhibits CXC chemokine production but augments neutrophil degranulation in whole blood stimulated with *Streptococcus pneumoniae*. *J Antimicrob Chemother* 46: 235–240

101 Hiratsuka T, Mukae H, Iiboshi H, Ashitani J, Nabeshima K, Minematsu T, Chino M, Ihi T, Kohno S, Nakazato M (2003) Increased concentrations of human beta-defensins in plasma and bronchoalveolar lavage fluid of patients with diffuse panbronchiolitis. *Thorax* 58: 425–430

102 Morikawa T, Kadota JI, Kohno S, Kondo T (2000) Superoxide dismutase in alveolar macrophages from patients with diffuse panbronchiolitis. *Respiration* 67: 546–551

103 Noma T, Ogawa N (2003) Roxithromycin enhances lymphocyte apoptosis in Dermatophagoides-sensitive childhood asthma. *J Allergy Clin Immunol* 111: 646–647

104 Spahn JD, Fost DA, Covar R, Martin RJ, Brown EE, Szefler SJ, Leung JY (2001) Clarithromycin potentiates glucocorticoid responsiveness in patients with asthma: results of a pilot study. *Ann Allergy Asthma Immunol* 87: 501–505

105 Enomoto F, Andou I, Nagaoka I, Ichikawa G (2002) Effect of new macrolides on the expression of adhesion molecules on neutrophils in chronic sinusitis. *Auris Nasus Larynx* 29:267–269

106 Iino Y, Sasaki Y, Kojima C, Miyazawa T (2001) Effect of macrolides on the expression of HLA-DR and costimulatory molecules on antigen-presenting cells in nasal polyps. *Ann Otol Rhinol Laryngol* 110: 457–463

107 Muhlestein JB, Anderson JL, Carlquist JF, Salunkhe K, Horne BD, Pearson RR, Bunch TJ, Allen A, Trehan S, Nielson C (2000) Randomized secondary prevention trial of azithromycin in patients with coronary artery disease. Primary clinical results of the ACADEMIC study. *Circulation* 102: 1755–1760

108 Seral C, Van Bambeke F, Tulkens PM (2003) Quantitative analysis of gentamicin, azithromycin, telithromycin, ciprofloxacin, moxifloxacin, and oritavancin (LY333328) activities against intracellular *Staphylococ-*

cus aureus in mouse J774 macrophages. *Antimicrob Agents Chemother* 47: 2283–2292

109 Toyoguchi T, Ebihara M, Ojima F, Hosoya J, Shoji T, Nakagawa Y (2000) Histamine release induced by antimicrobial agents and effects of antimicrobial agents on vancomycin-induced histamine release from rat peritoneal mast cells. J *Pharm Pharmacol* 52: 327–331

110 El Bekay R, Alvarez M, Carballo M, Martin-Nieto J, Monteseirin J, Pintado E, Bedoya FJ, Sobrino F (2002) Activation of phagocytic cell NADPH oxidase by norfloxacin: a potential mechanism to explain its bactericidal action. *J Leukoc Biol* 71: 255–261

111 Riesbeck K (2002) Immunomodulating activity of quinolones: Review. *J Chemother* 14: 3–12

112 Dalhoff A, Shalit I (2003) Immunomodulatory effects of quinolones. *Lancet Infect Dis* 3: 359–371

113 Ives TJ, Schwab UE, Ward ES, Hall IH (2003) *In-vitro* anti-inflammatory and immunomodulatory effects of grepafloxacin in zymogen A or *Staphylococcus aureus*-stimulated human THP-1 monocytes. *J Infect Chemother* 9: 134–143

114 Hall IH, Schwab UE, Ward ES, Ives TJ (2003) Effects of moxifloxacin in zymogen A or *S. aureus* stimulated human THP-1 monocytes on the inflammatory process and the spread of infection. *Life Sci* 73: 2675–2685

115 Holzheimer RG (2003) Oral antibiotic prophylaxis can influence the inflammatory response in aortic aneurysm repair: results of a randomized clinical study. *J Chemother* 15: 157–164

116 Ishikawa T, Okamura S, Oshimoto H, Kobayashi R, Mori M (2003) Metronidazole plus ciprofloxacin therapy for active Crohn's disease. *Intern Med* 42: 318–321

117 Labro MT (2002) Antibiotics as anti-inflammatory drugs. *Curr Op Investig Drugs* 3: 61–68

118 Yanagihara K, Kadoto J, Kohno S. (2001) Diffuse panbronchiolitis-pathophysiology and treatment mechanisms. *Int J Antimicrob Agents* 18 (Suppl 1): S83–S87

119 Hoyt JC, Robbins RA (2001) Macrolide antibiotics and pulmonary inflammation. *FEMS Microbiol Lett* 205: 1–7

120 Jaffé A, Bush A (2001) Anti-inflammatory effects of macrolides in lung disease. *Pediatr Pulmonol* 31: 464–473

121 Jaffé A, Rosenthal M (2002) Macrolides in the respiratory tract in cystic fibrosis. *J R Soc Med* 95 (Suppl 41): 27–31

122 Gaylor AS, Reilly JC (2002) Therapy with macrolides in patients with cystic fibrosis. *Pharmacother* 22: 327–335

123 Carey KW, Alwami A, Danziger LH, Rubinstein I (2003) Tissue reparative effects of macrolide antibiotics in chronic inflammatory sinopulmonary diseases. *Chest* 123: 261–265

124 Cervin A (2001) The anti-inflammatory effect of erythromycin and its derivatives, with special reference to nasal polyposis and chronic sinusitis. *Acta Otolaryngol* 121: 83–92

125 Equi A, Balfour-Lynn IM, Bush A, Rosenthal M (2002) Long term azithromycin in children with cystic fibrosis: a randomised, placebo-controlled crossover trial. *Lancet* 360: 978–984

126 Cazzola M, Salzillo A, Diamant F (2000) Potential role of macrolides in the treatment of asthma. *Monaldi Arch Chest* Dis 55: 231–236

127 Richeldi L, Ferrara G, Fabbri LM, Gibson PG (2002) Macrolides for chronic asthma. *Cochrane Database Syst Rev* 1: CD 002997

128 Komine M, Tamaki K (2000) An open trial of oral macrolide treatment for psoriasis vulgaris. *J Dermatol* 27: 508–512

129 Schilling F, Wagner AD (2000) Azithromycin: Eine antiinflammatorische Wirksamkeit im Einsatz bei der chronischen rekurrierenden multifokalen Osteomyelitis? Eine vorläufige Mitteilung. *Z Rheumatol* 59: 352–353

130 Cercek B, Shah PK, Noc M, Zahger D, Zeymer U, Matetzky S, Maurer G, Mahrer P; AZACS Investigators (2003) Effect of long-term treatment with azithromycin on recurrent ischaemic events in patients with acute coronary syndrome in the Azithromycin in Acute Coronary Syndrome (AZACS) trial: a randomised controlled trial. *Lancet* 361: 809–813

131 Stone AF, Mendall MM, Kaski JC, Edger TM, Risley P, Poloniecki J, Camm AJ, Northfield TC (2002) Effect of treatment for Chlamydia pneumoniae and Helicobacter pylori on markers of inflammation and cardiac events in patients with acute coronary syndromes: South Thames Trial of Antibiotics in Myocardial Infarction and Unstable Angina (STAMINA). *Circulation* 106: 219–223

132 Dunne MW (2000) Rationale and design of a secondary prevention trial of antibiotic use in patients after myocardial infarction: the WIZARD (weekly intervention with zithromax (azithromycin) for atherosclerosis

and its related disorders) trial. *J Infect Dis* 181 (Suppl 3): S572–S578
133 O'Connor CM, Dunne MW, Pfeffer MA, Muhlestein JB, Yao L, Gupta S, Benner RJ, Fisher MR, Cook TD; Investigators in the WIZARD Study (2003) Azithromycin for the secondary prevention of coronary heart disease events: the WIZARD study: a randomized controlled trial. *JAMA* 290: 1459–1466
134 Vammen S, Lindholt JS, Ostergaard L, Fasting H, Henneberg EW (2001) Randomized double-blind controlled trial of roxithromycin for prevention of abdominal aortic aneurysm expansion. *Br J Surg* 88: 1066–1072
135 Leiper K, Morris AI, Rhodes JM (2000) Open label trial of oral clarithromycin in active Crohn's disease. *Aliment Pharmacol Ther* 14: 801–806
136 Sakamoto M, Mikasa K, Majima T, Hamada K, Konishi M, Maeda K, Kita E, Narita N (2001) Anticachectic effect of clarithromycin for patients with unresectable non-small-cell lung cancer. *Chemother* 47: 444–451
137 Coleman M, Leonard J, Lyons L, Szelenyi H, Niesvizky R (2003) Treatment of Waldenström's macroglobulinemia with clarithromycin, low-dose thalidomide, and dexamethasone. *Semin Oncol* 30: 270–274
138 Sauter C, Blum S (2003) Regression of lung lesions in Hodgkin's disease by antibiotics: case report and hypothesis on the etiology of Hodgkin's disease. *J Clin Oncol* 26: 92–94
139 Bikowski JB (2003) Subantimicrobial dose doxycycline for acne and rosacea. *Skinmed* 2: 234–245
140 Alarcon GS (2000) Tetracyclines for the treatment of rheumatoid arthritis. *Expert Opin Investig Drugs* 9: 1491–1498
141 O'Dell JR, Blakely KW, Mallek JA, Eckoff PJ, Leff RD, Wees SJ, Sems KM, Fernandez AM, Palmer WR, Klassen LW et al (2001) Treatment of early seropositive rheumatoid arthritis: a two-year, double-blind comparison of minocycline and hydroxychloroquine. *Arthritis Rheum* 44: 2235–2241
142 Le CH, Morales A, Trentham DE (1998) Minocycline in early diffuse scleroderma. *Lancet* 352: 1755–1756
143 Robertson LP, Marshall RW, Hickling P (2003) Treatment of cutaneous calcinosis in limited systemic sclerosis with minocycline. *Ann Rheum Dis* 62: 267–269
144 Mosorin M, Javonen J, Biancari F, Satta J, Surcel HM, Leinonen M, Saikku P, Juvonen T (2001) Use of doxycycline to decrease the growth rate of abdominal aortic aneurysms: a randomized, double-blind, placebo-controlled pilot study. *J Vasc Surg* 34: 757–758
145 Nieman GF, Zerler BR (2001) A role for the anti-inflammatory properties of tetracyclines in the prevention of acute lung injury. *Curr Med Chem* 8: 317–325
146 Ahuja TS (2003) Doxycycline decreases proteinuria in glomerulonephritis. *Am J Kidney Dis* 42: 376–380
147 Caruso I (1997) Twenty years of experience with intra-articular rifamycin for chronic arthritides. *J Int Med Res* 25: 307–317
148 Caruso I, Cazzola M, Santandrea S (1992) Clinical improvement in ankylosing spondylitis with rifamycin SV infiltrations of peripheral joints. *J Int Med Res* 20: 171–181
149 Caruso I, Principi N, D'Urbino G, Santandrea S, Boccassini L, Montrone F, Sarzi Puttini PC, Bombaci A, Bozzato A, Azzolini V et al (1993) Rifamycin SV administered by intra-articular infiltrations shows disease modifying activity in patients with pauci or polyarticular juvenile rheumatoid arthritis. *J Int Med Res* 21: 243–256
150 Gui GP, Thomas PR, Tizard ML, Lake J, Sanderson JD, Hermon-Taylor J (1997). Two-year-outcomes analysis of Crohn's disease treated with rifabutin and macrolide antibiotics. *J Antimicrob Chemother* 39: 393–400
151 Gionchetti P, Rizello F, Venturi A, Ugolini F, Rossi M, Brigidi P, Johansson R, Ferrieri A, Poggioli G, Campieri M (1999) Review-antibiotic treatment in inflammatory bowel disease: rifaximin, a new possible approach. *Eur Rev Med Pharmacol Sci* 3: 27–30

Immunosuppressives in transplant rejection

C9

Henk-Jan Schuurman

Introduction

Suppression of immune reactivity can either be an undesirable effect or a situation which is specifically induced to the benefit of a patient. Examples of the first come from immunotoxicology, e.g., XENOBIOTICS or environmental factors causing IMMUNOSUPPRESSION. Virus infections, as exemplified by HIV, can cause severe immunodeficiency. Under clinical conditions suppression of the IMMUNE SYSTEM is specially indicated in two indications: autoimmunity and organ transplantation. AUTOIMMUNE DISEASES like rheumatoid arthritis are mainly treated by inhibition of the effector phase with antiinflammatory drugs like corticosteroids (see Chapters C.11, C.12, C.13). Immunosuppressants like cyclosporine, which have been mainly developed for transplantation, are increasingly used in AUTOIMMUNE DISEASES, and some, like leflunomide (see below), have been developed for rheumatoid arthritis as the first indication.

In contrast, in organ transplantation there is a principal need for interference with the initiation of an immune response which is induced by the grafted organ. Generally, high-dose IMMUNOSUPPRESSION is needed in the first period after transplantation (INDUCTION TREATMENT), or in the treatment of rejection episodes. To keep graft function stable so-called MAINTENANCE TREATMENT is given. Originally, when transplantation was introduced as a treatment of end-stage organ failure (the first kidney transplant was performed in the early fifties of the last century), there were few possibilities to prevent or treat allograft rejection. In the sixties and early seventies, this was mainly restricted to combinations of azathioprine, corticosteroids, and cyclophosphamide. Combinations of these drugs were effective, but associated with severe side-effects, mainly related to BONE MARROW depression (leukopenia, anemia). The only more specific reagent, anti-lymphocyte globulin (ALG), became available in 1966 and was used in induction treatment immediately after transplantation. Based on the complications working with these drugs, kidney transplantation developed slowly and the lack of appropriate immunosuppressive regimens slowed down the introduction of heart transplantation in clinical medicine after it was first performed in 1963.

The most widely used immunosuppressives at present are XENOBIOTICS, i.e., orally active drugs produced by microorganisms or chemically synthesized molecules (Tab. 1, Fig. 1; see selected reading for general references). A landmark in IMMUNOSUPPRESSION for transplantation was the introduction of cyclosporine in 1983. Using cyclosporine as a basic immunosuppressant in combination with azathioprine and corticosteroids, 1-year graft survival in kidney transplantation increased to 80–90% and heart transplantation also reached a 1-year patient survival exceeding 80%.

Cyclosporine is a so-called calcineurin inhibitor (see below): another calcineurin-inhibitor of more recent date is tacrolimus. In most transplant centres, INDUCTION IMMUNOSUPPRESSIVE TREATMENT during the first 2–4 weeks after transplantation nowadays includes either triple therapy, i.e., a calcineurin inhibitor, an inhibitor of cell proliferation like azathioprine or mycophenolate mofetil, and corticosteroids as an antiinflammatory agent, or in other centres, quadruple therapy is given, in which ANTIBODY OKT3 (muronomab-CD3) or another anti-lymphocyte ANTIBODY (ATG, anti-thymocyte globulin) are added to the immunosuppressive regimen. Many changes in the combination-drug treatment are currently proposed, in which one of the drugs is replaced by another one: as an example rapamycin

TABLE 1. MAIN XENOBIOTIC IMMUNOSUPPRESSANTS CURRENTLY ON THE MARKET OR IN ADVANCED CLINICAL DEVELOPMENT

Compound	Trade name	Mechanism of action
Cyclosporine	Neoral	Calcineurin inhibitor
Tacrolimus (FK506)	Prograf	Calcineurin inhibitor
Rapamycins		Inhibitor of growth factor-driven cell proliferation
sirolimus	Rapamune	
everolimus	Certican	
Cyclophosphamide		Inhibitor of cell proliferation
Methotrexate		Inhibitor of cell proliferation: anti-inflammatory
Azathioprine	Imuran	Inhibitor of cell proliferation
Mizoribine	Bredinin	Inhibitor of inosine monophosphate dehydrogenase
Mycophenolic acid derivatives		Inhibitor of inosine monophosphate dehydrogenase
mycophenolate mofetil	Cellcept	
mycophenolate sodium	Myfortic	
Leflunomide	Arava	Inhibitor of dihydroorotate dehydrogenase
Gusperimus (15-deoxyspergualin)	Spanidin	Inhibitor of cell differentiation
FTY720		Sphingosine 1-phosphate receptor agonist

is mentioned, which is an inhibitor of growth-factor induced cell proliferation, and can replace azathioprine. Most drugs, in particular those with a narrow therapeutic window between EFFICACY and side-effects, require regular monitoring of exposure on the basis of blood concentration [1]; for others, haematological parameters serve as a SURROGATE marker (like blood leukocyte counts in the case of azathioprine). Side-effects of calcineurin inhibitors primarily concern the kidney, and therefore in a number of kidney transplant centres patients are first treated with such drugs after a good kidney function is achieved. When patients show stable graft function, the INDUCTION TREATMENT is gradually converted into MAINTENANCE TREATMENT, in which lower immunosuppressive doses are used, or one component is gradually tapered down and eliminated from the regimen. Corticosteroids are mentioned as an example, because of the endocrinological side-effects (most visible are the Cushingoid features, but others include OSTEOPOROSIS, diabetes, hyperlipidemia, hypertension, hirsutism, and cataracts); in many transplant centres steroid-sparing regimens are the first aim in MAINTENANCE TREATMENT. Rejection crises (documented by histopathology of a graft biopsy, or biochemical markers in blood) are normally at first treated with high-dose intravenous corticosteroids (bolus injections on 3-5 successive days); so-called steroid-resistant rejections are to be treated by antibodies, either OKT3, ATG or, in severe forms, ALG.

FIGURE 1 (see next and following page)
Structures of cyclosporin A, tacrolimus (FK506), sirolimus (rapamycin), mycophenolate mofetil and mycophenolic acid, mizoribine, leflunomide and its active metabolite A 77 1726, brequinar, FTY720, gusperimus (15-deoxyspergualin), azathioprine, cyclophosphamide, and methotrexate.

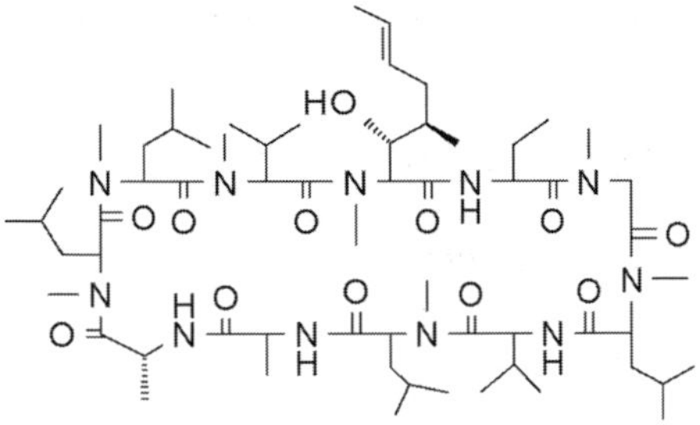

CYCLOSPORINE A

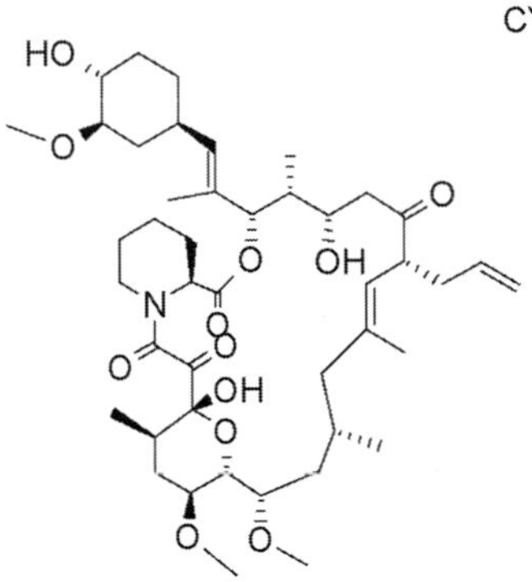

FK506

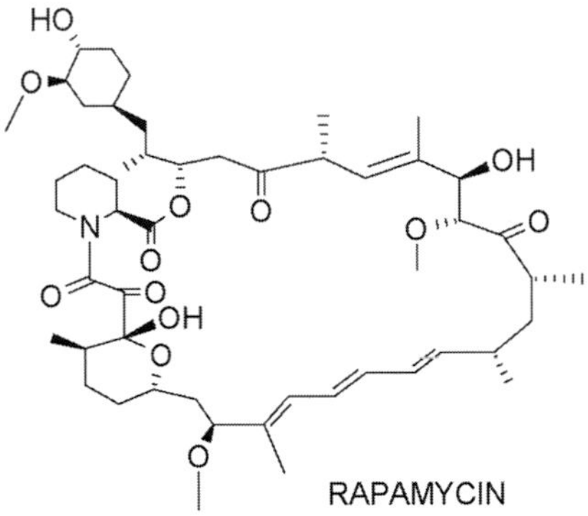

RAPAMYCIN

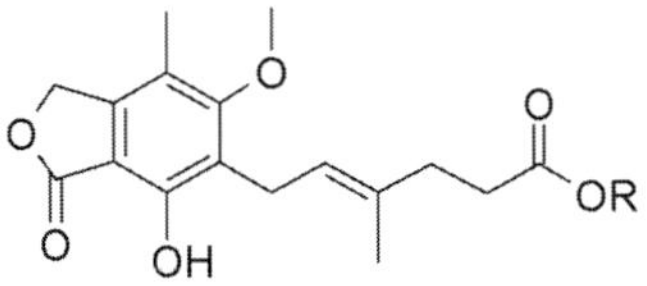

MYCOPHENOLIC ACID R = H

MYCOPHENOLATE MOFETYL R =

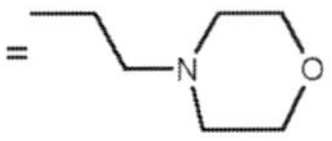

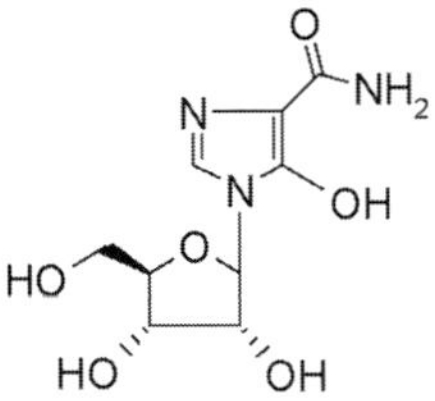

MIZORIBINE

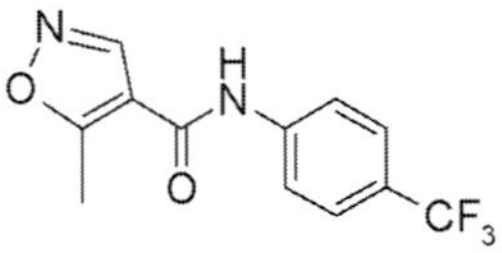

LEFLUNOMIDE

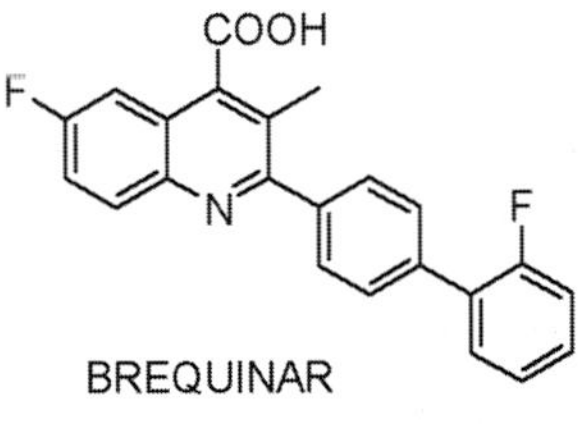

BREQUINAR

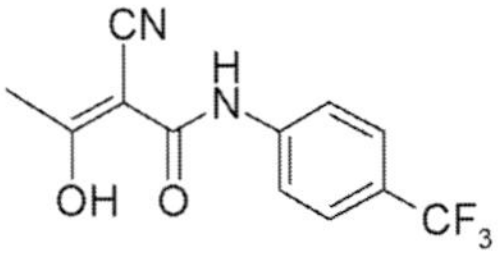

A 77 1726

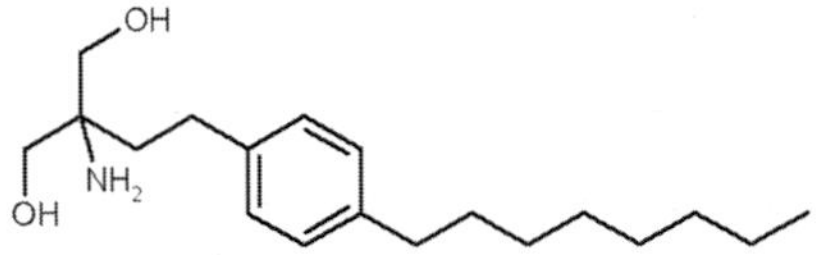

FTY720

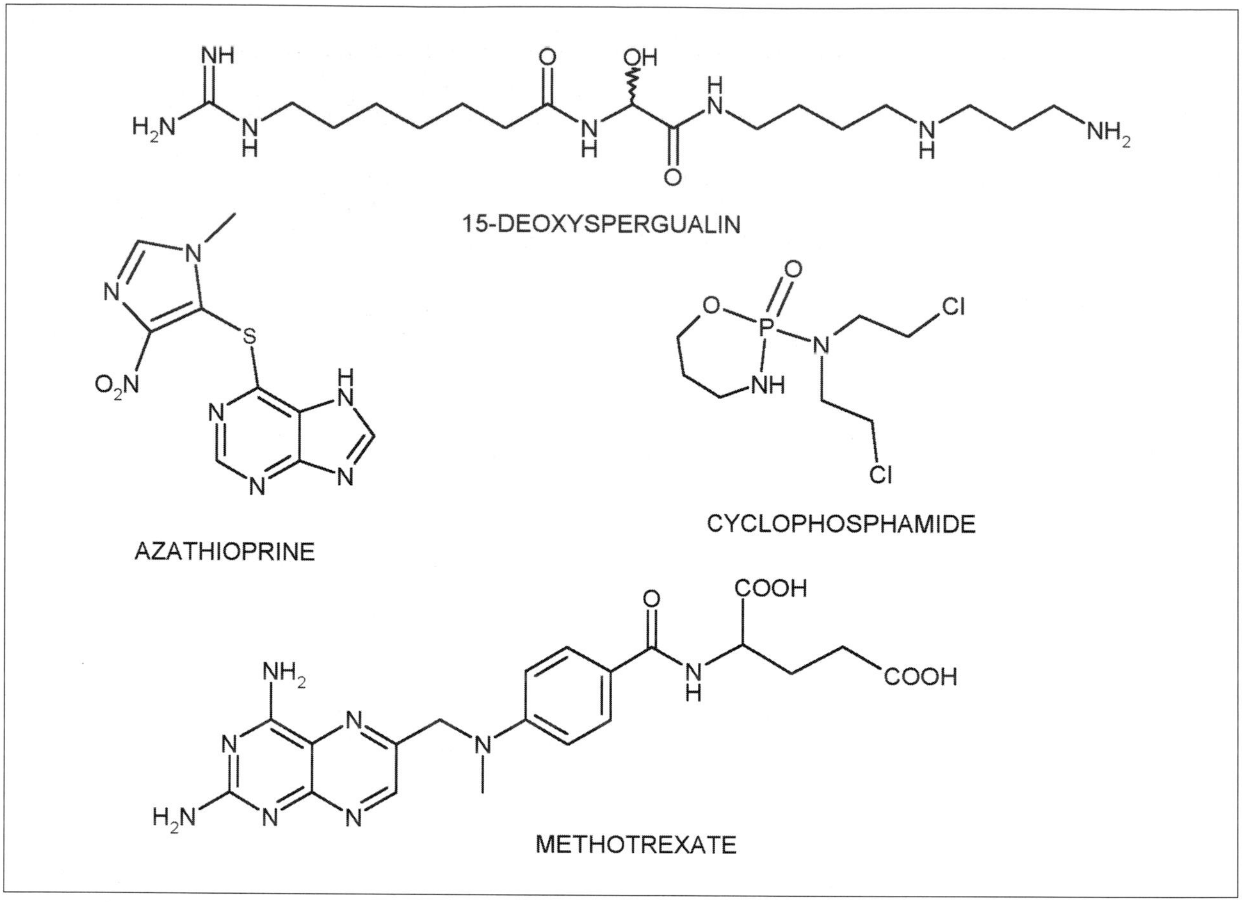

15-DEOXYSPERGUALIN

AZATHIOPRINE

CYCLOPHOSPHAMIDE

METHOTREXATE

Figure 1 (continued)

New immunosuppressants introduced during the last decade are tacrolimus (FK506), mycophenolate mofetil (RS-61443), leflunomide (HWA 486), gusperimus (15-deoxyspergualin), and mizoribine, and there are a number in an advanced stage of development or registration phase (new calcineurin inhibitors like ISA247; the rapamycin derivative RAD, everolimus; mycophenolate sodium, ERL 080; leflunomide analogues MNA 279 and MNA 715; the sphingosine 1-phosphate receptor agonist FTY720, and kinase inhibitors) (Tab. 1, Fig. 1; see selected reading for general references). Remarkably, most of these xenobiotic immunosuppressives originated from antiinfection or cancer drug development programmes; because of their failure in these indications or because IMMUNOSUPPRESSION was observed as a 'side-effect', the drugs were subsequently developed as immunosuppressants. Only tacrolimus was specifically developed as an immunosuppressant and turned out to have the same MECHANISM OF ACTION as cyclosporine. Some new immunosuppressants were first developed for other indications, like leflunomide for rheumatoid arthritis, and subsequently evaluated in clinical trials in transplantation. All these drugs work intracellularly, by inhibition of early or late events in intracellular signalling after lymphocyte activation, or by inhibition of cell proliferation (direct interference in DNA/RNA synthesis) following activation. As these intracellular pathways are not truly selective for LYMPHOCYTES, most drugs

TABLE 2. MAIN BIOLOGICALS CURRENTLY ON THE MARKET OR IN ADVANCED CLINICAL DEVELOPMENT

Compound	Trade name	Mechanism of action/specificity
Anti-lymphocyte globulin	Lymphoglobuline	Lymphocyte depletion
Anti-thymocyte globulin	Thymoglobulin, ATGAM	Lymphocyte depletion
Muronomab-CD3	Orthoclone-OKT3	Anti-CD3 mAb
Siplizumab (MEDI-507)		Anti-CD2 mAb
Clenoliximab (IDEC151)		Anti-CD4 mAb
Daclizumab	Zenapax	Anti-CD25 mAb
Basiliximab	Simulect	Anti-CD25 mAb
Rituximab	Rituxan	Anti-CD20 mAb, depletion of B cells
Alemtuzumab	Campath-1H	Anti-CD52 mAb, depletion of leukocytes
Odulinomab	Antilfa	Anti-CD11a (LFA-1) mAb
Efalizumab	Raptiva	Anti-CD11a (LFA-1) mAb
Natalizumab	Antegren	Anti-VLA-4 mAb
Infliximab	Remicade	anti-TNF-α mAb
Adalimumab	Humira	anti-TNF-α mAb
Etanercept	Enbrel	TNF R – IgG1 fusion protein
Anakinra	Kineret	IL-1 R antagonist
Denileukin diftitox	Ontak	IL-2 diphtheria toxin fusion protein
Alefacept	Amevive	LFA-3-Ig fusion protein
CTLA4-Ig/LEA29Y		Fusion proteins, inhibition of costimulation

have a quite narrow therapeutic window. Their side-effects can be reduced by combination treatment: administration at lower doses in combinations which are synergistic in IMMUNOSUPPRESSION can be associated with higher therapeutic windows. The different classes of drugs will be described further in the following sections.

BIOLOGICALS (i.e., poly- or mono-clonal Ab and fusion proteins generated by rDNA technology) have long been considered as a new class of promising innovative immunosuppressants (Tab. 2). This started with the introduction of ALG and the mouse mAb OKT3 in the eighties of the last century, and was followed by the introduction of humanized or chimeric anti-IL-2 receptor Ab for the indication transplantation (daclizumab, basiliximab). The main advantage of biologicals is their higher specificity resulting in a broadened therapeutic window: the main disavantage is their administration route (parenteral instead of oral for most low-molecular-weight XENOBIOTICS) and their potential immunogenicity, for instance the formation of anti-mouse Ab in case of a mouse mAb. In contrast to most low-molecular-weight XENOBIOTICS, most biologicals work extracellularly, i.e., they target cell surface molecules; their potential side-effects are therefore mediated by cross-reactivity to cell populations other than the prime target. In general, therapeutic Ab to cell surface molecules can affect the target cell in two mechanisms: either temporary blocking or down-regulating of surface molecules resulting in dysfunction or anergy, or lysis of the cell, for instance by the induction of apoptosis, complement-mediated effects or by Ab-dependent cell-mediated cytotoxicity (depleting Ab).

The field of biologicals is progressing nowadays and a number of new compounds have been launched during the last decade or are in advanced development (Tab. 2, see also selected reading for general references). In order to avoid the formation of anti-mouse Ab and to restore the effector function of the specific mAb, later generation Ab for clinical application are generated by genetic engineering. Two major approaches are followed: (1) chimeric Ab in which the constant part of Ig heavy and light chains in the mouse Ab molecule is replaced by human Ig sequences, and (2) humanized Ab in which the sequence encoding the CDR of the variable part of the mouse Ab is inserted in the sequence encoding human Ig. The second approach may lead to a loss of binding affinity as has been shown for the Ab daclizumab mentioned below. Engineered Ab have the advantage that the $t_{1/2}$ in the circulation can be substantially longer, e.g., from 24–48 hours for a mouse Ab to 2–4 weeks for an engineered Ab. Although murine antigenic determinants are still present in the variable part of light and heavy chains, albeit less so in the humanized Ab, the immunogenicity is strongly reduced.

With the present spectrum of immunosuppressants and extensive clinical experience with various drug combinations, the management of IMMUNOSUPPRESSION in patients with transplanted organs has markedly improved. However, major complications that are directly associated with IMMUNOSUPPRESSION still occur, such as drug toxicity, increased susceptibility to infection, and development of tumors, in particular the so-called posttransplant lymphoproliferative diseases [2]. On the other hand, long-term graft survival is hampered by graft dysfunction due to chronic rejection. This phenomenon not only relates to the immune response of the recipient to the graft, but also to intrinsic changes in the graft itself, mainly regarding the vasculature, so-called graft vessel disease or accelerated graft arteriosclerosis. This involves thickening of the intima of blood vessels of the graft and is ascribed to migration of smooth muscle cells to this site followed by cellular proliferation and extracellular matrix formation. With prolonged graft survival chronic rejection is now a major cause of graft loss in long-term surviving patients [3]. Drugs for the prevention or treatment of this condition are not yet available. Some new immunosuppressants (rapamycin, mycophenolate mofetil) are claimed to be effective, as these not only contribute to better IMMUNOSUPPRESSION (diminished host attack), but also inhibit the vascular response due to their inherent MECHANISM OF ACTION.

Since allograft rejection is a T CELL-mediated process, the main target for IMMUNOSUPPRESSION in transplantation is the T lymphocyte. But other cell types of the IMMUNE SYSTEM, like B LYMPHOCYTES and MACROPHAGES, are involved to a variable extent as well. The role of natural killer cells in graft rejection is still unclear. This might change in the future when xenotransplantation (transplantation of nonhuman organs into humans) will become available. The rejection of xenografts cannot be prevented by T CELL immunosuppressants since it not only involves T CELLS, but also B LYMPHOCYTES which are triggered in a T CELL-independent way. This is also relevant in allotransplantation because B CELL reactivity appears to be involved in chronic rejection as well. There are no specific B CELL drugs available as yet, but a number of immunosuppressants discussed below show both T and B CELL inhibitory activity.

Calcineurin inhibitors

Cyclosporine

Cyclosporine (Cyclosporin A), is a cyclic undecapeptide isolated from the fungus *Tolypocladium inflatum gams*. Its biological activity *in vivo* was first discovered in 1973 in a large microbiology screening programme, which included Ab formation to SRBC in mice. Subsequently, it showed EFFICACY in kidney allotransplantation in rats and pigs. This was followed by first clinical trials in human kidney transplant patients, and its introduction to clinical transplantation in 1983. Since cyclosporine is a highly lipophilic molecule, it is poorly soluble in water; for oral administration the first commercial formulation was an oil-based formulation with variable absorption. Since the drug shows a relatively narrow therapeutic window, the kidney being the first target for undesirable side-effects, drug-level monitoring proved necessary to control drug exposure in the therapeutic range. Instead of 16-h or 24-h trough levels, so-called C2 MONITORING (levels 2 hours after administration) has been introduced as an improved estimate of total exposure. At present, a microemulsion formulation (Neoral®) is marketed that shows improved absorption and far less inter- and intra-individual variation.

Cyclosporine shows immunosuppressive activity in a large spectrum of animal models of human immune-mediated diseases [4]. Its suppressive effect is mainly restricted to T LYMPHOCYTES. T CELL-independent B CELL responses are not affected. At the time of introduction the MECHANISM OF ACTION of cyclosporine was largely unknown. A first insight into the mode of action came from the observation that the compound inhibits the production of IL-2, one of the first CYTOKINES produced after T CELL activation. Subsequently it was demonstrated that cyclosporine inhibits IL-2 gene transcription by interfering with the Ca-dependent intracellular signalling mechanism (Fig. 2) [5]. A family of cytoplasmic proteins called cyclophilins (CYP (easily confused with cyt P450) in Fig. 2) has been identified which strongly binds to cyclosporine. CYP are enzymes catalysing cis-trans isomerisation of peptidyl-prolyl bonds (so-called proline isomerase or rotamase activity, important for proper folding of newly synthesized proteins *in vivo*). This inhibition of rotamase activity, however, does not cause the immunosuppressive effect which is actually mediated by the binding of the cyclosporine-immunophilin complex to the serine/threonine phosphatase calcineurin (CNA, CNB in Fig. 2), which plays a pivotal role in Ca-dependent intracellular signaling [6].

Activation of T CELLS via the TCR results in a cascade of events that, among others, involves the activation of the protein tyrosine kinases $p56^{lck}$, $p59^{fyn}$, and zeta chain-associated protein 70 (ZAP-70), followed by phosphorylation of phospholipase Cγ1, resulting in generation of second messengers phosphatidyl inositol (1,4,5) triphosphate and diacylglycerol, which in turn yield an increase in cytoplasmic free Ca^{2+} and activation of protein kinase C. Free Ca^{2+} upon complexing with calmodulin activates the phosphatase calcineurin. The cyclosporine-CYP complex upon binding to calcineurin inhibits its phosphatase activity. Calcineurin dephosphorylates the nuclear factor of activated T CELLS (NFAT) which is then translocated to the nucleus where it initiates together with other transcription factors (e.g., NF-κB and AP-1) expression of early T CELL activation genes, especially the gene encoding IL-2 [7]. Immunophilins and calcineurin are abundantly expressed in different cell types. The apparent T CELL selectivity of cyclosporine has therefore been related to the fact that Ca-dependent T CELL activation via the TCR uniquely involves the calmodulin-calcineurin pathway. There is as yet no unequivocal proof that potential side-effects, like damage to kidney tubules follow the same intracellular mechanism.

Following its MECHANISM OF ACTION, Ca-independent cell triggering is not affected by cyclosporine. For instance, T CELLS can be activated via the costimulatory CD28 molecule (Fig. 3) on the cell surface which in combination with activated protein kinase C (see above) can yield lymphokine gene transcription and T CELL activation in the absence of calcineurin activation. It has therefore been hypothesized that this pathway is involved in T CELL activation and allograft rejection that is resistant to cyclosporine treatment.

Tacrolimus (FK506)

Tacrolimus is a macrocyclic lactone isolated from the actinomycete *Streptomyces tsububaensis*. It was discovered in 1984 in an immunological screening programme specifically established to identify immunosuppressive compounds. Subsequently its immunosuppressive activity was demonstrated in various animal models of transplantation (rat, dog) [8]. The spectrum of immunosuppressive activity appeared to be identical to that of cyclosporine, but remarkably at much lower doses than cyclosporine, both *in vitro* and *in vivo*. Therapeutic drug levels in the circulation also appear to be much lower than those for cyclosporine. Tacrolimus is very lipophilic and poorly soluble in water, resulting in variable absorption necessitating regular drug monitoring; for oral adminstration, a solid dispersion formulation in hydroxypropylmethyl cellulose is used. Major side-effects are strikingly similar to those of cyclosporine but also involve the central nervous system as target organ. The difference with cyclosporine may at least in part be explained by differences in pharmacokinetics and tissue distribution, as the MECHANISM OF ACTION appears to be the same. This does not hold for other side-effects like the lack of gingiva hyperplasia and hirsutism of cyclosporine and the more pronounced diabetogenic effect of tacrolimus.

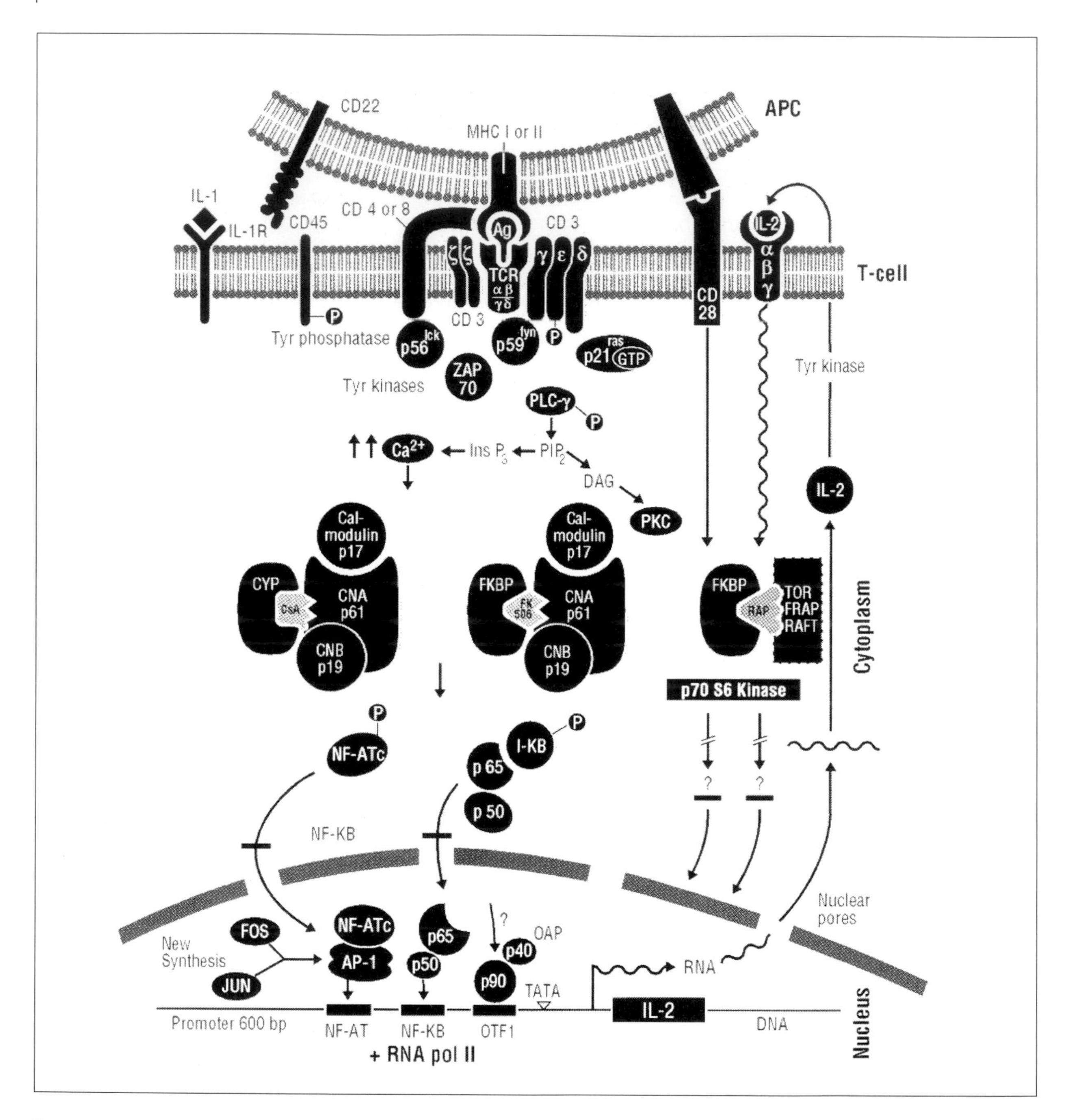

FIGURE 2

Intracellular signalling pathways and points of inhibition by cyclosporine (via binding to cyclophilin, CYP), tacrolimus (via binding to FKBP), and rapamycins (via binding to FKBP). Also cell surface molecules for biologicals (OKT3 binding to CD3; anti-IL-2-R Ab) or those currently considered as potential drug targets (CD4, CD28, CD45) are shown. For explanations see text.

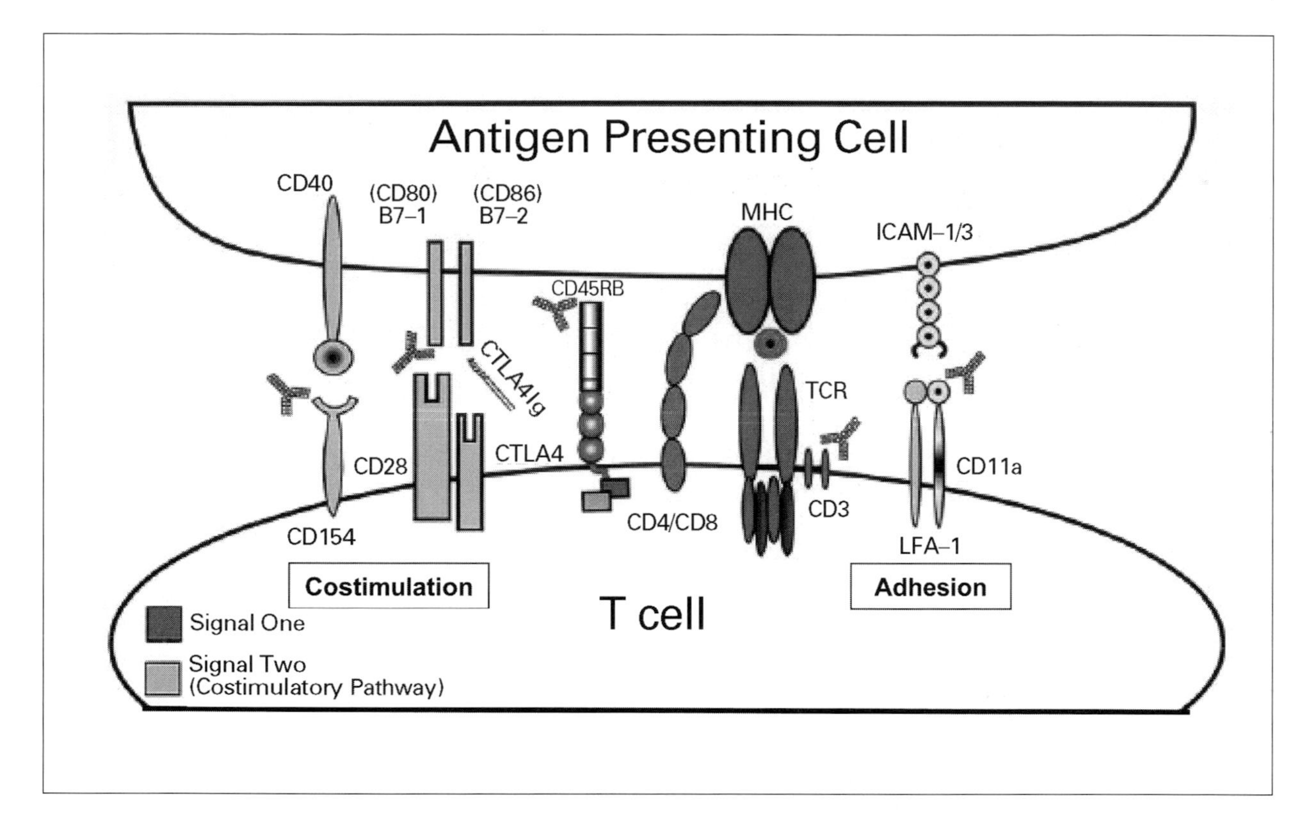

FIGURE 3

Target molecules on the surface of T cells and APC involved in the potential induction of T cell activation and immunological tolerance. T cell stimulation requires interaction between the TCR and the antigenic epitope in combination with costimulation via a number of cell surface interactions (depicted are those between CD40 and its ligand CD154, and between CD80/CD86 and CD28/CTLA-4). Inhibition of costimulation can result in an anergic state and subsequent unresponsiveness or tolerance. Inhibition of molecular interactions in cell adhesion can add to this inhibition of T cell activation (depicted is the interaction between LFA-1 and ICAM1/3 interaction, others include that between VLA-4 and ICAM/VCAM). Also depicted are presently available mAb used to interfere in the interactions. (CD45 is referred to in the text, but not explained here in the legend). Modified from [49].

The similarity between tacrolimus and cyclosporine regarding the immunosuppressive activity is based on the fact that both drugs inhibit calcineurin phosphatase activity and subsequent intranuclear events in T CELL activation (Fig. 2). However, in accord with the different molecular structure, tacrolimus does not bind to CYP as cyclosporine does. A family of immunophilins with rotamase activity has been identified, called FK506-binding proteins (FKBP). Of these, FKBP12 appears to be most relevant: upon binding to tacrolimus, the complex binds to calcineurin at the same site as the cyclosporine-CYP complex and thereby inhibits T CELL activation.

The EFFICACY of tacrolimus in humans was first shown in liver transplantation. In liver disease, e.g., liver allograft dysfunction, the absorption of tacrolimus is increased and its metabolism decreased, resulting in higher exposure, which is required in cases of rejection. Nowadays the drug is widely used in patients that poorly tolerate cyclosporine treatment; in addition to this conversion in immunosuppressive treatment regimen, the drug is also used in patients directly after transplantation (so-called *de novo* treatment).

Inhibitors of growth factor modulation

Sirolimus (rapamycin)

Sirolimus is a macrocyclic lactone, isolated from the actinomycete *Streptomyces hygroscopicus*, with a long history dating from the mid-1970s [9]. It was first discovered as an antifungal compound, which was not further developed as side-effects were encountered including involution of lymphoid tissue. Subsequently, antitumor effects were documented, as well as immunosuppressive activity in rat models of autoimmune disease. After structural similarities between sirolimus and tacrolimus were identified, studies on the EFFICACY of the compound in rat and mouse organ allograft models were initiated and first reported in 1989. Since then, the compound has been developed for clinical use in transplantation and was launched in 1999. A major complication was the development of a proper oral formulation with acceptable stability, BIOAVAILABILITY and predictability in absorption characteristics. The compound is very lipophilic and poorly soluble in water. In oily solution as well as in microemulsion, the compound appears readily absorbed after oral administration. A rapamycin derivative with improved physicochemical characteristics is in registration phase (everolimus) [10]. The relatively narrow therapeutic window requires regular drug exposure monitoring. Major side-effects are hyperlipidemia, followed by cholesterolemia and thrombocytopenia.

Compounds of the rapamycin class proved to be extremely potent immunosuppressants, affecting both T LYMPHOCYTES and Ab production by cells in the B lymphocyte lineage. This was demonstrated in a wide spectrum of experimental animal models and in clinical trials in renal transplantation. This is achieved at relatively low blood levels *in vivo*. In animal models, the drug shows synergistic action in combination with cyclosporine, which points to a difference in MECHANISM OF ACTION. Indeed, whereas cyclosporine and tacrolimus inhibit early events in T CELL activation, e.g., the expression of growth factors such as IL-2 in G0-G1 stage of the cell cycle (see above), rapamycin inhibits progression of G1 to S phase of the cell cycle. At the molecular level, it has been demonstrated that rapamycin binds, as FK506, to immunophilins of the FKBP family, in particular FKBP12 (Fig. 2). The rapamycin-FKBP complex does not bind to calcineurin, but to target molecules called mTOR (mammalian targets of rapamycin, alternative names in the literature are FRAP and RAFT, Fig. 2), proteins with a kinase activity [11]. This rapamycin-FKBP-TOR complex inhibits intracellular CYTOKINE-driven cell proliferation, presumably via p70 S6 kinase, which is involved in translational control (Fig. 2). Apparently, this pathway is particularly relevant in lymphoid cells, underlying the peculiar immunosuppressive characteristics of the drug.

Since growth factor-driven cell proliferation applies to other cell types as well, the therapeutic window of compounds in the rapamycin class is expected to be narrow. On the other hand, the antiproliferative action of the drug could be beneficial in chronic rejection of solid organ allografts. Indeed, it has been shown that rapamycin, in contrast to cyclosporine and tacrolimus, inhibits the proliferation of smooth muscle cells *in vitro*, and in animal transplantation models inhibits intima proliferation of blood vessels as observed in chronic rejection. There are first indications that this applies also to the clinical situation of patients with an organ allograft [12]. Recently, this beneficial effect on vasculopathies has opened up a new application in relation to stents placed in the vasculature of patients after balloon coronary angioplasty: such stents can show restenosis resulting in recurrence of the vessel occlusion. Either treatment of the patient with sirolimus, or bathing the stent before implantation in a rapamycin-containing medium reduces this complication [13]. The present understanding on the MECHANISM OF ACTION of compounds in the rapamycin class has renewed interest in mTOR inhibitors as potential anti-cancer drugs [14].

Cytotoxic drugs

The designation 'cytotoxic drugs' is used here to describe drugs that directly interfere with DNA/RNA synthesis and as such affect cell proliferation.

Cyclophosphamide

Cyclophosphamide is one of the first immunosuppressive drugs described. It is an alkylating agent that was originally used as an anticancer drug. The compound inhibits cells from entering the S phase of the cell cycle, which is subsequently blocked at the G2 phase. The drug was used in initial trials in clinical transplantation around 1970, in particular in patients with azathioprine toxicity. Severe side-effects were encountered, mainly BONE MARROW depression with severe leukopenia and anemia. Since the introduction of more selective T CELL immunosuppressants, the drug has barely been used any more because of these side-effects. However, cyclophosphamide is among the most powerful inhibitors of B LYMPHOCYTES, and has received renewed interest in experimental animal research in xenotransplantation.

Methotrexate

Methotrexate is a folate antagonist, which nowadays is mainly used in a low-dose treatment regimen (weekly administration) in subsets of patients with rheumatoid arthritis (see Chapter C.13). Its MECHANISM OF ACTION under these conditions is not completely resolved. It has been suggested that methotrexate at low dose is converted into a polyglutamate that inhibits transmethylation reactions resulting in an increased release of adenosine and decreased synthesis of guanosine. Although this condition affects purine metabolism, an antiinflammatory signal is delivered by the binding of adenosine to specific adenosine (A2) receptors. Thus, methotrexate at a low dose does not appear to be an immunosuppressant, but rather an antiinflammatory drug.

Azathioprine

Azathioprine has been used from the early days of clinical IMMUNOSUPPRESSION, being introduced as an immunosuppressant for transplantation in 1961. Its development followed the pioneering work on 6-mercaptopurine as an antileukemic agent in the 1940s. Nowadays, the drug is still in use in conjunction with baseline IMMUNOSUPPRESSION (e.g., cyclosporine) in transplantation: its dosing and dose adaptations are based on the initial side-effects, i.e., blood leukocyte counts. Despite its long clinical use, the exact MECHANISM OF ACTION is still not completely clear. The drug is converted by red blood cell glutathione to 6-mercaptopurine, which in turn is converted into a series of mercaptopurine-containing nucleotides which interfere with the synthesis of DNA and polyadenylate-containing RNA. One of the nucleotides formed is thioguanylic acid which can form thioguanosine triphosphate. This can be incorporated into nucleic acids and induce chromosome breaks as well as affect the synthesis of coenzymes. As a general inhibitor of cell proliferation, azathioprine affects both T and B lymphocyte reactivity.

Mizoribine

Mizoribine is an imidazole nucleoside, isolated as a potential antibiotic from the culture filtrate of the soil fungus *Eupenicillium brefeldianum*. Its immunosuppressive activity was demonstrated first by the inhibition of mouse lymphoma cell lines and subsequently by the inhibition of an Ab response in mice immunized with SRBC. It was subsequently shown that the drug is phosphorylated intracellularly to the active form, mizoribine 5'monophosphate, under the influence of adenosine kinase. This compound is a competitive inhibitor of the enzyme inosine monophosphate dehydrogenase (IMPDH), which is a rate-limiting enzyme in purine biosynthesis in lymphoid cells (Fig. 4). The drug has been in use in clinical transplantation since 1984, only in Japan, mainly as a replacement for azathioprine [15, 16].

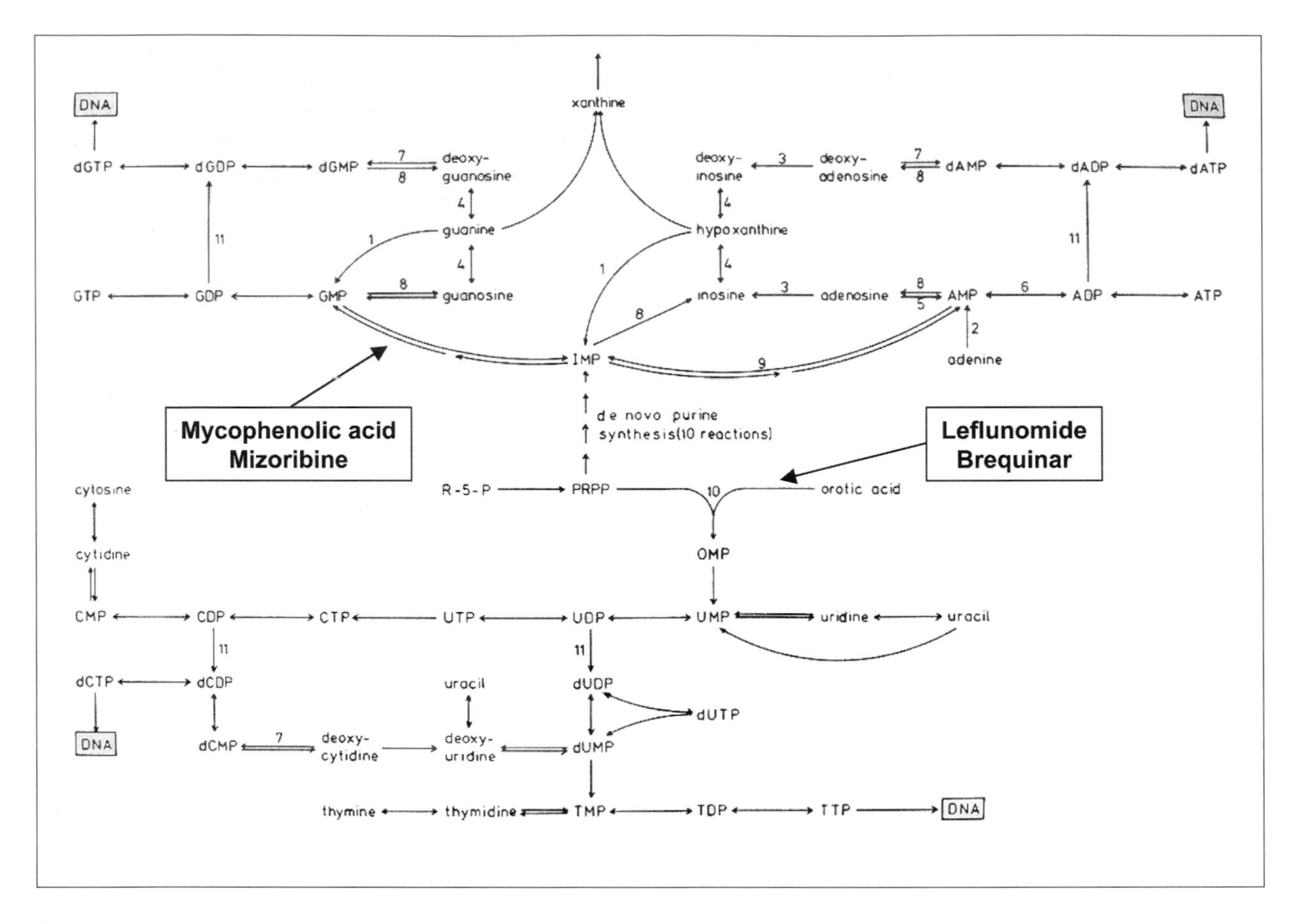

FIGURE 4

Purine/pyrimidine pathway and points of inhibition by mycophenolic acid, mizoribine, leflunomide, and brequinar. For purine metabolism two pathways are shown, the salvage pathway (conversion of guanine into guanosine monophosphate, GMP, by hypoxanthine-guanine phosphoribosyltransferase, HGPRT) and the de novo *pathway (conversion of inosine monophosphate into guanosine monophosphate, mediated by IMPDH). For pyrimidine metabolism the* de novo *pathway (conversion of orotic acid into uridine monophosphate, UMP) and the salvage pathway (conversion of uridine in UMP, and cytidine in cytidine monophosphate, CMP) are shown. Major enzymes involved are HGPRT (1), adenine phosphoribosyltransferase (2), adenosine deaminase (3), purine nucleoside phosphorylase (4), adenosine kinase (5), adenylate kinase (6), deoxycytidine kinase (7), 5'nucleotidase (8), adenosine monophosphate dehydrogenase (9), dihydroorotate dehydrogenase (10), and ribonucleotide reductase (11).*

Mycophenolic acid

Mycophenolate mofetil is the morpholinoester of mycophenolic acid (MPA) [16], a FERMENTATION product of various Penicillium species, orginally isolated and purified in the early 1910's. MPA was originally studied for its antibacterial and antifungal activity, and subsequently its antitumor activity in the late 1960s, but these activities were not further followed up in clinical development. The compound was specifically selected as a drug inhibiting IMPDH in the mid 1980's: this selection was based on the fact that MPA was not a nucleoside, that it failed to require phosphorylation to become active, and that it did not show unwanted side-effects of nucleosides, such as induction of chromosome breaks and inhibition of DNA repair enzymes. MPA is a noncompetitive reversible inhibitor of IMPDH (Fig. 4). The morpholinoester, mycophenolate mofetil (RS-61443), was developed as an immunosuppressant. It is rapidly hydrolysed by esterases to yield MPA. The drug was introduced to the market in 1995 for the indication transplantation, mainly as a replacement for azathioprine. Major side-effects are gastrointestinal intolerability and BONE MARROW depression, documented for a marketed formulation of the mycophenolate mofetil (Cellcept®): an enteric-coated formulation of mycophenolate sodium (Myfortic®) which has fewer gastrointestinal side-effects is now in registration phase.

The fact that inhibition of IMPDH causes quite selective IMMUNOSUPPRESSION is related to the relevance of different pathways in purine metabolism in different cell types (Fig. 4). Two pathways exist, the salvage pathway (conversion of guanine into guanosine monophosphate by hypoxanthine-guanine phosphoribosyltransferase) and the *de novo* pathway (conversion of inosine monophosphate into guanosine monophosphate, mediated by IMPDH). LYMPHOCYTES highly depend on the *de novo* pathway and do not use the salvage pathway. At the other end of the spectrum, cells of the central nervous system highly depend on the salvage pathway. Cell types like smooth muscle cells, fibroblasts, endothelial cells and epithelial cells can use both pathways for purine synthesis. Hence, inhibition of IMPDH results in a quite selective inhibition of purine biosynthesis in LYMPHOCYTES. For MPA, an additional selectivity has been documented for the two isoforms of IMDPH: the type I isoform is predominantly expressed in resting LYMPHOCYTES and the type II is strongly expressed in LYMPHOCYTES after activation. This type II isoform is 4-5 times more sensitive to inhibition by MPA than the type I isoform: hence MPA is a more potent inhibitor of activated LYMPHOCYTES. As both T and B LYMPHOCYTES are affected by IMPDH inhibition, MPA is an effective inhibitor of both T and B LYMPHOCYTES like mizoribine mentioned above. This has not only been demonstrated in rodent models, but also in pig-to-primate xenotransplantation models, where the suppression of xeno-Ab formation is a critical issue.

Apart from affecting RNA synthesis, IMPDH inhibition has other effects as well. IMDPH inhibition results in depletion of guanosine triphosphate (GTP) in LYMPHOCYTES (Fig. 4), which affects the transfer of fucose and mannose to glycoproteins, and this can affect ADHESION MOLECULES on the cell surface. Examples of ADHESION MOLECULES whose expression is inhibited are very late ANTIGEN-4 (VLA-4) and ligands for selectins. On the basis of this mechanism, IMPDH inhibitors could affect the recruitment of inflammatory cells into tissue, and effector-target cell interactions within tissues. This potential effect has been demonstrated by the inhibition of the adhesion of T LYMPHOCYTES to endothelial cells *in vitro* when either T CELLS or endothelial cells are pretreated with MPA.

Finally, as IMDPH inhibition can affect cell growth in other cell types besides LYMPHOCYTES, it might have an effect in chronic rejection. MPA inhibits proliferation of human smooth muscle cells *in vitro*, and is effective in a rat vessel transplantation model which mimicks chronic rejection in solid organ allografts.

Leflunomide

Leflunomide (HWA 486) is an isoxazole derivative, originally synthesized as part of an agriculture herbicide programme in the mid-1970s. Its antiinflammatory activity was demonstrated in animal models of adjuvant arthritis and experimental allergic encephalomyelitis; first studies on its immunosuppressive action in models of autoimmune disease were documented in 1990. Since then, the compound has been extensively investigated in animal models of solid organ allo- and xeno-transplantation (rodents, dog, nonhuman primates) and proved to be a potent immunosuppressant both for T and B LYMPHOCYTES. Leflunomide has been clinically developed for rheumatoid arthritis and was launched in 1999 [17]. Following this launch, a number of clinical trials have shown EFFICACY of the compound in transplantation. A major drawback to its development for the indication transplantation is the long $t_{1/2}$ in man (15–18 days): analogues (malononitrilamides MNA279 and MNA215) with a shorter $t_{1/2}$ are in clinical development [18].

Leflunomide is a prodrug, that under the influence of intestinal mucosa or liver is metabolized to the active isoxazole open-ring form (compound A 77 1726). The antiproliferative action (entry into S phase of the cell cycle), although still not completely understood, is based on its inhibition of two different intracellular pathways. The first involves the enzyme dihydroorotate dehydrogenase (DHODH), which is the fourth rate-limiting sequential enzyme in the *de novo* pyrimidine biosynthetic pathway (Fig. 4). This pathway is particularly relevant for the proliferative response of LYMPHOCYTES, as limited intracellular pools of substrates restrict the use of the salvage pathway by these cells. Also, the intracellular concentration of DHODH is relatively low in lymphoid cells, so that this pathway is easily inhibited. Inhibition of DHODH results in depletion of intracellular pyrimidine nucleotides, which have several vital cellular functions including synthesis of DNA, RNA, glycoproteins (adhesion proteins) and phospholipids. A second MECHANISM OF ACTION is inhibition of protein phosphorylation by inhibition of tyrosine kinase activity. The relevance of this mechanism for the immunosuppressive action of leflunomide is questionable as kinase inhibition generally requires higher concentrations in *in vitro* experiments than the inhibition of DHODH.

Like inhibition of IMPDH, inhibition of DHODH affects not only T CELLS but also B CELLS; hence leflunomide has been been shown to be a potent drug affecting T-dependent and T-independent Ab synthesis. There are reports that leflunomide affects B CELLS even more potently than T LYMPHOCYTES. There are also claims that the drug might be effective in chronic rejection, as demonstrated by the *in vitro* inhibition of rat smooth muscle cell proliferation, and the EFFICACY *in vivo* in rodent vessel transplantation, a model mimicking vascular pathology in chronic rejection of solid organ allografts. Apart from its immunosuppressive activity, the antiinflammatory action of the drug (inhibition of the production of inflammatory CYTOKINES) might be relevant for this indication.

Brequinar

Brequinar sodium (DUP 785) is another drug with a similar MECHANISM OF ACTION as leflunomide. The compound is a substituted 4-quinolinecarboxylic acid analogue produced by organic synthesis. It was originally developed as an anticancer drug and subsequently as an immunosuppressant for the indication transplantation. It is a potent inhibitor of DHODH; like leflunomide it also inhibits protein tyrosine kinase but only at relatively high concentrations. Its clinical development has been halted because of a narrow therapeutic window and side-effect profile (BONE MARROW depression, mainly thrombocytopenia).

Other immunosuppressive drugs

Gusperimus (15-deoxyspergualin)

Gusperimus is a synthetic derivative of spergualin, an antibiotic isolated from the soil bacterium *Bacillus laterosporus* in a screening programme for anticancer drugs in the early 1980s. Its immunosuppressive activity in a mouse skin transplantation model was demonstrated in 1985, and the drug was subsequently developed for the indication transplantation. It has been available commercially in Japan since 1994. Major drawbacks to its clinical application are the low oral BIOAVAILABILITY of the drug, which means that it has to be administered parenterally, and its instability in aqueous solution. Analogues have been prepared with improved stability and pharmacokinetics (tresperimus [19] and LF 15-0195).

Gusperimus is a potent immunosuppressive compound in animal models of transplantation and autoimmunity. B CELLS are inhibited equally well as T CELLS, and the drug also shows antiinflammatory activity. However, the MECHANISM OF ACTION is less well understood than that of the compounds mentioned above. The compound affects the differentiation of stimulated LYMPHOCYTES into effector cells, e.g., CTL or Ab-producing plasma cells, and the entry of cells from G0 or G1 into the S phase of the cell cycle. There are also indications that gusperimus blocks the maturation of DENDRITIC CELLS, which are important in ANTIGEN presentation: due to this blockade ANTIGEN presentation is inhibited. An inhibitory effect on macrophage activation has been claimed as well. At the molecular level, the drug blocks the nuclear transcription factor NF-κB, which appears to underlie its inhibitory activity in production of proinflammatory CYTOKINES and maturation of DENDRITIC CELLS. Also, binding of the compound to a cytosolic member of the heat-shock protein family, HSP70, has been described. Heat-shock proteins participate in the folding and unfolding of proteins and play a role in protein transport to intracellular organelles (the so-called chaperone function of these molecules in ANTIGEN presentation). The peptide-binding groove of this protein appears to be similar to that of MHC class I molecules. Therefore it has been suggested that gusperimus may compete for the peptide-binding groove and thereby inhibit ANTIGEN presentation.

FTY720

FTY720 has been recently identified as a novel IMMUNOMODULATORY agent which affects lymphocyte recirculation. It is a chemical derivative of myriocin, a metabolite of the ascomycete *Isaria sinclairii*. FTY720 prolongs allograft survival with high potency and EFFICACY in mice, rats, dogs and nonhuman primates, and is effective in other preclinical models of immune-mediated diseases as well [20, 21]. In these models a high level of synergy with other immunosuppressants like cyclosporine was documented. Also, in contrast to other low molecular weight xenobiotic immunosuppressants, the compound has a rather high therapeutic window. Remarkably, it does not affect protective immunity to, e.g., virus infections. Its MECHANISM OF ACTION has recently been elucidated: upon phosphorylation it is an agonist at four sphingosine 1-phosphate receptors that are expressed on LYMPHOCYTES and endothelial cells. The stimulation of such receptors results in the sequestration of LYMPHOCYTES in secondary LYMPHOID ORGANS. Thus, FTY720 treatment results in a depletion of T and B CELLS from the periphery so that they cannot participate in immune responses towards an allograft. The compound is presently in advanced clinical trials for the indication transplantation.

Biologicals

ALG/ATG/OKT3

Horse ALG and rabbit ATG [22] were originally introduced as immunosuppressants for INDUCTION TREATMENT or for treatment of rejection episodes. These reagents induce a severe but temporary depletion or inactivation of T CELLS (in the case of ATG) or LYMPHOCYTES (ALG) from the circulation. OKT3 is a mouse IgG2a mAb-specific for the CD3 ε-chain of the TCR complex. Upon OKT3 binding to CD3, the entire receptor complex is modulated (internalized) from the cell surface. This modulation results in depression of T CELL activity. The MECHANISM OF ACTION of OKT3 is not completely understood; there are claims that the ANTIBODY also induces APOPTOSIS of the cells, and hence has T CELL-depleting activity as well. Besides IMMUNOSUPPRESSION, there are indications that anti-CD3 antibodies might induce immune TOLERANCE as well [23].

OKT3 was the first mAb to be approved for clinical application in the mid-1980s. Its use in either INDUCTION TREATMENT, or treatment of steroid-resistant rejection, has shown the potential side-effects of this class of biologicals. A major side-effect is the so-called CYTOKINE-release syndrome that is related to the potent stimulatory activity of OKT3 (besides its depressing activity). This CYTOKINE-release syndrome can emerge quickly upon first dosing and results in malaise, fever, myalgia, rigors, headache and diarrhoea, in more severe cases hypotension, wheezing and/or pulmonary oedema. Also a temporary rise in serum creatinine is part of this CYTOKINE-release syndrome. A second side-effect is related to the fact that OKT3 is a mouse Ig, and thus can induce anti-mouse Ab formation. The presence of such Ab in the circulation reduces the EFFICACY of OKT3 in subsequent courses of treatment. There are a number of engineered CD3 Ab in development in which the original mouse Ab has been changed, for instance into a humanized version unable to bind FcγR [24].

Anti-IL-2 receptor Ab

The development of Ab to the α-chain of the IL-2R (anti-CD25 Ab) is based on the fact that this chain of the receptor is expressed on the surface of T CELLS only after activation; in peripheral blood CD25+ cells are present in quite low numbers. Thus, CD25 Ab are presumed to bind only activated T CELLS. Two Ab have been approved for the indication transplantation during the last decade: one is a humanized Ab (daclizumab [25]), and one is a chimeric molecule (basiliximab [26]). Both Ab have been shown to reduce the incidence of acute rejection after kidney transplantation. There are no relevant side-effects and CD25+ T CELLS are absent from the circulation as long as receptor-saturating ANTIBODY levels are maintained.

Other antibodies

There are a number of Ab that have been approved for other indications than transplantation, mainly for lymphoid malignancies, but that subsequently showed EFFICACY in transplantation. The best example is the humanized anti-CD52 mAb alemtuzumab recognizing a peptide linked to a membrane glycoprotein present on all LYMPHOCYTES, MONOCYTES and MACROPHAGES. This mAb has been successfully introduced for the treatment of B lymphoid malignancies. It causes a marked and persistent depletion of LYMPHOCYTES: within the T lymphocyte lineage this particularly affects the CD4+ population. Clinical trials in transplantation have revealed that a short course of treatment during induction at a much lower dose than used in the oncological indication results in a strongly reduced requirement for conventional IMMUNOSUPPRESSION (cyclosporine) during MAINTENANCE IMMUNOSUPPRESSION, a condition called 'almost' TOLERANCE [27, 28]. It is generally anticipated that this biological will get a solid position in INDUCTION IMMUNOSUPPRESSION after transplantation.

A variety of mAb to T CELLS other than anti-CD3 mAb are in various stages of clinical development: examples are the humanized anti-CD2 mAb siplizumab [29] and anti-CD4 mAb like the primatized mAb clenoliximab (IDEC151), and an anti-CD4 mAb with a mutated Fc portion called TRX1. Ab to the RB isoform of CD45 (Fig. 3) have received attention when it was shown that such antibodies can induce TOLERANCE in mice transplantation models and prolong survival in nonhuman primate transplantation models [30], and first clinical trials with a CD45RB mAb have been initiated. Similarly, Ab blocking costimulatory signals like anti-CD80 and anti-CD86 showed EFFICACY in a nonhuman primate transplantation model (these Ab had to be given together, as either CD80 or CD86 signalling only is sufficient in COSTIMULATION) [31] but it is not known whether these antibodies will be developed to a clinical application. Anti-CD154 mAb have been successfully applied in nonhuman primate transplantation models [32], and first clinical trials were initiated in transplantation: after thromboembolic side-effects were observed in these and other trials, the clinical development was put on hold. Anti-CD40 Ab have shown EFFICACY in nonhuman primates [33] and are currently in early development.

Therapeutic mAb to ADHESION MOLECULES have been generated (Fig. 3), either for the indication transplantation or for immune diseases like psoriaris. Anti-CD11a (anti-lymphocyte function antigen-1, LFA-1) antibodies received attention based on their EFFICACY in experimental transplantation models, including transplantation in nonhuman primates [34]. The murine anti-CD11a mAb odulinomab has been on the market for transplantation since 1997; it has not yet gained wide application, and its EFFICACY has not been proven unequivocally. Another humanized anti-CD11a mAb is efalizumab [35] that is on the market for psoriasis and is at present being evaluated in transplant patients as well. Other mAb target ADHESION MOLECULES, like VLA-4 or its ligands (VCAM, ICAM), are being studied, based on experimental animal data that the blockade of such interactions results in allograft survival: an example is the humanized anti-VLA-4 mAb natalizumab [36] at present in clinical development for multiple sclerosis and Crohn's disease.

Other biologicals

Recombinant DNA products are receiving increasing attention as immunosuppressives. One of the best examples is CTLA-4-Ig, a fusion protein between the CD28 antagonist CD152 (CTLA-4) in coreceptor blockade and an Ig Fc fragment, which showed EFFICACY in immune TOLERANCE induction in rodent transplantation models, and is nowadays in clinical development for psoriasis (Fig. 3). Clinical trials, amongst others in transplantation, are ongoing using a mutant form of this fusion protein (LEA29Y), that has an increased affinity in blocking the CD28-B7 interaction [37]. Soluble LFA-3-Ig fusion protein binds to CD2 on T CELLS and the FcR, thereby blocking T CELL activation and proliferation; such products have shown EFFICACY in transplantation models in rodents and nonhuman primates. The LFA-3-Ig fusion protein alefacept has recently been approved for the indication psoriasis [38], and is being considered for clinical evaluation in transplant patients.

Finally, fusion proteins incorporating a toxic moiety are to be mentioned, either including a diphtheria toxin or Pseudomonas exotoxin A. IL-2 fusion toxins were the first proteins of this class that were tested in clinics: denileukin diftitox is on the market for the treatment of T CELL lymphomas [39]. A main complication of this class of compounds is liver toxicity ascribed to the diphtheria toxin fragments in the molecule. Fusion proteins comprising an anti-CD3 Ab fragment are at present in early clinical trials: CD3-immunotoxins are considered as very effective T lymphocyte-depleting agents and could therefore be effective immunosuppressants in TOLERANCE-inducing protocols [40].

New approaches and perspectives

For the development of novel immunosuppressants, several aspects have to be considered:

First, the major problem of current baseline immunosuppressants is still toxicity (direct drug toxicity, increased susceptibility to infection, development of tumors). The development of drugs with fewer side-effects may be most likely achieved by selecting targets for pharmacological intervention that are specifically relevant for the immune response and not for other (cell) biological processes. Besides selectivity, a critical aspect in the selection of new drug targets is the redundancy of potential targets. Nonredundancy of targets is demonstrated most convincingly by the phenotype of human primary immunodeficiencies and knockout mice. Interference with such targets is expected to be very effective. Specific for the IMMUNE SYSTEM are different activation mechanisms which might be relevant for direct and indirect presentation of allograft ANTIGENS. Some of these new avenues in drug development will be discussed below: they concern either intracellular processes or cell surface interactions.

Second, for solid organ transplantation, chronic rejection is a major cause of graft failure in the long term after transplantation, and therefore there is a clinical need for drugs interfering with this process. Prevention or treatment of chronic rejection or graft vessel disease may require nonimmunological approaches in addition to improved novel immunosuppressants; these nonimmunological approaches are not considered here.

Third, specific drug targeting to the organ involved presents a possibility to cope with the systemic side-effects of immunosuppressants. An example of this approach comes from lung transplantation, which generally requires strong IMMUNOSUPPRESSION and hence is associated with a rather small therapeutic window between IMMUNOSUPPRESSION and side-effects. Using inhaled immunosuppressive therapy [41], for instance corticosteroids and cyclosporine, beneficial effects with more effective IMMUNOSUPPRESSION have been shown.

Fourth, the final goal in IMMUNOSUPPRESSION in transplantation and autoimmunity is the induction of TOLERANCE, often referred to as the 'Holy Grail', i.e., donor-specific unresponsiveness in the case of transplantation. Some examples of TOLERANCE induction are given below.

Finally, antisense technologies and GENE THERAPY have developed during the last few years to such an extent that first applications in IMMUNOSUPPRESSION become feasible.

Antigen presentation to T cells

As the interaction between T LYMPHOCYTES and APC and the subsequent activation of T CELLS are the central events leading to the initiation of an immune response (in the case of transplantation to graft rejection), current approaches focus on molecules specifically involved in this process. Of special interest are molecules involved in the costimulation of T CELLS, which according to the two-signal model are essential for T CELL activation (Fig. 3). Compounds preventing T CELL costimulation have the potential to induce T CELL anergy or unresponsiveness, which could lead to antigen-specific TOLERANCE.

One approach is to intervene at the cell surface with molecules intimately involved in ANTIGEN presentation using biologicals such as antagonistic mAb or soluble receptor antagonists. Some new biologicals are described above. Surface molecules of special interest for costimulatory blockade include on the T CELL, among others, CD2, CD4, CD28 and the related molecule CTLA-4, and CD154 (CD40-ligand). Regarding adhesion properties of T CELLS, the interaction between LFA-1 and ICAM-1, and that between VLA-4 and VCAM-1, are the subject of investigation. CD2 plays an important role not only in T CELL activation but also in T CELL adhesion, and anti-CD2 mAb have been shown to have EFFICACY in prevention of allograft rejection in animal models. CTLA-4-Ig has been shown to be efficacious in several rodent and primate transplantation models. More recently, anti-CD154 mAb have been introduced, which especially in combination with CTLA-4-Ig showed promising results in rodent transplantation models, including the prevention of chronic rejection. However, the clinical development of anti-CD154 mAbs has been halted because of adverse side-effects in some patients (thromboembolic phenomena).

On the APC side, mAb to ligands of T CELL costimulatory molecules like CD80 (B7.1), CD86 (B7.2) and CD40 have been used successfully in transplantation models. Blocking CD40 is especially attractive since CD40 is expressed not only on dendritic APC but also on MACROPHAGES and B CELLS. However, there is also expression to some extent on epithelial cells in inflammatory processes: non-stimulatory Ab described above appear not to bind to epithelium-expressed CD40 [42]. Further potential sites of intervention are at the level of ANTIGEN uptake, processing and assembly with MHC class II molecules.

Drug discovery programmes are aimed at the identification of small-molecular-weight-compounds which inhibit specific cell surface interactions. In contrast to the biologicals mentioned above, such inhibitors would have the advantage of being orally active. However, thus far it has proved difficult to identify small molecules which block the interactions of large protein surfaces, and to mimic the activity shown for larger-sized Ab molecules.

T-cell signal transduction

Present knowledge of T CELL transduction pathways (Chapter A.2) has revealed potentially interesting targets for drug development. The signalling pathway emanating from the TCR includes, among others, T CELL-selective molecules such as protein tyrosine kinases of the src family ($p56^{lck}$ and $p59^{fyn}$) and ZAP-70 (see also the section on calcineurin inhibitors above, Fig. 2). The pivotal and selective role of ZAP-70 for T CELL activation is documented by the severe combined immunodeficiency phenotype of humans who lack functional ZAP-70. CD28 coreceptor signalling is less well defined and involves components of several pathways like Janus activated kinase (JAK), PI3 kinase, $p21^{ras}$ and p38 MAP kinase. The relative contribution of these cascades to overall CD28 signalling may depend on the state of activation of the T CELLS and the level of CD28 activation. Interestingly, it has been reported that CD28 signalling may involve cyclosporine-resistant pathways (see above). All these kinases are potential targets for the development of new immunosuppressants. The major challenge here is to identify inhibitors with high selectivity which fail to inhibit other kinases, which are of critical importance in other cell types as well. Recently, a JAK3 inhibitor (CP-690,550) has been described that was highly efficacious in nonhuman primate transplantation without manifesting overt side-effects.

A number of molecules that do not exhibit catalytic function act as specific adaptor molecules by mediating the interaction between different components of signal transduction pathways. Generally these proteins contain domains that are important for protein-protein interactions such as SH2/SH3 domains. These interactions are being actively studied as potential drug targets as well. Here again, the major hurdle is to develop small-molecular-weight inhibitors which block the interaction of relatively large protein surfaces.

Cytokines

Immune cells communicate not only through direct contact of cell surface molecules, but also through CYTOKINES (see Chapter A.4). The cloning of CYTOKINES and CYTOKINE receptors and improved understanding of their function and elucidation of their signal transduction pathways has opened new possibilities for intervention (see anti-IL-2 receptor Ab described above). However, especially in the field of CYTOKINES, redundancy has to be considered. Of special interest is the paradigm of CYTOKINES associated with Th1/Th2 cells or immune deviation (see Chapter A.2). It is considered that transplant rejection may represent primarily a Th1 response, whereas transplantation TOLERANCE may be favored by Th2 cells. However, it is unclear whether this hypothesis applies to allograft rejection in humans. Nevertheless, purified proteins, engineered proteins (antagonistic mutants), domains of proteins (soluble CYTOKINE receptors) or antibodies are used to manipulate CYTOKINE responses. Certain CYTOKINES possessing intrinsic immunosuppressive properties, like IL-10 and TGF-β, have been administered as such to treat rejection. CYTOKINES promoting Th2 responses (e.g., IL-4) or blockade of CYTOKINES promoting Th1 responses (e.g., IL-12) might induce a beneficial deviation of the IMMUNE SYSTEM towards a Th2 response. Interference with CYTOKINES in the clinical setting so far has mainly been shown for the inflammatory CYTOKINE TNF-α: the mAb infliximab [43] is on the market since 1988 for the indication rheumatoid arthritis and Crohn's disease, and of more recent date is ETANERCEPT, a fusion protein comprising the extracellular portion of the 75kD TNF-R linked to the Fc component of human IgG1 [44]. Another antiinflammatory product on the market is the IL1-R antagonist anakinra [45].

Cell traffic/adhesion

The regulation of leukocyte trafficking by adhesion represents another important paradigm for the modulation of the IMMUNE SYSTEM. Adherence of LEUKOCYTES to the vascular endothelium is not only pivotal to an inflammatory process, but also plays a role in transplant rejection since the graft endothelium is the major contact site for immune cells of the host. Several cell surface molecules are involved in the adhesion process, which is subdivided into several steps, i.e., rolling, firm adhesion and extravasation. Members of the selectin family mediate the first contacts via carbohydrate structures. Members of the heterodimeric integrin superfamily like CD11a, LFA-1 and VLA-4 are involved in the later steps. Individual integrins are expressed on different cell types. They interact with their specific, widely expressed ligands that include not only surface molecules but also components of the complement system and the extracellular matrix, which all contribute to the activation of the graft endothelium. In addition to their function as ADHESION MOLECULES, VLA-1 and LFA-1 have a costimulatory activity for T LYMPHOCYTES (Fig. 3). All these features make these molecules attractive targets for the prevention of graft rejection, especially for chronic rejection. Recently, chemokine receptors CXCR3 received increased interest, after it was shown that such receptors are upregulated in allografts during acute rejection, and that animals lacking CXCR3 showed delayed or absent graft rejection [46]. Biologicals on the market or advanced clinical development have been described above: essentially the only low-molecular-weight compound affecting lymphocyte trafficking identified thus far is the sphingosine 1-phosphate receptor agonist FTY720.

Tolerance

Immune TOLERANCE, in other words unresponsiveness in transplant rejection or a status of allograft function and life-support without the necessity of chronic IMMUNOSUPPRESSION, can be relatively easily induced in rodent transplantation models, for instance in rats by using a short 2–3 week course of IMMUNOSUPPRESSION. This is not the case in humans or nonhuman primate models. In the clinic, blood transfusions (either from unrelated donors or donor-specific) have given a first indication that the IMMUNE SYSTEM can be 'modulated' in order to reduce alloreactivity towards the grafted organ. There are a number of approaches in induction of transplantation TOLERANCE that have shown promising results in nonhuman primates and hence could be developed towards a clinical application, including the following:

- Induction of anergy by blocking costimulatory pathways (Fig. 3). The main focus has been on the CD28-B7 and the CD40-CD154 interactions thus far, but other targets are under investigation as well. Some reagents have been described in the biologicals section above.
- Severe T CELL depletion using an immunotoxin targeting T CELLS. A chemical conjugate between an anti-CD3 mAb and a mutated diphtheria toxin yielded long-term allograft survival of rhesus monkey kidney or islet allografts when given in the peritransplant period [40]. This result is ascribed to the severe T CELL-depleting effect of the immunotoxin, that not only included T CELL-depletion from the blood circulation as observed with an anti-CD3 ANTIBODY, but also depletion from LYMPHOID ORGANS;
- BONE MARROW chimerism [47]. This includes the elimination of the functioning IMMUNE SYSTEM including STEM CELLS in BONE MARROW by high-dose chemotherapy and/or irradiation, followed by BONE MARROW transplantation. In this situation, a condition of 'mixed hematopoietic chimerism' is created, in which the individual accepts solid organ transplants from the BONE MARROW donor without the need of additional IMMUNOSUPPRESSION. This procedure has been shown to be effective in large animal transplantation models including pigs and nonhuman primates. Peripheral chimerism (in blood and/or BONE MARROW) during the peritransplant period is generally considered as crucial in establishing TOLERANCE; it can vanish later, when the graft itself is assumed to maintain a chimeric state. The induction of 'central TOLERANCE' (TOLERANCE at the level of precursors of LYMPHOCYTES) includes heavy IMMUNOSUPPRESSION, which make this approach not directly suitable for the general transplant setting. The regimen therefore needs optimization (e.g., in rodent models) with regard to minimizing the myeloablative regimen so that a clinical application may become feasible. Otherwise, transplant patients have been selected that are anyhow receiving such a heavy treatment as part of anticancer treatment. This is possible in, e.g., myeloma patients with end-stage renal failure. In a first series of such patients long-term TOLERANCE has been achieved [48].

Summary

A series of immunosuppressives is nowadays on the market or in advanced clinical development that are efficacious in prevention or treatment of rejection of a transplant in patients with end-stage organ failure. These can roughly be divided into low-molecular-weight XENOBIOTICS, orally active drugs produced by microorganisms or chemical synthesis, and biologicals, (monoclonal) antibodies or rDNA fusion proteins. Most XENOBIOTICS work intracellularly and affect different pathways in lymphocyte activation and/or proliferation; since such pathways are not truly selective for LYMPHOCYTES, most of these drugs show inherent side-effects and generally have a low therapeutic window. However, the availability of novel agents with a broader therapeutic window, and refinements in combination treatment have greatly added to improved tolerability of immunosuppressive regimens. Broadly acting cytotoxic drugs with severe side-effects are gradually being replaced by compounds with a more selective action towards lymphoid cells, and corticosteroid-sparing regimens are being applied to reduce the adverse side-effects of corticosteroids. Most biologicals work by binding to cell surface molecules, resulting in inactivation or depletion of the target cells. Progress in this area has not only resulted in improved IMMUNOSUPPRESSION, but also in potential approaches to induce a status of unresponsiveness, i.e., TOLERANCE to the transplant. This is achieved either by efficient depletion of reactive cells or by inhibition of second signals in coreceptor blockade in T CELL activation. New paradigms in IMMUNOSUPPRESSION involve interference with cell trafficking (lymphocyte recirculation) that appears possible by using either innovative XENOBIOTICS or biologicals binding to cell ADHESION MOLECULES.

Selected readings

Allison AC (2000) Immunosuppressive drugs: the first 50 years and a glance forward. *Immunopharmacology* 47: 63–83

Kanmaz T, Knechtle SJ (2003) Novel agents or strategies for immunosuppression after renal transplantation. *Curr Opin Organ Transplant* 8: 172–178

Lieberman R, Mukherjee A (eds) (1996) *Principles of Drug Development in Transplantation and Autoimmunity*. Chapman and Hall, New York

Matthews JB, Ramos E, Bluestone JA (2003) Clinical trials of transplant tolerance: slow but steady progress. *Am J Transplant* 3: 794–803

McAlister VC (2002) New immunosuppressants in clinical trial. *Curr Drug Targets Cardiovasc Haematol Disord* 2: 73–77

Schuurman H-J, Feutren G, Bach J-F (eds) (2001) *Modern Immunosuppressives*. Birkhäuser Verlag, Basel

Vilatoba M, Contreras JL, Eckhoff DE (2003) New immunosuppressive strategies in liver transplantation: balancing efficacy and toxicity. *Curr Opin Organ Transplant* 8: 139–145

Vincenti F (2002) What's in the pipeline? New immunosuppressive drugs in transplantation. *Am J Transpl* 2: 898–903

References

1 Kahan BD, Keown P, Levy GA, Johnston A (2002) Therapeutic drug monitoring of immunosuppressant drugs in clinical practice. *Clin Ther* 24: 330–350

2 Nalesnik MA (2002) Clinicopathologic characteristics of post-transplant lymphoproliferative disorders. *Recent Results Cancer Res* 159: 9–18

3 Allan JS, Madsen JC (2002) Recent advances in the immunology of chronic rejection. *Curr Opin Nephrol Hypertens* 11: 315–321

4 Borel JF (1990) Pharmacology of cyclosporine (Sandimmune). IV. Pharmacological properties *in vivo*. *Pharmacol Rev* 41: 259–371

5 Schreier MH (1997) Mechanism of action of cyclosporin. In: T Anke (ed): *Fungal Biotechnology*. Chapman and Hall, New York, 137–146

6 Rusnak F, Mertz P (2000) Calcineurin: form and function. *Physiol Rev* 80: 1483–1521

7 Lee JI, Burckart GJ (1998) Nuclear factor kappa B: important transcription factor and therapeutic target. *J Clin Pharmacol* 38: 981–993

8 Dumont FJ (2000) FK506, an immunosuppressant targeting calcineurin function. *Curr Med Chem* 7: 731–748

9 Kahan BD (2001) Sirolimus: a comprehensive review. *Expert Opin Pharmacother* 2: 1903–1917

10 Nashan B (2002) Review of the proliferation inhibitor everolimus. Expert Opin Investig Drugs 11: 1845–1857

11 Abraham RT (2002) Identification of TOR signaling complexes: more TORC for the cell growth engine. *Cell* 111: 9–12

12 Eisen HJ, Tuzcu EM, Dorent R, Kobashigawa J, Mancini D, Valantine-von Kaeppler HA, Starling RC, Sorensen K, Hummel M, Lind JM et al (2003) Everolimus for the prevention of allograft rejection and vasculopathy in cardiac-transplant recipients. *N Engl J Med* 349: 847–858

13 Indolfi C, Mongiardo A, Curcio A, Torella D (2003) Molecular mechanisms of in-stent restenosis and approach to therapy with eluting stents. *Trends Cardiovasc Med* 13: 142–148

14 Huang S, Houghton PJ (2002) Inhibitors of mammalian target of rapamycin as novel antitumor agents: from bench to clinic. *Curr Opin Invest Drugs* 3: 295–304

15 Yokota S (2002) Mizoribine: mode of action and effects in clinical use. *Pediatr Int* 44: 196–198

16 Ishikawa H (1999) Mizoribine and mycophenolate mofetil. *Curr Med Chem* 6: 575–597

17 Miceli-Richard C, Dougados M (2003) Leflunomide for the treatment of rheumatoid arthritis. *Expert Opin Pharmacother* 4: 987–997

18 Schorlemmer H, Bartlett R, Kurrle R (1998) Malononitrilamides: a new strategy of immunosuppression for allo- and xenotransplantation. *Transplant Proc* 30: 884–890

19 Simpson D (2001) Tresperimus: a new agent for transplant tolerance induction. *Expert Opin Investig Drugs* 10: 1381–1386

20 Brinkmann V, Lynch KR (2002) FTY720: targeting G-coupled receptors for sphingosine 1-phosphate in transplantation and autoimmunity. *Curr Opin Immunol* 14: 569–575

21 Brinkmann V, Pinschewer DD, Feng L, Chen S (2001) FTY720: altered lymphocyte traffic results in allograft protection. *Transplantation* 72: 764–769

22 Gaber AO, First MR, Tesi RJ, Gaston RS, Mendez R, Mulloy LL, Light JA, Gaber LW, Squiers E, Taylor RJ et al (1998) Results of the double-blind, randomized, multicentre, phase III clinical trial of Thymoglobulin *versus* ATGAM in the treatment of acute graft rejection episodes after renal transplantation. *Transplantation* 66: 29–37

23 Chatenoud L (2003) CD3-specific antibody-induced active tolerance: from bench to bedside. *Nat Rev Immunol* 3: 123–132

24 Renders L, Valerius T (2003) Engineered CD3 antibodies for immunosuppression. *Clin Exp Immunol* 133: 307–309

25 Vincenti F, Kirkman R, Light S, Bumgardner G, Pescovitz M, Halloran P, Neylan J, Wilkinson A, Ekberg H, Gaston R et al (1998) Interleukin-2-receptor blockade with daclizumab to prevent acute rejection in renal transplantation. *N Engl J Med* 338: 161–165

26 Nashan B, Moore R, Amlot P, Schmidt AG, Abeywickrama K, Soulillou JP (1997) Randomised trial of basiliximab *versus* placebo for control of acute cellular rejection in renal allograft recipients. *Lancet* 350: 1193–1198

27 Calne R, Moffat SD, Friend PJ, Jamieson NV, Bradley JA, Hale G, Firth J, Bradley J, Smith KG, Waldmann H (1999) Campath 1H allows low-dose cyclosporine monotherapy in 31 cadaveric renal allograft recipients. *Transplantation* 68: 1613–1616

28 Kirk AD, Hale DA, Mannon RB, Kleiner DE, Hoffmann SC, Kampen RL, Cendales LK, Tadaki DK, Harlan DM, Swanson SJ (2003) Results from a human renal allograft tolerance trial evaluating the humanized CD52-specific monoclonal antibody alemtuzumab (CAMPATH-1H). *Transplantation* 76: 120–129

29 Przepiorka D, Phillips GL, Ratanatharathorn V, Cottler-Fox M, Sehn LH, Antin JH, LeBherz D, Awwad M, Hope J, McClain JB (1998) A phase II study of BTI-322, a monoclonal anti-CD2 antibody, for treatment of steroid-resistant acute graft-*versus*-host disease. *Blood* 92: 4066–4071

30 Luke PP, O'Brien CA, Jevnikar AM, Zhong R (2001) Anti-CD45RB monoclonal antibody-mediated transplantation tolerance. Curr Mol Med 1: 533–543

31 Hausen B, Klupp J, Christians U, Higgins JP, Baumgartner RE, Hook LE, Friedrich S, Celnicker A, Morris RE (2001) Coadministration of either cyclosporine or steroids with humanized monoclonal antibodies against CD80 and CD86 successfully prolong allograft survival after life supporting renal transplantation in cynomolgus monkeys. *Transplantation* 72: 1128–1137

32 Kirk AD, Burkly LC, Batty DS, Baumgartner RE, Berning JD, Buchanan K, Fechner JH Jr, Germond RL, Kampen RL, Patterson NB et al (1999) Treatment with humanized monoclonal antibody against CD154 prevents acute renal allograft rejection in nonhuman primates. *Nat Med* 5: 686–693

33 Haanstra KG, Ringers J, Sick EA, Ramdien-Murli S, Kuhn EM, Boon L, Jonker M (2003) Prevention of kidney allograft rejection using anti-CD40 and anti-CD86 in primates. *Transplantation* 75: 637–643

34 Nakakura EK, Shorthouse RA, Zheng B, McCabe SM, Jardieu PM, Morris RE (1996) Long-term survival of solid organ allografts by brief anti-lymphocyte function-associated antigen-1 monoclonal antibody monotherapy. *Transplantation* 62: 547–552

35 Dedrick RL, Walicke P, Garovoy M (2002) Anti-adhesion antibodies efalizumab, a humanized anti-CD11a monoclonal antibody. *Transpl Immunol* 9: 181–186

36 Ghosh S (2003) Therapeutic value of alpha-4 integrin blockade in inflammatory bowel disease: the role of natalizumab. *Expert Opin Biol Ther* 3: 995–1000

37 Moreland LW, Alten R, van den Bosch F, Appelboom T, Leon M, Emery P, Cohen S, Luggen M, Shergy W, Nuamah I et al (2002) Costimulatory blockade in patients with rheumatoid arthritis: a pilot, dose-finding, double-blind, placebo-controlled clinical trial evaluating CTLA-4Ig and LEA29Y eighty-five days after the first infusion. *Arthritis Rheum* 46: 1470–1479

38 Ormerod AD (2003) Alefacept. Biogen. Curr Opin Investig Drugs 4: 608–613

39 Kreitman RJ (2003) Recombinant toxins for the treatment of cancer. *Curr Opin Mol Ther* 5: 44–51

40 Knechtle SJ, Hamawy MM, Hu H, Fechner JH Jr, Cho CS (2001) Tolerance and near-tolerance strategies in monkeys and their application to human renal transplantation. *Immunol Rev* 183: 205–213

41 Akamine S, Katayama Y, Higewnbottam T, Lock T (1998) Developments in inhaled immunosuppressive therapy for the prevention of pulmonary graft rejection. *Biodrugs* 9: 49–59

42 Boon L, Laman JD, Ortiz-Buijsse A, den Hartog MT, Hof-

fenberg S, Liu P, Shiau F, de Boer M (2002) Preclinical assessment of anti-CD40 Mab 5D12 in cynomolgus monkeys. *Toxicology* 174: 53–65
43 Nahar IK, Shojania K, Marra CA, Alamgir AH, Anis AH (2003) Infliximab treatment of rheumatoid arthritis and Crohn's disease. *Ann Pharmacother* 37: 1256–1265
44 Cohen SB, Rubbert A (2003) Bringing the clinical experience with anakinra to the patient. *Rheumatology* 42 (Suppl 2): ii36–ii40
45 Goffe B, Cather JC (2003) Etanercept: An overview. *J Am Acad Dermatol* 49 (2 Suppl): S105–S111
46 Haskell CA, Hancock WW, Salant DJ, Gao W, Csizmadia V, Peters W, Faia K, Fituri O, Rottman JB, Charo IF (2001) Targeted deletion of CX(3)CR1 reveals a role for fractalkine in cardiac allograft rejection. *J Clin Invest* 108: 679–688
47 Sykes M, Spitzer TR (2002) Non-myeloblative induction of mixed hematopoietic chimerism: application to transplantation tolerance and hematologic malignancies in experimental and clinical studies. *Cancer Treat Res* 110: 79–99
48 Bühler LH, Spitzer TR, Sykes M, Sachs DH, Delmonico FL, Tolkoff-Rubin N, Saidman SL, Sackstein R, McAfee S, Dey B et al (2002) Induction of kidney allograft tolerance after transient lymphohematopoietic chimerism in patients with multiple myeloma and end-stage renal disease. *Transplantation* 74: 1405–1409
49 Vincenti F (2002) What's in the pipeline? New immunosuppressive drugs in transplantation. *Am J Transpl* 2: 898–903

Cytotoxic drugs

Romano Danesi, Guido Bocci, Antonello Di Paolo and Mario Del Tacca

Background

Cytotoxic immunosuppressive agents have a longstanding important role in pharmacological IMMUNOSUPPRESSION. Azathioprine was among the first immunosuppressive drugs used in organ transplantation, and further development in this field has been landmarked by the introduction of ALKYLATING AGENTS (i.e., cyclophosphamide) and ANTIMETABOLITES (i.e., fludarabine, methotrexate and mycophenolic acid) in therapeutic regimens for the prevention of graft rejection (see chapter C8) and the treatment of AUTOIMMUNE DISEASES because of their well-documented lymphocytolytic effect.

Although the role of cytotoxic drugs is challenged by steroids, non-steroidal antiinflammatory agents (NSAIDs), poly- and MONOCLONAL ANTIBODIES, calcineurin and mTOR inhibitors, the reduction of the toxicity burden of IMMUNOSUPPRESSION is currently being investigated by the therapeutic use of cytotoxic drugs (i.e., azathioprine and prodrugs of mycophenolic acid) in combination with immunosuppressive antibodies with the aim of developing calcineurin- and steroid-free immunosuppressive regimens.

Azathioprine

Introduction

Azathioprine, an imidazolyl derivative of 6-mercaptopurine, was developed in the 1950s to improve the BIOAVAILABILITY of its parent drug, mercaptopurine [1]. Animal studies demonstrated that azathioprine had a higher THERAPEUTIC INDEX and was a better immunosuppressant than mercaptopurine. Azathioprine is now widely used in combination regimens with steroids and calcineurin inhibitors in patients receiving solid-organ transplants, as well as in rheumatology, dermatology and gastroenterology [1]. Its use as an immunosuppressive and a corticosteroid-sparing agent is challenged by mycophenolate mofetil, which is considered a safer and more effective agent, despite the fact that recent findings do not support this evidence [2].

Chemical structure

Azathioprine is an antimetabolite prodrug for 6-mercaptopurine, with an imidazolyl group attached to the SH group of 6-mercaptopurine (Fig. 1) to protect it from *in vivo* oxidation. In tissues, azathioprine is nonenzymatically converted to mercaptopurine.

Mechanism of action and pharmacological effect

Azathioprine is a cytotoxic antimetabolite inhibitor of nucleic acid synthesis. In particular, its metabolite 6-mercaptopurine is further activated intracellularly by ANABOLIC BIOTRANSFORMATION to 6-thioinosinic and 6-thioguanine acid which interfere with the metabolism of inosine monophosphate (IMP) to adenosine monophosphate (AMP) and triphosphate (ATP), thereby impairing the purine *de novo* biosynthetic pathway. Moreover, active metabolites are incorporated into RNA as well as DNA and its replication is inhibited. The drug suppresses the proliferation of T and B LYMPHOCYTES and reduces the number of cytotoxic T CELLS and plasma cells in circulation and peripheral organs, thereby decreasing the immunological reactivity of the host [3]. For these reasons,

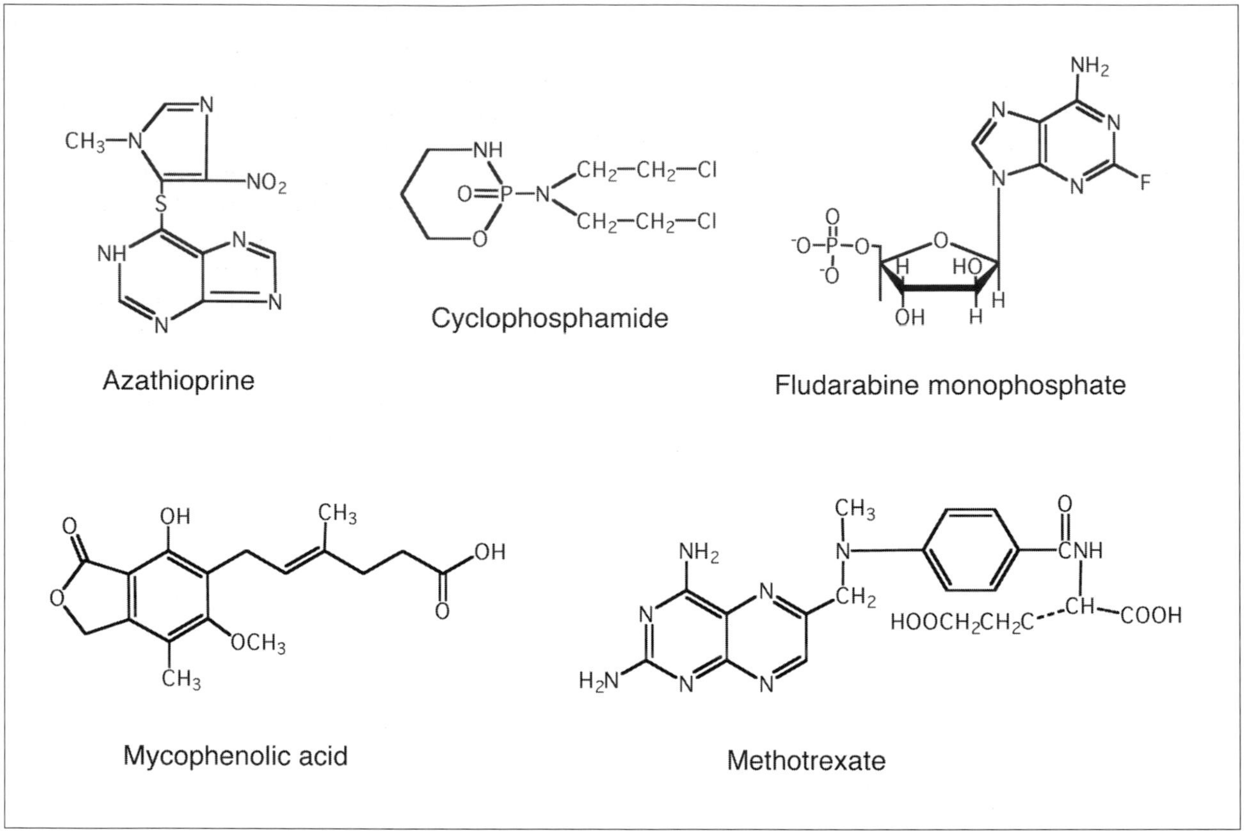

FIGURE 1. CHEMICAL STRUCTURES OF AZATHIOPRINE, CYCLOPHOSPHAMIDE, FLUDARABINE, METHOTREXATE AND MYCOPHENOLIC ACID

azathioprine may exert a modest anti-inflammatory effect.

Pharmacokinetics

Azathioprine is rapidly absorbed within 1–2 h after administration, and evenly distributed in all tissues, although the drug does not cross the blood–brain barrier. Azathioprine is rapidly converted in the liver and ERYTHROCYTES to 6-mercaptopurine and S-methyl-4-nitro-5-thioimidazole by sulfhydryl-containing compounds. Although it is generally recognized that 6-mercaptopurine is the active drug, a previous study has suggested that the imidazolyl moiety might have immunosuppressive activity on its own [4]. After standard oral doses, the terminal half-lives ($t_{1/2}$) of azathioprine and 6-mercaptopurine are 50 and 74 min, respectively. Azathioprine is mainly metabolized by xanthine oxidase, followed by thiopurine methyltransferase (TPMT) and hypoxanthine–guanine phosphoribosyltransferase (HGPRT). 6-Mercaptopurine is inactivated by xanthine oxidase to 6-thiouric acid and by the widely distributed enzyme TPMT to 6-methylmercaptopurine, with *S*-adenosylmethionine as the methyl donor. The anabolic pathway is dependent on the enzyme HGPRT of the purine salvage metabolic pathway, with subsequent multienzymatic steps leading to the formation of cytotoxic 6-thioguanine nucleotides. These active metabolites can accumulate in tissues, where they are catabolized or incorporated into nucleic acids. In addition to its action on 6-mercaptopurine, TPMT can also methylate metabolites of the HGPRT pathway, including thioinosine monophosphate, which is in turn a potent inhibitor of *de novo* purine synthesis. While xanthine oxidase is not believed to play a significant role in 6-mercaptopurine inactivation at the level of hematopoietic tissues, TPMT activity is the principal detoxification pathway for the cytotoxic thioguanine nucleotides in BONE MARROW. Hematopoietic toxicity is thus largely dependent on the activity of TPMT, a polymorphic ezyme with genetic variants characterized by low activity and increased risk of severe toxicity in patients [5]. Optimization of azathioprine treatment may thus be performed by TPMT genotyping as well as determination of erythrocyte 6-TGN levels.

Clinical indications

Azathioprine is an approved drug for renal transplantation and severe rheumatoid arthritis. Its EFFICACY has also been proven in the management of severe ulcerative colitis, and other autoimmune disorders, including bullous diseases in dermatology [6]. Azathioprine is also used as a steroid-sparing agent and has also been administered in cardiac transplantation. The initial dose of azathioprine is 3–5 mg/kg/day; the intravenous formulation may be used postoperatively followed by drug administration by oral route. In combination with cyclosporine and steroids, the dose may be lowered to 1–3 mg/kg/day as a maintenance level. During a 6-month treatment with azathioprine along with cyclosporine microemulsion and steroids in recipients of cadaveric kidney transplants, 35% had clinical rejections, compared with 12% over 15 additional months without steroids [2].

Adverse reactions

The dose of azathioprine should be reduced in patients administered allopurinol, because the inhibition of xanthine oxidase reduces the metabolic inactivation of azathioprine and increases the drug's adverse events. Overall, up to 20% of patients may discontinue treatment due to toxicity [7]. Hematopoietic toxicity, including neutropenia, anemia and thrombocytopenia, is the most common dose-limiting adverse effect, and it is frequently associated with low TPMT activity. Hepatic toxicity is the second most common adverse event and occurs independently of TPMT activity which might be correlated with the production of 6-methylmercaptopurine; therefore, therapeutic monitoring of 6-methylmercaptopurine levels may be useful in identifying patients at risk for hepatotoxicity. Gastrointestinal adverse events (mucositis, nausea, vomiting, abdominal pain, diarrhea, pancreatitis), neurotoxicity and photosensitive eruptions may occur. Severe hypersensitivity reactions with multi-organ involvement are uncommon. Finally, the incidence of squamous cell carcinoma of the skin and lymphoproliferative malignancies is increased [8].

Cyclophosphamide

Introduction

Cyclophosphamide is an antineoplastic and immunosuppressive agent used for the treatment of solid and hematological malignancies as well as severe AUTOIMMUNE DISEASES, including systemic lupus erythematosus, sclerodermia, and vasculitis. Although cyclophosphamide has been used in clinical practice since the 1950s, its therapeutic use is still widespread and only partially challenged by the introduction of newer drugs.

Chemical structure

Cyclophosphamide (Fig. 1) is a first-generation oxazaphosphorine alkylating agent. It belongs to the group of nitrogen mustards and, as do the other members of the family, has the property of becoming strongly electrophilic in body fluids and forming stable, covalent likages by alkylation of various nucleophilic moieties, particularly the N7 of guanine residues of DNA.

Mechanism of action and pharmacological effect

Cyclophosphamide acts via its principal active metabolite, phosphoramide mustard, through several mechanisms. At the molecular level, phosphoramide mustard is able to bind DNA [9] reacting with purine bases to form double strand adducts, but at higher doses cyclophosphamide may induce strand nicks by destabilizing purine-sugar bonds with the following loss of purine bases. At the cellular level, the drug is able to trigger APOPTOSIS and to induce a pronounced cytotoxic effect on mature LYMPHOCYTES with relative sparing of the respective precursor cells [10, 11]. The generation of reactive oxygen-free radicals may be considered another MECHANISM OF ACTION, leading to cell death by damaging DNA and inducing lipid peroxidation [12]. Moreover, acrolein – a cyclophosphamide metabolite – seems to be able to inhibit cell proliferation, to induce cell death by APOPTOSIS and to modulate expression of genes and transcription factors, because it reduces the activation of NUCLEAR FACTOR κB (NF-κ B) and ACTIVATOR PROTEIN 1 (AP-1). These effects could be increased by the depletion of cellular glutathione, which acts as a detoxifying molecule.

Pharmacokinetics

Cyclophosphamide is well absorbed after oral administration, with a BIOAVAILABILITY greater than 75%. The parent compound is widely distributed throughout the body with low plasma protein binding (20%). The half-life of cyclophosphamide is between 6 and 9 hours, and it is eliminated mainly in the urine as metabolites, even if 5–25% of an intravenous dose is excreted unchanged. Cyclophosphamide is quickly metabolised to active alkylating species by the mixed-function oxidase system of the smooth endoplasmic reticulum of hepatocytes, and plasma maximal concentrations of metabolites may be observed 2–3 h after an intravenous dose. Several cytochrome P450 (CYP) isoforms (CYP2A6, CYP2B6, CYP2C8, CYP2C9, and CYP3A4) are involved in the hydroxylation of the oxazaphosphorine ring of cyclophosphamide [13], leading to 4-OH-cyclophosphamide, which exists in equilibrium with the acyclic tautomer aldophosphamide. Aldophosphamide spontaneously releases acrolein and phosphoramide mustard, the former being a toxic byproduct. The involvement of cytochrome P-450 in cyclophosphamide metabolism explains enzyme induction (mainly of CYP2B, CYP3A4, CYP2C8, and CYP2C9 isoforms), which consists of increased cellular RNA and protein contents and associated catalytic activities following the exposure to cyclophosphamide itself. This phenomenon is responsible for increased CLEARANCE and shortened half-life of the parent drug, because it influences the rate of 4-hydroxylation.

Clinical indications

Cyclophosphamide is used for the treatment of systemic lupus erythematosus, vasculitis and other AUTOIMMUNE DISEASES. In systemic lupus erythematosus, pulse cyclophosphamide at the dose of 1 g/m^2 administered on a monthly schedule ensures a significant advantage in terms of survival and end-stage renal disease with respect to corticosteroids [14]. A remission rate of 75% was observed in patients affected by severe systemic lupus erythematosus and treated with high-dose cyclophosphamide (10–15 mg/kg) on a monthly schedule for 6 months, followed by quarterly pulses for 18 months. The same schedule has been adopted in lupus nephritis in children and adults, lowering the relapse rate to less than 10%. Positive results have been observed also in the treatment of optic neuritis associated with systemic lupus erythematosus. However, due to toxicity induced by the treatment, it has been proposed to use weekly low-dose pulses of 0.5 g until disease control is achieved, when it is switched to the monthly schedule and subsequently discontinued [15].

Churg-Strauss syndrome, a granulomatous, necrotizing vasculitis affecting small blood vessels, may be treated with daily doses of cyclophosphamide (2 mg) administered orally in combination with steroids, and the same schedule has been adopted for sclerodermia. Remission of Wegener's granulomatosis, a systemic necrotizing vasculitis affecting small and medium-size vessels, may be obtained with the use of cyclophosphamide, and daily low doses in combination with steroids are effective against the active disease [16]. Cyclophosphamide plus steroids is effective in the treatment of microscopic polyangiitis, a vasculitis that may be associated with severe pulmonary vasculitis and rapidly progressive glomerulonephritis, and polyarteritis nodosa.

It is noteworthy that the EFFICACY or tolerability of cyclophosphamide depends on the amount of phosphoramide mustard within cells which is controlled by two pathways: 1) aldehyde dehydrogenase which transforms aldophosphamide to carboxyphosphamide, the major urinary inactive metabolite, and 2) the isozymes CYP2B6 and CYP3A4, which catalyse the dechloroethylation of cyclophosphamide [17]. These mechanisms have been extensively investigated in the experimental and clinical setting. CYP2B6 POLYMORPHISM could be responsible for severe toxicities [18], because this CYP isoform catalyses the dechloroethylation of cyclophosphamide to 2- and 3-dechloroethylmetabolite and chloroacetaldehyde, the latter being a toxic by-product. Furthermore, the concentration of aldophosphamide is associated with the expression of aldehyde dehydrogenase isozymes [19]. Thus far, several polymorphic sites have been identified along the gene sequences of CYP and aldehyde dehydrogenase isoforms. Finally, O6-alkylguanine-DNA alkyltransferases are enzymes capable of protecting cells from the mutagenic effect of DNA alkylation, and therefore a low expression of these genes in target cells may be associated with a better response to cyclophosphamide therapy.

Adverse reactions

The use of cyclophosphamide is limited by the occurrence of moderate to severe side-effects, including gastrointestinal toxicity, alopecia, myelotoxicity, infertility, hemorrhagic cystitis and cardiotoxicity [15, 16]. Nausea and vomiting (10% of treated patients) require adequate prophylactic treatment, with steroids and 5-HT_3 antagonists. BONE MARROW toxicity (50–100% of cases) is commonly represented by leukopenia 7–14 days after the drug dose, whereas more severe side-effects are agranulocytosis and aplastic anemia. Herpes zoster and other OPPORTUNISTIC INFECTIONS occur in 37% of patients, multiple-organ involvement and lower trough leukocyte counts being additional risk factors of severity of infection. Ovarian failure and decreased sperm counts are associated with infertility (up to 100% of subjects), and their severity correlates with the duration of cyclophosphamide treatment and the patient's age. Adverse events may be mitigated by oral administration of the drug instead of using the intravenous route. The severity of hemorrhagic cystitis may be reduced when the drug is coadministered with the thiol-containing agent mesna which inactivates acrolein. Because of its MECHANISMS OF ACTION, cyclophosphamide is teratogenic and carcinogenic, the latter effect being more frequent for BONE MARROW (myeloproliferative disorders, 2% of patients) and bladder (transitional cell carcinoma, 2% of patients).

Fludarabine

Introduction

The marked lymphocytolytic activity of fludarabine, a deamination-resistant adenosine analogue, in indolent B-cell lymphoproliferative disorders has suggested its possible use as an immunosuppressive agent, although its EFFICACY and potential indications are still a matter of investigation and only limited data are available.

Chemical structure

The synthesis of fludarabine (9-β-Darabinosyl-2-fluoroadenine, F-ara-A, Fig. 1) was achieved by Montgomery and Hewson [20]. Because fludarabine is poorly soluble, the 5'-monophosphate derivative is used for human treatment. Fludarabine 5'-monophosphate is a prodrug that is converted metabolically by dephosphorylation to the active antimetabolite moiety by the 5'-nucleotidase activity present in most tissues, including ERYTHROCYTES and endothelial cells.

Mechanism of action and pharmacological effect

Fludarabine enters the cells by the nucleoside transport systems (Fig. 2), mostly through the human equilibrative nucleoside transporter-1 (hENT1) [21]. Fludarabine requires intracellular phosphorylation; the rate-limiting step of drug activation is catalyzed by deoxycytidine kinase (dCK), which phosphorylates fludarabine to F-ara-AMP, whereas 5'-nucleotidase (5'-NT) inactivates the monophosphate metabolite by dephosphorylation [22]. Being a deamination-resistant nucleoside analogue, cytidine deaminase (CdA) has no role in the CLEARANCE of fludarabine from the cells. Fludarabine triphosphate (F-ara-ATP) is the main active metabolite of fludarabine; F-ara-ATP is an alternative substrate that competes with the normal deoxynucleotide, deoxyadenosine 5'-triphosphate (dATP) and its principal action involves its incorporation into DNA, causing inhibition of RNA and DNA synthesis [22]. F-ara-ATP is an effective inhibitor of ribonucleotide reductase, thus resulting in self-potentiation of its activity by lowering deoxynucleotide pools. Enzymes of DNA synthesis and repair, including DNA polymerases α, β, γ and ε, DNA primase, DNA ligase I and the nucleotide excision repair system, have been shown to be inhibited by F-ara-ATP [22]. Once incorporated into DNA, F-ara-AMP is a poor substrate for subsequent DNA elongation by addition of deoxynucleotides and behaves as a chain terminator. Inactivation of DNA synthesis is followed by cellular APOPTOSIS by effector caspases (i.e., caspase-3). Finally, a unique characteristic of fludarabine is its ability to trigger APOPTOSIS in proliferating as well as in quiescent cells.

Pharmacokinetics

Fludarabine phosphate undergoes extensive first-pass metabolism due to the 5'-nucleotidase activities of ERYTHROCYTES, endothelial cells and parenchymal organs and rapidly disappears from plasma. Peak plasma concentrations (C_{max}) of fludarabine are observed in the first minutes after the end of the infusion; the standard dose of fludarabine (25 to 30 mg/m^2) results in a C_{max} of about 3 µmol/L. This concentration is adequate to generate intracellular levels of F-ara-ATP capable of triggering the cell death process in lymphoid cells.

Detailed pharmacokinetic data are available with fludarabine administered at doses ranging from 80 to 260 mg/m^2 as a rapid intravenous infusion of 2–5 min. Fludarabine displays a tri-exponential decay in plasma, with a $t_{1/2}\alpha$ of about 5 min, a $t_{1/2}\beta$ of 98 min and a $t_{1/2}\gamma$ ranging from 6.9 to 19.7 hours. The mean total body CLEARANCE (CL) of fludarabine is 4.08 L/h/m^2, the steady-state volume of distribution (Vss) is 44.2 L/m^2 and is independent of the drug dose but it is influenced by renal function, thus suggesting the need for dose reduction in patients with impaired kidney function [22].

Pharmacokinetic studies on fludarabine administered at 50, 70 or 90 mg and compared to a corresponding intravenous dose in patients with hematological malignancies, demonstrated that the oral BIOAVAILABILITY of the drug ranges from 55% to about 75% and it is independent of the dose [22].

Clinical indications

The use of fludarabine in IMMUNOSUPPRESSION for AUTOIMMUNE DISEASES is still limited. Preliminary clinical investigations indicate that the drug may be active in patients with lupus nephritis [23], psoriatic arthritis [24], inflammatory and paraproteinemic neuropathies [25], rheumatoid arthritis [26], and refractory dermatomyositis and polymyositis [27]. The schedule of administration is still under investigation. Fludarabine has been given as a single dose of 30 mg/m^2 over 30 min, every four weeks for four cycles in patients with psoriatic arthritis [24], at 20–30 mg/m^2/day for 3 consecutive days once a month for 6 months [26] or as a single infusion of 25 mg the first month, and then 25 mg for 3 consecutive days each month for 5 months [28] in patients with rheumatoid arthritis. Fludarabine was able to decrease the proliferative response of peripheral LYMPHOCYTES to mitogens, as well as the production of CYTOKINES by T CELLS (IL-2 and IFN-γ) and MONOCYTES (TNF-α and IL-10) [26].

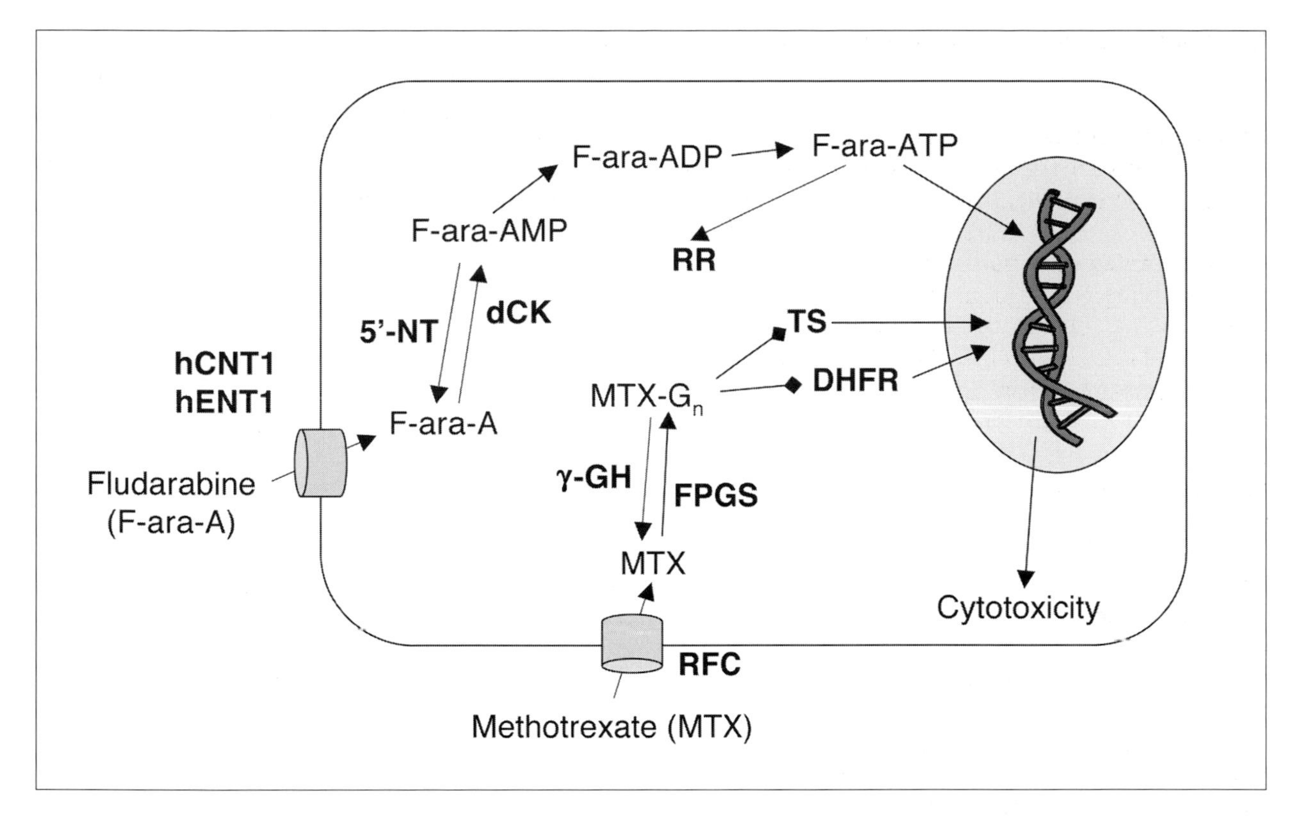

FIGURE 2. MEMBRANE TRANSPORT, METABOLISM AND INTRACELLULAR TARGETS OF FLUDARABINE AND METHOTREXATE
hCNT1, human concentrative nucleoside transporter-1; hENT1, human equilibrative nucleoside transporter-1; dCK, deoxycytidine kinase; 5'-NT, 5'-nucleotidase; CdA, cytidine deaminase; RR, ribonucleotide reductase; RFC, reduced folate carrier; FPGS, folylpolyglutamate synthase; γ-GH, γ-glutamyl hydrolase; TS, thymidylate synthase; DHFR, dihydrofolate reductase.

Adverse reactions

The incidence of drug-related toxicity is dose-dependent; while low doses of fludarabine are associated with a good tolerability profile [28], the administration of higher doses may be associated with substantial toxicity [26]. The most common adverse events in patients treated with fludarabine include myelosuppression (neutropenia, thrombocytopenia and anemia), fever and chills, nausea and vomiting [29]. The duration of clinically significant cytopenia in a few cases has ranged from 2 months to approximately 1 year. Other commonly reported events include malaise, fatigue, anorexia, and weakness. Serious OPPORTUNISTIC INFECTIONS have occurred in patients with chronic lymphocytic leukemia treated with fludarabine.

Methotrexate

Introduction

Methotrexate is a folate analogue that was introduced in the clinical practice more than 50 years ago. It is currently one of the most widely used disease-modifying antirheumatic drugs (DMARDs); its EFFICACY in rheumatoid arthritis has been confirmed in patients refractory or intolerant to other DMARDs or NSAIDs.

Chemical structure

Methotrexate (MTX, amethopterin, *N*-[4-[[(2,4-diamino-6-pteridinyl)methyl]methyl-amino]benzoyl]-*L*-glutamic acid) (Fig. 1) remains the only antifolate agent used for clinical IMMUNOSUPPRESSION to this date. It is a weak dicarboxylic organic acid, and is negatively charged at neutral pH, resulting in limited lipid solubility.

Mechanism of action and pharmacological effect

Methotrexate enters cells by the reduced folate carrier [30] and its long-chain polyglutamates inhibit the activity of the enzyme dihydrofolate reductase, involved in fundamental metabolic pathways such as *de novo* synthesis of purines, pyrimidines and polyamines; methotrexate also inhibits thymidylate synthase indirectly by diminishing levels of the enzyme cosubstrate 5,10-methylenetetrahydrofolate, while polyglutamated metabolites of methotrexate also directly bind and inhibit thymidylate synthase (Fig. 2). The use of high-dose methotrexate depletes the tumor cells of purine and pyrimidine precursors required for DNA and RNA synthesis, proliferation and division. Low-dose methotrexate has both immunosuppressive and antiinflammatory properties resulting in inhibition of proliferation of CD3 and CD4 LYMPHOCYTES, MONOCYTES-MACROPHAGES and neutrophils. Indeed, at low concentrations, methotrexate induces APOPTOSIS of activated T-cells from human peripheral blood *in vitro*. However, low-dose methotrexate does not act only as a cytotoxic drug against immunocompetent cells but also modulates CYTOKINE secretion from T-helper LYMPHOCYTES by increasing IL-4 and IL-10 and decreasing IFN-γ and IL-2. Intracellular methotrexate-polyglutamates inhibit the function of 5-amino-imidazole-4-carboxamide ribosyl-5-phosphate formyltransferase (AICAR-formyltransferase). The resulting high concentrations of AICAR lead to enhanced release of adenosine into the blood, activating A_{2a}, A_{2b} and A_3 extracellular receptors on MONOCYTES-MACROPHAGES. In this way, adenosine seems to promote the transcription of mRNA for an IL-1 receptor antagonist and increase the secretion of the potent anti-inflammatory CYTOKINE IL-10. Moreover, adenosine inhibits the production of TNF-α, IL-6 and IL-8 and the expression of E-SELECTIN on the cell surface [31].

Pharmacokinetics

After oral administration, active absorption of the drug occurs in the proximal jejunum with a saturable process and it decreases non-proportionally at increasing oral doses. The extent of absorption is highly variable between patients for doses higher than 10–15 mg/m^2, whereas only a moderate intra-individual pharmacokinetic variability has been described during long-term treatments with low methotrexate doses in patients with psoriasis and rheumatoid arthritis. Low doses of methotrexate are also administered parenterally to ensure compliance and uniform BIOAVAILABILITY. Indeed, the drug is absorbed more rapidly and reaches higher serum concentrations after intramuscular or subcutaneous administration compared with the oral route. Methotrexate may also be injected intra-articularly [32].

The volume of distribution of methotrexate corresponds to the intracellular distribution of the drug and, in blood, 30–70% of the drug is bound to albumin. Four hours after oral or intramuscular administration, the concentrations of methotrexate in the synovial fluid are equivalent to plasma concentrations. Methotrexate transport into cells occurs mainly by a carrier-mediated active transport system that methotrexate shares with folates. Once inside the cell, glutamate residues are progressively added to the drug by the folyl-polyglutamate synthetase enzyme. This intracellular accumulation of methotrexate polyglutamates allows a weekly bolus of the drug or the administration of the same dose equally divided in three doses, and relatively high concentrations of the drug are reached in the synovial membrane, cortical and trabecular bone. Methotrexate elimination has been described as biphasic or triphasic with a mean TERMINAL HALF-LIFE of 6–15 hours. A longer sampling interval is associated with longer TERMINAL HALF-LIFE estimates of the drug, because of intracellular methotrexate polyglutamylation and the slow release of the drug from cell to plasma.

Methotrexate can be metabolised by three different pathways: 1) in the gastrointestinal tract, intestinal bacteria could degrade the drug to 4-amino-deoxy-N10-methylpteroic acid, a metabolite that usually accounts for less than 5% of the administered dose; 2) in the liver, methotrexate is converted to 7-OH-methotrexate, 10-fold less potent at inhibiting dihydrofolate reductase but more nephrotoxic than the parent compound because of its precipitation in acidic urine; despite its extensive binding to serum albumin (>90%), 7-OH-methotrexate does not alter the protein binding of methotrexate; iii) inside the cells, the drug is converted to pharmacologically active long-chain methotrexate polyglutamates by folylpolyglutamate synthetase and inactivated by gamma glutamyl hydrolase, potentially contributing to drug resistance [33].

The main elimination route of methotrexate is by renal excretion. The drug is subjected to glomerular filtration and secretion/reabsorption by an active transport system across the renal tubules [34]. In addition, a variable amount of methotrexate is eliminated by active biliary excretion (10–30%), and undergoes enterohepatic recirculation.

Clinical indications

Methotrexate is indicated for the symptomatic control of severe and disabling psoriasis that is not adequately responsive to other forms of therapy and in the management of adult patients with severe, active, rheumatoid arthritis [35] or children with active polyarticular-course juvenile arthritis, who have had an insufficient therapeutic response to NSAIDs [36].

The influence of pharmacogenetics on both the immunosuppressive EFFICACY and toxicity of methotrexate in rheumatoid arthritis is starting to focus the interest of researchers [ ]. By affecting the intracellular folate pool, the drug influences the activity of the enzyme methylenetetrahydrofolate reductase (MTHFR), an important step in the generation of 5-methyl-tetrahydrofolate, which is the methyl donor for the conversion of homocysteine to methionine. Numerous POLYMORPHISMS have been described in the MTHFR gene and among them the C677T POLYMORPHISM has been associated with altered phenotypes and higher rates of adverse drug events, at least in oncology studies. The C677T variant of the MTHFR gene leads to alanine to valine substitution and a thermolabile MTHFR with decreased enzyme activity and increased plasma homocysteine levels.

Patients with rheumatoid arthritis receiving methotrexate have been assessed for toxicity, disease activity and the presence of the C677T POLYMORPHISM. Homozygous or heterozygous patients have revealed an increased risk of methotrexate discontinuation because of adverse events such as gastrointestinal symptoms (e.g., stomatitis, nausea, vomiting), hair loss, rash and hepatotoxicity (increase in transaminases). However, no relationship has been observed between the C677T POLYMORPHISM and the EFFICACY of methotrexate. Thus, the C677T POLYMORPHISM seems to make patients with rheumatoid arthritis more sensitive to methotrexate toxicity [37]. However, single nucleotide POLYMORPHISMS in other enzymes involved in the metabolic pathway of methotrexate, including dihydrofolate reductase and folylpolyglutamate synthase may be better predictors of methotrexate immunosuppressive EFFICACY and toxicity. Hence, further studies are needed to study POLYMORPHISMS in other enzymes of the folate pathway and their correlations with drug EFFICACY and toxicity in rheumatoid arthritis.

Adverse reactions

Methotrexate has the potential for severe toxicity, mostly related to dose or frequency of administration [35]. For this reason, strict monitoring of drug treatment is recommended as most adverse reactions are reversible if detected early. Severe toxicities are managed with leucovorin rescue and hemodialysis with a high-flux dialyzer. Abnormal liver function tests, nausea/vomiting, stomatitis, diarrhea, leukopenia, thrombocytopenia, dermatitis, alopecia and interstitial pneumonitis are adverse reactions observed in patients with rheumatoid arthritis treated with low-dose methotrexate (7.5–15 mg/week). With the exception of a higher incidence of alopecia, photosensitivity, and "burning of skin lesions", the

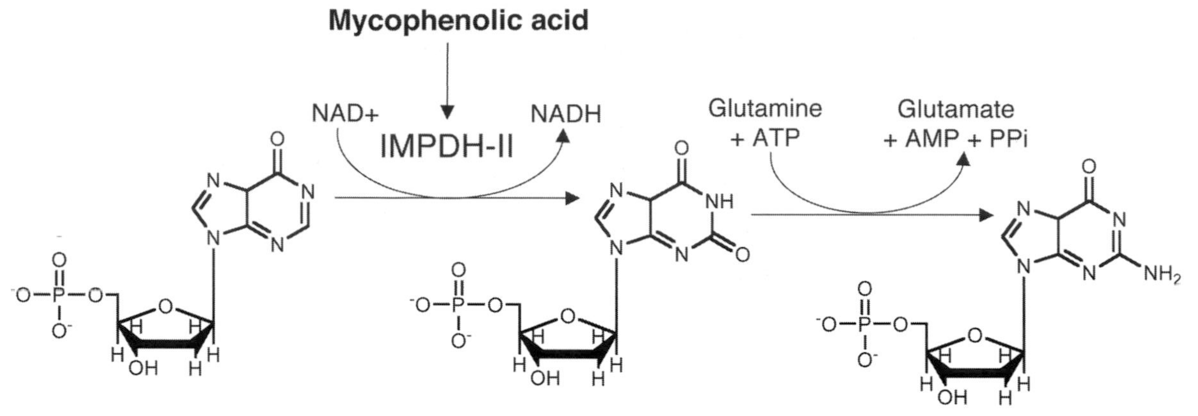

FIGURE 3. INHIBITION OF INOSINE MONOPHOSPHATE DEHYDROGENASE (IMPDH) TYPE II BY MYCOPHENOLIC ACID
PPi, pyrophosphate (inorganic)

adverse reaction rates in patients suffering from psoriasis are very similar to those with rheumatoid arthritis. In pediatric patients with juvenile arthritis treated with oral, weekly doses of methotrexate (5–20 mg/m^2/week or 0.1–0.65 mg/kg/week) most common adverse drug reactions are abnormal liver function tests, nausea, vomiting, diarrhea, stomatitis and leukopenia [36].

Mycophenolic acid

Introduction

The development of mycophenolic acid as an immunosuppressive agent was based on the observation that the proliferation of antigen-responsive T and B LYMPHOCYTES preferentially relies on the *de novo* purine synthesis, with a negligible contribution of the salvage pathway, which is in turn of primary importance for most cells [38]. The antiproliferative, anticancer activity of the drug is thus of secondary importance with respect to the lymphocytolytic effect which lead to further clinical characterization.

Chemical structure

Mycophenolic acid (Fig. 1) is a FERMENTATION product of several *Penicillium* species; it is the active moiety released by the prodrugs mycophenolate mofetil, a semisynthetic morpholinoethyl ester of mycophenolic acid, and mycophenolate sodium, which is administered as an enteric-coated formulation designed to prevent upper gastrointestinal tract absorption and reduce the gastrointestinal adverse events seen with mycophenolate mofetil [39].

Mechanism of action and pharmacological effect

Mycophenolic acid is a reversible, non-competitive inhibitor of inosine monophosphate dehydrogenase (IMPDH), and blocks *de novo* purine synthesis in T and B LYMPHOCYTES (Fig. 3), resulting in (1) inhibition of proliferation in response to antigenic stimuli and (2) initiation of apoptotic cascade. Two distinct isoforms of IMPDH have been identified, i.e., type I and type II. IMPDH type I is constitutively expressed mostly in non-replicating cells and IMPDH type II is the

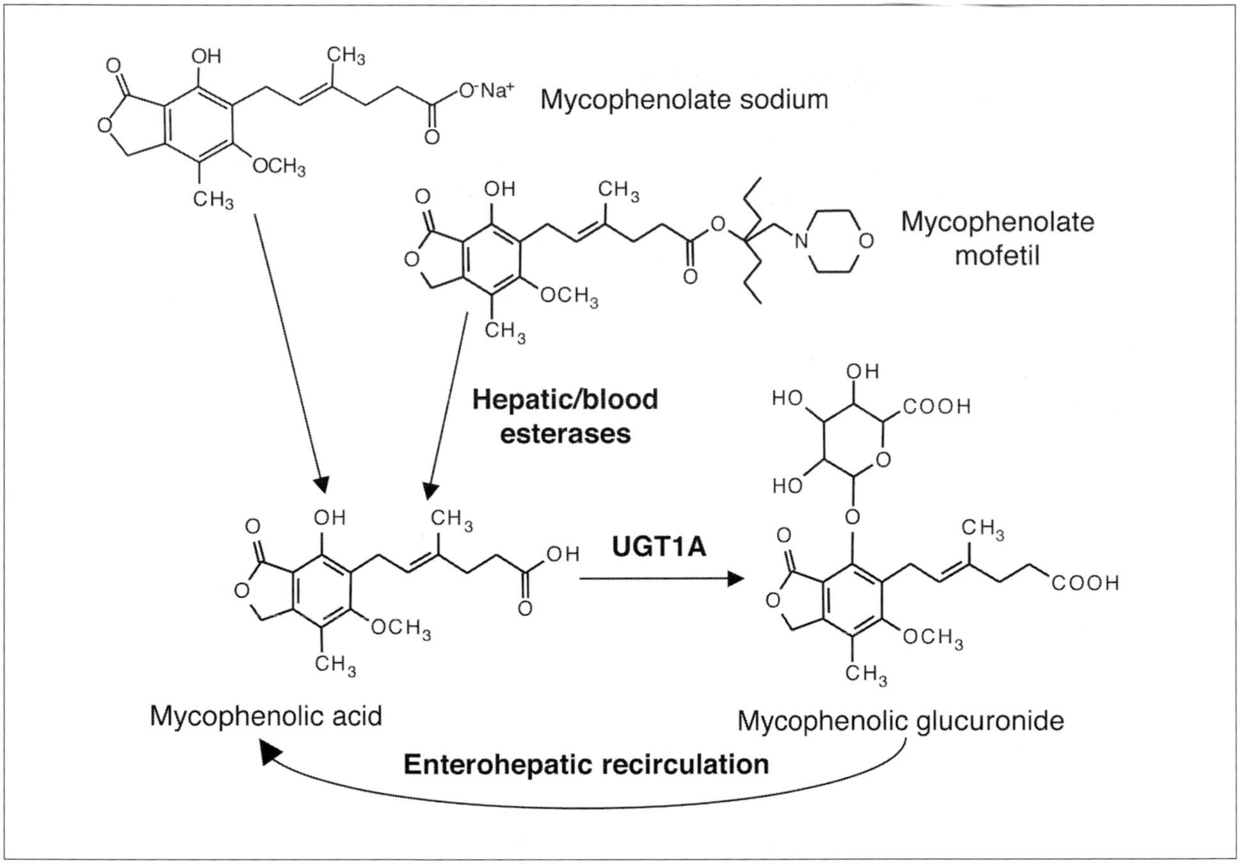

FIGURE 4
Metabolism of mycophenolate mofetil and mycophenolate sodium to the active moiety mycophenolic acid; UGT1-dependent metabolism yields mycophenolate glucuronide which undergoes enterohepatic recirculation.

inducible, predominant enzyme in activated LYMPHOCYTES. IMPDH type II is approximately five times more susceptible to inhibition by MPA than type I and this difference explains the unique susceptibility of proliferating LYMPHOCYTES to depletion of purine bases by mycophenolic acid. At variance with calcineurin inhibitors, mycophenolic acid has no effect on the production or release of CYTOKINES [40]. Additional pharmacological effects include suppression of ANTIBODY production, as a consequence of failure of B-cell activation, as well as anti-inflammatory effects resulting from decreased expression and activity of the inducible form of nitric oxide synthase [41].

Pharmacokinetics

Mycophenolate mofetil (Fig. 4) is absorbed in the stomach and the ester linkage is rapidly hydrolyzed by ubiquitous esterases to yield mycophenolic acid, the active immunosuppressive moiety [42]. Therefore, mycophenolate mofetil is undetectable in circulation, even after an intravenous administration. In contrast, enteric-coated mycophenolate sodium (Fig. 4) is mainly absorbed in the small intestine. The oral BIOAVAILABILITY of mycophenolic acid from mycophenolate mofetil and mycophenolate sodium is 94% and 71%, respectively, and the peak plasma concentration (C_{max}) occurs 1–2 h after oral administration [39, 43]. After a dose of 1 g of mycophenolate mofetil, the maximum plasma concentration ranges from 10 to 30 mg/L in patients with stable renal function and the elimination half-life averages 17 h, while pre-dose levels are approximately 1 mg/L [44]. The area under the curve (AUC) of mycophenolic acid is the most significant pharmacokinetic parameter for therapeutic drug monitoring; indeed, low mycophenolic acid AUC is significantly associated with an increased risk of acute rejection of kidney graft [42]. In patients given doses of mycophenolate mofetil tailored to achieve low (16.1), intermediate (32.2) or high (60.6 mg/h/L) total AUC of mycophenolic acid, a highly significant relationship was found between the AUC of the active metabolite and the incidence of rejection. In the low, intermediate and high AUC groups the incidence was 27.5%, 14.9% and 11.5%, respectively [45]. There was also a higher rate of premature withdrawal from the study as the AUC of mycophenolic acid increased: 7.8%, 23.4% and 44.2% for the three groups, respectively [45]. PHARMACOKINETIC MONITORING may be performed on the basis of AUC_{0-2h} of mycophenolic acid, this parameter being in good agreement with the AUC_{0-12h} measured across the complete dosing interval period.

Mycophenolic acid undergoes extensive hepatic glucuronidation by glucuronosyl-transferases (UGT) isoforms 1A8, 1A9 and 1A10 [46] to form the inactive metabolite mycophenolic acid glucuronide which is excreted into the bile (Fig. 4). The glucuronyl moiety of the metabolite is cleaved by enteric β-glucuronidases of intestinal bacteria to release mycophenolic acid which undergoes entero-hepatic recirculation producing a secondary peak plasma concentration about 6–12 h after administration. More than 90% of a dose of mycophenolate mofetil is excreted in the urine as mycophenolic acid glucuronide [47]. The comparison of pharmacokinetic parameters of mycophenolate sodium 720 mg and mycophenolate mofetil 1000 mg revealed that C_{max} and AUC_{0-24h} in the former group were consistently higher than in the latter group. Overall, there was a mean increase of 32% in systemic mycophenolic acid exposure over the length of the study in the mycophenolate sodium-treated patients [39].

Clinical indications

Mycophenolate mofetil is administered at 1–3 g/day orally or, less frequently, by the intravenous route and it is indicated for the prophylaxis of organ rejection in patients receiving allogeneic renal, cardiac or hepatic transplants mainly in combination with a calcineurin inhibitor (tacrolimus or cyclosporine) and corticosteroids. Clinical studies are underway to characterize the potential use of mycophenolate mofetil, administered in combination with immunosuppressive antibodies, in calcineurin- and steroid-sparing regimens, in order to reduce the toxicity burden of IMMUNOSUPPRESSION [48]. Mycophenolate sodium 720 mg twice daily has been compared with mycophenolate mofetil 1000 mg twice daily with respect to safety and EFFICACY. Both formulations demonstrated equivalent EFFICACY and safety in renal transplant recipients; in particular, the incidence of gastrointestinal adverse events was similar between the formulations despite the higher serum levels achieved with mycophenolate sodium [39]. Clinically important drug interactions of mycophenolate mofetil and mycophenolate sodium involve acyclovir/valacyclovir (increased hematological toxicity), cholestyramine, antacids containing aluminum hydroxide/magnesium hydroxide and cyclosporine (reduced BIOAVAILABILITY of mycophenolic acid), levonorgestrel (reduced AUC of the hormone) [39].

Adverse reactions

Overall, mycophenolate mofetil is better tolerated than azathioprine. The drug produces a lower incidence of leukopenia and fewer immunosuppressive-related malignancies compared with azathioprine and lacks the neurotoxicity and nephrotoxicity associated with calcineurin inhibitors. The most commonly observed adverse events associated with mycophenolate mofetil and mycophenolate sodium are gastrointestinal (i.e., nausea, vomiting, diarrhea, constipation, dyspepsia, flatulence, anorexia), hematological (leukopenia, thrombocytopenia, anemia), and OPPORTUNISTIC INFECTIONS. Patients experiencing gastrointestinal or hematological adverse events usually respond to dose fractionation or reduction; however, frequent dose changes have been associated with poorer outcomes, including a higher incidence of graft loss. A recent study, however, failed to confirm the advantages of mycophenolate mofetil over azathioprine, concerning both EFFICACY and safety [2].

Conclusions

The increasing understanding of the pathophysiology of AUTOIMMUNE DISEASE and graft rejection has revealed a number of potential targets that have been exploited for the design of potent immunosuppressive drugs, including MONOCLONAL ANTIBODIES and calcineurin-binding agents.

However, the occurrence of multiorgan toxicity associated with the use of calcineurin inhibitors, the severe metabolic adverse events induced by corticosteroids, the potential for long-term cancer risk associated with the use of selected MONOCLONAL ANTIBODIES and the occurrence of infectious diseases due to over-immunosuppression soon became key management issues that urged a reevaluation of immunosuppressive treatment schedules. In this context, cytotoxic drugs, particularly ANTIMETABOLITES, still play a crucial role in the control of IMMUNE SYSTEM activation, particularly in steroid- and calcineurin-inhibitor-resistant diseases, and are expected to be instrumental in the long term IMMUNOSUPPRESSION maintenance in steroid- and calcineurin-binding-sparing schedules.

Summary

Cytotoxic immunosuppressive drugs are a group of heterogeneous compounds characterized by their ability to damage immune cells by non-specific mechanisms, including nucleotide pool depletion, incorporation into DNA and nucleic acid alkylation. Their successful clinical use in IMMUNOSUPPRESSION for organ transplantation and AUTOIMMUNE DISEASES has been proven in a large number of clinical trials.

Azathioprine is an antimetabolite prodrug of 6-mercaptopurine with an imidazolyl group attached to the SH group of 6-mercaptopurine. Active metabolites of azathioprine are inhibitors of nucleic acid synthesis through impairment of the purine *de novo* biosynthetic pathway and incorporation into RNA and DNA. The drug suppresses the proliferation of T and B LYMPHOCYTES and reduces the number of cytotoxic T CELLS and plasma cells in circulation and peripheral organs. Azathioprine is mainly metabolized by xanthine oxidase; however, thiopurine methyltransferase (TPMT) is the principal detoxification pathway for the cytotoxic thioguanine nucleotides in BONE MARROW. Hematopoietic toxicity is dependent at least in part on the activity of TPMT, a POLYMORPHIC ENZYME with genetic variants characterized by low activity and increased risk of severe toxicity in patients. Azathioprine is used in renal transplantation, rheumatoid arthritis ulcerative colitis, and other skin autoimmune disorders. Hematopoietic toxicity, including neutropenia, anemia and thrombocytopenia, and hepatic toxicity are the most common adverse events and may be predicted by TPMT genotyping.

Cyclophosphamide is a first-generation oxazaphosphorine alkylating agent belonging to the group of nitrogen mustards, characterized by the ability to form stable, covalent likages by alkylation of the N7 of guanine residues of DNA via its active metabolite phosphoramide mustard. At the cellular level, the drug is able to trigger APOPTOSIS and to induce a pronounced cytotoxic effect on mature LYMPHOCYTES. Cyclophosphamide has a good oral BIOAVAILABILITY, and it is eliminated mainly in the urine as metabolites. Cyclophosphamide is transformed to active alkylating species by the mixed-function oxidase system of the smooth endoplasmic reticulum of hepatocytes. Several cytochrome P450 (CYP) isoforms (CYP2A6, CYP2B6, CYP2C8, CYP2C9, and CYP3A4) are involved in the hydroxylation of the oxazaphosphorine ring of cyclophosphamide, leading to 4-OH-cyclophosphamide, which exists in equilibrium with aldophosphamide. Aldophosphamide spontaneously releases acrolein and phosphoramide mustard, the former being a toxic by-product. The EFFICACY or tolerability of cyclophosphamide depends on the amount of phosphoramide mustard within cells. In this respect, key enzymes are aldehyde dehydrogenase, which transforms aldophosphamide to carboxyphosphamide, the major urinary inactive metabolite, and the isozymes CYP2B6 and CYP3A4, which catalyse the dechloroethylation of cyclophosphamide to 2- and 3-dechloroethyl-metabolite and chloroacetaldehyde, the latter being a neurotoxic by-product. The drug is used for the treatment of systemic lupus erythematosus, vasculitis and other AUTOIMMUNE DISEASES. The use of cyclophosphamide is limited by the occurrence of moderate to severe side-effects, including gastrointestinal toxicity, alopecia, myelotoxicity, infertility, hemorrhagic cystitis and cardiotoxicity.

Fludarabine, a deamination-resistant adenosine analogue, with marked lymphocytolitic effect, is a prodrug that requires intracellular phosphorylation; the rate-limiting step of drug activation is catalyzed by deoxycytidine kinase (dCK), which phosphorylates fludarabine to F-ara-AMP. Fludarabine triphosphate (F-ara-ATP) is the main active metabolite of fludarabine; F-ara-ATP is an alternative substrate that competes with the normal deoxynucleotide, deoxyadenosine 5'-triphosphate (dATP), and it is incorporated into DNA, causing inhibition of RNA and DNA synthesis. F-ara-ATP is an effective inhibitor of ribonucleotide reductase, thus resulting in self-potentiation of its activity by lowering deoxynucleotide pools. Inactivation of DNA synthesis is followed by cellular APOPTOSIS by effector caspases. A unique characteristic of fludarabine is its ability to trigger APOPTOSIS in proliferating as well as in quiescent cells. Fludarabine is administered by intravenous and oral routes as a phosphate prodrug and undergoes extensive metabolism due to the 5'-nucleotidase activities of ERYTHROCYTES, endothelial cells and parenchymal organs and rapidly disappears from plasma to release fludarabine. Fludarabine displays a tri-exponential decay in plasma, and drug CLEARANCE is influenced by renal function. The use of fludarabine in IMMUNOSUPPRESSION for AUTOIMMUNE DISEASES is still limited; however, preliminary clinical data indicate that the drug may be active in patients with lupus nephritis, psoriatic arthritis, inflammatory neuropathies, rheumatoid arthritis, and refractory dermatomyositis and polymyositis. The incidence of drug-related toxicity is dose-dependent; the most common adverse events being myelosuppression, fever, nausea and vomiting. Other commonly reported events include malaise, fatigue, anorexia, and weakness. Serious OPPORTUNISTIC INFECTIONS have occurred in patients with chronic lymphocytic leukemia treated with fludarabine.

Methotrexate, a folate analogue, is currently one of the most widely used disease-modifying antirheumatic drugs (DMARDs); its EFFICACY in rheumatoid arthritis has been confirmed in patients refractory or intolerant to other DMARDs. Methotrexate enters cells by the reduced folate carrier and its long-chain polyglutamates inhibit the enzyme dihydrofolate reductase, thereby blocking *de novo* synthesis of purines and pyrimidines; methotrexate also inhibits thymidylate synthase. Low-dose methotrexate has both immunosuppressive and anti-inflammatory properties resulting in inhibition of proliferation of CD3 and CD4 LYMPHOCYTES, MONOCYTES-MACROPHAGES and neutrophils. Oral BIOAVAILABILITY of methotrexate is predictable after administration of low doses, while higher doses should be administered parenterally. The main elimination route of methotrexate is by renal excretion; in addition, a variable amount is eliminated by active biliary excretion and undergoes enterohepatic recirculation. Methotrexate is indicated for the management of rheumatoid arthritis, juvenile arthritis and psoriasis. Methotrexate has the potential for severe toxicity, including hepatotoxicity, nausea/vomiting, stomatitis, diarrhea, leukopenia, thrombocytopenia, dermatitis, alopecia and interstitial pneumonitis. The homozygous or heterozygous condition of C677T POLYMORPHISM of the enzyme methylenetetrahydrofolate reductase (MTHFR) is associated with an increased risk of adverse events.

Finally, mycophenolic acid is an immunosuppressive agent that exerts a reversible, non-competitive inhibition of inosine monophosphate dehydrogenase (IMPDH), and blocks *de novo* purine synthesis in T and B LYMPHOCYTES. At variance with calcineurin inhibitors, mycophenolic acid has no effect on the production or release of CYTOKINES. Additional pharmacological effects include suppression of ANTIBODY production, as a consequence of failure of B-cell activation, as well as anti-inflammatory effects resulting from decreased expression and activity of the inducible nitric oxide synthase. The oral BIOAVAILABILITY of mycophenolic acid from mycophenolate mofetil and mycophenolate sodium is high and the drug undergoes extensive hepatic glucuronidation by glucuronosyl-transferases (UGT) isoforms 1A8, 1A9 and 1A10 to form the inactive metabolite mycophenolic acid glucuronide which is excreted into the bile. The drug is indicated for the prophylaxis of organ rejection in patients receiving allogeneic renal, cardiac or hepatic transplants mainly in combination with a calcineurin inhibitor and corticosteroids. Overall, mycophenolate mofetil is better tolerated than azathioprine. The drug produces a lower incidence of leukopenia and fewer IMMUNOSUPPRESSION-related malignancies compared with azathioprine and lacks the neurotoxicity and nephrotoxicity associated with calcineurin inhibitors. The most commonly observed adverse events associated with mycophenolate mofetil and mycophenolate sodium are gastrointestinal (i.e., nausea, vomiting, diarrhea, constipation, dyspepsia, flatulence, anorexia) and hematological (leukopenia, thrombocytopenia, anemia), as well as OPPORTUNISTIC INFECTIONS.

Selected readings

Cattaneo D, Perico N, Remuzzi G (2004) From pharmacokinetics to pharmacogenomics: a new approach to tailor immunosuppressive therapy. *Am J Transplant* 4: 299–310

Fischereder M, Kretzler M (2004) New immunosuppressive strategies in renal transplant recipients. *J Nephrol* 17: 9–18

Kuiper-Geertsma DG, Derksen RH (2003) Newer drugs for the treatment of lupus nephritis. *Drugs* 63: 167–180

Mueller XM (2004) Drug immunosuppression therapy for adult heart transplantation. Part 1: immune response to allograft and mechanism of action of immunosuppressants. *Ann Thorac Surg* 77: 354–362

References

1 Barshes NR, Goodpastor SE, Goss JA (2004) Pharmacologic immunosuppression. *Front Biosci* 9: 411–420

2 emuzzi G, Lesti M, Gotti E, Ganeva M, Dimitrov BD, Ene-Iordache B, Gherardi G, Donati D, Salvadori M, Sandrini S et al (2004) Mycophenolate mofetil *versus* azathioprine for prevention of acute rejection in renal transplantation (MYSS): a randomised trial. *Lancet* 364: 503–512

3 Mueller XM (2004) Drug immunosuppression therapy for adult heart transplantation. Part 1: immune response to allograft and mechanism of action of immunosuppressants. *Ann Thorac Surg* 77: 354–362

4 Sauer H, Hantke U, Wilmanns W (1988) Azathioprine lymphocytotoxicity. Potentially lethal damage by its imidazole derivatives. *Arzneimittelforschung* 38: 820–884

5 Evans WE. (2004) Pharmacogenetics of thiopurine S-methyltransferase and thiopurine therapy. *Ther Drug Monit* 26: 186–191

6 El-Azhary RA. (2003) Azathioprine: current status and future considerations. *Int J Dermatol* 42: 335–341

7 de Jong DJ, Goullet M, Naber TH (2004) Side effects of azathioprine in patients with Crohn's disease. *Eur J Gastroenterol Hepatol* 16: 207–212

8 Marcen R, Pascual J, Tato AM, Teruel JL, Villafruela JJ, Fernandez M, Tenorio M, Burgos FJ, Ortuno J (2003) Influence of immunosuppression on the prevalence of cancer after kidney transplantation. *Transplant Proc* 35: 1714–1716

9 Hengstler JG, Hengst A, Fuchs J, Tanner B, Pohl J, Oesch F (1997) Induction of DNA crosslinks and DNA strand lesions by cyclophosphamide after activation by cytochrome P450 2B1. *Mutat Res* 373: 215–223

10 Allison AC (2000) Immunosuppressive drugs: the first 50 years and a glance forward. *Immunopharmacol* 47: 63–83

11 Pette M, Gold R, Pette DF, Hartung HP, Toyka KV (1995) Mafosfamide induces DNA fragmentation and apoptosis in human T-lymphocytes. A possible mechanism of its immunosuppressive action. *Immunopharmacology* 30: 59–69

12 Sulkowska M, Sulkowski S, Skrzydlewska E, Farbiszewski R (1998) Cyclophosphamide-induced generation of reactive oxygen species. Comparison with morphological changes in type II alveolar epithelial cells and lung capillaries. *Exp Toxicol Pathol* 50: 209–220

13 Alan V, Boddy AV, Yule SM. (2000) Metabolism and pharmacokinetics of oxazaphosphorines. *Clin Pharmacokinet* 38: 291–304

14 Esdaile JM. (2002) How to manage patients with lupus nephritis. *Best Practice Res Clin Rheumatol* 16: 195–210

15 Mosca M, Ruiz-Irastorza G, Khamashta MA, Hughes

GRV (2001) Treatment of systemic lupus erythematosus. *Int Immunopharmacol* 1: 1065–1075

16 Langford CA (2001) Management of systemic vasculitis. *Best Practice Res Clin Rheumatol* 15: 281–297

17 Huang Z, Roy P, Waxman DJ (2000) Role of human liver microsomal CYP3A4 and CYP2B6 in catalyzing N-dechloroethylation of cyclophosphamide and ifosfamide. *Biochem Pharmacol* 59: 961–972

18 Jinno H, Tanaka-Kagawa T, Ohno A, Makino Y, Matsushima E, Hanioka N, Ando M (2003) Functional characterization of cytochrome P450 2B6 allelic variants. *Drug Metab Dispos* 31: 398–403

19 Giorgianni F, Bridson PK, Sorrentino BP, Pohl J, Blakley RL (2000) Inactivation of aldophosphamide by human aldehyde dehydrogenase isozyme 3. *Biochem Pharmacol* 60: 325–338

20 Montgomery JA, Hewson K (1969) Nucleosides of 2-fluoroadenine. *J Med Chem* 12: 498–504

21 Molina-Arcas M, Bellosillo B, Casado FJ, Montserrat E, Gil J, Colomer D, Pastor-Anglada M (2003) Fludarabine uptake mechanisms in B-cell chronic lymphocytic leukemia. *Blood* 101: 2328–2334

22 Gandhi V, Plunkett W (2002) Cellular and clinical pharmacology of fludarabine. *Clin Pharmacokinet* 41: 93–103

23 Kuiper-Geertsma DG, Derksen RH (2003) Newer drugs for the treatment of lupus nephritis. *Drugs* 63: 167–180

24 Takada K, Danning CL, Kuroiwa T, Schlimgen R, Tassiulas IO, Davis JC Jr, Yarboro CH, Fleisher TA, Boumpas DT, Illei GG (2003) Lymphocyte depletion with fludarabine in patients with psoriatic arthritis: clinical and immunological effects. *Ann Rheum Dis* 62: 1112–1115

25 Monaco S, Turri E, Zanusso G, Maistrello B (2004) Treatment of inflammatory and paraproteinemic neuropathies. *Curr Drug Targets Immune Endocr Metabol Disord* 4: 141–148

26 Davis JC Jr, Fessler BJ, Tassiulas IO, McInnes IB, Yarboro CH, Pillemer S, Wilder R, Fleisher TA, Klippel JH, Boumpas DT (1998) High dose *versus* low dose fludarabine in the treatment of patients with severe refractory rheumatoid arthritis. *J Rheumatol* 25: 1694–1704

27 Adams EM, Pucino F, Yarboro C, Hicks JE, Thornton B, McGarvey C, Sonies BC, Bartlett ML, Villalba ML, Fleisher T et al (1999) A pilot study: use of fludarabine for refractory dermatomyositis and polymyositis, and examination of endpoint measures. *J Rheumatol* 26: 352–360

28 Biasi D, Caramaschi P, Carletto A, Bambara LM. (2000) Unsuccessful treatment with fludarabine in four cases of refractory rheumatoid arthritis. *Clin Rheumatol* 19: 442–444

29 Bashey A (2002) Immunosuppression with limited toxicity: the characteristics of nucleoside analogs and anti-lymphocyte antibodies used in non-myeloablative hematopoietic cell transplantation. *Cancer Treat Res* 110: 39–49

30 Serra M, Reverter-Branchat G, Maurici D, Benini S, Shen JN, Chano T, Hattinger CM, Manara MC, Pasello M, Scotlandi K et al (2004) Analysis of dihydrofolate reductase and reduced folate carrier gene status in relation to methotrexate resistance in osteosarcoma cells. *Ann Oncol* 15: 151–160

31 Fraser AG (2003) Methotrexate: first-line or second-line immunomodulator? *Eur J Gastroenterol Hepatol* 15: 225–231

32 Grim J, Chládek J, Martínková J (2003) Pharmacokinetics and pharmacodynamics of methotrexate in non-neoplastic diseases. *Clin Pharmacokinet* 42: 139–151

33 Zhao R, Goldman ID (2003) Resistance to antifolates. *Oncogene* 22: 7431–7457

34 van Aubel RA, Smeets PH, Peters JG, Bindels RJ, Russel FG (2002) The MRP4/ABCC4 gene encodes a novel apical organic anion transporter in human kidney proximal tubules: putative efflux pump for urinary cAMP and cGMP. *J Am Soc Nephrol* 13: 595–603

35 Borchers AT, Keen CL, Cheema GS, Gershwin ME (2004) The use of methotrexate in rheumatoid arthritis. *Semin Arthritis Rheum* 34: 465–483

36 Ruperto N, Murray KJ, Gerloni V, Wulffraat N, de Oliveira SK, Falcini F, Dolezalova P, Alessio M, Burgos-Vargas R, Corona F et al (2004) A randomized trial of parenteral methotrexate comparing an intermediate dose with a higher dose in children with juvenile idiopathic arthritis who failed to respond to standard doses of methotrexate. *Arthritis Rheum* 50: 2191–2201

37 Ranganathan P, Eisen S, Yokoyama WM, McLeod HL (2003) Will pharmacogenetics allow better prediction of methotrexate toxicity and efficacy in patients with rheumatoid arthritis? *Ann Rheum Dis* 62: 4–9

38 Smak Gregoor PJ, van Gelder T, Weimar W (2000) Mycophenolate mofetil, Cellcept, a new immunosup-

pressive drug with great potential in internal medicine. *Neth J Med* 57: 233–246
39 Gabardi S, Tran JL, Clarkson MR (2003) Enteric-coated mycophenolate sodium. *Ann Pharmacother* 37: 1685–1693
40 Srinivas TR, Kaplan B, Meier-Kriesche HU (2003) Mycophenolate mofetil in solid-organ transplantation. *Expert Opin Pharmacother* 4: 2325–2345
41 Lui SL, Chan LY, Zhang XH, Zhu W, Chan TM, Fung PC, Lai KN (2001) Effect of mycophenolate mofetil on nitric oxide production and inducible nitric oxide synthase gene expression during renal ischaemia-reperfusion injury. *Nephrol Dial Transplant* 16: 1577–1582
42 Kelly P, Kahan BD (2002) Review: metabolism of immunosuppressant drugs. *Curr Drug Metab* 3: 275–287
43 Del Tacca M (2004) Prospects for personalized immunosuppression: pharmacologic tools – a review. *Transplant Proc* 36: 687–689
44 Holt DW (2002) Monitoring mycophenolic acid. *Ann Clin Biochem* 39: 173–183
45 Hale MD, Nicholls AJ, Bullingham RE, Hene R, Hoitsma A, Squifflet JP, Weimar W, Vanrenterghem Y, Van de Woude FJ, Verpooten GA (1998) The pharmacokinetic-pharmacodynamic relationship for mycophenolate mofetil in renal transplantation. *Clin Pharmacol Ther* 64: 672–683
46 Mackenzie PI (2000) Identification of uridine diphosphate glucuronosyl-transferases involved in the metabolism and clearance of mycophenolic acid. *Ther Drug Monit* 22: 10–13
47 Mele TS, Halloran PF (2000) The use of mycophenolate mofetil in transplant recipients. *Immunopharmacology* 47: 215–245
48 Shapiro R (2004) Low toxicity immunosuppressive protocols in renal transplantation. *Keio J Med* 53: 18–22

Corticosteroids

Ian M. Adcock and Kazuhiro Ito

Introduction

Cortisol secretion increases in response to any stress in the body, whether physical (such as illness, trauma, surgery, or temperature extremes) or psychological. However, this hormone is more than a simple marker of stress levels – it is necessary for the correct functioning of almost every part of the body. Excesses or deficiencies of this crucial hormone also lead to various physical symptoms and disease states [1]. Although cortisol is not essential for life per se it helps an organism to cope more efficiently with its environment, with particular metabolic actions on glucose production and protein and fat catabolism. Nevertheless, loss or profound diminishment of cortisol secretion leads to a state of abnormal metabolism and an inability to deal with stressors, which, if untreated, may be fatal [2].

The body's level of cortisol in the bloodstream displays a DIURNAL VARIATION, that is, normal concentrations of cortisol vary throughout a 24-hour period. Cortisol levels in normal individuals are highest in the early morning at around 8 am and are lowest just after midnight. This early morning dip in cortisol level often corresponds to increased symptoms of inflammatory diseases in man [3].

Increased levels of corticosteroids serve as potent suppressors of the immune and inflammatory systems. This is particularly evident when they are administered at pharmacological doses, but also is important in normal immune responses. As a consequence, corticosteroids are widely used as drugs to treat inflammatory conditions such as arthritis, asthma or dermatitis, and as adjunction therapy for conditions such as AUTOIMMUNE DISEASES. Synthetic corticosteroids may also be used in organ transplantation to reduce the chance of rejection. Thus, although the early effects of cortisol is to stimulate the IMMUNE SYSTEM, cortisol and synthetic corticosteroids predominantly repress the inflammatory response by decreasing the activity and production of IMMUNOMODULATORY and inflammatory cells [3].

The usefulness of corticosteroids in treating inflammatory diseases was exemplified by the early work of Kendall and Hench [4]. In a classic experiment, 100 mg of cortisone was injected into the muscle of a patient (Mrs G.) suffering from chronic rheumatoid arthritis on September 21st, 1948. 7 days later the patient was able to walk to the shops for the first time in years. Kendall and Hench were awarded the Nobel prize for this work in 1950 and it represented a new approach to therapy with natural hormones by utilizing pharmacological, rather than physiological, doses.

There are five main aspects of inflammation (1) the release of inflammatory mediators such as histamine, PROSTAGLANDINS, LEUKOTRIENES, CYTOKINES and CHEMOKINES; (2) increased blood flow in the inflamed area (erythema) caused by some of the released factors; (3) leakage of plasma from the vasculature into the damaged area (oedema) due to increased capillary permeability; (4) cellular infiltration signalled by chemoattractants, and (5) repair processes such as fibrosis. Corticosteroids can modify all of these processes [3].

Inflammation is a central feature of many chronic diseases including dermatitis, rheumatoid arthritis (RA), inflammatory bowel disease (IBD, Crohn's disease and ulcerative colitis), systemic lupus erythematosis (SLE), asthma and chronic obstructive pulmonary disease (COPD). The specific characteristics of the inflammatory response in each disease and the site of inflammation differ, but both involve the recruitment and activation of inflammatory cells and changes in the structural cells of the lung. All are

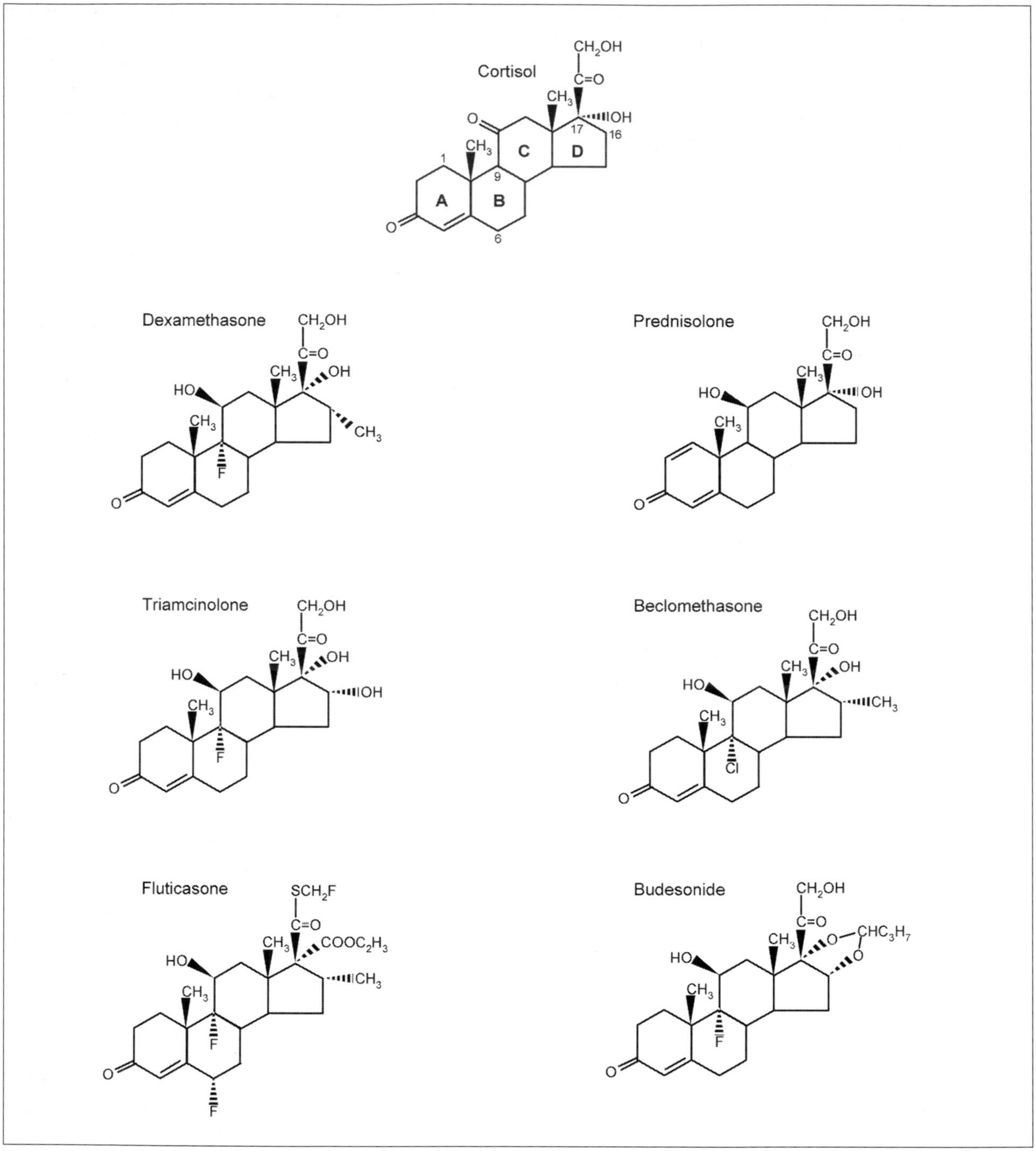

FIGURE 1

Structural modifications of cortisol that produce the clinically used corticosteroids dexamethasone, prednisolone, triamcinolone, beclomethasone, fluticasone and budesonide.

characterized by an increased expression of CYTOKINES, CHEMOKINES, growth factors, enzymes, receptors and ADHESION MOLECULES. The increased expression of these proteins is the result of enhanced gene transcription since many of the genes are not expressed in normal cells but are induced in a cell-specific manner during the inflammatory process [5].

Chemical structures

Corticosteroids are 21-carbon steroid hormones (Fig. 1) composed of 4 rings [6, 7]. The basic structure of the A ring is a 1α, 2β-half-chair, whatever the substitutions. Rings B and C are semi-rigid chairs with minimal structural influence by substituent groups. In contrast, the shape of the D ring depends on the nature and environment of the substituent groups. Modern TOPICAL corticosteroids are based on the cortisol structure with modification to enhance the anti-inflammatory effects such as insertion of a C=C double bond at C1,C2 or by the introduction of 6α-fluoro, 6α-methyl, 9α-fluoro and/or further substitutions with α-hydroxyl, α-methyl or β-methyl at the 16 position, for example in dexamethasone (Fig. 1). Lipophilic substituents, e.g., 16α-, 17α-acetals, 17α-esters or 21α-esters attached to the D-ring were found to further enhance receptor affinity, prolong local TOPICAL deposition and enhance hepatic metabolism, and are exemplified by the structures of budesonide and fluticasone, two of the most commonly used TOPICAL corticosteroids. However, the exact structural and lipophilic requirements to optimise corticosteroid pharmacokinetics, tissue retention and longevity of action are still unclear and corticosteroids with improved clinical characteristics are likely to be synthesized as our knowledge in this area increases.

Modes of action

Classically corticosteroids exert their effects by binding to a single 777 amino acid receptor (GR) that is localized to the cytoplasm of target cells. GRs are expressed in almost all cell types and their density varies from 200–30,000 per cell [8] with an affinity for cortisol of ~30nM, which falls within the normal range for plasma concentrations of free hormone. GR has several functional domains (Fig. 2). The corticosteroid ligand binding domain (LBD) is at the carboxyl terminus of the molecule and is separated from the DNA binding domain (DBD) by a hinge region. There is an N-terminal TRANSACTIVATION DOMAIN that is involved in activation of genes once binding to DNA has occurred. This region may also be involved in binding to other transcription factors. The inactive GR is part of a large protein complex (~300 kDa) that includes two subunits of the heat shock protein hsp90, which blocks the nuclear localisation of GR and one molecule of the immunophilin p59 [8].

Cortisteroids are thought to freely diffuse from the circulation into cells across the cell membrane and bind to cytoplasmic GR (Fig. 3). Once the cortisteroid binds to GR, hsp90 dissociates allowing the nuclear localisation of the activated GR-cortisteroid complex and its binding to DNA [8]. GR combines with another GR to form a dimer at consensus DNA sites termed glucocorticoid response elements (GREs) in the regulating regions of corticosteroid-responsive genes. This interaction allows GR to associate with a complex of DNA-PROTEIN MODIFYING AND REMODELLING PROTEINS including steroid receptor coactivator-1 (SRC-1) and CBP, which produce a DNA-protein structure that allows enhanced gene transcription [8]. The particular ligand and the number of GREs and their position relative to the transcriptional start site may be an important determinant of the magnitude of the transcriptional response to corticosteroids [8].

The GR complex can regulate gene products in at least 4 other ways. First, GR acting as a monomer can bind directly, or indirectly, with the transcription factors ACTIVATOR PROTEIN-1 (AP-1) and NUCLEAR FACTOR κB (NF-κB), which are up-regulated during inflammation, thereby inhibiting the pro-inflammatory effects of a variety of CYTOKINES [8]. Second, the GR dimer can bind to a GRE which overlaps the DNA binding site for a pro-inflammatory transcription factor or the start site of transcription [8]. Third, the GR dimer can induce the expression of the NF-κB inhibitor IκBα in certain cell types [9]. Last, corticos-

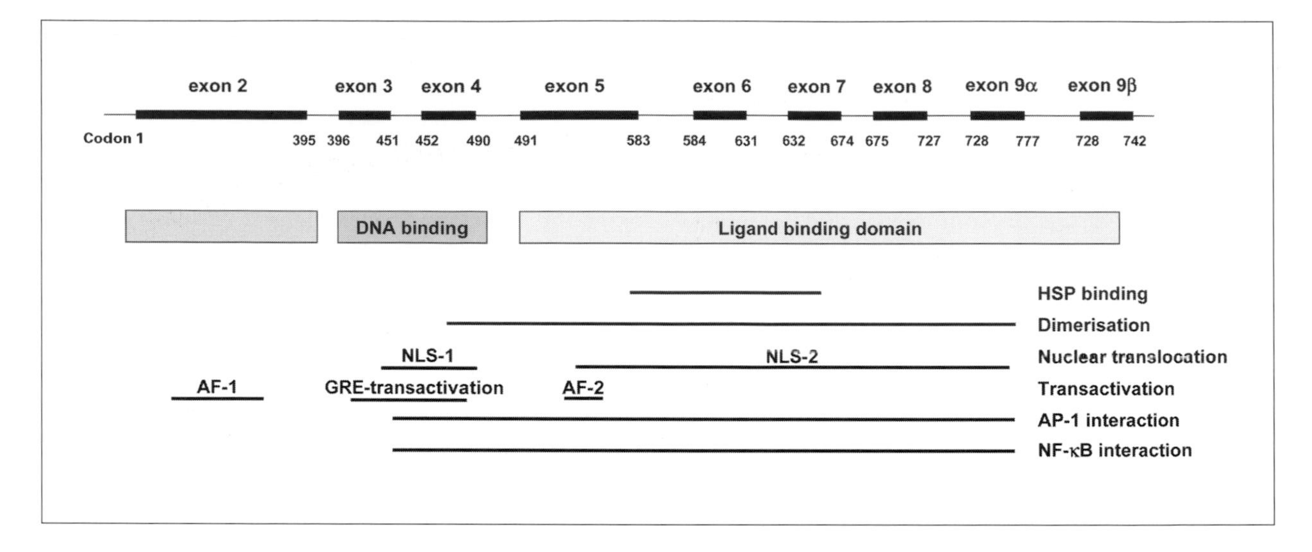

FIGURE 2. MODULAR STRUCTURE OF THE CORTICOSTEROID RECEPTOR (GLUCOCORTICOID RECEPTOR, GR)
The coding region of GR results from splicing together of exons 2–9 of the GR gene. The GRβ isoform of GR results from the use of the short 9β exon which removes the ligand binding domain seen in GRα. The modular design of GR enables distinct regions of the protein to function in isolation as ligand binding domains, dimerisation domains, nuclear localisation domains, transactivation and transrepression (AP-1 and NF-κB interacting) domains. NLS: nuclear localisation signal; AF-1/2: activating factor 1/2 and GRE: glucocorticoid response element (composed of two palindromic half sites (AGAACA) separated by three nucleotides).

teroids can increase the levels of cell ribonucleases and mRNA destabilizing proteins, thereby reducing the levels of mRNA [10] (Fig. 3). It is likely that the altered transcription of many different genes is involved in the anti-inflammatory action of corticosteroids in asthma, but the drugs' most important action may be to inhibit transcription of CYTOKINE and chemokine genes implicated in asthmatic inflammation [8].

Evidence for this has been described in a series of elegant experiments using mice expressing mutated GRs unable to dimerise and subsequently bind to DNA. Thus, Schutz and colleagues [11, 12] have confirmed a role for GR DNA binding as a dimer in the control of pro-opiomelanocortin (POMC) expression and T-cell APOPTOSIS but not in that of inflammatory genes regulated by AP-1 or NF-κB.

The traditional genomic theory of steroid action, whether directly interacting with DNA or involving cross-talk with other transcription factors, does not fully explain the rapid effects of hormonal steroids, and it is thought that the nongenomic actions are mediated by a distinct membrane receptor [13]. These receptors have distinctive hormone binding properties compared to the well-characterized cytoplasmic receptor and are probably linked to a number of intracellular signalling pathways acting through G-protein coupled receptors and a number of kinase pathways [13]. There are a number of reviews providing a summary of the evidence for these rapid effects which include rapid effects on actin structures, kinase activities and transmembrane currents [14]. In addition, the classical receptor is associated with a number of kinases and phosphatases within the inactive GR/hsp90 complex. These enzymes are released upon hormone binding and may also account for the rapid induction of tyrosine kinase activity seen in some cell types by glucocorticoids [15]. Evidence is also seen clinically whereby systemic doses of corticosteroids can lead to very rapid clinical improvement and inhibition of allergic/anaphylactic responses [14].

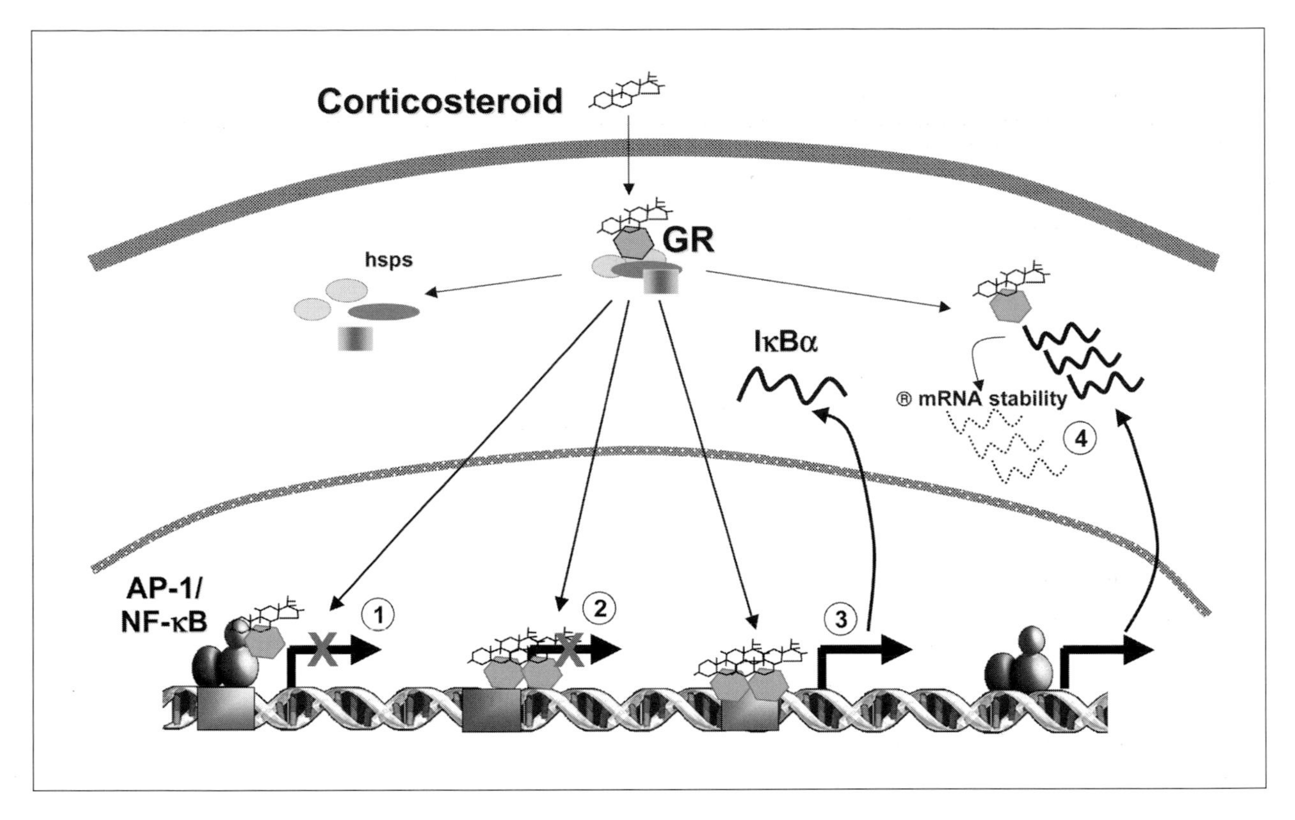

FIGURE 3. MECHANISMS OF GENE REPRESSION BY THE CORTICOSTEROID RECEPTOR (GR)
The corticosteroid can freely migrate across the plasma membrane where it associates with the cytoplasmic corticosteroid receptor (glucocorticoid receptor, GR). This results in activation of the GR and dissociation from the heat shock protein (hsp90) chaperone complex. Activated GR translocates to the nucleus where it can bind as a monomer either directly or indirectly with the transcription factors activator protein-1 (AP-1) and nuclear factor κB (NF-κB) preventing their ability to switch on inflammatory gene expression (1). Second, the GR dimer can bind to a GRE that overlaps the DNA binding site for a pro-inflammatory transcription factor or the start site of transcription to prevent inflammatory gene expression (2). Third, the GR dimer can induce the expression of the NF-κB inhibitor IκBα (3) and fourth, corticosteroids can increase the levels of cell ribonucleases and mRNA destabilizing proteins, thereby reducing the levels of mRNA (4).

Pharmacological effects

Effects on inflammation

Corticosteroids are the only therapeutic agents that clearly reverse the inflammation present in chronic diseases such as dermatitis, RA, Crohn's disease, SLE, asthma and COPD [3]. TOPICAL and systemic corticosteroids have the similar pharmacological effects with differences due to the dose delivered to the target organ and to an enhanced effect of systemic corticosteroids on the mobilisation and recruitment of inflammatory cells from the blood and BONE MARROW [16].

RA is a chronic systemic inflammatory disease of undetermined aetiology involving primarily the synovial membranes and articular structures of multiple joints. The disease is characterized by the inflammation of the membrane lining the joint, which causes pain, stiffness, warmth, redness and swelling and finally, later in the disease, joint destruction [17]. The inflamed joint lining, the synovium, can invade and damage bone and cartilage. Inflammatory cells release enzymes that may digest bone and cartilage. The involved joint can lose its shape and alignment, resulting in pain and loss of movement. Corticosteroids are effective in reducing joint pain, stiffness and swelling and the release of inflammatory mediators and tissue-digesting enzymes [18]. Evidence for slowing of radiographic progression of disease with corticosteroids has been available for >45 years [18]. Despite this, the use of corticosteroids in RA remains controversial, primarily because of possible undesirable side-effects [18].

In general, in all chronic inflammatory and immune diseases, corticosteroids cause a marked reduction in the number and activation of infiltrating cells including mast cells, MACROPHAGES, T LYMPHOCYTES, and eosinophils in the inflamed tissue [3]. Furthermore, TOPICAL and oral corticosteroids can have effects on tissue-resident cells and in asthma for example can reverse the shedding of epithelial cells, goblet-cell hyperplasia and basement-membrane thickening characteristically seen in biopsy specimens of bronchial epithelium from patients [3] (Fig. 4).

Cellular effects

Corticosteroids may have direct inhibitory effects on many of the cells involved in inflammation, including MACROPHAGES, T LYMPHOCYTES, eosinophils, mast cells, smooth muscle, endothelial and epithelial cells resulting in reduced mediator synthesis and release [3]. In general, corticosteroids substantially reduce the mast cell/eosinophil/lymphocyte driven processes, while leaving behind, or even augmenting, a neutrophil-mediated process [19]. For example, corticosteroids may enhance neutrophil function through increased leukotriene and superoxide production, in addition to inhibiting their APOPTOSIS [19]. Corticosteroids also decrease platelet CLEARANCE by the reticuloendothelial system, decrease ANTIBODY production and, by stabilising capillaries, decrease bleeding in patients with immune thrombocytopenic purpura (ITP) [20]. Furthermore, in autoimmune hepatitis, prednisolone prevents T-cell recognition of autoantigens, intrahepatic recruitment and proliferation of LYMPHOCYTES, IgG production and may enhance suppressor T-cell function in these patients [21].

In allergic diseases, corticosteroids also reduce the number of mast cells within the inflamed tissue; however, they do not appear to inhibit mediator release from these cells [3]. Treatment with TOPICAL corticosteroids also reduces the number of activated T LYMPHOCYTES ($CD25^+$ and HLA-DR^+) in the BAL and peripheral blood from asthmatic patients [22]. Corticosteroids are particularly effective against eosinophilic inflammation, possibly as a result of decreasing eosinophil survival by stimulating APOPTOSIS [23]. Interestingly, some patients with difficult-to-control asthma may develop exacerbations despite treatment with TOPICAL corticosteroids, and these often appear to have a T-cell- rather than eosinophil-dependent inflammatory mechanism [24, 25].

In addition to their suppressive effects on inflammatory cells, corticosteroids may also inhibit plasma exudation in most tissues and mucus secretion in inflamed airways [19]. There is an increase in vascularity in chronic inflammatory diseases [26] and high doses of TOPICAL systemic corticosteroids may reduce this [26] and the increased blood flow present in sites of inflammation [27].

Mediator effects

Corticosteroids block the generation of most pro-inflammatory CYTOKINES and CHEMOKINES, including INTERLEUKIN (IL)-1β, IL-4, IL-5, IL-8, granulocyte-macrophage COLONY-STIMULATING FACTOR (GM-CSF), TUMOR NECROSIS FACTOR-α (TNF-α), RANTES, and macrophage inflammatory protein-1α (MIP-1α) [19]. Despite the wide pleiotropy and redundancy in the CYTOKINE and chemokine families that exist and the subsequent inability to describe a precise role of most molecules in inflammatory disease pathogenesis, it is clear that these proteins are important mediators in chronic inflammation. The recent development of blocking antibodies and inhibitors to TNF-α has provided evidence that, in RA at least, this mole-

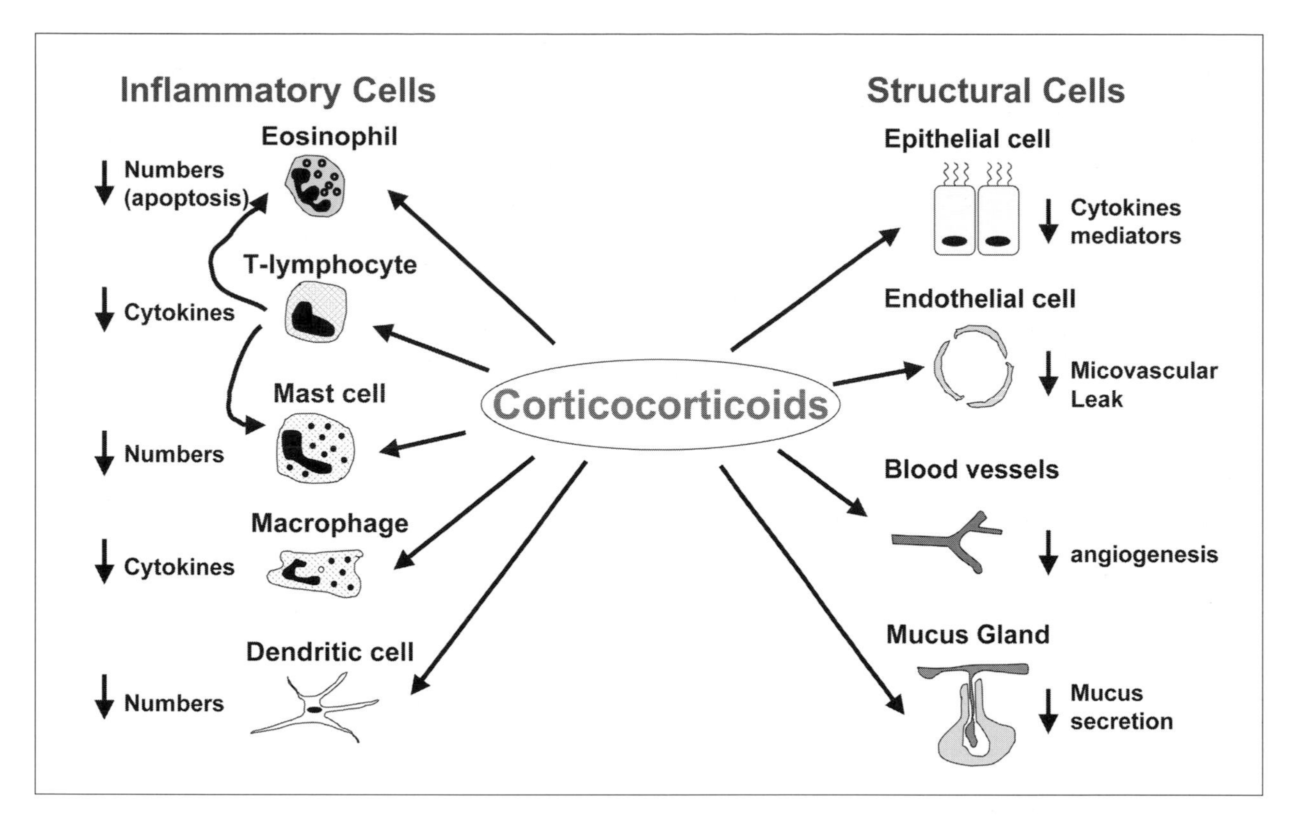

FIGURE 4
Corticosteroids act on most infiltrating and resident inflammatory cells to suppress inflammation. The activity (T lymphocytes and macrophages) and/or number of infiltrating cells (eosinophils, T lymphocytes, macrophages, mast cells and dendritic cells) are decreased by corticosteroids. Corticosteroids also have a suppressive effect on resident tissue cells and can reduce mediator release and adhesion molecule expression on epithelial and endothelial cells, microvascular leak from blood vessels, angiogenesis and both the numbers of mucus glands and release of mucus from these glands.

cule plays a key driving role in inflammation [28,29]. This does not appear to be the case with all inflammatory diseases, however [28–30]. Interestingly, corticosteroids can also enhance the expression of key anti-inflammatory molecules, such as IL-10 and IL-1ra, in some systems but again not all [3, 16]. For example, corticosteroids increase the production of IL-10, but not IL-1ra, from sites of inflammation in asthma [31,32].

Arachidonic acid metabolism via 5-lipoxygenase gives rise to a group of biologically active lipids known as LEUKOTRIENES: leukotriene B_4, which is a potent activator of leukocyte chemotaxis, and cysteinyl LEUKOTRIENES (leukotriene C_4, D_4 and E_4) which account for the spasmogenic activity previously described as a slow-reacting substance of anaphylaxis. LEUKOTRIENES, particularly cysteinyl LEUKOTRIENES, are thought to play a key role in some chronic and acute inflammatory diseases, however, these do not appear to be major targets for corticosteroids [33]. Thus, dexamethasone is ineffective in inhibiting the LTB_4/LTC_4 release from many cells and tissues [33]. Furthermore, systemic treatment with corticosteroids has no inhibitory effect on leukotriene release by whole-blood LEUKOCYTES, purified neutrophils, or MONOCYTES [33] and even increases *ex vivo* leukotriene biosynthesis by blood neutrophils [34]. Interestingly, *in vitro* corticosteroids accelerate LTC_4 catabolism by inducing activity of a LTC_4-degrading enzyme, γ-glutamyl transpeptidase-related enzyme (γ-GTPRE), in transformed human bronchial epithelial cells [35].

The inflammation in COPD [19] and in a high number of patients with IBD [36, 37] is scarcely suppressed by TOPICAL or oral corticosteroids, even at very high doses. Potential reasons for the failure of corticosteroids to function effectively in reducing inflammation in IBD and COPD include the fact that both diseases have a high oxidative load and oxidative stress may reduce corticosteroid receptor (GR) nuclear translocation [19, 36, 37], reduced GRα expression or altered GR-co-factor interactions within the nucleus [19]. Interestingly, cigarette smoke contains 10^{17} oxidant particles per puff and asthmatic subjects who smoke have a reduced responsiveness to both TOPICAL and oral corticosteroids [38] (see Chapter C.4).

Pharmacokinetics

The pharmacokinetics of many corticosteroids is well described. In general, plasma concentrations of corticosteroids vary considerably (up to 10-fold) after oral administration of the same dose by normal volunteers and patients [39]. Many compounds have elimination half-lives from plasma of 2–4 hours, whereas the biological half-lives range from 18–36 hours [40]. The pharmacokinetic properties of TOPICAL drugs depend upon a combination of tissue deposition/targeting, receptor binding, volume of distribution, tissue retention, and lipid conjugation. In addition, in order to achieve a good THERAPEUTIC INDEX, drugs need to possess a low oral BIOAVAILABILITY and small particle size, rapid metabolism, high CLEARANCE, high plasma protein binding and low systemic half-life. Furthermore, an ideal compound would be inactive at sites distal to the target organ [41, 42].

There are two main methods of reducing the systemic activity of TOPICAL corticosteroids (1) reducing gastrointestinal BIOAVAILABILITY and (2) prolonging TISSUE RESIDENCY. For example, oral administration of ileal release budesonide capsules for the treatment of Crohn's disease gives similar levels of systemic exposure to corticosteroid, corticosteroid availability and cortisol suppression in adults and children as seen with prednisolone but, importantly, no clinically relevant adverse side-effects were reported [43]. In asthma, changing the inhaler device can decrease oral delivery and subsequently gastrointestinal availability and enhance lung deposition by altering the particle size. Alternatively, for IBD the corticosteroid can be altered to reduce gastrointestinal absorption and/or enhance first-pass hepatic metabolism. Prolonged retention in the tissue can be achieved by increasing lipophilicity, as with fluticasone propionate (FP) and mometasone furoate, or by forming soluble intracellular esters, as with budesonide and ciclesonide [41, 42].

Lipophilicity generally correlates well with absorption characteristics. For example, fluticasone has high lipophilicity and binding affinity for the GR, resulting in a high volume of distribution and long plasma half-life. However, the systemic side-effects of fluticasone that arise from systemic absorption are limited due to its almost complete first-pass metabolism and, in the case of enteric delivery, low absorption from the gastrointestinal tract. In general, for TOPICAL corticosteroids, treatment EFFICACY and side-effects are directly related to tissue dose. This may also vary depending upon disease severity in the case of fluticasone but not, apparently, budesonide [41, 42]. The pharmacokinetic profile of TOPICAL corticosteroids therefore varies with the drug, the delivery mechanism and patient profile.

Clinical indications

Pharmacological control of inflammation may be obtained in most patients with varying doses of oral or TOPICAL corticosteroids depending upon the disease severity [3, 44]. In most patients adequate doses of TOPICAL corticosteroids allow systemic corticosteroids to be reduced or withdrawn completely [3]. As such, corticosteroids are standard therapy for disorders such as RA, asthma, connective tissue diseases, vasculitis, allergic reactions, IBD, psoriasis and eczema. However, when administered over a long period of time at reasonably high concentrations, their beneficial effects are overshadowed by a number of side-effects including OSTEOPOROSIS, skin atrophy and diabetes [44].

Overall, the duration, dosage and dosing regime, the particular corticosteroid used and its mode of application, along with a patient's individual susceptibility, appear to determine the incidence of adverse events [44, 45]. Prolonged TOPICAL application is also a high risk factor, whereas total dose is of secondary importance. Not surprisingly, side-effects are much more severe with systemic corticosteroid use, although even TOPICAL application can induce both local and system side-effects [46]. It is the presence of these side-effects that is the limiting factor for therapy and has driven research and development of new compounds [44, 47]. However, the newer potent TOPICAL steroids, such as mometasone and fluticasone, seem to carry a lower risk for adrenal suppression than older drugs [48].

Prednisone and prednisolone are highly effective in inducing remission of Crohn's disease [49, 50]. Crohn's disease causes inflammation in the ileum, although other parts of the gut may be involved. Crypt inflammation (cryptitis) and abscesses progress to focal ulcers which, in some cases, regress and, in others, the inflammatory process evolves with influx and proliferation of MACROPHAGES and other inflammatory cells, occasionally forming noncaseating granulomas with multinucleated giant cells. Transmural inflammation, deep ulceration, edema, muscular proliferation, and fibrosis cause deep sinus tracts and fistulas, mesenteric abscesses, and obstruction, which are the major local complications. As with other diseases, however, because they have systemic effects, their use has been associated with an increased incidence of adverse events [50].

Once it was demonstrated that corticosteroids were effective in inducing remission of Crohn's disease, oral drug delivery systems were developed to provide the EFFICACY of a corticosteroid without the associated corticosteroid toxicity. The choice of drug is based on the physicochemical and pharmacological properties of the corticosteroid, although this varies greatly between individual subjects [39]. However, TOPICAL corticosteroids are still mostly unproven clinically, possibly due to a lack of appropriate delivery systems [49, 50].

For example, local therapy of distal colitis has proved effective using retention enemas of hydrocortisone and betamethasone [49] and animal models have shown the feasibility of colonic delivery to increase local tissue concentrations and reduce adverse side-effects of corticosteroids [51]. Budesonide is the most commonly used TOPICAL corticosteroid in IBD and was initially used because of its extensive first-pass hepatic metabolism, which reduces oral BIOAVAILABILITY (10–15%) [52]. Budesonide also possesses a high rate of metabolism compared to other corticosteroids, such as beclamethasome and triamcinolone. Enteric-coated sustained-release pellets of budesonide (10 mg) are as effective as 40 mg prednisolone in controlling IBD endoscopic scores but did not affect plasma cortisol levels [50]. In an ongoing study of more than 4000 Crohn's disease patients treated with doses of up to 21 mg/day budesonide, some for more than 5 years, serious adverse events were experienced by only 3% of patients; most of these were gastrointestinal and unrelated to treatment [53, 54].

Immunosuppressive agents such as 6-mercaptopurine and azathioprine are also used to treat Crohn's disease. These drugs may cause side-effects like nausea, vomiting, and diarrhea and may lower a person's resistance to infection. However, when patients are treated with a combination of corticosteroids and immunosuppressive drugs, the immunosuppressives are CORTICOSTEROID-SPARING [50].

Corticosteroids are also the most widely used therapy in dermatology. Initial treatment involved hydrocortisone but great advances were seen following the introduction of halogenated compounds such as triamcinolone and the more potent drugs such as budesonide and fluticasone [44]. This inevitably led to over-prescription and adverse events becoming increasingly apparent and "steroid phobia" is still a considerable concern [44, 48] with over 70% of dermatology patients in the UK having worries about TOPICAL corticosteroid use and 24% being non-compliant as a result. Skin thinning was the major fear (35%) with 10% concerned about systemic absorption and effects on growth and development [48].

Other chronic immune and inflammatory diseases such as SLE, ITP and autoimmune hepatitis are also treated predominantly with systemic corticosteroids. Corticosteroids are generally taken orally with the dose varying with disease severity and organ involvement. Intramuscular methylprednisolone or hydrocortisone may be used during a mild flare characterised by fatigue or arthralgia in SLE for example; however, more aggressive therapy is often required [55]. Despite initially good response rates to corticosteroid therapy ranging from 65–78% for ITP for example, this soon drops to <20% in patients with chronic disease [56]. It is unclear whether this reflects a reduced EFFICACY of the drugs or a lack of PATIENT COMPLIANCE. Evidence that patients may respond to dexamethasone [20] suggests that the later may be a problem possibly due to the presence, or fear, of adverse events.

In asthma, corticosteroids consistently lessen airway hyperesponsiveness and the maximal response to a number of spasmogens and irritants [3, 57, 58]. Interestingly, the reduction in airway hyperesponsiveness may not be maximal until treatment has been given for several months. The magnitude of the reduction varies, and airway responsiveness can remain abnormal [3]. When corticosteroid therapy is discontinued airway responsiveness usually returns to pretreatment levels [59] (see Chapter C.4).

Side-effects

All currently available TOPICAL corticosteroids are absorbed into the systemic circulation and therefore inevitably have some systemic effect, although this is considerably less than those seen with oral corticosteroids (Tab. 1). The occurrence and severity of the side-effects seen depend upon the duration of use, dosage, dosing regime and specific drug used, along with individual patient variability [44]. However, the highest risk factor appears to be prolonged use. Side-effects of TOPICAL steroids include glaucoma, cataracts, tissue atrophy and wound healing, whilst at high doses there is an increased risk of infection, adrenal suppression and OSTEOPOROSIS. The growth retardation seen with oral corticosteroids does not appear to be a problem with modern TOPICAL corticosteroids, although there may be an initial reduction in growth velocity on starting therapy. Side-effects of oral corticosteroids include skin and muscle atrophy, delayed wound healing and increased risk of infection, OSTEOPOROSIS and bone necrosis, glaucoma and cataracts, behavioral changes, hypertension, peptic ulcers and gastrointestinal bleeding and Cushing's Syndrome and diabetes [44]. Interestingly, it appears that early skin atrophy induced by corticosteroid therapy is reversible, whereas major atrophy leading to striae formation is not [48].

These side-effects often occur together and this is exemplified by Cushing's Syndrome whose signs and symptoms include elevated blood pressure, development of diabetes, pink-to-purple stretch marks on the abdominal skin, fatigue, depression, moodiness, and accentuated fatty tissue on the face and upper back (Buffalo hump) [1, 44]. Women with Cushing's Syndrome often have irregular menstrual periods and develop new facial hair growth. Men may show a decrease in sex drive. Taken together, these side-effects seriously limit the value of corticosteroids in severe inflammation where the risk/benefit ratio is compromised. This has driven the need to develop novel agents with the anti-inflammatory capacity of corticosteroids but with reduced side-effects.

Whilst the major anti-inflammatory effects of corticosteroids are almost certainly due to TRANSREPRESSION, the underlying molecular mechanisms for the side-effects of corticosteroids are complex and not fully understood [44]. Certain side-effects, such as diabetes and glaucoma, are due to TRANSACTIVATION events, whilst others are due to TRANSREPRESSION (HPA suppression). In addition, the precise molecular events underlying corticosteroid induction of OSTEOPOROSIS is unclear but probably requires both gene induction and gene repression [44].

Despite this uncertainty, there has been a search for "dissociated" corticosteroids that selectively transrepress without significant TRANSACTIVATION, thus potentially reducing the risk of systemic side-effects. Several non-steroidal "selective corticosteroid receptor agonists" (SEGRA) have recently been reported that show dissociated properties in human cells and are now in clinical development where they show good separation between TRANSREPRESSION and TRANS-

TABLE 1. TISSUE/ORGAN-SPECIFIC SIDE-EFFECTS OF TOPICAL AND SYSTEMIC CORTICOSTEROIDS

Cardiovascular system	Hypertension
	Dyslipidemia
	Thrombosis
	Vasculitis
CNS	Disturbances in mood, behaviour, memory and cognition
	"Steroid psychosis", steroid dependence
	Cerebral atrophy
Endocrine system, metabolism, electrolytes	Cushing's syndrome
	Diabetes mellitus
	Adrenal atrophy
	Growth retardation
	Hypogonadism, delayed puberty
	Increased sodium retention and potassium excretion
Eye	Glaucoma
	Cataract
Gastrointestinal	Peptic ulcer
	Gastrointestinal bleeding
	Pancreatitis
Immune system	Increased risk of infection
	Re-activation of latent viruses
Skeleton and muscle	Muscle atrophy/myopathy
	Osteoporosis
	Bone necrosis
Skin	Atrophy, striae, distension
	Delayed wound healing
	Steroid acne, perioral dermatitis
	Erythema, teleangiectasia, petechia, hypertrichosis

ACTIVATION actions in the skin [8,47,60]. This suggests that the development of corticosteroids and SEGRA with a greater margin of safety is possible and may even lead to the development of oral compounds that have reduced adverse effects. Furthermore, the newer TOPICAL corticosteroids used today, such as fluticasone, mometasone and budesonide, appear to have more potent TRANSREPRESSION than TRANSACTIVATION effects, which may account, at least in part, for their selection as potent anti-inflammatory agents [8]. These new potent corticosteroids are particularly effective as TOPICAL agents and their use has overtaken that of oral/systemic corticosteroids for many diseases. For example, coated-enteric slow-release budesonide capsules are equally effective as prednisolone in Crohn's disease without the associated reduction in plasma cortisol seen with prednisolone [49, 61]. Similar results have been achieved with fluticasone, albeit in fewer well-controlled studies [62].

An alternative approach to obtain safer drugs is the use of soft drugs such as ciclesonide, which are only activated at the site of inflammation. Ciclesonide is a novel inhaled corticosteroid under clinical development for the treatment of asthma. Ciclesonide itself is inactive and needs to be cleaved by lung-specific esterases in order to bind to the corticosteroid receptor [63]. According to data from healthy volunteers, ciclesonide affected serum cortisol levels significantly less compared to beclomethasone dipropionate [63], suggesting that ciclesonide might have fewer systemic effects and hence a superior safety profile.

Other approaches to the production of 'safer' corticosteroids are the use of alternative agents that target other aspects of the inflammatory response and therefore act as steroid-sparing agents. These classes of drug include kinase inhibitors, IMMUNOMODULATORY agents such as cyclosporin and long-acting β-agonists [5].

Summary

Corticosteroids are the most effective therapy for chronic immune and inflammatory diseases in current use. Despite their success over the past 50 years, and especially since the advent of new potent halogenated compounds, worries about the detrimental side-effects of systemic corticosteroids has limited their effectiveness in severe disease. This has resulted in the increasing use of TOPICAL corticosteroids targeted to the site of inflammation rather than systemic administration. Improvements in risk/benefit ratios are likely to occur, as greater understanding of the role of chemical substitutions of the synthetic corticosteroid becomes clear, and more potent tissue selective drugs are developed. Drugs that target distinct aspects of corticosteroid function, switching on or off genes, are also under development and along with non-steroidal agents that target different aspects of the inflammatory response are likely to lead to safer drugs with a much reduced side-effect profile. However, until these become widely available, current systemic and TOPICAL corticosteroids are likely to remain the major treatment for most inflammatory diseases.

Selected readings

Eggert M, Schulz M, Neeck G (2001) Molecular mechanisms of glucocorticoid action in rheumatic autoimmune diseases. *J Steroid Biochem Mol Biol* 77: 185–191

Friend DR (1998) Review article: issues in oral administration of locally acting glucocorticosteroids for treatment of inflammatory bowel disease. *Aliment Pharmacol Ther* 12: 591–603

Hofer KN (2003) Oral budesonide in the management of Crohn's disease. *Ann Pharmacother* 37: 1457–1464

Leung DY, Bloom JW (2003) Update on glucocorticoid action and resistance. *J Allergy Clin Immunol* 111: 3–22

Saag KG, Koehnke R, Caldwell JR, Brasington R, Burmeister LF, Zimmerman B, Kohler JA, Furst DE (1994) Low dose long-term corticosteroid therapy in rheumatoid arthritis: an analysis of serious adverse events. *Am J Med* 96: 115–123

Schacke H, Docke WD, Asadullah K (2002) Mechanisms involved in the side-effects of glucocorticoids. *Pharmacol Ther* 96: 23–43

Recommended websites

National Heart, Lung and Blood Institute: *http://www.nhlbi.nih.gov/* (Accessed September 2005)

HealthLine plus Information. A service of the US National Library of Medicine and the National Institutes of Health: *http://www.nlm.nih.gov/medlineplus/druginfo/uspdi/202018.html* (Accessed September 2005)

A comprehensive information resource for patients with Crohn's Disease and Ulcerative Colitis: *http://ibd.patientcommunity.com/links.cfm?parentcat_id=109&cat_id=109* (Accessed September 2005)

American Academy of Allergy, Asthma and Immunology: *http://www.aaaai.org* (Accessed September 2005)

References

1 Magiakou MA, Chrousos GP (2002) Cushing's syndrome in children and adolescents: current diagnostic and therapeutic strategies. *J Endocrinol Invest* 25: 181–194

2 Kino T, Vottero A, Charmandari E, Chrousos GP (2002) Familial/sporadic glucocorticoid resistance syndrome and hypertension. *Ann NY Acad Sci* 970: 101–111

3 Barnes PJ, Adcock IM (2003) How do corticosteroids work in asthma? *Ann Intern Med* 139: 359–370

4 Raju TN (1999) The Nobel chronicles. 1950: Edward Calvin Kendall (1886–1972); Philip Showalter Hench (1896–1965); and Tadeus Reichstein (1897–1996). *Lancet* 353: 1370

5 Caramori G, Adcock I (2003) Pharmacology of airway inflammation in asthma and COPD. *Pulm Pharmacol Ther* 16: 247–277

6 Johnson M (1996) Pharmacodynamics and pharmacokinetics of inhaled glucocorticoids. *J Allergy Clin Immunol* 97: 169–176

7 Brattsand R, Thalen A, Roempke K, Kallstrom L, Gruvstad E (1982) Influence of 16 alpha, 17 alpha-acetal substitution and steroid nucleus fluorination on the topical to systemic activity ratio of glucocorticoids. *J Steroid Biochem* 16: 779–786

8 Adcock IM (2003) Glucocorticoids: new mechanisms and future agents. *Curr Allergy Asthma Rep* 3: 249–257

9 Auphan N, Didonato JA, Rosette C, Helmberg A, Karin M (1995) Immunosuppression by glucocorticoids: inhibition of NF-kappa B activity through induction of I kappa B synthesis. *Science* 270: 286–290

10 Shim J, Karin M (2002) The control of mRNA stability in response to extracellular stimuli. *Mol Cells* 14: 323–331

11 Reichardt HM, Tuckermann JP, Gottlicher M, Vujic M, Weih F, Angel P, Herrlich P, Schutz G (2001) Repression of inflammatory responses in the absence of DNA binding by the glucocorticoid receptor. *EMBO J* 20: 7168–7173

12 Reichardt HM, Kaestner KH, Tuckermann J, Kretz O, Wessely O, Bock R, Gass P, Schmid W, Herrlich P, Angel P et al (1998) DNA binding of the glucocorticoid receptor is not essential for survival. *Cell* 93: 531–541

13 Norman AW, Mizwicki MT, Norman DP (2004) Steroid-hormone rapid actions, membrane receptors and a conformational ensemble model. *Nat Rev Drug Discov* 3: 27–41

14 Buttgereit F, Scheffold A (2002) Rapid glucocorticoid effects on immune cells. *Steroids* 67: 529–534

15 Croxtall JD, Van Hal PT, Choudhury Q, Gilroy DW, Flower RJ (2002) Different glucocorticoids vary in their genomic and non-genomic mechanism of action in A549 cells. *Br J Pharmacol* 135: 511–519

16 Leung DY, Bloom JW (2003) Update on glucocorticoid action and resistance. *J Allergy Clin Immunol* 111: 3–22

17 Eggert M, Schulz M, Neeck G (2001) Molecular mechanisms of glucocorticoid action in rheumatic autoimmune diseases. *J Steroid Biochem* Mol Biol 77: 185–191

18 Morand EF (1998) Corticosteroids in the treatment of rheumatologic diseases. *Curr Opin Rheumatol* 10: 179–183

19 Adcock IM, Chung KF (2002) Overview: why are corticosteroids ineffective in COPD? *Curr Opin Investig Drugs* 3: 58–60

20 Huber MR, Kumar S, Tefferi A (2003) Treatment advances in adult immune thrombocytopenic purpura. *Ann Hematol* 82: 723–737

21 Czaja AJ, Ludwig J, Baggenstoss AH, Wolf A (1981) Corticosteroid-treated chronic active hepatitis in remission: uncertain prognosis of chronic persistent hepatitis. *N Engl J Med* 304: 5–9

22 Wilson JW, Djukanovic R, Howarth PH, Holgate ST (1994) Inhaled beclomethasone dipropionate downregulates airway lymphocyte activation in atopic asthma. *Am J Respir Crit Care Med* 149: 86–90

23 Schleimer RP, Bochner BS (1994) The effects of glucocorticoids on human eosinophils. *J Allergy Clin Immunol* 94: 1202–1213

24 Redington AE, Wilson JW, Walls AF, Madden J, Djukanovic R, Holgate ST, Howarth PH (2000) Persistent airway T-lymphocyte activation in chronic corticosteroid-treated symptomatic asthma. *Ann Allergy Asthma Immunol* 85: 501–507

25 in't Veen JC, Smits HH, Hiemstra PS, Zwinderman AE, Sterk PJ, Bel EH (1999) Lung function and sputum characteristics of patients with severe asthma during an induced exacerbation by double-blind steroid withdrawal. *Am J Respir Crit Care Med* 160: 93–99

26 Orsida BE, Li X, Hickey B, Thien F, Wilson JW, Walters EH (1999) Vascularity in asthmatic airways: relation to inhaled steroid dose. *Thorax* 54: 289–295

27 Brieva JL, Danta I, Wanner A (2000) Effect of an inhaled glucocorticosteroid on airway mucosal blood flow in mild asthma. *Am J Respir Crit Care Med* 161: 293–296

28 Arend WP (2002) The mode of action of cytokine inhibitors. *J Rheumatol* (Suppl) 65: 16–21

29 Taylor PC (2003) Anti-TNFalpha therapy for rheumatoid arthritis: an update. *Intern Med* 42: 15–20

30 Probert CS, Hearing SD, Schreiber S, Kuhbacher T, Ghosh S, Arnott ID, Forbes A (2003) Infliximab in moderately severe glucocorticoid resistant ulcerative colitis: a randomised controlled trial. *Gut* 52: 998-1002

31 John M, Lim S, Seybold J, Jose P, Robichaud A, O'Connor B, Barnes PJ, Chung KF (1998) Inhaled corticosteroids increase interleukin-10 but reduce macrophage inflammatory protein-1alpha, granulocyte macrophage colony-stimulating factor, and inter-

feron-gamma release from alveolar macrophages in asthma. *Am J Respir Crit Care Med* 157: 256–262

32 Sousa AR, Trigg CJ, Lane SJ, Hawksworth R, Nakhosteen JA, Poston RN, Lee TH (1997) Effect of inhaled glucocorticoids on IL-1 beta and IL-1 receptor antagonist (IL-1ra) expression in asthmatic bronchial epithelium. *Thorax* 52: 407–410

33 McMillan RM (2001) Leukotrienes in respiratory disease. *Paediatr Respir Rev* 2: 238–244

34 Thomas E, Leroux JL, Blotman F, Descomps B, Chavis C (1995) Enhancement of leukotriene A4 biosynthesis in neutrophils from patients with rheumatoid arthritis after a single glucocorticoid dose. *Biochem Pharmacol* 49: 243–248

35 Zaitsu M, Hamasaki Y, Aoki Y, Miyazaki S (2001) A novel pharmacologic action of glucocorticosteroids on leukotriene C4 catabolism. *J Allergy Clin Immunol* 108: 122–124

36 Hearing SD, Norman M, Probert CS, Haslam N, Dayan CM (1999) Predicting therapeutic outcome in severe ulcerative colitis by measuring *in vitro* steroid sensitivity of proliferating peripheral blood lymphocytes. *Gut* 45: 382–388

37 Norman M, Hearing SD (2002) Glucocorticoid resistance – what is known? *Curr Opin Pharmacol* 2: 723–729

38 Chalmers GW, Macleod KJ, Little SA, Thomson LJ, McSharry CP, Thomson NC (2002) Influence of cigarette smoking on inhaled corticosteroid treatment in mild asthma. *Thorax* 57: 226–230

39 Pickup ME (1979) Clinical pharmacokinetics of prednisone and prednisolone. *Clin Pharmacokinet* 4: 111–128

40 Swartz SL, Dluhy RG (1978) Corticosteroids: clinical pharmacology and therapeutic use. *Drugs* 16: 238–255

41 Colice GL (2000) Comparing inhaled corticosteroids. *Respir Care* 45: 846–853

42 Edsbacker S (1999) Pharmacological factors that influence the choice of inhaled corticosteroids. *Drugs* 58 (Suppl 4): 7–16

43 Lundin PD, Edsbacker S, Bergstrand M, Ejderhamn J, Linander H, Hogberg L, Persson T, Escher JC, Lindquist B (2003) Pharmacokinetics of budesonide controlled ileal release capsules in children and adults with active Crohn's disease. *Aliment Pharmacol Ther* 17: 85–92

44 Schacke H, Docke WD, Asadullah K (2002) Mechanisms involved in the side-effects of glucocorticoids. *Pharmacol Ther* 96: 23–43

45 Saag KG, Koehnke R, Caldwell JR, Brasington R, Burmeister LF, Zimmerman B, Kohler JA, Furst DE (1994) Low dose long-term corticosteroid therapy in rheumatoid arthritis: an analysis of serious adverse events. *Am J Med* 96: 115–123

46 Robertson DB, Maibach HI (1982) Topical corticosteroids. *Int J Dermatol* 21: 59–67

47 Schacke H, Hennekes H, Schottelius A, Jaroch S, Lehmann M, Schmees N, Rehwinkel H, Asadullah K (2002) SEGRAs: a novel class of anti-inflammatory compounds. *Ernst Schering Res Found Workshop* 357–371

48 Charman C, Williams H (2003) The use of corticosteroids and corticosteroid phobia in atopic dermatitis. *Clin Dermatol* 21: 193–200

49 Friend DR (1998) Review article: issues in oral administration of locally acting glucocorticosteroids for treatment of inflammatory bowel disease. *Aliment Pharmacol Ther* 12: 591–603

50 Sandborn WJ, Feagan BG (2003) Review article: mild to moderate Crohn's disease – defining the basis for a new treatment algorithm. *Aliment Pharmacol Ther* 18: 263–277

51 Cui N, Friend DR, Fedorak RN (1994) A budesonide prodrug accelerates treatment of colitis in rats. *Gut* 35: 1439–1446

52 Brattsand R, Linden M (1996) Cytokine modulation by glucocorticoids: mechanisms and actions in cellular studies. *Aliment Pharmacol Ther* 10 (Suppl 2): 81–90

53 Kane SV, Schoenfeld P, Sandborn WJ, Tremaine W, Hofer T, Feagan BG (2002) The effectiveness of budesonide therapy for Crohn's disease. *Aliment Pharmacol Ther* 16: 1509–1517

54 Hofer KN (2003) Oral budesonide in the management of Crohn's disease. *Ann Pharmacother* 37: 1457–1464

55 Ioannou Y, Isenberg DA (2002) Current concepts for the management of systemic lupus erythematosus in adults: a therapeutic challenge. *Postgrad Med* J 78: 599–606

56 Pizzuto J, Ambriz R (1984) Therapeutic experience on 934 adults with idiopathic thrombocytopenic purpura: Multicentric Trial of the Cooperative Latin American group on Hemostasis and Thrombosis. *Blood* 64: 1179–1183

57 Barnes PJ (1990) Effect of corticosteroids on airway hyperresponsiveness. *Am Rev Respir Dis* 141: S70–S76

58 van den BM, Kerstjens HA, Meijer RJ, de Reus DM, Koeter GH, Kauffman HF, Postma DS (2001) Corticosteroid-induced improvement in the PC20 of adenosine monophosphate is more closely associated with reduction in airway inflammation than improvement in the PC20 of methacholine. *Am J Respir Crit Care Med* 164: 1127–1132

59 Haahtela T, Jarvinen M, Kava T, Kiviranta K, Koskinen S, Lehtonen K, Nikander K, Persson T, Selroos O, Sovijarvi A et al (1994) Effects of reducing or discontinuing inhaled budesonide in patients with mild asthma. *N Engl J Med* 331: 700–705

60 Schacke H, Schottelius A, Docke WD, Strehlke P, Jaroch S, Schmees N, Rehwinkel H, Hennekes H, Asadullah K (2004) Dissociation of transactivation from transrepression by a selective glucocorticoid receptor agonist leads to separation of therapeutic effects from side-effects. *Proc Natl Acad Sci USA* 101: 227–232

61 Lofberg R, Danielsson A, Suhr O, Nilsson A, Schioler R, Nyberg A, Hultcrantz R, Kollberg B, Gillberg R, Willen R et al (1996) Oral budesonide *versus* prednisolone in patients with active extensive and left-sided ulcerative colitis. *Gastroenterology* 110: 1713–1718

62 de Kaski MC, Peters AM, Lavender JP, Hodgson HJ (1991) Fluticasone propionate in Crohn's disease. *Gut* 32: 657–661

63 Kanniess F, Richter K, Bohme S, Jorres RA, Magnussen H (2001) Effect of inhaled ciclesonide on airway responsiveness to inhaled AMP, the composition of induced sputum and exhaled nitric oxide in patients with mild asthma. *Pulm Pharmacol Ther* 14: 141–147

C12

Non-steroidal anti-inflammatory drugs

Regina M. Botting and Jack H. Botting

Introduction

Throughout history humans have experimented with herbal remedies to alleviate the symptoms of diseases. The active principles of some of these remedies are of proven value and have become established in modern therapeutics. None, however, has been more widely accepted nor as universally practised as the use of the extracts of certain plants for the treatment of the various symptoms of inflammatory conditions such as pain, swelling and fever.

The Egyptian Ebers papyrus records that 3500 years ago extracts of the dried leaves of myrtle applied to the abdomen and back were beneficial for rheumatic pains from the womb. A thousand years later, no less an authority than Hippocrates recommended the juice of willow bark to reduce fever and alleviate the pain of childbirth. Similar curative effects were attributed to decoctions from *Salix* and *Spiraea* species by early inhabitants of North America and South Africa.

The beneficial effects of willow bark were placed on a more scientific basis by the observations of a country parson, the Reverend Edward Stone of Chipping Norton in Oxfordshire. Stone gathered a pound of willow bark, dried it over a baker's oven and ground it to a fine powder. Doses of 1 dram (1.8g) were found to be successful in 50 patients with fever [1]. The restriction of the availability of willow bark due to the use of willows for manufacture of wickerware resulted in herbalists cultivating meadowsweet (*Spiraea ulmaria*) to provide treatment for the ague and similar conditions.

By the middle of the nineteenth century, advances in chemistry established that the common constituent of the plant extracts that reduced fever and inflammatory pain was salicylate. Salicylic acid was synthesized in Germany in 1860, and its ready availability lead to its widespread use in fever, rheumatism and as an external antiseptic. The value of salicylate as a medicine was limited by its unpleasant taste and tendency to produce nausea. In an attempt to make a more palatable preparation, Felix Hoffman, a chemist working for the Bayer Company, synthesised acetylsalicylate or aspirin. Bayer's Research Director, Dr. Heinrich Dreser, tested the effects of aspirin in animals and in the clinic and introduced it in 1899 as an antipyretic, anti-inflammatory and analgesic drug [2]. Aspirin has become perhaps the most widely used drug, and its value as an antipyretic and for the pain of rheumatoid- and osteoarthritis is well accepted. Sporadic reports of gastrotoxicity produced by aspirin were substantiated in 1938 by endoscopic studies which clearly demonstrated that aspirin produced erosions and even frank ulceration of the gastric mucosa [3].

The realisation that this valuable medicine could produce a serious and sometimes fatal gastrotoxicity stimulated a search for compounds with antipyretic, analgesic and anti-inflammatory actions wthout gastrotoxicity. Many compounds with differing chemical structure were produced and marketed from 1940 onwards, but all possessed to some degree the gastrotoxicity. As a group, these drugs were designated "non-steroid anti-inflammatory drugs" (NSAIDs).

Mode of action of non-steroid anti-inflammatory drugs

The fact that many compounds of diverse chemical structure not only possessed the same therapeutic actions but also shared identical toxic side-effects raised the intriguing possibility that a single biochemical action was responsible for all of the various actions of the NSAIDs. Many biochemical effects of NSAIDs were demonstrated: inhibition of dehydrogenases, aminotranferases, decarboxylases and several key enzymes involved in protein and RNA biosynthesis, as well as many others. However, there was no obvious correlation between these effects and the therapeutic and toxic actions of NSAIDs, and they were achieved only with concentrations well above those found in human plasma after therapy.

The enigma of the MECHANISM OF ACTION of NSAIDs was ultimately resolved by the elegantly simple pharmacological experiments of Vane (1971) [4], who showed that aspirin and some other NSAIDs inhibited, in a dose-dependent manner, the synthesis of the highly active lipid mediators PROSTAGLANDINS (PGs) from guinea pig lung homogenates. Vane hypothesised that both the therapeutic and side-actions of NSAIDs were due to inhibition of PG synthesis. PGs are formed from arachidonic acid mobilised from membrane phospholipids by a phospholipase enzyme. Arachidonic acid is acted upon by the microsomal enzyme cyclo-oxygenase (COX) to form the cyclic endoperoxides PGG_2 and PGH_2. These unstable endoperoxides are then isomerised, enzymatically or non-enzymatically, into various prostanoids such as thromboxane A_2 (TxA_2), prostacyclin (PGI_2) and PGD_2, E_2 and $F_{2\alpha}$ (see Fig. 1). Aspirin and other NSAIDs inhibit the cyclo-oxygenase enzyme, thus preventing the formation of the endoperoxide precursors of the various PGs.

PROSTAGLANDINS were known to be pyrogenic and were shown to be present in cerebrospinal fluid during fever. Similarly, PGE_2 and PGI_2 are vasodilator and present at inflammatory foci (such as the synovial fluid of arthritic joints), thus suggesting their involvement in the swelling typical of inflammatory conditions. The analgesic action of NSAIDs was initially less easy to explain, since PGs, unlike other mediators of INFLAMMATION such as bradykinin, were not pain-producing substances. However, they were subsequently shown to greatly potentiate the pain induced by other mediators; that is, they manifested a "hyperalgesic" effect. Clearly, the three therapeutic actions of NSAIDs could be expained by inhibition of PG synthesis.

That the common side-effects of NSAIDs, such as gastrotoxicity and nephrotoxicity, were also due to inhibition of PG synthesis was apparent when PGs were shown to be cytoprotective on the gastric mucosa, and could maintain renal blood flow when renal circulation was compromised. The decreased platelet reactivity observed, particularly after aspirin, was explained when the prostanoid TxA_2 was shown to be a potent inducer of platelet aggregation [5].

Cyclo-oxygenase-1 and cyclo-oxygenase-2

Implicit in the establishment of the MECHANISM OF ACTION of NSAIDs was the assumption that since all the pharmacological actions depended on the inhibition of COX, it would be impossible to separate the therapeutic effects from the toxic actions of these agents. However, there were some minor inconsistencies in this theory. Epidemiological and experimental studies showed that the severity of gastric toxic effects of different NSAIDs varied when the drugs were used at comparable anti-inflammatory doses. Ibuprofen, for example, causes less damage to the stomach than ketoprofen [6]. Similarly, the NSAID nimesulide was effective in models of inflammatory disease such as the carrageenan-injected rat paw, yet were poorly active in inhibiting PG synthesis in conventional systems [7].

The existence of multiple COX enzymes was a likely explanation for these inconsistencies, and over thirty years ago Flower and Vane had suggested that paracetamol, which lacks anti-inflammatory actions, exerted its antipyretic effect by inhibition of a distinct isoform of COX in brain tissue [8]. More recently a number of workers showed that PG synthesis could be upregulated in inflammatory conditions by induction of synthesis of more COX enzyme. For example Needleman and colleagues showed, using a model of the inflammatory condition of hydronephrosis in perfused rabbit kidney, that there was a marked increase in PGE_2 release following injection of bradykinin. This increased release could be prevented by prior treatment with the protein synthesis inhibitors cycloheximide or actinomycin D. Of

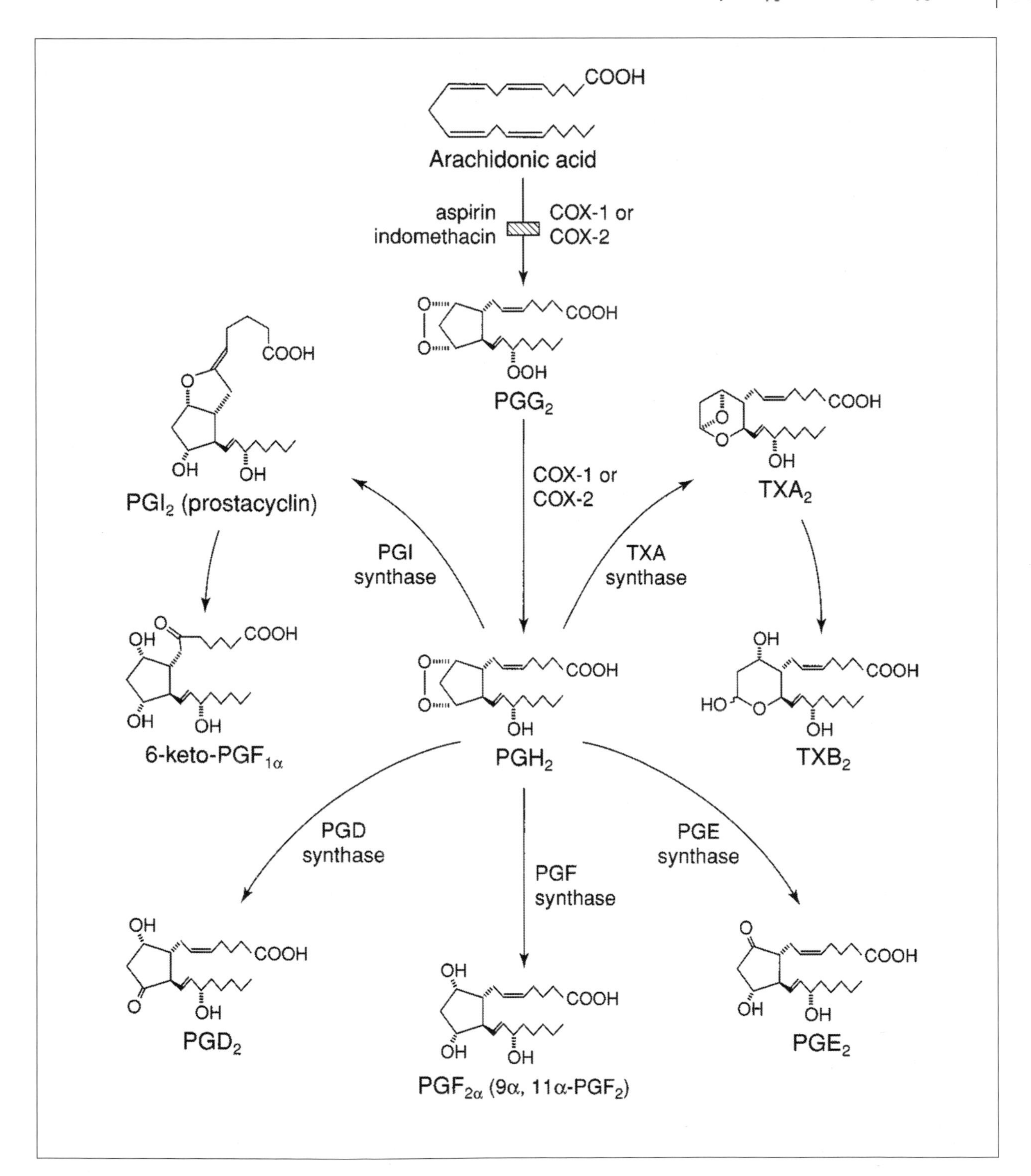

Figure 1. The arachidonic acid cascade

significance is that aspirin easily inhibited the release of PG during the early stages of perfusion, but became less effective as the experiment progressed. Needleman concluded that the increase in PG production in this model of INFLAMMATION was secondary to increased synthesis of COX enzyme [9].

Similar results were obtained from experiments using isolated cells stimulated with various inflammatory CYTOKINES or growth factors. Human umbilical vein endothelial cells so treated increased production of PGI_2, together with an increase in COX protein and mRNA [10], effects reduced by treatment with actinomycin D.

Needleman and his group [11–13] extended their earlier observations by demonstrating increased PG synthesis in human MONOCYTES *in vitro* and in mouse MACROPHAGES *in vivo* after activation with LPS. The increase in PG release was inhibited by prior treatment with dexamethasone, yet the basal level of COX activity was not affected. Needleman concluded that there may be various "pools" of COX within cells, one constitutive, another inducible, being upregulated by inflammatory stimuli [14]. The existence of such an inducible COX enzyme was firmly established by the discovery by Simmons and his colleagues at Brigham Young University of a novel PG synthase, encoded by an entirely different gene, induced in chick embryo fibroblasts by v-src, serum or phorbol esters [15, 16]. Subsequently, Herschman and his colleagues [17] found a similar gene in the mouse, as did Simmons et al. who, by cross hybridisation, cloned and characterised the murine homologue of the v-src-inducible COX [18] and showed the mRNA to be induced in an immediate early fashion by various mutagens The deduced protein structure of the inducible enzyme was found to be 60% homologous to constitutive COX, and the two enzymes were designated COX-1 (constitutive) and COX-2 (inducible) .

Although COX-1 is constitutive in most cells, very little COX-2 is present in resting cells. However, it is powerfully induced in fibroblasts, endothelial cells and vascular smooth muscle by mitogens or some CYTOKINES, and in monocytic cells by LPS. These observations have led to the general hypothesis that the constitutive COX-1 is a "housekeeping" enzyme involved in maintaining normal physiological processes. Thus COX-1 is responsible for synthesising thromboxane A_2 involved in platelet aggregation, PGE_2 and PGI_2 to protect the gastric mucosa and PGI_2 in the vascular endothelium to maintain dilation of blood vessels and to inhibit formation of inappropriate platelet thrombi (although recent work suggests that formation of PGI_2 from endothelial cells may be due to COX-2, being constantly induced in the cells by shear stress [19]).

The inducible COX-2, however, is believed to be primarily responsible for the production of PGs in pathological processes, for example the swelling and hyperalgesia associated with inflammatory disease (see Fig. 2). Certainly, increased COX-2 expression, which parallels an increase in PGE_2 synthesis, is demonstrable in models of inflammatory disease such as the murine air pouch, whilst COX-1 expression is unchanged [20]. Increased expression of COX-2 has also been demonstrated in synovial tissue taken from patients with rheumatoid arthritis, compared to tissues from patients with osteoarthritis or no arthritic pathology [21].

The implication that the COX-2 isoform was responsible for the synthesis of PGs at pathological foci and COX-1 for the synthesis of the beneficial, "housekeeping" PGs, raised the exciting possibility that selective inhibitors of COX-2 would exert the therapeutic actions of NSAIDs without, or with less of, the previously regarded inevitable side-effect of gastrotoxicity. This prompted many workers to undertake studies to establish the relative activity of established NSAIDs on COX-1 and COX-2 by comparison of the concentrations of NSAID necessary to inhibit the activity of each COX by 50% (IC_{50}). Although simple in concept, these measurements of selectivity were problematic in practice. Differences in source of enzyme, time of incubation, method of induction of COX-2 and protein concentration between various laboratories resulted in a variation in ratios and even differences in IC_{50} values. However, the human whole blood assay described by Patrignani et al. [22] and subsequently modified by Warner [6], provided a reliable method with similar incubation times, appropriate human target cells and performed in the presence of plasma proteins. In this method, the action of the NSAID on COX-1 is determined by the inhibition of production of the TXA_2 metabolite (TXB_2), where-

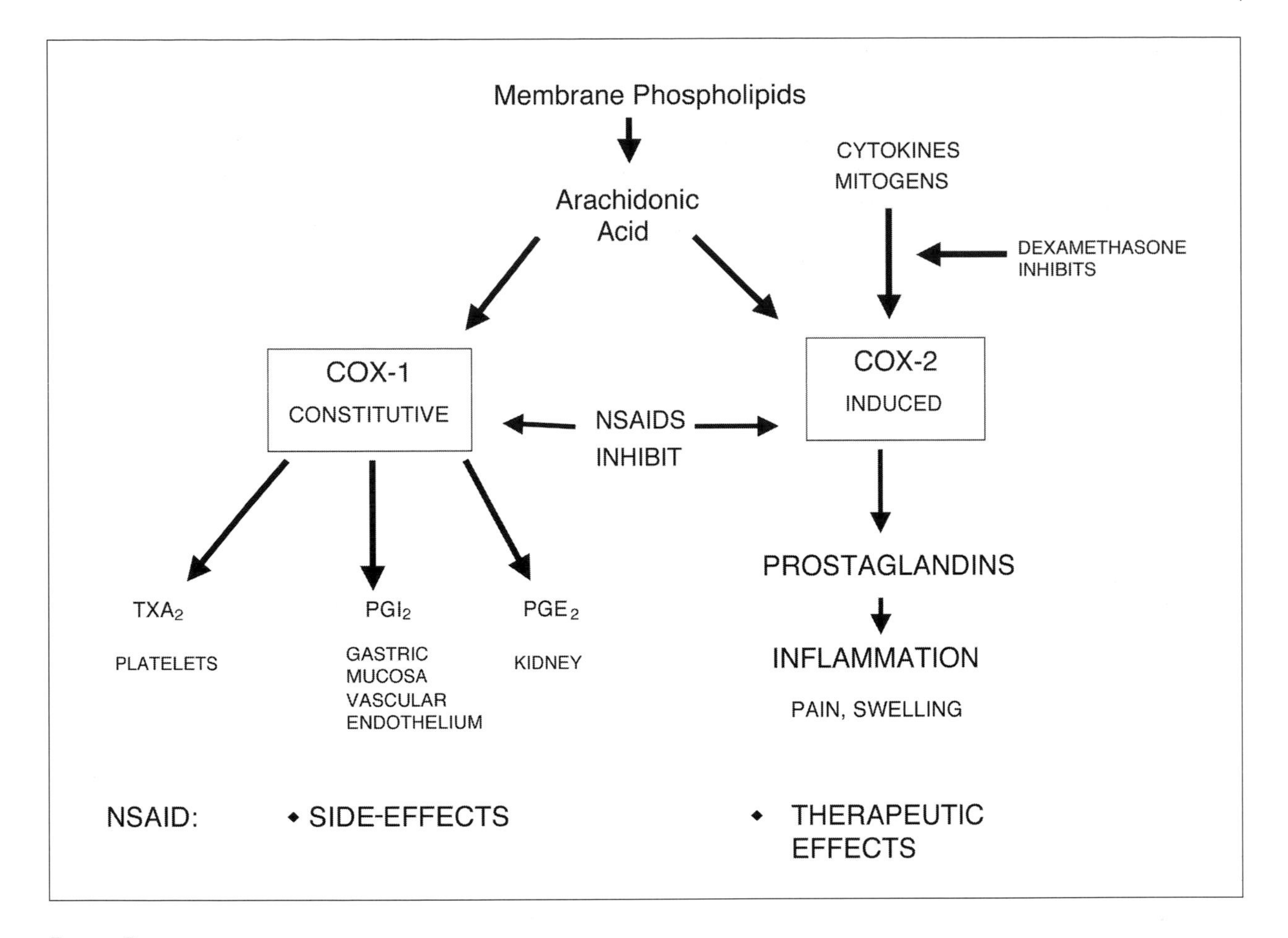

FIGURE 2
The side-effects of NSAIDs are caused by inhibition of the constitutive enzyme COX-1, which synthesises prostaglandins that serve essential physiological functions such as causing appropriate platelet aggregation, protection of the gastric mucosa, inhibition of thrombogenesis and maintenance of renal function. The therapeutic effects of NSAIDs are due to inhibition of COX-2, an enzyme induced by various factors released by bacteria, the vascular endothelium or other cells involved in the inflammatory response.

as the effect on COX-2 is assessed by the inhibition of PGE_2 formation from previously activated human A549 cells added to the whole blood. This method was reproducible and the selectivity of established NSAIDs towards inhibition of COX-2, rather than COX-1, was shown, in large-scale clinical trials and post-marketing surveillance studies, to be accompanied by relatively low gastrotoxicity. Thus nimesulide, for example, which is approximately 20 times as active on COX-2 compared to COX-1, has low gastrotoxicity, whereas ketorolac, which is many hundred times more selective for COX-1 has a relative risk of 25, compared to a value of 4–5 for general NSAID use.

Structural basis for COX-2 selectivity

The three-dimensional structures of both COX-1 [23] and COX-2 [24] have been determined and are remarkably similar. Each enzyme is composed of three distinct folding units: an epidermal growth factor-like domain, a membrane binding section and a C-terminal enzymatic domain containing both the cyclooxygenase and peroxidase active sites. Through the membrane binding domain the enzymes integrate into a single leaflet of the membrane lipid bilayer, thus positioning the cyclooxygenase site to allow access of the arachidonic acid substrate which is mobilised from adjacent membrane phospholipids.

Whereas NSAIDs generally act competitively to prevent access of arachidonic acid to the active site, aspirin acts in a unique manner to acetylate a serine residue at position 530 in COX-1, or the serine at an analogous position (516) in COX-2. Although the active sites of COX-1 and COX-2 are similar, there are differences which have been utilised by medicinal chemists to synthesise molecules which have a selective action on COX-2. Of crucial significance is position 523, which in COX-2 is valine and in COX-1 isoleucine. This difference of a single methyl group on an amino acid is sufficient to allow access of a potential inhibitor to a side pocket in the COX-2 enzymatic domain and, in addition, proximity to an arginine residue at position 513 (histidine in COX-1) which provides hydrogen bonding for an inhibitor of a structure that enables it to extend into the side pocket [25] (Fig. 3).

A second significant difference between COX-2 and COX-1 is at position 503, which is the aromatic amino acid phenylalanine in COX-1 but the relatively small, nonaromatic leucine in COX-2; this allows leucine at position 384 to reorient its methyl side chain away from the enzymatic site and thus extend the space available for a larger inhibitor molecule at the active site of the COX-2 enzyme [26].

Drugs specifically designed to inhibit COX-2 have entered the market and two such have undergone comprehensive clinical investigation. Celecoxib is the prototype of a number of highly selective COX-2 inhibitors produced by Monsanto/Searle. Celecoxib was shown to be an effective analgesic as assessed in patients after tooth extraction, and endoscopic studies after three month's chronic usage showed no more gastric damage than with placebo. In a large-scale clinical trial (the Celecoxib Long-term Arthritis Safety Study or CLASS, in 8000 patients) it was shown that, at six months, celecoxib produced less gastrotoxicity than the comparator drugs, ibuprofen and diclofenac [27]. Celecoxib is approved both for osteo- and rheumatoid arthritis in many countries.

Rofecoxib, produced by Merck, Sharp and Dohme, has also been approved by many regulatory authorities for arthritis and acute pain. As with celecoxib, rofecoxib in doses of 25-50mg was an effective analgesic for dental pain, dysmenorrhoea and osteoarthritis. In a 6-month endoscopic study, with a dose of rofecoxib 10-fold greater than that required for an effective anti-inflammatory effect, gastric damage was no greater than that seen with placebo. In a large-scale, 9-month trial involving 8000 patients (the Vioxx Gastrointestinal Outcome Research, VIGOR) rofecoxib was compared with naproxen [28]. As with celecoxib, rofecoxib was shown to produce substantially less gastrotoxicity than naproxen. Of possible significance is that in the VIGOR study the incidence of myocardial infarction in the rofecoxib group was 0.4%, compared with 0.1% in the naproxen-treated patients. This observation requires further study.

Additional uses for selective COX-2 inhibitors

The availability of NSAIDs with a clearly reduced toxicity raised the possibility of the chronic use of these drugs for additional indications. Thus far, selective COX-2 inhibitors have been postulated to have potential in some cancers, premature labour and Alzheimer's disease.

COX-2 inhibitors and cancers

There is persuasive epidemiological evidence for an association of chronic use of NSAIDs with reduced incidence of gastrointestinal carcinomas [29–35]. Regular use of aspirin and other NSAIDs can halve the risk of development of colon cancer. Clinical and experimental studies have provided further evidence to justify this beneficial effect of NSAIDs;

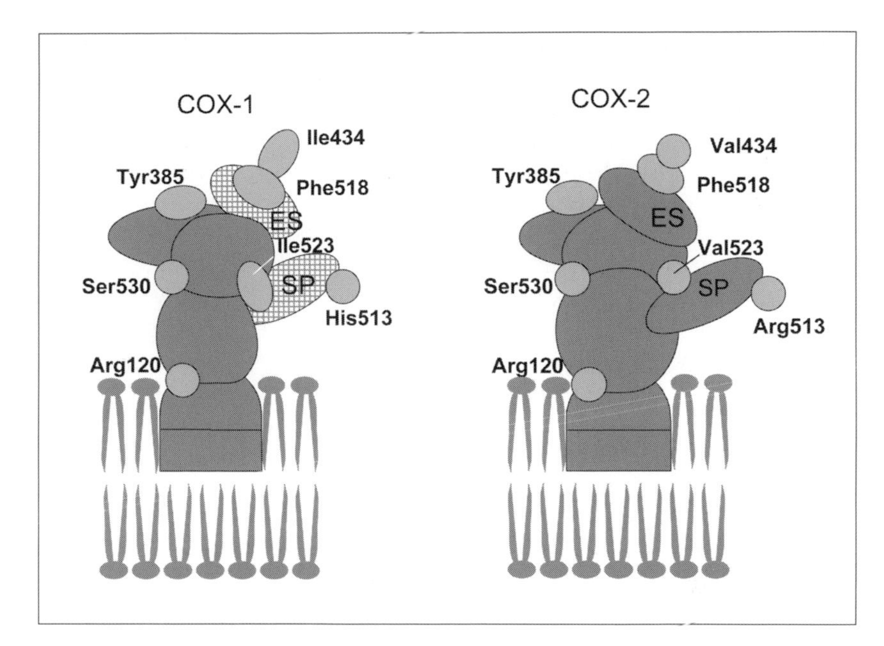

FIGURE 3. AMINO ACID SUBSTITUTIONS BETWEEN COX-1 AND COX-2
The hatched areas in COX-1 are those that are more accessible in COX-2 due to the amino acid substitutions. SP, side pocket; ES, extra space.

colonic tumour tissue has been shown to contain high levels of COX-2 protein but contiguous, normal tissue contains only COX-1 [36]. Patients with the inherited disorder of familial adenomatous polyposis (FAP), a condition characterised by a plethora of colonic polyps with eventual progression to tumour, manifest a marked reduction in polyp number on treatment with the NSAID sulindac [37,38]. Similarly, COX inhibitors reduce tumours in azoxymethane-induced colorectal carcinogenesis in rat [39] and selective COX-2 inhibitors reduce tumour or polyp numbers in animal models of FAP such as minimal intestinal neoplasias (min) mice or APC (adenomatous polyposis coli gene mutant) knockout mice. A clinical trial of celecoxib in patients with FAP has shown a 30% reduction in polyps and this use of the drug has been approved by the Food and Drug Administration, USA [40].

COX-2 is expressed in tumour tissue in many cancers other than colorectal carcinomas. Increased levels of COX-2 have been shown in human lung squamous cell and adenocarcinomas, squamous skin carcinomas and in tumours of the breast, liver, pancreas, prostate and cervical lymph nodes (see Ristimaki for review [41]).

Premature labour

The content of COX-2 mRNA in the uterus and the production of PGs increases just prior to and during labour, indicating an involvement of PGs (particularly $PGF_{2\alpha}$) in parturition. Premature labour, perhaps induced by inappropriate upregulation of COX-2 due to intrauterine infection, can be suppressed by COX inhibitors such as indomethacin, but inhibition of COX-1 will result in early closure of the ductus arteriosus which is maintained patent until birth by PGs synthesised by constitutive COX-1 [42]. Selective COX-2 inhibitors were thus thought to have a use in premature labour, producing a tocolytic effect without the problem of early closure of the ductus arteriosus. The COX-2-selective inhibitor nimesulide has been used with some success in premature labour [43] but has been shown subsequently to produce oligohydramnios in 50% of exposed fetuses [44]. The side-effects of nimesulide are reversible; however, it is accepted that the use of this drug for premature labour should be reserved for patients at high risk of early preterm delivery and where adequate patient surveillance can be maintained. Since some of the renal side-effects of nimesulide may be due to actions other than COX inhibition, such as calcium channel blockade, other selective COX-2 inhibitors are at present under trial for preterm labour [45].

Alzheimer's disease

Retrospective epidemiological studies suggest that long-term use of standard NSAIDs reduces the incidence of Alzheimer's disease (AD) [46], a benefit not seen with paracetamol or low-dose aspirin. Due to lack of suitable animal models there is little direct experimental confirmation, but it has been assumed that there is an inflammatory component in the pathogenesis of AD and postmortem studies have shown an increase of COX-2 expression in the cortex of AD patients [47]. Prolonged treatment with ibuprofen reduced the number of amyloid plaques in the brain of APP transgenic mice secondary to a lowered brain content of soluble $A\beta$ peptide, which on aggregation forms the amyloid plaques [48]. Large placebo-controlled trials of the highly selective COX-2 inhibitors celecoxib and rofecoxib have apparently shown no beneficial effect on the progress of AD [49, 50] which suggests that COX-2 may not be an appropriate target. It has been suggested that some NSAIDs may lower plaque formation through a mechanism independent of COX inhibition, possibly by inhibition of γ-secretase, which is the enzyme that cleaves the amyloid precursor protein into the plaque-forming $A\beta 42$ peptide [51]. Such a prophylactic effect of NSAIDs would be in accord with the Cache County epidemiological study, which showed that general NSAID therapy reduces incidence of AD if the drugs are used well before the onset of symptoms of dementia, whereas more recent exposure offered little protection [52].

Other cyclo-oxygenases

Although the discovery of COX-2 resolved the puzzle of why some NSAIDs manifested marked gastrotoxicity yet others more selective for COX-2, were relatively less toxic, some enigmas remained. As mentioned above, acetaminophen, like conventional NSAIDs, has marked antipyretic and analgesic activity yet lacks the typical anti-inflammatory and antiplatelet activities of this group of drugs. Twenty years ago it was suggested that acetaminophen might inhibit a cyclooxygenase in the brain, and that various, perhaps tissue-specific, cyclooxygenases may exist that are selectively inhibited by different NSAIDs [8].

Some light has recently been shed upon this field by Simmons and his colleagues who have identified a new COX-1 variant in dog and human brain which is sensitive to acetaminophen and which has been designated COX-3 [53]. COX-3 has the same structure as COX-1 except that the mRNA has retained intron-1 which is translated into a 30 amino acid extension to the enzyme protein. COX-3 was ectopically expressed in insect cells and found to be more sensitive to inhibition by acetaminophen than either COX-1 or COX-2. At least two other partial COX-1 mRNAs were characterised but only one, PCOX-1a, had retained intron-1 in transcription. The partial COX-1 proteins did not form prostanoids and their functions remain unknown.

Human COX-3 was found in highest concentrations in the cerebral cortex followed by the heart and skeletal muscle. The possible functions of COX-3 in organs other than the brain are as yet unknown. Simmons had previously found another COX isoform, sensitive to inhibition by acetaminophen, formed by a murine macrophage cell line (J774.2) treated with high concentrations of diclofenac [54]. However, unlike COX-3, this enzyme was insensitive to aspirin, and was considered to be a variant of COX-2, since a protein immunoreactive to COX-2 was co-induced together with the cyclooxygenase activity. Thus there may be yet other acetaminophen-sensitive cyclooxygenases derived from COX-2.

Summary

Since earliest times, extracts of salicylate-containing plants have been used by humans to alleviate the symptoms of inflammatory conditions, i.e., INFLAMMATION, pain and fever. Eventually, salicylic acid was synthesised chemically and became widely available for the treatment of INFLAMMATION. The more palatable derivative of salicylic acid, acetylsalicylic acid, was then synthesised in 1897 and became a popular medicine under the name of 'Aspirin'. But the MECHANISM OF ACTION of aspirin remained unknown until 1971 when John Vane discovered that aspirin reduced the synthesis of the ubiquitous lipid mediators, the PROSTAGLANDINS. Inhibition of the synthesis of PROSTAGLANDINS explained the anti-inflammatory, antipyretic and analgesic effects of aspirin, as release of PROSTAGLANDINS was the main cause of INFLAMMATION, fever and pain. After this discovery, many aspirin-like drugs such as ibuprofen, naproxen, piroxicam and others were developed by pharmaceutical companies and they all reduced the synthesis of PROSTAGLANDINS by inhibiting the activity of the cyclooxygenase enzyme which was responsible for their biosynthesis. These were collectively termed the non-steroid anti-inflammatory drugs or NSAIDs. However, all the NSAIDs to a lesser or greater extent shared the side-action of causing bleeding and ulceration of the stomach mucosa.

A major advance in the history of the NSAIDs took place in 1991 when Daniel Simmons and his colleagues found that there were 2 separate genes producing distinct cyclooxygenase enzymes. Thus, two cyclooxygenase enzymes were identified, cyclooxygenase-1 and cyclooxygenase-2. Cyclooxygenase-1 was the 'housekeeping' enzyme which maintained the integrity of the stomach mucosa, caused appropriate aggregation of platelets and maintained perfusion of the kidneys, whereas cyclooxygenase-2 was an inducible enzyme produced during INFLAMMATION by inflammatory stimuli such as mitogens, CYTOKINES and inflammatory mediators.

It was soon realised that to develop NSAIDs which would reduce INFLAMMATION without damaging the stomach would be a worthwhile goal and within ten years selective inhibitors of cyclooxygenase-2 were undergoing clinical trials for the treatment of arthritis. The first two selective COX-2 inhibitors, celecoxib and rofecoxib, were approved by the FDA in 1999 and were prescribed for millions of arthritic patients worldwide.

The extensive research into possible functions of cyclooxygenase-2 other than as a mediator of INFLAMMATION, revealed that the enzyme was overexpressed in cancer cells and in brain cells of Alzheimer patients. A major trial of rofecoxib in a precancerous state leading to colon cancer and familial adenomatous polyposis, and of celecoxib also for a possible beneficial effect in this condition, revealed some unexpected results. Whereas the clinical trials establishing EFFICACY and absence of gastrointestinal toxicity of the selective cyclooxygenase-2 inhibitors were completed in 6 to 9 months, in trials for an anti-cancer effect the drugs were administered for more than one year. Moreover, any adverse cardiovascular effects were carefully monitored. It became clear that with long-term chronic use of rofecoxib and to a lesser extent with celecoxib there was an increased incidence of cardiovascular toxic events, the most important of which was fatal myocardial infarction. As a result, rofecoxib was withdrawn from the market by Merck and Company and the treatment of arthritic patients with selective cyclooxygenase-2 inhibitors is being reassessed by the medical profession.

Selected readings

Vane JR (1971) Inhibition of prostaglandin synthesis as a mechanism of action for the aspirin-like drugs. *Nature* 231: 232–235

Xie W, Chipman JG, Robertson DL, Erikson RL, Simmons DL (1991) Expression of a mitogen-responsive gene encoding prostaglandin synthase is regulated by mRNA splicing. *Proc Natl Acad Sci USA* 88: 2692–2696

Silverstein FE, Faich G, Goldstein JL, Simon LS, Pincus T, Whelton A, Makuch R, Eisen G, Agrawal NM, Stenso WF et al (2000) Gastrointestinal toxicity with celecoxib vs nonsteroidal anti-inflammatory drugs for osteoarthritis and rheumatoid arthritis: the CLASS study: a randomised controlled trial. Celecoxib Long-term Arthritis Safety Study. *J Am Med Assoc* 284: 1247–1255

Bombardier C, Laine L, Reicin A, Shapiro D, Burgos-Vargas R, Davis B, Day R, Ferraz MB, Hawkey CJ, Hochberg MC et al; VIGOR Study Group (2000) Comparison of upper gastrointestinal toxicity of rofecoxib and naproxen in patients with rheumatoid arthritis. *N Eng J Med* 343: 1520–1528

Chandrasekharan NV, Dai H, Roos KL, Evanson NK, Tomsik J, Elton TS, Simmons DL (2002) COX-3, a cyclooxygenase-1 variant inhibited by acetaminophen and other analgesic/antipyretic drugs: cloning, structure, and expression. *Proc Natl Acad Sci USA* 99: 13926–13931

References

1 Vane JR, Botting RM (1996) The history of anti-inflammatory drugs and their mode of action. In: N Bazan, J Botting, J Vane (eds): *New Targets in Inflammation Inhibitors of COX-2 or Adhesion Molecules*. Kluwer Academic Publishers and William Harvey Press, London: 1–12

2 Dreser H (1899) Pharmacologisches über Aspirin (Acetylsalicylsäure). *Pflügers Arch* 76: 306–318

3 Douthwaite A, Lintott CA (1938) Gastroscopic observations of the effect of aspirin and certain other substances on the stomach. *Lancet* 2: 1222

4 Vane JR (1971) Inhibition of prostaglandin synthesis as a mechanism of action for the aspirin-like drugs. *Nature* 231: 232–235

5 Vane JR, Botting RM (1992) The Prostaglandins. In: JR Vane, RM Botting (eds): *Aspirin and other Salicylates*. Chapman & Hall, London, 17–34

6 Warner TD, Guiliano F, Vojnovic I, Bukasa A, Mitchell JA, Vane JR (1999) Nonsteroid drug selectivities for cyclo-oxygenase-1 rather than cyclo-oxygenase-2 are associated with human gastrointestinal toxicity: a full *in vitro* analysis. *Proc Natl Acad Sci USA*: 96: 7563–7568

7 Carr DP, Henn R, Green JR, Böttcher I (1986) Comparison of the systemic inhibition of thromboxane synthesis, anti-inflammatory activity and gastro-intestinal toxicity of non-steroidal anti-inflammatory drugs in the rat. *Agents and Actions* 19: 374–375

8 Flower RJ, Vane JR (1972) Inhibition of prostaglandin synthetase in brain explains the antipyretic activity of paracetamol (4-acetamidophenol). *Nature* 240: 410–411

9 Morrison AR, Moritz H, Needleman P (1978) Mechanism of enhanced renal biosynthesis of prostaglandins in ureter obstruction. Role of *de novo* protein biosynthesis. *J Biol Chem* 253: 8210–8212

10 Kawakami M, Ishibashi S, Ogawa H, Murase T, Takaku F, Shibata S (1986) Cachectin/TNF as well as interleukin-1 induces prostacyclin synthesis in cultured vascular cells. *Biochem Biophys Res Commun* 141: 482–487

11 Honda A, Raz A, Needleman P (1990) Induction of cyclo-oxygenase synthesis in human promyelocytic leukemia (HL-60) cells during monocytic or granulocytic differentiation. *Biochem J* 272: 259–262

12 Fu J-Y, Masferrer JL, Seibert K, Needleman P (1990) The induction and suppression of prostaglandin H_2 synthase (cyclooxygenase) in human monocytes. *J Biol Chem* 265: 16737-16740

13 Masferrer JL, Zweifel BS, Seibert K, Needleman P (1990) Selective regulation of cellular cyclooxygenase by dexamethasone and endotoxin in mice. *J Clin Invest* 86: 1375–1379

14 Masferrer JL, Zweifel BS, Seibert K, Needleman P (1990) Selective regulation of cellular cyclooxygenase by dexamethasone and endotoxin in mice. *J Clin Invest* 86: 1375–1379

15 Simmons DL, Levy DB, Yannoni Y, Erikson RL (1989) Identification of a phorbol ester-repressible v-src-inducible gene. *Proc Natl Acad Sci USA* 86: 1178–1182

16 Xie W, Chipman JG, Robertson DL, Erikson RL, Simmons DL (1991) Expression of a mitogen-responsive gene encoding prostaglandin synthase is regulated by mRNA splicing. *Proc Natl Acad Sci USA* 88: 2692–2696

17 Kujubu DA, Fletcher BS, Varnum BC, Lim RW, Herschman HR (1991) TIS10, a phorbol ester tumor promoter-inducible mRNA from Swiss 3T3 cells, encodes a novel prostaglandin synthase/cyclooxygenase homologue. *J Biol Chem* 266: 12866–12872

18 Simmonds DL, Xie W, Chipman J, Evett G (1991) Multiple cyclooxygenases: cloning of a mitogen-inducible form. In: M Bailey (ed): *Prostaglandins, Leukotrienes, Lipoxins and PAF*. Plenum Press, London, 67–68

19 McAdam BF, Catella-Lawson F, Mardini IA, Kapoor S, Fitzgerald GA (1999) Systemic biosynthesis of prostacyclin by cyclooxygenase (COX-2): The human pharmacology of a selective inhibitor of COX-2. *Proc Natl Acad Sci USA*: 96: 272–277

20 Willoughby DA, Tomlinson A, Gilroy D, Willis D (1996) Inducible enzymes with special reference to the inflammatory response. In: JR Vane, JH Botting, RM Botting (eds): Improved Non-steroid Anti-inflammatory Drugs. COX-2 Enzyme Inhibitors. Kluwer and William Harvey Press, London, 67–83

21 Sano H, Hla T, Maier JA, Crofford LJ, Case JP, Maciag T, Wilder RL (1992) *In vivo* cylooxygenase expression in synovial tissues of patients wth rheumatoid arthritis and osteoarthritis and rats with adjuvant and streptococcal cell wall arthritis. *J Clin Invest* 89: 97–108

22 Patrignani P, Panara MR, Greco A, Fusco O, Natoli C,

Iacobelli S, Cipollone F, Ganci A, Creminon C, Maclouf J et al (1994) Biochemical and pharmacological characterisation of the cyclooxygenase activity of human blood prostaglandin endoperoxide synthases. *J Pharmacol Exp Ther* 271: 1705–1710
23 Picot D, Loll PJ, Garavito RM (1994) The x-ray crystal structure of the membrane protein prostaglandin H_2 synthase-1. Nature 367: 243–249
24 Luong C, Miller A, Barnett J, Chow J, Ramesha C, Browner MF (1996) Flexibility of the NSAID binding site in the structure of human cyclooxygenase-2. *Nat Struct Biol* 3: 927–933
25 Wong E, Bayly C, Waterman HL, Reindeau D, Mancini JA (1997) Conversion of prostaglandin G/H synthase-1 into an enzyme sensitive to PGHS-2-selective inhibitors by a double His^{513} to Arg and Ile^{523} to Val mutation. *J Biol Chem* 272: 9280–9286
26 Browner MF (1998) The structure of human COX-2 and selective inhibitors. In: J Vane, J Botting J (eds): *Selective COX-2 Inhibitors. Pharmacology, Clinical Effects and Therapeutic Potential*. Kluwer and William Harvey Press, London, 19–26
27 Silverstein FE, Faich G, Goldstein JL, Simon LS, Pincus T, Whelton A, Makuch R, Eisen G, Agrawal NM, Stenson WF et al (2000) Gastrointestinal toxicity with celecoxib vs nonsteroidal anti-inflammatory drugs for osteoarthritis and rheumatoid arthritis: the CLASS study: a randomised controlled trial. Celecoxib Long-term Arthritis Safety Study. *J Am Med Assoc* 284: 1247–1255
28 Bombardier C, Laine L, Reicin A, Shapiro D, Burgos-Vargas R, Davis B, Day R, Ferraz MB, Hawkey CJ, Hochberg MC et al; VIGOR Study Group (2000) Comparison of upper gastrointestinal toxicity of rofecoxib and naproxen in patients with rheumatoid arthritis. *N Engl J Med* 343: 1520–1528
29 Giovannucci E, Egan KM, Hunter DJ (1995) Aspirin and the risk of colorectal cancer. *N Engl J Med* 333: 609–614
30 Greenberg ER, Baron JA, Freeman DH, Mandel JS, Haile R (1993) Reduced risk of large bowel adenomas among aspirin users. The Polyp Prevention Study Group. *J Natl Cancer Inst* 85: 912–916
31 Thun MJ, Namboodiri MM, Heath CWJ (1991) Aspirin use and reduced risk of coloc cancer. *N Engl J Med* 325: 1593–1596
32 Thun MJ, Namboodiri MM, Calle EE, Flandes WD, Heath CW (1993) Aspirin use and risk of fatal cancer. *Cancer Res* 53: 1322–1327
33 Peleg II, Maibach HT, Brown SH, Wilcox CM (1994) Aspirin and nonsteroidal antiinflammatory drug use and the risk of subsequent colorectal cancer. *Arch Int Med* 154: 394–399
34 Giovannucci E, Rimm EB, Stampfer MJ, Colditz GA, Ascherio A, Willett WC (1994) Aspirin use and the risk of colorectal cancer and adenoma in male health professionals. *Ann Intern Med* 121: 241–246
35 Eberhart CE, Coffey RJ, Radhika A (1994) Up-regulation of cyclooxygenase 2 gene expression in human colorectal adenomas and adenocarcinomas. *Gastroenterology* 107:1183–1188
36 Giardello FM, Hamilton SR, Krush AJ (1993) Treatment of colonic and rectal carcinomas with sulindac in familial adenomatous polyposis. *N Engl J Med* 328: 1313–1316
37 Waddell WR, Loughry RW (1983) Sulindac for polyposis of the colon. *J Surg Oncol* 24: 83–87
38 Waddell WR, Gasner GF, Cerise EJ, Loughry RW (1989) Sulindac for polyposis of the colon. *Am J Surg* 157:175–178
39 Reddy BS, Rao CV, Rivenson A, Kellof G (1993) Inhibitor effect of aspirin on azoxymethane-induced colon carcinogenesis in F344 rats. *Carcinogenesis* 14: 1493–1497
40 Steinbach G, Lynch PM, Phillips RK, Wallace MH, Hawk E, Gordon GB (2000) The effect of celecoxib, a cyclooxygenase-2 inhibitor, in familial adenomatous polyposis. *N Engl J Med* 342: 1946–1952
41 Ristimaki A, Narko K, Nieminen, Saukkonen K (2001) Role of cyclooxygenase-2 in carcinogenesis other than colorectal cancer. In: JR Vane, RM Botting (eds): *Therapeutic Roles of Selective COX-2 Inhibitors*. William Harvey Press, London, 418–441
42 Bennett PR, Elder MG (1992) The mechanisms of preterm labour: common genital tract pathogens do not metabolise arachidonic acid to prostaglandins or other eicosanoids. *Am J Obstet Gynecol* 166: 1541–1545
43 Sawdy R, Slater D, Fisk N, Edmonds DK, Bennett P (1997) Use of a cyclo-oxygenase type-2-selective non-steroidal anti-inflammatory agent to prevent preterm delivery. *Lancet* 350: 265–266
44 Peruzzi L, Gianoglio B, Porcellini MG, Coppo R (1999) Neonatal end-stage renal failure associated with

maternal ingestion of cyclo-oxygenase-type-1 selective inhibitor nimesulide as tocolytic. *Lancet* 354: 1615

45 Allport V, Bennett P (2001) Cyclooxygenase enzymes and human labour. In: JR Vane, RM Botting (eds): *Therapeutic Roles of Selective COX-2 Inhibitors*. William Harvey Press, London, 252–273

46 Stewart WF, Kawas C, Corrada M, Metter EJ (1997) Risk of Alzheimer's disease and duration of NSAID use. *Neurology* 48: 626–632

47 Pasinetti GM, Aisen PS (1998) Cyclooxygenase-2 expression is increased in frontal cortex of Alzheimer's disease brain. *Neuroscience* 87: 319–324

48 Lim GP, Yang F, Chu T, Chen P, Beech W, Teter B, Tran T, Ubeda O, Ashe KH, Frautschy SA et al (2000) Ibuprofen suppresses plaque pathology and inflammation in a mouse model for Alzheimer's disease. *J Neurosci* 20: 5709–5714

49 McGeer PL (2000) Cyclo-oxygenase-2 inhibitors: rationale and therapeutic potential for Alzheimer's disease. *Drugs Aging* 17: 1–11

50 Sainati SM, Ingram DM, Talwalker S, Geis GS (2000) Results of a double-blind, randomized, placebo-controlled study of celecoxib in the treatment of progression of Alzheimer's disease. *Abstracts of the Sixth International Stockholm/Springgfield Symposium on Advances in Alzheimer Therapy*. April 5–8, 2000

51 Weggen S, Eriksen JL, Das P, Sagi SA, Wang R, Pietrzik CU, Findlay KA, Smith TE, Murphy MP, Bulter T et al (2001) A subset of NSAIDs lower amyloidogenic Aβ42 independently of cyclooxygenase activity. *Nature* 414: 212–216

52 Zandi PP, Anthony JC, Hayden KM, Mehta K, Mayer L, Breitner JC (2002) Reduced incidence of AD with NSAID but not H2 receptor antagonists. The Cache County study. *Neurology* 59: 880–886

53 Chandrasekharan NV, Dai H, Roos KL, Evanson NK, Tomsik J, Elton TS, Simmons DL (2002) COX-3, a cyclooxygenase-1 variant inhibited by acetaminophen and other analgesic/antipyretic drugs: cloning, structure, and expression. *Proc Natl Acad Sci USA* 99: 13926–13931

54 Simmons DL, Botting RM, Robertson PM, Madsen ML, Vane JR (1999) Induction of an acetaminophen-sensitive cyclooxygenase with reduced sensitivity to nonsteroid antiinflammatory drugs. *Proc Natl Acad Sci USA* 96: 3275–3280

C13

Disease-modifying anti-rheumatic drugs

Clarissa Bachmeier, Peter M. Brooks, Richard O. Day, Garry G. Graham, Geoffrey O. Littlejohn, Eric F. Morand, Kevin D. Pile and Kenneth M. Williams

Introduction

Rheumatoid arthritis is the most common inflammatory rheumatic disease, and has a prevalence of about 1% [1]. The total incidence of this disease is low but its burden is significant. Untreated, rheumatoid arthritis usually leads to joint destruction, disability and increased mortality. Long-term management of patients is required, based not only on drug treatment, but also on non-pharmacological approaches such as physiotherapy and psychosocial support.

The pharmacological treatment of rheumatoid arthritis is still problematical, and as yet there are no reliably curative or disease-remitting therapies, although considerable gains have been made recently with the advent of biological therapies for this condition. However, rheumatoid arthritis often progresses to disability, and many patients eventually suffer from deformity and increased mortality. It is hoped that the full application of the old disease-modifying anti-rheumatic drugs (DMARD), together with the newer DMARD and biological agents, can improve this outlook. It should be noted that the principles of the use of DMARD in rheumatoid arthritis are paralleled in their treatment of less common inflammatory rheumatic diseases such as juvenile chronic arthritis.

First-line pharmacological therapy for all inflammatory arthritic diseases are the nonsteroidal inflammatory drugs (NSAID) (Chapter C.12). The NSAID are useful symptomatically but have no clear effect on the progression of rheumatoid arthritis. Virtually all patients require additional treatment with DMARD. This latter class of drugs is also known as slow-acting, anti-rheumatic drugs, as their onset of effect is generally delayed or slow. There has been debate about the effects of DMARD on the long-term effects of rheumatoid arthritis, but several DMARD are now known to slow joint damage in many patients with this disease. A recent change in the use of DMARD is that they are generally commenced as soon as a diagnosis of an inflammatory arthritis is made, in order to have the maximal effect on slowing the progress of the disease [2]. The rationale for early initiation of therapy is the evidence that many of the deleterious effects of rheumatoid arthritis, such as erosions of bone, occur within the first 1 to 2 years of disease. This approach contrasts with older regimens in which DMARD were only commenced on the appearance of bony erosions on joint X-rays.

Many patients with rheumatoid arthritis and other inflammatory arthritic diseases are also administered intra-articular and/or low-dose oral corticosteroids. They are frequently administered as bridging therapy until the effect of DMARD is fully established. In some patients, they are used long-term with DMARD, in order to satisfactorily suppress the disease.

A recent advance has been the introduction of biological drugs. These are proteins which are inhibitors of either TUMOR NECROSIS FACTOR (TNF) or INTERLEUKIN-1 (IL-1). These drugs produce rapid control of rheumatoid arthritis in a high percentage of cases and joint destruction may be halted and even reversed in some patients. There are, however, potential problems with the biological agents, particularly the activation of infections, as well as very high financial costs.

Pathophysiology of rheumatoid arthritis

The concept of a genetic predisposition and an environmental trigger has been applied to nearly all

AUTOIMMUNE DISEASES, including rheumatoid arthritis. However, despite an intensive search for transmissible agents, no infectious cause of rheumatoid arthritis has been proven. The worldwide distribution of rheumatoid arthritis has been interpreted as meaning that the environmental trigger is ubiquitous, or that multiple triggers are involved. Furthermore, genetic factors probably make a substantial contribution to rheumatoid arthritis.

The strongest genetic association in rheumatoid arthritis is with the MAJOR HISTOCOMPATIBILITY COMPLEX genes HLA-DR 1 and 4, which seem to associate with disease severity. The key function of the HLA-DR molecule is to present processed exogenous ANTIGENS to $CD4^+$ T-helper LYMPHOCYTES, supporting the hypothesis that T CELLS are important in the pathogenesis of rheumatoid arthritis. Against this hypothesis stands the failure of potent targeted anti-T cell MONOCLONAL ANTIBODIES in a carefully conducted clinical trial [3]. At present, the precise role of ANTIGEN-driven immune responses in the initiation of rheumatoid arthritis remains unresolved.

Despite the uncertainty relating to the initiating events, the ultimate pathology of rheumatoid arthritis, namely synovial INFLAMMATION (or synovitis), is well established. The normal synovium is a thin and delicate layer that reflects off the cartilage-periosteal border on to the underlying fibrous joint capsule. Synovium is composed of cells of fibroblast and macrophage origin, and it has two major functions: the provision of oxygen and nutrients to cartilage via the synovial fluid, and the production of lubricants, notably hyaluronic acid, that allow the articular surfaces to glide smoothly across each another.

In rheumatoid arthritis, the synovium is transformed into a chronically inflamed tissue. The normally thin synovial layer thickens dramatically due to accumulation of macrophage-like and fibroblast-like synoviocytes, and the subsynovial layer becomes oedematous, hypervascular and hypercellular with the accumulation of mononuclear cells, particularly $CD4^+$ T CELLS, B CELLS, and MACROPHAGES. The increased number of cells in the synovium in rheumatoid arthritis is believed to result from recruitment of blood-derived LEUKOCYTES, as well as increased proliferation and apoptosis. Neutrophils are abundant in rheumatoid synovial fluid but are sparse within the synovium.

Cellular expansion in the synovium is accompanied by significant overproduction of CYTOKINES such as TNF-α, IL-1, IL-6, IL-8, and macrophage migration inhibitory factor (MIF). Compelling evidence from *in vitro* and animal studies, as well as the effects of human treatment with CYTOKINE antagonists, suggests that these and other CYTOKINES drive continued cell recruitment and expansion, angiogenesis, and the production of pro-inflammatory mediators such as PROSTAGLANDINS, and reactive oxygen and nitrogen species [4] (Tab. 1). CYTOKINES induce cell activation by the activation of coordinated intracellular signal transduction cascades, which utilize molecules including NUCLEAR FACTOR κB (NF-κB) and mitogen-activated protein kinases. As CYTOKINES such as TNF-α, IL-1 and MIF can induce their own synthesis and secretion, a self-perpetuating inflammatory cycle of amplification ensues, leading to the chronic INFLAMMATION of rheumatoid arthritis. In concert with this inflammatory process, cartilage injury is caused by the generation of degradative enzymes including matrix metalloproteinases and aggrecanases. Bone injury is separately mediated by a process of osteoclast activation regulated by the receptor activator of the NF-κB-RANK-ligand system, again induced by macrophage and/or fibroblast production of CYTOKINES. The CYTOKINES and mediators operative in these processes constitute potential therapeutic targets in rheumatoid arthritis, and many are now known to be influenced by existing anti-rheumatic drugs.

Corticosteroids

The general chemistry, physiology and pharmacology of the corticosteroids are discussed in Chapter C.11 (Corticosteroids) of this volume.

Pharmacokinetics and metabolism

Prednisolone and prednisone are the most widely used corticosteroids for oral use in the treatment of rheumatoid arthritis and other inflammatory joint

TABLE 1. ACTIVITIES OF TNF-α AND IL-1 INVOLVED IN THE DEVELOPMENT OF RHEUMATOID ARTHRITIS

Activation of endothelial cells to produce adhesion molecules and subsequent transmigration of leukocytes
Increased activity of osteoclasts leading to destruction of bone
Suppressed synthesis of cartilage
Stimulated production of other inflammatory cytokines and chemokines
Increased phagocytic activity of leukocytes
Stimulated proliferation of fibroblasts and endothelial cells

diseases. Prednisolone is the active metabolite of the inactive pro-drug, prednisone, and is now often preferred to prednisone. The plasma half-life of prednisolone is short, on the order of 2 hours, but its pharmacological effects last for a longer time than the plasma half-life indicates. For example, the lymphocyte count in blood is decreased for about 12 hours following an intravenous dose (0.25 to 0.3 mg/kg) of prednisone [5]. As a result of the prolonged effect of prednisolone, it is usually only administered once a day. Other corticosteroids, such as betamethasone, have also been used for the treatment of rheumatic diseases, but their duration of action is generally longer with greater suppression of the pituitary-adrenal axis.

Corticosteroids are also administered by intra-articular injection. This provides a depot in the joint from which the corticosteroid dissolves slowly, thereby producing a prolonged anti-inflammatory effect in the joint [6]. Systemic effects are also seen due to absorption from the joint and may also be prolonged.

Mode of action

As outlined in Chapter C.11, the corticosteroids have multiple actions on the body with their major effects on specific genes. In relation to the rheumatic diseases, the most profound effects of corticosteroids are mediated through inhibition of the expression of pro-inflammatory genes, including many of the mediators listed above. This broad spectrum of effects is mediated by the interaction between the occupied glucocorticoid receptor and the NF-κB complex, resulting in inactivation of the transcriptional effects of NF-κB [7]. Although these effects of corticosteroids are very significant, corticosteroids also induce the expression of a number of anti-inflammatory genes, including annexin I (formerly known as lipocortin 1) and MAP kinase phosphatase 1, an endogenous inhibitor of MAP kinase activation and hence of cell activity in INFLAMMATION.

Clinical indications and efficacy

In addition to their wide use in the treatment of patients with rheumatoid arthritis, corticosteroids are used for all inflammatory arthritic states and conditions such as vasculitis and polymyalgia rheumatica. Oral or intra-articular corticosteroids significantly improve the clinical symptoms of rheumatoid arthritis and they are superior to the NSAID in this respect. The effect is rapid and most pronounced in the first weeks of administration. Patients with severe rheumatoid arthritis, vasculitis or active systemic lupus erythematosus may require high doses of oral corticosteroids, or if the patients are acutely ill, high doses of intravenous corticosteroids. Large single doses of corticosteroids (pulse therapy) have a beneficial effect for at least several weeks and may be useful while the effects of DMARD are developing [8].

The long-term use of corticosteroids in rheumatoid arthritis is controversial. A reduced rate of joint destruction has been reported but, in a recent study, no additional effect of low-dose prednisolone (7 mg daily) was found in patients treated with sulfasalazine [9]. Overall, the aims of treatment with the corticosteroids should always be to keep the dose and duration of therapy to the minimum which is compatible with good disease control.

Intra-articular corticosteroid injections are an important treatment for monoarticular inflammatory synovitis or for single joints that are difficult to control in the polyarthritic patient. The long-term effects of intra-articular therapy are unclear, with animal studies suggesting both beneficial and detrimental effects on articular cartilage. Generally, an individual joint may be injected up to 3 to 4 times per year, but no more in weight-bearing joints because of the risk of degeneration of cartilage.

Side-effects

The general side-effects of the corticosteroids are outlined in Chapter C.11 of this volume. In contrast to their anti-inflammatory effects, the side-effects of corticosteroids are mostly due to the induction of genes outside the IMMUNE SYSTEM. In part for this reason, corticosteroid side-effects are generally dose-related. The catabolic effect of corticosteroids on bone is strongest in the first 6 to 12 months after initiation of the drug. It is recommended that a dose equivalent to the physiological replacement dosage of cortisol (7.5 mg prednisolone per day) should not be exceeded, as increased OSTEOPOROSIS occurs with increasing doses [10]. Larger doses are often needed in other inflammatory rheumatic conditions such as polymyalgia rheumatica, but this limitation of dosage is often possible in the treatment of rheumatoid arthritis. In order to minimize the dose of corticosteroids, it is common to administer corticosteroids in conjunction with DMARD.

A major problem with the long-term use of corticosteroids is suppression of the hypothalamic-pituitary adrenal axis. This problem is reduced, but not eliminated, by administration in the morning. Alternate-day administration reduces the suppressive effect further, although it is often not practicable because of inadequate clinical response. In order for the hypothalamic-pituitary axis to recover, systemic corticosteroids must be withdrawn slowly, particularly when daily dosage is below the equivalent of 10 to 15 mg prednisolone and the systemic steroid has been used for more than about three weeks. Dosage of the systemic corticosteroids must be increased temporarily if patients become acutely unwell or if there is evidence of adrenal insufficiency.

Disease-modifying anti-rheumatic drugs (DMARD)

The most commonly used DMARD is methotrexate (Chapter C.10) and, consequently, most attention has been paid to this agent in this chapter. Other widely used DMARD are sulfasalazine and the antimalarial, hydroxychloroquine. The small molecule DMARD, leflunomide, and the biologicals, adalimumab, infliximab, etanercept and anakinra, are rapidly becoming more widely used. Azathioprine, cyclosporin, the gold complexes, penicillamine and the tetracyclines have little current use although they are clearly active. Clinical guidelines for the management of rheumatoid arthritis were recently published by the American College of Rheumatology [1] but there still is no universally accepted scheme for therapeutic progression in the management of rheumatoid arthritis. Most patients discontinue an individual agent because of inEFFICACY or toxicity [1]. The result is that patients with severe disease have often used several DMARD. The DMARD have well-described toxicity profiles, and there are guidelines for monitoring drug therapy in rheumatoid arthritis. An overview of DMARD used in the treatment of rheumatoid arthritis is shown in Appendix C.13.

Despite their long use, the mode of action of DMARD is not well understood. In part, this is due to the pathophysiology of rheumatoid arthritis which is still unclear.

Methotrexate

Methotrexate is a folate analogue originally developed in the 1940s as a cytotoxic drug for the treatment of various tumours (Chapter C.10). An earlier folate analogue, aminopterin, was shown to be useful in treatment of rheumatoid arthritis in 1951 [11] and methotrexate is now the major DMARD for the treatment of rheumatoid arthritis.

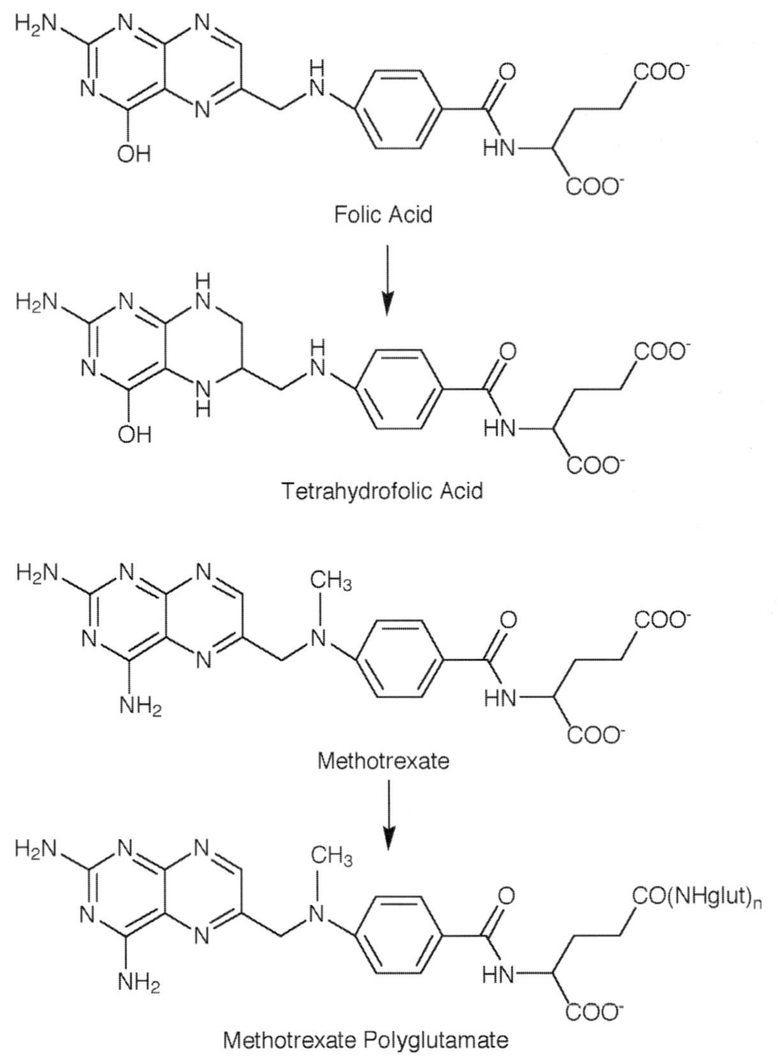

FIGURE 1 COMPARATIVE STRUCTURES OF METHOTREXATE AND FOLIC ACID
Folic acid is only active as cofactor for one-carbon transfers after reduction to tetrahydrofolate. Both folic acid and methotrexate form polyglutamates in cells, as is shown for methotrexate.

Chemistry

Methotrexate is a close analogue of folic acid (Fig. 1). Methotrexate is a hydrophilic ionized compound at physiological pH, indicating that it should not diffuse readily through cell membranes and may enter cells by carrier-mediated transport systems.

Pharmacokinetics and metabolism

Methotrexate is usually administered orally but may also be administered by subcutaneous or intramuscular injection if excessive nausea occurs when the drug is taken orally and the nausea is not controlled by folic acid treatment. Methotrexate is actively absorbed from the proximal jejunum, and may be taken regardless of meals. Urinary recovery of methotrexate indicates that the BIOAVAILABILITY is about 80% [12]. Following oral, subcutaneous or intramuscular dosage, the initial half-life of methotrexate is about 7 hours. This is followed by a very slow phase with a half-life of 2 to 3 days [12]. This slow elimination phase, possibly caused by the slow intracellular accumulation and loss of the active polyglutamate metabolites (see below), probably serves to prolong the effects of methotrexate and to allow its once a week dosage in rheumatoid arthritis. Although the data are limited, the peak and trough plasma concentrations of methotrexate are approximately 0.9 µmol/L and 0.8 nmol/L, respectively [12]. Methotrexate is about 50% unbound in plasma and the unbound peak and trough concentrations are therefore about 0.45 µmol/L and 0.4 nmol/L.

Methotrexate is primarily eliminated via the kidneys. Therefore, a lower dose of methotrexate should be used in patients with chronic renal impairment and temporary cessation of methotrexate treatment may be required at times of volume depletion (such as perioperatively). Dosage should also be decreased in older age because of decreasing renal function. Co-prescription of agents known to impair renal function, such as aminoglycosides and cyclosporin, should be undertaken with caution. It has also been reported that prolonged use of methotrexate itself may reduce renal function and hence its own CLEARANCE [13], a possible mechanism being increased plasma adenosine concentrations as a consequence of methotrexate activating A_1 receptors in the renal parenchyma, thereby diminishing renal blood flow and salt and water excretion [14].

Many patients taking low-dose methotrexate are also treated with NSAID in order to suppress the symptoms of INFLAMMATION, although toxicity from methotrexate in occasional patients has been attributed to this combination of drugs. Although renal blood flow and renal function can be decreased by NSAID, prospective studies do not indicate any NSAID-induced decrease in the renal CLEARANCE of methotrexate, except during treatment with high doses of aspirin [15]. Probenecid decreases the renal excretion of methotrexate and should be avoided [16]. Additionally, BONE MARROW suppression has occasionally been seen with the combination of cotrimoxazole and methotrexate, probably because cotrimoxazole has weak anti-folate activity.

Methotrexate contains a glutamate moiety and, after entering the cell, up to 6 glutamates are added (Fig. 1). This polyglutamation maintains a low intracellular concentration of methotrexate and a high accumulation of polyglutamated drug. This material cannot be transported extracellularly unless hydrolysed to the monoglutamate (i.e., methotrexate) by polyglutamate hydrolase. The polyglutamation of methotrexate effectively increases its intracellular life and enhances its enzyme inhibitory potency because the polyglutamates are active inhibitors of dihydrofolate reductase.

Mode of action

In the treatment of tumours, the major action of methotrexate is inhibition of dihydrofolate reductase, the result being the blockade of the intracellular production of reduced tetrahydrofolate which is important in the transfer of one-carbon units. These are necessary for the synthesis of some amino acids and nucleic acid bases. An action on dihydrofolate reductase is also indicated in rheumatoid arthritis because the trough concentration of unbound methotrexate (see above) exceeds the approximate dissociation constant of methotrexate from dihydrofolate reductase (0.01–0.2 nmol/L) [12].

Based on its actions on tumours, the MECHANISM OF ACTION of methotrexate in rheumatoid arthritis was postulated initially as the inhibition of the proliferation of activated LYMPHOCYTES. There is, however, no convincing evidence that lymphocyte proliferation is inhibited in rheumatoid arthritis patients. More recently, it been suggested that low-dose methotrexate may inhibit the recruitment of immature and inflammatory MONOCYTES into inflammatory sites and reduce their survival in the inflamed synovium, but with little or no effect on tissue-infiltrating MONOCYTES and resident MACROPHAGES [17].

Current hypotheses favour low-dose methotrexate having an anti-inflammatory action over an antiproliferative action. In general, low-dose methotrexate alters the CYTOKINE balance by inhibiting the pro-

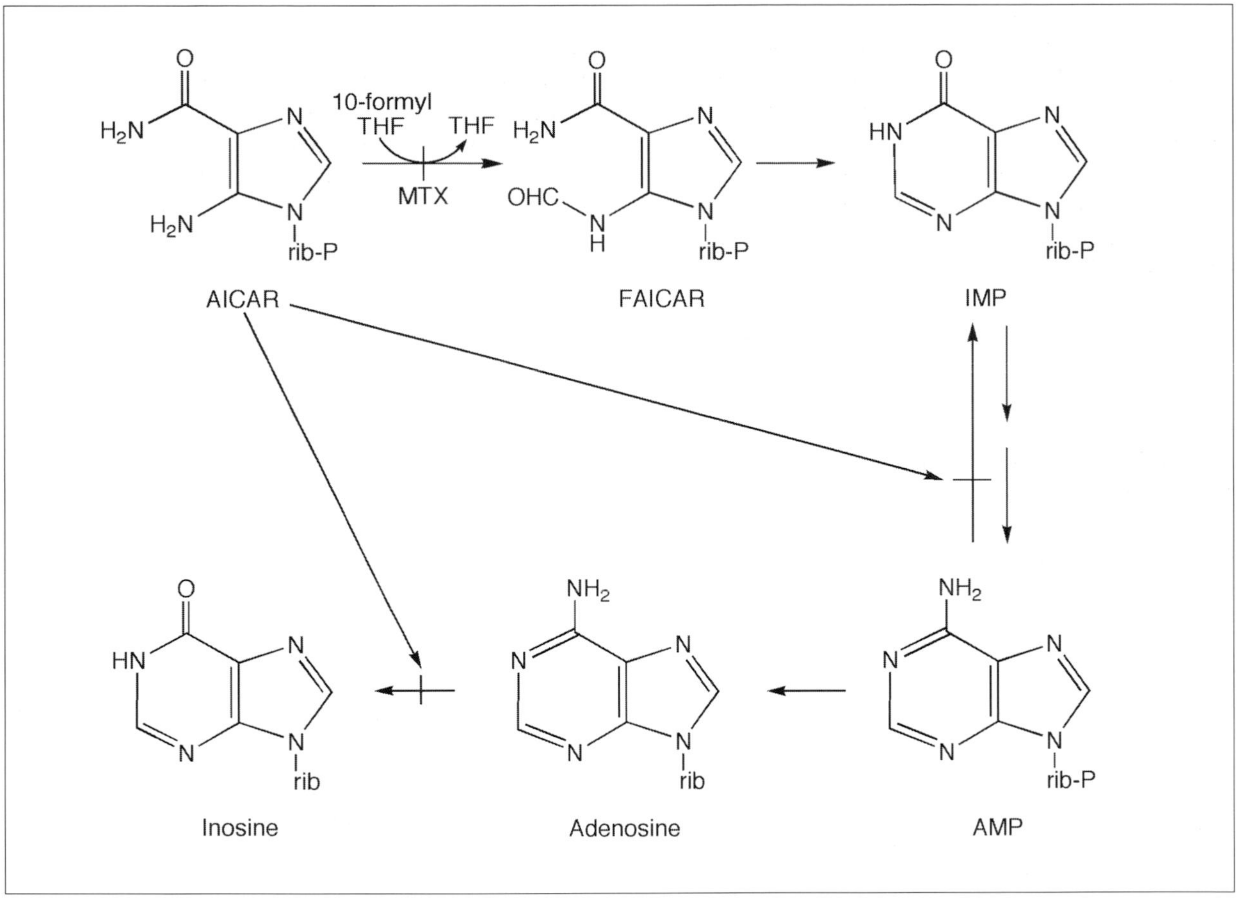

FIGURE 2

Proposed mode of action of methotrexate (MTX) based on inhibition of 5-aminoimidazole-4-carboxamide ribonucleotide (AICAR) transformylase which catalyzes the conversion of AICAR to N-formylAICAR (FAICAR). Inhibition of AICAR transformylase by methotrexate leads to the accumulation of AICAR which decreases the activity of intracellular enzymes which deaminate adenosine monophosphate (AMP) and adenosine. The extracellular concentrations of AMP and adenosine are increased consequently. The free adenosine and adenosine resulting from the extracellular hydrolysis of AMP then interact with adenosine receptors causing suppression of inflammation. Methotrexate may also inhibit a folate-dependent enzyme, glycinamide ribonucleotide (GAR) transformylase, which is earlier in the sequence of purine synthesis, but methotrexate inhibits this enzyme less potently than it inhibits AICAR transformylase. THF, tetrahydrofolate; rib, ribose; P, phosphate

duction of pro-inflammatory CYTOKINES (TNF, IL-6) and enhancing anti-inflammatory CYTOKINES (IL-1 receptor antagonist, IL-1ra).

The major anti-inflammatory effect of low-dose methotrexate may be inhibition of the enzyme AICAR transformylase, the result being the intracellular accumulation of adenosine monophosphate (AMP) and its conversion to adenosine in the extracellular space (Fig. 2). It is suggested that the higher levels of extracellular adenosine then bind to the transmembrane G-protein coupled adenosine cell surface receptors (A_1, $A_{2\alpha}$, $A_{2\beta}$, A_3) [14,18]. According to this hypothesis, methotrexate predominantly acts via ligation of the $A_{2\alpha}$ receptors that are present on neutrophils, MACROPHAGE-MONOCYTES, LYMPHOCYTES and BASOPHILS. Binding increases intracellular cAMP leading to IMMUNOSUPPRESSION by inhibition of phagocytosis, inhibition of secretion of TNF-α, IFN-γ, IL-2, IL-6, IL-8, and HLA expression, and increased secretion of IL-10, an anti-inflammatory CYTOKINE. Binding of adenosine to A_3 receptors on macrophage-MONOCYTES leads to inhibition of secretion of TNF-α, IL-12, IFN-γ, and IL-1ra [14,18]. Recent results on $A_{2\alpha}$ and A_3 knockout mice are consistent with adenosine mediating the anti-inflammatory effects of methotrexate because methotrexate does not have anti-inflammatory activity in mice lacking either receptor [19]. Furthermore, methotrexate increases adenosine concentrations in air pouch exudates, a model of INFLAMMATION. By contrast, methotrexate does not increase the blood concentration of adenosine, although changes at peripheral sites cannot be excluded [20].

Clinical indications and efficacy

Methotrexate is an established DMARD for rheumatoid arthritis, psoriatic arthropathy and other inflammatory joint diseases. Methotrexate is used at lower doses in the treatment of rheumatoid arthritis than when used as a cytotoxic drug. As outlined in Appendix C.13, the dose is slowly increased to a maintenance dose of 7.5–25 mg/wk. In individual patients, the dosage is increased up to a maximum of 25 mg/week or up to a level which produces satisfactory suppression of the activity of the disease with limited side-effects.

An outline of recent evaluations of methotrexate and other DMARD (Tab. 2) indicates the required quality of modern clinical trials in rheumatoid arthritis. Measures of symptoms, clinical signs and X-ray examination are all required. In terms of the results, these trials indicate little difference between methotrexate and other DMARD. The initial response is greater for both leflunomide and the biological, etanercept, than during treatment with methotrexate, but it is unclear whether this translates into a longer-term benefit. An important indication of the long-term therapeutic benefit of DMARD is their effect on the degree of damage to joints. Present evidence is that methotrexate retards, but does not entirely block, joint damage in many patients (Tab. 2).

The utility of methotrexate is seen from the high maintenance on treatment with about 60% to 70% of patients still taking methotrexate 6 years after initiation of the treatment [27, 28]. This retention rate is generally greater than seen with other DMARD although, of course, the continued use of methotrexate in individual patients does depend on the treatment strategies of the prescribers [28]. The value of methotrexate is also seen in its CORTICOSTEROID-SPARING effect.

The present view on methotrexate is that it should be considered for all patients at the time of diagnosis of rheumatoid arthritis. Individual factors such as pregnancy and alcohol intake (see sections on side-effects and monitoring below) may affect that decision but methotrexate needs to be considered. Whether the methotrexate regime should be used as the only DMARD or as part of a combination with other DMARD is unclear (see section on combination DMARD therapies below).

Cost

A cost-effectiveness analysis for the use of low-dose methotrexate indicates the utility of the drug. As with other drugs, the total cost of therapy of DMARD includes the cost of the drugs, the costs of monitoring treatment and the indirect costs incurred from lost productivity due to morbidity of the disease and its treatment. Methotrexate is one of the least expensive DMARD but has the highest costs associated with monitoring, including repeated clinical and laboratory monitoring for side-effects (see below). Methotrexate is a cost-saving option compared to no use of DMARD, but is very similar to sulfasalazine [29]. Leflunomide and etanercept were less cost-effective, although clearly therapeutically active.

Side-effects

Low-dose methotrexate produces a large number of adverse reactions (Appendix C.13). Long-term stud-

TABLE 2. COMPARATIVE RESULTS OF RECENT CLINICAL TRIALS ON METHOTREXATE AND OTHER DMARD

Treatments	Duration (months)	Methotrexate dose (mg week)	Folate supplement	ACR[1] 20% 50% 70%	Contrasts (ACR and radiographic progression)	Health assessment questionnaire (HAQ)[2]	Refs.
Methotrexate vs leflunomide vs placebo	12	7.5–15	Yes	46% 23% 9%	Methotrexate = leflunomide > placebo by ACR and X-ray	–0.26 methotrexate 0 placebo	[21]
Continuation of trial above	24	7.5–15	Yes	48% 28% 12%	Methotrexate = leflunomide by ACR and X-ray	–0.6 leflunomide greater improvement than with methotrexate –0.37	[22]
Methotrexate vs leflunomide	12	10–15	10%[3]	65% 44% 10%	Methotrexate > leflunomide by ACR but not by X ray	–0.7 all groups	[23]
Methotrexate vs sulfasalazine vs combination	12	7.5–15	No	59%	No differences		[24]
Methotrexate vs etanercept	12	19 (mean)	Yes	65% 42% 22%	Etanercept > methotrexate but only in first 6 months		[25]

[1]*ACR: Response score of the American College of Rheumatology. The percentage of patients achieving a 20%, 50% or 70% improvement in the ACR score during treatment with methotrexate is shown.*
[2]*A decrease in HAQ of 0.22 is considered to be the minimum clinically meaningful decrease [26].*
[3]*Folate usually started after an adverse reaction.*

ies (> 5 years) show that 10-30% of patients treated with methotrexate cease therapy due to toxicity. Oral ulceration, nausea and fatigue occur very frequently and are probably related to intracellular depletion of folates, resulting in increased adenosine and homocysteine. Hence the usefulness of supplementation with oral folic acid. Various doses of folic acid have been recommended, but recent reviewers advise that a single daily dose of 5 mg folic acid should be administered to all patients on the morning following the dose of methotrexate [30]. Supplementation at this level does not reduce the anti-rheumatic EFFICACY of low-dose methotrexate. Folinic acid contains the fully reduced form of folic acid and is used to treat methotrexate-induced haematotoxicity and overdose with the drug.

An unexpected side-effect of methotrexate is the accelerated formation of rheumatoid nodules, particularly around the fingers. Colchicine may prevent their formation [31]. Interstitial pneumonitis is a serious side-effect of methotrexate and occurs in 2–7% of patients. It is potentially fatal. Treatment consists of cessation of methotrexate, general supportive measures and high doses of corticosteroids. Although most patients with methotrexate-induced lung disease have a complete recovery, some have permanent lung damage. The strongest predictors for lung injury are age > 60 yrs, diabetes mellitus, rheumatoid pulmonary involvement, previous use of DMARD, and hypoalbuminemia [32]. Methotrexate should not be re-introduced after recovery from pneumonitis. Almost certainly, many reported cases of pneumonitis were the result of pulmonary infections which were not differentiated from methotrexate-induced pneumonitis.

As is the case during treatment with most DMARD, pregnancy should be avoided during treatment with low-dose methotrexate because of the high risk of teratogenic effects. Consequently, treatment with methotrexate, as well as with other DMARD, should be stopped before conception and not restarted until after delivery. Fortunately, the disease activity generally decreases during pregnancy but exacerbations can be treated with low-dose corticosteroids.

Monitoring

Methotrexate should be avoided in patients with significant pre-existing liver or lung disease. Therefore, at the start of therapy, a complete blood count, a chest radiograph and liver function tests, including measurements of aspartate aminotransferase, alanine aminotransferase, albumin and alkaline phosphatase, should be undertaken [1]. Tests for hepatitis B and C should be conducted in patients who are at risk of these diseases. Serum creatinine should be assayed as a measure of renal function. The complete blood count and tests of liver and renal function should be monitored every month for 6 months, then every 1 to 2 months subsequently. Treatment with methotrexate should be stopped in patients with transaminase concentrations persistently at twice the upper limit of normal or at three times the upper level of normal at any time. At this stage, measurement of plasma concentrations of methotrexate does not appear to be useful in predicting significant hepatotoxicity [33]. Liver biopsy is required only for those patients who need to continue methotrexate and who have sustained enzyme abnormalities.

Antimalarials (chloroquine and hydroxychloroquine)

Hydroxychloroquine and chloroquine were developed as antimalarial drugs. Their introduction for the treatment of rheumatoid arthritis followed the chance discovery of the value of an older antimalarial, mepacrine. Both hydroxychloroquine and chloroquine are now used in the treatment of rheumatoid arthritis but hydroxychloroquine is the more widely used. Both are relatively small-molecular-weight bases (Fig. 3).

Pharmacokinetics and metabolism

The BIOAVAILABILITY of hydroxychloroquine is very variable, ranging from below 20% up to 100%, but the BIOAVAILABILITY remains fairly constant within an individual [34]. The variable BIOAVAILABILITY may be responsible for much of the interpatient differences in the response to hydroxychloroquine. An important feature of the pharmacokinetics of hydroxychloroquine and chloroquine is their extremely long terminal half-lives of about 40 days. This means that steady-state concentrations may not be achieved until after 3 to 6 months of daily dosing. The use of loading regimens may decrease the time until the onset of effect of the antimalarials but this is not standard clinical practice.

Mode of action

The antimalarials produce a variety of *in vitro* effects but their clinical significance is unclear. A feature of both antimalarials is that they are weakly basic drugs that accumulate in acidic organelles, particularly lysosomes, where they raise the pH. This may affect the function of the acidic organelles, particularly lysosomal enzymes. One such enzyme is acidic sphingomyelinase which is located within the lipid membrane of lysosomes and is an important mediator in the signal transduction pathway between the TNF receptor on the cell surface and activation of transcription factor NF-κB in the nucleus [35]. Rais-

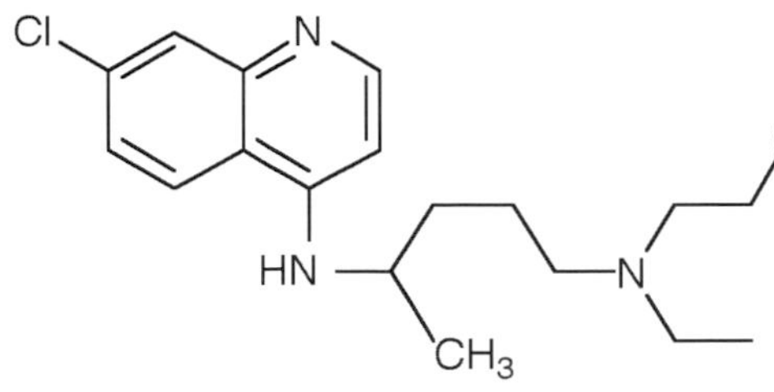

Figure 3. Structure of hydroxychloroquine
The base structure is shown but, at physiological pH values, the molecule is very largely present as a dication with both aliphatic amino groups ionized after taking up hydrogen ions. The structure of chloroquine is the same but does not have the hydroxyl group.

ing the pH of lysosomes inhibits acidic sphingomyelinase activity, consequently inhibiting NF-κB activity and proinflammatory gene expression. Antimalarials have also been reported to inhibit the activity of many other enzymes, including phospholipase A_2 and the production of IL-1.

Clinical indications and efficacy

Antimalarials are mild antirheumatic agents with low toxicity. This makes them attractive for use in the early stages of rheumatoid arthritis or in combination with other DMARD. However, in contrast to methotrexate and several other DMARD, the antimalarials do not appear to retard the damage to joints produced by rheumatoid arthritis [36].

Side-effects

The antimalarials are the least toxic DMARD. Most adverse effects (Appendix C.13) are transient and cessation of the drug is usually not required. Hydroxychloroquine and chloroquine have a very similar range of adverse effects, but hydroxychloroquine is associated with less toxicity than chloroquine. Of most concern is a rare, irreversible retinopathy, resulting in permanent visual loss. It is almost unknown with hydroxychloroquine, with most cases being due to chloroquine. This effect can be avoided by limiting the daily dosage of hydroxychloroquine relative to bodyweight (Appendix C.13).

Monitoring

It is recommended that there should be a baseline eye evaluation and examinations of the fundi and visual fields every 6 to 12 months. A relationship between the plasma concentrations of hydroxychloroquine and EFFICACY has been shown, indicating that there may be some value in measuring blood hydroxychloroquine concentrations to optimize dosing regimens [37]. However, adjusting the dosage of hydroxychloroquine after monitoring its plasma concentrations is still most uncommon.

Azathioprine

Azathioprine is a synthetic purine and anti-metabolite drug which is mostly used as an immunosuppressant to prevent transplant rejection (see Chapter C.10, Immunosuppressives in tissue rejection). It has limited use in the treatment of rheumatoid arthritis and ulcerative colitis.

Pharmacokinetics and metabolism

Azathioprine is cleaved non-enzymically to 6-mercaptopurine and the substituted thioimidazole after reaction with thiol compounds, such as glutathione (Fig. 4). 6-Mercaptopurine has been used for the same indications as azathioprine and undergoes several metabolic reactions which lead to both loss of activity and to its activation.

An inactive metabolite is 6-thiouric acid whose formation is catalyzed by the enzyme, xanthine oxidase (Fig. 4). Not surprisingly, treatment with the xanthine oxidase inhibitor, allopurinol, substantially decreases the formation of the inactive thiouric acid and the dosage of azathioprine and 6-mercaptop-

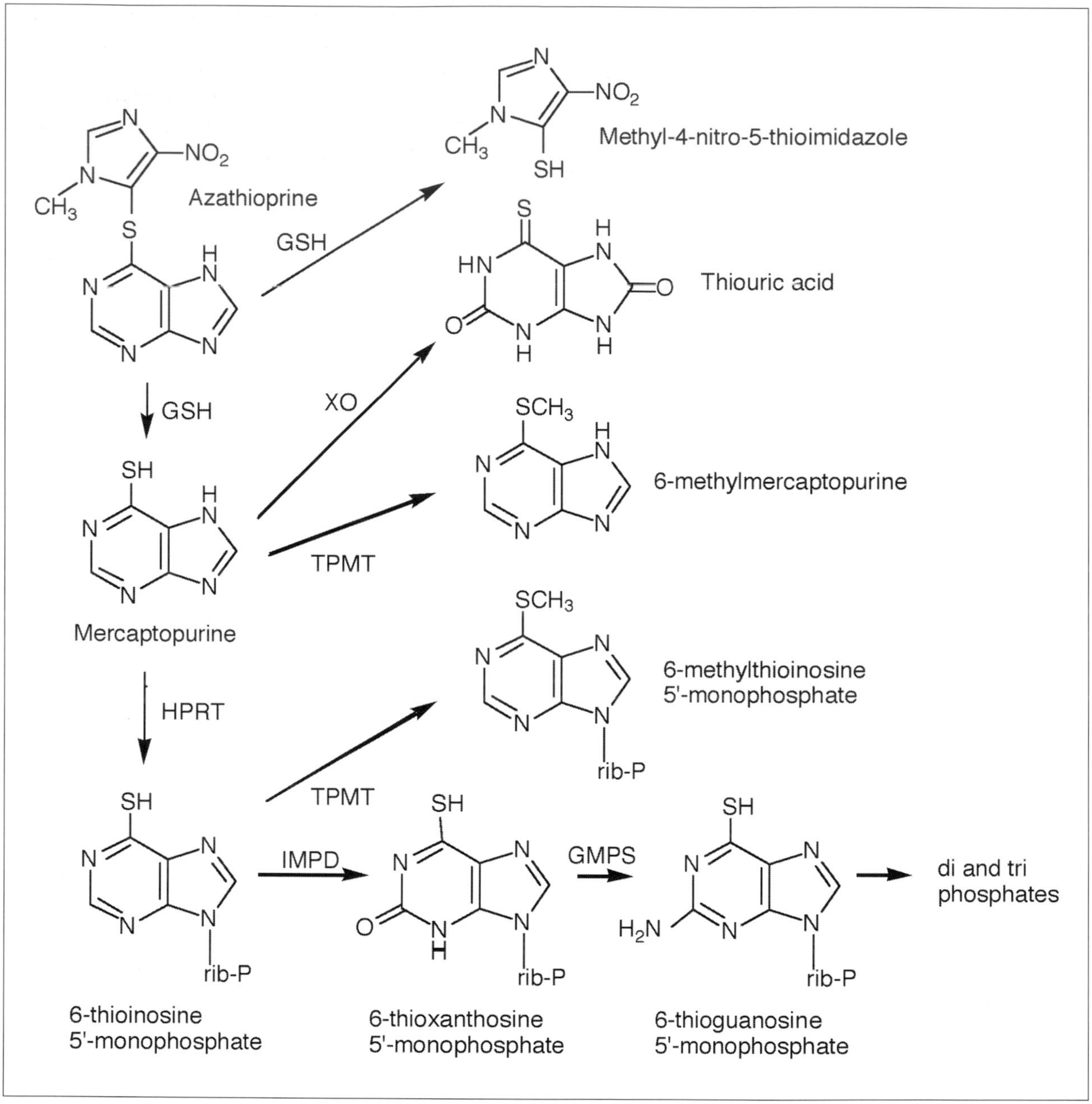

Figure 4 Structure and outline of the metabolism of azathioprine

Azathioprine is activated by its metabolism to 6-mercaptopurine and ultimately to thioguanine nucleotides which may exert their immunosuppressive activities after incorporation into DNA. On the other hand, the actions of xanthine oxidase (XO) and thiopurine methyltransferase (TPMT) largely cause the production of inactive products although 6-methylthioinosine 5'-monophosphate (methylthioinosinic acid) may cause feedback inhibition of purine production. HPRT, hypoxanthine phosphoribosyltransferase; IMPD, inosine monophosphate dehydrogenase; GMPS, guanosine monophosphate synthetase; rib-P, ribose-5-phosphate; GSH, glutathione.

urine should be reduced by about 75% in patients taking allopurinol. Optimal dosage of these drugs is also controlled by the activity of the thiopurine methyltransferase which is under genetic control [38]. About 85% of patients have normal activity of the enzyme, while about 1 in 300 individuals are homozygotes for the abnormal enzyme and have negligible inactivating capacity. Consequently, BONE MARROW toxicity is very common in the homozygotes for the abnormal enzyme due to the development of high concentrations of the active thioguanine nucleotides. If genetic analysis is available, these patients should not receive azathioprine or 6-mercaptopurine. The heterozygotes have intermediate activity of thiopurine methyltransferase and, optimally, should receive smaller than usual doses of azathioprine (Appendix C.13). The full significance of the metabolism of azathioprine is, however, not entirely clear and some patients who are homozygotes for the normal enzyme may produce low amounts of the thioguanine nucleotides and an overproduction of 6-methylmercaptopurine. These patients may not respond to azathioprine [39].

Mode of action

As is the case with most DMARD, the mode of action of azathioprine is unclear but, as outlined above, the major actions of azathioprine and 6-mercaptopurine are linked to their metabolism. The activities of these drugs are generally linked to the formation of the thioguanine nucleotides which are taken up into DNA. Other possibilities include feedback inhibition of *de novo* purine synthesis (particularly by 6-methylthioinosine 5'-monophosphate) and inhibition of inosine monophosphate dehydrogenase (Fig. 4) [40]. The overall effect appears to be a reduction in intracellular synthesis of purines and a decrease in numbers of circulating B and T LYMPHOCYTES, particularly CD8 cells.

Clinical indications and efficacy

Although azathioprine is not widely used for the treatment of rheumatoid arthritis, it has similar EFFICACY to several other DMARD such as parenteral GOLD, penicillamine and cyclosporin. It may not, however, be as effective as methotrexate. Azathioprine is generally reserved for use in the later stages of severe rheumatoid arthritis, particularly when other DMARD have failed. Azathioprine may be used alone, or in combination with other DMARD, for the management of rheumatoid arthritis and psoriatic arthritis. More studies are, however, required on the optimal use of azathioprine in COMBINATION THERAPY.

Side-effects

The most frequent toxicity that necessitates discontinuation of azathioprine is gastro-intestinal disturbance, though this is self-limiting. Significant blood dyscrasias may occur. Their occurence is related to both the dose of azathioprine and to the activity of thiopurine methyltransferase.

Monitoring

Regular blood counts are required during treatment with azathioprine. Infections require discontinuation of azathioprine. Hepatic transaminases should also be monitored but they usually normalize on dose reduction or cessation of treatment. There is considerable controversy about the induction of malignancies by azathioprine but careful monitoring of patients, particularly for lymphomas, is advised.

There are two approaches to the identification of homozygotes and heterozygotes with respect to the abnormal thiopurine methyltransferase: phenotyping and genotyping. Phenotypic detection of the abnormal enzyme is the less expensive [39]. At present, the detection of the abnormal enzyme appears to be cost-effective, although neither technique is in general use.

Cyclosporin

Cyclosporin is a fungal antimetabolite which has profound effects on the immune response. It widely used to prevent rejection of transplanted tissues (see

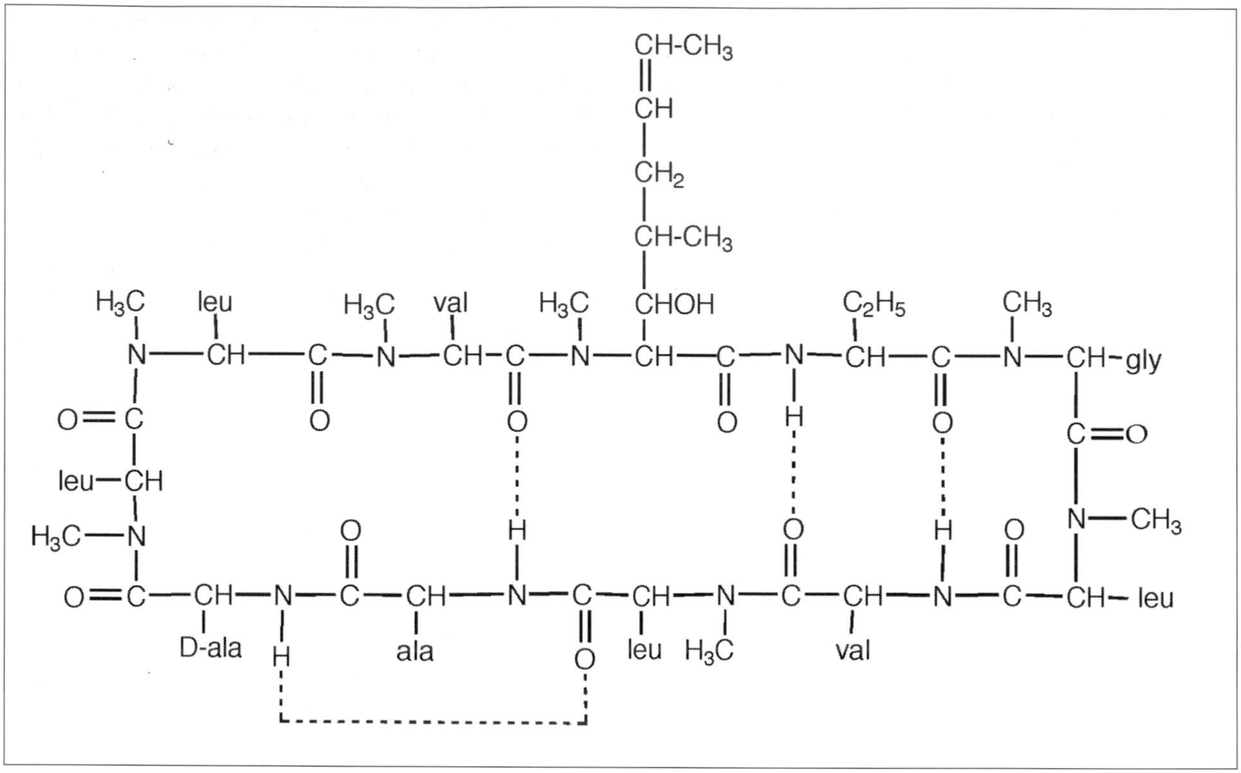

FIGURE 5 STRUCTURE OF CYCLOSPORIN

The hydrogen bonds (broken lines) in the polypeptide chain are internal or blocked by N-methylation of the amide nitrogens.

Chapter C.9, Immunosuppressives in tissue rejection). Cyclosporin is also efficacious in the treatment of rheumatoid arthritis but, as in all its clinical uses, has a low THERAPEUTIC INDEX.

Mode of action

Cyclosporin and similar drugs (rapamycin and tacrolimus) suppress the immune response by inhibiting key signal transduction pathways. This is achieved by intracellular binding to immunophilin receptors, the complexes thus formed inhibiting calcineurin. Blocking the actions of calcineurin leads to inhibition of the translocation to the nucleus of the cytosolic component of the NUCLEAR FACTOR of activated T CELLS (NF-AT). Thus, genes dependent on this transcription factor, such as IL-2 and IL-4, are not transcribed normally [41].

Pharmacokinetics and metabolism

Cyclosporin is a very hydrophobic polypeptide (Fig. 5) which is consequently difficult to formulate into a product which yields reliable absorption. The most recent changes in the formulation are liquid-filled capsules and a liquid preparation which produce an emulsion in water. These preparations are absorbed more consistently than the older formulations. The half-life of elimination is about 6 hours, but the drug is most commonly administered twice daily. Cyclosporin is metabolized by the hepatic cytochrome P-450 oxidase system (3A4 isoenzyme in particular). Consequently, its metabolism is inhibited by a range of drugs including ketoconazole and diltiazem, and induced by others, such as carbamazepine and phenytoin [42].

Clinical indications and efficacy

Cyclosporin therapy in rheumatoid arthritis is reserved currently for patients with refractory disease or severe extraarticular complications. Patients treated with cyclosporin at doses between 2.5 and 5.0 mg/kg/day show significant decreases in pain and in the number of inflamed joints, as well as improvements in the functional status of patients [43]. Doses of 10 mg/kg/day are more effective but nephrotoxicity now limits the dosage to 5 mg/kg/day. Cyclosporin retards further joint damage in previously involved joints and decreases the rate of new joint erosions in previously uninvolved joints in patients with early rheumatoid arthritis [44].

Side-effects

Cyclosporin causes a large variety of side-effects (Appendix C.13). Nephrotoxicity is the principal adverse effect of cyclosporin and its use should be avoided in patients with preexisting renal disease and it should be used carefully with other nephrotoxic agents or if creatinine CLEARANCE falls. Hypertension is also common, affecting up to one-third of cyclosporin-treated patients. Hypertension can usually be controlled with beta blockers or angiotensin converting enzyme (ACE) inhibitors or angiotensin blocking agents. However, ACE inhibitors or angiotensin blockers should not be used if hyperkalaemia has developed. Of concern are case reports of lymphoproliferative lesions and other malignancies.

Monitoring

A complete blood count, creatinine, uric acid and liver function tests, as well as the measurement of blood pressure, are recommended at baseline. For follow-up visits examination for oedema and measurement of blood pressure are suggested every 2 weeks until the cyclosporin dosage is stable, and monthly from then on. Serum creatinine should be checked every 2 weeks until the dosage is stable. Periodic complete blood counts, measurement of the concentrations of potassium in plasma and liver function tests are suggested. Monitoring whole blood concentrations of cyclosporin, not plasma concentrations, is uncommon in the treatment of rheumatoid arthritis but should be considered on starting or discontinuing any interacting drugs or if there is doubt about the compliance of the patients. Measurement of the concentrations in whole blood is recommended because of the temperature-dependent distribution of cyclosporin between red blood cells and plasma. Patients should be made aware of the inhibition of its metabolism by grapefruit juice and several other drugs.

Gold complexes

GOLD complexes were introduced into therapy of rheumatoid arthritis after attempts to use them for the treatment of infectious diseases. The first large-scale use was based on the mistaken belief that rheumatoid arthritis was a tubercular infection. Although the tubercular theory of rheumatoid arthritis has been disproven, the gold complexes are effective for rheumatoid and psoriatic arthritis.

Chemistry

Two forms are available, polymeric complexes, principally aurothiomalate, given by intramuscular injection (Fig. 6) and a monomeric complex, auranofin, which is given orally. Gold does not form simple salts and only complexes of gold are formed. Complexes can be prepared with gold principally in two oxidation states, I and III, but all the clinically used complexes contain gold (I). Gold in this oxidation state forms complexes with thiols, such as cysteine residues in proteins, and cyanide as described below. Complexes containing gold (III) are powerful oxidants and may be present transiently *in vivo*

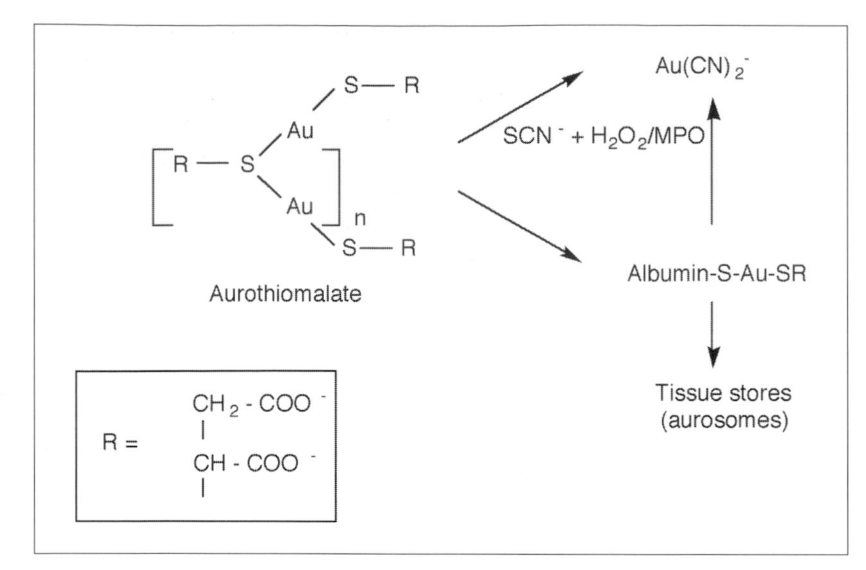

FIGURE 6. SCHEMATIC STRUCTURE AND METABOLISM OF AUROTHIOMALATE
Aurothiomalate is a polymeric anion consisting of about 8 to 10 units gold (I) thiomalate with the chain terminated by an additional thiomalate residue. Gold is present in plasma mainly as complexes with albumin and endogenous thiols (RS–). Aurothiomalate and albumin complexes are converted to aurocyanide ($Au(CN)_2^-$) by myeloperoxidase (MPO) during the oxidation of thiocyanate (SCN^-) by hydrogen peroxide (H_2O_2).

Pharmacokinetics and metabolism

Aurothiomalate is polymeric and highly water-soluble. Consequently, it is poorly absorbed orally and must be given by intramuscular injection which is administered once a week. The need for intramuscular injection makes the administration of aurothiomalate inconvenient and has limited its use in recent years because the newer DMARD, such as methotrexate, sulfasalazine and hydroxychloroquine, can be administered orally. Auranofin is much more lipid-soluble and is at least partly absorbed following oral administration.

GOLD complexes of albumin are the most significant complexes in plasma, and the total GOLD concentration is eliminated from plasma with an initial half-life of about 5 days [45]. After injection, much of the GOLD is excreted in urine, but some accumulates, particularly in the lysosomes of synovial lining cell aurosomes, where GOLD is present for many years following the last dose of GOLD. Much GOLD appears in faeces when oral GOLD is administered.

Aurocyanide is a metabolite of all GOLD (I) complexes. It is formed through the oxidation of thiocyanate (SCN^-) by myeloperoxidase, an important enzyme in the oxidative burst of neutrophils and MONOCYTES (Fig. 6). Thiocyanate is a normal body constituent which is oxidized, in part, to hydrogen cyanide which then reacts with GOLD complexes to yield aurocyanide which is a very stable complex of GOLD (I) [46].

Mode of action

Aurocyanide potently inhibits the oxidative burst of neutrophils and the proliferation of LYMPHOCYTES. Aurocyanide may thus mediate many of the anti-rheumatic and adverse effects of the GOLD complexes [46]. Unlike other GOLD complexes, aurocyanide enters cells readily. The intracellular site of action is unclear but may be proinflammatory transcription factors, such as AP-1 and NF-κB, which have cysteine residues within their DNA binding domains and bind GOLD [47]. GOLD(I) thioglucose also inhibits the IL-1 induced expression of NF-κB and AP-1-dependent genes and inhibits the DNA-binding activity of NF-κB *in vitro* [48].

Clinical indications and efficacy

The injectable GOLD compounds have useful anti-rheumatic activity with similar activity to sulfasalazine, methotrexate and penicillamine [49]. An excellent response occurs in about a third of the patients treated with intramuscular GOLD complexes within the first year. Patients receiving aurothiomalate show less progression of erosions and a higher rate of erosion repair than patients without GOLD treatment [36,50]. However, the long-term benefits of GOLD compounds are less impressive.

The EFFICACY of the orally administered complex, auranofin, during treatment for several months is well proven. However, the general conclusion is that auranofin is less efficacious than the injectable complex, aurothiomalate, and also less effective than methotrexate and penicillamine [49]. The initial clinical benefit is often not maintained and significant numbers of patients withdraw from treatment due to inefficacy, as well as side-effects.

Side-effects

Adverse effects of injected GOLD therapy are common, involving up to 40% of patients and can be severe (Appendix C.13). Toxicity, however, is generally minor and often manageable if dosing is adjusted early upon appearance of adverse effects, although adverse effects are a common cause of withdrawals from GOLD therapy over 2 to 4 years of therapy [49]. Toxicity does not correlate with the cumulative dose.

Auranofin has a lower incidence of the side-effects which are seen with the injectable GOLD complexes (Apppendix C.13). Diarrhoea is a common side-effect of auranofin but not the injectable complexes. Diarrhoea may decrease with continued dosage of auranofin and reduced dosage is often sufficient to stop it.

Monitoring

Before starting injectable GOLD therapy, a baseline evaluation should include a complete blood cell count, plasma creatinine and a qualitative test for urinary protein. The full blood cell count and urinary analysis should be repeated every 1 to 2 weeks for the first 20 weeks or at the time of each injection. Furthermore, the patients should be asked about skin itching and mouth ulcers before each injection. The 24-hour output of protein should be measured if the qualitative test for urinary protein is positive. Treatment should be ceased if the protein excretion is greater than 500 mg in 24 hours but can be restarted when urinary protein is no longer detectable.

For oral GOLD therapy, the recommended baseline assessment is the same as for injectable GOLD. The follow-up monitoring should include a full blood cell count and a urine analysis every 4 to 12 weeks.

Leflunomide

Leflunomide is a relatively new small molecule DMARD which is active after oral dosage. Chemically, it is unrelated to any other marketed DMARD.

Pharmacokinetics and metabolism

Leflunomide is rapidly and almost completely converted to an active open chain metabolite by first pass metabolism in the gut wall and liver (Fig. 7). The active metabolite of leflunomide has a long half-life of between 15–18 days because of its enterohepatic recirculation [51]. About 90% of a single dose of leflunomide is eliminated, about half in urine, primarily as metabolites, while about 50% is secreted in bile as the active metabolite and is ultimately excreted in faeces. Because the active metabolite relies heavily on biliary excretion for its CLEARANCE , and also given its risk of hepatotoxicity, leflunomide is contraindicated in patients with hepatic impairment.

The active metabolite of leflunomide may take 15 to 20 weeks to reach steady-state plasma concentrations. In order to achieve therapeutic concentrations rapidly, it is common to administer loading doses (Appendix C.13). In practice, the loading dose is often decreased or not used with the expectation of less "nuisance" problems with diarrhoea or nausea, both of which may influence early PATIENT COMPLIANCE. Anecdotally, clinical EFFICACY is maintained but delayed by a few weeks.

The active metabolite binds strongly to cholestyramine within the gastrointestinal tract. The result is that its plasma half-life is reduced to approximately one day and cholestyramine is used when rapid

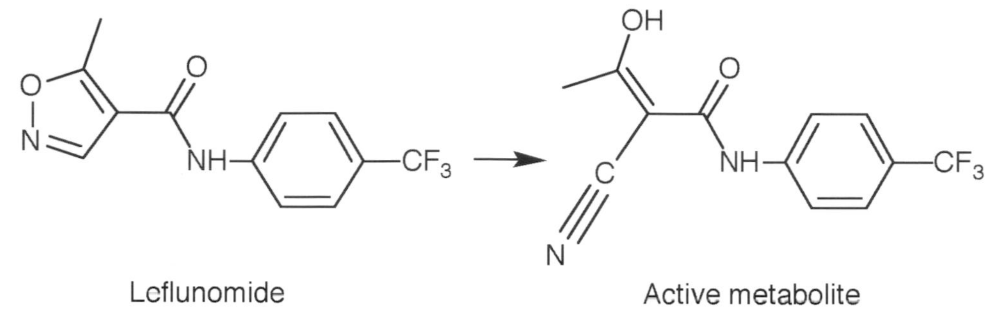

FIGURE 7. STRUCTURES OF LEFLUNOMIDE AND ITS ACTIVE METABOLITE
The active metabolite is the major form which circulates in plasma and is known by its code, A77 1726.

elimination of the active metabolite is required (see section on side-effects below).

Mode of action

Through its active metabolite, the primary mode of action of leflunomide is the selective and reversible inhibition of dihydroorotate dehydrogenase [52] (Fig. 8). This enzyme provides the rate-limiting step in the pyrimidine synthesis that is accelerated in the activated $CD4^+$ T CELLS that proliferate rapidly during the progression of rheumatoid arthritis. This anti-proliferative effect on activated LYMPHOCYTES is likely to be the key effect of leflunomide on the pathophysiology of rheumatoid arthritis. Additionally, leflunomide interferes with T CELL signalling and also has broad anti-inflammatory effects. For example, the active metabolite is a potent inhibitor of NUCLEAR FACTOR κB activation and causes a dose-dependent inhibition of CYTOKINE production (including TNF). Leflunomide also decreases the local production of synovial metalloproteinases that suggests a mechanism by which it acts to prevent joint destruction.

Clinical indications and efficacy

The EFFICACY of leflunomide has been examined in rheumatoid arthritis in six randomized clinical trials in over 2000 patients. These trials show that the EFFICACY of leflunomide is similar to that of methotrexate or sulfasalazine (Tab. 2) [53]. It not only decreases symptoms, and increases function and quality of life in rheumatoid arthritis but also retards radiographic joint damage [54]. Clinical improvement has been sustained for up to five years. COMBINATION THERAPY with leflunomide and methotrexate is also effective and well tolerated in patients inadequately responding to methotrexate alone [55].

Apart from its value in the treatment of rheumatoid arthritis, leflunomide is useful in the treatment of psoriatic arthritis and has been used in a small cohort of systemic lupus erythematosus patients with the drug appearing to be efficacious and safe [56].

Side-effects

The most common adverse events associated with leflunomide treatment are gastrointestinal but these tend to usually improve with time and/or dose reduction (Appendix C.13). Rash and reversible alopecia are also common.

The cytostatic effect of leflunomide may explain some of the side-effect profile, such as reversible alopecia and conversely the lack of OPPORTUNISTIC INFECTIONS. Most memory T CELLS circulate in the G_0 phase, and therefore do not require dihydroorotate

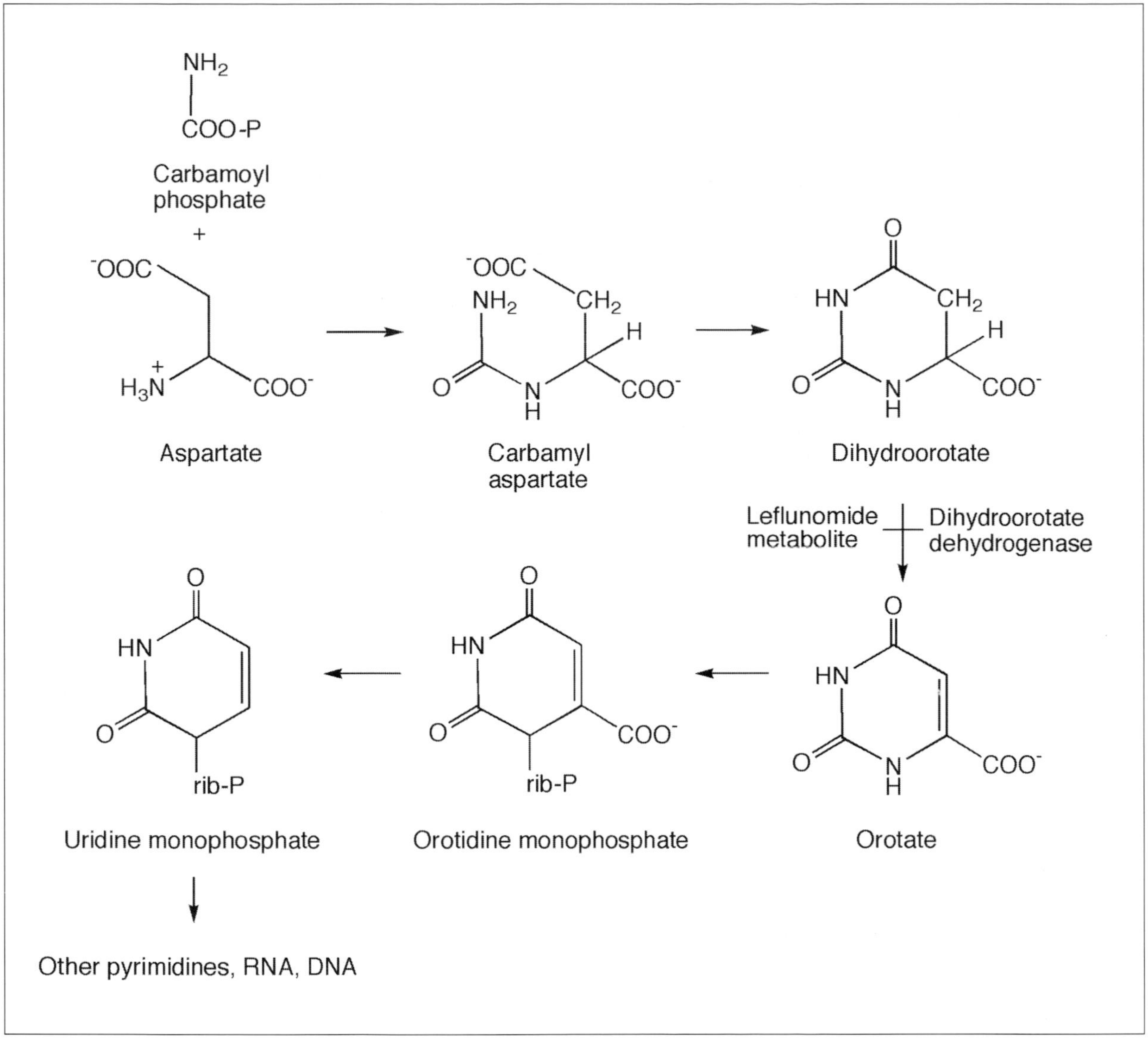

FIGURE 8
Synthesis of pyrimidines showing the site of action of the active metabolite of leflunomide on dihydroorotate dehydrogenase. rib, ribose; P, phosphate.

for any *de novo* pyrimidine synthesis, and are not susceptible to the anti-proliferative effect of leflunomide. In addition, because of the sparing of the salvage pathway, the replicating cells in the gastrointestinal tract and haematopoietic system are relatively unaffected, thus explaining the lack of mucositis or marrow toxicity.

When used as monotherapy in clinical trials, abnormal transaminase levels were noted in 5% to 15% of patients, but these effects were generally mild (less than 2-fold elevations) and reversible and usually resolved while continuing treatment. Post-marketing surveillance shows that almost all cases of hepatic dysfunction had other confounding factors present [57]. Leflunomide may increase plasma levels of cholesterol and low-density lipoproteins in a progressive manner but long-term effects of this are unknown.

Leflunomide is absolutely contraindicated in women who are or may become pregnant, because of its teratogenic effects (lymphomas) in animals. Termination of pregnancy is generally recommended if the patient has been on leflunomide, even though there have been a number of reported cases of delivery of full-term healthy infants [58]. Because of the prolonged half-life of the active metabolite, any woman taking leflunomide who is contemplating pregnancy should allow the plasma concentrations of the active metabolite to fall below 0.02 mg/L. This may take several months because of its long half-life. Alternatively, the elimination of the active metabolite can be accelerated by cholestyramine. The active metabolite of leflunomide diffuses into breast milk, although it is not known if the concentrations are sufficient to cause toxicity. At this stage, however, it is contraindicated in nursing mothers.

Any suspected toxicity may be further evaluated by use of a short course (1-2 days) of cholestyramine at a lower dose (4 g three times daily). This will often reverse the side-effect, be it rash or diarrhoea or other, quite quickly.

Monitoring

Baseline investigation should include hepatitis B and C serology and any persistent hepatic dysfunction should be further investigated. Patients should be advised to reduce alcohol consumption because of possible greater liver impairment. Liver function tests (including transaminases) should be monitored every four to six weeks for at least six months, and longer if patients are taking COMBINATION THERAPY with methotrexate or other hepatotoxic drugs. Thereafter, liver function tests should be repeated at least every three months, more frequently if transaminses have increased. Usually complete blood counts are performed at the same frequency. All women of childbearing age should be counselled to use effective forms of contraception and have a negative pregnancy test before beginning the drug.

Penicillamine

Penicillamine was introduced for the treatment of rheumatoid arthritis because of its ability to dissociate the disulphide bonds of immunoglobulin M (IgM) *in vitro* but it now seems probable that improvement in rheumatoid arthritis is unrelated to such an effect.

Chemistry

Penicillamine is a structural analogue of cysteine and a component of the penicillin molecule. It contains the D configuration, as opposed to the L configuration of the natural amino acids which are constituents of proteins (Fig. 9). Penicillamine is chemically reactive through its thiol (sulphydryl) group. It is oxidizable to disulphides as outlined below. It also forms chelates with several metals through the thiol group and, apart from its use in rheumatoid arthritis, is used for the treatment of lead, copper and arsenic poisoning and in Wilson's disease, in which there is excess copper in the body.

Pharmacokinetics and metabolism

Penicillamine is well absorbed orally but is subject to binding interactions in the gastrointestinal tract, particularly with iron salts and antacids. Administration on an empty stomach is therefore recommended. The elimination half-life of penicillamine is a few hours only. The main mechanism of CLEARANCE is via reactions with thiol molecules, particularly those found in albumin. The disulphide-bonded products, including penicillamine-disulphide, and penicillamine cysteine disulphide, act as a "sink" for penicillamine. A good example is the penicillamine-albumin complex which has a much longer half-life than the parent, penicillamine [59].

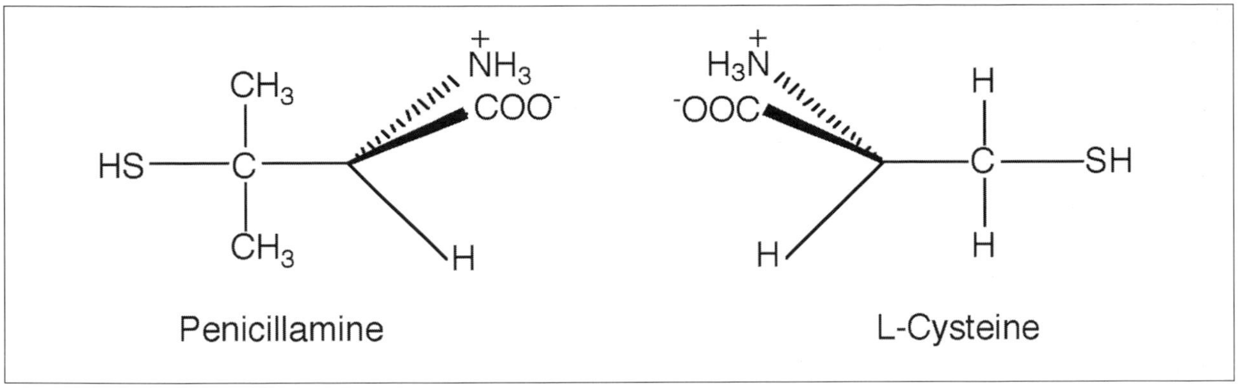

Figure 9
Structures of penicillamine and L-cysteine showing the contrast between the D configuration of penicillamine and the L structure of the naturally occurring amino acid, cysteine.

Mode of action

Penicillamine and the gold complexes have similar clinical properties in the treatment of rheumatoid arthritis, leading to the suggestion that they have a similar MECHANISM OF ACTION through their reactivities with endogenous thiols. Penicillamine should, however, not be administered with gold because it chelates gold and may interfere with its action.

Penicillamine inhibits AP-1 DNA binding in nuclear extracts in the presence of free radicals, presumably by forming disulphide bonds with the cysteine residues in the DNA-binding domains of Jun and Fos [60]. Conversely, some authors have sought to explain the mode of action of thiol drugs according to their reducing properties, whereas the argument put forward here, based on their oxidising action, is the exact opposite. Whatever the details of molecular interactions, it is considered that cysteine residues within the DNA-binding domains of proinflammatory transcription factors are potential targets for GOLD and penicillamine.

Clinical indications and efficacy

Penicillamine decreases clinical signs of rheumatoid arthritis as well as correcting abnormal laboratory parameters such as haemoglobin, erythrocyte sedimentation rate, rheumatoid factor, immunoglobulins and circulating, immune complexes. Joint destruction is not proven to be retarded by this DMARD. Otherwise, penicillamine seems to be as effective as GOLD and azathioprine [61]. Unfortunately, withdrawal rates from penicillamine are high (comparable to GOLD treatment) either due to inEFFICACY or side-effects.

Side-effects

The side-effects of penicillamine are similar to those of the gold complexes (Appendix C.13). Mild side-effects such as rash, stomatitis and metallic taste are common, whereas serious toxicity such as myelosuppression, proteinuria, nephrotic syndrome and autoimmune syndromes are rare. Most adverse effects occur in the first 6 months after commencement of the drug, involving about 50% of patients and resulting in a significant dropout rate from treatment.

Monitoring

The following baseline evaluations are recommended: complete blood cell count, creatinine and urine dipstick for protein. The complete blood cell count

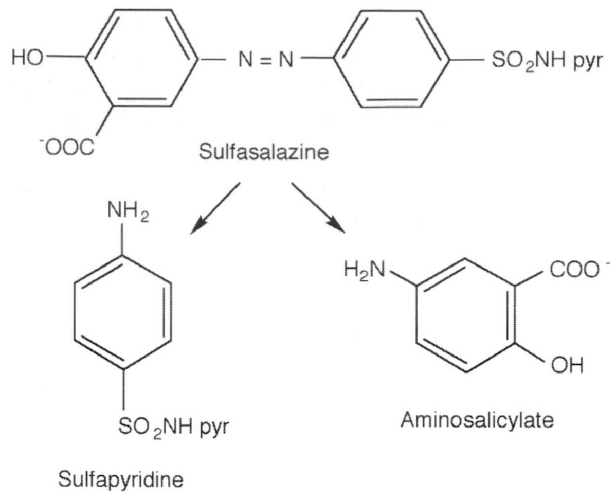

Figure 10. Structure and initial metabolism of sulfasalazine
Reduction of the azo bond of sulfasalazine within the large intestine yields sulphapyridine and aminosalicylate. Sulfapyridine is subsequently absorbed while the absorption of aminosalicylate is incomplete. pyr, pyridine residue

and test for urinary protein should be checked every 2 weeks until a stable dosage is achieved. After that, monitoring should be performed every 1 to 3 months.

Sulfasalazine

Sulfasalazine consists of mesalazine (aminosalicylate) and a sulphonamide, sulfapyridine, joined by an azo bond (Fig. 10). This drug was synthesized in the late 1930s and originally developed and used on the basis of a belief in an infectious cause of rheumatoid arthritis [62]. As discussed in the section on tetracyclines below, an infectious cause of rheumatoid arthritis is still intriguing although no organism has been definitely identified. Sulfasalazine is also widely used in the treatment of inflammatory bowel diseases, ulcerative colitis and Crohn's disease.

Pharmacokinetics and metabolism

Sulfasalazine is administered orally, and upon reaching the large bowel, the azo bond is reduced by colonic bacteria to yield sulfapyridine and mesalazine (Fig. 9). Sulfapyridine is absorbed, whereas mesalazine is poorly absorbed although low concentrations of the acetylated metabolite are found in plasma. Although metabolized in the large intestine, unchanged sulfasalazine also achieves substantial concentrations in plasma. During daily treatment with 2 g sulfasalazine, the average concentrations of the unchanged drug are about 4 mg/L (10 μM) [63].

Mode of action

The mode of action of sulfasalazine is unclear, although it has been reported to inhibit AICAR transformylase and therefore may act in a similar fashion to methotrexate through the accumulation of adenosine [64]. It is not known if sulfasalazine or its metabolites are therapeutically active. The apparent antirheumatic activity of sulfapyridine indicates that it may be the active metabolite [65]. On the other hand, both sulfasalazine and olsalazine appear active in the treatment of ankylosing spondylitis [66, 67], yet olsalazine is a dimer of mesalazine. Olsalazine is, like sulfasalazine, metabolized in the large bowel but, as is evident from its structure, olsalazine can yield only mesalazine. It is of note that sulfasalazine is used in both the treatment of rheumatoid arthritis and inflammatory bowel diseases. Furthermore, many patients with the bowel diseases also develop arthritic states of varying severity. These similarities indicate a commonality in the cause of the diseases and a common mode of action of sulfasalazine.

Clinical indications and efficacy

Sulfasalazine is widely used and effective in the treatment of rheumatoid arthritis as well as ankylosing spondylitis and HLA B27-related arthropathies. In particular, sulfasalazine slows the joint destruction of rheumatoid arthritis [68]. Cessation rates are similar to, or somewhat less than, those observed with gold and penicillamine, most commonly related to insufficient EFFICACY and less commonly to toxicity.

Side-effects

Although sulfasalazine can produce a wide range of side-effects, it is among the group of the best-tolerated DMARD (Appendix C.13), along with hydroxychloroquine, methotrexate and auranofin. Toxicity is most frequent in the first 2 to 3 months of usage, but its likelihood can be reduced by gradually increasing the dosage and the use of enteric coated formulations. Serious side-effects are rare, and most adverse effects are eliminated if the dose is reduced from the usual 1 g twice a day. Nausea and upper abdominal discomfort are the most frequent side-effects at the start of the therapy. Leukopenia is very uncommon but can develop rapidly. Its occurrence is most likely in the first 6 months of therapy but can develop later. The metabolite, sulfapyridine, can produce haemolysis in patients with a deficiency of glucose-6-phosphate dehydrogenase. Sulfasalazine should be avoided at around the time of conception and pregnancy, although no teratogenicity has been reported for this drug.

Monitoring

For the rapid detection of haematological side-effects, it is recommended that a complete blood cell count be performed before commencement of treatment. For the follow-up visits, complete blood cell counts every 2 to 4 weeks for the first 3 months are suggested, and at greater intervals subsequently. The haemolytic anaemia in patients with a deficiency of glucose-6-phosphate dehydrogenase will be detected in blood counts. Baseline measurement of hepatic transaminases is advised in patients with known or suspected liver disease.

Tetracyclines (minocycline)

There is a clear link between polyarthritis and infectious agents such as Parvovirus, Epstein-Barr virus, *Borrelia burgdorferi*, Hepatitis C, and Rubella. It is therefore not surprising that an infectious aetiology for rheumatoid arthritis was sought, and indeed the development of sulfasalazine [62] was based on the concept of rheumatoid arthritis being an infectious disease. Evidence of organisms in the joints of some patients, differences in the bowel flora of rheumatoid arthritis patients with and without erosions, and the ability of sulfasalazine to alter bowel flora remain intriguing links between rheumatoid arthritis and infection. The possible link between mycoplasma-like organisms and rheumatoid arthritis, together with the activity of tetracyclines in mycoplasma pneumonia, prompted the utilisation of tetracyclines in the treatment of this disease, with trials reported since the late 1960s. Several tetracyclines have been used in the treatment of rheumatoid arthritis but minocycline (Fig. 11) is the only tetracycline whose use in rheumatoid arthritis is supported by double-blind clinical trials [69, 70]. All the tetracyclines, particularly minocycline, chelate calcium and zinc and this activity may be related to an anti-rheumatic activity.

Pharmacokinetics

Minocycline is well absorbed after oral administration and has a half-life of elimination of about 16 hours, compared to about 11 hours for tetracycline. About 60% of an oral dose of tetracycline is excreted in urine but only about 10% of a dose of minocycline is excreted unchanged.

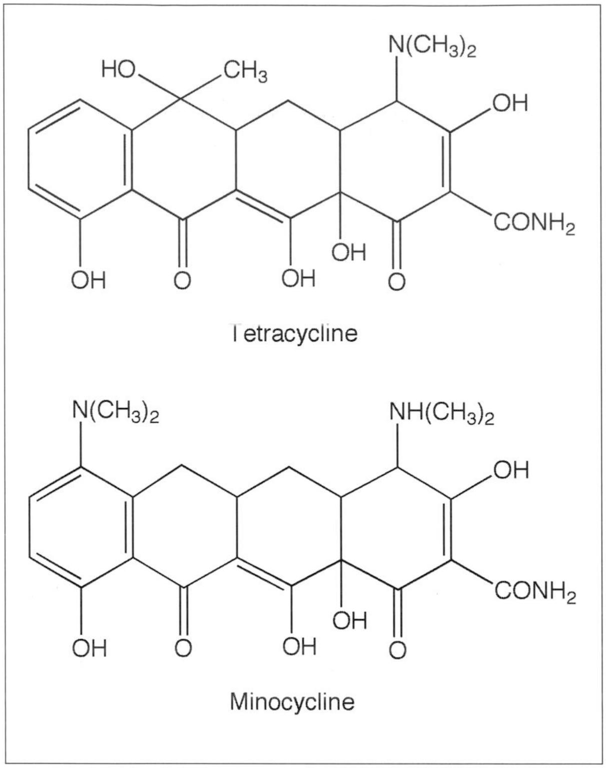

FIGURE 11. STRUCTURES OF TETRACYCLINE AND MINOCYCLINE

Mode of action

The mechanism of the anti-rheumatic effect of minocycline is not known, although several possibilities have been put forward. In addition to their known antimicrobial inhibition of protein synthesis, tetracyclines have anti-inflammatory and potentially IMMUNOMODULATORY effects, and the ability to inhibit angiogenesis. The chelating action is proposed as the basis of enzyme inhibition, particularly of the matrix metalloproteinases I and II. Minocycline reduces the production of phospholipase A_2 and consequent production of LEUKOTRIENES and PROSTAGLANDINS, thereby having effects similar to NSAID. Additionally tetracyclines may act as scavengers of reactive oxygen radicals, and have been shown to down-regulate type 2 nitric oxide synthase, inhibit PHAGOCYTE function, suppress B and T CELL function, and up-regulate IL-10 production [70]. IL-10 production is characteristic of a Th2 response and down-regulates the production of TNF-α.

Clinical indications and efficacy

Clinical trials indicate the usefulness of minocycline in the treatment of rheumatoid arthritis but minocycline is not widely used because of concerns about both its side-effects and development of resistant organisms resulting from its antibiotic activity. The usefulness of the tetracyclines is indicated from a meta-analysis of 10 studies, although only three studies were assessed to be high quality [70]. The small subject numbers and short duration of treatment in most studies did not allow a clear conclusion on the effects of the tetracyclines on the radiographic progression of the disease. However, two studies lasted for 48 weeks but showed no significant reduction in erosions or joint space narrowing.

In summary, minocycline is effective in the treatment of rheumatoid arthritis with a better response seen in early-onset seropositive disease. The cost of treatment and monitoring is low.

Side-effects

The tetracyclines produce a variety of side-effects (Appendix C.13). Tetracyclines bind to developing teeth and bone with consequent staining of teeth and inhibition of the maturation of teeth and bone [71]. Therefore, they should not be used from mid-pregnancy to term. The general use of DMARD during pregnancy is discussed above (see section on side-effects of methotrexate). There are also potential interactions with warfarin, oral contraceptives, antacids, iron preparations, and penicillins. Lupus-like manifestations including positive anti-nuclear antibodies have been described occasionally in patients receiving tetracyclines for the treatment of acne and testing may be required to differentiate between ongoing rheumatoid arthritis and polyarthritis or arthralgias from a lupus-like syndrome.

TABLE 3. PHARMACOLOGICAL PROPERTIES OF ANTI-RHEUMATIC BIOLOGICALS

Biological	Structure	Binding target	Half-life of elimination (approximate)
Adalimumab	Human – monoclonal antibody	TNF-α	10–14 days alone 15–19 days with methotrexate
Etanercept	Human TNF-α receptors covalently linked to Fc portion of human immunoglobulin G1	TNF-α, lymphotoxin	3–5 days
Infliximab	Chimeric antibody 25% mouse, 75% human	TNF-α	8–10 days
Anakinra	Non-glycosylated human protein	IL-1α, IL-1β	3 hours – longer in renal failure

Biological anti-rheumatic therapy

As outlined above in the section on the pathophysiology of rheumatoid arthritis, several CYTOKINES have major pro-inflammatory actions and are also involved with the systemic effects of rheumatoid arthritis. Two of these CYTOKINES, TNF-α and IL-1 (Tab. 3), are targeted by currently available antagonists. Three biologicals, infliximab, adalimumab and etanercept, bind TNF-α while anakinra binds to IL-1 receptors (Tab. 3). These anti-CYTOKINE therapies have revolutionized the treatment of rheumatoid arthritis and other inflammatory diseases. Indeed, their success is leading to an unprecedented phase of investigation of new drugs for the treatment of rheumatoid arthritis.

Chemistry

All four biological agents are proteins (Tab. 3). Infliximab is a chimeric ANTIBODY against TNF-α. It is a monoclonal ANTIBODY composed of a human ANTIBODY backbone with the region which binds TNF-α (IDIOTYPE) being derived from the mouse. Adalimumab is also an anti-TNF-α ANTIBODY but is a wholly "humanized" anti-TNF-α monoclonal ANTIBODY. Etanercept contains an immunoglobulin backbone and two TNF-α soluble receptors. It is thus a recombinant soluble TNF-α receptor fusion protein. Anakinra is very similar to IL-1ra, the naturally occurring antagonist at IL-1 receptors. The only differences are that anakinra is not glycosylated and it also contains a terminal methionine residue which is necessary for its biological production.

Pharmacokinetics

The biological agents, as proteins, require parenteral administration. Adalimumab, etanercept and anakinra are administered subcutaneously, but infliximab is administered by infusion, which makes its dosage less convenient (Appendix C.13). The three anti-TNF-α proteins have long half-lives. The half-lives of elimination of infliximab and adalimumab are one to two weeks [72–74] (Tab. 3). These long half-lives allow relatively infrequent dosage (Appendix C.13). The half-life of etanercept is somewhat shorter at about 3 to 5 days [75]. Consequently, once or twice a week dosage produces relatively constant plasma concentrations. Anakinra has the greatest CLEARANCE of all the present biological agents and has an apparent half-life of elimination of about 3 hours after the usual daily subcutaneous injection [76] (Tab. 3). The true half-life of elimination is shorter and the longer apparent half-life is due to prolonged absorption after subcutaneous injection. Anakinra is eliminated predominately by renal excretion and its dosage must therefore be reduced in patients with severe renal impairment.

Mode of action

The anti-TNF-α agents act by physically interacting with and thereby neutralising TNF-α. Infliximab and adalimumab bind to both soluble and cell-bound TNF-α, leading to a profile of immunological effects potentially different to that of etanercept, which binds only soluble TNF-α. This may in part explain the EFFICACY of infliximab in inflammatory bowel disease, in which etanercept is ineffective. Etanercept also binds the CYTOKINE, lymphotoxin, although the importance of this to its therapeutic effects is unclear. Anakinra binds to IL-1 receptors and thus prevents IL-1 interaction with its native receptor. As outlined above, TNF-α and IL-1 have multiple effects on cellular events operative in rheumatoid arthritis, and thus their neutralisation by these agents leads to profound inhibition of inflammatory activation in the joint.

Clinical indications and efficacy

The available anti-CYTOKINE biological agents are effective in the treatment of rheumatoid arthritis either when used alone or in combination with other DMARD, particularly methotrexate [77]. Because of its chimeric nature, infliximab is used in combination with methotrexate, which limits the development of neutralising anti-drug antibodies. These may be responsible for the tachyphylaxis observed in some patients treated with this agent. Although the other anti-CYTOKINES are not required to be administered with methotrexate, they often are used with methotrexate because the combination is more effective than anti-CYTOKINE treatment alone. The same consideration is applied to clinical trials where the biologicals are tested mostly in patients already taking older DMARD, particularly methotrexate.

Anti-cytokine therapies are markedly effective in rheumatoid arthritis although, in some studies, the benefit was modest. Overall, about two-thirds of patients respond with a clinically significant degree of disease control [78]. Compared to the other DMARD, the biologicals generally show a more rapid response, but this advantage may not be sustained. However, slowing of the radiographic progression of rheumatoid arthritis has been reported [79].

In addition to rheumatoid arthritis, infliximab and adalimumab have well-established roles in the treatment of inflammatory bowel disease, and an emerging role in the treatment of spondyloarthropathies and psoriasis. At this stage, anakinra has not been tested for inflammatory arthritic diseases other than rheumatoid arthritis.

Side-effects

The predominant toxicity profile of the biological anti-cytokine drugs relates to their profound effects on the IMMUNE SYSTEM and their major risk is infection (Appendix C.13). Reports of infections, presumably related to the immunosuppressive effects of these agents, especially relate to the reactivation of chronic infections such as tuberculosis. Screening for tuberculosis is therefore required before treatment [80]. In general, these agents should generally be avoided in patients with active or recurrent infections. In particular, concomitant therapy with corticosteroids may increase the susceptibility to infections. Combined treatment with anakinra and a TNF-α inhibitor may cause an increased risk of infection [81] and, at present, the use of this combination is contraindicated. Subcutaneous injection site reactions or intravenous infusion reactions are common but generally well tolerated. It may be thought that inhibitors of TNFa could lead to the development of tumours but, although tumours have occurred, it is difficult to determine if these were due to the anti-TNF-α activity. However, ongoing surveillance is required. A concern is the development of demyelinating conditions and, at this early stage, it is advised that the anti-TNF-α drugs should not be used in patients with multiple sclerosis [80].

Combination DMARD therapies

Despite maximization of dosage with individual DMARD, a large proportion of patients continue to have active rheumatoid arthritis. Combinations of DMARD are now used widely because they often show greater activity than single DMARD. COMBINATION THERAPY for the treatment of rheumatoid arthritis is advocated on two potential bases:

i) There may also be an additive or synergistic effect from combinations. Despite the lack of knowledge on mechanisms of DMARD action, it seems likely that these drugs work in different ways. Thus, modifying or inhibiting the disease processes at multiple sites may gain a greater degree of suppression of rheumatoid arthritis with less treatment resistance.

ii) Combinations of DMARD may allow a decreased dose of a particular agent leading to a concomitant decrease in its side-effects without diminished EFFICACY.

COMBINATION THERAPY can be delivered in at least three different styles. Multiple medications can be initiated together and combined dosage of DMARD maintained. Alternatively, one or more of the DMARD is gradually withdrawn. This is the "step down" approach. Alternatively the medications can be "stepped up", commencing with one agent and adding another DMARD if the desired outcome is not achieved.

A variety of combined treatments have been examined, particularly with methotrexate. Successful combinations which appear useful include methotrexate and hydroxychloroquine. This combination may produce less acute liver damage than methotrexate alone [82]. Triple therapy of methotrexate, sulfasalazine and hydroxychloroquine also appears useful, although it is difficult to make good comparisons of the various treatments [83], and, as outlined above, the biological preparations are often used with methotrexate.

An example of a step-up treatment is the addition of leflunomide to the treatment of patients treated with methotrexate. The response in patients taking the two drugs was better than in patients who continued methotrexate and placebo [84]. But is the outcome better than what you would expect by simply stopping methotrexate and commencing leflunomide? As there was no leflunomide-only arm, the answer is not attainable from that study.

Another step-up trial involved cyclosporin and methotrexate. Thus, the addition of cyclosporin (2.5–5 mg/kg/day) to the maximally tolerated dose of methotrexate increased the response to treatment in patients with active rheumatoid arthritis [85]. There was no difference in the reported toxicities. The basis for this improvement, however, may be more pharmacokinetic than an additive or synergistic response. Thus, cyclosporin (3 mg/kg/day) added to steady dose methotrexate (7.5–22.5 mg/wk) produced a 26% increase in mean peak plasma concentration of methotrexate and an 80% reduction in the metabolite 7-hydroxymethotrexate [86]. The metabolite is less efficacious than methotrexate in rat adjuvant arthritis, and 4- to 17-fold less cytotoxic in human cell culture. By altering the pharmacokinetic balance in favour of methotexate, the increased EFFICACY of the combination of methotrexate and cyclosporin may be explained.

Summary

The cause of rheumatoid arthritis is unknown but the synovial proliferation of the disease is accompanied by overproduction of CYTOKINES, such as TNF-α, IL-1, IL-6, IL-8 and macrophage inhibitory factor. Inflammatory and synovial cells, CYTOKINES and enzymes are important potential targets for the disease-modifying antirheumatic drugs (DMARDs). In contrast to previous recommendations, DMARDs are now prescribed in the early treatment of the disease with the goal of preventing damage to joints, but their use is still problematic. While DMARDs suppress the symptoms of the disease and most slow the joint destruction of rheumatoid arthritis in many patients, they do not stop the disease entirely and do not even produce a moderate response in all patients. Furthermore, their side-effects are considerable and, in rare instances, are life-threatening. However, DMARDs are more useful than the non-steroidal anti-inflammatory drugs (NSAIDs) which produce only mild to moderate suppression of the symptoms of rheumatoid arthritis and do not appear to alter the progress of the disease.

DMARDs are a very heterogenous group of drugs but can be divided into two major chemical groups: low-molecular-weight compounds, which are mainly administered orally, and polypeptides or biologicals which are administered systemically. DMARDs show EFFICACY over one to two years but their effects in the long-term is less clear. Low-molecular-weight DMARDs and their metabolites have been shown to affect multiple pathophysiological pathways, yet explanations for their full modes of action in their suppression of rheumatoid arthritis are still incomplete.

The most widely used low-molecular-weight DMARD is methotrexate, which slows the joint damage in many, but not all, patients with rheumatoid arthritis. It is used at weekly doses ranging from 7.5 up to 25 mg weekly, lower doses than those used in the treatment of tumours. Methotrexate is a folate analogue which is converted to polyglutamate derivatives in cells. In rheumatoid arthritis, a major mode of action of methotrexate is considered to be inhibition of 5-aminoimidazole-4-carboxamide ribonucleotide (AICAR) transformylase, an early step in the synthesis of purines. The subsequent accumulation of AICAR leads to inhibition of the deamination of adenosine and adenine ribnucleotide with the consequent extracellular accumulation of adenosine which has anti-inflammatory activity. Treatment with methotrexate also changes CYTOKINE balance by decreasing the production of pro-inflammatory CYTOKINES (TNF-α, IL-6) and enhancing anti-inflammatory CYTOKINES (IL-1 receptor antagonist, IL-1ra). Adverse events caused by methotrexate include diarrhoea, headache, rash, alopecia, hepatotoxicity, myelosuppression and the rare, but serious, pneumonitis. Oral ulcers, vomiting and fatigue occur frequently but are decreased by folic acid supplementation, for example, 5 mg on the morning following the weekly dose of methotrexate, without reduction in its anti-rheumatic activity.

Other widely used low-molecular-weight DMARDs are sulfasalazine, leflunomide and hydroxychloroquine. Unchanged sulfasalazine achieves substantial concentrations in plasma (peak approximately 10 μM) but it is also cleaved by colonic bacteria yielding its constituents, sulfapyridine which is absorbed and mesalazine (aminosalicylate) which is poorly absorbed. Sulfasalazine produces a range of adverse reactions but, with hydroxychloroquine, is the best-tolerated DMARD. Sulfasalazine reduces the radiological progression of rheumatoid arthritis in many patients. Cessation of treatment with sulfasalazine is still, however, common and is mostly due to inadequate therapeutic response.

Leflunomide is activated by its non-enzymatic conversion to a compound which has a very long half-life of elimination (about 15 to 18 days). Inhibition of the *de novo* synthesis of pyrimidines, via inhibition of dihydroorotate dehydrogenase, targets activated lymphocyte proliferation. Elevated hepatic transaminases have occurred commonly during treatment but, in many cases, are possibly caused by factors other than leflunomide. Leflunomide is absolutely contraindicated during pregnancy because of the foetal development of lymphomas in experimental animals. The active metabolite binds strongly to cholestyramine which rapidly removes the active metabolite by binding it within the gastrointestinal tract.

Hydroxychloroquine and chloroquine are both active in the treatment of rheumatoid arthritis but hydroxychloroquine is generally preferred because of its lesser ocular toxicity. Despite the low incidence of retinal toxicity during treatment with hydroxychloroquine, baseline and continuing ocular examination is required every 6 to 12 months. The greatest limitation of hydroxychloroquine is that it does not appear to retard the damage to joints and, for this reason, hydroxychloroquine is often used in mild rheumatoid arthritis or in combination with other DMARDs. The mode of action of hydoxychloroquine is unclear, but its accumulation in lysosomes and consequent inhibition of acidic sphingomyelinase which not only has enzymatic activity but also is a mediator in proinflammatory gene expression, may explain its anti-rheumatic effect.

Two types of gold complexes are used in the treatment of rheumatoid arthritis, although usage has declined rapidly in the last twenty years. Sodium aurothiomalate is the most widely used gold complex and has similar activity to methotrexate and sulfasalazine. Sodium aurothiomalate is a water-soluble polymeric complex which is administered by intramuscular injection once a week. Auranofin, an orally active gold complex, is also available but is less efficacious than the injectable complexes. The pharmacological effects of the gold complexes are probably mediated by their conversion to aurocyanide ($Au(CN)_2^-$). The cyanide is produced from the oxidation of thiocyanate (SCN^-) by myeloperoxidase, an enzyme in neutrophils and MONOCYTES. The effects of aurocyanide include inhibition of the oxidative burst of neutrophils and inhibition of proliferation of LYMPHOCYTES. The most common adverse effects of the gold complexes are blood dyscrasias, renal toxicity, and rash which can be very severe and lead to exfoliative dermatitis. Auranofin is a monomeric complex which is active orally. It has lesser systemic toxicity than aurothiomalate but its dosage is limited by diarrhoea.

Penicillamine is a thiol which decreases the clinical signs and abnormal laboratory clinical parameters, such as elevated erythrocyte sedimentation rate, but has not been proven to retard joint destruction. The thiol group is likely to be involved in the antirheumatic activity of penicillamine although the precise mode of action is unclear.

Several tetracycline antibiotics have also been used in the treatment of rheumatoid arthritis, but minocycline is the only tetracycline whose activity in rheumatoid arthritis is supported by double-blind clinical trials. However, minocycline is rarely used principally because of the risk of the development of resistant organisms. It also causes hyperpigmentation and lupus-like manifestations. It is contraindicated during pregnancy due to inhibition of foetal bone development

Azathioprine and cyclosporin are also effective but are generally reserved for the treatment of rheumatoid arthritis which is resistant to other treatments. Azathioprine is metabolized to 6-mercaptopurine which is further converted to thioguanine nucleotides which are taken up into DNA. Cyclosporin suppresses the immune response by binding to immunophilin receptors with consequent transcription inhibition of IL-2 and IL-4.

Several high-molecular-weight DMARDs (biologicals) are now available. Three, etanercept, adalimumab and infliximab, bind and inactivate TNF-α while anakinra blocks the INTERLEUKIN-1 receptor. The anti-TNF-α peptides have the longer half-lives of elimination and therefore require administration from twice weekly to two-monthly. Anakinra's half-life of about 3 hours (although longer in renal failure) requires daily administration. The recent introduction of these biologicals has provided a very substantial advance in the treatment of rheumatoid arthritis and other inflammatory arthritic diseases. They have also provided a critical proof of the concept that compounds which inhibit the actions of inflammatory CYTOKINES can be clinically useful in rheumatoid arthritis and other inflammatory rheumatic diseases. However, they are not effective in all patients and must be used carefully in patients with chronic infections. The need to inject the biologicals also limits their use. Present attention is being directed towards the development of antagonists of CYTOKINES, other than TNF-α and IL-1, and also to small-molecular-weight compounds which may block mechanisms and steps involved in the pathways of effects of CYTOKINES.

Corticosteroids are used widely in the treatment of rheumatoid arthritis. Their major actions are inhibition of the expression of a broad range of pro-inflammatory CYTOKINES and induced expression of anti-inflammatory genes, such as annexin-1 and MAP kinase 1. Because of their side-effects the dosage of the corticosteroids in rheumatoid arthritis should be kept as low as possible and for as short a time as possible.

Combined treatments with 2-3 DMARDs are used frequently, often in combination with methotrexate. There are two rationales to the combinations: better control of rheumatoid arthritis or reduced dosage of the DMARDs producing lesser side-effects. The biologicals and the corticosteroids are widely used in combinations with the low-molecular-weight DMARDs. One biological, infliximab, is recommended to be administered with methotrexate in order to prevent the development of neutralising antibodies. The other biologicals are often used with methotrexate because the combination is more effective than the biologicals alone.

Selected readings

RO Day, DE Furst, PL van Riel, B Bresnihan (eds) (2005) *Antirheumatic Therapy: Actions and Outcomes*, Birkhäuser, Basel

Olsen NJ, Stein CM (2004) Drug therapy: new drugs for rheumatoid arthritis *N Engl J Med* 350: 2167–2179

O'Dell JR (2004) Drug therapy: therapeutic strategies for rheumatoid arthritis *N Engl J Med* 350: 2591–2602

References

1 American College of Rheumatology Subcommittee on Rheumatoid Arthritis Guidelines (2002) Guidelines for the management of rheumatoid arthritis. *Arthritis Rheum* 46: 328–346

2 Nell VP, Machold KP, Eberl G, Stamm TA, Uffmann M, Smolen JS (2004) Benefit of very early referral and very early therapy with disease-modifying anti-rheumatic drugs in patients with early rheumatoid arthritis. *Rheumatology* 43: 906–914

3 Firestein GS, Zvaifler N (2002) How important are T cells in chronic rheumatoid synovitis?: II. T cell-independent mechanisms from beginning to end. *Arthritis Rheum* 46: 298–308

4 Smolen JS, Steiner, G. (2003) Therapeutic strategies for rheumatoid arthritis. *Nat Rev Drug Discov* 2: 473–488

5 Magee MH, Blum RA, Lates CD, Jusko WJ (2002) Pharmacokinetic/pharmacodynamic model for prednisolone inhibition of whole blood lymphocyte proliferation. *Br J Clin Pharmacol* 53: 474–484

6 Brady S, Day RO, Graham GG (1999) Are intra-articular steroids as effective. In: HA Bird and ML Snaith (eds): *Challenges in Rheumatoid Arthritis*. Blackwell Science, Oxford, 174–189

7 Barnes PJ (1998) Anti-inflammatory actions of glucocorticoids: molecular mechanisms. *Clin Sci* 94: 557–572

8 Smith MD, Ahern MJ, Roberts-Thompson PJ (1990) Pulse methylprednisolone therapy in rheumatoid arthritis: unproved therapy, unjustified therapy, or effective adjunctive treatment? *Ann Rheum Dis* 49: 265–267

9 Capell HA, Madhok R, Hunter JA, Porter D, Morrison E, Larkin J, Thomson EA, Hampson R, Poon FW (2004) Lack of radiological and clinical benefit over two years of low dose prednisolone for rheumatoid arthritis: results of a randomised controlled trial. *Ann Rheum Dis* 63: 797–803

10 Sambrook PN, Jones G (1995) Corticosteroid osteoporosis. *Br J Rheumatol* 34: 8–12

11 Gubner R, August S, Ginsberg V (1951) Therapeutic suppression of tissue reactivity. II. Effect of aminopterin in rheumatoid arthritis and psoriasis. *Am J Med Sci* 221: 176–182

12 Seideman P, Beck O, Eksborg S, Wennberg M (1993) The pharmacokinetics of methotrexate and its 7-hydroxy metabolite in patients with rheumatoid arthritis. *Br J Clin Pharmacol* 35: 409–412

13 Kremer JM, Petrillo GF, Hamilton RA (1995) Pharmacokinetics and renal function in patients with rheumatoid arthritis receiving a standard dose of oral weekly methotrexate: association with significant decreases in creatinine clearance and renal clearance of the drug after 6 months of therapy. *J Rheumatol* 22: 38–40

14 Cronstein BN (1996) Molecular therapeutics. Methotrexate and its mechanism of action. *Arthritis Rheum* 39: 1951–1960

15 Stewart CF, Fleming RA, Germain BF, Seleznick MJ, Evans WE (1991) Aspirin alters methotrexate disposition in rheumatoid arthritis patients. *Arthritis Rheum* 34: 1514–1520

16 Bannwarth B, Pehourcq F, Schaeverbeke T, Dehais J (1996) Clinical pharmacokinetics of low-dose pulse methotrexate in rheumatoid arthritis. *Clin Pharmacokinet* 30: 194–210

17 Cutolo M, Bisso A, Sulli A, Felli L, Briata M, Pizzorni C, Villaggio B (2000) Antiproliferative and antiinflammatory effects of methotrexate on cultured differentiating myeloid monocytic cells (THP-1) but not on synovial macrophages from patients with rheumatoid arthritis. *J Rheumatol* 27: 2551–2557

18 Cutolo M, Sulli A, Pizzorni C, Seriolo B, Straub RH (2001) Anti-inflammatory mechanisms of methotrexate in rheumatoid arthritis. *Ann Rheum Dis* 60: 729–735

19 Montesinos MC, Desai A, Delano D, Chen JF, Fink JS, Jacobson MA, Cronstein BN (2003) Adenosine A_{2A} or A_3 receptors are required for inhibition of inflammation by methotrexate and its analog MX-68. *Arthritis Rheum* 48: 240–247

20 Smolenska Z, Kaznowska Z, Zarowny D, Simmonds HA, Smolenski RT (1999) Effect of methotrexate on blood purine and pyrimidine levels in patients with rheumatoid arthritis. *Rheumatology* 38: 997–1002

21 Strand V, Cohen S, Schiff M, Weaver A, Fleischmann R, Cannon G, Fox R, Moreland L, Olsen N, Furst D et al (1999) Treatment of active rheumatoid arthritis with leflunomide compared with placebo and methotrexate. *Arch Intern Med* 159: 2542–2550

22 Cohen S, Cannon GW, Schiff M, Weaver A, Fox R, Olsen N, Furst D, Sharp J, Moreland L, Caldwell J et al (2001) Two-year, blinded, randomized, controlled trial of treatment of active rheumatoid arthritis with leflunomide

compared with methotrexate. *Arthritis Rheum* 44: 1984–1992

23 Emery P, Breedveld FC, Lemmel EM, Kaltwasser JP, Dawes PT, Gomor B, Van Den Bosch F, Nordstrom D, Bjorneboe O, Dahl R et al (2000) A comparison of the efficacy and safety of leflunomide and methotrexate for the treatment of rheumatoid arthritis. Rheumatology 39: 655–665

24 Dougados M, Combe B, Cantagrel A, Goupille P, Olive P, Schattenkirchner M, Meusser S, Paimela L, Rau R, Zeidler H et al (1999) Combination therapy in early rheumatoid arthritis: a randomised, controlled, double blind 52 week clinical trial of sulphasalazine and methotrexate compared with the single components. *Ann Rheum Dis* 58: 220–225

25 Bathon JM, Martin RW, Fleischmann RM, Tesser JR, Schiff MH, Keystone EC, Genovese MC, Wasko MC, Moreland LW, Weaver AL et al (2000) A comparison of etanercept and methotrexate in patients with early rheumatoid arthritis. *N Engl J Med* 343: 1586–1593

26 Guzman J, Maetzel A, Peloso P, Yeung M, Bombardier C (1996) Disability scores in DMARD trials : what is a clinically important change ? *Arthritis Rheum* 39: S208

27 Buchbinder R, Hall S, Sambrook PN, Champion GD, Harkness A, Lewis D, Littlejohn GO, Miller MH, Ryan PF (1993) Methotrexate therapy in rheumatoid arthritis: a life table review of 587 patients treated in community practice. *J Rheumatol* 20: 639–644

28 Hoekstra M, van de Laar MA, Bernelot Moens HJ, Kruijsen MW, Haagsma CJ (2003) Longterm observational study of methotrexate use in a Dutch cohort of 1022 patients with rheumatoid arthritis. *J Rheumatol* 30: 2325–2329

29 Choi HK, Seeger JD, Kuntz KM (2002) A cost effectiveness analysis of treatment options for methotrexate-naïve rheumatoid arthritis. *J Rheumatol* 29: 1156–1165

30 Whittle SL, Hughes RA (2004) Folate supplementation and methotrexate treatment in rheumatoid arthritis: a review. *Rheumatology* 43: 267–271

31 Abraham Z, Rozenbaum M, Rosner I (1999) Colchicine therapy for low-dose methotrexate-induced accelerated nodulosis in a rheumatoid arthritis patient. *J Dermatol* 26: 691–694

32 Alarcón GS, Kremer JM, Macaluso M, Weinblatt ME, Cannon GW, Palmer WR, St Clair EW, Sundy JS, Alexander RW, Smith GJ et al (1997) Risk factors for methotrexate-induced lung injury in patients with rheumatoid arthritis: A multicentre, case-control study. Methotrexate study group. *Ann Intern Med* 127: 356–364

33 Fathi NH, Mitros F, Hoffman J, Straniero N, Labreque D, Koehnke R, Furst DE (2002) Longitudinal measurement of methotrexate liver concentrations does not correlate with liver damage, clinical efficacy, or toxicity during a 3.5 year double blind study in rheumatoid arthritis. *J Rheumatol* 29: 2092–2098

34 McLachlan AJ, Tett SE, Cutler DJ, Day RO (1994) Bioavailability of hydroxychloroquine tablets in patients with rheumatoid arthritis. *Br J Clin Rheumatol* 33: 235–239

35 Wiegmann K, Schutze S, Machleidt T, Witte D, Kronke M (1994) Functional dichotomy of neutral and acidic sphingomyelinases in tumour necrosis factor signaling. *Cell* 78: 1005–1015

36 Sanders M (2000) A review of controlled clinical trials examining the effects of antimalarial compounds and gold compounds on radiographic progression in rheumatoid arthritis. *J Rheumatol* 27: 523–529

37 Tett SE, Day RO, Cutler DI (1993) Concentration-effect relationship of hydroxychloroquine in rheumatoid arthritis – a cross sectional study. *J Rheumatol* 20: 1874–1879

38 Weinshilboum R (2001) Thiopurine pharmacogenetics: clinical and molecular studies of thiopurine methylpurine transferase. *Drug Metab Disp* 29: 601–605

39 Seidman EG, Furst DE (2002) Pharmacogenetics for the individualization of treatment of rheumatological disorders using azathioprine. *J Rheumatol* 29: 2484–2487

40 Elion GB (1989) The purine path to chemotherapy. *Science* 244: 41–47

41 Ho S, Clipstone N, Timmermann L, Northrop J, Graef I, Fiorentino D, Nourse J, Crabtree GR (1996) The mechanism of action of cyclosporin A and FK506. *Clin Immunol Immunopathol* 80: S40–S45

42 Campana C, Regazzi MB, Buggia I, Molinaro M (1996) Clinically significant drug interactions with cyclosporin. An update *Clin Pharmacokinet* 30: 141–179

43 Tugwell P, Bombardier C, Gent M, Bennett KJ, Bensen WG, Carette S, Chalmers A, Esdaile JM, Klinkhoff AV, Kraag GR et al (1990) Low dose cyclosporin *versus* placebo in patients with rheumatoid arthritis. *Lancet* 335: 1051–1055

44 Pasero G, Priolo F, Marubini E, Fantini F, Ferraccioli G,

Magaro M, Marcolongo R, Oriente P, Pipitone V, Portioli I et al (1996) Slow progression of joint damage in early rheumatoid arthritis treated with cyclosporin A. *Arthritis Rheum* 39: 1006–1015

45 Gerber RC, Paulus HE, Bluestone R, Lederer M (1972) Kinetics of aurothiomalate in serum and synovial fluid. *Arthritis Rheum* 15: 625–629

46 Graham GG, Champion GD, Ziegler JB (1994) The cellular metabolism and effects of gold complexes. *Metal Based Drugs* 1: 395–404

47 Handel ML, Watts CKW, deFazio A, Day RO, Sutherland RL (1995) Inhibition of AP-1 binding and transcription by gold and selenium involving conserved cysteine residues in Jun and Fos. *Proc Natl Acad Sci USA* 92: 4497–4501

48 Yang JP, Merin JP, Nakano T, Kato T, Kitade Y, Okamoto T (1995) Inhibition of the DNA-binding activity of NF-KB by gold compounds *in vitro*. *FEBS Lett* 361: 89–96

49 Champion GD, Graham GG, Ziegler JB (1990) The gold complexes. In: P Brooks (ed): *Bailliere's Clinical Rheumatology, Slow Acting Anti-rheumatic Drugs and Immunosuppressives*. Bailliere, London, 4(3): 491–534

50 Buckland-Wright JC, Clarke GS, Chikanza IC, Grahame R (1993) Quantitative microfocal radiography detects changes in erosion area in patients with early rheumatoid arthritis treated with myocrisine. *J Rheumatol* 20: 243–247

51 Rozman B (2002) Clinical pharmacokinetics of leflunomide. *Clin Pharmacokinet* 41: 421–430

52 Davis JP, Cain GA, Pitts WJ, Magolda RL, Copeland RA (1996) The immunosuppressive metabolite of leflunomide is a potent inhibitor of human dihydroorotate dehydrogenase. *Biochemistry* 35: 1270–1273

53 Osiri M, Shea B, Robinson V, Suarez-Almazor M, Strand V, Tugwell P, Wells G (2003) Leflunomide for the treatment of rheumatoid arthritis: a systematic review and metaanalysis. *J Rheumatol* 30: 1182–1190

54 Kalden JR, Schattenkirchner M, Sorensen H, Emery P, Deighton C, Rozman B, Breedveld F (2003) The efficacy and safety of leflunomide in patients with active rheumatoid arthritis: a five-year followup study. *Arthritis Rheum* 48: 1513–1520

55 Weinblatt ME, Kremer JM, Coblyn JS, Maier AL, Helfgott SM, Morrell M, Byrne VM, Kaymakcian MV, Strand V (1999) Pharmacokinetics, safety, and efficacy of combination treatment with methotrexate and leflunomide in patients with active rheumatoid arthritis. *Arthritis Rheum* 42: 1322–1328

56 Remer CF, Weisman MH, Wallace DJ (2001) Benefits of leflunomide in systemic lupus erythematosus: a pilot observational study. *Lupus* 10: 480–483

57 Kaplan MJ (2001) Leflunomide Aventis Pharma. *Curr Opin Investig Drugs* 2: 222–230

58 Chakravarty EF, Sanchez-Yamamoto D, Bush TM (2003) The use of disease modifying antirheumatic drugs in women with rheumatoid arthritis of childbearing age: a survey of practice patterns and pregnancy outcomes. *J Rheumatol* 30: 241–246

59 Joyce DA, Day RO (1990) D-Penicillamine and D-penicillamine-protein disulphide in plasma and synovial fluid of patients with rheumatoid arthritis. *Br J Clin Pharmacol* 30: 511–517

60 Handel ML, Watts CK, Sivertsen S, Day RO, Sutherland RL (1996) D-Penicillamine causes free radical-dependent inactivation of activator protein-1 DNA binding. *Mol Pharmacol* 50: 501–505

61 Day RO, Paulus HE (1987) D-Penicillamine. In: HE Paulus, DE Furst, SH Dromgoole (eds): *Drugs for Rheumatic Disease*. Churchill Livingstone, New York, 85–112

62 Svartz N (1942) Salazopyrin, a new sulfanilamide preparation: a. therapeutic results in rheumatic polyarthritis; b. therapeutic results in ulcerative colitis; c. toxic manifestations in treatment with sulfanilamide preparations. Acta Med Scand 110: 577–98

63 Taggart AJ, McDermott BJ, Roberts SD (1992) The effect of age and acetylator phenotype on the pharmacokinetics of sulfasalazine in patients with rheumatoid arthritis. *Clin Pharmacokinet* 23: 311–320

64 Gadangi P, Longaker M, Naime D, Levin RI, Recht PA, Montesinos MC, Buckley MT, Carlin G, Cronstein BN (1996) The anti-inflammatory mechanism of sulfasalazine is related to adenosine release at inflamed sites. *J Immunol* 156: 1937–1941

65 Neumann VC, Taggart AJ, Le Gallez P, Astbury C, Hill J, Bird HA (1986) A study to determine the active moiety of sulphasalazine in rheumatoid arthritis. *J Rheumatol* 13: 285–287

66 Ferraz MB, Tugwell P, Goldsmith CH, Atra E (1990) Meta-analysis of sulfasalazine in ankylosing spondylitis. *J Rheumatol* 17: 1482–1486

67 Chapman CM, Zwillich SH (1994) Olsalazine in anky-

losing spondylitis: a pilot study. *J Rheumatol* 21: 1699–1701

68 van der Heijde DM, van Riel PL, Nuver-Zwart 1H, van de Putte LB (1990) Sulphasalazine *versus* hydroxychloroquine in rheumatoid arthritis: 3-year follow-up. *Lancet* 335: 539

69 O'Dell JR, Paulsen G, Haire CE, Blakely K, Palmer W, Wees S, Eckhoff PJ, Klassen LW, Churchill M, Doud D et al (1999) Treatment of early seropositive rheumatoid arthritis with minocycline: four-year followup of a double-blind, placebo-controlled trial. *Arthritis Rheum* 42: 1691–1695

70 Stone M, Fortin PR, Pacheco-Tena C, Inman RD (2003) Should tetracycline treatment be used more extensively for rheumatoid arthritis? Metaanalysis demonstrates clinical benefit with reduction in disease activity. *J Rheumatol* 30: 2112–2122

71 Capell HA, Madhok R (2003) Another DMARD option in rheumatoid arthritis? *J Rheumatol* 30: 2085–2087

72 den Broder A, van de Putte LB, Rau R, Schattenkirchner M, van Riel PL, Sander O, Binder C, Fenner H, Bankmann Y, Velagapudi R et al (2002) A single dose, placebo controlled study of the fully human anti-tumor necrosis factor-α antibody adalimumab (D2E7) in patients with rheumatoid arthritis. *J Rheumatol* 29: 2288–2298

73 Weisman MH, Moreland LW, Furst DE, Weinblatt ME, Keystone EC, Paulus HE, Teoh LS, Velagapudi RB, Noertersheuser PA, Granneman GR et al (2003) Efficacy, pharmacokinetic, and safety assessment of adalimumab, a fully human anti-tumor necrosis factor-alpha monoclonal antibody, in adults with rheumatoid arthritis receiving concomitant methotrexate: a pilot study. *Clin Ther* 25: 1700–1721

74 Kavanaugh A, St Clair EW, McCune WJ, Braakman T, Lipsky PE (2000) Chimeric anti-tumour necrosis factor-alpha monoclonal antibody treatment of patients with rheumatoid arthritis receiving methotrexate therapy. *J Rheumatol* 27: 841–850

75 Korth-Bradley JM, Rubin AS, Hanna RK, Simcoe DK, Lebsack ME (2000) The pharmacokinetics of etanercept in healthy volunteers. *Ann Pharmacother* 34: 161–164

76 Yang BB, Baughman S, Sullivan JT (2003) Pharmacokinetics of anakinra in subjects with different levels of renal function. *Clin Pharmacol Ther* 74: 85–94

77 Hochberg MC, Tracy JK, Hawkins-Holt M, Flores RH (2003) Comparison of the efficacy of the tumor necrosis factor alpha blocking agents adalimumab, etanercept, and infliximab when added to methotrexate in patients with active rheumatoid arthritis. *Ann Rheum Dis* 62 (Suppl 2): 13–16

78 Feldmann M, Maini RN (2001) Anti-TNFa therapy of rheumatoid arthritis: what have we learned? *Ann Rev Immunol* 19: 163–196

79 Pincus T, Ferraccioli G, Sokka T, Larsen A, Rau R, Kushner I, Wolfe F (2002) Evidence from clinical trials and long-term observational studies that disease-modifying anti-rheumatic drugs slow radiographic progression in rheumatoid arthritiis: updating a 1983 review. *Rheumatology* 41: 1346–1356

80 Mikuls TR, Moreland LW (2003) Benefit-risk assessment of infliximab in the treatment of rheumatoid arthritis. *Drug Saf* 26: 23–32

81 Genovese MC, Cohen S, Moreland L, Lium D, Robbins S, Newmark R, Bekker P (2004) Combination therapy with etanercept and anakinra in the treatment of patients with rheumatoid arthritis who have been treated unsuccessfully with methotrexate. *Arthritis Rheum* 50: 1412–1419

82 Fries JF, Singh G, Lenert L, Furst DE (1990) Aspririn, hydroxychloroquine, and hepatic enzyme abnormalities with methotrexate in rheumatoid arthritis. *Arthritis Rheum* 33:1611–1619

83 O'Dell JR, Leff R, Paulsen G, Haire C, Mallek J, Eckhoff PJ, Fernandez A, Blakely K, Wees S, Stoner J et al (2002) Treatment of rheumatoid arthritis with methotrexate and hydroxychloroquine, methotrexate and sulfasalazine, or a combination of the three medications: results of a two-year, randomized, double-blind, placebo-controlled trial. *Arthritis Rheum* 46: 1164–1170

84 Kremer JM, Genovese MC, Cannon GW, Caldwell JR, Cush JJ, Furst DE, Luggen ME, Keystone E, Weisman MH, Bensen WM (2002) Concomitant leflunomide therapy in patients with active rheumatoid arthritis despite stable doses of methotrexate: A randomized double-blind, placebo-controlled trial. *Ann Intern Med* 137: 726–733

85 Tugwell P, Pincus T, Yocum D, Stein M, Gluck O, Kraag G, McKendry R, Tesser J. Baker P, Wells G et al (19995) Combination therapy with cyclosporine and methotrexate in severe rheumatoid arthritis. *N Engl J Med* 333: 137–141

86 Fox RI, Morgan SL, Smith HT, Robbins BA, Choc MG,

Baggott JE (2003) Combined oral cyclosporin and methotrexate therapy in patients with rheumatoid arthritis elevates methotrexate levels and reduces 7-hydroxymethotrexate levels when compared with methotrexate alone. *Rheumatology* 42: 989–994

Perspectives of immunotherapy in the management of asthma and other allergic conditions

A.J. Frew

Introduction

In recent years, we have witnessed a gradual process of optimising the use of standard anti-asthma therapy, through the introduction of treatment guidelines, and the dissemination of specialist care throughout the medical community. It is now generally accepted that, at the population level, we have reached a plateau in terms of the level of control that can be achieved with standard drugs. At the individual patient level, there is still some scope to improve compliance and control, but it seems unlikely that we will ever be able to abolish all asthma symptoms simply by adjusting the doses and delivery devices used for standard drugs. Moreover, it is clear that our current therapies do not get rid of the disease, but they merely control the symptoms. Furthermore, while it is often asserted that long-term use of inhaled steroids may diminish the risk of complications of asthma, such as the development of fixed airways obstruction, we have only limited evidence for this. There is thus scope for us to use new therapies to control and suppress asthma.

We now know a great deal about the inflammatory and immunological basis of asthma. Over the past 20 years, many cells and molecules have been identified that can be implicated in one or more aspects of the pathogenesis of asthma [1]. Having identified these various "key players" and "key mediators", we have had to wait patiently for the development of effective inhibitors that can be used safely in humans, and hence allow us to assess whether these cells and mediators are really important in the process of asthma, or simply bystanders that just look suspicious or are merely accessories to the crime, rather than key players.

General aspects of immunotherapy in allergy and asthma

In the context of the immunopharmacology of asthma, the term IMMUNOTHERAPY generally refers to specific IMMUNOTHERAPY targeted against individual ALLERGENS, although increasingly this term is used to describe a range of immunological therapies that may influence allergic disease. Allergists and immunopharmacologists should not forget that the term IMMUNOTHERAPY is also used by oncologists and rheumatologists to describe a different range of IMMUNOMODULATORY treatments. For consistency it may be more appropriate to define IMMUNOTHERAPY as any form of treatment that seeks to modify the immune response in order to improve or alleviate immunological diseases. Within this general framework, one can identify non-specific treatments that are used to boost or suppress general immune functions, or alternatively, specific immunotherapies that aim to modify the immune response to specific external or internal ANTIGENS, without having any effect on the ability of the immune response to respond to unrelated ANTIGENS.

IMMUNOTHERAPY regimes for allergic diseases were mostly developed without a clear understanding of the fundamental biology. Consequently much effort has been expended in trying to understand the mechanisms by which immunostimulatory or immunosuppressive regimes may be effective. In some cases this exploration has led to improved regimes or novel approaches to IMMUNOTHERAPY.

Fundamental to any appraisal of IMMUNOTHERAPY is a coherent understanding of the immunological basis of the particular disease being studied. Thus, in attempting to manipulate the IMMUNE SYSTEM in allergic diseases such as asthma, it is crucial to establish which components of the disease have an immunological basis (as opposed to an inflammatory basis) and which of these immunological processes are necessary for the disease, but largely irreversible. For example, it is clear that an exaggerated tendency to develop IgE sensitisation is a substantial risk factor for the onset of asthma, but it is less clear whether allergic responses play a significant role in the maintenance of established asthma. It is thus entirely possible that a particular immunotherapeutic approach might be effective in preventing the onset of disease, but have no impact on the established condition. Conversely, in those patients where allergic reactions are relevant to the ongoing disease, specific or non-specific IMMUNOTHERAPY may be a useful way of reducing the effect of allergic trigger factors and may hence alleviate clinical symptoms.

A central thesis of IMMUNOTHERAPY is the plasticity of the immune response in man. For most of the past 15 years our thinking has been dominated by the Th1/Th2 concept. This is based on a view that T-helper cells regulate the immune and inflammatory response, through the production of IMMUNOMODULATORY CYTOKINES, and that once the T cell has differentiated into a Th1 or Th2 phenotype, its CYTOKINE pattern is stable. Most of the data showing that T CELLS acquire and retain CYTOKINE profiles were generated *in vitro* using mouse cells; similar general patterns are evident in human cells, but there are clear inter-species differences. Our thinking is now moving away from a simple Th1-Th2 spectrum towards a concept of contextual differentiation, in which the ANTIGEN-presenting cell plays a major role in determining the initial CYTOKINE response of the T CELL. We now believe that T CELLS are predisposed towards a particular phenotype, but whether they develop and proliferate appears to be regulated by specific T-REGULATORY CELLS, which use CYTOKINES such as IL-10 and TGF-β to control T-cell activity and function. Nonetheless, it is unclear at which point in this process the T CELL becomes irreversibly committed to a particular phenotype. This means we are unsure whether we can drive the response from Th2 to Th1 *in vivo* changing the phenotype of committed Th2 cells, or whether this shift requires the induction of previously uncommitted cells, which then acquire a Th1 or T regulatory phenotype.

Non-specific immunomodulatory therapies

Previous attempts to modulate the IMMUNE SYSTEM in asthma have drawn on a range of therapies that have been used for general IMMUNOSUPPRESSION in other disease areas. At one level, glucocorticosteroids (GCS) are IMMUNOMODULATORY: there is no doubt that they are effective agents in the control of asthma in most patients, and they have complex effects on airways biology as well as on the inflammatory processes of asthma. Other general immunosuppressive agents studied in asthma include cyclosporin A, methotrexate, azathioprine, gold, troleandomycin, hydroxychloroquine, and intravenous immunoglobulin. In general, these agents have been used mainly for their steroid-sparing properties, rather than for any particular aspect of their pharmacology. A more comprehensive review of the use of these agents in asthma lies elsewhere [2].

Specific mediator antagonists

A variety of newer treatments have been designed to target specific components of the inflammatory process found in asthmatic airways. These targets include IgE antibodies, specific CYTOKINES, CHEMOKINES and vascular ADHESION MOLECULES.

Anti-IgE

It is well established that IgE antibodies trigger release of inflammatory mediators from mast cells, via attachment to high-affinity IgE receptors and subsequent cross-linking. IgE also plays a role in ANTIGEN presentation and in the development of T and B LYMPHOCYTES.

The serum concentration of free IgE is significantly reduced by treatment with humanised monoclonal anti-IgE antibodies, as is the response to inhaled ANTIGEN [3, 4]. Initial clinical trials suggest that anti-IgE treatment may allow significant reduction of GCS doses [5]. It has also been suggested that anti-IgE may affect the way that ANTIGENS are captured and presented to LYMPHOCYTES, by disrupting the IgE-focussed presentation of ANTIGENS by DENDRITIC CELLS [6]. It remains to be proven whether this is a biologically important effect of anti-IgE. Treatment with anti-IgE (omalizumab) improves symptom scores and lung function in patients with moderate to severe asthma [7, 8], and reduces serum concentrations of IL-5 and IL-8, compared to pre-treatment values [9]. Anti-IgE therapy also reduces the expression of the high-affinity IgE receptor on mast cells in the human asthmatic airway [10]. Contrary to initial expectations, anti-IgE seems to work best when used as an add-on therapy in patients receiving higher doses of inhaled steroids, those who require frequent emergency treatment, and those with poorer lung function [11]. This can be interpreted as meaning that ALLERGY plays a more important role in this group than was previously believed, or else that anti-IgE has wider, non-specific effects which go well beyond chelation of free IgE molecules.

Anti-cytokine therapy

Several CYTOKINES have been implicated in the development and expression of allergic asthma. In particular, IL-4 plays a critical role in regulating B-cell switching to make IgE, but also modulates mucus production and endothelial ADHESION MOLECULE expression [12]. IL-5 has a key role [13] in the development, recruitment and activation of EOSINOPHILS and some have suggested it may be more relevant to asthma, whereas IL-4 may be the CYTOKINE relevant to ALLERGY. IL-13 can substitute for IL-4 in B CELL switching, but has also been implicated in some aspects of airways remodelling [14]. TUMOR NECROSIS FACTOR has also been implicated in airways remodelling [15–17].

Anti-IL-5 antibodies are very effective at blocking the development of eosinophilia and allergen-induced hyperresponsiveness in animal models [18–20]. In a clinical trial of monoclonal anti-IL-5 ANTIBODY, the sputum eosinophil response to allergen inhalation was reduced, but allergen-induced early and LATE-PHASE responses were not affected [21] This study was not really powered to address the clinical EFFICACY of anti-IL-5, but it has been widely cited as evidence that targeting IL-5 will not be an effective way of treating asthma. Subsequent work has shown that anti-IL-5 is more effective at reducing sputum and blood eosinophilia, than reducing tissue eosinophil numbers [22, 23]. An alternative anti-IL-5 ANTIBODY has also undergone initial trials in patients with severe asthma [24] but does not appear to have been followed through into more extensive clinical trials.

INTERLEUKIN-4 (IL-4) is regarded as a key CYTOKINE in asthma, due to its role in IgE class switching and its many other actions relevant to ALLERGY and asthma [25]. Inhalation of IL-4 causes increased sputum eosinophilia and increased bronchial responsiveness in patients with asthma [26]. Inhibition of IL-4 with nebulised synthetic soluble IL-4 receptor (sIL-4R) molecules blocked eosinophil recruitment and mucus production in animal models of asthma. An initial clinical trial was promising [27] and larger clinical studies have been performed, but are not yet published. Further use of anti-IL-5 appears to be stalled for commercial reasons, although the basic concept remains attractive. The same approach has been proposed to target IL-13, which may prove more effective, given the wider range of activity shown by IL-13, and a soluble IL-13 receptor (sIL-13R) is now in development for use in man.

TNF-α has been implicated in fibrosis and other aspects of airways remodelling. TNF-α may therefore be important in the chronicity of asthma, as distinct from the initial phases of allergic sensitisation, where IL-4 and IgE are more involved. This phase of the asthma process is relatively resistant to conventional therapies, such as inhaled steroids. An open study of a TNF-receptor antagonist has shown promise [28], and larger, double-blind, placebo-controlled studies are now in progress.

Cytokine therapy

Several CYTOKINES have "anti-allergic" properties, either by promoting the Th1 pattern of response, or by opposing the Th2 development. In particular IFN-γ opposes the actions of IL-4 while IL-12 works upstream, influencing ANTIGEN-PRESENTING CELLS so that they are biased towards the Th1 phenotype [29].

However, clinical trials of this approach have thus far been disappointing. Parenteral administration of IFN-γ to steroid-dependent asthmatic patients caused a reduction in blood eosinophil numbers, but had no effect on lung function or oral steroid requirements [30]. The pro-inflammatory CYTOKINE INTERLEUKIN-12 (IL-12) helps to regulate the balance between Th1 and Th2 LYMPHOCYTES and biases the IMMUNE SYSTEM away from a Th2 response. Patients with allergic asthma have reduced blood concentrations of IL-12 and their alveolar MACROPHAGES are deficient in IL-12 production [31, 32]. In animal models, exogenous IL-12 inhibits bronchial hyperresponsiveness and airway eosinophilia [33]. In mild asthma, administration of IL-12 resulted in a reduction in blood eosinophil count and in sputum eosinophil count following allergen challenge. However, airway hyperresponsiveness and the late asthmatic reaction were not affected. Further development seems unlikely as IL-12 administration was associated with significant side-effects in over 20% of patients [34]. Concern about the potential pro-inflammatory properties of IL-12 and IFN-γ may limit their potential use as IMMUNOMODULATORS in asthma. Moreover, IL-12 had little marginal benefit when used as a supplement to sublingual IMMUNOTHERAPY for house dust mite ALLERGY in children [35].

Specific immunotherapy

Specific allergen IMMUNOTHERAPY (SIT) involves the administration of allergen extracts to modify or abolish symptoms associated with atopic ALLERGY. The treatment is targeted at those ALLERGENS recognised by the patient and physician as responsible for symptoms. A decision to use SIT therefore demands a careful assessment of the patient's condition and the role of allergic triggers. IMMUNOTHERAPY was first developed in the late 19th century and many of the basic principles described in the initial publications remain valid today [36]. Usually, patients receive a course of injections, starting with a very low dose of allergen, and building up gradually until a plateau or maintenance dose is achieved. Maintenance injections are then given at 4–6 weekly intervals for 3 to 5 years. The updosing phase is normally given as a series of weekly injections, but several alternative induction regimes have been tried, some giving the whole series of incremental injections in a single day (rush protocol) while others give the doses in clusters of several doses on one day, then waiting a week before giving the next cluster of injections (semi-rush protocol). The main drawback to these rapid updosing regimes is the increased risk of adverse reactions compared to conventional protocols. On the other hand, full protection can be attained in a few days as compared to the three months required in the conventional regime. This may be particularly useful in patients being treated for life-threatening conditions such as anaphylaxis induced by bee and wasp stings.

Mechanisms of immunotherapy

Several mechanisms have been proposed to explain the beneficial effects of IMMUNOTHERAPY. These include induction of IgG (blocking) antibodies, long-term reductions in specific IgE, reduced recruitment of effector cells, altered T-cell CYTOKINE balance, induction of T-cell anergy and the induction of regulatory T CELLS. Until quite recently, SIT was thought to work by an effect on allergen-specific antibodies. Allergen-specific IgE levels rise temporarily during the initial phase of SIT, but fall back to pre-treatment levels during maintenance therapy [37]. The immediate skin test response can be reduced after SIT but this effect is relatively small compared to the degree of clinical benefit. In contrast, the LATE-PHASE skin test response is virtually abolished after successful SIT. Similar patterns are observed for LATE-PHASE nasal and airway responses [38]. SIT also induces allergen-specific IgG antibodies, which have been called "blocking antibodies". Current opinion is against this concept, partly because most mast cells are on the mucosal surfaces, and therefore meet the allergen before antibodies could interpose themselves, and partly because the rise in IgG follows rather than precedes clinical benefit. Moreover, there is a poor correlation between the amount of allergen-specific IgG and clinical protection. In reality, the main determinant of the allergen-specific IgG concentration is the total dose of allergen that has been given.

Allergen-specific T-cell responses are also affected by SIT. Both in the skin and in the nose, successful SIT is accompanied by a reduction in T-cell and eosinophil recruitment in response to allergen challenge. In parallel, there is a shift in the balance of Th1 and Th2 CYTOKINE expression in the allergen-challenged site. Th2 CYTOKINE expression is not affected but there is an increased proportion of T-cells expressing the Th1 CYTOKINES IL-2, IFN-γ and IL-12 [39–41]. SIT with inhalant ALLERGENS induces regulatory T CELLS, producing IL-10, as well as a shift in Th1:Th2 balance [42, 43]. Among other actions, IL-10 stimulates production of the IgG4 subclass, which may therefore be a marker of the immune response, rather that the precise mode of action conferring clinical benefit [44]. Taken together, these findings support the idea that SIT has a modulatory effect on allergen-specific T-cells, and this may be why the clinical and LATE-PHASE responses to allergen challenge are attenuated without there being a substantial change in allergen-specific IgE ANTIBODY levels.

SIT for asthma

IMMUNOTHERAPY has been widely used to treat allergic asthma, although the introduction of more effective inhaled therapies and the increased perception of side-effects have led to a gradual decline in the number of patients receiving specific IMMUNOTHERAPY for this indication. In some countries, most notably the UK, concern about adverse reactions has led to restrictions on the use of SIT for asthma, although this remains a valid indication for SIT in North America and continental Europe [45, 46].

The efficacy of SIT in adult asthma has been assessed in many trials over the last 50 years. Some of the earlier studies are difficult to interpret, because poor-quality allergen extracts were used or the studies were poorly designed. However, a meta-analysis of all trials published between 1954 and 1998 found clear evidence for the beneficial effects of SIT in asthma [46]. Symptom scores showed a small but definite improvement in those groups treated with mite SIT or pollen SIT, compared to the placebo-treated groups. There were parallel reductions in asthma medication usage and airways responses to inhaled allergen but no improvement in lung function measurements. Clinical trials have confirmed the EFFICACY of SIT in patients with grass pollen asthma, and in those with asthma caused by cat ALLERGY [47]. The benefits are greater for specific responses to allergen inhalation, than for NON-SPECIFIC AIRWAYS REACTIVITY. Arguments remain about the cost-effectiveness of SIT in asthma. While SIT does lead to improved peak flow rates and a reduction in anti-asthma medication, the overall medication cost (including SIT) goes up [48]. Assessment of cost: benefit therefore hinges on the perceived value of the clinical improvement.

Comparison of SIT with other types of treatment for asthma

The majority of clinical trials of SIT for asthma have compared SIT either with historical controls or with a matched placebo-treated group. To date, the effectiveness of specific SIT in asthma has rarely been compared with conventional management (with avoidance measures and conventional inhaled or oral drugs). One recent study assessed SIT in asthmatic children receiving conventional drug therapy and found no additional benefit in patients who were already receiving optimal drug therapy [49]. There are some significant criticisms of this study and further work of this type is urgently needed. It is also important that trials include analysis of cost-benefit and cost-effectiveness since purchasers of health care are increasingly demanding this evidence before agreeing to fund therapies.

Effects on natural history of allergic disease

It has been suggested that SIT may modify the natural history of asthma in children who are known to be atopic but have not yet developed asthma. Somewhere between 5 and 10% of atopic children and young adults may develop asthma symptoms each year, although these data are old and need updating [50]. Work from the 1950s and 1960s using mixed allergen extracts suggested that SIT may increase the rate of remission for children with asthma, and may also reduce the severity of symptoms in those who remain symptomatic [51]. Further evidence that SIT can modify the natural history of allergic disease has emerged in studies showing that the SIT reduces the probability of developing new sensitivities (to ALLERGENS other than the one used for therapy) [52]. An ongoing major multicentre study is assessing whether SIT is able to prevent allergic children aged 7–13 years from going on to develop asthma. After three years of therapy, 28% fewer children had asthma symptoms compared to the control group, and this difference has been maintained out to 5 years, suggesting that SIT really does affect the outcome of allergic sensitisation [53]. It remains to be seen whether SIT postpones the onset of asthma, or actually prevents it altogether; time will tell, but we will not know this for several years. If this effect is convincingly proven, then there will be a strong case to implement SIT in young patients with rhinitis who might not warrant SIT simply to control their nasal and eye symptoms.

Thus SIT is a valid but controversial treatment for asthma. While it seems entirely logical to try to treat allergic disorders by specifically suppressing the immune response to the triggering agents, it remains unclear whether a specific ALLERGY is really important in all asthmatic patients who are sensitised, as judged by skin tests. Moreover, we are unsure whether SIT in its present form is the best option for managing patients with allergic asthma. We await proper comparative studies of best current SIT, *versus* best current drug therapy, with robust endpoints including symptoms, objective measures of lung function, evaluation of cost:benefit ratios, safety, and quality of life. *In vitro* and *in vivo* measures such as skin test responses or allergen-specific IgG4 measurements are not sufficiently specific or sensitive to serve as SURROGATES for clinical EFFICACY. To date there have been relatively few well-controlled studies of SIT in asthma but there is increasing evidence that SIT is beneficial in mite-induced and pollen-induced asthma. The clinical EFFICACY of SIT in adult asthmatic patients sensitive to cats or moulds is less certain, and no comparative studies with conventional treatment have been performed. Further clinical trials are indicated, particularly in mild to moderate childhood asthma and also in patients with atopic disease who have not yet developed asthma but are at high risk of progression to asthma.

Future directions

There is still room to improve conventional SIT, through better allergen standardisation, and better patient selection. The development of recombinant ALLERGENS may allow better standardisation of allergen vaccines, as well as some fine tuning of vaccines for patients with unusual patterns of reactivity. Most allergic patients react to the same components of an allergen extract, the so-called major ALLERGENS, which are defined as those ALLERGENS recognised by over 50% of sera from a pool of patients with clinically significant ALLERGY to the material in question. However, not all patients recognise all major ALLERGENS and some patients only recognise ALLERGENS which are not recognised by the majority of allergic patient sera. This latter group may not respond to standard extracts, but might be better treated by a combination of ALLERGENS to which they are sensitive. Until the advent of molecular cloning, this has been impossible to achieve. The availability of recombinant ALLERGENS for SIT should thus lead to better characterisation of the range of sensitivities, and ultimately to better vaccine products. One area of uncertainty relates to the adjuvant effects of the non-allergen components of current allergen extracts. Most natural allergen extracts contain a variety of polysaccharide and lipid components that may be immunologically active, and once we have appropriate recombinant vaccines, it will be interesting to see whether these are as effective as natural extracts, or whether these extraneous elements are in fact contributing to the clinical EFFICACY of conventional extracts.

Recombinant molecular technology has also made it possible to develop novel forms of allergenic molecules. For example, one group has developed a recombinant trimer consisting of three covalently linked copies of the major birch pollen allergen, Bet v 1. This trimer exhibited profoundly reduced allergenic activity even though it contained the same B-CELL and T-CELL EPITOPES as the native molecule, and was able to induce Th1 CYTOKINE release. Interestingly, the rBet v 1 trimer induced IgG antibodies, analogous to the ANTIBODY response to standard SIT [54, 55].

Since the EPITOPES recognised by IgE molecules are usually three-dimensional whereas the EPITOPES recognised by T CELLS are short linear peptide fragments of the ANTIGEN, it should be possible to use peptide fragments of ALLERGENS to modulate T CELLS without risking anaphylaxis. Two distinct approaches have been tested. Either large doses of natural sequence peptides are given, deceiving the T-cell into high dose TOLERANCE [56], or else an altered peptide ligand can be given. Both approaches require consideration of the MHC type of the individual undergoing treatment. By sequential alteration of Der p peptides, it is possible to suppress proliferation of T-CELL clones recognising native Der p peptides, as well as suppressing their expression of CD40 ligand and their production of IL-4, IL-5 and IFN-γ. These anergic T CELLS do not provide help for B CELLS to switching class to IgE, and importantly this anergy cannot be reversed by providing exogenous IL-4 [57].

In an animal model, intranasal application of genetically produced hypoallergenic fragments of Bet v 1 produced mucosal TOLERANCE with significant reduction of IgE and IgG1 ANTIBODY responses, as well as reduced CYTOKINE production *in vitro* (IL-5, IFN-γ, IL-10). These reduced immunological responses were accompanied by inhibition of the cutaneous and airway responses that were seen with the complete Bet v 1 allergen. The mechanisms of IMMUNOSUPPRESSION seemed to be different for the allergen fragments and the whole molecule, in that TOLERANCE induced with the whole Bet v1 molecule was transferable with spleen cells whereas that induced by the fragments was not [58].

DNA vaccines may also prove useful. Two distinct approaches are in development. The first of these is a general approach, using CpG oligodeoxynucleotides (ODN). CpG ODN mimic bacterial DNA, and stimulate Th1-type CYTOKINE responses. In a mouse model of asthma, pre-administration of CpG ODN prevented both airways eosinophilia and bronchial hyperresponsiveness [59]. Moreover, these effects were sustained for at least six weeks after CpG ODN administration [60]. An alternative approach is to couple CpG ODN to the allergenic protein, which enhances immunogenicity in terms of eliciting a Th1-type response to the allergen, but reduces its allergenicity [61] and stimulates Th1 CYTOKINE expression in cultured human peripheral blood mononuclear cells [62]. Initial clinical trials have confirmed that the hybrid vaccine elicits a Th1-pattern response [63], and clinical trials in seasonal rhinitis have shown reductions in the nasal symptoms and the nasal inflammatory response, with a reduction in eosinophilia and Th2-CYTOKINE production, but an increase in Th1 CYTOKINES [64]. This benefit was sustained for at least one year after the course of therapy. A contrasting approach is to use allergen-specific naked DNA sequences as vaccines. This technology is still in its infancy, but preliminary data suggest that giving naked DNA leads to production of ALLERGENS from within the airways epithelial cells [65, 66]. Due to the different handling pathways for endogenous and exogenous ALLERGENS, it seems that the endogenously produced allergen elicits a Th1-type response and if this can be reproduced in allergic humans, it is hoped that this may overcome the existing Th2-pattern response and eliminate the ALLERGY. Initial animal studies have confirmed the concept and show that the DNA vaccine prevents sensitisation upon subsequent immunisation with the allergen [67]. However, the potential for generating a powerful Th1-type response to ubiquitous agents means that this approach will need further, careful evaluation in animal models before it can be pursued in man.

Conclusions

SIT has been in use for over a century, and is clinically effective in patients with rhinitis or asthma whose symptoms are clearly driven by allergic triggers. It is perhaps surprising that we are only now beginning to understand how SIT works, but the general view is that the vaccination protocol induces regulatory T CELLS that damp the response to allergen exposure in sensitised subjects. When used in appropriately selected patients, SIT is effective and safe, but care is needed to recognise and treat adverse reactions. As well as careful patient selection, appropriate training of allergists and SIT clinic support staff is essential. Future directions in SIT will include the development of better standardised vaccines, and the use of recombinant ALLERGENS, both of which should improve the safety profile of SIT. Some of these approaches may require individual tailoring of vaccines, which could present cost barriers. The emergence of more general IMMUNOMODULATORY therapies may prove more cost-effective, and would be particularly advantageous for patients sensitised to multiple ALLERGENS.

Summary

Current anti-asthma drugs offer a good level of control for most patients, but they do not cure the condition. Specific IMMUNOTHERAPY, targeted against individual ALLERGENS, can modify the immune response and improve the symptoms of rhinitis and asthma, but carries some definite risks, especially in those whose asthma is not well controlled. SIT appears to work by induction of T-REGULATORY CELLS, which downregulate the expression of the allergic response and the production of IgE. Other non-specific IMMUNOMODULATORY approaches have been used, mainly for their steroid-sparing properties. A number of novel therapies have been tested in recent years, including anti-IgE, and a variety of CYTOKINE and chemokine antagonists. Future developments are likely to involve a shift of focus away from CYTOKINES regulating the allergic pattern of response towards agents that prevent or reverse structural changes to the airway (airways remodelling).

Although we do not fully understand how SIT works, a number of attempts have been made to modulate the allergic response without the risks of side-effects seen with conventional SIT. These include using recombinant major ALLERGENS, peptide fragments, hybrid allergenic molecules and DNA vaccines (DNA sequences encoding ALLERGENS).

While SIT is effective and relatively safe, care is needed to recognise and treat adverse reactions, and there is scope for further improvement. Patient-specific vaccines care theoretically possible, but may prove too expensive for routine use. More general IMMUNOMODULATORY therapies may prove more cost-effective, particularly for patients sensitised to multiple ALLERGENS.

Selected readings

Jain VV, Kitagaki K, Kline JN (2003) CpG DNA and immunotherapy of allergic airway diseases. *Clin Exp Allergy* 33: 1330–1335

Norman PS (2004) Immunotherapy 1999–2004. *J Allergy Clin Immunol* 113: 1013–1023

Till SJ, Francis JN, Nouri-Aria K, Durham SR (2004) Mechanisms of immunotherapy. *J Allergy Clin Immunol* 113: 1024–1034

Wilson DR, Lima MT, Durham SR (2005) Sublingual immunotherapy for allergic rhinitis: systematic review and meta-analysis. *Allergy* 60: 4–12

References

1 Romagnani S (2002) Cytokines and chemoattractants in allergic inflammation. *Mol Immunol* 38: 881–885

2 Frew AJ, Plummeridge MJ (2001) Alternative agents for asthma. *J Allergy Clin Immunol* 108: 1–10

3 Boulet LP, Chapman KR, Cote J, Kalra S, Bhagat R, Swystun VA, Laviolette M, Cleland LD, Deschesnes F, Su JQ et al (1997) Inhibitory effects of an anti-IgE antibody E25 on allergen-induced early asthmatic response. *Am J Respir Crit Care Med* 155: 1835–1840

4 Fahy JV, Fleming HE, Wong HH, Liu JT, Su JQ, Reimann J, Fick RB Jr, Boushey HA (1997) The effect of an anti-IgE monoclonal antibody on the early and late-phase responses to allergen inhalation in asthmatic subjects. *Am J Respir Crit Care Med* 155: 1828–1834

5 Milgrom H, Fick RB, Su JQ, Reimann JD, Bush RK, Watrous ML, Metzger WJ (1999) Treatment of allergic asthma with monoclonal anti-IgE. *N Engl J Med* 341: 1966–1973

6 van Neerven RJJ, van Roomen CPA, Thomas WR, de Boer M, Knol EF, Davis FM (2001) Humanised anti-IgE mAb Hu-901 prevents the activation of allergen-specific T cells. *Int Arch Allergy Appl Immunol* 124: 400–402

7 Buhl R, Soler M, Matz J, Townley R, O'Brien J, Noga O, Champain K, Fox H, Thirlwell J, Della Cioppa G (2002) Omalizumab provides long-term control in patients with moderate-to-severe asthma. *Eur Respir J* 20: 73–78

8 Holgate ST, Chuchalin AG, Hebert J, Lotvall J, Persson GB, Chung KF, Bousquet J, Kerstjens HA, Fox H, Thirlwell J et al (2004) Efficacy and safety of a recombinant

anti-IgE antibody (omalizumab) in severe allergic asthma. *Clin Exp Allergy* 34: 632–638

9 Noga O, Hanf G, Kunkel G (2003) Immunological and clinical changes in allergic asthmatics following treatment with omalizumab. *Int Arch Allergy Immunol* 131: 46–52

10 Djukanovic R, Wilson SJ, Kraft M, Jarjour NN, Steel M, Chung KF, Bao W, Fowler-Taylor A, Matthews J, Busse WW et al (2004) The effects of anti-IgE (omalizumab) on airways inflammation in allergic asthma. *Am J Respir Crit Care Med*

11 Bousquet J, Wenzel S, Holgate S, Lumry W, Freeman P, Fox H (2004) Predicting response to omalizumab, an anti-IgE antibody, in patients with allergic asthma. *Chest* 125: 1378–1386

12 Dabbagh K, Takeyama K, Lee HM, Ueki IF, Lausier JA, Nadel JA (1999) IL-4 induces mucin gene expression and goblet cell metaplasia *in vitro* and *in vivo*. *J Immunol* 162: 6233–6237

13 Shi HZ, Chen YQ, Qin SM (1999) Inhaled IL-5 increased concentrations of soluble intracellular adhesion molecule-1 in sputum from atopic asthmatic subjects. *J Allergy Clin Immunol* 103: 463–467

14 Zhu Z, Homer RJ, Wang Z, Chen Q, Geba GP, Wang J, Zhang Y, Elias JA (1999) Pulmonary expression of IL-13 causes inflammation, mucus hypersecretion, subepithelial fibrosis, physiologic abnormalities and eotaxin production. *J Clin Invest* 103: 779–788

15 Thomas PS, Heywood G (2002) Effects of inhaled tumour necrosis alpha in subjects with mild asthma. *Thorax* 57: 774–778

16 Kanehiro A Kanehiro A, Lahn M, Makela MJ, Dakhama A, Fujita M, Joetham A, Mason RJ, Born W, Gelfand EW (2001) TNF-alpha negatively regulates airway hyperresponsiveness through gamma-delta T cells. *Am J Respir Crit Care Med* 164: 2229–2238

17 Benayoun L, Druilhe A, Dombret MC, Aubier M, Pretolani M (2003) Airway structural alterations selectively associated with severe asthma. *Am J Respir Crit Care Med* 167: 1360–1368

18 Mauser PJ, Pitman A, Witt A, Fernandez X, Zurcher J, Kung T, Jones H, Watnick AS, Egan RW, Kreutner W et al (1993) Inhibitory effect of the TRFK-5 anti-IL-5 antibody in a guinea pig model of asthma. *Am Rev Respir Dis* 148: 1623–1627

19 Mauser PJ, Pitman AM, Fernandez X, Foran SK, Adams GK 3rd, Kreutner W, Egan RW, Chapman RW (1995) Effects of an antibody to IL-5 in a monkey model of asthma. *Am J Respir Crit Care Med* 152: 467–472

20 Hamelmann E, Cieslewicz G, Schwarze J, Ishizuka T, Joetham A, Heusser C, Gelfand EW (1999) Anti-IL-5 but not anti-IgE prevents airway inflammation and airway hyperresponsiveness. *Am J Respir Crit Care Med* 160: 934–941

21 Leckie MJ, ten Brinke A, Khan J, Diamant Z, O'Connor BJ, Walls CM, Mathur AK, Cowley HC, Chung KF, Djukanovic R et al (2000) Effects of an IL-5-blocking monoclonal antibody on eosinophils, airway hyperresponsiveness and the late asthmatic response. *Lancet* 356: 2144–2148

22 Flood-Page P, Menzies-Gow AN, Kay AB, Robinson DS (2003) Eosinophil's role uncertain as anti-IL-5 only partially depletes numbers in asthmatic airway. *Am J Respir Crit Care Med* 167: 199–204

23 Gregory B, Kirchem A, Phipps S, Gevaert P, Pridgeon C, Rankin SM, Robinson DS (2003) Differential regulation of human eosinophil IL-3, IL-5, and GM-CSF receptor alpha-chain expression by cytokines: IL-3, IL-5, and GM-CSF down-regulate IL-5 receptor alpha expression with loss of IL-5 responsiveness, but up-regulate IL-3 receptor alpha expression. *J Immunol* 170: 5359–5366

24 Kips JC, O'Connor B, Langley SJ, Woodcock A, Kerstjens HAM, Postma DS, Danzig M Cuss F, Pauwels RA (2003) Effect of SCH55700, a humanised anti-human IL-5 antibody, in severe persistent asthma. A pilot study. *Am J Respir Crit Care Med* 167: 1655–1659

25 Larche M, Robinson DS, Kay AB (2003) The role of T lymphocytes in the pathogenesis of asthma. *J Allergy Clin Immunol* 111: 450–463

26 Shi H-Z, Deng JM, Nong ZX, Xiao CQ, Liu ZM, Qin SM, Jiang HX, Liu GN, Chen YQ (1998) Effect of inhaled IL-4 on airway hyperreactivity in asthmatics. *Am J Respir Crit Care Med* 157: 1818–1821

27 Borish L, Nelson HS, Lanz MJ, Claussen L, Whitmore JB, Agosti JM, Garrison L (1999) IL-4 receptor in moderate atopic asthma: a phase 1/2 randomised, placebo-controlled trial. *Am J Respir Crit Care Med* 160: 1816–1823

28 Babu K, Arshad SH, Howarth PH, Chauhan AJ, Bell EJ, Puddicombe S, Davies DE, Holgate ST (2003) Soluble TNFalpha receptor (Enbrel) as an effective therapeutic strategy in chronic severe asthma. *J Allergy Clin Immunol* 111: S277 (abst)

29 Wills-Karp M (2001) IL-12/IL-13 axis in allergic asthma. *J Allergy Clin Immunol* 107: 9–18

30 Boguniewicz M, Schneider LC, Milgrom H, Newell D, Kelly N, Tam P, Izu AE, Jaffe HS, Bucalo LR, Leung DY (1993) Treatment of steroid-dependent asthma with recombinant interferon-gamma. *Clin Exp Allergy* 23: 785–790
31 Tang C, Rolland JM, Ward C, Li X, Bish R, Thien F, Walters EH (1999) Modulatory effects of alveolar macrophages on CD4+ T-cell IL-5 responses correlate with IL-1beta, IL-6 and IL-12 production. *Eur Respir J* 14: 106–112
32 van der Pouw Kraan TC, Boeijc LC, de Groot ER, Stapel SO, Snijders A, Kapsenberg ML, van der Zee JS, Aarden LA (1997) Reduced production of IL-12 and IL-12-dependent IFN-gamma release in patients with atopic asthma. *J Immunol* 158: 5560–5565
33 Iwamoto I, Kumano K, Kasai M, Kurasawa K, Nakao A (1996) IL-12 prevents antigen-induced eosinophil recruitment into mouse airways. *Am J Respir Crit Care Med* 154: 1257–1260
34 Bryan SA, O'Connor BJ, Matti S, Leckie MJ, Kanabar V, Khan J, Warrington SJ, Renzetti L, Rames A, Bock JA et al (2000) Effects of recombinant human IL-12 on eosinophils, airway hyperresponsiveness and the late asthmatic response. *Lancet* 356 (9248): 2149–2153
35 Arikan C, Bahceciler NN, Deniz G, Akdis M, Akkoc T, Akdis CA, Barlan IB (2004) BCG-induced IL-12 did not additionally improve clinical and immunologic parameters in asthmatic children treated with sublingual immunotherapy. *Clin Exp Allergy* 34: 398–405
36 Freeman J (1914) Vaccination against hay fever: report of results during the first three years. *Lancet* 1: 1178
37 Creticos PS, Van Metre TE, Mardiney MR, Rosenberg GL, Norman PS, Adkinson NF (1984) Dose-response of IgE and IgG antibodies during ragweed immunotherapy. *J Allergy Clin Immunol* 73: 94–104
38 Iliopoulos O, Proud D, Adkinson NF, Creticos PS, Norman PS, Kagey-Sobotka A, Lichtenstein LM, Naclerio RM (1991) Effects of immunotherapy on the early, late and rechallenge nasal reaction to provocation with allergen: changes in inflammatory mediators and cells. *J Allergy Clin Immunol* 87: 855–866
39 Durham SR, Ying S, Varney VA, Jacobson MR, Sudderick RM, Mackay IS, Kay AB, Hamid QA (1996) Grass pollen immunotherapy inhibits allergen-induced infiltration of $CD4^+$ T-lymphocytes and eosinophils in the nasal mucosa and increases the number of cells expressing mRNA for interferon-gamma. *J Allergy Clin Immunol* 97: 1356–1365
40 McHugh SM, Deighton J, Stewart AG, Lachmann PJ, Ewan PW (1995) Bee venom immunotherapy induces a shift in cytokine responses from a Th2 to a Th1 dominant pattern: comparison of rush and conventional immunotherapy. *Clin Exp Allergy* 25: 828–838
41 Ebner C, Siemann U, Bohle B, Willheim M, Wiedermann U, Schenk S, Klotz F, Ebner H, Kraft D, Scheiner O (1997) Immunological changes during specific immunotherapy of grass pollen allergy: reduced lymphoproliferative responses to allergen and shift from Th2 to Th1 in T-cell clones specific for Phl p1, a major grass pollen allergen. *Clin Exp Allergy* 27: 1007–1015
42 Nasser SM, Ying S, Meng Q, Kay AB, Ewan PW (2001) IL-10 levels increase in cutaneous biopsies of patients undergoing wasp venom immunotherapy. Eur *J Immunol* 31: 3704–3713
43 Akdis CA, Blesken T, Akdis M, Wuthrich B, Blaser K (1998) Role of IL-10 in specific immunotherapy. *J Clin Invest* 102: 98–106
44 Akdis CA, Blaser K (1999) IL-10 induced anergy in peripheral T cells and reactivation by microenvironmental cytokines: two key steps in specific immunotherapy. *FASEB J* 13: 603–609
45 Bousquet J, Lockey RF, Malling HJ (1998) WHO position paper. Allergen immunotherapy: therapeutic vaccines for allergic disease. *Allergy* 53 (S44): 1–42
46 Abramson M, Puy R, Weiner J (1999) Immunotherapy in asthma: an updated systematic review. *Allergy* 54: 1022–1041
47 Lilja G, Sundin B, Graff-Lonnevig V, Hedlin G, Heilborn H, Norrlind K, Pegelow KO, Lowenstein H (1989) Immunotherapy with partially purified and standardised animal dander extracts. IV. Effects of 2 years of treatment. *J Allergy Clin Immunol* 83: 37–44
48 Creticos PS, Reed CE, Norman PS, Khoury J, Adkinson NF, Buncher CR, Busse WW, Bush RK, Gadde J, Li JT et al (1996) Ragweed immunotherapy in adult asthma. *N Engl J Med* 334: 501–506
49 Adkinson NF, Eggleston PA, Eney D, Goldstein EO, Schuberth KC, Bacon JR, Hamilton RG, Weiss ME, Arshad H, Meinert CL et al (1997) A controlled trial of immunotherapy for asthma in allergic children. *N Engl J Med* 336: 324–331
50 Horak F (1985) Manifestation of allergic rhinitis in

latent sensitised patients. A prospective study. *Arch Otorhinolaryngol* 242: 242–249

51 Johnstone DE, Dutton A (1968) The value of hyposensitization therapy for bronchial asthma in children. A 14 year study. *Pediatrics* 42: 793–802

52 Des Roches A, Paradis L, Menardo JL, Bouges S, Daures JP, Bousquet J (1997) Immunotherapy with a standardised *Dermatophagoides pteronyssinus* extract. VI. Specific immunotherapy prevents the onset of new sensitisations in children. *J Allergy Clin Immunol* 99: 450–453

53 Moller C, Dreborg S, Ferdousi HA, Halken S, Host A, Jacobsen L, Koivikko A, Koller DY, Niggemann B, Norberg LA et al (2002) Pollen immunotherapy reduces the development of asthma in children with seasonal rhinoconjunctivitis (the PAT-study). *J Allergy Clin Immunol* 109: 251–256

54 Vrtala S, Hirtenlehner K, Susani M, Akdis M, Kussebi F, Akdis CA, Blaser K, Hufnagl P, Binder BR, Politou A et al (2001) Genetic engineering of a hypoallergenic trimer of the major birch pollen allergen Bet v 1. *FASEB J* 15: 2045–2047

55 Vrtala S, Akdis CA, Budak F, Akdis M, Blaser K, Kraft D, Valenta R (2000) T cell epitope-containing hypoallergenic recombinant fragments of the major birch pollen allergen, Bet v 1, induce blocking antibodies. *J Immunol* 165: 6653–6659

56 O'Hehir RE, Yssel H, Verma S, de Vries JE, Spits H, Lamb JR (1991) Clonal analysis of differential lymphokine production in peptide and superantigen-induced T-cell anergy. *Int Immunol* 3: 819–826

57 Fasler S, Aversa G, de Vries JE, Yssel H (1998) Antagonistic peptides specifically inhibit proliferation, cytokine production, CD40L expression and help for IgE synthesis by Der p1-specific human T-cell clones. *J Allergy Clin Immunol* 101: 521–530

58 Wiedermann U, Herz U, Baier K, Vrtala S, Neuhaus-Steinmetz U, Bohle B, Dekan G, Renz H, Ebner C, Valenta R et al (2001) Intranasal treatment with a recombinant hypoallergenic derivative of the major birch pollen allergen Bet v 1 prevents allergic sensitization and airway inflammation in mice. *Int Arch Allergy Immunol* 126: 68–77

59 Kline JN, Waldschmidt TJ, Businga TR, Lemish JE, Weinstock JV, Thorne PS, Krieg AM (1998) Modulation of airway inflammation by CpG oligodeoxynucleotides in a murine model of asthma. *J Immunol* 160: 2555–2559

60 Sur S, Wild JS, Choudury BK, Sur N, Alam R, Klinman DM (1999) Long-term prevention of allergic lung inflammation in a mouse model of asthma by CpG oligodeoxynucleotides. *J Immunol* 162: 6284–6293

61 Tighe H, Takabayashi K, Schwartz D, van Nest G, Tuck S, Eiden JJ, Kagey-Sobotka A, Creticos PS, Lichtenstein LM, Spiegelberg HL et al (2000) Conjugation of immunostimulatory DNA to the short ragweed allergen Amb a1 enhances its immunogenicity and reduces its allergenicity. *J Allergy Clin Immunol* 106: 124–134

62 Marshall JD, Abtahi S, Eiden JJ, Tuck S, Milley R, Haycock F, Reid MJ, Kagey-Sobotka A, Creticos PS, Lichtenstein LM (2001) Immunostimulatory sequence DNA linked to the Amb a 1 allergen promotes Th1 cytokine expression while downregulating Th2 cytokine expression in PBMCs from human patients with ragweed allergy. *J Allergy Clin Immunol* 108: 191–197

63 Creticos PS, Eiden JJ, Broide D, Balcer-Whaley SL, Schroeder JT, Khattignavong A, Li H, Norman PP, Hamilton RG (2002) Immunotherapy with immunostimulatory oligonucleotides linked to purified ragweed Amb a 1 allergen: effects on antibody production, nasal allergen provocation and ragweed seasonal rhinitis. *J Allergy Clin Immunol* 109: 743–744

64 Tulic MK, Fiset PO, Christodoulopoulos P, Vaillancourt P, Desrosiers M, Lavigne F, Eiden J, Hamid Q (2004) Amb a1-immunostimulatory oligodeoxynucleotide conjugate immunotherapy decreases the nasal inflammatory response. *JACI* 113: 235–241

65 Hsu CH, Chua KY, Tao MH, Lai YL, Wu HD, Huang SK, Hsieh KH (1996) Immunoprophylaxis of allergen-induced IgE synthesis and airway hyperresponsiveness *in vivo* by genetic immunisation. *Nat Med* 2: 540–544

66 Hartl A, Kiesslich J, Weiss R, Bernhaupt A, Mostbock S, Scheiblhofer S, Ebner C, Ferreira F, Thalhamer J (1999) Immune responses after immunisation with plasmid DNA encoding Bet v 1, the major allergen of birch pollen. *J Allergy Clin Immunol* 103: 107–113

67 Hartl A, Hochreiter R, Stepanoska T, Ferreira F, Thalhamer J (2004) Characterisation of the protective and therapeutic efficiency of a DNA vaccine encoding the major birch pollen allergen Bet v1a. *Allergy* 590: 65–73

D

Immunotoxicology

Immunotoxicology

Joseph G. Vos, Jan Willem van der Laan, Henk van Loveren, Ruud Albers and Raymond Pieters

Introduction

Pharmaceuticals for human use comprise a very wide variety of product types. These include traditional products (i.e., chemically synthesized or plant-derived pharmaceuticals) as well as biological products (such as vaccines and blood products isolated from biological sources) and biotechnology-derived pharmaceuticals (such as peptide products manufactured by recombinant DNA techniques, monoclonal antibodies and gene therapy products). In the interest of the public, these medicinal products are subject to worldwide regulatory control by government authorities. The major objective of this regulation is to ensure that the benefit of the products to the patients are not outweighed by their adverse effects. To achieve this goal, the authorities carefully assess the balance between efficacy and safety. If this balance is positive, they allow marketing. To support applications for marketing authorisation, the pharmaceutical industry therefore has to submit scientific data which prove that their products are efficacious and acceptably safe in the proposed therapeutic indication. Furthermore, the pharmaceutical quality of the products applied for has to meet high standards.

Chemicals used for a variety of purposes can have adverse effects on the immune system of both animals and humans. In the case of drugs, this can be the result of pharmacological interference with the immune system, or an undesired reaction. One form of immunotoxicity is the direct toxicity of the compound to components of the immune system, which often leads to suppressed function. This may result in decreased resistance to infection, the development of certain types of tumors, or immune dysregulation and stimulation, thereby promoting ALLERGY or AUTOIMMUNITY. Other types or manifestations of immunotoxicity include ALLERGY or AUTOIMMUNITY in which the compound causes the immune system to respond as if the compound were an ANTIGEN or to respond to self-antigens that have been altered by the chemical.

Except for cancer patients on chemotherapy and organ transplant patients on long-term immunosuppressive therapy, there is little evidence that drugs are associated with undesired, clinically significant IMMUNOSUPPRESSION. However, only a few valid epidemiological studies of immunologically based diseases have been carried out [1], probably due to the complication of such studies by confounding factors such as (disease-associated) stress, nutritional status, lifestyle, (co)medication and genetics. Few conventional drugs have been shown to induce unexpected enhancement of immune competence. Unwanted immunostimulation has gained attention primarily through the introduction of new biotechnologically manufactured drugs such as CYTOKINES. Drug-induced HYPERSENSITIVITY reactions and autoimmune disorders are a major concern, and often the reason for withdrawing drugs from the market or restricting their use.

For the detection of chemical-induced direct immunotoxicity, animal models have been developed, and a number of these methods have been validated. Several compounds, including certain drugs, have been shown in this way to cause IMMUNOSUPPRESSION. Methods are also available for the detection of skin allergic responses, whereas no validated test is available to predict potential induction of AUTOIMMUNITY.

In this chapter, the various mechanisms of immunotoxicity by which pharmaceuticals affect different cell types and interfere with immune responses, ultimately leading to immunotoxicity, are introduced and discussed. Furthermore, procedures for preclinical testing of drugs are covered, comprising direct immunotoxicity as well as sensitizing capacity. This section is followed by consideration of procedures for clinical and epidemiological testing of drugs. Finally, regulatory aspects of immunotoxicity are discussed, including current guidelines and new developments in immunotoxicity assessment.

Mechanisms of immunotoxicity by pharmaceuticals

Effects on precursor stem cells

Precursor stem cells that are responsible for replenishing peripheral leukocytes reside in the bone marrow making it an organ that harbors many highly proliferating cells. All leukocyte lineages originate from these stem cells, but once distinct subsets of leukocytes are established, their dependence on replenishment from the bone marrow differs vastly. The short-lived neutrophils rely heavily on proliferation and new formation in the bone marrow, as each day more than 10^8 neutrophils enter and leave the circulation in a normal adult. In contrast, MACROPHAGES are long-lived and have little dependence on new formation of precursor cells [2]. The adaptive immune system, comprising ANTIGEN-specific T and B LYMPHOCYTES, is almost completely established around puberty and is therefore essentially bone marrow-independent in the adult.

As a consequence of their high proliferation rate, stem cells in the bone marrow are extremely vulnerable to antiproliferative cytostatic drugs such as the antineoplastic drugs cyclophosphamide (CY), and methotrexate (MTX), and the antirheumatic azathioprine (AZA) [1,3] (Fig. 1). This is particularly the case at high doses of these drugs, and lineages like neutrophils that are extremely bone marrow-dependent will be most vulnerable and are affected first by treatment with these drugs. After prolonged exposure, MACROPHAGES and T- or B CELLS of the adaptive immune system are also suppressed.

Effects on maturation of lymphocytes

After leaving the bone marrow, cells of both the T-CELL and the B-CELL lineages mature into ANTIGEN-specific lymphocytes. T LYMPHOCYTES mature in the thymus during a process referred to as thymocyte differentiation, which is a very complex selection process that takes place under the influence of the thymic microenvironment and ultimately generates an ANTIGEN-specific, host-tolerant population of mature T CELLS (see Chapters A2 and A3). Because this process involves cellular proliferation, gene rearrangement, apoptotic cell death, receptor up- and downregulation and ANTIGEN-presentation processes, it is very vulnerable to a number of chemicals, including pharmaceuticals (Fig. 1). Drugs may target different stages of T-CELL differentiation: bone marrow precursors (AZA), proliferating and differentiating thymocytes (AZA), ANTIGEN-presenting thymic epithelial cells and DENDRITIC CELLS (cyclosporin A, CsA) [4], and cell death processes (corticosteroids) [5] (Fig. 2).

In general, immunosuppressive drugs that affect the thymus cause a depletion of peripheral T CELLS, particularly after prolonged treatment and during early stages of life when thymus activity is high and important in establishing a mature T-CELL population.

After the bone marrow stage, B CELLS mature in the spleen. With the exception of certain monoclonal antibodies, there are no drugs that specifically affect B-CELL development, although some studies claim a more or less B-CELL-specific effect of CY and MTX. In general, suppression of the adaptive immune system at the antibody level is the result of an effect on T CELLS or their development.

Effects on initiation of immune responses

Once a mature immune system has been established, the innate and adaptive arms of the immune system cooperate to eliminate invading pathogens. Ideally, T CELLS tailor the responses to neutralise invaders with minimal damage to the host. After elimination of T CELLS with high affinity for self-antigens in the thymus, tolerance for AUTOANTIGENS is further maintained in the periphery by the two distinct signals that govern lymphocyte activation. Signal 1 is

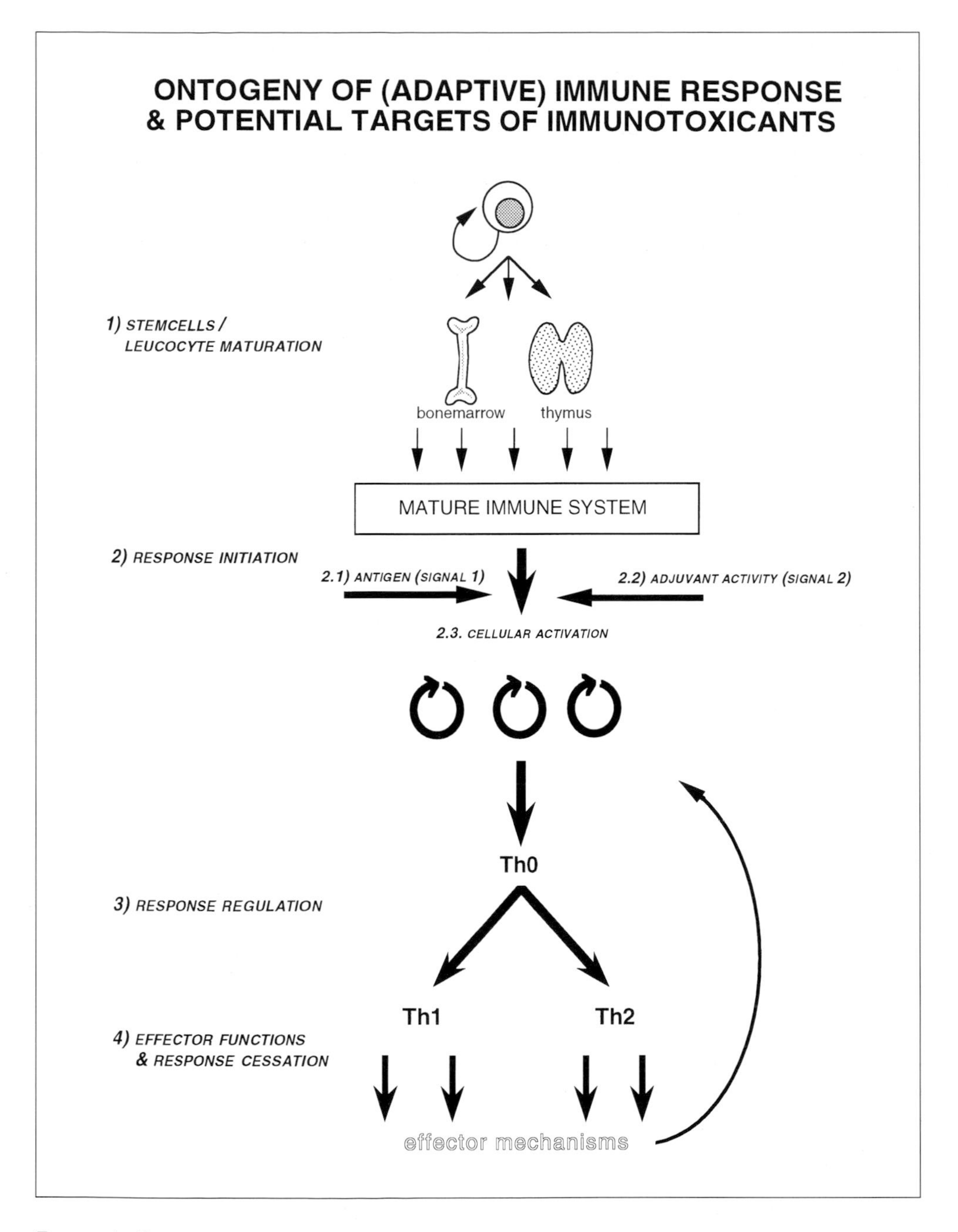

FIGURE 1. ONTOGENY OF THE IMMUNE RESPONSE AND TARGETS OF IMMUNOTOXIC PHARMACEUTICALS
This figure represents the different steps in the ontogeny of adaptive immune responses from stem cell to response cessation. It forms our conceptional framework to identify potential mechanisms of immunotoxicity. Indicated are effects of pharmaceuticals on stem cells and leucocyte maturation, and on the two signals that are essential for lymphocyte activation, and the resulting cellular activation. Also indicated are the regulation and cessation of immune responses as potential targets for immunotoxic pharmaceuticals.

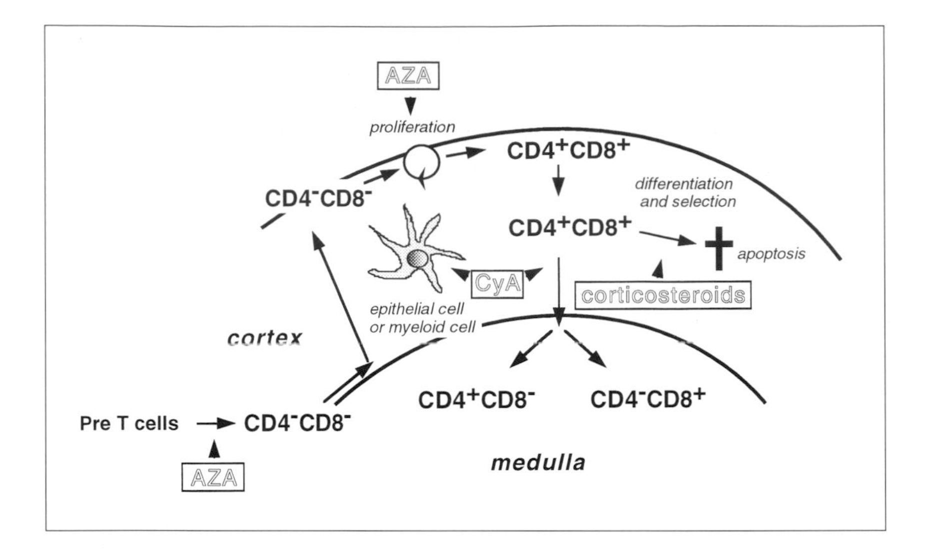

FIGURE 2. SCHEMATIC VIEW OF THE THYMUS SHOWING THE CORTICAL AND MEDULLARY REGION
Immature pre-T cells (CD4⁻CD⁻) enter the thymus at the cortico-medullary region and migrate to the subcapsular region where they show high proliferative activity and differentiate into CD4⁺CD8⁺ thymocytes. In the cortex, most thymocytes are CD4⁺CD8⁺ and at this stage thymoctes are selected under the influence of thymic epithelial cells and are prone to apoptotic cell death. After the CD4⁺CD8⁺ stage, cells differentiate either to CD4⁺ or CD8⁺ cells. Stages sensitive to pharmaceutical attack are indicated: AZA inhibits formation of pre-T cells and may inhibit immature thymocyte proliferation; CyA interferes with thymocyte selection, possibly through an effect on thymic dendritic cells; and corticosteroids stimulate apoptosis.

the specific recognition of ANTIGEN via clonally distributed ANTIGEN receptors. Signal 2 consists of ANTIGEN-nonspecific costimulation or "help" and involves interactions of various adhesive and signaling molecules [6]. It is imperative that lymphocytes receive both signals, as ANTIGEN recognition without costimulation induces tolerance, and lymphocytes are unresponsive to costimulation without an ANTIGEN-specific signal. The molecules transmitting signal 2 are thought to be expressed mainly in response to tissue damage, linking initiation of immune responses to situations of acute "danger" for the host [7]. This helps to aim immune responses at potentially dangerous microorganisms (non-self), while minimizing deleterious reactions to innocuous (non-self) ANTIGENS and to the host (self) (Chapter A1).

Certain pharmaceuticals and other XENOBIOTICS can interfere with the initiation of immune responses by forming complexes with self-proteins (e.g., HAPTEN-CARRIER complexes) or by releasing previously hidden self-antigens. By doing so these chemicals may provide signal 1 to neo-ANTIGEN-specific T CELLS. But probably more important is that chemicals may also induce signal 2 by triggering an inflammatory response. Finally, pharmaceuticals may directly affect cellular activation following occupation of the receptors involved in the two activation signals (Fig. 1).

Interference with antigen recognition (signal 1)

Large (protein) pharmaceuticals can be antigens

Large molecular weight pharmaceuticals (>4000 Da) can function as ANTIGENS and become targets of specific immune responses themselves (reviewed in [8]). This is particularly relevant for foreign biopharmaceuticals, as these can activate both T and B LYMPHOCYTES. The resulting specific immune responses may lead to formation of antibodies, and induce a specific memory which can lead to allergic responses to the substance. For example, passive immunization to tetanus toxin or snake venoms with serum from immunized horses causes the temporary formation of immune complexes with symptoms of fever, joint tenderness and proteinuria (serum sickness). Because serum proteins are given in large amounts and have a long half-life, SENSITIZATION and allergic reactions take place after a single dose. Similar immunotoxic effects due to immunogenicity may occur after repeated treatment with pharmaceuticals like porcine insulin, murine antibodies and biotechnologically engineered "novel proteins". In patients developing neutralizing antibodies, absence of response or reversal of clinical efficacy has been described [9]. The danger of immunotoxic effects due to immunogenicity is much lower when homologous recombinant human or "humanized" proteins are used as pharmaceuticals. Other factors that determine whether a certain biopharmaceutical raises an immune response are route of exposure, contaminants, and formulations [8].

Reactive pharmaceuticals can form haptens

Low molecular weight pharmaceuticals cannot function as ANTIGENS, because they are too small to be detected by T CELLS. Reactive drugs that bind to proteins, however, can function as HAPTENS and become immunogenic if epitopes derived from the CARRIER protein prime T CELLS, which in turn provide costimulation for HAPTEN-specific B CELLS. This effect is responsible for allergic responses to many new (neo)EPITOPES formed by chemical HAPTENS including pharmaceuticals (penicillin, penicillamine, cephalosporin, aspirin and many others), occupational contact sensitizers and respiratory sensitizers (Fig. 3). Other compounds require metabolic activation to form reactive metabolites that bind to proteins. The anesthetic halothane, for instance, is metabolized to alkyl halides by cytochrome P-450 in the liver. The alkyl halides bind to microsomal proteins including P-450, and the bound HAPTENS induce an immune response that causes so-called halothane hepatitis. Other compounds can be activated by extrahepatic metabolism, in particular by the myeloperoxidase system in phagocytic cells. For instance, activated MACROPHAGES and GRANULOCYTES can metabolize the antiarrhythmic procainamide to reactive metabolites that can bind to proteins, and immune responses to these HAPTENS seem to be responsible for procainamide-related agranulocytosis and lupus [10].

Induction of cross-reactivity by pharmaceuticals

Formation of neo-antigenic structures (either HAPTEN-CARRIER complexes, or non-covalent compound-protein structures) may cause stimulation of cross-reactive T CELLS. Cross-reactivity implies that T CELLS recognize not only the best fitting MAJOR HISTOCOMPATIBILITY COMPLEX (MHC)-(neo-)peptide complex but also completely different (neo-)peptides in the groove of unrelated MHC molecules (also termed allo-reactivity) (reviewed in [11]).

It has been found that certain drug-reactive T-CELL clones from patients were MHC-allele unrestricted and at the same time highly drug-specific, i.e., they did not respond to drug derivatives with small chemical alterations. Other drug-induced T-CELL clones appeared less stringent with respect to the structure of the drug they recognized but they were highly MHC allele-restricted. Still other drug-induced T CELLS responded to MHC-peptide complexes in the absence of the initiating drug, and in a MHC-allele unrestricted manner. So the specificity of drug-reactive T CELLS may range from highly drug-specific and non-MHC restricted to highly MHC-restricted and non-drug-specific. From these findings it can be inferred that drug-induced T CELLS may also react with auto-antigens through cross-reactivity.

Responses to haptens can spread to autoreactive responses

Chemical modification of AUTOANTIGENS can also lead to autoreactive responses to unmodified self-EPITOPES by a mechanism unrelated to cross-reactivity. ANTIGENS composed of neo- and self-EPITOPES (i.e.,

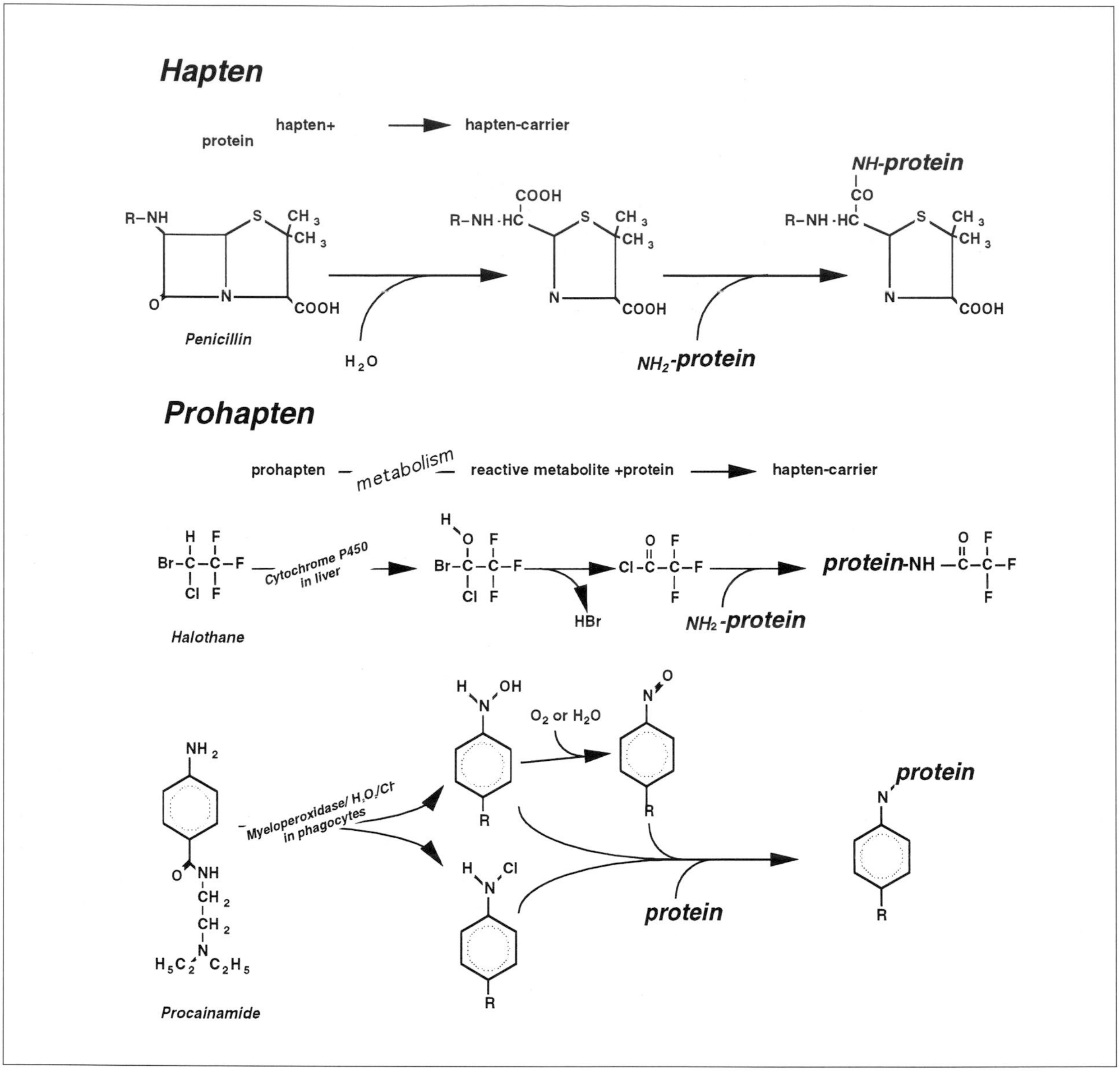

Figure 3. Haptens and prohaptens

Pharmaceuticals that are too small to attract a T-cell response can become antigenic when they bind as haptens to a protein carrier. In this case, T cells responding to chemically induced neo-epitopes on the carrier provide costimulation for B cells responding to the hapten. Prohaptens require metabolic activation to a reactive metabolite that can function as a hapten. Penicillin is a well-known example of a pharmaceutical that can form haptens by direct binding to proteins. In contrast, halothane itself does not form haptens, but cytochrome P450-mediated metabolism in the liver results in reactive metabolites that do bind to proteins. Procainamide can be metabolized by the myeloperoxidase/H_2O_2/Cl^- system of phagocytes. These metabolites are very reactive and can bind covalently to nucleophilic thiol and amino groups of proteins.

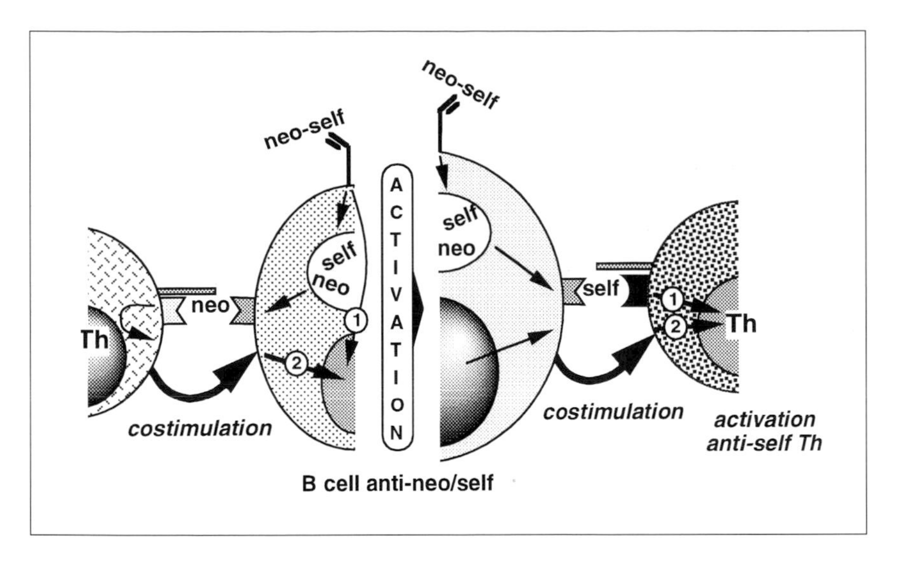

Figure 4. Determinant spreading
Immune responses to haptens can spread to autoreactive responses to the carrier protein. Haptenated autoantigens are recognized and internalized by specific B cells. After uptake and processing the B cells present a mixture of the neo- and self-EPITOPES complexed to distinct MHC-II molecules. Naive T cells do not respond to the self-EPITOPES, but neo-specific T cells provide costimulation for such B cells. This leads to activation of the B cell and production of antibodies. Moreover, when the B cells are activated, they can provide costimulation for naive T cells that recognize unmodified self-EPITOPES, leading to their activation and breaking of T-cell tolerance. 1, the antigen-specific or first signal; 2, the costimulatory or second signal; neo, neo-epitope.

haptenated AUTOANTIGENS) can be recognized and internalized by B CELLS specific to either the HAPTEN or to unmodified B-CELL EPITOPES on the AUTOANTIGEN. These cells subsequently present a mixture of neo- and self-EPITOPES complexed to distinct class II major histocompatibility (MHC-II) molecules on their surface (Fig. 4). Since T-CELL tolerance is obviously not established for the neo-EPITOPES, neo-specific Th cells provide signal 2 for the B CELL. This leads to production of either anti-hapten or anti-self antibodies, depending on the exact specificity of the B CELL. Moreover, once these B CELLS are activated, they can stimulate autoreactive Th cells recognizing unmodified self-EPITOPES. The underlying process is called epitope (determinant) spreading and causes the diversification of adaptive immune responses. Responses induced by injection of mercury salts, for instance, are initially directed only to unidentified chemically created neo-EPITOPES, but after 3–4 weeks include reactivity to unmodified self-EPITOPES [12]. The distinction between allergic and autoimmune responses induced by HAPTENS may therefore only be gradual, reflecting the relative antigenicity of the neo- and self-EPITOPES involved [13].

Pharmaceuticals can expose (epitopes of) autoantigens

Induction of self-tolerance involves specific recognition of AUTOANTIGEN leading to selective inactivation of autoreactive lymphocytes, and tolerance is therefore not established for (EPITOPES of) AUTOANTIGENS that are normally not available for immune recognition. Pharmaceuticals can expose such sequestered (EPITOPES of) AUTOANTIGENS by disrupting barriers between the ANTIGEN and the immune system (i.e., blood-brain barrier, blood-testis barrier, cell membranes). Tissue damage, cell death and

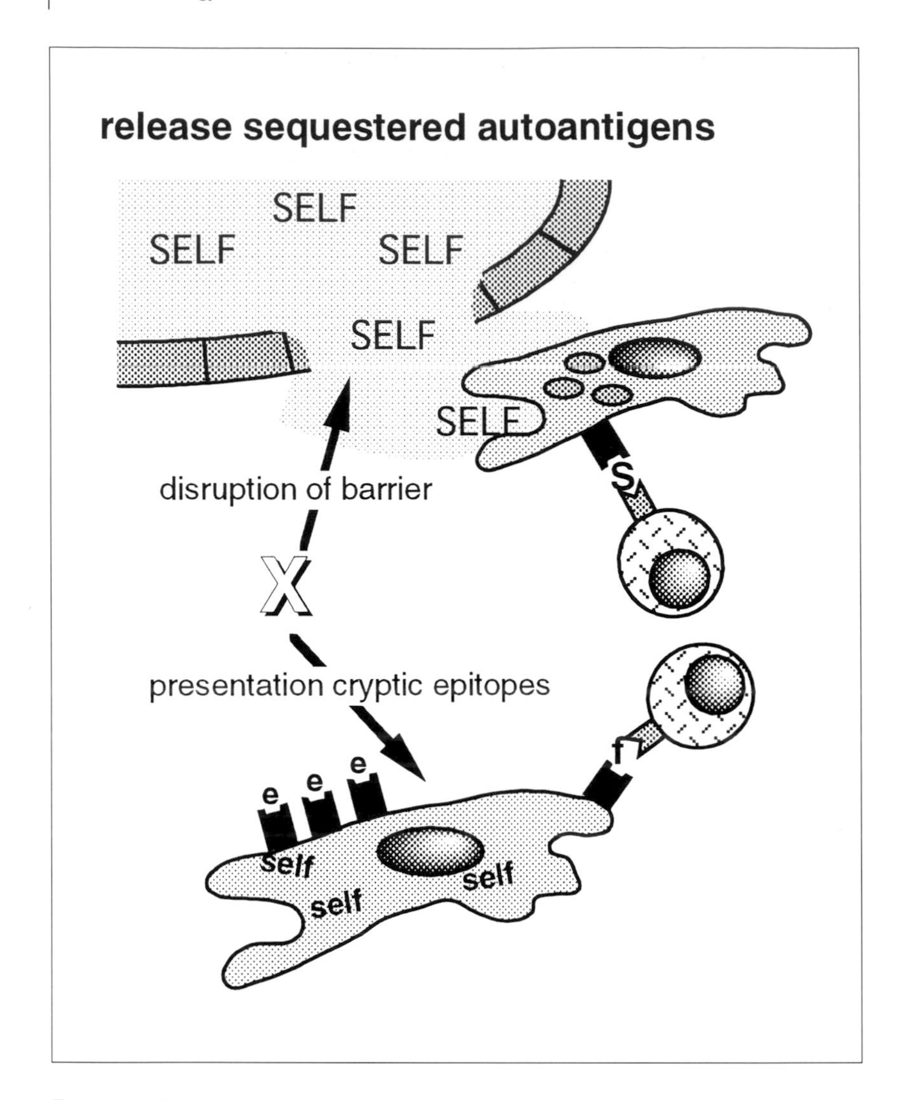

FIGURE 5. RELEASE OF SEQUESTERED SELF-EPITOPES

Pharmaceuticals can expose previously sequestered (epitopes of) self antigens by disrupting barriers between the antigen and the immune system (i.e., blood-brain barrier, blood-testis barrier, cell membranes). Similarly, augmented presentation of cryptic epitopes by altered antigen processing increases the availability of these epitopes for recognition by T cells.

protein denaturation induced or enhanced by pharmaceuticals can largely increase the availability of such (EPITOPES of) AUTOANTIGENS for immune recognition (Fig. 5). Moreover, altered ANTIGEN processing that augments presentation of EPITOPES that were previously not expressed increases the availability of these so-called subdominant or cryptic EPITOPES for recognition by T CELLS [14, 15]. It has been shown, for instance, that mice immunized with ANTIGEN in ADJUVANT respond exclusively to dominant EPITOPES, whereas preincubation with Au(III), the oxidized metabolite of the anti-rheumatic auranofin, elicits additional T-CELL responses to cryptic EPITOPES [13].

Interference with costimulation (signal 2)

It is important to stress that ANTIGEN recognition by itself does not lead to activation of lymphocytes (above and Chapter A3). In addition, costimulation (i.e., signal 2) is required in the initiation of immune responses [16–18]. Many XENOBIOTICS have the inherent capacity to induce this costimulation; they have intrinsic ADJUVANT activity. For instance, immunostimulatory responses in mice induced by D-penicillamine and phenytoin could be inhibited by blocking costimulatory interactions (i.e., mediated by CD40-CD154) with a specific monoclonal antibody to CD154 [19]. The underlying mechanisms are not always understood, but several mutually non-exclusive possibilities have been described.

Induction of inflammation

Cytotoxic pharmaceuticals or their reactive metabolites can induce tissue damage which leads to accumulation of tissue debris, release of proinflammatory CYTOKINES like TUMOR NECROSIS FACTOR-α (TNF-α), INTERLEUKIN-1 (IL-1), and IL-6, and attracts inflammatory cells like GRANULOCYTES and MACROPHAGES. CYTOKINES produced during this inflammatory response activate ANTIGEN-presenting cells. These present selected EPITOPES of ANTIGENS from the debris, and provide costimulation for Th cells, which leads to the initiation of an adaptive immune response [16]. All reactive and cytotoxic pharmaceuticals can conceivably have this effect to some extent. Side-effects reported after the therapeutic use of CYTOKINES have provided evidence that activation of the immune response may sometimes have deleterious consequences, such as flu-like reactions, vascular leak syndrome and CYTOKINE release syndrome. CYTOKINE-induced exacerbation of underlying autoimmune or inflammatory diseases may be other complications of concern [9]. The occurrence of CYTOKINE release syndrome has also been reported as a serious consequence of the administration of certain therapeutic monoclonal antibodies [20].

Noncognate T-B cooperation

Reactive XENOBIOTICS may also stimulate adaptive immune responses by disturbing the normal cooperation of Th- and B CELLS. Normally, B CELLS receive costimulation from Th cells that cognately recognize (EPITOPES of) the same ANTIGEN. As such, B-CELL tolerance for AUTOANTIGENS is a corollary of the T-CELL tolerance for such ANTIGENS. However, when Th cells respond to nonself EPITOPES on B CELLS, such B CELLS may be noncognately stimulated by the Th cell. This occurs during graft-*versus*-host responses following bone marrow transplantation, when Th cells of the host recognize nonself EPITOPES on B CELLS of the graft and *vice versa*. This leads to T- and B-CELL activation and results in production of AUTOANTIBODIES to distinct AUTOANTIGENS like DNA, nucleoli, nuclear proteins, erythrocytes and basal membranes. Drug-related lupus is characterized by a similar spectrum of AUTOANTIBODIES, and it has therefore been suggested that noncognate – graft-*versus*-host-like – T-B cooperation caused by T-CELL reactivity to HAPTENS on (autoreactive) B CELLS is one of the underlying mechanisms (Fig. 6) [21]. Chlorpromazine, hydralazine, phenytoine, isoniazid, α-methyldopa and procainamide are just a few of the pharmaceuticals that are associated with drug-related lupus [22].

Interference with cellular activation

Occupation of various lymphocyte receptors results in a cascade of molecular processes that eventually lead to production of growth factors and cellular proliferation and/or activation. Its complexity makes this cascade vulnerable to pharmaceuticals at numerous stages, although most of these chemicals target very crucial processes like purine metabolism as in the case of AZA. Most drugs that interfere with cellular activation are not cell type-selective at higher doses, as all living cells depend on the same basic molecular processes. At low doses, however, drugs like AZA, CY and MTX appear to have a more selective effect. AZA, for instance, is claimed to selectively suppress T-LYMPHOCYTE function at low doses, whereas CY or MTX preferentially affect B CELLS [1].

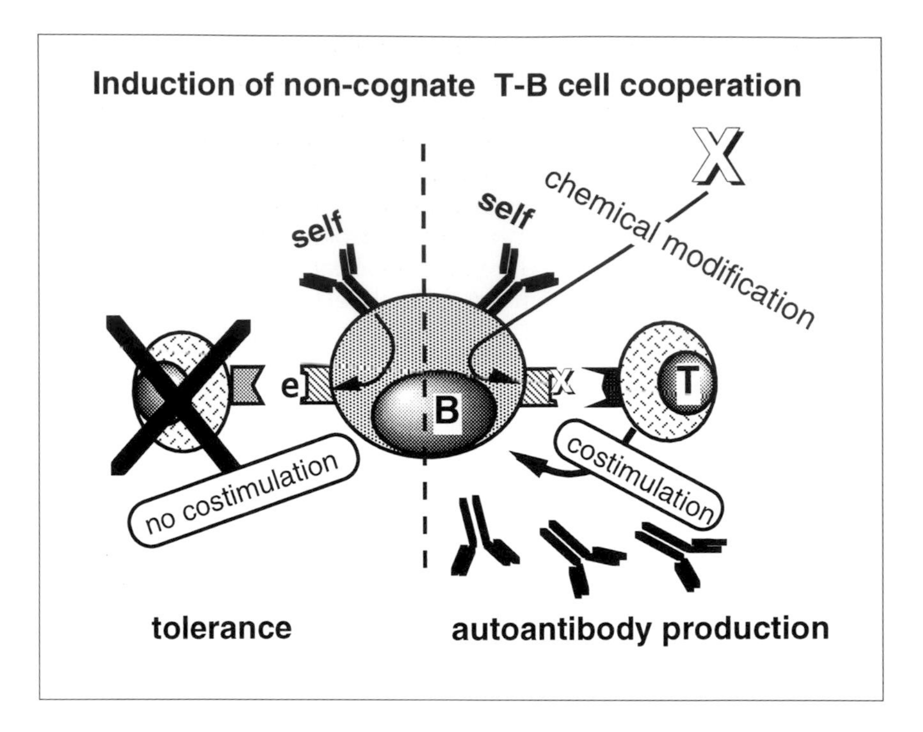

FIGURE 6. NON-COGNATE T-B CELL COOPERATION
*Normally, activation of B cells requires costimulation from activated T cells that cognately recognize epitopes of the same antigen. B cells with specificity for autoantigens do not receive this signal because T cells have learned to ignore autoantigens. However, certain pharmaceuticals can bind to B cell proteins and Th cells activated by the neo-epitopes created can non-cognately provide costimulation for the (e-specific) B cell (e, a part of self for which tolerance exists). This can bypass tolerance in the T-cell compartment and can lead to activation of autoreactive B cells and production of autoantibodies. This resembles stimulation of host B cells by graft T cells responding to the MHC molecules on the B cells during graft-*versus*-host reactions, and leads to a similar spectrum of autoantibodies.*

Suppressive effects of CsA and FK506, both interfering with the activation of the T-CELL-specific transcription factor NF-ATc, and rapamycin, preventing IL-2 receptor activation, are obviously more specific to the T LYMPHOCYTE.

Regulation of the immune response

Ongoing immune responses have to be carefully regulated in order to mount the most suitable defence (Fig. 1). Elimination of (intra)cellular targets like virally infected or neoplastic cells is most efficient by Th1-driven cellular responses using cytotoxic T CELLS and MACROPHAGES as effector mechanisms. Soluble targets, like extracellular bacteria and proteins, on the other hand, are most effectively eliminated by Th2-driven humoral responses, which rely on the formation of specific antibodies. The regulation of the type of immune response elicited and of the effector mechanisms activated is the result of a complex interplay of CYTOKINES produced by MACROPHAGES, DENDRITIC CELLS, MAST CELLS, GRANULOCYTES and LYMPHOCYTES (Chapter A4) and is influenced by a number of endo- and exogenous factors. Genetic make-up, in particular genes encoding for MHC molecules, but also gender (estrogens) are among the endogenous factors, whereas the type and dose of ANTIGEN, the route of exposure but also the type of (ongoing) costimulatory ADJUVANT activity are among the exogenous factors [23]. The role of the genetic makeup is illustrated by the model compound $HgCl_2$. This chemical is capable of inducing a Th1-dependent immunosuppressive state in an $H2^d$ strain of mice or an $RT1^l$ strain of rat, whereas it induces an autoreactive Th_2-dependent response in an $H2^s$ strain of mice or $RT1^n$ strain of rat [24, 25].

Within the same congeneic strain, however, different XENOBIOTICS can skew the response in opposing directions: $HgCl_2$ induces a Th2-like response in BALB/c mice, whereas the diabetogenic antitumor compound streptozotocin (STZ) induces a Th1 response in the same mouse strain [26]. Other examples are the ADJUVANTS complete Freund's ADJUVANT (CFA) and alum, stimulating the formation of immunoglobulin (Ig)G2a and IgG1/IgG2a isotypes of antibodies, respectively [23].

How chemicals exactly modulate the immune response is as yet unknown, but modulation of epitope selection by MHC molecules, selective activation of the innate immune system (e.g., MAST CELLS in the case of $HgCl_2$) and chemical-specific factors (e.g., macrophage activation by STZ) may be important factors.

Apart from regulation of the type of immune response, the occurrence of an immune response *per se* is tightly controlled. Recently, a number of regulatory T-CELL subsets have been identified that suppress auto-aggressive responses and ALLERGY [27]. A few recent studies indicate that regulatory cells are also involved in tolerance to orally encountered XENOBIOTICS. For instance, it has been demonstrated that $CD4^+CD25^+$ T CELLS prevent auto-antibody production by procainamide, gold sodium thiomalate or $HgCl_2$ in mice [28]. Similarly, induction of tolerance to adverse immune effects by D-penicillamine was at least partly mediated by T CELLS, possibly IFN-γ-producing $CD8^+$ T cells [29].

Effector functions and response cessation

To avoid unnecessary damage, the immune system has several feedback mechanisms to stop ineffective and obsolete responses (Fig. 1). The simplest feedback is the ANTIGEN itself, and complete degradation of the response-inducing ANTIGEN leads to response cessation. Pharmaceuticals that impair the activity of effector mechanisms delay ANTIGEN degradation and lead to accumulation of debris. It has been demonstrated, for instance, that several drugs, including D-penicillamine and procainamide, inhibit complement factor C4. This hampers the clearance of immune complexes and may therefore lead to their deposition and excessive tissue damage (reviewed in [30]).

Some drugs may also directly stimulate effector mechanisms, such as the complement system (contrast media) or effector cells, such as MAST CELLS and BASOPHILS. For instance the antibiotic vancomycin may induce histamine release and some non-steroidal anti-inflammatory drugs (NSAIDs) such as aspirin directly modulate the arachidonic acid pathway. In any case anaphylactoid clinical effects may occur without involvement of specific immune recognition. For those cases in which effector mechanisms are directly altered by drugs and clinically apparent allergic symptoms are seen, the term pseudo-ALLERGY is introduced (reviewed in [31]).

Clinical consequences of immunotoxicity of pharmaceuticals

In general, pharmaceuticals that inhibit cellular replication or activation induce IMMUNOSUPPRESSION which is dose-dependent. In particular, impaired activity of the first line of defense formed by the natural immune system can have disastrous consequences. These are generally not influenced by the genetic predisposition of the exposed individual, but actual outbreak of (opportunistic) infections or increased frequency of neoplasms may depend on the general immune status prior to exposition. This explains why immunosuppressive pharmaceuticals are most likely to have clinical consequences in immunocompromised individuals such as young children, the elderly, and transplant recipients.

Immunotoxic pharmaceuticals that somehow activate the immune system can lead to autoimmune or allergic diseases. Although on a drug-by-drug base immune-mediated adverse effects are mostly rare, overall it is estimated that 6–10% of all adverse drug reactions are immune-mediated and these immune-mediated adverse effects are the most frequent course of failure during clinical development [32].

Actual development of clinical symptoms is influenced by the route and duration of exposure, the dosage of the pharmaceutical, and by immunogenetic (MHC haplotype, Th1-type *versus* Th2-type responders) and pharmacogenetic (acetylator phenotype, sulfoxidizer, Ah receptor, etc.) predisposition of the exposed individual. From an immunological point of view it is clear that the polymorphic MHC molecules select the EPITOPES that are presented to T CELLS, and therefore influence all immune responses, including allergic and autoimmune responses induced by pharmaceuticals. Moreover, atopic individuals that tend to mount Th2 immune responses are more susceptible to anaphylaxis triggered by an IgE response to chemical HAPTENS than typical Th1 responders. Genetic variation in metabolism of pharmaceuticals is important as it determines the formation and clearance of immunotoxic metabolites. The slow acetylating phenotype, for instance, predisposes for drug-related lupus because reactive intermediates of phase I metabolism have an increased opportunity to bind proteins as they are only slowly conjugated.

Drugs known to interfere with the immune system are normally prescribed by well-trained physicians and are taken under more or less controlled conditions. As a result, adverse effects should be recognized as soon as they become apparent and measures can be taken before permanent harm is done. However, in the case of allergenic drugs, the immune system is sensitized, which may hamper future treatments with the same or a structurally related chemical (penicillin). Severe IMMUNOSUPPRESSION may cause permanent detrimental effects when neoplasms are formed, but also (opportunistic) infections may have serious consequences.

As the clinical consequence of exposure to immunotoxic pharmaceuticals ranges from immunodepressed conditions on the one hand to allergic and autoimmune diseases on the other hand, preclinical testing of pharmaceuticals in laboratory animals requires different approaches. In the following sections, procedures are covered comprising direct immunotoxicity as well as sensitizing capacity.

Procedures for preclinical testing of direct immunotoxicity

Testing in rodents by a tiered approach

Several laboratories have developed and validated a variety of methods to determine the effects of chemicals on the immune system of rats and mice [33]. Most employ a tier-testing system, whereas some investigators have advocated multiple testing in a single animal. The tier-testing approaches are similar in design in that the first tier is a screen for immunotoxicity with the second tier consisting of more specific or confirmatory studies, host resistance studies or in-depth mechanistic studies. At present, most information regarding these models comes from the model developed at the National Institute for Public Health and the Environment (RIVM) in Bilthoven, the Netherlands, and the model developed at the U.S. National Institute of Environmental Health Sciences National Toxicology Program (NIEHS-NTP). The RIVM tiered system [34,35] is based on the guideline 407 of the Organization for Economic Co-operation and Development (OECD), and performed in the rat using at least three dose levels, i.e., one resulting in overt toxicity, one aimed at producing no toxicity and one intermediate level. There is no immunization or challenge with an infectious agent. The first tier comprises general parameters including conventional hematology, serum immunoglobulin concentrations, bone marrow cellularity, weight and histology of LYMPHOID ORGANS [thymus, spleen, lymph nodes, MUCOSA-ASSOCIATED LYMPHOID TISSUE (MALT)], flow cytometric analysis of spleen cells and possibly immunophenotyping of tissue sections (Tab. 1). This approach has been used for the immunotoxic evaluation of pesticides [36] and pharmaceuticals [37].

The OECD guideline 407 includes weight of spleen and thymus, and histopathology of these organs, in addition to lymph nodes, PEYER'S PATCHES and bone marrow [38]. But, it should be borne in mind that in this guideline 407 some immunotoxic

TABLE 1. PANEL OF THE DUTCH NATIONAL INSTITUTE FOR PUBLIC HEALTH AND THE ENVIRONMENT FOR DETECTING IMMUNOTOXIC ALTERATIONS IN THE RAT

Parameters	Procedures
Tier 1	
Non-functional	- Routine hematology, including differential cell counting
	- Serum IgM, G, A, and E determination
	- Lymphoid organ weights (thymus, spleen, local and distant lymph nodes)
	- Histopathology of thymus, spleen, lymph nodes and mucosa-associated lymphoid tissue
	- Bone marrow cellularity
	- Analysis of lymphocyte subpopu-lations in spleen by flow cytometry
Tier 2 panel	
Cell-mediated immunity	- Sensitization to T-cell-dependent antigens (e.g., ovalbumin, tuberculin, *Listeria*), and skin test challenge
	- Lymphoproliferative responses to specific antigens (Listeria); mitogen responses (Con-A, PHA)
Humoral immunity	- Serum titration of IgM, IgG, IgA, IgE responses to T-cell-dependent antigens (ovalbumin, tetanus toxoid, *Trichinella spiralis*, sheep red blood cells) with ELISA;
	- Serum titration of T-cell-independent IgM response to LPS with ELISA
	- Mitogen response to LPS
Macrophage function	- *In vitro* phagocytosis and killing of *Listeria monocytogenes* by adherent spleen and peritoneal cells
	- Cytolysis of YAC-1 lymphoma cells by adherent spleen and peritoneal cells
NK cell function	- Cytolysis of YAC-1 lymphoma cells by non-adherent spleen and peritoneal cells
Host-resistance	- *Trichinella spiralis* challenge (muscle larvae counts and worm expulsion)
	- *Listeria monocytogenes* challenge (spleen clearance)
	- Rat cytomegalovirus challenge (clearance from salivary gland)
	- Endotoxin hypersensitivity
	- Autoimmune models (adjuvant arthritis, experimental allergic encephalomyelitis)

compounds may not be identified as such [39], for instance this concerned the opiate analgesic buprenorphine [40] and the long-acting b2-adrenoreceptor agonist salmeterol [41], which affected serum immunoglobulins in rats. Results of an interlaboratory validation study in the rat using cyclosporin A and hexachlorobenzene showed the importance of updating the 407 protocol with LYMPHOID ORGAN weights and serum immunoglobulin levels. The study concluded that histomorphological examination of lymphoid tissues resulted in the most reliable and sensitive data to be considered in regulatory toxicology and risk assessment [42].

In this OECD guideline 407 for testing toxicants, the immune system is not evaluated functionally. The inclusion of an *in vivo* ANTIGEN challenge test, e.g., with sheep red blood cells (SRBC), is currently considered to improve the sensitivity of the toxicity test. Results of a study [43] indicate that intravenous injection with SRBC during a 30- and 90-day toxicity study did not alter hematological and clinical chemistry parameters. With the expected exception of the spleen, administration of SRBC did not significantly alter the weights or morphology of routinely analysed tissues.

It should be noted that the array of tests currently included in the updated OECD guideline 407 is aimed at detecting potential immunotoxicity. Once immunotoxicity has been identified, further testing is required to confirm and extend the earlier findings. Further testing should include immune-function testing (Tab. 1). In addition to confirming functional implications of the immunotoxicity identified, functional tests are likely to provide information on no-adverse-effect levels, and are therefore valuable for the process of risk assessment. Caution is needed in determining the relevance of slight effects on immune parameters in view of the functional reserve capacity of the immune system. In those cases, infection models can be very helpful for risk assessment, as they are tools to elucidate the actual consequences of disturbances of immune function; effects observed using such infection models have surpassed the reserve capacity of the immune system. The fate of the pathogen and the associated host pathology may serve as indicators of the health implications of the immunotoxicity of the test chemical. Pathogens used in these host resistance models are chosen so that they are good models for human disease [35]. With some compounds, induction of immunotoxicity occurs especially during prenatal exposure. Yet, so far there are no immune parameters included in current OECD guidelines for developmental or reproductive toxicity testing [44].

The U.S. NTP has developed a tiered approach in mice that is linked closely to the standard protocol for chronic oral toxicity and carcinogenicity studies [45]. Exposure periods of 14–30 days have routinely been used at dose levels that have no effect on body weight or other toxicological end-points. In this way, compounds are identified for which the immune system represents the most sensitive target organ system. Tier 1 includes conventional hematology, LYMPHOID ORGAN weight, cellularity and histology of the spleen, thymus and lymph nodes, *ex vivo* splenic IgM-antibody plaque-forming cell assay following SRBC immunization, *in vitro* lymphocyte proliferation after stimulation with MITOGENS and allogeneic cells, and an *in vitro* assay for natural killer (NK) cell activity. In an adapted form of this approach, 51 different chemicals were evaluated, selected on the basis of structural relationships with previously identified immunotoxic chemicals [46]. The splenic SRBC IgM plaque-forming cell response and cell surface marker analyses showed the highest accuracy for identification of potential immunotoxicity.

Recently, the interlaboratory reproducibility of extended histopathology was studied by evaluating thymus, spleen and mesenteric lymph node of these past NTP studies performed in the mouse using 10 chemicals and 3 positive controls [47]. The consistency between 4 experienced toxicological pathologists with varied expertise in immunohistopathology was examined. Agreement between pathologists was highest in the thymus, in particular when evaluating thymus cortical cellularity, good in spleen follicular cellularity and in spleen and lymph node germinal center development, and poorest in spleen red pulp changes. The ability to identify histopathological change in lymphoid tissues was dependent upon the experience/training that the individual possesses in examining lymphoid tissue and the apparent severity of the specific lesion. In a further study, the accuracy of extended histopathology to detect immunotoxic chemicals was investigated [48]. While overall there was good agreement between histopathology and functional tests, the antibody forming cell (AFC) assay detected immune suppression in two instances where no changes in pathology were indicated. In contrast, the AFC assay failed to detect oxymetholone as IMMUNOTOXICANT, although extended histopathology indicated immunological changes. These data suggest that, while not as sensitive as functional tests, extended histopathology may provide a reasonable level of accuracy to identify immunotoxic chemicals. In the NTP protocol, in contrast to the OECD 407 guideline, the high dose was selected not to produce overt toxicity, thus limiting the likelihood of producing severe histopathological effects.

Immunotoxicity testing in nonrodent species

Various nonhuman primates, including *Macaca mulatta* (rhesus macaque), *Macaca nemestrina* (pig-

TABLE 2. ASSAYS RECOMMENDED FOR IMMUNOTOXICITY ASSESSMENT IN MAN

1. Complete blood count with differential count
2. Antibody-mediated immunity (one or more of following):
   - Primary antibody response to protein antigen (e.g., epitope-labelled influenza vaccine)
   - Immunoglobulin concentrations in serum (IgM, IgG, IgA, IgE)
   - Secondary antibody response to protein antigen (diphtheria, tetanus or polio)
   - Natural immunity to bloodgroup antigens (e.g., anti-A, anti-B)
3. Phenotypic analysis of lymphocytes by flow cytometry:
   - Surface analysis of CD3, CD4, CD8, CD20
4. Cellular Immunity:
   - Delayed-type hypersensitivity (DTH) skin testing
   - Primary DTH reaction to protein (KLH)
   - Proliferation to recall antigens
5. Autoantibodies and Inflammation:
   - C-reactive protein
   - Autoantibody titers to nuclei (ANA), DNA, mitochondria and IgE (rheumatoid factor)
   - IgE to allergens
6. Measure of non-specific immunity:
   - NK cell enumerations (CD56 or CD60) or cytolytic activity against K562 tumor cell line
   - Phagocytosis (NBT or chemiluminescence)
7. Clinical chemistry screen

   Proposal for all persons exposed to immunotoxicants

From: [30]

tailed macaque), *Macaca fascicularis* (cynomolgus monkey) and the marmoset, have been used in immunotoxicological studies. Virtually all of the immunotoxicology assays which are carried out in the mouse or rat can be and have been adapted for use with the nonhuman primates [49]. Phenotypic markers and functional assays in three different species of nonhuman primates were evaluated [50]. Functional assays included NK-cell activity, lymphocyte transformation and ANTIGEN presentation. The extensive phenotypic marker studies included the evaluation of over 20 markers or combinations of markers for each of the three monkey species. Otherwise, strategies and methods applied in studies in humans have been introduced in studies on nonhuman primates (see Tab. 2).

Also, other mammalian species have been used. While dogs are not the species of choice for immunotoxicological studies, they are one of the species predominantly used in toxicological safety assessments. Virtually all of the assays used for assessing immunotoxic potential have been adapted for use in the dog [33]. Among these are assay evaluation of basal immunoglobulin levels for IgA, IgG and IgM, ALLERGEN-specific serum IgE, mononuclear phagocyte function, NK-cell activity, cytotoxic T-CELL activity, and MITOGEN and cell-mediated immune responses.

Procedures for preclinical testing of sensitizing capacity

Structure-activity relationships

The intrinsic capacity of chemicals to exert adverse effects is linked to the structure of the compound. Structure-activity relationships with respect to direct toxicity of compounds to components of the immune system have received little attention. More attention has been given to structure-activity relationships with respect to the induction of ALLERGY. Here, structure-activity relationship models are directed towards a fuller understanding of the relationship between chemical structure and physicochemical properties and skin-sensitizing activity, in order ideally to derive quantitative structure-activity relationships (QSAR), linked perhaps to the development of expert, rule-based systems. In this context, parameters that appear to be of particular importance are protein reactivity and lipophilicity associated with the capacity to penetrate into the viable epidermis [51]. The correlation of the protein reactivity of chemicals with their skin SENSITIZATION potential is well established [52], so that it is now accepted that if a chemical is capable of reacting with a protein either directly or after appropriate (bio)chemical transformation, it has the potential to be a contact ALLERGEN, assuming of course that it can accumulate in the appropriate epidermal compartment. Each of the existing structure-activity relationship (SAR) models proposes structural alerts, i.e., moieties associated with sensitizing activity. In all cases, the structural alerts comprise electrophilic moieties, or moieties which can be metabolized into electrophilic fragments (proelectrophiles).

Testing for skin allergy

Guinea-pig models

The guinea pig was for many years the animal of choice for experimental studies of contact SENSITIZATION and several test methods were developed in this species (reviewed in [53]). The best-known and most widely applied are the Buehler test, the guinea-pig maximization test, and the guinea-pig optimization test, and have formed the basis of hazard assessment for many years. Both the guinea-pig maximization test and the Buehler test are now recommended according to an OECD guideline, accepted in 1992. While these tests differ with respect to procedural details, they are similar in principle. Guinea pigs are exposed to the test material, or to the relevant vehicle. In the Buehler test, both induction and challenge exposures are done topically. The Buehler test is sensitive, but false negatives are frequently observed. The test was improved by occluded application of the test compound. In the guinea-pig maximization test, induction involves intradermal and occluded epidermal exposure, and in the optimization test induction is done by intradermal, and challenge by intradermal and occluded epidermal exposure. ADJUVANT is employed in the guinea-pig maximization test and the optimization test to augment induced immune responses. Challenge-induced inflammatory reactions, measured as a function of erythema and/or edema, are recorded 24 and 48 h later. Classification of sensitizing activity in the guinea-pig tests is qualitative and not quantitative. It is based usually upon the percentage of test animals that display macroscopically detectable challenge reactions. Any compound inducing at least 30% positive animals in an ADJUVANT test is labeled as a sensitizer; in the case of a nonadjuvant test, 15% positivity is sufficient to classify the compound as a sensitizer.

Mouse models

In recent years, increased understanding of the cellular and molecular mechanisms associated with contact ALLERGY have derived largely from experimental investigations in the mouse [54, 55]. The most important test developed in mice is the local lymph node assay (LLNA) [56]. In contrast to guinea-pig assays described above, activity in the LLNA is measured by the primary T-CELL response in the draining lymph node following topical application to the mouse ear. Mice are treated daily, for 3 consecutive days, on the dorsum of both ears with the test material or with an equal volume of vehicle alone. Proliferative activity in draining lymph nodes (measured by the incorporation *in situ* of radiolabeled thymidine) is evaluated 5 days following the initiation of exposure. Currently, chemicals are classified as possessing sensitizing potential if with one test concentration a stimulation index, relative to vehicle-treated controls, of 3 or greater is induced. The method has the advantages of short duration and objective measurement of proliferation and minimal animal treatment. In contrast to guinea-pig assays, activity is measured as a function of events occurring during the induction, rather than elicitation phase, of contact SENSITIZATION.

Risk assessment of sensitizing chemicals requires, besides hazard identification, the assessment of potency. By using dose-response modelling (employment of a regression method that includes determination of the uncertainty margins) in the LLNA test, the potency of sensitizing chemicals can be determined, thus offering a possibility for classification [57]. The LLNA has been developed further to distinguish skin sensitizers from respiratory sensitizers based on the induction of $CD4^+$ T helper subsets (Th_1- *versus* Th_2-mediated responses) by the analysis of CYTOKINE profiles in draining lymph node cells [58]. Chemicals differ with respect to the types of HYPERSENSITIVITY they induce. Compounds that induce Th1 cells and mediate type IV delayed HYPERSENSITIVITY are skin sensitizers. Such responses are associated with the production by draining lymph node cells of IFN-γ. Compounds that induce Th2 cells and mediate type I immediate HYPERSENSITIVITY by the production of IgE and IgG1 are respiratory sensitizers, and are associated with the production by draining lymph node cells of high levels of IL-4. However, this is not true in all cases, as skin SENSITIZATION with some low molecular compounds such as picrylchloride [59] and toluene diisocyanate (TDI) [60] can induce respiratory HYPERSENSITIVITY with features of type IV HYPERSENSITIVITY in mice. Also, in humans, specific IgE is only detected in a minority of patients suffering from respiratory ALLERGY induced by TDI. In a recent study it is proposed that, by direct linkage of proliferation and CYTOKINE production in a dose-response manner, distinguishing contact ALLERGENS from respiratory ALLERGENS may be improved [61].

Testing for respiratory allergy

Most of the animal models that are used for studying specific respiratory tract HYPERSENSITIVITY were developed using high molecular weight ALLERGENS, notably proteins. Very few animal models have been developed as predictive tests for hazard identification and risk assessment in the area of chemical-induced respiratory ALLERGY [62]. The majority of these models are based upon antibody-mediated events. The models differ with regard to the following aspects: the animal species utilized, the route of administration of the agent, the protocol for both induction and elicitation of responses, type of response measured and judgment of significant response.

Guinea-pig models

The guinea pig has been used for decades for the study of anaphylactic shock and pulmonary HYPERSENSITIVITY. The guinea pig is similar to humans in that the lung is a major shock organ for anaphylactic responses to ANTIGENS. The guinea pig responds to histamine and can experience both immediate-onset and late-onset responses. Airway hyperreactivity and eosinophil influx and inflammation can also be demonstrated in this animal species. Mechanistic studies have been hampered by the lack of reagents needed to identify cells and mediators in respiratory ALLERGY. In addition, the major anaphylactic antibody is IgG1, whereas it is IgE in humans and other rodent species.

The guinea-pig model developed by Karol et al. [63] has proven to be valuable for low molecular weight chemical ALLERGENS. Guinea pigs sensitized by inhalation of free or protein-bound chemical ALLERGENS, such as TDI, will exhibit symptoms of pulmonary HYPERSENSITIVITY following subsequent inhalation challenge. HYPERSENSITIVITY reactions are measured usually as a function of challenge-induced changes in respiratory rate or alterations in other breathing parameters such as tidal volume. Changes in breathing patterns can also be provoked in dermally sensitized guinea pigs by inhalation challenge with the free chemical. In this approach it is not necessary to use hapten-protein conjugates.

A tiered approach to hazard assessment in guinea pigs proposed by Sarlo and Clark [64] comprises sequential analyses of physicochemical similarities with known ALLERGENS, the potential to associate covalently with protein, the ability to stimulate antibody responses and finally activity in a model of respiratory HYPERSENSITIVITY in which animals sensitized by subcutaneous injection are challenged by intratracheal instillation.

Mouse models

Models to investigate airway responses to sensitizing compounds have been developed in the mouse and comprise responses mediated by IgE [65] and non-IgE-mediated reactions [59, 60]. These models have not been used so far for predictive purposes.

As discussed earlier, analysis of the CYTOKINE profile in the LLNA in the mouse may provide information on whether a compound is a respiratory ALLERGEN. In the same series of investigations, it was found that topical administration to mice of chemical respiratory ALLERGENS stimulated a substantial increase in the serum concentration of total IgE, a response not seen with contact ALLERGENS considered to lack the ability to cause SENSITIZATION of the respiratory tract [58]. These observations suggested that it might be possible to identify chemical respiratory sensitizers as a function of induced changes in serum IgE concentration. The advantage of this approach, which forms the basis of the mouse IgE test, is that measurement of a serum protein is required, rather than of HAPTEN-specific antibody.

Investigations suggest that the mouse IgE test may provide a useful method for the prospective identification of chemical respiratory ALLERGENS [62]. It must be emphasized, however, that to date the assay has been evaluated only with a limited number of chemicals and that most of the analyses have been performed in a single laboratory. Difficulties arise from the assumption of IgE mediation of respiratory HYPERSENSITIVITY response in mice. As mentioned earlier, respiratory allergic responses, associated with increased reactivity of airways, may occur by a delayed type IV immune response-inducing compound [59]. For this reason, actual testing of lung functions *in vivo* seems prudent for those chemicals that are known to sensitize, but are unable to produce IgE responses.

Testing for autoimmunity

Autoimmune disease occurs when an individual's immune system attacks its own tissues or organs, resulting in functional impairment, inflammation and sometimes permanent damage. Diseases with multifactorial etiologies result from the loss of immune tolerance to self-antigens.

Induced and genetic models

For the detection of the potential of compounds to exacerbate induced or genetically predisposed AUTOIMMUNITY, animal models are available [66]. In induced models, a susceptible animal strain is immunized with a mixture of an ADJUVANT and an AUTOANTIGEN isolated from the target organ. Examples are ADJUVANT arthritis, experimental encephalomyelitis and experimental uveitis in the Lewis strain rat. Examples of spontaneous models of autoimmune disease are the BB-rat and the NOD-mouse that develop autoimmune pancreatitis and subsequently diabetes, and the $(NZBxNZW)F_1$ mouse or MRL/lpr mouse that develop pathology that resembles human systemic lupus erythematosus. These models are mainly used in the study of the pathogenesis of AUTOIMMUNITY and the preclinical evaluation of immunosuppressive drugs. Very few studies have addressed the potential of these models for assessment of whether a XENOBIOTIC exacerbates induced or congenital AUTOIMMUNITY.

Popliteal lymph node assay

Although currently no predictive assays have been developed and validated to identify, in the early phases of toxicity testing, the potential of drugs to induce systemic HYPERSENSITIVITY or autoimmune responses, it should be noted that available assays to identify contact sensitizers might also be helpful to identify systemic sensitizers. Clinical signs of systemic adverse immune-mediated effects usually become manifest only during advanced clinical development of drugs. The conditions used in routine preclinical toxicological screening are obviously not optimal for the detection of the immune-dysregulating potential of drugs and chemicals (e.g., small animal number, use of outbred animal strains, dynamics of disease development *versus* snapshot determinations, lack of predictive parameters).

AUTOIMMUNITY often results from the association of the compound with normal tisssue components, thereby rendering them immunogenic. A variety of chemicals and drugs, in particular the latter, have been found to induce autoimmune-like responses [67]. For the detection of chemicals that produce this type of reaction, the popliteal lymph node assay (PLNA) in mice is a promising tool [12]. The PLNA [68] is based upon the hyperplasia (increase in weight) of lymph nodes in graft-*versus*-host reactions or pseudo-graft-*versus*-host reactions, and has been modified to assess the immunomodulatory potential of drugs. The test substance is injected subcutaneously into one hind footpad, and the contralateral side is either untreated or inoculated with vehicle alone. Comparison of popliteal lymph nodes from both sides allows the effect of the test drug to be measured. Apart from differences in weight, histological evidence of *in vivo* immunostimulatory activity can be discerned. These pseudo-graft-*versus*-host reactions with follicular hyperplasia have been documented in mice for drugs such as diphenylhydantoin, D-penicillamine and streptozotocin. The assay appears to be appropriate to recognize sensitizing, i.e., ALLERGENic and autoimmunogenic chemicals, as well as nonsensitizing immunostimulating compounds, and has important advantages as it is a simple model based on local reactions that indicate direct immunostimulation with less interference of immunoregulatory mechanisms. So far, over 130 compounds (mainly pharmaceuticals and structural homologues) have been tested in the PLNA and outcomes (i.e., PLN-index: ratio of weight and cell numbers of PLN of compound-treated over vehicle-injected animals) appear to correlate well with documented adverse immune effects in humans [69]. Precaution has to be taken in the case of autoimmunogenic drugs, such as procainamide, that act as a prohapten. They are false-negative in the PLNA as such and require coinjection of metabolizing systems (S9 mix of GRANULOCYTES) to become positive. Results of a preliminary interlaboratory validation study indicate the potential predictive value of the PLNA in the rat as well [70]. Thus, the direct PLNA seems to be a versatile tool to recognize T-CELL activating drugs and chemicals, including autoimmunogenic chemicals, keeping in mind possible false-negative results with prohaptens. With the adoptive transfer PLNA, sensitized cells are used as probes to detect the formation *in vivo* of immunogenic metabolites of low molecular weight chemicals [12].

A recent modification of the PLNA uses bystander ANTIGENS to report the nature and type of immune response that is elicited by a given pharmaceutical [71]. In this assay, so-called reporter ANTIGENS (RA), either TNP-Ficoll or TNP-Ovalbumin (TNP-OVA), are injected together with the compound of interest and the IgG-response to the RA is measured. Here, it is important that unlike TNP-OVA, TNP-Ficoll cannot directly induce specific T-CELL activation. The IgG response to TNP-Ficoll is, however, susceptible to neoantigen specific T-CELL help, implying that an IgG response to TNP-Ficoll indicates that a co-injected compound has immunosensitizing potential. If a compound increases an IgG response to TNP-OVA and not to TNP-Ficoll, it can be concluded that the compound has the capacity to merely adjuvate an immune response. So, dependent on the type of RA, the IgG measured in this so-called RA-PLNA indicates whether compound-induced immunostimulation involves immunosensitization or proinflammatory ADJUVANT activity. The RA approach can also be applied in animal tests that use oral exposures to drugs [72].

It is important to note that the PLNA in any of its forms is a hazard identification test that can indicate whether a compound has the potency to induce adverse immune effects, including autoimmune derangements, in man.

Procedures for immunotoxicity testing in humans

Epidemiology design

It is obvious that many of the compounds causing direct immunotoxicity have been identified in rodent studies, as the database in humans is less complete and often inconclusive. The most common design used in immunotoxicity research in humans is the cross-sectional study, in which exposure parameters and effect parameters are assessed at the same time point [73]. The immune function of "exposed" subjects is compared with the immune function of "nonexposed" subjects by the measurement of various immunological parameters. For this reason, proper definition of exposure criteria in the exposed group is necessary. This group should include subjects at the upper end of exposure. Where possible, the study should incorporate individual estimates of exposure or actual measurements of the compound. In the broadest sense, biomarkers are measurements on biological specimens that will elucidate the relationship between environmental exposure and human diseases, so that exposure and diseases can be prevented. In clinical medicine, biomarkers are valued as a tool for the presence or absence of diseases or the course of the disease during therapeutic intervention. As such indicators are available for exposure, effect or susceptibility [74].

Markers of exposure

A biological marker of exposure is a XENOBIOTIC compound or its metabolite or the product of an interaction between the compound and some target cell or biomolecule. The most common markers of exposure are the concentration of the compound in urine, blood or target organ or tissue. Immune-specific biomarkers of exposure are antibodies or positive skin tests to the particular compound.

Markers of effect

A biomarker of effect is a measurable cellular or biochemical alteration within an organism that, depending on magnitude, can be recognized as an established or potential health impairment or disease. These range from markers of slight structural or functional changes to markers that are indicators of a subclinical stage of a disease or the manifestation of the disease itself. Functional changes in cells of the immune system by an immunotoxic chemical may be the first step in the process towards disease. For instance, longitudinal studies on asymptomatic individuals with low NK activity showed that they had an increased risk for upper respiratory infection and morbidity [75]. IMMUNOSUPPRESSION may lead to more subtle changes in resistance to infections, such as influenza or common cold, rather than opportunistic infection. Data in experimental animals also indicate that small changes in immune function could increase the likelihood of disease [46].

Markers of susceptibility

Markers of susceptibility, also called effect modifiers, can act at any point along the exposure-disease continuum. Important sources of variability are genetic, endocrine, age-related and environmental factors. Over the last 2 decades it has become clear that many immunological disorders are linked to alleles of the MHC gene complex. The products of MHC alleles in humans [human lymphocyte ANTIGENS (HLA)] have aroused interest at a clinical level as potential biomarkers of disease susceptibility. In some instances, there is a remarkable increase in relative risk of disease in individuals possessing particular alleles. Similar associations have been described in drug-induced immunological disorders. However, it should be noted that other genetic factors as well as environmental factors are also of importance. Stress of various types can also affect the immune system and influence the susceptibility to and recovery from infectious, autoimmune and neoplastic diseases. Age-related variability is shown by the developing fetus, which is more susceptible to immunotoxic effects than is the adult.

Assays for assessment of immune status

There is a plethora of tests developed to assess immunity in humans [76, 77], as described in laboratory manuals [78–80]. Many of these tests are nowadays commercially available as kits. A systematic approach to the evaluation of immune function, which is based on simple screening procedures followed by appropriate specialized tests of immune function, usually permits the definition of the immune alteration. This should include evaluation of the B-CELL system, of the T-CELL system and of nonspecific resistance (polymorphonuclear leukocytes, monocytes and MACROPHAGES, NK cells, the complement system).

Testing schemes for evaluation of individuals exposed to IMMUNOTOXICANTS are proposed, among others, by the Subcommittee on Immunotoxicology of the U.S. National Research Council [74] and by a task group of the World Health Organization (WHO) [33]. The panel proposed by the WHO is listed in Table 2, and is composed of assays that cover all major aspects of the immune system. Included are function assays to test for HUMORAL IMMUNITY, i.e., specific antibodies to tetanus or diphtheria (for which vaccination programs exist), and for cellular immunity using recall ANTIGENS. It should be mentioned that these tests were all developed for diagnostic purposes, but that in the context of immunotoxicity testing in humans they are to be used in an epidemiological setting. This means that distinctions found in parameters between an exposed group and a control group may have a different biological significance than an altered value in an individual. Whereas a decrease in a single immune parameter in an individual may not indicate increased susceptibility to disease, a subtle alteration in an immune biomarker in a population may indicate immunotoxicity.

Establishing immune changes in humans is considerably more complex than in animals, considering that noninvasive tests are limited, exposure levels to the agent (i.e., dose) are difficult to establish and responses in the population are extremely heterogeneous. Also, the normal population exhibits a wide range of immunological responses with no apparent health impact. In addition to this underlying population variability, certain host characteristics or common exposures may be associated with significant, predictable alterations in immunological parameters. If not recognized and effectively addressed in the study design or statistical analysis, these confounding factors may severely alter the results of population studies. Examples of factors associated with measurable alterations in immunological parameters include age, race, gender, pregnancy, acute stress and the behavioral ability to cope with stress, coexistent diseases or infections, nutritional status, lifestyle, tobacco smoking and some medications. Besides these variables, periodic (ranging from daily to seasonal) influences also exist. Some of these effects are relatively minor; other differences may be sufficiently large to exceed the expected effect from a low level of immunotoxic exposure. They are therefore of primary concern in large epidemiological studies. For a review of the influence of endogenous and environmental factors on vaccination responses see [81]. The importance of genetic factors on the response to a vaccine is shown by the role of CYTOKINE polymorphisms in the susceptibility of humans to ultraviolet B-induced modulation of immune responses after hepatitis B vaccination [82].

Predictive testing for allergy in humans

There are a variety of skin test procedures for the diagnosis of several types of allergic reactions, dealt with above. Basically, predictive tests in humans for skin ALLERGY are similar to diagnostic tests, but the aims are different. For diagnostic tests, the aim is to determine SENSITIZATION to chemicals to which there has been a prior exposure, whereas SENSITIZATION as a result of the procedure should be avoided. For predictive testing in humans, the aim is to show sensitizing capacity in individuals who have not been exposed previously to the compound.

For obvious reasons, predictive testing for respiratory SENSITIZATION is not done in humans. Occasionally, case reports may serve as an adequate hazard identification, but not as a risk estimate, because data on route and extent of exposure, and on the "population at risk" are usually missing. In the absence of case reports, it cannot be concluded that no potential for SENSITIZATION exists.

Immunotoxicity regulations

Regulatory guidance

There is great variation in the approaches adopted by regulatory agencies throughout the world to the control of human pharmaceuticals. Leading agen-

Box 1

For EMEA guidelines on immunotoxicity:

http://www.emea.eu.int/pdfs/human/swp/104299en.pdf Note for Guidance on Repeated Dose Toxicity, CPMP/SWP/1042/99, adopted July 2000

http://www.emea.eu.int/pdfs/human/swp/214500en.pdf Note for Guidance on Non-clinical Local Tolerance Testing of Medicinal Products. CPMP/SWP/2145/00, adopted February 2000.

http://www.emea.eu.int/pdfs/humans/swp/030295en.pdf Preclinical Safety Assessment of Biotechnology-Derived Pharmaceuticals. CPMP/ICH/302/95, adopted September 1997

For FDA guidelines on immunotoxicity:

http://www.fda.gov/cder/guidance/3010dft.pdf Guidance for Industry: Immunotoxicology Evaluation of Investigational New Drugs

For MHLW guideline on immunotoxicity:

a hyperlink is not available, as it is a draft guideline only.

cies involved in the regulation of pharmaceuticals for human use are the U.S. Food and Drug Administration (FDA), the Committee on Proprietary Medicinal Products (CPMP) of the European Agency for the Evaluation of Medicines (EMEA) in the European Union (EU), and the Ministry of Health, Labour and Welfare (MHLW) in Japan. The requirements the industry has to meet in order to gain marketing approval for its products have been laid down in official guidelines. These guidelines inform the industry about the data needed to demonstrate to the authorities the pharmaceutical quality of new pharmaceuticals to be marketed as well as their benefit and safety for the patient. The guidance given has a major impact on the development programmes adopted by the industry.

The regulations administered by government agencies are greatly influenced by the history, culture and legislations of the countries concerned. This still accounts for many national differences [83, 84]. However, worldwide harmonization of regulatory requirements is ongoing. In the 1990s, the International Conference on Harmonization of Technical Requirements for Registration of Pharmaceuticals for Human Use (ICH) has proved to be a success. In this international forum, government regulators and industry representatives of the three major regions of the world pharmaceutical market participated (i.e., United States, the EU, and Japan). A number of harmonized guidelines have been developed by the ICH. Most of these have already been adopted officially by the regulatory authorities mentioned. Existing national guidelines thus have been or will be replaced by new ones based on the ICH consensus.

Regulatory aspects of laboratory animal immunotoxicology

To identify potential target organs of toxicity in humans, the industry must screen the toxicity of pharmaceuticals in laboratory animals. Among targets such as liver and kidney, adverse effects on the immune system need to be assessed. As a rule, the regulatory authorities do not dictate how specific tests have to be conducted. The detailed technical requirements defined by OECD may or may not be followed. This approach allows deviations from routine protocol toxicity testing whenever justified. The study protocols may be adjusted in such a way that they provide the most relevant information depending on the nature and therapeutic indication of the pharmaceutical to be tested. Because immunotoxicology is a rather new discipline, guidelines are just recently (2000–2001) available in the EU, United States, and Japan which focus particularly on immunotoxicological issues regarding pharmaceuticals (see Box 1). In this regard we have to differentiate between the various aspects of immunotoxicity, i.e., IMMUNOSUPPRESSION, ALLERGY and AUTOIMMUNITY.

Unintended immunosuppression

Current predictive immunotoxicity testing is mainly done in the context of general toxicity, according to OECD guideline 407. Tests of the immune system comprise hematology, including differential cell counting, and histopathology of lymphoid tissues. Within OECD a debate as to functional testing, i.e., measurement of antibody responses to sheep red blood cells, is going on, in addition to discussions on the inclusion of measurement of NK activity and inclusion of FACS analysis of lymphocyte subpopulations.

The guidance documents from EU, USA and Japan embrace the concept of the tiered testing strategies described below, although there are differences in the sequence of tests recommended for the various tiers (see Tab. 3).

Both EMEA/CPMP, MHLW and FDA guidelines follow similar lines for immunotoxicity testing as the OECD. The major difference between the CPMP and the guidance in the other regions is that the CPMP has included functional testing in routine screening, mentioning explicitly the primary antibody response to a T CELL-dependent ANTIGEN. Measuring NK activity together with FACS analysis of lymphocyte subset populations might be used as an alternative approach. Such testing has been very instrumental in identifying compounds inducing unintended immune suppression. The FDA document advocates a cause-for-concern approach on the need for functional assays. A cause for concern is generally an adverse effect found in the toxicity endpoints, but might also be the intended use or the pharmacological activity of compounds (e.g., anti-HIV drugs, or anti-inflammatory activity). A list of additional studies mentioned in the various guidance documents is given in Table 4.

The MHLW draft guidance recommends lymphocyte subset phenotyping or spleen immunohistochemistry on a routine basis. Immunotoxicological findings in the repeated dose toxicity study require evaluation of the primary antibody response and the optional evaluation of NK-cell activity (Tier I). The Japanese document introduces 3 levels of screening; Tier I testing is required only when effects are observed in routine testing. Subsequently, Tier II testing is followed when effects are observed in Tier I. Currently only Europe requests functional assays in routine screening. Discussions within ICH tend to define a cause for concern approach, requiring functional testing as a result of defined concerns of potential immunotoxicity.

Sensitizing capacity

Regulatory HYPERSENSITIVITY testing generally focuses on locally applied compounds. Guidance documents focus predominantly on Type IV HYPERSENSITIVITY. The EMEA/CPMP Note for Guidance on Local Tolerance Testing refers to the OECD guidelines with regard to test methods. The OECD test 406 guideline on skin SENSITIZATION covers the guinea pig maximization test (GMPT) and the Buehler test (BT); OECD guideline 429 covers the LLNA. The FDA and CPMP regard the LLNA as a suitable stand-alone method for detecting HYPERSENSITIVITY potential. For the website on these guidelines see: *http://oecdpublications.gfi-nb.com/cgi-bin/OECDBookShop.storefront/*. The guidance document of the MHLW requests testing of all dermatological preparations in at least one skin SENSITIZATION study. The following tests are regarded as acceptable by the MHLW: ADJUVANT and Patch test, BT, Draize test, Freud's complete ADJUVANT test, GMPT, open cutaneous test, optimization test and the split ADJUVANT test. The most commonly used test in Japan is the GMPT, followed by the ADJUVANT Patch test or the BT. Table 5 provides an overview of the regional requirements.

Drugs intended for inhalation should be tested for their potential to induced Type I HYPERSENSITIVITY reaction according to the FDA. The Japanese and the

TABLE 3. PARAMETERS TESTING UNINTENDED IMMUNOSUPPRESSION IN THE (DRAFT) GUIDANCE OF THE EMEA/CPMP, THE FDA AND THE MHLW

		EMEA/CPMP	FDA	MHLW
Basic studies	Appearance and body weight	All	All	All
	White cell count and diff.	All	All	All
	Clinical chemistry	All	All	All
Organ weights	Spleen	All	All	All
	Thymus	All	All	All
	Adrenal	All	Suggested/all	All
	Lymph node, draining	All	–	–
	Lymph node, distant	All	–	–
Histopathology	Spleen	All	All	All
Tissues	Thymus	All	All	All
	Bone marrow smear	All	All	All
	Lymph node, mesenteric	All	If draining	-
	Lymph node, unspecified	All	All	All
	Lymph node, draining	All	All	-
	Peyer's patch	All	If draining	All
	MALT	–	If draining	-
Other studies	Lymphocyte subsets (A)	A or B		
	NK cell function (A)	All drugs		
	T-cell function (B)			

In the EMEA/CPMP Note for Guidance on Repeated Dose Toxicity the basic studies are called the „Initial screening phase" and the second round as „extended studies". In the guideline, the wording "Tiered" approach was avoided. The FDA Guidance describes "standard toxicology studies" and, if warranted, studies to determine drug effects on immune function should be considered.
The MHLW draft Guidance of Immunotoxicity Testing describes basis toxicity studies and all further steps are scheduled in Tier I and Tier II.

European authorities have not issued a non-clinical requirement for inhalation ALLERGY testing.

Autoimmunity

Because of the lack of suitable and validated animal models the potential risk for inducing AUTOIMMUNITY cannot be predicted. As regulatory guidance is following science, such guidance for AUTOIMMUNITY is not available. The FDA guidance document touches the issue of AUTOIMMUNITY mentioning the PLNA as a proposed test for AUTOIMMUNITY. It recognizes, however, that an extensive evaluation to support its use is lacking thus far. Biomarkers of T-CELL activation or markers of Th2 cell induction and autoantibody induction in experimental animals may be helpful, but their predictive value for determining the potential to induce autoimmune disease in humans has not been ascertained. Neither the EMEA nor the MHLW has released any guidance on AUTOIMMUNITY testing.

Immunotoxicology of biotechnological products

In 1997 ICH issued a guideline on "Preclinical Safety Assessment of Biotechnology-Derived Pharmaceuticals", which is adopted and implemented by all

Table 4. List of additional studies mentioned in the various guidance documents

Additional studies			
Delayed-type hypersensitivity[1]	EMEA/CPMP	FDA	MHLW
Lymphocyte proliferation (mitogen) responses/ blastogenesis	EMEA/CPMP	FDA	MHLW
Macrophage function/ phagocytosis activity	EMEA/CPMP	FDA	MHLW
Host resistance models	EMEA/CPMP	FDA	MHLW
Cytotoxic T-cell function		FDA	MHLW
Specific cytokine production		FDA	MHLW
Serum complement titer			MHLW
Popliteal lymph node responses[1]			MHLW
Immediate hypersensitivity[1]			MHLW
Autoantibodies			MHLW
Urine protein levels			MHLW

[1]Responses to other sensitizing antigens (not to the tested compound)

Table 5. Test requirements for topically applied drugs

Assays	EMEA	FDA	MHLW
GPMT	All[A]	All[A]	All[A]
BA	All[A]	All[A]	All[A]
LLNA	All[A]	All[A]	All[A]
MEST	All	All	All[A]
Adjuvant and Patch test	–	–	All[A]
Draize test	–	–	All[A]
FCA	–	–	All[A]
Open epicutaneous test	–	–	All[A]
Optimization test	–	–	All[A]
Spilt adjuvant test	–	–	All[A]
MIGET	–	Inhalation drugs	–

All Drugs, all topically applied drugs
All Drugs[A], test regarded as stand-alone assay

three regions (for the EMEA link see below). This document contains guidance on the assessment of immunogenicity as many products of this type are immunogenic in animals. Antibody responses should be characterised, and their appearance should be correlated with any pharmacological and/ or toxicological changes. The finding of antibodies against the product should, however, not be the sole criterion for the early termination of an animal study. Only in case of neutralisation of the response in the majority of the animals the study should be stopped. It should be kept in mind that antibody formation in animals is not predictive of a potential for antibody formation in humans.

Little specific information is present on general immunotoxicity. Inflammatory responses might be indicative of a stimulatory immune response, but may also be the result of toxic changes. Routine tiered testing approaches are not recommended for biotechnology-derived pharmaceuticals.

The ICH S6 document is the first one mentioning developmental immunotoxicity, but again the guidance is very general, indicating that the study design of the toxicity studies may be modified, e.g., in case of monoclonal antibodies with prolonged immunological effects. The more recent FDA guidance suggests the incorporation of immunotoxicological determinations in the ICH Stage C F reproductive toxicology study if a drug is expected to be used in pregnant women. The guidance documents from the other areas do not address developmental toxicity specifically.

The lack of detailed instructions is due to the fact that all three regions concerned have adopted a flexible, case-by-case, science-based approach to animal toxicity studies in this rapidly evolving scientific area. Consequently, the three regions involved in ICH are reluctant to formulate in-depth requirements which may turn out to be too strict. Such guidance might be valid for some biotechnology-derived products, but not for others.

Antigenicity

The Japanese Guidelines for Nonclinical Studies of Drugs do require antigenicity studies to be conducted on a case-by-case basis for conventional drugs. There is a lot of resistance to this request of the MHLW. Just recently, a representative of the Japanese government has indicated that they believe antigenicity data of conventional compounds are not predictive for any risks of these compounds, which promised to be a first step in deleting this issue from the guidance.

New developments

Under the auspices of the International Conference on Harmonization, a process has been started in November 2003 to harmonize the approach in screening human pharmaceuticals with regard to unintended IMMUNOSUPPRESSION. This process include the building of a database of pharmaceutical compounds being tested with respect to immune function. Until a harmonised document is available (expected no earlier than June 2006) the guidance documents of the individual regions remain in force.

Furthermore, the NIEHS and the NIOSH have started a discussion to develop a consensus document on the most appropriate experimental approaches and assays available to assess developmental immunotoxicity [85]. This initiative is also supported by the International Life Sciences Institute in Washington who organized a conference on this topic in 2003. Examples of pharmaceuticals as developmental IMMUNOTOXICANTS in rodents are acyclovir, cyclosporin A, cyclophosphamide, corticosteroids, benzodiazepines, azathioprine and tacrolimus. As mentioned above the FDA suggests incorporation of immunotoxicological determinations in reproductive toxicology studies.

In the future transgenic animals or adoptive transfer models might contribute to the insight in immunotoxic mechanisms of compounds *in vivo*. Another contribution in the screening for immunotoxicity can be expected from the toxicogenomics (microarray technology) and proteomics.

Conclusions

Preclinical testing of pharmaceuticals in laboratory animals requires different approaches to detect direct immunotoxicity, resulting in unwanted IMMUNOSUPPRESSION or immunostimulation, or to detect drug-induced HYPERSENSITIVITY and AUTOIMMUNITY. Tiered immunotoxicity-testing procedures have been developed and validated in the rat and mouse, and are being used successfully to detect drug-induced direct immunotoxicity.

Drug-induced HYPERSENSITIVITY and autoimmune reactions are of great clinical concern. For contact ALLERGY, routine contact SENSITIZATION testing in guinea pigs has been extended by the LLNA in the mouse as a stand-alone method for detecting HYPERSENSITIVITY potential. An important issue in contact ALLERGY is the development of quantitative measurements of the potency of ALLERGENS. No validated models are yet available to investigate the ability of drugs to induce respiratory SENSITIZATION. LLNA in mice or skin SENSITIZATION testing in guinea pigs should be recommended as a first screen.

Animal models are currently available to detect the potential of compounds to exacerbate induced or genetically predisposed AUTOIMMUNITY, but are seldom used in immunotoxicity studies. Models to investigate the ability of chemicals to induce AUTOIMMUNITY, as a result of an immune response to self-proteins modified by the chemical, are virtually limited to PLNA. As human data show that chemical agents, in particular drugs, can cause autoimmune diseases, new models should be developed.

In conclusion, immunotoxicology is a rapidly evolving field in the regulation of pharmaceuticals. This warrants updates of the guidelines available to date. The standard use of a large immunotoxicological test battery should not be recommended. Instead, a flexible approach on a case-by-case basis is needed, taking into account that some classes of drugs and some indications may be a greater cause of concern than others.

Summary

Immunotoxicity comprises direct toxicity to components of the immune system, resulting in suppressed function or inadvertent stimulation. Consequences may be reduced resistance to infections or tumours, or promotion of allergy or autoimmunity. Other types of immunotoxicity include allergy or autoimmunity by compounds that cause the immune system to respond as if the compound was an antigen or to respond to self-antigens that have been altered by the compound.

This chapter describes mechanisms of immunotoxicity by pharmaceuticals, procedures of preclinical testing of immunotoxicity of pharmaceuticals, procedures of immunotoxicity testing in humans, and immunotoxicity regulations.

Selected Readings

IPCS (1996) International Programme for Chemical Safety, Environmental Health Criteria Document 160. *Principles and methods for assessing direct immunotoxicity associated with chemical exposure*. Geneva: World Health Organization

Kimber I, Mitchell JA, Griffin AC (1986) Development of a murine local lymph node assay for the determination of sensitizing potential. *Food Chem Toxicol* 24: 585–586

Pieters R, Albers R (1999) Screening tests for autoimmune-related immunotoxicity. *Env Health Persp* 107 (Suppl 5): 673–677

Putman E, Van der Laan JW, Van Loveren H (2003) Assessing immunotoxicity: guidelines. *Fund Clin Pharmacol* 17: 615-626

Luster MI, Dean JH, Germolec DR (2003) Consensus workshop on methods to evaluate developmental immunotoxicity. *Environ Health Perspect* 111: 579–583

References

1 Descotes JG, Vial T (1994) Cytoreductive drugs. In: JH Dean, MI Luster, AE Munson, I Kimber (eds): *Immunotoxicology and Immunopharmacology*, 2nd edn. Raven Press, New York, 293–301

2 Broide DH (1991) Inflammatory cells:structure and function. In: DP Stites, AI Terr (eds): *Basic and Clinical Immunology*. Prentice Hall International, London, 141–153

3 Ryffel B, Car BD, Eugster H-P, Woerly G (1994) Transplantation Agents. In: JH Dean, MI Luster, AE Munson, I Kimber (eds): *Immunotoxicology and Immunopharmacology*, 2nd edn. Raven Press, New York, 267–292

4 Majoor GD, Wodzig WH, Vriesman PJCVB (1991) Cyclosporin-induced autoimmunity. In: MD Kendall, MA Ritter (eds): *Thymus Update: Thymus in Immunotoxicology*. Harwood Academic Publishers, London, 179–200

5 Cohen JJ (1992) *Glucocorticoid-induced Apoptosis in the Thymus. Seminars in Immunology*. Academic Press, New York

6 Clark EA, Ledbetter JA (1994) How B- and T-cells talk to each other. *Nature* 367: 425–428

7 Matzinger P (1994) Tolerance, danger and the extended family. *Annu Rev Immunol* 12: 991–1045

8 Crommelin DJA, Storm G, Verrijk R, de Leede L, Jiskoot W, Hennink WE (2003) Shifting paradigms: biopharmaceuticals *versus* low molecular weight drugs. *Int J Pharmaceutics* 266: 3–16

9 Vial T, Descotes J (1995) Immune-mediated side effects of cytokines in humans. *Toxicology* 105: 31–57

10 Uetrecht J (1990) Drug metabolism by leukocytes and its role in drug-induced lupus and other idiosyncratic drug reactions. *Crit Rev Toxicol* 20: 213–235

11 Kubicka-Muranyi M, Kremer J, Rottmann N, Lübben B, Albers R, Bloksma N, Lührmann R, Gleichmann E (1996) Murine systemic autoimmune disease induced by mercuric chloride: T helper cells reacting to self proteins. *Int Arch Allergy Immunol* 109: 11–20

12 Pichler WJ (2003) Drug-induced autoimmunity. *Curr Opin Allergy Clin Immunol* 3: 249–253

13 Griem P, Panthel K, Kalbacher H, Gleichmann E (1996) Alteration of a model antigen by Au (III) leads to T-cell sensitization to cryptic peptides. *Eur J Immunol* 26: 279–287

14 Bloksma N, Kubicka-Muranyi M, Schuppe H-C, Gleichmann E, Gleichmann H (1995) Predictive immunotoxicological test systems: suitability of the popliteal lymph node assay in mice and rats. *Crit Rev Toxicol* 25: 369–396

15 Griem P, Gleichmann E (1995) Metal ion induced autoimmunity. *Curr Opin Immunol* 7: 831–838

16 Janeway CA (1992) The immune system evolved to discriminate infectious nonself from noninfectious self. *Immunol Today* 13: 11–16

17 Cohen IR (1992) The cognitive paradigm and the immunological homunculus. *Immunol Today* 13: 490–494

18 Ibrahim MAA, Chain BM, Katz DR (1995) The injured cell: the role of the dendritic cell system as a sentinel receptor pathway. *Immunol Today* 16: 181–186

19 Nierkens S, Van Helden P, Bol M, Bleumink R, Van Kooten P, Ramdien-Murli S, Boon L, Pieters R (2002) Selective requirement for CD40-Cd154 in drug-induced type 1 *versus* type 2 responses to trinitrophenyl-ovalbumin. *J Immunol* 168: 3747–3754

20 Sgro C (1995) Side effects of a monoclonal antibody, muromonab CD3/orthoclone OKT3: bibliographic review. *Toxicology* 105: 23–29

21 Gleichmann E, Pals ST, Rolink AG, Radaszkiewicz T, Gleichmann H (1984) Graft-*versus*-host reactions: clues to the etiopathology of a spectrum of immunological diseases. *Immunol Today* 5: 324–332

22 Adams LE, Hess EV (1991) Drug-related lupus. Incidence, mechanisms and clinical implications. *Drug Safety* 6: 431–449

23 Janeway CA, Travers P (1994) *Immunobiology: The Immune System in Health and Disease*. Current Biology, Oxford

24 Druet P, Pelletier L, Rossert J, Druet E, Hirsch F, Sapin C (1989) Autoimmune reactions induced by metals. In: ME Kammuller, N Bloksma, W Seinen (eds): *Autoimmunity and Toxicology: Immune Disregulation Induced by Drugs and Chemicals*. Elsevier, Amsterdam, 349–361

25 Van Vliet E, Uhrberg M, Stein C, Gleichmann E (1993) MHC control of IL-4-dependent enhancement of B-cell Ia expression and Ig class switching in mice treated with mercuric chloride. *Int Arch Allergy Immunol* 101: 392–401

26 Albers R (1996) *Chemical-induced Autoimmunity – Immune Disregulation Assessed with Reporter Antigens*. Utrecht University, PhD thesis

27 Jonulit H, Schmitt E (2003) The regulatory T cell family: distinct subsets and their interrelations. *J Immunol* 171: 6323–6327

28 Layland LE, Wulferink M, Dierkes S, Gleichmann E (2004) Drug-induced autoantibody formation in mice: triggering by primed $CD4^+CD25^-$ T cells, prevention by primed $CD4^+CD25^+$ T cells. *Eur J Immunol* 34: 36–46

29 Mason MJ, Uetrecht J (2004) Tolerance induced by low dose D-penicillamine in the Brown Norway rat model of drug-induced autoimmunity is immune mediated. *Chem Res Toxicol* 17: 82–94

30 Coleman JW, Sim E (1994) Autoallergic responses to drugs: mechanistic aspects. In: JH Dean, MI Luster, AE Munson, I Kimber (eds): *Immunotoxicology and Immunopharmacology*, 2nd edn. Raven Press, New York, 553–572

31 Descotes J, Choquet-Kastylevsky G (2001) Gell and Coombs's classification: is it still valid? *Toxicol* 158: 43–49

32 Adkinson NF Jr, Essayan D, Gruchalla R, Haggerty H, Kawabata T, Sandler JD, Updyke L, Shear NH, Wierda D (2002) Task force report: future research needs for the prevention and management of immune-mediated drug hypersensitivity reactions. *J Allergy Clin Immunol* 109: S461–S478

33 IPCS (1996) International Programme for Chemical Safety, Environmental Health Criteria Document 160. *Principles and methods for assessing direct immunotoxicity associated with chemical exposure*. Geneva: World Health Organization

34 Vos JG (1980) Immunotoxicity assessment: screening and function studies. *Arch Toxicol* (Suppl) 4: 95–108

35 Van Loveren H, Vos JG (1989) Immunotoxicological considerations: a practical approach to immunotoxicity testing in the rat. In: AD Dayan, AJ Paine (eds): *Advances in Applied Toxicology*. Taylor and Francis, London, 143–165

36 Vos JG, Krajnc EI (1983) Immunotoxicity of pesticides. In: AW Hayes, RC Schnell, TS Miya (eds): *Developments in the Science and Practice of Toxicology*. Elsevier, Amsterdam, 229–240

37 De Waal EJ, Van Loveren H, Vos JG (1997) Practice of tiered testing for immunosuppression in rodents. *Drug Inf J* 31: 1317–1323

38 Koeter HBWM (1995) International harmonisation of immunotoxicity testing. *Hum Exp Toxicol* 14: 151–154

39 Van Loveren H, Vos JG, De Waal EJ (1996) Testing immunotoxicity of chemicals as a guide for testing approaches for pharmaceuticals. *Drug Inf J* 30: 275–279

40 Van Loveren H, Gianotten N, Hendriksen CFM, Schuurman H-J, Van Der Laan JW (1994) Assessment of immunotoxicity of buprenorphine. *Lab Animals* 28: 355–363

41 De Waal EJ, De Jong WH, Van Der Vliet H, Verlaan B, Van Loveren H (1996) An immunotoxicity screening study on salmeterol in rats. *Int J Immunopharmacol* 18: 523–528

42 Schulte A, Althoff J, Ewe S, Richter-Reichhelm HB (2002) Two immunotoxicity ring studies according to OECD TG 407-comparison of data on cyclosporin A and hexachlorobenzene. *Regul Toxicol Pharmacol* 36: 12–21

43 Ladics GS, Smith C, Heaps K, Ellioo GS, Slone TW, Loveless SE (1995) Possible incorporation of an immunotoxicological funtional assay for assessing humoral immunity for hazard identification purposes in rats on standard toxicology study. *Toxicology* 96: 225–238

44 Van Loveren H, Vos J, Putman E, Piersma A (2003) Immunotoxicological consequences of perinatal chemical exposures: a plea for inclusion of immune parameters in reproduction studies. *Toxicol* 185: 185–191

45 Luster MI, Munson AE, Thomas PT, Holsapple MP, Fenters JD, White KL, Lauer LD, Germolec DR, Rosenthal GJ, Dean JH (1988) Development of a testing battery to assess chemical-induced immunotoxicity: National Toxicology Program's guidelines for immunotoxicity evaluation in mice. *Fund Appl Toxicol* 10: 2–19

46 Luster MI, Portier C, Pait DG, Rosenthal GJ, Germolec DR, Corsini E, Blaylock BL, Pollock P, Kouchi Y, Craig W et al (1993) Risk assessment in immunotoxicology. II. Relationship between immune and host resistance tests. *Fund Appl Toxicol* 21: 71–82

47 Germolec DR, Nyska A, Kashon M, Kuper CF, Portier C, Kommineni C, Johnson KA, Luster MI (2004) Extended histopathology in immunotoxicity testing: interlaboratory validation studies. *Toxicol Sci* 78: 107–115

48 Germolec DR, Kashon M, Nyska A, Kuper CF, Portier C, Kommineni C, Johnson KA, Luster MI (2004) The accuracy of extended histopathology to detect immunotoxic chemicals. *Toxicol Sci* 82: 504–514

49 Bugelski PJ, Thiem PA, Solleveld HA, Morgan DG (1990) Effects of sensitization to dinitrochlorobenzene (DNCB) on clinical pathology parameters and mitogen-mediated blastogenesis in cynomolgus monkeys (Macaca fascicularis). *Toxicol Pathol* 18: 643–650

50 Ahmed-Ansari A, Brodie AR, Fultz PN, Anderson DC, Sell KW, McClure HM (1989) Flow microfluorometric analysis of peripheral blood mononuclear cells from nonhumam primates: Correlation of phenotype with immune function. *Amer J Primatol* 17: 107–131

51 Barratt MD, Basketter DA, Chamberlain M, Admans GD, Langowski JJ (1994) An expert system rulebase for identifying contact allergens. *Toxicol Vitro* 8: 1053–1060

52 Dupuis G, Benezra C (1982) *Contact Dermatitis to Simple Chemicals: a Molecular Approach*. Marcel Dekker, New York

53 Maurer T (1996) Predictive testing for skin allergy. In: JG Vos, M Younes, E Smith (eds): *Allergic Hypersensitivities Induced by Chemicals. Recommendations for Prevention*. CRC Press, Boca Raton, 237–259

54 Garssen J, Vandebriel RJ, Kimber I, Van Loveren H (1996) Hypersensitivity reactions: definitions, basic mechanisms, and localizations. In: JG Vos, M Younes, E Smith (eds): *Allergic Hypersensitivities Induced by Chemicals. Recommendations for Prevention*. CRC Press, Boca Raton, 19–58

55 Kimber I, Dearman RJ (1996) Contact hypersensitivity: immunological mechanisms. In: I Kimber, T Maurer (eds): *Toxicology of Contact Hypersensitivity*. Taylor and Francis, London, 14–25

56 Kimber I, Mitchell JA, Griffin AC (1986) Development of a murine local lymph node assay for the determina-

tion of sensitizing potential. *Food Chem Toxicol* 24: 585–586

57 Van Och FM, Slob W, de Jong WH, Vandebriel RJ, van Loveren H (2000) A quantitative method for assessing the sensitizing potency of low molecular weight chemicals using a local lymph node assay: employment of a regression method that includes determination of the uncertainty margins. *Toxicol* 20: 49–59

58 Dearman RJ, Mitchell JA, Basketter DA, Kimber I (1992) Differential ability of occupational chemical contact and respiratory allergens to cause immediate and delayed dermal hypersensitivity reactions in mice. *Int Arch Allergy Appl Immunol* 97: 315–321

59 Garssen J, Nijkamp FP, Van Der Vliet H, Van Loveren H (1991) T-cell mediated induction of airway hyperreactivity in mice. *Am Rev Resp Dis* 144: 931–938

60 Scheerens H, Buckley TL, Davidse EM, Garssen J, Nijkamp FP, Van Loveren H (1996) Toluene diisocyanate-induced *in vitro* tracheal hyperreactivity in the mouse. *Am J Respir Crit Care Med* 154: 858–865

61 Van Och FM, Van Loveren H, De Jong WH, Vandebriel RJ (2002) Cytokine production induced by low-molecular-weight chemicals as a function of the stimulation index in a modified local lymph node assay: an approach to discriminate contact sensitizers from respiratory sensitizers. *Toxicol Appl Pharmacol* 184: 46–56

62 Karol MH (1996) Predictive testing for respiratory allergy. In: JG Vos, M Younes, E Smith (eds): *Allergic Hypersensitivities Induced by Chemicals. Recommendations for Prevention*. CRC Press, Boca Raton, 125–137

63 Karol MH (1994) Animal models of occupational asthma. *Eur Respir J* 7: 555–568

64 Sarlo K Clark ED (1992) A tier approach for evaluating the respiratory allergenicity of low molecular weight chemicals. *Fund Appl Toxicol* 85: 55–58

65 Hessel EM, Van Oosterhout AJM, Hofstra CL, De Bie JJ, Garssen J, Van Loveren H, Verheyen AKCP, Savelkoul HFJ, Nijkamp FP (1995) Bronchoconstriction and airway hyperresponsiveness after ovalbumin inhalation in sensitized mice. *Eur J Pharmacol Env Tox Pharmacol Sect* 293: 401–412

66 Kammüller ME, Bloksma N, Seinen W (eds) (1989) *Autoimmunity and Toxicology: Immune Dysregulation Induced by Drugs and Chemicals*. Elsevier, Amsterdam

67 Gleichmann E, Vohr HW, Stringer C, Nuyens J, Gleichmann H (1989) Testing the sensitization of T-cells to chemicals. From murine graft-*versus*-host (GVH) reactions to chemical-induced GVH-like immunological diseases. In: ME Kammüller, N Bloksma, W Seinen (eds): *Autoimmunity and Toxicology: Immune Dysregulation Induced by Drugs and Chemicals*. Elsevier, Amsterdam, 364–390

68 Gleichmann E, Kind P, Schuppe HC, Merk H (1990) Tests for predicting sensitization to chemicals and their metabolites, with special reference to heavy metals. In: AD Dayan, RF Hertel, E Heseltine, G Kazantzis, EM Smith, Van Der Venne MT (eds): *Immunotoxicity of metals and Immunotoxicology*. Plenum Press, New York, 139–152

69 Pieters R, Albers R (1999) Screening tests for autoimmune-related immunotoxicity. *Env Health Persp* 107 (Suppl 5): 673–677

70 Vial T, Carleer J, Legrain B, Verdier F, Descotes J (1997) The popliteal lymph node assay: results of a preliminary interlaboratory validation study. *Toxicology* 122: 213–218

71 Albers R, Broeders A, ven der Pijl A, Seinen W, Pieters R (1997) The use of reporter antigens in the popliteal lymph node assay to assess immunomodulation by chemicals. *Toxicol Appl Pharmacol* 143: 102–109

72 Gutting BW, Updyke LW, Amacher DE (2002) BALB/c mice orally pretreated with diclofenac have augmented and accelerated PLNA responses to diclofenac. *Toxicol* 172: 217–230

73 Selgrade MJ, Cooper KD, Devlin RB, Van Loveren H, Biagini RE, Luster MI (1995) Immunotoxicity-Bridging the gap between animal research and human health effects. *Fund Appl Toxicol* 24: 13–21

74 National Research Council (1992) *Biological Markers in Immunotoxicology*. National Academy Press, Washington

75 Levy SM, Herberman RB, Lee J, Whiteside T, Beadle M, Heiden L, Simons A (1991) Persistantly low natural killer activity, age, and environmental stress as predictors of infectious morbidity. *Nat Immun Cell Growth Regul* 10: 289–307

76 Bentwich Z, Bianco N, Jager L, Houba V, Lambert PH, Knaap W, Rose N, Seligman M, Thompson R, Torrigiani G et al. (1982) Use and abuse of laboratory tests in clinical immunology: Critical considerations of eight widely used diagnostic procedures. Report of a Joint IUIS/WHO Meeting on Assessment of Tests Used in Clinical Immunology. *Clin Immunol Immunopathol* 24: 122–138

77 Bentwich Z, Beverley PCL, Hammarstrom L, Kalden JR, Lambert PH, Rose NR, Thompson RA (1988) Laboratory investigations in clinical immunology: Methods, pitfalls, and clinical indications. *Clin Immunol Immunopathol* 49: 478–497

78 Coligan JE, Kruisbeek AM, Margulies DH, Shevach EM, Strober W (eds) (1994) *Current Protocols in Immunology*, Vol. 1 and 2. John Wiley and Sons, New York

79 Lawlor GJ, Fischer TJ (eds.) (1988) *Manual of Allergy and Immunology. Diagnosis and Therapy*, 2nd edn. Little Brown, Boston

80 Miller LE, Ludke HR, Peacock JE, Tomar RH (eds.) (1991) *Manual of Laboratory Immunology*, 2nd edn. Lea and Fabiger, Philadelphia

81 Van Loveren H, Van Amsterdam JG, Vandebriel RJ, Kimman TG, Rumke HC, Steerenberg PS, Vos JG (2001) Vaccine-induced antibody responses as parameters of the influence of endogenous and environmental factors. *Environ Health Perspect* 109: 757–764

82 Sleijffers A, Yucesoy B, Kashon M, Garssen J, De Gruijl FR, Boland GJ, Van Hattum J, Luster MI, Van Loveren H (2003) Cytokine polymorphisms play a role in susceptibility to ultraviolet B-induced modulation of immune responses after hepatitis B vaccination. *J Immunol* 170: 3423–3428

83 Diggle GE (1993) Overview of regulatory agencies. In: B Ballantyne, T Marrs, P Turner (eds): *General and Applied Toxicology*. MacMillan, Basingstoke, 1071–1090

84 Putman E, Van der Laan JW, Van Loveren H (2003) Assessing immunotoxicity: guidelines. *Fund Clin Pharmacol* 17: 615-626

85 Luster MI, Dean JH, Germolec DR (2003) Consensus workshop on methods to evaluate developmental immunotoxicity. *Environ Health Perspect* 111: 579–583

Appendix

Appendix C1. Selected examples of vaccines registered in the USA

Generic name	Trade name, manufacturer	Antigens and antigen dose	Main excipients	How supplied	Recommended dosage and administration
Anthrax vaccine adsorbed	Biothrax, BioPort Corporation	Proteins from avirulent *Bacillus anthracis* strain	Aluminium hydroxide, benzethonium chloride, formaldehyde	Liquid, multidose vial	0.5 ml, 3 × s.c., given 2 weeks apart followed by 3 × s.c., given at 6, 12, and 18 months Risk groups only, individuals 18–65 years of age; recommendations in response to bioterrorism: http:/www.cdc.gov
Diphtheria and tetanus toxoids adsorbed, for paediatric use	(Generic), Aventis Pasteur	DT (6.7 Lf), TT (5 Lf)	Aluminium potassium sulphate, NaCl, phosphate buffer; multi-dose vial contains thiomersal	Liquid; single-dose vial or syringe, or multi-dose vial	0.5 ml, 2 × i.m. 4–8 weeks apart followed by 1 × 6–12 months after first injection For children 6 weeks–6 years of age
DTaP, adsorbed	Infanrix, GlaxoSmithKline Biologicals	DT (30 IU), TT (40 IU), 3 pertussis antigens (25 µg PT, 25 µg FHA, 8 µg pertactin)	Aluminium salts, NaCl, 2-phenoxyethanol	Liquid; single-dose vial	0.5 ml, 3 × deep i.m., at least 4 weeks between doses For children 2 months–6 years of age
DTaP-IPV-HepB, adsorbed	Pediarix, GlaxoSmithKline Biologicals	DT (25 Lf), TT (10 Lf), 3 pertussis antigens (25 µg PT, 25 µg FHA, 8 µg pertactin), HBsAg (10 µg), poliovirus types 1,2,3 (40, 8 and 32 DU, respectively)	Aluminium salts, NaCl, 2-phenoxyethanol	Liquid; single-dose vial or syringe	0.5 ml, 3 × i.m., at 6–8-week intervals Customary age for first dose: 2 months of age

Appendix C1. Selected examples of vaccines registered in the USA (continued)

Generic name	Trade name, manufacturer	Antigens and antigen dose	Main excipients	How supplied	Recom mended dosage and administration
HiB-HepB vaccine	COMVAX, Merck&Co	Haemophilus b conjugate (7.5 µg PRP, conjugated to meningococcal protein) and HBsAg (5 µg)	Aluminium hydroxy-phosphate sulphate, sodium borate, NaCl	Liquid, single-dose vial	0.5 ml, 3 × i.m., ideally at 2, 4, and 12–15 months of age
HiB vaccine	ActHIB, Aventis Pasteur	Haemophilus b conjugate (10 µg PRP, conjugated to 24 µg TT)	Sucrose, NaCl	Lyophilised, single-dose vial Diluent: 0.4% NaCl Reconstitution with DTP possible	0.5 ml, 4 × i.m., ideally at 2, 4, 6, and 15–18 months of age
HepA-HepB vaccine	TWINRIX, GlaxoSmithKline Biologicals	Inactivated HepA virus (720 ELISA units), HBsAg (20 µg)	Aluminium phosphate and hydroxide, amino acids, 2-phenoxyethanol, NaCl, phosphate buffer, Tween 80	Liquid, single-dose vial or syringe	1.0 ml, 3 × i.m. on a 0-, 1-, and 6-month schedule R sk groups only, individuals from 18 years of age
HepA vaccine	VAQTA, Merck&Co	Inactivated HepA virus (25 units, paedriatic; 50 units, adults)	aluminium hydroxy-phosphate sulphate, sodium borate, NaCl	Liquid, single-dose vial	2–18 years of age: 0.5 ml, 2 × i.m. on month 0 and 6–18 As of 19 years of age: 1.0 ml, 2 × i.m. on month 0 and 6–18
HepB vaccine	Engerix-B, GlaxoSmithKline Biologicals	HBsAg (10 µg)	Aluminium hydroxide, NaCl, phosphate buffer	Liquid, single-dose vial or syringe	0.5 ml, 3 × i.m. on a 0-, 1-, and 6-month schedule
Influenza virus vaccine, trivalent, types A and B	Fluzone, Aventis Pasteur	Haemagglutinin (3 × 15 µg) from 3 influenza virus strains	NaCl, phosphate buffer, thiomersal, gelatin	Liquid, single-dose vial or syringe	0.5 ml, 2 × i.m., at least 1 month apart Nct approved for infants < 6 months of age

Japanese encephalitis virus vaccine inactivated	JE-Vax	Inactivated virus	Sucrose, gelatin, thiomersal	Lyophilised, single-dose or multi-dose vial Diluent: water for injection	0.5 ml (1–3 years of age) or 1.0 ml (> 3 years of age), 3 × s.c. on day 0, 7, and 30; booster after 2 years
Lyme disease vaccine	LYMerix, GlaxoSmithKline Biologicals	Lipoprotein OspA (30 µg)	Aluminium hydroxide, NaCl, phosphate buffer, 2-phenoxyethanol	Liquid, single-dose vial or syringe	0.5 ml, 3x i.m. on a 0-, 1-, and 12-month schedule Indicated for individuals 15–70 years of age
Measles, mumps, and Rubella virus vaccine, live	M-M-R II, Merck & Co	Live attenuated measles, mumps and rubella viruses	Sorbitol, sodium phosphate, NaCl, hydrolised gelatin, human albumin	Lyophilised, single-dose vial Diluent: water for injection	0.5 ml, 2 × s.c. at 12–15 months of age and 4–6 years of age
Meningococcal polysaccharide vaccine, groups A, C, Y and W-135 Combined	Menomune-A/C/Y/W-135, Aventis Pasteur	Meningococcal polysaccharide groups A, C, Y, and W-135 (50 µg each)	Lactose, NaCl	Lyophilised, single-dose or multi-dose vial Diluent: water for injection (with thiomersal for multi-dose vial)	0.5 ml, s.c., for both adults and children, revaccination may be considered after 3–5 years
Pneumococcal vaccine, conjugate	Prevnar, Wyeth Lederle	7 pneumococcal polysaccharides (serotypes 4, 6B, 9V, 18C, 19F, and 23F (2 µg each, 4 µg of 6B) individually conjugated to diphtheria CRM_{197}	Aluminium phosphate	Liquid, single-dose vial	0.5 ml, 3 × i.m. separated by 1-month intervals for infants < 6 months of age, followed by second dose in 2nd year of age

Appendix C1. Selected examples of vaccines registered in the USA (continued)

Generic name	Trade name, manufacturer	Antigens and antigen dose	Main excipients	How supplied	Recom mended dosage and administration
Pneumococcal vaccine, polyvalent	Pneumovax 23, Merck & Co	23 pneumococcal polysaccharide serotypes (25 µg each)	NaCl, phenol	Liquid, single-dose or multi-dose vial	0.5 ml, s.c. or i.m. Not indicated for children < 2 years of age
Rabies vaccine	RabAvert, Chiron Behring	Inactivated virus (at least 2.5 IU of rabies antigen)	Buffered polygeline, potassium glutamate	Lyophilised, single-dose vial Diluent: water for injection	1.0 ml, 3 × i.m. on days 0, 7, and 21 or 28
Smallpox vaccine	Dryvax, Wyeth Lederle	Attenuated live vaccinia virus (ca. 10^8 infectious viruses/ml)	Glycerin, phenol	Lyophilised, multi-dose vial Diluent: 50% glycerin, 0.25% phenol in water for injection	0.25 ml, into superficial skin layers using a bifurcated needle Risk groups only; recommendations in response to bioterrorism: http:/www.cdc.gov
Typhoid vaccine Live Oral Ty21a	Vivotif, Berna Biotech	Live attenuated *Salmonella typhi* Ty21a (2–6*10^9 colony-forming units)	Sucrose, lactose, ascorbic acid, amino acids, magnesium stearate	Enteric-coated capsule	Capsule, 4 × orally, 1 h before a meal with a cold or luke-warm drink, on alternate days Risk groups only, > 6 years of age
Typhoid Vi Polysaccharide vaccine	TYPHIM Vi, Aventis Pasteur	Polysaccharide (25 µg) from *Salmonella typhi* strain Ty2	NaCl, phosphate buffer	Liquid, single-dose syringe	0.5 ml, 1 × i.m., reimmunisation every 2 years Risk groups only, > 2 years of age

Varicella vaccine	Varivax, Merck&Co	Live attenuated varicella virus (> 1350 plaque-forming unites)	Sucrose, hydrolysed gelatin, NaCl, sodium glutamate, phosphate buffer, KCl	Lyophilised, single-dose vial Diluent: water for injection	0.5 ml, 1–2 × s.c. Not indicated for children < 1 year of age
Yellow fever vaccine	YF-Vax, Aventis Pasteur	Live attenuated yellow fever virus (> $5.5*10^4$ plaque-forming units)	Sorbitol, gelatin, NaCl	Lyophilised, single-dose vial Diluent: isotonic NaCl	0.5 ml, 1 × s.c. Risk groups only, > 9 months of age

CRM_{197}, nontoxic variant of diphtheria toxin; DT, diphtheria toxoid; DTP, diphtheria-tetanus-pertussis combination vaccine; DU, D-antigen units; FHA, filamentous haemagglutinin; HBsAg, recombinant hepatitis B surface antigen; HepA, hepatitis A; HepB, hepatitis B; HiB, Haemophilus influenzae *type B; IU, international units; Lf, limits of flocculation (flocculation units); PRP, polyribosylribitol phosphate; PT, detoxified pertussis toxin; TT, tetanus toxoid*

Appendix C3. Marketed anti-allergic products

Chemical name	Trade name	Formulation	Usual dose	Pharmacokinetics	Indications	Contra-indications
Nedocromil	Tilade	Dose aerosol	2 mg/ inhalation 4 × 4 mg/day	Complete resorption from lungs. Not metabolised. Eliminated unchanged, 70% urine, 30% faeces. $T_{½el}$ = 1–2 h.	Prevention of allergic or exercise-induced asthma	
Nedocromil	Tilavist	Eye drops	20 mg/ml 1 drop 2 times/day /per eye	See above	Prevention of allergic conjunctivitis	
Disodium cromoglycate	Lomudal	Dose aerosol Inhalation powder Inhalation liquid	5 mg/ inhalation 4 × 10 mg/day 20 mg/capsule 4 × 20 mg/day 20 mg/2 ml 4 × 20 mg/day	Complete resorption from lungs. Not metabolised. Eliminated unchanged, 50% urine, 50% gal. $T_{½el}$ = 1.5–2 h.	Prevention of allergic or exercise-induced asthma	
Disodium cromoglycate	Lomusol	Nose drops	20 mg/ml 4–6 drops/day	See above	Prevention of allergic rhinitis	Preservative sensitivity
Disodium cromoglycate	Lomusol Otrivin Prevalin Vividrin Allergocrom	Nose spray	20 mg/ml 4–6 sprays/day	See above	Prevention of allergic rhinitis	Preservative sensitivity
Disodium cromoglycate	Opticrom Prevalin Vividrin Allergocrom	Eye drops	20 mg/ml 1–2 drops, 4–6 times/day	See above	Allergic conjunctivitis	Preservative sensitivity
Disodium cromoglycate	Nalcrom	Drink	200 mg 4 times/day	See above	Food allergy	

H_1-receptor antagonists						
Clemastine	Tavegil	Injection liquid tablet	1 mg/ml, 2 ml 1 mg	Well absorbed. T_{max} 5–7 h, duration 12 h. Metabolised in liver. $T_{½el}$ = 20–24 h.	Treatment of anaphylactic shock	Children under 1 year of age
Dimethindene	Fenistil	Capsule 1 mg	1–2 mg three times/day	Well absorbed. Metabolised in liver. $T_{½el}$ = 20–24 h.	Allergic conditions	Prostate hyperplasia, urine retention, glaucoma, new born babies
Dexchloro-feniramine	Polaramine	Tablet 2 mg Syrup 0.4 mg/ml	Max. 12 mg/day	Well absorbed. T_{max} 2 h. Metabolised in liver. $T_{½el}$ = 6 h.	Allergic conditions	Prostate hyperplasia, pylorus stenosis
Emedastine	Emadine	Eye drops 0.5 mg/ml	1 drop twice a day	Well absorbed. T_{max} 1–2 h. Metabolised in liver. $T_{½el}$ = 10 h.	Rhinitis/urticaria	Preservative sensitivity
Hydroxyzine	Atarax Navicalm	Tablet, 10, 25, 100 mg Syrup 2 mg/ml	50–100 mg/day	Well absorbed. T_{max} 2 h, duration 4–6h. Metabolised in liver. $T_{½el}$ = 20 h.	Pruritis/urticaria	
Mebhydroline	Mebhydro-line	Tablet 50 mg	50–100 mg, 2–3 times/day	Well absorbed. Metabolised in liver.	Allergic conditions	
Oxatomide	Tinset	Tablet 30 mg	30 mg twice a day	Well absorbed. T_{max} 2–4 h. Metabolised in liver. $T_{½el}$ = 14 h.	Allergic conditions	
Promethazine	Phenergan Prome-thazine	Injection liquid 2 mg/ml, 1 ml or 2 ml Tablet 25 mg Syrup 1 mg/ml	25–150 mg/day	Well absorbed. T_{max} 20 min, duration 6–12 h. Metabolised in liver. $T_{½el}$ = 10–14 h.	Allergic conditions / anaphylactic shock	Acute asthma attack, sensitivity to phenothiazines

Appendix C3. Marketed anti-allergic products

Chemical name	Trade name	Formulation	Usual dose	Pharmacokinetics	Indications	Contra-indications
Acrivastine	Semprex	Capsule 8 mg	8 mg/day	Well absorbed. T_{max} 2 h, duration 4–6 h. Metabolised in liver. $T_{½el}$ = 20h.	Allergic rhinitis / hayfever	Triprolidine sensitivity
Astemizole	Hismanal	Tablets 10 mg Suspension 10 mg/5 ml	10 mg/day		Urticaria	
Cetirizine	Reactine Zyrtec	Drink 1 mg/ml Tablet 10 mg	10 mg/day	Well absorbed. T_{max} 60–90 min. Metabolised in liver. $T_{½el}$ = 10 h.	Allergic rhinitis / conjunctivitis / urticaria	Preservative sensitivity
Fexofenadine	Telfast	Tablet 120 mg, 180 mg	120 mg/day rhinitis 180 mg/day urticaria	Well absorbed. T_{max} 1–3 h. Not metabolised. $T_{½el}$ = 11–15 h.	Allergic rhinitis / chron c urticaria	
Ketotifen	Zaditen	Tablet 1 mg, Syrup 0.2 mg/ml, drops 1 mg/ml	1–2 mg/day	Well absorbed. T_{max} 2–4 h, duration 4–6 h. Metabolised in liver. $T_{½el}$ = 22 h.	Allerg c rhinitis/ allergic skin conditions / prophylactic for asthma	Preservative sensitivity
Levocabastine	Livocab nose drops and eyedrops Livostin	0.5 mg/ml	Spray twice, twice a day 1 drop per eye Twice a day	Well absorbed. Not metabolised. $T_{½el}$ = 35–40h.	Allerg c rhinitis / conjunctivitis	Preservative sensitivity
Levocetirizine	Xyzal	Tablet 5 mg	5 mg/day	Well absorbed. T_{max} 1 h. Partially metabolised in liver. $T_{½el}$ = 6–10 h.	Allerg c rhinitis / chronic urticaria	Kidney insufficiency, lapp-lactase-deficiency, glucose-galactose malabsorption

Terfenadine	Triludan Terfenadine Seldane	Suspension 6 mg/ml Tablet 60 mg	60–120 mg/ day	Well absorbed. T_{max} 1–3 h. Not metabolised. $T_{½el}$ = 11–15 h.	Allergic rhinitis/ conjunctivitis/ allergic skin disorders	Disturbed liver or heart function
Azelastine	Allergodil Astelin	Nose spray 1 mg/ml Eye drops 0.5 mg/ml	Use spray twice a day Use drops twice a day	Well absorbed. Duration 12 h. Metabolised in liver. $T_{½el}$ = 20–45 h.	Allergic rhinitis / conjunctivitis	Preservative sensitivity
Desloratidine	Aerius	Tablet 5 mg Syrup 0.5 mg/ml	5 mg/day	Metabolised in liver. $T_{½el}$ = 27 h.	Allergic rhinitis	
Ebastine	Kestine Ebastel	Tablet 10 mg	10 mg/day	Well absorbed. T_{max} 1 h, duration > 48 h. $T_{½el}$ = 15–19h.	Allergic rhinitis / conjunctivitis	Liver insufficiency
Loratadine	Allerfre Claritin	Tablet 10 mg Syrup 1 mg/ ml	10 mg/day	Well absorbed. Metabolised in liver. $T_{½el}$ = 12 h.	Allergic rhinitis / conjunctivitis / chronic urticaria / pruritis	Preservative sensitivity
Mizolastine	Mistalin Mizollen	Tablet 10 mg	10 mg/day	Well absorbed. T_{max} 1–2 h, duration 24 h. Metabolised in liver. $T_{½el}$ = 13 h.	Allergic rhinitis / conjunctivitis / urticaria	Disturbed liver or heart function
Anti-IgE	Omali- zumab Xolair	Injection liquid	150–375 mg every 2–4 weeks		Moderate to severe asthma	Urticarial rash

Appendix C5. Licensed cytokines

Drug	Corporation	Target	Approval
Actimmune® IFN-γ-1b	InterMune Pharmaceuticals	Management of chronic granulomatous disease	December 90
Actimmune® IFN-γ-1b	InterMune Pharmaceuticals	Osteopetrosis	February 00
Alferon N Injection® IFN-α-n3 (human leukocyte derived)	Interferon Sciences	Genital warts	October 89
Betaseron® RECOMBINANT IFN-β-1b	Berlex Laboratories	Relapsing, remitting multiple sclerosis	July 93
BMP-2, bone morph. protein-2	Medtronic Sofamor Danek	Treatment of spinal degenerative disc disease	July 02
Enbrel® TNFR:Fc	Immunex	Moderate to severe active rheumatoid arthritis	November 98
Enbrel® TNFR:Fc	Immunex	Moderate to severe active juvenile rheumatoid arthritis	May 99
Enbrel® TNFR:Fc	Amgen	Active ankylosing spondylitis	July 03
EPOGEN® Epetin alfa (rEPO)	Amgen	Anemia caused by chemotherapy	April 93
EPOGEN® Epetin alfa (rEPO)	Amgen	Anemia, chronic renal failure, anemia in Retrovir® treated HIV-infected	June 89
EPOGEN® Epetin alfa (rEPO)	Amgen	Chronic renal failure, dialysis	November 99
EPOGEN® Epetin alfa (rEPO)	Amgen	Surgical blood loss	December 96
Interferon β-1a	Serono	Relapsing forms of multiple sclercsis	March 02
Infergen® IFN alfacon-1	Amgen	Treatment of chronic hepatitis C viral infection	October 97
Intron® A IFN-α-2b	Schering-Plough	AIDS-related Kaposi's sarcoma	November 88

Intron® A IFN-α-2b	Schering-Plough	Follicular lymphoma	November 97
Intron® A IFN-α-2b	Schering-Plough	Genital warts	June 88
Intron® A IFN-α-2b	Schering-Plough	Hairy cell leukemia	June 86
Intron® A IFN-α-2b	Schering-Plough	Hepatitis B	July 92
Intron® A IFN-α-2b	Schering-Plough	Hepatitis C	February 91
Intron® A IFN-α-2b	Schering-Plough	Malignant melanoma	December 95
Leukine™ sargramostim (GM-CSF)	Immunex	Allogeneic bone marrow transplantation	November 95
Leukine™ sargramostim (GM-CSF)	Immunex	Autologous bone marrow transplantation	March 91
Leukine™ sargramostim (GM-CSF)	Immunex	Neutropenia resulting from chemotherapy	September 95
Leukine™ sargramostim (GM-CSF)	Immunex	Peripheral blood progenitor cell mobilization	December 95
LFA-1/IgG1	Biogen	Moderate to severe chronic plaque psoriasis	January 03
Neumega®, oprelvekin rHu IL-11	Genetics Institute	Chemotherapy-induced thrombocytopenia	November 97
NEUPOGEN® Filgrastim (rG-CSF)	Amgen	Acute myelogenous leukemia	April 98
NEUPOGEN® Filgrastim (rG-CSF)	Amgen	Autologous or allogeneic bone marrow transplantation	
NEUPOGEN® Filgrastim (rG-CSF)	Amgen	Chemotherapy-induced neutropenia	February 91
NEUPOGEN® Filgrastim (rG-CSF)	Amgen	Chronic severe neutropenia	December 94
NEUPOGEN® Filgrastim (rG-CSF)	Amgen	Peripheral blood progenitor cell transplantation	December 95
Ontak Diptheria Tox – IL-2	Sevagen	CTCL	February 99

Appendix C5. Licensed cytokines (continued)

Drug	Corporation	Target	Approval
Proleukin® aldesleukin (IL-2)	Chiron	Metastatic melanoma	
Proleukin® aldesleukin (IL-2)	Chiron	Renal cell carcinoma	May 92
Rebetron™ ribavirin/IFN-α-2b	Schering-Plough	Chronic hepatitis C	June 98
Rebetron™ ribavirin/IFN-α-2b	Schering-Plough	Chronic hepatitis C, compensated liver disease	December 99
Roferon® IFN-α-2a	Hoffman-La Roche	AIDS-related Kaposi's sarcoma	November 88
Regranex®, Becaplermin, rHPDGF-BB	Ortho-McNeil	Diabetic neuropathy, foot ulcers	December 97
Roferon® IFN-α-2a	Hoffman-La Roche	Chronic myelogenous leukemia	November 95
Roferon® IFN-α-2a	Hoffman-La Roche	Hairy cell leukemia	June 86
Roferon® IFN-α-2a	Hoffman-La Roche	Hepatitis C	November 96
Stemgen (stem cell factor)	Amgen	Mobilization (Australia, New Zealand, Canada)	1997
Wellferon® IFN-n	Glaxo Wellcome	Treatment of hepatitis C in patients 18 years-of-age or older without decompensated liver disease	March 99

Appendix C6. Examples of Marketed Drugs

Cytokines	Cytokine inhibitors	Synthetic immunomodulators	Bacterial extracts
IFN-α-2a	Adalimumab	Imiquimod	Broncho-Munal
IFN-α-2b	Infliximab	Isoprinosine	Luivac
Peginterferon-α-2a	Etanercept		
Peginterferon-α-2b			
IFN-β-1a			
IFN-β-1b			
IFN-γ-1b			
Aldesleukin (IL-2)			
Filgrastim (G-CSF)			
Pegfilgrastim			
Lenograstim (G-CSF)			
Molgramostim (GM-CSF)			
Sargramostim (GM-CSF)			

Appendix C7. Commercially available plant and dietary immunomodulators

Chem. name	Trade name	Formulation	Usual dose	Pharmacokinetics	Indications	Contra-indications
Echinacea purpurea (purple cone-flower)	Echinacin, Echinagard, Echinaforce, Esberitox Mono	Extract as tincture or pastilles	60–80 g squeezed sap per 100 g	Alkylamides detected in blood after oral administration of extract	Adjuvant therapy of respiratory infections; wound healing	Hypersensitivity to Echinacea
Viscum album (mistle-toe)	Iscador, Lektinol	Extract for intra-cutaneous injection (Europe only)	1 ng/kg ML-I equiv. s.c.	ML-I crosses cell membrane to induce cytotoxicity	Adjuvant therapy of cancer	
Zinc sulphate	Solvezink	Tablets	Zinc sulfate 200 mg, equiv. to elemental zinc 45 mg	Zinc sulphate is partially absorbed from the gastrointestinal tract, distributed in the skeletal muscle, skin, bone, and pancreas, and excreted primarily through the duodenum and jejunum.	Zinc salts are primarily used to relieve minor eye irritations. They have been used to correct zinc-deficiency conditions (e.g., acrodermatitis enteropathica) and other diseases associated with zinc-depletion (e.g., anorexia nervosa, arthritis, diarrheas, eczema, recurrent infections, and recalcitrant skin problems).	Oral zinc preparations may cause gastrointestinal upset; ophthalmic zinc preparations should be used cautiously in persons with glaucoma.
	Zincol Zixol Oral-Z	Syrup oral	Zinc sulfate, 10 mg/5 ml Zinc sulfate, 1%			
	Colirio, Honfar	Drops ophthalmic	25 to 50 mg daily			
	Erectol	Capsules	Zinc sulfate, 220 mg		To prolong sexual activity and increase the production of semen	
Zinc gluconate	Zigg-eeze Zinc cold treatment Pharmezinc Amplified zinc	Lozenges Lozenges oral topical Tablets Tablets	Zinc gluconate equiv. to zinc element, 60 mg Zinc gluconate, 10 mg		Conditions requiring zinc supplementation	

Zinc picolinat	Zinc picoloinate	Tablets	25 mg			
Zinc mono-methio-nine	Optizinc	Capsules	30 mg of zinc per capsule	Independent human and animal studies show that zinc methionine exhibits a superior absorption rate when compared with many other forms of zinc. OptiZinc provides a superior source of bioavailable zinc. OptiZinc resists binding with dietary fibre and phytic acid.	Suitable for individuals with malabsorption and poor zinc status. Suitable for vegetarians and vegans.	Supplemental zinc can inhibit the absorption and availability of copper. Zinc monomethionine is not recommended for those persons showing hypersensitivity to zinc and/or methionine.
Sodium selenite	Selemun Selen Loges	Tablets film-coated Tablets oral unit dose liquids	46.7 mcg Tablets 300 mcg Se Oral unit dose liquids 100 mcg Se		Selenium deficiency	
Seleno-methio-nine	Sethotope				Contrast medium radiography	
Seleno-yeast	Selenium yeast	Tablets	50 mcg			

APPENDIX C7 (CONTINUED)

Chem. name	Trade name	Formulation	Usual dose	Pharmacokinetics	Indications	Contra-indications
Ascorbic acid (vitamin C)	Ascorbex, Ascorbic acid injection BP 2002, Ascorbic acid tablets BP 2002, Balanced C complex, C-Vit, C-Will, C-vimin, Cebiolon, Cebion, Cecap, Cecon, Cecrisina, Ceerexin, Celin, Cemina, Cereon, Cetebe, Cetrinets, Cevalin, Cevi-Bid, Cevi-Tablets, Cevi-drops, Cevita, Cevitol, Cewin, Citron, Citrovit, Dancimin-C, Delrosa, Demovit C, Energil C, Eriglobin, Eritropiu, Grad vitamine C, Hermes Cevitt, Hicee, Ido-C, Kendural C, Laroscorbine, Megafer, Megavit C 1000, Midy Vitamine C, Nutrol C, Poremax-C, Rovit C, Scorbex, Sidervim, Super C, Tetesept Vitamin C, Timed Release C, Upsavit C, Vicemex, Vicomin C	Tablets, effervescent, injections	In deficiency states: oral or intramuscular doses of 100 to 500 mg/day; for urine acid-ification: 3 to 12 g/day as divided doses every 4 h; prophylactic oral or intramuscular doses for pediatric patients: 30 to 60 mg/day (age-dependent); pediatric patients with scurvy:100 mg 3 times a day for 1 week and 100 mg daily thereafter until clinical symptoms abate.	Peak serum levels: 2 to 3 h after oral dosing; therapeutic serum levels: 0.4 to 1.5 mg/dL; without supplementation 75 mg of ascorbic acid is excreted in the urine daily, increasing to 400 mg within 24 h with the administration of 1 g day; hemodialysis and peritoneal dialysis remove significant amounts of the drug and supplementation is suggested following dialysis periods.	An essential vitamin for deficiency states. Also used in large doses as a urine acidifier for patients with urinary tract infections. Requirements for ascorbic acid may increase during infection, trauma, pregnancy, and lactation.	Adverse effects include hemolysis, esophagitis, intestinal obstruction, hyperoxaluria, renal failure, and injection site pain.

α-Tocopherol (vitamin E)	Alfa-E, Aquasol-E, Bio E, Bio-E-Vitamin, Biogenis Biopto-E, Biosan E 600 Burgerstein, Vitamin E Dalfarol, Dif-Vitamin E, E 200/400, E-Tab, E-Vitamin-ratiopharm, E-Vitamin, Malton E, Mediderm, Natural Made Vitamin E, One-A-Day Extras E, Optovit E, Optovit, Plenovit, Tocorell Vit. E, Togasan Vitamin E, Topher-E, Unique E	Capsules, injections	100 mg/kg/day orally; pediatric doses range from 1 mg/kg/day to 100 mg/day.	Oral absorption from the GI tract is variable; bile is necessary. Water-miscible forms are absorbed intramuscularly; oil preparations not as well. Vitamin E is distributed to all tissues and is predominately metabolized in the liver. The half-life after intramuscular injection is 44 h and after intravenous injection about 282 h.	Vitamin E is useful in many conditions including malabsorption diseases, hematological diseases, cardiovascular disease, and retrolental fibroplasia.	Vitamin E is generally well-tolerated and enhances the response to oral anticoagulants. Fatigue, weakness, headache, nausea, diarrhea, flatulence and abdominal pain, and delayed coagulation times in vitamin-K deficient patients may occur with large doses. Major signs of toxicity include fatigue and weakness. Topical vitamin E has rarely caused contact dermatitis and erythema multiforme.
Epigallocatechin gallate (EGCG)	Tea catechin	Tablets, capsules	As green tea extracts	EGCG is readily detected in human and rat plasma at ng/ml levels following oral ingestion	Lipid-lowering agent.	
Decaffeinated green tea polyphenol mixture	Polyphenon E	Ointment			Treatment of genital warts.	

Source: Lifecycle (IMS), Micromedex (International Health Series, Martindale, Index Nominum), PharmaProjects

Appendix C8. Antibacterial agents currently in use for their potential* immunomodulatory properties.

Erythromycin A
Clarithromycin
Azithromycin
Roxithromycin
Minocycline
Doxycycline
Rifamycin SV
Dapsone
Clofazimine

**The mechanism (immunomodulatory or antibacterial) underlying the therapeutic benefit still remains controversial for some clinical settings.*

Appendix C12/1: Drugs

Class I. Non-selective COX-1 and COX-2 inhibitors

Salicylates
Sodium salicylate
Aspirin
Diflunisal

Propionic acid derivatives
Ibuprofen
Naproxen
Fenbufen
Fenoprofen
Flurbiprofen
Ketoprofen
Tiaprofenic acid

Acetic acid derivatives
Indomethacin
Sulindac
Ketorolac
Tolmetin
Diclofenac
Aceclofenac

Pyrazolones
Azapropazone
Phenylbutazone

Oxicams
Piroxicam
Tenoxicam

Fenamic acids
Mefenamic acid
Meclofenamic acid

Nabumetone
6-methoxy-2-naphthylacetic acid (active metabolite)

Appendix C12/1. Chemical structures of the NSAIDs

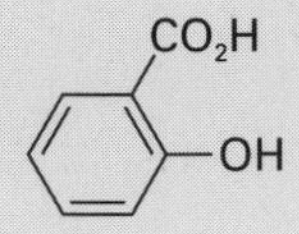

Salicylic acid

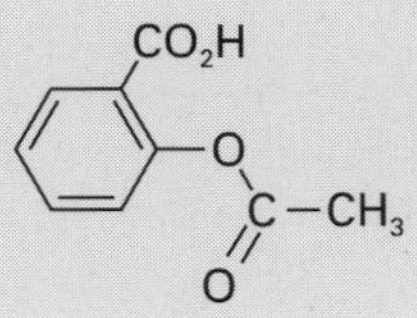

Aspirin
(Acetylsalicylic acid)

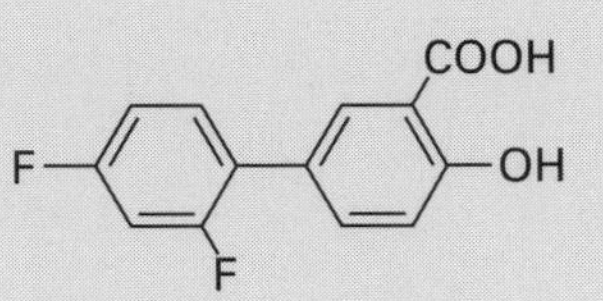

Diflunisal

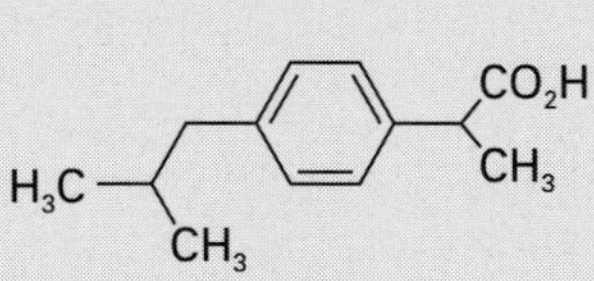

Ibuprofen

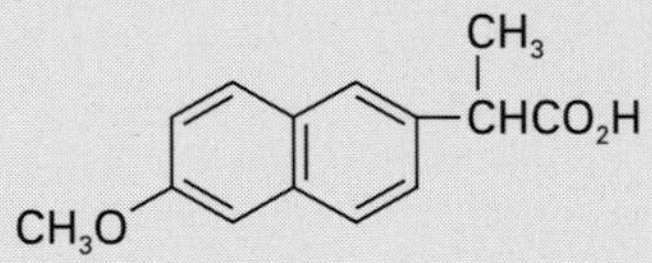

Naproxen

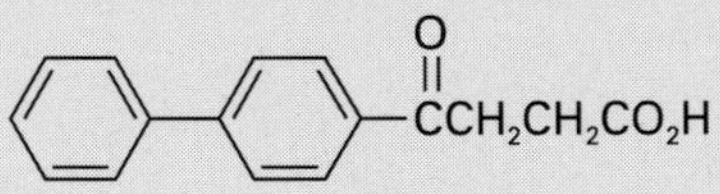

Fenbufen

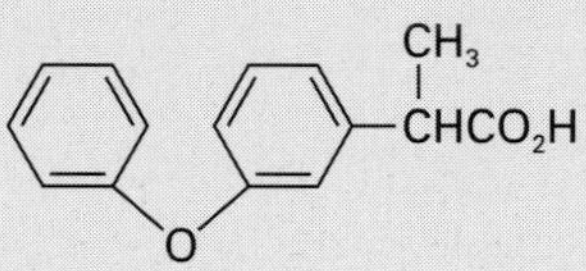

Fenoprofen

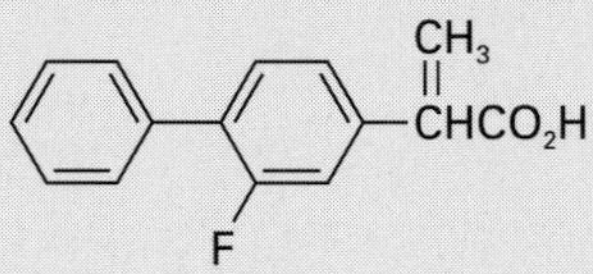

Flurbiprofen

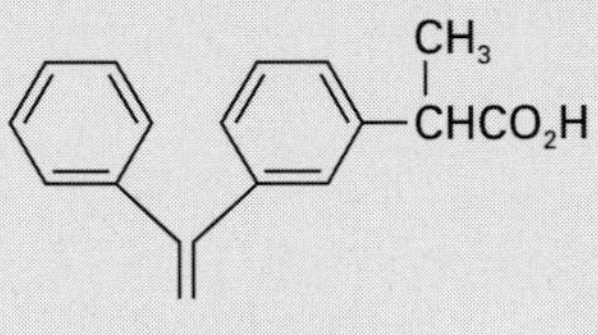

Ketoprofen

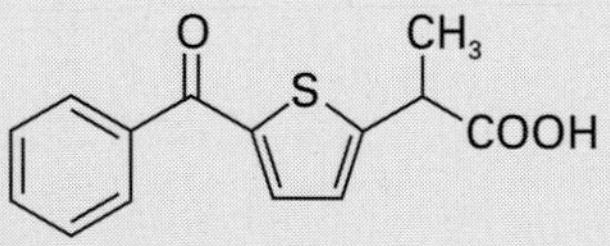

Tiaprofenic acid

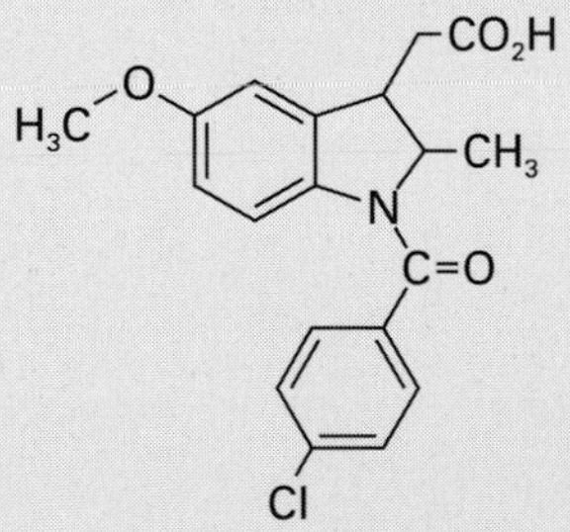

Indomethacin

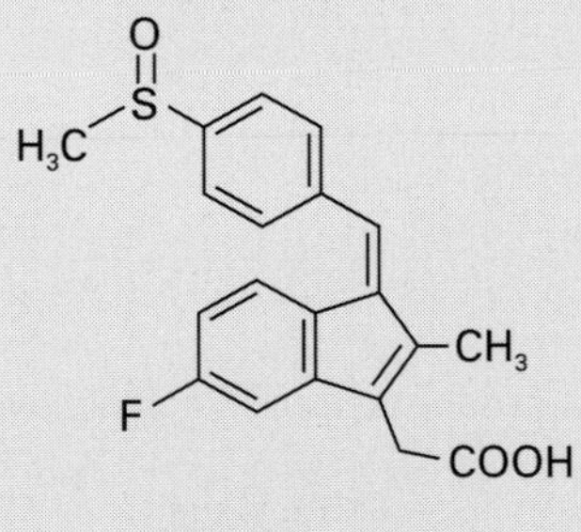

Sulindac

APPENDIX C12/1. CHEMICAL STRUCTURES OF THE NSAIDS (CONTINUED)

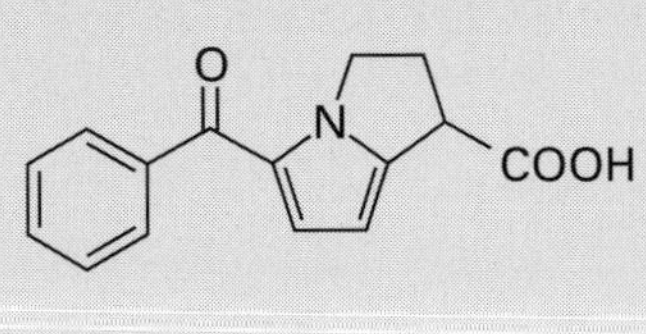

O
N
COOH

CH3
N
CH2CO2H
O CH3

CH2CO2H
NH
Cl Cl

Ketorolac 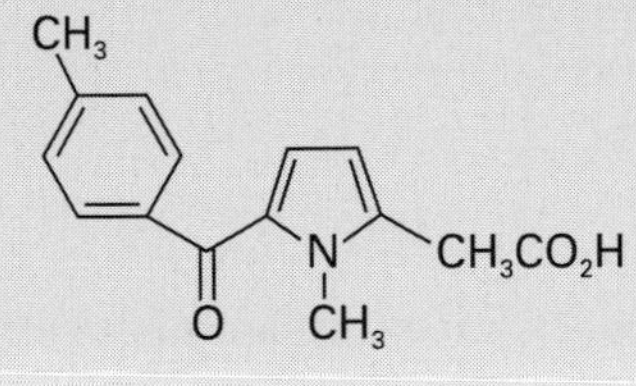 **Tolmetin** 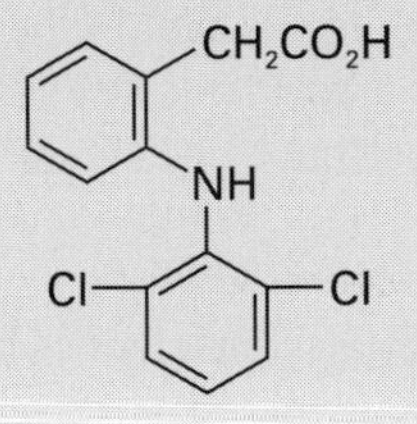 **Diclofenac**

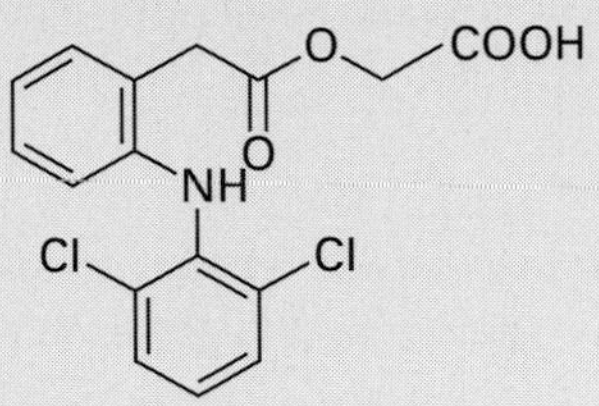

O COOH
O
NH
Cl Cl

O C3H7
H3C N N OH
N N(CH3)2

O
N CH3
N O

Aceclofenac 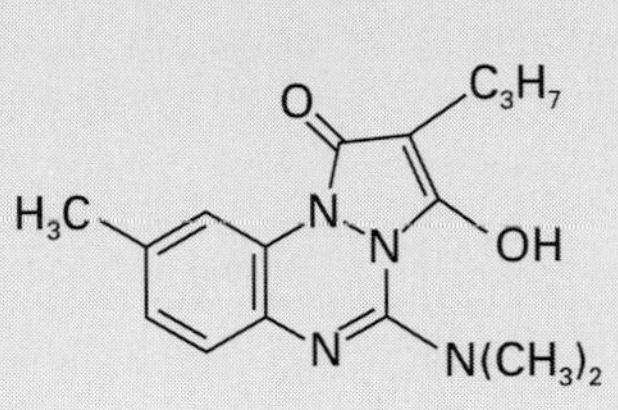 **Azapropazone** 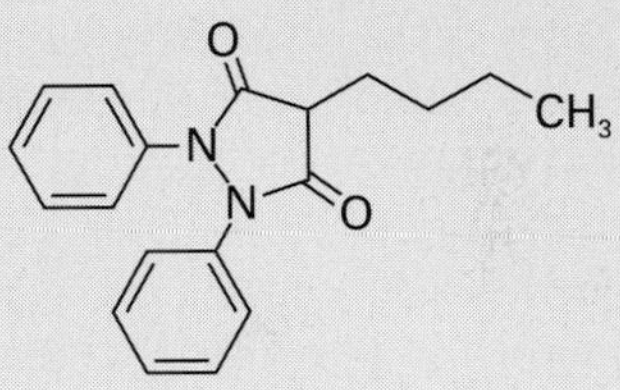 **Phenylbutazone**

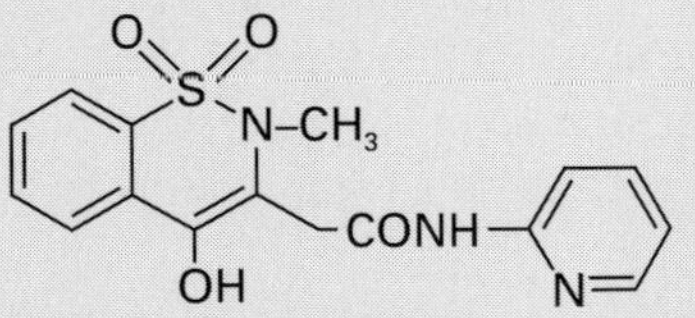

O O
S N–CH3
CONH
OH N

O O
S N CH3
H
S N
OH O N

CO2H H CH3
N CH3

Piroxicam 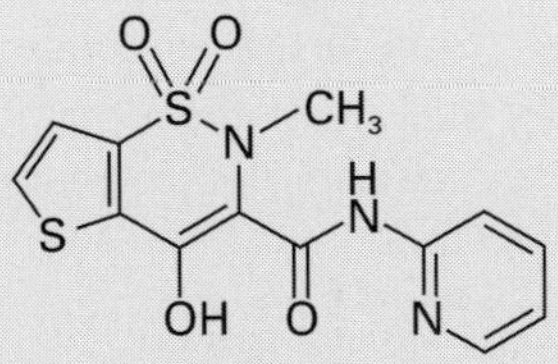 **Tenoxicam** 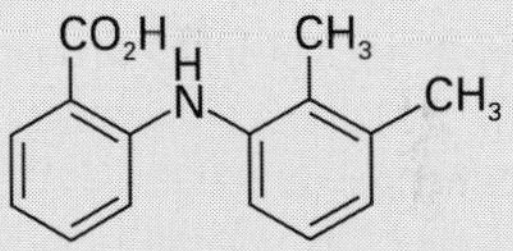 **Mefenamic acid**

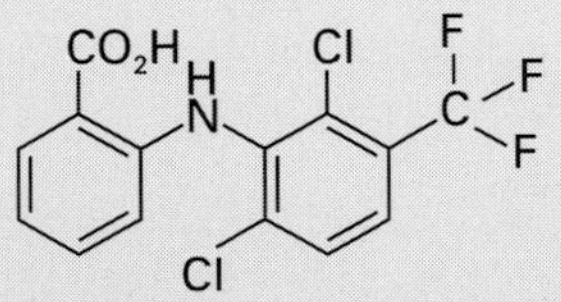

CO2H H Cl F
N C F
F
Cl

O
CH2CH2CCH3
CH3O

Mefenamic acid 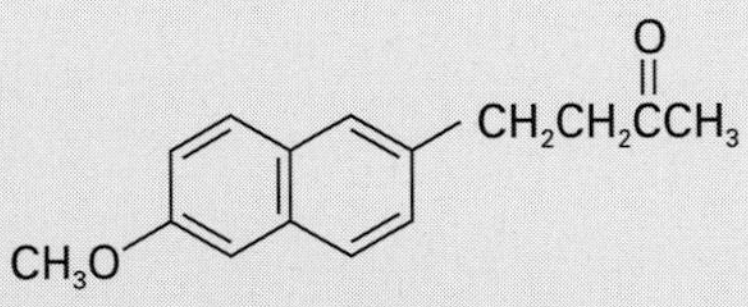 **Nabumetone**

Appendix C12/1: Drugs (continued) - Class I. Non-selective COX-1 and COX-2 inhibitors

Mode of action
These drugs all inhibit COX-1 and COX-2 non-selectively, but to different degrees, e.g., piroxicam and ketorolac inhibit COX-1 very much more potently than COX-2, but diclofenac and ibuprofen at therapeutic doses cause almost similar inhibition of both enzymes. They inhibit both COX-1 and COX-2 by competing with the substrate, arachidonic acid, for the active site on the enzyme. Aspirin is unusual in acetylating serine 530 in the active site and preventing access of arachidonic acid. The action of aspirin is irreversible and inhibition of COX-1 in platelets is responsible for the antithrombotic activity of aspirin.

Pharmacological effects
These drugs all have analgesic, antipyretic and anti-inflammatory effects.

Cellular pharmacokinetics
The inhibitory actions on COX enzymes are mainly peripheral as these drugs are acidic, polar compounds and cross the blood-brain barrier with difficulty. However, toxic doses may cause central symptoms, e.g., salicylate and aspirin in high doses cause dizziness and tinnitus. These acidic drugs are also highly bound to plasma proteins (95–99%) which leads to interactions with other highly bound drugs such as anticoagulants.

Clinical indications
The main therapeutic use is for pain and inflammation in rheumatic diseases such as rheumatoid arthritis and osteoarthritis; and also in various other musculoskeletal disorders. Daily low doses of aspirin are administered for their antiplatelet effects.

Side-effects
a) The major side-effect and disadvantage of the use of non-steroidal anti-inflammatory drugs is damage to the gastric mucosa resulting in ulceration and bleeding. This is due to removal of the gastroprotective prostaglandins synthesized by COX-1 in the stomach wall. Drugs which are highly selective for COX-1 over COX-2, such as piroxicam, cause greater incidence of bleeding than drugs such as diclofenac or ibuprofen which are almost equipotent for COX-1 and COX-2. Arthrotec is a formulation of diclofenac with the PGE_1 analogue, misoprostol, to prevent gastric mucosal damage.
b) COX-1 and COX-2 inhibitors other than aspirin have antithrombotic actions at therapeutic dose levels but aspirin is the most effective.
c) Most of the non-selective non-steroidal anti-inflammatory drugs precipitate an asthmatic episode in patients who suffer from aspirin-induced asthma.

APPENDIX C12/1: DRUGS (CONTINUED)

Class II. Selective COX-2 inhibitors

Low selectivity for COX-2
Etodolac
Meloxicam
Nimesulide

High selectivity for COX-2
Celecoxib
Rofecoxib
Etoricoxib
Valdecoxib

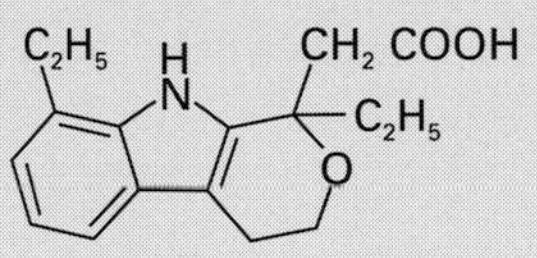

Etodolac 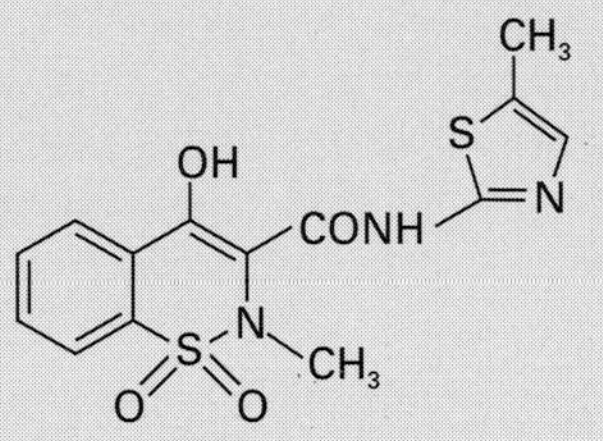 **Meloxicam** 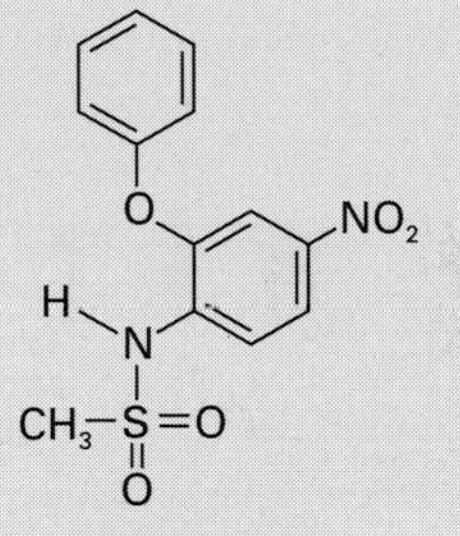 **Nimesulide**

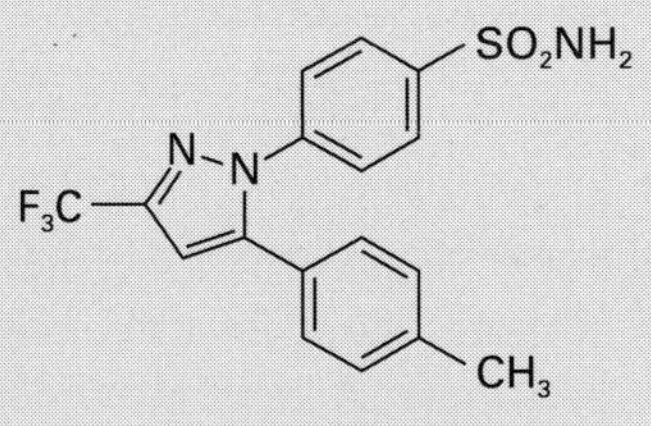

Celecoxib 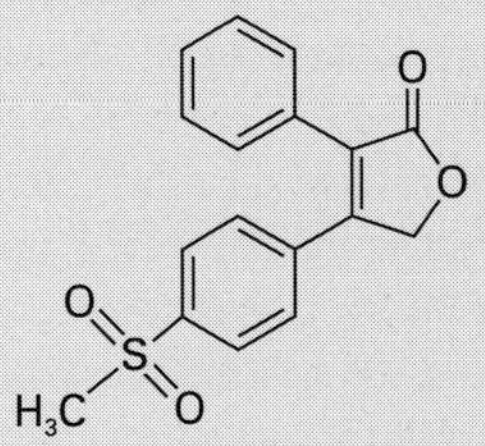 **Rofecoxib** 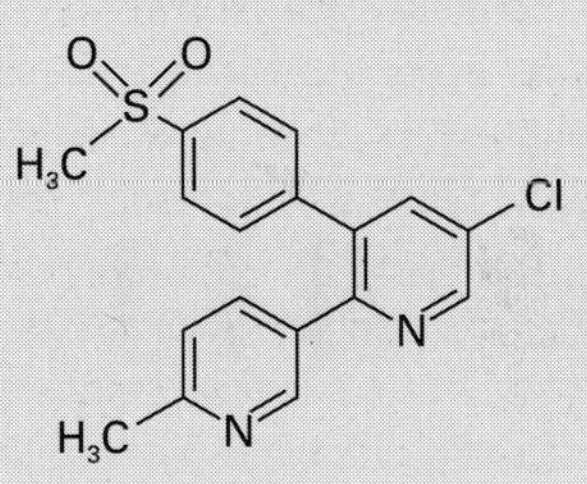 **Etoricoxib**

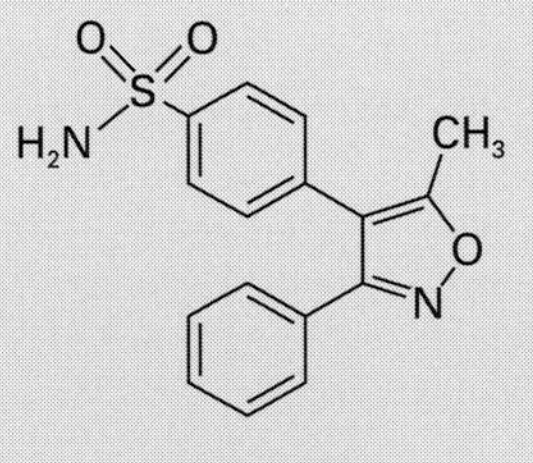

Valdecoxib

APPENDIX C12/1: DRUGS (CONTINUED) - II. Selective COX-2 inhibitors

Mode of action

Nimesulide, celecoxib, rofecoxib, etoricoxib and valdecoxib have sulphone anilide or sulphonamide structures and are made up of larger molecules than the non-selective anti-inflammatory drugs (listed above). The large molecules do not easily fit the COX-1 active site, but comfortably occupy the larger COX-2 active site with its side pocket. Etodolac and meloxicam also have large molecules which fit more easily into the active site of COX-2 than COX-1.

Etodolac, meloxicam and nimesulide inhibit COX-2, 5–20 times more selectively than COX-1 and celecoxib, rofecoxib, etoricoxib and valdecoxib are 10–100 times more selective for COX-2 than COX-1 depending on which pharmacological test is being applied.

Pharmacological effects

These drugs are analgesic, antipyretic and anti-inflammatory.

Cellular pharmacokinetics

The selective COX-2 inhibitors with sulphone anilide and sulphonamide structures cross the blood-brain barrier easily and may have central effects which have so far not been identified. They are bound to plasma proteins to a much lesser extent than the non-selective anti-inflammatory drugs (listed above) and interact to a lesser degree with other highly bound drugs such as anticoagulants.

Clinical indications

Their main therapeutic use is for the pain and inflammation of rheumatoid arthritis and osteoarthritis and also dysmenorrhoea and pain following dental surgery.

Side-effects

a) The main advantage of selective COX-2 inhibitors is the sparing effect on COX-1 of the gastric mucosa. Thus, these drugs cause a much lower incidence of stomach bleeding and ulceration than the anti-inflammatory drugs of Class I (listed above). However, if low-dose aspirin is administered with a selective COX-2 inhibitor, the gastroprotection of the COX-2 inhibitor is lost.
b) They are sparing of COX-1 in platelets and thus have no anti-thrombotic action.
c) The incidence of renal toxic actions is no greater than with the non-selective COX-1 and COX-2 inhibitors, but rofecoxib may cause a greater incidence of myocardial infarction than comparable non-selective inhibitors.
d) Selective COX-2 inhibitors do not precipitate asthma attacks in aspirin-sensitive asthma patients, confirming that aspirin-induced asthma is caused by inhibition of COX-1.

Appendix C12/1: Drugs (continued)

Class III. Para-aminophenol derivatives

Paracetamol (acetaminophen)

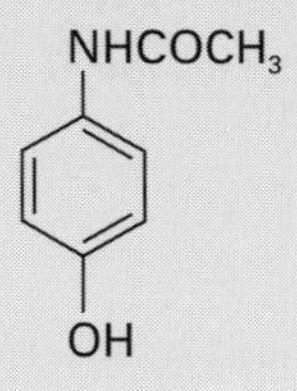

Paracetamol

Mode of action

In pharmacological tests, it is a weak inhibitor of both COX-1 and COX-2 and an inhibitory action on a third COX enzyme, COX-3, has been postulated.

Pharmacological effects

Paracetamol is an antipyretic analgesic with no anti-inflammatory actions in humans.

Cellular pharmacokinetics

Paracetamol enters the central nervous system easily and its analgesic and antipyretic actions are most likely by inhibition of COX-3 in the brain and spinal cord. Plasma protein binding is equivalent to approximately 20% of the therapeutic dose.

Clinical indications

Its therapeutic use is mainly for pain of various origins such as pain of rheumatic diseases and musculoskeletal pain. The use in fever is mostly for childhood fevers and occasionally for fever of cancer or stroke patients.

Side-effects

a) Paracetamol does not damage the gastric mucosa or prolong bleeding time in humans.
b) Paracetamol is safe to administer to aspirin-sensitive asthma patients.
c) The main disadvantage of paracetamol is the hepatotoxicity which can occur at doses only slightly greater than therapeutic doses. This is caused by a metabolite of paracetamol formed by cytochrome P450, N-acetyl-para-benzoquinonimine which depletes liver glutathione and rapidly produces renal tubular necrosis. N-acetyl-*para*-benzoquinonimine is inactivated by conjugation with glutathione and this reaction can be prevented by supplying sulfhydryl groups in the form of N-acetylcysteine. Thus, administration of N-acetylcysteine reverses the liver toxicity of paracetamol.

Appendix C12/2.

Drug	Chemical name	Trade name	Formulation	Usual dose	Pharmacokinetics	Indications	Contraindications
Aspirin	Acetylsalicylic acid	Angettes, Caprin, Nu-seals aspirin	Tablets	75 mg/day, or 300 mg 3 or 4 times daily	Highly bound in plasma, systemic action only at normal doses	Pain, rheumatism, antiplatelet therapy	Peptic ulcers, children under 12, asthma, anticoagulants, hypoglycaemic agents
Diflunisal	5-(2,4-difluorophenyl) salicylic acid	Dolobid	Tablets	500–1000 mg daily	Highly bound in plasma, systemic action only at normal doses	Pain, arthritis	Peptic ulcers, asthma, renal impairment, anticoagulants, pregnancy, lactation
Ibuprofen	2-(4-isobutylphenyl)-propionic acid	Brufen, Nurofen, Motrin, Ext: Ibugel, Fenbid gel.	Tablets, topical preparations	1200–1800 mg daily in divided doses	Highly bound in plasma, systemic action only at normal doses	Pain, pyrexia, arthritis	Peptic ulcer, asthma, renal impairment, anticoagulants, pregnancy, lactation
Naproxen	d-2-(6-methoxy-2-naphthyl)propionic acid	Naprosyn, Nycopren, Synflex	Tablets	250–500 mg twice daily	Highly bound in plasma, systemic action only at normal doses	Arthritis, musculoskeletal disorders, dys-menorrhoea	Peptic ulcer, asthma, renal impairment, anticoagulants
Fenoprofen	α-dl-2-(3-phenoxy-phenyl)propionic acid	Fenopron	Tablets	300–600 mg 3–4 times daily	Highly bound in plasma, systemic action only at normal doses	Pain, arthritis, musculoskeletal disorders	Peptic ulcer, asthma, renal impairment, anticoagulants, hypoglycaemics
Fenbufen	3-(4-biphenylyl-carbonyl)propionic acid	Lederfen	Tablets and capsules	450 mg twice daily	Highly bound in plasma, only systemic action at normal doses	Arthritis, mus-culoskeletal dis-orders	Peptic ulcer, asthma, renal impairment, anticoagulants

Flurbiprofen	2-(2-fluoro-4-biphenylyl)propionic acid	Froben, Ocufen, Streflam	Tablets and suppositories, eye drops, throat lozenges	150–200 mg daily, up to 300 mg daily	Highly bound in plasma, only systemic action at normal doses	Arthritis, musculoskeletal disorders, post-operative analgesia, sore throat	Peptic ulcer, asthma, renal impairment, anticoagulants, pregnancy, lactation
Ketoprofen	2(3-benzoylphenyl) propionic acid	Ketocid, Orudis, Oruvail, Oruvail gel, Powergel	Capsules, suppositories, injection or gel	100–200 mg once daily with food	Highly bound in plasma, only systemic action at normal doses	Arthritis, musculoskeletal disorders, after orthopaedic surgery, dysmenorrhoea	Peptic ulcer, asthma, renal impairment, anticoagulants, not children, pregnancy, lactation
Tiaprofenic acid	α-methyl-5-benzoyl-2-thienylacetic acid	Surgam	Tablets, capsules	600 mg daily in 2-3 divided doses	Highly bound in plasma, only systemic action at normal doses	Arthritis and other musculoskeletal disorders	Peptic ulcer, asthma, renal impairment, anticoagulants
Indomethacin	1-(*p*-chlorobenzoyl)-5-methoxy-2-methyl-3-indolylacetic acid	Indocid, Indomod, Flexin Continus, Indocid PDA	Capsules, suppositories, tablets, injection	50–200 mg daily, i.v. infusion over 20–30 mins	Highly bound in plasma, central effects, e.g., headache, vertigo, mental confusion, depression	Arthritis, musculoskeletal disorders, dysmenorrhoea, patent ductus arteriosus	Peptic ulcer, asthma, aspirin allergy, renal impairment, pregnancy, anticoagulants
Sulindac	Cis-5-fluoro-2methyl-1-[*p*-methylsulfinyl)benzylidene]indene-3-acetic acid	Clinoril	Tablets	200 mg twice daily	Highly bound in plasma, prodrug for sulindac sulphide	Arthritis, peri-articular disorders, musculoskeletal disorders	Peptic ulcer, asthma, renal impairment, anticoagulants, not children
Ketorolac	(+)-5-Benzoyl-2,3-dihydro-1*H*-pyrrolizine-1-carboxylic acid	Toradol, Acular,	Injection, eye drops	10–30 mg every 4–6 h	Highly bound in plasma, only systemic action	Post-operative pain	Peptic ulcer, asthma, renal impairment
Tolmetin	5-(*p*-toluoyl)-1-methylpyrrole-2-acetic acid	Reutol, Tolectin, Tolmene	Tablets	0.8–1.6 g per day	Highly bound in plasma, only systemic action	Rheumatoid and osteoarthritis	Peptic ulcer, anticoagulants, hypoglycaemics

APPENDIX C12/2. (CONTINUED)

Drug	Chemical name	Trade name	Formulation	Usual dose	Pharmacokinetics	Indications	Contraindications
Diclofenac	2-[(2,6-Dichloro-phenyl)amino]benzene acetic acid	Dicloflex, Diclomax, Motifene, Voltarol, Voltarol Emulgel, Volsaid Retard	Tablets, capsules, formulation with misoprostol (Arthrotec)	75-150 mg daily in 2 or 3 divided doses	Highly bound in plasma, only systemic action at therapeutic doses	Arthritis, musculoskeletal disorders, migraine, post-operative pain, dysmenorrhoea	Peptic ulcer, asthma, renal insufficiency, pregnancy, anticoagulants, hypoglycaemics
Aceclofenac	2-[(2,6-dichloro-phenyl)-amino]phenyl-actoxyacetic acid	Preservex	Tablets	100 mg twice daily	Highly bound in plasma, only systemic action	Rheumatoid and osteoarthritis	Peptic ulcer, renal impairment, pregnancy, antidiabetics
Azapropa-zone	5-(Dimethylamino)-9-methyl-2-propyl-1*H*-pyrazolo[1,2-α][1,2,4]benzotriazine-1,3(2*H*)-dione	Rheumox	Tablets	1–2 g daily in 2 or 4 doses	Highly bound in plasma, only systemic action	Rheumatoid arthritis, acute gout	Peptic ulcer, renal impairment, asthma, pregnancy
Piroxicam	4-Hydroxy-2methyl-N-2-pyridinyl-2*H*-1,2-benzothiazine-3-carboxamide 1,1-dioxide	Brexidol, Feldene, Feldene gel (ext.)	Tablets, capsules	20 mg per day	Highly bound in plasma, only systemic action	Arthritis, musculoskeletal disorders, post-operative pain	Peptic ulcer, renal impairment, asthma, anticoagulants
Tenoxicam	4-Hydroxy-2-methyl-N-2-pyridinyl-2*H*-thienol[2,3-e]-1,2-thiazine-3-carbox-amide 1,1-dioxide	Mobiflex	Tablets	20 mg per day	Highly bound in plasma, only systemic action	Arthritis, soft tissue injuries	Peptic ulcer, renal impairment, anticoagulants
Mefenamic acid	2-[(2,3-Dimethyl-phenyl)amino]benzoic acid	Ponstan	Tablets	500 mg 3 times daily	Highly bound in plasma	Arthritis, dysmenorrhea	Peptic ulcers, renal impairment, anticoagulants

Nabumetone	4-(6-Methoxy-2-naphthalenyl)-2-butanone	Relifex	Tablets	2 × 500 mg at bedtime	Prodrug for 6-methoxy-2-naphthylacetic acid	Rheumatoid and osteoarthritis	Peptic ulcer, renal impairment, pregnancy, anticoagulants
Etodolac	1,8-Diethyl-1,3,4,9-tetrahydropyrano-[3,4-b]indole-1-acetic acid	Lodine	Tablets	600 mg daily	Highly bound in plasma	Rheumatoid and osteoarthritis	Peptic ulcer, renal impairment, pregnancy, anticoagulants
Meloxicam	4-Hydroxy-2-methyl-N-(5-methyl-2-thiazolyl)-2*H*-1,2-benzothiazine-3-carboxamide 1,1-dioxide	Mobic	Tablets, suppositories	7.5–15 mg daily	Highly bound in plasma, CNS effects	Rheumatoid and osteoarthritis	Peptic ulcer, renal impairment, not children, anticoagulants
Nimesulide	N-(4-Nitro-2-phenoxy phenyl)methanesulphonamide	Aulin, Mesulid, Nimed	Tablets, powder	100 mg twice daily	Highly bound in plasma, short half-life	Osteoarthritis, musculoskeletal disorders, dysmenorrhea, post-operative pain	Peptic ulcer, liver insufficiency, asthma, allergic reactions to NSAIDs, anticoagulants
Celecoxib	(4-[5-(4-methyl-phenyl)-3-(trifluoro-methyl)-1*H*-pyrazol-1-yl]-benzenesulphonamide	Celebrex	Capsules	200–400 mg daily	Highly bound in plasma, crosses blood-brain barrier	Osteoarthritis, rheumatoid arthritis	Peptic ulcer, renal impairment, pregnancy, hypersensitivity
Rofecoxib	4-(4′-methylsulphonyl-phenyl)-3-phenyl-2,5*H*)-furanone	Vioxx	Tablets	12.5–25 mg once daily	Moderately bound in plasma, crosses blood-brain barrier	Osteoarthritis, rheumatoid arthritis, acute pain	Peptic ulcer, renal impairment, heart failure, pregnancy
Etoricoxib	(5-chloro-2-(6-methylpyridine-3-yl)-3-(4-methyl-sulphonylphenyl) pyridine	Arcoxia	Tablets	60–120 mg once daily	Highly bound in plasma, crosses blood-brain barrier	Osteoarthritis, rheumatoid arthritis, gouty arthritis	Peptic ulcer, renal impairment, pregnancy, heart failure

Appendix C12/2. (continued)

Drug	Chemical name	Trade name	Formulation	Usual dose	Pharmacokinetics	Indications	Contraindications
Valdecoxib	4-[5methyl-3-phenyl-oxisol-4-yl]-benzene sulphonamide	Bextra	Tablets	10–40 mg once daily	Interacts with inducers or inhibitors of cytochrome P450	Osteoarthritis, rheumatoid arthritis, dysmenorrhea	Peptic ulcer, renal impairment, hypersensitivity, anticoagulants
Parecoxib	(*N*-2[[(5-methyl-3-phenyloxisol-4yl)-phenyl]sul-phonyl]propane amide	Dynastat	Injection	40 mg i.v. or i.m.	Interacts with inducers or inhibitors of cytochrome P450	Post-operative pain	Peptic ulcer, renal impairment, pregnancy, anticoagulants
Paracetamol	N-(4-Hydroxyphenyl)-acetamide	Alvedon, Calpol, Infadrops, Medinol	Supposito-ries, suspen-sion, drops	60–250 mg	Crosses blood-brain barrier	Mild to moderate pain and pyrexia	Renal or hepatic impairment

Appendix C13. Overview of DMARDs used in the treatment of RA

Drug	Approximate time to benefit	Usual maintenance dose	Toxicity
Methotrexate	1–2 months	Oral initial 7.5-10 mg as a single dose once per week or split into 3 doses over 36 hours. Maintenance individualized dose 7.5–25 mg weekly.	Gastrointestinal (oral ulcers, nausea, vomiting, diarrhoea), CNS (headache, fatigue, fuzziness), stomatitis, rash, alopecia, rash, infections, infrequent myelosuppression, hepatotoxicity, rare but serious (even life-threatening) pneumonitis
Antimalarials (Hydroxychloroquine)	2–4 months	Oral 200 mg twice daily	Infrequent rash, diarrhoea, rare retinal toxicity
Azathioprine	2–3 months	Oral 50–150 mg daily	Myelosuppression, infrequent hepatotoxicity, early flu-like illness with fever, gastrointestinal symptoms, hepatotoxicity
Cyclosporin		1.25 mg/kg twice a day increasing to 2.5 mg/kg	Most common: nephrotoxicity, hypertension, nausea, vomiting, appetite loss Rare: anemia, gumhyperplasia, hirsutism, tremor, paresthesiae
Injectable gold complexes (Aurothiomalate)	3–6 months	Intramuscular initial dosage 10–25 mg once a week. Maintenance 25–50 mg every 2–4 weeks.	Rash, stomatitis, myelo-suppression, thrombocytopenia, proteinuria
Oral gold complex (Auranofin)	4–6 months	Oral 3 mg daily or twice daily	Frequent diarrhoea. Otherwise same as injectable gold complexes.
Leflunomide	1–4 months	Loading dose 100 mg daily for 3 days Maintenance dose 10–20 mg daily	Diarrhoea, dyspepsia, mild leukopenia, rash , alopecia, headache, dizziness. Possible severe hepatotoxicity
Penicillamine	3–6 months	Initial dose 125 mg daily for 4-8 weeks, dosage increased by 125 mg daily until improvement or a maximal daily dose of 750 mg.	Rash, stomatitis, dysgeusia, proteinuria, myelosuppression, infrequent but serious auto-immune disease

Appendix C13. Overview of DMARDs used in the treatment of RA (continued)

Drug	Approximate time to benefit	Usual maintenance dose	Toxicity
Sulfasalazine	1–2 months	Oral 1000 mg two or three times daily	Rash, infrequent myelo-suppression, GI intolerance
Tetracyclines Minocycline	3–12 months	Oral 100 to 200 mg daily	Most common: hyperpigmentation of skin nails and mucosa, rash, headache and dizziness, GI symptoms rare: pneumonitis, hepatitis, intracranial hypertension
Biologicals Adalimumab	1 month	Subcutaneous 40 mg fortnightly alone or with other DMARD	Injection site reactions Infections
Etanercept	1 month	Subcutaneous 25 mg once or twice a week Alone or with methotrexate only	Injection site reactions Infections
Infliximab	1 month	Intravenous maintenance dose 3 mg/kg every 8 weeks Only with methotrexate	Infusion reactions Infections
Anakinra	1 month	Subcutaneous 100 mg daily	Injection site reactions Infections

Appendix D1. EMEA/CPMP document "Guidance on Immunotoxicity"

Immunotoxicity concerns direct or indirect adverse effects on the immune system resulting from therapeutic exposure. It encompasses altered immunological events including immune dysregulation (suppression or enhancement), allergy, and autoimmunity. Tiered testing strategies have been developed to assess this direct type of immunotoxicity (i.e., suppression or enhancement). Signs of direct immunotoxicity in either exploratory or extended animal studies may trigger the incorporation of immunotoxicological endpoints into safety monitoring in clinical trials. This tiered approach is recommended for conventional medicinal products (cf. *Note for guidance on Preclinical safety evaluation of biotechnology-derived pharmaceuticals*, CPMP/ICH/302/95) or vaccines (cf. *Note for guidance on Preclinical pharmacological and toxicological testing of vaccines*, CPMP/465/95).

Initial screening phase

The initial screening phase consists predominantly of non-functional parameters. Addition of functional parameters will increase the predictive value of the screening phase and is therefore encouraged. The initial screening phase can be incorporated within at least one standard repeated dose toxicity study. Immmunotoxicity should be incorporated in a 28-days study ; if this is not feasible, 14-days' or 3-months' studies are acceptable. Rats or mice are the species of choice, unless another species are better justified.

The initial immunotoxicity screen consists of the following parameters: haematology (i.e., differential cell counting) lymphoid organ weights (i.e., thymus, spleen, draining and distant lymph nodes, Peyers' patches), bone marrow cellularity, distribution of lymphocyte subsets and NK-cell activity. If the latter two are unavailable the initial screening phase should be completed with the primary antibody response to a T-cell-dependent antigen (e.g., sheep red blood cells).

Changes in the above-mentioned parameters may be a trigger for additional testing.

Interpretation of the initial immunotoxicity screen should be based on an integrative view of changes in lymphoid tissues and immune cell populations as well as taking into account other types of toxicity and the health status of the test animal. For instance, involution of the thymus in the presence of overt systemic toxicity may be too easily explained as a secondary (stress-related) response to deterioration of the general health status. It may instead reflect a direct immunotoxic insult.

Extended studies

The primary aims of extended animal studies are to define the immunotoxicity and the dose-response relationship in order to facilitate risk assessment. Additionally they may indicate the target cell population(s) involved. The extended studies consist of functional assays; however, if data on lymphocyte subset distribution and NK cell activity are not available at this time, they should be considered as part of the extended studies. The design of extended animal studies will depend on the nature of the immunological changes observed in the initial screening phase. A scientifically motivated choice should be made from the following test parameters:

- delayed-type hypersensitivity (DTH)
- mitogen- or antigen-stimulated lymphocyte proliferative responses
- macrophage function
- primary antibody response to T-cell-dependent antigen
- *in vivo* models of host resistance, these may be employed to detect increased susceptibility to infectious agents (bacteria, parasites or viruses) or tumours. Such models are particularly important for risk assessment, as they may be tools to elucidate the actual consequences of disturbed immune function.

Glossary

ACCURATE: The accuracy of an assay is the degree to which it represents the true value. However, as commonly used it does not reflect the precision of the assay. Because of this some workers prefer the term bias. Accuracy is then a combination of the precision and bias of the assay.

ACQUISITION THRESHOLD: A boundary set during flow cytometric data acquisition, below which data will not be collected by the cytometer. The boundary can be set on the basis of light scatter (forward or side scatter), fluorescence, or a combination of fluorescence and light scatter. An acquisition threshold eliminates unwanted events (e.g., debris) that might otherwise obscure collection of relevant events and/or produce excessively large files.

ACTH, CORTICOTROPIN: Hormone of the anterior lobe of the hypophysis or pituitary gland; it governs the nutrition and growth of the adrenal cortex, stimulating it to release steroid hormones, and also possesses extra-adrenal adipokinetic activity; the hormone is a polypeptide containing 39 amino acids, but exact structure varies from one species to another.

ACTIVATOR PROTEIN-1 (AP-1): A key DNA binding protein that is required to activate the expression of many inflammatory and immune genes.

ACUTE REJECTION: Graft dysfunction of solid organs in the first posttransplant period, reflected histologically by infiltration of mononuclear cells and destruction of graft components by invading effector T lymphocytes (including CTL): in the histological definition there is no particular time period for the occurrence after transplantation.

ACUTE-PHASE REACTION: Increase in the circulating levels of certain serum proteins during infection, trauma or inflammatory reactions.

ADALIMUMAB: Fully human therapeutic monclonal antibody directed against human TNF-α

ADAPTIVE (ACQUIRED) IMMUNITY: Antigen-specific, lymphocyte-mediated defence mechanisms that take several days to become protective and are designed to remove the specific antigen.

ADHERENCE: Direct, ligand-mediated cell-cell contact allowing cells to interact.

ADHESION MOLECULE: A protein that enables cells to interact with each other.

ADJUVANT: A non-immunogenic material, co-administered with antigen that enhances the immune response to the antigen.

ADOPTIVE IMMUNOTHERAPY: Also known as passive immunotherapy. Therapeutic intervention consisting of the administration of antibodies or immune cells (typically T cells with predefined specificities, or NK lymphocytes).

AFFINITY: The strength with which an antibody molecule binds an epitope (= antigenic determinant).

AFFINITY MATURATION: The affinity of a particular antibody is increased as a result of somatic hypermutation and selection of the B-cell receptor by competition for pathogens.

AGAMMAGLOBULINEMIA: An immune disorder characterized by very low levels of protective immunoglobulins; affected people develop repeated infections.

AIRWAYS REMODELLING: Changes that occur in asthmatic airways, due to chronic inflammation, leading to stiffening of the airway and altered physiological responses.

ALKYLATING AGENT: A drug capable of interacting with macromolecules, mainly DNA, to form stable adducts.

ALLERGEN: A substance that induces an allergic immune response, typically involving IgE antibodies.

ALLERGY: Hypersensitivity caused by exposure to an exogenous antigen (allergen) resulting in a marked increase in reactivity and responsiveness to that antigen on subsequent exposure, resulting in adverse health effects.

ALLOGENEIC HEMATOPOIETIC STEM CELL TRANSPLANTATION: Therapeutic intervention by which the recipient's hematopoietic system is replaced by that of a donor who typically is matched for class I and II human leukocyte antigens. The therapy is aimed at eradication of cancer cells when given in the setting of (hematological) malignancies. Cytoreductive pretreatment is given to suppress the recipient's immune response against the donor cells and to reduce tumor cell burden.

ALLOGENEIC TRANSPLANT: A transplant involving organs from individuals of the same species that share different antigens.

ALTERNATIVE PATHWAY OF COMPLEMENT ACTIVATION: Pathway initiated by factor B activation at the surface of certain micro-organisms, involving complement factors C3, B, D, P, H and I, leading to the generation of an alternative pathway C3 convertase.

α-MSH, α-MELANOCORTIN STIMULATING HORMONE: Intermedin; a peptide hormone secreted by the intermediate lobe of the pituitary gland; it causes dispersion of melanin with melanophores (chromatophores), resulting in darkening of the skin, presumably by promoting melanin synthesis within melanocytes.

AMBIENT ANALYTE CONCENTRATION: The assay condition in which the ratio of total number of spotted ligands and the target concentration in the analyte are such that the latter is not the limiting factor. As a result the system is independent of sample volume and acts as a concentration sensor.

ANABOLIC BIOTRANSFORMATION: Metabolic pathway leading to active metabolites.

ANAKINRA: Therapeutic recombinant human interleukin-1 receptor antagonist.

ANALYTE SOLUTION: Sample that is analyzed on a microarray for the presence of a particular target.

ANALYTE: The compound to be analysed in the assay.

ANAPHYLATOXINS: Complement peptides (C3a, C4a, C5a) that cause histamine release from mast cells and basophils and smooth muscle contraction, leading to vasodilation.

ANAPHYLAXIS: A type of immune-mediated hypersensitivity reaction. Anaphylactic (type I) or immediate hypersensitivity reactions involve specific IgE antibodies.

ANGIOSTASIS: Cessation of angiogenesis.

ANTHROPOSOPHY: A philosophical school founded by Rudolf Steiner.

ANTIBACTERIAL AGENTS: Ansamycins; benzylpyrimidines; β-lactams; chloramphenicol; cyclines; fosfomycin; fusidic acid; gyrase B inhibitors; isoniazid; lincosamides; macrolides; peptides; quinolones; riminophenazines; sulfones/sulfonamides; other antibiotics (ethambutol, nitrofurans and minimally substituted imidazoles (metronidazole and tinidazole)).

ANTIBIOTICS: Therapeutic antimicrobial agents.

ANTIBODY (AB): Immunoglobulin molecule produced by B lymphocytes in response to immunization/sensitization with a specific antigen that specifically reacts with that antigen.

ANTIGEN (AG): Substance to which a specific immunological reaction mediated by either antibody or lymphocyte is directed.

ANTIGEN-PRESENTING CELL (APC): A cell that presents antigen to lymphocytes, enabling its specific recognition by receptors on the cell surface. In a more restricted way, this term is used to describe MHC class II-positive (accessory) cells that are able to present (processed) antigenic peptides complexed with MHC class II molecules to T-helper-inducer lymphocytes. These cells include macrophage populations (in particular Langerhans cells, and dendritic or interdigitating cells), B lymphocytes, activated T lymphocytes, certain epithelial and endothelial cells (after MHC-class II antigen induction by, e.g., interferon-γ).

ANTI-HISTAMINE: Drug that specifically antagonises receptors for the allergic mediator histamine; used most frequently for H1-receptor antagonists with anti-allergic activities.

ANTI-IGE: Antibody directed against immunoglobulin E; omalizumab is the first therapeutically used monoclonal anti-IgE antibody for the treatment of allergy and asthma.

ANTIMALARIALS: Group of heterogenous compounds with inhibitory actions on Plasmodium parasites, used in the treatment of malaria.

ANTIMETABOLITES: Substances (e.g., some cytotoxic drugs) that replace or inhibit the utilization of endogenous metabolites.

AORTA-GONAD-MESONEPHROS: Early-stage embryonic structure that gives rise to large blood vessels, gonads and kidney.

APOPTOSIS (PROGRAMMED CELL DEATH): A genetically determined process whereby the cell self-destructs after activation, by Ca^{2+}-dependent endonuclease-induced nuclear DNA (chromosomal) fragmentation into sections of about 200 base pairs. It can be spontaneous (as in neutrophils, explaining the short life span of these cells) or induced (as in tumour cells by interaction with cytotoxic lymphocytes or NK cells).

ASTHMA: Respiratory disease due to airway constriction,

associated with chronic inflammation and airway remodelling; either due to repeated inhaled exposure to allergen or to other environmental agent.

ASTHMATIC RESPONSE: Constriction of the bronchial tree due to an allergic reaction to inhaled antigen.

ATTENUATED VACCINE: Vaccine based on live bacteria or viruses that are made non-virulent, usually via serial passage *in vitro*. The mechanism by which mutations are introduced via this empirical approach is not well understood.

AUTOANTIBODIES: Immunoglobulins (antibodies) that are directed against endogenous molecules of the host (see: autoantigens). They circulate in the serum but may be also detectable in other body fluids or bound in target tissue structures. Autoantibodies may occur as a part of the natural immunoglobulin repertoire (natural antibodies) or are induced by different mechanisms (non-natural or pathological autoantibodies). A number of non-natural autoantibodies are diagnostic markers of defined autoimmune diseases regardless of their pathogenetic activity. They may be directed against conserved non-organ-specific autoantigens, organ-specific autoantigens or cell-specific autoantigens.

AUTOANTIGENS: Self constituents (antigens) of the organism, which may be targets of autoimmune responses mediated by autoreactive B cells (see: autoantibodies) or T cells; they include proteins, glycoproteins, nucleic acids, phospholipids and glycosphingolipids.

AUTOIMMUNE DISEASES: Diseases caused by antibodies or T cells targeting self-antigens.

AUTOIMMUNITY: A state of immune reactivity towards self constituents (see autoantigens) that may be either destructive or non-destructive. Destructive autoimmunity is associated with the development of autoimmune diseases.

AZATHIOPRINE: A cytotoxic drug used to treat autoimmune diseases.

B LYMPHOCYTE/CELL: Lymphocytes expressing immunoglobulin (antibody) surface receptors (on virgin B cells IgM and IgD) that recognize nominal antigen and after activation, proliferate and differentiate into antibody-producing plasma cells. During a T-cell-dependent process there is an immunoglobulin class switch (IgM into IgG, IgA, IgD, or IgE) with maintenance of the antigen-combining structure. For T-cell-independent antigens, cells differentiate only to IgM-producing plasma cells.

BACTERIAL EXTRACTS: Aqueous extracts of bacteria used as immunostimulants.

BASELINE IMMUNOSUPPRESSION: The standard immunosuppressive regimen given to transplant recipients.

BASILIXIMAB: Therapeutic humanized monoclonal antibody against the interleukin-2 receptor.

BASOPHILS: Granular leukocytes that bind basic dyes and contain granules that can digest microorganisms. Basophils, like other inflammatory granulocytes, accumulate at the site of inflammation or infection. There they discharge pro-inflammatory granule-associated mediators such as histamine, serotonin, bradykinin, heparin, cytokines, and newly formed mediators such as prostaglandins and leukotrienes. The mediators released by basophils also play an important role in some allergic responses.

BELL-SHAPED: symmetrical curve of a normal distribution of data, in reference to immune augmentation.

BIOACTIVITY: The effect of a given drug or biological, such as a vaccine or cytokine, on a living organism or tissue.

BIOAVAILABILITY: The degree to which an agent, such as a drug, becomes available to the target tissue after administration, depending on the degree of absorption of the agent into the circulating blood.

BIOLOGICAL RESPONSE MODIFIERS (BRMs): Natural, synthetic or engineered products that are used to boost, suppress, direct, or restore the body's ability to fight the disease.

BIOLOGICALS: Proteins, poly- or monoclonal antibodies and fusion proteins generated by recombinant DNA technology.

BIOPHARMACEUTICALS: Complex macromolecules created by the genetic manipulation of living organisms using biotechnological (gene cloning, recombinant DNA (gene splicing), or cell fusion) technologies rather than chemical manufacturing processes.

BIOTERRORISM: Hostile release into the environment of human pathological micro-organisms with the intent to cause disease and death.

BIOTHERAPEUTIC: Therapeutic micro-organism that has a beneficial effect because of its antagonistic activities against specific pathogens for the prevention and treatment of diseases.

BONE MARROW: Soft tissue in hollow bones, containing hematopoietic stem cells and precursor cells of all blood cell subpopulations (primary lymphoid organ). This is a major site of plasma cell and antibody production (secondary lymphoid organ).

BRADYKININ: Locally-acting, peptide inflammatory mediator, formed in the blood, with smooth muscle contracting, plasma exudation and pain receptor stimulating properties.

C1–C9: Components of the complement classical and lytic pathway responsible for mediating inflammatory reactions, opsonisation of particles and cell lysis.

C2 MONITORING: Measurement of drug concentration 2 hours after administration.

CALCINEURIN INHIBITOR: Immunosuppressants inhibiting the serine/threonine phosphatase calcineurin which plays a pivotal role in Ca-dependent intracellular signalling and results in blockade of interleukin-2 synthesis involved in T lymphocyte activation.

CANCER IMMUNITY: The complex of immune responses (cellular and humoral) that is elicited by cancer cells.

CANCER VACCINATION: Specific active immunotherapy of cancer (as opposed to adoptive or passive immunotherapy). This intervention consists of the immunization of patients against the antigens that are expressed in cancer cells with the goal of eradicating these cancer cells.

CAPSULE = VIRULENCE FACTOR: Protective structure outside the cell wall of a pathogen that can enable the pathogen directly or indirectly to cause disease.

CAPTURE MOLECULES (LIGANDS): Molecules immobilized on an array for high-affinity and high-specificity binding of the target from a sample.

CARRIER: An immunogenic macromolecule (usually protein) to which a hapten is attached, allowing this hapten to be immunogenic.

CATECHOLAMINES: Vasoactive substances (e.g., adrenaline, noradrenaline, dopamine) synthesized by neurones of the sympathetic and central nervous system and by the adrenal gland medulla.

CD: Cluster of differentiation, e.g., CD4, CD8, etc. This is a standard naming system for cell-surface proteins of the immune system. For example, CD4 and CD8 identify different subsets of T cells, and CD69 is a cell-surface protein induced upon short-term activation of T cells.

CD3: Molecule consisting of at least four invariant polypeptide chains, present on the surface of T lymphocytes associated with the T cell receptor, and thought to mediate transmembrane signalling (tyrosine phosphorylation) after antigen binding.

CD4: Glycoprotein of 55 kDa molecular weight on the surface of T lymphocytes and a proportion of monocytes/macrophages. On mature T cells, the presence is restricted to T helper (inducer) cells; the molecule has an accessory function to antigen binding by the T cell receptor, by binding to a non-polymorphic determinant of the MHC class II molecule.

CD8: Complex of dimers or higher multimers of 32 to 34 kDa molecular weight glycosylated polypeptides linked together by disulphide bridges, on the surface of T lymphocytes. On mature T cells, the presence of CD8 is restricted to T cytotoxic-suppressor cells; the molecule has an accessory function to antigen binding by the T cell receptor, by binding to a non-polymorphic determinant of the MHC class I molecule.

CD MARKERS: Cell surface molecules of leukocytes and platelets that are distinguishable with monoclonal antibodies and are used to differentiate cell populations.

cDNA ARRAYS: Microscope slide spotted with several hundreds or thousands of different chemically synthesized forms of DNA called complementary DNA (cDNA), which contains the coding part of gene sequences of interest, complementary to their corresponding messenger RNA (mRNA) transcripts. The immobilized DNA samples on the microarrays are hybridized. The main function of a microarray is to detect the level of mRNA transcript of genes of interest.

CELL ADHESION MOLECULES (CAMs): Group of proteins of the immunoglobulin supergene family involved in intercellular adhesion, including ICAM-1, ICAM-2, VCAM-1 and PECAM-1.

CELL-MEDIATED IMMUNITY: Immunological reactivity mediated by T lymphocytes.

CELLULAR THERAPY: Therapeutic measure involving the isolation of a patient's or control mononuclear cells, in vitro manipulation of these cells, and infusion of the modified cells back into the patient.

CHEMOKINES: Small-molecular-weight, pro-inflammatory peptide cytokines which attract cells of the immune system (chemotaxis) along a concentration gradient and activate them.

CHEMOTAXINS: Small molecules capable of inducing directed cell movement.

CHEMOTAXIS: Process of directed cell movement, usually to a site of infection in response to a gradient of chemotaxins.

CHIMERIC ANTIBODY: Antibody in which the constant parts of immunoglobulin heavy and light chains of the original antibody molecule are replaced by immunoglobulin sequences of another species.

CHIMERIC PROTEIN: A human-engineered protein that is encoded by a nucleotide sequence made by a splicing together of two or more complete or partial genes.

CHROMONES: Group of drugs, represented by sodium chromoglycate and nedocromil, with moderate anti-inflammatory activity used in the therapy of allegic rhinitis and mild asthma.

CHRONIC REJECTION: Graft dysfunction of solid organs late after transplantation, reflected histologically by changes in the vasculature, including remodelling of arteries and arterioles with intimal proliferation; there is no particular time period for its occurrence after transplantation.

CLASS SWITCH: The shift in a B cell or its progeny from the secretion of an immunoglobulin of one isotype or class to an immunoglobulin with the same V regions but a different heavy-chain constant region and, hence, a different isotype.

CLASSICAL PATHWAY OF COMPLEMENT ACTIVATION: Pathway initiated by immune complexes, e.g., microbes covered by antibodies, involving complement components C1, C2 and C4 and generating a classical C3 convertase.

CLEARANCE: The volume of blood or plasma freed of a drug during a specific time interval.

CLINICAL HYPOTHESIS: Theory(ies) to be tested during the clinical phase, such as required dose, frequency, duration, and route of administration, usually predetermined based on data from in vitro, animal, human safety, and dose-finding studies.

CLONE: A population of immunocompetent cells that emerges from one single precursor cell; within T or B lymphocytes these are cells with a fixed rearrangement of genes coding for T cell receptor or immunoglobulin.

CO- AND POST-TRANSLATIONAL MODIFICATION: Protein modification occurring during or after protein expression, covalent linking of a chemical moiety to residues of a protein, e.g., phosphorylation, nitrosylation, glycosylation.

COLOMBATION: An as yet still poorly understood process, whereby rapid evaporation of solvents during electrospray ionization results in the explosion of a supercharged micro-droplet, thus projecting some of its constituents into the gas phase as charged ions.

COLONY-STIMULATING FACTORS (CSF): Cytokines predominantly inducing the differentiation from bone marrow stem cells and the activation of non-lymphocytic leukocytes.

COMBINATION DMARD THERAPIES: Therapy of rheumatoid arthritis using a combination of usually two disease-modifying anti-rheumatic drugs (DMARDs); commonly involves administration of methotrexate with a biological.

COMBINATION THERAPY: Treatment incorporating two or more types of therapy, i.e., surgery, chemotherapy, radiation, hormonal, gene or drug therapy, etc. If two or more therapeutic measures/agents are taken together, the reduced response and increased resistance rates occurring when a single therapy is given over a prolonged period of time can be overcome.

COMBINATION VACCINE: Vaccine containing antigen derived from more than one pathogen. Examples include diphtheria-tetanus, diphtheria-pertussis-tetanus, diphtheria-pertussis-tetanus-polio, measles-mumps-rubella vaccines.

COMPLEMENT COMPONENTS B, P, D, H, I: Components of the alternative pathway of complement activation.

COMPLEMENT RECEPTORS (CR1–CR4): Set of four cell surface receptors for fragments of complement factor C3. CR3 and CR4 are integrins.

COMPLEMENT SYSTEM: Series of proteolytic enzymes in blood, capable of lysing microbes and enhancing the uptake of microbes by phagocytes.

COMPLEMENTARITY DETERMINING REGIONS: Three regions (CDR1; CDR2 and CDR3) of amino acid sequence in the immunoglobulin variable region that are highly divergent (hypervariable).

CORTICOSTEROID RECEPTOR (CR): The cytoplasmic protein to which corticosteroids bind to induce their effects.

CORTICOSTEROID SPARING: Therapeutic agent that when given with a corticosteroid drug, produces the same functional response as a higher dose of corticosteroid.

CORTICOSTEROIDS: Class of steroid hormones from the cortex of the adrenal gland and synthetic drugs with pronounced anti-inflammatory and immunomodulatory actions, metabolic and cardiovascular effects.

COSTIMULATION: Second signals generated to T lymphocytes after encounter with the antigen via the CD3-TCR complex, required to initiate T-cell activation. The absence of a costimulation signal can result in anergy (immune tolerance).

CRH, CORTICOTROPIN-RELEASING HORMONE: A peptide hormone released from the hypothalamus that stimulates the anterior pituitary to release ACTH.

CYCLO-OXYGENASE (COX): Enzyme catalysing the oxidative metabolism of arachidonic acid and other polyunsaturated fatty acids to the biologically active prostaglandins and other eicosanoids. COX-1 is a constitutive isoenzyme, COX-2 an isoform induced by inflammatory stimuli and COX-3 is a neuronal isoform, considered to be the target of the analgesic paracetamol.

CYCLOSPORIN A: Immunosuppressive drug inhibiting cytokine synthesis by T lymphocytes.

CYTOKINES: Proteins secreted by activated immunocompetent cells that act as intercellular mediators regulating cellular differentiation and activation, particularly within the immune system. They are produced by a number of tissue or cell types rather than by specialised glands and generally act locally in a paracrine or autocrine manner often with overlapping or synergistic actions.

CYTOTOXIC DRUGS: Drugs that directly interfere with DNA/RNA synthesis and as such affect cell proliferation.

CYTOTOXICITY: Induced cell death, either by binding of peptide mediators to specific death receptors, by insertion of membrane penetrating components (e.g. complement) or by products released by specific lymphocytes (NK cells, cytoxic T cells) or by granulocytes with cell-destroying properties.

DACLIZUMAB: Therapeutic humanized monoclonal antibody against the interleukin-2 receptor.

DE NOVO TREATMENT: Treatment with a (new) immunosuppressant starting at the beginning of transplantation.

DEFENSINS: Antibacterial peptides present in airway surface fluid that are produced by pulmonary epithelial cells, macrophages and neutrophils and form part of the innate immune defence of the lungs.

DEGRANULATION: Fusion of intracellular granules with the plasma membrane or with the phagosomal membrane, leading to release of granule contents into the extracellular space or into the phagosome, respectively.

DELETIONAL TOLERANCE: The process of tolerance induction by which reactive lymphocytes are removed from the repertoire (for instance by apoptosis).

DENDRITIC CELLS: Leukocytes in tissue that show elongations/protrusions of cytoplasm in the parenchyma, representing a specialized type of antigen-presenting cell derived from lymphocytes or monocytes.

DIAPEDESIS: Process of leukocyte movement from the blood through the endothelial blood vessel wall and basement membrane into the tissues.

DIURNAL VARIATION: Fluctuations that occur regularly within a 24 hour period.

DNA MICROARRAYS (DNA CHIPS): see "Microarrays"

DNA-PROTEIN MODIFYING AND REMODELLING PROTEINS: Proteins that can alter the number or type of side chains of DNA-associated proteins and thereby modify their interaction with DNA resulting in changes in DNA compaction and subsequent gene transcription.

DRUG RESISTANCE: Condition of lack of sensitivity to a drug that develops during its continued administration, resulting from increased metabolism or changes in target cells.

EFFICACY: Capacity or effectiveness of a drug to control or cure an illness. Efficacy should be distinguished from activity.

EICOSANOID: Metabolic product of arachidonic acid.

ENDOPEPTIDASE: An enzyme that catalyzes the splitting of proteins into smaller peptide fractions and amino acids by a process known as proteolysis.

ENDORPHINS: One of a family of opioid-like polypeptides originally isolated from the brain but now found in many parts of the body.

ENDOTOXIN: See "Lipopolysaccharide".

EOSINOPHILS: (Eosinocyte, eosinophilic leukocyte) are granular leukocytes stained by eosin that contain a typically bi-lobed nucleus and large specific granules. The eosinophils reside predominantly in submucosal tissue and normally low in blood. The cells participate in phagocytosis and inflammatory responses.

EPITOPE: The recognition site on an antigenic protein to which either a specific antibody or T-cell receptor binds.

ERYTHROCYTE: Red blood cell, involved in oxygen transport to tissue. Contains a nucleus in distinct avian species like chickens, but does not have a nucleus in mammals.

E-SELECTIN: A member of the family of cell-surface adhesion molecules of leukocytes and endothelial cells that bind to sugar moieties on specific glycoproteins with mucin-like features.

ETANERCEPT: Therapeutic fusion protein between the extracellular part of the human type 1 TNF-α receptor and human IgG.

FAB REGION: Region of an antibody that contains the antigen-combining site.

FC REGION: Region of an antibody responsible for binding to Fc receptors and the C1q component of complement.

FcεRI: High affinity receptor for immunoglobulin E antibodies present on effector cells such as mast cells and basophils.

FERMENTATION: An enzymatically controlled chemical transformation of an organic compound.

FERMOSERA: Protease-treated animal sera.

FICOLL: A solution of high-molecular-weight carbohydrate that is used as a density gradient for the isolation of mononuclear cells (lymphocytes and monocytes) from whole blood; also used for separation of viable from dead lymphocytes.

FILGRASTIM: Therapeutic recombinant human G-CSF with an additional structural methionine.

FIRST LINE OF DEFENCE: Initial cellular and humoral reaction to pathogen.

FLAVONOIDS: Secondary metabolites of the phenylpropanoid class with important functions as antioxidants.

FLOW CYTOMETRY: The analysis of fluorescence and light scatter properties of cells as they pass in suspension through a laser beam.

FLUORESCENCE-LABELED SAMPLES: Test reagents chemically labelled with a fluorescent marker.

FLUOROCHROME: An organic dye or protein that has fluorescent properties and thus can be conjugated to an antibody for use in flow cytometric staining assays.

FOLLICLES: Round to oval structures in lymphoid tissue, where B cells are lodged. Primary follicles only contain small-sized resting B cells: secondary follicles comprise a pale-stained germinal center, containing centrocytes and centroblasts being B lymphocytes in a state of activation/proliferation, macrophages and the stroma consisting of follicular dendritic cells. This germinal center is surrounded by a mantle (corona) with small-sized B lymphocytes.

FOLLICULAR DENDRITIC CELLS: Cells forming the stationary microenvironment of germinal centers of follicles in lymphoid tissue. They are elongated, often binucleated cells with long branches extending between germinal center cells and forming a labyrinth-like structure. The cells are linked together by desmosomes. The cells are of local parenchymal origin, presumably from pericytes surrounding blood vessels. Their main function is presentation of antigen, trapped as immune complex in the labyrinth, to B lymphocytes.

FORWARD SCATTER: The measurement of light deflected by a cell at narrow angles to the laser beam in a flow cytometer. This measurement correlates with cell size.

FULL GENOME ARRAYS: Array presentations of DNA expressing the whole of a species genome on a single DNA array.

GENE EXPRESSION PROFILES: The snapshot evaluation on DNA microarrays of the distribution of activated or inhibited genes from a selection of genes taken from cells under specific (e.g., inflammatory) conditions.

GENE THERAPY: Treatment correcting a genetic defect by the introduction of a normal gene into a cell.

GERMINAL CENTER: The pale-staining centre in follicles of lymphoid tissue, where B lymphocytes are activated by antigen in a T-lymphocyte-dependent manner and subsequently proliferate and differentiate, acquiring the morphology of centroblasts, centrocytes, and plasma cells. The germinal centre has a specialized microenvironment made up of follicular dendritic cells. Large macrophages are present, the so-called tingible-body macrophages (starry-sky macrophages).

GH, GROWTH HORMONE: Somatotropin. A 191-amino acid polypeptide hormone of the anterior lobe of the pituitary gland and cells of the immune system that is important in regulating growth.

GHRH, GROWTH HORMONE RELEASING HORMONE: Somatoliberin. A 44-amino acid polypeptide, produced in the neuronal bodies of the arcuate nucleus and cells of the immune system capable of stimulating the release of GH.

GLUCOCORTICOIDS OR GLUCOCORTICOSTEROIDS: Potent immunosuppresive and antiinflammatory drugs derived from the physiological steroid cortisol, the predominant mechanism of which is the inhibition of cytokine synthesis.

GLUTATHIONE PEROXIDASE(S): Selenium- and non-selenium-containing enzymes that convert lipid peroxides to hydroxyl moieties, thereby reducing their biological reactivity.

GOLD: Gold salts are one of the oldest specific antirheumatic therapies. The anti-inflammatory mechanism of these agents is still unclear.

GRAFT-*VERSUS*-TUMOR: Immune response to tumor cells by immune cells present in a donor's transplanted tissue, such as bone marrow or peripheral blood.

GRAM-POSITIVE BACTERIA: Bacteria with a thick outer cell wall consisting of peptidoglycan and teichoic acid that takes up the purple crystal violet Gram stain. In contrast to Gram-negative bacteria, the cell walls of Gram-positive bacteria do not contain lipopolysaccharide.

GRANULOCYTES: Granule-containing myeloid cells comprising neutrophilic granulocytes (neutrophils), eosinophilic granulocytes (eosinophils) and basophilic granulocytes (basophils).

GRE, GLUCOCORTICOID RESPONSE ELEMENT: Consensus glucocorticoid receptor DNA binding sequence found in the upstream regions of many corticosteroid genes.

HAPTEN: A non-immunogenic compound of low relative molecular mass which becomes immunogenic after conjugation with a carrier protein or cell and in this form induces immune responses. Antibodies, but not T cells, can bind the hapten alone in the absence of carrier.

HEMATOPOIESIS OR HAEMATOPOIESIS: The formation of blood cells: erythrocytes (red blood cells), thrombocytes (platelets), and leukocytes (white blood cells).

HEPATITIS B VIRUS: DNA virus specifically infecting the liver resulting in acute or chronic hepatitis.

HEPATITIS C VIRUS: RNA retrovirus specifically infecting the liver frequently resulting in chronic hepatitis.

HETEROGENEOUS: Commonly used to describe assays which require a step to separate the bound and free fractions.

HIGH-DOSE CHEMOTHERAPY: Type of chemotherapy treatment in which myeloablative doses are given.

HISTAMINE: Vasoactive amine released from mast cell and basophil granules.

HIV INFECTION, DESTRUCTION OF T CELLS: Human Immunodeficiency Virus targets the CD4 molecule on T lymphocytes leading to their destruction by cytotoxic T cells or by apoptosis.

HIV INFECTION, OPPORTUNISTIC INFECTIONS: Because of the suppression of the immune system by chronic human immunodeficiency virus infection, patients become susceptible to a variety of so-called opportunistic infections which "take advantage" of the compromised defences of the patient.

HIV INFECTION, REDUCED NUMBER OF T CELLS: Human immunodeficiency virus infection causes a gradual depletion of $CD4^+$ T lymphocytes which at a threshold of around 350 cells/ml ultimately leads to clinical symptoms of acquired immune deficiency syndrome (AIDS).

HIV INFECTION, TREATMENT: Drug treatment of HIV infection involves combination of viral reverse transcriptase inhibitors and inhibitors of viral proteases. A recently introduced drug (enfuvirtide) inhibits fusion of the virus with the host cell membrane.

HLA: See "Major histocompatibility complex".

HOMOGENEOUS: Commonly used to describe an assay which does not incorporate a step to remove the bound from the free label. However some assays that do not require a separation step still use include a solid phase such as a bead. These are sometimes referred to as non-separation heterogeneous assays. The term homogeneous assays is reserved for assays which entirely take place in the liquid phase.

HPA, HYPOTHALAMIC PITUITARY AXIS: A regulatory pathway including the hormones produced by the different pituitary cell types and the hypothalamic hormones that stimulate or inhibit their release.

HUMANIZED ANTIBODY: Therapeutic monoclonal antibody in which the sequence encoding the CDR of the variable part of the antibody is inserted into the sequence encoding immunoglobulin of another species.

HUMORAL IMMUNITY: Immunological reactivity mediated by antibodies.

HYPERSENSITIVITY: Abnormally increased, immunologically mediated response to a stimulus. Sometimes used loosely for any increased response or to describe allergy. The reaction can be mimicked by non-immunological mechanisms (e.g. chemical stimulation of mast cell degranulation).

ICAM-1 AND ICAM-2: Inter-cellular adhesion molecules on leukocytes and tissue cells that interact with β2-integrins and mediate binding of leukocytes to other cells.

IDIOSYNCRATIC: Refers to adverse effects of drugs for which an unexpected susceptibility of the host is presumed to be involved. The term covers either adverse effects reflecting a genetic predisposition, or those for which the mechanism has not been elucidated.

IDIOTYPE: Antigenic determinant of the region of the variable domain of antibody molecules or T cell receptor forming the paratope; that is, the site specifically recognizing a given antigenic determinant.

IGF-1, INSULIN-GROWTH FACTOR: A 70-amino acid polypeptide produced by many types of cells with potent mitogenic effects for a wide range of cells. The liver is a major site of production in response to growth hormone.

IMIQUIMOD: Immunomodulator that is an agonist at toll-like receptor-7 (TLR-7), stimulating release from leukocytes of cytokines; used topically for the treatment of external genital warts caused by human papilloma virus (HPV).

IMMUNE SYSTEM: A system including all aspects of host defense mechanisms against xenobiotics and pathogens that are encoded in the genes of the host. It includes barrier mechanisms, all organs of immunity, the innate (immediate, non-specific) immune response effectors (proteins, bioactive molecules and cells – mainly phagocytes) and the adaptive (delayed, specific) immune response effectors (T and B lymphocytes and their products). The two responses (specific and non-specific) act synergistically for a fully effective immune response.

IMMUNOALLERGIC: This term is used to describe immune-mediated hypersensitivity reactions.

IMMUNOGLOBULIN (IG): Immunoglobulins are synthesized by plasma cells. The basic subunit consists of two identical heavy chains (about 500 amino acid residues, organized into 4 homologous domains, for μ chain in IgM about 600 amino acid residues, organized in 5 homology domains) and two identical light chains (about 250 amino acid residues organized into 2 homologous domains), each consisting of a variable domain and 1-3 constant domains (in the μ chain 4 constant domains). The antigen-binding fragment (Fab) consists of variable domains of heavy and light chain (2 per basic subunit). Five classes of immunoglobulins exist which differ according to heavy chain type (constant domains): IgG (major Ig in blood), IgM (pentamer consisting of 5 basic units), IgA (major Ig in secretions, here present mainly as a dimeric Ig molecule), IgD (major function as receptor on B lymphocytes), and IgE. Effector functions after antigen binding are mediated by constant domains of the heavy chain (Fc part of the molecule) and include complement activation (IgG, IgM), binding to phagocytic cells (IgG), sensitization and antibody-dependent cell-mediated cytotoxicity (IgG), adherence to platelets (IgG), sensitization and degranulation of mast cells and basophils (IgE). IgA lacks these effector functions and acts mainly in immune exclusion (prevention of entry in the body) at secretory surfaces.

IMMUNOMODULATORY DRUG/IMMUNOMODULATOR: A drug capable of modifying or regulating, by enhancement or suppression, one or more immune functions.

IMMUNOPHENOTYPING: The use of fluorochrome-conjugated antibodies and flow cytometry to identify subsets of leukocytes in a sample such as peripheral blood. Such assays can be used to identify disease states such as leukemia/lymphoma.

IMMUNOSUPPRESSION: Defects in one or more components of the nonspecific/innate or specific/adaptive im-mune system resulting in inability to eliminate or neutralize non-self antigens. Congenital or primary immunodeficiencies are genetic or due to developmental disorders. Acquired or secondary immunodeficiencies develop as a consequence of immunosuppressive compounds, malnutrition, malignancies, radiation or infection. This may result in decreased resistance to infection, the development of certain types of tumors or immune dysregulation and stimulation, thereby promoting allergy or autoimmunity.

IMMUNOTHERAPY: Treatment of a disease by the artificial stimulation of the body's immune system to induce or suppress an immune response.

IMMUNOTOXICANT: Drug, chemical or other agent that is toxic to cells or other components of the immune system. One form of immunotoxicity is the direct toxicity of the compound to components of the immune system, which often leads to suppressed function. This may result in decreased resistance to infection, the development of certain types of tumors or immune dysregulation and stimulation, thereby promoting allergy or autoimmunity. Other types or manifestations of immunotoxicity include allergy or autoimmunity in which the compound causes the immune system to respond as if the compound were an antigen or to respond to self-antigens that have been altered by the chemical.

INDUCTION IMMUNOSUPPRESSION: The immunosuppressive regimen given at the beginning of transplantation.

INFLAMMATION: A complex biological and biochemical process involving cells of the immune system and a plethora of biological mediators (particularly cytokines); it may be defined as the normal response of living tissue to mechanical injury, chemical toxins, invasion by microorganisms, or hypersensitivity reactions. Excessive or chronic inflammation can have disastrous consequences for the host.

INFLAMMATORY CYTOKINES: Cytokines which primarily contribute to inflammatory reactions, including interferon-γ, interleukin-1, tumor necrosis factor, or chemokines.

INFLIXIMAB: Humanised monoclonal antibody (fusion protein between the variable portions of a murine antibody with the constant parts of human IgG) directed against human TNF-α.

INNATE IMMUNE SYSTEM/INNATE(NON-SPECIFIC) IMMUNITY: Non-adaptive, non-antigen-specific host defence system against pathogens and injurious stimuli, present at birth and consisting of phagocytes, natural killer cells and the complement system.

INTEGRINS: Family of heterodimeric cell surface molecules sharing in part a β chain (β1, β2, β3, about 750 amino acids long), each with a different α chain (about 1100 amino acids long), that mediate cell adhesion and migration by binding to other cell adhesion molecules (CAMs), complement fragments or extracellular matrix.

Based on strong structural and functional similarities, integrins form a protein family rather than a superfamily. Examples: leukocyte function-related antigen LFA-1 (αL/β1, CD11a/CD18; receptor for ICAM-1, ICAM-2 and ICAM-3); Mac-1 (αM/β2, CD11b/CD18; complement C3 receptor CR3); p150,95 (αX/β2, CD11c/CD18); very late antigens VLA-1 (α1/β1, CD49a/CD29; laminin, collagen receptor); VLA-2 (α2/β1, CD49b/CD29; laminin, collagen receptor); VLA-3 (α3/β1, CD49c/CD29; laminin, collagen, fibronectin receptor); VLA-4/LPAM-1 (α4/β1, CD49d/CD29; receptor for fibronectin and VCAM-1); VLA-5 (α5/β1, CD49e/CD29, fibronectin receptor); VLA-6 (α6/β1, CD49f/CD29; laminin receptor, and αV/β1, CD51/CD29; vitronectin receptor); LPAM-2 (α4/βp, CD49d/..., or α4/β7).

INTERFERONS (IFNS): Antiviral and immunoregulatory glycoproteins induced in different cell types by appropriate (mostly viral) stimuli, conferring resistance to infection with homologous or heterologous viruses.

INTERFERON-α 2A, 2B: Therapeutic recombinant interferons-α.

INTERFERON-α 2A, OR -α 2B PEGYLATED: Therapeutic recombinant interferons-α conjugated with monomethoxy-polyethyleneglycol (PEG), with a prolonged half-life in the circulation.

INTERFERON-β 1A, 1B: Therapeutic recombinant human interferons-β.

INTERFERON-β: Human β-interferons from fibroblasts.

INTERFERON-γ (IFN-γ): Cytokine produced by T cells. It induces an antiviral state and is cytostatic for tumor cells. It enhances MHC-class I and II expression on various cell types, is antagonistic with IL-4 in IgE/IgG1 synthesis, and stimulates IgG2a synthesis. It activates macrophages to become cytolytic, as well as inducing natural killer cell and lymphokine-activated killer activity.

INTERLEUKINS: Heterogenous group of immunoregulatory protein cytokines, also including lymphokines, monokines, interferons, acting as communication signals between cells..They generally have a low molecular weight (<80 kDa) and are frequently glycosylated; they regulate immune cell function and inflammation at picomolar concentrations by binding to specific cell surface receptors; they are transiently and locally produced; act in a paracrine or autocrine manner, with a wide range of overlapping functions.

INTERLEUKIN-1(IL-1): Multifunctional cytokine produced by several cell types with pro-inflammatory actions.

INTERLEUKIN-2 (IL-2): Cytokine generated by T helper type-1 cells.

INTERLEUKIN-10 (IL-10): An anti-inflammatory factor produced by Th2 helper-T cells, some B cells and LPS activated monocytes.

ISOAGGLUTININS: Specific antibodies agglutinating erythrocytes by reaction with blood-group-specific epitopes (isoantigens).

ISOELECTRIC FOCUSING: the technique used in electrophoresis that separates molecules on the basis of their different isoelectric points.

ISOELECTRIC POINT: The pH at which a molecule in solution will no longer move in an electric field because it no longer has a net electric charge.

ISOPRINOSINE: Complex salt of inosine used as an immunomodulator with stimulating effects on T cell proliferation.

ISOPROSTANES: Non-enzymatic, biologically active oxidation products of arachidonic acid.

ISOTYPE: Antigenic determinant that defines class or subclass of immunoglobulin molecules.

KILLER CELL IMMUNOGLOBULIN-LIKE RECEPTORS (KIRS): Also called killer cell inhibitory receptors: Class of receptors present on NK cells that bind to HLA molecules and inhibit cytotoxic reactions induced by NK cells in target cells.

LABEL: The moiety (enzyme, fluorochrome, radioisotope etc.) used in an immunoassay to quantitate the antigen:antibody interaction.

LATE PHASE: A delayed response to an allergic skin or inhalation challenge. Typically this is seen as swelling or broncho-constriction 3–8 hours after the initial allergic response.

LECTIN PATHWAY OF COMPLEMENT ACTIVATION: Pathway initiated by MBL that intersects with the classical pathway.

LECTINS: Unique group of proteins that have been found in plants, viruses, microorganisms and animals, but despite their ubiquity, their function in nature is unclear. Some lectins possess the ability to induce mitosis in cells (including lymphocytes) which normally are not dividing.

LEFLUNOMIDE: Synthetic inhibitor of dihydro-orotate dehydrogenase and pyrimidine synthesis with inhibitory activity on lymphocyte proliferation, used therapeutically to treat rheumatoid arthritis.

LENOGRASTIM: Therapeutic recombinant human G-CSF.

LENTINAN: Glucan polymer isolated from the shiitake mushroom (Lentinus edodes) with non-specific immunomodulating and tumour-suppressing activity.

LEUKOCYTES: Also called white blood cells, consisting of granulocytes (polymorphonuclear neutrophilic granulocytes, eosinophilic granulocytes and basophilic granulocytes), monocytes, and lymphocytes.

LEUKOTRIENES: A family of biologically active compounds derived by enzymatic oxidation from arachidonic acid. They participate in host defence reactions and pathophysiological conditions such as inflammation.

LEVAMISOLE: An oral immunomodulatory drug that was previously used for eliminating intestinal parasites in animals.

LIGAND: Molecule that binds to a receptor/target molecule.

LIPOPOLYSACCHARIDE (LPS): Product of some Gram-negative bacterial cell walls that binds to specific Toll-like receptors, to CD14 and to LPS-binding protein (LBP). It activates B lymphocytes, macrophages and neutrophils. Also referred to as endotoxin.

LIPOXIN: Biologically active products of arachidonic acid, formed through transmetabolism by two different lipoxygenases in different cells, with pro-apoptotic activity.

LYMPHOCYTES: Cells belonging to the lymphoid lineage of bone marrow-derived haematopoietic cells. A more restricted designation is that of a small resting or recirculating mononuclear cell in blood or lymphoid tissue, that measures about 7–8 µm and has a round nucleus containing densely aggregated chromatin, and little cytoplasm. Lymphocytes play a key role in immune reactions through specific recognition of antigens.

LYMPHOID ORGAN: Organ in which cells of the immune system, mainly lymphocytes, are lodged in an organised microenvironment, either in a resting stage or in a stage of activation/differentiation/proliferation. Lymphoid organs include bone marrow, thymus, lymph nodes, spleen, and mucosal-associated lymphoid tissue. Central (primary) lymphoid organs are those in which T and B lymphocytes develop and mature (bone marrow, thymus); peripheral (secondary) lymphoid organs are those where immunocompetent lymphocytes recognize antigen, and subsequently initiate immunologic reactions and produce effector elements of these reactions.

LYTIC PATHWAY: Complement pathway effected by complement fragments C5–C9, leading to lysis (destruction) of target cells (e.g cells to which antibody has bound).

MACROPHAGES: Large 12–20 µm mononuclear phagocytic and antigen-presenting cells, present in tissue (histiocyte), constituting the mononuclear phagocytic system that includes monocytes, macrophages, dendritic cells (in lymphoid organs), Langerhans cells (in skin) and Kupffer cells (in liver).

MACROSPOTS: Larger area on a microarray with a high total number of spotted ligands, resulting in assay conditions under which the concentration of target in the analyte is the limiting factor.

MAINTENANCE IMMUNOSUPPRESSION: The immunosuppressive regimen given when the transplant recipient has reached a stable situation after transplantation.

MAJOR HISTOCOMPATIBILITY COMPLEX (MHC): Set of genes that codes for tissue compatibility markers. The MHC complex of man is called HL-A (human leukocyte antigens), of mice H-2 and of rat RT-1. These markers are targets in the rejection of an allograft (matched grafted tissue or organ from a different individual) and hence determine the fate of allografts, play also a central role in control of cellular interactions during immunologic reactions. Tissue compatibility is coded by class I and class II gene loci. Class I MHC molecules are coded by the A, B, or C gene locus in the HLA complex, the K and D locus in the mouse H2 complex, in association with the β2-microglobulin molecule. These two-chain molecules are present on all nucleated cells. Class II MHC molecules are coded by the D (DR, DP, DQ) gene locus in the HLA complex, the I-A and I-E locus in the mouse H2 complex, and comprise an α and a β chain (intracellularly associated with an "invariant" chain). The two-chain molecules are present on B lymphocytes, activated T lymphocytes, monocytes/macrophages/interdigitating dendritic cells, some epithelial and endothelial cells (variable, dependent on species and state of activation) that are also called antigen-presenting cells. Genes within or closely linked to the MHC control certain complement components (MHC Class III genes). MHC-restriction is the phenomenon whereby immunological reactions can only occur in association with or parallel with recognition of the polymorphic determinant of a given MHC molecule, and not with that of another MHC molecule. This applies to T lymphocytes with an αβ T-cell receptor, that recognize antigenic peptides in association with the polymorphic determinant of MHC molecules and a fraction of the T-cell population with a γδ T-cell receptor.

MANNAN-BINDING LECTIN: Protein in blood that can bind to

mannan residues on certain micro-organisms and initiate complement activation via the lectin pathway.

MAST CELL: Tissue (mainly skin and mucosa)-associated cell activated by antigen/allergen bridging of surface bound IgE antibodies, releasing enzymes and vasoactive mediators, especially histamine.

MECHANISM OF ACTION: Complex process whereby a drug, which itself is not the triggering component of therapy; but which manipulates cellular and cytokine elements of the immune system that are the ultimate antitumor effectors.

MEDIUM-TO-LOW DENSITY: Microarrays that comprise a clinical number of probes (several hundred) in contrast to whole genome microarrays.

MEMBRANE ATTACK COMPLEX: Assembled complex of complement components C5b-C9 of the lytic pathway that is inserted into target cell membranes and causes cell lysis.

METHOTREXATE: Antagonist of folic acid used therapeutically as a cytostatic and as an antirheumatic drug.

METHYL INOSINE MONOPHOSPHATE (MIMP): Immunomodulator that preferentially augments T cell responses.

MICROARRAY: DNA display technology that allows the monitoring of the whole genome (or parts of it) on a single chip so that researchers can have a better simultaneous picture of the interactions among thousands of genes.

MICROSPOTS: Miniaturized area on a DNA microarray with a low total number of spotted ligands, thereby resulting in assay conditions under which the concentration of target in the analyte is not the limiting factor.

MICROVASCULAR LEAKAGE: Exudation of blood plasma resulting from the opening of gaps between endothelial cells lining the small vessels of the peripheral blood circulation under stimulation by inflammatory mediators.

MINOCYCLINE: Tetracycline antibacterial drug with additional anti-inflammatory activity, especially in inhibiting metalloproteinase activity and inflammatory cytokine release. Also used in the therapy of rheumatoid arthritis.

MITOGEN: Substance that activates resting cells to transform and proliferate.

MOLGRAMOSTIM: Therapeutic recombinant human GM-CSF.

MONOCLONAL ANTIBODIES: Identical copies of antibody with the same antigen specificity that consist of one heavy chain class and one light chain type. Typically, monoclonal antibodies are produced by a hybridoma, which is a transformed cell line grown in vivo or in vitro that is a somatic hybrid of 2 parent cell lines, one of which is a plasma cell originally producing the single antibody.

MONOCYTE: Large 10–15 μm non-differentiated mononuclear cell, present in the blood and in lymphatics, comprising the circulating component of the mononuclear phagocyte system.

mRNA: Messenger ribonucleic acid responsible during gene transcription for transferring information from DNA to ribosomes for protein synthesis.

MUCOSA-ASSOCIATED LYMPHOID TISSUE: Lymphoid tissue in immediate contact with the mucus-secreting mucosal layer in nasal cavity and nasopharynx (nasal-associated lymphoid tissue), airways (bronchus-associated lymphoid tissue), and intestinal tract (gut-associated lymphoid tissue). Serves as the immunological defence at secretory surfaces, to some extent independent of the systemic (internal) response.

MULTIPLE RESISTANT MICROORGANISMS: Micro-organisms (bacteria, viruses, fungi, parasites) which are no longer sensitive to the antimicrobial actions of a variety of therapeutic agents due to mutational changes.

MURABUTIDE: Synthetic muramyl peptide that selectively stimulates cytokine release from Th1 lymphocytes, enhancing host defence responses to viruses and bacteria.

MURAMYL DIPEPTIDES: Bacterial-derived peptides that stimulate host defence against bacterial infection by binding to receptors on macrophages causing them to release cytokines.

MUTEINS: Cytokine mutants, generated by gene technology, behaving mainly as receptor antagonists.

NADPH OXIDASE: Enzyme in phagocytes that generates superoxide, from which other bactericidal reactive oxygen species are derived.

NATURAL KILLER (NK) CELLS: Lymphocyte-like cells of the innate immune system capable of killing virus-infected and tumour-transformed cells in an antigen-independent manner.

NEO-ANGIOGENESIS: The growth of new capillaries from pre-existing vessels.

NEUROPEPTIDES: Peptides released from nerves that act as neurotransmitters and/or as mediators of inflammation.

NEUTROPHILS: These are highly specialized white blood cells characterized by a multi-lobed nucleus (polymorphonuclear) and a granular cytoplasm that is "neutral" to histological staining under the light microscope. Specialized constituents of the neutrophil membrane, cytoplasmic granules, and cytosol together mediate ingestion and killing of bacteria: after attachment and internalization of the microorganism into the phagocytic vacuole (phagosome), its destruction is mediated by the release of an array of antimicrobial polypeptides and reactive oxidant species (ROS).

NITRIC OXIDE SYNTHASE: Enzyme catalysing the synthesis from arginine of nitric oxide (NO). Occurs in a neuronal form (nNOS, NOS-1), endothelial cell (ecNOS, NOS-3) and inducible (iNOS, NOS-2) form.

NITRIC OXIDE: Gaseous local mediator with vasoactive, pro-inflammatory, bacteriocidal and neurotransmitter activities.

NON-SPECIFIC AIRWAYS REACTIVITY: Enhanced sensitivity to constriction of asthmatic airways in response to inhalation of non-specific pharmacological agonists such as histamine and methacholine.

NON-SPECIFIC IMMUNITY: (see "Innate immunity").

NON-SPECIFIC IMMUNOSTIMULATION: Antigen-independent enhancement of the sensitivity of to activation by a variety of antigens.

NUCLEAR FACTOR κB (NF-κB): Ubiquitous, inducible, transcription factor that binds to enhancer elements in most immune and inflammatory cells to stimulate inflammatory genes.

OLIGONUCLEOTIDE ARRAYS: Microscope slides spotted with several hundreds or thousands of different chemically synthesized forms of oligonucleotides (short sections of RNA or DNA with <20 base pairs), complementary to their corresponding messenger RNA (mRNA) or complementary DNA, which they are intended to detect.

OM-174: A soluble immunomodulator derived from *Escherichia coli* triacylated lipid A, used as an adjuvant in vaccination.

OPPORTUNISTIC INFECTIONS: Infections by micro-organisms that are usually harmless but can become pathogenic when host resistance to disease is impaired.

OPSONISATION: Covering of pathogenic particles and micro-organisms with antibody or complement proteins that enhance uptake of these by phagocytes.

OPTIMAL IMMUNOMODULATORY DOSE: Treatment dose that maximally induces the chosen immunomodulatory action with acceptable toxicity.

OSTEOPOROSIS: Metabolic disorder associated with fractures of the femoral neck, vertebrae, and distal forearm.

PASSIVE IMMUNOTHERAPY: Prophylactic or therapeutic treatment of disease by exogenous administration of antibody.

PATHOGEN-ASSOCIATED MOLECULAR PATTERNS (PAMPs): Protein, lipid or DNA moieties specifically expressed by microbes.

PATHOGENIC MICROORGANISMS: Micro-organisms that are capable of inducing disease in the host.

PATIENT COMPLIANCE: Voluntary cooperation of the patient in following a prescribed regimen.

PATTERN-RECOGNITION RECEPTORS: Molecules present on the surface of most immunocompetent cells, which recognize microbe-specific common pathogen-associated molecular patterns and produce an immediate defensive response against the invading microorganism.

PEGYLATION: Process whereby polyethylene glycol (PEG) is attached to a protein in order to extend its circulating half-life and thereby enhance its biological activity.

PENICILLAMINE: Thiol-containing drug with immunomodulating activity and copper-chelating activity; previously used as an antirheumatic drug. Also used to treat Wilson's disease.

PEROXYNITRITE: Highly reactive, tissue-damaging product of the oxidation of nitric oxide.

PEYER'S PATCH: Lymphoid tissue in the wall of the small intestine, separated from the gut lumen by a domed area and an epithelial layer; forms part of the mucosal-associated lymphoid tissue; main function is initiation of immue reactions towards pathogens entering through the dome epithelium.

PGG-GLUCAN: An immunomodulatory branched beta-(1,3) glucan purified from the cell walls of *Saccharomyces cerevisiae* which stimulates proliferation of leukocyte progenitor cells in the bone marrow and enhances leukocyte microbicidal activity without stimulating cytokine release.

PHAGOCYTES: Cells (etymologically "devouring cells") that are characterized by their ability to engulf via receptor-mediated endocytosis relatively large particles or micro-organisms (phagocytosis) into intracellular vacuoles by a process generally requiring actin polymerization. This property is esssential for the role of phagocytes in host defence. Two main phagocyte lineages exist: polymorphonuclear leukocytes (PMNL) comprising of neutrophils, eosinophils and basophils, and mononuclear phagocytes, comprising of monocytes, macrophages and dendritic cells.

PHAGOCYTOSIS: Process by which phagocytes bind and

engulf material >1 µm (e.g., microbes) in an Fc receptor-dependent manner, with accessory help of complement receptors. Phagocytosis occurs via a "zipper" mechanism, whereby the particle, opsonized (coated) with antibody or complement, becomes enclosed by the cell membrane of the phagocyte. The particle is then incorporated into a vacuole (phagosome) where it is degraded by proteases and an NADPH oxidase-mediated oxidative burst with formation of superoxide anion, peroxide anion, and hydroxyl radicals.

PHAGOSOME: Intracellular vacuole containing phagocytosed material.

PHARMACOKINETIC MONITORING: The monitoring of pharmacokinetic parameters of a drug or a drug metabolite in order to avoid drug-related toxicity, to maintain therapeutic concentrations or to adjust the administered doses.

PHYTATES: Phosphorus compounds found primarily in cereal grains, legumes, and nuts. They bind with minerals such as iron, calcium, and zinc and interfere with their absorption in the body. They also have beneficial heath effects, help disease prevention, and they can be considered as antioxidant compounds in foods.

PLATELET-ACTIVATING FACTOR (PAF): Molecule derived from phosphatidylcholine; released by immune cells and tissue cells; acts as a chemoattractant for and activator of phagocytes as well as an adhesion molecule.

PMNS: see "Polymorphonuclear granulocytes".

POLYMORPHIC ENZYME: An enzyme which displays variable activity as a result of the presence of several variants of the encoding gene.

POLYMORPHISM: The existence of a gene in several allelic forms in a single species.

POLYMORPHONUCLEAR GRANULOCYTES (POLYMORPHONUCLEAR LEUKOCYTES, PMNS): Leukocytes of bone marrow origin, with a lobular nucleus, involved in acute inflammatory reactions. Main subsets are basophilic, eosinophilic and neutrophilic granulocytes (different cytoplasmic granule colors under haematological staining). The cells contribute to bacterial killing and (acute) inflammatory reactions after attraction by specific (immune complex-mediated) or non-specific stimuli (including complement components); after activation, they release granules containing various hydrolytic enzymes.

POMC, PRO-OPIOMELANOCORTIN: The precursor protein for ACTH and endorphins.

PRECISE: The precision of an assay defines the reproducibility of multiple assay measurements. However a precise assay is not necessarily accurate.

PRL, PROLACTIN: A peptide hormone of the anterior lobe of the pituitary gland that stimulates the secretion of milk and possibly during pregnancy, breast growth. Also called lactogenic, mammotropic, or galactopoietic factor or hormone; lactotropin.

PROBIOTICS: Naturally occurring micro-organisms (e.g., *Lactobacillus*) used as dietary immunomodulators.

PROSTAGLANDINS (PGS): Biologically active products of the oxidative metabolism of polyunsaturated fatty acids by cyclo-oxygenase. Several different prostaglandins are formed by specific PG synthetases, including PGE_2 and PGD_2 from arachidonic acid, which are involved in inflammation.

PROTEOLYTIC FRAGMENTS: Peptide fragments resulting from proteolytic cleavage of a protein by an endopeptidase.

PROTEOME: The protein complement expressed by a genome.

PROTEOMICS: The scientific study of the proteome.

PSEUDO-ALLERGIC: Non-immune-mediated hypersensitivity reactions that bear clinical similarities to immune-mediated hypersensitivity reactions, because of the release of the same mediators. The term "anaphylactoid" is often used to depict pseudo-anaphylactic reactions in which non-immune-mediated hypersensitivity reactions are involved.

PURPLE CONEFLOWER (*ECHINACEA PURPUREA*): Indigenous North American plant an extract of which is used as a mild immunostimulant.

REACTIVE OXYGEN SPECIES (ROS): Unstable, highly reactive form of oxygen or oxygen-containing molecule (e.g. H_2O_2, superoxide anion, hydroxyl radical). That is formed by phagocyte mebrane NADPH oxidase and is involved in bacterial killing, protein degradation and acute inflammatory reactions.

RECEPTOR: Cell surface molecule that binds specifically to certain extracellular molecules.

RECOMBINANT: Protein produced in vitro from eukaryotic or prokaryotic cells as a result of alteration of the gene for the protein by mutation, addition, or deletion in the laboratory.

REPERTOIRE: The total spectrum of specific antigen-recognizing capacities (diversity) within the population of T or B lymphocytes.

RESPIRATORY BURST: Increase in oxygen consumption of phagocytes following cell activation by opsonised particles or soluble stimuli; leads to generation of superoxide and other reactive oxygen species (ROS).

RHEUMATOID ARTHRITIS: Systemic autoimmune disease mediated by autoantibodies and autoreactive lymphocytes, characterized by loss of bone density and painful inflammatory destruction of joint cartilage and bone, leading to deformation of the joints and limbs.

SAMPLE MATRIX: This refers to the type of sample (e.g., sample, plasma, urine, cell culture supernatant, etc.) to be analysed. These samples may contain components (e.g., binding proteins) which could potentially interfere in the antigen:antibody interaction.

SCURVY: Disease characterized by exhaustion and bleeding due to vitamin C deficiency.

SELECTINS: Group of three adhesion molecules on immune cells (L-selectin), platelets (P-selectin) and endothelial/epithelial cells (E-selectin, P-selectin) that bind to carbohydrate residues on molecules of opposing cells.

SELENOPROTEINS: Endogenous proteins with or without known functions into which selenium is incorporated as seleno-cysteine.

SENSITIVITY: The limit of detection of an assay. It is usually calculated by determining the mean and standard deviation from multiple measurements of a sample lacking any analyte. The value on the standard curve corresponding to two or three standard deviations above the mean is taken to be the assay sensitivity.

SENSITIZATION: Induction of specialized immunological memory in an individual by exposure to antigen.

SEROTONIN (5-HYDROXYTRYPTAMINE, 5-HT): Vasoactive amine mediator released by platelets, neurones and mast cells, with pro-inflammatory and smooth muscle contracting and neurotransmitter properties.

SERPINS: Serine protease inhibitors. Family of proteins with specific inhibitory action on serine proteases, located in all extracellular body fluids and intracellularly in many cell types.

SERUM THERAPY: Therapeutic treatment of disease by supply with heterologous (animal) serum containing antibody.

SHOCK: Acute hypotensive response resulting from massive systemic immunological activation of mast cells, with release of histamine (anaphylactic shock) or from systemic bacterial infection, with stimulation of inflammatory cytokine release by bacterial lipopolysaccharide (LPS; septic shock); the latter is often associated with extensive complement activation and inflammatory necrosis of essential organs (multiple organ failure, MOF).

SIDE SCATTER: The measurement of light deflected by a cell at wide angles to the laser beam in a flow cytometer. This measurement correlates with cell granularity.

SOMATOSTATIN (SOM): Somatotropin release-inhibiting factor; a tetradecapeptide capable of inhibiting the release of GH by the anterior lobe of the pituitary gland.

SOMATIC MUTATION: Small non-inherited changes in genes, resulting in alterations in protein amino acid sequences. In immunoglobulin molecules, such changes result in diversity of the antigen binding site (variable region).

SPECIFIC IMMUNITY: Immune responses directed towards specific antigen.

SPECIFICITY (OF AN IMMUNOASSAY): Defines whether or not compounds other than the analyte can interact with the binding sites of the antibodies used in the assay. This is determined by performing cross-reactivity studies with a series of structurally related compounds likely to be present in the biological samples under investigation.

STANDARD OPERATION PROCEDURE: Documented, standardized procedure ensuring reproducibility of a process.

STEM CELL: A multipotential self-renewing precursor cell of cells in a distinct lineage (for instance, the hematopoietic cell lineage).

STEM CELL TRANSPLANTATION: Infusion of infusing healthy stem cells into patients who have undergone high-dose chemotherapy for leukemia, immunodeficiency, lymphoma, anemias or metabolic disorders.

SUBSTRATE: The target molecule which is modified during the course of an enzyme catalyzed reaction.

SUBUNIT VACCINE: Vaccine containing (purified) Ag derived from pathogens.

SULFASALAZINE: Sulphonamide antibiotic used as an anti-inflammatory agent in inflammatory bowel disease.

SUPEROXIDE ANION: Reactive, higher energetic form of molecular oxygen generated enzymatically by phagocyte membrane NADPH oxidase.

SURROGATE: Biological marker that is considered likely to predict therapeutic benefit and is sufficiently correlated with the primary endpoint.

SYSTEMIC LUPUS ERYTHEMATOSUS (SLE): Human autoimmune disease, usually mediated by antinuclear antibodies.

T-CELL RECEPTOR (TCR): Heterodimeric molecule on the surface of the T lymphocyte that recognizes antigen. The polypeptide chains have a variable and a constant part, and can be an α, β, γ or δ chain. The $\alpha\beta$ T cell receptor occurs on most T cells and recognizes antigenic peptides in combination with the polymorphic determinant of MHC molecules (self-MHC restricted). The $\gamma\delta$ T cell receptor occurs on a small subpopulation, e.g., in mucosal epithelium, and can recognize antigen in a non-MHC restricted manner. The T-cell receptor occurs exclusively in association with the CD3 molecule that mediates transmembrane signalling.

T-CELL-DEPLETED: In patients who receive myeloablative (bone marrow replacement) therapy and an allogeneic stem cell transplant, removal of the T cells (T-cell depletion) from the administered stem cell product has been used to decrease the incidence and severity of graft-versus-host disease (GVHD). However, T-cell depletion has also been found to significantly delay immune reconstitution and increase the rate of graft failure and tumor relapse.

T LYMPHOCYTE/CELL: Thymus-derived lymphocytes that induce, regulate, and effect specific immunologic reactions stimulation by antigen, mostly in the form of processed antigen complexed with MHC product on an antigen-presenting cell. Most T lymphocytes recognize antigen by a heterodimeric α-β surface receptor molecule associated with CD3 molecule mediating transmembrane signalling. Subsets include helper-inducer (Th) and suppressor-cytotoxic (Tc) cells. Th1 and Th2 subpopulations exist: Th1 cells produce interleukins 2 and 3, interferon-γ, tumor necrosis factor-α and -β, and granulocyte/macrophage colony stimulating factor, and function in induction of delayed-type hypersensitivity, macrophage activation, and IgG2a synthesis. Th2 cells produce interleukins 3, 4 and 5, tumor necrosis factor-α and granulocyte/macrophage colony stimulating factor, and function in induction of IgG1, IgA and IgE synthesis, and induction of eosinophilic granulocytes. Cytotoxic T lymphocytes differentiate from precursor to effector cytotoxic cells and subsequently kill target cells.

TACROLIMUS: Immunosuppressive drug inhibiting transcription factor activation of cytokine synthesis in T lymphocytes.

TARGETS: Molecule of interest in an analyte.

T-CELL IMMUNITY: Immune responses involving activated T lymphocytes as helper or effector (cytotoxic) cells.

T-DEPENDENT ANTIGEN: Antigen for which antibody formation requires T cells.

TERMINAL HALF-LIFE: The time taken for the plasma concentration of a drug to fall by one-half during the elimination phase.

TH1 AND TH2 RESPONSES: Immune responses mediated by particular types of T lymphocytes. T lymphocytes are sub-divided into T-helper ($CD4^+$) and T-effector/cytotoxic ($CD8^+$) lymphocytes. $CD4^+$ T cells assure the regulation of immune responses via the release of different patterns of cytokines, primarily IL-2 and interferon-γ that characterize Th1 responses, and IL-4, IL-5 and IL-10 that characterize Th2 responses.

T-HELPER (TH) LYMPHOCYTES: Functional subset of T lymphocytes that can help to generate cytotoxic T lymphocytes and cooperates with B lymphocytes in the production of antibody.

THERAPEUTIC INDEX: The ratio of the largest dose producing no toxic symptoms to the smallest dose capable of exerting a therapeutic effect.

THERAPEUTIC STRATEGY: A pharmacologically based approach to the treatment of a disease based on the mechanism of action, pharmacokinetics, and toxicology, generally involving multiple modalities that integrate with clinical pathophysiological considerations.

THROMBOCYTE (PLATELET): Small cytoplasmic fragment in blood that is responsible for coagulation. Its main role is to block damaged vessel walls and prevent hemorrhage, by clumping and aggregation. Thrombocytes contain heparin and serotonin, which contribute after release to the acute vascular response in hypersensitivity reactions, and produce oxygen radicals.

THROMBOXANE: Biologically active product of the oxidative metabolism of arachidonic acid by cyclo-oxygenase and thromboxane synthetase with platelet activating and vasoconstricting activity.

THYROTROPIN STIMULATING HORMONE (TSH): Thyrotropin. A hormone of the anterior pituitary gland that stimulates thyroid hormone production.

TISSUE ENGINEERING: Development and manipulation of laboratory-grown molecules, cells, tissues, or organs to replace or support the function of defective or injured body parts.

TISSUE RESIDENCY: The amount of time a drug spends in a specific tissue – generally the target tissue of interest.

T LYMPHOCYTE SUBSETS: $CD8^+$ cytotoxic T cells recognize antigenic peptide in association with class I molecules of the major histocompatibility complex (MHC). $CD8^+$ T cells mediate their effector functions by killing the cells presenting the relevant antigenic peptide, or secreting effector cytokines. CD4+ T helper cells recognize antigenic peptide in association with class II MHC molecules. They mediate their effector functions by enhancing the persistence of antigen-stimulated T cells or through secretion of effector cytokines.

TOLERANCE: A state of unresponsiveness to antigenic stimulation, due to the absence of responding elements or the loss of capacity of existing elements to mount a reaction. Synonym for anergy.

TOLL-LIKE RECEPTORS: Family of pattern recognition receptors with extracellular leucine-rich repeats that bind pathogen-specific molecules and intracellular domains that resemble that of the IL-1 receptor, leading to activation of non-specific defence responses.

TOPICAL: Route of administration of a drug directly to the site of disease/inflammation.

TRACER: Used synonymously with label in an immunoassay.

TRANSACTIVATION DOMAIN: Region of a transcription factor that stimulates gene transcription.

TRANSACTIVATION: Stimulation of transcription by a transcription factor binding to DNA and activating adjacent proteins.

TRANSREPRESSION: Inhibition of transcription by a transcription factor binding to DNA and inhibiting adjacent proteins.

T-REGULATORY CELLS: T lymphocytes that regulate the activity and function of other T lymphocytes.

TRH, THYROTROPIN RELEASING HORMONE: Thyroid-stimulating hormone-releasing factor; a tripeptide hormone from the hypothalamus that stimulates the anterior pituitary to release thyrotropin.

TUMOR NECROSIS FACTOR (TNF): General mediator of inflammation and septic shock originally described as a tumor degrading factor induced by bacterial lipopolysaccharide. It comprises two forms, TNF-α and TNF-β, produced by monocytes/macrophages, TNF-β also by T lymphocytes and natural killer cells. Tumor necrosis factor has activity similar to IL-1, and acts synergistically with IL-1. It promotes an anti-viral state and is cytotoxic for tumor cells. It stimulates granulocytes and eosinophils, activates macrophages to IL-1 synthesis, stimulates B cells to proliferate and differentiate and T cells to proliferation, IL-2 receptor synthesis and IFN-γ synthesis. It induces fibroblasts to prostaglandin synthesis and proliferation, it induces fever and synthesis of acute-phase proteins. It reduces cytochrome p450 synthesis. It activates endothelium and promotes adherence of neutrophilic granulocytes to endothelium, and induces cell adhesion molecules like lymphocyte function-associated antigens LFA-1 and LFA-3, ICAM-1, and ELAM-1. It reduces lipoprotein lipase synthesis by adipocytes, and activates osteoclasts to bone resorption.

TUMOR-ASSOCIATED ANTIGENS: Antigens that are expressed by tumor cells and elicit specific B- or T-cell-mediated immune responses, irrespective of whether such immune responses affect tumor growth. Tumor-associated antigens include tissue-specific differentiation antigens, cancer-testis antigens, otherwise normal antigens that are overexpressed, fusion proteins, mutational antigens, virus-encoded antigens and minor histocompatibility antigens.

VACCINE: A substance or group of substances intended to induce an immune system to a tumor or to a microorganism, thus helping the body recognize and destroy cancer cells or microorganisms. It is a preparation of a weakened or killed pathogen, such as a bacterium or virus, or a fragment of the structure of a pathogen that upon administration stimulates antibody production or cellular immunity against the pathogen.

VACCINE DELIVERY SYSTEM: Colloidal carrier allowing multimeric antigen presentation (i.e., containing more than one antigen per particle), thereby increasing the immunogenicity of the antigen; it may also contain adjuvants.

WASH ASSAY: Shorthand for a flow cytometry assay that requires a washing and centrifugation step, as opposed to "no-wash" or homogeneous assays. Sensitivity is generally increased by washing, while homogeneous assays are more convenient.

XENOBIOTIC: Chemical or substance that is foreign to the biological system.

ZOOM GELS: 2DE gels with enhanced separation capacity or resolution. Several zoom gels with adjacent partially overlapping pH ranges can be combined *in silico* to generate even larger protein maps with a high number of proteins.

List of contributors

Ian M. Adcock, Airways Disease Section, National Heart & Lung Institute, Imperial College London, Guy Scadding Building, Dovehouse Street, London SW3 6LY, United Kingdom; E-Mail: ian.adcock@imperial.ac.uk

Ruud Albers, Skillbase leader immunology & inflammation, Unilever Health Institute, Olivier van Noortlaan 120, 3133 AT Vlaardingen, The Netherlands

Sergey G. Apasov, Laboratory of Immunology, National Institute of Allergy and Infectious Diseases, National Institutes of Health, Bldg10 Room 11N311, 10 Center Dr. MSC-1892, Bethesda, MD 20892-1892, USA; E-Mail: sapasov@niaid.nih.gov

Clarissa Bachmeier, Klinik für Rheumatologie & Klinische Immunologie/Allergologie, Inselspital, 3010 Bern, Switzerland; E-Mail: clarissa.bachmeier@insel.ch

Peter J. Barnes, Section of Airways Disease, National Heart and Lung Institute, Imperial College School of Medicine, Dovehouse St, London SW3 6LY, United Kingdom; E-Mail: p.j.barnes@imperial.ac.uk

J. Edwin Blalock, University of Alabama at Birmingham, Department of Physiology and Biophysics, 1918 University Blvd. MCLM 898, McCallum Building, Birmingham, AL 35294-0005, USA; E-Mail: Blalock@uab.edu

Guido Bocci, Division of Pharmacology and Chemotherapy, Department of Oncology, University of Pisa, 55 Via Roma, 56126 Pisa, Italy

Andreas Bosio, Miltenyi Biotec GmbH, Stöckheimer Weg 1, 50829 Köln, Germany; E-Mail: andreas.bosio@miltenyibiotec.de

Jack H. Botting, Department of Nephrology and Experimental Medicine, The John Vane Science Centre, St. Bartholomew's and the Royal London School of Medicine and Dentistry, Charterhouse Square, London, EC1M 6BQ, United Kingdom

Regina M. Botting, Department of Nephrology and Experimental Medicine, The John Vane Science Centre, St. Bartholomew's and the Royal London School of Medicine and Dentistry, Charterhouse Square, London, EC1M 6BQ, United Kingdom; E-Mail: r.m.botting@qmul.ac.uk

Peter M. Brooks, Faculty in Health Sciences, Royal Brisbane Hospital, Herston QLD 4006, Australia; E-Mail: P.Brooks@mailbox.uq.edu.au

Romano Danesi, Division of Pharmacology and Chemotherapy, Department of Oncology, University of Pisa, 55 Via Roma, 56126 Pisa, Italy; E-mail: r.danesi@med.unipi.it

Richard O. Day, Department of Physiology and Pharmacology, School of Medical Sciences, University of New South Wales, Sydney NSW 2052, Australia; and Department of Clinical Pharmacology, St Vincent's Hospital, Sydney NSW 2010, Australia

Reno Debets, Laboratory for Clinical and Tumor Immunology, Department of Internal Oncology, Erasmus MC - Daniel den Hoed, P.O. Box 5201, 3008 AE Rotterdam, The Netherlands

Mario Del Tacca, Division of Pharmacology and Chemotherapy, Department of Oncology, University of Pisa, 55 Via Roma, 56126 Pisa, Italy

Jacques Descotes, Poison Center and Pharmacovigilance Unit, 162 Avenue Lacassagne, 69424 Lyon Cedex 03, France; E-Mail: jacques-georges.descotes@chu-lyon.fr

Gerhard Dickneite, ZLB Behring GmbH, 35002 Marburg, Germany; E-Mail: Gerhard.Dickneite@zlbbehring.com

Antonello Di Paolo, Division of Pharmacology and Chemotherapy, Department of Oncology, University of Pisa, 55 Via Roma, 56126 Pisa, Italy

Julian D. Down, Harvard Institutes of Medicine, Boston, MA 02115, USA; E-Mail: jdown@rics.bwh.harvard.edu

John F. Dunne, BD Biosciences, 2350 Qume Drive, San Jose, CA 95131, USA; E-Mail: john_dunne@bd.com

A.J. Frew, Allergy and Inflammation Research Group, Division of Infection, Inflammation & Repair, School of Medicine, University of Southampton, Southampton General Hospital, MP 810, Southampton SO16 6YD, United Kingdom; E-Mail: lesley@soton.ac.uk

Garry G. Graham, Department of Physiology and Pharmacology, School of Medical Sciences, University of New South Wales, Sydney NSW 2052, Australia; and Department of Clinical Pharmacology, St Vincent's Hospital, Sydney NSW 2010, Australia; E-Mail: ggraham@stvincents.com.au

Jan W. Gratama, Laboratory for Clinical and Tumor Immunology, Department of Internal Oncology, Erasmus MC - Daniel den Hoed, P.O. Box 5201, 3008 AE Rotterdam, The Netherlands; E-Mail: j.w.gratama@erasmusmc.nl

Peter Gronski, ZLB Behring GmbH, 35002 Marburg, Germany; E-Mail: peter.gronski@zlbbehring.com

Tomasz J. Guzik, Chair of Pharmacology, Jagiellonian University School of Medicine,
16 Grzegorzecka, 31-531 Krakow, Poland; E-Mail: tguzik@well.ox.ac.uk

C. Erik Hack, Sanquin Research at CLB, and Landsteiner Laboratory, Academic Medical Centre, University of Amsterdam, Plesmanlaan 125, 1066 CX Amsterdam, The Netherlands; E-Mail: e.hack@crucell.com

Klaus Hermann, Technische Universität München, Klinik und Poliklinik für Dermatologie und Allergologie, Biedersteiner Str. 29, 80802 München, Germany

René Houtman, PamGene International, Nieuwstraat 30, 5211NL 's-Hertogenbosch, The Netherlands;
E-Mail: rhoutman@pamgene.com

Ian Humphery-Smith, Department of Pharmaceutical Proteomics, Faculty for Pharmaceutical Sciences, University of Utrecht, Sorbonnelaan 16, 3584CA Utrecht, The Netherlands

Kazuhiro Ito, Department of Thoracic Medicine, National Heart & Lung Institute, Imperial College London, Guy Scadding Building, Dovehouse Street, London SW3 6LY, United Kingdom

Wim Jiskoot, Department of Pharmaceutics, Utrecht Institute for Pharmaceutical Sciences (UIPS), P.O. Box 80082, 3508 TB Utrecht, The Netherlands; E-mail: w.jiskoot@pharm.uu.nl

Ernst-Jürgen Kanzy, ZLB Behring GmbH, Postfach 12 30, 35002 Marburg, Germany

Gideon F.A. Kersten, Unit Research and Development, Netherlands Vaccine Institute, Antonie van Leeuwenhoeklaan 11, 3720 BA Bilthoven, The Netherlands; E-Mail: gideon.kersten@nvi-vaccin.nl

Richard Korbut, Chair of Pharmacology, Jagiellonian University School of Medicine, 16 Grzegorzecka, 31-531 Krakow, Poland; E-Mail: mfkorbut@cyf-kr.edu.pl

Marie-Thérèse Labro, INSERM U479, CHU X. Bichat, 16 rue Henri Huchard, 75018 Paris, France;
E-Mail: labro@bichat.inserm.fr

Geoffrey O. Litttlejohn, Centre for Inflammatory Diseases, Monash Medical Centre, Clayton VIC 3168, Australia;
E-Mail: Geoff.Littlejohn@med.monash.edu.au

Holden T. Maecker, BD Biosciences, 2350 Qume Drive, San Jose, CA 95131, USA; E-Mail: holden_maecker@bd.com

K. Noel Masihi, Robert Koch Institute, Nordufer 20, 13353 Berlin, Germany; E-Mail: masihik@rki.de

Sue McKay, Department of Pharmacology and Pathophysiology, Faculty of Pharmaceutical Sciences, Utrecht University, Sorbonnelaan 16, 3584 CA Utrecht, The Netherlands; E-Mail: mc.schrader@tiscali.nl

Eric F. Morand, Centre for Inflammatory Diseases, Monash Medical Centre, Clayton VIC 3168, Australia;
E-Mail: eric.morand@med.monash.edu.au

Frans P. Nijkamp, Faculty of Pharmacy, Department of Pharmacology and Pathophysiology, Sorbonnelaan 16, 3584 CA Utrecht, The Netherlands; E-Mail: F.P.Nijkamp@pharm.uu.nl

Markus W. Ollert, Technische Universität München, Klinik und Poliklinik für Dermatologie und Allergologie, Biedersteiner Str. 29, 80802 München, Germany; E-Mail: ollert@lrz.tu-muenchen.de

Michael J. O'Sullivan, Amersham Biosciences, Amersham plc, The Maynard Centre, Whitchurch, Cardiff, United Kingdom; current address: 9 Firbanks Way, Talbot Grees, Pontyclun, Wales, UK, CF72 8LB; E-Mail: maddyos50@aol.com

Michael J. Parnham, PLIVA Research Institute Ltd., Prilaz baruna Filipovića 29, 10 000 Zagreb, Croatia;
E-Mail: michael.parnham@pliva.hr

Raymond Pieters, Institute for Risk Asessment Sciences, P.O. Box 80176, 3508 TD Utrecht, The Netherlands

Kevin D. Pile, Department of Medicine, James Cook University, Townsville QLD 4814, Australia;
E-Mail: kevin.pile@jcu.edu.au

Valerie F.J. Quesniaux, Experimental and Molecular Genetics (GEM, FRE2358), Transgenose Institute, CNRS, 45071 Orleans Cedex 2, France; E-Mail: quesinaux@cnrs-orleans.fr

Klaus Resch, Institut für Pharmakologie, Medizinische Hochschule Hannover, Carl-Neuberg-Straße 1, 30625 Hannover, Germany; Email: Resch.Klaus@MH-Hannover.de

Ger T. Rijkers, Department of Pediatric Immunology, Wilhelmina Children's Hospital, University Medical Center Utrecht, KC03.063.0, Lundlaan 6, 3584 EA Utrecht, The Netherlands; E-Mail: ger@rijkers.nl

Johannes Ring, Technische Universität München, Klinik und Poliklinik für Dermatologie und Allergologie, Biedersteiner Str. 29, 80802 München, Germany; E-Mail: Johannes.Ring@lrz.tu-muenchen.de

Dirk Roos, Sanquin Research at CLB, and Landsteiner Laboratory, Academic Medical Centre, University of Amsterdam, Plesmanlaan 125, 1066 CX Amsterdam, The Netherlands; E-Mail: d.roos@sanquin.nl

Barbara Schaffrath, Miltenyi Biotec GmbH, Stöckheimer Weg 1, 50829 Köln, Germany;
E-Mail: barbara.schaffrath@miltenyibiotech.com

Henk-Jan Schuurman, Numico Research B.V., 6704 PH Wageningen, The Netherlands;
E-Mail: henkjan.schuurman@numico-research.nl

Friedrich R. Seiler, Oberer Eichweg 10, 35001 Marburg, Germany

Michail V. Sitkovsky, Departments of Biology and Pharnaceutical Sciences, 134 Mugar Building, 360 Huntington Avenue, Boston, MA 02115, USA; E-Mail: m.sitkovsky@neu.edu

Harm Snippe, Eijkman-Winkler Center for Microbiology, Infectious Diseases and Inflammation, University Medical Center Utrecht, Utrecht University Hospital, HP G04.614, PO Box 85.500, 3508 GA Utrecht, The Netherlands; E-Mail: h.snippe@azu.nl

Hergen Spits, Department of Cell Biology and Histology, Academic Medical Centre, University of Amsterdam, Meibergdreef 9, 1105 AZ Amsterdam, The Netherlands; E-Mail: hergen.spits@amc.uva.nl

James E. Talmadge, Department of Pathology and Microbiology, University of Nebraska Medical Center, 987660 Nebraska Medical Center, Omaha, Nebraska 68198-7660, USA; E-Mail: jtalmadg@unmc.edu

Lisette van de Corput, Department of Pediatric Immunology, Wilhelmina Children's Hospital, University Medical Center Utrecht, KC03.063.0, Lundlaan 6, 3584 EA Utrecht, The Netherlands

Jan Willem van der Laan, Pharmacological-Toxicological Assessment Section, Centre for Biological Medicines and Medical Technology, National Institute of Public Health and the Environment, P.O. Box 1, 3720 BA Bilthoven, The Netherlands

Henk van Loveren, National Institute of Public Health and the Environment, Laboratory for Toxicology, P.O. Box 1, 3720 BA Bilthoven, The Netherlands; and Department of Health Risk Analysis and Toxicology, Maastricht University, P.O. Box 616, 6200 MD Maastricht, The Netherlands; E-Mail: h.van.loveren@rivm.nl

Antoon J.M. van Oosterhout, Laboratory of Allergology and Pulmonary Diseases, University Medical Center Groningen, Groningen University, P.O. Box 30.001, 9700 RB Groningen, The Netherlands: E-Mail: a.j.m.van.oosterhout@path.umcg.nl

Donatella Verbanac, PLIVA Research Institute Ltd., Prilaz baruna Filipovića 29, 10 000 Zagreb, Croatia;
E-Mail: donatella.verbanac@pliva.hr

Jan Verhoef, Eijkman-Winkler Center for Microbiology, Infectious Diseases and Inflammation, University Medical Center Utrecht, Utrecht University Hospital, HP G04.614, PO Box 85.500, 3508 GA Utrecht, The Netherlands; E-Mail: j.verhoef@azu.nl

Thierry Vial, Poison Center and Pharmacovigilance Unit, 162 Avenue Lacassagne, 69424 Lyon Cedex 03, France;
E-Mail: thierry.vial@chu-lyon.fr

Joseph G. Vos, National Institute of Public Health and the Environment, Lab. for Toxicology, Pathology and Genetics, P.O. Box 1, 3720 BA Bilthoven, The Netherlands; E-Mail: j.vos@rivm.nl

Douglas A. Weigent, University of Alabama at Birmingham, Department of Physiology and Biophysics, 1918 University Blvd. MCLM 898, McCallum Building, Birmingham, AL 35294-0005, USA; E-Mail: weigent@uab.edu

Ralph A. Willemsen, Laboratory for Clinical and Tumor Immunology, Department of Internal Oncology, Erasmus MC - Daniel den Hoed, P.O. Box 5201, 3008 AE Rotterdam, The Netherlands

Kenneth M. Williams, Department of Physiology and Pharmacology, School of Medical Sciences, University of New South Wales, Sydney NSW 2052, Australia; and Department of Clinical Pharmacology, St Vincent's Hospital, Sydney NSW 2010, Australia; E-Mail: ken.williams@unsw.edu.au

Index

absolute quantification (AQUA) 222
Access Allergy Diagnostic System 168
aceclofenac 613, 614
acellular pertussis vaccine 236
acetylator phenotype 570
acetylcholine 154
acquired immune deficiency syndrome (AIDS) 155
acrolein 468, 469
activator protein 1 (AP-1) 468, 485
acute-phase reaction 78
acyclovir 476
adalimumab 59, 535, 624
adaptive immunity 130
adjuvant 243, 566, 567
adoptive cellular immunotherapy 139
adoptive immunotherapy 131
adrenocorticotrophic hormone (corticotropin, ACTH) 150, 153, 155, 306
affinity maturation 38
agranulocytosis 469
Ah receptor 570
AlaSTAT liquid allergen technology 168
aldehyde dehyrogenase 469
aldophosphamide 468
alkylamide, immunomodulation by 392
alkylating agent 465
allergy 265, 270, 272, 277, 559
allogeneic hematopoietic stem cell transplantation (SCT) 140
allopurinol 467, 521
alopecia 469, 473
aluminium hydroxide/magnesium hydroxide 476
Alzheimer's disease 506
ambient analyte concentration 223
aminoglycoside 409
aminosalicylate (mesalazine) 532
5-amino-imidazole-4-carboxamide ribosyl-5-phosphate (AICAR)-formyltransferase 472
anakinra 59, 535, 624
analyte 171
analytical recovery 179
anaphylactic shock 117, 126
anaphylatoxin 77
anaphylaxis 165, 570
anemia 467, 471, 476
anergy 27
angiogenesis, effect of vitamin E 400
anorexia 476
ansamycin 410–412, 427, 429, 430
antibodies 132, 171
- allergy and 165
- autoimmune diseases and 165
- immunoadsorbent assay of 166
- microbial infection and 164
antibody
- detection by immunoblotting 166
- detection by immunocytochemistry 166
- detection method 165
- detection, clinical relevance of 164
- repertoire 13
- response, effect of vitamin E 400
- structure 164
- characterization of 164
antibody-dependent cellular cytotoxicity (ADCC) 40, 137
anti-CD11a 456
anti-CD52 mAb alemtuzumab 456
antigen 133
antigen-presenting cells (APC) 130, 547
antigen-recognition repertoire 8
antigen-specific tolerance 458
antigenic peptide 25
anti-idiotype antibody 240, 241
anti-IgE 275, 277, 546
anti-IL-2 receptor Ab 456
anti-inflammatory drug 59, 83
anti-lymphocyte globulin (ALG) 441, 455

antimalarials (chloroquine and hydroxychloroquine) 520, 623
antinociceptive 156
antisense oligodeoxynucleotide 154
antisense oligonucleotide 328
anti-thymocyte globulin (ATG) 441, 455
α1-antitrypsin 323
AP-1 325
AP50 77
aplastic anemia 469
apoptosis 71, 154
– green tea and 402
array data: acquisition, analysis and mining 205
ascorbic acid – *see* vitamin C
aspirin 87, 499, 612, 614
assay linearity 180
asthma 165, 545
astrocyte 152
atrial natriuretic peptide 300
attenuated bacterial vaccine 235
attenuated poliovaccine (Sabin) 233
attenuated viral vaccine 233
auranofin 525, 623
aurocyanide 526
aurosome 526
aurothiomalate 525, 623
autoantibody 567
autoantigen 563
autoimmunity 559
– clinical manifestation of 124
azapropazone 613, 614
azathioprine 58, 441, 451, 465–467, 521, 623
– adverse reactions to 467
– chemical structure of 465, 466
– clinical indication for 467
– pharmacokinetics of 466, 467
azurophil and specific granule 70

B lymphocyte (B cell) 30, 130, 150, 155, 266, 267, 269, 271
– receptor (BCR) complex 32
bacille Calmette-Guérin (BCG) 235, 358
bacterial extract 387
balloon coronary angioplasty 450
basiliximab 60, 456
bead-based immunoassay 187
benzylpyrimidine 414
bestatin 360
betalactam 414
bioavailability 490
biochips 197, 225
biological anti-rheumatic therapy 535
biotinylated antibody 174
Bishop, K. 399
bispecific monoclonal antibody (mAb) 139
bone marrow 38, 152
– chimerism 460
bradykinin 94
brequinar 454
bronchus-associated lymphoid tissue (BALT) 38
Bruton's tyrosine kinase (Btk) 32
Buehler test 574

C677T polymorphism 473
calcineurin 447, 465
– inhibitor 441, 446, 465, 475–477
calcitonin gene-related peptide (CGRP) 93, 155
calcium-activated potassium channel (KCa) 283
cancer 504
– testis antigen 133, 134
capillary electrophoresis 221
capsular polysaccharide 37, 237
capture molecule 222
caspase 470
catecholamine 100, 154
catechol-o-methyl transferase 281
cathepsin 322
CD4 – *see* T helper lymphocyte
CD8 – *see* T cytotoxic lymphocyte
CD19/CD21 37
CD22 41
CD28 36
CD40 34
CD40 ligand (CD40L) 34
CD45 41, 456
CD200 71
cDNA array 198
cDNA microarray 396
cefodizime 415, 426, 431
celecoxib 504, 615, 616
Celecoxib Long-term Arthritis Safety Study 504
A549 cell 503
cell culture supernatant 179
cell traffic/adhesion 459
Cellcept® 453

cellular proliferation 186
cellular response 92
cephalosporin 410
CH50 77
channel opener 300
Chediak-Higashi syndrome 71
chemokine 54, 66, 382, 488
- classes 93
- receptor 459
- CCR2 321
- CCR3 321
- CCR4 321
chemotaxin 64
chemotaxis 64
chemotherapeutic agent 6
chimeric Ab 446
chips 197
chloramphenicol 415
chlorpromazine 567
cholestyramine 476
chronic granulomatous disease 71
chronic injection 446
chronic rejection 450
Churg-Strauss syndrome 468
CLASS 504
classical vaccine component 240
clenoliximab 456
clindamycin 419, 431
clinical diagnostic test 171
clofazimine 425, 427
co- and post-translational modifications (PTMs) 213
coated well 175
collision-induced dissociation 218
colony-forming unit (CFU) 6
colony-stimulating factor (CSF) 47, 89, 91, 381
combination DMARD 536
combination inhaler 286
combination therapy 380
combination vaccine (MMR vaccine) 234
Committee on Proprietary Medicinal Products (CPMP) 580
competitive 171, 172
complement 63, 75
- activation product 78
- alternative pathway of activation 37, 76
- C1-esterase inhibitor 79
- C3b receptor 77
- classical pathway 76
- classical pathway deficiency 78
- receptor 69
complementarity determining region 30
conjugate vaccine 237
constant domain 30
contact allergen 574
Coomassie Blue 219
co-receptor 25
coronary artery disease 423, 427, 428
corticosteroid 300, 441, 512, 560
corticosteroid-sparing 491
corticotropin releasing hormone (CRH) 153, 156
- antisense oligodeoxynucleotide 156
co-stimulation 458
costimulatory pathway 460
cowpox vaccination 231
cyclo-oxygenase (COX), 82
- COX-1 82
- COX-2 82
- COX-3 506
C-reactive protein (CRP) 76
CSAID 59
CTLA-4-Ig 457
C-type lectin 75
CXCR antagonist 321
cyclic AMP 176, 283
cyclic nucleotide 171
cycline 416, 417, 427, 428, 431
cyclophilin 447
cyclophosphamide 441, 451, 467–469, 477, 560
- adverse reactions to 469
- chemical structure of 466, 467
- clinical indications for 468, 469
- pharmacokinetics 468
cyclosporin 441, 446, 467, 476, 523, 560, 623
cyclosporin A 15, 58, 313
cystic fibrosis (CF) 410, 427
cystic fibrosis transmembrane conductance regulator (CFTR) 411
cytidine deaminase (CdA) 470
cytidine-phosphate guanosine (CpG) 384
cytochrome P-450 (CYP) 468
- CYP2B6 polymorphism 469
cytokine 45, 92, 130, 132, 136, 149, 150, 152, 154, 157, 171, 187, 243, 378, 459, 488, 545
- biology 45

– immunomodulator 378
– inhibition 57, 58, 383
– network 73
– receptor 91
cytokine-release syndrome 455
cytotoxic drug 451

daclizumab 60, 456
dapsone 425, 427
data analysis and mining 206
DC 19, 138
de novo protein sequencing 218
decay-accelerating factor 77
defensin 385
degranulation 68, 69
dendritic cell 130, 152, 153, 378
deoxycytidine kinase (dCK) 470
dermatitis 473
dermatomyositis 470
developmental immunotoxicity 584
diapedesis 64
diarrhea 473, 474, 476
diclofenac 613, 614
diflunisal 612, 614
dihydrofolate reductase 472, 473
dihydroorotate dehydrogenase (DHODH) 454, 528
diphtheria toxin 457
disease-modifying anti-rheumatic drug (DMARD) 471, 511
doctrine of signatures 391
dose-response relationship349
D-penicillamine 569
Drug Rash with Eosinophilia and Systemic Symptoms (DRESS) 118, 123
drug screening 171
drug-induced autoimmune reaction 123
DtaP vaccine 239
DTP vaccine 236
DTP-Hib vaccine 237
dynamic range 172, 220
dyspepsia 476

ebselen 403
Echinacea purpurea 391, 606
– wound healing and 392
edible vaccine 241
efalizumab 456
eicosanoid 81, 86
electrospray ionization 215
Eleutherococcus senticosus 391
endopeptidase 214
endorphin 150, 155, 156
β-endorphin 151, 156
endothelin 176, 318
endotoxin 110, 111, 114
enzyme-linked immunoadsorbent assay (ELISA) 166
epigallocatechin gallate (EGCG) 396, 609
– immunomodulation by 402
epithelial cell 130
erythromycin A 410, 421, 431
erythropoietin 47
E-selectin 89
etanercept 59, 535, 624
etodolac 615, 616
etoricoxib 615, 616
European Medicines Evaluation Agency (EMEA) 580
Evans, H. 399
everolimus 450
extended histopathology 572

Fas-mediated apoptosis 24
fenbufen 612, 614
fenoprofen 612, 614
fermosera 254
filgrastim 47
FK-506 15, 568
flavonoid, immunomodulation by 400
flow cytometry 185
fludarabine 465, 469–471, 478
– adverse reactions to 471
– chemical structure of 466, 469
– clinical indications for 470
– pharmacokinetics 470
– triphosphate (F-ara-ATP) 470
fluorescence 174, 180, 188
fluorochrome 188
flurbiprofen 612, 614
follicle stimulating hormone (FSH) 153
folyl-polyglutamate synthetase 472, 473
formaldehyde 236
fosfomycin 418
FTY720 444, 455
functional assay 190

fungi 106, 109, 110
fusidic acid 419
fusion protein 135, 239–241
fyn 32

gamma glutamyl hydrolase 473
gastrotoxicity 499
gene
- expression profile 197
- transcription 485
- transfer 140
gene-rearrangement 30
genome 213
Ghrelin 152
glomerulonephritis 469
glucan 387
glucocorticoid 15, 58, 87
- receptor 301, 484
β-glucuronidase 476
glucuronosyl-transferase (UGT) 476
glutathione (GSH) 522
glutathione peroxidase (GSH-Px) 397
- phagocytes and 397
- selenium supply and 398
glycosylated lectin 394
gold 313, 623
- complex 525
G-protein 283
graft-*versus*-host disease (GVHD) 140
graft-*versus*-host reactions 577
graft-*versus*-tumor (GVT) effect 140
Gram-negative bacteria 110
Gram-negative shock 110, 111
Gram-positive bacteria 111, 112
granulocyte 51, 130
granulocyte-colony stimulating factor (G-CSF) 47
granulocyte-monocyte colony-stimulating factor (GM-CSF) 47, 354
growth factor 171
- inhibitor 450
growth hormone (GH) 150, 152, 153
- receptor 152
growth hormone releasing hormone (GHRH) 153
- antagonist 156
guinea-pig maximization test 576
gusperimus (15-deoxyspergualin) 444, 454
gyrase B inhibitor 419

Haemophilus influenzae type b (Hib) vaccine 237
Haemophilus influenzae type b conjugate vaccine 238
hair loss 473
halothane 563
hapten-carrier complex 562
heat-shock protein 455
hematopoiesis 3, 152, 156
hematopoietic stem cell 6
hemorrhagic cystitis 469
hepatitis B surface Ag (HBsAg) 237, 239, 242
hepatitis B virus 379
hepatitis C virus 380
hepatotoxicity 467
herd immunity 231
hereditary angio-oedema 79
herpes zoster 469
heterogenous 173
histamine 99, 265, 267–275, 277
- receptor 270, 273
histone deacetylase 291
homologous restriction factor 77
horseradish peroxidase 174
5-HPETE 84
12-HPETE 84
15-HPETE 84
human equilibrative nucleoside transporter-1 (hENT1) 470
human immunodeficiency virus (HIV)
- disease 185
- infection 381
- therapy 113
humanized Ab 446
hybrid resistance 75
hydralazine 567
hydrocortisone 305
5-hydroxytryptamine (5-HT) 99
- metabolism 100
- synthesis 100
- 5-HT2 receptor 100
hypersensitivity 165
- reaction 117, 119–123
hypervariable region 30
hypochlorous acid 71
hypoxanthine-guanine phosphoribosyltransferase (HGPRT) 466, 467

ibuprofen 610, 612

imiquimod 383
immobilized pH gradient (IPG) 214
immune reaction and inflammation 51
immune response modifier 415
immune surveillance 129
immune-stimulating complex (ISCOM) 243
immunoassay 164, 171
immunodeficiency virus (HIV) 155
immunoglobulin 29, 110
– class switching 13, 31, 34
– Fc receptor 40
– Fcγ receptor IIb 72
– IgE 165, 265–267, 269, 275–277
immunological synapse 26
immunological tolerance 26
immunomodulator 377, 388, 409
immunophenotyping 185
immunoprecipitation assay 165
immunoreceptor tyrosine-based activation motif (ITAM) 26, 32, 74
immunosuppression 546, 559
immunosuppressive drug 14
immunotherapy 545
– mechanism 548
immunotoxin 460, 572
in vitro 149, 152–154
– antibody response 154
in vivo 149, 152, 155, 156
inactivated polio vaccine (IPV) 232, 235
inactivated poliovirus vaccine (Salk) 233
indomethacin 612, 614
inducible costimulator (ICOS) 36
induction treatment 441
infertility 469
infliximab 59, 535, 624
inhibitory Fcγ receptor 72
innate immune system 63, 130
inosine monophosphate (IMP) 465
– dehydrogenase (IMPDH) 453, 474, 475
insulin 563
insulin-like growth factor-1 (IGF-1) 152, 154
integrin 65, 459
intercellular adhesion molecule-1 (ICAM-1) 66
interferon (IFN) 52, 89, 379
– α (IFN-α) 52, 149, 346, 379
– α 1a or 2b 52
– β (IFN-β) 52, 380
– β 1a or 1b 52
– γ (IFN-γ) 52, 73, 151, 327, 348, 380
– pegylated
interleukin (IL) 36, 48, 89
– IL-1 90, 150, 511, 535
– IL-2 48, 49, 129, 136, 150, 351
– IL-2 combination 381
– IL-4 48, 320, 547
– IL-5 48, 319
– IL-6 48, 150
– IL-8 66
– IL-9 320
– IL-10 177, 321
– IL-12 73, 322, 365, 548
– IL-13 48, 320, 547
– IL-15 49, 73
– IL-18 90, 156
International Conference on Harmonization of Technical Requirements for Registration of Pharmaceuticals for Human Use (ICH) 580
intracellular bacteria 109
intracrine 154
invasive bacteria 109
iodine-125 176
iodothyronine deiodinase 397
iontophoresis 241
ipatropium bromide 297
isoelectric focusing (IEF) 213
isoelectric point (pI) 213
isoniazide 419, 567
isoprinosine 384
isotope-coded affinity tagging (ICAT) 222
ITIM 71, 75

Janus activated kinase (JAK) 458
JAK3 inhibitor (CP-690,550) 458
Jenner, Edward 231, 233
juvenile arthritis 473, 474, 511

ketoprofen 612, 614
ketorolac 613, 614
killer cell immunoglobulin-like receptors (KIR) 75
kinin 93, 94
Koch, Robert 235

Langerhans cell 243
latency period 38

LEA29Y 457
leflunomide 441, 453, 527, 623
lenograstim 47
lentinan 387
leucovorin 473
leukemia/lymphoma 185
leukocyte adhesion deficiency (LAD) 66
leukocyte 149, 154, 155, 157, 399
leukopenia 469, 473, 474, 476
leukotriene (LT) 84, 171, 489
– cys-LT_1-receptor 310
– LTC_4 177
– LTD_4 177
– LTE_4 177
levamisole 360
levonorgestrel 476
LF 15-0195 454
LFA-3-Ig fusion protein 457
lincosamide 419
lipocortin 87
liposome 243
lipoxin 85
– receptor 85
5-lipoxygenase 310
live attenuated vaccine 233
live or inactivated organism 240
live vector 240
local lymph node assay (LLNA) 574
L-selectin 64, 89
luminescent 174, 180
luteotrophic hormone (LH) 153
– releasing hormone (LHRH) 153, 154
– releasing hormone (LHRH)-receptor 154
lymphocyte 151, 152, 154, 155
– effects of vitamin E on 399
– selenium and 398
– vitamin E content of 399
lymphoid organ 20, 149
lymphokine-activated killer (LAK) activity 139
lyn 32
lytic granule 24

macrolide 420, 421, 426, 427
– antibiotics 431
macrophage 130, 132, 151, 152
macrospot 221
magnesium sulphate 299
maintenance treatment 441
major histocompatibility complex (MHC) 34, 130, 151
malignancy-associated fusion protein 134
malignancy-associated minor histocompatibility gene 135
malignancy-associated point mutation of normal gene 135
malononitrilamide 454
mammalian target of rapamycin (mTOR) 450
mannan-binding lectin (MBL) 76, 78
MASP 76
mass coded abundance taging (MCT) 222
mass spectrometry (MS) 213
mass-over-charge ratio (m/z) 216
mast cell 265–271, 275–277, 284
matrix metalloproteinase 323
matrix-assisted laser desorption/ionization (MALDI) 215
meadowsweet 499
meclofenamic acid 613, 614
mefenamic acid 613, 614
α-melanocyte stimulating hormone (α-MSH) 151, 152, 156
meloxicam 87, 615, 616
membrane cofactor protein 77
membrane inhibitor of reactive lysis 77
membrane-bound immunoglobulin 32
memory B cell 39
meningococci 237
6-mercaptopurine 465–467, 521
6-methylmercaptopurine 467
8-mercaptoguanosine 43
mercury salt 565
mesna 469
methotrexate 312, 452, 465, 471–474, 478, 514, 560, 623
– adverse reactions to 473, 474
– chemical structure of 466, 472
– clinical indications for 473
– pharmacokinetics 472
methyl inosine monophosphate 385
methylenetetrahydrofolate reductase (MTHFR) 473
methylxanthine 289
α-methyldopa 567
microarray 197, 222
microbial immunomodulator 386
microbicidal protein 70
microcarrier 243
microglia 152
microscopic polyangiitis 469

microspot 223
microtitre plate 174, 175
Ministry of Health, Labour and Welfare (MHLW) 580
minocycline 533, 624
minor histocompatibility antigen 136
mistletoe (*Viscum album*) 393
– lectin, cytotoxicity of 393
mitogen-activated protein (MAP) kinase 34, 325
mizoribine 444, 451
molgramostim 47
monoamine oxidase (MAO) 282
monoclonal antibody (mAb) 129, 137, 465
– generation of 164, 174
monocyte 130, 153, 395
monocyte-colony stimulating factor (M-CSF) 47
mononuclear phagocyte 51
monooxygenase pathway (MOX) 86
monophosphoryl lipid A 386
mRNA amplification 204
mucoregulator 328
mucosal immunisation 242
multidimensional protein identification technology (MudPIT) 221
murabutide 386
muramyl dipeptide (MDP) 360
muscarinic receptor 296
mutational antigen 136
mutein 59
Mycobacterium bovis 235
mycophenolate mofetil 441, 465, 474–476
mycophenolate sodium 474–476
mycophenolic acid 452, 465, 474–478
– chemical structure of 466, 474
– clinical indications for 476
– pharmacokinetics 475, 476
myeloperoxidase 70, 526
Myfortic® 453
myrtle 499

nabumetone 613, 614
N-acetylcysteine 328
NADPH oxidase 70
naloxone 155
nanoparticle 243
naproxen 504, 612, 614
nasal or pulmonary delivery 243
natalizumab 456
natural killer (NK) cell 63, 72, 130, 395
– activity 572
– receptor 74
N-chloramine 71
nedocromil 309
negative selection 10
Neisseria infection 78
neo-antigenic structure 563
neurotransmitter 149
neutropenia 467, 471
neutrophil 152
– elastase 322
– phagocytic capacity, effect of vitamin E 400
new generation vaccine component 240
nimesulide 615, 616
nitric oxide (NO) 95-97
– metabolism 96
– synthesis 96
– synthase (NOS) 475
– eNOS 95, 97
– iNOS 96, 97, 319
– nNOS 95
non-invasive bacteria 109
non-isotopic 174
nonresonant ion 218
non-specific airways reactivity 548, 549
non-steroidal anti-inflammatory drug (NSAID) 465, 471, 473, 499, 511, 569
noradrenaline 281
normalisation 207
nuclear factor κB (NF-κB) 325, 455, 468, 485
nucleic acid vaccine 242
5'-nucleotidase 469, 470

O6-alkylguanine-DNA alkyltransferase 469
odulinomab 456
Organisation for Economic Co-operation and Development (OECD)
– guideline 407 570, 581
– test 406 guideline 581
oligonucleotide 361
– array 198
olsalazine 532
OM-174 386
omalizumab 313
opiate 154
– antagonist 155

opioid peptide 151
μ-opioid receptor 149, 156
opportunistic infection 570
opsonisation 69, 251
optic neuritis 468
oral polio vaccine (OPV) 232
overexpressed antigen 134
oxidative stress 318
oxymetholone 572

paracetamol 617
parasite 106, 107, 109, 110
parent ion gating 218
passivc immunothcrapy 262
Pasteur, Louis 231, 233
pathogen-associated molecular pattern (PAMP) 67
pathogenic microorganism 114
pattern recognition receptor (PRP) 67
PECAM-1 66
penicillamine 530, 623
penicillin G 410
Penicillium species 474
peptide hormone 149
peptide mass fingerprinting (PMF) 216
perforin 73
15-prostaglandin dehydrogenase (PGDH) 82
PGG-glucan 387
P-glycoprotein 413, 417, 422, 431
phagocyte 63
 – selenium and 398
phagocytosis 68, 69, 108, 109, 132, 251
 – effects of *Echinacea* on 392
pharmacoproteomics 225
phenylbutazone 613, 614
phenytoin 567
phosphodiesterase (PDE) 290
phosphoinositide-3-kinase 34
phospholipase Cγ2 34
phosphoramide mustard 468, 469
phytate 397
 – zinc absorption and 394
piroxicam 613, 614
plaque-forming cell assay 572
plasma cell 37
plasmid DNA 240
platelet-activating factor (PAF) 66, 87
 – synthesis and metabolism 88
 – receptor antagonist 89
pluripotent hematopoietic stem cell 3
pneumococci 237
pneumonitis 473
poliovirus 233
polyarteritis nodosa 469
polyclonal antibody 174, 465
polymorphonuclear leukocyte (PMNL) – *see* granulocyte
polymyositis 470
polyreactive antibody 13
polysaccharide antigen 39
polysaccharide-protein 42
 – conjugate vaccine 237
popliteral lymph node assay (PLNA) 577
positive selection 10
post-source decay 215
precision 172, 179
premature labour 505
primary (antibody) response 37
probiotics 386
procainamide 567
prolactin (PRL) 152–154
 – receptor 152
pro-opiomelanocortin (POMC) 149, 156
prostaglandin 171
prostanoid 82
protein tyrosine phosphate 40
proteolytic fragment 215
proteome 213
proteomics 213, 584
P-selectin 89
pseudo-graft-*versus*-host reaction 577
Pseudomonas exotoxin 457
psoriasis 472–474
psoriatic arthritis 470
pulmonary vasculitis 469
purine metabolism 453
purple coneflower – see *Echinacea*
pyogenic bacteria 78

quantitative structure-activity relationship (QSAR) 574
quadrupole coupled to TOF (QTOF) 218
quadrupole ion trap (QIT) 221
quinolone 410, 424, 432

radio frequencies voltage 218
radioactive iodine (iodine-125) 173

radio-allergo-sorbent test (RAST) 167
radioimmunoassay (RIA) 167
rapamycin 441
reactive oxygen species (ROS) 68, 70, 98
recombinant
– allergen 550
– hepatitis B vaccine 237
– live vaccine 239
– protein 345
– subunit vaccine 240
recombinase activating genes RAG-1 and RAG-2 30
regulation of immune response 55
repertoire 11
reporter antigen (RA) 577
reproductive toxicology 584
resistance, natural, innate immunity 105, 107, 114
resistance, specific immunity 107
resonant ion 218
respiratory allergen 575
respiratory burst 70
respiratory infection, Echinacea therapy of 392
respiratory tract hypersensitivity 575
rheumatoid arthritis 453, 470–474, 511
rifampicin 411, 429
riminophenazine 425
rofecoxib 504, 615, 616
rViscumin 394

salbutamol 281
salicylate 499, 611
Salmonella typhi vaccine 235
sample matrix 179
sandwich assay 172
Schwarz, Klaus 397
scintillation proximity 173
sclerodermia 467, 468
scurvy 398
SDS-PAGE 214
secondary antibody response 38
secretory IgA 242
selectin 459
selenium
– supplementation 398
– deficiency of 397
– immunomodulation by 396
selenomethionine 607
selenoprotein-P 397
sensitivity 173, 174
sensitization 563, 574
separation step 173
SEREX 133
seroconversion 238
serpin 71, 72, 323
serum therapy 254
shikimic acid 401
Siberian ginseng 391
silver staining 219
siplizumab 456
sirolimus – *see* rapamycin
skin, inflammation of 402
smallpox 231
sodium cromoglycate 308
sodium salicylate 611
sodium selenite 607
solid phase 172, 175
SOM 153, 154
somatic hypermutation 8, 13, 31
somatostatin 153
specificity 3, 174, 175
– rearrangement of genes 7
spergualin 454
sphingosine 1-phosphate receptor 455
– agonist FTY720 459
Spiraea 499
stable isotope labeling with amino acids during culture, or SILAC 221
standard 171, 175
standard curve fitting 175
Steiner, Rudolf 393
stem cell transplantation (SCT) 363
stomatitis 473, 474
streptavidin 174
structure-activity relationship (SAR) 574
substance P 155
substrate 172
subunit vaccine 232, 236, 240
sulfapyridine 532
sulfasalazine 532, 624
sulfonamide 425, 427
sulfoxidizer 570
sulindac 612, 614
surface/intracellular staining 193
Syk kinase 325
symptoms of inflammation 82

synthetic peptide vaccine 241
systemic lupus erythematosus 78, 468
Szent-Gyorgi, Albert 398, 400

tachykinin 93
tacrolimus 58, 441, 444, 476
tacrolimus (FK506) 447
tandem MS (MS/MS) 215
T cell receptor (TCR) 6–10, 19, 21, 22, 24–28, 130, 131, 353, 447, 455, 458
T cytotoxic lymphocyte (CTL, $CD8^+$) 19, 23, 129, 139
tea (*Camellia sinensis*), therapeutic effects of 401
tenoxicam 613, 614
terbutaline 281
terminal complement complex 77
terminal-deoxynucleotidyl transferase (TdT) 30
tetracycline 410, 624
T helper lymphocyte ($CD4^+$) 19, 185
- type-1 (Th1) 23, 154, 378
- type-2 (Th2) 23, 265, 378, 570
6-thioguanine acid 465
6-thioinosinic acid 465
thioinosine monophosphate 467
thiopurine methyltransferase (TPMT) 466, 467, 523
thioredoxin reductase (TR) 397
thrombocytopenia 467, 471, 473, 476
thrombopoietin 47
thymidylate synthase 472
thymosin-α1 380
thymulin 395
thymus 8
thyrotropin (thyroid stimulating hormone, TSH) 153, 154
thyrotropin releasing hormone 153
tiaprofenic acid 612, 614
tier-testing approach 570
time-of-flight (TOF) 216
tiotropium bromide 298
tissue residency 490
tissue-specific differentiation antigen 133, 134
T lymphocyte (T cell) 19, 23, 129, 130, 132, 150, 152, 153, 156, 395
- depletion 363
- signal transduction 458
T lymphocyte-independent antigen 37
α-tocopherol – *see* vitamin E
tolerance 14, 46
toll-like receptor (TLR) 67, 89, 112, 383
- signalling 69
tolmetin 613, 614
toxicogenomics 584
toxidemia 118
toxoid 236
tracer 173
transactivation 492
transcellular synthesis of lipoxin 85
transcription factor 325
transforming growth factor-α (TGF-α) 91
transforming growth factor-β (TGF-β) 129
transrepression 492
treatment guidelines 545
T-regulatory lymphocyte 24, 132, 546
tresperimus 454
trimethoprim 414
tritium tracer 174
tryptase 322
tumor necrosis factor (TNF) 53, 89, 150, 511, 547
- TNF-α 52, 90, 136, 137, 322, 535
- TNF-β 52, 90
tumor-associated antigen (TAA) 129, 132, 133
tumor-infiltrating lymphocyte (TIL) 129
two-dimensional gel electrophoresis (2DE) 213

U.S. Food and Drug Administration (FDA) 580

vaccination 137
vaccine 105, 113, 114, 231
- category 232
- classification 234
- delivery system 243
- efficacy 224, 238
- history 231, 233
vaccinia virus 239
valacyclovir 476
valdecoxib 615, 616
variable domain 30
variolation 231
vascular cell adhesion molecule-1 (VCAM-1) 326
vascular cell adhesion molecule-2 (VCAM-2) 66
vasculitis 467, 468
vasoactive intestinal polypeptide 300
vector 239, 243
Vioxx Gastrointestinal Outcome Research (VIGOR) 504
viral proteins expressed by malignant cells 135
virally encoded TAA 136

Virchow, Rudolph 81
virosome 243
virus 105–110, 112–114
viscumin 393
vitamin C (ascorbate) 396, 399, 608
 – immunomodulation by 398
vitamin E 396
 – deficiency of 400
 – immunomodulation by 399
VLA-4 326, 456

Wegener's granulomatosis 468
whole-cell pertussis vaccine 235, 238
willow bark 499
within-assay 179

xanthine oxidase 466, 467, 521
X-linked hyper-IgM syndrome 34

yeast glucan 387

ZAP-70 458
zinc 606
 – deficiency of 394
 – immunomodulation by 394
zoom gels 220